S0-AKZ-927

SpringerWienNewYork

Imme van den Berg
Vítor Neves (eds.)

The Strength of Nonstandard Analysis

SpringerWienNewYork

Imme van den Berg
Departamento de Matemática
Universidade de Évora, Évora, Portugal

Vítor Neves
Departamento de Matemática
Universidade de Aveiro, Aveiro, Portugal

Printed in Germany
SpringerWienNewYork is part of Springer Science + Business Media
springeronline.com

Typesetting: Camera ready by authors
Printing: Strauss GmbH, Mörlenbach, Germany
Cover image: Gottfried Wilhelm Freiherr von Leibniz (1646–1716), tinted lithograph by Josef Anton Seib. © ÖNB, Vienna
Printed on acid-free and chlorine-free bleached paper
SPIN 11923817

With 16 Figures

Library of Congress Control Number 2006938803

ISBN-10 3-211-49904-0 SpringerWienNewYork
ISBN-13 978-3-211-49904-7 SpringerWienNewYork

Foreword

Willst du ins Unendliche schreiten?
Geh nur im Endlichen nach alle Seiten!
Willst du dich am Ganzen erquicken,
So must du das Ganze im Kleinsten erblicken.

J. W. Goethe (*Gott, Gemüt und Welt, 1815*)

Forty-five years ago, an article appeared in the Proceedings of the Royal Academy of Sciences of the Netherlands Series A, 64, 432–440 and Indagationes Math. 23 (4), 1961, with the mysterious title "Non-standard Analysis" authored by the eminent mathematician and logician Abraham Robinson (1908–1974).

The title of the paper turned out to be a contraction of the two terms "Non-standard Model" used in model theory and "Analysis". It presents a treatment of classical analysis based on a theory of infinitesimals in the context of a non-standard model of the real number system $\mathbb{R}$.

In the Introduction of the article, Robinson states:

> "It is our main purpose to show that the models provide a natural approach to the age old problem of producing a calculus involving infinitesimal (infinitely small) and infinitely large quantities. As is well-known the use of infinitesimals strongly advocated by Leibniz and unhesitatingly accepted by Euler fell into disrepute after the advent of Cauchy's methods which put Mathematical Analysis on a firm foundation".

To bring out more clearly the importance of Robinson's creation of a rigorous theory of infinitesimals and their reciprocals, the infinitely large quantities, that has changed the landscape of analysis, I will briefly share with the reader a few highlights of the historical facts that are involved.

The invention of the "Infinitesimal Calculus" in the second half of the seventeenth century by Newton and Leibniz can be looked upon as the first fundamental new discovery in mathematics of revolutionary nature since the death of Archimedes in 212 BC. The fundamental discovery that the operations of differentiation (flux) and integration (sums of infinitesimal increments) are inverse operations using the intuitive idea that infinitesimals of higher order compared to those of lower order may be neglected became an object of severe criticism. In the "Analyst", section 35, Bishop G. Berkeley states:

> "And what are these fluxions? The velocities of evanescent increments. And what are these same evanescent increments? They are neither finite quantities, nor quantities infinitesimally small, nor yet nothing. May we call them ghosts of departed quantities?"

The unrest and criticism concerning the lack of a rigorous foundation of the infinitesimal calculus led the Academy of Sciences of Berlin, at its public meeting on June 3, 1774, and well on the insistence of the Head of the Mathematics Section, J. L. Lagrange, to call upon the mathematical community to solve this important problem. To this end, it announced a prize contest dealing with the problem of "Infinity" in the broadest sense possible in mathematics.The announcement read:

> "The utility derived from Mathematics, the esteem it is held in and the honorable name of 'exact science' par excellence, that it justly deserves, are all due to the clarity of its principles, the rigor of its proofs and the precision of its theorems. In order to ensure the continuation of these valuable attributes in this important part of our knowledge the prize of a 50 ducat gold medal is for:
>
> A clear and precise theory of what is known as 'Infinity' in Mathematics. It is well-known that higher mathematics regularly makes use of the infinitely large and infinitely small. The geometers of antiquity and even the ancient analysts, however, took great pains to avoid anything approaching the infinity, whereas today's eminent modern analysts admit to the statement 'infinite magnitude' is a contradiction in terms. For this reason the Academy desires an explanation why it is that so many correct theorems have been deduced from a contradictory assumption, together with a formulation of a truly clear mathematical principle that may replace that of infinity without, however, rendering investigations by its use overly difficult and overly lengthy. It is requested that the subject be treated in all possible generality and with all possible rigor, clarity and simplicity."

Twenty-three answers were received before the deadline of January 1, 1786. The prize was awarded to the Swiss mathematician Simon L'Huilier for his essay with motto:

> "The infinite is the abyss in which our thoughts are engulfed."

The members of the "Prize Committee" made the following noteworthy points: None of the submitted essays dealt with the question raised "why so many correct theorems have been derived from a contradictory assumption?" Furthermore, the request for clarity, simplicity and, above all, rigor was not met by the contenders, and almost all of them did not address the request for a newly formulated principle of infinity that would reach beyond the infinitesimal calculus to be meaningful also for algebra and geometry.

For a detailed account of the prize contest we refer the reader to the interesting biography of Lazare Nicolas M. Carnot (1753–1823), the father of the thermodynamicist Sadi Carnot, entitled "Lazare Carnot Savant" by Ch. C. Gillespie (Princeton Univ. Press, 1971), which contains a thorough discussion of Carnot's entry "Dissertation sur la théorie de l'infini mathématique", received by the Academy after the deadline. The above text of the query was adapted from the biography.

In retrospect, the outcome of the contest is not surprising. Nevertheless around that time the understanding of infinitesimals had reached a more sophisticated level as the books of J. L. Lagrange and L. N. Carnot published in Paris in 1797 show.

From our present state of the art, it seems that the natural place to look for a "general principle of infinity" is set theory. Not however for an "intrinsic" definition of infinity. Indeed, as Gian-Carlo Rota expressed not too long ago:

> "God created infinity and man, unable to understand it, had to invent finite sets."

At this point let me digress a little for further clarification about the infinity we are dealing with. During the early development of Cantor's creation of set theory, it was E. Zermelo who realized that the attempts to prove the existence of "infinite" sets, short of assuming there is an "infinite" set or a non-finite set as in Proposition 66 of Dedekind's famous "Was sind und was sollen die Zahlen?" were fallacious. For this reason, Zermelo in his important paper "Sur les ensembles finis et le principe de l'induction complete", Acta Math. 32 (1909), 185–193 (submitted in 1907), introduced an axiom of "infinity" by postulating the existence of a set, say A, non-empty, and that for each of its elements x, the singleton $\{x\}$ is an element of it.

Returning to the request of the Academy: To discover a property that all infinite sets would have in common with the finite sets that would facilitate their use in all branches of mathematics.What comes to mind is Zermelo's well-ordering principle. Needless to say that this principle and the manifold results and consequences in all branches of mathematics have had an enormous impact on the development of mathematics since its introduction. One may ask what has this to do with the topic at hand? It so happens that the existence of non-standard models depends essentially on it as well and consequently non-standard analysis too.

The construction of the real number system (linear continuum) by Cantor and Dedekind in 1872 and the Weierstrass ε-δ technique gradually replaced the use of infinitesimals. Hilbert's characterization in 1899 of the real number system as a (Dedekind) complete field led to the discovery, in 1907, by H. Hahn, of non-archimedian totally ordered field extensions of the reals. This development brought about a renewed interest in the theory of infinitesimals. The resulting "calculus", certainly of interest by itself, lacked a process of defining extensions of the elementary and special functions, etc., of the objects of classical analysis. It is interesting that Cantor strongly rejected the existence of non-archimedian totally ordered fields. He expressed the view that no actual infinities could exist other than his transfinite cardinal numbers and that, other than 0, infinitesimals did not exist. He also offered a "proof" in which he actually assumed order completeness.

It took one hundred and seventy-five years from the time of the deadline of the Berlin Academy contest to the publication of Robinson's paper "Non-standard Analysis". As Robinson told us, his discovery did not come about as a result of his efforts to solve Leibniz' problem; far from it. Working on a paper on formal languages where the length of the sentences could be countable, it occurred to him to look up again the important paper by T. Skolem "Über die Nichtcharakterisierbarkeit der Zahlenreihe mittels endlich oder abzählbar unendlich vieler Aussagen mit ausschliesslich Zahlenvariablen, Fund. Math. 23 (1934), 150–161*.

Briefly, Skolem showed in his paper the existence of models of Peano arithmetic having "infinitely large numbers". Nevertheless in his models the principle of induction holds only for subsets determined by admissible formulas from the chosen formal language used to describe Peano's axiom system. The non-empty set of the infinitely large numbers has no smallest element and so cannot be determined by a formula of the formal language and is called an external set; those that can were baptized as internal sets of the model.

Robinson, rereading Skolem's paper, wondered what systems of numbers would emerge if he would apply Skolem's method to the axiom system of the real numbers. In doing so, Robinson immediately realized that the real number

*See also: T. Skolem "Peano's Axioms and Models of Arithmetic", in Symposium on the Mathematical Interpretation of Formal Systems, North-Holland, Amsterdam 1955, 1–14.

system was a non-archimedian totally ordered field extension of the reals whose structure satisfies all the properties of the reals, and that, in particular, the set of infinitesimals lacking a least upper bound was an external set.

This is how it all started and the Academy would certainly award Robinson the gold medal.

At the end of the fifties at Caltech (California Institute of Technology) Arthur Erdelyi FRS (1908–1977) conducted a lively seminar entitled "Generalized Functions". It dealt with various areas of current research at that time in such fields as J. Mikusinski's rigorous foundation of the so-called Heaviside operational calculus and L. Schwartz' theory of distributions. In connection with Schwartz' distribution theory, Erdelyi urged us to read the just appeared papers by Laugwitz and Schmieden dealing with the representations of the Dirac-delta functions by sequences of point-functions converging to 0 pointwise except at 0 where they run to infinity. Robinson's paper fully clarified this phenomenon. Reduced powers of $\mathbb{R}$ instead of ultrapowers, as in Robinson's paper, were at play here. In my 1962 Notes on Non-standard Analysis the ultrapower construction was used, but at that time without using explicitly the Transfer Principle.

In 1967 the first International Symposium on Non-standard Analysis took place at Caltech with the support of the U.S. Office of Naval Research. At the time the use of non-standard models in other branches of mathematics started to blossom. This is the reason that the Proceedings of the Symposium carries the title: Applications of Model Theory to Algebra, Analysis and Probability.

A little anecdote about the meeting. When I opened the newspaper one morning during the week of the meeting, I discovered to my surprise that it had attracted the attention of the Managing Editor of the Pasadena Star News; his daily "Conversation Piece" read:

> "A Stanford Professor spoke in Pasadena this week on the subject 'Axiomatizations of Non-standard Analysis which are Conservative Extensions of Formal Systems for Standard Classical Analysis', a fact which I shall tuck away for reassurance on those days when I despair of communicating clearly."

I may add here that from the beginning Robinson was very interested in the formulation of an axiom system catching his non-standard methodology.

Unfortunately he did not live to see the solution of his problem by E. Nelson presented in the 1977 paper entitled "Internal Set Theory". A presentation by Nelson, "The virtue of Simplicity", can be found in this book.

A final observation. During the last sixty years we have all seen come about the solutions of a number of outstanding problems and conjectures, some centuries old, that have enriched mathematics. The century-old problem to create a rigorous theory of infinitesimals no doubt belongs in this category.

It is somewhat surprising that the appreciation of Robinson's creation was slow in coming. Is it possible that the finding of the solution in model theory, a branch of mathematical logic, had something to do with that?

The answer may perhaps have been given by Augustus de Morgan (1806–1871), who is well-known from De Morgan's Law, and who in collaboration with George Boole (1805–1864) reestablished formal logic as a branch of exact science in the nineteenth century, when he wrote:

> "We know that mathematicians care no more for logic than logicians for mathematics. The two eyes of exact sciences are mathematics and logic: the mathematical sect puts out the logical eye, the logical sect puts out the mathematical eye; each believing that it can see better with one eye than with two."

We owe Abraham Robinson a great deal for having taught us the use of both eyes.

This book shows clearly that we have learned our lesson well.

All the contributors are to be commended for the way they have made an effort to make their contributions that are based on the talks at the meeting "Nonstandard Mathematics 2004" as self-contained as can be expected. For further facilitating the readers, the editors have divided the papers in categories according to the subject. The whole presents a very rich assortment of the nonstandard approach to diverse areas of mathematical analysis.

I wish it many readers.

Wilhelmus A. J. Luxemburg
Pasadena, California
September 2006

Acknowledgements

The wide range of applicability of Mathematical Logic to classical Mathematics, beyond Analysis — as Robinson's terminology *Non-standard Analysis* might imply — is already apparent in the very first symposium on the area, held in Pasadena in 1967, as mentioned in the foreword.

Important indicators of the maturity of this field are the high level of foundational and pure or applied mathematics presented in this book, congresses which take place approximately every two years, as well as the experiences of teaching, which proliferated at graduate, undergraduate and secondary school level all over the world.

This book, made of peer reviewed contributions, grew out of the meeting Non Standard Mathematics 2004, which took place in July 2004 at the Department of Mathematics of the University of Aveiro (Portugal). The articles are organized into five groups, (1) Foundations, (2) Number theory, (3) Statistics, probability and measures, (4) Dynamical systems and equations and (5) Infinitesimals and education. Its cohesion is enhanced by many cross-overs.

We thank the authors for the care they took with their contributions as well as the three Portuguese Research Institutes which funded the production cost: the Centro Internacional de Matemática CIM at Coimbra, the Centro de Estudos em Optimização e Controlo CEOC of the University of Aveiro and the Centro de Investigação em Matemática e Aplicações CIMA-UE of the University of Évora. Moreover, NSM2004 would not have taken place were it not for the support of the CIM, CEOC and CIMA themselves, Fundação Luso-Americana para o Desenvolvimento, the Centro de Análise Matemática, Geometria e Sistemas Dinâmicos of the Instituto Superior Técnico of Lisbon and the Mathematics Department at Aveiro.

Our special thanks go to Wilhelmus Luxemburg of Caltech, USA, who, being one of the first to recognize the importance of Robinson's discovery of

Nonstandard Analysis, honoured us by writing the foreword. Last but not least, Salvador Sánchez-Pedreño of the University of Murcia, Spain, joined his expert knowledge of Nonstandard Analysis to a high dominion of the LaTeX editing possibilities in order to improve the consistency of the text and to obtain an outstanding typographic result.

Imme van den Berg
Vítor Neves
Editors

Contents

Part I

Foundations

1

The strength of nonstandard analysis

H. Jerome Keisler*

Abstract

A weak theory nonstandard analysis, with types at all finite levels over both the integers and hyperintegers, is developed as a possible framework for reverse mathematics. In this weak theory, we investigate the strength of standard part principles and saturation principles which are often used in practice along with first order reasoning about the hyperintegers to obtain second order conclusions about the integers.

1.1 Introduction

In this paper we revisit the work in [5] and [6], where the strength of nonstandard analysis is studied. In those papers it was shown that weak fragments of set theory become stronger when one adds saturation principles commonly used in nonstandard analysis.

The purpose of this paper is to develop a framework for reverse mathematics in nonstandard analysis. We will introduce a base theory, "weak nonstandard analysis" (*WNA*), which is proof theoretically weak but has types at all finite levels over both the integers and the hyperintegers. In *WNA* we study the strength of two principles that are prominent in nonstandard analysis, the standard part principle in Section 1.6, and the saturation principle in Section 1.9. These principles are often used in practice along with first order reasoning about the hyperintegers to obtain second order conclusions about the integers, and for this reason they can lead to the discovery of new results.

The standard part principle (*STP*) says that a function on the integers exists if and only if it is coded by a hyperinteger. Our main results show that in *WNA*, *STP* implies the axiom of choice for quantifier-free formulas (Theorem 17), *STP*+saturation for quantifier-free formulas implies choice for arithmetical formulas (Theorem 23), and *STP*+saturation for formulas with

*University of Wisconsin, Madison.
`keisler@math.wisc.edu`

first order quantifiers implies choice for formulas with second order quantifiers (Theorem 25). The last result might be used to identify theorems that are proved using nonstandard analysis but cannot be proved by the methods commonly used in classical mathematics.

The natural models of *WNA* will have a superstructure over the standard integers $\mathbb{N}$, a superstructure over the hyperintegers $^*\mathbb{N}$, and an inclusion map $j : \mathbb{N} \to {}^*\mathbb{N}$. With the two superstructures, it makes sense to ask whether a higher order statement over the hyperintegers implies a higher order statement over the integers. As is commonly done in the standard literature on weak theories in higher types, we use functional superstructures with types of functions rather than sets. The base theory *WNA* is neutral between the internal set theory approach and the superstructure approach to nonstandard analysis, and the standard part and saturation principles considered here arise in both approaches. For background in model theory, see [2, Section 4.4].

The theory *WNA* is related to the weak nonstandard theory $NPRA^\omega$ of Avigad [1], and the base theory RCA_0^ω for higher order reverse mathematics proposed by Kohlenbach [7]. The paper [1] shows that the theory $NPRA^\omega$ is weak in the sense that it is conservative over primitive recursive arithmetic (*PRA*) for Π_2 sentences, but is still sufficient for the development of much of analysis. The theory *WNA* is also conservative over *PRA* for Π_2 sentences, but has more expressive power. In Sections 1.11 and 1.12 we will introduce a stronger, second order Standard Part Principle, and give some relationships between this principle and the theories $NPRA^\omega$ and RCA_0^ω.

1.2 The theory PRA^ω

Our starting point is the theory *PRA* of primitive recursive arithmetic, introduced by Skolem. It is a first order theory which has function symbols for each primitive recursive function (in finitely many variables), and the equality relation $=$. The axioms are the rules defining each primitive recursive function, and induction for quantifier-free formulas. This theory is much weaker than Peano arithmetic, which has induction for all first order formulas.

An extension of *PRA* with all finite types was introduced by Gödel [4], and several variations of this extension have been studied in the literature. Here we use the finite type theory PRA^ω as defined in Avigad [1].

There is a rich literature on constructive theories in intuitionistic logic that are very similar to PRA^ω, such as the finite type theory HA^ω over Heyting arithmetic (see, for example, [9]). However, in this paper we work exclusively in classical logic.

We first introduce a formal object N and define a collection of formal objects called **types over** N.

(1) The **base type** over N is N.

(2) If σ, τ are types over N, then $\sigma \to \tau$ is a type over N.

We now build the formal language $L(PRA^\omega)$. $L(PRA^\omega)$ is a many-sorted first order language with countably many variables of each type σ over N, and the equality symbol $=$ at the base type N only. It has the usual rules of many-sorted logic, including the rule $\exists f \forall u\, f(u) = t(u, \ldots)$ where u, f are variables of type $\sigma, \sigma \to N$ and $t(u, \ldots)$ is a term of type N in which f does not occur.

We first describe the symbols and then the corresponding axioms.

$L(PRA^\omega)$ has the following function symbols:

- A function symbol for each primitive recursive function.
- The primitive recursion operator which builds a term $R(m, f, n)$ of type N from terms of type $N, N \to N$, and N.
- The definition by cases operator which builds a term $c(n, u, v)$ of type σ from terms of type N, σ, and σ.
- The λ operator which builds a term $\lambda v.t$ of type $\sigma \to \tau$ from a variable v of type σ and a term t of type τ.
- The application operator which builds a term $t(s)$ of type τ from terms s of type σ and t of type $\sigma \to \tau$.

Given terms r, t and a variable v of the appropriate types, $r(t/v)$ denotes the result of substituting t for v in r. Given two terms s, t of type σ, $s \equiv t$ will denote the infinite scheme of formulas $r(s/v) = r(t/v)$ where v is a variable of type σ and $r(v)$ is an arbitrary term of type N. $\equiv$ is a substitute for the missing equality relations at higher types.

The axioms for PRA^ω are as follows.

- Each axiom of PRA.
- The induction scheme for quantifier-free formulas of $L(PRA^\omega)$.
- Primitive recursion: $R(m, f, 0) = m$, $R(m, f, s(n)) = f(n, R(m, f, n))$.
- Cases: $c(0, u, v) \equiv u$, $c(s(m), u, v) \equiv v$.
- Lambda abstraction: $(\lambda u.t)(s) \equiv t(s/u)$.

The order relations $<, \leq$ on type N can be defined in the usual way by quantifier-free formulas.

In [1] additional types $\sigma \times \tau$, and term-building operations for pairing and projections with corresponding axioms were also included in the language, but as explained in [1], these symbols are redundant and are often omitted in the literature.

On the other hand, in [1] the symbols for primitive recursive functions are not included in the language. These symbols are redundant because they can be defined from the primitive recursive operator R, but they are included here for convenience.

We state a conservative extension result from [1], which shows that PRA^ω is very weak.

Proposition 1 *PRA^ω is a conservative extension of PRA, that is, PRA^ω and PRA have the same consequences in $L(PRA)$.*

The natural model of PRA^ω is the full functional superstructure $V(\mathbb{N})$, which is defined as follows. $\mathbb{N}$ is the set of natural numbers. Define $V_N(\mathbb{N}) = \mathbb{N}$, and inductively define $V_{\sigma\to\tau}(\mathbb{N})$ to be the set of all mappings from $V_\sigma(\mathbb{N})$ into $V_\tau(\mathbb{N})$. Finally, $V(\mathbb{N}) = \bigcup_\sigma V_\sigma(\mathbb{N})$. The superstructure $V(\mathbb{N})$ becomes a model of PRA^ω when each of the symbols of $L(PRA^\omega)$ is interpreted in the obvious way indicated by the axioms. In fact, $V(\mathbb{N})$ is a model of much stronger theories than PRA^ω, since it satisfies full induction and higher order choice and comprehension principles.

1.3 The theory $NPRA^\omega$

In [1], Avigad introduced a weak nonstandard counterpart of PRA^ω, called $NPRA^\omega$. $NPRA^\omega$ adds to PRA^ω a new predicate symbol $S(\cdot)$ for the standard integers (and S-relativized quantifiers $\forall^S, \exists^S$), and a constant H for an infinite integer, axioms saying that $S(\cdot)$ is an initial segment not containing H and is closed under each primitive recursive function, and a transfer axiom scheme for universal formulas. In the following sections we will use a weakening of $NPRA^\omega$ as a part of our base theory.

In order to make $NPRA^\omega$ fit better with the present paper, we will build the formal language $L(NPRA^\omega)$ with types over a new formal object $^*N$ instead of over N. The base type over $^*N$ is $^*N$, and if σ, τ are types over $^*N$ then $\sigma \to \tau$ is a type over $^*N$.

For each type σ over N, let $^*\sigma$ be the type over $^*N$ built in the same way. For each function symbol u in $L(PRA^\omega)$ from types $\vec{\sigma}$ to type τ, $L(NPRA^\omega)$ has a corresponding function symbol $^*u$ from types $^*\vec{\sigma}$ to type $^*\tau$. $L(NPRA^\omega)$ also has the equality relation $=$ for the base type $^*N$, and the extra constant symbol H and the standardness predicate symbol S of type $^*N$.

We will use the following conventions throughout this paper. When we write a formula $A(\vec{v})$, it is understood that $\vec{v}$ is a tuple of variables that contains all the free variables of A. If we want to allow additional free variables we write $A(\vec{v}, \ldots)$. We will always let:

- $m, n, \ldots$ be variables of type N,
- $x, y, \ldots$ be variables of type ${}^*N$,
- $f, g, \ldots$ be variables of type $N \to N$.

To describe the axioms of $NPRA^\omega$ we introduce the star of a formula of $L(PRA^\omega)$. Given a formula A of $L(PRA^\omega)$, a **star of** A is a formula ${}^*A$ of $L(NPRA^\omega)$ which is obtained from A by replacing each variable of type σ in A by a variable of type ${}^*\sigma$ in a one to one fashion, and replacing each function symbol in A by its star. The order relations on ${}^*N$ will be written $<$, $\leq$ without stars.

The axioms of $NPRA^\omega$ are as follows:

- The star of each axiom of PRA^ω.
- S is an initial segment: $\neg S(H) \wedge \forall x \forall y\, [S(x) \wedge y \leq x \to S(y)]$.
- S is closed under primitive recursion.
- Transfer: $\forall^S \vec{x}\ {}^*A(\vec{x}) \to \forall \vec{x}\ {}^*A(\vec{x})$, $A(\vec{m})$ quantifier-free in $L(PRA^\omega)$.

It is shown in [1] that if $A(m, n)$ is quantifier-free in $L(PRA)$ and $NPRA^\omega$ proves $\forall^S x \exists y\, {}^*A(x, y)$, then PRA proves $\forall m \exists n\, A(m, n)$. It follows that $NPRA^\omega$ is conservative over PRA for Π_2 sentences.

The natural models of $NPRA^\omega$ are the internal structures ${}^*V(\mathbb{N})$, which are proper elementary extensions of $V(\mathbb{N})$ in the many-sorted sense, with additional symbols S for $\mathbb{N}$ and H for an element of ${}^*\mathbb{N} \setminus \mathbb{N}$.

1.4 The theory *WNA*

We now introduce our base theory *WNA*, weak nonstandard analysis. The idea is to combine the theory PRA^ω with types over N with a weakening of the theory $NPRA^\omega$ with types over ${}^*N$, and form a link between the two by identifying the standardness predicate S of $NPRA^\omega$ with the lowest type N of PRA^ω. In this setting, it will make sense to ask whether a formula with types over ${}^*N$ implies a formula with types over N.

The language $L(WNA)$ of *WNA* has both types over N and types over ${}^*N$. It has all of the symbols of $L(PRA^\omega)$, all the symbols of $L(NPRA^\omega)$ except the primitive recursion operator ${}^*R$, and has one more function symbol j which goes from type N to type ${}^*N$.

We make the axioms of *WNA* as weak as we can so as to serve as a blank screen for viewing the relative strengths of additional statements which arise in nonstandard analysis.

The axioms of *WNA* are as follows:

- The axioms of PRA^{ω}.
- The star of each axiom of *PRA*.
- The stars of the Cases and Lambda abstraction axioms of PRA^{ω}.
- S is an initial segment: $\neg S(H) \wedge \forall x \forall y\, [S(x) \wedge y \leq x \rightarrow S(y)]$.
- S is closed under primitive recursion.
- j maps S onto $\mathbb{N}$: $\forall x\, [S(x) \leftrightarrow \exists m\, x = j(m)]$.
- Lifting: $j(\alpha(\vec{m})) = {}^*\alpha(j(\vec{m}))$ for each primitive recursive function α.

The star of a quantifier-free formula of $L(PRA)$, possibly with some variables replaced by H, will be called an **internal quantifier-free formula**. The stars of the axioms of *PRA* include the star of the defining rule for each primitive recursive function, and the induction scheme for internal quantifier-free formulas (which we will call **internal induction**).

The axioms of $NPRA^{\omega}$ that are left out of *WNA* are the star of the Primitive Recursion scheme, the star of the quantifier-free induction scheme of PRA^{ω}, and Transfer. These axioms are statements about the hyperintegers which involve terms of higher type.

Note that *WNA* is noncommittal on whether the characteristic function of S exists in type ${}^*N \rightarrow {}^*N$, while the quantifier-free induction scheme of $NPRA^{\omega}$ precludes this possibility.

In practice, nonstandard analysis uses very strong transfer axioms, and extends the mapping j to higher types. Strong axioms of this type will not be considered here.

Theorem 2 *$WNA + NPRA^{\omega}$ is a conservative extension of $NPRA^{\omega}$, that is, $NPRA^{\omega}$ and $WNA + NPRA^{\omega}$ have the same consequences in $L(NPRA^{\omega})$.*

Proof. Let M be a model of $NPRA^{\omega}$, and let M^S be the restriction of M to the standardness predicate S. Then M^S is a model of *PRA*. By Proposition 1, the complete theory of M^S is consistent with PRA^{ω}. Therefore PRA^{ω} has a model K whose restriction K^N to type N is elementarily equivalent to M^S. By the compactness theorem for many-sorted logic, there is a model M_1 elementarily equivalent to M and a model K_1 elementarily equivalent to K with an isomorphism $j : M_1^S \cong K_1^N$, such that $\langle K_1, M_1, j\rangle$ is a model of $WNA + NPRA^\omega$. Thus every complete extension of $NPRA^\omega$ is consistent with $WNA + NPRA^\omega$, and the theorem follows. □

Corollary 3 *WNA is a conservative extension of PRA for Π_2 formulas. That is, if $A(m,n)$ is quantifier-free in $L(PRA)$ and $WNA \vdash \forall m \exists n\, A(m,n)$, then $PRA \vdash \forall m \exists n\, A(m,n)$.*

Proof. Suppose $WNA \vdash \forall m \exists n\, A(m,n)$. By the Lifting Axiom, $WNA \vdash \forall^S x \exists^S y\, {}^*A(x,y)$. By Theorem 2, $NPRA^\omega \vdash \forall^S x \exists^S y\, {}^*A(x,y)$. Then $PRA \vdash \forall m \exists n\, A(m,n)$ by Corollary 2.3 in [1]. □

Each model of *WNA* has a $V(\mathbb{N})$ part formed by restricting to the objects with types over N, and a $V({}^*\mathbb{N})$ part formed by restricting to the objects with types over ${}^*N$. Intuitively, the $V(\mathbb{N})$ and $V({}^*\mathbb{N})$ parts of *WNA* are completely independent of each other, except for the inclusion map j at the zeroth level. The standard part principles introduced later in this paper will provide links between types $N \to N$ and $(N \to N) \to N$ in the $V(\mathbb{N})$ part and types ${}^*N$ and ${}^*N \to {}^*N$ in the $V({}^*\mathbb{N})$ part.

WNA has two natural models, the "internal model" $\langle V(\mathbb{N}), {}^*V(\mathbb{N}), j\rangle$ which contains the natural model ${}^*V(\mathbb{N})$ of $NPRA^\omega$, and the "full model" $\langle V(\mathbb{N}), V({}^*\mathbb{N}), j\rangle$ which contains the full superstructure $V({}^*\mathbb{N})$ over ${}^*\mathbb{N}$. In both models, j is the inclusion map from $\mathbb{N}$ into ${}^*\mathbb{N}$. The full natural model $\langle V(\mathbb{N}), V({}^*\mathbb{N}), j\rangle$ of *WNA* does not satisfy the axioms $NPRA^\omega$. In particular, the star of quantifier-free induction fails in this model, because the characteristic function of S exists as an object of type ${}^*N \to {}^*N$.

1.5 Bounded minima and overspill

In this section we prove some useful consequences of the *WNA* axioms.

Given a formula $A(x, \ldots)$ of $L(WNA)$, the **bounded minimum** operator is defined by

$$u = (\mu x < y)\, A(x, \ldots) \leftrightarrow \big[u \le y \wedge (\forall x < u) \neg A(x, \ldots) \wedge [A(u, \ldots) \vee u = y]\big],$$

where u is a new variable. By this we mean that the expression to the left of the $\leftrightarrow$ symbol is an abbreviation for the formula to the right of the $\leftrightarrow$ symbol. In particular, if z does not occur in A, $(\mu z < 1)\, A(\ldots)$ is the (inverted) characteristic function of A, which has the value 0 when A is true and the value 1 when A is false.

In *PRA*, the bounded minimum operator is defined similarly.

Lemma 4 *Let $A(m,\vec{n})$ be a quantifier-free formula of $L(PRA)$ and let $\alpha(p,\vec{n})$ be the primitive recursive function such that in PRA,*

$$\alpha(p,\vec{n}) = (\mu m < p)\, A(m,\vec{n}).$$

Then

(i) *WNA* $\vdash {}^*\alpha(y,\vec{z}) = (\mu x < y)\, {}^*A(x,\vec{z})$.

(ii) *In WNA, there is a quantifier-free formula $B(p,\ldots)$ such that*

$$(\forall m < p)\, A(m,\ldots) \leftrightarrow B(p,\ldots), \qquad (\forall x < y)\, {}^*A(x,\ldots) \leftrightarrow {}^*B(y,\ldots).$$

Similarly for $(\exists x < y)\, {}^*A(x,\ldots)$, and $u = (\mu x < y)\, {}^*A(x,\ldots)$.

Proof. (i) By the axioms of *WNA*, the defining rule for ${}^*\alpha$ is the star of the defining rule for α.

(ii) Apply (i) and observe that in *WNA*,

$$(\forall x < y)\, {}^*A(x,\ldots) \leftrightarrow y = (\mu x < y)\, \neg {}^*A(x,\ldots). \qquad \square$$

Let us write $\forall^\infty x\, A(x,\ldots)$ for $\forall x\, [\neg S(x) \rightarrow A(x,\ldots)]$ and $\exists^\infty x\, A(x,\ldots)$ for $\exists x\, [\neg S(x) \wedge A(x,\ldots)]$.

Lemma 5 *(Overspill) Let $A(x,\ldots)$ be an internal quantifier-free formula. In WNA,*

$$\forall^S x\, A(x,\ldots) \rightarrow \exists^\infty x\, A(x,\ldots) \quad \textit{and} \quad \forall^\infty x\, A(x,\ldots) \rightarrow \exists^S x\, A(x,\ldots).$$

Proof. Work in *WNA*. If $A(H,\ldots)$ we may take $x = H$. Assume $\forall^S x A(x,\ldots)$ and $\neg A(H,\ldots)$. By Lemma 4 (ii) we may take $u = (\mu x < H)\, \neg A(x,\ldots)$. Then $\neg S(u)$. Let $x = u - 1$. We have $x < u$, so $A(x,\ldots)$. Since S is closed under the successor function, $\neg S(x)$. $\square$

We now give a consequence of *WNA* in the language of *PRA* which is similar to Proposition 4.3 in [1] for $NPRA^\omega$. Σ_1**-collection** in $L(PRA)$ is the scheme

$$(\forall m < p)\exists n\, B(m,n,\vec{r}) \rightarrow \exists k (\forall m < p)(\exists n < k) B(m,n,\vec{r})$$

where B is a formula of $L(PRA)$ of the form $\exists q\, C$, C quantifier-free.

Proposition 6 *Σ_1-collection in $L(PRA)$ is provable in WNA.*

Proof. We work in *WNA*. By pairing existential quantifiers, we may assume that $B(m, n, \vec{r})$ is quantifier-free. Assume $(\forall m < p)\exists n\, B(m, n, \vec{r})$. Let $^*B$ be the formula obtained by starring each function symbol in B and replacing variables of type N by variables of type $^*N$.

By the Lifting Axiom and the axiom that S is an initial segment,

$$(\forall x < p)\exists^S y\, ^*B(x, y, j(\vec{r})).$$

Then

$$\forall^\infty w(\forall x < p)(\exists y < w)\, ^*B(x, y, j(\vec{r})).$$

By Lemma 4 and Overspill,

$$\exists^S w(\forall x < p)(\exists y < w)\, ^*B(x, y, j(\vec{r})).$$

By the Lifting Axiom again,

$$\exists k(\forall m < p)(\exists n < k)\, B(m, n, \vec{r}).$$ □

1.6 Standard parts

This section introduces a standard part notion which formalizes a construction commonly used in nonstandard analysis, and provides a link between the type $N \to N$ and the type $^*N$.

In type N let $(n)_k$ be the power of the k-th prime in n, and in type $^*N$ let $(x)_y$ be the power of the y-th prime in x. The function $(n, k) \mapsto (n)_k$ is primitive recursive, and its star is the function $(x, y) \mapsto (x)_y$.

Hereafter, when it is clear from the context, we will write t instead of $j(t)$ in formulas of $L(WNA)$.

Intuitively, we identify $j(t)$ with t, but officially, they are different because t has type N while $j(t)$ has type $^*N$. This will make formulas easier to read. When a term t of type N appears in a place of type $^*N$, it really is $j(t)$.

In the theory *WNA*, we say that x is **near-standard**, in symbols $ns(x)$, if $\forall^S z\, S((x)_z)$. Note that this is equivalent to $\forall n S((x)_n)$. We employ the usual convention for relativized quantifiers, so that $\forall^{ns} x\, B$ means $\forall x\, [ns(x) \to B]$ and $\exists^{ns} x\, B$ means $\exists x\, [ns(x) \wedge B]$. We write

$$x \approx y \text{ if } ns(x) \wedge \forall^S z\, (x)_z = (y)_z.$$

This is equivalent to $ns(x) \wedge \forall n\, (x)_n = (y)_n$. We write $f = {}^o x$, and say f is the **standard part of** x and x is a **lifting of** f, if

$$ns(x) \wedge \forall n f(n) = (x)_n.$$

Note that the operation $x \mapsto {}^o x$ goes from type ${}^*N$ to type $N \to N$. In nonstandard analysis, this often allows one to obtain results about functions of type $N \to N$ by reasoning about hyperintegers of type ${}^*N$.

Lemma 7 *In WNA, suppose that x is near-standard. Then*

(i) If $x \approx y$ then $ns(y)$ and $y \approx x$.

(ii) $(\exists y < H)\, x \approx y$.

Proof. (i) Suppose $x \approx y$. If $S(z)$ then $S((x)_z)$ and $(y)_z = (x)_z$, so $S((y)_z)$. Therefore $ns(y)$, and $y \approx x$ follows trivially.

(ii) Let β be the primitive recursive function $\beta(m,n) = \Pi_{i<m}\, p_i^{(n)_i}$. By Lifting and defining rules for β and ${}^*\beta$, $\forall x \forall u \forall z\, [z < u \to (x)_z = ({}^*\beta(u,x))_z]$. Therefore $\forall^\infty u\, \forall^S z\, (x)_z = ({}^*\beta(u,x))_z$, and hence $\forall^\infty u\, x \approx {}^*\beta(u,x)$. We have $\forall^S w\, w^w < H$, and by Overspill, there exists w with $\neg S(w) \wedge w^w < H$. Since x is near-standard, $\forall^S u\, [u \leq w \wedge (\forall z < u)\, p_z^{(x)_z} < w]$. By Overspill,

$$\exists^\infty u\, [u \leq w \wedge (\forall z < u)\, p_z^{(x)_z} < w].$$

Let $y = {}^*\beta(u,x)$. Then $x \approx y$. By internal induction,

$$\forall u\, [(\forall z < u)\, p_z^{(x)_z} < w \to {}^*\beta(u,x) < w^u].$$

Then $y \leq w^u \leq w^w < H$. □

We now state the Standard Part Principle, which says that every near-standard x has a standard part and every f has a lifting.

Standard Part Principle (*STP*):

$$\forall^{ns} x \exists f\, f = {}^o x \wedge \forall f \exists x\, f = {}^o x.$$

The following corollary is an easy consequence of Lemma 7.

Corollary 8 *In WNA, STP is equivalent to*

$$(\forall^{ns} x < H) \exists f\, f = {}^o x \wedge \forall f (\exists x < H)\, f = {}^o x.$$

The **Weak Koenig Lemma** is the statement that every infinite binary tree has an infinite branch. The work in reverse mathematics shows that many classical mathematical statements are equivalent to the Weak Koenig Lemma.

Theorem 9 *The Weak Koenig Lemma is provable in WNA + STP.*

Proof. Work in $WNA + STP$. Let $B(n)$ be the formula

$$(\forall m < n)\left[(n)_m < 3 \wedge (\forall k < m)\left[(n)_k = 0 \rightarrow (n)_m = 0\right]\right].$$

$B(n)$ says that n codes a finite sequence of 1's and 2's. Write $m \lhd n$ if

$$B(m) \wedge B(n) \wedge m < n \wedge (\forall k < m)\left[(m)_k > 0 \rightarrow (m)_k = (n)_k\right].$$

This says the sequence coded by m is an initial segment of the sequence coded by n. Suppose that $\{n : f(n) = 0\}$ codes an infinite binary tree T, that is,

$$\forall m \exists n\left[m < n \wedge f(n) = 0\right] \wedge \forall n\left[f(n) = 0 \rightarrow B(n) \wedge \forall m[m \lhd n \rightarrow f(m) = 0]\right].$$

The formulas $B(n)$ and $m \lhd n$ are PRA-equivalent to quantifier-free formulas, which have stars $^*B(y)$ and $z \,^*\!\lhd\, y$. By STP, f has a lifting x. By Lemma 4 and Overspill,

$$\exists^\infty y\left[^*B(y) \wedge (\forall z < y)\left[z \,^*\!\lhd\, y \rightarrow (x)_z = 0\right]\right].$$

Then $ns(y)$, and by the STP there exists $g = {}^o y$. It follows that g codes an infinite branch of T. □

The next proposition gives a necessary and sufficient condition for STP in $WNA + NPRA^\omega$. Let ϕ be a variable of type $^*N \rightarrow {}^*N$, and write $f \subset \phi$ for $\forall n f(n) = \phi(n)$.

Proposition 10 *In $WNA + NPRA^\omega$, STP is equivalent to*

$$\forall f \exists \phi f \subset \phi \wedge \forall \phi \exists f\left[\forall^S x\, S(\phi(x)) \rightarrow f \subset \phi\right].$$

Proof. Work in $WNA + NPRA^\omega$. Call the displayed sentence STP'.

Assume STP. Take any f. By STP, f has a lifting u. Since $(u, y) \mapsto (u)_y$ is primitive recursive, $\exists \phi \forall y \phi(y) = (u)_y$. Then $\forall n\, f(n) = (u)_n = \phi(n)$, so $f \subset \phi$.

Now take any ϕ and assume that $\forall^S x\, S(\phi(x))$. Using the star of the primitive recursion scheme in $NPRA^\omega$, there exists ψ such that $\forall x (\forall y < x)\, \phi(y) = (\psi(x))_y$. Let $u = \psi(H)$. We then have $(\forall y < H)\, \phi(y) = (u)_y$, so $\forall^S y\, \phi(y) = (u)_y$. It follows that u is near-standard, and by STP there exists f with $f = {}^o u$ and hence $f \subset \phi$.

Now assume STP'. Take any f. By STP' there exist ϕ with $f \subset \phi$. As before there exists ψ such that $\forall x (\forall y < x)\, \phi(y) = (\psi(x))_y$. Let $u = \psi(H)$. Then $\forall n\, (u)_n = \phi(n) = f(n)$, so u is a lifting of f.

Now let u be near-standard. Since $(u, y) \mapsto (u)_y$ is primitive recursive, $\exists \phi \forall y\, \phi(y) = (u)_y$. Then $\forall^S x\, S(\phi(x))$, so by STP' there exists f with $f \subset \phi$. Then $\forall n\, f(n) = (u)_n = \phi(n)$, so $f = {}^o u$. □

1.7 Liftings of formulas

In this section we will define some hierarchies of formulas with variables of type N and $N \to N$, and corresponding hierarchies of formulas with variables of type $^*N$. We will then define the lifting of a formula and show that liftings preserve the hierarchy levels and truth values of formulas.

In the following we restrict ourselves to formulas of $L(PRA^\omega)$ with variables of types N and $N \to N$. We now introduce a restricted class of terms, the basic terms, which behave well with respect to liftings.

By a **basic term over** N we mean a term of the form $\alpha(u_1, \dots, u_k)$ where α is a primitive recursive function of k variables and each u_i is either a variable n of type N or an expression of the form $f(n)$. These basic terms capture all primitive recursive functionals $\beta(\vec{m}, \vec{f})$ in the sense that there is a basic term $t(\vec{m}, \vec{f}, n)$ over N which gives the nth value in the computation of $\beta(\vec{m}, \vec{f})$ for each input $\vec{m}, \vec{f}, n$.

Let QF be the set of Boolean combinations of equations between basic terms over N.

If $A \in QF$, then $(\forall m < n)\, A$, $(\exists m < n)\, A$, and $u = (\mu x < y)\, A$ are PRA^ω-equivalent to formulas in QF.

The set $\Pi^1_0 = \Sigma^1_0$ of **arithmetical formulas** is the set of all formulas which are built from formulas in QF using first order quantifiers $\forall m$, $\exists m$ and propositional connectives.

For each natural number k, Π^1_{k+1} is the set of formulas of the form $\forall f A$ where $A \in \Sigma^1_k$, and Σ^1_{k+1} is the set of formulas of the form $\exists f A$ where $A \in \Pi^1_k$.

We observe that up to PRA^ω-equivalence, $\Pi^1_k \subseteq \Pi^1_{k+1} \cap \Sigma^1_{k+1}$, Π^1_k is closed under finite conjunction and disjunction, and that negations of sentences in Π^1_k belong to Σ^1_k (and vice versa).

In the following we restrict our attention to formulas with variables of type $^*N$. We build a hierarchy of formulas of this kind.

By a **basic term over** $^*N$ we mean a term of the form $^*\alpha(u_1, \dots, u_k)$ where α is a primitive recursive function of k variables and each u_i is either a variable of type $^*N$ or the constant symbol H. NQF is the set of finite Boolean combinations of equations $s = t$ and formulas $S(t)$ where s, t are basic terms over $^*N$. Note that the constant symbol H and the predicate symbol S are allowed in formulas of NQF, but the symbol j is not allowed.

The internal quantifier-free formulas are just the formulas $B \in NQF$ in which the symbol S does not occur.

Let $N\Pi^0_0 = N\Sigma^0_0$ be the set of formulas which are built from formulas in NQF using the relativized quantifiers $\forall^S$, $\exists^S$ and propositional connectives. Note that the relations $ns(x)$ and $x \approx y$ are definable by $N\Pi^0_0$ formulas.

For each natural number k, $N\Pi^0_{k+1}$ is the set of formulas of the form $\forall^{ns}x\, A$ where $A \in N\Sigma^0_k$. $N\Sigma^0_{k+1}$ is the set of formulas of the form $\exists^{ns}x\, A$ where $A \in N\Pi^0_k$.

Up to *WNA*-equivalence, $N\Pi^0_k \subseteq N\Pi^0_{k+1} \cap N\Sigma^0_{k+1}$, $N\Pi^0_k$ is closed under finite conjunction and disjunction, and negations of sentences in $N\Pi^0_k$ belong to $N\Sigma^0_k$ (and vice versa).

We now define the lifting mapping on formulas, which sends Π^1_k to $N\Pi^0_k$.

Definition 11 *Let $A(\vec{m}, \vec{f})$ be a formula in Π^1_k, where $\vec{m}, \vec{f}$ contain all the variables of A, both free and bound. The **lifting** $\overline{A}(\vec{z}, \vec{x})$ is defined as follows, where $\vec{z}$ and $\vec{x}$ are tuples of variables of type *N of the same length as $\vec{m}, \vec{f}$.*

- *Replace each primitive recursive function symbol in A by its star.*
- *Replace each m_i by z_i.*
- *Replace each $f_i(m_k)$ by $(x_i)_{z_k}$.*
- *Replace each quantifier $\forall m_i$ by $\forall^S z_i$, and similarly for $\exists$.*
- *Replace each quantifier $\forall f_i$ by $\forall^{ns} x_i$, and similarly for $\exists$.*

Lemma 12 *(Zeroth Order Lifting) For each formula $A(\vec{m}, \vec{f}) \in \Pi^1_0$, we have $\overline{A}(\vec{z}, \vec{x}) \in N\Pi^0_0$, and*

$$WNA \vdash {}^o\vec{x} = \vec{f} \to [A(\vec{m}, \vec{f}) \leftrightarrow \overline{A}(\vec{m}, \vec{x})].$$

Moreover, if $A \in QF$ then $\overline{A}(\vec{z}, \vec{x})$ is an internal quantifier-free formula.

Proof. It is clear from the definition that $\overline{A}(\vec{z}, \vec{x}) \in N\Pi^0_0$, and if $A \in QF$ then $\overline{A}(\vec{z}, \vec{x})$ is an internal quantifier-free formula. In the case that A is an equation between basic terms, the lemma follows from the Lifting Axiom. The general case is then proved by induction on the complexity of A, using the axiom that j maps $\mathbb{N}$ onto S. □

Lemma 13 *(First Order Lifting) For each formula $A(\vec{m}, \vec{f}) \in \Pi^1_k$, we have $\overline{A}(\vec{z}, \vec{x}) \in N\Pi^0_k$ and*

$$WNA + STP \vdash {}^o\vec{x} = \vec{f} \to [A(\vec{m}, \vec{f}) \leftrightarrow \overline{A}(\vec{m}, \vec{x})].$$

Proof. Zeroth Order Lifting gives the result for $k = 0$. The general case follows by induction on k, using STP. □

1.8 Choice principles in $L(PRA^\omega)$

In this section we state two choice principles in the language $L(PRA^\omega)$, and show that for quantifier-free formulas they are consequences of the Standard Part Principle. Given a function g of type $N \to N$, let $g^{(m)}$ be the function $g^{(m)}(n) = g(2^m 3^n)$.

In each principle, Γ is a class of formulas with variables of types N and $N \to N$, and $A(m, n, \ldots)$ denotes an arbitrary formula in Γ.

$(\Gamma, 0)$-**choice** $\forall m \exists n\, A(m, n, \ldots) \to \exists g \forall m\, A(m, g(m), \ldots)$.
$(\Gamma, 1)$-**choice** $\forall m \exists f\, A(m, f, \ldots) \to \exists g \forall m\, A(m, g^{(m)}, \ldots)$.

When Γ is the set of all quantifier-free formulas of PRA^ω, [7] calls these schemes $QF - AC^{0,0}$ and $QF - AC^{0,1}$ respectively. A related principle is

Γ-**comprehension** $\exists f \forall m\, f(m) = (\mu z < 1)\, A(m, \ldots)$.

Π^1_0-comprehension is called **Arithmetical Comprehension**. The following well-known fact is proved by pairing existential quantifiers.

Proposition 14 *In* PRA^ω:
$(\Pi^0_1, 0)$-*choice,* $(\Pi^1_0, 0)$-*choice, and Arithmetical Comprehension are equivalent;*
$(\Pi^0_k, 1)$-*choice is equivalent to* $(\Sigma^0_{k+1}, 1)$-*choice, and implies* $(\Sigma^0_{k+1}, 0)$-*choice;*
$(\Pi^1_k, 1)$-*choice is equivalent to* $(\Sigma^1_{k+1}, 1)$-*choice and implies* $(\Sigma^1_{k+1}, 0)$-*choice;*
$(\Sigma^1_{k+1}, 0)$-*choice implies* Π^1_k-*comprehension.*

In PRA^ω, one can define a subset of $\mathbb{N}$ to be a function f such that $\forall n\, f(n) \le 1$, and define $n \in f$ as $f(n) = 0$. With these definitions, $(\Pi^1_k, 1)$-choice implies Π^1_k-choice and Π^1_k-comprehension in the sense of second order number theory (see [8]).

Lemma 15 *For each internal quantifier-free formula* $A(x, y, \vec{z})$,

$$WNA \vdash \forall^S x \exists^S y\, A(x, y, \vec{z}) \to (\exists^{ns} y < H) \forall^S x\, A(x, (y)_x, \vec{z}).$$

Proof. Work in WNA. Assume that $\forall^S x \exists^S y\, A(x, y, \vec{z})$. By Lemma 4, there is a primitive recursive function α such that ${}^*\alpha(u, \vec{z}, w) = (\mu v < w)\, A(u, v, \vec{z})$. By internal induction there exists w such that $w^w < H \wedge \neg S(w)$. Then $\forall^S u\, S({}^*\alpha(u, \vec{z}, w))$ and

$$\forall^S x (\exists y < H)(\forall u < x)\, (y)_u = {}^*\alpha(u, \vec{z}, w).$$

By internal induction, there exists an x such that $\neg S(x)$ and a $y < H$ such that $(\forall u < x)\, (y)_u = {}^*\alpha(u, \vec{z}, w)$. It follows that y is near-standard, and by the definition of α, $(\forall u < x)\, A(u, (y)_u, \vec{z})$. Then

$$(\exists^{ns} y < H) \forall^S x\, A(x, (y)_x, \vec{z}). \qquad \square$$

Theorem 16 *$(QF, 0)$-choice is provable in WNA + STP.*

Proof. We work in *WNA* + *STP*. Let $A(m, n, \vec{r}, \vec{h}) \in QF$ and assume $\forall m \exists n$ $A(m, n, \vec{r}, \vec{h})$. Then $\overline{A}$ is an internal quantifier-free formula. By *STP*, $\vec{h}$ has a lifting $\vec{z}$. By First Order Lifting,

$$A(m, n, \vec{r}, \vec{h}) \leftrightarrow \overline{A}(m, n, \vec{r}, \vec{z}).$$

Then $\forall^S u \exists^S v\, \overline{A}(u, v, \vec{r}, \vec{z})$. By Lemma 15, there is a near-standard y such that $\forall^S u\, \overline{A}(u, (y)_u, \vec{r}, \vec{z})$. By *STP*, $\exists g\, g = {}^o y$. Then by First Order Lifting, $\forall m\, A(m, g(m), \vec{r}, \vec{h})$. □

Theorem 17 *$(QF, 1)$-choice is provable in WNA + STP.*

Proof. We use $(QF, 0)$-choice. Let $A(m, f, \vec{r}, \vec{h}) \in QF$. Assume for simplicity that the tuple $\vec{r}$ is a single variable r. Suppose that $\forall m \exists f\, A(m, f, r, \vec{h})$. By the definition of QF formulas, f occurs in A only in terms of the form $f(m)$ and $f(r)$. Then

$$A(m, f, r, \vec{h}) \leftrightarrow B(m, f(m), f(r), r, \vec{h})$$

where $B \in QF$. Hence $\forall m \exists k\, B(m, (k)_m, (k)_r, r, \vec{h})$. By $(QF, 0)$-choice,

$$\exists f \forall m\, B(m, (f(m))_m, (f(m))_r, r, \vec{h}).$$

Applying $(QF, 0)$-choice to the formula $\forall p\, \exists q\, q = (f((p)_0))_{(p)_1}$, we have

$$\exists g \forall p\, g(p) = (f((p)_0))_{(p)_1},$$

and since $(2^m 3^n)_0 = m$ and $(2^m 3^n)_1 = n$,

$$\exists g \forall m \forall n\, g^{(m)}(n) = g(2^m 3^n) = (f(m))_n.$$

Then $\forall m\, B(m, g^{(m)}(m), g^{(m)}(r), r, \vec{h})$, and $\forall m\, A(m, g^{(m)}, r, \vec{h})$. □

1.9 Saturation principles

We state two saturation principles which formalize methods commonly used in nonstandard analysis. In each principle, Γ is a class of formulas with variables of type $^*N$, and $A(x, y, \vec{u})$ denotes an arbitrary formula in the class Γ.

$(\Gamma, 0)$-**saturation** $\quad \forall^{ns} \vec{u}\, [\forall^S x \exists^S y\, A(x, y, \vec{u}) \rightarrow \exists y \forall^S x\, A(x, (y)_x, \vec{u})]$.
$(\Gamma, 1)$-**saturation** $\quad \forall^{ns} \vec{u}\, [\forall^S x \exists^{ns} y\, A(x, y, \vec{u}) \rightarrow \exists y \forall^S x\, A(x, (y)_x, \vec{u})]$.

Note that $(\Gamma, 1)$-saturation implies $(\Gamma, 0)$-saturation. $(N\Pi^0_k, 1)$-saturation is weaker than the $^*\Pi_k$-saturation principle in the paper [5]. $^*\Pi_k$-saturation is the same as $(N\Pi_k^0, 1)$-saturation except that the quantifiers $\forall^{ns}$, $\exists^{ns}$ are replaced by $\forall$, $\exists$.

In the rest of this section we prove some consequences of $(NQF, 0)$-saturation.

Proposition 18 *Let us write $w = st(v)$ for*

$$S(v) \to w = v \wedge \neg S(v) \to w = 0.$$

In WNA, $(NQF, 0)$-saturation implies that

$$\forall x \exists^{ns} y \forall^S z\, [(y)_z = st((x)_z)].$$

In WNA + STP, $(NQF, 0)$-saturation implies that

$$\forall x \exists f \forall m\, [f(m) = st((x)_m)].$$

Proof. Work in *WNA*. Note that $w = st(v)$ stands for a formula in NQF. Take any x. We have $\forall^S z \exists^S w\, w = st((x)_z)$. Then by $(NQF, 0)$-saturation, $\exists y \forall^S z\, (y)_z = st((x)_z)$, and it follows that $ns(y)$. The second assertion follows by taking $f = {}^o y$. □

Lemma 19
$WNA \vdash (\forall v > 1)(\exists w < v^{2vH})(\forall x < v)\, (w)_x = (\mu u < H)\, [(y)_{2^x 3^u} = 0]$.

Proof. Use internal induction on v. The result is clear for $v = 2$. Let $\alpha(x) = (\mu u < H)\, [(y)_{2^x 3^u} = 0]$. Assume the result holds for v, that is, $w < v^{2vH} \wedge (\forall x < v)\, (w)_x = \alpha(x)$. Let $z = w * p_v^{\alpha(v)}$. We have $p_v < v^2$ and $\alpha(v) < H$, so $z < v^{2vH} * v^{2H} \leq (v+1)^{2(v+1)H}$ and $(\forall x < v+1)\, (z)_x = \alpha(x)$. This proves the result for $v + 1$ and completes the induction. □

Lemma 20 *In WNA, $(NQF, 0)$-saturation implies that for every formula $A(x, \vec{u}) \in N\Pi_0^0$, $\forall^{ns} \vec{u}\, (\exists y < H) \forall^S x\, (y)_x = (\mu z < 1) A(x, \vec{u})$.*

Proof. Work in *WNA* and assume $(NQF, 0)$-saturation. Let Φ be the set of formulas $A(x, \vec{u})$ such that $\forall^{ns} \vec{u} (\exists y < H) \forall^S x\, (y)_x = (\mu z < 1) A(x, \vec{u})$. We prove that $N\Pi_0^0 \subseteq \Phi$ by induction on quantifier rank. Suppose first that $A \in NQF$. Let $C(x, w, \vec{u})$ be the formula $w = (\mu z < 1)\, A(x, \vec{u})$. Then C is a propositional combination of A, $w = 0$, and $w = 1$, so $C \in NQF$. We clearly have $\forall^{ns} \vec{u} \forall^S x\, \exists^S w\, C(x, w, \vec{u})$. By $(NQF, 0)$-saturation and Lemma 7 (ii),

$$\forall^{ns} \vec{u}\, (\exists y < H) \forall^S x\, C(x, (y)_x, \vec{u}),$$

so $A \in \Phi$.

It is clear that the set of formulas Φ is closed under propositional connectives. Suppose all formulas of $N\Pi_0^0$ of quantifier rank at most n belong to Φ, and $A(x,\vec{u}) = \exists^S w B(x,w,\vec{u})$ where $B(x,w,\vec{u}) \in N\Pi_0^0$ has quantifier rank at most n. There is a formula $D(v,\vec{u})$ with the same quantifier rank as B such that in *WNA*, $D(2^x 3^w,\vec{u}) \leftrightarrow B(x,w,\vec{u})$. Then $D \in \Phi$, so

$$\forall^{ns}\vec{u}\,(\exists t < H)\forall^S v\,(t)_v = (\mu z < 1)\,D(v,\vec{u}).$$

Then

$$\forall^{ns}\vec{u}\,(\exists t < H)\forall^S x \forall^S w\,(t)_{2^x 3^w} = (\mu z < 1)\,B(x,w,\vec{u}).$$

Assume that $ns(\vec{u})$ and take t as in the above formula. By Lemma 19 there exists s such that

$$(\forall x < H)\,(s)_x = (\mu w < H)\,[(t)_{2^x 3^w} = 0].$$

It is trivial that $\forall^S x \exists^S y\, y = (\mu z < 1)\,S((s)_x)$. By $(NQF,0)$-saturation and Lemma 7,

$$(\exists y < H)\forall^S x\,(y)_x = (\mu z < 1)\,S((s)_x).$$

Thus whenever $S(x)$,

$$(y)_x = 0 \text{ iff } S((s)_x) \text{ iff } \exists^S w\,[(t)_{2^x 3^w} = 0] \text{ iff } \exists^S w\, B(x,w,\vec{u}).$$

It follows that

$$\forall^{ns}\vec{u}\,(\exists y < H)\forall^S x\,(y)_x = (\mu z < 1)\exists^S w\, B(x,w,\vec{u}),$$

so $A \in \Phi$. □

Theorem 21 *In WNA, $(NQF,0)$-saturation implies $(N\Pi_0^0,0)$-saturation.*

Proof. We continue to work in *WNA* and assume $(NQF,0)$-saturation and $ns(\vec{u})$. Assume that $A(x,y,\vec{u}) \in N\Pi_0^0$ and $\forall^S x \exists^S y\, A(x,y,\vec{u})$. There is a formula $B(v,\vec{u}) \in N\Pi_0^0$ with $B(2^x 3^y,\vec{u})$ *WNA*-equivalent to $A(x,y,\vec{u})$. Applying Lemma 20 to B, we obtain w such that

$$\forall^S v\,(w)_v = (\mu z < 1)\,B(v,\vec{u}),$$

so

$$\forall^S x \forall^S y\,(w)_{2^x 3^y} = (\mu z < 1)\,A(x,y,\vec{u}).$$

By Lemma 19 there exists w' such that

$$(\forall x < H)\,(w')_x = (\mu y < H)\,[(w)_{2^x 3^y} = 0].$$

Then $\forall^S x\, A(x,(w')_x,\vec{u})$, and w is near-standard because $\forall^S x \exists^S y\, A(x,y,\vec{u})$. □

Theorem 22 *In WNA, $(NQF, 0)$-saturation implies that for every formula $A(\vec{u}) \in N\Pi^0_1$, there is a formula $B \in NQF$ such that*

$$\forall^{ns}\vec{u}\,[A(\vec{u}) \leftrightarrow \forall^{ns}x\exists^S y\, B(x,y,\vec{u})].$$

Proof. Work in *WNA* + $(NQF, 0)$-saturation. Suppose $A \in N\Pi^0_1$. Then there is a least k such that A is equivalent to a formula $\forall^{ns}x\, C$ where C is a prenex formula in $N\Pi^0_0$ of quantifier rank k. If C has the form $\forall^S y\, D$, the quantifier $\forall^S y$ can be absorbed into the quantifier $\forall^{ns}x$, contradicting the assumption that k is minimal. Suppose C has the form $\exists^S y \forall^S z\, D(x,y,z,\vec{u})$ and assume that $ns(\vec{u})$. Then $\neg C$ is equivalent to $\forall^S y\, \exists^S z\, \neg D(x,y,z,\vec{u})$. By $(N\Pi^0_0, 0)$-saturation, $\neg C$ is equivalent to $\exists^{ns} z \forall^S y\, \neg D(x,y,(z)_y,\vec{u})$. Then A is equivalent to $\forall^{ns}x\forall^{ns}z\exists^S y\, D(x,y,(z)_y,\vec{u})$. By combining the quantifiers $\forall^{ns}x\forall^{ns}z$, we contradict the assumption that k is minimal. Therefore C must have the form $\exists^S y\, B$ where $B \in NQF$, as required. □

1.10 Saturation and choice

In this section we prove results showing that in *WNA* + *STP*, saturation principles with quantifiers of type $^*N$ imply the corresponding choice principles with quantifiers of type $N \to N$.

Theorem 23 *In WNA+STP, $(NQF, 0)$-saturation implies Arithmetical Comprehension.*

Proof. Work in *WNA* + *STP* and assume $(NQF, 0)$-saturation. By Proposition 14, Arithmetical Comprehension is equivalent to $(\Pi^1_0, 0)$-choice. By Theorem 21, $(N\Pi^0_0, 0)$-saturation holds. Let $A(m,n,\vec{r},\vec{h})$ be an arithmetical formula such that $\forall m \exists n\, A(m,n,\vec{r},\vec{h})$. By *STP*, $\vec{h}$ has a lifting $\vec{u}$. By First Order Lifting, we have $\forall^S x \exists^S y\, \overline{A}(x,y,\vec{r},\vec{u})$, and $\overline{A} \in N\Pi^0_0$. By $(N\Pi^0_0, 0)$-saturation, there exists y such that $\forall^S x\, [S((y)_x) \wedge \overline{A}(x,(y)_x,\vec{r},\vec{u})]$. Then y is near-standard, and by *STP* there exists $g = {}^o y$.

By First Order Lifting again, $\forall m A(m, g(m), \vec{r}, \vec{h})$. □

We remark that the axioms of Peano Arithmetic are consequences of Arithmetical Comprehension, so $(NQF, 0)$-saturation implies Peano Arithmetic.

Theorem 24 *In WNA+STP, $(N\Pi^0_k, 0)$-saturation implies $(\Pi^1_k, 0)$-choice, and $(N\Sigma^0_k, 0)$-saturation implies $(\Sigma^1_k, 0)$-choice.*

Proof. Work in *WNA* + *STP*. For the Π^1_k case, assume $(N\Pi^0_k, 0)$-saturation. Let $A(m,n,\vec{r},\vec{h}) \in \Pi^1_k$ and suppose that $\forall m \exists n\, A(m,n,\vec{r},\vec{h})$. Now argue as in the proof of Theorem 23. The Σ^1_k case is similar. □

Theorem 25 *In WNA + STP, $(N\Pi_k^0, 1)$-saturation implies $(\Sigma_{k+1}^1, 1)$-choice.*

Proof. Work in $WNA + STP$ and assume $(N\Pi_k^0, 1)$-saturation. It suffices to prove $(\Pi_k^1, 1)$-choice.

Let $A(m, f, \vec{r}, \vec{h}) \in \Pi_k^1$ and suppose that $\forall m \exists f\, A(m, f, \vec{r}, \vec{h})$. By STP, $\vec{h}$ has a lifting $\vec{u}$. By First Order Lifting, $\forall^S x \exists^{ns} y\, \overline{A}(x, y, \vec{r}, \vec{u})$ and $\overline{A} \in N\Pi_k^0$. We may rewrite this as $\forall^S x \exists^{ns} y\, [ns(y) \wedge \overline{A}(x, y, \vec{r}, \vec{u})]$ and note that $ns(y) \wedge \overline{A} \in N\Pi_k^0$. By $(N\Pi_k^0, 1)$-saturation, there exists y such that

$$\forall^S x\, [ns((y)_x) \wedge \overline{A}(x, (y)_x, \vec{r}, \vec{u})].$$

Applying $(N\Pi_k^0, 0)$-saturation to the formula $\forall^S x \exists^S z\, z = ((y)_{(x)_0})_{(x)_1}$, we get a near-standard z such that $\forall^S x\, (z)_x = ((y)_{(x)_0})_{(x)_1}$. Then

$$\forall^S x \forall^S w\, (z)_{2^x 3^w} = ((y)_x)_w.$$

By STP, there exists $g = {}^o z$. Then for each m, n, $g^{(m)}(n) = g(2^m 3^n) = (z)_{2^m 3^n} = ((y)_m)_n$. Therefore $g^{(m)} = {}^o((y)_m)$ for each m. By First Order Lifting, we get the desired conclusion $\forall m\, A(m, g^{(m)}, \vec{r}, \vec{u})$. □

The literature in reverse mathematics shows that Π_1^1-comprehension is strong enough for almost all of classical mathematics (see [8]).

Let us work in $WNA + STP$ and aim for Π_1^1-comprehension. By Theorem 25, $(N\Pi_1^0, 1)$-saturation implies Π_1^1-comprehension. By Theorem 22, $(N\Pi_1^0, 1)$-saturation is equivalent to $(\Gamma, 1)$-saturation where Γ is the set of formulas of the form $\forall^{ns} v \exists^S w\, B$ with $B \in NQF$, so $(\Gamma, 1)$-saturation also implies Π_1^1-comprehension. By Theorem 25 at the next level, $(N\Pi_2^0, 1)$-saturation implies Π_2^1-comprehension, which is stronger than the methods used in most of classical mathematics.

1.11 Second order standard parts

In this section we introduce second order standard parts, which provide a link between the second level of $V(\mathbb{N})$ (type $(N \to N) \to N$), and the first level of $V({}^*\mathbb{N})$ (type ${}^*N \to {}^*N$). We will use $F, G, \ldots$ for variables of type $(N \to N) \to N$, and $\phi, \psi, \ldots$ for variables of type ${}^*N \to {}^*N$.

ϕ is called **near-standard**, in symbols $ns(\phi)$, if

$$\forall^{ns} x\, S(\phi(x)) \wedge \forall x \forall y\, [x \approx y \to \phi(x) = \phi(y)].$$

We write

$$\phi \approx \psi \text{ if } ns(\phi) \wedge \forall^{ns} x\, \phi(x) = \psi(x).$$

We write $G = {}^o\phi$, and say that G is the **standard part of** ϕ and that ϕ is a **lifting of** G, if

$$ns(\phi) \wedge \forall^{ns} x \forall f \, [{}^o x = f \rightarrow \phi(x) = G(f)].$$

Note that the operation $\phi \mapsto {}^o\phi$ goes from type ${}^*N \rightarrow {}^*N$ to type $(N \rightarrow N) \rightarrow N$. The following lemma is straightforward.

Lemma 26 *If $ns(\phi)$ and $\phi \approx \psi$ then $ns(\psi)$ and $\psi \approx \phi$.*

We now state the Second Order Standard Part Principle, which says that every near-standard ϕ has a standard part and every F has a lifting.

Second Order Standard Part Principle:

$$\forall^{ns} \phi \, \exists F \; F = {}^o\phi \; \wedge \forall F \exists \phi \; F = {}^o\phi.$$

By *WNA*+*STP*(2) we mean the theory *WNA* plus both the first and second order standard part principles.

We now take a brief look at the consequences of *STP*(2) in *WNA*+$NPRA^\omega$. Roughly speaking, in *WNA*+$NPRA^\omega$, the second order standard part principle imposes restrictions of the set of functionals which are reminiscent of constructive analysis. Besides the axioms of *WNA*, the only axiom of $NPRA^\omega$ that will be used in this section is the star of quantifier-free induction.

A functional G is **continuous** if it is continuous in the Baire topology, that is,

$$\forall f \exists n \forall h \left[[(\forall m < n) \, h(m) = f(m)] \rightarrow G(h) = G(f) \right].$$

Proposition 27 *WNA* + $NPRA^\omega$ + *STP*(2) $\vdash \forall G \, G$ *is continuous.*

Proof. Work in *WNA* + $NPRA^\omega$ + *STP*(2). Suppose G is not continuous at f. Then

$$\forall n \exists h \left[[(\forall m < n) \, h(m) = f(m)] \wedge G(f) \neq G(h) \right].$$

By *STP*(2) there are liftings ϕ of G and x of f. By Lemma 7 and *STP*,

$$\forall n (\exists y < H) \left[[(\forall m < n) \, (y)_m = (x)_m] \wedge \phi(x) \neq \phi(y) \right].$$

By the star of *QF* induction,

$$\exists^\infty w (\exists y < H) \left[[(\forall u < w) \, (x)_u = (y)_u] \wedge \phi(x) \neq \phi(y) \right].$$

But then $y \approx x$, contradicting the assumption that ϕ is near-standard. □

This result is closely related to Proposition 5.2 in [1], which says that in $NPRA^\omega$, every function $f \in \mathbb{R} \rightarrow \mathbb{R}$ is continuous.

The sentence

$$(\exists^2) = \exists G \forall f\, [G(f) = 0 \leftrightarrow \exists n f(n) = 0]$$

played a central role in the paper [7], where many statements are shown to be equivalent to $(\exists^2)$ in RCA_0^ω. Similar sentences are prominent in earlier papers, such as Feferman [3]. It is well-known that

$$PRA^\omega \vdash (\exists^2) \to \exists G\, G \text{ is not continuous.}$$

Corollary 28 $WNA + NPRA^\omega + STP(2) \vdash \neg(\exists^2)$.

1.12 Functional choice and $(\exists^2)$

In this section we obtain connections between WNA and two statements which play a central role in the paper of Kohlenbach [7], the statement $(\exists^2)$ and the functional choice principle $QF - AC^{1,0}$.

In [7], Kohlenbach proposed a base theory RCA_0^ω for higher order reverse mathematics which is somewhat stronger than PRA^ω, and is a conservative extension of the second order base theory RCA_0. Its main axioms are the axioms of PRA^ω and the scheme

$$QF - AC^{1,0}: \qquad \forall f \exists n\, A(f, n, \ldots) \to \exists G \forall f\, A(f, G(f), \ldots)$$

where $A(f, n, \ldots)$ is quantifier-free.

In [7], the formula A in the $QF - AC^{1,0}$ scheme is allowed to be an arbitrary quantifier-free formula in the language $L(PRA^\omega)$. Here we will make the additional restriction that $A(f, n, \ldots)$ is in the class QF as defined in Section 1.7, that is, $A(f, n, \ldots)$ is a Boolean combination of equations and inequalities between basic terms. These formulas only have variables of type N and $N \to N$, and do not have functional variables.

We show now that $QF - AC^{1,0}$ restricted in this way follows from WNA plus the standard part principles.

Theorem 29 $WNA + STP(2) \vdash QF - AC^{1,0}$.

Proof. Work in $WNA + STP(2)$. Assume $\forall f \exists n\, A(f, n, \vec{m}, \vec{h})$. By Zeroth Order Lifting, $\overline{A}(x, u, \vec{v}, \vec{z})$ is an internal quantifier-free formula, and

$$^o x = f \wedge {}^o\vec{z} = \vec{h} \to [A(f, n, \vec{m}, \vec{h}) \leftrightarrow \overline{A}(x, n, \vec{m}, \vec{z})].$$

By Lemma 4 there is a primitive recursive function α such that

$$^*\alpha(x, w, \vec{v}, \vec{z}) = (\mu u < w)\, \overline{A}(x, u, \vec{v}, \vec{z}).$$

By STP, there exists $\vec{z}$ such that $\vec{h} = {}^o\vec{z}$. By the Lambda Abstraction axiom,

$$\exists\phi\forall x\, \phi(x) = {}^*\alpha(x, H, \vec{m}, \vec{z}).$$

Then

$$\forall^{ns} x\, [S(\phi(x)) \wedge \overline{A}(x, \phi(x), \vec{m}, \vec{z})].$$

It follows that ϕ is near-standard. By $STP(2)$, there exists G such that $G = {}^o\phi$. Therefore $\forall f\, A(f, G(f), \vec{m}, \vec{h})$. □

One of the advantages of WNA over $NPRA^\omega$ is that one can add hypotheses which produce external functions and still keep the standard part principles. The simplest hypothesis of this kind is the following statement, which says that the characteristic function of S exists:

$$(1_S \text{ exists}): \qquad \exists\phi\forall y\, \phi(y) = (\mu z < 1)\, S(y).$$

It is clear that

$$NPRA^\omega \vdash \neg(1_S \text{ exists})$$

because by the star of quantifier-free induction, $\forall^S y\, \phi(y) = 0$ implies

$$\exists y\, [\neg S(y) \wedge \phi(y) = 0].$$

However, (1_S exists) is true in the full natural model $\langle V(\mathbb{N}), V({}^*\mathbb{N}), j\rangle$ of WNA. We now connect this principle with the statement $(\exists^2)$.

Theorem 30 *$WNA + STP(2) \vdash (1_S$ exists$) \to (\exists^2)$.*

Proof. Work in $WNA + STP(2)$. Let α be the primitive recursive function such that ${}^*\alpha(x, w) = (\mu u < w)\, (x)_u = 0$. Let ϕ be the function 1_S, so that $\forall y\, \phi(y) = (\mu z < 1)\, S(y)$. Then there exists ψ such that $\forall x\, \psi(x) = \phi({}^*\alpha(x, H))$. Observe that

$$\phi({}^*\alpha(x, H)) = 0 \leftrightarrow \exists^S u\, (x)_u = 0,$$

so $\psi(x) = 0 \leftrightarrow \exists^S u\ (x)_u = 0$. Moreover, $\forall x\, \psi(x) < 2$. We show that ψ is near-standard.

Suppose $ns(x)$ and $x \approx y$. We always have $S(\phi(x))$ since $\phi(x) < 2$. If $\psi(x) = 0$ then there exists u such that $S(u)$ and $(x)_u = 0$, so $(y)_u = 0$ and hence $\psi(y) = 0$. This shows that $ns(\psi)$. By $STP(2)$ there exists G such that $G = {}^o\psi$. Consider any f. By STP, f has a lifting x. Then $G(f) = 0$ iff $\psi(x) = 0$ iff $\exists^S u (x)_u = 0$ iff $\exists n\, f(n) = 0$, and thus $(\exists^2)$ holds. □

Let us now go back to Section 1.7 and redefine the set QF of formulas by allowing basic terms of the form $G_i(f_k)$ in addition to the previous basic terms, and redefining the hierarchy Π^1_k by starting with the new QF. Also redefine the set NQF and the hierarchy $N\Pi_k^0$ by allowing additional basic terms of the form $\phi_i(x_k)$. When $STP(2)$ is assumed, the lifting lemmas from Section 1.7 and the results of Section 1.9 can be extended to the larger classes of formulas just defined. The hierarchies Π_k^2 and $N\Pi_k^1$ at the next level can now be defined in the natural way. One can then obtain the following result, with a proof similar to the proofs in Section 1.9.

Theorem 31 *In $WNA + STP(2)$, $(N\Pi_k^1, 0)$-saturation implies $(\Pi_k^2, 0)$-choice, and $(N\Pi_k^1, 1)$-saturation implies $(\Pi_k^2, 1)$-choice.*

1.13 Conclusion

We have proposed weak nonstandard analysis, *WNA*, as a base theory for reverse mathematics in nonstandard analysis.

In $WNA + STP$, one can prove:

The Weak Koenig Lemma,

$(QF, 0)$-choice and $(QF, 1)$-choice,

$(NQF, 0)$-saturation implies $(\Pi_0^1, 0)$-choice.

$(N\Pi_k^0, i)$-saturation implies (Π_k^1, i)-choice, $i = 0, 1$.

In $WNA + STP(2)$ one can prove:

$QF - AC^{1,0}$,

$NPRA^\omega$ implies $\forall G\, G$ is continuous,

1_S exists implies $(\exists^2)$,

$(N\Pi_k^1, i)$-saturation implies (Π_k^2, i)-choice, $i = 0, 1$.

We envision the use of these results to calibrate the strength of particular theorems proved using nonstandard analysis. At the higher levels, this could give a way to show that a theorem cannot be proved with methods commonly used in classical mathematics.

Look again at the natural models of *WNA* discussed at the end of Section 1.4. Let ${}^*V(\mathbb{N})$ be an $\aleph_1$-saturated elementary extension of $V(\mathbb{N})$ in the model-theoretic sense, and consider the internal natural model $\langle V(\mathbb{N}), {}^*V(\mathbb{N}), j\rangle$ and the full natural model $\langle V(\mathbb{N}), V({}^*\mathbb{N}), j\rangle$. Both of these models satisfy the axioms of *WNA*, the *STP*, the statement $(\exists^2)$, and $(N\Pi_k^1, 1)$-saturation. In view of Corollary 28, in the internal natural model the axioms of $NPRA^\omega$ hold and $STP(2)$ fails, while in the full natural model $STP(2)$ holds and the axioms of $NPRA^\omega$ fail.

References

[1] J. Avigad, Weak theories of nonstandard arithmetic and analysis, to appear in *Reverse Mathematics*, ed. by S.G. Simpson.

[2] C.C. Chang and H.J. Keisler, *Model Theory*, Third Edition, Elsevier 1990.

[3] S. Feferman. Theories of finite type related to mathematical practice, in *Handbook of Mathematical Logic*, ed. by J. Barwise, 1977.

[4] K. Gödel. "Über eine bisher noch nicht benützte Erweiterung des finiten Standpunktes", Dialectica, **12** (1958) 280–287.

[5] C.W. Henson, M. Kaufmann and H.J. Keisler, "The strength of proof theoretic methods in arithmetic", J. Symb. Logic, **49** (1984) 1039–1058.

[6] C.W. Henson and H.J. Keisler, "The strength of nonstandard analysis", J. Symb. Logic, **51** (1986) 377–386.

[7] U. Kohlenbach, Higher order reverse mathematics, to appear in *Reverse Mathematics*, ed. by S.G. Simpson.

[8] S.G. Simpson, *Subsystems of Second Order Arithmetic. Perspectives in Mathematical Logic*, Springer-Verlag, 1999.

[9] A.S. Troelstra and D. van Dalen, *Constructivism in Mathematics. An Introduction*, Volume II, North-Holland, 1988.

2

The virtue of simplicity

Edward Nelson*

Part I. Technical

It is known that IST (internal set theory) is a conservative extension of ZFC (Zermelo-Fraenkel set theory with the axiom of choice); see for example the appendix to [2] for a proof using ultrapowers and ultralimits. But these semantic constructions leave one wondering what actually makes the theory work—what are the inner mechanisms of Abraham Robinson's new logic. Let us examine the question syntactically.

Notational conventions: we use x to stand for a variable and other lowercase letters to stand for a sequence of zero or more variables; variables with a prime $'$ range over finite sets; variables with a tilde $\tilde{}$ range over functions.

We take as the axioms of IST the axioms of ZFC together with the following, in which A is an internal formula:

(T) $\forall^{\mathrm{st}}\mathrm{t}\,[\forall^{\mathrm{st}}\mathrm{x}\mathrm{A} \to \forall\mathrm{x}\mathrm{A}]$, where A has free variables x and the variables of t,

(I) $\forall^{\mathrm{st}}\mathrm{y}'\exists\mathrm{x}\forall\mathrm{y}{\in}\mathrm{y}'\,\mathrm{A} \leftrightarrow \exists\mathrm{x}\forall^{\mathrm{st}}\mathrm{y}\,\mathrm{A}$,

(S) $\forall^{\mathrm{st}}\mathrm{x}\exists^{\mathrm{st}}\mathrm{y}\,\mathrm{A}(\mathrm{x},\mathrm{y}) \to \exists^{\mathrm{st}}\tilde{\mathrm{y}}\forall^{\mathrm{st}}\mathrm{x}\,\mathrm{A}\big(\mathrm{x},\tilde{\mathrm{y}}(\mathrm{x})\big)$.

We have written the standardization principle (S) in functional form and required A to be internal; we call this the *restricted* standardization principle. It can be shown that the general standardization principle is a consequence.

All functions must have a domain. There is a neat way, using the reflection principle of set theory, to ensure that $\tilde{\mathrm{y}}$ has a domain, but let me avoid discussion of this point.

We do not take the predicate symbol *standard* as basic, but introduce it by

$$x \text{ is standard } \leftrightarrow \exists^{\mathrm{st}} y[y = x].$$

In this way $\forall^{\mathrm{st}}$ and $\exists^{\mathrm{st}}$ are new *logical* symbols and (I), (S), (T) are *logical* axioms of Abraham Robinson's new logic.

*Department of Mathematics, Princeton University.
nelson@math.princeton.edu

For any formula A of IST we define a formula A^+, called the *partial reduction* of A. It will always be of the form $\forall^{st}u\exists^{st}vA^\bullet$ where $A^\bullet$ is internal. It is defined recursively as follows:

$$\begin{array}{rcl} \text{if A is internal, } A^+ & \text{is} & A \\ (\neg A)^+ & \text{is} & \forall^{st}\tilde{v}\exists^{st}u\neg A^\bullet(u,\tilde{v}(u)) \\ (A_1 \vee A_2)^+ & \text{is} & \forall^{st}u_1u_2\exists^{st}v_1v_2[A_1^\bullet \vee A_2^\bullet] \\ (\forall xA)^+ & \text{is} & \forall^{st}u\exists^{st}v'\forall x\exists v{\in}v'A^\bullet \\ (\forall^{st}xA)^+ & \text{is} & \forall^{st}xu\forall^{st}vA^\bullet. \end{array}$$

(We take $\neg$, $\vee$, and $\forall$ as the basic logical operators—the others can be defined in terms of them.) It is understood when forming $(A_1 \vee A_2)^+$ that a variant may be taken (bound variables changed) to avoid colliding variables. If z are the free variables of A, then the *reduction* of A, denoted by A°, is the internal formula

$$\forall u\exists v'\forall z\exists v{\in}v'A^\bullet.$$

This is the same as the partial reduction of the closure of A with $\forall^{st}$ and $\exists^{st}$ replaced by $\forall$ and $\exists$.

We need only show that if A is an axiom of IST, then A° is a theorem, and that for every rule of inference with premise A_1 (or premises A_1 and A_2) and conclusion B, if A_1° is a theorem (or A_1° and A_2° are theorems), then B° is a theorem. This turns out to be quite straightforward in the main, but there is one exception. When I spoke in Aveiro I thought I could present a truly simple syntactical proof of conservativity, but I was mistaken. This remains a desirable goal. So the first part of this paper celebrates the virtue of simplicity by its absence.

The complication lies with the rule of detachment, or modus ponens. First we need a purely internal lemma.

Lemma 1 *(Cross-section) Let A be internal. Then*

$$\exists\widetilde{v'}\forall u'\exists z\forall v{\in}\widetilde{v'}(u)\, A(u,v,z) \leftrightarrow \exists\tilde{v}\forall u'\exists z\forall u{\in}u'\, A\big(u,\tilde{v}(u),z\big).$$

Proof. The backward direction is trivial: let $\widetilde{v'}(u) = \{\tilde{v}(u)\}$. To prove the forward direction, fix $\widetilde{v'}$ and let

$$\Omega = \prod_u \widetilde{v'}(u).$$

Then Ω is the set of all cross-sections of $\widetilde{v'}$. Each $\widetilde{v'}(u)$ is a finite set; give it the discrete topology, so it is compact. Give Ω the product topology, so it is compact by Tychonov's theorem.

By hypothesis, for each u' there exists an element $\tilde{v}_{u'}$ of Ω such that we have $\exists z \forall u \in u' A\big(u, \tilde{v}(u), z\big)$ (let $\tilde{v}_{u'}$ be arbitrary outside u'). The u' are a directed set under inclusion, so $u' \mapsto \tilde{v}_{u'}$ is a net in Ω. Since Ω is compact, this net has a limit point $\tilde{v}$, which has the desired property. □

Corollary 2 *(Dual form of cross-section) Again let A be internal. Then*

$$\forall \widetilde{v'} \exists u' \forall z \exists v \in \widetilde{v'}(u)\, A(u, v, z) \leftrightarrow \forall \tilde{v} \exists u' \forall z \exists u \in u'\, A\big(u, \tilde{v}(u), z\big).$$

Theorem 3 *(Detachment) If* A° *and* $(A \to B)^\circ$ *are theorems, so is* B°.

Proof. Let y be the free variables common to A and B, let w be the remaining free variables of A, and let z be the remaining free variables of B. We shall derive a contradiction from A°, $(A \to B)^\circ$, and $\neg(B^\circ)$. These formulas are

(1) $\forall u_0 \exists v_0' \forall w_0 y_0 \exists v_0 \in v_0'\, A^\bullet(u_0, v_0, w_0, y_0)$

(2) $\forall \tilde{v}_1 r_1 \exists u_1' s_1' \forall w_1 y_1 z_1 \exists u_1 \in u_1' s_1 \in s_1'\, [\neg A^\bullet(u_1, \tilde{v}_1(u_1), w_1, y1) \vee$
$B^\bullet(r_1, s_1, y_1, z_1)]$

(3) $\exists r_2 \forall s_2' \exists y_2 z_2 \forall s_2 \in s_2'\, \neg B^\bullet(r_2, s_2, y_2, z_2)$.

Fix r_2 and let $r_1 = r_2$. (That is, delete $\exists r_2$ in (3), replace the variables r_2 by constants also denoted by r_2, delete $\forall r_1$ in (2) and replace each occurrence of r_1 by r_2.) Now apply choice to (1) to pull out v_0' as an existentially quantified function $\widetilde{v_0'}$ of u_0. Notice that (2) has the form of the right hand side of the dual form of the cross-section lemma, so replace it by the left hand side. In this way we obtain

(1′) $\exists \widetilde{v_0'} \forall u_0 z_0 \exists v_0 \in \widetilde{v_0'}(u_0)\, A^\bullet(u_0, v_0, w_0, y_0)$

(2′) $\forall \widetilde{v_1'} \exists u_1' s_1' \forall z_1 \exists u_i \in u_1' s_1 \in s_1' \forall v_1 \in \widetilde{v_1'}(u_1, s_1)\, [\neg A^\bullet(u_1, v_1, w_1, y_1) \vee$
$B^\bullet(r_2, s_1, y_1, z_1)]$

(3′) $\forall s_2' \exists z_2 \forall s_2 \in s_2'\, \neg B^\bullet(r_2, s_2, y_2, z_2)$.

Fix $\widetilde{v_0'}$; let $\widetilde{v_1'}$ be defined by $\widetilde{v_1'}(u, s) = \widetilde{v_0'}(u)$ for all u and s; fix u_1' and s_1'; let $s_2' = s_1'$; fix y_2 and z_2; let $y_1 = y_2$ and $z_1 = z_2$, and let w_1 be arbitrary; let $w_0 = w_1$ and $z_0 = z_2$; fix u_1 and s_1; let $u_0 = u_1$ and $s_2 = s_1$; fix v_0; let $v_1 = v_0$. Then we have

(1″) $A^\bullet(u_1, v_0, w_1, y_2)$

(2″) $\neg A^\bullet(u_1, v_0, w_1, y_2) \vee B^\bullet(r_2, s_1, y_2, z_2)$

(3″) $\neg B^\bullet(r_2, s_1, y_2, z_2)$,

which is a contradiction. □

I have sketched the main step in a syntactical proof of the conservativity of IST over ZFC. But a better argument is needed, one that gives a practical method for converting external proofs into internal proofs. This should be possible. Whenever one uses an ideal object, such as an infinitesimal or a finite set of unlimited cardinal, it depends on the free variables in only a finite way. I expect it to be possible to develop a syntactical procedure that examines the external proof and establishes this dependence in an internal fashion.

Part II. General

Much of mathematics is intrinsically complex, and there is delight to be found in mastering complexity. But there can also be an extrinsic complexity arising from unnecessarily complicated ways of expressing intuitive mathematical ideas. Heretofore nonstandard analysis has been used primarily to simplify proofs of *theorems*. But it can also be used to simplify *theories*. There are several reasons for doing this. First and foremost is the aesthetic impulse, to create beauty. Second and very important is our obligation to the larger scientific community, to make our theories more accessible to those who need to use them. To simplify theories we need to have the courage to leave results in simple, external form—fully to embrace nonstandard analysis as a new paradigm for mathematics.

Much can be done with what may be called *minimal nonstandard analysis*. Introduce a new predicate symbol *standard* applying *only to natural numbers*, with the axioms:

(1) 0 is standard,
(2) if n is standard then $n+1$ is standard,
(3) there exists a nonstandard number,
(4) if $\mathrm{A}(0)$ and if for all standard n whenever $\mathrm{A}(n)$ then $\mathrm{A}(n+1)$, then for all standard n, $\mathrm{A}(n)$.

A prime example of unnecessary complication in mathematics is, in my opinion, Kolmogorov's foundational work on probability expressed in terms of Cantor's set theory and Lebesgue's measure theory. A beautiful treatise using these methods is [1], but some probabilists find the alternate treatment in [3] more transparent. Please do not misunderstand what I am saying; these remarks are not polemical. Simplicity is not the only virtue in mathematics and I wish in no way to discount other approaches to the use of nonstandard analysis in probability. I just want to encourage a few others to explore the possibility of using minimal nonstandard analysis in probability theory, functional analysis, differential geometry, or whatever field engages your passion.

In this spirit I shall give a few examples from [3]. A *finite probability space* is a finite set Ω and a strictly positive function pr on Ω such that

$$\sum_{\omega\in\Omega} \mathrm{pr}(\omega) = 1.$$

(The set Ω is finite but we do not require its cardinal to be standard.) An *event* is a subset M of Ω, and its *probability* is

$$\mathrm{Pr}(M) = \sum_{\omega\in M} \mathrm{pr}(\omega).$$

A *random variable* is a function $x : \Omega \to \mathbb{R}$, and its *expectation* is

$$\mathbf{E}x = \sum_{\omega\in\Omega} x(\omega)\mathrm{pr}(\omega).$$

If $a \in \mathbb{R}$, we define

$$x^{(a)}(\omega) = \begin{cases} x(\omega), & |x(\omega)| \le a \\ 0, & \textit{otherwise}. \end{cases}$$

A random variable x *is* L^1 in case

$$\mathbf{E}|x - x^{(a)}| \simeq 0 \text{ for all } a \simeq \infty.$$

Theorem 4 *(Radon-Nikodym) A random variable x is L^1 if and only if we have $\mathbf{E}|x| \ll \infty$ and for all events M with* $\mathrm{Pr}(M) \simeq 0$ *we have* $\mathbf{E}|x|\chi_M \simeq 0$.

Proof. Suppose that x is L^1. We have $\mathbf{E}|x - x^{(a)}| \le 1$ for all $a \simeq \infty$, so by overspill this is true for some $a \ll \infty$. Then $\mathbf{E}|x| \le \mathbf{E}|x - x^{(a)}| + \mathbf{E}|x^{(a)}| \le 1 + a \ll \infty$. Now let $\mathrm{Pr}(M) \simeq 0$. Let $a \simeq \infty$ be such that $a\mathrm{Pr}(M) \simeq 0$—for example, let $a = 1/\sqrt{\mathrm{Pr}(M)}$. Then

$$\mathbf{E}|x|\chi_M \le \mathbf{E}|x^{(a)}|\chi_M + \mathbf{E}|x - x^{(a)}|\chi_M \le a\mathrm{Pr}(M) + \mathbf{E}|x - x^{(a)}| \simeq 0.$$

Conversely, suppose that $\mathbf{E}|x| \ll \infty$ and that for all M with $\mathrm{Pr}(M) \simeq 0$ we have $\mathbf{E}|x|\chi_M \simeq 0$. Let $a \simeq \infty$ and let $M = \{|x| > a\}$. Then we have $\mathrm{Pr}(M) \le \mathbf{E}|x|/a \simeq 0$, so that $\mathbf{E}|x|\chi_M \simeq 0$; that is, $\mathbf{E}|x - x^{(a)}| \simeq 0$. □

A property holds *almost everywhere* (a.e.) in case for all $\varepsilon \gg 0$ there in an event N with $\mathrm{Pr}(N) \le \varepsilon$ such that the property holds everywhere except possibly on N.

Theorem 5 *(Lebesgue) If x and y are L^1 and $x \simeq y$ a.e., then* $\mathbf{E}x \simeq \mathbf{E}y$.

Proof. Let $z = x - y$. Then $z \simeq 0$ a.e. For all $\lambda \gg 0$ we have $\Pr(\{|z| \geq \lambda\}) \leq \lambda$, so by overspill this holds for some infinitesimal λ. But then

$$|z| \leq |z|\chi_{\{|z|\geq\lambda\}} + \lambda$$

and since z is L^1, $\mathbf{E}|z| \simeq 0$ by the previous theorem. Hence $\mathbf{E}x \simeq \mathbf{E}y$. □

One final example, useful in probability theory but more general. Let I be a finite subset of $[0,1]$ of the form

$$0 = t_0 < t_1 \cdots < t_{\nu-1} < t_\nu = 1$$

such that $t_\mu \simeq t_{\mu+1}$ for all $0 \leq \mu < \nu$. To the naked eye, I looks just like $[0,1]$. Although I is finite, it is "uncountable" in the following sense:

Theorem 6 *(Cantor) For any sequence $x : \mathbb{N} \to I$ there exists $t \in I$ such that t is not infinitely close to any x_n with n standard.*

Proof. Construct t_0 by changing the nth decimal digit of x_n, so that $|t_0 - x_n| \geq 10^{-n}$ for all n. Let t be the greatest element of I that is less than t_0; then t is in I and has the desired property. □

References

[1] PATRICK BILLINGSLEY, *Convergence of Probability Measures*, second edition, Wiley & Sons, Inc., New York (1999).

[2] EDWARD NELSON, "Internal set theory: A new approach to nonstandard analysis", Bull. Amer. Math. Soc., **83** (1987) 1165–1198.

[3] EDWARD NELSON, *Radically Elementary Probability Theory*, Annals of Mathematics Studies #117, Princeton University Press, Princeton, New Jersey (1987). `http://www.math.princeton.edu/~nelson/books.html`

3

Analysis of various practices of referring in classical or non standard mathematics

Yves Péraire*

3.1 Introduction

The thesis underlying this text is that the various approaches of mathematics, both the conventional or the diverse non standard approaches, pure or applied, are characterized primarily by their mode of referring and in particular by the more or less important use of the *reconstructed* reference, the reference to the sets and collections, that I will distinguish from the *direct* reference, the reference to the world of the facts in a broad sense. The direct reference, in traditional mathematics as well as in non standard mathematics (for the main part) is ritually performed in the classical form of modelling, consisting in confronting the facts to a small paradise (a set) correctly structured. So the discourse on the model acts like a metaphor of the modelized reality.

I will defend another approach, the relative approach, which consists in using the mathematical languages like genuine languages of communication, referring directly to the facts, but accepting the usage of the metaphor of sets (revelead as such) too strongly culturally established. My thesis will be illustrated by the presentation of two articles

1. A mathematical framework for Dirac's calculus [14].

2. Heaviside calculus with no Laplace transform [15].

The first one starts with a semantical analysis of Dirac's article introducing the Delta function. Dirac said: "δ is improper, δ' is more improper". If we give it the meaning: "δ is not completely known, and δ' is less known", it works. So it is necessary to translate it in Relative Mathematics' language. We know that the *non relative* attitude consists in the formal affirmation that a model of the delta function is perfectly determined in a paradise, a space of generalized

*Département de Mathématiques, Université Blaise Pascal (Clermont II).
Yves.PERAIRE@math.univ-bpclermont.fr

functions. The continuation of the metaphor requires to provide the paradise with a topology. In the relative approach, we prefer to explore more deeply the concepts of a point, infinity, equality with *words* of relative mathematics. Finally the opposition

NON STANDARD/STANDARD

is replaced by the antagonism

RELATIVE/NON RELATIVE.

3.2 Généralités sur la référentiation

Cette conférence présente sous un autre angle, en les précisant, certaines des idées exposées au congrès PILM 2002 et qui paraîtront dans [13]. Les considérations générales concernant la sémantique des langues mathématiques, qui sous-tendent mon exposé, ne sont pas nouvelles. Toutefois, il est nécessaire de les réactualiser pour tenir compte des pratiques spécifiques des mathématiciens non standard.

On a dit quelquefois que la démarcation entre les diverses pratiques des mathématiques non standard, se ferait à partir du matériel linguistique mis en oeuvre. En gros, les différentes écoles non standard parleraient des même choses mais utiliseraient des dialectes différents.

Nous savons bien pourtant que le langage des mathématiques, classiques ou non standard, est contenu dans le langage du premier ordre du calcul des prédicats. Il est vrai cependant que les différentes écoles non standard n'utilisent pas la même partie de ce langage ; le seul prédicat binaire « $\cdot \in \cdot$ » pour les classiques, le prédicat « $\cdot \in \cdot$ » et le prédicat unaire « $\mathbf{st}\ \cdot$ » pour l'école Nelson-Reeb, j'utilise personnellement le prédicat « $\cdot \in \cdot$ » et un autre prédicat binaire, le prédicat de Wallet « $\cdot\ \mathcal{SR}\ \cdot$ ».

En réalité, ces différents choix induisent (autant qu'ils sont induits par) des pratiques sémantiques différenciées et c'est l'analyse de ce rapport au sens qui permet de comprendre la nature des différentes écoles mathématiques, standard ou non standard qu'elles se veuillent appliquées ou se prétendent pures.

Je distinguerai schématiquement trois mode d'attribution du sens, que je désignerai par les termes

- référence directe,
- référence reconstruite ou
- référence directe élargie.

Ce que j'appelle référence directe, c'est la référence aux faits au sens large, ne se limitant pas à la description des phénomènes, s'autorisant aussi la description des concepts. C'est en gros le mode de référence des physiciens.

La référence reconstruite, c'est la référence aux entités mentales stables engendrées par le discours *en langue mathématique* : les ensembles introduits formellement par le quantificateur $\exists$, les collections.

On pourrait dire que le référent des mathématiques pures consiste essentiellement en ces entités, éjectées à partir des axiomes de la théorie des ensembles. Dans les faits, les choses sont beaucoup plus mélangées. En effet, même si elle n'est pas mise en avant, la référence au monde des "faits élargis" aux concepts naturels, concept d'ensemble, concept naturel de nombre, à une certaine idée du fini et de l'infini, n'est pas absente de la pensée des mathématiciens purs. La pratique des mathématiques appliqués constitue un pas vers le 3ème mode d'attribution du sens ; la modélisation en est la forme la plus classique. Le référent est bien le mode des faits, mais le mode d'expression est systématiquement celui de la métaphore, prise dans l'univers reconstruit.

Les lignes précédentes illustrent ce que j'entend par référent : « ce à quoi le signe linguistique renvoie dans la "réalité extra-linguistique" soit l'univers "réel", soit l'univers imaginaire ». Mais de quels signes parlons-nous ? A quelle langue appartiennent-ils ? Il est tout à fait clair que c'est de la langue de communication, l'anglais ou le français, dont je parle et non pas des languages bien construites, de ZF de IST de RIST ou autre. En réalité ces langues sont utilisées comme matrices pour *produire* des entités structurées et également comme une écriture abrégée de la langue naturelle, une sorte de sténographie.

La pratique que je vais essayer de décrire maintenant, que j'appelle référence directe élargie, prend acte des pratiques sémantiques d'une partie du réseau Georges Reeb, les rend explicites et ouvre la voie à une généralisation. Ce qui caractérise cette approche, ce n'est pas la double référence — au monde des faits d'une part, au monde reconstruit de l'autre — qui est une pratique courante des mathématiques appliquées, mais le fait que ce processus de désignation du sens concerne non plus seulement la langue vernaculaire, mais aussi *la langue mathématique*, que l'on se met a pratiquer alors, comme le français ou l'anglais, en alternant style direct et style métaphorique en quantité variable selon ses préférences. Voici deux exemples de pratiques sémantiques utilisant une métaphore.

Considérons l'énoncé en français : « grand père est mort ». Il possède clairement une référence dans le monde des faits. On peut transcrire cet énoncé par un autre, métaphorique, « grand-père est au paradis ».

Admettons maintenant que ces deux énoncés disent la même chose et que l'on veuille à partir de l'un ou l'autre explorer plus finement le concept de la mort. On peut le faire de plusieurs manières. La méthode la plus directe consiste à utiliser le mot « mort » et à préciser encore le concept par l'introduction d'autre mots. On peut aussi tenter de « pousser la métaphore » un peu plus loin et proposer une description du paradis. Le second exemple provient des mathématiques de la physique.

Le point de départ est l'article de Paul Dirac dans [2], dans lequel est introduite la fonction δ. Si on ne regarde que les formules et les calculs qui figurent dans cet article, alors la définition donnée conduit sans nul doute à une contradiction. Il reste à expliquer pourquoi, malgré cela, ils ne conduisent pas à des absurdité physiques et semblent même avoir une certaine puissance explicative.

Le problème de la contradiction tel qu'il a été résolu par Laurent Schwartz [19] a consisté à parler de la fonction de Dirac, qui relie des grandeurs physiques, en termes d'un élément d'un paradis réputé sûr, l'espace des distributions. C'est une métaphore similaire à celle utilisée précédemment pour parler de la mort de grand-père, en effet dans les deux cas

(a) On veut rendre compte d'une relation entrée/sortie : vie/mort d'un côté, [valeur 0]/[valeur 1] de l'autre. Il y a une brève phase de transition de l'état initial à l'état final, que l'on ne peut pas décrire. On pense que l'état des choses pendant cette phase doit expliquer le changement d'état.
(b) Dans chaque exemple *l'indétermination* factuelle de la phase de transition est exprimée par une *détermination* dans un univers mythique, qui prétend expliquer le phénomène. Il faut remarquer toutefois que le mythe utilisé dans la démarche mathématique est d'une bien plus grande cohérence, peut-être parce que la langue mathématique, qui l'a engendré, est exempte d'ambiguïté.
(c) Il y a dans les deux cas une tentation de pousser trop loin la métaphore.

La métaphore de l'ensemble des distributions fonctionne assez bien, elle permet de faire correspondre à certaines formules posées par Dirac des transcriptions acceptables et utilisables mais on sait que certains faits ne sont pas pris en compte par ce modèle, la multiplication n'est pas permise, l'égalité au sens des distributions s'éloigne assez de la réalité de l'égalité physique... D'autres espaces abstraits, fonctions généralisées de Colombeau, ultrapuissances saturées... permettent de tenir un discours formellement cohérent sur la multiplication des fonctions de Dirac.

Cependant, malgré la réelle efficacité mathématique de ces modélisations, on peut souhaiter une approche plus directe, dans laquelle le champ sémantique serait moins encombré par les représentations ensemblistes. Il ne s'agit pas de renoncer complètement à ce type de méthodes, mais plutôt de briser l'automatisme du recours à ces dernières. Ma position sur ce point diverge donc de celle de Gilles Gaston Granger dans [4] pour qui ce qu'il appelle la *sortie de l'irrationnel* ne peut se faire *que par l'introduction* d'un espace fonctionnel. J'ai choisi une approche plus directe du même problème ; elle commence par une analyse sémantique du texte de Paul Dirac. Cette analyse permet de découvrir plusieurs faits, qu'il faut introduire dans la description en langue mathématique. En voici quelques uns.

– le point matériel à une épaisseur non nulle, physiquement infinitésimale,
– la fonction δ est partiellement indéterminée dans le point,

– l'égalité dans les formules de Dirac n' est pas l'égalité classique des ensembles.

Notre travail consistera donc à introduire, dans une langue bien construite, les mots pour exprimer ces faits (par une sorte de traduction de la langue naturelle). En ce qui concerne l'égalité de Dirac, l'analyse du texte fait apparaître que Dirac identifie à δ toute fonction nulle en dehors du point origine et d'intégrale égale à 1, je donnerais donc une définition de l'égalité qui intègre ce fait. Jusqu'à présent, le langage de la théorie relative des ensembles (voir [12]) m'a suffi pour fabriquer tout le lexique nécessaire.

Je propose d'appeller MATHÉMATIQUES RELATIVES la pratique que je viens de décrire. En conséquence les mathématiques classiques devront être qualifiées, pour une large part, de non relatives. En guise d'illustration je vais maintenant présenter les résultats et la philosophie de deux articles concernant les mathématiques de la physique.

1. A mathematical framework for Dirac's calculus [14].
2. Heaviside calculus with no Laplace transform [15].

Une partie importante de chacun de ces articles consiste à mettre en place une définition ad hoc de l'égalité.

3.3 Le calcul de Dirac. L'égalité de Dirac

Une description précise des notions évoquées plus bas, niveaux d'impropriété, dérivées observées, égalité de Dirac est disponible dans [14], chapitres 1 et 2 ainsi que de nombreux exemples de couples de fonctions Dirac-égales. Dans la discussion qui va suivre nous utiliserons des définitions incomplètes pour ne pas cacher l'essentiel, qui est la philosophie de ce travail. En particulier la définition de l'égalité de Dirac, Définition 4 de la section 2.4, est plus complexe.

Comment justifier les calculs de Dirac ? On a compris qu'il fallait reconnaître un sens physique à la notion de point, éclairer aussi la signification du signe « = » et exprimer cela en langue mathématique. L'étape initiale nécessaire pour y parvenir consiste à introduire des prédicats pour *dire l'indétermination.* Cette expression de l'indétermination est cachée dans les formulations de Dirac. En effet Considérons l'énoncé tiré de [2] :

> « *Strictly of course, $\delta(x)$ is not a proper function of x,... $\delta'(x)$, $\delta''(x)$ ... are even more discontinuous and less proper than $\delta(x)$ itself.* »

J'ignore quelle signification précise Dirac donnait au mot « improper », cependant si on l' interprète, en forçant un peu le sens, par « partiellement indéterminé », la transcription fonctionne. Par contre, la traduction par le mot « anormal » qui a été donnée dans [3], peut provoquer un bloquage.

Voici rapidement comment, dans l'article [14], j'ai exprimé les choses.

1. J'ai utilisé le prédicat de standardicité relative $\cdot SR \cdot$ pour définir les niveaux d'impropriété. On pourra trouver dans [14], première section, la définition précise des niveaux d'impropriété. Rapidement, nous dirons que plus un objet est impropre moins il est standard.
2. A tout nombre p j'associe le point analysé $\wr p \wr$, c'est un halo dont le degré d'infinitésimalité est d'autant plus petit que p est impropre.
3. Je définis la *fonction de Dirac principale* δ de la manière suivante :
   (a) Je fixe un infinitésimal h_1 strictement positif et impropre,
   (b) je pose $\delta = \dfrac{1}{2h_1}\mathrm{Ind}_{[-h_1,h_1]}$.

   On obtient pour δ le graphe suivant, que l'on peut trouver aussi dans les livres de physique. J'ai représenté sur le même graphique la *fonction échelon de Heaviside*, elle vaut 1 pour les valeurs positives de la variable et 0 partout ailleurs.

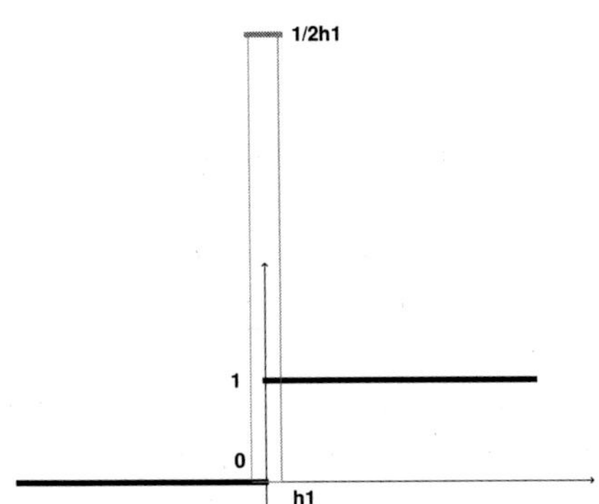

Figure 3.1: La fonction de Heaviside et sa dérivée observée, la fonction de Dirac

   Le réflexe de tout mathématicien sera d'observer que ma définition de δ n'est pas indépendante de la valeur de h_1. La réponse à cette objection nécessite une meilleure investigation de la notion d'égalité.

4. Je définis la Dirac-égalité, « $\stackrel{D}{=}$ » sur un ensemble de fonctions $\mathcal{C}^\infty$ par morceaux, standard ou pas, de telle sorte que toute fonction ayant toutes ses dérivées infinitésimales pour les valeurs appréciables et d'intégrale infiniment proches de 1 sur tout intervalle $[x, y]$ contenant 0 et ayant des bornes appréciables, soit Dirac-égale à la fonction de Dirac principale.

5. J'ai défini la dérivée observée d'une fonction, standard ou pas, *pouvant présenter des discontinuités* du premier ordre par une formule de la forme

$$f^{]'[}(x) = \frac{f(x+h) - f(x-h)}{2h}$$

qui, à première vue, dépend de h, ou h est infiniment petit d'un ordre de petitesse adapté au niveau d'impropriété de f. Après avoir posé cette définition,

il est préférable de résister à la tentation de chercher une limite quelque-part quand h tend vers 0.

On repousse ainsi l'intrusion dans le champ sémantique d'un objet encombrant. La dérivée observée de la fonction de Heaviside est précisément notre fonction de Dirac principale. On montre facilement que la valeur de $f^{]'[}$ est *Dirac-indépendante* de h :

deux valeurs distinctes de h donnent des valeurs Dirac-égales de la dérivée observée.

Cela répond à l'objection évoquée à l'item 3 et explique aussi pourquoi dans la notation de la dérivée observée je n'ai pas fait apparaître l'accroissement h. On se convaincra facilement que cette approche donne une description plus proche de la réalité physique.

Le chapitre 3 présente les propriétés de base de la dérivation observée. Parmi celles-ci nous avons :

Compatibilité avec la dérivée classique.

$$\text{Si } f \text{ est dérivable } f' \stackrel{D}{=} f^{]'[}.$$

Linéarité au sens de Dirac.

Si f et g ont le même niveau d'impropriété, alors

$$(f+g)^{]'[} \stackrel{D}{=} f^{]'[} + g^{]'[}.$$

Formule de Leibniz au sens de Dirac.

Sous la même condition sur le niveau d'impropriété, alors

$$(f \times g)^{]'[} \stackrel{D}{=} f \times g^{]'[} + f^{]'[} \times g.$$

On trouvera dans [14], section 3 théorème 4, une formulation d'écriture plus complexe mais qui s'applique aux cas ou les fonctions f et g ont des niveaux d'impropriété distincts.

On remarquera que ces égalités ne sont pas testées sur des espaces fonctionnels comme chez Schwartz ou Colombeau et que toutes les multiplications sont autorisées. Cela permet de légitimer des formules de physiques où interviennent des carrés de fonction delta, comme celle-ci dans laquelle δ est la fonction de Dirac principale :

$$\frac{(-1)^{p+1}}{(p+1)!}\, x_+^{p+1}\, \delta^{]p+2[} + \delta^2 \stackrel{D}{=} \frac{p+2}{2}\, \delta^{]'[}.$$

Il existe une formule analogue dans [1] dans laquelle δ est représentée dans l'espace des fonctions généralisées de Colombeau et où l'égalité est une égalité faible. La formule doit être modifiée si on utilise une autre fonction de Dirac que la fonction de Dirac principale.

Remarques

J'ai qualifié d'égalité la relation $\stackrel{D}{=}$. On peut préférer voir en cette relation une indiscernabilité ou une approximation de "*quelque chose quelque part*". Je prétends pourtant que la relation de Dirac mérite, autant que l'égalité classique des ensembles, le nom d'égalité. En effet que dit la formule précédente ? En dehors du point origine, supposé d'épaisseur radicalement indétectable, et pour toute valeur accessible de la variable, les fonctions de part et d'autre du signe $\stackrel{D}{=}$ ainsi que leur dérivées, à tout ordre accessible, prennent des valeurs indistinguables à 10^{-n} près quelle que soit la valeur de n à laquelle on puisse accéder. Les intégrales itérées à tout ordre matériellement possible, avec des bornes d'intégrations de part et d'autre du point $\wr 0 \wr$ sont également indistinguables. On peut dire que deux fonctions Dirac-égales représentent des phénomènes physiquement identiques. En revanche, si on tourne le regard vers le monde des entités, la relation $\stackrel{D}{=}$ n'est pas une égalité, c'est une relation d'équivalence ; mais on sait bien que ce qui est présenté comme égalité, dans le monde reconstruit, n'est qu'un avatar de l'équivalence logique.

On remarquera que je n'ai pas hésité, dans ma description du calcul de Dirac, à faire usage de l'ensemble des nombres réels, pour parler des grandeurs physiques. Je n'ai commencé à renoncer à la métaphore ensembliste qu'à partir du moment où elle m'a semblé gênante. Kinoshita dans [6] et Grenier dans [5] ont préféré pour leur part utiliser un "*continuum discret*".

Chacun est libre d'utiliser la quantité et le type de représentations ensemblistes qui lui plait, chacun a droit à son style propre.

On remarquera que les formules précédentes parlent *en premier lieu* de faits concernant des grandeurs physiques, pas des ensembles. Il vaut mieux éviter, bien que cela reste possible, et même souhaitable, pour d'autres travaux, d'attribuer un statut ontologique dans la "réalité mathématique" aux infinitésimaux de différents niveaux dont on fait usage, cela risque de devenir vite désagréable. L'enrichissement du vocabulaire que nous avons réalisé *ici* avait pour but de rendre plus précis le discours en langue mathématique... mais si on tournait les yeux vers les ensembles on pourrait y voir un monde d'entités de plus en plus idéales, un paradis étrange d'idéalités stratifiées...

La fonction de Dirac principale est celle que l'on obtient, prenant au sérieux notre ignorance ce qu'il se passe dans le point $\wr 0 \wr$, quand on applique la définition de la dérivée observée à la fonction échelon de Heaviside.

Dans la section suivante, j'utiliserai des fonctions de Dirac à droite de zéro et de classe $\mathcal{C}^\infty$. Cela peut sembler beaucoup de précisions, pour une fonction mal connue à l'origine. Cela peut se justifier de la manière suivante : dans l'article que je vais présenter le but n'est pas de *décrire* une partie de la réalité physique mais de donner une plus ample "*justification*" pour le concept de Heaviside d'opérateur de dérivation, ayant pour inverse un opérateur d'intégration.

3.4 Calcul de Heaviside sans transformée de Laplace. L'égalité de Laplace

Comme pour la section précédente, j'inviterai le lecteur à se reporter à un article complet, l'article [15]. C'est la philosophique sous-jacente à cet article que je veux présenter ici, je devrai donc passer sous silence les démonstrations, et même donner des définitions incomplètes.

Le calcul opérationnel, quand on l'applique formellement, sans vérifier l'existence d'une transformation de Laplace, donne les bonnes solutions. Ces solutions elles même n'ont le plus souvent pas d'intégrale de Laplace, du moins dans les cas que j'ai pu traiter. D'autre part on a le sentiment que ces méthodes ne devraient pas dépendre de la convergence de l'intégrale de Laplace. Aussi ai-je tenté (avec succès) de mettre en place une notion d'*image généralisée* qui opère même quand les fonctions en jeu ne sont pas Laplace-transformables. Bien sur, ce n'est pas la première tentative dans ce sens. Après le premier travail de Vigneaux publié en 1929 dans [20] d'autres auteurs dont récemment, Komatsu dans [7], G.Lumer et F. Neubrander dans [8, 9] se sont attaqués à ce problème par des méthodes classiques.

L'approche relative permet d'obtenir une solution complète et beaucoup plus directe. Les résultats obtenus sont les suivants.

1. Les difficultés dues à la divergence de l'intégrale de Laplace disparaissent.
2. Les dérivations par rapport aux paramètres sont toujours permises.
3. On peut utiliser des séries divergentes comme images généralisées.
4. Les calculs peuvent faire intervenir des fonctions de Dirac, des peignes de Dirac, etc. sans qu'il soit nécessaire d'introduire des espaces fonctionnels.

5. L'application des méthodes de ce calcul de Heaviside rejustifié, aboutit à une description plus fine des solutions trouvées. En particulier, si il apparaît de l'indétermination dans une équation au dérivées partielles, à cause de la présence d'une fonction delta, alors on retrouve la trace de cette indétermination dans les solutions (Voir l'exemple 1 plus bas, à titre d'illustration).

Indiquons maintenant plus précisément le chemin suivi. Notre calcul va s'appliquer à une classe de bonnes fonctions, par exemple de classe C^∞, mais pas nécessairement intégrables.

Définition de l'égalité de Laplace

Contrairement à ma définition de l'égalité de Dirac, dont la description s'inspirait de l'observation de l'usage que faisait Dirac de l'égalité, ma définition de ce que j'appelle égalité de Laplace n'est pas justifiée par une pratique de Laplace. La définition utilise deux niveaux d'impropriétés définis au moyen du prédicat $\cdot\mathcal{SR}\cdot$. Cette classification permet de définir des ordres de grandeur relatifs dans $\mathbb{R}$.

Nous connaissons les définitions suivantes.

Un nombre x est infinitésimal et on écrit $x \sim 0$ si

$$\forall^{st}\varepsilon > 0 \quad |x| < \varepsilon. \tag{1}$$

Le nombre x est infiniment grand et on écrit $x \sim +\infty$ si

$$\forall^{st} l \quad |x| > l. \tag{1'}$$

Nous dirons maintenant qu'un nombre x est un *infinitésimal relatif*, et nous écrirons $x \approx 0$, si

$$\forall^{Imp}\varepsilon > 0 \quad |x| < \varepsilon. \tag{2}$$

et que x est un *infiniment grand relatif* si

$$\forall^{Imp} l > 0 \quad |x| > l \tag{2'}$$

ce qui s'écrira $x \approx +\infty$. ($\forall^{Imp}x\ F(x)$ est une abréviation pour $\forall x\ (Imp(x) \Rightarrow F(x))$, ce qui s'oralise « pour tout x impropre, $F(x)$ »).

On peut montrer que tout nombre relativement infinitésimal est infinitésimal. Un nombre non relativement infiniment grand est dit *relativement limité*.

On fixe ensuite une fois pour toute :
– un nombre $\bowtie$ relativement infiniment grand,
– une *fonction de Dirac relative* à droite de 0, δ.

Cela signifie que δ reste une fonction de Dirac même pour un observateur idéal capable de "voir" des fonctions de Dirac impropres. Cette fonction δ est très impropre, elle est nulle en dehors d'un intervalle $[0, h]$ avec $h \approx 0$.

J'ai défini ensuite dans [15] une collection $\mathcal{N}$ de fonctions négligeables de classe $\mathcal{C}^\infty$ de deux variables x et t. La généralisation à un nombre plus grand de variables ne pose pas de problème. *On peut même étendre nos résultats aux cas où on a un nombre infiniment grand, impropre, de variables.* Cela peut être utile si on a besoin de faire intervenir des paramètres qui sont des fonctions standard, représentées par un ensemble fini impropre contenant toutes leurs valeurs standard. La classe $\mathcal{N}$ vérifie les propriétés suivantes.

- Si $\alpha(x,t) \in \mathcal{N}$ alors $\alpha(x,t) \approx 0$ pour tout x relativement limité et tout t relativement appréciable.
- $t \cdot \mathcal{N} \subset \mathcal{N}$.

Si $\star$ désigne le produit de convolution des fonctions, alors

- $\mathcal{N} \star \mathcal{N} \subset \mathcal{N}$.
- $\underline{f} \star \mathcal{N} \subset \mathcal{N}$ pour toute bonne fonction f.
- $\dfrac{\partial^{n+m}\mathcal{N}}{\partial x^m \partial t^n} \subset \mathcal{N}$, pour tous $n \in \mathbb{N}$ et $m \in \mathbb{Z}$, limités

On remarque que dans la dernière propriété m peut être négatif. La dérivée d'ordre négatif $-k$ d'une fonction α s'obtient en calculant l'intégrale

$$\underbrace{\int_0^{t_k=t} \cdots \int_0^{t_2} \int_0^{t_1}}_{\text{On intègre } k \text{ fois}} \alpha(s)\, ds\, dt_1 \cdots dt_{k-1}.$$

Remarque. La classe $\mathcal{N}$, "n'est pas" un ensemble pas plus que $t \cdot \mathcal{N}$, $\mathcal{N} \star \mathcal{N}$, $\underline{f} \star \mathcal{N}$ ou $\frac{\partial^{n+m}\mathcal{N}}{\partial x^m \partial t^n}$. Les inclusions précédentes sont des inclusions de collections.

On définit alors l'égalité de Laplace $\overset{\mathcal{L}}{=}$ en posant

$$F \overset{\mathcal{L}}{=} G \overset{Def}{\Leftrightarrow} \exists \alpha \in \mathcal{N}\ F - G = \mathcal{L}\alpha$$

où $\mathcal{L}\alpha$ est la transformée de Laplace classique de α, qui doit donc exister.

On pose ensuite

$$(\mathbf{L}f)(p) \overset{Def}{=} \left[\int_0^{\bowtie} e^{-pt} f(t)\, dt\right] \cdot (\mathcal{L}\delta)(p).$$

La fonction $(\mathcal{L}\delta)(p)$ n'est pas égale à 1, comme pour la distribution de Dirac. On trouve parfois dans les livres de physique, concernant les *fonctions* de Dirac, l'égalité $(\mathcal{L}\Delta)(p) = 1$. En réalité pour une fonction de Dirac Δ non standard, on a $(\mathcal{L}\delta)(p) \sim 1$ pour les valeurs limitées de p. Notre fonction δ, fixée plus haut, vérifie quant à elle

$$\text{pour tout } p > 0 \text{ relativement limité, } (\mathcal{L}\delta)(p) \approx 1.$$

Démontrons ce dernier point. La deuxième formule de la moyenne donne, pour chaque $p > 0$,

$$(\mathcal{L}\delta)(p) = e^{p\theta h} \int_0^h \delta(t)\, dt, \quad \theta \in [0,1].$$

$h \approx 0$ implique $\theta h \approx 0$, $p\theta h \approx 0$ pour p relativement limité et donc, $e^{p\theta h} \approx 1$. Comme $\int_0^h \delta(t)\, dt \approx 1$, on en déduit que $(\mathcal{L}\delta)(p) \approx 1$ pour tout p relativement limité.

Notation. On écrira $f \sqsupset\!\square\; F$ si $F \overset{\mathcal{L}}{=} \mathbf{L}f$ et $f \sqsupset F$ si f admet F comme transformation de Laplace.

Quelques résultats obtenus

Pour toutes bonnes fonctions f et g standard ou impropre

1. $\mathbf{L}f$ est *Laplace-indépendant* de δ. Des choix distincts de la fonction δ donneraient des transformations Laplace-égales.
2. Si $f \sqsupset\!\square\; F$ et $g \sqsupset\!\square\; F$, alors $f = g$.
3. Si f est de type exponentiel, $f \sqsupset F \Rightarrow f \sqsupset\!\square\; F$.
   Ce qui précède implique que toutes les fonctions qui figurent dans les tables de transformations de Laplace sont des images généralisées.
4. $(\mathbf{L}f')(p) \overset{\mathcal{L}}{=} p(\mathbf{L}f)(p) - f(0+) \cdot (\mathcal{L}\delta)(p)$.
5. $\left(\mathbf{L} \int_0^t f(s)\, ds\right)(p) \overset{\mathcal{L}}{=} \dfrac{1}{p}(\mathbf{L}f)(p)$.
6. $\mathbf{L}(f \star g) \overset{\mathcal{L}}{=} (\mathbf{L}f) \cdot (\mathbf{L}g)$.
7. Si $f(t) = \sum_{n=0}^{+\infty} a_n t^n$ alors $\mathbf{L}F \overset{\mathcal{L}}{=} \sum_{n=0}^{\bowtie} a_n \dfrac{n!}{p^{n+1}}$, même si la série $\sum_{n=0}^{+\infty} a_n \dfrac{n!}{p^{n+1}}$ diverge.
8. $\mathbf{L}\dfrac{\partial f}{\partial x} \overset{\mathcal{L}}{=} \dfrac{\partial \mathbf{L}f}{\partial x} \ldots$ etc.

3.5 Exemples

Exemple 1.
Recherche de la solution de l'équation aux dérivées partielles définies pour $x > 0,\ t > 0$.

$$\frac{\partial^2}{\partial x^2}\, u(x,t) - \frac{\partial}{\partial t}\, u(x,t) = e^{t^2}$$

$$u(0_+, t) = \Delta(t), \quad u(x, 0_+) = 0 \text{ pour tout } x.$$

Δ est une fonction impropre de Dirac. L'application du calcul formel généralisé donne la solution "exacte"

$$u(x,t) = \left[\Delta(t) - H(t) \star e^{t^2}\right] \star \frac{x}{2\sqrt{\pi t}} e^{\frac{-x^2}{4t}} + H(t) \star e^{t^2}.$$

Cependant LA RÉPONSE PHYSIQUE CORRECTE au problème est la suivante : Pour tous $x > 0$ et $t > 0$,

(a) Si t est appréciable la solution est indiscernable de

$$H(t) \star e^{t^2} - \left[H(t) \star e^{t^2}\right] \star \frac{x}{2\sqrt{\pi t}} e^{\frac{-x^2}{4t}} + \frac{x}{2\sqrt{\pi t}} e^{\frac{-x^2}{4t}}.$$

(b) On ne connaît pas avec précision le comportement de la solution quand t est très petit. Cette imprécision est héritée du celle de la fonction Δ.

Exemple 2.
La fonction e^{-t^2} a une transformation de Laplace et admet un développement en série de rayon infini, cependant sa transformée de Laplace n'est pas égale à la somme — qui n'est pas définie — des transformées de Laplace des termes de la série. On a cependant

$$e^{-t^2} \sqsupset \sum_{2n < \bowtie} (-1)^n \frac{(2n)!}{n! p^{n+1}}.$$

Pour la fonction e^{t^2} il n'y a pas de transformée de Laplace mais on peut écrire une image généralisée :

$$e^{t^2} \sqsupset \sum_{2n < \bowtie} \frac{(2n)!}{n! p^{n+1}}.$$

En conclusion, je dirai que l'examen systématique des questions de sémantique peut accroître considérablement l'efficacité de l'outil mathématique. Nous venons de le constater pour les mathématiques de la physique.

Références

[1] B. DAMYANOV, "Multiplication of Schwartz distributions and Colombeau Generalized functions", Journal of Appl. Anal., **5** (1999), 249–60.

[2] P.A.M. DIRAC, "The physical interpretation of the Quantum Dynamics", Proc. of the Royal Society, section A, **113** (1926-27) 621–641.

[3] P.A.M. DIRAC, *Les principes de la mécanique quantique* (1931), Jacques Gabay, Paris, 1990.

[4] G.G. GRANGER, *L'irrationnel*, Editions Odile Jacob, 1998.

[5] J.P. GRENIER, "Représentation discrète des distributions standard", Osaka J. of Math., **32** (1995) 799–815.

[6] M. KINOSHITA, "Non-standard representations of distributions", Osaka J. Math., I **25** (1988) 805–824 and II **27** (1990) 843–861.

[7] H. KOMATSU, "Laplace transform of Hyperfunctions a new foundation of the Heaviside Calculus", J. Fac. Sci. Tokyo, Sect. IA. Math., **34** (1987) 805–820.

[8] G. LUMER and F. NEUBRANDER, *The Asymptotic Laplace Transform and evolution equations*, Advances in Partial differential equations, Math. Topics Vol. 16, Wiley-VCH 1999.

[9] G. LUMER and F. NEUBRANDER, Asymptotic Laplace Transform and relation to Komatsu's Laplace Transform of Hyperfunctions, preprint 2000.

[10] J. MIKUSINSKI, Bull. Acad. Pol. Ser. Sci. Math. Astron. Phys., **43** (1966) 511–13.

[11] E. NELSON, "Internal set theory : a new approach to nonstandard analysis", Bull. A.M.S., **83** (1977) 1165–1198.

[12] Y. PÉRAIRE, "Théorie relative des ensembles internes", Osaka J. Math., **29** (1992) 267–297.

[13] Y. PÉRAIRE, "Le replacement du référent dans les pratiques de l'analyse issues de Edward Nelson et de Georges Reeb", Archives Henri Poincaré, Philosophia Scientiae (2005), à paraître.

[14] Y. PÉRAIRE, "A mathematical framework for Dirac's calculus", Bulletin of the Belgian Mathematical Society Simon Stevin, (2005), à paraître.

[15] Y. PÉRAIRE, Heaviside calculus with no Laplace transform, Integral Transform and Special Functions, à paraître.

[16] C.RAJU, "Product and compositions with the Dirac delta function", J. Phys. A : Marh. Gen., **15** (1982) 381–96.

[17] K.D. STROYAN and W.A.J. LUXEMBURG, *Introduction to the theory of infinitesimals*, Academic Press, New-York, 1976.

[18] L. SCHWARTZ, "Sur l'impossibilité de la multiplication des distributions", C. R. Acad. Sci. Paris, **239** (1954) 847–848.

[19] L. SCHWARTZ, *Théorie des distributions*, Hermann, Paris (1966)

[20] J.C. VIGNEAUX, "Sugli integrali di Laplace asintotici", Atti Accad. Naz. Lincei, Rend. Cl. Sci. Fis. Math., **6** (29) (1939) 396–402.

4

Stratified analysis?

Karel Hrbacek*

It is now over forty years since Abraham Robinson realized that "*the concepts and methods of Mathematical Logic are capable of providing a suitable framework for the development of the Differential and Integral Calculus by means of infinitely small and infinitely large numbers*" (Robinson [29], Introduction, p. 2). The magnitude of Robinson's achievement cannot be overstated. Not only does his framework allow rigorous paraphrases of many arguments of Leibniz, Euler and other mathematicians from the classical period of calculus; it has enabled the development of entirely new, important mathematical techniques and constructs not anticipated by the classics. Researchers working with the methods of nonstandard analysis have discovered new significant results in diverse areas of pure and applied mathematics, from number theory to mathematical physics and economics.

It seems fair to say, however, that acceptance of "nonstandard" methods by the larger mathematical community lags far behind their successes. In particular, the oft-expressed hope that infinitesimals would now replace the notorious ε-δ method in teaching calculus remains unrealized, in spite of notable efforts by Keisler [20], Stroyan [31], Benci and Di Nasso [4], and others. Sociological reasons — the inherent conservativity of the mathematical community, the lack of a concentrated effort at proselytizing — are often mentioned as an explanation. There is also the fact that "nonstandard" methods, at least in the form in which they are usually presented, require heavier reliance on formal logic than is customary in mathematics at large. While acknowledging much truth to all of the above, here I shall concentrate on another contributing difficulty. At the risk of an overstatement, it is this: while it is undoubtedly possible to do calculus by means of infinitesimals in the Robinsonian framework, it does *not* seem possible to do calculus *only* by means of infinitesimals in it. In particular, the promise to replace the ε-δ method by the use of infinitesimals cannot be carried out in full.

*Department of Mathematics, The City College of CUNY, New York, NY 10031.
`khrbacek@ccny.cuny.edu`

In Section 4.1 I examine this shortcoming in detail, review earlier relevant work, and propose a general plan for extending the Robinsonian framework with the goal of remedying this problem and — possibly — diminishing the need for formal logic as well. Section 4.2 contains a few examples intended to illustrate how mathematical arguments can be conducted in this extended framework. Section 4.3 presents an axiomatic system in which the techniques of Section 4.2 can be formalized, and discusses the motivation and prospects for its further extension.

4.1 The Robinsonian framework

Here and in the rest of the paper, by *Robinsonian framework* I mean any presentation of "nonstandard" methods that postulates a fixed hierarchy of standard, internal and, in most cases, also external sets. Thus the original type-theoretic foundations of Robinson [29], the superstructure method of Robinson and Zakon [30] (also Chang and Keisler [6]), and direct use of ultrafilters à la Luxemburg [21], as well as axiomatic nonstandard set theories like **HST** [13, 18, 19] or Nelson's **IST** [23], are covered by the term, and the discussion in this section applies to all of them. I present the arguments in the "internal picture" employed in **IST** and **HST**; that is, for example, $\mathbf{R}$ denotes the set of all (internal) real numbers, and is referred to as *the standard set of reals*; if needed, $^\circ\mathbf{R}$ denotes the external set of all standard reals. Superstructure afficionados would use $^*\mathbf{R}$ for $\mathbf{R}$ and $\mathbf{R}$ for $^\circ\mathbf{R}$. The same conventions apply to the standard set of natural numbers $\mathbf{N}$ and other standard sets.

The paradigmatic example below, the familiar *nonstandard definition of continuity*, illustrates the difficulty I am concerned about.

Definition 4.1.1 *Let $f : \mathbf{R} \to \mathbf{R}$ be a standard function, and $x \in \mathbf{R}$ a standard real number.*

(i) f is continuous at x *iff for all infinitesimal h, $f(x+h) - f(x)$ is infinitesimal.*

(ii) f is (pointwise) continuous *iff for all standard $x \in \mathbf{R}$, f is continuous at x. Explicitly: $(\forall^{\mathrm{st}} x \in \mathbf{R})(\forall$ infinitesimal $h)$ $(f(x+h)-f(x)$ is infinitesimal$)$.*

It is a basic and useful fact of nonstandard analysis that the notion of continuity of a standard function at a standard point defined above can be extended in a natural way to the notion of continuity at a nonstandard point, and that a standard continuous function $f : \mathbf{R} \to \mathbf{R}$ is continuous at *all* $x \in \mathbf{R}$, even the nonstandard ones. But, what precisely does this mean in the Robinsonian framework, and how do we know that it is true? Certainly not by transfer! In the Robinsonian framework, for a statement about standard objects to be transferable from the standard to the internal universe, all of its quantifiers have to range over standard sets; formally, it has to be of the form φ^{st} where φ is an $\in$-formula (internal formula). Definition 4.1.1 is not of this form; the quantifier ($\forall$ infinitesimal h) ranges over internal sets. Briefly, **Definition 4.1.1 is not transferable.** A naive attempt to transfer 4.1.1(ii) will likely produce something along the lines of

$$(\forall x \in \mathbf{R})(\forall \text{ infinitesimal } h)\ (f(x+h) - f(x) \text{ is infinitesimal }).$$

As is well known, this statement is equivalent to *uniform* continuity of f (for standard f).

How then do we arrive at our "basic and useful fact"? Every treatment of elementary nonstandard analysis has to answer this question somehow. Moreover, similar difficulties appear with derivatives, integrals —in fact, with every concept defined by nonstandard methods. There seems to be little explicit attention paid to this issue in the literature. Important exceptions are the writings of Péraire [25]-[27], Gordon [11, 12], and Andreev's thesis [1]; discussions of a number of points considered in this paper can be found there. I realized the crucial importance of this issue for teaching of nonstandard analysis during O'Donovan's talk in Aveiro. While describing his experiences with the nonstandard definition of derivative, O'Donovan recounted some questions his students typically ask: "Can we use this formula when x is not standard? When f is not standard?" The answer of course is **NO** —but what then are they supposed to use? After all, a standard function like $\sin x$ does have a derivative at all x![1]

Three implicit responses applicable in Robinsonian framework can be discerned.

Response I (Robinson [29], Goldblatt [10]).

Although Definition 4.1.1 is not expressible by an $\in$-formula, it is *equivalent* to an $\in$-formula, namely, to the *standard definition of continuity.*

Definition 4.1.2 *Let $f : \mathbf{R} \to \mathbf{R}$ be a standard function, and $x \in \mathbf{R}$ standard. f is* continuous at x *iff* $(\forall^{\text{st}}\varepsilon > 0)(\exists^{\text{st}}\delta > 0)(\forall^{\text{st}}y \in \mathbf{R})(\,|y - x| < \delta \Rightarrow |f(y) - f(x)| < \varepsilon)$.

The formula on the right side of Definition 4.1.2 *is* transferable, and yields a natural notion of continuity for all $f : \mathbf{R} \to \mathbf{R}$ and all $x \in \mathbf{R}$ that agrees with Definition 4.1.1 for standard f and x.

[1] I am grateful to R. O'Donovan for many subsequent email exchanges that have been extremely helpful in further clarification of the difficulties with using infinitesimals to teach calculus.

Definition 4.1.3 *Let* $f : \mathbf{R} \to \mathbf{R}$ *be any (internal) function, and* $x \in \mathbf{R}$ *any real.* f *is* continuous at x *iff*

$$(\forall \varepsilon > 0)(\exists \delta > 0)(\forall y \in \mathbf{R})(|y - x| < \delta \Rightarrow |f(y) - f(x)| < \varepsilon).$$

The problem is resolved, but at the cost of a relapse to the usual ε-δ definition of continuity, at the internal level. This response also illustrates one of the chief reasons why nonstandard methods have to rely so heavily on formal logic. In Definitions 4.1.1(i) and 4.1.2 we have two equivalent formulas, of which one is transferable and the other is not. Transferability is an attribute of formulas; it is a logical, metamathematical concept.

Response II (Nelson [23]).

First, we use Definition 4.1.1(i) to define continuity for standard f, x. Then we let

$C := {}^s\{\langle f, x\rangle : f : \mathbf{R} \to \mathbf{R},\ x \in \mathbf{R},\ f, x \text{ standard},\ f \text{ continuous at } x\}$
(C is a standard set), and finally, for any (internal) f and x, define

Definition 4.1.4 f *is* continuous at x *iff* $\langle f, x\rangle \in C$.

If f is standard and continuous at all standard $x \in \mathbf{R}$ then $(\forall^{\mathrm{st}} x)(\langle f, x\rangle \in C)$, this formula is transferable, and gives $(\forall x)(\langle f, x\rangle \in C)$, i.e., f is continuous at all $x \in \mathbf{R}$, as desired.

The problem here is that this ("somewhat implicit" [23]) definition of continuity is completely divorced from the usual intuition (captured for standard f, x, in different ways, both by the standard ε-δ definition and by the nonstandard definition using infinitesimals): a function f is continuous at x if arguments "near" x yield values "near" $f(x)$. According to 4.1.4, the implicit meaning of the statement "f is continuous at x" for standard f and unlimited x is "$x \in {}^s\{y \in \mathbf{R} : y \text{ standard}, f \text{ continuous at } y\}$"; it is not related to the behavior of f near x in the sense of the order topology on $\mathbf{R}$. Similarly, continuity of $f(x) := x^\nu$, where $\nu \in \mathbf{N}$ is unlimited, translates to "$\nu \in {}^s\{n \in \mathbf{N} : f_n(x) := x^n \text{ is continuous }\}$"; i.e., x^ν is continuous "because" x^n is a continuous function for all finite n! Definition 4.1.4 can be decoded via Nelson's reduction algorithm; it then gives the usual ε-δ definition, at the internal level, as in Response I. We note that here too there are two equivalent formulas, one of which transfers and the other does not.

Response III.

This is a feasible response in an approach based on superstructures, even though I found no discussion of it in the literature. While considering it I switch to the asterisk notation.

We recall that the definition of (pointwise) continuity in terms of infinitesimals can be generalized to standard $f : T_1 \to T_2$, where T_1 and T_2 are arbitrary standard topological spaces. The internal open sets of $^*\mathbf{R}$ form a base for an (external) topology on $^*\mathbf{R}$, the *Q-topology*. The meaning of "f is continuous at x" for internal f and $x \in {}^*\mathbf{R}$ can be given by "f is continuous at x in the Q-topology on $^*\mathbf{R}$." As noted above, this concept has a nonstandard definition, although applying it requires working with $^{\circledast}(^*\mathbf{R})$ in a "second-order" enlargement[2] of the superstructure that contains $^*\mathbf{R}$. Yet the difficulty is not resolved. In order to prove that the two definitions of continuity are equivalent for standard f and x, one needs to apply transfer to the (equivalent) standard definitions in terms of open neighborhoods, i.e., fall back on the ε-δ method, in topological disguise. The Robinsonian framework does not provide for direct transfer of nonstandard definitions. This response also begs the question, how do we know that $^{\circledast}(^*f)$ is continuous, and so on. Clearly, an infinite sequence of consecutive enlargements would be needed.

It seems that every attempt to define continuity ultimately has to be grounded on the ε-δ method. As remarked above, the same difficulty appears with derivatives, integrals —in fact, with all standard concepts introduced by nonstandard methods. I see it as a serious problem for the Robinsonian framework, if not as a research tool, surely as a teaching tool and, fundamentally, as a satisfactory answer to the question about the place of infinitesimals, and nonstandard objects in general, in mathematics.

What is to be done?

Contemplation of the three responses suggests some ideas. First, we need to abandon the fixed distinction between standard and nonstandard and be able to treat any (internal) object as if it were standard, and in this capacity subject to application of nonstandard definitions and theorems. This is the idea of **relativization of standardness**. Second, we need to be able to transfer properties described by *arbitrary* (external) formulas, not just $\in$-formulas; for emphasis, I refer to this facility as **general transfer**.

Both ideas have some history in the literature of nonstandard analysis. A definition of relative standardness seems to appear first in Cherlin and Hirschfeld [7], although its model-theoretic roots can be discerned in [6]; but the subsequent development occurred mostly in the axiomatic setting. Gordon [11] defined two notions of relative standardness in **IST** (one of them is essentially the same as in [7]). Wallet [24] proposed to use a binary relative

[2]For some applications of "second-order" enlargements see Molchanov [22]. In the alpha-theory of Benci and DiNasso [4], the $*$-embedding is defined for all sets, but $^*(^*\mathbf{R})$ is only ω_1-saturated, and monads in the Q-topology on $^*\mathbf{R}$ are trivial.

standardness predicate as a primitive in an axiomatic treatment, an idea that was developed systematically by Péraire in [25]. The notion of relative standardness *stratifies* the universe into *levels of standardness*. Both Gordon and Péraire give a nonstandard definition of continuity applicable to all f and x (see Definition 4.2.2), and numerous other examples (see also [1, 12, 26, 27, 28]). Gordon's approach does not work as smoothly for concepts whose definitions involve shadows, such as derivative, because standardization does not hold in full. A sufficiently strong (for this purpose) standardization does hold in Péraire's theory **RIST**. Another difference is that, unlike Gordon's, Péraire's relative standardness predicate is a *total* preordering. Yet another stratified nonstandard set theory (not employing a binary relative standardness predicate) was put forward by Fletcher [9].

None of [9, 11, 25] give an explicit formulation of transfer for more than just $\in$-formulas. To my knowledge, the idea of (more) general transfer appears first in the work of Benninghofen and Richter [5]. The main result of [5] is a transfer theorem for a certain (complicated) class of $\in$-st-formulas; Cutland [8] gives some simpler special cases. Although the class of transferable formulas is limited, it has led to interesting applications (see the proof of l'Hôpital rule in Section 4.2, and [5, 32, 33]). However, the idea of relative standardness is not explicit in [5]. Further discussion of the mutual relationship of these various approaches can be found in [16].

In my opinion, the decisive step needed to resolve the difficulty discussed above is to combine relative standardness with general transfer. This step was taken by Péraire in [26] with his proof (in **RIST**) of "polytransfer," essentially, transfer for all formulas that do not quantify over levels of standardness. Many standard concepts have satisfactory nonstandard, transferable definitions in **RIST**. Nevertheless, there are situations (see Example 4 in Section 4.2) where quantification over levels of standardness is both natural and necessary. Moreover, the need to single out the special classes of formulas to which principles of **RIST** apply increases reliance of the framework on formal logic.

It is my belief that for a theory of the "nonstandard" to be fully satisfactory, both foundationally and practically, *all* of its *principles need to apply uniformly at all levels, and to all formulas.* In addition to a complete resolution of the difficulty that is the subject of this discussion, such framework would also diminish the need for appeals to formal logic in practical work: all formulas would be transferable, and equivalent formulas would have equivalent transfers. The axiomatic system **FRIST** presented in Section 4.3 achieves these objectives for internal sets. The examples in the next section show some of the power of internal methods extended by relativized standardness and general transfer.

4.2 Stratified analysis

This section gives several examples intended to illustrate "nonstandard" mathematics in stratified framework. The presentation is informal; an axiomatic system **FRIST** in which all of the arguments in this section can be formalized is described fully in Section 4.3.

Our basic assumption is that the universe of mathematical objects (= sets) is stratified by a binary relation $\sqsubseteq$ into "levels of standardness." The notation $x \sqsubseteq y$ is to be read "*x is standard relative to y*" or "*x is y-standard*", and it is a dense total preordering with a least element 0. The 0-standard sets are called simply *standard*; they form the lowest level of standardness. The class of x-standard sets is denoted $\mathbb{S}_x$.

Let α be a set; *relativization of standardness* to level α consists in regarding $\mathbb{S}_\alpha$, rather than $\mathbb{S}_0$, as the lowest level of standardness. A statement about standard sets is *relativized* to level α by replacing all references to "standard" with "α-standard" (more explicitly, by replacing "*x is y-standard*" with "*x is $\langle y, \alpha \rangle$-standard*").

The key principle that governs the stratified universe is **general transfer**: *All valid statements about standard sets remain valid when relativized to any level α.*

Mathematical practice proceeds by enriching the language with new definitions. We make it a *general rule* that, whenever some standard notion (a new predicate or function) is defined for standard x in terms of some property of x, the definition of the notion is extended to all x by relativizing the defining statement to level x (the definition has to be *fully relativized* —see Section 4.3).

In addition, we make the familiar assumptions that

- all the usual mathematical operations preserve standardness (*the class $\mathbb{S}_0$ of all standard sets satisfies* **ZFC**);
- given any property of x and any standard set A, there is a standard set B, whose standard elements are precisely those standard $x \in A$ having that property (*standardization*);
- for every level $\alpha \sqsupset 0$ there exist α-standard unlimited natural numbers (a much stronger *idealization* is available — see Section 4.3 — but this suffices for calculus).

The examples that follow illustrate the formulation of relativizations and the general rule.

Definition 4.2.1

(i) $r \in \mathbf{R}$ *is* α-limited *iff* $|r| < n$ *for some α-standard* $n \in \mathbf{N}$.

(ii) $h \in \mathbf{R}$ *is* α-infinitesimal *iff* $|h| < \frac{1}{n}$ *for all* α*-standard* $n \in \mathbf{N}$, $n \neq 0$.

(iii) $r \approx^\alpha s$ *iff* $r - s$ *is* α*-infinitesimal.*

(iv) For α*-limited* $r \in \mathbf{R}$*, the* α-shadow *of* r, $\mathbf{sh}_\alpha(r)$, *is the unique* α*-standard* $s \in \mathbf{R}$ *such that* $r \approx^\alpha s$.

(Remark. *As usual, every limited (i.e., 0-limited) real has a unique 0-shadow, by standardization. By general transfer, every* α*-limited real has a unique* α*-shadow.)*

EXAMPLE 1. Continuity (Gordon [11, 12], Péraire [25, 27]).

Definition 4.1.1 relativizes as follows:

Definition 4.2.2 *Let* $f : \mathbf{R} \to \mathbf{R}$ *and* $x \in \mathbf{R}$.

(i) f *is* continuous at x *iff for all* $\langle f, x\rangle$*-infinitesimal* h, $f(x+h) - f(x)$ *is* $\langle f, x\rangle$*-infinitesimal.*

(ii) f *is (pointwise)* continuous *iff it is continuous at all* f*-standard* x*. Explicitly: for all* f*-standard* x *and all* $\langle f, x\rangle$*-infinitesimal* h, $f(x+h)-f(x)$ *is* $\langle f, x\rangle$*-infinitesimal.*

The closure of $\langle f, x\rangle$-standard sets under set-theoretic operations (due to transfer of this property from $\mathbb{S}_0$) implies that f and x are $\langle f, x\rangle$-standard. In particular, for f-standard x, "$\langle f, x\rangle$-standard" is equivalent to "f-standard", and we have:

• f is *continuous* iff for all f-standard x and all f-infinitesimal h, $f(x+h) - f(x)$ is f-infinitesimal.

If f is α-standard, transfer gives:

f is *continuous* iff it is continuous at all α-standard $x \in \mathbf{R}$.

But every $x \in \mathbf{R}$ is α-standard for some $\alpha \sqsupseteq f$, so we have also:

f is *continuous* iff it is continuous at all $x \in \mathbf{R}$.

This reasoning is a special case of a general *global transfer principle* (Section 4.3, Proposition 4.3.1).

In contrast with •, relativized definition of uniform continuity is

Definition 4.2.3 *Let* $f : \mathbf{R} \to \mathbf{R}$. f *is* uniformly continuous *iff for all* x *and all* f*-infinitesimal* h, $f(x+h) - f(x)$ *is* f*-infinitesimal.*

EXAMPLE 2. Derivative (Péraire [27]).

Relativization of the usual nonstandard definition of derivative gives

Definition 4.2.4 *Let $f : \mathbf{R} \to \mathbf{R}$ and $x \in \mathbf{R}$. f is* differentiable at x *iff there is an $\langle f, x\rangle$-standard $L \in \mathbf{R}$ such that $\frac{f(x+h)-f(x)}{h} - L$ is $\langle f, x\rangle$-infinitesimal, for all $\langle f, x\rangle$-infinitesimal $h \neq 0$.*

If this is the case, $f'(x) := L = \mathbf{sh}_{\langle f,x\rangle}\left(\frac{f(x+h)-f(x)}{h}\right)$.

We next give two proofs of an elementary result from calculus, in order to illustrate two styles of work that are supported by the stratified framework.

Proposition 4.2.5 *If f is differentiable at x then f is continuous at x.*

Proof 1. By Definition 4.2.4, for any $\langle f, x\rangle$-infinitesimal h, $f(x + h) - f(x) = Lh + kh$ where k is $\langle f, x\rangle$-infinitesimal. The usual arguments show that the product of an $\langle f, x\rangle$-limited real and an $\langle f, x\rangle$-infinitesimal is an $\langle f, x\rangle$-infinitesimal, and that the sum of two $\langle f, x\rangle$-infinitesimals is $\langle f, x\rangle$-infinitesimal. □

Proof 2. For any standard f and x and any infinitesimal h, $f(x+h)-f(x) = Lh+kh$ and $Lh+kh$ is infinitesimal. Hence a standard function f differentiable at a standard x is continuous at x. By transfer, any function f differentiable at any x is continuous at x. □

The style of the first proof is to give the argument uniformly for all f, x. The advantage is that there is no explicit evocation of transfer; the disadvantage is the need to keep track of the level $\langle f, x\rangle$ throughout the argument.

The second proof is Robinson's, i.e., for standard f, x; followed by transfer of the result to all f, x. The formula being transferred is not an $\in$-formula, but it is easily seen that our general transfer applies to it. This is another instance of the global transfer principle from Proposition 4.3.1 in Section 4.3: any statement that invokes relative standardness only via previously defined standard notions and is valid for all standard sets remains valid for all (internal) sets.

One can, if one so chooses, work in stratified analysis exactly as one would in the Robinsonian framework; that is, give the nonstandard definitions and proofs for standard arguments only. This is what we do in the remaining examples. But in the stratified framework, all such definitions and proofs automatically transfer to definitions and proofs that are meaningful and natural for *all* arguments (as long as no essential use is made of external sets).

Stratified analysis also provides opportunities for proofs and constructions that are not readily available in the Robinsonian framework. They have not been much explored as yet; two examples of what is possible are given below. First we list some general results about infinitesimals.

Lemma 4.2.6

(*a*) *If $x \in \mathbf{R}$ is α-infinitesimal and $\beta \sqsubseteq \alpha$ then x is β-infinitesimal.*

(*b*) *Every* α*-limited natural number is* α*-standard.*

(*c*) *If* y *is infinitesimal then there is an infinitesimal* x *such that* y *is* x*-infinitesimal.*

Proof. (a) is trivial from transitivity of $\sqsubseteq$, and (b) is just transfer (to level α) of the well-known fact that every limited natural number is standard.

(c) is a consequence of (b) and density of $\sqsubseteq$. Let y be a positive infinitesimal and let $\nu \in \mathbf{N}$ be such that $\nu \leq \frac{1}{y} < \nu + 1$. Then $0 \sqsubset \nu$; we fix α such that $0 \sqsubset \alpha \sqsubset \nu$. By (b), ν is α-unlimited and so $y < \frac{1}{n}$ for all α-standard n. So y is α-infinitesimal, hence x-infinitesimal for any α-standard infinitesimal x. □

EXAMPLE 3. l'Hôpital Rule (Benninghofen and Richter [5]; see also [12]).

Proposition 4.2.7 (*l'Hôpital Rule*)
If $\lim_{x \to a} |g(x)| = \infty$ *and* $lim_{x \to a} \frac{f'(x)}{g'(x)} = d \in \mathbf{R}$ *then* $lim_{x \to a} \frac{f(x)}{g(x)} = d$.

Proof. It suffices to prove the proposition for standard f, g, a, d. Also, w.l.o.g. we can let $a = 0$ (replace x by $x - a$). Let x be infinitesimal and y be x-infinitesimal. By Cauchy's Theorem, there is η between x and y (hence, η is infinitesimal) such that $\frac{f(y)-f(x)}{g(y)-g(x)} = \frac{f'(\eta)}{g'(\eta)} \approx d$. Now factor

$d \approx \frac{f(y)-f(x)}{g(y)-g(x)} = \frac{f(y)-f(x)}{g(y)} \times \frac{g(y)}{g(y)-g(x)}$ and observe that $\frac{f(x)}{g(y)} \approx 0$, $\frac{g(x)}{g(y)} \approx 0$.

($\lim_{x \to 0} |g(x)| = \infty$ implies that for all infinitesimal z, $g(z)$ is unlimited. By transfer to x-level, for all x-infinitesimal z, $g(z)$ is x-unlimited. As y is x-infinitesimal, $\frac{f(x)}{g(y)}$ and $\frac{g(x)}{g(y)}$ are x-infinitesimal.)

It follows that the first factor is infinitely close to $\frac{f(y)}{g(y)}$ and the second to 1. From properties of infinitesimals we conclude that $\frac{f(y)}{g(y)} \approx d$.

By Lemma 4.2.6(c), every infinitesimal y is x-infinitesimal for some infinitesimal x. Hence $\frac{f(y)}{g(y)} \approx d$ holds for *every* infinitesimal y, and we are done.□

Remark. The assumption of density of levels allows for a simpler argument than that of [5, 12], but it is not essential.

EXAMPLE 4. Higher derivatives.

We assume that f, x are standard and $f'(y)$ exists for all $y \approx x$. If $f''(x) = L \in \mathbf{R}$ exists, then $L \approx \frac{f(x+2h)-2f(x+h)+f(x)}{h^2}$ holds for all $h \approx 0$, $h \neq 0$. However, the converse of this statement is false; existence of a standard $L \in \mathbf{R}$ with the above property does not imply that $f''(x)$ exists.

In the stratified framework, we can give a description of $f''(x)$ in terms of values of f, analogous to Definition 4.2.4, using two levels of standardness.

Proposition 4.2.8 *Assume that f and x are standard and $f'(y)$ exists for all $y \approx x$. Then $f''(x)$ exists iff there is a standard $L \in \mathbf{R}$ such that*

$$L \approx \frac{f(x+h_0+h_1) - f(x+h_0) - f(x+h_1) + f(x)}{h_0 h_1}$$

for all $h_0 \approx 0,\ h_1 \approx^{h_0} 0,\ h_0, h_1 \neq 0$.

If this is the case, $f''(x) = L$.

As usual in the stratified framework, the proposition holds for arbitrary f and x provided $\approx$ is replaced by $\approx^{\langle f,x\rangle}$ and "standard" by "$\langle f,x\rangle$-standard."

Proof. We assume that $f'(y)$ exists for all $y \approx x$; in particular, $f'(x)$ and $f'(x+h_0)$ exist. Hence

$$\tfrac{f(x+h_0+h_1)-f(x+h_0)}{h_1} = f'(x+h_0) + k_1, \qquad \tfrac{f(x+h_1)-f(x)}{h_1} = f'(x) + k_2,$$

where $k_1, k_2 \approx^{h_0} 0$.

From this we get

$$Q := \tfrac{f(x+h_0+h_1)-f(x+h_0)-f(x+h_1)+f(x)}{h_0 h_1} = \tfrac{f'(x+h_0)-f'(x)}{h_0} + \tfrac{(k_1-k_2)}{h_0}.$$

We note that $(k_1 - k_2) \approx^{h_0} 0$ and hence $\frac{(k_1-k_2)}{h_0} \approx^{h_0} 0$, by transfer to the level h_0 of the fact that a quotient of an infinitesimal by a standard real $\neq 0$ is infinitesimal. In particular, $\frac{(k_1-k_2)}{h_0} \approx 0$.

If $L := f''(x)$ exists, we have $\frac{f'(x+h_0)-f'(x)}{h_0} = L + k$ for $k \approx 0$ and hence $Q \approx L$.

Conversely, if $Q \approx L \in {}^{\circ}\mathbf{R}$ for all h_0, h_1 as above, we have $\frac{f'(x+h_0)-f'(x)}{h_0} \approx L$ for all $h_0 \approx 0, h_0 \neq 0$, and hence $f''(x)$ exists and equals to L. □

By induction we get a characterization of $f^{(n)}(x)$ valid for any *standard* $n \in \mathbf{N}$.

Proposition 4.2.9 *Assume that f and x are standard and $f^{(n-1)}(y)$ exists for all $y \approx x$. Then $f^{(n)}(x)$ exists iff there is a standard $L \in \mathbf{R}$ such that*

$$L \approx \frac{1}{h_0 \ldots h_{n-1}} \sum_i (-1)^{i_0+\ldots+i_{n-1}} f(x + h^{i_0} + \ldots + h^{i_{n-1}})$$

holds for all $\langle h_0, \ldots, h_{n-1}\rangle$, where $i = \langle i_0, \ldots, i_{n-1}\rangle \in \{0,1\}^n$, $h^{i_k} := h_k$ if $i_k = 0$, $h^{i_k} := 0$ if $i_k = 1$; $h_0 \approx 0$, $h_k \approx^{h_{k-1}} 0$ for $0 < k < n$, and all $h_k \neq 0$.

If this is the case, $f^{(n)}(x) = L$.

General transfer implies that Proposition 4.2.9 holds for **all** n, even the unlimited (hyperfinite) ones, provided $\approx$ is replaced everywhere by $\approx^n$ and "standard L" by "n-standard L". As there are functions that have derivatives of all orders, *there have to exist "strongly decreasing" sequences of infinitesimals of any hyperfinite length n: $\langle h_0, \ldots, h_{n-1}\rangle$ where each h_k is h_{k-1}-infinitesimal.* Such constructions are not available in **RIST** or in any other framework for nonstandard mathematics.

4.3 An axiomatic system for stratified set theory

The theory **FRIST** (*Fully Relativized Internal Set Theory*) presented here is formalized in first-order logic with equality and two primitive binary predicates, $\in$ (membership) and $\sqsubseteq$ (relative standardness). It postulates the axioms of **ZFC** (the schemata of separation and replacement for $\in$-formulas only), and $(\forall x)(x \sqsubseteq x)$, $(\forall x, y, z)(x \sqsubseteq y \wedge y \sqsubseteq z \rightarrow x \sqsubseteq z)$, $(\forall x, y)(x \sqsubseteq y \vee y \sqsubseteq x)$, $(\forall x)(\varnothing \sqsubseteq x)$, $(\exists x)(\neg\, x \sqsubseteq \varnothing)$. Thus $\sqsubseteq$ is a nontrivial total preordering with a least element $0 = \varnothing$; $x \sqsubseteq y$ reads "*x is standard relative to y*" or "*x is y-standard.*" We also write $x \sqsubset y$ for $x \sqsubseteq y \wedge \neg\, y \sqsubseteq x$, and postulate that $\sqsubseteq$ is *dense*: $(\forall x, y)(x \sqsubset y \rightarrow (\exists z)(x \sqsubset z \sqsubset y))$.

Let α be a set; we let $\mathbb{S}_\alpha := \{x : x \sqsubseteq \alpha\}$ be the class of all α-standard sets; in particular $\mathbb{S} := \mathbb{S}_0 = \{x : x \sqsubseteq 0\}$ is the class of standard sets. If $\bar{\alpha}$ is a list $\alpha_1, \ldots, \alpha_n$, the statement that x is $\bar{\alpha}$-standard is shorthand for "x is $\langle \alpha_1, \ldots, \alpha_n\rangle$-standard"; it is easy to see that (in **FRIST**) this is equivalent to "x is β-standard" for $\beta := \max\{\alpha_1, \ldots, \alpha_n\}$.

For any α let $x \sqsubseteq_\alpha y \equiv (x \sqsubseteq \alpha \wedge y \sqsubseteq \alpha) \vee (x \sqsubseteq y)$. This "relativized" relative standardness predicate treats $\mathbb{S}_\alpha$ as the "standard" (i.e. "level 0") universe, and keeps the higher levels unchanged.

Let Φ be any $\in$-$\sqsubseteq$-formula. Φ^α denotes the *relativization of Φ to level* α, the formula obtained from Φ by replacing each occurrence of $\sqsubseteq$ by $\sqsubseteq_\alpha$. Informally, each occurrence of "x is y-standard" is replaced by "x is $\langle y, \alpha\rangle$-standard." Clearly $x \sqsubseteq_0 y$ is equivalent to $x \sqsubseteq y$, and Φ^0 to Φ.

We now state the principal axioms of **FRIST**.

Transfer: For all α, $(\forall \bar{x} \in \mathbb{S}_0)(\Phi^0(\bar{x}) \leftrightarrow \Phi^\alpha(\bar{x}))$.

Standardization: For all $\bar{x}$,
$(\forall x \in \mathbb{S}_0)\ (\exists y \in \mathbb{S}_0)\ (\forall z \in \mathbb{S}_0)\ (z \in y \leftrightarrow z \in x \wedge \Phi^0(z, x, \bar{x}))$.

Idealization: For all $0 \sqsubset \alpha$, $A, B \in \mathbb{S}_0$ and $\bar{x}$,
$(\forall a \in A^{\text{fin}} \cap \mathbb{S}_0)(\exists x \in B)(\forall y \in a)\ \Phi^\alpha(x, y, \bar{x}) \leftrightarrow$
$(\exists x \in B)(\forall y \in A \cap \mathbb{S}_0)\ \Phi^\alpha(x, y, \bar{x})$.

Transfer captures the idea of *full relativization*: whatever is true about the standard universe $\mathbb{S}_0$ is also true about each relativized standard universe $\mathbb{S}_\alpha$. Precisely, for any α and any statement Φ of the $\in$-$\sqsubseteq$-language, if **FRIST** $\vdash \Phi$ then also **FRIST** $\vdash \Phi^\alpha$. In particular, it follows that we can replace 0 by any α in standardization, and by any $\beta \sqsubset \alpha$ in the other two schemata. It is also easy to show that each $\mathbb{S}_\alpha$ satisfies **ZFC**. Details of these and other results about **FRIST**, as well as a discussion of its relationship to **RIST** and **IST**, can be found in [15].

We prove a version of transfer from the standard universe to the entire internal universe. Let $\Phi(\bar{x})$ be an $\in$-$\sqsubseteq$-formula. We define a new predicate $P(\bar{x})$ by postulating $P(\bar{x}) \equiv \Phi^{\bar{x}}(\bar{x})$, and say that the definition of P is *fully relativized* if it has this form. We note that for any α and any $\bar{x} \in \mathbb{S}_\alpha$, $P(\bar{x}) \leftrightarrow \Phi^\alpha(\bar{x})$; in particular, $P(\bar{x}) \leftrightarrow \Phi(\bar{x})$ for standard $\bar{x}$.

Let the definitions of $P_1, \ldots, P_n$ be fully relativized. Given any formula $\Psi(\bar{x})$ in the $\in$-$\bar{P}$-language, we denote the formula obtained by restricting all quantifiers in Ψ to $\mathbb{S}_\alpha$ by $\mathbb{S}_\alpha \vDash \Psi(\bar{x})$.

Proposition 4.3.1 (*Global Transfer*) $(\forall \bar{x} \in \mathbb{S}_\alpha)[(\mathbb{S}_\alpha \vDash \Psi(\bar{x})) \leftrightarrow \Psi(\bar{x})]$.

Proof. If Ψ is a formula in the $\in$-$\bar{P}$-language, we let $\Psi^{(\alpha)}$ be the formula obtained from Ψ by replacing (first) each occurrence of $(\exists x)$ [$(\forall x)$, resp.] in Ψ by $(\exists x \in \mathbb{S}_\alpha)$ [$(\forall x \in \mathbb{S}_\alpha)$, resp.], and (then) each occurrence of $P_i(\bar{x})$ by $\Phi_i^\alpha(\bar{x})$.

We prove that for all α and all $\bar{x} \in \mathbb{S}_\alpha$, $(\mathbb{S}_\alpha \vDash \Psi(\bar{x})) \leftrightarrow \Psi^{(\alpha)}(\bar{x}) \leftrightarrow \Psi(\bar{x})$, by induction on complexity of Ψ.

If $\Psi(\bar{x})$ is $P_i(\bar{x})$, the claim follows from the definitions and remarks above. If $\Psi(\bar{x})$ is "$x_i \in x_j$" or "$x_i = x_j$" the assertion is trivial, as are the induction steps corresponding to logical connectives.

Assume that, for all β and $\bar{x}, y \in \mathbb{S}_\beta$, $(\mathbb{S}_\beta \vDash \Psi(\bar{x}, y)) \leftrightarrow \Psi^{(\beta)}(\bar{x}, y) \leftrightarrow \Psi(\bar{x}, y)$.

Let $\bar{x} \in \mathbb{S}_\alpha$; we then have $(\mathbb{S}_\alpha \vDash (\forall y)\Psi(\bar{x}, y)) \leftrightarrow (\forall y \in \mathbb{S}_\alpha)(\mathbb{S}_\alpha \vDash \Psi(\bar{x}, y)) \leftrightarrow (\forall y \in \mathbb{S}_\alpha)\Psi^{(\alpha)}(\bar{x}, y) \leftrightarrow [(\forall y)\Psi(\bar{x}, y)]^{(\alpha)}$.

By transfer of the penultimate statement to level β we get its equivalence with $(\forall \beta \sqsupseteq \alpha)(\forall y \in \mathbb{S}_\beta)[\Psi^{(\alpha)}(\bar{x}, y)]^\beta$. It is easily seen that, for $\alpha \sqsubseteq \beta$, $[\Psi^{(\alpha)}]^\beta \leftrightarrow \Psi^{(\beta)}$; hence the last statement is further equivalent to $(\forall \beta \sqsupseteq \alpha)$ $(\forall y \in \mathbb{S}_\beta)\Psi^{(\beta)}(\bar{x}, y) \leftrightarrow (\forall \beta \sqsupseteq \alpha)(\forall y \in \mathbb{S}_\beta)\Psi(\bar{x}, y) \leftrightarrow (\forall y)\Psi(\bar{x}, y)$. (For the last step, note that for every y there is $\beta \sqsupseteq \alpha$ such that $y \in \mathbb{S}_\beta$; e.g. $\beta = \langle \alpha, y \rangle$.) $\square$

Corollary 4.3.2 *For any* $A, \bar{p}$, $\{x \in A : \Psi(x, A, \bar{p})\}$ *is a set in* $\mathbb{S}_{\langle A, \bar{p} \rangle}$.

Proof. Fix α so that $A, \bar{p} \in \mathbb{S}_\alpha$. Standardization into $\mathbb{S}_\alpha$ yields a set $B \in \mathbb{S}_\alpha$ such that $\mathbb{S}_\alpha \vDash (\forall x)(x \in B \leftrightarrow x \in A \wedge \Psi(x, A, \bar{p}))$. By global transfer, $(\forall x)(x \in B \leftrightarrow x \in A \wedge \Psi(x, A, \bar{p}))$. $\square$

This corollary shows that we can use fully relativized predicates in definitions of sets without fear of encountering external sets. For example, f' is a standard function whenever f is, even though we use infinitesimals to define $f'(x)$.

The consistency of **FRIST** is established by the following theorem, whose proof can be found in [15], Theorems 4.7, 5.1, 5.2.

Theorem 4.3.3 ***FRIST*** *is a conservative extension of* ***ZFC****.*

In fact, ***FRIST*** *has an interpretation in* ***ZFC****, in which the class of standard sets is (definably) isomorphic to the class of all (****ZFC****) sets.*

An interpretation of **FRIST** in **ZFC** involves a new method of iterating ultrapowers, where the stages of the iteration used to obtain the final universe $^*\mathbb{V}$ are not indexed by any a priori given linear ordering $\langle \Lambda, \leq \rangle$, but by $\langle {}^*\Lambda, \leq^* \rangle$, a linear ordering constructed from $\langle \Lambda, \leq \rangle$ simultaneously with $^*\mathbb{V}$.

FRIST differs from the theory of the same name presented in [15] in two ways.

1. Idealization has been weakened to "bounded idealization" (see **FRBST** of [15]). Foundational reasons for this move are discussed at length in [16] in the context of **IST** vs. **BST**. Without some such weakening of idealization, the second part of Theorem 4.3.3 does not hold.

2. The postulate of density of levels has been added. This property holds in models from [15] based on $\Lambda = \mathbf{Q}$ or any other dense total ordering. There are foundational reasons that seem to favor this assumption, and it also has practical uses (see the proof of l'Hôpital rule).

Like **IST**, **FRIST** is a theory of internal sets. Nelson [23] exhibited a *reduction algorithm* for **IST**; it takes every (bounded) $\in$-st-formula to an $\in$-formula that is equivalent to it for all standard values of the free variables. The question arises, whether such algorithm is also available for **FRIST**. On one level, the answer is *YES*; such algorithm is provided by the interpretation of **FRIST** in **ZFC** from the proof of Theorem 4.3.3. On the other hand, $\in$-formulas yielded by this algorithm are far too complicated to be helpful with practical work. It is possible that a simpler, more natural reduction algorithm can be given, but this matter is still under investigation.

As demonstrated by Nelson and other adherents of **IST**, "internal" methods, cleverly used, suffice for a large area of "nonstandard" applications. Nevertheless, I have always maintained [13, 14, 16] that a truly comprehensive nonstandard set theory has to incorporate external sets as well. There are numerous constructs (nonstandard hulls, Loeb measures,...) where external sets are not just a convenient bookkeeping device but the object of interest per se; a foundational framework that does not allow these constructs can hardly be universally acceptable to practitioners of "nonstandard" methods. The foundational aspirations of nonstandard set theory also require external sets; "they are there," and have to be accounted for. The guiding "maxim" of this paper, to wit, that all principles should apply uniformly at all levels and to all formulas, gives yet another reason why external sets are necessary. In **FRIST**, transfer, standardization and idealization do indeed satisfy it, but the axioms of **ZFC** do not! In particular, separation applies in **FRIST** only to $\in$-formulas. As soon as we attempt to extend it to all formulas, we introduce external sets (e.g. the set of standard integers $\{n \in \mathbf{N} : n \sqsubseteq 0\}$).

A thorough discussion of various axiomatic nonstandard set theories for external sets and of the difficulties they face can be found in Kanovei and Reeken's monograph [19] and in the survey article [16]. Perhaps the only researcher who considered an extension of the idea of relativization of standardness to external sets in an axiomatic framework was David Ballard. In [3], Ballard proposed an axiomatic system **EST**, where external sets are allowed and any set can be regarded as "standard." **EST** was inspired by Fletcher's [9], and employs neither the binary standardness predicate nor general transfer. After the paper [15] was written, Ballard and I concurrently started to develop an extension of its ideas to external sets. David Ballard died unexpectedly in May 2004 in the midst of his work. My research on this topic is still in progress [17]; results obtained so far indicate that an extension of **FRIST** to a theory that incorporates external sets is both possible and natural.

References

[1] P.V. Andreev, The notion of relative standardness in axiomatic systems of nonstandard analysis, Thesis, Dept. Math. Mech., State University of N.I. Lobachevskij in Nizhnij Novgorod, 2002, 96 pp. (Russian).

[2] P.V. Andreev, "On definable standardness predicates in internal set theory", Mathematical Notes, **66** (1999) 803 - 809 (Russian).

[3] D. Ballard, *Foundational Aspects of "Non" standard Mathematics,* Contemporary Mathematics, vol. 176, American Mathematical Society, Providence, R.I., 1994.

[4] V. Benci and M. Di Nasso, "Alpha-theory: an elementary axiomatics for nonstandard analysis", Expositiones Math., **21** (2003) 355-386.

[5] B. Benninghofen and M.M. Richter, "A general theory of superinfinitesimals", Fund. Math., **123** (1987) 199–215.

[6] C.C. Chang and H.J. Keisler, *Model Theory,* 3rd edition, North-Holland Publ. Co., 1990.

[7] G. Cherlin and J. Hirschfeld, Ultrafilters and ultraproducts in non-standard analysis, in *Contributions to Non-standard Analysis*, ed. by W.A.J. Luxemburg and A. Robinson, North Holland, Amsterdam 1972.

[8] N.J. Cutland, "Transfer theorems for π-monads", Ann. Pure Appl. Logic, **44** (1989) 53–62.

[9] P. Fletcher, "Nonstandard set theory", J. Symbolic Logic, **54** (1989) 1000–1008.

[10] R. Goldblatt, *Lectures on the Hyperreals: An Introduction to Nonstandard Analysis*, Graduate Texts in Math. 188, Springer-Verlag, New York, 1998.

[11] E.I. Gordon, "Relatively nonstandard elements in the theory of internal sets of E. Nelson", Siberian Mathematical Journal, **30** (1989) 89–95 (Russian).

[12] E.I. Gordon, *Nonstandard Methods in Commutative Harmonic Analysis*, American Mathematical Society, Providence, Rhode Island, 1997.

[13] K. Hrbacek, "Axiomatic foundations for nonstandard analysis", Fundamenta Mathematicae, **98** (1978) 1–19; *abstract* in J. Symbolic Logic, **41** (1976) 285.

[14] K. Hrbacek, "Nonstandard set theory", Amer. Math. Monthly, **86** (1979) 1–19.

[15] K. Hrbacek, Internally iterated ultrapowers, in *Nonstandard Models of Arithmetic and Set Theory*, ed. by A. Enayat and R. Kossak, Contemporary Math. 361, American Mathematical Society, Providence, R.I., 2004.

[16] K. Hrbacek, Nonstandard objects in set theory, in *Nonstandard Methods and Applications in Mathematics*, ed. by N.J. Cutland, M. Di Nasso and D.A. Ross, Lecture Notes in Logic 25, Association for Symbolic Logic, Pasadena, CA., 2005, 41 pp.

[17] K. Hrbacek, Relative Set Theory, work in progress.

[18] V. Kanovei and M. Reeken, "Internal approach to external sets and universes", Studia Logica, Part I, **55** (1995) 227–235; Part II, **55** (1995) 347–376; Part III, **56** (1996) 293–322.

[19] V. Kanovei and M. Reeken, *Nonstandard Analysis, Axiomatically*, Springer-Verlag, Berlin, Heidelberg, New York, 2004.

[20] H.J. KEISLER, *Calculus: An Infinitesimal Approach,* Prindle, Weber and Scmidt, 1976, 1986.

[21] W.A.J. LUXEMBURG, A general theory of monads, in: W.A.J. Luxemburg, ed., *Applications of Model Theory to Algebra, Analysis and Probability,* Holt, Rinehart and Winston 1969.

[22] V.A. MOLCHANOV, "On applications of double nonstandard enlargements to topology", Sibirsk. Mat. Zh., **30** (1989) 64–71.

[23] E. NELSON, "Internal set theory: a new approach to Nonstandard Analysis", Bull. Amer. Math. Soc., **83** (1977) 1165–1198.

[24] Y. PÉRAIRE and G. WALLET, "Une théorie relative des ensembles intérnes", C.R. Acad. Sci. Paris, Sér. I, **308** (1989) 301–304.

[25] Y. PÉRAIRE, "Théorie relative des ensembles intérnes", Osaka Journ. Math., **29** (1992) 267–297.

[26] Y. PÉRAIRE, "Some extensions of the principles of idealization transfer and choice in the relative internal set theory", Arch. Math. Logic, **34** (1995) 269–277.

[27] Y. PÉRAIRE, "Formules absolues dans la théorie relative des ensembles intérnes", Rivista di Matematica Pura ed Applicata, **19** (1996) 27–56.

[28] Y. PÉRAIRE, "Infinitesimal approach of almost-automorphic functions", Annals of Pure and Applied Logic, **63** (1993) 283–297.

[29] A. ROBINSON, *Non-standard Analysis,* Studies in Logic and the Foundations of Mathematics, North-Holland, Amsterdam, 1966.

[30] A. ROBINSON and E. ZAKON, A set-theoretical characterization of enlargements, in W.A.J. Luxemburg, ed., *Applications of Model Theory to Algebra, Analysis and Probability,* Holt, Rinehart and Winston 1969.

[31] K.D. STROYAN, *Foundations of Infinitesimal Calculus*, 2nd ed., Academic Press 1997.

[32] K.D. STROYAN, B. BENNINGHOFEN and M.M. RICHTER, "Superinfinitesimals in topology and functional analysis", Proc. London Math. Soc., **59** (1989) 153–181.

[33] K.D. STROYAN, Superinfinitesimals and inductive limits, in *Nonstandard Analysis and its Applications*, ed. by N. Cutland, Cambridge University Press, New York, 1988.

5

ERNA at work

C. Impens* and S. Sanders**

Abstract

Elementary Recursive Nonstandard Analysis, in short ERNA, is a constructive system of nonstandard analysis proposed around 1995 by Chuaqui, Suppes and Sommer. It has been shown to be consistent and, without standard part function or continuum, it allows major parts of analysis to be developed in an applicable form. We briefly discuss ERNA's foundations and use them to prove a supremum principle and provide a square root function, both up to infinitesimals.

5.1 Introduction

Hilbert's Program, proposed in 1921, called for an axiomatic formalization of mathematics, together with a proof that this axiomatization is consistent. The consistency proof itself was to be carried out using only what Hilbert called *finitary* methods. The special character of finitary reasoning then would justify classical mathematics. In due time, many characterized Hilbert's informal notion of 'finitary' as that which can be formalized in Primitive Recursive Arithmetic (PRA), proposed in 1923 by Skolem. In PRA one finds (a) an absence of explicit quantification, (b) an ability to define primitive recursive functions, (c) a few rules for handling equality, e.g., substitution of equals for equals, (d) a rule of instantiation, and (e) a simple induction principle.

By Gödel's second incompleteness theorem (1931) it became evident that only *partial* realizations of Hilbert's program are possible. The system proposed by Chuaqui and Suppes is such a partial realization, in that it provides an axiomatic foundation for basic analysis, with a PRA consistency proof ([1], p. 123 and p. 130). Sommer and Suppes's improved system allows definition by recursion (which does away with a lot of explicit axioms)

*Department of Pure Mathematics and Computer Algebra, University of Gent, Belgium. ci@cage.ugent.ac.be

**sasander@cage.ugent.be

and still has a PRA proof of consistency ([2], p. 21). This system is called *Elementary Recursive Nonstandard Analysis*, in short ERNA. Its consistency is proved via Herbrand's Theorem (1930), which is restricted to quantifier-free formulas $Q(x_1, \ldots, x_n)$, usually containing free variables. Alternatively, one might say it is restricted to universal sentences

$$(\forall x_1) \ldots (\forall x_n) Q(x_1, \ldots, x_n),$$

obtained by closing the open quantifier-free formulas by means of universal quantifiers. Herbrand's theorem states that, if a collection of such formulas resp. sentences is consistent, it has a simple 'Herbrand' model and, if it is not, its inconsistency will show up in some finite procedure.

Herbrand's theorem requires that ERNA's axioms be written in a quantifier-free form. As a result, some axioms definitely look artificial; fortunately, theorems don't suffer from the quantifier-free restriction.

Calculus applications of ERNA have been, so far, scarce and sketchy. Thus, [3] contains an outline of an existence theorem for first-order ordinary differential equations, relying on the property, stated without proof, that a continuous function on a compact interval is bounded. As part of a less anecdotical approach we will provide an ERNA version of the supremum principle and deduce from it a square root function. Both results hold *up to infinitesimals*; as ERNA has no standard part function, it is intrinsically impossible to do better.

5.2 The system

The system we are about to describe was first presented in [2], and all our undocumented results are quoted from that paper. The foundations are also exposed, in a more informal manner, in [3].

Notation 5.2.1 $\mathbb{N} = \{0, 1, 2, \ldots\}$ *consists of the (finite) integers.*

Notation 5.2.2 $\vec{x}$ *stands for some finite (possibly empty) sequence* $(x_1, \ldots, x_k)$.

Notation 5.2.3 $\tau(\vec{x})$ *denotes a term in which* $\vec{x} = (x_1, \ldots, x_k)$ *is the list of the distinct free variables.*

5.2.1 The language

- connectives: $\wedge, \neg, \vee, \rightarrow, \leftrightarrow$
- quantifiers: $\forall, \exists$

- an infinite set of variables
- relation symbols:[1]
  - binary $x = y$
  - binary $x \leq y$
  - unary $\mathcal{I}(x)$, read as 'x is infinitesimal', also written '$x \approx 0$'
  - unary $\mathcal{N}(x)$, read as 'x is hypernatural'.
- individual constant symbols:
  - 0
  - 1
  - ε (The Axiom 3 (6) of 5.2.2 shows that ε denotes a positive infinitesimal.)
  - ω (The axioms 3 (7) and 2 (4) of 5.2.2 show that $\omega = 1/\varepsilon$ denotes an infinite hypernatural.)
  - $\uparrow$, read as 'undefined'.

Notation 5.2.4 *'x is defined' stands for '$x \neq \uparrow$'.* (Examples: $1/0$ is undefined, $1/0 = \uparrow$.)

- function symbols:[2]
  - (unary) 'absolute value' $|x|$, 'ceiling' $\lceil x \rceil$, 'weight' $\|x\|$. (For the meaning of $\|x\|$, see Theorem 5.2.3.)
  - (binary) $x + y$, $x - y$, $x \cdot y$, x/y, $x\hat{\ }y$. (Axiom set 6 and Axiom 12 (4) of 5.2.2 show that $x\hat{\ }n = x^n$ for hypernatural n, else undefined.)
  - for each $k \in \mathbb{N}$, k k-ary function symbols $\pi_{k,i}$ ($i = 1, \ldots, k$). (The Axiom schema 7 of 5.2.2 shows that $\pi_{k,i}(\vec{x})$ are the projections of the k-tuple $\vec{x}$.)
  - for each formula φ with $m + 1$ free variables, without quantifiers or terms involving min, an m-ary function symbol $\min_\varphi$. (For the meaning of which, see Theorem 5.2.6 and Theorem 5.2.7.)
  - for each triple $(k, \sigma(x_1, \ldots, x_m), \tau(x_1, \ldots, x_{m+2}))$ with $0 < k \in \mathbb{N}$, σ and τ terms not involving min, an $(m+1)$-ary function symbol $\mathrm{rec}^k_{\sigma\tau}$. (Axiom schema 9 of 5.2.2 shows that this is the term obtained from σ and τ by recursion, after the model $f(0, \vec{x}) = \sigma(\vec{x})$, $f(n+1, \vec{x}) = \tau(f(n, \vec{x}), n, \vec{x})$, if terms are defined and don't weigh too much.)

[1] For better readibility we express the relations in x or in (x, y), according to arity.

[2] We denote the values as computed in x or (x, y) according to the arity.

5.2.2 The axioms

Axiom set 1 (Logic). *Axioms of first-order logic.*

Axiom set 2 (Hypernaturals).

1. *0 is hypernatural;*
2. *if x is hypernatural, so is $x+1$;*
3. *if x is hypernatural, then $x \geq 0$;*
4. *ω is hypernatural.*

Definition 5.2.1 *'x is infinite' stands for '$x \neq 0 \wedge 1/x \approx 0$'; '$x$ is finite' stands for 'x is not infinite'; 'x is natural' stands for 'x is hypernatural and finite'.*

Axiom set 3 (Infinitesimals).

1. *if x and y are infinitesimal, so is $x+y$;*
2. *if x is infinitesimal and y is finite, xy is infinitesimal;*
3. *an infinitesimal is finite;*
4. *if x is infinitesimal and $|y| \leq x$, then y is infinitesimal;*
5. *if x and y are finite, so is $x+y$;*
6. *ε is infinitesimal;*
7. *$\varepsilon = 1/\omega$.*

Axiom set 4 (Ordered field). *Axioms expressing that the elements, with $\uparrow$ excluded, constitute an ordered field of characteristic zero with absolute value function. These include (quantifier-free)*

- *if x is defined, then $x+0=0+x=x$;*
- *if x is defined, then $x+(0-x)=(0-x)+x=0$;*
- *if x is defined and $x \neq 0$, then $x \cdot (1/x) = (1/x) \cdot x = 1$.*

Axiom set 5 *(Archimedean). If x is defined, $\lceil x \rceil$ is a hypernatural and $\lceil x \rceil - 1 < x \leq \lceil x \rceil$.*

Theorem 5.2.1 *If x is defined, then $\lceil x \rceil$ is the least hypernatural $\geq x$.*

Theorem 5.2.2 *x is finite iff there is a natural n such that $|x| \leq n$.*

Proof. The statement is trivial for $x = 0$. If $x \neq 0$ is finite, so is $|x|$ because, assuming the opposite, $1/|x|$ would be infinitesimal and so would $1/x$ be by axiom (4) of set 3. By axiom (5) of the same set, the hypernatural $\lceil |x| \rceil < |x|+1$ is then also finite. Conversely, let n be natural and $|x| \leq n$. If $1/|x|$ were infinitesimal, so would $1/n$ be by axiom (4) of set 3, and this contradicts the assumption that n is finite. □

Corollary 5.2.1 *$x \approx 0$ iff $|x| < 1/n$ for all natural $n \geq 1$.*

Axiom set 6 (Power).

1. *if $x \neq \uparrow$, then $x\hat{\ }0 = 1$;*
2. *if $x \neq \uparrow$ and n is hypernatural, then $x\hat{\ }(n+1) = (x\hat{\ }n) \cdot x$.*

Axiom schema 7 (Projection).

If $x_1, \ldots, x_n$ are defined, then $\pi_{n,i}(x_1, \ldots, x_n) = x_i$ for $i = 1, \ldots, n$.

Axiom set 8 (Weight).

1. *If $\|x\|$ is defined, then $\|x\|$ is a nonzero hypernatural.*
2. *If $|x| = m/n \leq 1$ (m and $n \neq 0$ hypernaturals), then $\|x\|$ is defined, $\|x\|.|x|$ is hypernatural and $\|x\| \leq n$.*
3. *If $|x| = m/n \geq 1$ (m and $n \neq 0$ hypernaturals), then $\|x\|$ is defined, $\|x\|/|x|$ is hypernatural and $\|x\| \leq m$.*

Definition 5.2.2 *A* hyperrational *is of the form $\pm p/q$, with p and $q \neq 0$ hypernatural.*

Theorem 5.2.3

1. *If x is not a hyperrational, then $\|x\| = \uparrow$.*
2. *If x is a hyperrational, say $x = \pm p/q$ with p and $q \neq 0$ relatively prime hypernaturals, then*

$$\| \pm p/q\| = \max\{|p|, |q|\}.$$

Remark. In both statements of this theorem, the antecedent can be expressed in a quantifier-free way, but the whole sentence cannot. (This explains why it is a theorem and not part of the axioms.) For instance, $\mathcal{N}(p) \to \neg\mathcal{N}(p|x|)$ expresses 'x is not hyperrational'.

Theorem 5.2.4

1. *$\|0\| = 1$;*

2. *if* $n \geq 1$ *is hypernatural,* $\|n\| = n$*;*

3. *if* $\|x\|$ *is defined, then* $\|1/x\| = \|x\|$ *and* $\|\lceil x \rceil\| = |\lceil x \rceil| \leq \|x\|$*;*

4. *if* $\|x\|$ *and* $\|y\|$ *defined,* $\|x+y\|$*,* $\|x-y\|$*,* $\|xy\|$ *and* $\|x/y\|$ *are at most equal to* $(1+\|x\|)(1+\|y\|)$*, and* $\|x\hat{\ }y\|$ *is at most* $(1+\|x\|)\hat{\ }(1+\|y\|)$.

Notation 5.2.5 *For any* $0 < n \in \mathbb{N}$ *we write*

$$\|(x_1,\ldots,x_n)\| = \max\{\|x_1\|,\ldots,\|x_n\|\}.$$

Notation 5.2.6 *For any* $0 < n \in \mathbb{N}$ *we write*

$$2_n^x := \underbrace{2\hat{\ }(\ldots 2\hat{\ }(2\hat{\ }(2\hat{\ }x)))}_{n\ 2\text{'s}}.$$

Theorem 5.2.5 *If* $\tau(\vec{x})$ *is a term not involving* ω*,* ε*,* rec *or* min*, then there exists a* $0 < k \in \mathbb{N}$ *such that*

$$\|\tau(\vec{x})\| \leq 2_k^{\|\vec{x}\|}.$$

Axiom schema 9 (Recursion) *For* $0 < k \in \mathbb{N}$*,* σ *and* τ *not involving* min*:*

$$\mathrm{rec}_{\sigma\tau}^k(0,\vec{x}) = \begin{cases} \sigma(\vec{x}) & \text{if this is defined, and has weight } \leq 2_k^{\|\vec{x}\|}, \\ \uparrow & \text{if } \sigma(\vec{x}) = \uparrow, \\ 0 & \text{otherwise.} \end{cases}$$

$$\mathrm{rec}_{\sigma\tau}^k(n+1,\vec{x}) = \begin{cases} \tau(\mathrm{rec}_{\sigma\tau}^k(n,\vec{x}),n,\vec{x}) & \text{if defined, with weight } \leq 2_k^{\|\vec{x},n+1\|}, \\ \uparrow & \text{if } \tau(\mathrm{rec}_{\sigma\tau}^k(n,\vec{x}),n,\vec{x}) = \uparrow, \\ 0 & \text{otherwise.} \end{cases}$$

If σ is constant, the list $\vec{x}$ is empty, and the weight requirements mentioned in this axiom schema are void.

A few words concerning the restrictions included in this axiom schema. One of ERNA's main advantages over the Chuaqui-Suppes system is, that it allows some form of recursion while preserving a finitary consistency proof. In achieving this, a crucial role is played by the weight function, introduced axiomatically but given explicitly in theorem 5.2.3. Recursion is an essential feature of PRA, and it is therefore impossible to prove inside PRA the consistency of a system that has unrestricted recursion. ERNA's axiom schema 9 restricts recursion by truncating objects outgrowing the preset weight standard. In view of the huge bounds allowed, it seems unlikely that access to calculus applications will suffer from this restriction; computing weights is the price to be paid in practice.

Axiom schema 10 (Internal minimum). *For any quantifier-free formula $\varphi(y, \vec{x})$ not involving* min *or* $\mathcal{I}$ *we have*

1. $\min_\varphi(\vec{x})$ *is a hypernatural number;*
2. *if* $\min_\varphi(\vec{x}) > 0$*, then* $\varphi(\min_\varphi(\vec{x}), \vec{x})$*;*
3. *if* n *is a hypernatural and* $\varphi(n, \vec{x})$*, then*

$$\min_\varphi(\vec{x}) \leq n \quad \textit{and} \quad \varphi(\min_\varphi(\vec{x}), \vec{x}).$$

Theorem 5.2.6 *If the quantifier-free formula* $\varphi(y, \vec{x})$ *does not involve* $\mathcal{I}$ *or* min*, and if there are hypernatural* n*'s such that* $\varphi(n, \vec{x})$*, then* $\min_\varphi(\vec{x})$ *is the least of these. If there are none,* $\min_\varphi(\vec{x}) = 0$.

Corollary 5.2.2 *Proofs by hypernatural induction.*

Example 5.2.1 *The sum of two hypernaturals is a hypernatural.*

Proof. Fix any hypernatural x. If the theorem is wrong, there exists at least one y with $\mathcal{N}(y) \wedge \neg\mathcal{N}(x+y)$. By Theorem 5.2.6, there is a least number with these properties, say y_0. Then $y_0 \neq 0$ since $x + 0 = x$ (field axiom) and $\mathcal{N}(x)$ (assumption). From $y_0 \neq 0$, $\mathcal{N}(y_0 - 1)$ (hypernatural axiom). By leastness, $\mathcal{N}(x + (y_0 - 1))$. Hence (field axiom) $\mathcal{N}((x + y_0) - 1)$ and finally $\mathcal{N}(x + y_0)$ (hypernatural axiom). This contradiction proves the theorem. □

Axiom schema 11 (External minimum). *For any quantifier-free formula $\varphi(y, \vec{x})$ not involving* min, ω *or* ε *we have*

1. $\min_\varphi(\vec{x})$ *is a hypernatural number;*
2. *if* $\min_\varphi(\vec{x}) > 0$*, then* $\varphi(\min_\varphi(\vec{x}), \vec{x})$*;*
3. *if* n *is a natural number,* $\|x\|$ *is finite and* $\varphi(n, \vec{x})$*, then* $\min_\varphi(\vec{x}) \leq n$ *and* $\varphi(\min_\varphi(\vec{x}), \vec{x})$.

Remark. $\mathcal{I}$ is allowed in φ.

Theorem 5.2.7 *Let* $\varphi(y, \vec{x})$ *a quantifier-free formula not involving* min, ω *or* ε*. If* $\|\vec{x}\|$ *is finite and if there are natural* n*'s such that* $\varphi(n, \vec{x})$*, then* $\min_\varphi(\vec{x})$ *is the least of these. If there are none,* $\min_\varphi(\vec{x}) = 0$.

Corollary 5.2.3 *Proofs by natural induction.*

Axiom schema 12 ((Un)defined terms).

1. 0, 1, ω, ε *are defined;*

2. $|x|$, $\lceil x \rceil$, $\|x\|$ *are defined iff* x *is;*
3. $x+y$, $x-y$, xy *are defined iff* x *and* y *are;* x/y *is defined iff* x *and* y *are and* $y \neq 0$;
4. $x\hat{\ }y$ *is defined iff* x *and* y *are and* y *is hypernatural;*
5. $\pi_{k,i}(x_1, \ldots, x_k)$ *is defined iff* $x_1, \ldots, x_k$ *are;*
6. *if* x *is not a hypernatural,* $\mathrm{rec}^k_{\sigma\tau}(x, \vec{y})$ *is undefined;*
7. $\min_\varphi(x_1, \ldots, x_k)$ *is defined iff* $x_1, \ldots, x_k$ *are.*

Theorem 5.2.8 (Hypernatural induction) *Let* $\varphi(x)$ *be a quantifier-free formula not involving* min *or* $\mathcal{I}$, *such that*

1. $\varphi(0)$ *holds,*
2. *the implication* $(\mathcal{N}(n) \wedge \varphi(n)) \rightarrow \varphi(n+1)$ *holds.*

Then $\varphi(n)$ *holds for all hypernatural* n.

Proof. Suppose, on the contrary, that there is a hypernatural n such that $\neg\varphi(n)$. By Theorem 5.2.6, there is a least hypernatural n_0 such that $\neg\varphi(n_0)$. By our assumption (1), $n_0 > 0$. Consequently, $\varphi(n_0 - 1)$ does hold. But then, by our assumption (2), so would $\varphi(n_0)$. This contradiction proves the theorem. □

Example 5.2.2 *The sum of two hypernaturals is a hypernatural.*

Proof. Fix any hypernatural N and consider the formula $\mathcal{N}(N+x)$. Both $\mathcal{N}(N+0)$ and $\mathcal{N}(N+n) \rightarrow \mathcal{N}(N+n+1)$ are included in axiom set 5.2.2. Hence $\mathcal{N}(N+n)$ for every hypernatural n. □

Example 5.2.3 *Let* $\varphi(n)$ *be a quantifier-free formula not involving* min *or* $\mathcal{I}$. *If* $n_0 < n_1$ *are hypernaturals such that* $n_0 \leq n \leq n_1 - 1 \rightarrow \varphi(n) = \varphi(n+1)$, *then* $\varphi(n_0) = \varphi(n_1)$.

Proof. The formula $\varphi(n_0) = \varphi(n_0 + x)$ holds for $x = 0$. If $\varphi(n_0) = \varphi(n_0 + n)$ for any hypernatural $n_0 + n \leq n_1 - 1$, then also $\varphi(n_0) = \varphi(n_0 + n + 1)$ by assumption. Hence $\varphi(n_0) = \varphi(n_0 + n)$ for $0 \leq n \leq n_1 - n_0$. □

For further use we collect here some definable functions, being terms of the language that (provably in ERNA) have the properties of the function.

1. The identity function $id(x) = x$ is $\pi_{1,1}$.

2. For each closed term τ and each arity k, the constant function

$$C_{k,\tau}(x_1,\ldots,x_k)=\tau,$$

   is $\pi_{k+1,k+1}(x_1,\ldots,x_k,\tau)$.

3. The hypersequence

$$r(n)=\begin{cases}0 & \text{if } n=0\\ 1 & \text{if } n\geq 1\end{cases}$$

   is $\mathrm{rec}^k_{\sigma\tau}$ with $k=1$, $\sigma=0$, $\tau=C_{2,1}$.

4. The function

$$\zeta(x)=\begin{cases}1 & \text{if } x=0\\ x & \text{otherwise}\end{cases}$$

   is $1+x-r(\lceil |x| \rceil)$.

5. The functions

$$h(x)=\begin{cases}1 & \text{if } x>0\\ 0 & \text{otherwise}\end{cases} \quad\text{and}\quad H(x)=\begin{cases}1 & \text{if } x\geq 0\\ 0 & \text{otherwise}\end{cases}$$

   are $\frac{x+|x|}{2\zeta(x)}$ and $\frac{1}{2}+\frac{\zeta(|x|)}{2\zeta(x)}$, respectively.

6. The function

$$1_{(a,b]}(x)=\begin{cases}1 & \text{if } a<x\leq b\\ 0 & \text{otherwise}\end{cases}$$

   is $h(x-a)H(b-x)$. Likewise for the characteristic function of any other interval.

7. For constants a,b and terms ρ,σ,τ, the function

$$d(x)=\begin{cases}\sigma(x) & \text{if } a<x\leq b \text{ and } \rho(x)>0\\ \tau(x) & \text{otherwise}\end{cases}$$

   is $1_{(a,b]}(x)(h(\rho(x))\,\sigma(x)+(1-h(\rho(x)))\,\tau(x))$. Likewise for any other type of interval in $a<x\leq b$ and/or any other inequality in $\rho(x)>0$. Any such construction will be called a *definition by cases.* If no interval is specified, the terms ρ,σ,τ and the resulting function can have more than one free variable.

The next theorem is to be considered as an ERNA version of the supremum principle for a set of type $\{x \mid f(x)<0\}$.

Notation 5.2.7 *We write $a \ll b$ if $a < b$ and $a \not\approx b$.*

Theorem 5.2.9 *Let $b < c$ be constants such that $d := c - b$ is finite. Further, let $f(x)$ be a term not involving $\mathcal{I}$ or* min*, such that $f(x)$ is never undefined for $b \leq x \leq c$. If*

i. $f(c) \geq 0$,

ii. $f(b) < 0$,

then there is a constant γ with the following properties:

iii. $f(\gamma) \geq 0$,

iv. for every natural number $n \geq 1$ there are $x > \gamma - 1/n$ such that $f(x) < 0$.

If $f(x)$ has the extra property

$$(f(x) < 0 \wedge b < y < x) \to f(y) < 0 \tag{5.1}$$

then γ is, up to infinitesimals, the only constant $> b$ with the properties (iii) and (iv).

Proof. In order to apply recursion, we choose c as our term σ and use definition by cases to obtain the term

$$\tau(t, n) = \begin{cases} t - d/2^n & \text{if } f(t - d/2^n) \geq 0 \\ t & \text{otherwise.} \end{cases}$$

Note that 'otherwise' is equivalent here to 'if $f(t - d/2^n) < 0$', because we have excluded undefined values for $f(x)$. ERNA's unary function symbol $\text{rec}^1_{\sigma\tau}$ for this particular σ and τ will be shortened to rec. Its properties can be stated simply as

$$\text{rec}(0) = c \quad \text{and} \quad \text{rec}(n+1) = \tau(\text{rec}(n), n)$$

because undefined terms cannot occur, and there are no weight requirements because τ has arity two.

If we prove that for any hypernatural n the two properties

$$f(\text{rec}(n)) \geq 0 \tag{5.2}$$

$$f(\text{rec}(n) - d/2^{n-1}) < 0 \quad (n \geq 1) \tag{5.3}$$

hold, we are done. It suffices to take $\gamma = \text{rec}(\omega)$ and to note that, because $d/2^{\omega-1} \approx 0$,

$$\text{rec}(\omega) - \frac{1}{n} < \text{rec}(\omega) - d/2^{\omega-1}$$

for any natural number $n \geq 1$. We prove (5.2) by hypernatural induction. For $n = 0$ the requirement (5.2) is identical with the assumption (i). Now let n be a hypernatural for which (5.2) holds. If $f(\text{rec}(n) - d/2^n) \geq 0$, the definition of τ implies that $\text{rec}(n+1) = \tau(\text{rec}(n), n) = \text{rec}(n) - d/2^n$, which translates the assumption into $f(\text{rec}(n+1)) \geq 0$. Otherwise, $\text{rec}(n+1) = \tau(\text{rec}(n), n) = \text{rec}(n)$, making the induction hypothesis identical with the requirement $f(\text{rec}(n+1)) \geq 0$.

Next we consider (5.3). Our proof demands that $n = 1$ be treated separately. We have $\text{rec}(1) = \tau(\text{rec}(0), 0) = \tau(c, 0)$, and this is simply b since $f(c - d) = f(b) < 0$. Therefore, the property (5.3) is identical with the assumption (ii). Now the proof for any hypernatural $N \geq 2$. We consider the formula

$$\mathcal{N}(n) \ \wedge \ n \leq N-2 \ \wedge \ \text{rec}(N-n) \neq \text{rec}(N-n-1) - d/2^{N-n-1} \tag{5.4}$$

and consider two possibilities. First possibility: there are no hypernaturals n satisfying (5.4). This means that

$$\text{rec}(N-n) - d/2^{N-n-1} = \text{rec}(N-n-1) - d/2^{N-n-2}$$

for $0 \leq n \leq N-2$, and by example 5.2.3 it follows that

$$\text{rec}(N) - d/2^{N-1} = \text{rec}(1) - d = c. \tag{5.5}$$

As $f(c) < 0$, we conclude that (5.3) holds for our N. Second possibility: there are hypernaturals n satisfying (5.4). If so, let n_0 be the smallest one, as provided by theorem 5.2.6. Then $n_0 \leq N-2$ and

$$\text{rec}(N-n_0) \neq \text{rec}(N-n_0-1) - d/2^{N-n_0-1}$$

i.e.

$$\tau(\text{rec}(N-n_0-1), N-n_0-1) \neq \text{rec}(N-n_0-1) - d/2^{N-n_0-1}.$$

The definition of $\tau(t, n)$ shows that then, inevitably,

$$\tau(\text{rec}(N-n_0-1), N-n_0-1) = \text{rec}(N-n_0-1),$$

meaning that

$$f(\text{rec}(N-n_0-1) - d/2^{N-n_0-1}) \equiv f(\text{rec}(N-n_0) - d/2^{N-n_0-1}) < 0. \tag{5.6}$$

By the leastness of n_0,

$$\text{rec}(N-n) - d/2^{N-n-1} = \text{rec}(N-n-1) - d/2^{N-n-2}$$

for $0 \leq n \leq n_0 - 1$. Hence

$$\text{rec}(N) - d/2^{N-1} = \text{rec}(N - n_0) - d/2^{N-n_0-1}$$

by example 5.2.3. Substituting in (5.6) gives

$$f(\text{rec}(N) - d/2^{N-1}) < 0,$$

as was to be proved.

Finally, assume the extra property (5.1). If $\gamma' > b$ is another constant with the properties (iii) and (iv), we cannot have $\gamma' \ll \gamma$, as property (iv) for γ would imply that there are $x > \gamma'$ satisfying $f(x) < 0$, which by (5.1) leads to $f(\gamma') < 0$ and contradicts the property (iii) for γ'. Likewise for the possibility $\gamma \ll \gamma'$. Therefore $\gamma' \approx \gamma$. □

This theorem allows us to equip ERNA with a *square root up to infinitesimals* function.

Example 5.2.4 *For every finite constant* $p > 0$, ERNA *provides a constant* $\gamma > 0$, *unique up to infinitesimals, such that* $\gamma^2 \approx p$.

Proof. It follows from the properties of an ordered field that the term $f(x) = x^2 - p$ and the constants $b = 0, c = 1 + p$ satisfy the requirements of theorem 5.2.9, including the extra requirement (5.1). If γ is the constant resulting from the theorem, then $\gamma^2 \geq p$ and for every natural $n \geq 1$ there are $x > \gamma - 1/n$ with $x^2 < p$. Moreover, $x < 1 + p$ by the properties of the ordered field. Hence $\gamma^2 < x^2 + 2x/n + 1/n^2 < p + 2(1+p)/n + 1/n^2$. By corollary 5.2.1, we conclude that $0 \leq \gamma^2 - p \approx 0$. □

References

[1] R. Chuaqui and P. Suppes, "Free-Variable Axiomatic Foundations of Infinitesimal Anaysis: A Fragment with Finitary Consistency Proof", J. Symb. Logic, **60** (1995) 122–159.

[2] R. Sommer and P. Suppes, "Finite Models of Elementary Recursive Nonstandard Analysis", Notas de la Sociedad Matematica de Chile, **15** (1996) 73–95.

[3] R. Sommer and P. Suppes, "Dispensing with the Continuum", J. Math. Psychology, **41** (1997) 3–10.

[4] P. Suppes and R. Chuaqui, A finitarily consistent free-variable positive fragment of Infinitesimal Anaysis, Proceedings of the IXth Latin American Symposium on Mathematical Logic, Notas de Logica Mathematica, **38** (1993) 1–59, Universidad Nacional del Sur, Bahia Blanca, Argentina.

6

The Sousa Pinto approach to nonstandard generalised functions

R. F. Hoskins*

Abstract

Nonstandard Analysis suggests several ways in which the standard theories of distributions and other generalised functions could be reformulated. This paper reviews the contributions of José Sousa Pinto to this area up to his untimely death four years ago. Following the original presentation of nonstandard models for the Sebastião e Silva axiomatic treatment of distributions and ultradistributions he worked on a nonstandard theory of Sato hyperfunctions, using a simple ultrapower model of the hyperreals. (This in particular allows nonstandard representations for generalised distributions, such as those of Roumieu, Beurling, and so on.) He also considered a nonstandard theory for the generalised functions of Colombeau, and finally turned his attention to the hyperfinite representation of generalised functions, following the work of Kinoshita.

6.1 Introduction

José Sousa Pinto of the University of Aveiro, Portugal, died in August 2000 after a prolonged and debilitating illness. His interest in nonstandard methods, particularly in their application to the study of generalised functions, was of long standing and he will be especially remembered for his part in the organisation of the highly successful **International Colloquium of Nonstandard Mathematics** held at Aveiro [2] in 1994. A most modest and unassuming mathematician, his contributions to NSA are less well known than their value deserves and this paper is concerned to report his work and to stand as some tribute to his memory. From a personal point of view I would also wish to take

*Department of Electronic and Electrical Engineering, Loughborough University, Loughborough, UK.
royhosk@aol.com

this opportunity to acknowledge the value and pleasure I have had in working with him over many years.

6.1.1 Generalised functions and N.S.A.

The theory of generalised functions is a subject of major importance in modern analysis and one that has gone through many changes since the original presentation [18] of the theory of distributions in the form given to it by Laurent Schwartz in the 1950s. Various alternative approaches to the theory have been explored over the years, and the subject has been expanded (and complicated) by the introduction of generalised distributions of several types, ultradistributions, hyperfunctions and so on. In recent years the development of a unifying and simplifying treatment of the whole subject area has become possible through the use of Nonstandard Analysis (N.S.A.) The application of nonstandard methods to distributions and other generalised functions was already considered by Abraham Robinson in his classic text [15] on N.S.A. in 1966, and various workers have extended and developed this approach since then. It was Sousa Pinto who first considered the possibility of developing a nonstandard realisation of the Sebastiao e Silva axioms for Schwartz distributions [6], and later for ultradistributions [7]. His further work on Sato hyperfunctions remained unpublished at the time of his death and an outline of this forms the main part of the present paper.

The first section of the paper briefly reviews standard material on distributions, ultradistributions and Sato hyperfunctions, and summarises the earlier nonstandard re-formulation of that material on which the subsequent development is based.

6.2 Distributions, ultradistributions and hyperfunctions

6.2.1 Schwartz distributions

We recall first some basic facts about distributions. A distribution, in the sense of Schwartz, is a continuous linear **functional** on the space $\mathcal{D} \equiv \mathcal{D}(\mathbb{R})$ of all infinitely differentiable functions of compact support, equipped with an appropriate topology. That is to say, a distribution is simply a member μ of the topological dual $\mathcal{D}'(\mathbb{R})$ of that space. The **distributional derivative** of $\mu \in \mathcal{D}'$ is the distribution $D\mu$ defined by

$$< D\mu, \phi >=< \mu, -\phi' >, \quad \forall \phi \in \mathcal{D}.$$

It follows from this definition that all distributions are infinitely differentiable in this sense. Moreover, it can be shown that every distribution is locally a finite-order derivative of a continuous function. Those distributions which are *globally* representable as finite-order derivatives of continuous functions are called, not unnaturally, **finite-order distributions**. The space of all such finite-order distributions is denoted by $\mathcal{D}'_{fin}(\mathbb{R})$.

Every locally integrable function f defines a so-called **regular** distribution μ_f according to

$$< \mu_f, \phi >= \int_{-\infty}^{+\infty} f(x)\phi(x)dx, \qquad \text{for all } \phi \in \mathcal{D}.$$

$\mathcal{D}'$ contains elements other than such regular distributions, so that $\mathcal{D}'$ is a proper extension of the space $\mathcal{L}_{loc}(\mathbb{R})$ of all locally integrable functions. In this sense distributions may be legitimately described as *generalised functions.* However there is no direct sense in which a distribution can be said to have a value at a point. This becomes particularly clear in the case of those distributions which are not regular. The delta function is the prime example of such a **singular** distribution, being defined simply as that functional δ (obviously linear and continuous) which maps each function $\phi \in \mathcal{D}$ into the number $\phi(0)$. It is a finite-order distribution since it is the second derivative of the continuous function $x_+(t) \equiv tH(t)$, where H denotes the Heaviside unit step.

6.2.2 The Silva axioms

The definition of distributions as equivalence classes of nonstandard internal functions in a nonstandard universe was already made explicit in Abraham Robinson's original text [15] on N.S.A. Several other nonstandard models for $\mathcal{D}'(\mathbb{R})$ have since appeared. In particular such a model was presented by Hoskins and Pinto [6] in 1991, based on the axiomatic treatment of distributions given by Sebastião e Silva [19] in 1956. The Silva axioms for finite-order distributions on an interval $I \subset \mathbb{R}$ can be stated as follows:

Silva axioms for finite order distributions

Distributions are elements of a linear space $\mathcal{E}(I)$ for which two linear maps are defined: $\iota : \mathcal{C}(I) \to \mathcal{E}(I)$ and $D : \mathcal{E}(I) \to \mathcal{E}(I)$, such that

S1 ι is the injective identity, (every $f \in \mathcal{C}$ is a distribution).

S2 To each $\nu \in \mathcal{E}$ there corresponds $D\nu \in \mathcal{E}$ such that, if $\nu = \iota(f) \in \mathcal{C}^1(I)$ then $D\nu = \iota(f')$

S3 For $\nu \in \mathcal{E}$ there exists $f \in \mathcal{C}$ and $r \in \mathbb{N}_0$ such that $\nu = D^r\iota(f)$.

S4 Given $f, g \in \mathcal{C}$ and $r \in \mathbb{N}_0$, the equality $D^r\iota(f) = D^r\iota(g)$ holds if and only if $(f-g)$ is a polynomial of degree $< r$.

Silva gives an abstract model for this set of axioms as follows: define an equivalence relation $\diamond$ on $\mathbb{N}_0 \times \mathcal{C}$ by

$$(r, f) \diamond (s, g) \leftrightarrow \exists m \in \mathbb{N}_0 \left\{m \geq r, s \wedge (\mathcal{I}_a^{m-r} f - \mathcal{I}_a^{m-s} g) \in \Pi_m\right\}$$

where Π_m is the set of all complex-valued polynomials of degree less than m and $\mathcal{I}_a^k$ is the kth iterated indefinite integral operator with origin at $a \in I$. Now write

$$\mathcal{C}_\infty \equiv \mathcal{C}_\infty(I) = \mathbb{N}_0 \times \mathcal{C}/\diamond$$

Then $\mathcal{C}_\infty$ is a model for the Silva axioms S1-S4, and every model for the Silva axioms is isomorphic to $\mathcal{C}_\infty$. In particular $\mathcal{D}'_{fin}(\mathbb{R})$ is isomorphic to $\mathcal{C}_\infty(\mathbb{R})$. The extension to global distributions of arbitrary order is straightforward. See, for example, the exposition of the Silva approach to distribution theory given in Campos Ferreira, [3].

A nonstandard model for $\mathcal{C}_\infty$

A nonstandard model for these axioms was presented by Hoskins and Pinto [6], using a simple ultrapower model ${}^*\mathbb{R} = \mathbb{R}^\mathbb{N}/\sim$ for the hyperreals. It may be summarised as follows:

The internal set ${}^*\mathcal{C}^\infty(\mathbb{R})$ is the nonstandard extension of the standard set $\mathcal{C}^\infty(\mathbb{R})$ of all infinitely differentiable functions on $\mathbb{R}$,

$${}^*\mathcal{C}^\infty(\mathbb{R}) = \{F = [(f_n)_{n\in\mathbb{N}}] : \ f_n \in \mathcal{C}^\infty(\mathbb{R}) \quad \text{for nearly all } n \in \mathbb{N}\}.$$

This set is a differential algebra. We denote by ${}^S\mathcal{C}(\mathbb{R})$ the (external) set of all functions $F \in {}^*\mathcal{C}^\infty(\mathbb{R})$ which are finite-valued and S-continuous at each point of ${}^*\mathbb{R}_b$. An internal function $F \in {}^*\mathcal{C}^\infty(\mathbb{R})$ is then said to be a **predistribution** if it is a finite-order $*$derivative of a function in ${}^S\mathcal{C}(\mathbb{R})$. The set of all such pre-distributions is given by,

$${}^*D^\infty\{{}^S\mathcal{C}(\mathbb{R})\} \equiv \bigcup_{r\geq 0} {}^*D^r\{{}^S\mathcal{C}(\mathbb{R})\}$$

$$= \{F \in {}^*\mathcal{C}^\infty(\mathbb{R}) : \ F = {}^*D^r\Phi \text{ for some } \Phi \in {}^S\mathcal{C}(\mathbb{R}) \text{ and some } r \in \mathbb{N}_0\}.$$

We then have the following (strict) inclusions:

$${}^S\mathcal{C}(\mathbb{R}) \subset {}^*D^\infty\{{}^S\mathcal{C}(\mathbb{R})\} \subset {}^*\mathcal{C}^\infty(\mathbb{R}).$$

The members of ${}^*D^\infty\{{}^S\mathcal{C}(\mathbb{R})\}$ are the nonstandard representatives of finite order distributions on $\mathbb{R}$. Given two such pre-distributions F and G, we say that they are **distributionally equivalent**, and write $F\Xi G$, if and only if there exists an integer $m \in \mathbb{N}_0$ and a polynomial p_m of degree m (with coefficients in ${}^*\mathbb{R}$) such that

$$ {}^*\mathcal{I}_a^m(F-G) \approx p_m $$

where ${}^*\mathcal{I}_a^m$ denotes the mth-order *indefinite integral operator from $a \in {}^*\mathcal{R}_b$. Then for any $F \in {}^*\mathcal{D}^\infty\{{}^S\mathcal{C}(\mathbb{R})\}$ we denote by $\mu_F = {}^\Xi[F]$ the equivalence class containing F and call it a **finite order Ξdistribution**. The set of all such equivalence classes is denoted by

$$ {}^\Xi\mathcal{C}_\infty(\mathbb{R}) \equiv {}^*\mathcal{D}^\infty\{{}^S\mathcal{C}(\mathbb{R})\}/\Xi. $$

${}^\Xi\mathcal{C}_\infty(\mathbb{R})$ is a nonstandard model for the Silva axioms and is isomorphic with $\mathcal{D}'_{fin}(\mathbb{R})$.

6.2.3 Fourier transforms and ultradistributions

For the classical Fourier transform of sufficiently well-behaved functions we have

$$ \tilde{f}(x) = \int_{-\infty}^{+\infty} f(y)e^{-ixy}dy \quad ; \quad f(y) = \frac{1}{2\pi}\int_{-\infty}^{+\infty} \tilde{f}(x)e^{ixy}dx $$

and the Parseval relation

$$ \int_{-\infty}^{+\infty} f(x)\tilde{g}(x)dx = \int_{-\infty}^{+\infty} \tilde{f}(y)g(y)dy. $$

To extend the definition of Fourier transform to distributions Schwartz used a generalised form of Parseval relation to define $\tilde{\mu}$ as the functional satisfying

$$ <\tilde{\mu}, \phi> = <\mu, \tilde{\phi}> . $$

The difficulty here is that if $\phi \in \mathcal{D}$ then its Fourier transform $\tilde{\phi}$ belongs not to $\mathcal{D}$ but to another space $\mathcal{Z} \equiv \mathcal{Z}(\mathbb{R})$ which comprises all those functions ψ such that $\psi(z)$ is defined on $\mathbb{C}$ as an entire function satisfying an inequality of the form

$$ |z^k\psi(z)| \leq C_k \exp(a|y|), \quad a > 0, \quad k = 0, 1, 2, \ldots . $$

Since $\mathcal{D} \cap \mathcal{Z} = \emptyset$ it follows that although $\tilde{\mu}$ is well defined as a linear continuous functional on the space $\mathcal{Z}$ it is not necessarily defined on $\mathcal{D}$ and may therefore not be a distribution. The members of $\mathcal{Z}'(\mathbb{R})$ are called **ultradistributions**, and constitute another class of generalised functions. Although there are functionals which are both distributions and ultradistributions there exist distributions which are not ultradistributions and ultradistributions which are not distributions.

Nonstandard representation of ultradistributions

In [7] a slight modification of the argument presented in [6] showed that every Ξdistribution may be represented by an internal function in ${}^*\mathcal{D}(\mathbb{R})$. Accordingly we can redefine ${}^{\Xi}\mathcal{C}_\infty(\mathbb{R})$ as follows:

$$^{\Xi}\mathcal{C}_\infty(\mathbb{R}) = {}^*D^\infty({}^S\mathcal{D})/\Xi = \bigcup_{r\geq 0}\{{}^*D^r({}^S\mathcal{D})\}/\Xi$$

where ${}^S\mathcal{D}$ is the ${}^*\mathbb{C}_b$ submodule of ${}^S\mathcal{C}$ comprising all infinitely *differentiable functions of hypercompact support which are finite-valued and S-continuous on ${}^*\mathbb{R}_b$.

If $\tilde{F} \equiv [(\tilde{f}_n)_{n\in\mathbb{N}}]$ is any internal function in ${}^*\mathcal{D}$ then its inverse Fourier transform $F = {}^*\mathcal{F}^{-1}\{\tilde{F}\}$ is defined in the obvious way as $F \equiv [(f_n)_{n\in\mathbb{N}}] = [(\mathcal{F}^{-1}\{f_n\})_{n\in\mathbb{N}}]$ and it follows readily that $\tilde{F} \in {}^*\mathcal{D}$ if and only if $F \equiv \mathcal{F}^{-1}\{\hat{F}\} \in {}^*\mathcal{Z}$. Not every internal function in ${}^*\mathcal{D}$ represents a Ξdistribution and similarly not every internal function in ${}^*\mathcal{Z}$ represents an ultradistribution. However the following result was established in [7].

Let $\mathcal{H}(\mathbb{C})$ denote the space of all standard complex-valued functions which can be extended into the complex plane as entire functions. For each entire function $A(z) = \sum_{n=0}^\infty a_n z^n$ in $\mathcal{H}(\mathbb{C})$ define the ∞*-order operator* $\mathbf{A} : \mathcal{Z} \to \mathcal{Z}$ by setting, for each $\phi \in \mathcal{Z}$,

$$\mathbf{A}[\phi(t)] = \sum_{n=0}^{\infty} a_n(-iD)^n\phi(t) = \sum_{n=0}^{\infty}(-i)^n\phi^{(n)}(t).$$

Then:

Theorem 1 *The inverse Fourier transform of a Ξdistribution in ${}^{\Xi}\mathcal{C}_\infty(\mathbb{R})$ is representable as a finite sum of (standard) ∞-order derivatives of internal functions in ${}^*\mathcal{Z}$ whose standard parts are continuous functions of polynomial growth.*

6.2.4 Sato hyperfunctions

Another approach to the required generalisation of the Fourier transform stems from the work of Carlemann [4]. He observed that if a function f, not necessarily in $\mathcal{L}^1(\mathbb{R})$, satisfies a condition of the form

$$\int_0^x |f(y)|dy = 0(|x|^\kappa) \qquad \text{for some natural number } \kappa,$$

and if we write

$$g_1(z) = \int_{-\infty}^{0} f(y)e^{-izy}dy, \quad \text{and} \quad g_2(z) = -\int_0^{+\infty} f(y)e^{-izy}dy$$

then $g_1(z)$ is analytic for all $\Im(z) > 0$ and $g_2(z)$ is analytic for all $\Im(z) < 0$. Moreover, for $\beta > 0$, the function

$$g(x) \equiv g_1(x + i\beta) - g_2(x - i\beta)$$

is the classical Fourier transform of the function $e^{-\beta|t|}f(t)$. The original function f can be recovered by taking the inverse Fourier transform of g and multiplying by $e^{\beta|t|}$. This suggested that a route to a generalisation of the Fourier transform could be found by associating with f a pair of functions $f_1(z)$ and $f_2(z)$ analytic in the upper and lower half-planes respectively. This idea forms the basis of the theory of **hyperfunctions** developed by M. Sato [17] in 1959/60, (although it was anticipated by several other mathematicians). In order to give a brief sketch of this theory it is convenient to introduce the following notation:

$\mathcal{H}(\mathbb{C}\backslash\mathbb{R})$ = the space of all functions analytic outside the real axis.

$\mathcal{H}(\mathbb{C})$ = subspace of all functions in $\mathcal{H}(\mathbb{C}\backslash\mathbb{R})$ which are entire.

$\mathcal{H}^{p,loc}(\mathbb{C}\backslash\mathbb{R})$ = space of all functions θ in $\mathcal{H}(\mathbb{C}\backslash\mathbb{R})$ which are of arbitrary growth to infinity but locally of polynomial growth to the real axis (that is, such that for each compact $K \subset \mathbb{R}$ there exists $C_K > 0$ and $r_K \in \mathbb{N}_0$ such that

$$|\theta(z)| \leq C_K |\Im(z)|^{r_K}$$

for all $z \in \mathbb{C}$ with $\Re(z) \in K$ and sufficiently small $\Im(z) \neq 0$).

Definition 2 *The* ***hyperfunctions*** *of Sato are the members of the quotient space $\mathcal{H}_S(\mathbb{R}) = \mathcal{H}(\mathbb{C}\backslash\mathbb{R})/\mathcal{H}(\mathbb{C})$, that is, the set of all equivalence classes $[\theta]$, where $\theta(z)$ is defined and analytic on $\mathbb{C}\backslash\mathbb{R}$ and $\theta_1 \sim \theta_2$ iff $\theta_1 - \theta_2$ is entire.*

Sato hyperfunctions constitute a genuine extension of Schwartz distributions. This is shown by the following crucial result established by Bremermann [1], in 1965:

Theorem 3 (Bremmerman) *If μ is any distribution in $\mathcal{D}'$ then there exists a function $\mu^0(z)$ defined and analytic in $\mathbb{C}\backslash\mathbb{R}$ such that*

$$< \mu, \phi >= \lim_{\varepsilon \to 0} \int_{-\infty}^{+\infty} \{\mu^0(x + i\varepsilon) - \mu^0(x - i\varepsilon)\}\phi(x)dx.$$

Conversely, if $\theta \in \mathcal{H}^{p,loc}(\mathbb{C}\backslash\mathbb{R})$ then there exists $\mu \in \mathcal{D}'$ such that $\theta(z) = \mu^0(z)$. Identifying $\mu \in \mathcal{D}'$ with $[\theta] \in \mathcal{H}_S(\mathbb{R})$ gives an embedding of $\mathcal{D}'$ such that the mapping $S : \mathcal{D}' \to \mathcal{H}^{p,loc}(\mathbb{C}\backslash\mathbb{R})$ is a topological isomorphism.

Remark 4 Note that if μ has compact support then $\mu^0(z)$ is given explicitly by

$$\mu^0(z) = \frac{1}{2\pi i} < \mu, (x-z)^{-1} > .$$

For example, $\delta^0(z) = \frac{1}{2\pi i} < \delta, (x-z) >= -\frac{1}{2\pi i z}$ and we have

$$< \delta, \phi >= \lim_{\varepsilon \to 0} \frac{\varepsilon}{\pi} \int_{-\infty}^{+\infty} \frac{1}{x^2 + \varepsilon^2} \phi(x)\, dx = \phi(0).$$

6.2.5 Harmonic representation of hyperfunctions

For each Sato hyperfunction $[\theta]$ in $\mathcal{H}_S(\mathbb{R})$ we can choose some specific function $\theta^0 \in [\theta]$ as the **defining function** of the hyperfunction and write

$$\theta_\pi(x, y) = \theta^0(x + iy) - \theta^0(x - iy).$$

Then θ_π maps the half-plane $\Pi^+ \equiv \mathbb{R} \times \mathbb{R}^+$ into $\mathbb{C}$, and is harmonic on Π^+. Define $\mathbf{H}(\Pi^+)$ to be the linear space of all (real or complex-valued) functions defined and harmonic on Π^+, and let $\Gamma : \mathcal{H}(\mathbb{C} \backslash \mathbb{R}) \to \mathbf{H}(\Pi^+)$ denote the map given by

$$\theta^0 \in \mathcal{H}(\mathbb{C} \backslash \mathbb{R}) \to \Gamma(\theta^0) = \theta_\pi(x, y).$$

Then we have the result

Theorem 5 (Li Bang-He, [13]) Γ *is an onto map and if* $\Gamma(\theta^0) = \Gamma(\nu^0)$ *then* $\theta^0 - \nu^0$ *is a complex constant.*

Now suppose that $[\theta]$ is the null hyperfunction, so that θ^0 belongs to $\mathcal{H}(\mathbb{C})$. It is easily shown that $\theta_\pi(x, y)$ is an entire function in both variables (that is, can be extended into $\mathbb{C} \times \mathbb{C}$ as an entire function), is odd in the variable y and such that

$$\lim_{y \to 0} \theta_\pi(x, y) = 0.$$

On the other hand we have immediately from the above theorem,

Corollary 6 *Let* $\theta_\pi \in \mathbf{H}(\Pi^+)$ *be an harmonic function entire in both variables and odd in the variable* y. *Then there exists an entire function* $\theta \in \mathcal{H}(\mathbb{C})$ *such that*

$$\theta_\pi(x, y) = \theta(x + iy) - \theta(x - iy)$$

for all $(x, y) \in \Pi^+$.

Denote by $\mathbf{H}_0(\Pi^+)$ the linear subspace of $\mathbf{H}(\Pi^+)$ comprising all functions which extend into $\mathbb{C} \times \mathbb{C}$ as entire functions in both variables and which are odd in the second variable. Then we have

$$\mathcal{H}_S(\mathbb{R}) \sim \mathbf{H}(\Pi^+)/\mathbf{H}_0(\Pi^+)$$

and every equivalence class $[\theta_\pi(x,y)] \in \mathbf{H}(\Pi^+)/\mathbf{H}_0(\Pi^+)$ is a representation by harmonic functions of the corresponding hyperfunction $[\theta] \in \mathcal{H}_S(\mathbb{R})$. In the sequel we will use analytic or harmonic representation for hyperfunctions as occasion demands.

6.3 Prehyperfunctions and predistributions

Nonstandard representation of hyperfunctions

In the present context $\omega \in {}^*\mathbb{R}$ denotes the infinite hypernatural number defined by $[(n)_{n\in\mathbb{N}}]$ in ${}^*\mathbb{N}_\infty$. Then to each harmonic function $\nu_\pi \in \mathbf{H}(\Pi^+)$ there corresponds an internal function $F_{\{\nu\}} : {}^*\mathbb{R} \to {}^*\mathbb{C}$ defined by

$$F_{\{\nu\}}(x) = {}^*\nu_\pi(x,\varepsilon)$$

where $\varepsilon = \omega^{-1} \in mon(0)$ and ${}^*\nu_\pi \equiv {}^*(\nu_\pi)$ is the nonstandard extension of ν_π. The internal function $F_{\{\nu\}}$ clearly belongs to ${}^*\mathcal{C}^\infty(\mathbb{R})$. If we define a map $\hat{\omega} : \mathbf{H}(\Pi^+) \to {}^*\mathcal{C}^\infty(\mathbb{R})$ by setting $\hat{\omega}(\nu_\pi) = F_{\{\nu\}}$ then we have

$${}^\omega\mathcal{H}_S(\mathbb{R}) \equiv \hat{\omega}(\mathbf{H}(\Pi^+)) \subset {}^*\mathcal{C}^\infty(\mathbb{R})$$

where the inclusion is strict.

The set ${}^\omega\mathcal{H}_S \equiv \hat{\omega}(\mathbf{H}(\Pi^+))$ is an external subset of ${}^*\mathcal{C}^\infty(\mathbb{R})$; however it can be embedded into an internal subset as follows: Consider the nonstandard extension ${}^*\mathbf{H}(\Pi^+)$ of $\mathbf{H}(\Pi^+)$ comprising all *harmonic functions on ${}^*\Pi^+$ and then define,

$${}^\omega\mathbf{H}(\mathbb{R}) \equiv \{F \in {}^*\mathcal{C}^\infty(\mathbb{R}) : [\exists\Theta \in {}^*\mathbf{H}(\Pi^+) : F(x) = \Theta(x,\varepsilon), \forall_{x\in{}^*R}]\}.$$

Then we have

$${}^\omega\mathcal{H}_S \equiv \hat{\omega}(\mathbf{H}(\Pi^+)) \subset {}^\omega\mathbf{H}(\mathbb{R}) \subset {}^*\mathcal{C}^\infty(\mathbb{R})$$

where ${}^\omega\mathbf{H}(\mathbb{R})$ contains elements which are infinitely close to members of ${}^\omega\mathcal{H}_S \equiv \hat{\omega}(\mathbf{H}(\Pi^+))$ and also elements which are far from any internal function in that space. The members of ${}^\omega\mathbf{H}(\mathbb{R})$ will generally be called **prehyperfunctions**. Prehyperfunctions which are near some element of ${}^\omega\mathcal{H}_S(\mathbb{R})$ are said to be **nearstandard prehyperfunctions**, and the others may be called **remote prehyperfunctions**. The set of all such nearstandard prehyperfunctions will be denoted by

$${}^\omega\mathbf{H}_{ns}(\mathbb{R}) \supset {}^\omega\mathcal{H}_S \equiv \hat{\omega}(\mathbf{H}(\Pi^+)).$$

The elements of ${}^\omega\mathbf{H}(\mathbb{R})$ enjoy an important property which is not shared by all internal functions in ${}^*\mathcal{C}^\infty(\mathbb{R})$, namely:

Theorem 7 *Every prehyperfunction in* ${}^{\omega}\mathbf{H}(\mathbb{R})$ *on the line may be extended into the hypercomplex plane as a* **analytic function in the infinitesimal strip*

$$\Omega_{\varepsilon} = \{z \in {}^{*}\mathbb{C} : \ |Im(z)| < \varepsilon\}.$$

Proof. Any function is analytic at the centre of an open disc on which it is harmonic. Hence any internal function $F(x)$, $x \in {}^{*}R$ in ${}^{\omega}\mathbf{H}(\mathbb{R})$ extends into the hypercomplex plane $z = \xi + i\eta$ and is *analytic on the disc

$$(\xi - x)^2 + \eta^2 < \varepsilon^2.$$

For $\xi = x$ we have $-\varepsilon < \eta < \varepsilon$ and since x may be any hyperreal it follows that the internal function $F_{\{\nu\}}(x)$ extends as a *analytic function into the infinitesimal strip in ${}^{*}\mathbb{C}$ defined by $|Im(z)| < \varepsilon$. □

The converse does not hold: not every *analytic internal function in the infinitesimal strip Ω_{ε} is a prehyperfunction on the line. The product of two prehyperfunctions, for example, is a *analytic function on the strip Ω_{ε} but need not itself be a prehyperfunction: ${}^{\omega}\mathbf{H}(\mathbb{R})$ is a linear space over ${}^{*}\mathbb{C}$ but not an algebra. Now let $\mathcal{A}(\Omega_{\varepsilon})$ denote the set of all internal functions in ${}^{*}\mathcal{C}^{\infty}(\mathbb{R})$ which may be extended *analytically into the strip Ω_{ε}. $\mathcal{A}(\Omega_{\varepsilon})$ is a differential subalgebra of ${}^{*}\mathcal{C}^{\infty}(\mathbb{R})$ with respect to the usual operations of addition, multiplication and *differentiation. Moreover we have,

$${}^{\omega}\mathcal{H}_S(\mathbb{R}) \subset {}^{\omega}\mathbf{H}_{ns}(\mathbb{R}) \subset {}^{\omega}\mathbf{H}(\mathbb{R}) \subset \mathcal{A}(\Omega_{\varepsilon}) \subset {}^{*}\mathcal{C}^{\infty}(\mathbb{R}).$$

The product of any two prehyperfunctions makes sense within the algebra $\mathcal{A}(\Omega_{\varepsilon})$ although such a product will not generally be a prehyperfunction. In view of the above inclusions it seems appropriate to call the members of $\mathcal{A}(\Omega_{\varepsilon})$ **generalised prehyperfunctions**.

Finally let $\mathcal{O}(\Omega_{\varepsilon})$ be the subset of all internal functions $\Theta \in \mathcal{A}(\Omega_{\varepsilon})$ such that ${}^{*}D^k\Theta(x) \approx 0$ for all $x \in {}^{*}\mathbb{R}_b$ and for all $k \in \mathbb{N}_0$. $\mathcal{O}(\Omega_{\varepsilon})$ is a linear ${}^{*}\mathbb{C}_b$-submodule (but not an ideal) of $\mathcal{A}(\Omega_{\varepsilon})$ and therefore

$$\mathcal{A}(\Omega_{\varepsilon})/\mathcal{O}(\Omega_{\varepsilon})$$

is a module over ${}^{*}\mathbb{C}_b$: its elements might conveniently be called **generalised** ${}^{\Xi}$**-hyperfunctions**. Every such generalised ${}^{\Xi}$-hyperfunction which may be represented by an internal function in ${}^{\omega}\mathbf{H}_{ns}(\mathbb{R})$ is called a **standard** ${}^{\Xi}$**-hyperfunction** or simply a ${}^{\Xi}$**-hyperfunction**.

The set of all ${}^{\Xi}$-hyperfunctions on the line is defined by

$${}^{\Xi}\mathcal{H}_S(\mathbb{R}) \equiv {}^{\omega}\mathcal{H}_S(\mathbb{R})/\mathcal{O}(\Omega_{\varepsilon})$$

and we have the isomorphism

$$\mathcal{H}_S(\mathbb{R}) \sim {}^{\Xi}\mathcal{H}_S(\mathbb{R}).$$

6.4 The differential algebra $\mathcal{A}(\Omega_\varepsilon)$

6.4.1 Predistributions of finite order

Although not every continuous function f on $\mathbb{R}$ may be continued analytically into the complex plane, every such function does admit an analytic representation in the sense that there exists a unique hyperfunction $[f_\pi(x,y)]$ in $\mathcal{H}_S(\mathbb{R})$ such that $f_\pi(x,y) \to f(x)$ as $y \downarrow 0$ uniformly on compacts. Then the internal function $F(x) = {}^*f_\pi(x,\varepsilon)$ belongs to ${}^\omega\mathcal{H}_S(\mathbb{R}) \subset {}^\omega\mathbf{H}_{ns}(\mathbb{R}) \subset \mathcal{A}(\Omega_\varepsilon)$ and is such that

$$F(x) \approx {}^*f(x), \qquad \text{for all } x \in {}^*\mathbb{R}_b.$$

$F(x)$ is S-continuous at every point (standard and nonstandard) of ${}^*\mathbb{R}_b$; that is,

$$\forall x, y \in {}^*\mathbb{R}_b\ [x \approx y \Rightarrow F(x) \approx F(y)].$$

Reciprocally, every internal function $F \in \mathcal{A}(\Omega_\varepsilon)$ which is finite and S-continuous at every point $x \in {}^*\mathbb{R}_b$ is infinitely close to a (standard) continuous function f defined on $\mathbb{R}$. We denote by ${}^S\mathcal{C}(\Omega_\varepsilon)$ the subalgebra of all functions in $\mathcal{A}(\Omega_\varepsilon)$ which are finite and S-continuous on ${}^*\mathbb{R}_b$.

An internal function $F : {}^*\mathbb{R} \to \mathbb{C}$ is said to be **finitely *differentiable** at $x \in {}^*\mathbb{R}$ if it is *differentiable at x and, in addition, ${}^*DF(x)$ is a bounded number.

Theorem 8 *An internal function $F \in \mathcal{A}(\Omega_\varepsilon)$ which is finitely *differentiable on ${}^*\mathbb{R}_b$ belongs to ${}^S\mathcal{C}(\Omega_\varepsilon)$.*

Definition 9 *For any internal function F in ${}^S\mathcal{C}(\Omega_\varepsilon)$ the standard function $f = st(F)$ will be called the* ***shadow*** *of F while the regular distribution $\nu_F = st_{\mathcal{D}}(F) \in \mathcal{D}'(\mathbb{R})$ generated by F will be called the* ***$\mathcal{D}$-shadow*** *of F.*

Now let Φ be an arbitrary function in ${}^S\mathcal{C}(\Omega_\varepsilon)$. Since ${}^S\mathcal{C}(\Omega_\varepsilon)$ is not a differential algebra the derivative, ${}^*D\Phi$, will not necessarily belong to ${}^S\mathcal{C}(\Omega_\varepsilon)$. However it is easy to see that the functional $\mu_{{}^*D\Phi} : \mathcal{D}(\mathbb{R}) \to \mathbb{C}$ defined by

$$< \mu_{{}^*D\Phi}, \phi >= st(< \Phi, -{}^*\phi' >)$$

is a well defined distribution. Hence ${}^*D\Phi$ has a well-defined $\mathcal{D}$-shadow. If, in addition, ${}^*D\Phi$ itself belongs to ${}^S\mathcal{C}(\Omega_\varepsilon)$ then $\nu_{{}^*D\Phi}$ is the (regular) distribution generated by the standard function $f' = st({}^*D\Phi)$ which is the standard derivative of the function $f = st(\Phi)$. Let ${}^*D^0\{{}^S\mathcal{C}(\Omega_\varepsilon)\}$ denote the set of derivatives of functions in ${}^S\mathcal{C}(\Omega_\varepsilon)$ and for each $k \in \mathbb{N}_0$ define

$${}^*D^k\{{}^S\mathcal{C}(\Omega_\varepsilon)\} = {}^*D\left\{{}^*D^{k-1}\{{}^S\mathcal{C}(\Omega_\varepsilon)\}\right\}$$

where ${}^*D^0\{{}^S\mathcal{C}(\Omega_\varepsilon)\} \equiv {}^S\mathcal{C}(\Omega_\varepsilon)$. We can now define the following (external) subset of ${}^*\mathcal{H}(\Omega_\varepsilon)$:

$$ {}^S\mathcal{C}_\infty(\Omega_\varepsilon) = \bigcup_{k=0}^{\infty} {}^*D^k\{{}^S\mathcal{C}(\Omega_\varepsilon)\}. $$

Then for each $F \in {}^S\mathcal{C}_\infty(\Omega_\varepsilon)$ there exists an S-continuous function $\Phi \in {}^S\mathcal{C}(\Omega_\varepsilon)$ and an integer $r \in \mathbb{N}_0$ such that $F = {}^*D^r\Phi$. The functional $\mu_F : \mathcal{D}(\mathbb{R})$, defined by

$$ <\mu_F, \phi>= st\left((-1)^r < \Phi, {}^*\phi^{(r)} >\right) = st\left({}^*\!\!\int_{{}^*K_\phi} {}^*\phi(x)F(x)dx\right) $$

is a distribution. We call μ_F the **D-shadow** of the internal function F and write $\mu_F = st_\mathcal{D}(F)$. That is to say, we extend $st_\mathcal{D}$ into ${}^S\mathcal{C}_\infty(\Omega_\varepsilon)$ as a mapping with values in $\mathcal{D}'(\mathbb{R})$. An internal function $F \in {}^*\mathcal{H}(\Omega_\varepsilon)$ is said to be **$^*S_\mathcal{D}$differentiable** if, for every $\phi \in \mathcal{D}$ there exists a standard number b_ϕ such that

$$ <\tau^{-1}\{F(x+\tau) - F(x)\}, {}^*\phi > \approx b_\phi $$

for all $\tau \approx 0$, $\tau \neq 0$. As is easily confirmed, every function F in ${}^S\mathcal{C}_\infty(\Omega_\varepsilon)$ is $^*S_\mathcal{D}$differentiable, and we define the **^{S}differential order** of F to be the number ${}^So(F)$ defined by

$$ {}^So(F) = \min\{j \in \mathbb{N}_0 : \ F = {}^*D^j\Phi, \quad \text{for some } \Phi \in {}^S\mathcal{C}(\Omega_\varepsilon)\}. $$

Accordingly we call ${}^S\mathcal{C}_\infty(\Omega_\varepsilon)$ the set of all **predistributions of finite ^{S}differential order**.

Replacing the (standard) concept of **distribution of finite order** by the (nonstandard) concept of **predistribution of finite ^{S}differential order** it is clear that ${}^S\mathcal{C}_\infty(\Omega_\varepsilon)$ with the ${}^*D$ operator constitutes a natural (nonstandard) model for the axiomatic definition of distributions of finite order given by J.S. Silva.

Distributional equivalence

We may now glue together all internal functions in ${}^S\mathcal{C}_\infty(\Omega_\varepsilon)$ which have the same distributional shadow. That is to say, we define the following equivalence relation on ${}^S\mathcal{C}_\infty(\Omega_\varepsilon)$:

$F, G \in {}^S\mathcal{C}_\infty(\Omega_\varepsilon)$ are **distributionally equivalent**, written $F \Xi G$, if and only if they have the same distributional shadow. The quotient space

$$ {}^\Xi\mathcal{C}_\infty(\Omega_\varepsilon) = {}^S\mathcal{C}_\infty(\Omega_\varepsilon)/\Xi $$

is a ${}^*\mathbb{C}_b$-module which is isomorphic to the space $\mathcal{C}_\infty(\mathbb{R})$ of J.S. Silva distributions of finite order.

6.4.2 Predistributions of local finite order

Finite order predistributions in ${}^S\mathcal{C}_\infty(\mathbb{R})$ are not the only prehyperfunctions which have a distributional shadow. Let $F \in {}^\omega\mathcal{H}_S(\mathbb{R})\backslash{}^S\mathcal{C}_\infty(\Omega_\varepsilon)$ be a prehyperfunction such that for every compact $K \subset \mathbb{R}$ there exists an integer $r_K \in \mathbb{N}_0$ and an internal function $\Psi_K \in {}^\omega\mathcal{H}_S(\mathbb{R})$ which is S-continuous on some *-neighbourhood of $^*K$ so that

$$F(x) = {}^*D^{r_K}\Phi_K(x)$$

for all $x \in {}^*K \subset \mathbb{R}_b$. The smallest such r_K will be called the ^{S}differential order of F on $^*K$, and denoted ${}^So_K(F)$. If $\phi \in \mathcal{D}(\mathcal{R})$ has support contained in K then

$$< F, {}^*\pi >=< \Phi_K, (-1)^{r_K}\, {}^*\phi^{(r_K)} >$$

is a bounded number and so F has a distributional shadow in $\mathcal{D}_K(\mathbb{R}) \subset \mathcal{D}(\mathbb{R})$. Since K may be any compact in $\mathcal{R}$ it follows that F has a shadow in $\mathcal{D}(\mathcal{R})$ and so $\mu = st_D(F)$ will be a well defined (standard) distribution in $\mathcal{D}'(\mathcal{R})$.

Denote by ${}^S\mathcal{C}_\pi(\mathbb{R})$ the subset of all prehyperfunctions which have a distributional shadow. Then we have the inclusion

$${}^S\mathcal{C}_\infty(\mathbb{R}) \subset {}^S\mathcal{C}_\pi(\mathbb{R})$$

Moreover ${}^S\mathcal{C}_\pi(\mathbb{R})\backslash{}^S\mathcal{C}_\infty(\mathbb{R})$ is not empty since it contains, for example, the internal function $F : {}^*\mathbb{R} \to {}^*\mathbb{C}$ defined by

$$F(x) = \sum_{i=0}^{+\infty} {}^*D^i \left\{ \frac{1}{\pi} \frac{\omega}{1 + \omega^2(x-i)^2} \right\}$$

The members of ${}^S\mathcal{C}_\pi(\mathbb{R})$ will be called **predistributions of local finite order** or, more simply, **predistributions**.

6.4.3 Predistributions of infinite order

Let Φ be any internal function in ${}^S\mathcal{C}(\Omega_\varepsilon)$ and suppose that there exists an harmonic function $g \in \mathbf{H}(\Pi^+)$ such that $\Phi(x) = {}^*g(x, \varepsilon)$. Let also $r \equiv [r_n]$ be an arbitrary infinite hypernatural number. For every $n \in \mathbb{N}$ the function

$$\frac{\partial^{r_n} g}{\partial x^{r_n}}(x, y), \qquad (x, y) \in \Pi^+$$

is again harmonic on Π^+, and so ${}^*D^r\Phi$ is a generalised prehyperfunction in ${}^*\mathcal{H}(\Omega_\varepsilon)$. The internal function ${}^*D^r\Phi$ is locally bounded by $r!\,\omega^r$; that is to say, for each compact $K \subset \mathbb{R}$ there exists a bounded constant C_K such that for all $x \in {}^*K$ we have

$$|{}^*D^r\Phi(x)| \leq C_K r!\,\omega^r.$$

In general the infinite order derivative ${}^*D^r\Phi$ may have neither an ordinary shadow nor a distributional shadow, and $st_{\mathcal{D}}({}^*D^r\Phi)$ may have no meaning. On the other hand, for any $\phi \in \mathcal{D}(\mathbb{R})$, we have

$$< {}^*D^r\Phi, {}^*\phi >= (-1)^r < \Phi, {}^*\phi^r > .$$

But, since there exist test functions in $\mathcal{D}(\mathbb{R})$ whose derivatives may grow arbitrarily with the order, then

$$< {}^*D^r\Phi, {}^*\phi >\equiv \left[\left(< g(x,1/n), (-1)^{r_n}\phi^{(r_n)}(x) >\right)_{n\in N}\right]$$

will not in general be a bounded hypercomplex number. Hence $st_{\mathcal{D}}({}^*D^r\Phi)$ may have no meaning. However we can define a family of standard part maps which allow us to attach a type of shadow to derivatives of the form ${}^*D^r\Phi$ for infinite $r \in {}^*\mathbb{N}_\infty$ and internal functions Φ in ${}^S\mathcal{C}(\Omega_\varepsilon)$. These standard part maps are defined on certain subspaces of $\mathcal{D}(\mathbb{R})$, for example on those of so-called **Roumieu type** which we recall very briefly as follows.

Spaces of Roumieu type [16]

Let $\mathcal{M}$ denote the set of all positive real sequences $(M_p)_{p\in\mathbb{N}_0}$ such that

(a) $(M_p)^2 \leq M_{p-1}M_{p+1}, \quad p = 0, 1, \ldots,$

(b) $M_p \leq Ah^p \min_{0\leq q\leq p}\{M_pM_{p-q}\}$, $p = 0, 1, \ldots,$ for some positive constants A and h,

(c) $\sum_{p=0}^{+\infty}(M_p)^{-1/p} < +\infty$. Further, let $\mathcal{D}^{(M_p)}(\mathbb{R})$ be the subset of $\mathcal{D}(\mathbb{R})$ comprising all functions ϕ whose derivatives satisfy

$$|\phi^{(p)}(x)| \leq Ah^pM_p, \qquad p = 0, 1, \ldots,$$

for some sequence $(M_p) \in \mathcal{M}$, where A and h are positive constants (generally dependent on ϕ).

It can be shown that $\mathcal{D}^{(M_p)}(\mathbb{R})$ is not empty for every sequence $\{M_p\}_{p\in\mathbb{N}_0} \in \mathcal{M}$. In particular, if there exists $p \in \mathbb{N}_0$ such that $M_p = +\infty$ for all $p \geq p_0$ then $\{M_p\}_{p\in\mathbb{N}_0}$ belongs to $\mathcal{M}$ and $\mathcal{D}^{(M_p)}(\mathbb{R}) \equiv \mathcal{D}(\mathbb{R})$. The space $\mathcal{D}^{(M_p)}(\mathbb{R})$ is the union of the family of spaces

$$\left\{\mathcal{D}^{(M_p)}_{K,h}(\mathbb{R})\right\}_{K\subset\mathbb{R},\,h>0}$$

where K runs over the set of all compact subsets of $\mathbb{R}$ and h runs over all positive numbers. For each real number $h > 0$ and compact $K \subset \mathbb{R}$, $\mathcal{D}^{(M_p)}_{K,h}(\mathbb{R})$ contains all functions $\phi \in \mathcal{D}^{(M_p)}(\mathbb{R})$ with support contained in K and satisfying the above inequality for that particular value of $h > 0$. Each space $\mathcal{D}^{(M_p)}_{K,h}$ is a Banach space with respect to the norm

$$\|\phi\|_{\mathcal{D}^{(M_p)}_{K,h}} = \sup_{p \geq 0} \left\{ \frac{1}{h^p M_p} \sup_{x \in K} |\phi^{(p)}(x)| \right\},$$

and $\mathcal{D}^{(M_p)}(\mathbb{R})$ is provided with the inductive limit topology.

$\mathcal{D}'^{(M_p)}$ denotes the topological dual of $\mathcal{D}^{(M_p)}(\mathbb{R})$ and its elements are sometimes called **generalised distributions in the sense of Roumieu**. A linear functional is in $\mathcal{D}^{(M_p)}(\mathbb{R})$ if and only if it is continuous on that space for every $h > 0$ and compact $K \subset \mathbb{R}$.

6.5 Conclusion

The above outline of Sousa Pinto's nonstandard treatment of hyperfunctions is reported in greater detail in [10]. Sousa Pinto's later work, developing the hyperfinite approach to distributions initiated by Kinoshita, is given in [8] and [9], but most comprehensively in his last publication, the book [20], which is now available in an English translation.

References

[1] H.J. BREMMERMANN, *Distributions, Complex Variables and Fourier Transforms*, Massachusetts, 1965.

[2] N.J. CUTLAND ET AL., *Developments in Nonstandard Mathematics*, Longman, Essex, 1995.

[3] J. CAMPOS FERREIRA, *Introduçao a Teoria das Distribuiçoes*, Calouste Gulbenkian, Lisboa 1993.

[4] T. CARLEMANN, *L'intégrale de Fourier et questions qui s'y ratachent*, Uppsala, 1944.

[5] B. FISHER, "The product of the distributions x^{-r} and $\delta^{(r-1)}(x)$", Proc. Camb. Phil. Soc., **72** (1972) 201–204.

[6] R.F. HOSKINS and J. SOUSA PINTO, "A nonstandard realisation of the J.S. Silva axiomatic theory of distributions", Portugaliae Mathematica, **48** (1991) 195–216.

[7] R.F. Hoskins and J. Sousa Pinto, "A nonstandard definition of finite order ultradistributions", Proc. Indian Acad. Sci. (Math. Sci.), **109** (1999) 389–395.

[8] R.F. Hoskins and J. Sousa Pinto, "Hyperfinite representation of distributions", Proc. Indian Acad. Sci. (Math. Sci.), **110** (2000) 363–377.

[9] R.F. Hoskins and J. Sousa Pinto, "Sampling and Π-sampling expansions", Proc. Indian Acad. Sci. (Math. Sci.), **110** (2000) 379–392.

[10] R.F. Hoskins and J. Sousa Pinto, *Theories of Generalised Functions*, Horwood Publishing Ltd., 2005.

[11] Kinoshita Moto-o, "Nonstandard representations of distributions, I", Osaka J. Math., **25** (1988), 805-824.

[12] Kinoshita Moto-o, "Nonstandard representations of distributions, II", Osaka J. Math., **27** (1990) 843–861.

[13] Li Bang-He, "Non-standard Analysis and Multiplication of Distributions", Scientia Sinica, **XXI-5** (1978) 561–585.

[14] Li Bang-He, "On the harmonic and analytic representation of distributions", Scientia Sinica (Series A), **XXVIII-9** (1985) 923–937.

[15] A. Robinson, *Nonstandard Analysis*, North Holland, 1966.

[16] C. Roumieu, "Sur quelques extensions de la notion de distribution", Ann. Scient. Ec. Norm. Sup., 3e série (1960) 41–121

[17] M. Sato, "Theory of Hyperfunctions", Part I, J. Fac. Sci. Univ. Tokyo, Sect. I, **8** (1959) 139–193; Part II, Sect. I, **8** (1960) 387–437.

[18] L. Schwartz, *Theorie des Distributions I, II*, Hermann, 1957/59.

[19] J.S. Silva, Sur l'axiomatique des distributions et ses possible modeles, *Obras de Jose Sebastiao e Silva*, vol III, I.N.I.C., Portugal.

[20] J. Sousa Pinto, *Metodos Infinitesimais de Analise Matematica*, Lisboa, 2000. English edition: *Infinitesimal Methods of Mathematical Analysis*, Horwood Publishing Ltd., 2004

7

Neutrices in more dimensions

Imme van den Berg*

Abstract

Neutrices are convex subgroups of the nonstandard real number system, most of them are external sets. They may also be viewed as modules over the external set of all limited numbers, as such non-noetherian. Because of the convexity and the invariance under some translations and multiplications, the external neutrices are appropriate models of orders of magnitude of numbers. Using their strong algebraic structure a calculus of *external numbers* has been developped, which includes solving of equations, and even an analysis, for the structure of external numbers has a property of completeness. This paper contains a further step, towards linear algebra and geometry. We show that in $\mathbb{R}^2$ every neutrix is the direct sum of two neutrices of $\mathbb{R}$. The components may be chosen orthogonal.

7.1 Introduction

7.1.1 Motivation and objective

Consider the problem to specify a mathematical model for the intuitive notion of "order of magnitude". Orders of magnitude have some intrinsic vagueness, haziness or superficiality. They are bounded and invariant under at least some additions, or alternatively, translations. Classically, orders of magnitude are modelled with the O- or o-notation, which can be seen as additive groups of real functions in one variable [19]. One may argue that there is some friction between the intuitive notion of order of magnitude, which is about numbers, and its model, which concerns functions. If one wishes to preserve the properties of boundedness and invariance under some additions in the real number system, one enters into conflict with the archimedian property, a conflict also

*University of Évora, Portugal.
ivdb@uevora.pt

known as the *Sorites* paradox [28, 35]. However, this difficulty can be circumvented, if one models within the real number system of nonstandard analysis.

Nonstandard analysis disposes of so-called external sets, which do not correspond to sets of classical analysis. External sets of real numbers may be convex and bounded without having an infimum and supremum, and they are invariant under at least some additions. It may be shown that such an external set E has a group property: There exists $\varepsilon > 0$ such that whenever $x \in E$ one has $x + l\varepsilon \in E$ for limited (i.e., bounded by a standard integer) real numbers l. So it is natural to consider convex (external) additive subgroups of $\mathbb{R}$, which have been called *neutrices* in [29] and [30]. The term is borrowed from Van der Corput [20], who uses it to designate groups of functions, which may be more general than Oh's and oh's. Within the nonstandard real number system there exists a rich variety of neutrices; simple examples are $\oslash$, the set of all infinitesimals, and $\pounds$, the set of all limited real numbers, for more intricate examples see [29, 30] and also section 7.3.2. *External numbers* are the sum of a (nonstandard) real number and a neutrix. One could define an *order of magnitude* simply to be a convex (external) subset of the nonstandard reals; then it is in fact an *external interval*, bounded by two external numbers (see [5] for a proof).

The external numbers satisfy a calculus, which is rich enough to include addition, subtraction, multiplication, division, order, solution of equations and calculation of integrals. All in all, this calculus of orders of magnitude resembles closely the calculus of the reals. We refer to [29, 30] for definitions, results and notations. We recall here a notation for two orders of magnitudes which are not neutrices. The set of positive *appreciable* numbers @ corresponds to the external interval $(\oslash, \pounds]$ and the set $\not\infty$ of positive unlimited numbers corresponds to the external interval $(\pounds, \mathbb{R}]$.

The neutrices and external numbers have been applied in various settings (the terminology not being explicit in earlier papers): singular perturbation theory ([2, this paper describes the discovery of the "canard" phenomenon, the set of parameters for which it occurs is an external interval)], [22], [10], [11, papers on exponentially small thicknesses of boundary layers], [12, thickness of transitions of boundary layers]), asymptotics [5, sets of numbers having the same (nonstandard) asymptotic expansion, and domains of validity of asymptotic approximations], probability theory [6, modelling of mass and queue of probability distributions] and psychology [7, inperfect knowledge of maximal utility]. Special mention has to be made of the work of Bosgiraud, who applies the external numbers nontrivially to problems of modelling and calculation of insecure statistical events in a series of papers [13, 14, 15, 16, 17, 18].

This paper makes a step towards an external calculus in more variables. The main result (7.2.2, decomposition theorem) states that every neutrix in the two-dimensional real space is the direct sum of two neutrices in the one-dimensional nonstandard real line. The neutrices are uniquely determined, and correspond to orthogonal directions. So in a sense the neutrices in the plane possess a dimension, too, which may be interpreted as "length" times "width". In a second paper [8] we extend the decomposition theorem to $\mathbb{R}^k$ for arbitrary standard k, and also show that the decomposition fails if k is unlimited. One reason for dividing the publication of the result into two separate papers is that the proofs of the two-dimensional case and the k-dimensional case are both rather lengthy. It is to be noted that the proof of the k-dimensional case uses in an essential way the result in two dimensions, but is by no means an extension of its proof by some form of external induction.

For standard dimension, the result answers in part a conjecture by Georges Reeb, who suggested that one should be able to recognize the dimension of a space on its external subsets. He also conjectured that there should be a relatively easy nonstandard proof of the topological dimension theorem, or the invariance of domain theorem, but this remains unsettled (some progress has been made by Reveilles [34]).

One may define an *external point* to be the sum of a (nonstandard) vector and a neutrix. On the basis of the decomposition theorem it could be interesting to develop fragments of linear algebra, matrix-calculus, geometry and multivariate analysis and statistics, in order to model, in the case of more variables, approximate qualifications ("small", "enormous", "somewhat", "good"), approximate phenomena ("mistbanks", "stains", "spheres of influence", "superficiality"), and approximate calculus and reasoning.

7.1.2 Setting

Roughly spoken, an internal set is a set defined by a formula of classical analysis (which may contain parameters), and an external set cannot be defined this way.

In k-dimensional space, except for its linear subspaces, every neutrix is an external set. The class of external sets differs from one nonstandard model to another, or alternatively, from one nonstandard axiomatics to another. However, the particular external sets met in applications are to a large extent the same. Indeed, usually they reduce either to $\varnothing$ or to $\pounds$; these two sets have essentially the same properties through all common nonstandard models and axiomatics.

The setting of this paper is the axiomatic system Internal Set Theory (IST) of Nelson [32, 33], and we refer to [24] and [23] for up-to-date presentations and terminology. The language of this system contains two primitive symbols $\in$ and *st* (standard). It differs from model-theoretic approaches in the way that nonstandard elements are already contained in infinite standard sets, instead of extensions of such sets. It has the advantage of simplicity, of saturation in all (standard) cardinals, and of full validity of the "Fehrele principle" [5]: no *galactic* formula, i.e. a Σ_1-formula, starting with the "external quantifier" $\exists x\,(st\,x \wedge \cdots)$ is equivalent to a *halic* formula, i.e. a Π_1-formula, starting with the "external quantifier" $\forall x\,(st\,x \longrightarrow \cdots)$ unless they are equivalent to an internal formula. These are exactly the two characteristics that enable to prove the classification theorem of halflines (theorem 7.3.23), which has as a direct consequence that every order of magnitude is an external interval, and constitutes a crucial step in the proof of the decomposition theorem for neutrices in two dimensions (theorem 7.3.42).

Formally, Nelson's axiomatics does not regard external sets, but there are no major problems if we consider "external sets" which have only internal elements, and are defined by an external formula in which all "external quantifiers" range over standard sets. This is supposedly the case for neutrices, and for sets reduced to such. Axiomatics which enable to deal with this kind of external sets are given by [31, 27, 1, 25, 26].

7.1.3 Structure of this article

In section 7.2 we define formally the notion of neutrix and state the decomposition theorem for neutrices in 2-dimensional space.

The decomposition theorem is proved in section 7.3. The actual proof, which is contained in section 7.3.3, needs some elementary external geometry (section 7.3.1) and algebra (section 7.3.2). In particular we study two kinds of division for neutrices in one dimension, and their relation to certain geometric properties of neutrices in two dimensions. We give special attention to practical aspects, like the calculation of the divisions, and present many examples.

In general, the proof is neither fully algebraic, nor fully analytic. Instead, it tends to be a mixture of algebraic and analytic arguments, where typically algebraic operations are adapted to the order of magnitude of the quantities involved.

7.2 The decomposition theorem

Definition 7.2.1 *Let $k \in \mathbb{N}$, $k \geq 1$ be standard. A* neutrix *is a convex additive subgroup of $\mathbb{R}^k$.*

A convex (external) subset N of $\mathbb{R}$ is a subgroup if and only if it is symmetric with respect to 0 and whenever $x \in N$ one has $2x \in N$. Then by external induction $nx \in N$ for all standard $n \in \mathbb{N}$; by convexity, one has $lx \in$

$\mathbb{N}$ for all limited $l \in \mathbb{R}$. This indicates an alternative way to define neutrices: they should be *modules* over $\pounds$, the external ring of all limited real numbers: a subset N of $\mathbb{R}$ is a neutrix if and only if $\pounds \cdot N = N$.

The only internal neutrices of $\mathbb{R}^k$ are its linear subspaces. Thus the only internal neutrices of $\mathbb{R}$ are $\{0\}$ and $\mathbb{R}$ itself. Two obvious external neutrices in $\mathbb{R}$ are $\pounds$ itself and $\oslash$, the external set of all infinitesimals. The neutrices of the form $\varepsilon\pounds$ for some positive $\varepsilon \in \mathbb{R}$ ("ε-galaxies") are isomorphic to $\pounds$ and the neutrices of the form $\varepsilon\oslash$ for some positive $\varepsilon \in \mathbb{R}$ ("ε-halos") are isomorphic to $\oslash$. Every neutrix $N \neq \{0\}$ is non-noetherian in an external sense: there exists always a strictly ascending chain of subneutrices $(N_n)_{n\in\mathbb{N}}$ with $N_1 \subsetneq N_2 \subsetneq \cdots \subsetneq N_n \subsetneq N_{n+1} \subsetneq \cdots$ for standard indices n. Indeed, let $\omega \in \mathbb{R}$ be positive unlimited, and put $N_n = \omega^n\pounds$. Then $N_n \not\subseteq N_{n+1}$ for all indices n. Consider ω^ω which is not an element of N_n for all standard indices n. Let $\varepsilon \in N$ be sufficiently small such that also $\varepsilon\omega^\omega \in N$. Then $(\varepsilon N_n)_{st\,n}$ is a strictly ascending chain of $\pounds$-submodules of N. It may be proved [5] that for any strictly ascending chain $(N_n)_{st\,n}$ of neutrices, the union $\cup_{st\,n}N_n$ is neither isomorphic (for internal homomorphisms) to $\pounds$ nor to $\oslash$. This suggests that there is a rich variety of external neutrices in $\mathbb{R}$, non-isomorphic with respect to internal homomorphisms. Still, as it is tacitly understood that a neutrix is defined by a bounded formula ϕ of Internal Set Theory, a neutrix has a simple logical form, for ϕ may be supposed to be internal, galactic or halic [5].

The main theorem of this paper asserts that in a sense augmenting the dimension to two does not generate entirely new types of neutrices, for any neutrix in $\mathbb{R}^2$ may be decomposed into two neutrices of $\mathbb{R}$. We adopt the notation $Nx \equiv \{nx \mid n \in N\}$ for the neutrix of all multiples of some vector x with coefficients in some neutrix $N \subset \mathbb{R}$.

Theorem 7.2.2 (Decomposition theorem) *Let $N \subset \mathbb{R}^2$ be a neutrix. Then there are neutrices $N_1 \supset N_2$ in $\mathbb{R}$ and orthonormal vectors u_1, u_2 such that*

$$N = N_1u_1 \oplus N_2u_2$$

Moreover, if $M_1 \supset M_2$ in $\mathbb{R}$ are neutrices and v_1, v_2 are orthonormal vectors with $N = M_1v_1 \oplus M_2v_2$, it holds that

$$M_1 = N_1\ ,\ M_2 = N_2.$$

7.3 Geometry of neutrices in $\mathbb{R}^2$ and proof of the decomposition theorem

The present section is divided into various subsections, conform the various stages of the proof of the decomposition theorem.

The first subsection contains the definitions of the notions of thickness, width (smallest thickness) and length (largest thickness) of neutrices, which are the basic ingredients of the proof. We prove an important theorem, called the "sector-theorem", expressing convexity of the thicknesses of neutrices in two dimensions on not too large sectors. One of the consequences is that the thicknesses in most directions are minimal. This implies that the width of a neutrix is realized in some direction, so in the decomposition represented in the main theorem we obtained already the neutrix N_2. Other important notions used in the proof are near-orthogonality and near-parallelness.

The most important step in the proof of the decomposition theorem in $\mathbb{R}^2$ consists in establishing a direction, which realizes the length of the neutrix; then up to a rotation, the neutrix will simply be "length times width".

This part of the proof uses a form of euclidean geometry in which the points and lines may have non-zero thickness, in fact such a thickness takes the form of a neutrix. We have to adapt some definitions, operations and theorems of ordinary, exact euclidean geometry to this geometry of "clouds" or "mistbanks". This is done in the first subsection, while the second subsection contains an algebraic tool: the "division" of a neutrix by another, the result of which may be calculated through a "subtraction", after taking logarithms.

The final subsection establishes the existence of a direction which realizes the length. It uses an argument of "external analysis": every (external) lower halfline has a supremum, which is an *external number*, i.e., the sum of an ordinary real number and a neutrix (theorem 7.3.23). In a sense, the external set of directions which realize the length is the supremum of an external set of directions which do *not* realize the length.

7.3.1 Thickness, width and length of neutrices

Definition 7.3.1 *Let $N \subset \mathbb{R}^2$ be a neutrix. We call N* square *in case there exists a neutrix $M \subset \mathbb{R}$ such that $N = M \times M$.*

Definition 7.3.2 *Let $N \subset \mathbb{R}^2$ be a neutrix and $r \in \mathbb{R}^2$ be a non-zero vector. The* thickness *of N in the direction of r is the neutrix T_r of $\mathbb{R}$ defined by*

$$T_r = \left\{ x \in \mathbb{R} \,\middle|\, x \cdot \frac{r}{\|r\|} \in N \right\}.$$

The width *W of N is defined by*

$$W = \bigcap_{r \in S^1} T_r,$$

and its length *L by*

$$L = \bigcup_{r \in S^1} T_r.$$

In an obvious way, if X is a subset of $\mathbb{R}^2$ one may define $\|X\| = \{\|x\| \mid x \in X\}$. So, if r is unitary, we may write the thickness in the direction of r alternatively as $T_r = \pm \|N \cap \mathbb{R}r\|$. As an example, consider the neutrix $\pounds \times \varnothing \subset \mathbb{R}^2$. Let $r = \binom{\cos\phi}{\sin\phi}$. Then $T_r = \pounds$ if $\phi \simeq 0 \ (\operatorname{mod} \pi)$, otherwise $T_r = \varnothing$. Hence $W = \varnothing$ and $L = \pounds$. It will be shown later on that for every neutrix the width is assumed in some direction (in fact most of the directions) and the same holds for the length (in only few directions). For square neutrices $N = M^2$ all thicknesses are equal to M, hence their width and length are also equal to M.

Definition 7.3.3 *Let $N \subset \mathbb{R}^2$ be a neutrix and $x, y \in \mathbb{R}^2$. We call x and y* nearly orthonormal *if $\|x\| = \|y\| = 1$ and $< x, y > \simeq 0$. We call x and y* nearly orthogonal *if $\frac{x}{\|x\|}$ and $\frac{y}{\|y\|}$ are nearly orthonormal.*

Let $\varepsilon \simeq 0$. Then the vectors $\binom{1}{0}$ and $\binom{\varepsilon}{\sqrt{1-\varepsilon^2}}$ are nearly orthonormal. The vectors $\binom{1/\varepsilon}{0}$ and $\binom{\varepsilon}{1}$ are nearly orthogonal. We introduce now a notion of near-orthogonality with respect to neutrices.

Definition 7.3.4 *Let $N \subset \mathbb{R}^2$ be a non-square neutrix and $x, y \in \mathbb{R}^2$ be non-zero vectors. We call a line $\mathbb{R}y$* nearly parallel *to N if $T_y > W$. We call x a* near-normal *vector of N if x is nearly orthogonal to all $y \in S^1$ such that y is nearly parallel to N.*

As an example, consider the neutrix $N \equiv \pounds \times \varnothing \subset \mathbb{R}^2$. The vector $\binom{0}{1}$ is nearly orthogonal to N. Indeed unitary vectors y_ε such that y is nearly parallel, i.e. $T_{y_\varepsilon} = \pounds > \varnothing$, are all of the form $\binom{\varepsilon}{\sqrt{1-\varepsilon^2}}$ with $\varepsilon \simeq 0$. Then $\left\langle \binom{0}{1}, \binom{\sqrt{1-\varepsilon^2}}{\varepsilon} \right\rangle \simeq 0$. Note that all vectors of the form $\binom{\alpha}{1+\beta}$ with $\alpha, \beta \simeq 0$ are also nearly orthogonal to N.

The notions of near-parallelness and near-normality will be justified for neutrices N in two dimensions by theorem 7.3.6. In fact it is a consequence of the next simple, but important theorem, based on the convexity of N, that most directions have the same thickness. This proves a fortiori the existence of a direction which realizes the width.

Theorem 7.3.5 (Sector-theorem) *Let $N \subset \mathbb{R}^2$ be a neutrix and a and b two unitary vectors which make an angle θ with $0 \leq \theta \lessapprox \pi$. Let c be a unitary vector, which makes an angle γ with a such that $0 \leq \gamma \leq \theta$. Then*

$$T_c \geq \min(T_a, T_b).$$

Proof. Without restriction of generality, we may assume that $T_b \geq T_a > 0$ and that $a = \binom{1}{0}$. Let $x \in \mathbb{R}a \cap N$. Consider the change of scale $M = N/\|x\|$.

The image of x is $a \in M$. Notice that also $b \in M$, for $T_b \geq T_a$. Let ζ be the intersection of the lines ab and $\mathbb{R}c$. Then $\zeta \in M$ by convexity and $\|\zeta\|$ is appreciable. Because $\pounds\zeta \in M$ it holds that $\zeta/\ \|\zeta\| \in M$. Put $z = \|x\| \cdot \zeta/\ \|\zeta\|$. Then $z \in N$. Because $\|z\| = \|x\|$ we conclude that $T_c \geq T_a$. □

Theorem 7.3.6 *Let $N \subset \mathbb{R}^2$ be a neutrix. Then there exists an unitary vector u such that $T_u = W$. In fact the set of directions corresponding to an unitary vector r with $T_r = W$ contains an interval of the form $(\alpha + \varnothing, \alpha + \pi + \varnothing)$.*

Proof. Let $M = \min\left(T_{\binom{1}{0}}, T_{\binom{0}{1}}\right)$. By the sector-theorem $T_u \geq M$ for all unitary u in the first quadrant. But $T_{\binom{1}{0}} = T_{\binom{-1}{0}}$ and $T_{\binom{0}{1}} = T_{\binom{0}{-1}}$ hence $T_u \geq M$ for all unitary u in every quadrant. Hence $M = W$.

If N is square, one has $T_r = W$ for all unitary vectors r. If not, we may rotate N in order to obtain that $T_{\binom{1}{0}} = T_{\binom{-1}{0}} > W$. Let r be an unitary vector which makes an angle θ with the horizontal axis such that $0 \lessapprox \theta \lessapprox \pi$. If $T_r > W$, one should have $T_{\binom{0}{1}} > W$, by the sector-theorem applied to $[\theta, \pi]$ or $[0, \theta]$, depending to whether $\theta \leq \frac{\pi}{2}$ or $\theta \geq \frac{\pi}{2}$. This implies a contradiction, hence $T_r = W$. Thus we proved the second assertion,with $\alpha = 0$. □

Corollary 7.3.7 *Let $N \subset \mathbb{R}^2$ be a neutrix with width W. Let u, v be two orthonormal vectors. Then $T_u = W$ or $T_v = W$.*

The proof of the decomposition theorem for a neutrix of $\mathbb{R}^2$ is easy, once we know that its length is realized in some direction.

Theorem 7.3.8 *Let $N \subset \mathbb{R}^2$ be a neutrix with length L and width W. Assume there exists a unitary vector u such that $N \cap \mathbb{R}u = Lu$. Let v be unitary such that $u \perp v$. Then $N = Lu \oplus Wv$. Moreover, if u', v' are orthonormal and $N_1, N_2 \subset \mathbb{R}$ are neutrices with $N_1 \supset N_2$ such that $N = N_1u' \oplus N_2v'$, one has $N_1 = L$ and $N_2 = W$.*

Proof. By corollary 7.3.7 it holds that $T_v = W$. Because N is a neutrix, we have $Lu \oplus Wv \subset N$.

Conversely, let $n \in N$. Let p be the orthogonal projection of n on $\mathbb{R}u$. Because $\|n\| \in L$ and $\|p\| \leq \|n\|$ one has $\|p\| \in L$, so $p \in N$. Then $n - p \in N$, and because $T_v = W$, it follows that $n - p \in Wv$. Hence $n = p + (n - p) \in Lu \oplus Wv$ and $N \subset Lu \oplus Wv$. We conclude that $N = Lu \oplus Wv$.

We prove now the uniqueness part. The neutrix N cannot contain vectors larger than its length, so $N_1 \subset L$. If there exists $\lambda \in L \setminus N_1$, all vectors n in N satisfy $\|n\| < |\lambda|$, so L cannot be the length of N. So $L \supset N_1$, from which we conclude that $L = N_1$. By corollary 7.3.7, one has $N_2 = W$. □

It is now almost straightforward to prove the decomposition theorem for neutrices in $\mathbb{R}^2$ if their length is of the form $\lambda\pounds$.

Proposition 7.3.9 *Let $\lambda \in \mathbb{R}, \lambda > 0$. Let N be a neutrix with length $L = \lambda\pounds$ and width W. Then there are orthonormal vectors u and v such that $N = Lu \oplus Wv$.*

Proof. By rescaling if necessary we may assume that $\lambda = 1$. Let $u \in N$ be unitary. Then $\pounds u \in N$ because N is a neutrix, and $N \cap \mathbb{R}u \subset \pounds u$ because the length of N is $\pounds$. Hence $N \cap \mathbb{R}u = \pounds u = Lu$.

Let v be unitary such that $u \perp v$. Then $N = Lu \oplus Wv$ by theorem 7.3.8.□

The decomposition theorem is also easily proved for subneutrices of a given neutrix with length less than the length of this neutrix. Indeed, we have the following definition and proposition.

Definition 7.3.10 *Let $N \subset \mathbb{R}^2$ be a neutrix with length L and width W. Let $M \subset \mathbb{R}$ be a neutrix with $W \subset M \subset L$. We define*

$$N_M = \{n \in N \mid \|n\| \in M\}.$$

Clearly N_M is a neutrix with length M. Its length is realized in any direction for which N contains a vector u with $T_u \geq M$. Then the next proposition is a direct consequence of theorem 7.3.8.

Proposition 7.3.11 *Let $N \subset \mathbb{R}^2$ be a neutrix with length L and width W. Let $M \subset \mathbb{R}$ be a neutrix with $W \subset M \subset L$. Assume there is a unitary vector u such that $T_u \geq M$. Let v be a unitary vector v such that (u, v) is orthonormal. Then $N_M = Mu \oplus Wv$.*

Definition 7.3.12 *Let $N \subset \mathbb{R}^2$ be a neutrix. We call N* lengthy *if it is not square, and if its length is not of the form $L = \varepsilon\pounds$ for some $\varepsilon \in \mathbb{R}$, $\varepsilon > 0$.*

So the two-dimensional decomposition theorem will follow once we have proved that lengthy neutrices assume there length. Note that the proof of proposition 7.3.9 does not work, because for every $\lambda \in L$ there exist $n \in N$ such that $\|n\| = \not\!\!\infty \cdot \lambda$. In order to prove that a lengthy neutrix assumes its length too, we work "from outside in". We consider lines with unitary directions u such that $T_u < L$, with u in an appropriate segment of the unit circle, in such a way that all lines "leave to the right". We divide the segment into two (external) classes: directory vectors for lines which are "leaving upside" and directory vectors for lines which are "leaving downside". It follows from the completion argument mentioned earlier that the two classes do not entirely fill up the segment, just as two disjoint open intervals within some closed interval omit at least one point. The unitary vectors left out are then exactly those directions v such that $T_v = L$.

We introduce an appropriate scaling and orientation for lengthy neutrices, and make also precise what we understand by "leaving to the right upside" and "leaving to the right downside".

Definition 7.3.13 *Let $N \subset \mathbb{R}^2$ be a lengthy neutrix. We call N* appropriately scaled *if $W \subsetneqq \oslash$ and $L \supsetneqq \pounds$.*

Proposition 7.3.14 *A lengthy neutrix $N \subset \mathbb{R}^2$ is homothetic to an appropriately scaled neutrix.*

Proof. Let $\lambda, \omega \in L \setminus W$ be positive such that $\lambda/\omega \simeq 0$. Consider $M = N/\omega$. Its width is $W/\omega \subset W/\lambda \subset \oslash$ and its length is L/ω which contains at least some unlimited elements. If $W/\lambda \subsetneqq \oslash$ we are done. If $W/\lambda = \oslash$ we have $W/\omega \subsetneqq W/\lambda$, so clearly $W/\omega \subsetneqq \oslash$. Hence M is appropriately scaled. □

Definition 7.3.15 *Let $N \subset \mathbb{R}^2$ be a lengthy appropriately scaled neutrix with length L. We call N* appropriately oriented *if*

$$N_{\pounds} = \pounds \begin{pmatrix} 1 \\ 0 \end{pmatrix} \oplus W \begin{pmatrix} 0 \\ 1 \end{pmatrix}.$$

Lemma 7.3.16 *Let $N \subset \mathbb{R}^2$ be a lengthy appropriately scaled neutrix. There exists a rotation ρ of the plane such that $\rho(N)$ is appropriately oriented.*

Proof. Let $W \subsetneqq \oslash$ be the width of N. Then by proposition 7.3.11 there are orthonormal vectors u and v such that $N_{\pounds} = \pounds u \oplus W v$. So let ρ be a rotation such that $\rho(u) = \binom{1}{0}$. Then $\rho(N)_{\pounds} = \pounds \binom{1}{0} \oplus W \binom{0}{1}$. □

Lemma 7.3.17 *Let $N \subset \mathbb{R}^2$ be a lengthy appropriately scaled and oriented neutrix with length L and width W. Let $\lambda \in L, \lambda \gtrsim 0$. Then there exist orthonormal vectors $u \simeq \binom{1}{0}$ and $v \simeq \binom{0}{1}$ such that $N_{\pounds\lambda} = \pounds\lambda u \oplus W v$.*

Proof. If λ is limited, the property follows from definition 7.3.15. If λ is unlimited, any element $n \in N$ with $\|n\| = \lambda$ is of the form $n = \binom{\xi}{\varepsilon\xi}$ with $\xi \simeq \infty$ and $\varepsilon \simeq 0$. Take $u = \frac{1}{\sqrt{1+\varepsilon^2}} \binom{1}{\varepsilon}$ and $v = \frac{1}{\sqrt{1+\varepsilon^2}} \binom{-\varepsilon}{1}$. Then the proposition follows from lemma 7.3.16. □

In the final part of this section we consider points and lines close to a lengthy appropriately scaled and oriented neutrix.

Definition 7.3.18 *Let $N \subset \mathbb{R}^2$ be a lengthy appropriately scaled and oriented neutrix with and length L and width W. Let $q \in \mathbb{R}^2$ be such that $\|q\| \in L$. Then q is said to be* infinitely close *to N if $q \simeq n$ for some element $n \in N$. Let $u \simeq \binom{1}{0}$, $v \simeq \binom{0}{1}$ be orthonormal vectors such that $N_{\pounds\lambda} = \pounds\lambda u \oplus Wv$ for some unlimited λ with $\lambda \geq \|q\|$. Then $q = \xi u + \eta v$ with $\eta \simeq 0$. It is called a* lower *point if $\eta < W$ and an* upper *point if $\eta > W$. Let x be a unitary vector, with $x \simeq \binom{1}{0}$. We say that the nearly parallel line $\mathbb{R}x$ is* downward *if it contains an infinitely close lower point, and* upward *if it contains an infinitely close upper point.*

As an example, consider $N = \omega\varnothing \times \frac{1}{\omega^2}\pounds$. The line containing the vector $\binom{1}{1/\omega}$ is nearly parallel upward, and the line containing the vector $\binom{1}{-1/\omega}$ is nearly parallel downward. By convexity, a nearly parallel line cannot be both downward and upward with respect to a neutrix N. By the next proposition, if its intersection with N is not maximal, it should be either one.

Proposition 7.3.19 *Let $N \subset \mathbb{R}^2$ be a lengthy appropriately scaled and oriented neutrix with length L. Let $x \simeq \binom{1}{0}$ be a unitary vector. Assume that $T_x < L$. Then the nearly parallel line x is either downward or upward with respect to N.*

Proof. Let W be the width of N. Let λ be unlimited such that $T_x < \lambda < L$. By proposition 7.3.9 there exist orthonormal vectors $u \simeq \binom{1}{0}$ and $v \simeq \binom{0}{1}$ such that $N_{\pounds l} = \pounds\lambda \oplus Wv$. By continuity $x \cap \pounds\lambda \oplus \varnothing v$ contains some point $y = \xi u + \eta v$ with $\xi \in L^+$ and $|\eta| > W$. If $\eta < W$ the line x is downward, and $\eta > W$ the line x is upward. □

Definition 7.3.20 *Let $N \subset \mathbb{R}^2$ be a lengthy appropriately scaled and oriented neutrix with length L and width W. Let $x \notin N$ be an infinitely close lower point and $y \notin N$ be an infinitely close upper point, with $\|x\| = \|y\|$. Let $u \simeq \binom{1}{0}$ and $v \simeq \binom{0}{1}$ be orthonormal vectors such that $u \perp xy$. If for some unlimited λ with $|\lambda| \geq \|x\|$ it holds that $N_{\pounds\lambda} = \pounds\lambda u \oplus Wv$, the points x and y are called* opposite *with respect to N. Also, the lines $\mathbb{R}x$ and $\mathbb{R}y$ are called* opposite *with respect to N.*

The final proposition of this section states that opposite lines generate the same thicknesses.

Proposition 7.3.21 *Let $N \subset \mathbb{R}^2$ be a lengthy appropriately scaled and oriented neutrix. Let $a, b \simeq \binom{1}{0}$ be unitary such that $\mathbb{R}a$ is a nearly parallel downward line, and $\mathbb{R}b$ an opposite nearly parallel upward line. Then $T_a = T_b$.*

Proof. Let L be the length of N and W be its width. Let $x \in \mathbb{R}a$ be an infinitely close lower point and y be its opposite infinitely close upper point on $\mathbb{R}b$. Let $\lambda \in L^+, |\lambda| \geq \|x\|$ be such that $N_{\pounds\lambda} = \pounds\lambda u \oplus Wv$, where $u \simeq \binom{1}{0}$ and $v \simeq \binom{0}{1}$ are orthonormal vectors such that $u \perp xy$. Let $\alpha > 0$ be such that $T_a < \alpha < \|x\|$. Then $\alpha a \notin N$, so there exist $\xi \leq \|x\|$, $\eta > W$ such that $\alpha a = \xi u - \eta v$. Because $N_{\pounds\lambda} = \pounds\lambda u \oplus Wv$, it holds that $\alpha b = \xi u + \eta v \notin N$, so $\alpha \notin T_b$. Hence $T_b \subset T_a$. In a symmetric manner we prove that $T_a \subset T_b$. We conclude that $T_a = T_b$. □

7.3.2 On the division of neutrices

Let $N \subset \mathbb{R}^2$ be a neutrix with width W and length L. We assume that N is of the form $N = L\binom{1}{0} \oplus W\binom{0}{1}$. Consider all lines $\mathbb{R}x$ with x of the form $x = \binom{1}{y}$, $y \in \mathbb{R}$.

The following sets appear to be of interest:

1. $R = \left\{ y \mid \mathbb{R}\binom{1}{y} \cap L\binom{1}{0} \oplus W^C\binom{0}{1} = \emptyset \right\}$
2. $S = \left\{ y \mid \mathbb{R}\binom{1}{y} \cap L^C\binom{1}{0} \oplus W\binom{0}{1} \neq \emptyset \right\}$ $\quad (L \subsetneq \mathbb{R})$.

Clearly, if $y \in R$ or $y \in S$ the neutrix realizes its length in the direction x, i.e. one has $T_x = L$. If $y \in S$, the line $\mathbb{R}\binom{1}{y}$ leaves N on its "small" side. On the other hand, if $y \in R^C$, the line $\mathbb{R}\binom{1}{y}$ leaves N on its "large" side. But N may have "corners", i.e., there may exist lines $\mathbb{R}x$ which after leaving N enter into the set $L^C\binom{1}{0} \oplus W^C\binom{0}{1}$. Then $S \subsetneq R$. An example of such a neutrix is $N = \pounds\binom{1}{0} \oplus \varepsilon\pounds\binom{0}{1}$, with $\varepsilon \simeq 0$, $\varepsilon > 0$. An example of a neutrix without such "corners" is given by $N = \pounds\binom{1}{0} \oplus \oslash\binom{0}{1}$, see also [5].

If $W \subsetneq L$, the sets R and S are neutrices, in fact they result from an algebraic operation applied to W and L. The first one is the well-known division operator on ideals or modules, commonly written ":" [36]. We recall here its definition, in the context of neutrices.

Definition 7.3.22 *Let $M, N \subset \mathbb{R}$ two neutrices. We write*

$$M : N = \{x \in \mathbb{R} \mid (\forall n \in N)(nx \in M)\}.$$

One can also abbreviate by

$$M : N = \{x \in \mathbb{R} \mid Nx \subset M)\}.$$

Since $N(\pounds x) = (\pounds N)x = Nx \subset M$, the set $M : N$ is a neutrix. Notice that

$$N(M : N) \subset M$$

and that $M : N$ is the maximal set X which satisfies the property

$$N \cdot X \subset M. \tag{7.1}$$

As such, we call $M : N$ the *solution* of the equation (7.1).

The second one is also a sort of division, that we note M/N. This division is based on a sort of inverse. Its definition needs more knowledge on neutrices, and will be postponed.

The study of divisions is highly related, but distinct from earlier work on the division of neutrices by Koudjeti [29] (see also [30]), mainly in the sense that our definitions are of algebraic or analytic nature, instead of set-theoretic. We use many of his tools and results, sometimes in a slightly modified form.

Following Koudjeti, the argumentation becomes more simple and intrinsic, if instead of multiplications of neutrices we study additions of lower halflines.

The transformation from neutrices N to lower halflines is done by the *symmetrical logarithm* $\log_s(N) = \log(N^+ \setminus \{0\})$; formally we define $\log_s\{0\} = \emptyset$. We transform halflines G back by the *symmetrical exponential* $\exp_s(G) = [-\exp(G), \exp(G)]$; formally we define $\exp_s \emptyset = \{0\}$.

Below we recall some fundamental properties of halflines and neutrices. A neutrix I is *idempotent* if $I \cdot I = I$. A lower halfline H is *idempotent* if $H + H = H$. A lower halfline is idempotent if and only if it is of the form $(-\infty, N)$ or $(-\infty, N]$, where N is a neutrix. A lower halfline H is idempotent if and only if $\exp_s H$ is an idempotent neutrix. A neutrix N is idempotent if and only if $\log_s N$ is an idempotent lower halfline.

Let $\varepsilon \simeq 0$, $\varepsilon > 0$. Examples of idempotent neutrices are, in increasing order $\{0\}$, $\pounds \cdot e^{-@/\varepsilon}$, $\pounds \cdot \varepsilon^{\not\infty}$, $\oslash$, $\pounds$, $\pounds e^{(1/\varepsilon)^{@}}$ and $\mathbb{R}$. The neutrices $\pounds \cdot e^{-(1/\varepsilon)^2 - @/\varepsilon}$, $(1/\varepsilon)^{1/\varepsilon} \cdot \pounds \cdot \varepsilon^{\not\infty}$, $(1/\varepsilon) \cdot \oslash$, $\varepsilon\pounds$ and $\pounds e^{\Gamma(1/\varepsilon) + (1/\varepsilon)^{@}}$ are not idempotent. They are all of the form $a \cdot I$ where I is an idempotent neutrix, and a is a real number. In fact every neutrix can be written is this form. This is a consequence of the following classification theorem of lower half-lines, which with successive generalizations has been proved in [9, 3, 5]:

Theorem 7.3.23 *Every lower half-line $H \subset \mathbb{R}$ has a representation either of the form $H = (-\infty, r + N)$ or $H = (-\infty, r + N]$, where N is a neutrix, which is unique, and r is a real number, determined up to the neutrix N.*

We see that a lower half-line H may be written in the form $H = r + K$, where K is of the form $(-\infty, N)$ or $(-\infty, N]$, i.e. the lower halfline K is idempotent. Then $\exp_s H = e^r \exp_s K$, and we obtained as a consequence that every neutrix is the product of a real number and an idempotent neutrix.

With respect to the above theorem we recall some notation. Halflines of the form $(-\infty, r + N)$ may be called *open*, and halflines of the form $(-\infty, r + N]$

closed. The external set $r + N$ is called an *external number*; it has been shown [29], [30] that many algebraic laws valid for the real number system continue to be valid for the external numbers. Certain analytic laws, too, on behalf of the above theorem. For instance, it may be justified to call the external number $r + N$ the *supremum* $\sup H$ of H. We call $\overline{H} \equiv (-\infty, \sup H]$ the *closure* of H, and $\underline{H} \equiv (-\infty, \sup H)$ the *interior* of H.

The pointwise addition of lower halflines H and K, satisfies

$$H + K = \begin{cases} K & \sup H \subsetneqq \sup K, \text{ or } \sup H = \sup K \text{ and } K \text{ is open} \\ H & \sup H \supsetneqq \sup K, \text{ or } \sup H = \sup K \text{ and } K \text{ is closed.} \end{cases} \tag{7.2}$$

Similar definitions and rules hold for upper halflines, working with infimums instead of supremums. Since lower halflines may be translated into idempotent lower halflines, and neutrices may be rescaled to idempotent neutrices, in defining algebraic operations we may restrict ourselves to the idempotent cases. The extension to the general case is straightforward and will be briefly addressed to at the end of this section.

We turn first to the problem of defining subtractions. The first operation $\div$ will correspond to the division :, and the second operation $\cdot\cdot$, which will correspond to the division /, is defined through inverses.

Definition 7.3.24 *Let H, K be idempotent lower halflines. We define $H \div K$ by*

$$H \div K = \{x \,|\, (\forall k \in K)(k + x \in H)\}\,.$$

Notice that $K + (H \div K) \subset H$ and that we have a maximality property similar to the division :, i.e. $H \div K$ may be called the (maximal) *solution* of the equation $K + X = H$.

Definition 7.3.25 *Let H be an idempotent lower halfline. We define the* symmetrical inverse $(-H)_s$ *of H by*

$$(-H)_s = \begin{cases} \overline{H} & H \text{ open} \\ \underline{H} & H \text{ closed} \end{cases}$$

If H is open ("boundary to the left of zero"), its symmetric reciprocal is a idempotent halfline, which is closed ("boundary to the right of zero"); in a sense the "distance" of the "boundaries" of H and $(-H_s)$ to zero is equal. So there is some geometric justification in calling the reciprocal symmetric. Formally it holds that $(-\emptyset)_s = \mathbb{R}$ and $(-\mathbb{R})_s = \emptyset$.

Definition 7.3.26 *Let H, K be idempotent lower halflines. We define $H \cdot\cdot K$ by*

$$H \cdot\cdot K = H + (-K)_s.$$

Notice that $H \cdot\cdot K$ is an idempotent lower halfline, too.

Proposition 7.3.27 *Let H, K be lower halflines. Let $S = \sup H$ and $T = \sup K$. Then*

1. $(-H)_s = -H^C$.
2. $H \cdot\cdot K = H - K^C$.
3. $H \cdot\cdot K = \begin{cases} (-K)_S & S \subsetneq T, \text{ or } S = T \text{ and } H \text{ is closed} \\ H & S \supsetneq T, \text{ or } S = T \text{ and } H \text{ is open.} \end{cases}$

The proofs are straightforward, using formula (7.2) in 3.

Proposition 7.3.28 *Let H, K be two idempotent lower half-lines of $\mathbb{R}$. Let $S = \sup H$ and $T = \sup K$. Then*

1. $(H \div K) = (H^C - K)^C$.
2. $H \div K = \begin{cases} (-K)_s & S \subsetneq T, \text{ or } S = T \text{ and } K \text{ is open} \\ H & S \supsetneq T, \text{ or } S = T \text{ and } K \text{ is closed.} \end{cases}$

Proof.

1. Let $x \in H \div K$. Suppose $x \in H^C - K$. Then there exist $y > H$, $k \in K$ such that $x = y - k$. Thus $x + k > H$, so $x \notin H \div K$, a contradiction. Then $x \in (H^C - K)^C$, hence $H \div K \subset (H^C - K)^C$. Conversely, let $x \in (H^C - K)^C$. Suppose $x \notin H \div K$. Then there exists $k \in K$, $z \in H^C$ such that $k + x = z$. Thus $x \in H^C - K$, a contradiction. So $x \in (H^C - K)^C$ and $(H^C - K)^C \subset H \div K$. We conclude that $H \div K = (H^C - K)^C$.
2. Straightforward, from 1 and formula (7.2). □

The following theorem is a direct consequence of propositions 7.3.27.3 and 7.3.28.2.

Theorem 7.3.29 *Let H, K be two idempotent lower half-lines of $\mathbb{R}$. Whenever $H \neq K$, it holds that*

$$H \cdot\cdot K = H \div K,$$

but

$$H \cdot\cdot H = \underline{H} \subsetneq \overline{H} = H \div H.$$

So, generically, the subtraction $\cdot\cdot$, defined through reciprocals, and the subtraction $\div$, defined through solutions, yields the same result; thus the equation $K+X=H$ can be solved through reciprocals. In the exceptional case of the subtraction of two identical halflines the outcomes are strongly interrelated, for $H \cdot\cdot H$ is the interior of $H \div H$, and $H \div H$ the closure of $H \cdot\cdot H$.

We will now consider divisions. We start by defining the symmetric inverse. We relate this notion to the symmetric reciprocal.

Definition 7.3.30 *Let $N \subset \mathbb{R}$ be an idempotent neutrix. The* symmetric inverse $(N^{-1})_s$ *is defined by*

$$(N^{-1})_s = \exp_s(-\log_s N)_s .$$

Notice that $(N^{-1})_s$ is an idempotent neutrix, for it is the exponential of an idempotent lower halfline.

Below we calculate the symmetric inverse for some familiar neutrices. We have always $\varepsilon \simeq 0$, $\varepsilon > 0$.

$\mathbf{N}$	$\log_s\mathbf{N}$	$(-\log_s\mathbf{N})_s$	$(\mathbf{N}^{-1})_s$
$\{0\}$	$\emptyset$	$\mathbb{R}$	$\mathbb{R}$
$\pounds e^{-@/\varepsilon}$	$(-\infty, \oslash/\varepsilon)$	$(-\infty, \oslash/\varepsilon]$	$\pounds e^{\oslash/\varepsilon}$
$\pounds \varepsilon^{\not\propto}$	$(-\infty, \pounds \log \varepsilon)$	$(-\infty, \pounds \log 1/\varepsilon]$	$\pounds (1/\varepsilon)^{@}$
$\oslash$	$(-\infty, \pounds)$	$(-\infty, \pounds]$	$\pounds$
$\pounds$	$(-\infty, \pounds]$	$(-\infty, \pounds)$	$\oslash$
$\pounds e^{(1/\varepsilon)^{@}}$	$(-\infty, \pounds/\varepsilon]$	$(-\infty, \pounds/\varepsilon)$	$\pounds e^{-\not\propto/\varepsilon}$
$\mathbb{R}$	$\mathbb{R}$	$\emptyset$	$\{0\}$

Table 7.1: Inverses of some idempotent neutrices.

In [29] the symmetric inverse of a (convex) set N is defined to be $\frac{1}{N^C} \cup \{0\}$.

The next theorem states that, if N is a neutrix, the two definitions are equivalent.

Proposition 7.3.31 *Let $N \subset \mathbb{R}$ be a neutrix. Then*

$$(\mathbf{N}^{-1})_s = \frac{1}{N^C} \cup \{0\}.$$

Proof. The equality holds formally for $N = \{0\}$ and $N = \mathbb{R}$. Let $0 \subsetneqq N \subsetneqq \mathbb{R}$. We prove only that $(\mathbf{N}^{-1})_s \subset \frac{1}{N^C} \cup \{0\}$. Clearly $0 \in \frac{1}{N^C} \cup \{0\}$. Now assume that x is a nonzero element of $(\mathbf{N}^{-1})_s$, that we may suppose to be positive by reasons of symmetry. Then

$$\begin{aligned}\log x &\in (-\log_s N)_s \\ &= -(\log_s N)^C \\ &= -(\log(|N| \setminus \{0\})^C \\ &= -\log(\mathbb{R}^+ \setminus N) \\ &= \log \tfrac{1}{\mathbb{R}^+ \setminus N}.\end{aligned}$$

So $x \in 1/N^C$. Hence $(\mathbf{N}^{-1})_s \subset \frac{1}{N^C} \cup \{0\}$. □

We define the symmetric division through multiplication by the symmetric inverse, and relate it to the symmetric subtraction.

Definition 7.3.32 *Let $M, N \subset \mathbb{R}$ be two idempotent neutrices. The* symmetric division *M/N of M and N is defined by*

$$M/N = M \cdot (N^{-1})_s.$$

Proposition 7.3.33 *Let M, N be idempotent neutrices. Then*

$$M/N = \exp_s(\log_s M \cdot\cdot \log_s N).$$

Proof. We have, using the algebraic relations $M = \exp_s \log_s M$ and $\exp_s H \cdot \exp_s K = \exp_s(H + K)$,

$$\begin{aligned}M/N &= M \cdot (N^{-1})_s \\ &= \exp_s \log_s M \cdot \exp_s(-\log_s N)_s \\ &= \exp_s \left(\log_s M + (-\log_s N)_s\right) \\ &= \exp_s \left(\log_s M \cdot\cdot \log_s N\right).\end{aligned}$$

□

We refrain from giving a general formula for M/N and point out that it can be calculated with the help of the propositions 7.3.27 and 7.3.33. However in some special cases M/N may be readily calculated.

1. $M \subsetneqq N \subset \oslash:\ M/N = M,\ N/M = (M^{-1})_s$.
2. $£ \subset M \subsetneqq N:\ M/N = (N^{-1})_s,\ N/M = N$.
3. $M \subset \oslash:\ M/M = M$.
4. $M \supset £:\ M/M = (M^{-1})_s$.

Observe that always $M/M \subset \varnothing$.

We consider now the division $M : N$, of definition 7.3.22. It bears the following relation to the subtraction $\dot{-}$:

Proposition 7.3.34 *Let M, N be idempotent neutrices. Then*

$$M : N = \exp_s(\log_s M \dot{-} \log_s N).$$

Proof. We prove only that $M : N \subset \exp_s(\log_s M \dot{-} \log_s N)$. Formally, one has $0 \in \exp_s(\log_s M \dot{-} \log_s N)$. Let $x \in M : N, x \neq 0$. For reasons of symmetry, we may suppose that $x > 0$. Because $xN \subset M$ one has $\log x + \log_s N \subset \log_s M$. So $\log x \in \log_s M \dot{-} \log_s N$, hence $x \in \exp_s(\log_s M \dot{-} \log_s N)$. We conclude that $M : N \subset \exp_s(\log_s M \dot{-} \log_s N)$. □

As a consequence of theorem 7.3.29 and propositions 7.3.33 and 7.3.34 we obtain that $M/N = M : N$ whenever $M \neq N$. If $M = N$ we have

$$M : M = \exp_s(\log_s M \dot{-} \log_s M) = \exp_s \overline{\log_s M} = \begin{cases} (M^{-1})_s & M \subset \varnothing \\ M & M \supset £. \end{cases}$$

Notice that always $M : M \supset £$. Because $M/M \subset \varnothing$, we obtain

$$M/M \subsetneq M : M.$$

The division has the following set-theoretic characterization.

Proposition 7.3.35 *Let $M, N \subset \mathbb{R}$ be two neutrices. Then*

$$M : N = \left(\frac{M^C}{N \setminus \{0\}}\right)^C.$$

Proof. The proof is very similar to the proof of proposition 7.3.28.1. □

It follows from the results above that, analogously to the subtraction $\dot{-}$, the division $M : N$ can in practice be calculated through inverses. See also the table and the special cases we presented earlier.

We indicate briefly how subtractions of non-idempotent halflines can be reduced to subtractions of idempotentent halflines, and consider also the analogous reduction for divisions of neutrices.

Let F, G be two lower halflines. By theorem 7.3.23 there exist real numbers f and g and idempotent halflines H and K such that

$$F = f + H \quad , \quad G = g + K.$$

We define

$$F \cdot\cdot G = f - g + H \cdot\cdot K \quad , \quad F \div G = f - g + H \div K.$$

In the same manner, let M, N be neutrices. Let $m, n \in \mathbb{R}$ and I, J be idempotent neutrices such that

$$M = mI \quad , \quad N = nJ.$$

We define

$$M/N = \frac{m}{n} I/J.$$

As regards to the operation $M : N$, the relation

$$M : N = \frac{m}{n}(I : J) \tag{7.3}$$

readily follows from definition 7.3.22. It is a matter of straightforward verification to show that the above formulae do not depend on the choice of f, g, m and n, and that, mutatis mutandis, the properties considered earlier in this section continue to hold.

We state three useful properties of the division $:$. We recall that a neutrix N is *linear* if there exists $\varepsilon > 0$ such that $N = \varnothing\varepsilon$ or $N = \pounds\varepsilon$, else N is *nonlinear*. Nonlinear neutrices have the property that $N = \omega \cdot N$ for at least some $\omega \simeq +\infty$ (see [5]).

Proposition 7.3.36 *Let $M, M_1, M_2, N, N_1, N_2 \subset \mathbb{R}$ be neutrices.*

1. *If $M_1 \subset M_2$, it holds that $M_1 : N \subset M_2 : N$.*
2. *If $N_1 \subset N_2$, it holds that $M : N_1 \supset M : N_2$.*
3. *If $M \subsetneqq N$, one has $M : N = \varnothing$ if and only if there exists $\varepsilon > 0$ such that $M = \varnothing\varepsilon$ and $N = \pounds\varepsilon$, otherwise $M : N \subsetneqq \varnothing$.*

Proof. We only prove 3. Let $\varepsilon, \eta > 0$. As regards to linear neutrices we have the following table

$:$	$\varnothing\varepsilon$	$\pounds\varepsilon$
$\varnothing\eta$	$\pounds\varepsilon/\eta$	$\varnothing\varepsilon/\eta$
$\pounds\eta$	$\pounds\eta/\varepsilon$	$\pounds\eta/\varepsilon$

So the only possibility for such neutrices M, N to obtain $M : N = \varnothing$, is when $M = \varnothing\varepsilon$ and $N = \pounds\eta$, with ε/η appreciable, i.e., when $N = \pounds\varepsilon$. Else the condition $M \subsetneqq N$ ensures that $\varepsilon/\eta \simeq 0$, respectively $\eta/\varepsilon \simeq 0$, which implies that $M : N \subsetneqq \varnothing$.

Assume M is nonlinear. Let $\omega \simeq +\infty$ be such that $M/\omega = M$. Then

$$M : N = \frac{M}{\omega} : N = \frac{1}{\omega}(M : N) \subset \frac{1}{\omega}\varnothing \subsetneqq \varnothing.$$

The case that N is nonlinear is similar. This concludes the proof. □

Finally we establish the relation between the two divisions and the families of directions in the plane R and S.

Theorem 7.3.37 *Let $N \subset \mathbb{R}^2$ be a neutrix with width W and length L, of the form $N = Lu \oplus Wv$, where u and v are orthonormal vectors. Let*

$$R = \left\{ y \,\middle|\, \mathbb{R}\left(\begin{smallmatrix}1\\ y\end{smallmatrix}\right) \cap Lu \oplus W^C v = \emptyset \right\}$$

and, if $L \subsetneqq \mathbb{R}$, let

$$S = \left\{ y \,\middle|\, \mathbb{R}\left(\begin{smallmatrix}1\\ y\end{smallmatrix}\right) \cap L^C u \oplus Wv \neq \emptyset \right\}.$$

Then $R = W : L$ and $S = W/L$.

Proof. Without restriction of generality, we may assume that $u = \left(\begin{smallmatrix}1\\ 0\end{smallmatrix}\right)$ and $v = \left(\begin{smallmatrix}0\\ 1\end{smallmatrix}\right)$. First, let $y \in R$. If $y = 0$, clearly, $y \in W : L$. Assume $y \neq 0$. Then there exist $\lambda \in L, \lambda \neq 0$ such that $\left(\begin{smallmatrix}\lambda\\ \lambda y\end{smallmatrix}\right) \notin L\left(\begin{smallmatrix}1\\ 0\end{smallmatrix}\right) \oplus W^C\left(\begin{smallmatrix}0\\ 1\end{smallmatrix}\right)$, so

$$y = \frac{\lambda y}{\lambda} \in \left(\frac{W^C}{L \setminus \{0\}}\right)^C = W : L.$$

Hence $R \subset W : L$. Conversely, let $y \in W : L$. If $y = 0$, clearly $y \in R$. Assume $y \neq 0$. Suppose there is $\lambda \in L$ such that $\left(\begin{smallmatrix}\lambda\\ \lambda y\end{smallmatrix}\right) \in L\left(\begin{smallmatrix}1\\ 0\end{smallmatrix}\right) \oplus W^C\left(\begin{smallmatrix}0\\ 1\end{smallmatrix}\right)$. Then $y \in \frac{W^C}{L \setminus \{0\}}$, which means that $y \notin W : L$, a contradiction. So $y \in R$, hence $W : L \subset R$. We conclude that $R = W : L$.

Second, let $y \in S$. Then there exists $\mu \in L^C$ such that $\left(\begin{smallmatrix}\mu\\ \mu y\end{smallmatrix}\right) \in L^C\left(\begin{smallmatrix}1\\ 0\end{smallmatrix}\right) \oplus W\left(\begin{smallmatrix}0\\ 1\end{smallmatrix}\right)$. So $\mu y \in W$ and $y \in \frac{W}{L^C} = W/L$. Hence $S \subset W/L$. Conversely, let $y \in W/L$. Then there is $\mu \in L^C$ and $\eta \in W$ such that $y = \frac{\eta}{\mu}$. So $\left(\begin{smallmatrix}\mu\\ \mu y\end{smallmatrix}\right) \in L^C\left(\begin{smallmatrix}1\\ 0\end{smallmatrix}\right) \oplus W\left(\begin{smallmatrix}0\\ 1\end{smallmatrix}\right)$, which means that $y \in S$. Hence $W/L \subset S$. We conclude that $S = W/L$. □

7.3.3 Proof of the decomposition theorem

Let us consider a neutrix in $\mathbb{R}^2$. We know already that, if it is square or not lengthy, it can be decomposed into two neutrices of $\mathbb{R}$. For lengthy neutrices, we sketch here the remaining part of the proof of the decomposition theorem.

A lengthy neutrix may be assumed to be appropriately scaled and oriented. By theorem 7.3.8 it suffices to look for a direction in the plane with maximal thickness. The set of directions with maximal thickness will be obtained as the complement of the directions with nonmaximal thickness, where by the sector-theorem we may restrain ourselves to directions nearly parallel to the neutrix. By proposition 7.3.19 and 7.3.21 they are divided into two "equal" opposite parts. The set of directions with maximal thicknesses will then be the supremum, in the sense of theorem 7.3.23, of the set of downward nearly parallel directions, or alternatively, the infimum of the set of upward nearly parallel directions. In fact, the "gap" between the two opposite families of directions is of the form $\binom{1}{x+W:L}$, where W is the width of the neutrix, and L its length.

We turn now to the proper proof, and start with some terminology.

Definition 7.3.38 *Let $N \subset \mathbb{R}^2$ be a lengthy appropriately scaled and oriented neutrix. We write*

$$\begin{aligned} D &= \left\{ y \simeq 0 \,\middle|\, \tbinom{1}{y} \text{ is nearly parallel downward} \right\} \\ U &= \left\{ y \simeq 0 \,\middle|\, \tbinom{1}{y} \text{ is nearly parallel upward} \right\} \\ S_D &= \left\{ x \in \mathbb{R} \,\middle|\, D + x = D \right\} \\ S_U &= \left\{ x \in \mathbb{R} \,\middle|\, U + x = U \right\}. \end{aligned}$$

For $y \simeq 0$ we write

$$I_y = \left\{ x \in \mathbb{R} \,\middle|\, T_{\binom{1}{y+x}} = T_{\binom{1}{y}} \right\},$$

and we define

$$I = \bigcap_{y \simeq 0} I_y.$$

Theorem 7.3.39 *Let $N \subset \mathbb{R}^2$ be a lengthy appropriately scaled and oriented neutrix with length L and width W. Then*

$$I = W : L \subsetneq \varnothing.$$

Proof. It follows from proposition 7.3.36.3 that $W : L \subsetneq \varnothing$. Let $x \in I$. Suppose $x \notin W : L$. By proposition 7.3.35 there are $\eta \in W^C$ and $\lambda \in L$ such that $x = \eta/\lambda$. By proposition 7.3.9 there are orthonormal vectors $u \simeq \binom{1}{0}$ and $v \simeq \binom{0}{1}$ such that $N_{\pounds\lambda} = \pounds\lambda u \oplus Wv$; up to a rotation we may assume that $u = \binom{1}{0}$ and $v = \binom{0}{1}$. So $\binom{\lambda}{\eta} \notin N$. Hence $T_{\binom{1}{x}} \subset T_{\binom{1}{\eta/\lambda}} \subsetneq \pounds\lambda \subset T_{\binom{1}{0}}$. Hence $x \notin I_u \supset I$, a contradiction. This implies that $x \in W : L$, which means that $I \subset W : L$.

Conversely, let $x \in W : L$. Let $y \simeq 0$. By proposition 7.3.11 there are orthonormal vectors u, v such that $N_{T\binom{1}{y}} = T\binom{1}{y}u \oplus Wv$. It follows from theorems 7.3.37 and 7.3.8 that $T_{\binom{1}{y+x}} = T_{\binom{1}{y}}$ for all $x \in W : T_{\binom{1}{y}} \supset W : L$, so $x \in I_y$. Because y is arbitrary, it holds that $x \in I$. Hence $W : L \subset I$. We conclude that $I = W : L$. □

Theorem 7.3.40 *Let $N \subset \mathbb{R}^2$ be a lengthy appropriately scaled and oriented neutrix with length L and width W. Then*

1. *For all $y, z \in D$ such that $y \leq z$ one has $T_{\binom{1}{y}} \leq T_{\binom{1}{z}}$.*
2. *For all $y, z \in U$ such that $y \leq z$ one has $T_{\binom{1}{y}} \geq T_{\binom{1}{z}}$.*
3. *For all $\varepsilon > W : L$ there is $y \in D, z \in U$ such that $z - y \leq \varepsilon$.*
4. *$S_D = S_U = W : L$.*
5. *There exists $x \simeq 0$ such that $D = [\varnothing, x + W : L)$ and $U = (x + W : L, \varnothing]$.*

Proof.

1. Suppose $T_{\binom{1}{y}} > T_{\binom{1}{z}}$. By proposition 7.3.21 there is $x \in U$ such that $T_{\binom{1}{x}} = T_{\binom{1}{y}}$. So $y < z < x$, $y \simeq x$, while $T_{\binom{1}{y}} > T_{\binom{1}{z}} < T_{\binom{1}{x}}$. This contradicts the sector-theorem. Hence $T_{\binom{1}{y}} \leq T_{\binom{1}{z}}$.
2. Analogous to 1.
3. Because $W : L$ is a neutrix, one has $\varepsilon/2 > W : L$. By proposition 7.3.35 there exist $\lambda \in L, \eta > W$ such that $\eta/\lambda \leq \varepsilon/2$. Let u, v be orthonormal such that $N_{\pounds\lambda} = \pounds\lambda u \oplus Wv$. We see that $\lambda u + \eta v, \lambda u - \eta v \notin N$, so $-\eta/\lambda \in D$ and $\eta/\lambda \in U$, while $\eta/\lambda - (-\,\eta/\lambda) \leq \varepsilon$.
4. From 3 we derive that $S_D \subset W : L$. Conversely, let $y \in D$. Then it follows from theorem 7.3.39 that $T_{\binom{1}{y+W:L}} = T_{\binom{1}{y}} < L$. This implies that $y + x \in D$ for all $x \in W : L$. Hence $W : L \subset S_D$. We conclude that $S_D = W : L$. The proof that $S_U = W : L$ is analogous.
5. By theorem 7.3.23 and 4 the set D is either of the form $D = [\varnothing, x + W : L)$ or $D = [\varnothing, x + W : L]$. We show that the second possibility is absurd. Then the only way to satisfy 3 is when $U = (x + W : L, \varnothing]$. By 7.3.39 and 1 one has $T_{\binom{1}{y}} \leq T_{\binom{1}{x}}$ for all $y \in D$. Then also $T_{\binom{1}{z}} \leq T_{\binom{1}{x}}$ for all $z \in U$, by proposition 7.3.21. Because $D \cup U = \varnothing$, one has $T_{\binom{1}{y}} \leq T_{\binom{1}{x}}$ for all $y \simeq 0$. By theorem 7.3.6 all other thicknesses are equal to W. Hence $T_r \leq T_{\binom{1}{x}} < L$ for all unitary vectors r. Then L cannot be the length of N, a contradiction. Hence $D = [\varnothing, x + W : L)$. The proof that $U = (x + W : L, \varnothing]$ is analogous. □

Theorem 7.3.41 *Let $N \subset \mathbb{R}^2$ be a lengthy appropriately scaled and oriented neutrix with length L and width W. Then there exists $x \simeq 0$ such that* $\left\{y \middle| T_{\binom{1}{y}} = L\right\} = x + W : L$.

Proof. By theorem 7.3.40.5 there exists $x \simeq 0$ such that $\varnothing \setminus (D \cup U) = x + W : L$. By theorem 7.3.39 we have $T_{\binom{1}{y}} = T_{\binom{1}{x}}$ for all $y \in x + W : L$. Suppose $T_{\binom{1}{x}} < L$. By proposition 7.3.19 either $x \in D$ or $x \in U$, a contradiction. Hence $T_{\binom{1}{x}} \geq L$, in fact $T_{\binom{1}{y}} = L$ for all $y \in x + W : L$. □

Theorem 7.3.42 (Two-dimensional decomposition theorem) *Let $N \subset \mathbb{R}^2$ be a neutrix. Then there are neutrices L, W with $W \subset L \subset \mathbb{R}$, and orthonormal vectors u, v such that*

$$N = Lu \oplus Wv.$$

Moreover, the neutrix L is the length of N, and W is its width.

Proof. Let L be the length of N and W its width. If $L = W$ one has $N = L\binom{1}{0} \oplus W\binom{0}{1}$. If $L = \pounds\lambda$ for some $\lambda \in \mathbb{R}$, the theorem follows from proposition 7.3.9. In the remaining cases we may assume that N is appropriately scaled and oriented. By theorem 7.3.41 there exist a unitary vector u such that $T_u = L$. Let v be unitary such that u, v are orthonormal. By theorem 7.3.8 we have $N = Lu \oplus Wv$. The last part of the theorem also follows from theorem 7.3.8. □

References

[1] D. Ballard, *Foundational aspects of "non" standard mathematics*, Contemporary Mathematics 176, American Mathematical Society, Providence, RI, 1994.

[2] E. Benoit, J.-L. Callot, F. Diener, and M. Diener, "Chasse au Canard", Collectanea Mathematica, **31** 1-3 (1983) 37–119.

[3] I.P. van den Berg, "Un principe de permanence général", Astérisque, **110** (1983) 193–208.

[4] I.P. van den Berg, "Un point-de-vue nonstandard sur les développements en série de Taylor", Astérisque, **110** (1983) 209–223.

[5] I.P. van den Berg, *Nonstandard Asymptotic Analysis*, Lecture Notes in Mathematics 1249, Springer-Verlag, 1987.

[6] I.P. van den Berg, An external probability order theorem with some applications, in F. and M. Diener eds., *Nonstandard Analysis in Practice*, Universitext, Springer-Verlag, 1995.

[7] I.P. van den Berg, An external utility function, with an application to mathematical finance, in H. Bacelar-Nicolau, ed., *Proceedings 32nd European Conference on Mathematical Psychology*, National Institute of Statistics INE, Portugal (2001).

[8] I.P. van den Berg, A decomposition theorem for neutrices, IWI preprint 2004-2-01, Univ. of Groningen, submitted.

[9] I.P. van den Berg and M. Diener, "Diverses applications du lemme de Robinson en analyse nonstandard", C. R. Acad. Sc. Paris, Sér. I, Math., **293** (1981) 385–388.

[10] A. Bohé, "Free layers in a singularly perturbed boundary value problem", SIAM J. Math. Anal., **21** (1990) 1264–1280.

[11] A. Bohé, "The existence of supersensitive boundary-value problems", Methods Appl. Anal., **3** (1996) 318–334.

[12] A. Bohé, "The shock location for a class of sensitive boundary value problems", J. Math. Anal. Appl., **235** (1999) 295–314.

[13] J. Bosgiraud, "Exemple de test statistique non standard", Ann. Math. Blaise Pascal, **4** (1997) 9–13.

[14] J. Bosgiraud, "Exemple de test statistique non standard, risques externes", Publ. Inst. Statist. Univ. Paris, **41** (1997) 85–95.

[15] J. Bosgiraud, "Tests statistiques non standard sur des proportions", Ann. I.S.U.P., **44** (2000) 3–12.

[16] J. Bosgiraud, "Tests statistiques non standard pour les modèles à rapport de vraisemblance monotone I", Ann. I.S.U.P., **44** (2000) 87–101.

[17] J. Bosgiraud, "Tests statistiques non standard pour les modèles à rapport de vraisemblance monotone II", Ann. I.S.U.P., **45** (2001) 61–78.

[18] J. Bosgiraud, Nonstandard likely ratio test in exponential families, this volume.

[19] N.G. de Bruijn, *Asymptotic Analysis*, North Holland, 1961.

[20] J.G. van der Corput, Neutrix calculus, neutrices and distributions, MRC Tecnical Summary Report, University of Wisconsin (1960).

[21] F. DIENER, "Sauts des solutions des équations $\varepsilon\ddot{x} = f(t, x, \dot{x})$", SIAM J. Math. Anal., **17** (1986) 533–559.

[22] F. DIENER, "Équations surquadratiques et disparition des sauts", SIAM J. Math. Anal., **19** (1988) 1127–1134.

[23] F. DIENER, M. DIENER (eds.), *Nonstandard Analysis in Practice*, Universitext, Springer-Verlag, 1995.

[24] F. DIENER and G. REEB, *Analyse Non Standard*, Hermann, Paris, 1989.

[25] V. KANOVEI and M. REEKEN, "Internal approach to external sets and universes I. Bounded set theory", Studia Logica, **55** (1995) 229–257.

[26] V. KANOVEI and M. REEKEN, "Internal approach to external sets and universes II. External universes over the universe of bounded set theory", Studia Logica, **55** (1995) 347–376.

[27] T. KAWAI, An axiom system for nonstandard set theory, Rep. Fac. Sci. Kagoshima Univ., **12** (1979) 37–42.

[28] R. KEEFE, *Theories of Vagueness*, Cambridge University Press, 2000.

[29] F. KOUDJETI, *Elements of External Calculus with an aplication to Mathematical Finance*, thesis, Labyrinth publications, Capelle a/d IJssel, The Netherlands, 1995.

[30] F. KOUDJETI and I.P. VAN DEN BERG, Neutrices, external numbers and external calculus, in F. and M. Diener eds., *Nonstandard Analysis in Practice*, Universitext, Springer-Verlag, 1995.

[31] R. LUTZ and M. GOZE, *Nonstandard Analysis: a practical guide with applications*, Lecture Notes in Mathematics 881, Springer-Verlag, 1981.

[32] E. NELSON, "Internal Set Theory, an axiomatric approach to nonstandard analysis", Bull. Am. Math. Soc., **83** (1977) 1165–1198.

[33] E. NELSON, "The syntax of nonstandard analysis", Ann. Pure Appl. Logic, **38** (1988) 123–134.

[34] J. P. REVEILLES, "Une définition externe de la dimension topologique", C. R. Acad. Sci. Paris, Sér. I, Math., **299** (1984) 707–710.

[35] R.M. SAINSBURY, *Paradoxes*, Cambridge University Press, 2nd ed., 1995.

[36] B.L. VAN DER WAERDEN, *Moderne Algebra II*, Springer-Verlag, 1931.

Part II

Number theory

8

Nonstandard methods for additive and combinatorial number theory. A survey

Renling Jin*

8.1 The beginning

In this article my research on the subject described in the title is summarized. I am not the only person who has worked on this subject. For example, several interesting articles by Steve Leth [21, 22, 23] were published around 1988. I would like to apologize to the reader that no efforts have been made by the author to include other people's research.

My research on nonstandard analysis started when I was a graduate student in the University of Wisconsin. A large part of my thesis was devoted towards solving the problems posed in [19]. By the time when my thesis was finished, many of the problems had been solved. However, some of them were still open including [19, Problem 9.13]. It took me another three years to find a solution to [19, Problem 9.13]. Before this my research on nonstandard analysis was mainly focused on foundational issues concerning the structures of nonstandard universes. After I told Steve about my solution to [19, Problem 9.13], he immediately informed me how it could be applied to obtain interesting results in combinatorial number theory. This opened a stargate in front of me and lead me into a new and interesting field.

For nonstandard analysis we use a superstructure approach. We fix an $\aleph_1$-saturated nonstandard universe $^*\mathbb{V}$. For each standard set A we write $^*A$ for the nonstandard version of A in $^*\mathbb{V}$.

*Department of Mathematics, College of Charleston, Charleston, SC 29424. jinr@cofc.edu

8.2 Duality between null ideal and meager ideal

Given an ordered measure space Ω such as the Lebesgue measure space on the real line with the natural order, one can discuss the relationship between measurable sets and open sets[1]. The null ideal on Ω is the collection of null sets, i.e. the sets with measure zero, and the meager ideal is the collection of all meager sets[2]. The sets in an ideal are often considered to be small. The duality between null ideal and meager ideal means that there exists a meager set with positive measure, i.e. the smallness in terms of null ideal is incomparable with the smallness in terms of meager ideal. However, it is usually true that if the space also has an additive structure, then the sum of two set with positive measure may not be meager. See Corollary 8.2.2 for example. What can we say about a Loeb space?

Let H be a hyperfinite integer and let $[0, H]$ be an interval of integers. The term $[a, b]$ in this article always means the interval of integers between a and b including a and b if they are also integers. On $[0, H]$ one can construct a Loeb measure generated by the normalized counting measure. By a Loeb space we always mean the hyperfinite set $[0, H]$ with the Loeb measure generated by the normalized counting measure[3]. On $[0, H]$ there is also a natural order and an additive structure. However, the order topology on $[0, H]$ is discrete, therefore uninteresting. In [19] a U-topology is introduced for each cut $U \subseteq [0, H]$ that gives meaningful analogy of the order topology on the real line. An infinite initial segment U of non-negative integers is called a cut if it is closed under addition. For example $\mathbb{N}$, the set of all standard non-negative integers, is the smallest cut. We often write $x > U$ for a positive integer x and a cut U if $x \notin U$. Let $U \subseteq [0, H]$ be a cut. A set $A \subseteq [0, H]$ is called U-open if for every $x \in A$, there exists a positive integer $y > U$ such that $[x - y, x + y] \cap [0, H] \subseteq A$. A U-topology is the collection of all U-open sets and a U-meager set is a meager set in terms of U-topology. In [19] it was proven that for any cut $U \subseteq [0, H]$ there is always a U-meager set of Loeb measure one in $[0, H]$. The question 9.13 in [19] asked whether the sum (modulo $H + 1$) of two sets in $[0, H]$ with positive Loeb measure can be U-meager for some cut $U \subseteq [0, H]$. For two sets A and B and a binary operation $\circ$ between A and B, we write $A \circ B$ for the set $\{a \circ b : a \in A \text{ and } b \in B\}$. For a number k, we write kA for the set $\{ka : a \in A\}$. In [11] we prove the following theorem.

[1] An order on a space can generate a topology called order topology on the space so that a set is open if it is the union of open intervals.

[2] A set A in a topological space is nowhere dense if every non-empty open set O contains another non-empty open set R disjoint from A. A meager set is the union of finitely many or countably many nowhere dense sets. A meager set is also called a set of the first category.

[3] The Loeb space here is often called the hyperfinite uniform Loeb space.

Theorem 8.2.1 *Let H be a hyperfinite integer and $U \subseteq [0, H]$ be any cut. If $A, B \subseteq [0, H]$ are two internal sets with positive Loeb measure, then $A \oplus_H B$ is not U-nowhere dense, where $\oplus_H$ is the usual addition modulo $H + 1$.*

Note that Theorem 8.2.1 yields a negative answer to [19, Problem 9.13]. Also note that the theorem is still true if $\oplus_H$ is replaced by the usual addition $+$ and the sumset $A+B$ is considered to live in $[0, 2H]$. Theorem 8.2.1 has several corollaries in the standard world. If one lets U be the cut $\bigcap_{n\in\mathbb{N}}[0, \frac{H}{n}]$, then Theorem 8.2.1 implies the following well known non-trivial fact.

Corollary 8.2.2 *If A and B are two sets of reals with positive Lebesgue measure, then $A + B$ must contain a non-empty open interval of reals.*

Corollary 8.2.2 was credited to Steinhaus in [21].

Let $A \subseteq \mathbb{N}$ be infinite. The upper Banach density $BD(A)$ of A is defined by

$$BD(A) = \lim_{k\to\infty} \sup_{n\in\mathbb{N}} \frac{|A \cap [n, n+k]|}{k+1}.$$

A set $C \subseteq \mathbb{N}$ is called piecewise syndetic if there is a positive integer k such that $C + [0, k]$ contains arbitrarily long sequence of consecutive numbers. The definition of upper Banach density, syndeticity, and piecewise syndeticity can be found in [1, 7]. If one let $U = \mathbb{N}$, then Theorem 8.2.1 implies the following result.

Corollary 8.2.3 *Let $A, B \subseteq \mathbb{N}$. If $BD(A) > 0$ and $BD(B) > 0$, then $A + B$ is piecewise syndetic.*

Corollary 8.2.3 was pointed out to me by Steve Leth. By choosing other cuts U one can have more corollaries. These corollaries also have their own corollaries. The reader can find more of them in [11, 18].

Corollary 8.2.2 and Corollary 8.2.3 can also be proven using standard methods [13]. However, Theorem 8.2.1 does not have a standard version. The generality of Theorem 8.2.1 shows the advantages of the nonstandard methods. Theorem 8.2.1 reveals a universal phenomenon, which says that if two sets are large in terms of "measure", then $A + B$ must not be small in terms of "order-topology".

8.3 Buy-one-get-one-free scheme

Excited by the results such as Corollary 8.2.3, I was eager to let people know what I had obtained. After a talk I gave at a meeting in 1997, I was informed by a member in the audience that Corollary 8.2.3 had probably already been proven in [1] or in [7]. This made me rush to the library to check out the book and the paper; I was anxious to see whether my efforts were a waste of time. Fortunately, they weren't; in fact, Corollary 8.2.3 complemented a theorem in [7] which says that if a set $A \subseteq \mathbb{N}$ has positive upper Banach density, then $A - A$ is syndetic. From [1, 7] I also learned of terms such as *upper Banach density*, *syndeticity*, *piecewise syndeticity*, etc. the first time.

One thing which caught my eye when I read [1, 7] was the use of Birkhoff Ergodic Theorem. It is natural for a nonstandard analyst to think what one can achieve if Birkhoff Ergodic Theorem is applied to some problems in a Loeb measure space setting. With that in mind, I derived Theorem 8.3.1 and Theorem 8.3.2 as lemmas in [12].

Given a set $A \subseteq \mathbb{N}$, the lower asymptotic density $\underline{d}(A)$ of A is defined by

$$\underline{d}(A) = \liminf_{n \to \infty} \frac{|A \cap [1, n]|}{n},$$

where $|X|$ means the cardinality of X when X is finite. Later, we will use $|X|$ representing the internal cardinality of X when X is a hyperfinite set. For a set A and a number x we often write $A \pm x$ for $A \pm \{x\}$ and write $x \pm A$ for $\{x\} \pm A$.

Theorem 8.3.1 *Suppose $A \subseteq \mathbb{N}$ with $BD(A) = \alpha$. Then there is an interval of hyperfinite length $[H, K]$ such that for almost all $x \in [H, K]$ in terms of the Loeb measure on $[H, K]$, we have $\underline{d}((^*A - x) \cap \mathbb{N}) = \alpha$. On the other hand, if $A \subseteq \mathbb{N}$ and there is a positive integer x such that $\underline{d}((^*A - x) \cap \mathbb{N}) \geqslant \alpha$, then $BD(A) \geqslant \alpha$.*

Given a set $A \subseteq \mathbb{N}$, the Shnirel'man density $\sigma(A)$ of A is defined by

$$\sigma(A) = \inf_{n \geqslant 1} \frac{|A \cap [1, n]|}{n}.$$

Theorem 8.3.2 *Suppose $A \subseteq \mathbb{N}$ with $BD(A) = \alpha$. Then there is a positive integer x such that $\sigma((^*A - x) \cap \mathbb{N}) = \alpha$.*

Theorem 8.3.1 is [12, Lemma 2] and Theorem 8.3.2 is the combination of [12, Lemma 3, Lemma 4 and Lemma 5]. It is often the case that a result involving Shnirel'man density is obtained first. Then people explore possible generalizations to some results involving lower asymptotic density. The behaviors of these two densities are quite similar. From Theorem 8.3.1 and Theorem 8.3.2 we can see that the behavior of upper Banach density is also similar to the behavior of lower asymptotic density or Shnirel'man density. We can now claim that there is a theorem involving upper Banach density parallel to each existing theorem involving lower asymptotic density or Shnirel'man density. This is the scheme that can be called *buy-one-get-one-free* because we can get a parallel theorem involving upper Banach density for free as soon as a theorem involving lower asymptotic density or Shnirel'man density is obtained. I would now like to briefly describe how this works.

Given a set $A \subseteq \mathbb{N}$ with $BD(A) = \alpha$, there is a positive integer x (may be nonstandard) such that $\underline{d}((^*A - x) \cap \mathbb{N}) = \alpha$ (or $\sigma((^*A - x) \cap \mathbb{N}) = \alpha$). This means that in $x + \mathbb{N}$, a copy of $\mathbb{N}$ above x, the set $^*A$ has lower asymptotic density (or Shnirel'man density) α. Now apply the existing theorem involving $\underline{d}$ (or σ) to the set $^*A \cap (x+\mathbb{N})$ to obtain a result about $^*A$. Finally, pushing down the result to the standard world, one can obtain a parallel theorem involving upper Banach density. Corollary 8.3.3 and Corollary 8.3.4 below are the results obtained using this scheme.

The first one is a corollary parallel to Mann's Theorem. Mann's Theorem says that if two sets $A, B \subseteq \mathbb{N}$ both contain 0, then $\sigma(A + B) \geqslant \min\{\sigma(A) + \sigma(B), 1\}$. Mann's Theorem is an important theorem; in [20] it is referred as one of three pearls in number theory. We can now easily prove the following corollary of Theorem 8.3.2.

Corollary 8.3.3 *$BD(A + B + \{0, 1\}) \geqslant \min\{BD(A) + BD(B), 1\}$ for all $A, B \subseteq \mathbb{N}$.*

The addition of $\{0, 1\}$, which substitutes the condition $0 \in A \cap B$ in Mann's Theorem, is necessary because without it, Corollary 8.3.3 is no longer true. For example if A and B both are the set of all even numbers, then $BD(A) = BD(B) = BD(A + B) = \frac{1}{2}$.

The second corollary is parallel to Plünnecke's Theorem [24, p. 225] which says that if $B \subseteq \mathbb{N}$ is a basis of order h, then $\sigma(A + B) \geqslant \sigma(A)^{1-\frac{1}{h}}$ for every set $A \subseteq \mathbb{N}$. A set $B \subseteq \mathbb{N}$ is called a basis of order h if every non-negative integer n is the sum of at most h non-negative integers (repetition is allowed) from B. In the upper Banach density setting we can define piecewise basis of order h. A set $B \subseteq \mathbb{N}$ is called a piecewise basis of order h if there is a sequence of intervals $[a_k, b_k] \subseteq \mathbb{N}$ with $\lim_{k\to\infty}(b_k - a_k) = \infty$ such that every integer $n \in [0, b_k - a_k]$ is the sum of at most h integers (repetition allowed) from $(B - a_k) \cap [0, b_k - a_k]$. If B is a basis of order h, then it must also be a piecewise basis of order h because one can take $b_k = k$ and $a_k = 0$.

Corollary 8.3.4 *If B is a piecewise basis of order h, then $BD(A + B) \geqslant BD(A)^{1-\frac{1}{h}}$ for every set $A \subseteq \mathbb{N}$.*

Corollary 8.3.3 and Corollary 8.3.4 can also be proven using the standard methods [13] from Ergodic Theory. For more results similar to Corollary 8.3.3 and Corollary 8.3.4, see [12].

8.4 From Kneser to Banach

In the last section we didn't use the full power of Theorem 8.3.1. Suppose $A \subseteq \mathbb{N}$ has upper Banach density α. We use only one x such that $\underline{d}((^*A - x) \cap \mathbb{N}) = \alpha$ while there are almost all x in an interval $[H, K]$ of hyperfinite length such that $\underline{d}((^*A - x) \cap \mathbb{N}) = \alpha$. We can take this advantage and prove a theorem involving upper Banach density parallel to Kneser's Theorem [9]. For two sets $A, B \subseteq \mathbb{N}$ we write $A \sim B$ if $(A \setminus B) \cup (B \setminus A)$ is a finite set. Kneser's Theorem[4] says that for all $A, B \subseteq \mathbb{N}$, if $\underline{d}(A + B) < \underline{d}(A) + \underline{d}(B)$, then there is a positive integer g and a set $G \subseteq [0, g-1]$ such that $A + B \subseteq G + g\mathbb{N}$, $A + B \sim G + g\mathbb{N}$, and $\underline{d}(A + B) = \frac{|G|}{g} \geqslant \underline{d}(A) + \underline{d}(B) - \frac{1}{g}$.

Kneser's Theorem was motivated by Mann's Theorem. Can Mann's Theorem be true if one replaces Shnirel'man density with lower asymptotic density? There are obvious counterexamples. Let d, k, and k' be positive integers. Suppose $G = \{0, d, 2d, \ldots, (k-1)d\}$ and $G' = \{0, d, 2d, \ldots, (k'-1)d\}$. Suppose also that $g > (k + k' - 2)d$, $A = G + g\mathbb{N}$, and $B = G' + g\mathbb{N}$. Then $\underline{d}(A + B) = \frac{k+k'-1}{g} = \underline{d}(A) + \underline{d}(B) - \frac{1}{g}$. Roughly speaking, Kneser's Theorem says that the only kind of counterexamples which make the inequality false in Mann's Theorem with σ replaced by $\underline{d}$ are similar to the one just described.

In [13] a parallel theorem [13, Theorem 3.8] was obtained. Let $A, B \subseteq \mathbb{N}$ with $BD(A) = \alpha$ and $BD(B) = \beta$. Then there are intervals $[a_n, b_n]$ and $[c_n, d_n]$ such that $\lim_{n\to\infty}(b_n - a_n) = \infty$, $\lim_{n\to\infty}(d_n - c_n) = \infty$, $\lim_{n\to\infty} \frac{|A\cap[a_n,b_n]|}{b_n - a_n + 1} = \alpha$, and $\lim_{n\to\infty} \frac{|B\cap[c_n,d_n]|}{d_n - c_n + 1} = \beta$. We hope to characterize the structure of $A + B$ inside the intervals $[a_n + c_n, b_n + d_n]$. However, [13, Theorem 3.8] when restricted to the addition of two sets, only characterized the structure of $A+B$ on a very small part of $\mathbb{N}$. The reason for this is because we used only one x and one y with $\underline{d}((^*A - x) \cap \mathbb{N}) = \alpha$ and $\underline{d}((^*B - y) \cap \mathbb{N}) = \beta$ in the proof.

During the summer of 2003 my undergraduate research partner Prerna Bihani and I conducted an undergraduate research project funded by the College of Charleston to work on theorems parallel to Kneser's Theorem. The work done during the summer and the following year produced the paper [2], which contains Theorem 8.4.1. To avoid some technical difficulties we considered only the sum of two copies of the same set in [2].

[4]The version of Kneser's Theorem in [9] is about the addition of multiple sets. We stated the version here for the addition of two sets just for simplicity.

Theorem 8.4.1 *Let A be a set of non-negative integers such that $BD(A) = \alpha$ and $BD(A + A) < 2\alpha$. Let $\{[a_n, b_n] : n \in \mathbb{N}\}$ be a sequence of intervals such that $\lim_{n\to\infty}(b_n - a_n) = \infty$ and $\lim_{n\to\infty} \frac{|A\cap[a_n,b_n]|}{b_n-a_n+1} = \alpha$. Then there are $g \in \mathbb{N}$, $G \subseteq [0, g - 1]$, and $[c_n, d_n] \subseteq [a_n, b_n]$ for each $n \in \mathbb{N}$ such that*

(1) $\lim_{n\to\infty} \frac{d_n-c_n}{b_n-a_n} = 1$,

(2) $A + A \subseteq G + g\mathbb{N}$,

(3) $(A + A) \cap [2c_n, 2d_n] = (G + g\mathbb{N}) \cap [2c_n, 2d_n]$ *for all* $n \in \mathbb{N}$,

(4) $BD(A + A) = \frac{|G|}{g} \geqslant 2\alpha - \frac{1}{g}$.

Note that (1) above shows that the structure of $A + A$ is characterized on a large portion of $[2a_n, 2b_n]$. Note also that we cannot replace $[2c_n, 2d_n]$ with $[2a_n, 2b_n]$ in (3) because all conditions for A still hold if we delete any elements from $A \cap ([a_n, b_n] \setminus [c_n, d_n])$.

The proof of Theorem 8.4.1 can be described in several steps. Given a hyperfinite integer N, we know that for almost all $x, y \in [a_N, b_N]$ we have $\underline{d}((^*A-x)\cap\mathbb{N}) = \underline{d}((^*A-y)\cap\mathbb{N}) = \alpha$. We can also assume that $\underline{d}((x-{}^*A)\cap\mathbb{N}) = \underline{d}((y-{}^*A)\cap\mathbb{N}) = \alpha$. Step one: characterize the structure of $^*(A+A)$ in $x+y+\mathbb{Z}$ using Kneser's Theorem, where $\mathbb{Z}$ is the set of all standard integers. Step two: show that the structures of $^*(A+A)\cap(x+y+\mathbb{Z})$ for almost all $x, y \in [a_N, b_N]$ are consistent with one another so that these structures can be combined into one structure. Hence we can characterize the structure of $^*(A + A)$ in $[2c_N, 2d_N]$, where $\frac{d_N-c_N}{b_N-a_N} \approx 1$. Step three: prove that for different hyperfinite integers N and N', the structure of $^*(A + A)$ in $[2a_N, 2b_N]$ and the structure of $^*(A + A)$ in $[2a_{N'}, 2b_{N'}]$ are consistent so that these structures of $^*(A+A)$ in $[2a_N, 2b_N]$ for all hyperfinite integers N can be combined into one structure of $^*(A + A)$ in $\bigcup\{[a_N, b_N] : N \text{ is hyperfinite}\}$. Step five: pushing down the structure of $^*(A + A)$ to the standard world results Theorem 8.4.1.

The methods developed in [13] do not seem to be enough for proving Theorem 8.4.1. So it is interesting to see whether one can produce a reasonably nice and short standard proof of the theorem.

In [3] the structure of A was characterized when $\underline{d}(A)$ is very small and $\underline{d}(A + A) \leqslant c\underline{d}(A)$ for some constant $c \geqslant 2$. It is also interesting to see how one can characterize the structure of $A + A$ when $BD(A + A) \leqslant cBD(A)$ for some constant $c \geqslant 2$.

8.5 Inverse problem for upper asymptotic density

In January of 2000, I was invited to give a talk at the DIMACS workshop "Unusual Applications of Number Theory". One of the workshop organizers was Melvyn Nathanson to whom I am grateful for being the first number theorist to express an interest in my research on number theory not to mention his continued encouragement. During the workshop I had a chance to meet another number theorist G. A. Freiman who is well-known for his work on inverse problems in additive and combinatorial number theory. He gave me a preprint of his list of open problems [5]. This list and the book [24] have since gotten me interested in the inverse problems.

Inverse problems study the properties of A when $A+A$ satisfies certain conditions. Freiman discovered a phenomenon that if $A+A$ is small, then A must have some arithmetic structure. In fact Kneser's Theorem and Theorem 8.4.1 can be viewed as two examples of the phenomenon. One can characterize the arithmetic structure of A from the structure of $A+A$ in Theorem 8.4.1 and characterize the structure of A and the structure of B from the structure of $A+B$ in Kneser's Theorem (see [2] for details). In this section we characterize the structure of A when the upper asymptotic density of $A+A$ is small. Given $A \subseteq \mathbb{N}$, the upper asymptotic density $\bar{d}(A)$ of A is defined by

$$\bar{d}(A) = \limsup_{n\to\infty} \frac{|A \cap [1,n]|}{n}.$$

Without loss of generality we always assume $0 \in A$ in this section. We can also assume that $\gcd(A) = 1$ because if $\gcd(A) = d > 1$, then we can recover the structure of A from the structure of A', where $A' = \{a/d : a \in A\}$. When $0 \in A$ and $\gcd(A) = 1$, one can easily prove, using Freiman's result (1) at the beginning of the next section, that $\bar{d}(A+A) \geqslant \frac{3}{2}\bar{d}(A)$ if $\bar{d}(A) \leqslant \frac{1}{2}$ and $\bar{d}(A+A) \geqslant \frac{1+\bar{d}(A)}{2}$ if $\bar{d}(A) \geqslant \frac{1}{2}$. The following two examples show that the lower bounds above are optimal.

Example 8.5.1 *For every real number* $0 \leqslant \alpha \leqslant 1$, *let*

$$A = \{0\} \cup \bigcup_{n=1}^{\infty} [\lceil (1-\alpha)2^{2^n} \rceil, 2^{2^n}].$$

Then $\bar{d}(A) = \alpha$, $\bar{d}(A+A) = \frac{1+\alpha}{2}$ *if* $\alpha \geqslant \frac{1}{2}$, *and* $\bar{d}(A+A) = \frac{3}{2}\alpha$ *if* $\alpha \leqslant \frac{1}{2}$.

Example 8.5.2 *Let* $k, m \in \mathbb{N}$ *be such that* $k \geqslant 4$ *and* 0, m, $2m$ *are pairwise distinct modulo* k. *Let* $A = k\mathbb{N} \cup (m + k\mathbb{N})$. *Then* $\bar{d}(A) = \frac{2}{k} = \alpha \leqslant \frac{1}{2}$ *and* $\bar{d}(A+A) = \frac{3}{k} = \frac{3}{2}\alpha$. *It is easy to choose* k, m *such that* $\gcd(A) = 1$.

So we can say that $\bar{d}(A+A)$ is small when $\bar{d}(A+A) = \min\{\frac{3}{2}\bar{d}(A), \frac{1+\bar{d}(A)}{2}\}$ and we need to characterize the structure of A when $\bar{d}(A+A)$ is small.

Clearly the characterization of the structure of A should cover the cases in both Example 8.5.1 and Example 8.5.2. We hope to show that A must have the structure described in one of the examples above when $\bar{d}(A+A)$ is small. However, some variations of the examples are unavoidable. If the set A is replaced by $A' \subseteq A$ with $\bar{d}(A') = \alpha$ in both Example 8.5.1 and Example 8.5.2, then $\bar{d}(A'+A')$ is also small. Furthermore, if $A = A' \cup A''$ where

$$A' = \{0\} \cup \bigcup_{n=1}^{\infty} [\lceil (1-\alpha)2^{2^{2n}} \rceil, 2^{2^{2n}}]$$

and A'' is an arbitrary subset of

$$\bigcup_{n=1}^{\infty} [\lceil (1-\alpha)2^{2^{2n+1}} \rceil, 2^{2^{2n+1}}],$$

then again $\bar{d}(A+A)$ is small. This example shows that we can only hope to characterize the structure of A along the increasing sequence h_n such that $\lim_{n\to\infty} \frac{|A\cap[1,h_n]|}{h_n} = \alpha$.

It was a long journey for me to arrive at the most recent result in [16] due to the technical difficulties of the proof. First the structure of A was characterized in [14] when $\bar{d}(A+A+\{0,1\})$ is small. Later the structure of A was characterized in [15] when $\bar{d}(A+A)$ is small and A contains two consecutive numbers. Finally in [16] the following theorem was proven.

Theorem 8.5.3 *Let* $\bar{d}(A) = \alpha > 0$.

Part I: Assume $\alpha > \frac{1}{2}$. *Then* $\bar{d}(A+A) = \frac{1+\alpha}{2}$ *implies that for every increasing sequence* $\{h_n : n \in \mathbb{N}\}$ *with* $\lim_{n\to\infty} \frac{|A\cap[0,h_n]|}{h_n+1} = \alpha$, *we have*

$$\lim_{n\to\infty} \frac{|(A+A)\cap[0,h_n]|}{h_n+1} = \alpha.$$

Part II: Assume $\alpha < \frac{1}{2}$ *and* $\gcd(A) = 1$. *Then* $\bar{d}(A+A) = \frac{3}{2}\alpha$ *implies that either (a) there exist* $k > 4$ *and* $c \in [1, k-1]$ *such that* $\alpha = \frac{2}{k}$ *and* $A \subseteq k\mathbb{N} \cup (c + k\mathbb{N})$ *or (b) for every increasing sequence* $\{h_n : n \in \mathbb{N}\}$ *with* $\lim_{n\to\infty} \frac{|A\cap[0,h_n]|}{h_n+1} = \alpha$, *there exist two sequences* $0 \leqslant c_n \leqslant b_n \leqslant h_n$ *such that*

$$\lim_{n\to\infty} \frac{|A\cap[b_n,h_n]|}{h_n - b_n + 1} = 1,$$

$$\lim_{n\to\infty} \frac{c_n}{h_n} = 0,$$

and $[c_n+1, b_n-1] \cap A = \emptyset$ *for every* $n \in \mathbb{N}$.

Part III: Assume $\alpha = \frac{1}{2}$ *and* $\gcd(A) = 1$. *Then* $\bar{d}(A+A) = \frac{3}{2}\alpha$ *implies that either (a) there exists* $c \in \{1,3\}$ *such that* $A \subseteq 4\mathbb{N} \cup (c+4\mathbb{N})$ *or (b) for every increasing sequence* $\{h_n : n \in \mathbb{N}\}$ *with* $\lim_{n\to\infty} \frac{|A\cap[0,h_n]|}{h_n+1} = \alpha$, *we have*

$$\lim_{n\to\infty} \frac{|(A+A)\cap[0,h_n]|}{h_n+1} = \alpha.$$

I would like to make some remarks here on Theorem 8.5.3. First, the proof of Part I is easy; the most difficult part is Part II. Second, Part I and (b) of Part III cannot be improved so that set A has the structure similar to the structure described in (b) of Part II. For example, if one lets

$$A = \{0\} \cup \bigcup_{n=1}^{\infty} \left([3\cdot 2^{2^n-3}, 4\cdot 2^{2^n-3}] \cup [5\cdot 2^{2^n-3}, 2^{2^n}]\right),$$

then $\bar{d}(A) = \frac{1}{2}$ and $\bar{d}(A+A) = \frac{1+\bar{d}(A)}{2}$. Clearly A does not have the structure described in (b) of Part II.

The main ingredient of the proof of Theorem 8.5.3 is the following lemma in nonstandard analysis. For an internal set $A \subseteq [0,H]$ and a cut $U \subseteq [0,H]$ we define the lower U-density $\underline{d}_U(A)$ by

$$\underline{d}_U(A) = \sup\left\{\inf\left\{st\left(\frac{|A\cap[0,n]|}{n+1}\right) : n \in U \smallsetminus [0,m]\right\} : m \in U\right\},$$

where st means the standard part map. Note that if $U = \mathbb{N}$ and $A \subseteq \mathbb{N}$, then $\underline{d}(A) = \underline{d}_U({}^*A)$. A set $I = \{a, a+d, a+2d, \ldots\}$ is called an arithmetic progression with difference d. An arithmetic progression can be finite (hyperfinite) or infinite. If an arithmetic progression is finite (hyperfinite), then its cardinality (internal cardinality) is its length. A set $I \cup J$ is called a bi-arithmetic progression if both I and J are arithmetic progressions with the same difference d and $I+I$, $I+J$, and $J+J$ are pairwise disjoint. A finite (hyperfinite) bi-arithmetic progression $I\cup J$ has its length $|I|+|J|$. Let U be a cut. A bi-arithmetic progression $B \subseteq U$ is called U-unbounded if both I and J are upper unbounded in U.

Lemma 8.5.4 *Let* H *be hyperfinite and* $U = \bigcap_{n\in\mathbb{N}}[0, \frac{H}{n}]$. *Suppose* $A \subseteq [0,H]$ *be such that* $0 < \underline{d}_U(A) = \alpha < \frac{2}{3}$. *If* $A \cap U$ *is neither a subset of an arithmetic progression of difference greater than* 1 *nor a subset of a* U*-unbounded bi-arithmetic progression, then there is a standard positive real number* $\gamma > 0$ *such that for every* $N > U$, *there is a* $K \in A$, $U < K < N$, *such that*

$$\frac{|(A+A)\cap[0,2K]|}{2K+1} \geqslant 3\frac{|A\cap[0,K]|}{2K+1} + \gamma.$$

Lemma 8.5.4 is motivated by Kneser's Theorem. It basically says that either $A+A$ is large in an interval $[0, 2K]$ with $K > \frac{H}{n}$ for some standard n or A has desired arithmetic structure in an interval $[0, K]$ with $K > \frac{H}{n}$ for some standard n. The proof uses the fact that U is an additive semi-group. This can be done only in a nonstandard setting. It is interesting to see whether this lemma can be replaced by a standard argument with a reasonable length.

Recently G. Bordes [4] generalized Part II of Theorem 8.5.3 for sets A with small upper asymptotic density. He characterized the structure of A when $\bar{d}(A) \leqslant \alpha_0$ for some small positive number α_0 and $\bar{d}(A+A) < \frac{5}{3}\bar{d}(A)$. It is interesting to see whether one can replace α_0 by a relatively large value, say $\frac{2}{5}$, in Bordes' Theorem.

8.6 Freiman's $3k-3+b$ conjecture

After Theorem 8.5.3 was proven, I realized that the same methods used there could also be used to advance the existing results towards the solution of Freiman's $3k-3+b$ conjecture [5]. This is important because the conjecture is about the inverse problem for the addition of *finite sets*. Let A be a finite set of integers with cardinality $k > 0$. It is easy to see that $|A+A| \geqslant 2k-1$. On the other hand, if $|A+A| = 2k-1$, then A must be an arithmetic progression. In the early 1960s, Freiman obtained the following generalizations [6].

(1) Let $A \subseteq \mathbb{N}$. Suppose $k = |A|$, $0 = \min A$, and $n = \max A$. Suppose also $\gcd(A) = 1$. Then $|A+A| \geqslant 3k-3$ if $n \geqslant 2k-3$ and $|A+A| \geqslant k+n$ if $n \leqslant 2k-3$.

(2) If $k > 3$ and $|2A| = 2k-1+b < 3k-3$, then A is a subset of an arithmetic progression of length at most $k+b$

(3) If $k > 6$ and $|2A| = 3k-3$, then either A is a subset of an arithmetic progression of length at most $2k-1$ or A is a bi-arithmetic progression.

In [6] a result was also mentioned without proofs for characterizing the structure of A when $k > 10$ and $|A+A| = 3k-2$. In [10] an interesting generalization of (3) above was obtained by Hamidoune and Plagne, where the condition $|2A| = 3k-3$ is replaced by $|A+tA| = 3k-3$ for every integer t. However, no further progress of this kind had been made for a larger value of $|A+A|$ before my recent work. In fact, Freiman made the following conjecture in [5] five years ago.

Conjecture 8.6.1 *There exists a natural number K such that for any finite set of integers A with $|A| = k > K$ and $|A+A| = 3k-3+b < \frac{10}{3}k-5$ for* *some $b \geqslant 0$, A is either a subset of an arithmetic progression of length at most $2k-1+2b$ or a subset of a bi-arithmetic progression of length at most $k+b$.*

Note that the conclusion of Conjecture 8.6.1 could be false if one allows $|A+A| = \frac{10}{3}k - 5$. Simply let A be the union of three intervals $[0, a-1]$, $[b, b+a-1]$, and $[2b, 2b+a-1]$, where $k = 3a$ and b is a sufficiently large integer. Clearly $|A+A| = \frac{10}{3}k - 5$. Since b can be as large as we want, we can choose a b so that set A is neither a subset of an arithmetic progression of a restricted length nor a subset of a bi-arithmetic progression of a restricted length.

Using nonstandard methods such as Lemma 8.5.4, I was able to prove the following theorem in [17].

Theorem 8.6.2 *Suppose $f : \mathbb{N} \mapsto \mathbb{N}$ is a function with $\lim_{n\to\infty} \frac{f(n)}{n} = 0$. There exists a natural number K such that for any finite set of integers A with $|A| = k$, if $k > K$ and $|A+A| = 3k-3+b$ for some $0 \leqslant b \leqslant f(k)$, then A is either a subset of an arithmetic progression of length at most $2k-1+2b$ or a subset of a bi-arithmetic progression of length at most $k+b$.*

Theorem 8.6.2 gives a new result even for $f(x) \equiv 2$. However, we still have a long way before solving Conjecture 8.6.1. It is already interesting to see whether we can obtain the same result with $f(x) = \alpha x$ for some positive real number α.

The ideas for proving Theorem 8.6.2 are similar to the proof of Theorem 8.5.3, but much more technical. Suppose Theorem 8.6.2 is not true; then one can find a sequence of counterexamples A_n such that $|A_n| \to \infty$. Given a hyperfinite integer N, let $A = A_N$. Without loss of generality, we can assume that $0 = \min A$, $H = \max A$, $\gcd(A) = 1$, and $\alpha \approx \frac{|A|}{H+1} \gg 0$. Note that $\frac{|A+A|-3|A|+3}{H} \approx 0$. Hence $|A+A|$ is almost the same as $3|A|-3$ from the nonstandard point of view. Using the case-by-case argument, we can show that if $\frac{|A|}{H+1} \ll \frac{1}{2}$, then A is a subset of a bi-arithmetic progression. If $\frac{|A|}{H+1} \approx \frac{1}{2}$ and $b = |A+A| - 3|A| + 3$, then we can show that $H+1 \leqslant 2|A| - 1 + 2b$ when A is not a subset of a bi-arithmetic progression. The proof for the case $\frac{|A|}{H+1} \approx \frac{1}{2}$ is much harder than the proof for the case $\frac{|A|}{H+1} \ll \frac{1}{2}$ although the former depends on the latter. In both cases Lemma 8.5.4 was used to get structural information of A on an interval with length longer than $\frac{H}{n}$ for some standard positive integer n.

There are some similarities between our methods and analytic methods. In order to detect some structural properties of $A \subseteq [0, n]$, one may need to show that either A is uniformly distributed on $[0, n]$ or A has a greater density on a well formed subset of $[0, n]$. The analytic methods usually look for a large Fourier coefficient $\bar{A}(r)$ (cf. [8, Corollary 2.5]) or a large exponential sum $\sum_{i=0}^{n-1} A(i) e^{\frac{2\pi i}{n}}$ (cf. [24, Theorem 2.9]) to detect the greater density on a well formed subset of $[0, n]$ when n is a prime number. When n is not a prime number then one needs to replace it with a prime number $p > n$ and consider A in $[0, p]$ instead. This replacement may not work well for Conjecture 8.6.1 as the structure of A needs to be very precise. In our methods we look for the greater density of $A \subseteq [0, H]$ on an interval $[0, K]$ for some $K > U$ by checking the value of $\underline{d}_U(A)$, where $U = \bigcap_{n \in \mathbb{N}} [0, \frac{H}{n}]$. If $\underline{d}_U(A) \geqslant \frac{2}{3}$, then the density of A on $[0, K]$ for some $K > U$ is significantly greater than $|A|/H$, which will lead to a contradiction that $|A + A|$ is almost the same as $3k - 3$. If $\underline{d}_U(A) = 0$, then the density of A on $[K, H]$ is significantly greater than $|A|/H$, which will again lead to a contradiction. Otherwise either $|(A + A) \cap [0, 2K]|$ is large, which is impossible by the fact that $\frac{|A+A|}{2H+1} \lessapprox \frac{3}{2}\alpha$, or A has very nice structural properties on $[0, K]$ following Lemma 8.5.4, which will force A to have the structure we hope for.

References

[1] V. Bergelson, Ergodic Ramsey theory—an update, in *Ergodic theory of $\mathbb{Z}^d$ actions (Warwick, 1993-1994)*, London Mathematical Society Lecture Note Ser. 228, Cambridge University Press, Cambridge, 1996.

[2] P. Bihani and R. Jin, Kneser's Theorem for upper Banach density, submitted.

[3] Y. Bilu, "Addition of sets of integers of positive density", Journal of Number Theory, **64** (1997) 233–275.

[4] G. Bordes, "Sum-sets of small upper density", Acta Arithmetica, to appear.

[5] G. A. Freiman, Structure theory of set addition. II. Results and problems, in *Paul Erdös and his mathematics, I* (Budapest, 1999), Bolyai Soc. Math. Stud., **11**, János Bolyai Math. Soc., Budapest, 2002.

[6] G. A. Freiman, *Foundations of a structural theory of set addition*, Translated from the Russian. Translations of Mathematical Monographs, Vol. 37, American Mathematical Society, Providence, R. I., 1973.

[7] H. Furstenberg, *Recurrence in Ergodic Theory and Combinatorial Number Theory*, Princeton University Press, 1981.

[8] T. Gowers, "A new proof of Szemerédi's theorem", Geometric and Functional Analysis, **11** (2001) 465–588.

[9] H. Halberstam and K. F. Roth, *Sequences*, Oxford University Press, 1966.

[10] Y. O. HAMIDOUNE and A. PLAGNE, "A generalization of Freiman's $3k-3$ theorem", Acta Arith., **103** (2002) 147–156.

[11] R. JIN, "Sumset phenomenon", Proceedings of American Mathematical Society, **130** (2002) 855–861.

[12] R. JIN, "Nonstandard methods for upper Banach density problems", The Journal of Number Theory, **91** (2001) 20–38.

[13] R. JIN, Standardizing nonstandard methods for upper Banach density problems, in the DIMACS series, *Unusual Applications of Number Theory*, edited by M. Nathanson, Vol. 64, 2004.

[14] R. JIN, "Inverse problem for upper asymptotic density", Transactions of American Mathematical Society, **355** (2003) 57–78.

[15] R. JIN, "Inverse problem for upper asymptotic density II", to appear.

[16] R. JIN, "Solution to the inverse problem for upper asymptotic density", to appear.

[17] R. JIN, Freiman's $3k-3+b$ conjecture and nonstandard methods, preprint.

[18] R. JIN and H. J. KEISLER, "Abelian group with layered tiles and the sumset phenomenon", Transactions of American Mathematical Society, **355** (2003) 79–97.

[19] H. J. KEISLER and S. LETH, "Meager sets on the hyperfinite time line", Journal of Symbolic Logic, **56** (1991) 71–102.

[20] A. I. KHINCHIN, *Three pearls of number theory*, Translated from the 2d (1948) rev. Russian ed. by F. Bagemihl, H. Komm, and W. Seidel, Rochester, N.Y., Graylock Press, 1952.

[21] S. C. LETH, "Applications of nonstandard models and Lebesgue measure to sequences of natural numbers", Transactions of American Mathematical Society, **307** (1988) 457–468.

[22] S. C. LETH, "Sequences in countable nonstandard models of the natural numbers", Studia Logica, **47** (1988) 243–263.

[23] S. C. LETH, "Some nonstandard methods in combinatorial number theory", Studia Logica, **47** (1988) 265–278.

[24] M. B. NATHANSON, *Additive Number Theory—Inverse Problems and the Geometry of Sumsets*, Springer-Verlag, 1996.

9

Nonstandard methods and the Erdős-Turán conjecture

Steven C. Leth*

9.1 Introduction

A major open question in combinatorial number theory is the Erdős-Turán conjecture which states that if $A = \langle a_n \rangle$ is a sequence of natural numbers with the property that $\sum_{n=1}^{\infty} 1/a_n$ diverges then A contains arbitrarily long arithmetic progressions [1]. The difficulty of this problem is underscored by the fact that a positive answer would generalize Szemerédi's theorem which says that if a sequence $A \subset \mathbb{N}$ has positive upper Banach Density then A contains arbitrarily long arithmetic progressions. Szemerédi's theorem itself has been the object of intense interest since first conjectured, also by Erdős and Turán, in 1936. First proved by Szemerédi in 1974 [9], the theorem has been re-proved using completely different approaches by Furstenberg in 1977 [2, 3] and Gowers in 1999 [4], with each proof introducing powerful new methods.

The Erdős-Turán conjecture immediately implies that the primes contain arbitrarily long arithmetic progressions, and it was thought by many that a successful proof for the primes would be the result of either a proof of the conjecture itself or significant progress toward the conjecture. However, very recently Green and Tao were able to solve the question for the primes without generalizing Szemerédi's result in terms of providing weaker density conditions on a sequence guaranteeing that it contain arithmetic progressions.

In this paper we outline some possible ways in which nonstandard methods might be able to provide new approaches to attacking the Erdős-Turán conjecture, or at least other questions about the existence of arithmetic progressions. Heavy reference will be made to results in [7] and [8], and the proofs for all results quoted but not proved here appear in those two sources.

*Department of Mathematical Sciences, University of Northern Colorado, Greeley, CO 80639.
steven.leth@unco.edu

9.2 Near arithmetic progressions

We begin with some definitions that first appear in [8].

Definition 9.2.1 *Let $A \subset \mathbb{N}$, and let $I = [a, b]$ be an interval in $\mathbb{N}$. We will write* $\boldsymbol{l(I)}$ *for the length of I (i.e. $l(I) = b-a+1$) and we will write* $\boldsymbol{\delta(A, I)}$ *or* $\boldsymbol{\delta(A, [a, b])}$ *for the density of the set A on the interval I. Thus $\delta(A, I) = \frac{|A \cap I|}{l(I)}$.*

Definition 9.2.2 *Let t, d and w be in $\mathbb{N}$, and let $\alpha \in \mathbb{R}$ with $0 < \alpha < 1$. For $A \subset \mathbb{N}$ and I an interval in $\mathbb{N}$ of length $l(I)$ we say that A contains a t-termed α-homogeneous cell of distance d and width w in I or simply a* $\boldsymbol{<t, \alpha, d, w>}$ *cell in I iff there exists $b \in I$ with $b + (t-1)d + w$ also in I such that for each $\nu, \xi = 0, 1, 2, \ldots, t-1$:*

$$\delta\big(A, [b+\xi \cdot d, b+\xi \cdot d+w]\big) \geq (1-\alpha)\delta\big(A, [b+\nu \cdot d, b+\nu \cdot d+w]\big) \geq (1-\alpha)^2\delta(A, I).$$

If each $\delta(A, [b+\xi \cdot d, b+\xi \cdot d+w])$ is simply nonzero, i.e. the intervals are nonempty, then we say that A contains a $\boldsymbol{<t, d, w>}$ *cell.*

For $\beta > 0$ and $0 \leq u \leq w$ we will say that a $< t, \alpha, d, w >$ cell is $\boldsymbol{u, \beta}$ ***uniform*** *if for each $\nu = 0, 1, 2, \ldots, t-1$, and all x such that $u \leq x \leq w$:*

$$(1-\beta)\delta(A, J_\nu) \leq \delta(A, [b+\nu \cdot d, b+\nu \cdot d+x]) \leq (1+\beta)\delta(A, J_\nu),$$

where J_ν denotes the interval $[b+\nu \cdot d, b+\nu \cdot d+w]$.

It is clear that an actual arithmetic progression of length t and distance d is an example of a $< t, \alpha, w, d >$ cell with $w = 0$ and α any non-negative number. Furthermore, this cell is u, β uniform for $u = 0$ and any non-negative β. We could view the existence of a $<t, \alpha, d, w>$ cell inside a sequence A as a weak form of an actual arithmetic progression inside A. These cells are "near" arithmetic progressions in some (perhaps rather weak) sense, and intuitively are "nearer" to arithmetic progressions as the size of w decreases. In some of the results that we look at w will be "small" in the sense that the ratio of w to d will be small compared to the ratio of d to the length of the interval I. In other results w will be "small" by actually being bounded by a finite number while d gets arbitrarily large.

Definition 9.2.3 *Let I be an interval in $\mathbb{N}$, and $A \subset I$, with $r > 1 \in \mathbb{R}$ and $m \in \mathbb{N}$. We say that A has the* $\boldsymbol{m, r}$ ***density property*** *on I iff for any interval $J \subset I$, if $l(J) \geq \frac{l(I)}{m}$, then $\delta(A, J) \leq r\delta(A, I)$.*

Theorem 9.2.1 below gives a condition for the existence of "near" arithmetic progressions for any sequence on any interval I in which the density does not drastically increase as the size of the subinterval decreases. More specifically, it provides an absolute constant such that whenever the density of a sequence does not increase beyond a fixed ratio for any subinterval of size greater than the length of I divided by that fixed constant, then the sequence will contain a $<t,\alpha,w,d>$ cell with some relative "smallness" conditions for w.

A complete proof of this theorem appears in [8], but we will outline the proof here, as it provides the clearest illustration of how the use of the non-standard model provides us with a new set of tools for questions of this type.

Theorem 9.2.1 *Let $h(x)$ be any increasing real valued function such that $h(x) > 0$ whenever $x > 0$, and let $g(x)$ be any real valued function which approaches infinity as x approaches infinity. For all real $\alpha > 0$, $r > 1$ and $j, t \in \mathbb{N}$ there exists a standard natural number m such that for all $n > m$, whenever I is an interval of length n and any nonempty set $A \subset I$ has the m, r density property on I then A contains a u, β uniform $< t, \alpha, d, w >$ cell with $\frac{u}{w} < h(\frac{w}{d})$, $\frac{w}{d} < h(\frac{d}{n})$, $\beta < h(\frac{d}{n})$ and $\frac{n}{g(m)} < d < \frac{n}{j}$. Furthermore, we may take w and d to be powers of 2.*

Proof. (Sketch only). Suppose $h(x), g(x), \alpha, j, r, t$ are given as in the statement above and that no such m exists. By "overspill" there exists an M, N in $^*\mathbb{N} - \mathbb{N}$ with $M < N$ and a hyperfinite internal set A such that A has the M, r density property on an interval of length N but A contains no $< t, \alpha, d, w >$ cell on this interval with the required properties. Since the conditions are translation invariant we may assume that the interval is $[0, N-1]$. We now define a standard function $f : [0,1] \longrightarrow [0,1]$ by:

$$f(x) = st\left(\frac{|A \cap [0, xN]|}{|A \cap [0, N]||}\right).$$

Using the fact that A has the M, r density property on $[0, N]$ it is not difficult to show that $f(x)$ satisfies a Lipschitz condition with constant r. Thus, the function f is absolutely continuous, differentiable almost everywhere and equal to the integral of its derivative. Since $f(1) = 1$, $f(0) = 0$ and f is the integral of its derivative, it must be that the Lebesgue measure of $\{x : f'(x) \geq (1 - \frac{\alpha}{4})\}$ is nonzero. Thus, there exists a real number $c \geq 1$ such that the Lebesgue measure of the set

$$E = \left\{x : c - \frac{\alpha}{4} \leq f'(x) \leq c\right\}$$

is nonzero.

By using the Lebesgue density theorem it is straightforward to show that any set of positive measure contains arbitrarily long arithmetic progressions, and that, in fact, these progressions may have arbitrarily small differences between elements. This allows us to obtain a $<t, \alpha, D, W>$ cell, with $D, W \in {}^*\mathbb{N} - \mathbb{N}$ with the property that there exists $B \in {}^*\mathbb{N} - \mathbb{N}$ such that

$$st\left(\frac{B}{N}\right), st\left(\frac{B+D}{N}\right), st\left(\frac{B+2D}{N}\right), \ldots, st\left(\frac{B+(t-1)D}{N}\right)$$

forms an arithmetic progression in E. The α homogeneity follows from the definition of E. The fact that f is differentiable at each point in E allows us to obtain the uniformity condition, and allows us to take U, D and W arbitrarily small but not infinitesimal to N. This, in turn, allows us the freedom to make those quantities powers of 2.

We are thus able to obtain a U, β uniform $< t, \alpha, D, W >$ cell for A in $[0, N-1]$ with all the properties required in the theorem, contradicting our assumption. □

Definition 9.2.4 *For A a sequence of positive integers we define the* ***upper Banach Density*** *of A or* ***BD(A)*** *by:*

$$BD(A) = \inf_{x \in \mathbb{N}-\{0\}} \max_{a \in \mathbb{N}} \frac{|A \cap [a+1, a+x]|}{x}.$$

Upper Banach density is often simply called Banach density, and is sometimes referred to in the literature as strong upper density, with notation $d^*(A)$ in place of $BD(A)$. That notation is used in [7].

The theorem allows us to obtain some results about the existence of uniform $<t, \alpha, d, w>$ cells in sequences that are relatively sparse (certainly too sparse to necessarily contain actual arithmetic progressions). The theorem below, also proved in [8], is of this type.

Theorem 9.2.2 *Let $\alpha > 0$ and $t > 2 \in \mathbb{N}$ be given, $h(x)$ be any continuous real valued function such that $h(x) > 0$ whenever $x > 0$ and let A be a sequence in $\mathbb{N}$ with the property that for all $\varepsilon > 0$, $|A \cap [0, n-1]| > n^{1-\varepsilon}$ for sufficiently large n. Then for sufficiently large n, A contains a u, β uniform $<t, \alpha, d, w>$ cell on $[0, n-1]$ with w and d powers of 2 and such that:*

$$\frac{u}{w} < h\left(\frac{w}{d}\right), \ \frac{w}{d} < h\left(\frac{\log d}{\log n}\right), \ \textit{and} \ \ \beta < h\left(\frac{\log d}{\log n}\right).$$

The condition that $\frac{w}{d} < h\left(\frac{\log d}{\log n}\right)$ is not very strong, and it would be desirable to improve this weak "smallness" condition. The theorem below shows that, even for $t = 3$, if we keep the density condition on A as above then we cannot improve this smallness condition to $\frac{w}{d} < \left(\frac{d}{n}\right)^{\alpha}$ for any $\alpha > 0$, even if we do not insist on any homogeneity or uniformity conditions.

Theorem 9.2.3 *Let $\alpha > 0$. There exist constants $r > 0$ such that for arbitrarily large n there are subsets A of $[0, n-1]$ such that*

$$\left|A \cap [0, n-1]\right| > \frac{n}{2^{r \log\log n \sqrt{\log n}}},$$

and yet A contains no $<3, d, w>$ cell in $[0, n-1]$ satisfying $\frac{w}{d} < \left(\frac{d}{n}\right)^{\alpha}$, with w and d a power of 2.

Here we recall that a $<t, d, w>$ cell is merely a collection of t intervals in arithmetic progression on which A is nonempty.

Proof. Since the statement is strictly stronger as α decreases, we will assume that $\alpha \leq 1$. In [5, p. 98] it is shown that there exists a constant $c > 0$ and a sequence A satisfying

$$\left|A \cap [0, n-1]\right| > \frac{n}{e^{c\sqrt{\log n}}} \text{ for sufficiently large } n$$

that contains no 3-term arithmetic progression. This result is due to Behrend. For convenience we will adjust the constant and use log base 2 here, and also replace e with 2. By adjusting the constant if necessary (and using $2A$) we may assume that A contains no two consecutive numbers. We may also translate so that $0 \in A$ without changing the density condition. Thus, we begin with a sequence A which contains 0 and no two consecutive numbers and satisfies

$$\left|A \cap [0, n-1]\right| > \frac{n}{2^{c\sqrt{\log n}}} \text{ for sufficiently large } n.$$

Let $N \in {}^*\mathbb{N} - \mathbb{N}$ and let

$$\beta = (1/2)^{2/\alpha}; \quad m_0 = N; \quad m_1 = \text{the largest power of 2 less than } \beta^{(1+\alpha/2)}N$$

$$m_{k+1} = \text{the smallest power of 2 greater than } \left(\frac{m_k}{N}\right)^{\alpha} m_k$$

$$L = \text{the smallest number such that } m_{L+1} \leq 1.$$

We now wish to show by induction that

$$m_k \leq \beta^{(1+\alpha/2)^k} N. \tag{9.2.1}$$

To see this we note that for $k = 1$ the definition of m_1 guarantees this. By the construction we have

$$\left(\frac{m_k}{N}\right)^{\alpha} m_k < m_{k+1} \leq 2\left(\frac{m_k}{N}\right)^{\alpha} m_k.$$

so that, assuming the induction hypothesis,

$$\begin{aligned} m_{k+1} &\leq 2\Big(\frac{m_k}{N}\Big)^{\alpha} m_k \\ &\leq 2\Big(\beta^{(1+\alpha/2)^k}\Big)^{\alpha} \beta^{(1+\alpha/2)^k} N \\ &= 2\Big(\beta^{\alpha(1+\alpha/2)^k}\Big) \beta^{(1+\alpha/2)^k} N \\ &= 2\Big(\beta^{\alpha/2(1+\alpha/2)^k}\Big)\Big(\beta^{\alpha/2(1+\alpha/2)^k}\Big) \beta^{(1+\alpha/2)^k} N \\ &\leq 2\beta^{\alpha/2}\beta^{\alpha/2(1+\alpha/2)^k+(1+\alpha/2)^k} N \\ &\leq \beta^{\alpha/2(1+\alpha/2)^k+(1+\alpha/2)^k} N \\ &= \beta^{(1+\alpha/2)^{k+1}} N, \end{aligned}$$

completing the induction step, and establishing 9.2.1 above.

We will define a subset B of $[0, N-1]$ with the property that it contains no $<3, d, w>$ cell in $[0, N-1]$ satisfying $\frac{w}{d} < \left(\frac{d}{N}\right)^{\alpha}$, with w and d a power of 2, and such that

$$\big|B \cap [0, N-1]\big| > \frac{N}{2^{cL\sqrt{\log N}}}$$

by essentially using block copies of initial segments of $^*A$. More specifically we let

$$B_1 = [0, m_1 - 1]$$

and, for $1 \leq k < L$

$$i \in B_{k+1} \text{ iff } \left\lfloor \frac{i'}{m_{k+1}} \right\rfloor \in {}^*A, \text{ where } i' \text{ is the remainder of } i \text{ mod } m_k.$$

When k is finite, the fact that $0 \in {}^*A$ means that B_k intersected with $B_1 \cap \ldots \cap B_{k-1}$ has cardinality at least a noninfinitesimal multiple of that of $B_1 \cap \ldots \cap B_{k-1}$. When k is in $^*\mathbb{N} - \mathbb{N}$ the density condition on A guarantees that at least a $\frac{1}{2^{c\sqrt{\log n}}}$ portion of B_k intersects with $B_1 \cap \ldots \cap B_{k-1}$, where $n < N$. Thus

$$B_1 \cap \ldots \cap B_k \text{ has cardinality at least } \frac{N}{2^{ck\sqrt{\log N}}}. \tag{9.2.2}$$

We now let

$$B = B_1 \cap \ldots \cap B_L.$$

Now suppose that B contains a $<3, d, w>$ cell on $[0, N-1]$ with $\frac{w}{d} < \left(\frac{d}{N}\right)^{\alpha}$ where both w and d are powers of 2. We show that this forces an actual arithmetic progression of length 3 in $^*A$ and thus in A (by transfer), contradicting our assumption about A.

To see this, we let i be such that $m_{i+1} \leq d < m_i$. Then since ${}^*A$ contains no two consecutive numbers, the $<3, d, w>$ cell on $[0, N-1]$ must be completely contained inside one of the blocks of length m_i, i.e. inside some $[\nu m_i, (\nu+1)m_i]$. But

$$w \leq \left(\frac{d}{N}\right)^{\alpha} d < \left(\frac{m_i}{N}\right)^{\alpha} m_i \leq m_{i+1},$$

and since w, d and the m_k's are powers of 2

$$w | m_{i+1} \quad \text{and} \quad m_i | d,$$

so that there exist 3 intervals of length m_{i+1} inside $[\nu m_i, (\nu+1)m_i]$ which contain elements of B and are in arithmetic progression. By the construction this means that ${}^*A$ contains an arithmetic progression of length 3, and then by transfer, so does A.

It remains to estimate L in terms of N. From 9.2.1 we see that

$$m_k \leq 1 \text{ when } \beta^{(1+\alpha/2)^k} N \leq 1 \quad \text{i.e. } N \leq (1/\beta)^{(1+\alpha/2)^k}$$

so that

$$\log N \leq (1+\alpha/2)^k \log(1/\beta)$$

or

$$\log\log N \leq k \log(1+\alpha/2) + \log\log(1/\beta).$$

Thus

$$m_k \leq 1 \text{ whenever } k \geq \frac{\log\log N - \log\log(1/\beta)}{\log(1+\alpha/2)}.$$

This and the definition of L now yield

$$L + 1 \leq \frac{\log\log N - \log\log(1/\beta)}{\log(1+\alpha/2)}.$$

The above inequality, the definition of B and 9.2.2 imply that for any $r > c/\log(1+\alpha/2)$

$$\left|B \cap [0, N-1]\right| > \frac{N}{2^{r \log\log N \sqrt{\log N}}}.$$

For such an r the result now follows by transfer. □

We note that the density condition given in the theorem above is stronger than simply being greater than $n^{1-\varepsilon}$ for sufficiently large n, since $n^{1-\varepsilon}$ is of the form

$$\frac{n}{2^{c \log N}} < \frac{n}{2^{\log\log n \sqrt{\log n}}} \quad \text{for large } n.$$

Thus, weak as theorem 9.2.2 is in its smallness conditions, there are clear limits to how much it can be strengthened for sets of this relative sparseness.

It appears to be more promising to look at denser sequences in the hope of maintaining a stronger "smallness" condition. The theorem below is proved in [8], and provides just one possible example of conditions like this that might provide a means for approaching deep questions about arithmetic progressions. The proof of theorem 9.2.4 below is similar to that of the proof of theorem 3 given above. In particular these proofs illustrate how "smallness" conditions that may not seem very strong can be used to show that somewhat denser sets contain actual arithmetic progressions.

Theorem 9.2.4 *The Erdős-Turán conjecture follows if we can show that for fixed t and constant $c > 0$, there exists n_0 such that for all $n > n_0$, whenever the sequence A satisfies*

$$\left|A \cap [0, n-1]\right| > \frac{n}{(c \log n)^{2 \log \log n}}$$

then A contains a $<t, d, w>$ cell on $[0, n-1]$ with $\frac{w}{d} < \frac{d}{n}$ where both w and d are powers of 2.

9.3 The interval-measure property

The conditions given below are natural from the nonstandard perspective and not at all so from the standard perspective. They might provide another means of attacking questions about arithmetic progressions. These definitions first appear in [7].

Definition 9.3.1 *Let A be an internal subset of $^*\mathbb{N}$, $y, z \in {}^*\mathbb{N}$ with $z - y \in {}^*\mathbb{N} - \mathbb{N}$. Then we say that A has the **IM (interval-measure)** property on $[y, z]$ iff for every real, standard $\beta > 0$ there is a real, standard $\alpha > 0$ such that whenever $[u, v] \subset [y, z]$ with $v - u \in {}^*\mathbb{N} - \mathbb{N}$ and the largest gap of A on $[u, v]$ is $\leq \alpha(v - u)$ then*

$$\lambda\left(st\left\{\left(\frac{a-y}{z-y}\right) : a \in A \cap [y, z]\right\}\right) \geq 1 - \beta.$$

*If A is a standard subset of $\mathbb{N}$, we say that A has the **SIM (standard interval-measure)** property iff $^*A$ has the IM property on every interval $[y, z] \subset {}^*\mathbb{N}$ with $z - y \in {}^*\mathbb{N} - \mathbb{N}$, and*

$$\lambda\left(st\left\{\left(\frac{a-y}{z-y}\right) : a \in {}^*A \cap [y, z]\right\}\right) > 0 \text{ on some such interval } [y, z].$$

Here λ is used to denote Lebesgue measure.

A somewhat cumbersome but standard equivalent definition for the SIM property is given in [7]. Through the use of the standard equivalent it is easy to see that if A has the SIM property then given $\beta > 0$ there exists a fixed $\alpha > 0$ that works for every infinite interval. The theorem below follows immediately from theorem 3.2 in that same work.

Theorem 9.3.1 *Let $t \in \mathbb{N}$, $0 < \beta < 1/t$, and suppose that $A \subset \mathbb{N}$ has the SIM property, with α corresponding to the given β. Then there exists a constant $j \in \mathbb{N}$ such that whenever A contains a $\boldsymbol{<t, d, w>}$ cell consisting of the intervals $[b + \xi \cdot d, b + \xi \cdot d + w]$ for $0 \le \xi \le t - 1$ in which the largest gap of A on each of these intervals is $\le \alpha w$ then A contains $\boldsymbol{<t, d, j>}$ subcell, i.e. consisting of intervals of the form $[b' + \xi \cdot d, b' + \xi \cdot d + j] \subset [b + \xi \cdot d, b + \xi \cdot d + w]$.*

This result is significant in that a fixed constant size to the intervals is a much stronger "smallness" condition than was achieved in previous results. However, the assumption that A is a SIM set is a strong condition. It is shown in [7] that any sequence $A = \langle a_n \rangle$ in which $\lim_{n \to \infty}(a_{n+1} - a_n) = \infty$ does not have the SIM property, so no pure density condition weaker than positive Banach density can imply that A contains a SIM set. On the other hand, sets may certainly have the SIM property without having positive Banach density. Even the question of whether or not positive Banach density is sufficient for a sequence to contain a SIM set is still open. A positive solution to either of the conjectures below would be a major step toward establishing the SIM condition as a useful tool for questions of this type. Since the two conjectures together imply Szemerédi's theorem, at least one of them is certain to be quite difficult.

Conjecture 9.3.1 *Let $A \subset \mathbb{N}$ have positive Banach density. Then A contains a subset B with the SIM property.*

Conjecture 9.3.2 *Every set $A \subset \mathbb{N}$ with the SIM property contains arbitrarily long arithmetic progressions.*

References

[1] P. Erdős and P. Turán "On some sequences of Integers", J. London Math. Soc., **11** (1936) 261–264.

[2] H. Furstenberg, "Ergodic behavior of diagonal measures and a theorem of Szemerédi on arithmetic progressions", J. Anal. Math. Soc., **31** (1977) 204–256.

[3] H. Furstenberg, *Recurrence in Ergodic Theory and Combinatorial Number Theory*, Princeton University Press, 1981.

[4] T. GOWERS, "A new proof of Szemerédi's theorem", GAFA, **11** (2001) 465–588.

[5] R. GRAHAM, B. ROTHSCHILD and J. SPENCER, *Ramsey Theory*, 2nd ed. Wiley, 1990.

[6] B. GREEN and T. TAO, The Primes Contain Arbitrarily Long Arithmetic Progressions, Preprint, 8 Apr 2004. `http://arxiv.org/abs/math.NT/0404188`

[7] S. LETH, "Some nonstandard methods in combinatorial number theory", Studia Logica, **XLVII** (1988) 85–98.

[8] S. LETH, "Near arithmetic progressions in sparse sets", Proc. of the Amer. Math. Soc., **134** (2005) 1579-1589.

[9] E. SZEMERÉDI, "On sets of integers containing no k elements in arithmetic progression", Acta Arith., **27** (1975) 199–245.

Part III

Statistics, probability and measures

10

Nonstandard likelihood ratio test in exponential families

Jacques Bosgiraud*

Abstract

Let $(p_\theta)_{\theta\in\Theta}$ be an exponential family in $\mathbb{R}^k$. After establishing nonstandard results about large deviations of the sample mean $\overline{X}$, this paper defines the nonstandard likelihood ratio test of the null hypothesis $H_0 : \theta \in \text{hal}(\widetilde{\Theta}_0)$, where $\widetilde{\Theta}_0$ is a standard subset of Θ and $\text{hal}(\widetilde{\Theta}_0)$ its halo. If α is the level of the test, depending on whether $\frac{\ln \alpha}{n}$ is infinitesimal or not we obtain different rejection criteria. We calculate risks of the first and second kinds (external probabilities) and prove that this test is more powerful than any "regular" nonstandard test based on $\overline{X}$.

10.1 Introduction

10.1.1 A most powerful nonstandard test

In a preceding paper [9], we proved that the nonstandard likelihood ratio test (NSLRT) is more powerful than the nonstandard chi-squared test. Our purpose now is to generalize this result to classical exponential families in $\mathbb{R}^k$ (where k is standard): the NSLRT is more powerful than any nonstandard test issuing from a (family of) standard test(s) with rejection criterion $\overline{X} \in R$, where $\overline{X}$ is the sample mean and R a sufficiently regular set. The standard likelihood ratio test is not so powerful (see §10.5). We hope that viewing the problem from a general perspective will lead to a clearer understanding of its structure and simpler and better proofs. In §10.3, we establish some results about large deviations of the mean, more or less similar to classical results, before studying the NSLRT in §10.4 and comparing it to nonstandard "regular" tests based on $\overline{X}$ in §10.5.

*Universíté Paris VIII, Paris.
jacques.bosgiraud@univ-paris8.fr

This paper follows also papers about nonstandard tests for monotone likelihood ratio families ([5], [6], [8]) and we shall maintain the same definition for a nonstandard test in this paper. For notation and definitions of nonstandard analysis, external calculus, and external probabilities we refer to [15], [11], [14], [4].

10.2 Some basic concepts of statistics

10.2.1 Main definitions

A *statistical family* is a triplet $(\Omega, \mathcal{F}, \mathcal{P})$ where Ω is a set, $\mathcal{F}$ a σ-field of subsets of Ω, $\mathcal{P}$ a family of probabilities on $(\Omega, \mathcal{F})$: $\mathcal{P} := \{P_\theta : \theta \in \Theta\}$. In the following, we suppose that Ω is a subset of $\mathbb{R}^d$ (where d is a standard integer), that Θ is a subset of $\mathbb{R}^k$ (where k is a standard integer) and that there exists a positive mesure μ defined on $(\Omega, \mathcal{F})$ such that for each $\theta \in \Theta$, P_θ is absolutely continuous with respect to μ: we note $P_\theta := p_\theta \mu$ where p_θ is a $\mathcal{F}$-measurable function defined on Ω.

So, $X : \Omega \to \mathbb{R}^d$, $x \to x$ is a random variable with distribution P_θ.

Let n be a (standard or nonstandard) integer; denoting $\mathcal{X} := \Omega^n$, $P_\theta^n := P_\theta^{\otimes n}$ is a probability defined on $(\mathcal{X}, \mathcal{F}^{\otimes n})$ and we can write $P_\theta^n = p_\theta^n \mu^n$, where $\mu^n := \mu^{\otimes n}$ and $p_\theta^n(x_1, \ldots, x_n) = \prod_{i=1}^n p_\theta(x_i)$. A *n-sample* $(x_1, \ldots, x_n)$ is an element of $\mathcal{X}$. For $i = 1, \ldots, n$ we define $X_i : \mathcal{X} \to \mathbb{R}^d$ by $X_i(x_1, \ldots, x_n) = x_i$; then $X_1, \ldots, X_n$ are independent identically distributed (i.i.d.) random variables with distribution P_θ. $(X_1, \ldots, X_n)$ is a *n-sampling* of X. Note that some authors do not distinguish X_i and x_i, sample and sampling.

A *statistic* is a measurable function

$$T : \mathcal{X} \to \mathbb{R}^m, \quad (x_1, \ldots, x_n) \to T(x_1, \ldots, x_n).$$

For example, the sample mean $\overline{X} := \frac{1}{n}\sum_{i=1}^n X_i$ is a statistic (here $m = d$). We note $E_\theta T$ the expectation of T for the probability P_θ^n.

10.2.2 Tests

Let Θ_0 a nonempty proper subset of Θ, $\Theta_1 := \Theta_0^{\mathrm{C}}$ its complement and $\alpha \in]0,1[$. A *level-α test* of $H_0 : \theta \in \Theta_0$ against $H_1 : \theta \in \Theta_1$ is a statistic $\varphi : \mathcal{X} \to [0,1]$ such that $\forall \theta \in \Theta_0$, $E_\theta \varphi \leq \alpha$.

$H_0 : \theta \in \Theta_0$ is called the null hypothesis; $1 - \varphi$ is the probability of accepting H_0.

$H_1 : \theta \in \Theta_1$ is the alternative hypothesis and φ is the probability of rejecting H_0.

Also, $\sup_{\theta\in\Theta_0} E_\theta\varphi$ is called the size of the test. $\{(x_1,\ldots,x_n) \in \mathcal{X}, \varphi(x_1,\ldots,x_n) = 1\}$ is called the *rejection set* and $\{(x_1,\ldots,x_n) \in \mathcal{X}, \varphi(x_1,\ldots,x_n) = 0\}$ is called the *acceptation set*; if $\varphi(x_1,\ldots,x_n) \in]0,1[$ the test is said to be randomized.

Very often, φ is defined through a statistic T and $\varphi(x_1,\ldots,x_n) = 1 \Leftrightarrow T(x_1,\ldots,x_n) \in R$ (in fact, we shall write $\varphi = 1 \Leftrightarrow T \in R$) where R is a subset of $\mathbb{R}^m$. R is also called the *rejection set* and "$T \in R$" is the *rejection criterium* (e.g. if $m = 1$, $T > t_0$ is a rejection criterium; t_0 is a constant depending on Θ_0 and α).

For $\theta \in \Theta_0$, $E_\theta\varphi$ is the risk of the first kind (at θ); for $\theta \in \Theta_1$, $E_\theta(1-\varphi)$ is the risk of the second kind (at θ); for $\theta \in \Theta$, $E_\theta\varphi$ is the power of φ (at θ).

For testing a given null hypothesis, there are generally a lot of level-α tests (for example the constant test $\varphi := \alpha$). A level-α test ϕ is said *uniformly the most powerful* (U.M.P.) if for any level-α test φ, ϕ is uniformly more powerful than ψ, i.e. $\forall\theta \in \Theta_1, E_\theta\phi \geq E_\theta\varphi$ (in fact "more powerful" means "at least as powerful"). U.M.P. tests only exist in particular cases: for example for 1-dimensional exponential families if $H_0 : \theta \leq \theta_0$ (where $\theta_0 \in \Theta$, an interval of $\mathbb{R}$). So, some more sophisticated notions are used to compare the power of two tests: for exemple the relative efficiency (cf. §10.5). It is not possible to summarize this notion in some words; so we suggest to refer to [1], [2] or [12].

10.3 Exponential families

10.3.1 Basic concepts

The following classical results and more information about exponential families can be found in [10] or [13]. Let k be a standard integer. We denote by $(x \mid y) = \sum_{j=1}^k x_j y_j$ the scalar product of x and y, vectors of $\mathbb{R}^k$. Let μ be a probability mesure on $\Omega := \mathbb{R}^k$ (the σ-field is the field of borelian sets) and let

$$\Theta := \left\{\theta \in \mathbb{R}^k : \int \exp(\theta \mid x)\mu(dx) < \infty\right\}.$$

The set Θ is convex and for $\theta \in \Theta$, let

$$\psi(\theta) = \ln \int \exp(\theta \mid x)\mu(dx).$$

The function ψ is convex and continuous on Θ^0 (the interior of Θ).

The statistical family $\{P_\theta : \theta \in \Theta\}$ defined by $P_\theta := p_\theta\mu$ where

$$p_\theta(x) = \exp\big((\theta \mid x) - \psi(\theta)\big)$$

is the (full) *exponential family associated to* μ. A lot of classical statistical families (e.g. multinomial distributions, multidimensional normal distributions,...) are exponential families, generally after reparametrization. This reparametrization can be chosen such that Θ^0 is nonempty. Let

$$\Theta' := \{\theta \in \Theta : \; E_\theta \|X\| < \infty\}$$

where $\|\|$ denotes the Euclidean norm.

For $\theta \in \Theta'$ we define $\lambda(\theta) := E_\theta X$; this mapping is 1-1 from Θ' onto $\Lambda := \lambda(\Theta')$: Θ' contains Θ^0 and λ is a 1-1 diffeomorphism from Θ^0 onto Λ^0 (cf. [10, pp.74,75]). To see this, notice that for $\theta \in \Theta^0$, $E_\theta X = \nabla\psi(\theta)$ and all derivatives of ψ exist at θ. For $\theta_1, \theta_2 \in \Theta^0$, $\theta_1 \neq \theta_2$, ψ is strictly convex on the line joining θ_1 and θ_2 and then $\big(\theta_1 - \theta_2 \mid \lambda(\theta_1) - \lambda(\theta_2)\big) > 0$.

10.3.2 Kullback-Leibler information number

For $\theta_0 \in \Theta$ and $\theta \in \Theta'$, the Kullback-Leibler information number is given by

$$I(\theta, \theta_0) = \psi(\theta_0) - \psi(\theta) + \big(\theta - \theta_0 \mid \lambda(\theta)\big).$$

If Θ_0 is a proper subset of Θ, let

$$I(\theta, \Theta_0) := \inf\left\{I(\theta, \theta_0) : \; \theta_0 \in \Theta_0\right\}.$$

For $(\xi, \xi_0) \in \Lambda^2$, we set

$$J(\xi, \xi_0) := I\big(\lambda^{-1}(\xi), \lambda^{-1}(\xi_0)\big),$$

and for $A \subset \Lambda$ and $\xi \in \Lambda$, let

$$J(\xi, A) := \inf\left\{J(\xi, a) : \; a \in A\right\},$$

$$J(A, \xi) := \inf\left\{J(a, \xi) : \; a \in A\right\}.$$

The classical likelihood ratio test of $H_{\Theta_0} : \theta \in \Theta_0$ against $\theta \in \Theta \setminus \Theta_0$ is based on the statistic

$$R_{\Theta_0} := \frac{1}{n}\left(\ln \sup_{\theta \in \Theta} \prod_{i=1}^{n} p_\theta(X_i) - \ln \sup_{\theta_0 \in \Theta_0} \prod_{i=1}^{n} p_{\theta_0}(X_i)\right)$$

where $(X_i)_{1 \leq i \leq n}$ is a n-sampling of X (see § 10.2.1). Here, denoting $\overline{X} = \frac{1}{n}\sum_{i=1}^{n} X_i$

$$\begin{aligned} R_{\Theta_0} &= \sup_{\theta \in \Theta}\left\{(\theta \mid \overline{X}) - \psi(\theta)\right\} - \sup_{\theta_0 \in \Theta_0}\left\{(\theta_0 \mid \overline{X}) - \psi(\theta_0)\right\} \\ &= \inf_{\theta_0 \in \Theta_0} \sup_{\theta \in \Theta}\left\{(\theta - \theta_0 \mid \overline{X}) - \psi(\theta) + \psi(\theta_0)\right\}. \end{aligned}$$

If $\overline{X} \in \Lambda$, then $\lambda^{-1}(\overline{X})$ is the maximum likelihood estimator of θ and so

$$R_{\Theta_0} = \inf_{\theta_0 \in \Theta_0} I(\lambda^{-1}(\overline{X}), \theta_0) = I(\lambda^{-1}(\overline{X}), \Theta_0) = J(\overline{X}, \Lambda_0)$$

where $\Lambda_0 := \lambda(\Theta_0)$.

$\overline{\Lambda}$ (the closure of Λ) is the closure of the convex hull of the support of μ (cf. [10]); so, in any case, $\overline{X} \in \overline{\Lambda}$. As $\sup\{(\theta \mid \xi) - \psi(\theta) : \theta \in \Theta\}$ is lower semi-continuous with respect to ξ, we define (as in [10]), for $\xi \in \partial\Lambda$ (the boundary of Λ) and $\xi_0 \in \Lambda$, $J(\xi, \xi_0)$ by

$$J(\xi, \xi_0) := \liminf_{\xi' \to \xi,\, \xi' \in \Lambda} J(\xi', \xi_0).$$

So, if $\overline{X} \notin \Lambda$ (then $\overline{X} \in \partial\Lambda$) it remains possible to write

$$R_{\Theta_0} = J(\overline{X}, \Lambda_0).$$

In the following, we shall suppose that, for each $\xi_0 \in \Lambda$, $J(\cdot, \xi_0)$ is continuous on $\overline{\Lambda}$. This assumption is verified by classical exponential families, but it is possible to build counter-examples (cf. exercise 7.5.6 in [10]). Note that if $\xi_0 \in \Lambda \setminus \text{hal}(\partial\Lambda)$ is limited, this assumption implies the S-continuity of $J(\cdot, \xi_0)$ on any limited subset of $\overline{\Lambda}$. Indeed, if ξ_0 is standard, it is obvious. If ξ_0 is non-standard, and if ξ is limited, let $\theta_0 := \lambda^{-1}(\xi_0)$; as λ is a diffeomorphism from Θ^0 onto Λ^0, then ${}^o\theta_0 = \lambda^{-1}({}^o\xi_0)$ and so, as ψ is continuous on Θ^0, we can write

$$J(\xi, \xi_0) - J(\xi, {}^o\xi_0) = \psi(\theta_0) - \psi({}^o\theta_0) + ({}^o\theta_0 - \theta_0 \mid \xi) \simeq 0.$$

The S-continuity of $J(\cdot, \xi_0)$ is then deduced from the S-continuity of $J(\cdot, {}^o\xi_0)$.

In the following, if A is a subset of $\mathbb{R}^k$, A^0 will denote its interior, $\overline{A}$ its closure, ∂A its boundary for the euclidian topology of $\mathbb{R}^k$; its shadow ${}^oA$ is a closed standard set (cf. [11, p. 63]). If A is a subset of $\overline{\Lambda}$, $\widehat{\partial} A$ will denote its boundary for the induced topology of $\overline{\Lambda}$.

10.3.3 The nonstandard test

In this paper n will be supposed unlimited, so we shall use classical asymptotic properties of the likelihood ratio test (cf. §10.4). Let $\widetilde{\Theta}_0$ be a standard subset of Θ, such that $\text{hal}(\widetilde{\Theta}_0)$ — its halo — is included in a standard compact K included in Θ^0.

Let $\mathcal{F} := \{\Theta_0 \text{ internal} : \widetilde{\Theta}_0 \subset \Theta_0 \subset \text{hal}(\widetilde{\Theta}_0)\}$ and let Φ_{Θ_0} be the likelihood ratio test of $H_{\Theta_0} : \theta \in \Theta_0$ against $\theta \notin \Theta_0$ of size α. Recall that Φ_{Θ_0} is defined in the following way: there exists a number $d := d(\alpha, \Theta_0)$ such that

$$\sup_{\theta \in \Theta_0} P_\theta^n(R_{\Theta_0} > d) \leq \alpha \leq \sup_{\theta \in \Theta_0} P_\theta^n(R_{\Theta_0} \geq d).$$

Consequently, if $R_{\Theta_0} < d$ then $\Phi_{\Theta_0} = 0$, if $R_{\Theta_0} > d$ then $\Phi_{\Theta_0} = 1$ and one randomizes if $R_{\Theta_0} = d$.

Definition 10.3.1 *The nonstandard likelihood ratio test (NSLRT) of the nonstandard null hypothesis* (H_0) : $\theta \in hal(\widetilde{\Theta}_0)$ *of level* α *is defined by* $\Phi_0 \equiv \inf \{\Phi_{\Theta_0} : \Theta_0 \in \mathcal{F}\}$.

We prove in this paper that, except for a case studied in §10.4.3 where Φ_0 is randomized, Φ_0 is equal to 0 or 1:

- either $\forall \Theta_0 \in \mathcal{F}, \Phi_{\Theta_0} = 1$ and then $\Phi_0 = 1$ (if H_{Θ_0} is rejected for each $\Theta_0 \in \mathcal{F}$ then H_0 is rejected),
- or $\exists \Theta_0 \in \mathcal{F}, \Phi_{\Theta_0} = 0$ and then $\Phi_0 = 0$ (if H_{Θ_0} is accepted for at least one $\Theta_0 \in \mathcal{F}$ then H_0 is accepted).

At the beginning of §10.4 we shall explain that α has to be infinitesimal. We shall prove in §10.5 that this NSLRT is uniformly more powerful than any "regular" (this term will be defined later) nonstandard test based on $\overline{X}$.

Convention: if E is an external event, and $PR(E)$ its external probability and if η is an external number, "the probability of E is equal to η" means $PR(E) = \eta$ and "the probability of E is exactly equal to η" means $PR(E) \equiv \eta$.

10.3.4 Large deviations for $\overline{X}$

In exponential families, $J(\cdot, \lambda(\theta_0))$ can be regarded as the Cramér transform of P_{θ_0}. More generally, classical literature establishes that

$$\lim_{n \to \infty} \frac{1}{n} \ln P^n(\overline{X}_n \in A) = - \inf_{x \in A} C(x)$$

where A is a Borel set of $\mathbb{R}^k$ such that $A \subset \overline{A^0}$ and $\overline{X}_n$ the mean of n i.i.d. random variables taking values in $\mathbb{R}^k$, C the Cramér transform of their distribution (see [16], section 4: the proof is based on the minimax theorem for compact convex sets). Even more generally, if the i.i.d. random variables take values in some topological vector space E, classical literature (cf. [3]) establishes the same result if A is a finite union of convex Borel sets of E. For exponential families, it is possible to establish specific proofs by using half-spaces (cf. [10], chap. 7 and exercises 7.5.1 to 7.5.6). In a more general setting, half-spaces are also used in [3], section 2.

Our propositions 10.3.2 and 10.3.3 give nonstandard results which imply (by transfer) the classical results. If the first part of the proofs is based on the law of large numbers as in the classical literature, the second part, based on infinitesimal pavings, seems to be original.

10.3.5 n-regular sets

Definition 10.3.2 *Let A be a subset of $\overline{\Lambda}$ and $a \in \partial A$. We shall say that A is n-regular in a iff $\exists b \in hal(a)$, $\exists \rho \in \left]\frac{\mathcal{L}}{\sqrt{n}}, \emptyset\right]$ such that $\rho \|\lambda^{-1}(b)\| \simeq 0$ and $B(b,\rho) \subset A$ (where $B(b,\rho)$ is the open ball of centrum b and radius ρ).*

Proposition 10.3.1 *Let A be a limited subset of $\overline{\Lambda}$ such that ${}^{o}A = \overline{({}^{o}A)^{0}}$ and $hal(\partial A) = hal(\partial\, {}^{o}A)$; then A is n-regular in any point $a \in \partial A$.*

Proof. We first claim that $A \setminus \text{hal}(\partial A) = {}^{o}A \setminus \text{hal}(\partial\, {}^{o}A)$. Indeed, if $x \in A \setminus \text{hal}(\partial A)$ then ${}^{o}x \in {}^{o}A$ and ${}^{o}x \notin \text{hal}(\partial A) = \text{hal}(\partial\, {}^{o}A)$ and so ${}^{o}x \in {}^{o}A \setminus \text{hal}(\partial\, {}^{o}A)$; consequently, ${}^{o}x \in ({}^{o}A)^{0}$ and so $x \in ({}^{o}A)^{0}$; finally, as $x \notin \text{hal}(\partial A) = \text{hal}(\partial\, {}^{o}A)$, $x \in {}^{o}A \setminus \text{hal}(\partial\, {}^{o}A)$. Conversely, if $x \in {}^{o}A \setminus \text{hal}(\partial\, {}^{o}A)$, there exists $y \in A$ such that $x \simeq y$; then $x \in A$ for if not $x \in \text{hal}(\partial A) = \text{hal}(\partial\, {}^{o}A)$.

If $a \notin \text{hal}(\partial\Lambda)$, choose an infinitesimal ρ such that $\rho > \frac{\mathcal{L}}{\sqrt{n}}$. Then for each $b \in A \setminus \text{hal}(\partial A)$, $B(b,\rho) \subset A$: it is obvious if b is standard and if b is nonstandard, with ${}^{o}b$ its shadow, there exists a standard ρ_0 such that $B({}^{o}b, \rho_0) \subset A$ and then $B(b,\rho) \subset B({}^{o}b, \rho_0) \subset A$. So

$$\forall \varepsilon \gtrapprox 0,\ \exists b \in A,\ d(a,b) < \varepsilon \text{ and } B(b,\rho) \subset A.$$

Then the internal set $\{\varepsilon \in \mathbb{R} : \exists b \in A, d(a,b) < \varepsilon \wedge B(b,\rho) \subset A\}$ contains all appreciable ε. Cauchy's principle yields that

$$\exists \varepsilon \simeq 0,\ \exists b \in A,\ d(a,b) < \varepsilon \text{ and } B(b,\rho) \subset A;$$

as $a \notin \text{hal}(\partial\Lambda)$ then $b \notin \text{hal}(\partial\Lambda)$ and so $\lambda^{-1}(b)$ is limited (because λ is a standard diffeomorphism from Θ^0 onto Λ^0). Finally $\rho \|\lambda^{-1}(b)\| \simeq 0$.

If $a \in \text{hal}(\partial\Lambda)$, fix $\rho = n^{-1/4}$; if $b \in A \setminus (\text{hal}(\partial\Lambda) \cup \text{hal}(\partial A))$ then $B(b,\rho) \subset A$ and, as $\lambda^{-1}(b)$ is limited $\rho \|\lambda^{-1}(b)\| < n^{-1/8}$ (for example). So the internal set

$$\left\{\varepsilon \in \mathbb{R} :\ \exists b \in A,\ d(a,b) < \varepsilon\ \wedge\ B(b,\rho) \subset A\ \wedge\ \rho \|\lambda^{-1}(b)\| < n^{-1/8}\right\}$$

contains all appreciable ε. Cauchy's principle yields that

$$\exists \varepsilon \simeq 0,\ \exists b \in A,\ d(a,b) < \varepsilon \wedge B(b,\rho) \subset A \wedge \rho \|\lambda^{-1}(b)\| < n^{-1/8}. \qquad \square$$

For example, a standard set A such that $A \subset \overline{A^0}$ (e.g. an open standard set) is n-regular in any boundary point. It is possible to prove that a limited convex set A such that ${}^{o}A$ has a nonempty interior is also n-regular in any boundary point (we shall not use this result in the following).

Lemma 10.3.1

(i) Let $(\xi_0, \xi_1) \in \Lambda^2$ *and let* $\xi \in]\xi_0, \xi_1[$ *(the segment between* ξ_0 *and* ξ_1*). Then* $J(\xi, \xi_1) < J(\xi_0, \xi_1)$.

(ii) Let A *be a nonempty subset of* Λ *and* $\xi_1 \in \Lambda^0 \setminus A$. *Then* $J(A, \xi_1) = J(\overline{A}, \xi_1) = J(\hat{\partial} A, \xi_1)$.

(iii) Let E *and* F *be subsets of* Λ *such that* $\overline{E}$ *is a compact set included in* F^0*; let* $\xi \in \Lambda^0 \setminus \overline{E}$. *Then* $J(F, \xi) < J(E, \xi)$ *and* $J(\xi, F) < J(\xi, E)$.

Proof. *(i)* As Λ is convex, $\xi \in \Lambda$. We set $\xi_0 - \xi =: \alpha(\xi - \xi_1)$ where $\alpha > 0$; denoting $\theta = \lambda^{-1}(\xi)$ and $\theta_1 := \lambda^{-1}(\xi_1)$, we have (using corollary 2.5 in [10])

$$J(\xi_0, \xi_1) - J(\xi, \xi_1) = (\theta - \theta_1 \mid \xi_0 - \xi) + J(\xi_0, \xi) = \alpha(\theta - \theta_1 \mid \xi - \xi_1) + J(\xi_0, \xi) > 0.$$

(ii) Let $\xi_0 \in A$. If $\xi_0 \notin \hat{\partial} A$, let $\xi \in \hat{\partial} A \cap]\xi_0, \xi_1[$. Using *(i)*, we can write $J(\xi, \xi_1) \leq J(\xi_0, \xi_1)$. So $J(\widehat{\partial} A, \xi_1) \leq J(A, \xi_1)$. Conversely, using the continuity of $J(\cdot, \xi_1)$, we have

$$J(\widehat{\partial} A, \xi_1) \geq J(\overline{A}, \xi_1) = J(A, \xi_1).$$

(iii) If $\xi \in \overline{F}$, then $J(F, \xi) = J(\xi, F) = 0$ and $J(E, \xi) \neq 0$, $J(\xi, E) \neq 0$ because $\xi \notin \overline{E}$.

We suppose now that $\xi \notin F$. Let $\xi_E \in \widehat{\partial} E$ be such that $J(E, \xi) = J(\xi_E, \xi)$; there exists at least one $\xi_F \in \widehat{\partial} F \cap]\xi_E, \xi[$ and then, using *(i)*, $J(\xi_F, \xi) < J(\xi_E, \xi)$ and so $J(F, \xi) < J(E, \xi)$.

Let now $\xi'_E \in \widehat{\partial} E$ be such that $J(\xi, E) = J(\xi, \xi'_E)$; there exists at least one $\xi'_F \in \widehat{\partial} F \cap]\xi'_E, \xi[$. We set $\theta_E := \lambda^{-1}(\xi'_E)$ and $\theta_F := \lambda^{-1}(\xi'_F)$. Then

$$\begin{aligned} J(\xi, \xi'_E) - J(\xi, \xi'_F) = \\ &= \psi(\theta_E) - \psi(\theta) + (\theta - \theta_E \mid \xi) - \psi(\theta_F) + \psi(\theta) - (\theta - \theta_F \mid \xi) \\ &= \psi(\theta_E) - \psi(\theta_F) + (\theta_F - \theta_E \mid \xi) \\ &= \psi(\theta_E) - \psi(\theta_F) + (\theta_F - \theta_E \mid \xi'_F) + (\theta_F - \theta_E \mid \xi - \xi'_F) \\ &= I(\theta_F, \theta_E) + (\theta_F - \theta_E \mid \xi - \xi'_F). \end{aligned}$$

Then setting $\xi - \xi'_F =: \beta(\xi'_F - \xi'_E)$ where $\beta > 0$, we can write

$$J(\xi, \xi'_E) - J(\xi, \xi'_F) = J(\xi'_F, \xi'_E) + \beta(\theta_F - \theta_E \mid \xi'_E - \xi'_F) > 0,$$

according to corollary 2.5 in [10]. □

Proposition 10.3.2 *Let A be a limited subset of $\Lambda \setminus hal(\partial\Lambda)$, $\xi_1 = \lambda(\theta_1)$ be a limited point of $\Lambda \setminus hal(\partial\Lambda)$ and $\xi_0 \in \hat{\partial} A$ such that $J(\xi_0, \xi_1) \simeq J(A, \xi_1)$. If A is n-regular in ξ_0, then*

$$\frac{1}{n} \ln P^n_{\theta_1}(\overline{X} \in A) \simeq -J(A, \xi_1),$$

and if furthermore $d(\xi_1, A) \not\simeq 0$, then

$$-\frac{1}{n} \ln P^n_{\theta_1}(\overline{X} \in A) = J(A, \xi_1)(1 + \emptyset) = @.$$

Proof. For all $(\theta_1, \theta_2) \in \Theta^2$, denoting $x = (x_1, \ldots, x_n)$ and $\overline{x} = \frac{1}{n}\sum_{j=1}^n x_j$ (where $x_j = (x_{j,i})_{1 \le i \le k} \in \mathbb{R}^k$) we can write

$$P^n_{\theta_1}(dx) = \exp\Big(n\big((\theta_1 - \theta_2 \mid \overline{x}) - \psi(\theta_1) + \psi(\theta_2)\big)\Big) P^n_{\theta_2}(dx).$$

Let $\theta_0 := \lambda^{-1}(\xi_0)$ and let $\xi_2 =: \lambda(\theta_2)$ and $\rho > \frac{\mathcal{L}}{\sqrt{n}}$ be such that $\xi_2 \in \mathrm{hal}(\xi_0)$ and $B(\xi_2, \rho) \subset A$. As λ is a standard diffeomorphism from Θ^0 onto Λ^0, $\theta_0, \theta_1, \theta_2$ are limited. Then

$$\begin{aligned}
P^n_{\theta_1}(\overline{X} \in A) &\ge P^n_{\theta_1}(\overline{X} \in B(\xi_2, \rho)) \\
&\ge \int_{\overline{X} \in B(\xi_2,\rho)} \exp\Big(n\big((\theta_1 - \theta_2 \mid \overline{X}) - \psi(\theta_1) + \psi(\theta_2)\big)\Big)\, dP^n_{\theta_2} \\
&\ge \exp\Big(n \int_{\overline{X} \in B(\xi_2,\rho)} \big((\theta_1 - \theta_2 \mid \overline{X}) - \psi(\theta_1) + \psi(\theta_2)\big)\, dP^n_{\theta_2}\Big)
\end{aligned}$$

by Jensen's inequality.

According to (8.4) in [4], $P^n_{\theta_2}(\overline{X}_i \in [\xi_{2,i} - \rho/2, \xi_{2,i} + \rho/2]) = 1 + \emptyset$ for each $i = 1 \ldots k$. Then the nonstandard law of large numbers yields $P^n_{\theta_2}(\overline{X} \in B(\xi_2, \rho)) = 1 + \emptyset$ and we can write (denoting by $\mathcal{L}^k$ the external set of limited vectors of $\mathbb{R}^k$)

$$\begin{aligned}
\frac{1}{n} \ln P^n_{\theta_1}(\overline{X} \in A) &\ge (\theta_1 - \theta_2 \mid \xi_2 + \mathcal{L}^k \rho) - \psi(\theta_1) + \psi(\theta_2) + \frac{1}{n}\ln(1 + \emptyset) \\
&= (\theta_1 - \theta_2 \mid \xi_2) + \mathcal{L}\rho - \psi(\theta_1) + \psi(\theta_2) + \frac{\emptyset}{n} \\
&= -I(\theta_2, \theta_1) + \emptyset.
\end{aligned}$$

As

$$I(\theta_2, \theta_1) - I(\theta_0, \theta_1) = (\theta_0 - \theta_1 \mid \xi_2 - \xi_0) + I(\theta_2, \theta_0),$$

with $\|\theta_0 - \theta_1\|$ limited, $\|\xi_2 - \xi_0\| \simeq 0$ and $I(\theta_2, \theta_0) = @\|\theta_2 - \theta_0\| = @\|\xi_2 - \xi_0\|$ (cf. lemma 3.2.2 in [13]), we have

$$-I(\theta_2, \theta_1) = -I(\theta_0, \theta_1) + \emptyset = -J(A, \xi_1) + \emptyset,$$

and so

$$\frac{1}{n}\ln P^n_{\theta_1}(\overline{X}\in A) \geq -J(A,\xi_1)+\emptyset.$$

Conversely, the limited set A is included in a hypercube $[-p,p]^k$ where p is a standard integer; pave it with $n^k(2p)^k$ hypercubes (T_l) with side $\delta = \frac{1}{n}$. Among these hypercubes, eliminate the ones which do not intersect A and for the others choose $t_l \in A\cap T_l$. In the aim of simplicity, we shall denote again the selected hypercubes by $(T_l)_{1\leq l\leq N}$. Let $\theta_l := \lambda^{-1}(t_l)$, θ_l is limited because t_l is limited and $t_l \notin \mathrm{hal}(\partial\Lambda)$. From the relation $N < n^k(2p)^k$ we deduce $\frac{\ln N}{n} = \frac{\ln n}{n}\mathcal{L}$. Then, we can write

$$\begin{aligned}
P^n_{\theta_1}(\overline{X}\in A) &\leq \sum_{l=1}^{N}\int_{\{\overline{X}\in T_l\}} dP^n_{\theta_1}\\
&= \sum_{l=1}^{N}\int_{\{\overline{X}\in T_l\}} \exp\Big(n\big((\theta_1-\theta_l \mid \overline{X})-\psi(\theta_1)+\psi(\theta_l)\big)\Big)\, dP^n_{\theta_l}\\
&\leq \sum_{l=1}^{N}\exp\left(n\left(\left(\theta_1-\theta_l \mid t_l+\frac{\mathcal{L}^k}{n}\right)-\psi(\theta_1)+\psi(\theta_l)\right)\right)\\
&\leq N\max_{l=1\dots N}\exp\left\{n\left(\left(\theta_1-\theta_l \mid t_l+\frac{\mathcal{L}^k}{n}\right)-\psi(\theta_1)+\psi(\theta_l)\right)\right\}.
\end{aligned}$$

Thus

$$\begin{aligned}
\frac{1}{n}\ln P^n_{\theta_1}(\overline{X}\in A) &\leq \frac{1}{n}\ln N+\max_{l=1\dots N}\left\{\left(\theta_1-\theta_l \mid t_l+\frac{\mathcal{L}^k}{n}\right)-\psi(\theta_1)+\psi(\theta_l)\right\}\\
&\leq \frac{1}{n}\ln N+\max_{l=1\dots N}\{(\theta_1-\theta_l\mid t_l)-\psi(\theta_1)+\psi(\theta_l)\}+\frac{\mathcal{L}}{n}\\
&\leq \frac{1}{n}\ln N+\max_{l=1\dots N} -I(\theta_l,\theta_1)+\frac{\mathcal{L}}{n}\\
&\leq -J(A,\xi_1)+\emptyset.
\end{aligned}$$

Finally,

$$\frac{1}{n}\ln P^n_{\theta_1}(\overline{X}\in A) = -J(A,\xi_1)+\emptyset.$$

As we said previously, according to lemma 3.2.2 in [13], if a and ξ_1 are limited elements of $\Lambda\setminus\mathrm{hal}(\Lambda)$, then $J(a,\xi_1) = @d(a,\xi_1)$. Consequently, if $d(A,\xi_1)=@$ then $J(A,\xi_1)=@$. Thus

$$-\frac{1}{n}\ln P^n_{\theta_1}(\overline{X}\in A) = J(A,\xi_1)(1+\emptyset) = @. \qquad \square$$

In fact, the continuity of $J(\cdot,\xi_1)$ allows us to generalize this result to limited subsets of $\overline{\Lambda}$:

Proposition 10.3.3 *Let A be a limited subset of $\overline{\Lambda}$, ξ_1 be a limited point of $\Lambda \setminus hal(\partial\Lambda)$ and let $\xi_0 \in \hat{\partial} A$ such that $J(\xi_0, \xi_1) \simeq J(A, \xi_1)$. If A is n-regular in ξ_0 then*

$$\frac{1}{n} \ln P^n_{\theta_1}(\overline{X} \in A) \simeq -J(A, \xi_1).$$

Proof. Let $\xi_2 =: \lambda(\theta_2) \in \text{hal}(\xi_0)$ and $\rho > \frac{\mathcal{L}}{\sqrt{n}}$ be such that $B(\xi_2, \rho) \subset A$ and $\rho \|\theta_2\| \simeq 0$. As in the proof of proposition 10.3.2, we can write

$$\begin{aligned} \frac{1}{n} \ln P^n_{\theta_1}(\overline{X} \in A) &\geq (\theta_1 - \theta_2 \mid \xi_2 + \mathcal{L}^k \rho) - \psi(\theta_1) + \psi(\theta_2) + \frac{1}{n} \ln(1 + \emptyset) \\ &= (\theta_1 - \theta_2 \mid \xi_2) + \mathcal{L}\rho + \emptyset - \psi(\theta_1) + \psi(\theta_2) + \emptyset \\ &= -J(\xi_2, \xi_1) + \emptyset = -J(\xi_0, \xi_1) + \emptyset = -J(A, \xi_1) + \emptyset \end{aligned}$$

because $\|\xi_2 - \xi_0\| \simeq 0$ and $J(\cdot, \xi_1)$ is S-continuous.

Conversely, as in the proof of proposition 10.3.2, use a paving $(T_l)_{1 \leq l \leq N}$ with side $\delta = \frac{1}{n}$ where each T_l is closed and such that $T_l \cap A \neq \varnothing$, but choose t_l in T_l (maybe outside A) such that:

$$\begin{aligned} &\text{if } \theta_{1,i} - \theta_{l,i} > 0 \text{ then } t_{l,i} \text{ is maximal in } T_l \\ &\text{if } \theta_{1,i} - \theta_{l,i} \leq 0 \text{ then } t_{l,i} \text{ is minimal in } T_l \end{aligned}$$

So, for $\overline{X} \in T_l$ we can write $(\theta_1 - \theta_l \mid \overline{X}) \leq (\theta_1 - \theta_l \mid t_l)$ and then

$$\begin{aligned} P^n_{\theta_1}(\overline{X} \in A) &\leq \sum_{l=1}^{N} \int_{\{\overline{X} \in T_l\}} dP^n_{\theta_1} \\ &\leq \sum_{l=1}^{N} \int_{\{\overline{X} \in T_l\}} \exp\Big(n\big((\theta_1 - \theta_l \mid \overline{X}) - \psi(\theta_1) + \psi(\theta_l)\big)\Big)\, dP^n_{\theta_l} \\ &\leq \sum_{l=1}^{N} \exp\Big(n\big((\theta_1 - \theta_l \mid t_l) - \psi(\theta_1) + \psi(\theta_l)\big)\Big). \end{aligned}$$

Thus

$$\begin{aligned} \frac{1}{n} \ln P^n_{\theta_1}(\overline{X} \in A) &\leq \frac{1}{n} \ln N + \max_{l=1\ldots N} \{(\theta_1 - \theta_l \mid t_l) - \psi(\theta_1) + \psi(\theta_l)\} \\ &\leq \frac{1}{n} \ln N + \max_{l=1\ldots N} \{-J(t_l, \xi_1)\}. \end{aligned}$$

As $J(\cdot, \xi_1)$ is S-continuous, $J(t_l, \xi_1) = J({}^o t_l, \xi_1) + \emptyset$ and then, as ${}^o t_l \in {}^o A$

$$\frac{1}{n} \ln P^n_{\theta_1}(\overline{X} \in A) \leq -J({}^o A, \xi_1) + \emptyset.$$

We now claim that $J({}^oA,\xi_1) \simeq J(A,\xi_1)$. Indeed, if $a \in A$, then ${}^oa \in {}^oA$ and $J(a,\xi_1) \simeq J({}^oa,\xi_1)$; so $J({}^oA,\xi_1) \lesssim J(A,\xi_1)$. Conversely, if $a \in {}^oA$, there exists $a' \in A$ such that $a \simeq a'$ and thus such that $J(a,\xi_1) \simeq J(a',\xi_1)$; so $J(A,\xi_1) \lesssim J({}^oA,\xi_1)$.

Finally $\frac{1}{n}\ln P^n_{\theta_1}(\overline{X} \in A) \leq -J(A,\xi_1) + \emptyset$. □

10.3.6 *n*-regular sets defined by Kullback-Leibler information

Let Θ_0 be an internal subset of Θ, included in a standard compact K included in the interior of Θ.

Let $c_0 := c_0(\Theta_0) := \sup\{I(\theta,\Theta_0) : \theta \in \Theta\} \leq \infty$. For $c \in]0, c_0[$, let $A_c := \{\xi \in \overline{\Lambda} : J(\xi,\Lambda_0) \leq c\}$ where $\Lambda_0 := \lambda(\Theta_0)$. We shall see in lemma 10.3.2 that $J(\cdot,\Lambda_0)$ is continuous; then if $c < c' < c_0$, A_c is a proper subset of $A_{c'}$ (for if not, $\{\xi \in \overline{\Lambda} : J(\xi,\Lambda_0) \leq \frac{c+c'}{2}\} = \{\xi \in \overline{\Lambda} : J(\xi,\Lambda_0) < \frac{c+c'}{2}\}$ is a connected component of $\overline{\Lambda}$ which is a connected set.)

Lemma 10.3.2 *If $c \leq c_0$ is limited, then $\hat{\partial} A_c$ is a limited compact set.*

Proof. $\hat{\partial} A_c = \{\xi \in \overline{\Lambda} : J(\xi,\Lambda_0) = c\}$ is closed in $\overline{\Lambda}$ since the function $\xi \to J(\xi,\Lambda_0)$ is continuous on $\overline{\Lambda}$ as we prove now. By transfer, we just have to prove this continuity for a standard Λ_0: if ξ_1 and ξ_2 are such that $\|\xi_1 - \xi_2\| \simeq 0$, if $\xi_0 \in \Lambda_0$ is such that $J(\xi_1,\xi_0) \simeq J(\xi_1,\Lambda_0)$ then, as $J(\cdot,\xi_0)$ is S-continuous, $J(\xi_2,\xi_0) \simeq J(\xi_1,\xi_0)$ and so $J(\xi_2,\Lambda_0) \lesssim J(\xi_1,\Lambda_0)$; similarly $J(\xi_1,\Lambda_0) \lesssim J(\xi_2,\Lambda_0)$.

We prove now that if ξ is unlimited, then $J(\xi,\Lambda_0)$ is unlimited. For each $\xi_0 \in \Lambda_0$, ξ_0 belongs to the standard compact K; then $\|\xi - \xi_0\| \simeq \infty$ and so $J(\xi,\xi_0) \simeq \infty$ (cf. [10, p. 177]). Thus $J(\xi,\Lambda_0)$ is unlimited.

Therefore $\{\xi \in \overline{\Lambda} : J(\xi,\Lambda_0) = c\}$ is limited. □

Proposition 10.3.4 *Let $c \leq c_0$ be limited and θ_1 be a limited point of Θ' such that $\xi_1 := \lambda(\theta_1) \notin hal(\partial\Lambda)$.*

(i) If $\xi_1 \notin A_c$, then $\frac{1}{n}\ln P^n_{\theta_1}(\overline{X} \in A_c) = \frac{1}{n}\ln P^n_{\theta_1}(\overline{X} \in A^0_c) \simeq -J(\hat{\partial} A_c,\xi_1)$.

(ii) If $\xi_1 \in A_c$, then $\frac{1}{n}\ln P^n_{\theta_1}(\overline{X} \in A_c) \simeq 0$.

(iii) If $\xi_1 \notin A^{\mathbf{C}}_c$, then $\frac{1}{n}\ln P^n_{\theta_1}(\overline{X} \in A^{\mathbf{C}}_c) = \frac{1}{n}\ln P^n_{\theta_1}(\overline{X} \in \overline{A^{\mathbf{C}}_c}) \simeq -J(\hat{\partial} A_c,\xi_1)$.

(iv) If $\xi_1 \in A^{\mathbf{C}}_c$, then $\frac{1}{n}\ln P^n_{\theta_1}(\overline{X} \in A^{\mathbf{C}}_c) \simeq 0$.

Proof. We first prove the both similar results *(i)* and *(iii)*, and then the both similar results *(ii)* and *(iv)* which are obtained in a same way.

(i) If Θ_0 and c are standard, A_c is a standard compact set such that $A_c = \overline{A_c^0}$. Indeed, $A_c^0 = \{\xi \in \Lambda : J(\xi, \Lambda_0) < c\}$. If ξ is such that $J(\xi, \Lambda_0) = c$, let $\lambda_0 \in \overline{\Lambda}_0$ (a compact set) be such that $J(\xi, \Lambda_0) = J(\xi, \overline{\Lambda}_0) = J(\xi, \lambda_0) = c$. According to lemma 10.3.1 (i), for any $\xi' \in]\xi, \lambda_0[$, $J(\xi', \lambda_0) < J(\xi, \lambda_0)$ and so $J(\xi', \Lambda_0) < c$. Thus $]\xi, \lambda_0[\in A_c^0$ and then $\xi \in \overline{A_c^0}$. So, according to proposition 10.3.1, A_c and A_c^0 are n-regular in any point of ∂A_c and then, according to proposition 10.3.3, we can write $\frac{1}{n} \ln P_{\theta_1}^n(\overline{X} \in A_c) \simeq -J(A_c, \xi_1)$ and $\frac{1}{n} \ln P_{\theta_1}^n(\overline{X} \in A_c^0) \simeq -J(A_c^0, \xi_1)$. Finally, lemma 10.3.1 (ii) yields $J(A_c, \xi_1) = J(A_c^0, \xi_1) = J(\widehat{\partial} A_c, \xi_1)$.

Now, if c or Θ_0 are not standard, the continuity of $J(\xi, \cdot)$ on $\lambda(K)$ implies ${}^oA_c(\Theta_0) = A_{{}^oc}({}^o\Theta_0)$ and the hypothesis of proposition 10.3.1 is verified since

$$\mathrm{hal}(\partial A_c) = \mathrm{hal}(\widehat{\partial} A_c) \cup \left\{\xi \in \partial\Lambda : \ J(\xi, \Lambda_0) \lesssim c\right\}$$

and

$$\mathrm{hal}(\partial\, {}^oA_c) = \mathrm{hal}(\widehat{\partial}\, {}^oA_c) \cup \left\{\xi \in \partial\Lambda : \ J(\xi, {}^o\Lambda_0) \lesssim {}^oc\right\}.$$

Indeed, on one hand, the continuity of $J(\xi, \cdot)$ implies

$$\left\{\xi \in \partial\Lambda : J(\xi, \Lambda_0) \lesssim c\right\} = \left\{\xi \in \partial\Lambda : J(\xi, {}^o\Lambda_0) \lesssim {}^oc\right\},$$

and on the other hand, this continuity also implies

$$\begin{aligned} \mathrm{hal}(\hat{\partial} A_c) &= \left\{\xi \in \overline{\Lambda} : J(\xi, \lambda(\Theta_0)) \simeq c\right\} \\ &= \left\{\xi \in \overline{\Lambda} : J(\xi, \lambda({}^o\Theta_0)) \simeq {}^oc\right\} = \mathrm{hal}(\hat{\partial}\, {}^oA_c). \end{aligned}$$

Then we can conclude as before, using proposition 10.3.1, proposition 10.3.3 and lemma 10.3.1.

(iii) In the same way, $A_c^{\mathbf{C}}$ is n-regular in any point of $\widehat{\partial} A_c^{\mathbf{C}} = \widehat{\partial} A_c$, but $A_c^{\mathbf{C}}$ is not always limited. Meanwhile, $\widehat{\partial} A_c$ is limited and so the proof (in proposition 10.3.3) of $\frac{1}{n} \ln P_{\theta_1}^n(\overline{X} \in A_c^{\mathbf{C}}) \geq -J(\hat{\partial} A_c, \xi_1) + \emptyset$ remains valid.

Suppose now that $\xi_1 \in A_c$. A_c is included in a standard hypercube $U_p := [-p, p]^k$ (where p is a standard integer). If we set $S_{p,i} := \left\{x \in \mathbb{R}^k : x_i > p\right\}$ and $S'_{p,i} := \left\{x \in \mathbb{R}^k : x_i < -p\right\}$ then $U_p^{\mathbf{C}} = \bigcup_{i=1}^k (S_{p,i} \cup S'_{p,i})$. It is clear that

$$P_{\theta_1}^n(\overline{X} \in A_c^{\mathbf{C}}) = P_{\theta_1}^n(\overline{X} \in U_p \setminus A_c) + P_{\theta_1}^n(\overline{X} \in U_p^{\mathbf{C}}).$$

On one hand, as $U_p \setminus A_c$ is limited and n-regular in any boundary point, we can write

$$\frac{1}{n} \ln P_{\theta_1}^n(\overline{X} \in U_p \setminus A_c) \simeq -J(U_p \setminus A_c, \xi_1) = -J(\widehat{\partial} U_p \cup \widehat{\partial} A_c, \xi_1).$$

Using lemma 10.3.1(iii) with $E = \widehat{\partial} U_p$ and $F = A_c^{\mathbf{C}}$, we have $J(\widehat{\partial} U_p \cup \widehat{\partial} A_c, \xi_1) = J(\widehat{\partial} A_c, \xi_1)$ and so $\frac{1}{n} \ln P_{\theta_1}^n(\overline{X} \in U_p \setminus A_c) \simeq -J(\widehat{\partial} A_c, \xi_1)$.

On the other hand,

$$P_{\theta_1}^n(\overline{X} \in U_p^{\mathbf{C}}) \leq \sum_{i=1}^{k} (P_{\theta_1}^n(\overline{X} \in S_{p,i}) + P_{\theta_1}^n(\overline{X} \in S'_{p,i})).$$

We know that $\frac{1}{n} \ln P_{\theta_1}^n(\overline{X} \in S_{p,i}) \simeq -J(S_{p,i}, \xi_1)$ and $\frac{1}{n} \ln P_{\theta_1}^n(\overline{X} \in S'_{p,i}) \simeq -J(S'_{p,i}, \xi_1)$ (it is a classical result concerning half-spaces: cf. chap. 7 in [10]); so

$$\frac{1}{n} \ln P_{\theta_1}^n(\overline{X} \in U_p^{\mathbf{C}}) \leq \max_{i=1\ldots k} \max(-J(S_{p,i}, \xi_1), -J(S'_{p,i}, \xi_1)) + \emptyset$$

which is less than $-J(\widehat{\partial} A_c, \xi_1)$ according to lemma 10.3.1(i). Then $\frac{1}{n} \ln P_{\theta_1}^n(\overline{X} \in A_c^{\mathbf{C}}) \leq -J(\widehat{\partial} A_c, \xi_1) + \emptyset$.

Finally, $\frac{1}{n} \ln P_{\theta_1}^n(\overline{X} \in A_c^{\mathbf{C}}) \simeq -J(\widehat{\partial} A_c, \xi_1)$.

(ii) If $\xi_1 \in \mathrm{hal}(\widehat{\partial} A_c)$, then exists $\xi_0 \in \widehat{\partial} A_c$ such that $J(\xi_0, \xi_1) \simeq J(A, \xi_1) \simeq 0$ and it remains possible to use proposition 10.3.1, proposition 10.3.3 and lemma 10.3.1 as for the proof of *(i)*.

If $\xi_1 \notin \mathrm{hal}(\widehat{\partial} A_c)$, then $\xi_1 \in {}^oA_c \setminus \mathrm{hal}(\widehat{\partial} A_c)$ according to the proof of *(i)* and so there is a standard number ρ such that the ball $B(\xi, \rho)$ is included in ${}^oA_c \setminus \mathrm{hal}(\widehat{\partial} A_c)$ and consequently included in A_c. Then the nonstandard law of large numbers yields $P_{\theta_1}^n(\overline{X} \in A_c) = 1 + \emptyset$ and finally $\ln P_{\theta_1}^n(\overline{X} \in A_c) \simeq 0$.

(iv) is obtained in a similar way. □

Definition 10.3.3 *Let $\theta \in \Theta'$. Let h_θ be the function defined on $[0, c_0[$ by $h_\theta(c) := J(A_c^{\mathbf{C}}, \lambda(\theta))$ and g_θ the function defined on $[0, c_0[$ by $g_\theta(c) := J(A_c, \lambda(\theta))$.*

As $c \to A_c$ is increasing, g_θ is a non-increasing function and h_θ is a non-decreasing function. Let $\gamma(\theta) := I(\theta, \Theta_0)$; $\lambda(\theta) \in A_c$ iff $c \geq \gamma(\theta)$ and so $g_\theta(c) = 0$ if and only if $c \in [\gamma(\theta), c_0[$ and $h_\theta(c) = 0$ if and only if $c \in]0, \gamma(\theta)]$.

Proposition 10.3.5

(i) The functions $(\theta, c) \to g_\theta(c)$ and $(\theta, c) \to h_\theta(c)$ are S-continuous on $\{\theta \in \Theta \setminus \mathrm{hal}(\partial\Theta) : \theta \text{ limited}\} \times ([0, c_0[\cap \mathcal{L})$.

(ii) The function g_θ is decreasing on $[0,\gamma(\theta)]$ and h_θ is increasing on $[\gamma(\theta), c_0]$.

Proof. Let θ be standard and $\theta' \simeq \theta$, then $\lambda(\theta) \simeq \lambda(\theta')$; let c and c' be such that $c' \simeq c$. Then ${}^oA_{c'} = {}^oA_c$ and

$$J(A_c, \lambda(\theta')) \simeq J(A_c, \lambda(\theta)) \simeq J({}^oA_c, \lambda(\theta)).$$

Similarly, $J(A_{c'}, \lambda(\theta')) \simeq J({}^oA_{c'}, \lambda(\theta))$.

The monotonicity of g_θ is deduced from lemma 10.3.1 (iii) which shows that if $c' < c$ and $\lambda(\theta) \notin A_{c'}$ then $J(A_{c'}, \lambda(\theta)) > J(A_c, \lambda(\theta))$.

The continuity and monotonicity of h_θ are proved in the same way. □

10.4 The nonstandard likelihood ratio test

Recall that $\mathcal{F} := \{\Theta_0 \text{ internal } : \widetilde{\Theta}_0 \subset \Theta_0 \subset \text{hal}(\widetilde{\Theta}_0)\}$, that the NSLRT is defined by $\Phi_0 :\equiv \inf\{\Phi_{\Theta_0} : \Theta_0 \in \mathcal{F}\}$, that Φ_{Θ_0} is defined through R_{Θ_0} and that, for exponential families, $R_{\Theta_0} = J(\overline{X}, \Lambda_0)$ where $\Lambda_0 := \lambda(\Theta_0)$. So, the relation

$$\sup_{\theta \in \Theta_0} P_\theta^n(R_{\Theta_0} > d) \le \alpha \le \sup_{\theta \in \Theta_0} P_\theta^n(R_{\Theta_0} \ge d)$$

(see §10.2.1) can be written

$$\sup_{\theta \in \Theta_0} P_\theta^n(\overline{X} \in A_d^{\mathbf{C}}(\Theta_0)) \le \alpha \le \sup_{\theta \in \Theta_0} P_\theta^n(\overline{X} \in \overline{A_d^{\mathbf{C}}(\Theta_0)}).$$

Consequently, $A_d^{\mathbf{C}}$ is the rejection set (relatively to $\overline{X}$), A_d^0 the acceptation set and the test Φ_{Θ_0} is randomized if $\overline{X} \in \hat{\partial} A_c$.

As n is illimited, if $d < c_0(\Theta_0)$ and if d is limited then, for $\theta \in \Theta_0$, proposition 10.3.4 yields

$$\frac{1}{n} \ln P_\theta^n(\overline{X} \in A_d^{\mathbf{C}}) \simeq \frac{1}{n} \ln P_\theta^n(\overline{X} \in \overline{A_d^{\mathbf{C}}}) \simeq -J(\hat{\partial} A_d, \lambda(\theta)).$$

As $\sup_{\theta \in \Theta_0} -J(\hat{\partial} A_d, \lambda(\theta)) = -d$ by definition of A_d, we have

$$d(\alpha, \Theta_0) \simeq -\frac{\ln \alpha}{n}.$$

This is the nonstandard form of a classical result of Bahadur ([1], [2]). We shall study the NSLRT according as these $d(\alpha, \Theta_0)$ are infinitesimal or appreciable. In this latter case, we shall have to distinguish the cases $d(\alpha, \Theta_0) < c_0(\Theta_0)$ and $d(\alpha, \Theta_0) \ge c_0(\Theta_0)$.

In the following, $\widetilde{c}_0$ stands for $c_0(\widetilde{\Theta}_0)$, $\widetilde{\Lambda}_0$ for $\lambda(\widetilde{\Theta}_0)$ and $R_0 = J(\overline{X}, \widetilde{\Lambda}_0)$ for $R_{\widetilde{\Theta}_0}$; A_c will denote $A_c(\widetilde{\Lambda}_0)$; the notations $h_\theta(c)$ and $g_\theta(c)$ will also refer to these $A_c := A_c(\widetilde{\Lambda}_0)$. Note that if $\Theta_0 \in \mathcal{F}$, then the S-continuity of λ on K implies

$$R_{\Theta_0} \simeq R_0.$$

The following lemmas will be used for the calculation of the risks of the first and the second kind. They are valid not only for exponential families, but also for any statistical family. Here Θ'' is a standard subset of Θ.

Lemma 10.4.1 *Let T be a statistic, $\mathcal{S}$ a (perhaps external) subset of Θ'' and $F : D \to \mathbb{R}_+, (\theta, c) \to F_\theta(c)$ a standard continuous function defined on a domain $D = \{(\theta, c) \in \Theta'' \times \mathbb{R}_+ : d_\theta \le c \le d'_\theta\}$ such that*

- *the functions $\theta \to d_\theta$ and $\theta \to d'_\theta$ are continuous,*
- *for each $\theta \in \mathcal{S}$, F_θ is decreasing on $[d_\theta, d'_\theta]$,*
- *for each θ in $\mathcal{S}$ and c nearly standard in $]d_\theta, d'_\theta[$,*

$$\frac{1}{n} \ln P^n_\theta(T \ge c) \simeq F_\theta(c).$$

Then, for each θ nearly standard in $\mathcal{S}$ and C_0 standard in $[d_\theta, d'_\theta[$

$$PR^n_\theta(T \overset{>}{\not\approx} C_0) \equiv \mathcal{L}e^{-n@+nF_\theta(C_0)}.$$

Proof. $\underline{PR^n_\theta}(T \overset{>}{\not\approx} C_0) \equiv \sup_{w=@} P^n_\theta(T > C_0 + w)$. Now write

$$\frac{1}{n} \ln P^n_\theta(T \ge C_0 + w) \simeq F_\theta(C_0 + w) \simeq {}^oF_\theta(C_0 + w).$$

So, as ${}^oF_\theta = F_{\theta_0}$ (where $\theta_0 := {}^o\theta$) is a standard continuous decreasing function, we have

$$\frac{1}{n} \ln \underline{PR^n_\theta}(T \overset{>}{\not\approx} C_0) \equiv \left] -\infty, \sup_{w=@} F_\theta(C_0 + w) \right[\equiv]-\infty, F_\theta(C_0) + \emptyset[\,.$$

Similarly,

$$\frac{1}{n} \ln \overline{PR^n_\theta}(T \overset{>}{\not\approx} C_0) \equiv]-\infty, \; {}^oF_\theta(C_0) + \emptyset[\equiv]-\infty, F_\theta(C_0) + \emptyset[\,,$$

for if not, there exists $\eta < 0$, $\eta \simeq 0$ such that

$$\forall \varepsilon \in \mathrm{hal}^+(0), \; \frac{1}{n} \ln P^n_\theta(T \ge C_0 + \varepsilon) > {}^oF_\theta(C_0) + \eta$$

where $\mathrm{hal}^+(0) := \{\varepsilon \in \mathrm{hal}(0) : \varepsilon \ge 0\}$.

Then $\left\{\varepsilon \in \mathbb{R}_+ : \frac{1}{n}\ln P_\theta^n(T \geq C_0+\varepsilon) > {}^oF_\theta(C_0)+\eta\right\}$ is an internal set including hal$^+(0)$ and consequently (by Cauchy's principle) including an interval $[0, w_0]$ where w_0 is standard. But

$$\frac{1}{n}\ln P_\theta(T \geq C_0 + w_0) \simeq F_\theta(C_0 + w_0) \simeq {}^oF_\theta(C_0 + w_0)$$

with ${}^oF_\theta(C_0 + w_0) < {}^oF_\theta(C_0)$ which is a contradiction, because both these numbers are standard. Finally

$$\underline{PR}_\theta^n(T \gtrapprox C_0) \equiv \overline{PR}_\theta^n(T \gtrapprox C_0) \equiv \mathcal{L}e^{-n@+nF_\theta(C_0)}. \qquad \square$$

Lemma 10.4.2 *Let T be a statistic, $\mathcal{S}$ a (perhaps external) subset of Θ'' and $G : E \to \mathbb{R}_+, (\theta, c) \to G_\theta(c)$ a standard continuous function defined on a domain $E = \{(\theta, c) \in \Theta'' \times \mathbb{R}_+ : e_\theta \leq c \leq e'_\theta\}$ such that*

- *the functions $\theta \to e_\theta$ and $\theta \to e'_\theta$ are continuous,*
- *for each $\theta \in \mathcal{S}$, G_θ is increasing on $[e_\theta, e'_\theta]$,*
- *for each θ in $\mathcal{S}$ and c nearly standard in $]e_\theta, e'_\theta[$,*

$$\frac{1}{n}\ln P_\theta^n(T \leq c) \simeq G_\theta(c).$$

Then, for each θ nearly standard in $\mathcal{S}$ and C_0 standard in $[e_\theta, e'_\theta[$

$$PR_\theta^n(T \lessapprox C_0) \equiv \mathcal{L}e^{nG_\theta(C_0)+n\emptyset}.$$

Proof. $\overline{PR}_\theta^n(T \lessapprox C_0) \equiv \inf_{t=@} P_\theta^n(T \leq C_0 + t)$. Now write

$$\frac{1}{n}\ln P_\theta^n(T \leq C_0 + t) \simeq G_\theta(C_0 + t) \simeq {}^oG_\theta(C_0 + t).$$

So, as ${}^oG_\theta$ is a standard continuous increasing function, we have

$$\frac{1}{n}\ln \overline{PR}_\theta^n(T \lessapprox C_0) \equiv \left]-\infty, \inf_{t=@} G_\theta(C_0 + t)\right] \equiv \left]-\infty, G_\theta(C_0) + \emptyset\right].$$

As in lemma 10.4.1, one establishes that $\overline{PR}_\theta^n(T \lessapprox C_0) \equiv \underline{PR}_\theta^n(T \lessapprox C_0)$ and finally

$$PR_\theta^n(T \lessapprox C_0) \equiv \mathcal{L}e^{nG_\theta(C_0)+n\emptyset}. \qquad \square$$

10.4.1 $\dfrac{\ln\alpha}{n}$ infinitesimal

Proposition 10.4.1 *For* $\frac{\ln\alpha}{n}\simeq 0$,

(i) if $R_0\simeq 0$ *then* $\Phi_0\equiv 0$ *otherwise* $\Phi_0\equiv 1$*;*

(ii) for $\theta\in hal(\widetilde{\Theta}_0)$*, the risk of the first kind is exactly equal to* $\mathcal{L}e^{-n@-nh_\theta(0)}$*;*

(iii) for a limited $\theta\in\Theta\setminus(hal(\partial\Theta)\cup hal(\widetilde{\Theta}_0))$*, the risk of the second kind is exactly equal to* $\mathcal{L}e^{-ng_\theta(0)+n\emptyset}$*.*

Proof. *(i)* If $R_0\not\simeq 0$ then $\forall\Theta_0\in\mathcal{F}$, $R_{\Theta_0}\not\simeq 0$ and so $\forall\Theta_0\in\mathcal{F}$, $\Phi_{\Theta_0}=1$ because $d(\alpha,\Theta_0)\simeq-\frac{\ln\alpha}{n}\simeq 0$ and finally $\Phi_0\equiv 1$.

If $R_0\simeq 0$, $\overline{X}\in\text{hal}(\widetilde{\Lambda}_0)$ (for if not, $R_0=J(\overline{X},\widetilde{\Lambda}_0)$ is appreciable) and so $\lambda^{-1}(\overline{X})\in\widetilde{\Theta}_0\cup\{\lambda^{-1}(\overline{X})\}\in\mathcal{F}$. Then, $\Phi_{\Theta_0}=0$ for this $\Theta_0:=\widetilde{\Theta}_0\cup\{\lambda^{-1}(\overline{X})\}$. Finally, $\Phi_0\equiv 0$.

(ii) According to proposition 10.3.4 (iii) and (iv), for a limited $c\le\widetilde{c}_0$, we have $\frac{1}{n}\ln P^n_\theta(R_0\ge c)\simeq -h_\theta(c)$. Let Θ'' a standard compact set such that $\text{hal}(\widetilde{\Theta}_0)\subset\Theta''\subset\Theta\setminus\text{hal}(\partial\Theta)$ (for example $\Theta''=K$). Then according to proposition 10.3.5, $(\theta,c)\to -h_\theta(c)$ is continuous on $\Theta''\times[0,c_0(\Theta'')]$ and for $\theta\in\Theta''$, $-h_\theta$ is decreasing on $[\gamma(\theta),c_0(\Theta'')]$.

We can now apply 10.4.1 with $D=\{(\theta,c)\in\Theta''\times\mathbb{R}_+:\gamma(\theta)\le c\le c_0(\Theta'')\}$, $T=R_0$, $\mathcal{S}=\text{hal}(\widetilde{\Theta}_0)$, $F_\theta=-h_\theta$, $C_0=0$ to obtain

$$PR^n_\theta(R_0 \underset{\not\simeq}{>} 0)\equiv\mathcal{L}e^{-n@-nh_\theta(0)}.$$

(iii) According to proposition 10.3.4 (i) and (ii), for a limited $c\le\widetilde{c}_0$, we have $\frac{1}{n}\ln P^n_\theta(R_0\le c)\simeq -g_\theta(c)$. For any limited $\theta\in\Theta\setminus(\text{hal}(\partial\Theta)\cup\text{hal}(\widetilde{\Theta}_0))$, there exists a standard compact set $\Theta''\subset\Theta\setminus\text{hal}(\partial\Theta)$, such that $\theta\in\Theta''$. According to proposition 10.3.5, $(\theta,c)\to -g_\theta(c)$ is continuous on $\Theta''\times[0,c_0(\Theta'')]$ and for $\theta\in\Theta''$, $-g_\theta$ is increasing on $[0,\gamma(\theta)]$.

We can now apply lemma 10.4.2 with $D=\{(\theta,c)\in\Theta''\times\mathbb{R}_+:0\le c\le\gamma(\theta)\}$, $\mathcal{S}=\Theta''\setminus\text{hal}(\widetilde{\Theta}_0)$, $T=R_0$, $G_\theta=-g_\theta$, $C_0=0$, to obtain $PR^n_\theta(R_0\lesssim 0)\equiv\mathcal{L}e^{-ng_\theta(0)+n\emptyset}$. □

Remark 10.4.1 *Let* $\theta\in hal(\widetilde{\Theta}_0)$*; the nonstandard version of the law of large numbers yields*

$$PR^n_\theta(\overline{X}\in hal(\lambda(\theta)))=1+\emptyset\Rightarrow PR^n_\theta(\overline{X}\in hal(\widetilde{\Lambda}_0))=1+\emptyset.$$

As the reject criterion is $\overline{X}\notin hal(\widetilde{\Lambda}_0)$*, it is not logical to take* α *appreciable, but infinitesimal.*

10.4.2 $0 \lnapprox \frac{1}{n}|\ln\alpha| \lnapprox \widetilde{c}_0$

If $\widetilde{c}_0 = \infty$, we shall suppose that $\frac{|\ln\alpha|}{n}$ is limited because some preliminary results (in §10.3) are no longer available (for example A_c is no longer limited if $c \simeq \infty$).

Proposition 10.4.2 *For* $0 \lnapprox \frac{|\ln\alpha|}{n} \lnapprox \widetilde{c}_0$ *and* $\frac{|\ln\alpha|}{n}$ *limited, put* $C := {}^o\!\left(-\frac{\ln\alpha}{n}\right)$.

(i) If $R_0 \gnapprox C$ *then* $\Phi_0 \equiv 1$ *otherwise* $\Phi_0 \equiv 0$.

(ii) For $\theta \in hal(\widetilde{\Theta}_0)$, *the risk of the first kind is exactly equal to* $\mathcal{L}e^{-n@-nh_\theta(C)}$.

(iii) For a limited $\theta \in \Theta \setminus ((hal(\partial\Theta) \cup hal(\widetilde{\Theta}_0))$, *if* $\gamma(\theta) \gnapprox C$ *then the risk of the second kind is exactly equal to* $\mathcal{L}e^{-ng_\theta(C)+n\emptyset}$; *and if* $\gamma(\theta) \lesssim C$ *then the risk of the second kind is exactly equal to* $1 + \mathcal{L}e^{-n@-nh_\theta(C)}$.

Proof. *(i)* If $R_0 \gnapprox C$ then $\forall\Theta_0 \in \mathcal{F}, R_{\Theta_0} \gnapprox -\frac{\ln\alpha}{n}$, so $R_{\Theta_0} > d(\alpha, \Theta_0)$ and $\Phi_{\Theta_0} = 1$. Finally, $\Phi_0 \equiv 1$.

If $R_0 \simeq 0, \Phi_0 \equiv 0$, as in proposition 10.4.1 (i).

If $0 \lnapprox R_0 \lesssim C$, then $\overline{X} \notin \text{hal}(\widetilde{\Lambda}_0)$; we prove that $\exists\Theta_0 \in \mathcal{F}$, $\Phi_{\Theta_0} = 0$. If not, the external set $\mathcal{F}$ is included in the internal set $\{\Theta_0 : \widetilde{\Theta}_0 \subset \Theta_0 \subset K \wedge \Phi_{\Theta_0} \neq 0\}$, and using the Cauchy's principle, there exists an internal set $\Theta_0 \supset \text{hal}(\widetilde{\Theta}_0)$ such that $\Phi_{\Theta_0} \neq 0$, i.e. $R_{\Theta_0} \geq d(\alpha, \Theta_0) \simeq C$. Denoting by $\widehat{\Theta}_0$ the shadow of this Θ_0 we have $R_{\widehat{\Theta}_0} \simeq R_{\Theta_0} \gtrsim C$.

$\widehat{\Theta}_0$ and $\widetilde{\Theta}_0$ are two distinct relatively compact standard sets such that $\partial\widetilde{\Theta}_0 \subset \widehat{\Theta}_0^0$ and then such that $\partial\widetilde{\Lambda}_0 \subset \widehat{\Lambda}_0^0$ (where $\widehat{\Lambda}_0 := \lambda(\widehat{\Theta}_0)$). According to lemma 10.3.1 (iii), $J(\xi, \widehat{\Lambda}_0) < J(\xi, \widetilde{\Lambda}_0)$ for each $\xi \notin \widetilde{\Lambda}_0$ and then

$$R_{\widehat{\Theta}_0} = J(\overline{X}, \widehat{\Lambda}_0) \simeq J({}^o\overline{X}, \widehat{\Lambda}_0) < J({}^o\overline{X}, \widetilde{\Lambda}_0) \simeq J(\overline{X}, \widetilde{\Lambda}_0) = R_0.$$

As $J({}^o\overline{X}, \widehat{\Lambda}_0)$ and $J({}^o\overline{X}, \widetilde{\Lambda}_0)$ are both standard numbers, this contradicts $R_{\widehat{\Theta}_0} \gtrsim C$ and $R_0 \lesssim C$.

(ii) The proof is similar to the proof of proposition 10.4.1 (ii), with $C_0 = C$ instead of $C_0 = 0$. We note that the risk of the first kind calculated in §10.4.1 is a particular case of this result.

(iii) For the first result, the proof is similar to the proof of proposition 10.4.1 (iii), with $C_0 = C$ instead of $C_0 = 0$ and $\mathcal{S} = \{\theta \in \Theta'' \setminus \text{hal}(\widetilde{\Theta}_0) : \gamma(\theta) \gnapprox C\}$ instead of $\mathcal{S} = \Theta'' \setminus \text{hal}(\widetilde{\Theta}_0)$. For the second, we choose a standard compact set $\Theta'' \subset \Theta \setminus \text{hal}(\partial\Theta)$, such that $\theta \in \Theta''$. In fact, the proof of (ii) is valid with $\mathcal{S} = \{\theta \in \Theta'' : \gamma(\theta) \lesssim C\}$ and one obtains

$$PR_\theta^n(R_0 \gnapprox C) \equiv \mathcal{L}e^{-n@-nh_\theta(C)} \qquad \square$$

10.4.3 $\frac{1}{n}|\ln\alpha| \gtrapprox \widetilde{c}_0$

The number $\widetilde{c}_0$ is standard because $\widetilde{\Theta}_0$ is standard and $\widetilde{c}_0 \neq \infty$; Λ is a limited set for if not $J(\xi, \widetilde{\Lambda}_0)$ is unlimited because ξ is unlimited (see the proof of lemma 10.3.2) and then $c_0(\widetilde{\Theta}_0) \geq \sup_{\xi\in\Lambda} J(\xi, \widetilde{\Lambda}_0)$ is unlimited. We shall suppose here that for each subset Θ_0 of Θ, the function $\theta \to I(\theta, \Theta_0)$ is continuous on Θ. This assumption is verified by classical exponential families. Then for any Θ_0, $c_0(\Theta_0) = \sup_{\xi\in\Lambda} J(\xi, \lambda(\Theta_0))$.

For any $\Theta_0 \subset K$, if $c_0(\Theta_0) \lnapprox \frac{1}{n}|\ln\alpha|$ then $d(\alpha, \Theta_0) = c_0(\Theta_0)$ and so $\Phi_{\Theta_0} < 1$ because $A_d = \overline{\Lambda}$. If $\overline{X} \in \Lambda^0$, then $\Phi_{\Theta_0} = 0$ and if $\overline{X} \in \partial\Lambda$, either $\Phi_{\Theta_0} = 0$, or $0 < \Phi_{\Theta_0} < 1$ (there is a randomization); the problem is not simple, as the example of multinomial distributions shows (cf. [9]). We note that Kallenberg avoids this case in theorem 3.3.2 of [13].

Proposition 10.4.3 *For* $\frac{1}{n}|\ln\alpha| \gtrapprox \widetilde{c}_0$,

(i) $\Phi_0 < 1$,

(ii) if $\overline{X} \in \Lambda^0$ *then* $\Phi_0 \equiv 0$.

Proof. As Φ_0 is a non-increasing function of α, we just have to give proofs for $\frac{1}{n}|\ln\alpha| \simeq c_0$.

(i) We prove that $\exists\Theta_0 \in \mathcal{F}$, $\Phi_{\Theta_0} < 1$. If not, $\forall\Theta_0 \in \mathcal{F}$, $\max\Phi_{\Theta_0} = 1$ (which means that Φ_{Θ_0} takes value 1), and by Cauchy's principle $\exists\widehat{\Theta}_0 \supset \mathrm{hal}(\widetilde{\Theta}_0)$, such that $\max\Phi_{\widehat{\Theta}_0} = 1$. According to lemma 10.3.1 (iii), for any $\xi \in \Lambda$, $J({}^o\xi, {}^o\widehat{\Lambda}_0) < J({}^o\xi, {}^o\widetilde{\Lambda}_0)$ because $\mathrm{hal}(\widetilde{\Lambda}_0) \subset \widehat{\Lambda}_0 := \lambda(\widehat{\Theta}_0)$. Both these numbers being standard, the relation

$$J(\xi, \widehat{\Lambda}_0) \simeq J({}^o\xi, {}^o\widehat{\Lambda}_0) < J({}^o\xi, {}^o\widetilde{\Lambda}_0) \simeq J(\xi, \widetilde{\Lambda}_0)$$

implies $J(\xi, \widehat{\Lambda}_0) \lnapprox J(\xi, \widetilde{\Lambda}_0)$ and then $c_0(\widehat{\Theta}_0) \lnapprox c_0(\widetilde{\Theta}_0)$. So $c_0(\widehat{\Theta}_0) \lnapprox \frac{|\ln\alpha|}{n}$ which contradicts $\max\Phi_{\widehat{\Theta}_0} = 1$.

(ii) Let $\overline{X} \in \Lambda^0$; if $\forall\Theta_0 \in \mathcal{F}, \Phi_{\Theta_0} \neq 0$, by Cauchy's principle $\exists\widehat{\Theta}_0 \supset \mathrm{hal}(\widetilde{\Theta}_0)$ such that $\Phi_{\widehat{\Theta}_0} \neq 0$. But $c_0(\widehat{\Theta}_0) \lnapprox \widetilde{c}_0 \simeq \frac{|\ln\alpha|}{n}$ and so, as $\overline{X} \in \Lambda^0$, $\Phi_{\widehat{\Theta}_0} = 0$, which contradicts $\Phi_{\widehat{\Theta}_0} \neq 0$. □

Remark 10.4.2 *Let* θ *be a limited element of* $\Theta \setminus (hal(\partial\Theta) \cup (\widetilde{\Theta}_0))$*; the nonstandard version of the law of large numbers yields* $PR_\theta^n(\overline{X} \in hal(\lambda(\theta))) = 1 + \emptyset$ *and thus* $PR_\theta^n(\overline{X} \in \Lambda^0) = 1 + \emptyset$*. So, for this* θ*, the risk of the second kind is equal to* $1 + \emptyset$*: if* $\frac{1}{n}|\ln\alpha| \gtrsim \widetilde{c}_0$*, the NSLRT is not consistent (in the classical* *theory, a sequence of tests* $(\varphi_n)_{n\geq n_0}$ *[where n is the size of the sample] is consistent if for any* $\theta \in \Theta_1$, $\lim_{n\to\infty} E_\theta(1-\varphi_n) = 0$*: then, for* $n \simeq \infty$*, the risk of the second kind is infinitesimal).*

Remark 10.4.3 *As the example of multinomial distributions shows (cf. [9]), it seems impossible to give general results about the value of* Φ_0 *if* $\overline{X} \in \partial\Lambda$*. We can only say that* Φ_0 *is randomized if* $\overline{X}$ *belongs to a subset of* $\partial\Lambda$ *and takes the value 0 if* $\overline{X}$ *belongs to the complement of this subset. This subset can be empty (as shown in [9]), and then* $\Phi_0 \equiv 0$ *in any case.*

10.5 Comparison with nonstandard tests based on $\overline{X}$

In the classical theory, for testing $\theta \in \Theta_0$ against $\theta \in \Theta_1$, one fixes the level α and one tries to find a test that has maximum power in any $\theta \in \Theta_1$. However, such uniformly most powerful (U.M.P.) tests exist only in a few exceptional cases: for example for 1-dimensional exponential families, if $\Theta_0 := \{\theta \in \Theta : \theta \leq \theta_0\}$. If no U.M.P. exists, one studies an asymptotic approach, for sample sizes tending to infinity. One of the tools to compare asymptotically the power of two tests of the same hypothesis is the *Bahadur relative efficiency.* In a more general framework (not only for exponential families) Bahadur ([1] or [2], see also [12]) has proved that the likelihood ratio test is at last as "efficient" as any test based on a rejection criterion of the form $T > c$ (with randomization if $T = c$ if necessary), where T is a statistic built with the n-sampling. However, it may be that to get the same power in a point $\theta \in \Theta \setminus \Theta_0$ the size of the sampling for the likelihood ratio test has to be larger than for the test based on T: thus was introduced the notion of "Bahadur deficiency" (cf. [13]).

We shall prove that the NSLRT is uniformly the most powerful in a large family of nonstandard tests if $\alpha \lesssim\!\!\!\!\not\simeq 1$ and $\frac{\ln\alpha}{n} \lesssim\!\!\!\!\not\simeq \widetilde{c}_0$. The standard likelihood ratio test is not as powerful in the corresponding standard family.

Let (φ_{Θ_0}) be a family of tests of level α (where φ_{Θ_0} tests $H_{\Theta_0} : \theta \in \Theta_0$ against $\theta \notin \Theta_0$). The corresponding level-α nonstandard test φ_0 is defined by

$$\varphi_0 :\equiv \inf\{\varphi_{\Theta_0} : \Theta_0 \in \mathcal{F}\}$$

and when we shall refer to a "nonstandard test", it will be defined thusly.

10.5.1 Regular nonstandard tests

Suppose that for each Θ_0, φ_{Θ_0} is a test of size α defined by a rejection set $R := R(\alpha, \Theta_0)$ such that

$$\sup_{\theta\in\Theta_0} P_\theta^n(\overline{X} \in R) \leq \alpha \leq \sup_{\theta\in\Theta_0} P_\theta^n(\overline{X} \in R \cup \widehat{\partial} R).$$

If R is n-regular in any point of $\widehat{\partial}R$ and if $\widehat{\partial}R$ is limited, φ_{Θ_0} will be called "regular". The most famous example of such a test is the chi-squared test since the relative frequency F is in fact the sample mean for the multivariate distribution. In order to get $R \neq \emptyset$ we shall suppose that $\frac{1}{n}|\ln\alpha| \lessapprox \widetilde{c}_0$ and so $\frac{1}{n}|\ln\alpha| \lessapprox c_0(\Theta_0)$ for $\Theta_0 \in \mathcal{F}$. If φ_{Θ_0} is regular for any $\Theta_0 \in \mathcal{F}$, φ_0 will be said to be "regular". We prove now that Φ_0 is more powerful than any regular level-α nonstandard test φ_0 in the following sense: $\Phi_0 \geq \varphi_0$. (For standard tests Φ and φ, it is clear that if $\Phi \geq \varphi$, then Φ is more powerful than φ —in the classical sense— because $E_\theta\Phi \geq E_\theta\varphi$ for any θ.)

Proposition 10.5.1 *For $\alpha \lessapprox 1$ and $\frac{1}{n}|\ln\alpha| \lessapprox \widetilde{c}_0$, if φ_0 is a regular level-α nonstandard test, then $\Phi_0 \geq \varphi_0$.*

Proof. As n is illimited and as R is n-regular at any point of $\widehat{\partial}R$ which is limited, we can write

$$\frac{1}{n}\ln P_\theta^n(\overline{X} \in R) \simeq \frac{1}{n}\ln P_\theta^n(\overline{X} \in \overline{R}) \simeq -J(\widehat{\partial}R, \lambda(\theta)),$$

so $\inf_{\theta\in\Theta_0} J(\hat{\partial}R, \lambda(\theta)) =: J(\hat{\partial}R, \lambda(\Theta_0))$ verifies $J(\hat{\partial}R, \lambda(\Theta_0)) \simeq -\frac{\ln\alpha}{n}$.

Put $c := {}^o\!\left(\frac{|\ln\alpha|}{n}\right)$. The region where $\varphi_0 \neq 0$ is included in the external subset $R_{\varphi_0} := \bigcap_{\Theta_0\in\mathcal{F}} R(\alpha, \Theta_0)$. We prove that $R_{\varphi_0} \subset \{\xi \in \overline{\Lambda}, J(\xi, \widetilde{\Lambda}_0) \gtrapprox c\} \equiv \{\Phi_0 \equiv 1\}$ which will prove that $\Phi_0 \geq \varphi_0$ (we remember that Φ_0 takes only values 0 and 1).

Let $\xi_0 \in \overline{\Lambda}$ such that $J(\xi_0, \widetilde{\Lambda}_0) \lesssim c$; we now claim that $\exists\Theta_0 \in \mathcal{F}$, $\xi_0 \notin R(\alpha, \Theta_0)$. Indeed, suppose that $\forall\Theta_0 \in \mathcal{F}$, $\xi_0 \in R(\alpha, \Theta_0)$. Then, by Cauchy's principle $\exists\widehat{\Theta}_0 \supset \mathrm{hal}(\widetilde{\Theta}_0)$ such that $\xi_0 \in R(\alpha, \widehat{\Theta}_0)$. Denote $\theta_0 := \lambda^{-1}(\xi_0)$.

• If $\xi_0 \in \mathrm{hal}(\widetilde{\Lambda}_0)$, then $\theta_0 \in \widehat{\Theta}_0$ since λ is a standard diffeomorphism from Θ^0 onto Λ^0.

Suppose $\rho_0 := d(\xi_0, \hat{\partial}R(\alpha, \widehat{\Theta}_0)) > \frac{\mathcal{L}}{\sqrt{n}}$ and so $B(\xi_0, \rho_0) \cap \overline{\Lambda} \subset R(\alpha, \widehat{\Theta}_0)$. As $\overline{\Lambda}$ contains the support of P_{θ_0}, the nonstandard law of large numbers yields $P_{\theta_0}^n(\overline{X} \in R(\alpha, \hat{\Theta}_0)) = 1 + \emptyset$ which is impossible because $P_{\theta_0}^n(\overline{X} \in R(\alpha, \hat{\Theta}_0)) \leq \alpha$.

Suppose then $\rho_0 \leq \frac{\mathcal{L}}{\sqrt{n}}$. Let $\xi_1 \in \hat{\partial}R(\alpha, \widehat{\Theta}_0) \cap \mathrm{hal}(\xi_0)$; the n-regularity at ξ_1 implies that $\exists\, \xi_2 \in \mathrm{hal}(\xi_1)$, $\exists\rho \in \left]\frac{\mathcal{L}}{\sqrt{n}}, \emptyset\right]$ such that $B(\xi_2, \rho) \subset R(\alpha, \widehat{\Theta}_0)$. As $d(\xi_0, \xi_2) \simeq 0$, $\xi_2 \in \lambda(\widehat{\Theta}_0)$ and $\theta_2 := \lambda^{-1}(\xi_2) \in \widehat{\Theta}_0$. As previously, $P_{\theta_2}^n(\overline{X} \in R(\alpha, \hat{\Theta}_0)) = 1 + \emptyset$ which is impossible because $P_{\theta_2}^n(\overline{X} \in R(\alpha, \hat{\Theta}_0)) \leq \alpha$.

• If $\xi_0 \notin \mathrm{hal}(\widetilde{\Lambda}_0)$, we can write

$$\begin{aligned} J(\xi_0, \lambda(\widehat{\Theta}_0)) &\simeq J(\xi_0, \lambda({}^o\widehat{\Theta}_0)) \simeq J({}^o\xi_0, \lambda({}^o\widehat{\Theta}_0)) \\ &< J({}^o\xi_0, \lambda(\widetilde{\Theta}_0)) \simeq J(\xi_0, \lambda(\widetilde{\Theta}_0)) \\ &\lesssim c. \end{aligned}$$

But $J({}^o\xi_0, \lambda({}^o\widehat{\Theta}_0))$ and $J({}^o\xi_0, \lambda(\widetilde{\Theta}_0)) \neq 0$ are both standard numbers; as $\lambda(\widetilde{\Theta}_0) \subset (\lambda({}^o\widehat{\Theta}_0))^0$ we have $J({}^o\xi_0, \lambda({}^o\widehat{\Theta}_0)) < J({}^o\xi_0, \lambda(\widetilde{\Theta}_0))$ and so $J(\xi_0, \lambda(\hat{\Theta}_0)) \lesssim c$.

On the other hand $J(R(\alpha, \hat{\Theta}_0), \lambda(\hat{\Theta}_0)) \simeq -\frac{\ln \alpha}{n} \simeq c$: this contradicts $\xi_0 \in R(\alpha, \hat{\Theta}_0)$. □

10.5.2 Case when $\widetilde{\Theta}_0$ is convex

For any $\Theta_0 \in \mathcal{F}$ and any $\theta_1 \in \Theta \setminus \mathrm{hal}(\widetilde{\Theta}_0)$, there is a level-$\alpha$ test $\Psi_{\Theta_0}^{\theta_1}$ which is the most powerful at θ_1 (for any level-α test φ testing $\theta \in \Theta_0$ against $\theta \in \Theta_1$, $E_{\theta_1}\varphi \leq E_{\theta_1}\Psi_{\Theta_0}^{\theta_1}$). This test is defined in the following way (see [13, p. 49]) let

$$f(x) := \int_{\overline{\Theta}_0} \exp\Big(n\big((\theta_0 - \theta_1 \mid x) - \psi(\theta_0) + \psi(\theta_1)\big)\Big)\tau(d\theta_0)$$

where τ is the least favorable distribution. Then $\Psi_{\Theta_0}^{\theta_1} = 1$ if $f(\overline{X}) < C$ and $\Psi_{\Theta_0}^{\theta_1} = 0$ if $f(\overline{X}) > C$ where C is a constant depending on α, Θ_0, θ_1. So, the rejection set (relatively to $\overline{X}$) is $\big\{\xi \in \overline{\Lambda} : f(\xi) < C\big\}$.

Definition 10.5.1 *For* $\theta_1 \in \Theta \setminus \mathrm{hal}(\widetilde{\Theta}_0)$, *let* $\Psi_0^{\theta_1} :\equiv \inf\big\{\Psi_{\Theta_0}^{\theta_1} : \Theta_0 \in \mathcal{F}\big\}$.

We might think that $\Psi_0^{\theta_1}$ is "the most powerful at θ_1 nonstandard level-α test of H_0"; but in fact, in some cases, Φ_0 is more powerful than $\Psi_{\Theta_0}^{\theta_1}$:

Proposition 10.5.2 *If* $\widetilde{\Theta}_0$ *is convex, then for any* $\theta_1 \in \Theta \setminus \mathrm{hal}(\widetilde{\Theta}_0)$, $\Phi_0 \geq \Psi_{\Theta_0}^{\theta_1}$.

Proof. Let $\theta_1 \in \Theta \setminus \mathrm{hal}(\widetilde{\Theta}_0)$. We prove that the rejection sets

$$A := \big\{\xi \in \overline{\Lambda} :\ f(\xi) < C\big\}$$

are n-regular and then the proposition 10.5.1 yields the result.

The set $\widetilde{\Theta}_0$ is convex and $d(\theta_1, \widetilde{\Theta}_0) \not\simeq 0$ and so exists a standard cone $\Delta \subset \mathbb{R}^k$ and a standard positive number $\varepsilon > 0$ such that if $u \in \Delta$ and $\|u\| = 1$ then $(\theta_0 - \theta_1 \mid u) < -\varepsilon$ for any $\theta_0 \in \widetilde{\Theta}_0$. Therefore, for any $\Theta_0 \in \mathcal{F}$, for any $\theta \in \Theta_0$ and for any $u \in \Delta \setminus \{0\}$, $(\theta_0 - \theta_1 \mid u) < 0$.

Choose $\Theta_0 \in \mathcal{F}$. If $u \in \Delta \setminus \{0\}$,

$$f(x+u) := \int_{\overline{\Theta}_0} \exp\big(n(\theta_0 - \theta_1 \mid u)\big)\cdot$$
$$\cdot \exp\Big(n\big((\theta_0 - \theta_1 \mid x) - \psi(\theta_0) + \psi(\theta_1)\big)\Big)\tau(d\theta_0) < f(x).$$

Let now $a \in \partial A$, then $f(a) = C$. For any $u \in \Delta \setminus \{0\}$, $f(a+u) < C$, so $a + (\Delta \setminus \{0\}) \subset A$. It is now possible to choose a point $b \in a + \Delta$ such that $d(a,b) = \frac{1}{n}$ (for example) and such that $B(b, \frac{1}{n^{3/2}}) \subset a + \Delta$. If $a \notin \text{hal}(\partial\Lambda)$, this proves that A is n-regular in a. If $a \in \text{hal}(\partial\Lambda)$, we have to use Cauchy's principle like at the end of the proof of proposition 10.3.1 □

References

[1] R. R. BAHADUR, An optimal property of likelihood ratio statistic, in *Proc. Fifth Berkeley Symp. Math. Statist. Prob.*, Vol. 1, 1965.

[2] R. R. BAHADUR *Some limit theorems in statistics*, SIAM, Philadelphia, 1971.

[3] R.R. BAHADUR and S.L. ZABELL, "Large deviations of the sample mean in general vector spaces", Annals of Probability, **31** (1979) 587–621.

[4] I. VAN DEN BERG, An external probability order theorem with applications, in F. and M. Diener editors, *Non Standard Analysis in Practice*, Springer-Verlag, Universitext, 1995.

[5] J. BOSGIRAUD, "Exemple de test statistique non standard, risque externe", Pub. Inst. Stat. Paris, **41** (1997) 85–95.

[6] J. BOSGIRAUD, "Tests statistiques non standard sur des proportions", Pub. Inst. Stat. Paris, **44** (2000) 3–12.

[7] J. BOSGIRAUD, "Tests statistiques non standard pour les modèles à rapport de vraisemblance monotone (première partie)", Pub. Inst. Stat. Paris, **44** (2000) 87–101

[8] J. BOSGIRAUD, "Tests statistiques non standard pour les modèles à rapport de vraisemblance monotone (seconde partie)", Pub. Inst. Stat. Paris, **45** (2001) 61–78.

[9] J. BOSGIRAUD, "Nonstandard chi-squared test", Journal of Information & Optimization Sciences, **26** (2005) 443–470.

[10] L.D. BROWN, *Fundamentals of Statistical Exponential Families*, Institute of Mathematical Statistics, Hayward, California, 1986.

[11] F. DIENER and G. REEB, *Analyse non standard*, Hermann, Paris, 1989.

[12] P. GROENEBOOM and J. OOSTERHOFF, "Bahadur efficiency and probabilities of large deviations", Statistica Neerlandica, **31** (1977) 1–24.

[13] W.C.M. KALLENBERG, *Asymptotic Optimality of likehood ratio tests in exponential families*, Mathematisch Centrum, Amsterdam, 1978.

[14] F. KOUDJETI and I. VAN DEN BERG, Neutrices, external numbers and external calculus, in F. et M. Diener editors: *Non Standard Analysis in Practice*, Springer-Verlag, Universitext, 1995.

[15] E. NELSON, *Radically Elementary Probability Theory*, Princeton University Press, 1987.

[16] S.R.S. VARADHAN, *Large Deviations and Applications*, SIAM, Philadelphia, 1984.

11

A finitary approach for the representation of the infinitesimal generator of a markovian semigroup

Schérazade Benhabib*

Abstract

This work is based on Nelson's paper [1], where the central question was: under suitable regularity conditions, what is the form of the infinitesimal generator of a Markov semigroup?

In the elementary approach using IST [2], the idea is to replace the continuous state space, such as $\mathbb{R}$ with a finite state space X possibly containing an unlimited number of points. The topology on X arises naturally from the probability theory. For $x \in X$, let $\mathcal{I}_x$ be the set of all $h \in \mathcal{M}$ vanishing at x where $\mathcal{M}$ is the multiplier algebra of the domain $\mathcal{D}$ of the infinitesimal generator. To describe the structure of the semigroup generator A, we want to split $Ah(x) = \sum_{y \in X \setminus \{x\}} a(x,y)\ h(y)$ so that the contribution of the external set F_x of the points far from x appears separately. A definition of the quantity $\alpha_{ah}(x) = \sum_{y \in F} a(x,y)\ h(y)$ is given using the least upper bound of the sums on all internal sets W included in the external set F. This leads to the characterization of the global part of the infinitesimal generator.

11.1 Introduction

Let X be a finite state space possibly containing an unlimited number N of points. For all x in X, let $a(x,y)$ be a real number for each y on X such that $a(x,y)$ is positive if $y \neq x$ and $\sum_y a(x,y) = 0$. Then $A = (a(x,y))$ is a finite $N \times N$ matrix, with positive off-diagonal elements and row sums 0. Thus $P^t = \sum_n \frac{t^n A^n}{n!}$ exists and is a markovian semigroup with infinitesimal generator A.

*EIGSI, 17041 La Rochelle, France.
scherazade.benhabib@eigsi.fr

Then we can define externally the domain of definition of A and its multiplier algebra as follows.

Definition 11.1.1 *The domain $\mathcal{D}$ of definition of the infinitesimal generator A is the external set*

$$\mathcal{D} = \{f \in \mathbb{R}^X / \ f \text{ limited and } Af \text{ limited}\} \tag{11.1.1}$$

And its multiplier algebra is the external set

$$\mathcal{M} = \{f \in \mathcal{D} / \ \forall g \in \mathcal{D} \ fg \in \mathcal{D}\} \tag{11.1.2}$$

The topology on X arises naturally from the probability theory and a proximity relation is defined on X, more particulary from the way the functions in the multiplier algebra $\mathcal{M}$ of the domain $\mathcal{D}$ of the infinitesimal generator act on X.

Definition 11.1.2 *Let u and v be in X, we say that u is close to v ($u \simeq v$) if and only if*

$$\forall h \in \mathcal{M} \ \ h(u) \simeq h(v)$$

Thus, the elements of $\mathcal{M}$ have a macroscopic property of continuity formally analogous to the S-continuity.

With this relation, we define the external sets

$$C_x = \{y \in X : \ y \simeq x\} \tag{11.1.3}$$

of points close to x and

$$F_x = \{y \in X : \ y \not\simeq x\} \tag{11.1.4}$$

of those far from x.

For each x in X, $(a(x,y))$ are positive real numbers if $x \neq y$ and can take unlimited values. Let $\mathcal{I}_x$ be the set of all h in $\mathcal{M}$ vanishing at x, and for h in $\mathcal{I}_x$ let us define

$$Ah(x) = \sum_{y \in X \setminus \{x\}} a(x,y) \ h(y) \tag{11.1.5}$$

To describe the structure of the semigroup generator, we want to split $Ah(x)$ so that the contribution of the points far from x in X, appears separately.

One is tempted to define directly the quantity

$$\alpha_{ah}(x) = \sum_{y \in F_x} a(x,y)h(y) \tag{11.1.6}$$

which alas has no meaning because of the external character of the set of indices F_x

Nevertheless, we will show in Section 11.2 how we can attach a definite meaning to it. That leads in Section 11.3 to a definition for an α_x-integrable function. Theorem 11.3.1 establishes the representation of $Ah(x)$ for functions in $\mathcal{M}$ which have a zero at x of the third order. This is the global or integral part of A. Theorem 11.3.2 and its corollary extend the notion of α_x-integrable to functions which have a zero of the first or the second order.

11.2 Construction of the least upper bound of sums in IST

Lemma 11.2.1 *Let E be an external subset of X and f a positive function defined on X. If there exists a standard natural number n_0 such that for all internal sets W contained in E we have*

$$\sum_{y\in W} f(y) \leq n_0 \tag{11.2.1}$$

then there exists a unique up to an infinitesimal least number $\alpha(f)$ such that

$$\sum_{y\in W} f(y) \lesssim \alpha(f) \tag{11.2.2}$$

for all internal sets W contained in E.

Proof. Let n_0 be a standard natural number such that for all internal sets W contained in E we have (11.2.1).

By external induction, there is a least n_0 such that (11.2.1) holds. Now for each standard natural k there is a largest natural number j with $0 \leq j \leq 2^k$ such that for all such W

$$\sum_{y\in W} f(y) \leq n_0 - \frac{j}{2^k} \tag{11.2.3}$$

Let $\alpha(k) = n_0 - \frac{j}{2^k}$. Then for each standard k we have a standard $\alpha(k)$, so there is a standard sequence $k \to \alpha(k)$. This sequence is decreasing, therefore it has a standard limit $\alpha(f)$. Moreover, for all standard k and all such W, $\left(\alpha(f) - \sum_{y\in W} f(y)\right) \leq \frac{1}{2^k}$. □

This proof uses very elementary tools. It can easily be formulated in the minimal non-standard analysis introduced by Nelson in [3] in which "standard"

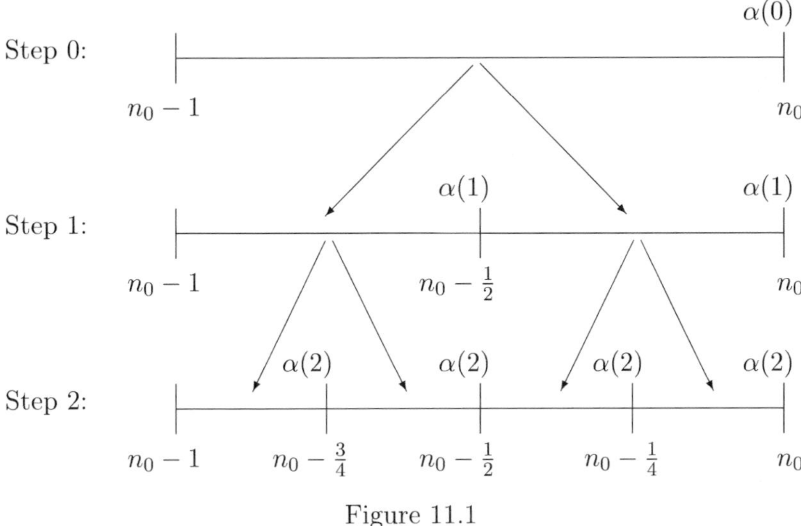

Figure 11.1

applies only to natural numbers. On the contrary, if all the force of IST is used, one could check that

$$\alpha(f) = \sup{}^{S}\left\{x \in \mathbb{R}^+;\ \exists^{int} W \subset E \sum_{y \in W} f(y) \lesssim x\right\}.$$

Figure 11.1 exhibits the way the sequence is built.

Definition 11.2.1

1. *If f is a positive function defined on X, and if the conditions required for Lemma 11.2.1 are verified the quantity*

$$\alpha_f = \sum_{y \in E} f(y) \tag{11.2.4}$$

is defined to be the standard limit $\alpha(f)$.

2. *If the conditions for Lemma 11.2.1 are not verified, let α_f be the formal symbol ∞ and say that (11.2.2) is not defined.*

3. *In the general case, let $f = f^+ - f^-$ be the decomposition of f into the difference of two positive functions. Say that α_f is defined in case both α_{f^+} and α_{f^-} are defined, and let α_f be their difference:*

$$\alpha_f = \alpha_{f^+} - \alpha_{f^-}$$

Notation 11.2.1 *For each x in X, if f depends on x, write $\alpha_f(x)$ instead of α_f.*

11.3 The global part of the infinitesimal generator

Definition 11.3.1 *For $x \in X$ and $h \in \mathcal{M}$, let $f = a(x, \cdot)h$, then h is said to be α_x-integrable if and only if $\alpha_f(x)$ is defined.*

Notation 11.3.1 *In that case write $\alpha_{ah}(x)$ instead of $\alpha_f(x)$.*

Definition 11.3.2

1. *h is a function in $\mathcal{I}_x$ if and only if $h \in \mathcal{M}$ and $h(x) = 0$.*

2. *h is a function in $\mathcal{I}_x^2$ if and only if $h \in \mathcal{M}$ and $h = \sum_{k=1}^{p} e_k\, g_k$ with p limited and e_k, g_k in $\mathcal{I}_x$.*

3. *h is a function in $\mathcal{I}_x^3$ if and only if $h \in \mathcal{M}$ and is of the form $h = \sum_1^p e_k f_k g_k$ with $e_k, f_k, g_k \in \mathcal{I}_x$ and p limited.*

4. *We say that h has a zero at x of the first order when $h \in \mathcal{I}_x$, of the second order when $h \in \mathcal{I}_x^2$, of the third order when $h \in \mathcal{I}_x^3$.*

Theorem 11.3.1 *If h has a zero at x of the third order, then h is α_x-integrable and $Ah(x)$ is infinitely close to $\alpha_{ah}(x)$.*

Proof. The proof goes through the following three steps:

1. Let h be in $\mathcal{I}_x^3$, therefore it is of the form $h = \sum_1^q f_k^2\, g_k$ with q limited and f_k, g_k in $\mathcal{I}_x$.

2. If $f, g \in \mathcal{I}_x$ then $f^2 g$ is α_x-integrable and so is h.

3. $A(f^2 g)(x) \simeq \alpha_{a(f^2 g)}(x)$ thus $Ah(x) \simeq \alpha_{ah}(x)$.

These steps are proved as follows:

1. Let $h \in \mathcal{I}_x^3$, as

$$efg = \frac{1}{4}\left(e + f\right)^2 g - \frac{1}{2}e^2 g - \frac{1}{2}f^2 g,$$

therefore h is of the mentioned form.

2. Let $f^2 g = \left(f^2 g^+\right) - \left(f^2 g^-\right)$ and W be an internal subset of F_x

$$\left|\sum_W a(x,y) f^2(y) g^+(y)\right| \leq ||g|| A(f^2)(x)$$

As $||g||$ and $A(f^2)$ are limited, they are both infinitely close to a standard number. Taking the integer part of $°(||g||A(f^2)(x))$ plus 1, we get the existence of the natural number n_0 required by Lemma 11.2.1.

The same argument applies to f^2g^-. Thus $\alpha_{a(f^2g^+)}(x)$ and $\alpha_{a(f^2g^-)}(x)$ exist and as $\alpha_{a(f^2g)}(x) = \alpha_{a(f^2g^+)}(x) - \alpha_{a(f^2g^-)}(x)$, then f^2g is α_x-integrable, and so is h.

3. Let $\beta \gg 0$. Since $g \in \mathcal{I}_x$ there is an internal set W of F_x such that $g^+(y) \leq \beta$ for $y \in W^C$. We have

$$\begin{aligned}
\left|A(f^2g^+)(x) - \alpha_{a(f^2g^+)}\right| &= \left|\sum_{y\in X\backslash\{x\}} a(x,y)f^2(y)g^+(y) - \alpha_{a(f^2g^+)}\right| \\
&\leq \sum_{y\in X\backslash\{x\}} a(x,y)f^2(y)g^+(y) - \sum_{y\in W} a(x,y)f^2(y)g^+(y) \\
&= \sum_{y\in W^C\backslash\{x\}} a(x,y)f^2(y)g^+(y) \\
&\leq \beta \sum_{y\in W^C\backslash\{x\}} a(x,y)f^2(y) \\
&\leq \beta \sum_{y\in X\backslash\{x\}} a(x,y)f^2(y) \\
&= \beta A(f^2)(x)
\end{aligned}$$

Since $A(f^2)$ is limited and $\beta \gg 0$ is arbitrary, $A(f^2g^+) \simeq \alpha_{a(f^2g^+)}(x)$. The same applies to f^2g^-.

Then $A(f^2g^-)(x) \simeq \alpha_{a(f^2g^-)}(x)$ and $\alpha_{a(f^2g)} = \alpha_{a(f^2g^+)}(x) + \alpha_{a(f^2g^-)}(x)$. Thus $A(f^2g)(x) \simeq \alpha_{a(f^2g)}(x)$. Therefore $A(h)(x) \simeq \alpha_{(ah)}(x)$. □

Theorem 11.3.2 *If h is positive and has a zero at x of the first order, then h is α_x-integrable and α_h is less or infinitely close to $Ah(x)$.*

Proof. Let W be any internal subset of F_x and h a positive function in $\mathcal{I}_x^+$, so

$$0 \leq \sum_W a(x,y)h(y) \leq A(h)(x)$$

As $A(h)(x)$ is limited, we argue as in Theorem 11.3.1 and apply Lemma 11.2.1. One concludes that h is α_x-integrable.

As the inequality above holds for all internal subset W of F_x, we get that $\alpha_{ah}(x)$ is less or infinitely close to $A(h)(x)$. □

Corollary 11.3.1 *If h has a zero at x of the second order, then h is α_x-integrable.*

Proof. Let h be a function in $\mathcal{I}_x^2$, as

$$fg = \frac{1}{4}(e+g)^2 - \frac{1}{4}(e-g)^2$$

h is also of the form $h = \sum_{k=1}^{q} \beta_k f_k^2$ with $\beta_k = \pm 1$, q limited and f_k^2 in $\mathcal{I}_x^+$. Consequently we can apply Theorem 11.3.2 and h is α_x-integrable. □

11.4 Remarks

The next goal will be to find an IST construction of the pure diffusion part of the generator of the semi-group and to complete a work already in progress concerning the rescaling of the time parameter for the markovian semigroup P^t.

Acknowledgment *I would like to thank Professor E.Nelson for the initial suggestion of this problem and numerous enlightening discussions. I would also like to thank Professor G. Wallet of the University of la Rochelle for his support.*

References

[1] E. NELSON, "Representation of a Markovian Semi-group and Its Infinitesimal Generator", Journal of Mathematics and Mechanics, **7** (1958) 997–988.

[2] E. NELSON, "Internal Set Theory: a new approach to nonstandard analysis", Bulletin Amer. Math. Soc., **83** (1977) 1165–1198.

[3] E. NELSON, *Radically Elementary Probability Theory*, Princeton University Press, 1987

[4] E. B. DYNKIN, *Markov Processes*, Springer-Verlag, 2 vols., 1965.

[5] W. FELLER, *An Introduction to Probability Theory and Its Applications*, John Wiley & Sons.

[6] I. VAN DEN BERG, *Non Standard Asymptotic Analysis*, Lectures Notes in Mathematics, 1249, Springer-Verlag, 1987.

12

On two recent applications of nonstandard analysis to the theory of financial markets

Frederik S. Herzberg*

Abstract

Suitable notions of "unfairness" that measure how far an empirical discounted asset price process is from being a martingale are introduced for complete and incomplete-market settings. Several limit processes are involved each time, prompting a nonstandard approach to the analysis of this concept. This leads to an existence result for a "fairest price measure" (rather than a martingale measure) for an asset that is simultaneously traded on several stock exchanges. This approach also proves useful when describing the impact of a currency transaction tax.

12.1 Introduction

Nonstandard analysis is most successful when it concerns itself with mathematical concepts that involve several limit processes in a non-trivial way. One such example is the quantification of a stochastic process's distance from being a martingale.

Given the well-known correspondence between the existence of a martingale measure for a discounted asset price process and the non-existence of arbitrage opportunities, there is a natural economic interpretation to the comparison of two positive stochastic processes in regard to "how far" they actually are from being a martingale. We will render two economic questions related to such-conceived "unfairness" mathematical: (1) Is there a "fairest price measure" (rather than a martingale measure) for an asset that is simultaneously traded at multiple stock exchanges? (2) Does a currency transaction tax imposed on a currency that is subject to herd behaviour among traders have a fairness-

*Mathematical Institute, University of Oxford, Oxford OX1 3LB, United Kingdom. herzberg@maths.ox.ac.uk / herzberg@wiener.iam.uni-bonn.de enhancing impact on the currency exchange rates preceding a financial crash? Both questions will be shown to allow for an affirmative answer.

Techniques from nonstandard analysis have already been successfully applied to mathematical finance in the work of Cutland, Kopp and Willinger [5]. This contribution is based on two more recent papers by the author [8, 9].

12.2 A fair price for a multiply traded asset

In this Section, we will *en passant* motivate notions of "unfairness" for both complete-market as well as incomplete-market settings.

Theorem 12.2.1 *Let $n \in \mathbb{N}$, $p > 0$ and $c, N > 0$. Consider an adapted probability space $(\Gamma, \mathcal{G}, \mathbb{P})$ and $\mathcal{B}[0,1] \otimes \mathcal{G}_1$-measurable geometric Brownian motions with constant multiplicative drift $\tilde{g}_i : \Gamma \times [0,1] \to \mathbb{R}^{d \cdot n}$, $i \in \{1, \ldots, n\}$. Then there exist processes g_i, $i \in \{1, \ldots, n\}$ on an adapted probability space $(\Omega, \mathcal{F}, \mathbb{Q})$, equivalent to the processes $\tilde{g}_i$ on Γ (in the sense of adapted equivalence [7]), such that there is a probability measure $\mathbb{M}_m$ on Ω in the class of measures*

$$\mathcal{C}(\Omega, g) := \left\{ Q : \mathcal{F}_1 \to [0,1] : \begin{array}{rl} & Q \text{ probability measure}, \\ \forall A \in \mathcal{F}_1 & \frac{1}{N} \cdot \mathbb{Q}(A) \leq Q(A) \leq N \cdot \mathbb{Q}(A), \\ \forall i \neq j \in \{1, \ldots, n\} & \int_0^1 \frac{\mathrm{Cov}_Q\left((g_i)_s, (g_j)_s\right)}{\mathbb{E}_Q\left|(g_i g_j)_s\right|} ds \geq c \end{array} \right\}$$

minimising

$$Q \mapsto m_\Omega(Q, g) := \sum_{i=1}^n \int_0^1 \int_s^1 \int_\Omega \left| (g_i)_s - \mathbb{E}_Q \left[(g_i)_t \,|\mathcal{F}_s \right] \right|^p dQ \; dt \; ds$$

in the class of Loeb extensions of finitely additive measures from $({}^*\mathcal{C})(\Omega, G)$ (G being an arbitrary lifting of g).

Analogously, there is a probability measure $\mathbb{M}_n$ minimising

$$Q \mapsto n_\Omega(Q, g) = \int_0^1 \mathbb{E}_\mathbb{Q} \left| \frac{1}{g_t} \frac{d}{du}\Big|_{u=0} \mathbb{E}\left[g_{t+u} \mid \mathcal{G}_t \right] \right| dt$$

(where $n_\Omega(Q, g)$ is defined to be $+\infty$ if the derivative in 0 in this definition does not a.s. exist as a continuous function in t) in the class of Loeb extensions of finitely additive measures from $({}^*\mathcal{C})(\Omega, G)$. Moreover,

$$\inf_{\mathcal{C}(\Omega, g)} m_\Omega(\cdot, g) \leq \inf_{\mathcal{C}(\Gamma, \tilde{g})} m_\Gamma(\cdot, \tilde{g})$$

as well as

$$\inf_{\mathcal{C}(\Omega, g)} n_\Omega(\cdot, g) \leq \inf_{\mathcal{C}(\Gamma, \tilde{g})} n_\Gamma(\cdot, \tilde{g}) .$$

Remark 12.2.1 *The original paper [8] does not state the Theorem and the subsequent Lemmas as precisely as it is done here, although the exact result becomes clear from the proofs in that paper.*

The proof for this Theorem can be split into the following Lemmas which might also be interesting in their own right.

Lemma 12.2.1 *Using the notation of the previous Theorem 12.2.1, for any hyperfinite adapted space Ω [12],*

$$\inf_{\mathcal{C}(\Omega,g)} m_\Omega(\cdot,g) \leq \inf_{\mathcal{C}(\Gamma,\tilde{g})} m_\Gamma(\cdot,\tilde{g}).$$

Lemma 12.2.2 *Under the assumptions of Theorem 12.2.1 and choosing Ω to be any hyperfinite adapted space, the infimum of $m_\Omega(\cdot,g)$ in the class of Loeb extensions of finitely additive measures from $({}^*\mathcal{C})(\Omega,G)$ is attained by some measure $\mathbb{M}_m \in \mathcal{C}(\Omega,g)$.*

Proofs for both of these Lemmas can be found in a recent paper by the author [8].

Although they look very similar, it is technically slightly more demanding to prove the following two Lemmas (which in turn obviously entail the second half of the Theorem, i.e. the assertions concerned with the map n).

Lemma 12.2.3 *Using the notation of the previous Theorem 12.2.1, for any hyperfinite adapted space Ω [12],*

$$\inf_{\mathcal{C}(\Omega,g)} n_\Omega(\cdot,g) \leq \inf_{\mathcal{C}(\Gamma,\tilde{g})} n_\Gamma(\cdot,\tilde{g}).$$

Lemma 12.2.4 *Under the assumptions of Theorem 12.2.1 and choosing Ω to be any hyperfinite adapted space, the infimum of $n_\Omega(\cdot,g)$ on $\mathcal{C}(\Omega,g)$ is attained by some measure $\mathbb{M}_n \in \mathcal{C}(\Omega,g)$.*

Easy results are

Lemma 12.2.5 *A semimartingale x is a P-martingale on Ω if and only if $m_\Omega(P,x)=0$. The function $m_\Omega(P,\cdot)^{\frac{1}{p}}$ on the space of measurable processes of Ω satisfies the triangle inequality and is 1-homogeneous. For $p=2$, it defines an inner product on the space*

$$\mathcal{E} := \left\{x : \Omega \times [0,1] \to \mathbb{R}^d \ : \ x \text{ measurable},\ m(P,x) < +\infty\right\}$$

which becomes a Hilbert space by this construction.

Lemma 12.2.6 *A semimartingale x is a P-martingale on Ω if and only if $n_\Omega(P,x) = 0$. The function $n_\Omega(P,\cdot)$ on the space of measurable processes of Ω remains unchanged when multiplying the argument by constants* (scaling invariance).

We can generalise this to the following

Definition 12.2.1 *Let Ω be an adapted probability space. Define*

$$\mathcal{L}(\Omega,\mathbb{R}^d) := \left\{x : \Omega\times[0,1]\to\mathbb{R}^d \ : \ x \textit{ measurable}\right\}.$$

A function $\Upsilon : \mathcal{L}\left(\Omega,\mathbb{R}^d\right) \to [0,+\infty]$ is an incomplete-market notion of unfairness *if and only if it satisfies the triangle inequality, is 1-homogeneous, and assigns 0 to a semimartingale y if and only if y is a martingale. Υ is said to be a* complete-market notion of unfairness *if and only if it remains unchanged under multiplication by constants and Υ vanishes exactly for those semimartingales that are in fact martingales.*

By Lemmas 12.2.5 and 12.2.6 respectively, the function $m_\Omega(Q,\cdot)^{\frac{1}{p}}$ (for a fixed probability measure Q) is an example for an incomplete-market notion of unfairness, whereas similarly $n_\Omega(Q,\cdot)$, again for a fixed probability measure Q, is an example for a complete-market notion of unfairness.

Remark 12.2.2 *The distinction between notions of unfairness for complete and incomplete markets can be justified by the following reasoning: If it is, under assumption of completeness, possible to buy as much of an asset as one intends to, multiplication of the discounted price process by a constant does not enhance the arising arbitrage opportunities at all; therefore, for complete markets, a suitable notion of unfairness should be scaling invariant in that it does not change under multiplication of the argument — which is conceived as being a discounted price process — by constants.*

12.3 Fairness-enhancing effects of a currency transaction tax

In this Section, we shall analyse the empirical price of an asset that is subject to herd-behaviour preceding a financial crash and how the resulting distortion can be mitigated through imposing a transaction tax. This is of particular relevance if the crashing asset in question is a currency, given the massive (usually destabilising and therefore adverse) macroeconomic consequences of such financial crises.

We shall restrict our attention to a model of currency prices where extrapolation from the observed behaviour of other traders, up to white noise and constant inflation, completely accounts for the evolution of the currency price. This is to say that, in analogy to a Nash equilibrium, we assume that every agent is acting in such a manner that he gains most if all other agents follow his pattern. In our model this pattern will consist in using some average value of past currency prices as a "sunspot", that is a proxy for a general perception that depreciation or appreciation of the particular currency in question is due.

In order to introduce the model for the pre-crash discounted currency price, consider $\alpha : \mathbb{R} \to \mathbb{R}$, a piecewise constant function, such that

$$\alpha = \alpha(1) > 0 \text{ on } (0, +\infty) \quad , \quad \alpha = \alpha(-1) < 0 \text{ on } (-\infty, 0).$$

We assume that the logarithmic discounted price process $x^{(v)}$ is governed, given some initial condition that without loss of generality can be taken to be $x^{(v)} = 0$, by the stochastic differential equation

$$\begin{aligned} dx_t^{(v)} &= \alpha \left(x_t^{(v)} - \sum_{u \in I} p_u \cdot x_{(t-u)\vee 0}^{(v)} \right) \chi_{\left\{ \left| x_t^{(v)} - \sum_{u \in I} p_u \cdot x_{(t-u)\vee 0}^{(v)} \right| \geq v \right\}} dt \\ &\quad + \sigma \cdot db_t - \frac{\sigma^2}{2} dt, \end{aligned} \tag{12.3.1}$$

where $r > 0$ is the logarithmic discount rate (assumed to be constant), b is the one-dimensional Wiener process, $(p_u)_{u \in I}$ is a convex combination — i.e. $I \subset (0, +\infty)$ is finite, $\forall u \in I \quad p_u > 0$ and $\sum_{u \in I} p_u = 1$. The parameter v depends very much on the tax rate we assume. If ρ is the logarithmic tax rate and T the expected time during which one will hold the asset (that is the expected duration of the upward or downward tendency of the stock price), one can compute v as follows:

$$v = T \cdot \rho.$$

The equation (12.3.1), is of course first of all only a formal equation that — thanks to the boundedness of α — can be made rigorous using the theory of stochastic differential equations developed by Hoover and Perkins [11] or Albeverio et al. [2] for instance.

We will introduce the following abbreviation:

$$\psi^{(v)} := \chi_{\{|\cdot| \geq v\}} \cdot \alpha.$$

12.4 How to minimize "unfairness"

Intuitively, the map $n_\Omega(P, \cdot)$ (for a fixed probability measure P on a probability space Ω) when applied to a discounted asset price process measures how often (in terms of time and probability) and how much it will be the case that one may expect to obtain a multiple (or a fraction) of one's portfolio simply by selling or buying the stock under consideration.

For the following, we will drop the first component of n (indicating the probability measure) if no ambiguity can arise; for internal hyperfinite adapted spaces and Loeb hyperfinite adapted spaces, we will assume that the canonical measure (the internal uniform counting measure and its Loeb measure, respectively) on the space is referred to.

First of all we derive a formula that will make explicitly computing n easier in our specific setting:

Lemma 12.4.1 *If $x^{(v)}$ satisfies (12.3.1) for some $v > 0$ on an adapted probability space $(\Gamma, \mathcal{G}, \mathbb{Q})$, then the discounted price process $\left(\exp\left(x_t^{(v)}\right) : t \geq 0\right)$ is of finite unfairness. More specifically,*

$$
\begin{aligned}
n_\Gamma\left(\exp\left(x^{(v)}\right)\right) &= \int_0^1 \mathbb{E}\left[\left|\psi^{(v)}\left(x_t^{(v)} - \sum_{i\in I} p_i x^{(v)}_{(t-i)\vee 0}\right)\right|\right] dt \\
&= \int_0^1 \mathbb{E}\left[\left|\frac{d}{du}\right|_{u=0} \mathbb{E}\left[x^{(v)}_{t+u}\middle|\mathcal{G}_t\right] + \frac{\sigma^2}{2}\right|\right] dt
\end{aligned}
$$

Proof. The proof is more or less a formal calculation, provided one is aware of the path-continuity of our process and the fact that the filtrations generated by b and $x^{(v)}$ are identical. For this implies that, given $t > 0$, the value

$$
\psi^{(v)}\left(x^{(v)}_{t+u} - \sum_{i\in I} p_i x^{(v)}_{(t+u-i)\vee 0}\right)(\omega)
$$

does not change within sufficiently small times u — almost surely for all those paths ω where $x_t^{(v)}(\omega) - \sum_{i\in I} p_i x^{(v)}_{(t-i)\vee 0}(\omega) \notin \{\pm v\}$, this condition itself being satisfied with probability 1. Now, using this result and the martingale property of the quotient of the exponential Brownian motion and its exponential bracket, we can deduce that for all $t > 0$ almost surely:

$$
\begin{aligned}
&\frac{1}{\exp\left(x_t^{(v)}\right)} \frac{d}{du}\bigg|_{u=0} \mathbb{E}\left[\exp\left(x^{(v)}_{t+u}\right)\middle|\mathcal{G}_t\right] \\
&\qquad = \frac{d}{du}\bigg|_{u=0} \mathbb{E}\left[\exp\left(x^{(v)}_{t+u} - x_t^{(v)}\right)\middle|\mathcal{G}_t\right] \\
&\qquad = \frac{d}{du}\bigg|_{u=0} \exp\left(\psi^{(v)}\left(x_t^{(v)} - \sum_{i\in I} p_i x^{(v)}_{(t-i)\vee 0}\right) u\right) \cdot \\
&\qquad\qquad \cdot \mathbb{E}\left[\exp\left(\sigma b_{t+u} - \frac{\sigma^2}{2}(t+u) - \left(\sigma b_t - \frac{\sigma^2}{2} t\right)\right)\middle|\mathcal{G}_t\right]
\end{aligned}
$$

$$= \frac{d}{du}\bigg|_{u=0} \exp\left(\psi^{(v)}\left(x_t^{(v)} - \sum_{i\in I} p_i x_{(t-i)\vee 0}^{(v)}\right)u\right)\cdot$$
$$\cdot \mathbb{E}\left[\exp\left(\sigma b_{t+u} - \frac{\sigma^2}{2}(t+u)\right)\bigg|\mathcal{G}_t\right]\exp\left(-\sigma b_t + \frac{\sigma^2}{2}t\right)$$
$$= \frac{d}{du}\bigg|_{u=0} \exp\left(\psi^{(v)}\left(x_t^{(v)} - \sum_{i\in I} p_i x_{(t-i)\vee 0}^{(v)}\right)u\right)\cdot 1$$
$$= \psi^{(v)}\left(x_t^{(v)} - \sum_{i\in I} p_i x_{(t-i)\vee 0}^{(v)}\right)$$

Analogously, one may prove the second equation in the Lemma: for, one readily has almost surely

$$\frac{d}{du}\bigg|_{u=0} \mathbb{E}\left[x_{t+u}^{(v)}\big|\mathcal{G}_t\right]$$
$$= \frac{d}{du}\bigg|_{u=0}\left(\mathbb{E}\left[\psi^{(v)}\left(x_t^{(v)} - \sum_{i\in I} p_i x_{(t-i)\vee 0}^{(v)}\right)u\bigg|\mathcal{G}_t\right] - \frac{\sigma^2}{2}u + x_t^{(v)}\right)$$
$$= \psi^{(v)}\left(x_t^{(v)} - \sum_{i\in I} p_i x_{(t-i)\vee 0}^{(v)}\right) - \frac{\sigma^2}{2}$$

In order to proceed from these pointwise almost sure equations to the assertion of the Theorem, one will apply Lebesgue's Dominated Convergence Theorem, yielding

$$n_\Gamma\left(\exp\left(x^{(v)}\right)\right) = \int_0^1 \mathbb{E}\left[\left|\frac{d}{du}\bigg|_{u=0}\mathbb{E}\left[x_{t+u}^{(v)}\big|\mathcal{G}_t\right] + \frac{\sigma^2}{2}\right|\right]dt. \qquad \square$$

For finite hyperfinite adapted probability spaces, an elementary proof for the main Theorem of this Section can be contrived:

Lemma 12.4.2 *For any hyperfinite number H we will let $X^{(v)}$ for all $v > 0$ denote the solution to the hyperfinite initial value problem*

$$\begin{aligned}
&X_0^{(v)} = 0,\\
&\forall t \in \left\{0,\ldots,1-\frac{1}{H!}\right\}\\
&X_{t+\frac{1}{H!}}^{(v)} - X_t^{(v)} =\\
&= \alpha\left(X_t^{(v)} - \sum_{u\in I} p_u\cdot X_{(t-u)\vee 0}^{(v)}\right)\chi_{\left\{\left|X_t^{(v)} - \sum_{u\in I} p_u\cdot X_{(t-u)\vee 0}^{(v)}\right|\geq v\right\}}\cdot\frac{1}{H!}\\
&\quad + \sigma\cdot\pi_{t+\frac{1}{H!}}\cdot\frac{1}{(2H!)^{1/2}} - \frac{\sigma^2}{2}\frac{1}{H!}
\end{aligned} \qquad (12.4.1)$$

(where $\pi_{\ell/H!} : \Omega = \{\pm 1\}^{H!} \to \{\pm 1\}$ is for all hyperfinite $\ell \leq H!$ the projection to the ℓ-th coordinate) which is just the hyperfinite analogue to (12.3.1). Using this notation, and considering a hyperfinite adapted probability space of mesh size $H!$, one has for all $k < H!$,

$$\sum_{k<H!} \mathbb{E}\left| X^{(\tau)}_{k/H!} - \mathbb{E}\left[X^{(\tau)}_{\frac{k+1}{H!}} \middle| \mathcal{F}_{k/H!} \right] - \frac{\sigma^2}{2}\frac{1}{H!} \right|$$
$$\leq \sum_{k<H!} \mathbb{E}\left| X^{(v)}_{k/H!} - \mathbb{E}\left[X^{(v)}_{\frac{k+1}{H!}} \middle| \mathcal{F}_{k/H!} \right] - \frac{\sigma^2}{2}\frac{1}{H!} \right|$$

for all $\tau \geq v$. As a consequence, $m\left(X^{(\cdot)}, \Omega\right)$ is monotonely decreasing for all hyperfinite adapted probability spaces $\Omega = \{\pm 1\}^{H!}$.

Proof of Lemma 12.4.2. It suffices to prove the result for finite (rather than merely hyperfinite) adapted spaces. By transfer to the nonstandard universe, we will obtain the same result for infinite hyperfinite H as well. The proof of this Lemma for finite H relies on exploiting the assumption that α is piecewise constant, since $\alpha = \alpha(1)\chi_{(0,+\infty)} + \alpha(-1)\chi_{(-\infty,0)} + \alpha(0)\chi_{\{0\}}$ yields

$$\forall k < H! \quad \forall v > 0$$
$$\mathbb{E}\left[X^{(v)}_{\frac{k+1}{H!}} \middle| \mathcal{F}_{k/H!} \right] - X^{(v)}_{\frac{k}{H!}} + \frac{\sigma^2}{2}\frac{1}{H!}$$
$$= \frac{1}{H!}\alpha\left(X^{(v)}_{k/H!} - \sum_{u\in I} p_u \cdot X^{(v)}_{\left(\frac{k}{H!}-u\right)\vee 0} \right) \chi_{\left\{ \left| X^{(v)}_{k/H!} - \sum_{u\in I} p_u \cdot X^{(v)}_{\left(\frac{k}{H!}-u\right)\vee 0} \right| \geq v \right\}}$$
$$= \frac{1}{H!}\left(\begin{array}{c} \alpha(1)\chi_{\left\{ X^{(v)}_{k/H!} - \sum_{u\in I} p_u \cdot X^{(v)}_{\left(\frac{k}{H!}-u\right)\vee 0} \geq v \right\}} \\ +\alpha(-1)\chi_{\left\{ X^{(v)}_{k/H!} - \sum_{u\in I} p_u \cdot X^{(v)}_{\left(\frac{k}{H!}-u\right)\vee 0} \leq -v \right\}} \end{array} \right),$$

which immediately follows from the construction of Anderson's random walk [3] $B_\cdot = \frac{1}{\sqrt{2H!}}\pi_\cdot$ and the recursive difference equation defining the process $X^{(v)}$. However, this last equation implies

$$\forall k < H! \quad \forall v > 0$$
$$\mathbb{E}\left| \mathbb{E}\left[X^{(v)}_{\frac{k+1}{H!}} \middle| \mathcal{F}_{k/H!} \right] - X^{(v)}_{\frac{k}{H!}} + \frac{\sigma^2}{2}\frac{1}{H!} \right|$$
$$= \frac{1}{H!}\mathbb{E}\left| \begin{array}{c} \alpha(1)\chi_{\left\{ X^{(v)}_{k/H!} - \sum_{u\in I} p_u \cdot X^{(v)}_{\left(\frac{k}{H!}-u\right)\vee 0} \geq v \right\}} \\ +\alpha(-1)\chi_{\left\{ X^{(v)}_{k/H!} - \sum_{u\in I} p_u \cdot X^{(v)}_{\left(\frac{k}{H!}-u\right)\vee 0} \leq -v \right\}} \end{array} \right|$$

$$= \frac{1}{H!}\left(\begin{array}{c} |\alpha(1)|\,\mathbb{P}\left\{X^{(v)}_{k/H!} - \sum\limits_{u\in I} p_u \cdot X^{(v)}_{(\frac{k}{H!}-u)\vee 0} \geq v\right\} \\ + \,|\alpha(-1)|\,\mathbb{P}\left\{X^{(v)}_{k/H!} - \sum\limits_{u\in I} p_u \cdot X^{(v)}_{(\frac{k}{H!}-u)\vee 0} \leq -v\right\} \end{array}\right).$$

Now all that remains to be shown is that

$$\sum_{k<H!} \mathbb{P}\left\{X^{(v)}_{k/H!} - \sum_{u\in I} p_u \cdot X^{(v)}_{(\frac{k}{H!}-u)\vee 0} \geq v\right\}$$

and

$$\sum_{k<H!} \mathbb{P}\left\{X^{(v)}_{k/H!} - \sum_{u\in I} p_u \cdot X^{(v)}_{(\frac{k}{H!}-u)\vee 0} \leq -v\right\}$$

are monotonely decreasing in v. The former statement is a consequence of the following assertion:

$$\begin{aligned} \forall v' \leq v\ \forall \ell < H!\quad & \sum_{k\leq\ell} \mathbb{P}\left\{X^{(v)}_{k/H!} - \sum_{u\in I} p_u \cdot X^{(v)}_{(\frac{k}{H!}-u)\vee 0} \geq v\right\} \\ & \leq \sum_{k\leq\ell} \mathbb{P}\left\{X^{(v')}_{k/H!} - \sum_{u\in I} p_u \cdot X^{(v')}_{(\frac{k}{H!}-u)\vee 0} \geq v'\right\} \end{aligned} \tag{12.4.2}$$

One can prove this estimate by considering the minimal ℓ such that the pointwise inequality

$$\sum_{k\leq\ell} \chi_{\left\{X^{(v)}_{k/H!} - \sum\limits_{u\in I} p_u \cdot X^{(v)}_{(\frac{k}{H!}-u)\vee 0} \geq v\right\}} \leq \sum_{k\leq\ell} \chi_{\left\{X^{(v')}_{k/H!} - \sum\limits_{u\in I} p_u \cdot X^{(v')}_{(\frac{k}{H!}-u)\vee 0} \geq v'\right\}} \tag{12.4.3}$$

fails to hold. Then one has an $\omega \in \Omega$ such that

$$\begin{aligned} \forall k < \ell\quad & \chi_{\left\{X^{(v)}_{k/H!} - \sum\limits_{u\in I} p_u \cdot X^{(v)}_{(\frac{k}{H!}-u)\vee 0} \geq v\right\}}(\omega) \\ & = \chi_{\left\{X^{(v')}_{k/H!} - \sum\limits_{u\in I} p_u \cdot X^{(v')}_{(\frac{k}{H!}-u)\vee 0} \geq v'\right\}}(\omega), \end{aligned} \tag{12.4.4}$$

$$1 = \chi_{\left\{X^{(v)}_{\ell/H!} - \sum\limits_{u\in I} p_u \cdot X^{(v)}_{(\frac{\ell}{H!}-u)\vee 0} \geq v\right\}}(\omega), \tag{12.4.5}$$

$$0 = \chi_{\left\{X^{(v')}_{\ell/H!} - \sum\limits_{u\in I} p_u \cdot X^{(v')}_{(\frac{\ell}{H!}-u)\vee 0} \geq v'\right\}}(\omega). \tag{12.4.6}$$

But equation (12.4.4) implies, via the difference equation for $X^{(\cdot)}$ (12.4.1), inductively in k the relation

$$\forall k < \ell \quad X^{v}_{k}(\omega) = X^{v'}_{k}(\omega).$$

If one combines this with

$$1 = \chi_{\left\{X^{(v)}_{\ell/H!} - \sum\limits_{u\in I} p_u \cdot X^{(v)}_{\left(\frac{\ell}{H!}-u\right)\vee 0} \geq v\right\}}(\omega)$$

(which is equation (12.4.5)) and $v \leq v'$, one can derive — again via the recursive difference equation (12.4.1) — that $X^{(v')}_{\ell/H!} \geq X^{(v)}_{\ell/H!}$ applied in $t = \frac{\ell}{H!}$ as well as the estimates

$$\begin{aligned}
\chi_{\left\{X^{(v')}_{\ell/H!} - \sum\limits_{u\in I} p_u \cdot X^{(v')}_{\left(\frac{k}{H!}-u\right)\vee 0} \geq v'\right\}}(\omega) &= \chi_{\left\{X^{(v')}_{\ell/H!} - \sum\limits_{u\in I} p_u \cdot X^{(v)}_{\left(\frac{\ell}{H!}-u\right)\vee 0} \geq v'\right\}}(\omega) \\
&\geq \chi_{\left\{X^{(v)}_{\ell/H!} - \sum\limits_{u\in I} p_u \cdot X^{(v)}_{\left(\frac{\ell}{H!}-u\right)\vee 0} \geq v'\right\}}(\omega) \\
&\geq \chi_{\left\{X^{(v)}_{\ell/H!} - \sum\limits_{u\in I} p_u \cdot X^{(v)}_{\left(\frac{\ell}{H!}-u\right)\vee 0} \geq v\right\}}(\omega) \\
&= 1.
\end{aligned}$$

This contradicts equation (12.4.6). Hence, the estimate (12.4.3) has been established for all $k < H!$, leading to (12.4.2).

Similarly, one can prove

$$\begin{aligned}
\forall v' \leq v \, \forall \ell < H! \quad & \sum_{k\leq \ell} \mathbb{P}\left\{X^{(v)}_{k/H!} - \sum_{u\in I} p_u \cdot X^{(v)}_{\left(\frac{k}{H!}-u\right)\vee 0} \leq -v\right\} \\
& \leq \sum_{k\leq \ell} \mathbb{P}\left\{X^{(v')}_{k/H!} - \sum_{u\in I} p_u \cdot X^{(v')}_{\left(\frac{k}{H!}-u\right)\vee 0} \leq -v'\right\}
\end{aligned}$$

which entails that $\sum_{k<H!} \mathbb{P}\left\{X^{(v)}_{k/H!} - \sum\limits_{u\in I} p_u \cdot X^{(v)}_{\left(\frac{k}{H!}-u\right)\vee 0} \leq -v\right\}$ must be monotonely decreasing in v. □

Using nonstandard analysis and the model theory of stochastic processes as developed by Keisler and others [10, 12, 7], we can prove the following result:

Theorem 12.4.1 *Suppose $\left(y^{(v)} : v > 0\right)$ is a family of stochastic processes on an adapted probability space Γ such that $y^{(v)}$ solves the stochastic differential equation (12.3.1) formulated above for all $v > 0$. Then the function $\sigma \mapsto n_\Gamma\left(y^{(\sigma)}\right)$ attains its minimum on $[0, S]$ in S.*

Proof. By the previous Lemmas 12.4.2 and the formula for n from Lemma 12.4.1, the assertion of the Theorem holds true internally for hyperfinite adapted spaces Γ, if we replace $y^{(\sigma)}$ by a lifting $Y^{(\sigma)}$ and if we let n_Ω when applied to internal processes denote the hyperfinite analogue of the standard n_Ω. Now, according to results by Hoover and Perkins [11] as well as Albeverio et al. [2], the solution $X^{(\upsilon)}$ of the hyperfinite initial value problem (12.4.1) is a lifting for the solution $x^{(\upsilon)}$ of (12.3.1) on a hyperfinite adapted space for any $\upsilon \geq 0$. Now, $y \mapsto n_\Omega(y)$ is the expectation of a conditional process in the sense of Fajardo and Keisler [7]. Therefore, due to the Adapted Lifting Theorem [7], we must have

$$\forall \upsilon \geq 0 \quad {}^\circ n_\Omega\left(X^{(\upsilon)}\right) = n_\Omega\left(x^{(\upsilon)}\right)$$

(where we identify n with its internal analogue when applied to internal processes). Since the internal equivalent of the Theorem's assertion holds for internal hyperfinite adapted space, the previous equation implies that it is also true for Loeb hyperfinite adapted spaces.

Now let $\left(y^{(\upsilon)} : \upsilon > 0\right)$ be a family of processes on some (not necessarily hyperfinite) adapted probability space Γ with the properties as in the Theorem. Because of the universality of hyperfinite adapted spaces [7, 12], we will find a process $x^{(\upsilon)}$ on any hyperfinite adapted space Ω such that x and y are automorphic to each other. This implies [2] that $x^{(\upsilon)}$ satisfies (12.3.1) as well. Furthermore, as one can easily see using Lebesgue's Dominated Convergence Theorem,

$$\forall \upsilon > 0 \quad n_\Omega\left(x^{(\upsilon)}\right) = n_\Gamma\left(y^{(\upsilon)}\right).$$

Due to our previous remarks on the solutions of (12.3.1) on hyperfinite adapted spaces, this suffices to prove the Theorem. □

Acknowledgements. The author gratefully acknowledges funding from the German National Academic Foundation (*Studienstiftung des deutschen Volkes*) and a pre-doctoral research grant of the German Academic Exchange Service (*Doktorandenstipendium des Deutschen Akademischen Austauschdienstes*). He is also indebted to his advisor, Professor Sergio Albeverio, as well as an anonymous referee for helpful comments on the first version of this paper.

References

[1] S. Albeverio, Some personal remarks on nonstandard analysis in probability theory and mathematical physics, in *Proceedings of the VIIIth International Congress on Mathematical Physics — Marseille 1986*, eds. M. Mebkhout, R. Sénéor, World Scientific, Singapore, 1987.

[2] S. Albeverio, J. Fenstad, R. Høegh-Krohn and T. Lindstrøm, *Nonstandard methods in stochastic analysis and mathematical physics*, Academic Press, Orlando, 1986.

[3] R. M. Anderson, "A nonstandard representation for Brownian motion and Itô integration", Israel Journal of Mathematics, **25** (1976) 15–46.

[4] A. Corcos et al., "Imitation and contrarian behaviour: hyperbolic bubbles, crashes and chaos", Quantitative Finance, **2** (2002) 264–281.

[5] N. Cutland, E. Kopp and W. Willinger, "A nonstandard treatment of options driven by Poisson processes", Stochastics and Stochastics Reports, **42** (1993) 115–133.

[6] M. Drehmann, J. Oechssler and A. Roider, Herding with and without payoff externalities — an internet experiment, mimoe 2004.

[7] S. Fajardo and H.J. Keisler, *Model theory of stochastic processes*, Lecture Notes in Logic 14, A. K. Peters, Natick (Mass.), 2002.

[8] F.S. Herzberg, "The fairest price of an asset in an environment of temporary arbitrage", International Journal of Pure and Applied Mathematics, **18** (2005) 121–131.

[9] F.S. Herzberg, On measures of unfairness and an optimal currency transaction tax, *arXiv* Preprint math. PR/0410543. `http://www.arxiv.org/pdf/math.PR/0410543`

[10] D.N. Hoover and H.J. Keisler, "Adapted probability distributions", Transactions of the American Mathematical Society, **286** (1984) 159–201.

[11] D.N. Hoover and E. Perkins, "Nonstandard construction of the stochastic integral and applications to stochastic differential equations I, II", Transactions of the American Mathematical Society, **275** (1983) 1–58.

[12] H.J. Keisler, Infinitesimals in probability theory, in *Nonstandard analysis and its applications*, ed. N. Cutland, London Mathematical Society Student Texts 10, Cambridge University Press, Cambridge, 1988.

[13] P.A. Loeb, "Conversion from nonstandard to standard measure and applications in probablility theory", Transactions of the American Mathematical Society, **211** (1974) 113–122.

[14] E. Nelson, *Radically elementary probability theory*, Annals of Mathematics Studies 117, Princeton University Press, Princeton (NJ), 1987.

[15] E. Perkins, Stochastic processes and nonstandard analysis, in *Nonstandard analysis–recent developments*, ed. A.E. Hurd, Lecture Notes in Mathematics 983, Springer-Verlag, Berlin, 1983.

[16] K.D. Stroyan and J.M. Bayod, *Foundations of infinitesimal stochastic analysis*, Studies in Logic and the Foundations of Mathematics 119, North-Holland, Amsterdam, 1986.

13

Quantum Bernoulli experiments and quantum stochastic processes

Manfred Wolff*

Abstract

Based on a $W^\star$-algebraic approach to quantum probability theory we construct basic discrete internal quantum stochastic processes with independent increments. We obtain a one-parameter family of (classical) Bernoulli experiments as linear combinations of these basic processes.

Then we use the nonstandard hull of the internal GNS-Hilbert space $\mathcal{H}_\tau$ corresponding to the chosen state τ (the underlying quantum probability measure) in order to derive nonstandard hulls of our internal processes. Finally continuity requirements lead to the specification of a certain subspace $\mathcal{L}$ of $\widehat{\mathcal{H}}_\tau$ to which the nonstandard hulls of our internal processes can be restricted and which turns out to be isomorphic to the Loeb-Guichardet space introduced by Leitz-Martini [10]. A subspace of $\mathcal{L}$ then is shown to be isomorphic to the symmetric Fock space $\mathcal{F}_+\left(L^2([0,1],\lambda)\right)$ and our basic processes agree with the processes of Hudson and Parthasarathy on this subspace.

13.1 Introduction

In the early 1980's Hudson and Parthasarathy [5], Hudson and Lindsay [4], and Hudson and Streater [6] started the theory of quantum stochastic processes, in particular of quantum Brownian motion. In [12] Parthasarathy showed one way how to make the passage from quantum random walk to diffusion. All these topics are extensively treated in P.A. Meyer's compendium [11]. For another general introduction to this field see [13] where an extensive motivation from quantum physics may be found. The Hudson-Parthasarathy approach is based on the classical symmetric Fock space over $L^2(\mathbb{R}_+, \lambda)$ where

*Mathematisches Institut, Eberhard-Karls-Univ. Tübingen.
manfred.wolff@uni-tuebingen.de

λ denotes the Lebesgue measure (see section 13.6). Guichardet [3] gave an interesting representation of this Fock space as an L^2-direct sum of spaces $L^2_n = L^2(X_n)$ where X_n is the space of all subsets of the unit interval $[0,1]$ having exactly n elements. The measure on X_n is defined via the corresponding Lebesgue measure on $\mathbb{R}^n_+$. This representation was extensively used by Maassen developing his kernel approach to general quantum stochastic processes.

Journé [7] invented the so called toy Fock spaces (bébé Fock in French) as discrete approximations of the symmetric Fock space. Let $\Omega_0 = \{0,1\}$ be equipped with the uniform distribution μ_0. Then the toy Fock space of order n is the space $L^2(\Omega_0^n, \mu_0^{\otimes n})$. A rigorous discrete approximation of the Guichardet space and the basic quantum stochastic processes defined on it which uses the space $L^2(\Omega_0^{\mathbb{N}}, \mu_0^{\otimes \mathbb{N}})$ as a toy Fock space is given by Attal [2].

There is another approach based also on the Guichardet space to prove such approximation theorems using nonstandard analysis which was developed by Leitz-Martini [10]. He discretized the Guichardet space in the following manner: let $T = \{\frac{k}{N} : k \in {}^*\mathbb{Z}, 0 \leq k \leq N\}$ be the hyperfinite time line where $N \in {}^*\mathbb{N}$ is infinitely large. Let Γ_n be the set of all internal subsets of T of internal cardinality n and set $\Gamma = \bigcup_{k=0}^{N} \Gamma_k$. Now let M be an internal subset of Γ and set

$$m(M) = \sum_{n \leq N} \frac{|M \cap \Gamma_n|}{N^n}$$

where $|A|$ denotes the internal cardinality of A. Then m is an internal measure on Γ with $m(\Gamma) \approx e$. The space $L^2(\Gamma, L_m)$, where L_m denotes the Loeb measure associated to m, contains the Guichardet space as a Banach sublattice. In fact $L^2(\Gamma, L_m)$ is the nonstandard version of the Guichardet space, so to speak the *Loeb–Guichardet space* over the time interval $[0,1]$. Among other things Leitz-Martini constructed all the relevant discrete quantum stochastic processes and showed that their nonstandard hulls exist in a precise manner and form the basic quantum stochastic processes: the time process, the creation and annihilation processes, and the number process.

All approaches mentioned so far start with some kind of discrete approximation of the *symmetric Fock space* or the isomorphic Guichardet space over $\mathbb{R}_+$, $[0,1]$ respectively.

In contrast to these approaches we present here a new one based on the more natural *$W^\star$-algebraic* foundation of quantum stochastic as described e.g. in [8]. In [13] the author sketched a different though similar approach in the finite-dimensional setting but he was not able to make the transition from the finite-dimensional to the continuous infinite-dimensional case. Our access is exactly the noncommutative version of Anderson's way [1] to Brownian motion. In particular we are able to *justify* the use of the symmetric Fock space (or equivalently the Guichardet space) in the context of quantum stochastic processes. We believe that by our approach the applications of nonstandard analysis as given by Leitz-Martini [10] are also better understandable.

We begin with the discrete quantum Bernoulli experiment modelled in the von Neumann algebra of all $2^n \times 2^n$-matrices over $\mathbb{C}$. Then we prove a Moivre-Laplace type theorem for quantum Bernoulli experiments, i.e. we show that these discrete stochastic processes converge to quantum Brownian motion in the same sense in which the classical approximation of the usual Brownian motion is treated by Anderson [1]. Furthermore we construct the basic quantum stochastic processes out of the corresponding discrete versions in the case of a vacuum state. Finally we show how the symmetric Fock space approach fits into our setting.

Applications to the theory of stochastic processes are under preparation.

13.2 Abstract quantum probability spaces

Recall that almost all information about a given probability space (Ω, Σ, μ) is to be found in the pair $(\mathcal{A}, \mu)$ with $\mathcal{A} = L^\infty(\Omega, \Sigma, \mu)$. Let us denote by $[f]$ the equivalence class modulo negligible functions in which the essentially bounded measurable function f is contained. Then $[f] \to \int_\Omega f(\omega) d\mu(\omega) =: \mu(f)$ defines a positive, linear, order continuous normalized functional on $\mathcal{A}$.

Analogously a **quantum probability space** is a pair $(\mathcal{A}, \tau)$ where $\mathcal{A}$ denotes a $W^\star$-algebra and τ is a positive, linear order continuous normalized functional on $\mathcal{A}$, or a **normal state** for short. The reader not familiar with the abstract theory of $W^\star$-algebras may look at them as subalgebras of the algebra of all bounded operators on an appropriate Hilbert space which are closed under involution $^\star : a \to a^\star$ and which are also closed with respect to the strong operator topology. In this context order continuity means the following: whenever a downward directed net (a_α) of nonnegative selfadjoint operators from $\mathcal{A}$ converges to the operator 0 with respect to the strong operator topology then $\lim_\alpha \tau(a_\alpha) = 0$ holds. In particular one may interpret $L^\infty(\Omega, \Sigma\mu)$ as the $W^\star$–algebra of all multiplication operators $M_f : g \to M_f(g) = fg$ $(f \in L^\infty(\Omega, \Sigma, \mu))$ on the Hilbert space $L^2(\Omega, \Sigma, \mu)$.

Let $(\mathcal{A}, \tau)$ be a quantum probability space. Let $\mathcal{A}_1, \mathcal{A}_2$ be $W^\star$-subalgebras generating the $W^\star$-subalgebra $\mathcal{A}_3$; moreover let τ_k be the restriction of τ to $\mathcal{A}_k$ $(k = 1, 2, 3)$. Then $\mathcal{A}_1$ and $\mathcal{A}_2$ are called **stochastically independent** if $(\mathcal{A}_3, \tau_3)$ is isomorphic to $(\mathcal{A}_1 \overline{\otimes} \mathcal{A}_2, \tau_1 \overline{\otimes} \tau_2)$ where $\overline{\otimes}$ denotes the $W^\star$-tensor product.

Remark: Let us point out that this notion of independence agrees with the usual one in the classical (commutative) case $\mathcal{A} = L^\infty(\Omega, \Sigma, \mu)$ but that there are other notions of independence in the quantum stochastic setting also generalizing the classical one (e.g. free independence).

Two probability spaces $(\mathcal{A}, \tau)$ and $(\mathcal{A}', \tau')$ are called **equivalent** if there exists an order continuous algebraic isomorphism $\varphi : \mathcal{A} \to \mathcal{A}'$ with $\tau' = \tau \circ \varphi$.

Now let us construct the L^2-space corresponding to the probability space $(\mathcal{A}, \tau)$. It is nothing else than the so called GNS-space (after Gelfand, Naimark, and Segal). To this end we introduce the $\mathbb{C}$-valued mapping $(a|b) = \tau(a^\star b)$ on $\mathcal{A} \times \mathcal{A}$. It is sesquilinear, that means, it is linear in the second argument, antilinear in the first argument, and it satisfies $(a|a) \geq 0$ for all $a \in \mathcal{A}$. The space $L_\tau = \{a : (a|a) = 0\}$ is a left ideal, and on $\mathcal{A}/L_\tau$ there is well-defined a scalar product by $(\hat{a}|\hat{b}) = (a|b)$ where $\hat{a} = a + L_\tau$ denotes the equivalence class in which a is contained. $\mathcal{H}_\tau$ is the completion of $\mathcal{A}/L_\tau$ with respect to the associated scalar product norm $\|\hat{a}\| = \sqrt{(\hat{a}|\hat{a})} = \sqrt{\tau(a^\star a)}$. The representation $\pi_\tau : \mathcal{A} \to \mathcal{L}(\mathcal{H}_\tau)$ is given by $\pi_\tau(a)(b + L_\tau) = ab + L_\tau$ which is well-defined since L_τ is a left ideal. The representation is injective whenever $\mathcal{A}$ is simple (i.e. has no nontrivial closed ideals), e.g. $\mathcal{A} = M_n(\mathbb{C})$, the algebra of all $n \times n$-matrices.

In case $\mathcal{A} = L^\infty(\Omega, \mu)$ the space $\mathcal{H}_\tau$ turns out indeed to be the space $L^2(\Omega, \mu)$ and in this way $L^\infty(\Omega, \mu)$ is represented as the algebra of operators of multiplication by L^∞-functions (see above).

13.3 Quantum Bernoulli experiments

Let us begin with the easiest example of a quantum probability space: Let $\mathcal{A} = M_2(\mathbb{C})$ be the $W^\star$-algebra of all complex 2×2 matrices $a = (a_{ik})$ and choose $\lambda \in]1/2, 1]$. Then $\tau_\lambda(a) = \lambda a_{11} + (1 - \lambda) a_{22}$ defines a normal state on $\mathcal{A}$. As in the case of classical probability we will often write $\mathbb{E}_\lambda(a)$ in place of $\tau_\lambda(a)$, and we call it the **expectation of a**.

Now consider the following model of tossing a fair coin: choose $\mathcal{A}_0$ as the commutative $W^\star$-algebra $\mathbb{C}^{\{0,1\}}$ and let $\mu_0 : \mathcal{A}_0 \to \mathbb{C}$ be given by $\mu_0(f) = \frac{1}{2}(f(0) + f(1))$. The underlying classical probability space is to be rediscovered in $\mathcal{A}_0$ as the set of two idempotents $1_{\{0\}}$ and $1_{\{1\}}$ where 1_A denotes the indicator function of a subset A of a given set. $\mathcal{A}_0$ is spanned also by the elements

$$X = -1_{\{0\}} + 1_{\{1\}} \quad \text{and} \quad 1 = 1_{\{0\}} + 1_{\{1\}} = 1_{\{0,1\}}.$$

We now construct all injective $\star$-algebra homomorphisms $\pi : \mathcal{A}_0 \to \mathcal{A}$ satisfying $\tau_\lambda \circ \pi = \mu_0$. In the sense of section 13.2 this means that we will find all abstract probability spaces $(\mathcal{B}, \tau_{\lambda_{|\mathcal{B}}})$ which are equivalent to $(\mathcal{A}_0, \mu_0)$. An easy calculation shows that they are given by $\pi := \pi_z$ with

$$\pi(1_{\{1\}}) = \frac{1}{2}\begin{pmatrix} 1 & z \\ \bar{z} & 1 \end{pmatrix} =: P_z \quad , \quad \pi(1_{\{0\}}) = \frac{1}{2}\begin{pmatrix} 1 & -z \\ -\bar{z} & 1 \end{pmatrix} =: Q_z$$

and linear extension, where $z \in \mathbb{C}$ is a parameter with $|z| = 1$. $\pi_z(X) = P_z - Q_z = \left(\begin{smallmatrix} 0 & z \\ \bar{z} & 0 \end{smallmatrix}\right) =: X_z$.

In summary we have seen that the simplest quantum probability space $(\mathcal{A}, \tau_\lambda)$ contains a one-parameter family $(\mathcal{A}_z, \tau_\lambda|_{\mathcal{A}_z})$ of models of classical coin tossing.

For $w = e^{is}$, $z = e^{it}$ $(-\pi < s, t \leq \pi)$ the commutator $[X_w, X_z]. = X_w X_z - X_z X_w$ is easily determined:

$$[X_w, X_z] = 2\,i\,\sin(s-t)\begin{pmatrix} 1 & 0 \\ 0 & -1 \end{pmatrix}.$$

So X_w and X_z commute if $w = z$ or $w = -z$.

We obtain $\mathbb{E}_\lambda([X_w, X_z]) = 2\,i\,(2\lambda - 1)\sin(s-t)$. The absolute value of it attains its maximum at $s - t = \pm\pi/2$. The pair (X_w, X_z) with $z = -iw$ is called a **quantum coin tossing**. It is unique up to automorphisms. More precisely this means the following: choose $w = e^{it}$ $(t \in]-\pi, \pi])$, and consider the automorphism φ on $\mathcal{A}$ given by $\varphi(a) = u^\star a u$ where

$$u = \begin{pmatrix} e^{-it/2} & 0 \\ 0 & e^{it/2} \end{pmatrix}.$$

Then $\varphi(X_1) = X_w$, $\varphi(X_{-i}) = X_{-iw}$ and $\tau_\lambda \circ \varphi = \tau_\lambda$.

So in the following we need only to consider the quantum coin tossing (X_1, X_{-i}). Notice that $X_1 = \sigma_x$, $X_{-i} = \sigma_y$ and $[X_w, X_{-iw}]x = 2i\sigma_z$, where $\sigma_x, \sigma_y, \sigma_z$ denote the Pauli matrices. The following important matrices can be constructed by (σ_x, σ_y):

$$b^- = \begin{pmatrix} 0 & 1 \\ 0 & 0 \end{pmatrix} = \frac{1}{2}(\sigma_x + i\sigma_y), \tag{13.3.1}$$

$$b^+ = \begin{pmatrix} 0 & 0 \\ 1 & 0 \end{pmatrix} = \frac{1}{2}(\sigma_x - i\sigma_y), \tag{13.3.2}$$

$$b^\circ = \begin{pmatrix} 0 & 0 \\ 0 & 1 \end{pmatrix} = \mathbf{1} - \frac{1}{2i}[\sigma_x, \sigma_y]. \tag{13.3.3}$$

Obviously the following formulae hold:

$$\sigma_x = b^+ + b^-,\ \sigma_y = i \cdot (b^+ - b^-),\ [b^-, b^+] = \sigma_z = \mathbf{1} - 2b^\circ,\ X_w = \overline{w}b^+ + wb^-.$$

These formulae show that the *random variables* b^+ and b^- are in a certain sense more basic than X_w though they are not selfadjoint.

As in the classical case we describe the experiment of n $(\in \mathbb{N})$ independent coin tossings by the n-fold tensor product $(\mathcal{A}^{\otimes n}, \tau_\lambda^{\otimes n})$. The algebra corresponding to the k-th trial is described by the elements

$$\underbrace{\mathbf{1} \otimes \cdots \otimes \mathbf{1}}_{(k-1)\text{ times}} \otimes x \otimes \underbrace{\mathbf{1} \otimes \cdots \otimes \mathbf{1}}_{(n-k)\text{ times}} =: x_k$$

where $x \in \mathcal{A}$ is arbitrary. Describing k times repetition of our quantum coin tossing we obtain $S_{w,k} := \sum_{j=1}^{k} X_{w\,j}$. We have

$$[S_{w,k}, S_{-iw,k}] = 2i \sum_{j=1}^{k} \sigma_{z,\,k}.$$

The pair $(S_{w,k}, S_{-iw,k})$ is called a **quantum Bernoulli experiment**.

Setting

$$\mathcal{A}_k := \mathcal{A}^{\otimes k} \otimes \underbrace{\mathbf{1} \otimes \cdots \otimes \mathbf{1}}_{n-k \text{ times}} = \{a \otimes \underbrace{\mathbf{1} \otimes \cdots \otimes \mathbf{1}}_{n-k \text{ times}} : \; a \in \mathcal{A}^{\otimes k}\}$$

we obtain a natural filtering $\mathbb{C} \cdot \mathbf{1} \subset \mathcal{A}_1 \subset \cdots \subset \mathcal{A}_n = \mathcal{A}^{\otimes n}$. Moreover there also exists a conditional expectation $\mathbb{E}_k$ from $\mathcal{A}_n$ onto $\mathcal{A}_k$ which is given by $\mathbb{E}_k(a \otimes b) = \tau_\lambda^{\otimes(n-k)}(b) \cdot a$, where $b \in \mathcal{A}^{\otimes(n-k)}$, and linear extension.

13.4 The internal quantum processes

Now we consider a polysaturated model ${}^*V(\mathbb{R})$ of the full structure $V(\mathbb{R})$ over $\mathbb{R}$. As usual we replace the standard integer n above by an infinitely large integer N. Then all considerations in the previous section remain true for internal objects.

We use the discrete time interval $T = \{\frac{k}{N} : 0 \le k < N\}$ denoting its elements by $\underline{s}, \underline{t}, \underline{u}, \ldots$ Moreover we rescale also the family $(\mathcal{A}_k)$ by setting $\mathcal{A}_{\underline{t}} := \mathcal{A}_{N\underline{t}}$.

Set $\rho = \frac{1}{\sqrt{N}}$. Then we introduce

$$a_{\underline{t}}^{\pm} = \rho b_{\underline{t}}^{\pm}, \tag{13.4.1}$$

$$a_{\underline{t}}^{\circ} = b_{\underline{t}}^{\circ}, \tag{13.4.2}$$

$$a_{\underline{t}}^{\bullet} = \rho^2 \begin{pmatrix} 1 & 0 \\ 0 & 0 \end{pmatrix}. \tag{13.4.3}$$

These elements are the internal increments by which we construct the basic internal stochastic processes

$$A_{\underline{t}}^{\sharp} = \Sigma_{0 \le \underline{s} < \underline{t}}\, a_{\underline{s}}^{\sharp}, \; \sharp \in \{+, -, 0, \bullet\} =: I.$$

$(A_{\underline{t}}^{+})_{\underline{t}}$ is called the internal **creation process**, $(A_{\underline{t}}^{-})_{\underline{t}}$ the **annihilation process**, $(A_{\underline{t}}^{\bullet})_{\underline{t}}$ the **time process** and finally $(A_{\underline{t}}^{\circ})_{\underline{t}}$ the **number process**.

The internal Brownian motion corresponding to the parameter $w \in \mathbb{C}$ then is given by $B_{w,\underline{t}} = \bar{w} A_{\underline{t}}^{+} + w A_{\underline{t}}^{-}$ $(|w| = 1)$. It satisfies the internal difference equation

$$dB_{w,\underline{t}} = \overline{w} a_{\underline{t}}^{+} + w a_{\underline{t}}^{-}.$$

Like in the previous section we consider the state $\tau_\lambda^{\otimes N} =: \tau_{\lambda,N}$ and we obtain

$$\mathbb{E}_{\lambda,N}(B_{w,\underline{t}}) := \tau_{\lambda,N}(B_{w,\underline{t}}) = 0, \quad \mathbb{E}_{\lambda,N}(B^2_{w,\underline{t}}) = \underline{t}. \tag{13.4.4}$$

The second equation follows from the facts that

$$\tau_{\lambda,N}\left((\bar{w}a^+_{\underline{s}} + wa^-_{\underline{s}})(\bar{w}a^+_{\underline{u}} + wa^-_{\underline{u}})\right) = 0$$

for $\underline{s} \neq \underline{u}$, but $\tau_{\lambda,N}\left((\bar{w}a^+_{\underline{s}} + wa^-_{\underline{s}})^2\right) = \tau_{\lambda,N}(1/N) = 1/N$.

The commutator is

$$\left[B_{w,\underline{t}}, B_{-w,\underline{t}}\right] = 2i\rho\Sigma_{0\leq\underline{s}<\underline{t}}\,\sigma_{z,\underline{s}}.$$

The so-called internal Itô-table for the increments is easily computed:

	$a^\bullet_{\underline{s}}$	$a^+_{\underline{s}}$	$a^\circ_{\underline{s}}$	$a^-_{\underline{s}}$
$a^\bullet_{\underline{s}}$	$\rho^2 a^\bullet_{\underline{s}}$	0	0	$\rho^2 a^-_{\underline{s}}$
$a^+_{\underline{s}}$	$\rho^2 a^+_{\underline{s}}$	0	0	$\rho^2 a^\circ_{\underline{s}}$
$a^\circ_{\underline{s}}$	0	$a^+_{\underline{s}}$	$a^\circ_{\underline{s}}$	0
$a^-_{\underline{s}}$	0	$a^\bullet_{\underline{s}}$	$a^-_{\underline{s}}$	0

Let $0 \neq v \in {}^*\mathbb{C}$ be arbitrary. Then the process $(P_{\underline{t}})$ given by

$$P^v_{\underline{t}} = A^\circ_{\underline{t}} + |v|B_{\frac{v}{|v|},\underline{t}} + |v|^2 A^\bullet_{\underline{t}}$$

is called the **Poisson Process with parameter** v.

Now we compute the **characteristic operator functions** of these processes. Whenever $V = (V_{\underline{t}})_{\underline{t}}$ is a stochastic process its characteristic operator function is defined by $f_V(u,\underline{t}) = \exp(iuV_{\underline{t}})$, where $u \in {}^*\mathbb{R}$. Obviously $\dot{f}_V(u,\underline{t})\,|_{u=0} = iV_{\underline{t}}$ and $f_V(u,\underline{t})$ is unitary iff $V_{\underline{t}}$ is selfadjoint.

The processes $V = (V_{\underline{t}})$ we considered so far are all of the form

$$V_{\underline{t}} = \sum_{0\leq\underline{s}<\underline{t}} d_{\underline{s}},$$

where

$$d_{\underline{s}} = \underbrace{1\otimes\cdots\otimes 1}_{N\underline{s}-1 \text{ times}} \otimes d \otimes \underbrace{1\otimes\cdots\otimes 1}_{N-N\underline{s} \text{ times}},$$

and $d \in \mathcal{A}$ is appropriately chosen. Since $d_{\underline{u}}$, $d_{\underline{v}}$ commute for $\underline{u} \neq \underline{v}$ we obtain

$$f_V(u,\underline{t}) = \bigotimes_{0\leq\underline{s}<\underline{t}} (\exp(iud))_{\underline{s}}.$$

The most important formulas are the following ones:

$$\exp(iuB_{w,\underline{t}}) = \bigotimes_{0\le\underline{s}<\underline{t}} (\cos(\rho u)\underline{\mathbf{1}} + i\sin(u)X_w)_{\underline{s}}$$

$$\exp(iuP_{\underline{t}}) = \bigotimes_{0\le\underline{s}<\underline{t}} Q_{\underline{s}}(u)$$

where $Q(u) = \frac{1}{\xi}\begin{pmatrix} 1+\rho^2|v|^2e^{iu\xi} & \rho v(e^{iu\xi}-1) \\ \rho\bar{v}(e^{iu\xi}-1) & \rho^2|v|^2+e^{iu\xi} \end{pmatrix}$ with $\xi = 1+\rho^2|v|^2$.

13.5 From the internal to the standard world

The operator norm topology on $\mathcal{A}$ is too fine in order to obtain reasonable nonstandard hulls. Moreover the processes discussed so far are internally bounded but not necessarily S-bounded. For example $B_{w,k/N}$ is selfadjoint, hence its operator norm is equal to its spectral radius. This number is easily computed as $\frac{k}{\sqrt{N}}$, or in other words $\|B_{w,\underline{t}}\| = \sqrt{N}\cdot\underline{t}$.

In the following we only consider τ_λ for $\lambda = 1$. The other case will be treated elsewhere. We set $\tau_1 =: \tau$ and we construct the Hilbert space $\mathcal{H}_\tau$ corresponding to $(\mathcal{A},\tau)$ ($\mathcal{A}$ the algebra of 2×2-matrices, see section 13.2). To this end we denote the columns of the 2×2-matrix a by $a = (a_1^\downarrow, a_2^\downarrow)$. Then $\tau(a^*b) = (a_1^\downarrow \mid b_1^\downarrow)$ where the latter is the canonical scalar product on $\mathbb{C}^2$. Hence $L_\tau = \{a : a_2^\downarrow = 0\}$ and $\pi_\tau(a)(b+L_\tau) = ab_1^\downarrow + L_\tau$. In particular $\mathcal{H}_\tau \cong \mathbb{C}^2$, where the isometric isomorphism is given by $a + L_\tau \to a_1^\downarrow$.

We have previously introduced the random variable $X_1 = \sigma_x$. Its equivalence class in $\mathcal{H}_\tau$ is denoted by $\widetilde{X_1} = X_1 + L_\tau$ We set $e_0 = b_1^{\circ\downarrow} = \binom{1}{0} = \widetilde{X_1^0}$ and $e_1 = b_1^{-\downarrow} = \binom{0}{1} = \widetilde{X}_1$.

Now we consider $\tau_N = \tau^{\otimes N}$. Obviously $\mathcal{H}_{\tau_N} \cong ({}^*\mathbb{C}^2)^{\otimes N}$ holds. Let $\omega \in \{0,1\}^T$ be an arbitrary internal function. Then we define e_ω by $e_\omega = \otimes_{k=0}^{N-1} e_{\omega(k/N)}$. The set $\{e_\omega : \omega \in \{0,1\}^T\}$ forms an orthonormal basis in $\mathcal{H}_{\tau_N}$.

$\mathcal{H}_{\tau_N}$ is an internal Hilbert space, so its nonstandard hull $\widehat{\mathcal{H}}_{\tau_N}$ is a Hilbert space.

The formulae (13.3.1) up to (13.4.3) above lead to the following equations, which result from the fact that $\mathcal{H}_{\tau_N}$ is the GNS-Hilbert space of $(\mathcal{A}_N, \tau_N)$. In order to make the formulae as simple as possible we denote the addition mod 2 by $\oplus$, i.e. we consider $\Omega = \{0,1\}^T$ as the cartesian product of the group $(\mathbb{Z}_2, \oplus)$. Then we set $\omega^+ = 1_T \oplus \omega$ and $\omega^- = \omega$ for $\omega \in \{0,1\}^T$. Moreover we write $1_{\underline{t}}$ in place of $1_{\{\underline{t}\}}$.

With all theses conventions we obtain

$$a_{\underline{t}}^{\pm} e_\omega = \frac{\omega^{\pm}(\underline{t})}{\sqrt{N}} e_{\omega \oplus 1_{\underline{t}}}, \tag{13.5.1}$$

$$a_{\underline{t}}^{\bullet} e_\omega = \frac{\omega^{+}(\underline{t})}{N} e_\omega = a_{\underline{t}}^{-} a_{\underline{t}}^{+} e_\omega, \tag{13.5.2}$$

$$a_{\underline{t}}^{\circ} e_\omega = \omega^{-}(\underline{t}) e_\omega = N \cdot a_{\underline{t}}^{+} a_{\underline{t}}^{-} e_\omega. \tag{13.5.3}$$

13.5.1 Brownian motion

First of all we show that our internal Brownian motion $(B_{w,\underline{t}})$ leads to the standard Brownian motion on a subspace of $\widehat{\mathcal{H}}_{\tau_N}$. To this end we recall that the standard Brownian motion (β_t) on $L^2(\Omega, L_\mu)$ (with $\Omega = \{0,1\}^T$, $\mu = \mu_N = \mu_0^{\otimes N}$, L_μ the corresponding Loeb measure) can be viewed as a family of multiplication operators $f \mapsto \beta_t f$ on $L^2(\Omega, L_\mu)$ which are selfadjoint but unbounded (cf. the end of section 13.2). From equation (13.4.4) it follows that $B_{w,\underline{t}}$ is S-bounded in $\mathcal{H}_N$. We claim that classical Brownian motion (or better to say: Anderson's Brownian motion) is equivalent (in the sense of section 13.2) to our $(\widehat{B}_{w,{}^\circ\underline{t}})$ where $(\widehat{B}_{w,{}^\circ\underline{t}})$ denotes the family of selfadjoint operators on a subspace of $\widehat{\mathcal{H}}_{\tau_N}$ coming from the family $(B_{w,\underline{t}})$. To this end let$|w| = 1$ and

$$P_1 = \frac{1}{2}\begin{pmatrix} 1 & w \\ \overline{w} & 1 \end{pmatrix} \quad , \quad P_0 = \frac{1}{2}\begin{pmatrix} 1 & -w \\ -\overline{w} & 1 \end{pmatrix}.$$

For $\omega \in \Omega$ we set $P_\omega = \otimes_{0 \le k < N} P_{\omega(k/N)}$. It follows $\tau_N(P_\omega) = 2^{-N} = \mu(\{\omega\})$. We denote the internal $W^\star$-algebra generated by $\{P_\omega : \omega \in \{0,1\}^T\}$ by $\mathcal{B}(w)$. The formula

$$b_{w,t} := \overline{w} a_{\underline{t}}^{+} + w a_{\underline{t}}^{-} = \frac{\rho}{2}(P_{1_{\underline{t}}^{-}} - P_{1_{\underline{t}}^{+}})$$

shows that $(B_{w,t}) = \sum_{0 \le \underline{s} < \underline{t}} b_{w,\underline{t}}$ is contained in $\mathcal{B}(w)$.

The internal L^∞-space $L^\infty(\Omega, \mu)$ is nothing else than $L^\infty(\{0,1\}, \mu_0)^{\otimes N}$. Thus it is not very hard to see that $\pi_N := \pi_w^{\otimes N}$ (see section 13.2) maps $L^\infty(\Omega, \mu)$ onto $\mathcal{B}(w)$ with $\tau_N \circ \pi_N|_{\mathcal{B}(w)} = \mu$. Now let $D = \{f \in L^\infty(\Omega, \mu) :$ f is S-bounded$\}$. Then

$$\begin{aligned} \pi_N(D) &\subset \{a \in \mathcal{A}_N : \|a\| \text{ is } S\text{-bounded}\} \\ &\subset \{x \in \mathcal{H}_{\tau_N} : \|x\|_2 = \sqrt{\tau(x^* x)} \text{ is } S\text{-bounded}\}. \end{aligned}$$

Moreover for f, $g \in D$ we have

$$\int_\Omega {}^\circ\bar{f}\, {}^\circ g \, dL_\mu \approx \mu(\bar{f} g) = \tau_N\left(\pi_N(\bar{f} g)\right) = (\pi_N(f) | \pi_N(g))$$

because of $\pi_N(\bar{f}g) = \pi_N(f)^\star \pi_N(g)$. Since $L^2(\Omega, L_\mu)$ is the completion of $\{{}^\circ f : f \in D\}$ the mapping ${}^\circ f \to \widehat{\pi_N(f)}$ can be extended to a linear isometry U from $L^2(\Omega, L_\mu)$ onto a subspace $\mathcal{M}$, say, of $\mathcal{H}_{\mathcal{T}N}$.

For $t \in [0,1]$ set $\tilde{t} = \min\{\underline{s} \in T : \underline{s} > t\}$. We show that $U(\beta_t) = \widehat{B_{w,\tilde{t}}}$. Define

$$y_{\underline{t}}(\omega) = \begin{cases} 1 & \omega(\underline{t}) = 1 \\ -1 & \omega(\underline{t}) = 0 \end{cases}$$

and $Y_{\underline{t}} = \frac{1}{\sqrt{N}} \sum_{0 \le \underline{s} < \underline{t}} y_{\underline{s}}$. Then $\beta_t(\omega) = {}^\circ Y_{\tilde{t}}(\omega)$. Moreover $\pi_N(Y_{\underline{t}}) = B_{w,\underline{t}}$. Set $M_n = \{\omega : |Y_{\tilde{t}}(\omega)| \le n\}$. Then for $n \in \mathbb{N}$ we have $M_n = \{\omega : |\beta_t(\omega)| \le n\}$. This implies that $\lim_{n\to\infty} \beta_t 1_{M_n} = \beta_t$ holds in $L^2(\Omega, L_\mu)$. Therefore for every standard $\varepsilon > 0$ there exists a standard n_0 with $\|\beta_t(1_\Omega - 1_{M_n})\|_2 < \varepsilon$ for $n \ge n_0$ standard. But this in turn implies $\|Y_{\tilde{t}}(1_\Omega - 1_{M_n})\|_2 < \varepsilon$ for n standard, $n \ge n_0$. Since π_N is an isometry also for the internal L^2-norm on $\mathcal{H}_{\mathcal{T}N}$ we obtain $\|B_{w,\tilde{t}} \pi_N(1_\Omega - 1_{M_n})\|_2 < \varepsilon$ for these $n \ge n_0$. $\|B_{w,\tilde{t}} \pi_N(1_{M_n})\|_2 = \|Y_{\tilde{t}} 1_{M_n}\|_2 \le n$ implies that $B_{w,\tilde{t}} \pi_N(1_{M_n}) \in \mathcal{H}_{\mathcal{T}N,\mathrm{fin}}$ and $U(\beta_t 1_{M_n}) = \widehat{B_{w,\tilde{t}} \pi_N(1_{M_n})}$. But this in turn gives $U(\beta_t) = \widehat{B_{w,\tilde{t}}}$. That the operator of multiplication by β_t is mapped onto the operator of multiplication by $\widehat{B_{w,\tilde{t}}}$ follows from an easy additional argument.

13.5.2 The nonstandard hulls of the basic internal processes

The deeper problem is to find a joint subspace $\mathcal{K}$ of $\widehat{\mathcal{H}}_{\mathcal{T}N}$ such that all processes have standard parts as closable operators densely defined in $\mathcal{K}$. We begin the construction with some more notations: For $\omega \in \{0,1\}^T =: \Omega$ we set $M_\omega = \{\underline{t} : \omega(\underline{t}) = 1\}$, and $|\omega| := |M_\omega|$, that is the (internal) cardinality of M_ω. Using these notations we obtain the following formulae by means of (13.4.4)–(13.5.2). Let $y = \sum_{\omega\in\Omega} y_\omega e_\omega$ be arbitrary. Then

$$A^\bullet_{\underline{t}} y = \sum_{\omega\in\Omega} \left(\frac{1}{N} \sum_{0\le\underline{s}<\underline{t}} \omega^+(\underline{s}) \right) y_\omega e_\omega \tag{13.5.4}$$

$$A^\circ_{\underline{t}} y = \sum_{\omega\in\Omega} \left(\sum_{0\le\underline{s}<\underline{t}} \omega(\underline{s}) \right) y_\omega e_\omega \tag{13.5.5}$$

$$A^+_{\underline{t}} y = \sum_{\varphi\in\Omega} \left(\frac{1}{\sqrt{N}} \sum_{0\le\underline{s}<\underline{t}} y_{\varphi\oplus 1_{\underline{s}}} \varphi^-(\underline{s}) \right) e_\varphi \tag{13.5.6}$$

$$A^-_{\underline{t}} y = \sum_{\varphi\in\Omega} \left(\frac{1}{\sqrt{N}} \sum_{0\le\underline{s}<\underline{t}} y_{\varphi\oplus 1_{\underline{s}}} \varphi^+(\underline{s}) \right) e_\varphi \tag{13.5.7}$$

Consider now the projection $\mathbb{E}_{\underline{t}}$ from $\mathcal{H}$ onto $\mathcal{H}_{\underline{t}}$ which is the Hilbert space coming from $\mathcal{A}_{\underline{t}}$ (see section 13.3). $\mathbb{E}_{\underline{t}}$ is given by

$$\mathbb{E}_{\underline{t}}(e_\omega) = \begin{cases} e_\omega & \omega(\underline{s}) = 0 \text{ for all } \underline{s} \geq t \\ 0 & \text{else} \end{cases}$$

Then we obtain $\mathbb{E}_{\underline{t}} A^\natural_{\underline{t}} = A^\natural_{\underline{t}} \mathbb{E}_{\underline{t}}$ for each $\sharp \in \{\bullet, \circ, +, -\}$ or in other words *the family* $(A^\sharp_{\underline{t}})_{\underline{t}<1}$ *is adapted to* $(\mathcal{H}_{\underline{t}})_{\underline{t}<1}$.

We prove now some continuity properties. To this end we introduce

$$\mathcal{H}^{(n)} := \left\{ \sum_{|\omega|=n} y_\omega e_\omega : y_\omega \in {}^*\mathbb{C} \right\}$$

and we set $\mathcal{H} := \mathcal{H}_{\tau_N}$ for short.

Then $\mathcal{H} = \perp_{n=0}^N \mathcal{H}^{(n)}$ where $\perp$ denotes the orthogonal direct sum. Moreover $A^\bullet_{\underline{t}}(\mathcal{H}^{(n)}) \subset \mathcal{H}^{(n)}$, $A^\circ_{\underline{t}}(\mathcal{H}^{(n)}) \subset \mathcal{H}^{(n)}$, $A^+_{\underline{t}}(\mathcal{H}^{(n)}) \subset \mathcal{H}^{n+1}$, and finally $A^-_{\underline{t}}(\mathcal{H}^{(n)}) \subset \mathcal{H}^{(n-1)}$ where $\mathcal{H}^{(-1)} := \{0\} =: \mathcal{H}^{(N+1)}$.

For $y = \sum_{|\omega|=n} y_\omega e_\omega$ and $0 \leq \underline{s} < \underline{t}$ we obtain

$$\begin{aligned} \|A^\bullet_{\underline{t}} y - A^\bullet_{\underline{s}} y\|^2 &= \sum_{|\omega|=n} \left| \frac{1}{N} \sum_{\underline{s} \leq \underline{u} < \underline{t}} \omega^+(s) \right|^2 |y_\omega|^2 \\ &\leq \sum_{|\omega|=n} \left| \frac{1}{N} \sum_{\underline{s} \leq \underline{u} < \underline{t}} 1 \right|^2 |y_\omega|^2 = (\underline{t} - \underline{s})^2 \|y\|^2. \end{aligned} \tag{13.5.8}$$

It follows that $\|A^\bullet_{\underline{t}} - A^\bullet_{\underline{s}}\| \leq (\underline{t} - \underline{s})$, in particular the mapping $\underline{t} \mapsto A^\bullet_{\underline{t}}$ is S-continuous from T to $\mathcal{L}(\mathcal{H})$ with respect to the operator norm.

Let $y = \sum_{|\omega|=n} y_\omega \in \mathcal{H}^{(n)}$, and $0 \leq \underline{s} < \underline{t}$

$$\|A^\circ_{\underline{t}} y - A^\circ_{\underline{s}} y\|^2 = \sum_{|\omega|=n} \Big(\sum_{\underline{s} \leq \underline{u} < \underline{t}} \omega(\underline{s}) \Big)^2 |y_\omega|^2.$$

It follows that $\|A^\circ_{\underline{t}}|_{\mathcal{H}^{(n)}}\| \leq n$, but the mapping $\underline{t} \mapsto A^\circ_{\underline{t}} y$ is not S-continuous in general. So finally we have to choose an appropriate subspace in order to get S-continuity of $\underline{t} \to A^\circ_{\underline{t}} y$.

Next we show the following inequality:

Proposition 1 *For* $0 \leq \underline{s} < \underline{t}$

$$\|(A^+_{\underline{t}} - A^+_{\underline{s}})|_{\mathcal{H}^{(n)}}\|^2 \leq (n+1)(\underline{t} - \underline{s}) \tag{13.5.9}$$

holds.

Proof. Let $y = \sum_{|\omega|=n} y_\omega e_\omega \in \mathcal{H}^{(n)}$ be arbitrary. Then

$$(A_{\underline{t}}^+ - A_{\underline{s}}^+)y = \frac{1}{\sqrt{N}} \sum_{|\varphi|=n+1} \left(\sum_{\underline{s} \le \underline{u} < \underline{t}} y_{\varphi \oplus 1_{\underline{u}}} \varphi(\underline{u}) \right) e_\varphi$$

holds. If $\varphi(u) = 0$ then $|\varphi \oplus 1_{\underline{u}}| = n+1$ hence $y_{\varphi \oplus 1_{\underline{u}}} = 0$. It follows

$$\begin{aligned} \|(A_{\underline{t}}^+ - A_{\underline{s}}^+)y\|^2 &= \frac{1}{N} \sum_{|\varphi|=n+1} \left(\left| \sum_{\underline{s} \le \underline{u} < \underline{t}} y_{\varphi \oplus 1_{\underline{u}}} \right| \right)^2 \\ &\le \frac{n+1}{N} \sum_{|\varphi|=n+1} \sum_{\underline{s} \le \underline{u} < \underline{t}} |y_{\varphi \oplus 1_{\underline{u}}}|^2 \end{aligned}$$

Let $\omega \in \Omega$ be arbitrary with $|\omega| = n$. Then there exist at most $N(\underline{t} - \underline{s}) - |\omega \cdot 1_{[\underline{s},\underline{t}[}|$ different φ with $\varphi \oplus 1_{\underline{u}} = \omega$ and $\underline{s} \le \underline{u} < \underline{t}$. This gives $\|(A_{\underline{t}}^+ - A_{\underline{s}}^+)y\|^2 \le \frac{N(\underline{t}-\underline{s})}{N}(n+1)\|y\|^2$, and the asserted inequality follows. □

Corollary 1

$$\|(A_{\underline{t}}^- - A_{\underline{s}}^-)|_{\mathcal{H}^{(n)}}\|^2 \le (\underline{t} - \underline{s})n. \tag{13.5.10}$$

Proof. $A_{\underline{t}}^-$ is the adjoint of $A_{\underline{t}}^+$.

By construction we have $\mathcal{H}^{(n)} \perp \mathcal{H}^{(m)}$ for $n \ne m$. So the subspace $\mathcal{K}^\infty$ of $\hat{\mathcal{H}}$, generated by all the spaces $\hat{\mathcal{H}}^{(n)}$ $(n \in \mathbb{N})$ is nothing else than the orthogonal direct sum $\mathcal{K}^\infty = \perp_{n \in \mathbb{N}_0} \hat{\mathcal{H}}^{(n)}$. Recall the definition $\tilde{t} = \min\{\underline{u} \in T : \underline{u} > \underline{t}\}$. Then from the inequality (13.5.8) it follows that by

$$\hat{A}_{t,n}^\bullet \hat{y} = (A_{\tilde{t},n}^\bullet y)^\wedge$$

there is uniquely defined an operator on $\hat{\mathcal{H}}^{(n)}$ bounded by t. So we obtain the operator $\hat{A}_t^\bullet = \perp_{n \in \mathbb{N}_0} \hat{A}_{t,n}$ on $\mathcal{K}^\infty$ which is selfadjoint and bounded by t. Moreover $\|\hat{A}_t^\bullet - \hat{A}_s^\bullet\| \le t - s$ holds. □

In order to define the nonstandard hull of one of the other operators $A_{\underline{t}}^\sharp$ $(\sharp \in \{+, -, 0\})$ we choose the following joint domain of definition:

$$\mathcal{K}_0^\infty := \left\{ y \in \mathcal{K}^\infty : y = \perp_{n \in \mathbb{N}_0} \hat{y}_n, \hat{y}_n \in \hat{\mathcal{H}}^{(n)}, \sum_{n=0}^\infty n\|\hat{y}_n\|_2^2 < \infty \right\}$$

which is dense in $\mathcal{K}^\infty$. From the inequalities (13.5.9) to (13.5.10) we obtain that by

$$\hat{A}_t^\circ(\perp_{n \in \mathbb{N}_0} \hat{y}_n) = \perp_{n \in \mathbb{N}_0} (A_{\tilde{t}}^\circ y_n)^\wedge$$

$$\hat{A}_t^\pm(\perp_{n \in \mathbb{N}_0} \hat{y}_n) = \perp_{n \in \mathbb{N}_0} (A_{\tilde{t}}^\pm y_n)^\wedge$$

the operators $\hat{A}_t^\sharp : \mathcal{K}_0^\infty \to \mathcal{K}^\infty$ are uniquely defined. Moreover the following proposition holds:

Proposition 2 *$\hat{A}_t^\circ$ is selfadjoint and $\hat{A}_t^+$ and $\hat{A}_t^-$ are adjoint to each other.*

Proof. Set $\hat{A}_{t,n}^\sharp = \hat{A}_t^\sharp \mid \hat{\mathcal{H}}_n$. These operators are bounded and $\hat{A}_{t,n}^\circ$ is selfadjoint, whereas $\hat{A}_{t,n}^+$ and $\hat{A}_{t,(n+1)}^-$ are adjoint to each other. Now the assertion is obvious. □

The final problem we have to solve is to find an appropriate filtration with the necessary conditions of continuity. To this end we calculate the difference $\mathbb{E}_{\underline{t}} - \mathbb{E}_{\underline{s}}$ for $0 \leq \underline{s} < \underline{t}$. For $y \in \mathcal{H}^{(n)}$ we obtain

$$(\mathbb{E}_{\underline{t}} - \mathbb{E}_{\underline{s}})y = \sum_{|\omega|=n,\, \omega \leq 1_{[\underline{s},\underline{t}]}} y_\omega e_\omega .$$

Let $\underline{t} = k/N$, $\underline{s} = l/N$. It follows

$$\begin{aligned} \|\mathbb{E}_{\underline{t}} y - \mathbb{E}_{\underline{s}} y\|^2 &= \sum_{|\omega|=n,\, \omega \leq 1_{[\underline{s},\underline{t}[}} |y_\omega|^2 \\ &\leq \max_{|\omega|=n} |y_\omega|^2 \binom{k-l}{n} \\ &= \max_{|\omega|=n} |y_\omega|^2 N^n (\underline{t} - \underline{s})^n \cdot \left(1 - \frac{1}{k-l}\right) \cdots \left(1 - \frac{n-1}{k-l}\right). \end{aligned}$$

So the mapping $y \to \mathbb{E}_{\underline{t}} y$ will be S-continuous if $\max_{|\omega|=n} |y_\omega|^2 N^n$ will be finite.

This consideration leads to the internal measure m on $\Omega = \{0,1\}^T$ given by $m(\{\omega\}) = \frac{1}{N^{|\omega|}}$. Then (Ω, m) is the direct sum of $(\Omega_n, m_n)_{n \leq N}$ with $\Omega_n = \{\omega \in \Omega : |\omega| = n\}$ and $m_n = m|_{\Omega_n}$. Notice that $m(\Omega_n) = \binom{N}{n} \cdot \frac{1}{N^n}$, so that $m(\Omega) = \sum_{k=0}^N \binom{N}{k} \cdot \frac{1}{N^k} = \left(1 + \frac{1}{N}\right)^N \approx e$. Moreover $\sum_{k=L}^N m(\Omega_k) \approx 0$ for all $L \approx \infty$.

Let (Ω, L_m) be the Loeb measure space associated to (Ω, m). Then $L_m(\bigcup_{k \in \mathbb{N}} \Omega_k) = e$ and $\Omega \setminus (\bigcup_{k \in \mathbb{N}} \Omega_k)$ is an L_m- null set. Finally $L^2(\Omega, L_m) \cong \perp_{n \in \mathbb{N}} L^2(\Omega_n, L_{m_n})$ where L_{m_n} is the Loeb measure associated to m_n.

Consider the external subspace X of all S-bounded internal *complex valued functions on Ω. Then for $f \in X$ the complex-valued function ${}^\circ f : \omega \to {}^\circ(f(\omega))$ is in $L^2(\Omega, L_m)$ and the space $\{{}^\circ f : f \in X\}$ is dense in $L^2(\Omega, L_m)$. Moreover ${}^\circ f = \sum_{k \in \mathbb{N}_0} {}^\circ f_k$ in the L^2-sense, where $f_k = f \mid_{\Omega_k}$.

Now we map X onto a subspace Y of $\mathcal{H}$ by the mapping V defined as follows:

$$V(f) = \sum_{\omega \in \Omega} f(\omega) N^{-|\omega|/2} e_\omega,$$

and set

$$W(f) = \widehat{V(f)}.$$

Then $\int_\Omega |f|^2 dm = \|V(f)\|^2$. Let $\mathcal{L}$ be the closure of $\{W(f) : f \in X\} =: \mathcal{L}^{(b)}$ in $\mathcal{K}^\infty$. Then the last equation shows $L^2(\Omega, L_m) \cong \mathcal{L}$. Moreover $X_k = \{f \in X : f(\Omega_\ell) = 0 \text{ for } \ell \neq k\}$ is mapped onto a subspace of $\mathcal{H}^{(k)}$ (k standard). Let $\mathcal{L}_k$ be the closure in $\mathcal{K}^\infty$ of $\{W(f) : f \in X_k\} =: \mathcal{L}_k^{(b)}$. Then $\mathcal{L} = \perp_{k\in\mathbb{N}} \mathcal{L}_k$ and $\mathcal{L}_k \cong L^2(\Omega_k, L_{m_k})$.

The following facts are easily seen to hold:

Theorem 1

$$\begin{aligned}
\mathcal{L}_k^{(b)} &\subset \mathcal{K}_0^\infty \textit{ for } k \in \mathbb{N}_0 \\
\hat{A}_{\underline{t}}^{\bullet} \mathcal{L}_k^{(b)} &\subset \mathcal{L}_k \\
\hat{A}_{\underline{t}}^{\circ} \mathcal{L}_k^{(b)} &\subset \mathcal{L}_k \\
\hat{A}_{\underline{t}}^{+} \mathcal{L}_k^{(b)} &\subset \mathcal{L}_{k+1}^{(b)} \\
\hat{A}_{\underline{t}}^{-} \mathcal{L}_k^{(b)} &\subset \mathcal{L}_{k-1}^{(b)}.
\end{aligned}$$

Moreover $\mathcal{L} \cap \mathcal{K}_0^\infty$ is dense in $\mathcal{L}$ and it is the joint domain of definition of the operators $\hat{A}_t^\sharp$, $\sharp \in \{\bullet, \circ, +, -\}$ which are closable and essentially selfadjoint in case $\sharp = \bullet, \circ$, and adjoint to each other in case $\sharp = +, -$.

Remark: We map an internal set M of T onto its indicator function $1_M \in \{0,1\}^T = \Omega$. This mapping ψ is bijective and therefore we obtain an isometric isomorphism U of $L^2(\Omega, m)$ onto the internal Guichardet space of Leitz-Martini. This shows that our considerations concerning the continuity of the filtering leads in a very natural manner to the Loeb-Guichardet space for the basic quantum stochastic processes (cf. section 13.1).

13.6 The symmetric Fock space and its embedding into $\mathcal{L}$

Let $\mathcal{H}$ be a Hilbert space and let $\mathcal{H}^{\otimes n}$ be the nth tensor product of $\mathcal{H}$. We denote the full symmetric group of n elements by $\S_n$. Then by

$$P_n(x_1 \otimes \cdots \otimes x_n) = \frac{1}{n!} \sum_{\pi \in \S_n} x_{\pi(1)} \otimes \cdots \otimes x_{\pi(n)}$$

there is uniquely defined an orthogonal projection, the range of which is called the ***n*th symmetric tensor product** $\mathcal{H}^{\odot n}$. The **symmetric Fock space**

$\mathcal{F}_+(\mathcal{H})$ **over** $\mathcal{H}$ is the orthogonal sum

$$\mathcal{F}_+(\mathcal{H}) = \perp_{n=0}^{\infty} \mathcal{H}^{\bigodot n}$$

where as usual $\mathcal{H}^{\bigodot 0} = \mathbb{C}$.

For the special case $\mathcal{H} = L^2([0,1],\lambda)$ (λ: Lebesgue measure) the tensor product $\mathcal{H}^{\otimes n}$ is nothing else than $L^2([0,1]^n,\lambda^n)$ where λ^n denotes the n-dimensional Lebesgue measure, and the projection P_n is given by

$$P_n f(t_1,\ldots t_n) = \frac{1}{n!}\sum_{\pi\in S_n} f(t_{\pi(1)},\ldots t_{\pi(n)}).$$

Now we give the more or less classical definition of the basis quantum processes. To this end we set

$$f_1 \circ \cdots \circ f_n := P_n(f_1 \otimes \cdots \otimes f_n).$$

For $0 \le t < 1$ the definition of the creation operator is given by

$$\mathcal{A}_t^+((f_1 \circ \cdots \circ f_n) := 1_{[0,t[} \circ f_1 \circ \cdots \circ f_n \tag{13.6.1}$$

and linear extension to the largest possible domain.

Similarly the annihilation operator is given by

$$\mathcal{A}_t^-(f_1 \circ \cdots \circ f_n) = \sum_{j=1}^{n} \left(1_{[0,t[}|f_j\right) f_1 \circ \cdots \circ \hat{f}_j \circ \cdots \circ f_n, \tag{13.6.2}$$

where the expression $\hat{f}_j$ indicates that this factor is omitted.

The number operator is given by

$$\mathcal{A}_t^\circ(f_1 \circ \cdots \circ f_n) = 1_{[0,t[}\cdot f_1 \circ \cdots \circ f_n + \cdots + f_1 \circ f_2 \circ \cdots \circ f_{n-1} \circ 1_{[0,t[}\cdot f_n, \tag{13.6.3}$$

whereas the time-operator is

$$\mathcal{A}_t^\bullet(f_1 \circ \cdots \circ f_n) = t f_1 \circ \cdots \circ f_n. \tag{13.6.4}$$

Remark: These operators are those ones introduced by Hudson and Parthasarathy [5] as the basic quantum processes.

Consider now the set $X_n = \{(t_1,t_2,\ldots,t_n) : 0 \le t_1 < \cdots < t_n \le 1\}$. It is a measurable subset of $[0,1]^n$. Let f be an element of $L^2(X_n,\lambda^n)$. Setting

$$\tilde{f}(t_1,\ldots,t_n) = \begin{cases} f(t_1,\ldots,t_n) & (t_1,\ldots,t_n) \in X_n \\ 0 & \text{otherwise} \end{cases}$$

we obtain an element $\tilde{f}$ of $L^2([0,1]^n, \lambda^n)$. Then $P_n(\tilde{f}) \in L^2([0,1],\lambda)^{\odot n}$ and $\|P_n(\tilde{f})\|^2 = \int_{X_n} |f|^2 d\lambda^n$, i.e. $\|f\| = \|P_n(\tilde{f})\|$ holds. On the other hand let $f \in L^2([0,1],\lambda)^{\odot n}$ be arbitrary. Then $f = n!\, P_n(\widetilde{f_{|X_n}})$ (equality in the L^2-sense). It follows that the mapping $f \to P_n(\tilde{f})$ is a linear isometry from $L^2(X_n, \lambda^n)$ onto $L^2([0,1],\lambda)^{\odot n}$. Its inverse is given by $g \to n!\, g_{|X_n} =: U_n(g)$.

Now we embed $L^2(X_n, \lambda^n)$ isometrically into the space $\mathcal{L}_n$ constructed at the end of the previous section. To this end let $\omega \in \Omega_n$ be arbitrary. Set $t_1(\omega) = \min(\underline{s} : \omega(\underline{s}) = 1)$ and by induction $\underline{t}_{k+1}(\omega) = \min(\underline{s} > \underline{t}_k : \omega(\underline{s}) = 1)$. Then $\vec{\underline{t}}(\omega) := (\underline{t}_1(\omega), \ldots, \underline{t}_n(\omega)) \in {}^*X_n \cap T^n$. Moreover if $\vec{\underline{t}} := (k_1/N, \ldots, k_n/N) \in {}^*X_n \cap T^n$ then $\vec{\underline{t}} = \vec{\underline{t}}(\omega)$ for $\omega = 1_M$ where $M = \{\underline{t}_1(\omega), \ldots, \underline{t}_n(\omega)\}$.

We consider the dense subspace C_n of $L^2(X_n, \lambda^n)$ consisting of the restrictions to X_n of continuous functions on the closure $\overline{X_n}$. For $f \in C_n$ we set $S(f)(\omega) = {}^*f(\vec{\underline{t}}(\omega))$. This is an element of $L^2(\Omega_n, m_n)$ satisfying $\|f\|_2 \approx \|S(f)\|_2$. It follows that C_n is linearly and isometrically embedded into $L^2(\Omega_n, L_{m_n})$, and this embedding can obviously be extended to the whole of $L^2(X_n, \lambda^n)$. Since $L^2(\Omega_n, L_{m_n})$ is linearly and isometrically embedded in $\mathcal{L}_n$, we obtain that the symmetric tensor product $L^2([0,1],\lambda)^{\odot n}$ can be embedded in $\mathcal{L}_n$. An obvious extension then yields an embedding denoted by U of the symmetric Fock space $\mathcal{F}_+(L^2([0,1],\lambda)$ into $\mathcal{L}$. Call $\mathcal{G}$ the image of this embedding and set $\mathcal{G}_0 = G \cap \mathcal{K}_0^\infty$.

Theorem 2 *$\mathcal{G}_0$ is dense in $\mathcal{G}$. Moreover the restrictions of $A^\sharp$ to $\mathcal{G}_0$ yield closed densely defined operators which are selfadjoint for $\sharp \in \{\circ, \bullet\}$ and adjoint to each other in the other cases and which moreover satisfy*

$$\mathcal{A}_t^\sharp = U^{-1} A_t^\sharp U$$

on $U^{-1}(\mathcal{G}_0)$ where $\mathcal{A}^\sharp$ are the operators given by equations (13.6.1) up to (13.6.4).

The proof is obvious.

References

[1] R. M. Anderson, "A nonstandard representation for Brownian motion and Itô integral", Israel J. Mathem., **25** 15–46.

[2] S. Attal, Approximating the Fock space with the toy Fockspace, in J. Azéma, M. Émery, M. Ledoux, M. Yor (eds.), *Séminaire de Probabilités XXXVI*, Lecture Notes in Mathematics 1801, Springer-Verlag, Berlin New York, 2003.

[3] A. GUICHARDET, *Symmetric Hilbert spaces and related topics*, Lecture Notes in Mathematics 261, Springer-Verlag, Berlin, 1972.

[4] R. L. HUDSON and J. M. LINDSAY, "A noncommutative martingale representation theorem for non-Fock- Brownian motion", J. Funct. Anal., **61** (1985) 202–221.

[5] R. L. HUDSON and K. R. PARTHASARATHY, "Quantum Itô's formula and stochastic evolutions", Comm. Math. Physics, **93** (1984) 311–323.

[6] R. L. HUDSON and R. F. STREATER, "Itô's formula is the chain rule with Wick ordering", Phys. Letters, **86** (1982) 277–279.

[7] J. L. JOURNÉ, "Structure des cocycles markoviens sur l'espace de Fock", Prob. Theory Re. Fields, **75** (1987) 291–316.

[8] B. KÜMMERER and H. MAASSEN, "Elements of quantum probability", Quantum Probab. Communications, QP-PQ, **10** (1998) 73–100.

[9] H. MAASSEN, Quantum Markov processes on Fock space described by integral kernels, in L. Accardi, W. von Waldenfels, *Quantum probability and applications II, Proceedings Heidelberg 1984*, Lecture Notes in Mathematics 1136, Springer-Verlag, Berlin, New York, 1985.

[10] M. LEITZ-MARTINI, *Quantum stochastic calculus using infinitesimals*, Ph.-D. Thesis, University of Tübingen, 2001.

[11] P.A. MEYER, *Quantum probability for probabilists*, Lecture Notes in Mathematics 1538, Springer-Verlag, Berlin, New York, 1993.

[12] K. R. PARTHASARATHY, The passage from random walk to diffusion in quantum probability, *Celebration Volume in Appl. Probability* (1988), Applied Probability Trust, Sheffield, 231–245.

[13] K. R. PARTHASARATHY, *An introduction to quantum stochastic calculus*, Birkhäuser, Basel, 1992.

14

Applications of rich measure spaces formed from nonstandard models

Peter Loeb*

Abstract

We review some recent work by Yeneng Sun and the author. Sun's work shows that there are results, some used for decades without a rigourous foundation, that are only true for spaces with the rich structure of Loeb measure spaces. His joint work with the author uses that structure to extend an important result on the purification of measure valued maps.

14.1 Introduction

In 1975 [11], the author constructed a class of standard measure spaces formed on nonstandard models. These spaces, now called "Loeb spaces" in the literature, are very close to the underlying internal spaces, and are rich in structure. (See the author's three chapters and Osswald's two chapters in [12] for background.) We will briefly review here some recent work by Yeneng Sun that uses these measure spaces. Sun has shown that there are results, some used in applications for decades without a rigorous foundation, that are only true for spaces with the rich structure of Loeb measure spaces. We will follow the description of Sun's work with a detailed proof of a special case of the recent result in [14] by Yeneng Sun and the author on the purification of measure-valued maps. A counterexample shows that the result we give is false if the Loeb space we use is replaced by the unit interval with Lebesgue measure. We note in passing here that the result in [13] by Osswald, Sun, Zhang and the author presents yet another example needing rich measure spaces.

*Department of Mathematics, University of Illinois, Urbana, IL 61801.
`loeb@math.uiuc.edu`

14.2 Recent work of Yeneng Sun

We begin with some work in the late 1990's of Yeneng Sun ([16], [17], [18], and Chapter 7 of [12]). That work, Sun's alone, is published elsewhere, so what is said here is an invitation to read, and not a substitute for, the original articles. To set the background we consider the following question: Does it make sense to speak of an infinite number of independent individuals or random variables indexed by points in a uniform probability space? Can one, for example, reasonably consider an infinite number of independent tosses of a fair coin with the tosses indexed by a uniform probability space, and if so, does it make sense to say that half the tosses should be heads? Of course, there is no problem in speaking of independent random variables indexed by the first n integers with each integer having probability $1/n$. An infinite index set with a uniform probability measure, however, must be uncountable; there is no uniform probability measure on the full set of natural numbers. The problem thus evolves to finding the possible meaning of an uncountable family of independent random variables. Whatever way one approaches this problem, the usual measure-theoretic tools fail.

A natural attempt to generalize coin tossing replaces the natural numbers with points in the interval $[0,1]$, and thus replaces sequences of 1's and -1's with functions from $[0,1]$ to the two-point set $\{-1,1\}$. This is a reasonable model for an uncountable family of independent coin tosses. In 1937, Doob [2] exhibited a problem with this and similar spaces of functions when standard techniques are applied. To see the problem, let Ω denote the set of $\{-1,1\}$-valued functions on $[0,1]$, and let P be the product measure on Ω constructed from the measure taking the values $1/2$ at 1 and $1/2$ at -1. In the usual construction of an appropriate σ-algebra on Ω, measurable sets are formed from the algebra of cylinder sets, with functions in each cylinder set restricted at only a finite number of elements of $[0,1]$ to take either the value 1 or -1. It follows that each measurable set in Ω is determined in a way described below by a countable subset of $[0,1]$, and so the following result holds.

Proposition 14.2.1 *Fix any $h \in \Omega$. Set*

$$M_h := \{\omega \in \Omega : \omega(t) = h(t) \text{ except for countably many } t \in [0,1]\}.$$

Then M_h has P-outer measure 1.

Proof. For each measurable $B \subseteq \Omega$, there is a countable set $C \subset [0,1]$ such that for all α, β in Ω, if $\alpha(t) = \beta(t)$ for all $t \in C$, then $\alpha \in B$ if and only if $\beta \in B$. Suppose $M_h \subseteq B$. Given $\omega \in \Omega$, Let ω' agree with ω on C and agree with h on $[0,1] \setminus C$. Then $\omega' \in M_h$, so $\omega' \in B$. It then follows that $\omega \in B$. Thus $B = \Omega$, so the outer measure $P^*(M_h) = 1$. □

Remark 14.2.2 Note that if $\omega \in M_h$, then since Lebesgue measure λ of a countable set is 0, $\omega = h$ λ-a.e. on $[0,1]$. This is true if h is nonmeasurable, or if $h \equiv 1$, or if $h \equiv -1$. Now outer measure when applied to the intersection of measurable sets with M_h is finitely additive, hence countably additive. Since the outer measure of M_h is 1, one can trivially extend P to a measure $\overline{P}$ with $\overline{P}(M_h) = 1$. Thus, no matter what h might be, one can claim that $\overline{P}$ almost every function is equal to h at λ-almost very point of $[0,1]$. This, and other highly questionable arguments have been used for decades to work around the measure theoretic problem indicated by Doob's example.

Another approach to representing a continuum of independent random variables is to consider a function $f(i,\omega)$, called a process, where i is an index from an uncountable probability space called the parameter space (the probability measure need not be uniform) and ω is taken from a second probability space called the sample space. The question then is whether it makes sense to work with the usual product of these two probability spaces. Sun has shown in Proposition 7.33 of [12] that no matter what kind of measure spaces, even Loeb measure spaces, one might take as the parameter space and sample space of a process, independence and joint measurability with respect to the classical measure-theoretic product, i.e., formed using measurable rectangles as in [15], are never compatible with each other except for a trivial case. Here is the exact statement of that proposition.

Proposition 14.2.3 (Sun) *Let $(I,\mathcal{I},\mu)$ and $(X,\mathcal{X},\nu)$ be any two probability spaces. Form the classical, complete product probability space $(I \times X, \mathcal{I} \otimes \mathcal{X}, \mu \otimes \nu)$. Let f be a function from $I \times X$ to a separable metric space. If f is jointly measurable on the product probability space, and for $\mu \otimes \mu$-almost all $(i_1, i_2) \in I \times I$, f_{i_1} and f_{i_2} are independent (call this **almost sure pairwise independence**), then, for μ-almost all $i \in I$, $f(i,\cdot)$ is a constant function on X.*

Sun notes that Proposition 14.2.3 is still valid when μ has an atom A. The almost sure pairwise independence condition implies the essential constancy of the random variables f_i for almost all $i \in A$.

In the articles cited at the beginning of this section, Sun has shown that a construction overcoming these measure-theoretic problems is obtained by forming the internal product of internal factors, and then taking not just the Loeb space of each factor, but also the Loeb space of the internal product. This yields a rich extension of the usual product σ-algebra formed from the Loeb factors, one that still has the Fubini Property equating the integral over the product space to the iterated integrals over the factor spaces forming that product. Here in more detail is that construction.

Sun's starts with internal spaces $(T, \mathcal{T}, \lambda)$ and $(\Omega, \mathcal{A}, P)$. The space T may be a hyperfinite set with λ given by uniform weights. He then forms the Loeb spaces $\big(T, L_\lambda(\mathcal{T}), \widehat{\lambda}\big)$ and $\big(\Omega, L_\mu(\mathcal{A}), \widehat{P}\big)$. He lets $\lambda \otimes P$ denote the internal product measure, while $\mathcal{T} \otimes \mathcal{A}$ denotes the internal product σ-algebra, and $L_\lambda(\mathcal{T}) \otimes L_P(\mathcal{A})$ denotes the classical product σ-algebra formed from $L_\lambda(\mathcal{T})$ and $L_P(\mathcal{A})$ as in [15]. On the other hand, forming the Loeb space from the internal product $\mathcal{T} \otimes \mathcal{A}$ produces a larger σ-algebra $L_{\lambda\otimes P}(\mathcal{T} \otimes \mathcal{A})$ on $T \times \Omega$. (See the following propositions.) Here are some properties of what we shall call the big product space $\big(T \times \Omega, L_{\lambda\otimes P}(\mathcal{T} \otimes \mathcal{A}), \widehat{\lambda \otimes P}\big)$.

Proposition 14.2.4 (Keisler-Sun [7]) *The big product space depends only on the Loeb factor spaces* $\big(T, L_\lambda(\mathcal{T}), \widehat{\lambda}\big)$ *and* $\big(\Omega, L_P(\mathcal{A}), \widehat{P}\big)$.

Proposition 14.2.5 (Anderson [1]) *If* $E \in L_\lambda(\mathcal{T}) \otimes L_P(\mathcal{A})$, *then* $E \in L_{\lambda\otimes P}(\mathcal{T} \otimes \mathcal{A})$ *and* $\widehat{\lambda \otimes P}(E) = \widehat{\lambda} \otimes \widehat{P}(E)$.

Proposition 14.2.6 (Hoover (first example), Sun [17]) *The inclusion of* $L_\lambda(\mathcal{T}) \otimes L_P(\mathcal{A})$ *in* $L_{\lambda\otimes P}(\mathcal{T} \otimes \mathcal{A})$ *is strict if and only if both* $\widehat{\lambda}$ *and* $\widehat{P}$ *have non-atomic parts.*

Proposition 14.2.7 (Keisler [6]) *A Fubini Theorem holds for the big product space.*

In his articles, cited above, Sun shows that to work with independence in a continuum setting, as has been attempted in informal mathematical applications without rigor, and for decades, one needs a rich product σ-algebra such as the σ-algebra of a big product space. Here is that general result as stated in Proposition 7.4.1 of [12] and proved in Theorem 6.2 of [17].

Proposition 14.2.8 (Sun) *Let* X *be a complete, separable and metrizable topological space. Let* $\mathcal{M}(X)$ *be the space of Borel probability measures on* X, *where* $\mathcal{M}(X)$ *is endowed with the topology of weak convergence of measures. Let* μ *be any Borel probability measure on the space* $\mathcal{M}(X)$. *If both* $\widehat{\lambda}$ *and* $\widehat{P}$ *are atomless, then there is a process* f *from* $(T\times\Omega, L_{\lambda\otimes P}(\mathcal{T}\otimes\mathcal{A}), \widehat{\lambda \otimes P})$ *to* X *such that the random variables* $f_t = f(t, \cdot)$ *are almost surely pairwise independent (i.e., for* $\widehat{\lambda \otimes \lambda}$*-almost all* $(t_1, t_2) \in T\times T$, f_{t_1} *and* f_{t_2} *are independent), and the probability measure on* $\mathcal{M}(X)$ *induced by the function* $\widehat{P} f_t^{-1}$ *from* T *to* $\mathcal{M}(X)$ *is the given measure* μ.

Remark 14.2.9 Here, for any given $t \in T$, $\widehat{P} f_t^{-1}$ is the probability measure on X induced by the random variable $f_t : \Omega \to X$. It is the measurable mapping from T to $\mathcal{M}(X)$ taking the value $\widehat{P} f_t^{-1}$ at each $t \in T$ that induces the measure μ on $\mathcal{M}(X)$.

Remark 14.2.10 A consequence of the proposition is that when the space $(T, L_\lambda(\mathcal{T}), \widehat{\lambda})$ is a hyperfinite Loeb counting probability space and $\widehat{\mu}$ is a non-atomic Loeb probability measure, it makes sense to use the big product space as the underlying product space for an infinite number of equally weighted, independent random variables or agents.

As part of his work with large product spaces, Sun has extended the law of large numbers. The usual strong law states that if random variables X_i, $i \in \mathbb{N}$, are independent with the same distribution and finite mean m, then $\frac{1}{n}\Sigma_{i=1}^n X_i$ tends almost surely to the constant random variable m. That is, for almost all samples ω, the value of the sequence at ω tends to the constant m.

Theorem 14.2.11 (Sun) *Let f be a real-valued integrable process on the big product space. If the random variables $f_t := f(t, \cdot)$ are almost surely pairwise independent, then for almost all samples $\omega \in \Omega$, the mean of the sample function $f_\omega := f(\cdot, \omega)$ on the parameter space T is the mean of f viewed as a random variable on the big product space. There is no requirement of identical distributions.*

Another facet of this same work deals with independence. It is well known that for a finite collection of random variables, pairwise independence is strictly weaker than mutual independence. Sun has shown that for processes on big product spaces, pairwise independent and mutual independent coalesce, and they coalesce with other notions of independence that are distinct for a finite number of random variables. This implies asymptotic results for finite families of random variables where the families are ordered by containment and have increasing cardinality.

14.3 Purification of measure-valued maps

We now turn to a special case of the joint work of Yeneng Sun with the author in [14]. That special case generalizes the following celebrated theorem of Dvoretzky, Wald and Wolfowitz (see [3], [4], [5]).

Theorem 14.3.1 *Let A be a finite set and $\mathcal{M}(A)$ the space of probability measures on A. Let $(T, \mathcal{T})$ be a measurable space, and μ_k, $k = 1, \cdots, m$, finite, atomless signed measures on $(T, \mathcal{T})$. Given $f : T \to \mathcal{M}(A)$ so that for each $a \in A$, $f(\cdot)(\{a\})$ is $\mathcal{T}$-measurable, there is a $\mathcal{T}$-measurable map $g : T \to A$ so that for each $a \in A$, and each $k \leq m$,*

$$\int_T f(t)(\{a\}) d\mu_k(t) = \mu_k(\{t \in T : g(t) = a\}).$$

This theorem justifies the elimination, i.e., purification, of randomness in some settings. For example, in games, T represents information available to the players, and A represents the actions players may choose, given $t \in T$. Every player's objective is to maximize her own expected payoff, but that payoff depends on the actions chosen by all the players.

For each player, a mapping from T to A is called a pure strategy. A mapping from T to $\mathcal{M}(A)$ is called a mixed strategy; in this case, each player chooses a "lottery on A". A Nash equilibrium is achieved if every player is satisfied with her own choice of strategy given the choices of the other players.

In quite general settings, such an equilibrium can be achieved using mixed strategies. When results such as Theorem 14.3.1 apply, those mixed strategies can be purified yielding an equilibrium with the same expected payoffs. We refer the reader to [14] for more information on the game-theoretic consequences of Theorem 14.3.1 and its extension. We will concentrate here on the proof of that extension, a proof simpler than that of the more general result in [14].

We extend the DWW Theorem from a finite set A to a complete, separable metric space. There always exists a Borel bijection from such a space to a compact metric space. That is, if A is uncountable, then it follows from Kuratowski's theorem (see [15], p. 406) that there is a Borel bijection from A to $[0, 1]$. On the other hand, for a countable set A, one can use a bijection from A to $\{0, 1, 1/2, \dots, 1/n, \dots\}$. Therefore, it suffices to let A be a compact metric space. For simplicity we will work with a finite set of measures, but may extend to a countable collection.

Theorem 14.3.2 *Let K be a finite set, and let A be a compact metric space. For each $k \in K$, let μ_k be a nonatomic, finite, signed Loeb measure on a Loeb measurable space $(T, \mathcal{T})$. If f is a $\mathcal{T}$-measurable mapping from T to $\mathcal{M}(A)$, then there is a $\mathcal{T}$-measurable mapping g from T to A such that for each $k \in K$ and for all Borel sets B in A, $\int_T f(t)(B)\mu_k(dt) = \mu_k\left(g^{-1}[B]\right)$. This is equivalent to the condition that for each $k \in K$ and for any bounded Borel measurable function θ on A,*

$$\int_T \int_A \theta(a) f(t)(da)\mu_k(dt) = \int_T \theta(g(t))\mu_k(dt).$$

Example 14.3.3 Sun and the author show by a counter example in [14] that the Loeb measures of the theorem cannot be replaced with measures absolutely continuous with respect to Lebesgue measure on $[0, 1]$. In their example, the compact set A is the interval $[-1, 1]$, and the measure-valued map is given by $f(t) := (\delta_t + \delta_{-t})/2$ for each $t \in [0, 1]$. The example uses just two measures, $\mu_1 = \lambda$ on $[0, 1]$, and $\mu_2 = 2t\lambda$ on $[0, 1]$. If there is a function g that works, then it is not hard to see using the functions on A given by $\theta(a) = |a|$ and $\theta(a) = a^2$

that $\int_0^1 (t - |g(t)|)^2 \lambda(dt) = 0$. That is, $g(t)$ must take the value t or $-t$ λ-a.e. on $[0,1]$. It is also not hard to see that this is impossible. The moral is that a Lebesgue measurable g can't switch values on $[0,1]$ fast enough to work.

To set up the proof of the theorem, we let $\widetilde{\mu}_k$ be the signed internal measure generating μ_k for each $k \in K$. Also, we set $P_0 = \frac{1}{c}\Sigma_{k\in K} |\widetilde{\mu}_k|$, where $|\widetilde{\mu}_k|$ is the internal total variation of $\widetilde{\mu}_k$ and $c = \Sigma_{k\in K} |\widetilde{\mu}_k|(T)$. Then P_0 is an internal probability measure on T, and for each $k \in K$, $\widetilde{\mu}_k << P_0$. Let $P = \frac{1}{\mathrm{st}(c)}\Sigma_{k\in K} |\mu_k|$. The probability measure P is the Loeb measure generated by P_0, and for each $k \in K$, $\mu_k << P$. Let $\widetilde{\beta}_k$ be the internal Radon-Nikodym derivative of $\widetilde{\mu}_k$ with respect to P_0. Since $\mu_k << P$, $\widetilde{\beta}_k$ is S-integrable with respect to P_0 and $\beta_k := \mathrm{st}\,\widetilde{\beta}_k$ is the Radon-Nikodym derivative of μ_k with respect to P. We let $\mathcal{B}$ denote the collection of Borel subsets of A.

The following lemma is the only part of the proof that needs nonstandard analysis.

Lemma 14.3.4 *Let $\{\phi_i : i \in \mathbb{N}\}$ be a countable, dense (with respect to the sup-norm topology) subcollection of the continuous real-valued functions on A. Assume that there is a sequence of $\mathcal{T}$-measurable mappings $\{g_n,\ n \in \mathbb{N}\}$ from T to A such that for each $i \in \mathbb{N}$ and $k \in K$, the sequence $\int_T \phi_i(g_n(t))\beta_k(t)P(dt)$ converges; let $c_{i,k} \in \mathbb{R}$ denote the limit. Then, there is a $\mathcal{T}$-measurable mapping g from T to A such that for each $i \in \mathbb{N}$ and $k \in K$,*

$$\int_T \phi_i(g(t))\beta_k(t)P(dt) = c_{i,k}.$$

Proof. For each $n \in \mathbb{N}$, let $h_n : T \to {}^*A$ be a $\mathcal{T}_0$-measurable lifting of g_n with respect to the internal measure P_0. Then for each n and $i \in \mathbb{N}$ and each $k \in K$,

$$\int_T \phi_i(g_n(t))\beta_k(t)P(dt) \simeq \int_T {}^*\phi_i(h_n(t))\widetilde{\beta}_k(t)P_0(dt),$$

and so

$$\lim_{\substack{n\in\mathbb{N}\\ n\to\infty}} \left(\mathrm{st}\left|\int_T {}^*\phi_i(h_n(t))\widetilde{\beta}_k(t)P_0(dt) - c_{i,k}\right|\right) = 0.$$

Using $\aleph_1$-saturation, we may extend the sequence h_n to an internal sequence and choose an unlimited integer $H \in {}^*\mathbb{N}$ so that for every $i \in \mathbb{N}$ and each $k \in K$,

$$\int_T {}^*\phi_i(h_H(t))\widetilde{\beta}_k(t)P_0(dt) \simeq c_{i,k}.$$

The desired function g is obtained by setting $g(t) := \mathrm{st}(h_H(t)) \in A$ at each $t \in T$. □

Now for the proof of Theorem 14.3.2, we note first that since $\mathcal{M}(A)$ is a compact metric space under the Prohorov metric ρ (which induces the topology of weak convergence of measures), there is a sequence of simple functions $\{f_n\}_{n=1}^{\infty}$ from $(T, \mathcal{T})$ to $\mathcal{M}(A)$ such that

$$\forall t \in T, \lim_{n\to\infty} \rho(f_n(t), f(t)) = 0.$$

Assume we know that for each $n \in \mathbb{N}$, there is a $\mathcal{T}$-measurable mapping g_n from T to A such that for each $k \in K$, and any bounded Borel measurable function θ on A,

$$\int_T \int_A \theta(a) f_n(t)(da)\mu_k(dt) = \int_T \theta(g_n(t))\mu_k(dt).$$

Let $\mathcal{G} = \{\phi_i : i \in \mathbb{N}\}$ be a countable dense subset of the continuous real-valued functions on A with the sup-norm topology. For a given $\phi_i \in \mathcal{G}$ and each $t \in T$, $\int_A \phi_i(a) f_n(t)(da) \to \int_A \phi_i(a) f(t)(da)$. Moreover, each integral is bounded by the maximum value of $|\phi_i|$ on A, so by the bounded convergence theorem, for each $k \in K$,

$$\begin{aligned}\int_T \phi_i(g_n(t))\beta_k(t)P(dt) &= \int_T \phi_i(g_n(t))\mu_k(dt)\\ &= \int_T \int_A \phi_i(a) f_n(t)(da)\mu_k(dt)\\ &\to \int_T \int_A \phi_i(a) f(t)(da)\mu_k(dt)\end{aligned}$$

By the lemma, there is a $\mathcal{T}$-measurable mapping g from T to A such that for each $k \in K$,

$$\begin{aligned}\int_T \phi_i(g(t))\mu_k(dt) &= \int_T \phi_i(g(t))\beta_k(t)P(dt)\\ &= \int_T \int_A \phi_i(a) f(t)(da)\mu_k(dt).\end{aligned}$$

Since this is true for each ϕ_i in $\mathcal{G}$, it is true with ϕ_i replaced by an arbitrary bounded Borel measurable function θ on A. Therefore, without loss of generality, we may assume that $f : T \to \mathcal{M}(A)$ is simple.

Next, we fix a sequence of Borel measurable, finite partitions $\mathcal{P}^m = \{A_1^m, \ldots, A_{l_m}^m\}$ of A such that the diameter of each set in $\mathcal{P}_m$ is at most $1/2^m$, and $\mathcal{P}_{m+1}$ is a refinement of $\mathcal{P}_m$. Also for each l, $1 \le l \le l_m$, we pick a point $a_l^m \in A_l^m$.

Since f is a simple function from T to $\mathcal{M}(A)$, there is a $\mathcal{T}$-measurable partition $\{S_j\}_{j=1}^N$ of T such that $f \equiv \gamma_j \in \mathcal{M}(A)$ on S_j. It follows that for each $B \in \mathcal{B}$, and each $k \in K$, $\int_T f(t)(B)\mu_k(dt) = \sum_{j=1}^N \gamma_j(B)\mu_k(S_j)$. By Lyapunov's Theorem, each S_j can be decomposed by a finite, $\mathcal{T}$-measurable partition $\{T_1^{j,m}, \ldots, T_{l_m}^{j,m}\}$ so that for every $l \leq l_m$ and $k \in K$, $\mu_k(T_l^{j,m}) = \gamma_j(A_l^m)\mu_k(S_j)$, whence

$$\int_T f(t)(A_l^m)\mu_k(dt) = \sum_{j=1}^N \gamma_j(A_l^m)\mu_k(S_j)$$
$$= \sum_{j=1}^N \mu_k\left(T_l^{j,m}\right).$$

(As an alternative to the above use of the Lyapunov theorem, one can with small modifications of the proof here use the author's Lyapunov theorem [10], but then all of the simple functions must be modified on a P-null set T_0 so that for each of them, the corresponding partition sets S_j of T are internal.)

Now for each $m \geq 1$, we define a $\mathcal{T}$-measurable mapping $g_m : T \to A$ so that for each $l \leq l_m$ and each $j \leq N$, $g_m(t) \equiv a_l^m$ on $T_l^{j,m}$. For each continuous, real-valued ψ on A, for each $m \geq 1$ and each $k \in K$,

$$\sum_{l=1}^{l_m} \psi(a_l^m)\left(\int_T f(t)(A_l^m)\mu_k(dt)\right) = \sum_{l=1}^{l_m} \psi(a_l^m)\left(\sum_{j=1}^N \gamma_j(A_l^m)\mu_k(S_j)\right)$$
$$= \sum_{l=1}^{l_m}\sum_{j=1}^N \left(\psi(a_l^m)\gamma_j(A_l^m)\right)\mu_k(S_j)$$
$$= \sum_{l=1}^{l_m}\sum_{j=1}^N \psi(a_l^m)\mu_k\left(T_l^{j,m}\right)$$

approximates as $m \to \infty$ the integral

$$\int_T\left(\int_A \psi(a)f(t)(da)\right)\mu_k(dt) = \sum_{l=1}^{l_m}\int_T\left(\int_{A_l^m}\psi(a)f(t)(da)\right)\mu_k(dt)$$
$$= \sum_{l=1}^{l_m}\sum_{j=1}^N \int_{S_j}\left(\int_{A_l^m}\psi(a)\gamma_j(da)\right)\mu_k(dt)$$
$$= \sum_{l=1}^{l_m}\sum_{j=1}^N \left(\int_{A_l^m}\psi(a)\gamma_j(da)\right)\mu_k(S_j)$$

so

$$\int_T \psi(g_m(t))\beta_k(t)P(dt) = \int_T \psi(g_m(t))\mu_k(dt)$$
$$= \sum_{l=1}^{l_m}\sum_{j=1}^{N} \psi(a_l^m)\mu_k\left(T_l^{j,m}\right)$$
$$\to \int_T \left(\int_A \psi(a)f(t)(da)\right)\mu_k(dt)$$

It follows from the lemma that there is a $\mathcal{T}$-measurable mapping g from T to A such that for each $k \in K$ and for each continuous real-valued function θ on A from a countable dense set of such functions, and therefore for each bounded Borel measurable function θ on A,

$$\int_T \theta(g(t))\mu_k(dt) = \int_T \int_A \theta(a)f(t)(da)\mu_k(dt). \qquad \square$$

References

[1] R.M. ANDERSON, "A nonstandard representation of Brownian motion and Itô integration", Israel J. Math., **25** (1976) 15–46.

[2] J.L. DOOB, "Stochastic processes depending on a continuous parameter", Trans. Amer. Math. Soc., **42** (1937) 107–140.

[3] A. DVORETSKY, A. WALD and J. WOLFOWITZ, "Elimination of randomization in certain problems of statistics and of the theory of games", Proc. Nat. Acad. Sci. USA, **36** (1950) 256–260.

[4] A. DVORETSKY, A. WALD and J. WOLFOWITZ, "Relations among certain ranges of vector measures", Pac. J. Math., **1** (1951) 59–74.

[5] A. DVORETSKY, A. WALD and J. WOLFOWITZ, "Elimination of randomization in certain statistical decision problems in certain statistical decision procedures and zero-sum two-person games", Ann. Math. Stat., **22** (1951) 1–21.

[6] H.J. KEISLER, *An infinitesimal approach to stochastic analysis*, Memoirs Amer. Math. Soc. 48, 1984.

[7] H.J. KEISLER and Y.N. SUN, "A metric on probabilities, and products of Loeb spaces", Jour. London Math. Soc., **69** (2004) 258–272.

[8] M.A. Khan, K. P. Rath and Y. N. Sun, "The Dvoretzky-Wald-Wolfowitz theorem and purification in atomless finite-action games", International Journal of Game Theory, **34** (2006) 91–104.

[9] M.A. Khan and Y. N. Sun, "Non-cooperative games on hyperfinite Loeb spaces", J. Math. Econ., **31** (1999) 455–492.

[10] P.A. Loeb, "A combinatorial analog of Lyapunov's Theorem for infinitesimally generated atomic vector measures", Proc. Amer. Math. Soc., **39** (1973) 585–586.

[11] P.A. Loeb, "Conversion from nonstandard to standard measure spaces and applications in probability theory", Trans. Amer. Math. Soc., **211** (1975) 113–122.

[12] P.A. Loeb and M. Wolff, eds., *Nonstandard Analysis for the Working Mathematician*, Kluwer Academic Publishers, Amsterdam, 2000.

[13] P.A. Loeb, H. Osswald, Y. Sun and Z. Zhang, "Uncorrelatedness and orthogonality for vector-valued processes", Tran. Amer. Math. Soc., **356** (2004) 3209–3225.

[14] P.A. Loeb and Y. Sun, "Purification of measure-valued maps", Doob Memorial Volume of the Illinois Journal of Mathematics, **50** (2006) 747–762.

[15] H.L. Royden, *Real Analysis*, third edition, Macmillan, New York, 1988.

[16] Y. Sun, "Hyperfinite law of large numbers", Bull. Symbolic Logic, **2** (1996) 189–198.

[17] Y. Sun, "A theory of hyperfinite processes: the complete removal of individual uncertainty via exact LLN", J. Math. Econ., **29** (1998) 419–503.

[18] Y. Sun, "The almost equivalence of pairwise and mutual independence and the duality with exchangeability", Probab. Theory and Relat. Fields, **112** (1998) 425–456.

15

More on S-measures

David A. Ross*

15.1 Introduction

In their important (but often overlooked) paper [1], C. Ward Henson and Frank Wattenberg introduced the notion of *S-measurability*, and showed that S-measurable functions are "approximately standard" (in a sense made precise in the next section).

In a recent paper ([3]), the author used this machinery to transform a well-known nonstandard plausibility argument for the Radon-Nikodým theorem into a correct and complete nonstandard proof of the theorem. The problem of finding an "essentially nonstandard" proof for Radon-Nikodým had been one of long standing. Although Luxemburg gave a nonstandard proof as long ago as 1972 ([2]), he obtained the result as a consequence of another equally-deep theorem in analysis due to Riesz. (Beate Zimmer [6] has recently proved vector-valued extensions of Radon-Nikodym starting from the same plausibility argument, though using different standardizing machinery than that in [3] or the present paper.)

In this paper I use an S-measure argument very like the one in [3] to give an intuitive nonstandard proof of the Riesz result used by Luxemburg. Along the way I give a new proof for the main technical result from [1] on S-measures, a new nonstandard proof for Egoroff's Theorem, and a nonstandard proof for the existence of the conditional expectation operator which is more elementary than that in [3].

15.2 Loeb measures and S-measures

I will assume that we work in a nonstandard model in the sense of Robinson, and that this model is as saturated as it needs to be to carry out all constructions; in particular, it is an enlargement.

*Department of Mathematics, University of Hawaii, Honolulu, HI 96822.
ross@math.hawaii.edu

Suppose X a set and $\mathcal{A}$ is an algebra on X. There are two natural algebras on ${}^*X$: $\mathcal{A}_0 = \{{}^*A : A \in \mathcal{A}\}$ (the algebra of *standard* subsets of ${}^*X$), and ${}^*\mathcal{A}$. Note that except in the simplest cases, (i) neither $\mathcal{A}_0$ nor ${}^*\mathcal{A}$ are σ-algebras; (ii) $\mathcal{A}_0$ is external; and (iii) ${}^*\mathcal{A}$ is vastly larger than $\mathcal{A}_0$.

These leads to two distinct σ-algebras:

1. $\mathcal{A}_S$ = the smallest σ-algebra containing $\mathcal{A}_0$
2. $\mathcal{A}_L$ = the smallest σ-algebra containing ${}^*\mathcal{A}$

Recall that if μ is a (finitely- or countably-additive) finite measure on $(X, \mathcal{A})$ then ${}^*\mu$ maps ${}^*\mathcal{A}$ to ${}^*[0, \infty)$, and in particular is not normally a measure, unless the range of μ is finite. However, ${}^{\circ *}\mu$ takes its values in $[0, \infty)$, and so $({}^*X, {}^*\mathcal{A}, {}^{\circ *}\mu)$ is an external, standard, finitely-additive finite measure space.

The following was first noticed by Loeb. It is an immediate consequence of $\aleph_1$-saturation and the Carathéodory Extension Theorem, or can be proved directly; see [4] for details.

Theorem 1 *${}^{\circ *}\mu$ extends to a σ-additive measure μ_L on $({}^*X, \mathcal{A}_L)$. Moreover, $\forall E \in \mathcal{A}_L$*

$$\mu_L(E) = \inf\{\mu_L(A) \;:\; E \subseteq A \in {}^*\mathcal{A}\} \tag{15.2.1}$$
$$= \sup\{\mu_L(A) \;:\; A \subseteq E,\ A \in {}^*\mathcal{A}\} \tag{15.2.2}$$

Of course, the measure μ_L remains σ-additive when restricted to any σ-algebra contained in $\mathcal{A}_L$, in particular $\mathcal{A}_S$; this corresponds to replacing ${}^*\mathcal{A}$ by $\mathcal{A}_0$ as the generating algebra. However, it is not so obvious that the approximation properties (15.2.1) and (15.2.2) hold with this replacement as well. The next lemma, which asserts that they do, guarantees that a set $E \subseteq {}^*X$ has "*S-measure* 0" in the sense of Henson and Wattenberg [1] precisely when $E \subseteq D$ for some $D \in \mathcal{A}_S$ with $\mu_L(D) = 0$. For this lemma, and the duration of the paper, it will be convenient to assume that $\mathcal{A}$ is a σ-algebra.

Lemma 1 $\forall E \in \mathcal{A}_S$,

$$\mu_L(E) = \inf\{\mu(A) :\ E \subseteq {}^*A,\ A \in \mathcal{A}\} \tag{15.2.3}$$
$$= \sup\{\mu(A) :\ {}^*A \subseteq E,\ A \in \mathcal{A}\} \tag{15.2.4}$$

Proof. Let

$$\begin{aligned} \mathcal{B} &= \{E \in \mathcal{A}_S :\ \forall \varepsilon > 0\ \exists A, B \in \mathcal{A} \text{ such that } {}^*A \subseteq E \subseteq {}^*B \\ &\qquad \text{and } \mu(B \setminus A) < \varepsilon\} \\ &= \{E \in \mathcal{A}_S :\ \forall \varepsilon > 0\ \exists A, B \in \mathcal{A} \text{ such that } {}^*A \subseteq E \subseteq {}^*B \\ &\qquad \text{and } \mu_L({}^*B) < \mu_L(E) + \varepsilon \text{ and } \mu_L({}^*A) > \mu_L(E) - \varepsilon\}. \end{aligned}$$

Evidently $^*A \in \mathcal{B}$ for every $A \in \mathcal{A}$. It suffices to show that $\mathcal{B}$ is a σ-algebra, so that $\mathcal{A}_S \subseteq \mathcal{B}$. If $A \subseteq E \subseteq B$ then $B^{\complement} \subseteq E^{\complement} \subseteq A^{\complement}$ and $\mu_L(B^{\complement} \setminus A^{\complement}) = \mu_L(A \setminus B)$, so $\mathcal{B}$ is closed under complements. It remains to show that $\mathcal{B}$ is closed under countable unions. Let $E = \bigcup_n E_n$ where $E_n \in \mathcal{B}$ ($n \in \mathbb{N}$) increases to E, and let $\varepsilon > 0$. For some N, $\mu_L(E_N) > \mu_L(E) - \varepsilon/2$, and for some $A \in \mathcal{A}$, $^*A \subseteq E_N$ and $\mu_L(^*A) > \mu_L(E_N) - \varepsilon/2$; it follows that $^*A \subseteq E$ and $\mu_L(^*A) > \mu_L(E) - \varepsilon$. For the exterior approximation, there exists $B_n \in \mathcal{A}$ with $E_n \subseteq {}^*B_n$ and $\mu_L(^*B_n \setminus E_n) < \varepsilon 2^{-(n+1)}$. Put $B = \bigcup_n B_n$, then $E = \bigcup_n E_n \subseteq \bigcup_n {}^*B_n \subseteq {}^*B$, and

$$\begin{aligned}\mu(B) = \mu_L(^*B) &\leq \mu_L\Big(^*B \setminus \bigcup_n {}^*B_n\Big) + \mu_L\Big(\Big(\bigcup_n {}^*B_n\Big) \setminus E\Big) + \mu_L(E) \\ &< 0 + \sum_n \varepsilon 2^{-(n+1)} + \mu_L(E) \leq \mu_L(E) + \varepsilon.\end{aligned}$$

□

Call a function $f : {}^*X \to \mathbb{R}$ *approximately standard* provided:

1. f is $\mathcal{A}_S$-measurable;
2. the restriction $g = f|_X$ of f to X is an $\mathcal{A}$-measurable function from X to $\mathbb{R}$; and
3. $f \approx {}^*g$ almost everywhere (with respect to μ_L).

For example, suppose $h : X \to \mathbb{R}$ is a bounded $\mathcal{A}$-measurable function. For $r \in \mathbb{R}$, ${}^{\circ *}h^{-1}(-\infty, r] = \bigcap_{n \in \mathbb{N}} {}^*h^{-1}(-\infty, r + 1/n)$, so ${}^{\circ *}h$ is $\mathcal{A}_S$-measurable. For $x \in X$, $h(x) = {}^*h(x) = {}^{\circ *}h(x)$, so $h = ({}^{\circ *}h)|_X$. It follows that ${}^{\circ *}h$ is approximately standard.

The main result of this section is the theorem of Henson and Wattenberg, that *all* $\mathcal{A}_S$-measurable functions are approximately standard. The proof here differs from theirs, and makes it possible to prove some new results about integrability.

Denote by **APS** the set of all approximately standard functions. The following enumerates some useful properties of **APS**.

Lemma 2

1. **APS** *is closed under finite linear combinations.*
2. *If* $f, g \in$ **APS** *then* $\min\{f, g\} \in$ **APS** *and* $\max\{f, g\} \in$ **APS**.
3. *If* $f_n \in$ **APS**, $n \in \mathbb{N}$, *and* f_n *monotonically increases pointwise to* f, *then* $f \in$ **APS**.

4. *Let $h : X \to \mathbb{R}$ be an $\mathcal{A}$-measurable function, and put $E = \{x \in {}^*X : |{}^*h(x)| < \infty\}$. Then (i) $E \in \mathcal{A}_S$, (ii) $\mu_L({}^*X \setminus E) = 0$, and (iii) ${}^\circ({}^*h\chi_E) \in$* **APS** *(where χ_E is the characteristic function of E).*

Proof.

1. Follows immediately from the observation that if $f, g : {}^*X \to \mathbb{R}$ and $\alpha, \beta \in \mathbb{R}$ then $(\alpha f + \beta f)|_X = (\alpha f|_X + \beta f|_X)$.

2. As in (1), note that $\max\{f, g\}|_X = \max\{f|_X, g|_X\}$ and $\min\{f, g\}|_X = \min\{f|_X, g|_X\}$.

3. Let $g_n = f_n|_X$, $A_n = \{x \in {}^*X : {}^*g_n(x) \not\approx f_n(x)\}$, and $A = \bigcup_{n \in \mathbb{N}} A_n$. Since $\mu_L(A_n) = 0$ for each n, $\mu_L(A) = 0$. Let $g = f|_X$ and note $g = \sup_n g_n$. Fix a standard $\varepsilon > 0$, and for $n \in \mathbb{N}$ let $E_n = \{x \in X : g_n(x) > g(x) - \varepsilon\}$. Since $X = \bigcup_{n \in \mathbb{N}} E_n$, $\mu_L({}^*E_n) = \mu(E_n)$ increases to $\mu(X) = \mu_L({}^*X)$, so $\mu_L(A \cup ({}^*X \setminus \bigcup_{n \in \mathbb{N}} {}^*E_n)) = 0$. Suppose $x \in ({}^*X \setminus A) \cap \bigcup_{n \in \mathbb{N}} {}^*E_n$. For some $n, m \in \mathbb{N}$, $x \in {}^*E_n$ and $f_m(x) > f(x) - \varepsilon$. It follows that ${}^*g(x) < {}^*g_n(x) + \varepsilon \approx f_n(x) + \varepsilon \leq f(x) + \varepsilon < f_m(x) + 2\varepsilon \approx {}^*g_m(x) + 2\varepsilon \leq {}^*g(x) + 2\varepsilon$. Since ε was arbitrary, ${}^*g(x) \approx f(x)$.

4. Put $E_n = \{x \in X : |h(x)| < n\}$, so $E = \bigcup_{n \in \mathbb{N}} {}^*E_n \in \mathcal{A}_S$. $X = \bigcup_{n \in \mathbb{N}} E_n$, so $\mu_L({}^*E_n) = \mu(E_n)$ increases to $\mu(X) = \mu_L({}^*X)$, and $\mu_L({}^*X \setminus E) = 0$. The proof of (iii) is now like the example preceding the statement of this lemma. □

Theorem 2 *Every $\mathcal{A}_S$-measurable function is approximately standard.*

Proof. Note that if $A \in \mathcal{A}$ then $\chi_{{}^*A} = {}^*\chi_A = {}^{\circ *}\chi_A$ is in **APS** by Lemma 2 (4). By (1) and (3) of Lemma 2 and a suitable version of the Monotone Class Theorem (for example, Theorem 3.14 of [5]), every bounded $\mathcal{A}_S$-measurable function is in **APS**. If $f \geq 0$ is $\mathcal{A}_S$-measurable then $f = \sup_n \max\{f, n\}$, so is in **APS** by (2) and (3) of Lemma 2. Any general $\mathcal{A}_S$-measurable f can be written $f = \max\{f, 0\} - \max\{-f, 0\}$ so is in APS by (1) and (2) of Lemma 2. □

For example, suppose $E \in \mathcal{A}_S$; denote by $S(E)$ the set $X \cap E$ of standard elements of E. If f is the characteristic function of E then $g = f|_X$ is the characteristic function of $S(E)$. By the theorem, $S(E)$ is $\mathcal{A}$-measurable and $\mu(S(E)) = \mu_L({}^*S(E)) = \int {}^*g \, d\mu_L = \int f \, d\mu_L = \mu_L(E)$.

Corollary 1 *Let $f : {}^*X \to \mathbb{R}$ be approximately standard, and $p > 0$. Put $g = f|_X$. Then $f \in \mathfrak{L}^p(\mu_L)$ if and only if $g \in \mathfrak{L}^p(\mu)$, in which case $\int f^p d\mu_L = \int g^p d\mu$ and $\|f\|_p = \|g\|_p$.*

Proof. Without loss of generality, $f \geq 0$. Let $s_n : {}^*X \to \mathbb{R}$, $n \in \mathbb{N}$, be simple functions increasing pointwise to f^p. Put $t_n = s_n|_X$, then t_n is a sequence of simple functions on X increasing pointwise to g^p. Moreover, if $s_n = \sum_{k=1}^m \alpha_k \chi_{E_k}$ then $t_n = \sum_{k=1}^m \alpha_k \chi_{S(E_k)}$, and $\int s_n d\mu_L = \sum_{k=1}^m \alpha_k \mu_L(E_k) = \sum_{k=1}^m \alpha_k \mu(S(E_k)) = \int t_n d\mu$. Let $n \to \infty$, and it follows that $\int f^p d\mu_L$ and $\int g^p d\mu$ are either both infinite or are both finite and equal. □

The next result lets us extend the notion of approximate standardness to completion-measurable functions.

Lemma 3 *Let $f : {}^*X \to \mathbb{R}$; the following are equivalent:*

(i) f is measurable with respect to the μ_L-completion of $\mathcal{A}_S$;

(ii) $f|_X$ is measurable with respect to the μ-completion of $\mathcal{A}$, and $f \approx {}^*(f|_X)$ off a μ_L-nullset in $\mathcal{A}_S$.

Proof. Recall (without proof) the standard fact that a function is completion measurable if and only if it agrees with a measurable function off a nullset.

($i \Rightarrow ii$) There is an $\mathcal{A}_S$-measurable f' and a μ_L-nullset $A \in \mathcal{A}_S$ such that $f = f'$ off A. Let $g = f|_X$ and $g' = f'|_X$, then g' is $\mathcal{A}$-measurable, ${}^*g' \approx f'$ off a μ_L-nullset $B \in \mathcal{A}_S$, and $\{x \in X : g(x) \neq g'(x)\} \subseteq S(A)$. Since $\mu_L(A) = 0$, $\mu(S(A)) = 0$ by the example above, and so g' is measurable with respect to the μ-completion of $\mathcal{A}$. Note also that ${}^*\mu(S(A)) = 0$, so $\mu_L(A \cup B \cup {}^*S(A)) = 0$. For $x \in {}^*X, x \notin A \cup B \cup {}^*S(A), f(x) = f'(x) \approx g'(x) = g(x)$, proving (ii).

($ii \Rightarrow i$) Let $A \in \mathcal{A}_S$ be a μ_L-nullset with $f \approx {}^*(f|_X)$ off A. Let $g = f|_X$, g' $\mathcal{A}$-measurable such that $g = g'$ off a μ-nullset $B \in \mathcal{A}$, $E = \{x \in {}^*X : |{}^*g'(x)| < \infty\}$, and $f' = {}^\circ({}^*g'\chi_E)$. By Lemma 2, f' is in **APS**. For $x \notin E \cup {}^*B \cup A$, $f'(x) \approx {}^*g'(x) = {}^*g(x) \approx f(x)$, so in fact $f'(x) = f(x)$. Since f' is $\mathcal{A}_S$-measurable, f is completion measurable. □

15.3 Egoroff's Theorem

The most striking application [1] made of the S-measure construction was a proof of Egoroff's Theorem. Their proof relied on a nonstandard condition shown by Robinson to be equivalent to approximate uniform convergence.

Here I use the machinery above to give an alternate proof of the theorem not depending on Robinson's condition.

Theorem 3 *Let $(X, \mathcal{A}, \mu)$ be a finite measure space, $f, f_n : X \to \mathbb{R}$ measurable, $f_n \to f$ a.e. Then*

$$\forall \varepsilon > 0\ \exists A \in \mathcal{A}\ \mu(A) > \mu(X) - \varepsilon\ \&\ f_n \to f \textit{ uniformly on } A.$$

Proof. For $n \in \mathbb{N}$ put $g_n = \inf_{m>n} f_m$ and $h_n = \sup_{m>n} f_m$. Note that for almost all x, $h_n(x) - g_n(x)$ is nonnegative and nondecreasing in n with $\lim_{n\to\infty} h_n(x) - g_n(x) = 0$. Put $\hat{g}_n = {}^{\circ *}g_n$, $\hat{h}_n = {}^{\circ *}h_n$. Let $E = \{x \in {}^*X : \lim_{n\to\infty} \hat{h}_n(x) - \hat{g}_n(x) \neq 0\}$. Note $E \in \mathcal{A}_S$, so $\mu_L(E) = \mu(S(E)) = \mu(\emptyset) = 0$. It follows that for some $A \in \mathcal{A}$, ${}^*A \subseteq E^{\mathsf{C}}$ and $\mu(A) > \mu(X) - \varepsilon$. It remains to show that f_n converges to f uniformly on A; equivalently, that $h_n - g_n$ converges to 0 uniformly on A.

Fix $\delta > 0$. For $x \in {}^*A$ there is a $k \in \mathbb{N}$ such that $\hat{h}_k(x) - \hat{g}_k(x) < \delta$. Note ${}^*h_k(x) - {}^*g_k(x) < \delta$ as well. Let $\phi(x)$ be the least k such that ${}^*h_k(x) - {}^*g_k(x) < \delta$. The function $\phi : {}^*A \to \mathbb{N}$ is internal and finite-valued, so has a standard upper bound $N \in \mathbb{N}$. Then ${}^*h_k - {}^*g_k < \delta$ on ${}^*A$ for every $k \geq N$, so $h_k - g_k < \delta$ on A; this completes the proof. □

15.4 A Theorem of Riesz

Theorem 4 *Let T be a continuous linear real functional on $\mathfrak{L}^2(X, \mathcal{A}, \mu)$; then there is a g such that for every $f \in \mathfrak{L}^2(X, \mathcal{A}, \mu)$, $T(f) = \int fg\, d\mu$.*

After the proof we show that the function g is actually in $\mathfrak{L}^2(X, \mathcal{A}, \mu)$.

Proof. By saturation let $\widehat{\mathcal{A}}$ be a *-finite algebra with $\mathcal{A}_0 \subseteq \widehat{\mathcal{A}} \subseteq {}^*\mathcal{A}$. There is an internal *-partition Π of ${}^*X$ which corresponds to $\widehat{\mathcal{A}}$ in the sense that the latter is the internal closure of the former under hyperfinite unions.

Now, define

$$\hat{\gamma}(x) := \begin{cases} \dfrac{{}^*T(\chi_p)}{{}^*\mu(p)}, & x \in p \in \Pi \text{ and } {}^*\mu(p) > 0; \\ 0, & \text{otherwise.} \end{cases}$$

Let $a : \Pi \to {}^*X$ be an internal choice function, that is, $a_p \in p$ for $p \in \Pi$. For any bounded, measurable $f : X \to \mathbb{R}$,

$$T(f) = \sum_p {}^*T({}^*f\chi_p) \tag{15.4.1}$$

$$\approx \sum_p {}^*f(a_p){}^*T(\chi_p) \tag{15.4.2}$$

$$= \sum_p {}^*f(a_p)\hat{\gamma}(a_p){}^*\mu(p) \tag{15.4.3}$$

$$= \sum_p \int_p {}^*f(a_p)\hat{\gamma}(a_p)\, {}^*d\mu \tag{15.4.4}$$

$$= \sum_p \int_p {}^*f(a_p)\hat{\gamma}(x)\,{}^*d\mu \tag{15.4.5}$$

$$\approx \sum_p \int_p {}^*f(x)\hat{\gamma}(x)\,{}^*d\mu \tag{15.4.6}$$

$$= \int {}^*f\hat{\gamma}\,{}^*d\mu \tag{15.4.7}$$

The steps from (15.4.1) to (15.4.2) and (15.4.5) to (15.4.6) follow from boundedness of f and the definition of $\widehat{\mathcal{A}}$. For (15.4.2) to (15.4.3) note that by continuity of T, if $p \in \Pi$ and ${}^*\mu(p) = 0$ then ${}^*T(\chi_p) = 0$.

Moreover, if ${}^*f$ is replaced in the above by any internal function $h : {}^*X \to {}^*\mathbb{R}$ with the property that h is constant on each $p \in \Pi$, then all but the first equality in the above still holds, and we obtain ${}^*T(h) = \int h\hat{\gamma}\,d^*\mu$. In particular, if $h = \chi_A$ for some $A \in \widehat{\mathcal{A}}$ then ${}^*T(\chi_A) = \int_A \hat{\gamma}\,d^*\mu$.

The proof now proceeds as follows: (i) Show $\hat{\gamma}$ is finite almost everywhere (and therefore $\gamma =^\circ \hat{\gamma}$ exists almost everywhere). (ii) Show that $\hat{\gamma}$ is S-integrable (and therefore $\gamma =^\circ \hat{\gamma}$ is integrable). (iii) Put $G = \mathbb{E}[\gamma|\mathcal{A}_S]$ (the conditional expectation of γ). (iv) Let g be the restriction of G to X; by Theorem 2 g is $\mathcal{A}$-measurable. (v) Show that g works.

For (i), write $[\hat{\gamma} < n] = \{x \in {}^*X : \hat{\gamma}(x) < n\}$ $(n \in {}^*\mathbb{N})$, and $[\hat{\gamma} < \infty] = \bigcup_{n\in\mathbb{N}}[\hat{\gamma} < n]$. Suppose (for a contradiction) that $\mu_L([\hat{\gamma} < \infty]) < 1 - r$ for some standard $r > 0$. Then $\mu([\hat{\gamma} < n]) < 1 - r$ for each standard $n \in \mathbb{N}$, so $\mu([\hat{\gamma} < H]) < 1 - r$ for some infinite H. But then $\infty > T(1) \approx {}^*\!\int 1\hat{\gamma}d^*\mu$ (by the note above) $\geq {}^*\!\int_{[\hat{\gamma}\geq H]}\hat{\gamma}d^*\mu \geq H^*\mu([\hat{\gamma} \geq H]) > rH$, which is infinite, a contradiction.

For (ii), the reader is referred to [4] for a discussion of S-integrability. In particular, to show that $\hat{\gamma}$ is S-integrable, it suffices to show that $\forall H$ infinite, ${}^*\!\int_{[\hat{\gamma}>H]}\hat{\gamma}d^*\mu \approx 0$. (Indeed, this can be adopted as the *definition* of S-integrability.)

So, fix such an H. Note $[\hat{\gamma} > H] \in \widehat{\mathcal{A}}$; it follows that ${}^*\!\int_{[\hat{\gamma}>H]}\hat{\gamma}d^*\mu = {}^*T(\chi_{[\hat{\gamma}>H]})$. Suppose (for a contradiction) that ${}^*T(\chi_{[\hat{\gamma}>H]}) > r > 0$, r standard. By (i), ${}^*\mu([\hat{\gamma} > H]) \approx 0$. It follows from transfer that for every standard $n \in \mathbb{N}$ there is a $B_n \in \mathcal{A}$ with $T(\chi_{B_n}) > r$ and $\mu(B_n) < 1/2^{-n}$. Put $B^n := \bigcup_{m\geq n} B_m$. Then $\mu(B^n) < \sum_{m\geq n} 2^{-m} = 2^{-n+1}$, but $T(\chi_{B^n}) > r$. As $n \to \infty$, $\chi_{B^n} \to 0$ in $\mathfrak{L}^2$ but $T(\chi_{B^n}) > r$, contradicting continuity of T.

It follows from S-integrability of $\hat{\gamma}$ that $\gamma =^\circ \hat{\gamma}$ is integrable, so its conditional expectation with respect to $\mathcal{A}_S$, $\mathbb{E}[\gamma|\mathcal{A}_S]$, exists; put $G = \mathbb{E}[\gamma|\mathcal{A}_S]$. (Conditional expectation is defined in the next section, where a simple nonstandard proof for its existence is given. The reader is referred to chapter 9 of [5] for properties of the conditional expectation.)

Finally, let $g = G|_X$, which (as noted above) is $\mathcal{A}$-measurable by Theorem 2. It remains to show that g satisfies the conclusion of Theorem 4.

Let $f \in \mathcal{L}^2(X, \mathcal{A}, \mu)$, and suppose first that f is bounded. This ensures that $({}^*f)\hat{\gamma}$ is S-integrable, and $({}^{\circ *}f)\gamma \in \mathcal{L}^1({}^*X, \mathcal{A}_L, \mu_L)$.

Then

$$T(f) \approx \int {}^*f\hat{\gamma}\, d^*\mu \tag{15.4.8}$$

$$\approx \int ({}^{\circ *}f)\gamma\, d\mu_L \tag{15.4.9}$$

$$= \int \mathbb{E}[({}^{\circ *}f)\gamma|\mathcal{A}_S]\, d\mu_L \tag{15.4.10}$$

$$= \int ({}^{\circ *}f)\mathbb{E}[\gamma|\mathcal{A}_S]\, d\mu_L \tag{15.4.11}$$

$$= \int ({}^{\circ *}f)G\, d\mu_L \tag{15.4.12}$$

$$= \int fg\, d\mu \tag{15.4.13}$$

The step from (15.4.8) to (15.4.9) is by S-integrability, (15.4.10) to (15.4.11) is a standard property of conditional expectation (using the fact that ${}^{\circ *}f$ is $\mathcal{A}_S$-measurable), and (15.4.12) to (15.4.13) is Corollary 1 with $p = 1$.

For unbounded $f \in \mathcal{L}^2(X, \mathcal{A}, \mu)$ and $n \in \mathbb{N}$, let $f_n = \max\{-n, \min\{f, n\}\}$. By the result above, $T(f_n) = \int f_n g\, d\mu$. By continuity of T and Lebesgue's Dominated Convergence Theorem ([5], Theorem 5.9), $T(f) = \int fg\, d\mu$. □

The Riesz Theorem is usually stated in the following nominally stronger form.

Corollary 2 *Let T be a continuous linear real functional on $\mathcal{L}^2(X, \mathcal{A}, \mu)$; then there is a $g \in \mathcal{L}^2(X, \mathcal{A}, \mu)$ such that for every $f \in \mathcal{L}^2(X, \mathcal{A}, \mu)$, $T(f) = \int fg\, d\mu$.*

Proof. It suffices to show that any g satisfying the conclusion of Theorem 4 is already in $\mathcal{L}^2(X, \mathcal{A}, \mu)$.

Since T is continuous, it is bounded, so there is a constant C such that for any $f \in \mathcal{L}^2(X, \mathcal{A}, \mu)$, $T(f) \leq C\|f\|_2$.

Put $g_n = \min\{g, n\}$. Then $\|g_n\|_2^2 = \int g_n^2\, d\mu \leq \int g_n g\, d\mu = T(g_n) \leq C\|g_n\|_2$. Divide both sides of this inequality by $\|g_n\|_2$ and square, obtain $\int g_n^2\, d\mu = \|g_n\|_2^2 \leq C^2$. The result now follows by Fatou's Lemma ([5], Lemma 5.4). □

15.4.1 Conditional expectation

Theorem 5 *Suppose* $(X, \mathcal{A}, \mu)$ *is a probability measure, that* $\mathcal{B} \subseteq \mathcal{A}$ *is another* σ*-algebra, and that* $f \in \mathcal{L}^1(X, \mathcal{A}, \mu)$*. There is a* $\mathcal{B}$*-measurable function* $g : X \to \mathbb{R}$ *such that for any* $B \in \mathcal{B}$*,* $\int_B f\, d\mu = \int_B g\, d\mu$*.*

The function g is called the *conditional expectation* of f on $\mathcal{B}$, and denoted by $\mathbb{E}[f|\mathcal{B}]$.

To avoid circularity, it is necessary to show that $\mathbb{E}[f|\mathcal{B}]$ exists without use of the Riesz Theorem (or related results, such as the Radon-Nikodým Theorem). Such a proof appears in [3]. This section presents a modification of that proof which is even more elementary, in that it does not require the Hahn decomposition.

Without loss of generality $f \geq 0$. Put $\mathcal{G} = \{g \in \mathcal{L}^1(X, \mathcal{B}, \mu) : \forall B \in \mathcal{B}, \int_B g\, d\mu \leq \int_B f\, d\mu\}$. Note if $g_1, g_2 \in \mathcal{G}$ then $\max\{g_1, g_2, 0\} \in \mathcal{G}$. Let $r = \sup\{\int_X g d\mu : g \in \mathcal{G}\}$, and for $n \in \mathbb{N}$ let $g_n \in \mathcal{G}$ with $\int_X g_n d\mu > r - 1/2^n$; we may assume that $0 \leq g_1 \leq g_2 \leq \cdots$. Put $g = \sup_n g_n$. Note that if $B \in \mathcal{B}$, $\int_B g\, d\mu = \sup_n \int_B g_n\, d\mu \leq \int_B f\, d\mu$, so $g \in \mathcal{G}$. Moreover, $\int_X g\, d\mu = r$. It remains to show that for any $B \in \mathcal{B}$, $\int_B f\, d\mu = \int_B g\, d\mu$.

Suppose not; then there is a $B \in \mathcal{B}$ and $\varepsilon > 0$ with $\int_B f\, d\mu - \int_B g\, d\mu = \varepsilon$.

For some $\delta > 0$, $\varepsilon' = \int_B (f - \hat{g}) d\mu > 0$ where $\hat{g} = g + \delta\chi_B$. $\hat{g}$ *almost* witnesses a contradiction; the perturbation by δ needs to be localized to a slightly smaller set.

As in the proof of Theorem 4 let $\widehat{\mathcal{B}}$ be a *-finite algebra with $\{^*B : B \in \mathcal{B}\} \subseteq \widehat{\mathcal{B}} \subseteq {}^*\mathcal{B}$, and let Π be the internal hyperfinite *-partition of $^*X$ corresponding to $\widehat{\mathcal{B}}$. Let $B^+ = \bigcup\{b \in \Pi : \int_b {}^*(f - \hat{g}) d^*\mu > 0\}$. Put $s = {}^\circ\int_{B^+} {}^*(f - \hat{g}) d^*\mu$.

Note that for every $C \in \mathcal{B}$, if $C \subseteq B$ then $\int_C (f - \hat{g}) d\mu = \int_{^*C} {}^*(f - \hat{g}) d^*\mu \leq s$; in particular, $\varepsilon' \leq s$.

For $n \in \mathbb{N}$ consider the statement,

$$\exists B_n \in \mathcal{B},\ B_n \subseteq B,\ \int_{B_n} (f - \hat{g}) d\mu > s - 2^{-n}$$

As this holds in the nonstandard model (with B^+ for B_n), it holds by transfer in the standard model. For $m \in \mathbb{N}$ put $B^m = \bigcup_{n>m} B_n$, and put $B_\infty = \bigcap_m B^m$. Note $B_\infty \in \mathcal{B}$ and $B_\infty \subseteq B$, so $\int_{B_\infty} (f - \hat{g}) d\mu \leq s$. On the other hand, since always $\int_{B_n} (f - \hat{g}) d\mu > s - 2^{-n}$, $\int_{B^m} (f - \hat{g}) d\mu > s - \sum_{n>m} 2^{-n} = s - 2^{-m}$, therefore $\int_{B_\infty} (f - \hat{g}) d\mu \geq s$.

Put $g' = g + \delta\chi_{B_\infty}$. If $C \in \mathcal{B}$,

$$\int_C (f - g')\, d\mu = \int_{C \setminus B_\infty} (f - g)\, d\mu + \int_{C \cap B_\infty} (f - \hat{g})\, d\mu\,.$$

The first term in this sum is nonnegative since $g \in \mathcal{G}$. The second is nonnegative since otherwise $\int_{B_\infty \setminus C}(f - \hat{g})d\mu > s$. It follows that $\int_C (f - g')d\mu \geq 0$, so $g' \in \mathcal{G}$. Since $s > 0$, $\mu(B_\infty) > 0$, so $\int g'\, d\mu = \int g\, d\mu + \delta\mu(B_\infty) > r$, a contradiction.

References

[1] C. Ward Henson and Frank Wattenberg, "Egoroff's theorem and the distribution of standard points in a nonstandard model", Proc. Amer. Math. Soc., **81** (1981) 455–461.

[2] W. A. J. Luxemburg, On some concurrent binary relations occurring in analysis, in *Contributions to non-standard analysis (Sympos., Oberwolfach, 1970)*, Studies in Logic and Found. Math., Vol. 69, North-Holland, Amsterdam, 1972.

[3] David A. Ross, Nonstandard measure constructions — solutions and problems, in *Nonstandard methods and applications in mathematics*, Lecture Notes in Logic, 25, A.K. Peters, 2006.

[4] David A. Ross, Loeb measure and probability, in *Nonstandard analysis (Edinburgh, 1996)*, NATO Adv. Sci. Inst. Ser. C. Math. Phys. Sci., vol. 493, Kluwer Acad. Publ., Dordrecht, 1997.

[5] David Williams, *Probability with martingales*, Cambridge Mathematical Textbooks, Cambridge University Press, Cambridge, 1991.

[6] Beate Zimmer, "A unifying Radon-Nikodým theorem through nonstandard hulls", Illinois Journal of Mathematics, **49** (2005) 873-883.

16

A Radon-Nikodým theorem for a vector-valued reference measure

G. Beate Zimmer*

Abstract

The conclusion of a Radon-Nikodým theorem is that a measure μ can be represented as an integral with respect to a reference measure such that for all measurable sets A, $\mu(A) = \int_A f_\mu(x)\, d\lambda$ with a (Bochner or Lebesgue) integrable derivative or density f_μ. The measure λ is usually a countably additive σ-finite measure on the given measure space and the measure μ is absolutely continuous with respect to λ. Different theorems have different range spaces for μ, which could be the real numbers, or Banach spaces with or without the Radon-Nikodým property. In this paper we generalize to derivatives of vector valued measures with respect a vector-valued reference measure. We present a Radon-Nikodým theorem for vector measures of bounded variation that are absolutely continuous with respect to another vector measure of bounded variation. While it is easy in settings such as $\mu << \lambda$, where λ is Lebesgue measure on the interval $[0, 1]$ and μ is vector-valued to write down a nonstandard Radon-Nikodým derivative of the form $\varphi : {}^*[0,1] \to \text{fin}({}^*E)$ by $\varphi_\mu(x) = \sum_{i=1}^{H} \frac{{}^*\mu(A_i)}{{}^*\lambda(A_i)} 1_{A_i}(x)$, a vector valued reference measure does not allow this approach, as the quotient of two vectors in different Banach spaces is undefined. Furthermore, generalizing to a vector valued control measure necessitates the use of a generalization of the Bartle integral, a bilinear vector integral.

16.1 Introduction and notation

For nonstandard notions and notations not defined here we refer for example to the book by Albeverio, Fenstad, Hoegh-Krøhn and Lindstrøm [1] and the survey on nonstandard hulls by Henson and Moore [6] or the introduction

*Department of Mathematics and Statistics, Texas A&M University - Corpus Christi, Corpus Christi, TX 78412.

to nonstandard analysis in [7]. Plenty of information about vector measures can be found in [5].

Throughout, Ω is a set and Σ is a σ-algebra of subsets of Ω. E, F and G denote Banach spaces. $\mu : \Sigma \to E$ and $\nu : \Sigma \to F$ are countably additive vector measures of bounded variation. The variation of ν is defined as $|\nu|(A) = \sup \sum_{\pi \in \Pi} \|\nu(A_\pi)\|$ where $A \in \Sigma$ and Π is a finite partition of A into sets in Σ. By Proposition I.1.9 in [5] the variation $|\nu|$ of a countably additive measure ν is also a countably additive measure on Σ.

We think of the nonstandard model in terms of superstructures $V(X)$ and $V({}^*X)$ connected by the monomorphism $* : V(X) \to V({}^*X)$ and call an element $b \in V({}^*X)$ internal, if it is an element of a standard entity, i.e. if there is an $a \in V(X)$ with $b \in {}^*a$. We assume that the nonstandard model is at least $\aleph$-saturated, where $\aleph$ is an uncountable cardinal number such that the cardinality of the σ-algebra Σ of $|\nu|$-measurable subsets of Ω is less than $\aleph$.

Let $({}^*\Omega, L_{|\nu|}({}^*\Sigma), \widehat{|\nu|})$ denote the Loeb space constructed from the nonstandard extension of the measure space $(\Omega, \Sigma, |\nu|)$. This measure space is obtained by extending the measure ${}^{\circ *}|\nu|$ from ${}^*\Sigma$ to the σ-algebra generated by ${}^*\Sigma$. The completion of this σ-algebra is denoted by $L_{|\nu|}({}^*\Sigma)$. The Loeb measure $\widehat{|\nu|}$ is a standard countably additive measure.

The functions we work with take their values in a Banach space E or its nonstandard hull $\widehat{E}$. The nonstandard hull of a Banach space E is defined as $\widehat{E} = \text{fin}({}^*E)/\approx$, the quotient of the elements of bounded norm by elements of infinitesimal norm. The nonstandard hull is a standard Banach space and contains E as a subspace. We denote the quotient map from $\text{fin}({}^*E)$ onto $\widehat{E}$ by π or π_E. Whenever we encounter products of Banach spaces, we equip them with the ℓ_∞ norm: $\|e \times f\| = \max(\|e\|, \|f\|)$.

By a result of Loeb in [8], there exists an internal *-finite partition of ${}^*\Omega$ consisting of sets $A_1, \ldots, A_H \in {}^*\Sigma$ $(H \in {}^*\mathbb{N})$ such that the partition is finer than the image under $*$ of any finite partition of Ω into sets in Σ. The proof uses a concurrent relation argument. From this fine partition we discard all partition sets of ${}^*|\nu|$-measure zero. The remaining sets still form an internal collection, which we also denote by $A_1, \ldots, A_H$.

16.2 The existing literature

Ross gives a nice detailed nonstandard proof of the Radon-Nikodým theorem for two real-valued σ-finite measures $\mu << \nu$ on a measurable space in [9]. Previously we have studied nonstandard Radon-Nikodým derivatives of vector measures $\mu : \Sigma \to E$ with respect to Lebesgue measure on $[0, 1]$. We have shown that through nonstandard analysis a unifying approach to

Radon-Nikodým derivatives independent of the Radon-Nikodým property or lack thereof of E can be found. The generalized derivatives we constructed are not necessarily essentially separably valued and therefore not always Bochner integrable. However, a generalization of the Bochner integral in [10] for functions with values in the nonstandard hull of a Banach space allows one to integrate the generalized derivatives. In [11] with methods similar to those used by Ross [9] or Bliedtner and Loeb in [3], and with the use of local reflexivity we "standardize" the generalized derivatives from maps ${}^*\Omega \to \widehat{E}$ to maps $\Omega \to E''$ with values in the second dual of the original Banach space E.

The vector measures book [5] by Diestel and Uhl is still considered as the authoritative work on Radon-Nikodým theorems. The standard literature yields very little on vector-vector derivatives; only Bogdan [4] together with his student Kritt has made an initial foray into this field. In their article they assume that they have two vector measures $\mu : \Sigma \to E$ and $\nu : \Sigma \to F$ with $\mu << |\nu|$, and they assume that the range space of μ has the Radon-Nikodým property and the range space of ν is uniformly convex. They then define a derivative and integral in the form $\mu(A) = \int_A \frac{d\mu}{d|\nu|}(\omega) \cdot \frac{d|\nu|}{d\nu}(\omega)\, d\nu$, where $\frac{d|\nu|}{d\nu}$ is a function from Ω into F', the dual space of the range of ν. The integrand is then of the form $\frac{d\mu}{d\nu}(\omega) = e \cdot f'$, a function from Ω into $E \times F'$. A trilinear product on $E \times F \times F'$ with values in E can be defined as $e \cdot f' \cdot f = f'(f) \cdot e$. This is needed to integrate simple functions of the form $\sum_{i=1}^n e_i \cdot f_i' \cdot 1_{A_i} : \Omega \to E \times F'$ with respect to the F-valued measure ν as follows:

$$\int_\Omega \sum_{i=1}^n e_i \cdot f_i' \cdot 1_{A_i}\, d\nu = \sum_{i=1}^n e_i \cdot f_i' \cdot \nu(A_i) = \sum_{i=1}^n e_i \cdot f_i'(\nu(A_i)) \in E.$$

The main difficulty is the definition of $\frac{d|\nu|}{d\nu}$, the derivative of a real-valued measure with respect to a vector-valued measure, the opposite setting from the ordinary Radon-Nikodým theorems. Bogdan uses his assumption that F is uniformly convex to find this derivative. Since $\nu << |\nu|$ and since uniform convex spaces are reflexive and hence have the Radon-Nikodým property, there is a Bochner integrable function $\frac{d\nu}{d|\nu|} : \Omega \to F$ such that for all measurable sets A, the vector measure can be written as a Bochner integral $\nu(A) = \int_A \frac{d\nu}{d|\nu|}\, d|\nu|$. A standard theorem (e.g. Theorem II.2.4 in [5]) about vector measures states that the norm of the Radon-Nikodým derivative is a Radon-Nikodým derivative for the total variation of the vector measure, i.e. for all $A \in \Sigma$

$$|\nu|(A) = \int_A \left\| \frac{d\nu}{d|\nu|} \right\| d|\nu|.$$

This implies that $|\nu|$-almost everywhere on Ω, $\left\| \frac{d\nu}{d|\nu|} \right\| = 1$. For uniformly convex spaces F there is a continuous map $g : S_F \to S_{F'}$, where S_X denotes the unit sphere of X, such that $\langle f, g(f)\rangle = 1$ for all $f \in S_F$. In this case, the composition $g \circ \frac{d\nu}{d|\nu|}$ is a map from Ω into the unit sphere of F'. The derivative of μ with respect to ν is defined as $\frac{d\mu}{d\nu} = \frac{d\mu}{d|\nu|} \cdot \left(g \circ \frac{d\nu}{d|\nu|}\right) : \Omega \to E \times F'$, or, if one regards the product $e \times f'$ as a rank one continuous linear operator from F to E, the derivative can be considered a map: $\frac{d\mu}{d\nu} : \Omega \to L(F, E)$. Since $\frac{d\mu}{d|\nu|}$ and $\frac{d\nu}{d|\nu|}$ are both Bochner integrable and g is continuous and $g \circ \frac{d\nu}{d|\nu|}$ is almost everywhere of norm one, it is easy to show that the result is integrable with respect to the vector measure ν with an integral that uses approximation by simple functions.

16.3 The nonstandard approach

With nonstandard analysis we can significantly weaken the assumptions on the Banach spaces E and F. We use Bogdan's idea of writing the derivative as $\frac{d\mu}{d|\nu|} \cdot \frac{d|\nu|}{d\nu} = \frac{d\mu}{d|\nu|} \cdot \left(g \circ \frac{d\nu}{d|\nu|}\right)$.

The derivatives $\frac{d\mu}{d|\nu|}$ and $\frac{d\nu}{d|\nu|}$ are derivatives of a vector measure with respect to a real valued measure. They can be found and integrated with the methods described in [10], [11] or [12].

In [10] we defined a Banach space $M(\widehat{|\nu|}, \widehat{E})$ of extended integrable functions as the set of equivalence classes under equality $\widehat{|\nu|}$-almost everywhere of functions $f : {}^*\Omega \to \widehat{E}$ for which there is an internal, $*$-simple, S-integrable $\varphi : {}^*\Omega \to {}^*E$ such that $\pi_E \circ \varphi = f$ $\widehat{|\nu|}$-almost everywhere on ${}^*\Omega$. Such a φ is called a lifting of f. On $M(\widehat{|\nu|}, \widehat{E})$ we defined an integral by setting $\int_A f \, d\widehat{|\nu|} = \pi\left(\int_A \varphi \, d^*|\nu|\right)$, for all $A \in {}^*\Sigma$, where the integral of the internal simple function φ is defined in the obvious way. This integral can be extended to sets in the Loeb σ-algebra and generalizes the Bochner integral in the sense that $M(\widehat{|\nu|}, \widehat{E})$ contains $L_1(\widehat{|\nu|}, \widehat{E})$ and the integrals agree on that subspace. However, $M(\widehat{|\nu|}, \widehat{E})$ also contains functions which fail to be essentially separably valued and hence fail to be measurable. Lemma 1 in [11] translates to this situation as:

Lemma 16.3.1 *Let E be any Banach space and let $\mu : \Sigma \to E$ be a $|\nu|$-absolutely continuous countably additive vector measure of bounded variation. Then the internal $*$-simple function $\varphi_\mu : {}^*\Omega \to {}^*E$ defined by*

$$\varphi_\mu(\omega) = \sum_{i=1}^{H} \frac{{}^*\mu(A_i)}{{}^*|\nu|(A_i)} \, 1_{A_i}(\omega)$$

is S-integrable, where $A_1, \ldots, A_H$ is the fine partition of ${}^*\Omega$ introduced above.

S-integrability implies that the function is $\widehat{|\nu|}$-almost everywhere fin$(^*E)$-valued. This allows us compose an S-integrable internal function with the quotient map $\pi_E : \text{fin}(^*E) \to \widehat{E}$ to make it a nonstandard hull valued function defined on the Loeb space $\left(^*\Omega, L_{|\nu|}(^*\Sigma), \widehat{|\nu|}\right)$. Define $f_\mu \in M(\widehat{|\nu|}, \widehat{E})$ by

$$f_\mu = \pi_E \circ \varphi_\mu.$$

Theorem 4 in [11] asserts that if the vector measure μ has a Bochner integrable Radon-Nikodým derivative, then the generalized derivative "is" the Radon-Nikodým derivative in the following sense:

Theorem 16.3.2 *Let $\mu : \Sigma \to E$ be a countably additive vector measure with a Bochner integrable Radon-Nikodým derivative $f : \Omega \to E$. Define the generalized Radon-Nikodým derivative $f_\mu : {}^*\Omega \to \widehat{E}$ as above. Then*

$$\pi_E \circ {}^*f = f_\mu \quad \textit{on each set } A_i \textit{ in the fine partition of } {}^*\Omega.$$

Of course, the same arguments also work for finding a $\widehat{F}$-valued generalized derivative $f_\nu \in M(\widehat{|\nu|}, \widehat{F})$ of the vector measure $\nu : \Sigma \to F$ with respect to its total variation $|\nu|$. Differentiating ν with respect to its total variation gives one extra feature of the derivative:

Lemma 16.3.3 *Let F be a Banach space and let $\nu : \Sigma \to F$ be a countably additive vector measure of bounded variation. Define*

$$\varphi_\nu(\omega) = \sum_{i=1}^{H} \frac{{}^*\nu(A_i)}{{}^*|\nu|(A_i)} 1_{A_i}(\omega).$$

Then $f_\nu = \pi_F \circ \varphi_\nu$ is $\widehat{|\nu|}$-almost everywhere of norm one.

Proof. Since $\nu << |\nu|$, the S-integrability of f_ν follows from Lemma 16.3.1. The norm condition follows from an adaptation of a basic standard result (Theorem II.2.4 in [5]): if f is Bochner integrable and a vector measure F is defined by $F(A) = \int_A f\, d|\nu|$, then $|F|(A) = \int_A \|f\| d|\nu|$. By construction of the generalized derivative,

$${}^*\nu(A) \approx \int_A \varphi_\nu = \int_A \sum_{i=1}^{H} \frac{{}^*\nu(A_i)}{{}^*|\nu|(A_i)} 1_{A_i}(\omega)\, d^*|\nu|$$

for all $A \in {}^*\Sigma$. In this situation it implies that

$${}^*|\nu|(A) = {\int_A}^* \left\| \frac{d\nu}{d|\nu|} \right\| d^*|\nu|$$

and hence ${}^*\|\varphi_\nu(\omega)\| = 1$ for $\widehat{|\nu|}$-almost every $\omega \in {}^*\Omega$. □

16.4 A nonstandard vector-vector integral

The next theorem allows us to generalize the generalized Bochner integral to a vector-vector integral which for any simple function obviously agrees with the Bartle integral developed in [2]. For the definition of the Bartle integral, E, F and G are Banach spaces equipped with a continuous bilinear multiplication $E \times F \to G$ and $\nu : \Sigma \to F$ is a countably additive vector measure of bounded variation. Such a product can arise by taking $E = F = G$ to be a C^*-algebra, or by taking $E = F'$ and $G = \mathbb{R}$, or by taking $E = L(F, G)$. A function $f : \Omega \to E$ is Bartle integrable if there is a sequence of simple functions s_n that converges to f almost everywhere and for which the sequence $\lambda_n(A) = \int_A s_n \, d\nu$ converges in the norm of G for all $A \in \Sigma$. Then $\int_A f \, d\nu = \lim_{n\to\infty} \lambda_n(A)$ exists in norm uniformly for all $A \in \Sigma$. Essentially bounded measurable functions are Bartle-integrable. The Bartle integral is a countably additive set function.

Theorem 16.4.1 *Let E, F and G be Banach spaces equipped with a continuous bilinear multiplication $E \times F \to G$. Assume that $\nu : \Sigma \to F$ is a countably additive vector measure of bounded variation and $f \in M(\widehat{|\nu|}, \widehat{E})$. The integrand f has an internal lifting φ_f and there is an internal $*$-simple lifting φ_ν of the generalized derivative of ν with respect to $|\nu|$. Then for all $A \in {}^*\Sigma$, the integral $\int_A f \, d\widehat{\nu}$ defined by*

$$\int_A f \, d\widehat{\nu} = \pi_G \left(\int_A \varphi_f \cdot \varphi_\nu \, d^*|\nu| \right)$$

is a well-defined countably additive $\widehat{G}$-valued integral.

Proof. The function $f \in M(\widehat{|\nu|}, \widehat{E})$ has a internal, S-integrable and $*$-simple lifting $\varphi_f : {}^*\Omega \to \text{fin}({}^*E)$. By Lemma 16.3.3 $\varphi_\nu : {}^*\Omega \to \text{fin}({}^*F)$ is internal, $*$-simple, S-integrable and of norm one almost everywhere. Continuity of the multiplication $E \times F \to G$ implies that the product $\varphi_f \cdot \varphi_\nu$ is an S-integrable and internal $*$-simple function from ${}^*\Omega \to \text{fin}({}^*G)$. This means that it is a lifting of an extended Bochner integrable function in $M(\widehat{|\nu|}, \widehat{G})$ and hence the $\widehat{G}$-valued generalized integral with respect to $\widehat{|\nu|}$ applies and gives a well-defined vector-vector integral. Since the extended Bochner integral is countably additive, so is this integral. □

The use of nonstandard analysis simplifies standard arguments here, since in the standard setting this argument would only work if the Banach space F had the Radon-Nikodým property.

16.5 Uniform convexity

Bogdan needed the assumption that F is uniformly convex to convert the F-valued derivative $\frac{d\nu}{d|\nu|}$ into the F'-valued derivative $\frac{d|\nu|}{d\nu}$. He composed the derivative $\frac{d\nu}{d|\nu|}$ with a continuous map g from the unit sphere of F to the unit sphere of its dual space F' such that for all $f \in S_F$, $\langle f, g(f)\rangle = 1$. The resulting map $g \circ \frac{d\nu}{d|\nu|} : \Omega \to F'$ is $|\nu|$ measurable and inherits the integrability of $\frac{d\nu}{d|\nu|}$.

If the Banach space F happens to be uniformly convex, we can compose the nonstandard extension of g with φ_ν. In this case, $\langle f, g(f)\rangle = 1$ for all $f \in S_F$ translates into $\left\langle \frac{{}^*\nu(A_i)}{{}^*|\nu|(A_i)}, {}^*g\left(\frac{{}^*\nu(A_i)}{{}^*|\nu|(A_i)}\right)\right\rangle = 1$ for $i = 1, \ldots, H$ i.e.

$$\left\langle {}^*\nu(A_i), {}^*g\left(\frac{{}^*\nu(A_i)}{{}^*|\nu|(A_i)}\right)\right\rangle = {}^*|\nu|(A_i).$$

Theorem 16.5.1 *Let E be any Banach space and let F be a uniformly convex vector space. If the vector measure $\mu : \Sigma \to E$ is absolutely continuous with respect to the total variation of the vector measure $\nu : \Sigma \to F$, then there is an extended Bochner integrable function $f : {}^*\Omega \to \widehat{E \times F'}$ such that for all $A \in \Sigma$*

$$\mu(A) = \int_{{}^*A} f \, d\widehat{\nu}.$$

Proof. Define $\varphi_f = \varphi_\mu \cdot ({}^*g \circ \varphi_\nu)$, where φ_μ is a lifting of the generalized derivative $\frac{d\mu}{d|\nu|}$, φ_ν is a lifting of the generalized derivative $\frac{d\nu}{d|\nu|}$ and g is the continuous map from S_F to $S_{F'}$. This lifting is internal, S-integrable, $*$-simple and takes its values in $\mathrm{fin}({}^*E) \times {}^*S_{F'}$. Define $f = \pi_{E\times F'} \circ \varphi_f$. Then

$$\begin{aligned}
\int_{{}^*\Omega} f \, d\widehat{\nu} &= \pi_E\left(\int_{{}^*\Omega} \varphi_\mu(\omega) \cdot {}^*g\left(\varphi_\nu(\omega)\right) d^*\nu\right) \\
&= \pi_E\left(\sum_{i=1}^{H} \frac{{}^*\mu(A_i)}{{}^*|\nu|(A_i)} \cdot {}^*g\left(\frac{{}^*\nu(A_i)}{{}^*|\nu|(A_i)}\right) {}^*\nu(A_i)\right) \\
&= \pi_E\left(\sum_{i=1}^{H} \frac{{}^*\mu(A_i)}{{}^*|\nu|(A_i)} \cdot {}^*|\nu|(A_i)\right) \\
&= \pi_E\left({}^*\mu({}^*\Omega)\right) \\
&= \mu(\Omega),
\end{aligned}$$

and similarly for $\mu(A)$, where (A_i) is replaced by $(A_i \cap {}^*A)$ throughout. Since the fine partition refines the standard partition of Ω into A and its complement, $A_i \cap {}^*A$ is either the empty set or A_i. □

If E has the Radon-Nikodým property, then this lifting coincides with the image of the derivative found by Bogdan in the nonstandard hull of the space of vector-vector integrable functions on Ω.

16.6 Vector-vector derivatives without uniform convexity

We can weaken the assumption on F: it follows from the Hahn-Banach theorem that for each nonzero $f \in F$ there is an $f' \in F'$ with $\|f'\| = 1$ and $f'(f) = \|f\|$. Without uniform convexity, the choice of f' is not necessarily unique, and the map $f \mapsto f'$ need not be continuous on the unit sphere of F. We can forego the use of a function g defined on all of S_F. All we need is a derivative on each set A_i in the fine partition of ${}^*\Omega$.

Theorem 16.6.1 *Let F be any Banach space and let $\nu : \Sigma \to F$ be a countably additive vector measure of bounded variation. Then there is a function $g_{|\nu|} \in M(\widehat{|\nu|}, \widehat{F'})$ such that*

$$\int_{^*A} g_{|\nu|}\, d\widehat{\nu} = {}^\circ({}^*|\nu|({}^*A)) = |\nu|(A)$$

for all $A \in \Sigma$.

Proof. As the collection of sets A_i is internal, we can use the axiom of choice to get an internal collection of norm one functionals $g_i \in {}^*S_{F'}$ with

$$g_i\left({}^*\nu(A_i)\right) = {}^*|\nu|(A_i) \quad \text{for } i = 1, \ldots, H.$$

Define an internal $*$-simple S-integrable function. $\psi_{|\nu|} : {}^*\Omega \to {}^*S_{F'}$ by

$$\psi_{|\nu|} = \sum_{i=1}^{H} g_i \cdot 1_{A_i}.$$

This function is a lifting of an extended integrable function $g_{|\nu|} \in M(\widehat{|\nu|}, \widehat{F'})$. We can define a continuous bilinear multiplication $F' \times F \to \mathbb{R}$ by $f' \cdot f = f'(f)$ and then use the integral from theorem 16.4.1. In this case the quotient map π_G becomes the standard part map $^\circ : \mathrm{fin}({}^*\mathbb{R}) \to \mathbb{R}$. By the choice of the fine partition, $A_i \cap {}^*A$ is either the empty set or equals A_i.

$$\begin{aligned}
\int_{^*A} g_{|\nu|}\, d\widehat{\nu} &= {}^\circ\!\left(\int_{^*A} \psi_{|\nu|}\, d^*\nu\right) \\
&= {}^\circ\!\left(\int_{^*A} \sum_{i=1}^{H} g_i \cdot 1_{A_i}\, d^*\nu\right)
\end{aligned}$$

$$
\begin{aligned}
&= {}^{\circ}\left(\sum_{i=1}^{H} g_i \cdot {}^*\nu(A_i \cap {}^*A)\right) \\
&= {}^{\circ}\left(\sum_{i=1}^{H} {}^*|\nu|(A_i \cap {}^*A)\right) \\
&= {}^{\circ}({}^*|\nu|({}^*A))
\end{aligned}
$$

for any set $A \in \Sigma$. This finishes the proof, as for all $A \in \Sigma$, $|\nu|(A) = {}^{\circ}({}^*|\nu|({}^*A))$.
□

The last theorem constructed a derivative for any countably additive vector measure of bounded variation with respect to its total variation as an element of $M(\widehat{|\nu|}, \widehat{F'})$. This was the main difficulty in differentiation with respect to a vector measure. Now we can differentiate one vector measure with respect to another vector measure.

Theorem 16.6.2 *Let E and F be any Banach spaces, let $\mu : \Sigma \to E$ and $\nu : \Sigma \to F$ be a countably additive vector measures of bounded variation such that $\mu << |\nu|$. Then there is a function $h \in M(\widehat{|\nu|}, \widehat{E \times F'})$ such that for all $A \in \Sigma$*

$$
\mu(A) = \int_{{}^*A} h \, d\widehat{\nu}.
$$

Proof. The continuous trilinear multiplication $E \times F' \times F \to E$ defined by $e \times f' \times f = f'(f) \cdot e$ extends to a continuous bilinear multiplication $\widehat{E \times F'} \times \widehat{F} \to \widehat{E}$ defined by $\pi(e \times f') \times \pi(f) = \pi_E(f'(f) \cdot e)$. The extended integrable function $h : {}^*\Omega \to \widehat{E \times F'}$ is defined through its lifting $\varphi = \varphi_\mu \cdot \psi_{|\nu|}$, the product of the lifting φ_μ of $\frac{d\mu}{d|\nu|} \in M(\widehat{|\nu|}, \widehat{E})$ and the lifting $\psi_{|\nu|}$ of $\frac{d|\nu|}{d\nu} \in M(\widehat{|\nu|}, \widehat{F'})$ found in theorem 16.6.1. Setting

$$
\varphi = \varphi_\mu \cdot \psi_{|\nu|} = \sum_{i=1}^{H} \frac{{}^*\mu(A_i)}{{}^*|\nu|(A_i)} \cdot g_i \cdot 1_{A_i},
$$

we obtain a $*$ simple internal function with values in $\text{fin}({}^*E) \times {}^*S_{F'}$. The S-integrability of φ follows from the S-integrability of φ_μ and Lemma 16.3.3 which asserts that ${}^*\|\psi_{|\nu|}\| \approx 1$ almost everywhere. Then for all $A \in \Sigma$

$$
\begin{aligned}
\mu(A) &= \pi_E\left({}^*\mu({}^*A)\right) \\
&= \pi_E\left(\sum_{i=1}^{H} {}^*\mu(A_i \cap {}^*A)\right) \\
&= \pi_E\left(\sum_{i=1}^{H} \frac{{}^*\mu(A_i \cap {}^*A)}{{}^*|\nu|(A_i \cap {}^*A)} \cdot {}^*|\nu|(A_i \cap {}^*A)\right)
\end{aligned}
$$

$$
\begin{aligned}
&= \pi_E\left(\sum_{i=1}^{H} \frac{{}^*\mu(A_i \cap {}^*A)}{{}^*|\nu|(A_i \cap {}^*A)} \cdot (g_i \cdot 1_{A_i \cap {}^*A}) \cdot {}^*\nu(A_i \cap {}^*A)\right) \\
&= \pi_E\left(\int_{{}^*A} \varphi_\mu \cdot \psi_{|\nu|}\, d^*\nu\right) \\
&= \int_{{}^*A} h\, d\widehat{\nu},
\end{aligned}
$$

and $h : {}^*\Omega \to \widehat{E \times F'}$ is an extended integrable generalized vector-vector derivative $\frac{d\mu}{d\nu}$. □

16.7 Remarks

All the constructions above depend on the choice of a fine partition $A_1, \ldots, A_H$ of ${}^*\Omega$ that refines all finite standard partitions of Ω into measurable sets. A different choice of the partition would yield different derivatives, especially in the case of a non-uniformly convex Banach space F. However, for the standard analyst there is no noticeable difference between different choices of fine partitions, as the integrals over any standard set will be the same, no matter which fine partition is used.

Representing a vector measure $\nu : \Sigma \to F$ as an integral of its generalized derivative $\frac{d\nu}{d|\nu|}$ simplifies some of the results on Loeb completions of internal measures [13] by Živaljević.

References

[1] S. Albeverio, J.E. Fenstad, R. Høegh-Krohn and T. Lindstrøm, *Nonstandard methods in stochastic analysis and mathematical physics*, Academic Press, Orlando, FL, 1986.

[2] R.G. Bartle, "A general bilinear vector integral", Studia Math., **15** (1956) 337–352.

[3] J. Bliedtner and P.A. Loeb, "The Optimal Differentiation Basis and Liftings of L^∞", Trans. Amer. Math. Soc., **352** (2000) 4693–4710.

[4] W.M. Bogdanovicz and B. Kritt, "Radon–Nikodým Differentiation of one Vector-Valued Volume with Respect to Another", Bulletin De L'Academie Polonaise Des Sciences, Série des sciences math. astr. et phys., **XV** (1967) 479–486.

[5] J. Diestel and J.J. Uhl, *Vector Measures*, Mathematical Surveys 15, American Mathematical Society, Providence, RI, 1977.

[6] C.W. Henson and L.C. Moore, Nonstandard Analysis and the theory of Banach spaces, in A.E. Hurd (ed.) *Nonstandard Analysis—Recent Developments*, Springer Lecture Notes in Mathematics 983, 1983.

[7] A.E. Hurd and P.A. Loeb, *An Introduction to Nonstandard Real Analysis*, Academic Press, Orlando, FL, 1985.

[8] P.A. Loeb, "Conversion from nonstandard to standard measure spaces and applications to probability theory", Trans. Amer. Math. Soc., **211** (1975) 113–122.

[9] D. A. Ross, Nonstandard Measure Constructions — Solutions and Problems submitted to NS2002, Nonstandard Methods and Applications in Mathematics, June 10-16 2002, Pisa, Italy.

[10] G.B. Zimmer, "An extension of the Bochner integral generalizing the Loeb-Osswald integral", Proc. Cambr. Phil. Soc., **123** (1998) 119–131.

[11] G. B. Zimmer, "A unifying Radon-Nikodým Theorem through Nonstandard Hulls", Illinois Journal of Mathematics, **49**, (2005) 873-883.

[12] G. B. Zimmer, *Nonstandard Vector Integrals and Vector Measures*, Ph.D. Thesis, University of Illinois at Urbana-Champaign, October 1994.

[13] R. T. Živaljević, "Loeb completion of internal vector-valued measures", Math. Scand., **56** (1985) 276–286.

17

Differentiability of Loeb measures

Eva Aigner*

Abstract

We introduce a general definition of S-differentiability of an internal measure and compare different special cases. It will be shown how S-differentiability of an internal measure yields differentiability of the associated Loeb measure. We give some examples.

17.1 Introduction

In this paper we present some new results about differential properties of Loeb measures. The theory of differentiable measures was suggested by Fomin [9] as an infinite dimensional substitute for the Sobolev-Schwartz theory of distributions and has extended rapidly to a strong field of research. In particular during the last ten years it has become the foundation for many applications in different fields such as quantum field theory (see e.g. [10] or [14]) or stochastic analysis (see e.g. [4], [5], [6] and [15]).

There are many different notions of measure differentiability. The following two are most common (see e.g. [3], [6], [9], [13] and [15]). Let ν be a Borel measure on a locally convex space E and y an element of E.

1) The measure ν is called *Fomin-differentiable along* y if for all Borel subsets $B \subset E$ the limit

$$\lim_{r\to 0,\, r\in\mathbb{R}} \frac{\nu(B+ry)-\nu(B)}{r}$$

exists.

2) The measure ν is called *Skorohod-differentiable with respect to* y, if there exists another Borel measure ν' on E, such that for all continuous real-

*Ludwig-Maximilians-Universität,München, Germany.

valued bounded functions g on E

$$\lim_{r\to 0,\, r\in\mathbb{R}} \frac{\int g(x-ry)d\nu(x) - \int g(x)d\nu(x)}{r} = \int g(x)d\nu'(x).$$

The relationships between these two approaches were studied by Averbukh, Smolyanov and Fomin [3] and Bogachev [6].

A very general definition of differentiability of a curve of measures is given by Weizsäcker in [16].

In this paper we define measure differentiability as follows.

Definition 17.1.1 *Let $(\Omega, \mathcal{F}, \nu)$ be a measure space, $(\nu_r)_{-\varepsilon<r<\varepsilon}$, $\varepsilon \in \mathbb{R}^+$, a curve of nonnegative finite σ-additive measures on Ω such that $\nu = \nu_0$ and $\tilde{\mathcal{C}}$ a set of $\mathcal{F}$-measurable real-valued bounded functions on Ω. We say that ν is* differentiable with respect to the set $\tilde{\mathcal{C}}$ *if there exists a signed finite σ-additive measure ν' on Ω, such that for all functions g of $\tilde{\mathcal{C}}$*

$$\lim_{r\to 0} \frac{\int g(\omega)d\nu_r(\omega) - \int g(\omega)d\nu(\omega)}{r} = \int g(\omega)d\nu'(\omega).$$

The measure ν' is called a derivative of ν.

Note that Definition 17.1.1 covers the cases mentioned above, since for a locally convex space Ω and a fixed vector $y \in \Omega$ a curve $(\nu_r)_{-\varepsilon<r<\varepsilon}$ can be defined by $\nu_r(B) = \nu(B + ry)$ for all Borel subsets $B \subset \Omega$. When choosing $\tilde{\mathcal{C}} = \{1_B : B \in \mathcal{F}\}$, where 1_B is the indicator function, we obtain Fomin-differentiability. When choosing $\tilde{\mathcal{C}}$ as the set of all continuous bounded functions we obtain Skorohod-differentiability.

Differentiability (in the above sense) for Loeb measures has not been studied previously as far as we are aware. Hence the aim of this paper is to present the foundation and basic results. Since we want to obtain results for Loeb measures, in the first part we provide and discuss natural and very general assumptions for the underlying internal measures. The arising results for the Loeb measures — in particular a powerful theorem for the case of Fomin differentiability — are presented and discussed in the second part. Short and simple examples will illustrate the results.

A more complex and detailed description of differential properties of Loeb measures will be given in the author's thesis ([1]). A main topic of that thesis is also the application of the basic results to nonstandard representations of abstract Wiener spaces (see [7], [8] and [12]), which yields new insights also in standard mathematics.

The reader should be familiar with the basic results on nonstandard analysis and the Loeb measure construction, presented e.g. in [2], [7] and [12].

17.2 S-differentiability of internal measures

Standing Assumption

Throughout this paper let Ω be an internal set, $\mathcal{A}$ an internal ${}^\star\sigma$-field on Ω and $\mu \geq 0$ an internal ${}^\star\sigma$-additive S-bounded measure on $\mathcal{A}$. Let $(\mu_t)_{t\in J}$ be an internal curve of nonnegative ${}^\star\sigma$-additive S-bounded measures on $\mathcal{A}$. Since we want to obtain an external curve of Loeb measures, we assume that for some $\varepsilon \in \mathbb{R}^+$ the internal parameter set J is either an interval of ${}^\star\mathbb{R}$ containing the standard interval $I =]-\varepsilon, \varepsilon[$ or J is a discrete interval $\{\frac{-k}{H}, \frac{-k+1}{H}, \ldots, \frac{k-1}{H}, \frac{k}{H}\}$ with $H \in {}^\star\mathbf{N} \setminus \mathbf{N}$, $k \in \{1, \ldots, H\}$ and $\varepsilon \leq \frac{k}{H}$. Moreover, the internal curve $(\mu_t)_{t\in J}$ shall be S-continuous in the following sense. If $t, s \in J$ with $t \approx s$, then $\mu_t(A) \approx \mu_s(A)$ for each $A \in \mathcal{A}$. Finally we assume that $\mu = \mu_0$.

We now introduce S-differentiability for internal measures.

Definition 17.2.1 *Suppose we have a (not necessarily internal) set $\mathcal{C}$ of internal ${}^\star\mathbb{R}$-valued functions on Ω, each being $\mathcal{A}$-measurable and S-bounded. We say that the internal measure μ is S-differentiable with respect to the set $\mathcal{C}$ if there exists an internal S-bounded signed measure μ' on $\mathcal{A}$ so that for all $f \in \mathcal{C}$ and for all infinitesimals $t \in J$, $t \neq 0$*

$$\frac{\int f(\omega)\, d\mu_t(\omega) - \int f(\omega)\, d\mu(\omega)}{t} \approx \int f(\omega)\, d\mu'(\omega).$$

We call μ' an (internal) derivative of μ.

Note that a derivative is not uniquely determined by the above definition. If μ is S-differentiable with respect to a set $\mathcal{C}$ and if for some infinitesimal $t \in J$ the internal measure $\frac{\mu_t - \mu}{t}$ has limited values, then $\frac{\mu_t - \mu}{t}$ is a derivative of μ.

In this section we will regard and compare S-differentiability for different sets $\mathcal{C}$ of functions. Following the standard literature we define S-Fomin-differentiability.

Definition 17.2.2 *The internal measure μ is called S-Fomin-differentiable if the differentiability is with respect to $\mathcal{C} = \{1_A : A \in \mathcal{A}\}$ where 1_A is the indicator function.*

Note that in the case of S-Fomin-differentiability each measure $\frac{\mu_t - \mu}{t}$ with $t \approx 0$, $t \in J \setminus \{0\}$, is a derivative of μ. The following proposition shows the power of S-Fomin-differentiability.

Proposition 17.2.1 *If μ is S-Fomin-differentiable and μ' is a derivative of μ, then μ is S-differentiable with respect to the set $\mathcal{C}$ of all S-bounded ${}^\star\mathbb{R}$-valued $\mathcal{A}$-measurable functions. The Fomin-derivative μ' is also a derivative with respect to $\mathcal{C}$.*

Proof. Let μ be S-Fomin-differentiable and let μ' be a derivative of μ. Let $t \neq 0$ be an infinitesimal of J and set $\tilde{\mu} := \frac{\mu_t - \mu}{t}$. Since $\mu' \approx \tilde{\mu}$ on $\mathcal{A}$ we obtain $\int f(\omega) d\mu'(\omega) \approx \int f(\omega) d\tilde{\mu}(\omega)$ for all S-bounded $\mathcal{A}$-measurable functions $f : \Omega \to {}^\star\mathbb{R}$. □

When considering measures with Lebesgue densities, then the following lemma is useful.

Lemma 17.2.1 *Let $\Omega = {}^\star\mathbb{R}$, $\mathcal{A}$ the internal field of Borel subsets, y a fixed element of ${}^\star\mathbb{R}$, μ an internal S-bounded measure on $\mathcal{A}$. Take $J = {}^\star\mathbb{R}$ and for all $t \in J$ define $\mu_t(A) = \mu(A + ty)$ for all $A \in \mathcal{A}$. Assume that μ has an internal Lebesgue density f satisfying the following three conditions:*

1) $f \geq 0$, $f = 0$ only on a set of internal Lebesgue measure 0.

2) f is *differentiable in the direction of y (with derivative $f'_y(x) := f'(x) \cdot y$).

3) If $t \approx 0$, then for all $x \in {}^\star\mathbb{R}$ with $f(x) \neq 0$

$$\frac{1}{t}\left(\frac{f(x+ty)}{f(x)} - 1\right) \approx \frac{f'_y(x)}{f(x)}.$$

Now if the internal function $\beta^y_\mu : {}^\star\mathbb{R} \to {}^\star\mathbb{R}$, defined by

$$\beta^y_\mu(x) = \begin{cases} \dfrac{f'_y(x)}{f(x)} & \text{if } f(x) \neq 0 \\ 0 & \text{if } f(x) = 0, \end{cases}$$

is S_μ-integrable, then μ is S-Fomin-differentiable and if μ' is a derivative, then for all $A \in \mathcal{A}$

$$\mu'(A) \approx \int_A \beta^y_\mu(x)\, d\mu(x).$$

Proof. Since β^y_μ is S_μ-integrable, $A \mapsto \int_A \beta^y_\mu(x) d\mu(x)$ defines an S-bounded measure. Now let $N = \{x \in {}^\star\mathbb{R} : f(x) = 0\}$. Then $\lambda(N) = 0$, where λ is the internal Lebesgue measure. If $t \approx 0$, $t \neq 0$, then

$$\begin{aligned} \left| \frac{\mu_t(A) - \mu(A)}{t} - \int_A \beta^y_\mu(x) d\mu(x) \right| &\leq \int_A \left| \frac{f(x+ty) - f(x)}{t} - \beta^y_\mu(x) \cdot f(x) \right| d\lambda(x) \\ &= \int_{A \setminus N} \left| \frac{1}{t}\left(\frac{f(x+ty)}{f(x)} - 1\right) - \frac{f'_y(x)}{f(x)} \right| d\mu(x) \\ &\approx 0. \end{aligned}$$

Here we have used condition 3) and the fact that μ is S-bounded. □

Example 17.2.1 Let $\Omega = {}^\star\mathbb{R}$, $\mathcal{A}$ be the internal field of Borel subsets and μ defined by $\mu(A) = \int_A \frac{1}{1+x^2}\, d\lambda(x)$. Fix an element $y \in {}^\star\mathbb{R}$ and define the curve by $\mu_t(A) = \int_{A+ty} \frac{1}{1+x^2}\, d\lambda(x)$, $t \in {}^\star\mathbb{R}$. If y is limited, then it's easy to see that the assumptions of Lemma 17.2.1 are satisfied. Hence the measure μ is S-Fomin-differentiable and if μ' is a derivative, then $\mu'(A) \approx \int_A \frac{-2xy}{1+x^2}\, d\mu(x)$ for each $A \in \mathcal{A}$. If y is an unlimited element of ${}^\star\mathbb{R}$, then μ is not S-Fomin-differentiable, because for $t = \frac{1}{y}$ and for the internal interval $A = {}^\star[0,1] \subset {}^\star\mathbb{R}$ the value $\frac{\mu_t(A)-\mu(A)}{t}$ is unlimited.

Remark Note that Lemma 17.2.1 is also true for $\Omega = {}^\star\mathbb{R}^L$, $L \in {}^\star\mathbf{N}$. In the author's thesis [1] this is used to show the S-Fomin-differentiability of a nonstandard representation of the Wiener measure introduced by Cutland and Ng in [8].

We now turn to other forms of S-differentiability. Again, following the standard literature, we define S-Skorohod-differentiability.

Definition 17.2.3 *Let Ω be a subset of ${}^\star M$ where M is a metric space and let $\mathcal{A}$ be an internal ${}^\star\sigma$-field on Ω. The measure μ is called S-Skorohod-differentiable if it is S-differentiable with respect to the set of all S-bounded, $\mathcal{A}$-measurable functions $f : \Omega \to {}^\star\mathbb{R}$ that are S-continuous. This means, $f(\omega) \approx f(\tilde{\omega})$ for $\omega, \tilde{\omega} \in \Omega$, $\omega \approx \tilde{\omega}$.*

If Ω is as described in Definition 17.2.3, then, as a consequence of Proposition 17.2.1, S-Fomin-differentiability implies S-Skorohod-differentiability. The following example shows that the converse is not true.

Example 17.2.2 Fix a natural number $H \in {}^\star\mathbf{N} \setminus \mathbf{N}$ and let $\Omega \subset {}^\star\mathbb{R}$, $\Omega = \{\frac{1}{H} \cdot z : z \in {}^\star\mathbf{Z}\}$. The measure μ is the counting measure defined on the field of internal subsets of Ω, i.e.

$$\mu(A) = \frac{\left|A \cap \left[0, \frac{H-1}{H}\right]\right|}{H}.$$

Here $\left[0, \frac{H-1}{H}\right] = \{0, \frac{1}{H}, \ldots, \frac{H-1}{H}\}$ and $|A|$ is the internal number of elements. Now let $J \subset \Omega$, $J = \left[-\frac{l}{H}, \frac{l}{H}\right]$ with $l \in {}^\star\mathbf{N}$, $\frac{l}{H} \approx \frac{1}{2}$ and let $(\mu_t)_{t \in J}$ be defined by $\mu_t(A) = \mu(A+t)$. Note that for any $k \in {}^\star\mathbf{N}$ with $\frac{k}{H} \in J$ we obtain:

$$\frac{\mu_{\frac{k}{H}} - \mu}{\frac{k}{H}}(A) = \frac{1}{k}\left(\left|A \cap \left[\frac{-k}{H}, \frac{-1}{H}\right]\right| - \left|A \cap \left[\frac{H-k}{H}, \frac{H-1}{H}\right]\right|\right)$$

and

$$\frac{\mu_{\frac{-k}{H}} - \mu}{\frac{-k}{H}}(A) = \frac{1}{k}\left(\left|A \cap \left[0, \frac{k-1}{H}\right]\right| - \left|A \cap \left[\frac{H}{H}, \frac{H+k-1}{H}\right]\right|\right).$$

Now choose $k \in {}^\star\mathbf{N}$ with $\frac{k}{H} \in J$ and $\frac{k}{H} \approx 0$ and set $A = [0, \frac{k-1}{H}]$. Then

$$\frac{\mu_{\frac{k}{H}} - \mu}{\frac{k}{H}}(A) = 0 \quad \text{and} \quad \frac{\mu_{\frac{-k}{H}} - \mu}{\frac{-k}{H}}(A) = 1.$$

Hence μ is not S-Fomin-differentiable. It is not hard to check that μ is S-Skorohod-differentiable with internal derivatives $\frac{\mu_t - \mu}{t}$ for all infinitesimals $t \in J \setminus \{0\}$. We will give a standard application of this example in Example 17.3.1.

Finally we consider S-differentiability with respect to *continuous functions. We will see that this is equivalent to S-Fomin-differentiability. The following proposition can be easily shown by transfer of Urysohn's Lemma.

Proposition 17.2.2 *Let Ω be an internal *normal space, $\mathcal{A}$ the internal field of Borel subsets. If $(\mu_t)_{t \in J}$ is a curve of *regular measures, then μ is S-differentiable with respect to the set $\mathcal{C}$ of all internal *continuous S-bounded functions if and only if μ is S-Fomin-differentiable.*

17.3 Differentiability of Loeb measures

In this section we show how S-differentiability of an internal measure yields differentiability (in the sense of Definition 17.1.1) of the corresponding Loeb measure. Recall the Standing Assumption on page 240.

Now we can define a curve of Loeb measures in a unique way. Let $\varepsilon \in \mathbb{R}^+$ as described in the Standing Assumption (page 240), $I =]-\varepsilon, \varepsilon[$. For each $r \in I$ choose $t \in J$ such that $t \approx r$ and set $\mu_r := \mu_t$. Let us denote the associated Loeb spaces by $(\Omega, L_{\mu_r}(\mathcal{A}), (\mu_r)_L)$. Since the Loeb σ-fields $L_{\mu_r}(\mathcal{A})$ are not necessarily identical we choose a joint σ-field $\mathcal{F} \subset \bigcap_{r \in I} L_{\mu_r}(\mathcal{A})$. We now define the curve $((\mu_L)_r)_{r \in I}$ of measures on $\mathcal{F}$ by $(\mu_L)_r := (\mu_r)_L$ restricted to $\mathcal{F}$.

Let us mention some obvious connections between S-differentiability and differentiability.

Lemma 17.3.1 *Let μ be S-differentiable with respect to a set $\mathcal{C}$ of internal $^\star\mathbb{R}$-valued, $\mathcal{A}$-measurable and S-bounded functions on Ω and let μ' be an internal derivative of μ.*

1) Then for all $f \in \mathcal{C}$

$$\lim_{r \to 0} \frac{{}^\circ(\int f(\omega) d\mu_r(\omega)) - {}^\circ(\int f(\omega) d\mu(\omega))}{r} = {}^\circ\left(\int f(\omega) d\mu'(\omega)\right).$$

The convergence is uniform, if $\mathcal{C}$ is internal.

2) Suppose

$$\tilde{\mathcal{C}} := \{g : \Omega \to \mathbb{R} : \text{ there is an } f \in \mathcal{C} \text{ with } {}^\circ(f(\omega)) = g(\omega) \text{ for all } \omega \in \Omega\}$$

and

$$\mathcal{F}_{\mu'} := \Big(\bigcap_{r \in I} L_{\mu_r}(\mathcal{A})\Big) \cap L_{\mu'}(\mathcal{A}).$$

Then the Loeb measure μ_L, restricted to $\mathcal{F}_{\mu'}$, is differentiable with respect to the set $\tilde{\mathcal{C}}$ and the Loeb extension $(\mu')_L$ of ${}^\circ(\mu')$, restricted to $\mathcal{F}_{\mu'}$, is a derivative $(\mu_L)'$ of μ_L. This means that for all $g \in \tilde{\mathcal{C}}$:

$$\lim_{r \to 0} \frac{\int g(\omega) d(\mu_L)_r(\omega) - \int g(\omega) d\mu_L(\omega)}{r} = \int g(\omega) d(\mu')_L(\omega).$$

Note that, if μ' and γ' are two internal derivatives of μ, then $\int g(\omega) d(\mu')_L(\omega) = \int g(\omega) d(\gamma')_L$ for all $g \in \tilde{\mathcal{C}}$, but the Loeb measures $(\mu')_L$ and $(\gamma')_L$ on $\mathcal{A}$ and hence also the σ-fields $L_{\mu'}(\mathcal{A})$ and $L_{\gamma'}(\mathcal{A})$ may be different.

Example 17.3.1 In Example 17.2.2 we defined an internal counting measure μ. Since μ is S-Skorohod-differentiable with internal derivatives $\frac{\mu_t - \mu}{t}$, $t \in J \setminus \{0\}$, $t \approx 0$, we can apply Lemma 17.3.1. But as we have seen in Example 17.2.2, for $k \in {}^\star\mathbf{N}$ with $\frac{k}{H} \in J$ and $\frac{k}{H} \approx 0$ and $A = [0, \frac{k-1}{H}]$

$$\left(\frac{\mu_{\frac{k}{H}} - \mu}{\frac{k}{H}}\right)_L (A) = 0 \text{ and } \left(\frac{\mu_{\frac{-k}{H}} - \mu}{\frac{-k}{H}}\right)_L (A) = 1.$$

Nevertheless, there exists a σ-field, on which the Loeb measures $\left(\frac{\mu_t - \mu}{t}\right)_L$ coincide for all infinitesimals $t \in J \setminus \{0\}$. Let $\mathcal{B}(\mathbb{R})$ be the (standard) σ-field of all Borel subsets of $\mathbb{R}$. For $B \in \mathcal{B}(\mathbb{R})$ let $st^{-1}[B] = \{\omega \in \Omega : {}^\circ\omega \in B\}$. According to the usual approach to standard Lebesgue measure by a nonstandard counting measure (see [7] or [12]), the external set $st^{-1}[B]$ is an element of $L_{\mu_r}(\mathcal{A})$ for all $r \in I$ and $(\mu_r)_L(st^{-1}[B]) = \nu_r(B)$, where $\nu_r(B) = \int_{B+r} 1_{[0,1]}(x) d\lambda(x)$ with standard Lebesgue measure λ. But $st^{-1}[B]$ is also an element of $L_{\frac{\mu_t - \mu}{t}}(\mathcal{A})$ for all infinitesimals $t \in J \setminus \{0\}$ and

$$\left(\frac{\mu_t - \mu}{t}\right)_L (st^{-1}[B]) = 1_B(1) - 1_B(0).$$

Hence the Loeb measures $\left(\frac{\mu_t - \mu}{t}\right)_L$, restricted to the σ-field $\{st^{-1}[B] : B \in \mathcal{B}(\mathbb{R})\}$, coincide for all infinitesimals $t \in J \setminus \{0\}$. If we define a measure ν' on $\mathcal{B}(\mathbb{R})$ by $\nu'(B) = 1_B(1) - 1_B(0)$, then the S-differentiability of the internal counting measure μ yields the Skorohod-differentiability (with respect to $1 \in \mathbb{R}$) of the standard measure ν with derivative ν'.

We will now see that in the case of S-Fomin-differentiability the σ-field $\mathcal{F}$ doesn't depend on the chosen internal derivative and the derivative $(\mu_L)'$ of μ_L is uniquely determined. Moreover, the differentiability of μ_L is true not only with respect to standard parts of internal functions, but also with respect to all $\mathcal{F}$-measurable S-bounded real-valued functions on Ω.

We gather these facts together into the following, which is the main theorem of this paper.

Theorem 17.3.1 *Let μ be S-Fomin-differentiable and $\mathcal{F} = \bigcap_{r\in I} L_{\mu_r}(\mathcal{A})$. Then μ_L is differentiable on $\mathcal{F}$ with respect to the set $\tilde{\mathcal{C}} = \{1_B : B \in \mathcal{F}\}$ and the differentiability is uniform on $\tilde{\mathcal{C}}$. The derivative $(\mu_L)'$ is uniquely determined and is absolutely continuous with respect to μ_L. If μ' is an internal derivative of μ, then the Loeb extension $(\mu')_L$ is defined on $\mathcal{F}$ and coincides with $(\mu_L)'$. In particular this is true for all internal measures $\frac{\mu_t-\mu}{t}$, where $t \in J \setminus \{0\}$ is infinitesimal.*

Proof. Let μ' be a derivative of μ. We will show the following statements:

(A) For all internal sets $A \in \mathcal{A}$ the limit

$$\lim_{r\to 0} \frac{(\mu_L)_r(A) - \mu_L(A)}{r}$$

exists and is equal to $(\mu')_L(A)$. The convergence is uniform on the internal field $\mathcal{A}$.

(B) If $N \in \mathcal{F}$ is a μ_L-nullset, then

$$\lim_{r\to 0} \frac{(\mu_L)_r(N) - \mu_L(N)}{r} = 0.$$

The convergence is uniform for all μ_L-nullsets of $\mathcal{F}$. Any μ_L-nullset of $\mathcal{F}$ is also a $(\mu')_L$-nullset of $\mathcal{F}$.

(C) If $B \in \mathcal{F}$ and if $A \in \mathcal{A}$ is μ_L-equivalent to B, then

$$\lim_{r\to 0} \frac{(\mu_L)_r(B) - \mu_L(B)}{r} = \lim_{r\to 0} \frac{(\mu_L)_r(A) - \mu_L(A)}{r},$$

in particular the left limit exists. The convergence is uniform on $\mathcal{F}$.

(D) The Loeb extension $(\mu')_L$ is defined on $\mathcal{F}$ and for all $B \in \mathcal{F}$

$$(\mu')_L(B) = \lim_{r\to 0} \frac{(\mu_L)_r(B) - \mu_L(B)}{r}.$$

(A) follows from Lemma 17.3.1.

To prove *(B)* let N be a μ_L-nullset of $\mathcal{F}$, i.e. there exists a sequence $(N_{\frac{1}{n}})_{n\in\mathbf{N}} \subset \mathcal{A}$ so that for all $n \in \mathbf{N}$ we have $N \subseteq N_{\frac{1}{n}}$, $N_{\frac{1}{n+1}} \subseteq N_{\frac{1}{n}}$ and

$\mu(N_{\frac{1}{n}}) \leq \frac{1}{n}$. Let $\tilde{N} := \bigcap_{n=1}^{\infty} N_{\frac{1}{n}}$. Then $\tilde{N} \in \mathcal{F}$, $\mu_L(\tilde{N}) = 0$ and we obtain for all $r \in I$

$$(\mu_L)_r(N) \leq (\mu_L)_r(\tilde{N}).$$

Since $\mu_L(N) = 0$ we get

$$\left| \frac{(\mu_L)_r(N) - \mu_L(N)}{r} \right| = \left| \frac{(\mu_L)_r(N)}{r} \right| \leq \left| \frac{(\mu_L)_r(\tilde{N})}{r} \right| .$$

Therefore it is sufficient to show that $\lim\limits_{r \to 0} \frac{(\mu_L)_r(\tilde{N})}{r} = 0$. Now since $\mu_L(\tilde{N}) = 0$,

$$\begin{aligned} \frac{(\mu_L)_r(\tilde{N})}{r} &= \frac{(\mu_L)_r(\tilde{N}) - \mu_L(\tilde{N})}{r} \\ &= \frac{\lim_{n\to\infty} (\mu_L)_r(N_{\frac{1}{n}}) - \lim_{n\to\infty} \mu_L(N_{\frac{1}{n}})}{r} \\ &= \lim_{n\to\infty} \frac{(\mu_L)_r(N_{\frac{1}{n}}) - \mu_L(N_{\frac{1}{n}})}{r}. \end{aligned}$$

It follows from *(A)*, that for each $n \in \mathbf{N}$ the limit $\lim_{r\to 0} \frac{(\mu_L)_r(N_{\frac{1}{n}}) - \mu_L(N_{\frac{1}{n}})}{r}$ exists and is equal to $(\mu')_L(N_{\frac{1}{n}})$. Since $(\mu')_L$ is defined on the smallest σ-field containing $\mathcal{A}$, the limit $\lim_{n\to\infty} (\mu')_L(N_{\frac{1}{n}}) = (\mu')_L(\tilde{N})$ also exists. Because of the uniform convergence stated in *(A)* and since $(N_{\frac{1}{n}})_{n\in\mathbf{N}} \subset \mathcal{A}$ we can exchange the limits as follows:

$$\begin{aligned} \lim_{r\to 0} \frac{(\mu_L)_r(\tilde{N})}{r} &= \lim_{r\to 0} \lim_{n\to\infty} \frac{(\mu_L)_r(N_{\frac{1}{n}}) - \mu_L(N_{\frac{1}{n}})}{r} = \\ \lim_{n\to\infty} \lim_{r\to 0} \frac{(\mu_L)_r(N_{\frac{1}{n}}) - \mu_L(N_{\frac{1}{n}})}{r} &= \lim_{n\to\infty} (\mu')_L(N_{\frac{1}{n}}) = (\mu')_L(\tilde{N}). \end{aligned}$$

So the measure μ_L is differentiable at $\tilde{N}$ and the value of the derivative is $(\mu')_L(\tilde{N})$. It remains to show that $(\mu')_L(\tilde{N}) = 0$. To this end we use an argument due to Smolyanov and Weizsäcker [15]. Let us consider the function

$$f : I \longrightarrow \mathbb{R}, r \mapsto (\mu_L)_r(\tilde{N})$$

Then f is differentiable at $t = 0$ and $f'(0) = (\mu')_L(\tilde{N})$. Since f is nonnegative and $f(0) = 0$, the first derivative of f in 0 must be 0. Hence $(\mu')_L(\tilde{N}) = 0$. The uniformity of the convergence can be seen by using again the uniform convergence on $\mathcal{A}$. Of course $(\mu')_L(N) = 0$ since $N \subset \tilde{N}$. Hence the μ_L-nullsets of $\mathcal{F}$ are also $(\mu')_L$-nullsets of $\mathcal{F}$.

(C) Here we show the differentiability for an arbitrary element of $\mathcal{F}$. So let $B \in \mathcal{F}$, $A \in \mathcal{A}$ μ_L-equivalent to B and $r \in I$. Since $\mu_L(B) = \mu_L(A)$, it is sufficient to show that

$$\lim_{r \to 0} \frac{(\mu_L)_r(B) - (\mu_L)_r(A)}{r} = 0.$$

Since μ_L is nonnegative the following estimate is easy to verify.

$$|(\mu_L)_r(B) - (\mu_L)_r(A)| \leq (\mu_L)_r(A \triangle B),$$

where $A \triangle B$ denotes the symmetric difference $(A \setminus B) \cup (B \setminus A)$. Now *(B)* yields

$$\lim_{r \to 0} \left| \frac{(\mu_L)_r(B) - (\mu_L)_r(A)}{r} \right| \leq \lim_{r \to 0} \frac{(\mu_L)_r(A \triangle B)}{|\, r \,|} = 0.$$

The uniform convergence follows from *(A)* and *(B)*. Hence *(C)* is proved.

(D) follows from *(A)*, *(B)* and *(C)*. □

Example 17.3.2 Let μ be a measure with internal Lebesgue density satisfying the assumptions of Lemma 17.2.1 for a fixed $y \in {}^\star\mathbb{R}$. Recall that the internal curve is given by $\mu_t(A) = \mu(A + ty)$ for all Borel subsets $A \subset {}^\star\mathbb{R}$. Let $\varepsilon \in \mathbb{R}^+$, $I =]-\varepsilon, \varepsilon[$. Since μ is S-Fomin-differentiable, we can apply Theorem 17.3.1. Hence μ_L is differentiable on $\mathcal{F} = \bigcap_{r \in I} L_{\mu_r}(\mathcal{A})$ with respect to $\tilde{\mathcal{C}} = \{1_B : B \in \mathcal{F}\}$. Moreover, for the uniquely determined derivative $(\mu_L)'$ we obtain:

$$(\mu_L)'(B) = \int_B {}^\circ\beta_\mu^y(x)\, d\mu_L(x)$$

for all $B \in \mathcal{F}$, where β_μ^y is defined in Lemma 17.2.1

The power of Fomin-differentiability is also shown in the last result.

Corollary 17.3.1 *If μ is S-Fomin-differentiable with an internal derivative μ' and $\mathcal{F}$ is defined as in Theorem 17.3.1, then the Loeb measure μ_L, restricted to $\mathcal{F}$, is differentiable with respect to the set $\tilde{\mathcal{C}}$ of all $\mathcal{F}$-measurable real-valued bounded functions on Ω. The Loeb measure $(\mu')_L$, restricted to $\mathcal{F}$, is the derivative of $(\mu)_L$.*

Proof. This is routine integration theory using the uniform differentiability on $\mathcal{F}$ and the approximation of measurable functions by simple functions. □

Remark An application of Fomin-differentiability is given in [15] by Smolyanov and Weizsäcker. There, a nonnegative measure ν on a locally convex space

E is considered which is Fomin-differentiable along all elements of a Hilbert subspace of E. Using this measure differentiability, Smolyanov and Weizsäcker introduce an operator on a subspace of $\mathcal{L}^2(E, \nu)$. In the Gaussian case this operator is the derivative operator of the Malliavin calculus (see [11]).

The question of the assumptions that are needed for differential properties of Loeb measures to yield such a derivative operator is the subject of ongoing investigation.

References

[1] E. Aigner, *Differentiability of Loeb measures and applications*, PhD thesis, Universität München, (in prep.)

[2] S. Albeverio, J.E. Fenstad, R. Høegh-Krohn and T. Lindstrøm, *Nonstandard Methods in Stochastic Analysis and Mathematical Physics*, Academic Press, New York, 1986.

[3] V.I. Averbukh, O.G. Smolyanov and S.V. Fomin, "Generalized functions and differential equations in linear spaces, I. Differentiable measures", Trudi of Moscow Math. Soc., **24** (1971) 140–184.

[4] V.I. Bogachev, "Differential properties of measures on infinite dimensional spaces and the Malliavin calculus", Acta Univ. Carolinae, Math. Phys., **30** (1989) 9–30.

[5] V.I. Bogachev, "Smooth Measures, the Malliavin Calculus and Approximations in Infinite Dimensional Spaces", Acta Univ. Carolinae, Math. Phys., **31** (1990) 9–23.

[6] V.I. Bogachev, "Differentiable Measures and the Malliavin Calculus", Journal of Mathematical Sciences, **87** (1997) 3577–3731.

[7] N. Cutland, *Loeb Measures in Practice: Recent Advances*, Lecture Notes in Mathematics 1751, Springer-Verlag, 2000.

[8] N. Cutland and S.-A. Ng, A nonstandard approach to the Malliavin calculus, in *Applications of Nonstandard-Analysis to Analysis, Functional Analysis Propability Theory and Mathematical Physics* (eds. S. Albeverio, W.A.J. Luxemburg and M. Wolff), D.Reidel-Kluwer, Dordrecht, 1995.

[9] S.V. Fomin, Differential measures in linear spaces, in *Proc. Int. Congr. of Mathematicians. Sec.5*, Izd. Mosk. Univ., Moscow, 1966.

[10] A.V. KIRILLOV, "Infinite-dimensional analysis and quantum theory as semimartingale calculus", Russian Math. Surveys, No. 3 (1994) 41–95.

[11] D. NUALART, *The Malliavin Calculus and Related Topics*, Probability and its Applications, Springer-Verlag, 1995.

[12] H. OSSWALD, *Malliavin calculus in abstract Wiener spaces. An introduction*, Book manuscript, 2001.

[13] A.V. SKOROHOD, *Integration in Hilbert Spaces*, Ergebnisse der Mathematik, Springer-Verlag, Berlin, New-York, 1974.

[14] O.G. SMOLYANOV and H.V. WEIZSÄCKER, "Formulae with logarithmic derivatives of measures related to the quantization of infinite-dimensional Hamilton systems", Russian Math. Surveys, **551** (1996) 357–358.

[15] O.G. SMOLYANOV and H.V. WEIZSÄCKER, "Smooth probability measures and associated differential operators", Inf. Dim. Analysis, Quantum Prob. and Rel. Topics, **2** (1999) 51–78.

[16] H.V. WEIZSÄCKER, *Differenzierbare Maße und Stochastische Analysis*, 6 Vorträge am Graduiertenkolleg 'Stochastische Prozesse und probabilistische Analysis'. Januar/Februar, Berlin, 1998.

Part IV

Differential systems and equations

18

The power of Gâteaux differentiability

Vítor Neves*

Abstract

The search for useful non standard minimization conditions on C^1 functionals defined on Banach spaces lead us to a very simple argument which shows that if a C^1 function $f : E \to F$ between Banach spaces is actually Gâteaux differentiable on finite points along finite vectors, then it is uniformly continuous on bounded sets if and only if it is lipschitzian on bounded sets. The following is a development of these ideas starting from locally convex spaces.

18.1 Preliminaries

This section consists of an informal description of tools to frame the discourse and therefore contains only a small number of theorems and few proofs. Careful foundational treatments may be found in [3], [9], [14] or [1]; we shall also use Nelson's quantifiers $\forall^{st}$ and $\exists^{st}$ as in [8].

We assume the existence of two set-theoretical structures $\mathcal{A}$ and $\mathcal{B}$ and a function $^*(\cdot) : \mathcal{A} \to \mathcal{B}$ satisfying the properties we proceed to describe. Elements of either of the structures which are sets shall be called **entities**[1].

(P1) $\mathcal{A}$ *is a model of the relevant Analysis* in the sense that any object of Classical Mathematical Analysis as well as the mathematical structures under study have an interpretation in $\mathcal{A}$ and theorems about them are true in $\mathcal{A}$. In particular the complete ordered field $\mathbb{R}$ of real numbers or any other classical space are elements of $\mathcal{A}$.

---

*Departamento de Matemática, Universidade de Aveiro.
vneves@mat.ua.pt
Work for this article was partially supported by FCT via both the grant POCTI\MAT\41683\01 and funds from the R&D unit CEOC.

[1] We shall also admit the existence of *atoms* i.e. objects without elements which are not the empty set; for instance numbers are atoms.

The image by $^*(\cdot)$ of any given element $a \in \mathcal{A}$ will be denoted $^*a$ and be called **standard** as well as a itself; elements of standard sets in $\mathcal{B}$ shall be called **internal**; non internal sets in $\mathcal{B}$ are called **external**.

A set of sets verifies the **Finite Intersection Property**, abbreviated f.i.p., if *all its finite subsets have non-empty intersection.* Denote the cardinal (number of elements) of a set S by $card(S)$ and its power set by $\mathcal{P}(S)$.

(P2) Polysaturation[2]

Given $E \in \mathcal{A}$ and $\mathcal{C} \subseteq {}^*\mathcal{P}(E)$, if $\mathcal{C}$ verifies the f.i.p. and $card(\mathcal{C}) < card(\mathcal{A})$ then $\bigcap \mathcal{C} \neq \emptyset$

There is an encompassing formal first order language $\mathcal{L}$ with equality $=$ and $\in$ (to be read as the membership relation), with the usual connectives $\neg$ for negation, $\wedge$ for conjunction, $\vee$ for disjunction, $\Rightarrow$ for implication, $\Leftrightarrow$ for equivalence, universal $\forall$ and existential $\exists$ quantifiers and enough constants, predicate and function symbols to denote any elements, relations and functions under consideration either in $\mathcal{A}$ or in $\mathcal{B}$.

Say that a formula of $\mathcal{L}$ is **bounded** if its quantified subformulae, hereby including the formula itself, are of the form

$$\forall x\ [x \in a \Rightarrow \phi] \qquad \text{or} \qquad \exists x\ [x \in a \wedge \phi]$$

for some constant a and formula ϕ; a **sentence** is a formula without free variables. $\mathcal{B}$ contains a *formal* copy of $\mathcal{A}$ in the following sense.

(P3) Transfer

If $\phi(a_1, \cdots, a_n)$ is a bounded sentence with occurrences of constants a_i $(1 \leq i \leq n)$ and no more constants, then $\phi(a_1, \cdots, a_n)$ is true in $\mathcal{A}$ iff $\phi({}^*a_1, \cdots, {}^*a_n)$ is true in $\mathcal{B}$[3].

It is always useful to keep in mind theorems 18.1.1 and 18.1.2 below. Say that a formula of $\mathcal{L}$ is **standard** (resp. **internal**) if it is bounded and all its constants denote standard (resp. internal) elements of $\mathcal{B}$.

Theorem 18.1.1 (Principles of Standard and of Internal Definition) *A set $B \in \mathcal{B}$ is standard (resp. internal) iff there exists $a \in \mathcal{A}$ and a standard (resp. internal) formula ϕ such that $B = \{x \in {}^*a | \ \phi(x)\}$ or, more informally, a set in $\mathcal{B}$ is standard or internal iff it is definable by a respectively standard or internal formula.*

[2] Actually this is stronger than needed for our purposes, but it is powerful and easy to formulate.

[3] $\mathcal{B}$ is a kind of elementary extension of $\mathcal{A}$ with embedding $^*(\cdot)$.

Call **monad** any intersection $\bigcap_{C\in\mathcal{C}} {}^*C$ with $\emptyset \neq \mathcal{C} \in \mathcal{A}$.

Theorem 18.1.2 (Cauchy's Principle) *Any internal set which contains a monad* $\bigcap_{C\in\mathcal{C}} {}^*C$ *also contains one of the standard sets* ${}^*C$.

Hyper-real numbers $\lambda \in {}^*\mathbb{R}$ are classified the following way

$$\begin{aligned} \lambda \textit{ is } \textbf{finite} \quad &:\equiv \quad \exists r \in \mathbb{R}\ |x| \leq {}^*r \\ \lambda \textit{ is } \textbf{infinitesimal} \quad &:\equiv \quad \forall r \in \mathbb{R}\ [r > 0 \Rightarrow |x| \leq {}^*r] \\ \lambda \textit{ is } \textbf{infinite} \quad &:\equiv \quad \lambda \textit{ is not finite}. \end{aligned}$$

$\mathcal{O}$ and μ denote respectively the set of finite and infinitesimal hyper-real numbers.

For any given (real) locally convex space S, let Γ_S denote a gauge, i.e. a *directed set*, of semi-norms defining the topology of S; $\mathbf{fin}({}^*S)$ and $\mu({}^*S)$ shall denote respectively the sets of finite and infinitesimal elements in ${}^*S$, i.e., for any given $x \in {}^*S$,

$$x \in \mathbf{fin}({}^*S) \quad :\equiv \quad \forall \gamma \in \Gamma_S\ \gamma(x) \in \mathcal{O} \tag{18.1.1}$$

$$x \in \mu({}^*S) \quad :\equiv \quad \forall \gamma \in \Gamma_S\ \gamma(x) \in \mu \tag{18.1.2}$$

When Γ_S is unbounded, in the sense that

$$\forall x \in S \quad \Gamma_S(x) \ := \ \{\gamma(x) |\ \gamma \in \Gamma_S\} \quad \text{is unbounded}, \tag{18.1.3}$$

it so happens that

$$\forall x \in {}^*S \quad \big[x \in \mu({}^*S) \ \Leftrightarrow \ \forall^{st}\gamma \in \Gamma_S\ \gamma(x) \leq 1\big]. \tag{18.1.4}$$

Condition (18.1.4) has the syntactical advantage of dispensing with a quantifier; as we will present strongly syntactical reasoning,

we assume from now on that condition (18.1.3) *is always verified.*

Denote ${}^\sigma C$ the set of **standard** elements of ${}^*C$, i.e.,

$${}^\sigma C \ := \ \{{}^*x |\ x \in C\}.^4$$

If E and F are locally convex spaces, $L^{(k)}(E, F)$ denotes the space of continuous k-linear maps $l : E^k \to F$; a map $l \in {}^*L^{(k)}(E, F)$ is said to be **finite** (resp. **infinitesimal**) if $l(\mathbf{fin}({}^*E)^k) \subseteq \mathbf{fin}({}^*F)$ (resp. $l(\mathbf{fin}({}^*E)^k) \subseteq \mu({}^*F)$).

It is useful to keep in mind the following results ([4] is a most complete source; also see [14, chap. 10]).

[4]We sometimes identify C with ${}^\sigma C$ for the sake of simplifying notation.

Theorem 18.1.3 *In a locally convex space S,*

1. $x \in \mu({}^*S)$ *if and only if* $\forall \lambda \in \mathcal{O}\ \lambda x \in \mu({}^*S) \qquad (x \in {}^*S)$.
2. $x \in \mathbf{fin}({}^*S)$ *if and only if* $\forall \lambda \in \mu\ \lambda x \in \mu({}^*S) \qquad (x \in {}^*S)$.
3. $\forall x \in \mu({}^*S)\ \exists \lambda \in {}^*\mathbb{R} \setminus \mathcal{O}\ \lambda x \in \mu({}^*S)$, *therefore it is also true that*

$$\forall x \in {}^*S\ [x \in \mu({}^*S) \Leftrightarrow \exists \lambda \in {}^*\mathbb{R} \setminus \mathcal{O}\ \lambda x \in \mu({}^*S)]$$

Let $x \approx y$ mean that x is **infinitely near** y, i.e., $x - y \in \mu({}^*S)$; also let **st** denote **standard part** whenever appropriate, i.e.,

$$y = \mathbf{st}(x) \ :\equiv\ y \in {}^\sigma S \ \&\ x \approx y;$$

$\mathbf{ns}({}^*S)$ will denote the set of **near-standard vectors** of ${}^*S$, i.e.,

$$\mathbf{ns}({}^*S) := {}^\sigma S + \mu({}^*S).$$

A function $f : {}^*E \to {}^*F$ is **S-continuous on** $A \subseteq {}^*E$ if for all $x, y \in A$, $f(x) \approx f(y)$, whenever $x \approx y$; by definition, any standard function f is S-continuous if ${}^*f$ is. Theorem 18.1.2 implies that S-continuity of internal functions is nothing else than continuity measured with standard tolerances; this and condition (18.1.3) imply:

Lemma 18.1.1 *Suppose $A \in {}^*\mathcal{P}(E)$. An internal function $f : A \to {}^*F$ is S-continuous at $x \in A$ iff*

$$\forall \nu \in {}^\sigma\Gamma_F\, \forall \varepsilon \in {}^\sigma\mathbb{R}\ \exists \gamma \in {}^\sigma\Gamma_E\ \exists \delta \in {}^\sigma\mathbb{R}\ \forall y \in A$$
$$[\gamma(x - y) < \delta \Rightarrow \nu(f(y) - f(x)) < \varepsilon].$$

In particular

Theorem 18.1.4 *A function $f : A \subseteq E \to F$ between locally convex spaces E and F is*

1. *continuous (on A) iff it is S-continuous on* ${}^*A \cap \mathbf{ns}({}^*E)$;
2. *uniformly continuous (on A) iff it is S-continuous on* ${}^*A$.

And

Theorem 18.1.5 *Given locally convex spaces E and F and an internal ${}^*k$-linear function $l : {}^*E^k \to {}^*F$, the following conditions are equivalent*

1. *l is S-continuous.*

2. *$l(\mathbf{fin}({}^*E)) \subseteq \mathbf{fin}({}^*F)$ (i.e. the internal S-continuous maps are the internal finite maps).*

3. *$l(\mu({}^*E)) \subseteq \mu({}^*F)$.*

It might also be interesting to notice the following

Theorem 18.1.6 *Let E and F be locally convex spaces, A be a subset of ${}^*E$ and $\phi : (A \times \mathbf{fin}({}^*E)) \to {}^*F$ be an internal function which is linear in the second coordinate. If ϕ is S-continuous in the sense that*

$$\forall x, y \in A\ \forall u, v \in \mathbf{fin}({}^*E)\ [x \approx y\ \&\ u \approx v\ \Rightarrow\ \phi(x,u) \approx \phi(y,v)],$$

then, for all $x \in A$, $\phi(x, \cdot)$ is a finite map.

Proof. Assume $\phi : (A \times \mathbf{fin}({}^*E)) \to {}^*F$ is S-continuous, $x \in A \subseteq {}^*E$ and that $u \in \mathbf{fin}({}^*E)$; for all $t \in \mu$, $tu \approx 0$, therefore $t\phi(x,u) = \phi(x,tu) \approx \phi(x,0) = 0$, hence, by theorem 18.1.3, $\phi(x,u) \in \mathbf{fin}({}^*F)$ as required. □

Observe that, when S is normed, an element $x \in {}^*S$ is infinitesimal (resp. finite) iff $\|x\| \in \mu$ (resp. $\|x\| \in \mathcal{O}$) and the following holds.

Corollary 18.1.1 *Given normed spaces E and F and a k-linear function $l : E^k \to F$, the following conditions are equivalent*

1. *l is continuous*

2. *$\forall x \in {}^*S\ [\, \|x\| \approx 0\ \Rightarrow\ \|l(x)\| \approx 0\,]$*

3. *$\forall x \in {}^*S\ [\, \|x\| \in \mathcal{O}\ \Rightarrow\ \|l(x)\| \in \mathcal{O}\,]$.*

Of course theorem 18.1.5 and corollary 18.1.1 essentially say that, for linear maps *continuity is equivalent to continuity at zero.*

18.2 Smoothness

Let E and F be complete locally convex spaces, A be an internal subset of ${}^*E$, $f : A \to {}^*F$ and $l_{(\cdot)} : A \to {}^*L(E,F)$ be internal and write

$$f(x+tv) = f(x) + tl_x(v) + t\iota \quad (x \in A,\ v \in {}^*E,\ t \in {}^*\mathbb{R},\ \iota \in {}^*F). \tag{18.2.1}$$

f is **GS-differentiable** at x with **GS-derivative** l_x, if $\iota \approx 0$ whenever $v \in \mathbf{fin}({}^*E)$ and $t \approx 0$; f is **uniformly differentiable** at $a \in {}^*E$ with **uniform derivative** $l_{(\cdot)}$ if f is GS-differentiable at x and l_x is finite, for all $x \approx a$. We shall also call $l : A \times \mathbf{fin}({}^*E) \to {}^*F$ the derivative of f and define

$$df(x,v) := l(x,v) := l_x(v) := df_x(v) \qquad (x, v \in {}^*E).$$

Observe that, when one takes x and v in E within equation (18.2.1), the *GS*-derivative $df(x,v)$ is just the Gâteaux derivative (**G-derivative** for short) of f at x along v. The following theorem is well known too (see [13]).

Theorem 18.2.1 *If E and F are Banach spaces, the function $f : E \to F$ has G-derivative $df(x,y)$, for all $x \in E$, and $x \mapsto df(x,\cdot) : E \to L(E,F)$ is continuous for the uniform topology on $L(E,F)$, then $x \mapsto df(x,\cdot)$ is actually a Fréchet derivative.*

The following is a simple generalization of proposition 2.4 in [15].

Theorem 18.2.2 *If the internal map $f : {}^*E \to {}^*F$ is uniformly differentiable at all points of an internal set $A \in {}^*\mathcal{P}(E)$ with derivative df, then*

1. *f is S-continuous on A*
2. *$df : A \times \mathbf{fin}({}^*E) \to {}^*F$ verifies*
   (a) *$df(A \times \mathbf{fin}({}^*E)) \subseteq \mathbf{fin}({}^*F)$*
   (b) *df is S-continuous in the sense that*

$$\forall x, y \in A\ \forall u, v \in \mathbf{fin}({}^*E)\ [x \approx y\ \&\ u \approx v \ \Rightarrow\ df_x(u) \approx df_y(v)].$$

Proof. Borrowing from [15]: suppose that $f : {}^*E \to {}^*F$ is internal and uniformly differentiable at all points of the internal set A, that $x, y \in A$ and that $x \approx y$; take $\lambda \in {}^*\mathbb{R} \setminus \mathcal{O}$ such that $\lambda(x-y) \in \mu({}^*E)$ (theorem 18.1.3) and define $t := \frac{1}{\lambda}$; t is infinitesimal and, for some $\iota \in \mu({}^*F)$,

$$\begin{aligned} f(x) - f(y) &= f(y + t\lambda(x-y)) - f(y) \\ &= t df_y(\lambda(x-y)) + t\iota \\ &= df_y(x-y) + t\iota \in \mu({}^*F) \end{aligned}$$

by theorem 18.1.5, and 1 is proven.

2 (a) is an immediate consequence of the fact that the maps df_x are finite and this is useful in proving S-continuity of df in (b): take $x, y \in A$ and $u, v \in \mathbf{fin}({}^*E)$ such that $x \approx y$ & $u \approx v$; observe that

$$df(x,u) - df(y,v) = df_x(u-v) + df_x(v) - df_y(v).$$

As the maps df_x are finite, by theorem 18.1.5, $df(x, u-v) \in \mu({}^*F)$ and it is enough to show that, under the present assumptions, $df_x(v) - df_y(v) \approx 0$.

Now, taking λ and t as above and $v \in \mathbf{fin}({}^*E)$, there exist infinitesimal vectors $\iota, \eta, \delta \in {}^*F$ such that

$$\begin{aligned} f(x) - f(y) &= df_y(x-y) + t\iota & (18.2.2)\\ f(x+tv) - f(x) &= tdf_x(v) + t\eta & (18.2.3)\\ f\big(y + t(v + \lambda(x-y))\big) - f(y) &= tdf_y(v) + df_y(x-y) + t\delta. & (18.2.4) \end{aligned}$$

Summing (18.2.2)+(18.2.3)-(18.2.4), we obtain

$$t(df_x(v) - df_y(v)) = t(\iota + \eta - \delta)$$

and $df_x(v) - df_y(v) \approx 0$ as required. □

Moreover Stroyan also showed in [15] that uniform differentiability is a very good generalization of G-differentiability in that, among other properties, it overcomes the impossibility of topologizing $L(E,F)$ in such a way that the evaluation map $(x,y) \mapsto l(x) : L(E,F) \times E \to F$ be continuous when E and F are not normable locally convex spaces[5]; we established the equivalence between uniform and C_{qb} differentiability of standard functions on complete locally convex spaces, and discuss relations with a weaker kind of differentiability when $\mathbf{fin}({}^*E) = \mathbf{ns}({}^*E)$, in [12]. The main results in this context read as follows.

Assume that U is a non-empty open subset of E, $f : U \to F$, $a \in {}^\sigma U + \mu({}^*E)$ and $k \in \mathbb{N}$.

Definition 18.2.1 (Stroyan [15]) *The function f is k-uniformly differentiable at a if there are finite maps $d^{(i)} f_{(\cdot)} : U \to L^{(i)}(E,F)$, $(1 \le i \le k)$ — the* ***derivatives*** *of f — such that, with $d := d^{(1)}$, for all $x \in \mathbf{fin}({}^*E)$ and all $t \in \mu({}^*\mathbb{R})$, there exist $\eta \in \mu({}^*F)$ and infinitesimal maps η_i $(2 \le i \le k)$, so that*

$$\begin{aligned} f(a+tx) &= f(a) + tdf_a(x) + t\eta & (18.2.5)\\ d^{(i)} f_{a+tx}(\cdot) &= d^{(i)} f_a(\cdot) + td^{(i+1)} f_a(x,\cdot) + t\eta_i(\cdot)\ (1 \le i < k). & (18.2.6) \end{aligned}$$

f is k-uniformly differentiable in U if it is k-uniformly differentiable at all $a \in {}^\sigma U + \mu({}^*E)$. When f is 1-uniformly differentiable we just say that it is ***uniformly*** *differentiable.*

The following may also be found in [15].

Theorem 18.2.3 *When f is k-uniformly differentiable, not only f itself is S-continuous but also the derivatives $d^{(i)} f : \mathbf{ns}({}^*E) \times \mathbf{fin}({}^*E) \to {}^*F$ are S-continuous:*

$$\forall x, y \in \mathbf{ns}({}^*E)\ \forall u, v \in \mathbf{fin}({}^*E)\quad [x \approx u\ \&\ y \approx v\ \Rightarrow\ d^{(i)} f_x(u) \approx d^{(i)} f_y(v)] \quad (1 \le i \le k).$$

[5] A quite simple and clear explanation of this can be found in [7, page 2].

Theorem 18.2.4 (Taylor's formula [15]) *f is k-uniformly differentiable in U if there are maps $df^{(i)} : U \to L^{(i)}_{(\cdot)}(E, F)$ $(1 \le i \le k)$, such that, whenever $a \in {}^{\sigma}U + \mu({}^*E)$, $x \in \mathbf{fin}({}^*E)$, $t \in \mu$, there exists $\eta \in \mu({}^*F)$ such that*

$$f(a + tx) = f(a) + \sum_{i=1}^{k} \frac{t^i}{i!} df_a^{(i)}(x, \cdots, x) + t^k \eta.$$

Relations with F-differentiability on Banach spaces are actually studied in [14, chap. 5]. On what regards standard definitions of smoothness we take from [6], whereto we refer the reader for details, namely on **convergence structures**.

Definition 18.2.2 *Let $\mathcal{N}$ denote the filter of neighborhoods of zero in $\mathbb{R}$, $\mathcal{B}$ be a filter on E and, for any filter $\mathcal{F}$ in a (real) vector space, $\mathcal{NF}$ denote the filter generated by $\{\{rb|\ r \in N \ \&\ b \in F\}|\ N \in \mathcal{N} \ \&\ F \in \mathcal{F}\}$; also, for $k \in \mathbb{N}$, let $\mathcal{B}^k$ be the filter in E^k generated by $\{B_1 \times \cdots \times B_k|\ B_1, \cdots B_k \in \mathcal{B}\}$ and, for any filter $\mathcal{F}$ in $L^{(k)}(E, F)$, $F(\mathcal{B}^k)$ be generated by $\{\bigcup_{l \in F} l(B)|\ F \in \mathcal{F} \ \&\ B \in \mathcal{B}\}$; finally, if ϕ is a function and $\mathcal{F}$ is a filter on the domain of ϕ, $\phi(\mathcal{F})$ is generated by $\{\phi(F)|\ F \in \mathcal{F}\}$.*

1. *$\mathcal{B}$ is **quasi-bounded** if $\mathcal{NB}$ converges to zero; Λ_{qb} denotes the convergence structure of quasi-bounded convergence; when a filter $\mathcal{F}$ converges to $x \in E$ with respect to Λ_{qb}, we write $\mathcal{F} \in \Lambda_{qb}(x)$.*

2. *A filter $\mathcal{F}$ in $L^{(k)}(E, F)$ qb-converges to zero if $\mathcal{F}(\mathcal{B}^k)$ converges to zero in F whenever $\mathcal{B}$ is quasi-bounded in E.*

3. *A function $l : U \to L^{(k)}(E, F)$ is **qb-continuous** if $l(\mathcal{F}) \in \Lambda_{qb}(l(a))$ whenever $a \in U$ and $\mathcal{F}$ is a filter in E converging to a.*

4. *Recall that U is an open subset of E and $f : U \to F$. We say that f is **of class C^k_{qb}** if*

   (a) f is qb-continuous.

   (b) There exist maps $d^i f_{(\cdot)} : U \to L^{(i)}(E, F)$ $\quad (1 \le i \le k)$ such that

   i. $d^i f_{(\cdot)}$ is qb-continuous, for all $i = 1, \cdots, k$

   ii. For all $a \in U$, all $x \in E$ and all $i = 0, \cdots, k-1$

$$\lim_{t \to 0} \frac{1}{t} \left(d^i f_{a+tx} - d^i f_a\right) = d^{i+1} f_a(x, \cdot)$$

   for the topology of pointwise convergence on $L^i(E, F)$.

Theorem 18.2.5 ([12]) *The function f is of class C^k_{qb}, with derivatives $d^i f_{(\cdot)}$, $(1 \leq i \leq k)$ iff it is k-uniformly differentiable with the same derivatives.*

The following is also a consequence (for example see [6]) either of this theorem (18.2.5) or of theorem 18.2.4.

Theorem 18.2.6 *The function $f : E \to F$, between Banach spaces E and F, is of Fréchet class C^k, with derivatives $d^i f_{(\cdot)}$, $(1 \leq i \leq k)$ iff it is k-uniformly differentiable with the same derivatives.*

18.3 Smoothness and finite points

Recall that E and F denote real locally convex spaces.

Lemma 18.3.1 *Suppose $A \in {}^*\mathcal{P}(E)$. The following conditions are equivalent for any internal map $f : A \to {}^*F$ which is GS-differentiable at all $x \in A \cap \mathbf{fin}({}^*E)$, with GS-derivative $df : (A \cap \mathbf{fin}({}^*E)) \times \mathbf{fin}({}^*E) \to {}^*F$.*

1. *f is uniformly differentiable at all $x \in A \cap \mathbf{fin}({}^*E)$*
2. *f is S-continuous on $A \cap \mathbf{fin}({}^*E)$*
3. *$df((A \cap \mathbf{fin}({}^*E)) \times \mathbf{fin}({}^*E)) \subseteq \mathbf{fin}({}^*F)$*

Proof. For the sake of simplicity, we assume $A = {}^*E$; the proof of the general case is an easy adaptation thereof.

Suppose that $df : \mathbf{fin}({}^*E) \times \mathbf{fin}({}^*E) \to {}^*F$ is the GS-derivative of the internal map $f : {}^*E \to {}^*F$.

$(1 \Rightarrow 2)$ This is 1 in theorem 18.2.2 above.

$(2 \Rightarrow 3)$ Assume that f is S-continuous on $\mathbf{fin}({}^*E)$, let $df_{(\cdot)} : \mathbf{fin}({}^*E) \to {}^*L(E, F)$ be the GS-derivative of f. Pick x and v in $\mathbf{fin}({}^*E)$; we must show that $df_x(v)$ is finite; suppose this is not the case, take a semi-norm $\gamma \in \Gamma_F$ such that $\gamma(df_x(v)) \notin \mathcal{O}$ and let $t := \frac{1}{\gamma(df_x(v))}$; t is infinitesimal and, by hypothesis, there exists $\iota \in \mu({}^*F)$ such that $f(x + tv) = f(x) + tdf_x(v) + t\iota$; but then

$$1 = \gamma(tdf_x(v)) = \gamma\left(f(x + tv) - f(x) - t\iota\right) \approx 0$$

which is impossible; it follows that $df_x(v)$ must be finite as required.

$(3 \Rightarrow 1)$ In the presence of GS-differentiability, 3 completes the definition of uniform differentiability. □

Say that a standard function $f : E \to F$ is GS, or uniformly, differentiable at $x \in {}^*E$ with derivative df_x if respectively ${}^*f$ is GS, or uniformly differentiable, at x with derivative ${}^*df$.

Theorem 18.3.1 *If a standard function $f : E \to F$ is GS-differentiable with standard derivative df_x at all $x \in \mathbf{fin}({}^*E)$, the following conditions are equivalent*

1. *f is uniformly differentiable on bounded sets with derivative df.*
2. *f is uniformly continuous on bounded sets.*
3. *For any bounded subset B of E^2, $df(B)$ is bounded in F.*
4. *df is uniformly continuous on bounded subsets of E^2.*

As we said before, the proof is easy. Start by keeping in mind that

Theorem 18.3.2 *A subset B of the locally convex space S is bounded iff ${}^*B \subseteq \mathbf{fin}({}^*S)$; in particular, if E and F are normed and $L(E;F)$ is endowed with the usual uniform norm, a subset B of $L(E,F)$ is bounded iff*

$$\forall \phi \in {}^*B \ \forall x \in \mathbf{fin}({}^*E) \quad \phi(x) \in \mathbf{fin}({}^*F). \tag{18.3.1}$$

Proof of thm. 18.3.1. Equivalences $(1 \Leftrightarrow 2 \Leftrightarrow 3)$ are immediate consequences of lemma 18.3.1 and theorem 18.3.2 (note that, by theorem 18.3.2 above, ${}^*B \cap \mathbf{fin}({}^*E) = {}^*B$ if and only if B is bounded).

$(3 \Rightarrow 4)$ Follows from theorems 18.3.2 and 18.2.2.

$(4 \Rightarrow 2)$ If we show that condition 4 implies that df_x is a finite map whenever B is a bounded subset of E and $x \in {}^*B$, we are done; but this follows from theorem 18.1.6. □

As an application to Banach spaces:

Corollary 18.3.1 *Let E and F be Banach spaces and $f : E \to F$ be a S-continuous and GS-differentiable function at all $x \in \mathbf{fin}({}^*E)$ with standard derivative df, then*

1. *df is actually the Fréchet derivative of f and f is of class C^1.*
2. ***(Mean Value)** Denoting $[x,y]$ the line segment from x to y,*

$$\forall x, y \in E \ \exists z \in [x,y] \quad \|f(x) - f(y)\| \leq \|df_z\| \cdot \|x - y\|; \tag{18.3.2}$$

*therefore f is **Lipschitzian** on bounded sets, i.e., for each bounded subset B of E, there exists $K \in \mathbb{R}$ such that*

$$\forall x, y \in B \quad \|f(x) - f(y)\| \leq K\|x - y\|. \tag{18.3.3}$$

3. *For all infinitely near finite vectors* $x, y, z \in {}^*E$, *there exists an infinitesimal* $\iota \in {}^*F$ *such that*

$$f(x) - f(y) \;=\; df_z(x-y) + \|x-y\|\iota \qquad (18.3.4)$$

Proof. 1. As f is S-continuous on $\mathbf{fin}({}^*E)$, it is uniformly continuous on bounded sets and hence, by theorem 18.3.1, of Fréchet class C^1.

2. Condition (18.3.2) is true in view of part 1 above; therefore condition (18.3.3) is true (by theorems 18.3.1 and 18.3.2) because $\mathbf{fin}({}^*E)$ is *convex.

3. We borrow from [14, page 97]: the Fréchet derivative $x \mapsto df_x$ is S-continuous at finite points by 2b in theorem 18.2.2 and we may apply Transfer of the Mean Value condition (18.3.2) to the function

$$g \;\equiv\; v \mapsto f(v) - df_z(v-z),$$

whose derivative $v \mapsto dg_v = df_v - df_z$ is infinitesimal whenever $v \approx z$, and therefore verifies, for some $c \in [x, y]$, hence $c \approx z$,

$$\frac{\|f(x) - f(y) - df_z(x-y)\|}{\|x-y\|} = \|g(x) - g(y)\| \le \left\| dg_c\left(\frac{x-y}{\|x-y\|}\right)\right\| \approx 0. \quad \square$$

18.4 Smoothness and the nonstandard hull

This section evolves along main ideas due to Manfred Wolff.

Recall that $\cdot \approx \cdot$ is an equivalence relation on ${}^*S$, with equivalence classes $\hat{x} := x + \mu({}^*S)$ and let $\hat{S}$ be the **nonstandard hull** of ${}^*S$ with semi-norms $\hat{\gamma}$ or

$$\hat{S} := \mathbf{fin}({}^*S)_{/\approx} \quad \& \quad \hat{\gamma}(\hat{x}) := \mathbf{st}(\gamma(x)) \quad (x \in \mathbf{fin}({}^*S)).$$

Nonstandard hulls are complete spaces, but may vary with the chosen model of analysis $\mathcal{A}$; actually Banach spaces with invariant nonstandard hulls are finite dimensional, but this is not the case with locally convex non-normable ones; spaces with invariant non-standard hulls were characterized in [5] and were named **HM-spaces** by Keith Stroyan (relevant data is summarized in detail in [14, chap. 10]; also see [10, sec. 3.9] for nonstandard hulls of metric spaces).

Any internal function between locally convex spaces, $f : A \subseteq {}^*E \to {}^*F$ that is S-continuous on $\mathbf{fin}({}^*E) \cap A$ and such that $f(\mathbf{fin}({}^*E) \cap A) \subseteq \mathbf{fin}({}^*F)$ has a natural **non-standard hull** $\hat{f} : \hat{A} \subseteq \hat{E} \to \hat{F}$ too defined by

$$\begin{aligned} \hat{A} &:= \{\hat{x} |\ x \in \mathbf{fin}({}^*E) \cap A\} \\ \hat{f}(\hat{x}) &:= \widehat{f(x)} \quad (x \in A \cap \mathbf{fin}({}^*E)); \end{aligned}$$

note that $\hat{f}$ *is not* a standard function in the sense we have been considering, for it certainly is not *a priori* an element of the model of analysis $\mathcal{A}$, nevertheless an application of Nelson's Algorithm ([8], [11]) will provide us with definitions of differentiability for functions without reference to nonstandard extensions.

From now on $f : {}^*E \to {}^*F$ denotes an internal GS-differentiable function with derivative df such that

$$f(\mathbf{fin}({}^*E)) \subseteq \mathbf{fin}({}^*F) \tag{18.4.1}$$

and

$$d\hat{f}(\hat{x}, \hat{v}) := \widehat{df(x, v)} \tag{18.4.2}$$

whenever this is a good definition, namely when df is S-continuous on $\mathbf{fin}({}^*E)$ and $x, v \in \mathbf{fin}({}^*E)$. For any GS-differentiable function $h : E \to F$

$$\Delta(h, x, v, t) := \left(\frac{1}{t}\Big(h(x + tv) - h(x) - dh(x, v)\Big)\right) \quad (x, v \in {}^*E;\ t \in {}^*\mathbb{R} \setminus \{0\}).$$

18.4.1 Strong uniform differentiability

Definition 18.4.1 *An* ***internal*** *function* $f : {}^*E \to {}^*F$ *is* ***Strongly Uniformly Differentiable*** *(**SUD** for short) if it is uniformly differentiable on* $\mathbf{fin}({}^*E)$ *and*

$$f(\mathbf{fin}({}^*E)) \subseteq \mathbf{fin}({}^*F).$$

Let $\mathcal{P}$ and $\mathcal{Q}$ denote respectively gauges of semi-norms for E and F. In these terms, f is SUD when the two following conditions are simultaneously verified

$$\forall (x, v) \in {}^*E^2 \quad \big[(x, v) \in \mathbf{fin}({}^*E)^2 \Rightarrow df(x, v) \in \mathbf{fin}({}^*F)\big]$$
$$\forall (x, v) \in {}^*E^2\ \forall t \in {}^*\mathbb{R} \quad \big[(x, v) \in \mathbf{fin}({}^*E)^2 \wedge 0 \neq t \in \mu \Rightarrow \Delta(f, x, v, t) \in \mu({}^*F)\big];$$

these expand respectively to

$$\begin{aligned}\forall (x, v) \in {}^*E^2\ \big[\forall^{st} p \in {}^*P\, \exists^{st} m \in {}^*\mathbb{N}\ p(x) + p(v) \leq m \\ \Rightarrow \forall^{st} q \in {}^*Q \exists^{st} n \in {}^*\mathbb{N}\ q\big(df(x, v)\big) \leq n\big]\end{aligned}$$

$$\begin{aligned}\forall (x, v) \in {}^*E^2\ \forall t \in {}^*\mathbb{R} \Big[\forall^{st} p \in {}^*P\, \exists^{st} m \in {}^*\mathbb{N}\ p(x) + p(v) \leq m \\ \wedge \forall^{st} n \in {}^*\mathbb{N}\ 0 < |t| \leq \frac{1}{n} \Rightarrow \forall^{st} q \in {}^*Q\ q(\Delta(f, x, v, t)) \leq 1\Big];\end{aligned}$$

which reduce respectively to the following, where we leave the domains implicit,

$$\forall (x, v)\ \forall^{st} q\ \exists^{st} (p, n)\ \forall^{st} m\ \big[p(x) + p(v) \leq m \Rightarrow\ q\big(df(x, v)\big) \leq n\big]$$

$$\begin{aligned}\forall (x, v, t)\, \forall^{st} q\, \exists^{st} (p, n)\, \forall^{st} m \Big[p(x) + p(v) \leq m \\ \wedge\, 0 < |t| \leq \frac{1}{n} \Rightarrow q(\Delta(f, x, v, t)) \leq 1\Big];\end{aligned}$$

Nelson's algorithm then gives

$$\forall^{st} q\ \forall^{st} \widetilde{m}\ \exists^{st\ fin} P \times N\ \forall (x,v)\ \exists (p,n) \in P \times N \\ \big[p(x) + p(v) \leq \widetilde{m}(p,n)\ \Rightarrow\ q\big(df(x,v)\big) \leq n\big] \tag{18.4.3}$$

$$\forall^{st} q\ \forall^{st} \widetilde{m}\ \exists^{st\ fin} P \times N\ \forall (x,v,t)\ \exists (p,n) \in P \times N \\ \Big[p(x) + p(v) \leq \widetilde{m}(p,n)\ \wedge\ 0 < |t| \leq \frac{1}{n} \Rightarrow q(\Delta(f,x,v,t)) \leq 1\Big]. \tag{18.4.4}$$

Formulas (18.4.3) and (18.4.4) together define strong uniform differentiability of an internal function and we actually proved the following

Theorem 18.4.1 *An internal GS-differentiable function* $f : {}^*E \to {}^*F$ *is* ***SUD*** *iff*

$$\forall^{st} q\ \forall^{st} \widetilde{m}\ \exists^{st\ fin} P \times N\ \forall (x,v,t)\ \exists (p,n) \in P \times N \\ \Big[\Big[p(x) + p(v) \leq \widetilde{m}(p,n)\ \wedge\ 0 < |t| \leq \frac{1}{n}\Big] \Rightarrow \\ \big[q\big(df(x,v)\big) \leq n\ \wedge\ q(\Delta(f,x,v,t)) \leq 1\big]\Big]; \tag{18.4.5}$$

When f is standard, transfer of (18.4.5) provides the definition, 18.4.2 below, of SU differentiability for functions between locally convex spaces in *any* "universe" $\mathcal{A}$:

Definition 18.4.2 *A* ***standard*** *function* $h : S \to T$ *between locally convex spaces* S *and* T*, with unbounded gauges of semi-norms respectively* Σ *and* Θ*, is* ***Strongly Uniformly Differentiable****, with derivative* $dh_{(\cdot)} : S \to L(S,T)$*, (****SUD*** *for short) when, leaving implicit that* $\sigma \in \Sigma$*,* $\tau \in \Theta$*,* $\widetilde{m} : \Sigma \times \mathbb{N} \to \mathbb{N}$*,* $x \in S$*,* $v \in S$ *and* $t \in \mathbb{R} \setminus \{0\}$*,*

$$\forall \tau\ \forall \widetilde{m}\ \exists^{fin} C \times N \subseteq \Sigma \times \Theta\ \forall (x,v,t)\ \exists (p,n) \in C \times N \\ \Big[\Big[\sigma(x) + \sigma(v) \leq \widetilde{m}(\sigma,n)\ \wedge\ 0 < |t| \leq \frac{1}{n}\Big] \Rightarrow \\ \big[\tau\big(df(x,v)\big) \leq n\ \wedge\ \tau(\Delta(h,x,v,t)) \leq 1\big]\Big]. \tag{18.4.6}$$

NB: *variants of this formula where any inequality* $\leq$ *is taken to be strict, i.e., to be* $<$*, are equivalent.*

18.4.2 The non-standard hull

Theorem 18.4.2 *Let E and F be locally convex spaces and $f : {}^*E \to {}^*F$ be an internal GS-differentiable function with derivative df such that*

$$f(\mathbf{fin}({}^*E)) \subseteq \mathbf{fin}({}^*F). \tag{18.4.7}$$

f is SUD iff $\hat{f} : \hat{E} \to \hat{F}$ is uniformly differentiable with derivative $d\hat{f} : \hat{E}^2 \to \hat{F}$ given by

$$d\hat{f}(\hat{x}, \hat{v}) := \widehat{df(x, v)}.$$

Proof. Let us first assume that the internal function $f : {}^*E \to {}^*F$ is SUD with derivative df, and takes finite vectors of ${}^*E$ into finite vectors of ${}^*F$.

Define

$$\widehat{\mathcal{P}} := \{\hat{p}|\ p \in \mathcal{P}\} \quad , \quad \widehat{\mathcal{Q}} := \{\hat{q}|\ q \in \mathcal{Q}\};$$

although there might exist continuous semi-norms not of the type $\hat{\gamma}$, $\hat{\mathcal{P}}$ and $\hat{\mathcal{Q}}$ are unbounded directed families of semi-norms, so that condition (18.1.4) still holds with obvious adaptations. Also recall that

$$\hat{f}(\hat{x}) = \widehat{f(x)}$$

and observe that equation

$$d\hat{f}(\hat{x}, \hat{v}) := \widehat{df(x, v)}$$

defines $d\hat{f}$ well, in view of theorems 18.1.5 and 18.1.6 and lemma 18.3.1.

Take $\hat{q} \in \hat{\mathcal{Q}}$ and $\widetilde{m} : \hat{\mathcal{Q}} \times \mathbb{N} \to \mathbb{N}$. Define

$$\widetilde{M}(p, n) := \widetilde{m}(\hat{p}, n) \quad (p \in \mathcal{P},\ n \in \mathbb{N});$$

$\widetilde{M}$ is a standard function, therefore we may deduce from (18.4.5) that there exists a standard finite set $P \times N \subseteq {}^*\mathcal{P} \times {}^*\mathbb{N}$ such that

$$\begin{aligned}&\forall (x, v, t) \in {}^*E^2 \times {}^*\mathbb{R}\ \exists (p, n) \in P \times N \\ &\Big[p(x) + p(v) < \widetilde{M}(p, n)\ \wedge\ 0 < |t| \leq \frac{1}{n}\ \Rightarrow \\ &\qquad\qquad q(df(x, v)) \leq n\ \wedge\ q(\Delta(f, x, v, t)) < 1\Big].\end{aligned}$$

P and N being standard and finite, define

$$\hat{N} := N \quad , \quad \hat{P} := \{\hat{p}|\ p \in P\},$$

and take $(\hat{x}, \hat{v}, t)$; whatever the representatives x and v, there is a corresponding pair $(p, n) \in P \times N$, which may be vary with x and v, but always verifies (18.4.3); suppose that

$$\hat{p}(\hat{x}) + \hat{p}(\hat{v}) < \widetilde{M}(\hat{p}, n) \ \wedge \ 0 < |t| \leq \frac{1}{n}; \tag{18.4.8}$$

all the numbers involved here are standard, $\hat{p}(\hat{x}) \approx p(x)$ & $\hat{p}(\hat{v}) \approx p(v)$ as well as $\hat{q}(\widehat{df(x,v)}) \approx q(df(x,v))$; in particular, it follows

$$\begin{array}{ccccc} p(x) + p(v) & < & \widetilde{m}(\hat{p}, n) & = & \widetilde{M}(p, n) \\ & \text{so that} & & & \\ q(df(x,v)) & < & n & & \\ & \text{an thus} & & & \\ \hat{q}(\widehat{df(x,v)}) & \leq & n & & \\ & \text{that is} & & & \\ \hat{q}(d\hat{f}(\hat{x}, \hat{v})) & \leq & n & & \end{array}$$

and the first factor in the consequent of (18.4.5) follows for $\hat{f}$ as required; for the remaining factor of the consequent, observe that, actually

$$p(x) + p(v) < \widetilde{M}(p, n) \ \wedge \ 0 < |t| \leq \frac{1}{n};$$

implies

$$q\left(\frac{1}{t}\Big(f(x+tv) - f(x) - df(x,v)\Big)\right) \ = \ q(\Delta(f, x, v, t)) \ < \ 1$$

too; t and n are standard, $0 \neq t$ and f as well as df are S-continuous by lemma 18.3.1, thus

$$\frac{1}{t}\Big(f(x+tv) - f(x) - df(x,v)\Big) \ \approx \ \frac{1}{t}\Big(\hat{f}(\hat{x}+t\hat{v}) - \hat{f}(\hat{x}) - \widehat{df(x,v)}\Big);$$

finally

$$\hat{q}(\Delta(\hat{f},, x, \hat{v}, t)) \leq 1.$$

The proof that $\hat{f}$ is SUD with derivative $d\hat{f}$ is finished.

Now suppose that $\hat{f}$ is itself uniformly differentiable with derivative $d\hat{f}$: $\hat{E}^2 \to \hat{F}$. Take $q \in {}^{\sigma}\mathcal{Q}$ and a standard function $\widetilde{m} : {}^*(\mathcal{P} \times \mathbb{N}) \to {}^*\mathbb{N}$; since $\widehat{(\cdot)}$ is 1-1

$$\widetilde{M}(\hat{p}, n) := \widetilde{m}(p, n)$$

defines a function from $\widehat{\mathcal{P}} \times \mathbb{N}$ into $\mathbb{N}$ and we may apply (18.4.6) appropriately written in terms of $\hat{f}$, and obtain the corresponding finite set $\hat{P} \times N \subseteq \widehat{\mathcal{P}} \times \mathbb{N}$; pick $x, v \in E,\ t \in \mathbb{R} \setminus \{0\}$ and use (18.4.6), with $<$ for $\leq$, in order to obtain $(\hat{p}, n) \in \hat{P} \times N$ so that

$$\Big[\hat{p}(\hat{x}) + \hat{p}(\hat{v}) \leq \widetilde{M}(\hat{p}, n)\ \wedge\ \ 0 < |t| \leq \frac{1}{n}\Big] \Rightarrow \\ \big[\hat{q}\big(d\hat{f}(\hat{x}, \hat{v})\big) < n\ \wedge\ \hat{q}(\Delta(\hat{f}, \hat{x}, \hat{v}, t)) < 1\big]; \tag{18.4.9}$$

therefore, if

$$p(x) + p(v) < \widetilde{m}(p, n)\ \wedge\ \ 0 < |t| \leq \frac{1}{n} \tag{18.4.10}$$

then

$$\hat{p}(\hat{x}) + \hat{p}(\hat{v}) \leq \widetilde{M}(\hat{p}, n)\ \wedge\ \ 0 < |t| \leq \frac{1}{n},$$

hence

$$\hat{q}\big(d\hat{f}(\hat{x}, \hat{v})\big) < n\ \wedge\ \hat{q}(\Delta(\hat{f}, \hat{x}, \hat{v}, t)) < 1$$

or

$$\hat{q}\big(\widehat{df(x, v)}\big) < n\ \wedge\ \hat{q}(\Delta(\hat{f}, \hat{x}, \hat{v}, t)) < 1.$$

It immediately follows that

$$q\big(df(x, v)\big)\ \leq\ n,$$

because both $\hat{q}\big(\widehat{df(x, v)}\big)$ and the n above are standard; moreover $\widehat{(\cdot)}$ is linear and when $t \not\approx 0$

$$q(\Delta(f, x, v, t))\ \leq\ 1; \tag{18.4.11}$$

it so happens that $t \overset{\mathcal{I}}{\mapsto} q(\Delta(f, x, v, t))$ is always an internal function and we actually have shown that

$$\forall q \in {}^{\sigma}\mathcal{Q}\ \exists n \in {}^{\sigma}\mathbb{N}\ \left[0 < |t| < \frac{1}{n} \Rightarrow q(\Delta(f, x, v, t))\ \leq\ 1\right]$$

which is the same as

$$\forall q \in {}^{\sigma}\mathcal{Q}\ \ \big[0 \neq t \approx 0\ \Rightarrow\ q(\Delta(f, x, v, t)) \approx 0\big]$$

and thus (18.4.11) also holds when $t \approx 0$. Again because $\widehat{(\cdot)}$ is 1-1, we may define

$$P\ :=\ \{p |\ \hat{p} \in \hat{P}\}$$

and we have shown that (18.4.5) holds. □

References

[1] S. Albeverio, J-E. Fenstad, R. Høegh-Krohn and T. Lindstrøm (eds.), *Nonstandard Methods in Stochastic Analysis and Mathematical Physics*, Academic Press, 1986.

[2] Leif O. Arkeryd, Nigel J. Cutland and C. Ward Henson (eds.), *Nonstandard Analysis, Theory and Applications*, Kluwer, 1997.

[3] C. Ward Henson, A Gentle Introduction to Nonstandard Extensions, in *Nonstandard Analysis, Theory and Applications*, Arkeryd, Cutland & Henson (eds.), Kluwer, 1997.

[4] C. Ward Henson and L.C. Moore, "The Nonstandard theory of topological vector spaces", Trans. of the AMS, **172** (1972) 405–435.

[5] C. Ward Henson and L.C. Moore, "Invariance of The Nonstandard Hulls of Locally Convex Spaces", Duke Math. J., **40** 93–205.

[6] H.H. Keller, *Differential Calculus in Locally Convex Spaces*, Lec. Notes in Math 417, Springer-Verlag, 1974.

[7] Andreas Kriegl and Peter W. Michor, The Convenient setting of Global Analysis, Math. Surveys and Monographs 53, AMS (1997) 405–435.

[8] Francine Diener and Keith Stroyan, Syntactical Methods in Infinitesimal Analysis, in Cutland (ed.), *Nonstandard Analysis and its Applications*, CUP, 1988.

[9] Tom Lindstrøm, An invitation to Nonstandard Analysis, in Cutland (ed.), *Nonstandard Analysis and its Applications*, CUP, 1988.

[10] Peter Loeb and Manfred Wolff (eds.), *Nonstandard Analysis for the Working Mathematician*, Kluwer, 2000.

[11] E. Nelson, "Internal Set Theory: a new approach to Nonstandard Analysis", Bull. AMS, **83** (1977).

[12] Vítor Neves, "Infinitesimal Calculus in HM spaces", Bull. Soc. Math. Belgique, **40** (1988) 177–198.

[13] Jacob T. Schwartz, *Nonlinear Functional Analysis*, Gordon & Breach, 1969.

[14] Keith D. Stroyan and W.A.J. Luxemburg: *Introduction to the Theory of Infinitesimals*, Academic Press, 1976.

[15] Keith D. Stroyan, "Infinitesimal Calculus in Locally convex Spaces: 1. Fundamentals", Trans. Amer. Math. Soc., **240** (1978), 363–383.

19

Nonstandard Palais-Smale conditions

Natália Martins* and Vítor Neves**

Abstract

We present nonstandard versions of the Palais-Smale condition (**PS**) below, some of them generalizations, but still sufficient to prove Mountain Pass Theorems, which are quite important in Critical Point Theory.

19.1 Preliminaries

For notation, other preliminaries and basic references on Nonstandard Analysis we suggest section 1 in article [13] in this volume, namely we assume that we have two set-theoretical structures $\mathcal{A}$ and $\mathcal{B}$ — the former a model of "the relevant" Analysis — and a 1-1 function $^*(\cdot) : \mathcal{A} \to \mathcal{B}$ which satisfies the Transfer and Polysaturation Principles. Specific notation and results follow. Recall that $^*\mathbb{R}$ denotes the set of ***hyperreal numbers*** and $\mathcal{O}$ the set of finite hyperreal numbers; if $x \in {}^*\mathbb{R}$ is infinitesimal we write $x \approx 0$. If $x, y \in {}^*\mathbb{R}$ are such that $x - y \approx 0$, we say that x is ***infinitely close*** to y and write $x \approx y$. Also remember that finite hyperreal numbers have a standard part:

Theorem 19.1.1 (Standard Part Theorem) *If $x \in \mathcal{O}$ there exists a unique $r \in \mathbb{R}$ such that $x \approx r$; r is called the **standard part** of x and is denoted by $\mathbf{st}(x)$ or ${}^\circ x$. Moreover, for all $x, y \in \mathcal{O}$, $st(x+y) = st(x) + st(y)$, $st(xy) = st(x)st(y)$ and if $x \leq y$ then $st(x) \leq st(y)$.*

---

*Departamento de Matemática, Universidade de Aveiro.
nataliam@mat.ua.pt

Work for this article was partially supported by the R&D unit Center for Research in Optimization and Control (CEOC) of the University of Aveiro and grant POCTI/MAT/41683/2001 of the Portuguese Foundation for Science and Technology (FCT) via FEDER.

**vneves@mat.ua.pt

Next we restrict some definitions to normed spaces; just recall that a normed space is a locally convex space whose topology is defined by a single norm and also recall that, for any set $A \in \mathcal{A}$,

$$^{\sigma}A := \{^{*}a : a \in A\},$$

and often A and $^{\sigma}A$ are identified for simplicity.

Definition 19.1.1 *Let $(E, \|\cdot\|)$ be a real normed space and $x \in {}^{*}E$.*

1. $x \in \mathbf{fin}({}^{*}E)$, *i.e., x is* ***finite*** *if $\|x\| \in \mathcal{O}$;*

2. $x \in \mu({}^{*}E)$, *i.e., x is* ***infinitesimal*** *if $\|x\| \approx 0$ in ${}^{*}\mathbb{R}$; $x \approx 0$ means $x \in \mu({}^{*}E)$;*

3. $x \in \mathbf{ns}({}^{*}E)$, *i.e., x is* ***near-standard*** *if there exists $a \in {}^{\sigma}E$ such that $x \approx a$, which means $x - a \approx 0$;*

4. $x \in \mathbf{pns}({}^{*}E)$, *i.e., x is* ***pre-near-standard*** *if for all positive $\varepsilon \in \mathbb{R}$ there exists $a \in {}^{\sigma}E$ such that $\|x - a\| < \varepsilon$.*

It follows from the above definitions that $\mathbf{ns}({}^{*}E) \subseteq \mathbf{pns}({}^{*}E)$ and $\mathbf{ns}({}^{*}E) \subseteq \mathbf{fin}({}^{*}E)$. In general, $\mathbf{ns}({}^{*}E) \neq \mathbf{pns}({}^{*}E)$ and $\mathbf{ns}({}^{*}E) \neq \mathbf{fin}({}^{*}E)$.

Theorem 19.1.2 *Let $(E, \|\cdot\|)$ be a real normed space.*

1. *E is complete if and only if $\mathbf{pns}({}^{*}E) = \mathbf{ns}({}^{*}E)$;*

2. *E is finite dimensional if and only if $\mathbf{fin}({}^{*}E) = \mathbf{ns}({}^{*}E)$;*

3. *$A \subseteq E$ is compact if and only if ${}^{*}A \subseteq \mathbf{ns}({}^{*}E)$ and $\mathbf{st}({}^{*}A) = A$.*

The nonstandard extension of $\mathbb{N}$, ${}^{*}\mathbb{N}$, is the set of ***hypernatural numbers*** and ${}^{*}\mathbb{N}_{\infty}$ denotes the set of infinite hypernatural numbers. It is also useful to keep in mind the following properties of sequences and functions in a real normed space $(E, \|\cdot\|)$.

Proposition 19.1.1 *Suppose $(x_n)_{n \in \mathbb{N}}$ is a sequence in E. Then*

1. *$(x_n)_{n \in \mathbb{N}}$ is bounded if and only if $\forall n \in {}^{*}\mathbb{N}_{\infty}\ x_n \in \mathbf{fin}({}^{*}E)$;*

2. *$(x_n)_{n \in \mathbb{N}}$ converges to $a \in E$ if and only if $\forall n \in {}^{*}\mathbb{N}_{\infty}\ x_n \approx a$;*

3. *$(x_n)_{n \in \mathbb{N}}$ has a convergent subsequence if and only if $\exists m \in {}^{*}\mathbb{N}_{\infty}$, $x_m \in \mathbf{ns}({}^{*}E)$.*

Theorem 19.1.3 *Let E and F be two real normed spaces and $f : E \to F$. The function f is continuous on $a \in E$ if and only if*

$$\forall x \in {}^{*}E\ [x \approx a \Rightarrow f(x) \approx f(a)].$$

19.2 The Palais-Smale condition

Many results in Critical Point Theory involve the following condition, originally introduced in 1964 [10] by Palais and Smale:

Definition 19.2.1 *If E is a real Banach space, the C^1 functional $f : E \to \mathbb{R}$ satisfies the **Palais-Smale condition** if for all $(x_n)_{n\in\mathbb{N}} \in E^{\mathbb{N}}$,*

$$\textbf{(PS)} \qquad (f(x_n))_{n\in\mathbb{N}} \text{ is bounded and } \lim_{n\to\infty} f'(x_n) = 0 \Rightarrow$$
$$(x_n)_{n\in\mathbb{N}} \text{ has a convergent subsequence.}$$

Suppose $(E, \|\cdot\|)$ is a real Banach space with continuous dual E' and duality pairing $\langle\cdot,\cdot\rangle$. Let $C^1(E,\mathbb{R})$ denote the set of continuously Fréchet differentiable functionals defined on E; a is ***critical point*** (resp. ***almost critical point***) of f if $f'(a) = 0$ (resp. $f'(a) \approx 0$), i.e., if $\langle f'(a), v\rangle = 0$ (resp. $\langle f'(a), v\rangle \approx 0$) for all $v \in E$ (resp. $v \in \mathbf{fin}({}^*E)$); $f(a)$ is a ***critical value*** of f if $f^{-1}(f(a))$ contains critical points.

If $f : E \to \mathbb{R}$ is Fréchet differentiable we will denote by K the set of all critical points of f, that is,

$$K = \{x \in E : \ f'(x) = 0\}$$

and, for each $c \in \mathbb{R}$, K_c will denote the set of critical points with value c, that is,

$$K_c = \{x \in E : \ f'(x) = 0 \ \wedge \ f(x) = c\} = K \cap f^{-1}(c).$$

The following is easy to prove.

Proposition 19.2.1 *Suppose that $f \in C^1(E,\mathbb{R})$ satisfies* **(PS)**. *Then*

1. *If f is bounded, K is a compact set.*

2. *For each $a, b \in \mathbb{R}$ such that $a \leq b$,*

$$\{u \in E : \ a \leq f(u) \leq b \ \wedge \ f'(u) = 0\} = f^{-1}([a,b]) \cap K$$

   is a compact set.

Note that

$$f \text{ satisfies } \textbf{(PS)} \text{ and } K \text{ is compact } \not\Rightarrow \ f \text{ is bounded}$$

and

$$K \text{ is compact and } f \text{ is bounded } \not\Rightarrow \ f \text{ satisfies } \textbf{(PS)},$$

as can be seen with the following two examples.

Example 19.2.1 The real function $f(x) = x^3$ ($x \in \mathbb{R}$) satisfies (**PS**), $K = \{0\}$ is compact and f is not bounded.

Example 19.2.2 Let

$$f(x) = \begin{cases} 2 - x^2 & \text{if } x \in [-1, 1] \\ \frac{1}{x^2} & \text{if } x \notin [-1, 1] \end{cases}$$

f is bounded, $K = \{0\}$ but f does not satisfies (**PS**).

Example 19.2.3 The real functions $\exp(x)$, $\cos(x)$, $\sin(x)$ and all the constant functions defined on $\mathbb{R}$ do not satisfy (**PS**).

Now we will present an important class of functionals that satisfies (**PS**).

Proposition 19.2.2 *If E is a finite dimensional real Banach space and $f \in C^1(E, \mathbb{R})$ is coercive (i.e. $f(x) \to +\infty$ whenever $\|x\| \to +\infty$), then f satisfies* (**PS**).

Proof. Let $(x_n)_{n\in\mathbb{N}} \subseteq E$ be such that $(f(x_n))_{n\in\mathbb{N}}$ is bounded and $\lim_{n\to\infty} f'(x_n) = 0$. Then, for all $n \in {}^*\mathbb{N}_\infty$, $f(x_n) \in \mathcal{O}$ (Proposition 19.1.1). Since f is coercive, for all $n \in {}^*\mathbb{N}_\infty$, $x_n \in \mathbf{fin}({}^*E)$. E is finite dimensional then, by Theorem 19.1.2, $\mathbf{fin}({}^*E) = \mathbf{ns}({}^*E)$ and therefore, for all $n \in {}^*\mathbb{N}_\infty$, $x_n \in \mathbf{ns}({}^*E)$. Finally, Proposition 19.1.1 implies that $(x_n)_{n\in\mathbb{N}}$ has a convergent subsequence. □

19.3 Nonstandard Palais-Smale conditions

The following definition often shortens statements

Definition 19.3.1 *Suppose $f \in C^1(E, \mathbb{R})$. We say that a sequence $(u_n)_{n\in\mathbb{N}}$ is a* ***Palais-Smale sequence for*** f *if*

$$(f(u_n))_{n\in\mathbb{N}} \text{ is bounded and } \lim_{n\to\infty} f'(u_n) = 0.$$

Therefore,

f satisfies the ***Palais-Smale condition*** (**PS**) *if every Palais-Smale sequence for f has a convergent subsequence.*

Suppose E is a real Banach space and $f \in C^1(E,\mathbb{R})$. Next we present some nonstandard variations of (**PS**):

$$(PS0)\qquad \begin{array}{c} (u_n)_{n\in\mathbb{N}} \text{ is a Palais-Smale sequence} \\ \Downarrow \\ \exists m \in {}^*\mathbb{N}_\infty \; u_m \in \mathbf{ns}({}^*E) \end{array}$$

$$(PS1)\qquad f(u) \in \mathcal{O} \;\wedge\; f'(u) \approx 0 \Rightarrow u \in \mathbf{ns}({}^*E)$$

$$(PS2)\qquad \begin{array}{c} f(u) \in \mathcal{O} \;\wedge\; f'(u) \approx 0 \\ \Downarrow \\ u \in \mathbf{fin}({}^*E) \;\wedge\; \mathbf{st}(f(u)) \text{ is a critical value of } f \end{array}$$

$$(PS3)\qquad \begin{array}{c} (u_n)_{n\in\mathbb{N}} \text{ is a Palais-Smale sequence} \\ \Downarrow \\ (u_n)_{n\in\mathbb{N}} \text{ is bounded} \;\wedge\; \forall n \in {}^*\mathbb{N}_\infty \; \mathbf{st}(f(u_n)) \text{ is a critical value of } f \end{array}$$

$$(PS4)\qquad \begin{array}{c} (u_n)_{n\in\mathbb{N}} \text{ is a Palais-Smale sequence} \\ \Downarrow \\ (u_n)_{n\in\mathbb{N}} \text{ is bounded} \;\wedge\; \exists n \in {}^*\mathbb{N}_\infty \; \mathbf{st}(f(u_n)) \text{ is a critical value of } f \end{array}$$

Proposition 19.3.1 *If* $f \in C^1(E,\mathbb{R})$ *then*

$$(PS1) \Leftrightarrow \Big[f(u) \in \mathcal{O} \wedge f'(u) \approx 0 \Rightarrow \\ u \in \mathbf{ns}({}^*E) \;\wedge\; \mathbf{st}(f(u)) \textit{ is a critical value of } f \Big].$$

Proof. The implication $\Leftarrow$ is trivial. For the proof of the other implication, suppose that $u \in {}^*E$ is such that $f(u) \in \mathcal{O}$ and $f'(u) \approx 0$. By $(PS1)$, there exists $a \in {}^\sigma E$ such that $u \approx a$ and, therefore, from the continuity of f and f' it follows that (Theorem 19.1.3)

$$f(a) \approx f(u) \quad \text{and} \quad f'(a) \approx f'(u) \approx 0$$

too, so that $f(a) = \mathbf{st}(f(u))$ and $f(a)$ is a critical value of f. □

Nonstandard versions of (**PS**) and (**PS**) itself are related as follows.

Theorem 19.3.1 *For any real Banach space E we have*

$$(PS1) \Rightarrow (\mathbf{PS}) \Leftrightarrow (PS0) \Rightarrow (PS2) \Leftrightarrow (PS3) \Rightarrow (PS4).$$

Proof. It is clear that $(\mathbf{PS}) \Leftrightarrow (PS0)$, $(PS1) \Rightarrow (PS0)$ and $(PS3) \Rightarrow (PS4)$, therefore

$$(PS1) \Rightarrow (\mathbf{PS}) \Leftrightarrow (PS0) \quad \text{and} \quad (PS3) \Rightarrow (PS4)$$

are true.

First we prove that $(PS0) \Rightarrow (PS2)$. Let $u \in {}^*E$ such that $f(u) \in \mathcal{O}$ and $f'(u) \approx 0$. Fix $M \in \mathbb{R}^+$ such that $|f(u)| < M$. Suppose $u \notin \mathbf{fin}({}^*E)$. For each $n \in \mathbb{N}$, define the standard set

$$H_n := \Big\{x \in E : \ |f(x)| < M \ \wedge \ \|f'(x)\| < \frac{1}{n} \ \wedge \ \|x\| > n\Big\}.$$

Since, for each $n \in \mathbb{N}$,

$$u \in {}^*H_n = \Big\{x \in {}^*E : \ |f(x)| < M \ \wedge \ \|f'(x)\| < \frac{1}{n} \ \wedge \ \|x\| > n\Big\},$$

we conclude that ${}^*H_n \neq \emptyset$ and the Transfer Principle says that $H_n \neq \emptyset$.

For each $n \in \mathbb{N}$, take $x_n \in H_n$. Then $(x_n)_{n\in\mathbb{N}}$ is a Palais-Smale sequence but, for all $n \in {}^*\mathbb{N}_\infty$, $x_n \notin \mathbf{fin}({}^*E)$, and hence, for all $n \in {}^*\mathbb{N}_\infty$, $x_n \notin \mathbf{ns}({}^*E)$. This is a contradiction with $(PS0)$ and therefore $u \in \mathbf{fin}({}^*E)$.

Now we will prove that if $f(u) \in \mathcal{O}$ and $f'(u) \approx 0$, then $\mathbf{st}(f(u))$ is a critical value of f.

Let $\alpha = \mathbf{st}(f(u))$ and, for each $n \in \mathbb{N}$, define

$$F_n := \Big\{x \in E : \ |f(x)| < M \ \wedge \ \|f'(x)\| < \frac{1}{n} \ \wedge \ |\alpha - f(x)| < \frac{1}{n}\Big\}.$$

Since, for each $n \in \mathbb{N}$,

$$u \in {}^*F_n = \Big\{x \in {}^*E \ : \ |f(x)| < M \ \wedge \ \|f'(x)\| < \frac{1}{n} \ \wedge \ |\alpha - f(x)| < \frac{1}{n}\Big\}$$

then, ${}^*F_n \neq \emptyset$ and therefore $F_n \neq \emptyset$.

For each $n \in \mathbb{N}$ take $x_n \in F_n$. Then $(x_n)_{n\in\mathbb{N}}$ is a Palais-Smale sequence and from $(PS0)$ we conclude

$$\exists a \in {}^\sigma E \ \exists m \in {}^*\mathbb{N}_\infty \ x_m \approx a.$$

Since $f \in C^1(E, \mathbb{R})$,

$$\alpha \approx f(x_m) \approx f(a) \ \text{ and } \ 0 \approx f'(x_m) \approx f'(a).$$

Therefore $\alpha = f(a)$, $f'(a) = 0$, α is a critical value of f and $(PS0) \Rightarrow (PS2)$ is proven.

Lets prove that $(PS2) \Rightarrow (PS3)$. Let $(u_n)_{n\in\mathbb{N}}$ be a Palais-Smale sequence. Then,

$$\forall n \in {}^*\mathbb{N}_\infty \; [\, f(u_n) \in \mathcal{O} \;\wedge\; f'(u_n) \approx 0\,].$$

From $(PS2)$ we conclude that

$$\forall n \in {}^*\mathbb{N}_\infty \; [\, u_n \in \mathbf{fin}({}^*E) \;\wedge\; \mathbf{st}(f(u_n)) \text{ is a critical value of } f\,]$$

so we may conclude that $(u_n)_{n\in\mathbb{N}}$ is bounded and

$$\forall n \in {}^*\mathbb{N}_\infty \; \mathbf{st}(f(u_n)) \text{ is a critical value of } f.$$

Finally we will prove that $(PS3) \Rightarrow (PS2)$. Let $u \in {}^*E$ such that $f(u) \in \mathcal{O}$ and $f'(u) \approx 0$. Let $M \in \mathbb{R}^+$ such that $|f(u)| < M$. Suppose $u \notin \mathbf{fin}({}^*E)$; we construct a Palais-Smale sequence $(x_n)_{n\in\mathbb{N}}$ using the sets H_n, as in the proof of the first part of the proof that $(PS0) \Rightarrow (PS2)$, in such a way that $(x_n)_{n\in\mathbb{N}}$ is not bounded, which contradicts $(PS3)$.

Suppose now that $f(u) \in \mathcal{O}$ and $f'(u) \approx 0$. We need to prove that $\alpha = \mathbf{st}(f(u))$ is a critical value of f. Just as in the second part of the proof of $(PS0) \Rightarrow (PS2)$, we can build a Palais-Smale sequence $(x_n)_{n\in\mathbb{N}}$ such that for all $n \in {}^*\mathbb{N}_\infty$, $f(x_n) \approx \alpha$. From $(PS3)$ we conclude that for all $n \in {}^*\mathbb{N}_\infty$, $\mathbf{st}(f(x_n)) = \alpha$ is a critical value of f. □

More can be said when E is separable:

Theorem 19.3.2 *When E is a separable Banach space,* $(\mathbf{PS})$ *and* $(PS1)$ *are equivalent; in other words: if E is a separable Banach space, a C^1 function $f : E \to \mathbb{R}$ verifies the Palais-Smale condition if and only if almost critical points where f is finite are near-standard.*

Proof. Suppose f satisfies $(\mathbf{PS})$ and $u \in {}^*E$ is such that $f(u) \in \mathcal{O}$ and $f'(u) \approx 0$. If $u \notin \mathbf{ns}({}^*E)$, it follows from Theorem 19.1.2 that $u \notin \mathbf{pns}({}^*E)$, that is,

$$\exists \varepsilon \in \mathbb{R}^+ \; \forall y \in {}^\sigma E \;\; \|u - y\| > \varepsilon.$$

Let $V := \{v_p : p \in \mathbb{N}\}$ be dense in E. We will construct a Palais-Smale sequence $(x_n)_{n\in\mathbb{N}}$ in E such that for all $N \in {}^*\mathbb{N}_\infty$, $x_N \notin \mathbf{ns}({}^*E)$ which contradicts $(\mathbf{PS})$.

Let $M \in \mathbb{R}^+$ be such that $|f(u)| < M$. For each $n \in \mathbb{N}$, define

$$C_n := \Big\{ x \in E : |f(x)| < M \wedge \|f'(x)\| < \frac{1}{n} \wedge \forall p \in \mathbb{N}\, [\, p \le n \Rightarrow \|x - v_p\| > \varepsilon\,] \Big\}.$$

Since, for each $n \in \mathbb{N}$, $u \in {}^*C_n$, we conclude that ${}^*C_n \neq \emptyset$ and therefore $C_n \neq \emptyset$.

For each $n \in \mathbb{N}$, take $x_n \in C_n$. Let $N \in {}^*\mathbb{N}_\infty$ and $v \in {}^\sigma E$. Since $\overline{V} = E$, there exists $p_0 \in \mathbb{N}$ such that $\|v - v_{p_0}\| < \frac{\varepsilon}{2}$. Since

$$\|x_N - v\| \geq \|x_N - v_{p_0}\| - \|v_{p_0} - v\| > \frac{\varepsilon}{2}$$

we conclude that $x_N \notin \mathbf{ns}({}^*E)$ which contradicts (**PS**). □

It is not clear whether this equivalence is true in general.

The following result is consequence of the fact that if E is finite dimensional, then E is separable and $\mathbf{fin}({}^*E) = \mathbf{ns}({}^*E)$.

Theorem 19.3.3 *If E is finite dimensional*

$$(PS1) \Leftrightarrow (\mathbf{PS}) \Leftrightarrow (PS0) \Leftrightarrow (PS2) \Leftrightarrow (PS3) \Leftrightarrow (PS4).$$

Another easy observation:

Proposition 19.3.2 *Any C^1 functional in a Banach space which verifies $(PS4)$ and admits a Palais-Smale sequence, has at least one critical point.*

Proof. Suppose that $(u_n)_{n\in\mathbb{N}}$ is a Palais-Smale sequence for the functional $f \in C^1(E, \mathbb{R})$. Then there exist $m \in {}^*\mathbb{N}_\infty$ such that $\mathbf{st}(f(u_m))$ is a critical value of f. Hence, there exists $a \in E$ such that $f(a) = \mathbf{st}(f(u_m))$ and $f'(a) = 0$. □

Next we present an example which shows that

$$(PS2) \not\Leftarrow (PS1).$$

Example 19.3.1 Let H be an infinite dimensional Hilbert space and define

$$\begin{array}{rcl} f: H & \rightarrow & \mathbb{R} \\ x & \mapsto & f(x) = g(\|x\|^2 - 1) \end{array}$$

where $g : \mathbb{R} \rightarrow \mathbb{R}$ is given by

$$g(t) = \begin{cases} 0 & \text{if } \ t \leq 0 \\ t^2 \exp^{-\frac{1}{t^2}} & \text{if } \ t > 0 \end{cases}.$$

Observe that g is a C^1 function and

$$g'(t) \approx 0 \Leftrightarrow [t \leq 0 \ \vee \ t \approx 0]. \tag{19.3.1}$$

Also,

$$\begin{array}{rcl} h: H & \rightarrow & \mathbb{R} \\ x & \mapsto & \|x\|^2 - 1 \end{array}$$

is a C^1 functional and

$$\forall a \in H\ \forall x \in H\ \langle h'(a), x\rangle = 2a \bullet x$$

where $\cdot \bullet \cdot$ denotes the inner product in H. Therefore, f is a C^1 functional. We will prove that f does not satisfy $(PS1)$ but satisfies $(PS2)$.

By Theorem 19.1.2 we can take $u \in {}^*H$ such that $u \in \mathbf{fin}({}^*H) \setminus \mathbf{ns}({}^*H)$ and $\|u\| = 1$. Hence, $f(u) = 0$ and $f'(u) = 0$, which shows that f does not satisfy $(PS1)$.

Note that

$$v \notin \mathbf{fin}({}^*H) \Rightarrow f(v) \notin \mathcal{O}$$

and

$$\begin{aligned} f'(v) \approx 0 \quad &\Leftrightarrow \quad \|f'(v)\| \approx 0 \\ &\Leftrightarrow \quad \forall x \in \mathbf{fin}({}^*H)\ \langle f'(v), x\rangle = (2v \bullet x) g'(\|v\|^2 - 1) \approx 0. \end{aligned} \tag{19.3.2}$$

Next we will prove that

$$f'(v) \approx 0 \ \Rightarrow\ [\,\|v\| \leq 1 \ \vee\ \|v\| \approx 1\,]. \tag{19.3.3}$$

If $f'(v) \approx 0$ and $v \neq 0$, by (19.3.2) either $2v \bullet \frac{v}{\|v\|} \approx 0$, and thus $\|v\| \approx 0$, or $g'(\|v\|^2 - 1) \approx 0$, so that, by (19.3.1), $\|v\| \leq 1$ or $\|v\| \approx 1$; in all possible cases (19.3.3) holds.

Since

$$[\,\|v\| \leq 1 \ \vee\ \|v\| \approx 1\,] \ \Rightarrow\ f(v) \approx 0$$

and 0 is a critical value of f, we may conclude that

$$f(v) \in \mathcal{O} \ \wedge\ f'(v) \approx 0 \ \Rightarrow\ v \in \mathbf{fin}({}^*H) \ \wedge\ \mathbf{st}(f(v)) = 0 \text{ is a critical value of } f$$

proving that f does satisfy $(PS2)$.

Remark 19.3.1 Other consequences of Example 19.3.1*:*

1. If we assume H to be separable, Theorem 19.3.2 shows that

$$(PS2) \not\Rightarrow (\mathbf{PS}).$$

2. Moreover, the fact that $f \in C^1(E, \mathbb{R})$ and satisfies $(PS2)$, does not imply that the sets

$$f^{-1}([a, b]) \cap K \qquad (a \leq b)$$

are compact (see Proposition 19.2.1).

It follows easily from Proposition 19.1.1 that condition $(PS3)$ is equivalent to

$$(u_n)_{n\in\mathbb{N}} \text{ is a Palais-Smale sequence}$$
$$\Downarrow$$
$$(u_n)_{n\in\mathbb{N}} \text{ is bounded } \wedge \text{ all convergent subsequences of } (f(u_n))_{n\in\mathbb{N}} \text{ converge to a critical value of } f$$

and $(PS4)$ is equivalent to

$$(u_n)_{n\in\mathbb{N}} \text{ is a Palais-Smale sequence}$$
$$\Downarrow$$
$$(u_n)_{n\in\mathbb{N}} \text{ is bounded } \wedge \text{ there is a subsequence of } (f(u_n))_{n\in\mathbb{N}} \text{ which converges to a critical value of } f.$$

Moreover, in the finite dimensional case, these two standard conditions are equivalent.

19.4 Palais-Smale conditions per level

In this section we present a weaker *compactness condition* for C^1 functionals introduced in 1980 [3] by Brézis, Coron and Nirenberg. In the survey book [7] the reader can find more variants of the (**PS**) condition.

Definition 19.4.1 *Suppose $f \in C^1(E,\mathbb{R})$ and $c \in \mathbb{R}$. We say that $(u_n)_{n\in\mathbb{N}}$ is a **Palais-Smale sequence of level** c (for f) if*

$$\lim_{n\to\infty} f(u_n) = c \quad \textit{and} \quad \lim_{n\to\infty} f'(u_n) = 0.$$

*f satisfies the **Palais-Smale condition of level** c, $(\mathbf{PS})_c$, if every Palais-Smale sequence of level c has a convergent subsequence.*

Remark 19.4.1 Suppose that $f \in C^1(E,\mathbb{R})$. Then

1. If f satisfies (**PS**), then f satisfies $(\mathbf{PS})_c$ for all $c \in \mathbb{R}$.

2. If f satisfies $(\mathbf{PS})_c$, then the set of critical points of value c, K_c, is compact.

Example 19.4.1 The function $\exp(x) : \mathbb{R} \to \mathbb{R}$ satisfies $(\mathbf{PS})_c$ for all c except for $c = 0$. The real functions $\sin(x)$ and $\cos(x)$ defined in $\mathbb{R}$ satisfy $(\mathbf{PS})_c$ for all c except for $c = 1$ and $c = -1$.

19.5 Nonstandard variants of Palais-Smale conditions per level

As above, E is a real Banach space, $f \in C^1(E, \mathbb{R})$ and $c \in \mathbb{R}$.

Obvious adaptations of conditions $(PS0)$, $(PS1)$ and $(PS2)$ provide the following conditions *per* level:

$$(PS0)_c \qquad \begin{array}{c} (u_n)_{n\in\mathbb{N}} \text{ is a Palais-Smale sequence of level } c \\ \Downarrow \\ \exists m \in {}^*\mathbb{N}_\infty \; u_m \in \mathbf{ns}({}^*E) \end{array}$$

$$(PS1)_c \qquad f(u) \approx c \;\wedge\; f'(u) \approx 0 \Rightarrow u \in \mathbf{ns}({}^*E)$$

$$(PS2)_c \qquad \begin{array}{c} f(u) \approx c \;\wedge\; f'(u) \approx 0 \\ \Downarrow \\ u \in \mathbf{fin}({}^*E) \;\wedge\; c = \mathbf{st}(f(u)) \text{ is a critical value of } f \end{array}$$

Note that if $(u_n)_{n\in\mathbb{N}}$ is a Palais-Smale sequence of level c, then

$$\forall n \in {}^*\mathbb{N}_\infty \; \mathbf{st}(f(u_n)) = c$$

hence, the adaptation of conditions $(PS3)$ and $(PS4)$ to this context are equivalent to the standard condition:

$$\begin{array}{c} (u_n)_{n\in\mathbb{N}} \text{ is a Palais-Smale sequence of level } c \\ \Downarrow \\ (u_n)_{n\in\mathbb{N}} \text{ is bounded } \wedge\; c \text{ is a critical value of } f. \end{array}$$

The variants of Theorems 19.3.1, 19.3.2 and 19.3.3 can be easily proven.

Theorem 19.5.1 *For any real Banach space E we have*

$$(PS1)_c \Rightarrow (\mathbf{PS})_c \Leftrightarrow (PS0)_c \Rightarrow (PS2)_c.$$

Theorem 19.5.2 *Suppose E is a real separable Banach space and $f \in C^1(E, \mathbb{R})$. Then f satisfies $(\mathbf{PS})_c$ if and only if f satisfies $(PS1)_c$.*

Theorem 19.5.3 *If E has finite dimension, then*

$$(PS1)_c \Leftrightarrow (\mathbf{PS})_c \Leftrightarrow (PS0)_c \Leftrightarrow (PS2)_c.$$

Remark 19.5.1 Example 19.3.1 also shows that condition $(PS2)_c$ does generalize $(\mathbf{PS})_c$ and $(PS1)_c$ when $c = 0$.

19.6 Mountain Pass Theorems

Some important results in Critical Point Theory, such as the Mountain Pass Theorems and some variants of Ekeland's Variational Principle, can be obtained using a *deformation technique.* This technique was introduced in 1934 by Lusternik and Schnirelman [8] and consists in deforming a given C^1 functional outside the set of critical points. In 1983 Willem proved in [14] the Quantitative Deformation Lemma; a very technical lemma that involves the concept of pseudo-gradient vector field (see [14] or [9]).

For $S \subseteq E$, $\alpha \in \mathbb{R}^+$ and $c \in \mathbb{R}$, we use the following notations

$$f^c := \{x \in E : \ f(x) \leq c\} \quad \text{and} \quad S_\alpha := \{x \in E : \ dist(x, S) \leq \alpha\}$$

where

$$dist(x, S) = \inf\{\|x - y\| \ : \ y \in S\}.$$

Lemma 19.6.1 (Quantitative Deformation Lemma) *Let $f \in C^1(E, \mathbb{R})$, $S \subseteq E$, $c \in \mathbb{R}$, $\varepsilon, \delta \in \mathbb{R}^+$ be such that*

$$\forall y \in f^{-1}([c - 2\varepsilon, c + 2\varepsilon]) \cap S_{2\delta} \ \left[\ \|f'(y)\| \geq \frac{8\varepsilon}{\delta} \ \right].$$

Then there exists $\eta \in C([0,1] \times E, E)$ such that

1. $\eta(0, y) = y$;
2. $\eta(t, y) = y$ *if* $y \notin f^{-1}([c - 2\varepsilon, c + 2\varepsilon]) \cap S_{2\delta}$;
3. $\eta(1, f^{c+\varepsilon} \cap S) \subseteq f^{c-\varepsilon}$;
4. *for all* $t \in [0, 1]$, $\eta(t, \cdot) : E \to E$ *is a homeomorphism;*
5. $\forall t \in [0, 1], \forall y \in E \ \ \|\eta(t, y) - y\| \leq \delta$;
6. *for each* $y \in E$, $f(\eta(\cdot, y))$ *is non increasing;*
7. $\forall t \in]0, 1], \forall y \in f^c \cap S_\delta \ \ f(\eta(t, y)) < c$.

An easy consequence of the Quantitative Deformation Lemma is the following variant of Ekeland's Variational Principle (see [9, page 14]).

Corollary 19.6.1 *Let $f \in C^1(E, \mathbb{R})$ be bounded from below. Then, for any $\varepsilon \in \mathbb{R}^+$, there exists $u \in E$ such that*

$$f(u) \leq \inf_{x \in E} f(x) + \varepsilon \quad \text{and} \quad \|f'(u)\| < \sqrt{\varepsilon}.$$

Applying the Transfer Principle to Corollary 19.6.1 we obtain

Corollary 19.6.2 *Let $f \in C^1(E, \mathbb{R})$ be bounded from below. Then there exists a point $u \in {}^*E$ such that*

$$f(u) \approx \inf_{x \in E} f(x) \quad \text{and} \quad f'(u) \approx 0.$$

Now we can deduce

Theorem 19.6.1 *Let $f \in C^1(E, \mathbb{R})$ be bounded below and $c = \inf_{x \in E} f(x)$. If f satisfies $(PS2)_c$ then c is a critical value of f.*

Proof. By Corollary 19.6.2 we conclude that there exists $u \in {}^*E$ such that $f(u) \approx c$ and $f'(u) \approx 0$. Since f satisfies $(PS2)_c$, $c = \mathbf{st}(f(u))$ is a critical value of f. □

Theorem 19.6.1 is a generalization of the classical result (see [7, page 16]):

Theorem 19.6.2 *Let $f \in C^1(E, \mathbb{R})$ be bounded below and $c = \inf_{x \in E} f(x)$. If f satisfies $(\mathbf{PS})_c$ then c is a critical value of f.*

We now state the Mountain Pass Theorem introduced by Ambrosetti and Rabinowitz in 1973 [1].

Theorem 19.6.3 (Mountain Pass Theorem, Ambrosetti-Rabinowitz) *Let E be a real Banach space and $f \in C^1(E, \mathbb{R})$. Suppose that*

1. *there exists $e \in E$ and $r \in \mathbb{R}^+$ such that $\|e\| > r$ and*

$$\max\{f(0), f(e)\} \; < \; b := \inf_{\|x\|=r} f(x);$$

2. *$\Gamma = \{\gamma \in C([0,1], E) : \gamma(0) = 0 \wedge \gamma(1) = e\}$ and $c := \inf_{\gamma \in \Gamma} \max_{t \in [0,1]} f(\gamma(t))$;*

3. *f satisfies $(\mathbf{PS})$.*

Then $c \geq b$ and c is a critical value of f.

We usually say that if $f : E \to \mathbb{R}$ satisfies condition *1* of Theorem 19.6.3, then f satisfies the **mountain pass geometry** (with respect to 0 and e).

Remark 19.6.1 Conditions *1* and *2* of Theorem 19.6.3 are not enough to imply that c is a critical value of f as we can see with the following example (see [7, page 36]). The function $f : \mathbb{R}^2 \to \mathbb{R}$ defined by $f(x, y) = x^2 + (x+1)^3 y^2$ satisfies the mountain pass geometry with $0 = (0,0)$, $e = (-2, 3)$ and $r = \frac{1}{2}$. $(0,0)$ is a strict local minimizer and is the only critical point of f. Therefore, there is no $z \in \mathbb{R}^2$ such that $f(z) = c > 0$ and $f'(z) = 0$.

In 1980 Brézis, Coron and Nirenberg obtained in [3] a generalization of the Mountain Pass Theorem of Ambrosetti-Rabinowitz for functionals satisfying the $(\mathbf{PS})_c$ condition:

Theorem 19.6.4 (Mountain Pass Theorem, Brézis-Coron-Nirenberg) *Let E be a real Banach space and $f \in C^1(E,\mathbb{R})$. Suppose that*

1. *there exists $e \in E$ and $r \in \mathbb{R}^+$ such that $\|e\| > r$ and*
$$\max\{f(0), f(e)\} \; < \; b := \inf_{\|x\|=r} f(x);$$
2. *$\Gamma = \{\gamma \in C([0,1],E) : \; \gamma(0) = 0 \wedge \gamma(1) = e\}$ and $c := \inf_{\gamma\in\Gamma} \max_{t\in[0,1]} f(\gamma(t))$;*
3. *f satisfies $(\mathbf{PS})_c$.*

Then $c \geq b$ and c is a critical value of f.

This result and the Mountain Pass Theorem of Ambrosetti-Rabinowitz are easy consequences of the Quantitative Deformation Lemma, since it is possible to prove that for every C^1 functional that satisfies conditions *1* and *2* of both theorems, there exists a Palais-Smale sequence of level c (see [9, page 19]).

We now deduce the following generalization of the Mountain Pass Theorem of Brézis-Coron-Nirenberg.

Theorem 19.6.5 *Let E be a real Banach space and $f \in C^1(E,\mathbb{R})$. Suppose that*

1. *there exists $e \in E$ and $r \in \mathbb{R}^+$ such that $\|e\| > r$ and*
$$\max\{f(0), f(e)\} \; < \; b := \inf_{\|x\|=r} f(x);$$
2. *$\Gamma = \{\gamma \in C([0,1],E) : \gamma(0) = 0 \wedge \gamma(1) = e\}$ and $c := \inf_{\gamma\in\Gamma} \max_{t\in[0,1]} f(\gamma(t))$;*
3. *f satisfies $(PS4)_c$.*

Then $c \geq b$ and c is a critical value of f.

Proof. From the Quantitative Deformation Lemma there exists a Palais-Smale sequence of level c. Since f satisfies $(PS4)_c$, c is a critical value of f. □

References

[1] A. AMBROSETTI and P. RABINOWITZ, "Dual vatiational methods in critical point theory and applications", J. Funct. Anal., **14** (1973) 349–381.

[2] LEIF O. ARKERYD, NIGEL J. CUTLAND and C. WARD HENSON (eds.), *Nonstandard Analysis, Theory and Applications*, Kluwer, 1997.

[3] H. BRÉZIS, J.M. CORON and L. NIRENBERG, "Free vibrations for a nonlinear wave equation and a theorem of P. Rabinowitz", Comm. Pure Appl. Math., **33** (1980) 667–684.

[4] NIGEL J. CUTLAND (ed.), *Nonstandard Analysis and its Applications*, CUP, 1988.

[5] NIGEL J. CUTLAND, VÍTOR NEVES, FRANCO OLIVEIRA and JOSÉ SOUSA PINTO (eds.), *Developments in nonstandard mathematics*, Longman, 1995.

[6] IVAR EKELAND "Nonconvex minimization Problems", Bulletin (New Series) of the American Mathematical Society, **1** (1979).

[7] MARIA DO ROSÁRIO GROSSINHO and STEPAN AGOP TERSIAN, *An Introduction to Minimax Theorems and their Applications to Differential Equations*, Kluwer Academic Publishers, 2001.

[8] L. LUSTERNIK and L. SCHNIRELMANN, *Méthodes topologiques dans le problèmes variationnels*, Hermann, Paris, 1934.

[9] JEAN MAWHIN, *Critical Point Theory and Applications to Nonlinear Differential Equations*, VIGRE Minicourse on Variational Methods and Nonlinear PDE, University of Utah, 2002.

[10] R.S. PALAIS and S. SMALE, "A generalized Morse theory", Bull. Amer. Math. Soc., **70** (1964) 165–172.

[11] PAUL H. RABINOWITZ, *Minimax Methods in Critical Point Theory with Applications to Differential Equations*, CBMS Regional Conference, 65, AMS, Providence, Rhode Island, 1986.

[12] K.D. STROYAN and W.A.J. LUXEMBURG, *Introduction to the theory of Infinitesimals*, Academic Press, 1976.

[13] V. NEVES, The power of Gâteaux differentiability, this volume.

[14] M. WILLEM, *Lectures on Critical Point Theory*, Trabalho de Matemática 199, Fund. Univ. Brasilia, Brasilia, 1983.

20

Averaging for ordinary differential equations and functional differential equations

Tewfik Sari[*]

Abstract

A nonstandard approach to averaging theory for ordinary differential equations and functional differential equations is developed. We define a notion of perturbation and we obtain averaging results under weaker conditions than the results in the literature. The classical averaging theorems approximate the solutions of the system by the solutions of the averaged system, for Lipschitz continuous vector fields, and when the solutions exist on the same interval as the solutions of the averaged system. We extend these results to perturbations of vector fields which are uniformly continuous in the spatial variable with respect to the time variable and without any restriction on the interval of existence of the solution.

20.1 Introduction

In the early seventies, Georges Reeb, who learned about Abraham Robinson's *Nonstandard Analysis* (NSA) [29], was convinced that NSA gives a language which is well adapted to the study of perturbation theory of differential equations (see [6, p. 374] or [25]). The axiomatic presentation *Internal Set Theory* (IST) [26] of NSA given by E. Nelson corresponded more to the Reeb's dream and was in agreement with his conviction *"Les entiers naïfs ne remplissent pas* $\mathbb{N}$*"*. Indeed, no formalism can recover exactly all the actual phenomena, and *nonstandard objects* which may be considered as a formalization of *non-naïve objects* are already elements of our usual (standard) sets. We do not need any use of *stars* and *enlargements*. Thus, the Reebian school adopted IST. For more informations about Reeb's dream and convictions see the Reeb's preface of Lutz and Goze's book [25], Stewart's book [40] p. 72, or Lobry's book [23].

The Reebian school of *nonstandard perturbation theory of differential equations* produced various and numerous studies and new results as attested by a

[*]Université de Haute Alsace, Mulhouse, France.
T.Sari@uha.fr

lot of books and proceedings (see [2, 3, 4, 7, 8, 9, 10, 12, 23, 25, 30, 37, 42] and their references). It has become today a well-established tool in asymptotic theory, see, for instance, [17, 18, 20, 24, 39] and the five-digits classification 34E18 of the 2000 Mathematical Subject Classification. Canards and rivers (or Ducks and Streams [7]) are the most famous discoveries of the Reebian school.

The classical perturbation theory of differential equations studies deformations, instead of perturbations, of differential equations (see Section 2.1). Classically the phenomena are described asymptotically, when the parameter of the deformation tends to some fixed value. The first benefit of NSA is a natural and useful notion of perturbation. A perturbed equation becomes a simple nonstandard object, whose properties can be investigated directly. This aspect of NSA was clearly described by E. Nelson in his paper *Mathematical Mythologies* [30], p. 159, when he said *"For me, the most exciting aspect of nonstandard analysis is that concrete phenomena, such as ducks and streams, that classically can only be described awkwardly as asymptotic phenomena, become mythologized as simple nonstandard objects."*

The aim of this paper is to present some of the basic nonstandard techniques for averaging in Ordinary Differential Equations (ODEs), that I obtained in [32, 36], and their extensions, obtained by M. Lakrib [19], to Functional Differential Equations (FDEs). This paper is organized as follows. In Section 20.2 we define the notion of *perturbation* of a vector field. The main problem of perturbation theory of differential equations is to describe the behavior of trajectories of perturbed vector fields. We define a standard topology on the set of vector fields, with the property that f is a perturbation of a standard vector field f_0 if and only if f is infinitely close to f_0 for this topology. In Section 20.3 we present the Stroboscopic Method for ODEs and we show how to use it in the proof of the averaging theorem for ODEs. In Section 20.4 we present the Stroboscopic Method for FDEs and we show how to use it in the proofs of the averaging theorem for FDEs. The nonstandard approaches of averaging are rather similar in structure both in ODEs and FDEs. It should be noticed that the usual approaches of averaging make use of different concepts for ODEs and for FDEs: compare with [5, 31] for averaging in ODEs and [13, 14, 15, 22] for averaging in FDEs.

20.2 Deformations and perturbations

20.2.1 Deformations

The classical *perturbation theory of differential equations* studies families of differential equations

$$\dot{x} = F(x, \varepsilon), \tag{20.2.1}$$

where x belongs to an open subset U of $\mathbb{R}^n$, called *phase space*, and ε belongs to a subset B of $\mathbb{R}^k$, called *space of parameters.*

The family (20.2.1) of differential equations is said to be a *k-parameters deformation* of the vector field $F_0(x) = F(x, \varepsilon_0)$, where ε_0 is some fixed value of ε. The main problem of the perturbation theory of differential equations is to investigate the behavior of the vector fields $F(x, \varepsilon)$ when ε tends to ε_0.

The intuitive notion of a *perturbation* of the vector field F_0 which would mean any vector field which is *close to* F_0 does not appear in the theory. The situation is similar in the theory of *almost periodic functions* which, classically, do not have *almost periods.* The nonstandard approach permits to give a very natural notion of almost period (see [16, 28, 33, 41]). The *classical perturbation theory of differential equations* considers *deformations* instead of *perturbations* and would be better called *deformation theory of differential equations.* Actually the vector field $F(x, \varepsilon)$ when ε is sufficiently close to ε_0 is called a perturbation of the vector field $F_0(x)$. In other words, the differential equation

$$\dot{x} = F_0(x) \tag{20.2.2}$$

is said to be the *unperturbed equation* and equation (20.2.1), for a fixed value of ε, is called the *perturbed equation.* This notion of *perturbation* is not very satisfactory since many of the results obtained for the family (20.2.1) of differential equations take place in all systems that are close to the unperturbed equation (20.2.2). Noticing this fact, V. I. Arnold (see [1], footnote page 157) suggested to study a neighborhood of the unperturbed vector field $F_0(x)$ in a suitable function space. For the sake of mathematical convenience, instead of neighborhoods, one considers deformations. According to V. I. Arnold, the situation is similar with the historical development of variational concepts, where the directional derivative (Gâteaux differential) preceded the derivative of a mapping (Fréchet differential). Nonstandard analysis permits to define a notion of perturbation. To say that a vector f is a *perturbation* of a standard vector field f_0 is equivalent to say that f is *infinitely close* to f_0 is a suitable function space, that is f *is in any standard neighborhood* of f_0. Thus, studying perturbations in our sense is nothing than studying neighborhoods, as suggested by V. I. Arnold.

20.2.2 Perturbations

Let X be a standard topological space. A point $x \in X$ is said to be infinitely close to a standard point $x_0 \in X$, which is denoted by $x \simeq x_0$, if x is in any standard neighborhood of x_0. Let A be a subset of X. A point $x \in X$ is said to be *nearstandard in* A if there is a standard $x_0 \in A$ such that $x \simeq x_0$.

Let us denote by

$$^{NS}A = \{x \in X :\ \exists^{st} x_0 \in A\ x \simeq x_0\},$$

the *external-set* of nearstandard points in A [34]. Let E be a standard uniform space. The points $x \in E$ and $y \in E$ are said to be infinitely close, which is denoted by $x \simeq y$, if (x, y) lies in every standard entourage. If E is a standard metric space, with metric d, then $x \simeq y$ is equivalent to $d(x, y)$ infinitesimal.

Definition 1 *Let X be a standard topological space X. Let E be a standard uniform space. Let D and D_0 be open subsets of X, D_0 standard. Let $f : D \to E$ and $f_0 : D_0 \to E$ be mappings, f_0 standard. The mapping f is said to be a perturbation of the mapping f_0, which is denoted by $f \simeq f_0$, if $^{NS}D_0 \subset D$ and $f(x) \simeq f_0(x)$ for all $x \in {}^{NS}D_0$.*

Let $\mathcal{F}_{X,E}$ be the set of mappings defined on open subsets of X to E :

$$\mathcal{F}_{X,E} = \{(f, D) :\ D \text{ open subset of } X \text{ and } f : D \to E\}.$$

Let us consider the topology on this set defined as follows. Let $(f_0, D_0) \in \mathcal{F}_{X,E}$. The family of sets of the form

$$\{(f, D) \in \mathcal{F}_{X,E} :\ K \subset D\ \forall x \in K\ (f(x), f_0(x)) \in U\},$$

where K is a compact subset of D_0, and U is an entourage of the uniform space E, is a basis of the system of neighborhoods of (f_0, D_0). Let us call this topology the *topology of uniform convergence on compacta.* If all the mappings are defined on the same open set D, this topology is the usual topology of uniform convergence on compacta on the set of functions on D to E.

Proposition 1 *Assume X is locally compact. The mapping f is a perturbation of the standard mapping f_0 if and only if f is infinitely close to f_0 for the topology of uniform convergence on compacta.*

Proof. Let $f : D \to E$ be a perturbation of $f_0 : D_0 \to E$. Let K be a standard compact subset of D_0. Let U be a standard entourage. Then $K \subset D$ and $f(x) \simeq f_0(x)$ for all $x \in K$. Hence $(f(x), f_0(x)) \in U$. Thus $f \simeq f_0$ for the topology of uniform convergence on compacta. Conversely, let f be infinitely close to f_0 for the topology of uniform convergence on compacta. Let $x \in {}^{NS}D_0$. There exists a standard $x_0 \in D_0$ such that $x \simeq x_0$. Let K be a standard compact neighborhood of x_0, such that $K \subset D_0$ (such a neighborhood exists since X is locally compact). Then $x \in K \subset D$ and $(f(x), f_0(x)) \in U$ for all standard entourages U, that is $^{NS}D_0 \subset D$ and $f(x) \simeq f_0(x)$ on $^{NS}D_0$. Hence f is a perturbation of f_0. □

The notion of perturbation can be used to formulate Tikhonov's theorem on slow and fast systems whose fast dynamics has asymptotically stable equilibrium points [24], and Pontryagin and Rodygin's theorem on slow and fast systems whose fast dynamics has asymptotically stable cycles [39]. In the following section we use it to formulate the theorem of Krilov, Bogolyubov and Mitropolski of averaging for ODEs. All these theorems belong to the singular perturbation theory. In this paper, by a solution of an Initial Value Problem (IVP) associated to an ODE we mean a maximal (i.e. noncontinuable) solution. The fundamental nonstandard result of the regular perturbation theory of ODEs is called the Short Shadow Lemma. It can be stated as follows [36, 37].

Let $g : D \to \mathbb{R}^d$ and $g_0 : D_0 \to \mathbb{R}^d$ be continuous vector fields, $D, D_0 \subset \mathbb{R}_+ \times \mathbb{R}^d$. Let a_0^0 and a^0 be initial conditions. Assume that g_0 and a_0^0 are standard. The IVP

$$dX/dT = g(T, X), \quad X(0) = a^0 \tag{20.2.3}$$

is said to be a perturbation of the standard IVP

$$dX/dT = g_0(T, X), \quad X(0) = a_0^0, \tag{20.2.4}$$

if $g \simeq g_0$ and $a^0 \simeq a_0^0$. To avoid inessential complications we assume that equation $dX/dT = g_0(T, X)$ has the uniqueness of the solutions. Let ϕ_0 be the solution of the IVP (20.2.4). Let I be its maximal interval of definition. Then, by the following theorem any solution of problem (20.2.3) also exist on I and is infinitely close to ϕ_0.

Theorem 1 (Short Shadow Lemma) *Let problem (20.2.3) be a perturbation of problem (20.2.4). Every solution ϕ of problem (20.2.3) is a perturbation of the solution ϕ_0 of problem (20.2.4), that is, for all nearstandard t in I, $\phi(t)$ is defined and satisfies $\phi(t) \simeq \phi_0(t)$.*

Let us consider the restriction ψ of ϕ to ${}^{NS}I$. By the Short Shadow Lemma, for standard $t \in I$, it takes nearstandard values $\psi(t) \simeq \phi_0(t)$. Thus its shadow, which is the unique standard mapping which associate to each standard t the standard part of $\psi(t)$, is equal to ϕ_0. In general the shadow of ϕ is not equal to ϕ_0. Thus, the Short Shadow Lemma describes only the "short time behaviour" of the solutions.

20.3 Averaging in ordinary differential equations

The method of averaging is well-known for ODEs. The fundamental result of this theory asserts that, for small $\varepsilon > 0$, the solutions of a nonautonomous system

$$\dot{x} = f\left(t/\varepsilon, x, \varepsilon\right), \quad \text{where} \quad \dot{x} = dx/dt, \tag{20.3.1}$$

are approximated by the solutions of the averaged autonomous system

$$\dot{y} = F(y), \quad \text{where} \quad F(x) = \lim_{T\to\infty} \frac{1}{T} \int_0^T f(t, x, 0)\, dt. \tag{20.3.2}$$

The approximation of the solutions of (20.3.1) by the solutions of (20.3.2) means that if $x(t, \varepsilon)$ is a solution of (20.3.1) and $y(t)$ is the solution of the averaged equation (20.3.2) with the same initial condition, which is assumed to be defined on some interval $[0, T]$, then for $\varepsilon \simeq 0$ and for all $t \in [0, T]$, we have $x(t, \varepsilon) \simeq y(t)$.

The change of variable $z(\tau) = x(\varepsilon\tau)$ transforms equation (20.3.1) into equation

$$z' = \varepsilon f(\tau, z, \varepsilon), \quad \text{where} \quad z' = dz/d\tau. \tag{20.3.3}$$

Thus, the method of averaging can be stated for equation (20.3.3), that is, if ε is infinitesimal and $0 \le \tau \le T/\varepsilon$ then $z(\tau, \varepsilon) \simeq y(\varepsilon\tau)$.

Classical results were obtained by Krilov, Bogolyubov, Mitropolski, Eckhaus, Sanders, Verhulst (see [5, 31] and the references therein). The theory is very delicate. The dependence of $f(t, x, \varepsilon)$ in ε introduces many complications in the formulations of the conditions under which averaging is justified. In the classical approach, averaging is justified for systems (20.3.1) for which the vector field f is Lipschitz continuous in x. Our aim in this section is first to formulate this problem with the concept of perturbations of vector fields and then to give a theorem of averaging under hypothesis less restrictive than the usual hypothesis. In our approach, averaging is justified for all perturbations of a continuous vector field which is continuous in x uniformly with respect to t. This assumption is of course less restrictive than Lipschitz continuity with respect to x.

20.3.1 KBM vector fields

Definition 2 *Let U_0 be an open subset of $\mathbb{R}^d$. The continuous vector field $f_0 : \mathbb{R}_+ \times U_0 \to \mathbb{R}^d$ is said to be a Krilov-Bogolyubov-Mitropolski (KBM) vector field if it satisfies the following conditions*

1. *The function $x \to f_0(t, x)$ is continuous in x uniformly with respect to the variable t.*

2. *For all $x \in U_0$ the limit $F(x) = \lim_{T\to\infty} \frac{1}{T} \int_0^T f_0(t, x)\, dt$ exists.*

3. *The averaged equation $\dot{y}(t) = F(y(t))$ has the uniqueness of the solution with prescribed initial condition.*

Notice that, in the previous definition, conditions (1) and (2) imply that the function F is continuous, so that the averaged equation considered in condition (3) is well defined. In the case of non autonomous ODEs, the definition of a perturbation given in Section 20.2 must be stated as follows.

Definition 3 *Let U_0 and U be open subsets of $\mathbb{R}^d$. A continuous vector field $f : \mathbb{R}_+ \times U \to \mathbb{R}^d$ is said to be a perturbation of the standard continuous vector field $f_0 : \mathbb{R}_+ \times U_0 \to \mathbb{R}^d$ if U contains all the nearstandard points in U_0, and $f(s, x) \simeq f_0(s, x)$ for all $s \in \mathbb{R}_+$ and all nearstandard x in U_0.*

Theorem 2 *Let $f_0 : \mathbb{R}_+ \times U_0 \to \mathbb{R}^d$ be a standard KBM vector field and let $a_0 \in U_0$ be standard. Let $y(t)$ be the solution of the IVP*

$$\dot{y}(t) = F(y(t)), \qquad y(0) = a_0, \tag{20.3.4}$$

defined on the interval $[0, \omega[$, $0 < \omega \leq \infty$. Let $f : \mathbb{R}_+ \times U \to \mathbb{R}^d$ be a perturbation of f_0. Let $\varepsilon > 0$ be infinitesimal and $a \simeq a_0$. Then every solution $x(t)$ of the IVP

$$\dot{x}(t) = f\left(t/\varepsilon, x(t)\right), \qquad x(0) = a, \tag{20.3.5}$$

is a perturbation of $y(t)$, that is, for all nearstandard t in $[0, \omega[$, $x(t)$ is defined and satisfies $x(t) \simeq y(t)$.

The proof, in the particular case of almost periodic vector fields, is given in Section 20.3.4. The proof in the general case is given in Section 20.3.5.

20.3.2 Almost solutions

The notion of almost solution of an ODE is related to the classical notion of *ε-almost solution.*

Definition 4 *A function $x(t)$ is said to be an almost solution of the standard differential equation $\dot{x} = G(t, x)$ on the standard interval $[0, L]$ if there exists a finite sequence $0 = t_0 < \cdots < t_{N+1} = L$ such that for $n = 0, \cdots, N$ we have*

$$t_{n+1} \simeq t_n, \quad x(t) \simeq x(t_n) \ \ \textit{for} \ \ t \in [t_n, t_{n+1}],$$

$$\textit{and} \quad \frac{x(t_{n+1}) - x(t_n)}{t_{n+1} - t_n} \simeq G(t_n, x(t_n)).$$

The aim of the following result is to show that an almost solution of a standard ODE is infinitely close to a solution of the equation. This result which was first established by J. L. Callot (see [11, 27]) is a direct consequence of the nonstandard proof of the existence of solutions of continuous ODEs [26].

Theorem 3 *If $x(t)$ is an almost solution of the standard differential equation $\dot{x} = G(t, x)$ on the standard interval $[0, L]$, $x(0) \simeq y_0$, with y_0 standard, and the IVP $\dot{y} = G(t, y)$, $y(0) = y_0$, has a unique solution $y(t)$, then $y(t)$ is defined at least on $[0, L]$ and we have $x(t) \simeq y(t)$, for all $t \in [0, L]$.*

Proof. See [11, 36] □

Let us apply this theorem to obtain an averaging result for an ODE which does not satisfy all the hypothesis of Theorem 2.

Consider the ODE (see [11, 27, 36])

$$\dot{x}(t) = \sin \frac{tx}{\varepsilon}. \tag{20.3.6}$$

The conditions (2) and (3) in Definition 2 are satisfied with $F(x) = 0$. Thus, the solutions of the averaged equation are constant. But condition (1) of the definition is not satisfied, since the function $f(t, x) = \sin(tx)$ is not continuous in x uniformly with respect to t. Hence Theorem 2 does not apply. In fact the solutions of (20.3.6) are not nearly constant and we have the following result:

Proposition 2 *If $\varepsilon > 0$ is infinitesimal then, in the region $t \geq x > 0$ the solutions of (20.3.6) are infinitely close to hyperbolas $tx =$ constant. In the region $x > t \geq 0$, they are infinitely close to the solutions of the ODE*

$$\dot{x} = G(t, x), \quad \textit{where} \quad G(t, x) = \frac{\sqrt{x^2 - t^2} - x}{t}. \tag{20.3.7}$$

Proof. The isocline curves $I_k = \{(t, x) : tx = 2k\pi\varepsilon\}$ and $I'_k = \{(t, x) : tx = (2k + \frac{3}{2})\pi\varepsilon\}$ define, in the region $t \geq x > 0$, tubes in which the trajectories are trapped. Thus for $t \geq x > 0$ the solutions are infinitely close to the hyperbolas $tx =$ constant. This argument does not work for $x > t \geq 0$. In this region, we consider the microscope

$$T = \frac{t - t_k}{\varepsilon}, \qquad X = \frac{x - x_k}{\varepsilon},$$

where (t_k, x_k) are the points where a solution $x(t)$ of (20.3.6) crosses the curve I_k. Then we have

$$\frac{dX}{dT} = \sin(x_k T + t_k X + \varepsilon T X), \qquad X(0) = 0.$$

By the Short Shadow Lemma (Theorem 1), $X(T)$ is infinitely close to a solution of $dX/dT = \sin(x_k T + t_k X)$. By straightforward computations we have

$$\frac{x_{k+1} - x_k}{t_{k+1} - t_k} \simeq G(t_k, x_k).$$

Hence, in the region $x > t \geq 0$, the function $x(t)$ is an almost solution of the ODE (20.3.7). By Theorem 3, the solutions of (20.3.6) are infinitely close to the solutions of (20.3.7). □

20.3.3 The stroboscopic method for ODEs

In this section we denote by $G : \mathbb{R}_+ \times D \to \mathbb{R}^d$ a standard continuous function, where D is a standard open subset of $\mathbb{R}^d$. Let $x : I \to \mathbb{R}^d$ be a function such that $0 \in I \subset \mathbb{R}_+$.

Definition 5 *We say that x satisfies the Strong Stroboscopic Property with respect to G if there exists $\mu > 0$ such that for every positive limited $t_0 \in I$ with $x(t_0)$ nearstandard in D, there exists $t_1 \in I$ such that $\mu < t_1 - t_0 \simeq 0$, $[t_0, t_1] \subset I$, $x(t) \simeq x(t_0)$ for all $t \in [t_0, t_1]$, and*

$$\frac{x(t_1) - x(t_0)}{t_1 - t_0} \simeq G(t_0, x(t_0)).$$

The real numbers t_0 and t_1 are called *successive instants of observation* of the stroboscopic method.

Theorem 4 (Stroboscopic Lemma for ODEs) *Let $a_0 \in D$ be standard. Assume that the IVP $\dot{y}(t) = G(t, y(t))$, $y(0) = a_0$, has a unique solution y defined on some standard interval $[0, L]$. Assume that $x(0) \simeq a_0$ and x satisfies the Strong Stroboscopic Property with respect to G. Then x is defined at least on $[0, L]$ and satisfies $x(t) \simeq y(t)$ for all $t \in [0, L]$.*

Proof. Since x satisfies the Strong Stroboscopic Property with respect to G, it is an almost solution of the ODE $\dot{x} = G(t, x)$. By Theorem 3 we have $x(t) \simeq y(t)$ for all $t \in [0, L]$. The details of the proof can be found in [36]. □

The Stroboscopic Lemma has many applications in the perturbation theory of differential equations (see [11, 32, 35, 36, 38, 39]). Let us use this lemma to obtain a proof of Theorem 2.

20.3.4 Proof of Theorem 2 for almost periodic vector fields

Suppose that f_0 is an almost periodic in t, then any of its translates $f_0(s + \cdot, x_0)$ is a nearstandard function, and f_0 has an average F which satisfies [16, 28, 33, 41]

$$F(x) = \lim_{T \to \infty} \frac{1}{T} \int_s^{s+T} f_0(t, x)\, dt,$$

uniformly with respect to $s \in \mathbb{R}_+$. Since F is standard and continuous, we have

$$F(x) \simeq \frac{1}{T} \int_s^{s+T} f_0(t, x)\, dt, \tag{20.3.8}$$

for all $s \in \mathbb{R}_+$, all $T \simeq \infty$ and all nearstandard x in U_0. Let $x : I \to U$ be a solution of problem (20.3.5). Let t_0 be an instant of observation: t_0 is limited in I, and $x_0 = x(t_0)$ is nearstandard in U_0. The change of variables

$$X = \frac{x(t_0 + \varepsilon T) - x_0}{\varepsilon},$$

transforms (20.3.5) into

$$dX/dT = f(s+T, x_0 + \varepsilon X), \quad \text{where} \quad s = t_0/\varepsilon.$$

By the Short Shadow Lemma (Theorem 1), applied to $g(T,X) = f(s + T, x_0 + \varepsilon X)$ and $g_0(T,X) = f_0(s + T, x_0)$, for all limited $T > 0$, we have $X(T) \simeq \int_0^T f_0(s + r, x_0)dr$. By Robinson's Lemma this property is true for some unlimited T which can be chosen such that $\varepsilon T \simeq 0$. Define $t_1 = t_0 + \varepsilon T$. Then we have

$$\frac{x(t_1) - x(t_0)}{t_1 - t_0} = \frac{X(T)}{T} \simeq \frac{1}{T}\int_0^T f_0(s+r, x_0)\, dr = \frac{1}{T}\int_s^{s+T} f_0(t, x_0)\, dt \simeq F(x_0).$$

Thus x satisfies the Strong Stroboscopic Property with respect to F. Using the Stroboscopic Lemma for ODEs (Theorem 4) we conclude that $x(t)$ is infinitely close to a solution of the averaged ODE (20.3.4).

20.3.5 Proof of Theorem 2 for KBM vector fields

Let f_0 be a KBM vector field. From condition (2) of Definition 2 we deduce that for all $s \in \mathbb{R}_+$, we have $F(x) = \lim_{T\to\infty} \frac{1}{T}\int_s^{s+T} f_0(t,x)\, dt$, but the limit is not uniform on s. Thus for unlimited positive s, the property (20.3.8) does not hold for all unlimited T, as it was the case for almost periodic vector fields. However, using also the uniform continuity of f_0 in x with respect to t we can show that (20.3.8) holds for some unlimited T which are not very large. This result is stated in the following technical lemma [36].

Lemma 1 *Let $g : \mathbb{R}_+ \times \mathcal{M} \to \mathbb{R}^d$ be a standard continuous function where $\mathcal{M}$ is a standard metric space. We assume that g is continuous in $m \in \mathcal{M}$ uniformly with respect to $t \in \mathbb{R}_+$ and that g has an average $G(m) = \lim_{T\to\infty} \frac{1}{T}\int_0^T g(t,m)\, dt$. Let $\varepsilon > 0$ be infinitesimal. Let $t \in \mathbb{R}_+$ be limited. Let m be nearstandard in $\mathcal{M}$. Then there exists $\alpha > \varepsilon$, $\alpha \simeq 0$ such that, for all limited $T \geq 0$ we have*

$$\frac{1}{S}\int_s^{s+TS} g(r,m)\, dr \simeq TG(m), \quad \textit{where} \quad s = t/\varepsilon, \quad S = \alpha/\varepsilon.$$

The proof of Theorem 2 needs another technical lemma whose proof can be found also in [36].

Lemma 2 *Let $g : \mathbb{R}_+ \times \mathbb{R}^d \to \mathbb{R}^d$ and $h : \mathbb{R}_+ \to \mathbb{R}^d$ be continuous functions. Suppose that $g(T, X) \simeq h(T)$ holds for all limited $T \in \mathbb{R}_+$ and all limited $X \in \mathbb{R}^d$, and $\int_0^T h(r)\,dr$ is limited for all limited $T \in \mathbb{R}_+$. Then, any solution $X(T)$ of the IVP $dX/dT = g(T, X)$, $X(0) = 0$, is defined for all limited $T \in \mathbb{R}_+$ and satisfies $X(T) \simeq \int_0^T h(r)\,dr$.*

Proof of Theorem 2. Let $x : I \to U$ be a solution of problem (20.3.5). Let $t_0 \in I$ be limited, such that $x_0 = x(t_0)$ is nearstandard in U_0. By Lemma 1, applied to $g = f_0$, $G = F$ and $m = x(t_0)$, there is $\alpha > 0$, $\alpha \simeq 0$ such that for all limited $T \geq 0$ we have

$$\frac{1}{S}\int_s^{s+TS} f_0(r, x_0)\,dr \simeq TF(x_0), \quad \text{where } s = t_0/\varepsilon,\ S = \alpha/\varepsilon. \tag{20.3.9}$$

The change of variables

$$X(T) = \frac{x\,(t_0 + \alpha T) - x_0}{\alpha}$$

transforms (20.3.5) into

$$dX/dT = f(s + ST, x_0 + \alpha X).$$

By Lemma 2, applied to $g(T, X) = f(s + ST, x_0 + \alpha X)$ and $h(T) = f_0(s + ST, x_0)$, and (20.3.9), for all limited $T > 0$, we have

$$X(T) \simeq \int_0^T f_0(s + Sr, x_0)\,dr = \frac{1}{S}\int_s^{s+TS} f_0(r, x_0)\,dr \simeq TF(x_0).$$

Define the successive instant of observation of the stroboscopic method t_1 by $t_1 = t_0 + \alpha$. Then we have

$$\frac{x(t_1) - x(t_0)}{t_1 - t_0} = X(1) \simeq F(x_0).$$

Since $t_1 - t_0 = \alpha > \varepsilon$ and $x(t) - x(t_0) = \alpha X(T) \simeq 0$ for all $t \in [t_0, t_1]$, we have proved that the function x satisfies the Strong Stroboscopic Property with respect to F. By the Stroboscopic Lemma, for any nearstandard $t \in [0, \omega[$, $x(t)$ is defined and satisfies $x(t) \simeq y(t)$. □

20.4 Functional differential equations

Let $\mathcal{C} = \mathcal{C}([-r, 0], \mathbb{R}^d)$, where $r > 0$, denote the Banach space of continuous functions with the norm $\|\phi\| = \sup\{\|\phi(\theta)\| : \theta \in [-r, 0]\}$, where $\|\cdot\|$ is a norm of $\mathbb{R}^d$. Let $L \geq t_0$. If $x : [-r, L] \to \mathbb{R}^d$ is continuous, we define $x_t \in \mathcal{C}$ by setting $x_t(\theta) = x(t + \theta)$, $\theta \in [-r, 0]$ for each $t \in [0, L]$. Let $g : \mathbb{R}_+ \times \mathcal{C} \to \mathbb{R}^d$, $(t, u) \mapsto g(t, u)$, be a continuous function. Let $\phi \in \mathcal{C}$ be an initial condition. A Functional Differential Equation (FDE) is an equation of the form

$$\dot{x}(t) = g(t, x_t), \qquad x_0 = \phi.$$

This type of equation includes differential equations with delays of the form

$$\dot{x}(t) = G(t, x(t), x(t - r)),$$

where $G : \mathbb{R}_+ \times \mathbb{R}^d \times \mathbb{R}^d \to \mathbb{R}^d$. Here we have $g(t, u) = G(t, u(0), u(-r))$.

The method of averaging was extended [13, 22] to the case of FDEs of the form

$$z'(\tau) = \varepsilon f(\tau, z_\tau), \tag{20.4.1}$$

where ε is a small parameter. In that case the averaged equation is the ODE

$$y'(\tau) = \varepsilon F(y(\tau)), \tag{20.4.2}$$

where F is the average of f. It was also extended [14] to the case of FDEs of the form

$$\dot{x}(t) = f(t/\varepsilon, x_t). \tag{20.4.3}$$

In that case the averaged equation is the FDE

$$\dot{y}(t) = F(y_t). \tag{20.4.4}$$

Notice that the change of variables $x(t) = z(t/\varepsilon)$ does not transform equation (20.4.1) into equation (20.4.3), as it was the case for ODEs (20.3.3) and (20.3.1), so that the results obtained for (20.4.1) cannot be applied to (20.4.3). In the case of FDEs of the form (20.4.1) or (20.4.3), the classical averaging theorems require that the vector field f is Lipschitz continuous in x uniformly with respect to t. In our approach, this condition is weakened and we only assume that the vector field f is continuous in x uniformly with respect to t. Also in the classical averaging theorems it is assumed that the solutions $z(\tau, \varepsilon)$ of (20.4.1) and $y(\tau)$ of (20.4.2) exist in the same interval $[0, T/\varepsilon]$ or that the solutions $x(t, \varepsilon)$ of (20.4.3) and $y(t)$ of (20.4.4) exist in the same interval $[0, T]$. In our approach, we assume only that the solution of the averaged equation is defined on some interval and we give conditions on the vector field f so that, for ε sufficiently small, the solution $x(t, \varepsilon)$ of the system exists at least on the same interval.

20.4.1 Averaging for FDEs in the form $z'(\tau) = \varepsilon f(\tau, z_\tau)$

We consider the IVP, where ε is a small parameter

$$z'(\tau) = \varepsilon f(\tau, z_\tau), \qquad z_0 = \phi.$$

The change of variable $x(t) = z(t/\varepsilon)$ transforms this equation in

$$\dot{x}(t) = f(t/\varepsilon, x_{t,\varepsilon}), \qquad x(t) = \phi(t/\varepsilon), \quad t \in [-\varepsilon r, 0], \tag{20.4.5}$$

where $x_{t,\varepsilon} \in \mathcal{C}$ is defined by $x_{t,\varepsilon}(\theta) = x(t + \varepsilon\theta)$ for $\theta \in [-r, 0]$.

Let $f : \mathbb{R}_+ \times \mathcal{C} \to \mathbb{R}^d$ be a standard continuous function. We assume that

(H1) The function $f : u \mapsto f(t, u)$ is continuous in u uniformly with respect to the variable t.

(H2) For all $u \in \mathcal{C}$ the limit $F(u) = \lim_{T\to\infty} \frac{1}{T}\int_0^T f(t, u)\,dt$ exists.

We identify $\mathbb{R}^d$ to the subset of constant functions in $\mathcal{C}$, and for any vector $c \in \mathbb{R}^d$, we denote by the same letter, the constant function $u \in \mathcal{C}$ defined by $u(\theta) = c$, $\theta \in [-r, 0]$. Averaging consists in approximating the solutions $x(t, \varepsilon)$ of (20.4.5) by the solution $y(t)$ of the averaged ODE

$$\dot{y}(t) = F(y(t)), \qquad y(0) = \phi(0). \tag{20.4.6}$$

According to our convention, $y(t)$, in the right-hand side of this equation, is the constant function $u^t \in \mathcal{C}$ defined by $u^t(\theta) = y(t)$, $\theta \in [-r, 0]$. Since F is continuous, this equation is well defined. We assume that

(H3) The averaged ODE (20.4.6) has the uniqueness of the solution with prescribed initial condition.

(H4) The function f is quasi-bounded in the variable u uniformly with respect to the variable t, that is, for every $t \in \mathbb{R}_+$ and every limited $u \in \mathcal{C}$, $f(t, u)$ is limited in $\mathbb{R}^d$.

Notice that conditions (H1), (H2) and (H3) are similar to conditions (1), (2) and (3) of Definition 2. In the case of FDEs we need also condition (H4). In classical words, the *uniform quasi boundedness* means that for every bounded subset B of $\mathcal{C}$, $f(\mathbb{R}_+ \times B)$ is a bounded subset of $\mathbb{R}^d$. This property is strongly related to the continuation properties of the solutions of FDEs (see Sections 2.3 and 3.1 of [15]).

Theorem 5 *Let $f : \mathbb{R}_+ \times \mathcal{C} \to \mathbb{R}^d$ be a standard continuous function satisfying the conditions (H1)-(H4). Let ϕ be standard in $\mathcal{C}$. Let $L > 0$ be standard and let $y : [0, L] \to \mathbb{R}^d$ be the solution of (20.4.6). Let $\varepsilon > 0$ be infinitesimal. Then every solution $x(t)$ of the problem (20.4.5) is defined at least on $[-\varepsilon r, L]$ and satisfies $x(t) \simeq y(t)$ for all $t \in [0, L]$.*

20.4.2 The stroboscopic method for ODEs revisited

In this section we give another formulation of the stroboscopic method for ODEs which is well adapted to the proof of Theorem 5. Moreover, this formulation of the Stroboscopic Method will be easily extended to FDEs (see Section 20.4.4). We denote by $G : \mathbb{R}_+ \times \mathbb{R}^d \to \mathbb{R}^d$, a standard continuous function. Let $x : I \to \mathbb{R}^d$ be a function such that $0 \in I \subset \mathbb{R}_+$.

Definition 6 *We say that x satisfies the Stroboscopic Property with respect to G if there exists $\mu > 0$ such that for every positive limited $t_0 \in I$, satisfying $[0, t_0] \subset I$ and $x(t)$ is limited for all $t \in [0, t_0]$, there exists $t_1 \in I$ such that $\mu < t_1 - t_0 \simeq 0$, $[t_0, t_1] \subset I$, $x(t) \simeq x(t_0)$ for all $t \in [t_0, t_1]$, and*

$$\frac{x(t_1) - x(t_0)}{t_1 - t_0} \simeq G\left(t_0, x(t_0)\right).$$

The difference with the Strong Stroboscopic Property with respect to G considered in Section 20.3.3 is that now we assume that the successive instant of observation t_1 exists only for those values t_0 for which $x(t)$ is limited for all $t \in [0, t_0]$. In Definition 5, in which we take $D = \mathbb{R}^d$, we assumed the stronger hypothesis that t_1 exists for all limited t_0 for which $x(t_0)$ is limited.

Theorem 6 (Second Stroboscopic Lemma for ODEs) *Let $a_0 \in D$ be standard. Assume that the IVP $\dot{y}(t) = G\left(t, y(t)\right)$, $y(0) = a_0$, has a unique solution y defined on some standard interval $[0, L]$. Assume that $x(0) \simeq a_0$ and x satisfies the Stroboscopic Property with respect to G. Then x is defined at least on $[0, L]$ and satisfies $x(t) \simeq y(t)$ for all $t \in [0, L]$.*

Proof. Since x satisfies the Stroboscopic Property with respect to G, it is an almost solution of the ODE $\dot{x} = G(t, x)$. By Theorem 3 we have $x(t) \simeq y(t)$ for all $t \in [0, L]$. The details of the proof can found in [19] or [21]. □

Proof of Theorem 5. Let $x : I \to \mathbb{R}^d$ be a solution of problem (20.4.5). Let $t_0 \in I$ be limited, such that $x(t)$ is limited for all $t \in [0, t_0]$. By Lemma 1, applied to $g = f$, $G = F$ and the constant function $m = x(t_0)$, there is $\alpha > 0$, $\alpha \simeq 0$ such that for all limited $T \geq 0$ we have

$$\frac{1}{S} \int_s^{s+TS} f(r, x(t_0))\, dr \simeq TF(x(t_0)), \quad \text{where } s = t_0/\varepsilon,\ S = \alpha/\varepsilon. \tag{20.4.7}$$

Using the uniform quasi boundedness of f we can show (for the details see [19] or [21]) that $x(t)$ is defined and limited for all $t \simeq t_0$. Hence the function

$$X(\theta, T) = \frac{x(t_0 + \alpha T + \varepsilon\theta) - x(t_0)}{\alpha}, \quad \theta \in [-r, 0], \quad T \in [0, 1],$$

is well defined. In the variable $X(\cdot,T)$ system (20.4.5) becomes

$$\frac{\partial X}{\partial T}(0,T) = f(s+ST, x(t_0)+\alpha X(\cdot,T)).$$

Using assumptions (H1) and (H4) together with (20.4.7), we obtain after some computations that for all $T \in [0,1]$, we have

$$X(0,T) \simeq \int_0^T f(s+Sr, x(t_0))\,dr = \frac{1}{S}\int_s^{s+TS} f(r, x(t_0))\,dr \simeq TF(x(t_0)).$$

Define the successive instant of observation of the stroboscopic method t_1 by $t_1 = t_0 + \alpha$. Then we have

$$\frac{x(t_1)-x(t_0)}{t_1-t_0} = X(0,1) \simeq F(x(t_0)).$$

Since $t_1 - t_0 = \alpha > \varepsilon$ and $x(t) - x(t_0) = \alpha X(0,T) \simeq 0$ for all $t \in [t_0, t_1]$, we have proved that the function x satisfies the Stroboscopic Property with respect to F. By the Second Stroboscopic Lemma for ODEs, for any $t \in [0,L]$, $x(t)$ is defined and satisfies $x(t) \simeq y(t)$. □

20.4.3 Averaging for FDEs in the form $\dot{x}(t) = f\left(t/\varepsilon, x_t\right)$

We consider the IVP, where ε is a small parameter

$$\dot{x}(t) = f\left(t/\varepsilon, x_t\right), \qquad x_0 = \phi, \tag{20.4.8}$$

We assume that f satisfies conditions (H1), (H2) and (H4) of Section 20.4.1. Now, the averaged equation is not the ODE (20.4.6), but the FDE

$$\dot{y}(t) = F\left(y_t\right), \qquad y_0 = \phi. \tag{20.4.9}$$

Averaging consists in approximating the solutions $x(t,\varepsilon)$ of (20.4.8) by the solution $y(t)$ of the averaged FDE (20.4.9). Condition (H3) in Section 20.4.1 must be restated as follows

(H3) The averaged FDE (20.4.9) has the uniqueness of the solution with prescribed initial condition.

Theorem 7 *Let $f : \mathbb{R}_+ \times \mathcal{C} \to \mathbb{R}^d$ be a standard continuous function satisfying the conditions (H1)-(H4). Let ϕ be standard in $\mathcal{C}$. Let $L > 0$ be standard and let $y : [0,L] \to \mathbb{R}^d$ be the solution of problem (20.4.9). Let $\varepsilon > 0$ be infinitesimal. Then every solution $x(t)$ of the problem (20.4.8) is defined at least on $[-r,L]$ and satisfies $x(t) \simeq y(t)$ for all $t \in [-r,L]$.*

20.4.4 The stroboscopic method for FDEs

Since the averaged equation (20.4.9) is an FDE, we need an extension of the stroboscopic method for ODEs given in Section 20.4.2. In this section we denote by $G : \mathbb{R}_+ \times \mathcal{C} \to \mathbb{R}^d$, a standard continuous function. Let $x : I \to \mathbb{R}^d$ be a function such that $[-r, 0] \subset I \subset \mathbb{R}_+$.

Definition 7 *We say that x satisfies the Stroboscopic Property with respect to G if there exists $\mu > 0$ such that for every positive limited $t_0 \in I$, satisfying $[0, t_0] \subset I$ and $x(t)$ and $G(t, x_t)$ are limited for all $t \in [0, t_0]$, there exists $t_1 \in I$ such that $\mu < t_1 - t_0 \simeq 0$, $[t_0, t_1] \subset I$, $x(t) \simeq x(t_0)$ for all $t \in [t_0, t_1]$, and*

$$\frac{x(t_1) - x(t_0)}{t_1 - t_0} \simeq G\left(t_0, x_{t_0}\right).$$

Notice that now we assume that the successive instant of observation t_1 exists for those values t_0 for which both $x(t)$ and $G(t, x_t)$ are limited for all $t \in [0, t_0]$. In the limit case $r = 0$, the Banach space $\mathcal{C}$ is identified with $\mathbb{R}^d$ and the function x_t is identified with $x(t)$ so that, $G(t, x_t)$ is limited, for all limited $x(t)$. Hence the "Stroboscopic Property with respect to G" considered in the previous definition is a natural extension to FDEs of the "Stroboscopic Property with respect to G" considered in Definition 6.

Theorem 8 (Stroboscopic Lemma for FDEs) *Let $\phi \in \mathcal{C}$ be standard. Assume that the IVP $\dot{y}(t) = G(t, y_t)$, $y_0 = \phi$, has a unique solution y defined on some standard interval $[-r, L]$. Assume that the function x satisfies the Stroboscopic Property with respect to G and $x_0 \simeq \phi$. Then x is defined at least on $[-r, L]$ and satisfies $x(t) \simeq y(t)$ for all $t \in [-r, L]$.*

Proof. Since x satisfies the Stroboscopic Property with respect to G, it is an almost solution of the FDE $\dot{x} = G(t, x_t)$. For FDEs, we have to our disposal an analog of Theorem 3. Thus $x(t) \simeq y(t)$ for all $t \in [0, L]$. The details of the proof can found in [19] or [21]. □

Proof of Theorem 7. Let $x : I \to \mathbb{R}^d$ be a solution of problem (20.4.8). Let $t_0 \in I$ be limited, such that both $x(t)$ and $F(x_t)$ are limited for all $t \in [0, t_0]$. From the uniform quasi boundedness of f we deduce that $x(t)$ is S-continuous on $[0, t_0]$. Thus x_t is nearstandard for all $t \in [0, t_0]$. By Lemma 1, applied to $g = f$, $G = F$ and $m = x_{t_0}$, there is $\alpha > 0$, $\alpha \simeq 0$ such that for all limited $T \geq 0$ we have

$$\frac{1}{S} \int_s^{s+TS} f(r, x_{t_0})\, dr \simeq TF\left(x_{t_0}\right), \quad \text{where } s = t_0/\varepsilon,\ S = \alpha/\varepsilon. \tag{20.4.10}$$

Using the uniform quasi boundedness of f we can show (for the details see [19] or [21]) that $x(t)$ is defined and limited for all $t \simeq t_0$. Hence the function

$$X(\theta, T) = \frac{x(t_0 + \alpha T + \theta) - x(t_0 + \theta)}{\alpha}, \quad \theta \in [-r, 0], \quad T \in [0, 1],$$

is well defined. In the variable $X(\cdot, T)$ system (20.4.8) becomes

$$\frac{\partial X}{\partial T}(0, T) = f(s + ST, x_{t_0} + \alpha X(\cdot, T)).$$

Using assumptions (H1) and (H4) together with (20.4.10), we obtain that for all $T \in [0, 1]$, we have

$$X(0, T) \simeq \int_0^T f\left(s + Sr, x_{t_0}\right) dr = \frac{1}{S} \int_s^{s+TS} f\left(r, x_{t_0}\right) dr \simeq TF\left(x_{t_0}\right).$$

Define the successive instant of observation of the stroboscopic method t_1 by $t_1 = t_0 + \alpha$. Then we have

$$\frac{x(t_1) - x(t_0)}{t_1 - t_0} = X(0, 1) \simeq F\left(x_{t_0}\right)$$

Since $t_1 - t_0 = \alpha > \varepsilon$ and $x(t) - x(t_0) = \alpha X(0, T) \simeq 0$ for all $t \in [t_0, t_1]$, we have proved that the function x satisfies the Stroboscopic Property with respect to F. By the Stroboscopic Lemma for FDEs, for any $t \in [0, L]$, $x(t)$ is defined and satisfies $x(t) \simeq y(t)$. □

References

[1] V.I. Arnold (ed.), *Dynamical Systems V*, Encyclopedia of Mathematical Sciences, Vol. 5, Springer-Verlag, 1994.

[2] H. Barreau and J. Harthong (éditeurs), *La mathématique Non Standard*, Editions du CNRS, Paris, 1989.

[3] E. Benoît (Ed.), *Dynamic Bifurcations*, Proceedings, Luminy 1990, Springer-Verlag, Berlin, 1991.

[4] I.P. van den Berg, *Nonstandard Asymptotic Analysis*, Lectures Notes in Math. 1249, Springer-Verlag, 1987.

[5] N.N. Bogolyubov and Yu.A. Mitropolski, *Asymptotic Methods in the Theory of Nonlinear Oscillations*, Gordon and Breach, New York, 1961.

[6] J.W. Dauben, *Abraham Robinson, The Creation of Nonstandard Analysis, A Personal and Mathematical Odyssey*, Princeton University Press, Princeton, New Jersey, 1995.

[7] F. Diener and M. Diener (eds.), *Nonstandard Analysis in Practice*, Universitext, Springer-Verlag, 1995.

[8] F. Diener and G. Reeb, *Analyse non standard*, Hermann, 1989.

[9] M. Diener and C. Lobry (éditeurs), *Analyse non standard et représentation du réel*, OPU (Alger), CNRS (Paris), 1985.

[10] M. Diener and G. Wallet (éditeurs), *Mathématiques finitaires et analyse non standard*, Publication mathématique de l'Université de Paris 7, Vol. 31-1, 31–2, 1989.

[11] J.L. Callot and T. Sari, Stroboscopie et moyennisation dans les systèmes d'équations différentielles à solutions rapidement oscillantes, in *Mathematical Tools and Models for Control, Systems Analysis and Signal Processing*, vol. 3, CNRS Paris, 1983.

[12] A. Fruchard and A. Troesch (éditeurs), *Colloque Trajectorien à la mémoire de G. Reeb et J.L. Callot*, Strasbourg-Obernai, 12-16 juin 1995, Prépublication de l'IRMA, Strasbourg, 1995.

[13] J.K. Hale, "Averaging methods for differential equations with retarded arguments and a small parameter", J. Differential Equations, **2** (1966) 57–73.

[14] J.K. Hale and S.M. Verduyn Lunel, "Averaging in infinite dimensions", J. Integral Equations Appl., **2** (1990) 463–494.

[15] J.K. Hale and S.M. Verduyn Lunel, *Introduction to Functional Differential Equations*, Applied Mathematical Sciences 99, Springer-Verlag, New York, 1993.

[16] L.D. Klugler, Nonstandard Analysis of almost periodic functions, in *Applications of Model Theory to Algebra, Analysis and Probability*, W.A.J. Luxemburg ed., Holt, Rinehart and Winston, 1969.

[17] M. Lakrib, "The method of averaging and functional differential equations with delay", Int. J. Math. Sci., **26** (2001) 497–511.

[18] M. Lakrib, "On the averaging method for differential equations with delay", Electron. J. Differential Equations, **65** (2002) 1–16.

[19] M. Lakrib, *Stroboscopie et moyennisation dans les équations différentielles fonctionnelles à retard*, Thèse de Doctorat en Mathématiques de l'Université de Haute Alsace, Mulhouse, 2004.

[20] M. Lakrib and T. Sari, "Averaging results for functional differential equations", Sibirsk. Mat. Zh., **45** (2004) 375–386; translation in Siberian Math. J., **45** (2004) 311–320.

[21] M. Lakrib and T. Sari, Averaging Theorems for Ordinary Differential Equations and Retarded Functional Differential Equations. `http://www.math.uha.fr/ps/200501lakrib.pdf`

[22] B. Lehman and S.P. Weibel, "Fundamental theorems of averaging for functional differential equations", J. Differential Equations, **152** (1999) 160–190.

[23] C. Lobry, *Et pourtant... ils ne remplissent pas* $\mathbb{N}$*!*, Aleas Editeur, Lyon, 1989.

[24] C. Lobry, T. Sari and S. Touhami, "On Tykhonov's theorem for convergence of solutions of slow and fast systems", Electron. J. Differential Equations, **19** (1998) 1–22.

[25] R. Lutz and M. Goze, *Nonstandard Analysis: a practical guide with applications*, Lectures Notes in Math. 881, Springer-Verlag, 1982.

[26] E. Nelson, "Internal Set Theory", Bull. Amer. Math. Soc., **83** (1977) 1165–1198.

[27] G. Reeb, Équations différentielles et analyse non classique (d'après J. L. Callot), in *Proceedings of the 4th International Colloquium on Differential Geometry* (1978), Publicaciones de la Universidad de Santiago de Compostela, 1979.

[28] A. Robinson, "Compactification of Groups and Rings and Nonstandard Analysis", Journ. Symbolic Logic, **34** (1969) 576–588.

[29] A. Robinson, *Nonstandard Analysis*, American Elsevier, New York, 1974.

[30] J.-M. Salanskis and H. Sinaceur (eds.), *Le Labyrinthe du Continu*, Colloque de Cerisy, Springer-Verlag, Paris, 1992.

[31] J.A. Sanders and F. Verhulst, *Averaging Methods in Nonlinear Dynamical Systems*, Applied Mathematical Sciences 59, Springer-Verlag, New York, 1985.

[32] T. SARI, "Sur la théorie asymptotique des oscillations non stationnaires", Astérisque **109–110** (1983) 141–158.

[33] T. SARI, Fonctions presque périodiques, in *Actes de l'école d'été Analyse non standard et representation du réel*, Oran-Les Andalouses 1984, OPU Alger – CNRS Paris, 1985.

[34] T. SARI, General Topology, in *Nonstandard Analysis in Practice*, F. Diener and M. Diener (Eds.), Universitext, Springer–Verlag, 1995.

[35] T. SARI, Petite histoire de la stroboscopie, in *Colloque Trajectorien à la Mémoire de J. L. Callot et G. Reeb*, Strasbourg-Obernai 1995, Publication IRMA, Univ. Strasbourg (1995), 5–15.

[36] T. SARI, Stroboscopy and Averaging, in *Colloque Trajectorien à la Mémoire de J.L. Callot et G. Reeb*, Strasbourg-Obernai 1995, Publication IRMA, Univ. Strasbourg, 1995.

[37] T. SARI, Nonstandard Perturbation Theory of Differential Equations, Edinburgh, invited talk in *International Congres in Nonstandard Analysis and its Applications*, ICMS, Edinburgh, 1996.
`http://www.math.uha.fr/sari/papers/icms1996.pdf`

[38] T. SARI, Averaging in Hamiltonian systems with slowly varying parameters, in *Developments in Mathematical and Experimental Physics, Vol. C, Hydrodynamics and Dynamical Systems*, Proceedings of the First Mexican Meeting on Mathematical and Experimental Physics, El Colegio Nacional, Mexico City, September 10-14, 2001, Ed. A. Macias, F. Uribe and E. Diaz, Kluwer Academic/Plenum Publishers, 2003.

[39] T. SARI and K. YADI, "On Pontryagin-Rodygin's theorem for convergence of solutions of slow and fast systems", Electron. J. Differential Equations, **139** (2004) 1–17.

[40] I. STEWART, *The Problems of Mathematics*, Oxford University Press, 1987.

[41] K.D. STROYAN and W.A.J. LUXEMBURG, *Introduction to the theory of infinitesimals*, Academic Press, 1976.

[42] IIIe *rencontre de Géométrie du Schnepfenried, Feuilletages, Géométrie symplectique et de contact, Analyse non standard et applications*, Vol. 2, 10-15 mai 1982, Astérisque 109-110, Société Mathématique de France, 1983.

21

Path-space measure for stochastic differential equation with a coefficient of polynomial growth

Toru Nakamura*

Abstract

A σ-additive measure over a space of paths is constructed to give the solution to the Fokker-Planck equation associated with a stochastic differential equation with coefficient function of polynomial growth by making use of nonstandard analysis.

21.1 Heuristic arguments and definitions

Consider a stochastic differential equation,

$$dx(t) = f\big(x(t)\big)\,dt + db(t), \tag{21.1.1}$$

where $f(x)$ is a real-valued function and $b(t)$ a Brownian motion with variance $2Dt$ for a time interval t. We wish to construct a measure over a space of paths for (21.1.1). To my knowledge, the coefficient function has been assumed to have at most linear growth, that is $|f(x)| \le \text{const}\cdot|x|$ for sufficiently large x, otherwise some paths explode to infinity in finite times. As an example, let $f(x) = |x|^{1+\delta}$ ($\delta > 0$) and define *explosion time* $\mathfrak{e}$ for each continuous path $x(t)$ by $\lim_{t\to\mathfrak{e}-0} x(t) = \pm\infty$. Then, it is proved that $P(\mathfrak{e} = \infty) < 1$ and more strongly $P(\mathfrak{e} = \infty) = 0$. For general case, see Feller's test for explosion in [1].

Despite the explosion, we shall consider $f(x)$ of polynomial growth of an arbitrary order and define a measure over a space of paths. We use nonstandard analysis because it has a very convenient theory, Loeb measure theory [2, 3], which enables us to construct a standard σ-additive measure in a simple way.

*Department of Mathematics, Sundai Preparatory School, Kanda-Surugadai, Chiyoda-ku, Tokyo 101-0062, Japan.

Let us interpret (21.1.1) as a law for a particle momentum $x = x(t)$ at time t with the force $f\big(x(t)\big)$ for drift and the random force $db(t)/dt$ acting on the particle. By a time $t > 0$ some particles may disappear to infinity along the exploding paths, but others still exist with finite momentum so that they should make up a "probability" density of a particle momentum. We wish to use the word "probability" though its total value may be less than 1 because of the disappearance of particles to infinity.

Especially when $f\big(x(t)\big)$ is a repulsive force, that is, its sign is the same as that of $x(t)$, a particle can get larger momentum compared to the case where $f\big(x(t)\big)$ is absent. However, once it gets very large momentum, it can hardly come back to the former one because the drift force $f\big(x(t)\big)$ acts on the particle so as to increase its momentum. Thus, the particles which have gone to infinity by a time $t > 0$ could not contribute to the probability density at t. This consideration suggests that we can introduce a cutoff at an infinite number in momentum space if it is necessary in order to define a measure over a space of paths so that the probability density should be constructed by a path integral with respect to the measure.

Eq. (21.1.1) gives the forward Fokker-Planck equation for the probability density $U(t,x)$ of a particle momentum x at time t,

$$\frac{\partial}{\partial t}U(t,x) = D\frac{\partial^2}{\partial x^2}U(t,x) - \frac{\partial}{\partial x}\{f(x)U(t,x)\}. \tag{21.1.2}$$

We assume that the drift coefficient $f(x)$ and the initial function $U(0,x)$ satisfy the following conditions:

(A1) For some natural number $n \in \mathbb{N}$, $|f(x)| \leq \text{const}\cdot|x|^n$ for sufficiently large x.

(A2) $f(x) \in C^2(\mathbb{R})$.

(A3) $f(x)^2/(4D) + f'(x)/2$ is bounded from below. We denote the bound by

$$c = \min\{f(x)^2/(4D) + f'(x)/2 \mid x \in \mathbb{R}\}.$$

(A4) $U(0,x) \in C^2(\mathbb{R})$ and its support is a bounded set.

Rewrite (21.1.2) into a difference equation with infinitesimal time-spacing ε and momentum-spacing $\delta = \sqrt{2D\varepsilon}$ using a forward difference quotient

$$\frac{\partial}{\partial t}U(t,x) \Rightarrow \frac{1}{\varepsilon}\big\{U(t+\varepsilon,x) - U(t,x)\big\}$$

for time-derivative, and central ones

$$\frac{\partial}{\partial x}U(t,x) \Rightarrow \frac{1}{2\delta}\big\{U(t,x+\delta) - U(t,x-\delta)\big\}$$

and

$$\frac{\partial^2}{\partial x^2}U(t,x) \Rightarrow \frac{1}{\delta^2}\{U(t,x+\delta)+U(t,x-\delta)-2U(t,x)\}$$

for space-derivative. The result is

$$\begin{aligned} U(t+\varepsilon,x) &= \frac{1}{2}\Big\{1+f(x-\delta)\sqrt{\varepsilon/(2D)}\Big\}U(t,x-\delta) \\ &+\frac{1}{2}\Big\{1+f(x+\delta)\big(-\sqrt{\varepsilon/(2D)}\big)\Big\}U(t,x+\delta), \end{aligned} \tag{21.1.3}$$

which indicates that the coefficients $\frac{1}{2}\{1+f(x-\delta)\sqrt{\varepsilon/(2D)}\}$ should be assigned to the infinitesimal line-segment with end points $\big(t,\,x(t)\big)=\big(t,\,x(t+\varepsilon)-\delta\big)$ and $\big(t+\varepsilon,\,x(t+\varepsilon)\big)$ of a $*$-polygonal path $x(s)$, and $\frac{1}{2}\{1+f(x+\delta)(-\sqrt{\varepsilon/(2D)})\}$ to the segment with end points $\big(t,\,x(t)\big)=\big(t,\,x(t+\varepsilon)+\delta\big)$ and $\big(t+\varepsilon,\,x(t+\varepsilon)\big)$.

Taking into account the approximation

$$\begin{aligned} \log\big[1+f\big(x(t)\big)\big(\pm\sqrt{\varepsilon/(2D)}\big)\big] &\simeq \frac{1}{2D}f\big(x(t)\big)\big(\pm\sqrt{2D\varepsilon}\big)-\frac{1}{4D}f\big(x(t)\big)^2\varepsilon \\ &= \frac{1}{2D}f\big(x(t)\big)\{x(t+\varepsilon)-x(t)\}-\frac{1}{4D}f\big(x(t)\big)^2\varepsilon, \end{aligned}$$

let us interpret the coefficients as

$$\begin{aligned} &\frac{1}{2}\Big\{1+f\big(x(t)\big)\big(\pm\sqrt{\varepsilon/(2D)}\big)\Big\} \\ &\qquad\simeq \frac{1}{2}\exp\left[\int_t^{t+\varepsilon}\frac{1}{2D}f\big(x(s)\big)\,db(s)-\int_t^{t+\varepsilon}\frac{1}{4D}f\big(x(s)\big)^2\,ds\right]. \end{aligned}$$

The first integral on the right-hand side is the Ito-integral. Thus, we define $*$-path ω, $*$-measure μ for each ω, and $\mathcal{U}(t,x)$ by

Definition 1

(1) Let ν be $[t/\varepsilon]$ with Gauss' parenthesis. For each internal function $\alpha:\{0,1,\cdots,\nu-1\}\to\{-1,1\}$ and $y\in\mathbb{R}$, define x_k by $x_k=y+\sum_{i=0}^{k-1}\alpha(i)\delta$, and ω by the $*$-polygonal path with vertices $(0,y)$, (ε,x_1), $\cdots$, $(\nu\varepsilon,x_\nu)$.

(2) Define $*$-measure μ by

$$\mu(\omega)=\frac{1}{2^\nu}\exp\left[\int_0^t\frac{1}{2D}f\big(\omega(s)\big)\,db(s)-\int_0^t\frac{1}{4D}f\big(\omega(s)\big)^2\,ds\right] \tag{21.1.4}$$

with the first integral in the exponent being the Ito integral, and

$$\mathcal{U}(t,x)=\sum_\omega U\big(0,\omega(0)\big)\mu(\omega) \tag{21.1.5}$$

where the sum is taken over all ω satisfying $\omega(\nu\varepsilon)=x$.

We note that the $*$-measure (21.1.4) corresponds to the Girsanov formula [4].

21.2 Bounds for the $*$-measure and the $*$-Green function

Making use of the Ito formula

$$\int_0^t \frac{1}{2D} f\big(\omega(s)\big)\, db(s) = \frac{1}{2D}\Big\{F\big(\omega(t)\big) - F\big(\omega(0)\big)\Big\} - \frac{1}{2}\int_0^t f'\big(\omega(s)\big)\, ds$$

where $F'(x) = f(x)$, we obtain a bound for μ as

$$\begin{aligned} \mu(\omega) &= \frac{1}{2^\nu} \exp\Big[\frac{1}{2D}\Big\{F\big(\omega(t)\big) - F\big(\omega(0)\big)\Big\} - \int_0^t \Big\{\frac{1}{4D} f\big(\omega(s)\big)^2 + \frac{1}{2} f'\big(\omega(s)\big)\Big\} ds\Big] \\ &\le \frac{1}{2^\nu} \exp\Big[\frac{1}{2D}\Big\{F\big(\omega(t)\big) - F\big(\omega(0)\big)\Big\} - ct\Big]. \end{aligned} \tag{21.2.1}$$

The constant c in the last line was given in the assumption (A3).

Let us fix the end points of ω at finite numbers y and x, and consider a space of $*$-paths,

$$P(t, x : 0, y) = \big\{\,\omega \mid \omega(0) = y \quad \text{and} \quad \omega\big(\varepsilon[t/\varepsilon]\big) = x\,\big\},$$

and define the $*$-Green function for the interval $[0,\, t]$ by

$$\mathcal{G}(t, x : 0, y) = \frac{1}{2\delta} \sum_{\omega \in P(t,x:0,y)} \mu(\omega). \tag{21.2.2}$$

Then, $\mathcal{U}(t, x)$ defined in (21.1.5) is written as

$$\mathcal{U}(t, x) = \sum_y U(0, y) \mathcal{G}(t, x : 0, y) 2\delta. \tag{21.2.3}$$

The infinitesimal spacing corresponding to dy is not δ but 2δ in (21.2.3) because only the paths that start every other point y could reach the end point x. Since

$$\sum_{\omega \in P(t,x:0,y)} \frac{1}{2^\nu} = \frac{2\delta}{(4\pi D t)^{1/2}} \exp\Big[-\frac{(x-y)^2}{4Dt}\Big]\Big(1 + \mathcal{O}(\varepsilon^{1/2})\Big), \tag{21.2.4}$$

the $*$-Green function is bounded as

$$\begin{aligned} \mathcal{G}(t, x : 0, y) \le \exp\Big[\frac{F(x)}{2D} - \frac{F(y)}{2D} - ct\Big] \\ \times \frac{1}{(4\pi D t)^{1/2}} \exp\Big[-\frac{(x-y)^2}{4Dt}\Big]\Big(1 + \mathcal{O}(\varepsilon^{1/2})\Big), \end{aligned} \tag{21.2.5}$$

and hence

$$\begin{aligned}\big|\mathcal{U}(t,x)\big| \leq \sum_y \big|U(0,y)\big| \exp\Big[\frac{F(x)}{2D} - \frac{F(y)}{2D} - ct\Big] \\ \times \frac{1}{(4\pi Dt)^{1/2}} \exp\Big[-\frac{(x-y)^2}{4Dt}\Big]\Big(1+\mathcal{O}(\varepsilon^{1/2})\Big)2\delta.\end{aligned} \tag{21.2.6}$$

Since the support of $U(0,y)$ is assumed in (A4) to be a bounded set, the right-hand side of (21.2.6) is near-standard.

To define a standard function U as the standard part of $\mathcal{U}$, we introduce two time scales of different orders, one for the Brownian motion, ε, and the other, τ, for the changes in the drift term in (21.1.1); we choose $\varepsilon = \mathcal{O}(\tau^3)$, for example. The spacing ε is finer and stands for the time-spacing of $*$-random walks, and τ is long enough to cover many steps of the $*$-random walks, yet short enough for the change in the drift term to be small. Define a standard function $U(t,x)$ as the standard part of the value of $\mathcal{U}$ at the coarse-grained lattice point of time $\underline{t}$:

$$U(t,x) = \mathrm{st}\,\mathcal{U}(\underline{t},x) \quad \text{where} \quad \underline{t} = \tau\big[t/\tau\big], \tag{21.2.7}$$

which we expect to be the solution to the Fokker-Planck equation (21.1.2).

21.3 Solution to the Fokker-Planck equation

In order to prove that $U(t,x)$ in (21.2.7) is the solution to (21.1.2), or more concretely to estimate I_1 in (21.3.7) below, we should truncate the $*$-paths at an infinite number A as

$$P_A(\underline{t},x:0,y) = \big\{\omega \in P(\underline{t},x:0,y) \mid \forall s \in \big[0,\,\underline{t}\big] \ \ |\omega(s)| < A\big\}.$$

The magnitude of A will be determined later in (21.3.5). Then the corresponding $*$-Green function $\mathcal{G}_A$ and $\mathcal{U}_A$ are defined by

$$\mathcal{G}_A(\underline{t},x:0,y) = \frac{1}{2\delta} \sum_{\omega\in P_A(\underline{t},x:0,y)} \mu(\omega) \tag{21.3.1}$$

and

$$\mathcal{U}_A(\underline{t},x) = \sum_y U(0,y)\mathcal{G}_A(\underline{t},x:0,y)2\delta. \tag{21.3.2}$$

Let us first calculate the difference between (21.2.2) and (21.3.1). Consider a $*$-path $\omega \in P(\underline{t},x:0,y)\setminus P_A(\underline{t},x:0,y)$ and put $\mathfrak{t}(\omega) = \min\{s \mid |\omega(s)| = A\}$. Define a $*$-path ω' by turning upside down the section of the path $\omega(s)$ for the interval $0 \leq s \leq \mathfrak{t}(\omega)$ so that ω' should start $2A - y$ or $-2A - y$ at time $s = 0$. The section of $\omega'(s)$ for $\mathfrak{t}(\omega) \leq s \leq t$ is that of $\omega(s)$ for the same interval. In this way, we can define a one-to-one correspondence from $\omega \in P(\underline{t}, x : 0, y) \setminus P_A(\underline{t}, x : 0, y)$ to $\omega' \in P(\underline{t}, x : 0, 2A-y) \cup P(\underline{t}, x : 0, -2A-y)$. Then by (21.2.4),

$$\sum_{\omega \in P(\underline{t},x:0,y)\setminus P_A(\underline{t},x:0,y)} \frac{1}{2^\nu} = \frac{2\delta}{(4\pi Dt)^{1/2}} \Big\{ \exp\Big[-\frac{(2A-y-x)^2}{4Dt}\Big] + \\ + \exp\Big[-\frac{(-2A-y-x)^2}{4Dt}\Big] \Big\} \times \Big(1 + \mathcal{O}(\varepsilon^{1/2})\Big),$$

and hence

$$\big|\mathcal{G}(\underline{t}, x : 0, y) - \mathcal{G}_A(\underline{t}, x : 0, y)\big| \leq \exp\Big[\frac{F(x)}{2D} - \frac{F(y)}{2D} - ct\Big] \frac{1}{(4\pi Dt)^{1/2}} \times \\ \Big\{ \exp\Big[-\frac{(2A-y-x)^2}{4Dt}\Big] + \exp\Big[-\frac{(-2A-y-x)^2}{4Dt}\Big] \Big\} \times \Big(1 + (\varepsilon^{1/2})\Big), \tag{21.3.3}$$

which implies

$$\big|\mathcal{U}(\underline{t}, x) - \mathcal{U}_A(\underline{t}, x)\big| = \mathcal{O}\Big(\exp\Big[-\frac{A^2}{Dt}\Big]\Big) \tag{21.3.4}$$

for any finite x.

Now, we are ready to prove that $U(t, x)$ is the solution to (21.1.2). We wish to evaluate the standard part of

$$\frac{1}{\sigma}\Big\{\mathcal{U}(t+\sigma, x) - \mathcal{U}(t, x)\Big\}$$

for an infinitesimal $\sigma = k\tau$ ($k \in {}^*\mathbb{N}$) given arbitrarily. Choose the truncation parameter A as

$$A = (D/\beta)^{1/2} \big|\log \beta\sigma\big| \tag{21.3.5}$$

where $\beta > 0$, a standard constant, is just introduced to make the argument of logarithm dimensionless. Then by (21.3.4),

$$\mathcal{U}(\underline{t}, x) - \mathcal{U}_A(\underline{t}, x) = o(\sigma^n) \quad \text{and} \quad \mathcal{U}(\underline{t}+\sigma, x) - \mathcal{U}_A(\underline{t}+\sigma, x) = o(\sigma^n) \tag{21.3.6}$$

for any standard natural number n, meaning that $\mathcal{U}_A(\underline{t}, x)$ can be identified with $\mathcal{U}(\underline{t}, x)$ up to negligible error. Therefore we shall hereafter deal with the truncated $\mathcal{U}_A$ instead of $\mathcal{U}$.

Then the difference quotient we should calculate is

$$\frac{1}{\sigma}\Big\{\mathcal{U}_A(\underline{t}+\sigma, x) - \mathcal{U}_A(\underline{t}, x)\Big\} = \frac{1}{\sigma}(I_1 + I_2) + o(1)$$

where

$$I_1 = \sum_{\xi} \mathcal{U}_A(\underline{t}, x+\xi) \times \Big\{ \mathcal{G}_A(\underline{t}+\sigma, x : \underline{t}, x+\xi) - \frac{1}{(4\pi D\sigma)^{1/2}} \exp\Big[-\frac{\xi^2}{4D\sigma}\Big]\Big\} 2\delta, \tag{21.3.7}$$

$$I_2 = \sum_{\xi} \Big\{ \mathcal{U}_A(\underline{t}, x+\xi) - \mathcal{U}_A(\underline{t}, x)\Big\} \frac{1}{(4\pi D\sigma)^{1/2}} \exp\Big[-\frac{\xi^2}{4D\sigma}\Big] 2\delta. \tag{21.3.8}$$

In order to calculate the sum in I_1, we wish to expand the summand as power series in ξ except for the exponential function, which is possible if ξ is sufficiently small. Since

$$\mathcal{G}_A(\underline{t}+\sigma, x : \underline{t}, x+\xi) = \frac{1}{2\delta} \exp\Big[\frac{1}{2D}\{F(x) - F(x+\xi)\}\Big] \times \sum_{\omega \in P_A(\underline{t}+\sigma, x : \underline{t}, x+\xi)} \frac{1}{2^{\nu'}} \exp\Big[-\int_{\underline{t}}^{t+\sigma} \Big\{\frac{1}{4D} f\big(\omega(s)\big)^2 + \frac{1}{2} f'\big(\omega(s)\big)\Big\} ds\Big] \tag{21.3.9}$$

satisfies

$$\mathcal{G}_A(\underline{t}+\sigma, x : \underline{t}, x+\xi) \leq \exp\Big[\frac{1}{2D}\{F(x) - F(x+\xi)\} - c\sigma\Big] \times \frac{1}{(4\pi D\sigma)^{1/2}} \exp\Big[-\frac{\xi^2}{4D\sigma}\Big], \tag{21.3.10}$$

the summand in (21.3.7) contains $\exp\big[-\frac{\xi^2}{4D\sigma}\big]$ as a factor which enables us to restrict ξ to be sufficiently small, $|\xi| = \mathcal{O}(\sigma^{1/2})$, in rough estimation. In reality, it is restricted to

$$|\xi| < (D/\beta)^{\frac{1}{2}} (\beta\sigma)^{\frac{1}{2}-a}$$

for some small a such as $1/10$ as shown in the following. Note that

$$\exp\Big[\frac{1}{2D}\{F(x) - F(x+\xi) - ct\}\Big] = \mathcal{O}\big(e^{|\log \beta\sigma|^m}\big) \tag{21.3.11}$$

for some $m \in \mathbb{N}$ by the truncation at $A = \mathcal{O}(|\log \beta\sigma|)$ and the assumption (A1). Then by (21.3.10),

$$\Big| \mathcal{U}_A(\underline{t}, x+\xi)\Big\{\mathcal{G}_A(\underline{t}+\sigma, x : \underline{t}, x+\xi) - \frac{1}{(4\pi D\sigma)^{1/2}} \exp\Big[-\frac{\xi^2}{4D\sigma}\Big]\Big\}\Big| \leq \text{const}\cdot e^{|\log \beta\sigma|^m} \frac{1}{(4\pi D\sigma)^{1/2}} \exp\Big[-\frac{\xi^2}{4D\sigma}\Big].$$

If $|\xi| \geq (D/\beta)^{\frac{1}{2}}(\beta\sigma)^{\frac{1}{2}-a}$, the infinitely large number $e^{|\log\beta\sigma|^m}$ in the last line can be controlled by the half of the factor $e^{-\xi^2/(4D\sigma)}$. In fact,

$$\exp\Big[-\frac{\xi^2}{8D\sigma}\Big] \leq \exp\Big[-\frac{1}{8(\beta\sigma)^{2a}}\Big] = o\big(e^{-|\log\beta\sigma|^m}\big).$$

Hence, the sum $\sum_\xi$ in (21.3.7) for such ξ is estimated as

$$\text{const}\cdot \int\limits_{|\xi|\geq(D/\beta)^{1/2}(\beta\sigma)^{1/2-a}} \frac{1}{(4\pi D\sigma)^{1/2}}\exp\Big[-\frac{\xi^2}{8D\sigma}\Big]d\xi = \mathcal{O}\Big(\sigma^a e^{-\frac{1}{8(\beta\sigma)^{2a}}}\Big) = o(\sigma^n)$$

for any $n\in\mathbb{N}$, so that it can be neglected. Now, we have only to consider ξ of infinitesimal order equal or less than $\sigma^{\frac{1}{2}-a}$, so that we can expand

$$F(x) - F(x+\xi) = -\xi f(x) - \frac{\xi^2}{2!}f'(x) + \mathcal{O}(\sigma^{\frac{3}{2}-3a}). \tag{21.3.12}$$

Next, we wish to replace the integral in (21.3.9) by

$$\frac{1}{4D}\int_t^{t+\sigma} f\big(\omega(s)\big)^2 ds + \frac{1}{2}\int_t^{t+\sigma} f'\big(\omega(s)\big)\,ds \simeq \frac{\sigma}{4D}f(x)^2 + \frac{\sigma}{2}f'(x). \tag{21.3.13}$$

It is justifiable if paths going very far at some time $s\in[\underline{t},\,\underline{t}+\sigma]$ can be neglected. Let us take an infinitesimal

$$A' = (D/\beta)^{\frac{1}{2}}(\beta\sigma)^{\frac{1}{2}-a'}$$

where a' is small but $a' > a$, for example, $a' = 2/10$ if we take $a = 1/10$. Then, the sum over ω which satisfies $|\omega(s)-x| \geq A'$ for some $s\in[t,\,t+\sigma]$ is bounded by

$$\begin{aligned}\frac{\text{const}}{(4\pi D\sigma)^{1/2}}\Big\{\exp\Big[-\frac{(2A'-\xi)^2}{4D\sigma}\Big] + \exp\Big[-\frac{(2A'+\xi)^2}{4D\sigma}\Big]\Big\} &= \\ \mathcal{O}\big(\sigma^{-1/2}e^{-\frac{1}{(\beta\sigma)^{2a}}}\big) &= o(\sigma^n)\end{aligned} \tag{21.3.14}$$

for any $n\in\mathbb{N}$ in the same way as (21.3.3), and hence negligible. Here, we have used the fact $\xi = \mathcal{O}(\sigma^{\frac{1}{2}-a}) = o(A')$ because $a' > a$. Thus, we have only to consider ω satisfying

$$|\omega(s) - x| \leq (D/\beta)^{\frac{1}{2}}(\beta\sigma)^{\frac{1}{2}-a'}$$

for all $s \in [t,\, t+\sigma]$, and hence

$$
\begin{aligned}
&\exp\Big[-\int_{\underline{t}}^{\underline{t}+\sigma}\Big\{\frac{1}{4D}f\big(\omega(s)\big)^2+\frac{1}{2}\int_{\underline{t}}^{\underline{t}+\sigma} f'\big(\omega(s)\big)\Big\}ds\Big] = \\
&= \exp\Big[-\frac{\sigma}{4D}f(x)^2-\frac{\sigma}{2}f'(x)\Big]\times \\
&\quad\times \exp\Big[-\frac{1}{4D}\int_{\underline{t}}^{\underline{t}+\sigma}\big\{f\big(\omega(s)\big)^2-f(x)^2\big\}\,ds \\
&\qquad -\frac{1}{2}\int_{\underline{t}}^{\underline{t}+\sigma}\big\{f'\big(\omega(s)\big)ds - f'(x)\big\}ds\Big] = \\
&= \exp\Big[-\frac{\sigma}{4D}f(x)^2-\frac{\sigma}{2}f'(x)\Big]\Big(1+\mathcal{O}\big(\sigma^{\frac{3}{2}-a'}\big)\Big).
\end{aligned}
\tag{21.3.15}
$$

Putting (21.3.12), (21.3.15) and

$$
\sum_{\omega\in P_A(\underline{t}+\sigma,x:\underline{t},x+\xi)}\frac{1}{2^{\nu'}}=\frac{2\delta}{(4\pi D\sigma)^{1/2}}\exp\Big[-\frac{\xi^2}{4D\sigma}\Big]\Big(1+\mathcal{O}(\sigma^{3/2})\Big)
$$

into (21.3.9), we obtain

$$
\begin{aligned}
&\mathcal{G}_A(\underline{t}+\sigma,x:\underline{t},x+\xi)-\frac{1}{(4\pi D\sigma)^{1/2}}\exp\Big[-\frac{\xi^2}{4D\sigma}\Big] = \\
&=\Big\{-\frac{\xi}{2D}f(x)-\frac{\xi^2+2D\sigma}{4D}f'(x)+\frac{\xi^2-2D\sigma}{8D^2}f(x)^2+o(\sigma)\Big\}\times \\
&\qquad\qquad\times\frac{1}{(4\pi D\sigma)^{1/2}}\exp\Big[-\frac{\xi^2}{4D\sigma}\Big].
\end{aligned}
\tag{21.3.16}
$$

Lastly, we use the expansion

$$
\mathcal{U}_A(\underline{t},x+\xi)=\mathcal{U}_A(\underline{t},x)+\xi D_\delta\mathcal{U}_A(\underline{t},x)+\mathcal{O}(\sigma^{1-2a})
\tag{21.3.17}
$$

with

$$
D_\delta\mathcal{U}_A(\underline{t},x)=\frac{1}{\delta}\Big\{\mathcal{U}_A(\underline{t},x+\delta)-\mathcal{U}_A(\underline{t},x)\Big\}
$$

for I_1. For I_2, we use the expansion taken one step further,

$$
\mathcal{U}_A(\underline{t},x+\xi)=\mathcal{U}_A(\underline{t},x)+\xi D_\delta\mathcal{U}_A(\underline{t},x)+\frac{\xi^2}{2!}D_\delta^2\mathcal{U}_A(\underline{t},x)+o(\sigma)
\tag{21.3.18}
$$

with

$$
D_\delta^2\mathcal{U}_A(\underline{t},x)=\frac{1}{\delta^2}\Big\{\mathcal{U}_A(\underline{t},x+\delta)+\mathcal{U}_A(\underline{t},x-\delta)-2\mathcal{U}_A(\underline{t},x)\Big\},
$$

because (21.3.16) contains a factor of order $\mathcal{O}(\sigma^{\frac{1}{2}-a})$, whereas there is no such factor in I_2. We shall not give the proof of these expansions, for it requires elementary but a little long estimations. Estimations similar to this case can be found in [5, 6, 7].

Putting (21.3.16) and (21.3.17) into (21.3.7) and replacing the sum by integral which is permitted because $\delta = o(\tau^{3/2})$, we obtain

$$\begin{aligned}
I_1 = & \int\limits_{|\xi|\leq(D/\beta)^{1/2}(\beta\sigma)^{1/2-a}} \Big\{-\frac{\xi}{2D}f(x) - \frac{\xi^2+2D\sigma}{4D}f'(x) + \frac{\xi^2-2D\sigma}{8D^2}f(x)^2 + o(\sigma)\Big\} \\
& \times \Big\{\mathcal{U}_A(\underline{t},x) + \xi D_\delta \mathcal{U}_A(\underline{t},x)\Big\}\frac{1}{(4\pi D\sigma)^{1/2}}\exp\Big[-\frac{\xi^2}{4D\sigma}\Big]d\xi \\
= & \int\limits_{|\xi|\leq(D/\beta)^{1/2}(\beta\sigma)^{1/2-a}} \Big\{\Big(-\frac{\xi}{2D}f(x) - \frac{\xi^2+2D\sigma}{4D}f'(x) + \frac{\xi^2-2D\sigma}{8D^2}f(x)^2\Big)\mathcal{U}_A(\underline{t},x) \\
& -\frac{\xi^2}{2D}f(x)D_\delta\mathcal{U}_A(\underline{t},x) + o(\sigma)\Big\}\frac{1}{(4\pi D\sigma)^{1/2}}\exp\Big[-\frac{\xi^2}{4D\sigma}\Big]d\xi.
\end{aligned}$$

Since the domain of integration can be extended to ${}^*\mathbb{R}$ by the factor $\frac{1}{(4\pi D\sigma)^{1/2}}\exp\big[-\frac{\xi^2}{4D\sigma}\big]$,

$$\begin{aligned}
I_1 = & \int\limits_{{}^*\mathbb{R}} \Big\{\Big(-\frac{\xi}{2D}f(x) - \frac{\xi^2+2D\sigma}{4D}f'(x) + \frac{\xi^2-2D\sigma}{8D^2}f(x)^2\Big)\mathcal{U}_A(\underline{t},x) \\
& -\frac{\xi^2}{2D}f(x)D_\delta\mathcal{U}_A(\underline{t},x) + o(\sigma)\Big\}\frac{1}{(4\pi D\sigma)^{1/2}}\exp\Big[-\frac{\xi^2}{4D\sigma}\Big]d\xi \\
= & -\sigma\Big\{f'(x)\mathcal{U}_A(\underline{t},x) + f(x)D_\delta\mathcal{U}_A(\underline{t},x)\Big\} + o(\sigma).
\end{aligned} \tag{21.3.19}$$

Similarly, putting (21.3.18) into (21.3.8), we obtain

$$I_2 = D\sigma D_\delta^2\mathcal{U}_A(\underline{t},x) + o(\sigma). \tag{21.3.20}$$

Therefore,

$$\begin{aligned}
& \frac{1}{\sigma}\Big\{\mathcal{U}_A(\underline{t}+\sigma,x) - \mathcal{U}_A(\underline{t},x)\Big\} = \\
& DD_\delta^2\mathcal{U}_A(\underline{t},x) - \Big\{f'(x)\mathcal{U}_A(\underline{t},x) + f(x)D_\delta\mathcal{U}_A(\underline{t},x)\Big\} + o(1)
\end{aligned} \tag{21.3.21}$$

holds for any infinitesimal $\sigma = k\tau$, which means that $U(t,x) = \mathrm{st}\,\mathcal{U}_A(\underline{t},x)$ is differentiable with respect to t and its t-derivative is the standard part of the right-hand side of (21.3.21). Though we omit the proof, it is proved that $U(t,x)$ is twice differentiable with respect to x and their values are

$$\frac{\partial}{\partial x}U(t,x) = \mathrm{st}\, D_\delta \mathcal{U}_A(\underline{t},x) \quad \text{and} \quad \frac{\partial^2}{\partial x^2}U(t,x) = \mathrm{st}\, D_\delta^2 \mathcal{U}_A(\underline{t},x).$$

Taking the standard part of the both-hand sides of (21.3.21), we obtain

$$\frac{\partial}{\partial t}U(t,x) = D\frac{\partial^2}{\partial x^2}U(t,x) - \frac{\partial}{\partial x}\{f(x)U(t,x)\}, \tag{21.3.22}$$

namely $U(t,x)$ is the solution to the Fokker-Planck equation (21.1.2).

Let us finally note that, by Loeb measure theory, a standard σ-additive measure over a space of paths can be derived from the $*$-measure μ we have constructed in this paper so far.

References

[1] H.P. McKean, Jr., *Stochastic Integrals*, Academic Press, New York and London, 1969, (§3.6).

[2] P. A. Loeb, "Conversion from nonstandard to standard measure spaces and applications in probability theory", Trans. Amer. Math. Soc., **211** (1975) 113–122.

[3] K. D. Stroyan and J. M. Bayod, *Foundations of infinitesimal stochastic analysis*, Studies in Logic and the Foundations of Mathematics, North-Holland, Amsterdam, 1986.

[4] I. V. Gigrsanov, "On transforming a certain class of stochastic processes by absolutely continuous substitution of measures", Theor. Prob. Appl., **5** (1960) 285–301.

[5] T. Nakamura, "A nonstandard representation of Feynman's path integrals", J. Math. Phys., **32** (1991) 457–463.

[6] T. Nakamura, "Path space measures for Dirac and Schrödinger equations: Nonstandard analytical approach", J. Math. Phys., **38** (1997) 4052–4072.

[7] T. Nakamura, "Path space measure for the 3+1-dimensional Dirac equation in momentum space", J. Math. Phys., **41** (2000) 5209–5222.

22

Optimal control for Navier-Stokes equations

Nigel J. Cutland* and Katarzyna Grzesiak**

Abstract

We survey recent results on existence of optimal controls for stochastic Navier-Stokes equations in 2 and 3 dimensions using Loeb space methods.

22.1 Introduction

In this paper we give a brief survey of recent results[1] concerning the existence of optimal controls for the stochastic Navier-Stokes equations (NSE) in a bounded domain D in 2 and 3 space dimensions; that is, $D \subset \mathbb{R}^d$ with $d = 2$ or 3. The controlled equations in their most general form are as follows (see the next section for details):

$$\begin{aligned} u(t) = u_0 + &\int_0^t \{-\nu A u(s) - B(u(s)) + f(s, u(s), \theta(s, u))\}\, ds \\ &+ \int_0^t g(s, u(s))\, dw(s) \end{aligned} \tag{22.1.1}$$

Here the evolving velocity field $u = u(t, \omega)$ is a stochastic process with values in the Hilbert space $\mathbf{H} \subseteq \mathbf{L}^d(D)$ of divergence free functions with domain D; this gives the (random) velocity $u(t, x, \omega) \in \mathbb{R}^d$ of the fluid at any time t and point $x \in D$. The most general kind of control θ that we consider acts through the external forcing term f, and takes the form $\theta : [0, T] \times \mathcal{H} \to M$ where $\mathcal{H}$ is the space of paths in $\mathbf{H}$ and the control space M is a compact metric space.

*Mathematics Department, University of York, UK.
nc507@york.ac.uk

**Department of Finance, Wyzsza Szkola Biznesu, National-Louis University, Nowy Sacz, Poland.
jermakowicz@wsb-nlu.edu.pl

[1] The results reported here are developed from the second author's PhD thesis [16] written under the supervision of the first author. Full details may be found in the papers [9] and [10].

In certain settings however it is necessary to restrict to controls of the form $\theta : [0,T] \to M$ that involve no feedback, or those where the feedback only takes account of the instantaneous velocity $u(t)$.

The terms νA, B in the equations are the classical terms representing the effect of viscosity and the interaction of the particles of fluid respectively; the term B is quadratic in u and is the cause of the difficulties associated with solving the Navier-Stokes equations (even in the deterministic case $g = 0$). The final term in the equation represents noisy external forces, with w denoting an infinite dimensional Wiener process.

The methods involve the Loeb space techniques that were employed in [4, 6] to solve the stochastic Navier-Stokes equations with general force and multiplicative noise (that is, with noise $g(s, u(s))$ involving feedback of the solution u), combined with the nonstandard ideas used earlier in the study of optimal control of finite dimensional equations (see [7] for example). For the 3-d case it is necessary to utilize the idea of approximate solutions developed in [11] for the study of attractors.

We assume a fixed time horizon for the problem, and then the aim is to establish the existence of an optimal control that minimizes a general cost J that has a running component and a terminal cost, modelled by

$$\mathcal{J}(\theta, u) = \mathbb{E}\left(\int_0^T h\left(t, u(t), \theta(t, u)\right) dt + \bar{h}(u(T))\right). \tag{22.1.2}$$

The existence of an optimal control (in some cases, a generalized or *relaxed* control) can be established in a number of settings. The results for $d = 2$ are somewhat stronger than for $d = 3$, which is a reflection of the well-known distinction between these two cases even for the uncontrolled equations: in dimension $d = 2$ there is uniqueness of solutions and the solutions are strong (essentially this means that the field $u(t, x)$ is differentiable in x for each t) whereas for $d = 3$ the solutions that exist for all time are weak and uniqueness is a major open problem[2]. Consequently even the formulation of the optimal control problem is more difficult, and optimal controls are obtained using our methods for more restricted classes of controls compared to the results for $d = 2$.

The plan of the paper is as follows. First (Section 22.2) we provide some details of the Hilbert space setup for the equations and their solution, and then recall the basic nonstandard ideas concerning controls. The results for dimension $d = 2$ are then outlined in Section 22.3, and in the final section we do the same for $d = 3$. We omit proofs, referring the interested reader to [9] and [10], although in some cases where new ideas are introduced we provided sketches.

Alternative approaches to control theory for stochastic Navier-Stokes equations have been studied in [3, 13, 18], where the controls are assumed to be

[2]In fact this is one of the Millennium problems. Uniqueness *is* known for strong solutions but for $d = 3$ such solutions can only be obtained for short times.

stochastic processes

$$\theta : [0, T] \times \Omega \to \mathbf{H}$$

adapted to the information filter given by complete observations of the fluid velocity. In [3] for example, the author assumes that there is a constant $\rho > 0$ such that $|\theta(t, \omega)| < \rho$ for all $t \in [0, T]$ and $\omega \in \Omega$, while in [13, 18] it is assumed that the controls satisfy the condition

$$\mathbb{E}\left(\int_0^T |\theta(t)|^2 dt\right) < \infty.$$

In these papers the control is subject to the action of a linear operator. The stochastic minimum principle is derived providing a necessary condition for an optimal control, and a dynamic programming approach is presented to give a sufficient condition. It is shown that the minimum value function is a viscosity solution to the Hamilton-Jacobi-Bellman equation associated with the problem. In [13] the authors obtain smooth solutions to the HJB equation which justifies the dynamic programming approach.

22.2 Preliminaries

22.2.1 Nonstandard analysis

For optimal control theory, nonstandard analysis is a natural tool to use because one can always find a "nonstandard" optimal control — this is simply θ_N for any infinite N, where $(\theta_n)_{n\in\mathbb{N}}$ is a minimizing sequence of controls. The task then is to see whether θ_N can be transformed into a *standard* optimal control of some kind. For finite dimensional DEs and SDEs this idea was developed in [7] and related papers.

In the study of the Navier-Stokes equations (particularly the *stochastic* version) Loeb space techniques have proved very powerful, providing for example the first general existence proof for the stochastic Navier-Stokes equations in dimensions up to 4 (see [6]) and more recently new results concerning the existence of attractors (see [5], [11] and [12]).

In the work reported here the nonstandard techniques used in these two areas are combined. We work in a standard universe $\mathbb{V} = \mathbb{V}(S)$ where S is a base set that contains all the objects of interest, and take an $\aleph_1$-saturated extension ${}^*\mathbb{V}(S) \subset \mathbb{V}({}^*S)$. For ease of reference we gather together in an Appendix the most important facts about the nonstandard representation of the spaces used in the study of the NSE equations.

22.2.2 The stochastic Navier-Stokes equations

The classical form of the uncontrolled stochastic Navier-Stokes equations with zero boundary condition on a time interval $0 \le t \le T$ is as follows:

$$\begin{cases} du = \{\nu\Delta u - <u, \nabla> u - \nabla p + f(t,u)\}dt + g(t,u)dw(t) \\ \operatorname{div} u = 0 \\ u|_{\partial D} = 0 \\ u(0,x) = u_0(x) \end{cases}$$

These equations are considered in a bounded domain $D \subset \mathbb{R}^d$ ($d = 2, 3$) which is fixed throughout the paper, with the boundary ∂D of class $\mathcal{C}^2$. Here $u : [0,T] \times D \times \Omega \to \mathbb{R}^3$ is the random velocity field, ν is the viscosity, p is the pressure, and f represents external forces; u_0 is the initial condition. The diffusion term g together with the driving Wiener process w represents additional *random* external forces, or noise. Underlying this model is a filtered probability space $\mathbf{\Omega} = (\Omega, \mathcal{F}, (\mathcal{F}_t)_{t\geq 0}, P)$.

For the conventional Hilbert space formulation, write $\mathbf{L}^2(D) = (L^2(D))^d$ and let

$$\mathcal{V} = \{u \in \mathcal{C}_0^\infty(D, \mathbb{R}^d) \colon \operatorname{div} u = 0\}.$$

Then $\mathbf{H}$ is the closure of $\mathcal{V}$ in $\mathbf{L}^2(D)$ with the norm given by $|u|^2 = (u,u)$, where

$$(u,v) = \sum_{i=1}^{d} \int_D u^i(x) v^i(x) dx.$$

and $\mathbf{V}$ is the closure of $\mathcal{V}$ in the norm $|u| + \|u\|$ where $\|u\|^2 = ((u,u))$ and

$$((u,v)) = \sum_{j=1}^{d} \left(\frac{\partial u}{\partial x_j}, \frac{\partial v}{\partial x_j} \right).$$

$\mathbf{H}$ and $\mathbf{V}$ are real Hilbert spaces, $\mathbf{V}$ dense in $\mathbf{H}$. The dual space to $\mathbf{V}$ is denoted by $\mathbf{V}'$ with the duality extending the scalar product in $\mathbf{H}$ and

$$\mathbf{V} \subset \mathbf{H} \equiv \mathbf{H}' \subset \mathbf{V}'.$$

Write A for the Stokes operator on $\mathbf{H}$ (the self-adjoint extension of the projection of $-\Delta$) which is densely defined in $\mathbf{H}$; it can be extended to $A : \mathbf{V} \to \mathbf{V}'$ by $Au[v] = ((u,v))$ for $u, v \in \mathbf{V}$. The operator A has an orthonormal basis of eigenfunctions $\{e_k\}_{k\in\mathbb{N}} = \mathcal{E} \subset \mathbf{H}$ with eigenvalues $0 < \lambda_k \nearrow \infty$. For $u \in \mathbf{H}$ write $u = \sum u_k e_k$. Write $\mathbf{H}_n$ for the finite dimensional subspace $\mathbf{H}_n = \operatorname{span}\{e_1, e_2, \dots, e_n\}$ and Pr_n for the projection onto $\mathbf{H}_n$.

A family of spaces $\mathbf{H}^r$ for $r \in \mathbb{R}$ is defined as follows: for $r \geq 0$

$$\mathbf{H}^r = \left\{ u \in \mathbf{H} : \sum_{k=1}^{\infty} \lambda_k^r u_k^2 < \infty \right\}$$

with the norm given by

$$|u|_r^2 = \sum_{k=1}^{\infty} \lambda_k^r u_k^2.$$

and $\mathbf{H}^{-r}$ is the dual of $\mathbf{H}^r$. We may represent $\mathbf{H}^{-r}$ by

$$\mathbf{H}^{-r} = \left\{ (u_k)_{k \in \mathbb{N}} : \sum_{k=1}^{\infty} \lambda_k^{-r} u_k^2 < \infty \right\}$$

In terms of this family we have $\mathbf{H}^0 = \mathbf{H}$, $\mathbf{H}^1 = \mathbf{V}$, and $\mathbf{H}^{-1} = \mathbf{V}'$ with the norms $|u| = |u|_0$ and $\|u\| = |u|_1$.

The quadratic function B is given by $\big(B(u), y\big) = b(u, u, y)$, where

$$b(u, v, y) = \sum_{i,j=1}^{d} \int_D u^i(x) \frac{\partial v^j}{\partial x_i}(x) y^j(x)\, dx$$

whenever the integral is defined. This describes the nonlinear inertia term in the equation. The trilinear form b has many properties [19] including the following that we need here.

$$b(u, v, v) = 0, \tag{22.2.1}$$

$$|b(u, v, y)| \leq c|u|^{\frac{1}{4}} \|u\|^{\frac{3}{4}} |v|^{\frac{1}{4}} \|v\|^{\frac{3}{4}} \|y\| \tag{22.2.2}$$

There are additional properties that hold only in dimension $d = 2$.

The inequality (22.2.2) gives the second of the following properties.

Proposition 22.2.1 *For $u \in \mathbf{V}$*

$$|Au|_{V'} = \|u\|$$
$$|B(u)|_{V'} \leq c|u|^{\frac{1}{2}} \|u\|^{\frac{3}{2}}$$

In the above setting the evolution form of the Navier-Stokes equations (without explicit control) in the space $\mathbf{V}'$ (the vector $\nabla p = 0$ in this space) is given by

$$\begin{aligned} u(t) = u_0 &+ \int_0^t \{-\nu A u(s) - B\,(u(s)) + f\,(s, u(s))\}\, ds \\ &+ \int_0^t g\,(s, u(s))\, dw(s) \end{aligned} \tag{22.2.3}$$

with the initial condition $u_0 \in \mathbf{H}$. The first integrals are Bochner integrals in $\mathbf{V}'$. The driving noise process w is an $\mathbf{H}$-valued Wiener process with covariance Q, a fixed non-negative trace class operator (see [6, 14] for details), and the stochastic integral is the extension of the Itô integral to Hilbert spaces due to Ichikawa [17].

In the study of the Navier-Stokes equation there are several types of solution. The most general is a *weak solution*, but in the case $d = 2$ we also have *strong solutions*. The definitions are as follows (see [6]):

Definition 22.2.2

(a) An adapted process $u : [0,T] \times \Omega \to H$ is a ***weak solution*** to the stochastic Navier-Stokes equations (22.2.3) if

(i) for P-a.a. ω the path $u(\cdot,\omega)$ has

$$u(\cdot,\omega) \in L^\infty([0,T];\mathbf{H}) \cap L^2([0,T];\mathbf{V}) \cap C([0,T];\mathbf{H}_{\text{weak}}),$$

(ii) for all $t \in [0,T]$ the equation (22.2.3) holds as an identity in V',

(iii) u satisfies the energy inequality

$$\mathbb{E}\left(\sup_{t\in[0,T]} |u(t)|^2 + \int_0^T \|u(t)\|^2 dt\right) < \infty. \tag{22.2.4}$$

By *solution* henceforth in this paper we mean *weak solution.*

(b) A weak solution is ***strong*** if

(i) for P-a.a. ω the path $u(\cdot,\omega)$ has

$$u(\cdot,\omega) \in L^\infty([0,T];\mathbf{V}) \cap L^2([0,T];\mathbf{H}^2) \cap C([0,T];\mathbf{V}_{\text{weak}}), \tag{22.2.5}$$

(which implies that $u(\cdot,\omega) \in C([0,T];H)$);

(ii) for P-a.a. ω the path $u(\cdot,\omega)$ has

$$\sup_{t\in[0,T]} \|u(t)\|^2 + \int_0^T Au(t)^2 dt < \infty. \tag{22.2.6}$$

When controls are introduced into the forcing term f the equation (22.2.3) takes the form

$$\begin{aligned} u(t) = u_0 &+ \int_0^t \{-\nu Au(s) - B(u(s)) + f(s,u(s),\theta(s,u))\}\, ds \\ &+ \int_0^t g(s,u(s))\, dw(s). \end{aligned}$$

In the next section we discuss the types of control that we consider.

The following general existence result for solutions to the sNSE was first proved in [4] (see also [6] for an exposition), using Loeb space methods.

First define $K_m = \{v : \|v\| \leq m\} \subseteq \mathbf{V}$, with the strong topology of $\mathbf{H}$. In the theorem below, continuity on each K_m turns out to be the appropriate condition for the coefficients f, g; this is weaker than continuity on $\mathbf{V}$ in either the $\mathbf{H}$-norm or the weak topology of $\mathbf{V}$.

Theorem 22.2.3 *Suppose that* $u_0 \in \mathbf{H}$ *and*

$$f : [0, \infty) \times \mathbf{V} \to \mathbf{V}', \qquad g : [0, \infty) \times \mathbf{V} \to L(\mathbf{H}, \mathbf{H})$$

are jointly measurable functions with the following properties

(i) $f(t, \cdot) \in C(K_m, \mathbf{V}'_{\text{weak}})$ *for all* m,

(ii) $g(t, \cdot) \in C(K_m, L(\mathbf{H}, \mathbf{H})_{\text{weak}})$ *for all* m,

(iii) $|f(t, u)|_{\mathbf{V}'} + |g(t, u)|_{\mathbf{H},\mathbf{H}} \leq a(t)(1 + |u|)$ *where* $a \in L^2(0, T)$ *for all* T.

Then equation (22.2.3) has a solution u *on a filtered Loeb space.*

In dimension $d = 2$ a stronger existence theorem (and uniqueness, given Lipschitz coefficients) has been established — this will be noted when required in Section 22.3.

22.2.3 Controls

The simplest controls considered in this paper are those with no feedback, as follows. As is customary in optimal control theory, it is often necessary to extend this class to its natural completion, which is the class of generalized or *relaxed* controls. These, and the topology on them are as defined thus (as in [7]). The compact metric space M is fixed for the whole paper.

Definition 22.2.4

(i) The class C of ***ordinary controls*** is the set of measurable functions $\theta : [0, T] \to M$, where M is a fixed compact metric space.

(ii) *The class* D of ***relaxed controls*** is the set of measurable functions $\varphi : [0, T] \to M_1(M)$, where $M_1(M)$ is the set of probability measures on M. (We regard $C \subseteq D$ by identifying $a \in M$ with the Dirac measure δ_a).

The *weak* or *narrow* topology on $\mathfrak{C}$ is given by means of the set $\mathcal{K}$ of bounded measurable functions $z : [0, T] \times M \to \mathbb{R}$ with $z(t, \cdot)$ continuous for all $t \in [0, T]$. The action of $\theta \in \mathfrak{C}$ on $z \in \mathcal{K}$ is defined by

$$\theta(z) = \int_0^T z(t, \theta(t))dt.$$

Then the topology on $\mathfrak{C}$ is defined by specifying as subbase of open neighbourhoods the sets

$$\{\theta : |\theta(z)| < \varepsilon\}_{\varepsilon>0,\, z\in\mathcal{K}}.$$

The effect of a relaxed control on a function $z \in \mathcal{K}$ is obtain by first extending the domain of each $z \in \mathcal{K}$ to $[0,T] \times \mathcal{M}_1(M)$ as follows: for a probability measure $\mu \in \mathcal{M}_1(M)$ define

$$z(t, \mu) = \int_M z(t, m)d\mu(m).$$

The weak (or narrow) topology is then extended to $\mathfrak{D}$ by defining the action of $\varphi \in \mathfrak{D}$ on $z \in \mathcal{K}$ to be

$$\varphi(z) = \int_0^T z\big(t, \varphi(t)\big)dt = \int_0^T \left(\int_M z(t,m)d\varphi_t(m)\right) dt$$

where for brevity we use $\varphi_t := \varphi(t)$.

The fundamental results about controls that we shall use are summarized below; for details consult for example [7], [15], [2] or [20] (noting that a relaxed control is a particular case of a *Young measure*).

Theorem 22.2.5 *The set of relaxed controls $\mathfrak{D}$ is compact.*

This means that for an internal ("nonstandard") control Φ (*ordinary or *relaxed) there is a well defined *standard part* $^\circ\Phi \in \mathfrak{D}$. (In [7] it is shown how this can be defined explicitly using Loeb measures.) Note that for a control $\varphi \in \mathfrak{D}$ we have $^\circ({}^*\varphi) = \varphi$.

For the next result we define a *uniform step control* to be an ordinary control θ such that there is a partition of $[0,T]$ into intervals of constant length with θ constant on each partition interval.

Theorem 22.2.6 *The set of uniform step controls $\mathfrak{C}^S$ is dense in $\mathfrak{D}$.*

To handle controls in a nonstandard setting the next definition and result is basic.

Definition 22.2.7 $Z : {}^*[0,T] \times {}^*M \to {}^*R$ is a **bounded uniform lifting** of $z \in \mathcal{K}$ if Z is an internal *measurable function such that

(i) Z is finitely bounded

(ii) for a.a. $\tau \in {}^*[0,T]$, for all $\alpha \in {}^*M$: $Z(\tau,\alpha) \approx z({}^\circ\tau, {}^\circ\alpha)$.

Theorem 22.2.8 ([7]) *Let Φ be an internal control with $\varphi = {}^\circ\Phi$ and Z be a bounded uniform lifting of z. Then $\Phi(Z) \approx \varphi(z)$, i.e.*

$$\int_0^T Z(\tau, \hat{\varphi}_\tau)d\tau \approx \int_0^T z(t, \varphi_t)dt$$

In later sections we will define wider classes of controls of the form $\theta = \theta(t,u)$ for $u \in \mathbf{H}$ or even $\theta = \theta(t, u(\cdot))$ where $u(\cdot)$ denotes the trajectory $\{u(s) : 0 \le s \le t\}$ up to time t.

22.3 Optimal control for $d = 2$

Throughout this section we take $d = 2$ and utilize the stronger existence and uniqueness results that obtain in this case.

22.3.1 Controls with no feedback

First we consider the simplest kind of optimal control problem, taking the set of controls to be the no-feedback controls $\mathfrak{C}$ and its generalization $\mathfrak{D}$. The controlled equation then takes the form

$$\begin{aligned} u(t) = u_0 &+ \int_0^t \{-\nu Au(s) - B(u(s)) + f(s, u(s), \varphi(s))\}\, ds \\ &+ \int_0^t g(s, u(s))\, dw(s) \end{aligned} \tag{22.3.1}$$

for a control in $\mathfrak{D}$. Provided f and g are suitably Lipschitz then solutions to this equation are unique. Theorem 22.2.3 is refined as follows.

Theorem 22.3.1 *Let $d = 2$. Consider the following conditions on the functions f, g: there exist constants $c > 0$ and $a(\cdot) \in L^2(0,T)$ such that:*

(a) $f : [0,T] \times \mathbf{H} \times M \to \mathbf{V}'$ *is jointly measurable and*

- **(i)** $|f(t,u,m) - f(t,v,m)|_{\mathbf{V}'} \le c|u - v|$
- **(ii)** $f(t, u, \cdot)$ *is continuous*
- **(iii)** *for a.a.* $t \in [0,T]$

$$|f(t,u,m)|_{\mathbf{V}'} \le a(t)(1 + |u|)$$

for all $u, v \in \mathbf{H}$ and $m \in M$.

(b) $g : [0,T] \times \mathbf{H} \to L(\mathbf{H}, \mathbf{H})$ *is jointly measurable and*

(i) $|g(t,u) - g(t,v)|_{\mathbf{H},\mathbf{H}} \leq c|u - v|$

(ii) *for a.a.* $t \in [0,T]$

$$|g(t,u)|_{\mathbf{H},\mathbf{H}} \leq a(t)(1 + |u|)$$

for all $u, v \in \mathbf{H}$.

There is a filtered probability space $\mathbf{\Omega} = (\Omega, \mathcal{F}, (\mathcal{F}_t)_{t\geq 0}, P)$ *carrying a Wiener process* w*, with the universal property that for any* f*,* g *satisfying the above conditions (for any* $a(t)$ *and* c*) and any ordinary or relaxed control the equation (22.3.1) has a unique solution on* $\mathbf{\Omega}$ *satisfying the energy inequality*

$$\mathbb{E}\left(\sup_{t\in[0,T]} |u(t)|^2 + \int_0^T \|u(t)\|^2 dt \right) < E \tag{22.3.2}$$

with the constant E *uniform over the set of controls; in fact* $E = E\left(u_0, a(\cdot)\right)$.

The uniqueness of solutions requires the property

$$|b(u,v,y)| \leq c\, |u|^{\frac{1}{2}} \|u\|^{\frac{1}{2}}\, |v|^{\frac{1}{2}}\, \|v\|^{\frac{1}{2}}\, \|y\|$$

of the form b which is only valid in dimension 2.

We now fix a space $\mathbf{\Omega}$ as given by this theorem. This may be a Loeb space (as in [6, Theorem 6.4.1 and 6.6.2]), but for the current purpose this is not essential[3]: we simply assume that there is such a space $\mathbf{\Omega}$ in the basic standard universe of discourse and do not care about its provenance. (In Section 22.3.11 it will be necessary to specify the space $\mathbf{\Omega}$ more precisely.) For a given control φ and initial condition u_0 we define

$$u^\varphi = \text{the unique solution for the control } \varphi \text{ with } u(0) = u_0$$

(Strictly we should write $u^\varphi_{\mathbf{\Omega}}$ to denote the underlying space, but where not mentioned this will be $\mathbf{\Omega}$.)

22.3.2 Costs

To formulate an optimal control problem it is necessary to consider the cost of a control. Here we consider a general cost comprising a running cost and a terminal cost, defined as follows.

[3]This is in contrast to Section 22.3.11 and also when working in $d = 3$, where it will be necessary to specify the space $\mathbf{\Omega}$ more precisely.

Definition 22.3.2 We assume given (and fixed for the paper) two functions $h : [0,T] \times H \times M \to R$ and $\bar{h} : H \to R$ with the following properties.

(i) h is jointly measurable, non-negative, continuous in the second and the third variables and satisfies the following growth condition: for a.a. $t \in [0,T]$, all $u \in H$ and $m \in M$

$$|h(t,u,m)| \leq a(t)(1 + |u|).$$

(ii) $\bar{h} : H \to R$ is non-negative, continuous and has linear growth: that is,

$$|\bar{h}(u)| \leq c(1 + |u|)$$

for all $u \in H$.

The ***cost*** for a control φ (ordinary or relaxed) is $J(\varphi)$ defined by

$$\mathcal{J}(\varphi) = \mathbb{E}\left(\int_0^T h\left(t, u^{\varphi}(t), \varphi(t)\right) dt + \bar{h}\left(u^{\varphi}(T)\right)\right)$$

(Strictly we should write $J_{\mathbf{\Omega}}(\varphi)$ to denote the underlying space, but if the space is not mentioned then we mean Ω.)

The minimal cost for ordinary controls is

$$\mathcal{J}_0 = \inf\{\mathcal{J}(\theta) : \theta \in \mathfrak{C}\}$$

and the minimum cost for relaxed controls is

$$\hat{\mathcal{J}}_0 = \inf\{\mathcal{J}(\varphi) : \varphi \in \mathfrak{D}\}$$

22.3.3 Solutions for internal controls

In the present context we are not concerned with proving existence of solutions to the stochastic Navier-Stokes equations (sNSE) — either by Loeb space methods or any other technique: we are simply assuming the basic existence result above. However, for the purposes of obtaining an optimal control we wish to transfer this result to give the existence of an *internal* solution U^{Φ} to the internal equation on the internal space ${}^*\mathbf{\Omega}$ controlled by an internal control Φ. The idea then is to take the standard part to give a standard solution to the equation controlled by the standard part ${}^{\circ}\Phi$; of course the solution will live on the adapted Loeb space $L({}^*\mathbf{\Omega})$ rather than the original space $\mathbf{\Omega}$, and later we see how we can come back to $\mathbf{\Omega}$.

The proof of the following result, making the first part of the above procedure precise, is almost identical to that involved in the proof of existence — the main difference being that in the existence proof the internal process U lives in $\mathbf{H}_N$ whereas here the process U^{Φ} lives in ${}^*\mathbf{H}$. Here is the result needed.

Theorem 22.3.3 *Let $u_0 \in \mathbf{H}$ and f, g, h and $\bar{h}$ satisfy the conditions of Theorem 22.3.1 and Definition 22.3.2, and suppose that Φ is an internal control. Let U^{Φ} be the unique solution (which exists by transfer of Theorem 22.3.1) on ${}^*\boldsymbol{\Omega}$ to the internal equation *(22.3.1) with control Φ; that is*

$$U^{\Phi}(\tau) = {}^*u_0 + \int_0^{\tau} \left\{-\nu\, {}^*AU^{\Phi}(\sigma) - {}^*B\left(U^{\Phi}(\sigma)\right) + {}^*f\left(\sigma, U^{\Phi}(\sigma), \Phi(\sigma)\right)\right\} d\sigma + \\ + \int_0^{\tau} {}^*g\left(\sigma, U^{\Phi}(\sigma)\right) dW(\sigma) \tag{22.3.3}$$

and satisfying the internal energy inequality

$$\mathbb{E}_{{}^*P}\left(\sup_{\tau\in{}^*[0,T]} |U^{\Phi}(\tau)|^2 + \int_0^T \|U^{\Phi}(\tau)\|^2 d\tau\right) < E. \tag{22.3.4}$$

Then U^{Φ} has a standard part $u = {}^{\circ}U^{\Phi}$, which is a solution on $L({}^*\boldsymbol{\Omega})$ to the standard equation (22.3.1) with control ${}^{\circ}\Phi$, i.e. $u = U^{{}^{\circ}\Phi}$ and

$${}^*\mathcal{J}_{{}^*\Omega}(\Phi) \approx \mathcal{J}_{L({}^*\Omega)}({}^{\circ}\Phi). \tag{22.3.5}$$

22.3.4 Optimal controls

The main theorem for controls with no feedback is now easy to derive.

Theorem 22.3.4 *Let $u_0 \in \mathbf{H}$ and f, g, h and $\bar{h}$ satisfy the conditions of Theorem 22.3.1 and Definition 22.3.2. Then there is an optimal relaxed control φ_0, i.e. such that $\mathcal{J}(\varphi_0) = \mathcal{J}_0$*

Proof. Take a sequence of controls $(\theta_k)_{k\in\mathbb{N}} \subset \mathfrak{C}$ such that

$$\lim_{k\to\infty} \mathcal{J}(\theta_k) = \mathcal{J}_0 \quad \text{and} \quad \mathcal{J}(\theta_k) \searrow \mathcal{J}_0.$$

Fix an infinite $K \in {}^*\mathbb{N}$; then ${}^*\mathcal{J}_{{}^*\Omega}(\theta_K) \approx \mathcal{J}_0$ for the internal ordinary control $\theta_K : {}^*[0,T] \to {}^*M$. Corresponding to this control is the unique internal solution U^{θ_K} to the internal controlled Navier-Stokes equation (22.3.3) which obeys the energy inequality (22.3.4).

Let $\varphi_0 := {}^{\circ}\theta_K : [0,T] \to \mathcal{M}_1(M)$ and apply Theorem 22.3.3 to the control $\Phi = \theta_K$ and to the solution U^{θ_K}. Then ${}^{\circ}U^{\theta_K} = u^{\varphi_0}_{L({}^*\boldsymbol{\Omega})}$ and

$${}^*\mathcal{J}_{{}^*\boldsymbol{\Omega}}(\theta_K) \approx \mathcal{J}_{L({}^*\boldsymbol{\Omega})}(\varphi_0)$$

so

$$\mathcal{J}_{L({}^*\boldsymbol{\Omega})}(\varphi_0) = \mathcal{J}_0.$$

It only remains to show that $\mathcal{J}_{L(^*\mathbf{\Omega})}(\varphi_0) = \mathcal{J}(\varphi_0)$. First notice that $\mathcal{J}(\varphi_0)$ is a real number so $\mathcal{J}(\varphi_0) = {}^*\mathcal{J}_{^*\mathbf{\Omega}}({}^*\varphi_0)$. Apply Theorem 22.3.3 to the internal control ${}^*\varphi_0$ and the internal solution $u^{{}^*\varphi_0}$ on ${}^*\mathbf{\Omega}$ giving

$$\mathcal{J}(\varphi_0) = {}^*\mathcal{J}_{^*\mathbf{\Omega}}({}^*\varphi_0) \approx \mathcal{J}_{L(^*\mathbf{\Omega})}({}^\circ({}^*\varphi_0)) = \mathcal{J}_{L(^*\mathbf{\Omega})}(\varphi_0) = \mathcal{J}_0$$

so φ_0 is optimal. □

Actually the above theorem is easily derived from the following more general consequence of Theorem 22.3.3.

Theorem 22.3.5 *The function $\mathcal{J}$ is continuous with respect to the narrow topology on $\mathfrak{D}$.*

Proof. This is routine using the nonstandard criterion for continuity, together with Theorem 22.3.3 and the argument in the final part of the above proof. □

This result, together with the density of $\mathfrak{C}$ in $\mathfrak{D}$ (Theorem 22.2.6) means that $\hat{\mathcal{J}}_0 = \mathcal{J}_0$ and so we have

Theorem 22.3.6 *The optimal relaxed control φ_0 is also optimal in the class of relaxed controls.*

As usual in optimal control theory, if the force term f is suitably convex then from an optimal relaxed control it is easy to obtain an optimal ordinary control.

22.3.5 Hölder continuous feedback controls ($d = 2$)

The natural generalization of the control problem for the stochastic 2D Navier-Stokes equation is to allow θ to depend on the instantaneous fluid velocity as well as on time:

$$\begin{aligned} u(t) = u_0 + &\int_0^t \{-\nu Au(s) - B\,(u(s)) + f\big(s, u(s), \theta(s, u(s))\big)\}\, ds + \\ &+ \int_0^t g\,(s, u(s))\, dw(s). \end{aligned} \tag{22.3.6}$$

It is straightforward to extend the approach outlined above for no-feedback controls to the special situation when the controls considered are *Hölder continuous feedback controls* with uniform constants, as follows.

Definition 22.3.7 For a given set of constants $(\alpha, \beta, \gamma, c_1, c_2)$ with $\alpha, \beta \in (0,1)$ and $\gamma, c_1, c_2 > 0$ the set of ***Hölder continuous feedback controls*** $C(\alpha, \beta, \gamma, c_1, c_2)$ is the set of jointly measurable functions

$$\theta : [0,T] \times \mathbf{H} \to M$$

which are locally Hölder continuous in both variables with these constants; that is:

$$|\theta(t,u) - \theta(s,v)| \le c_1|t-s|^\alpha + c_2|u-v|^\beta \tag{22.3.7}$$

for all $t, s \in [0,T]$ such that $|t-s| \le \gamma$ and all $u, v \in H$ such that $|u-v| \le \gamma$.

We will see that this definition guarantees the existence of an optimal ordinary control; the Hölder continuity allows us to construct the standard part of an internal control as an ordinary control without the need to consider measure-valued (i.e. relaxed) controls.

The cost of a control is as before:

$$\mathcal{J}(\theta) = \mathbb{E} \int_0^T h(t, u^\theta(t), \theta(t, u^\theta(t)))\, dt + \bar{h}(u^\theta(T)) \tag{22.3.8}$$

where h and $\bar{h}$ satisfy the conditions of Definition 22.3.2. The minimal cost is now defined as

$$\mathcal{J}_0 = \inf\{\mathcal{J}(\theta) :\ \theta \in \mathfrak{C}(\alpha, \beta, \gamma, c_1, c_2)\}$$

(of course, strictly this should be $\mathcal{J}_0(\alpha, \beta, \gamma, c_1, c_2)$).

The standard part of an internal control is defined in the obvious way:

Definition 22.3.8 Suppose that $\Theta \in {}^*\mathfrak{C}(\alpha, \beta, \gamma, c_1, c_2)$, so $\Theta : {}^*[0,T] \times {}^*\mathbf{H} \to {}^*M$. Define $\theta \in \mathfrak{C}(\alpha, \beta, \gamma, c_1, c_2)$ by

$$\theta(t,u) = {}^\circ\Theta(t,u) = {}^\circ\Theta(\tau, U)$$

for any $\tau \approx t$ and $U \approx u$ in $\mathbf{H}$ (because of the condition (22.3.7)). Write $\theta = {}^\circ\Theta$.

Now we have the counterpart of Theorem 22.3.3 for Hölder continuous controls.

Theorem 22.3.9 *Let $\Theta \in {}^*\mathfrak{C}(\alpha, \beta, \gamma, c_1, c_2)$ be an internal control and f, g satisfy the conditions of Theorem 22.3.1. Let U^Θ be the unique solution on ${}^*\mathbf{\Omega}$ to the internal equation with control Θ:*

$$\begin{aligned} U^\Theta(\tau) &= {}^*u_0 + \\ &+ \int_0^\tau \{-\nu^*AU^\Theta(\tau) - {}^*B\left(U^\Theta(\tau)\right) + {}^*f\left(\tau, U^\Theta(\tau), \Theta(\tau, U^\Theta(\tau))\right)\} d\tau \\ &+ \int_0^\tau {}^*g\left(\tau, U^\Theta(\tau)\right) dW(\tau) \end{aligned} \tag{22.3.9}$$

and satisfying the internal energy inequality

$$\mathbb{E}_{^*P}\left(\sup_{\tau\in^*[0,T]}|U^{\Theta}(\tau)|^2+\int_0^T\|U^{\Theta}(\tau)\|^2d\tau\right)<E.$$

Then its standard part $u = {}^{\circ}U^{\Theta}$ is a solution on $L(^*\boldsymbol{\Omega})$ to the standard equation (22.3.6) with control ${}^{\circ}\Theta$ (i.e. $u = u^{{}^{\circ}\Theta}$), and

$$^*\mathcal{J}_{^*\Omega}(\Theta)\approx\mathcal{J}_{L(^*\Omega)}({}^{\circ}\Theta).$$

This gives the following with proof almost identical to that of Theorem 22.3.4.

Theorem 22.3.10 *Let f, g satisfy the conditions of Theorem 22.3.1. There exists an optimal admissible control $\theta_0 \in \mathfrak{C}(c_1, c_2, \alpha, \beta, \gamma)$, i.e. such that $\mathcal{J}(\theta_0) = \mathcal{J}_0$.*

22.3.6 Controls based on digital observations ($d = 2$)

The most general results for optimal control of 2D stochastic Navier-Stokes equations involves feedback controls that depend on the entire past of the path of the process u through digital observations made at a fixed finite number of points of time. This model of control was discussed in [1] and [7] for finite dimensional stochastic equations; in the latter paper digital observations were made at random times, but we do not include that feature here.

The controlled system takes the form

$$\begin{aligned}du(t) = &\{-\nu Au(t) - B(u(t),u(t)) + f(t,u,\theta(y(u),t))\}\,dt\\ &+ g(t,u)dw(t)\end{aligned} \tag{22.3.10}$$

with initial condition $u(0) = u_0 \in \mathbf{H}$. In this equation $u = u(\cdot)$ denotes the path of the process $u(t)$ up to the present time (which will be clear from the context). The function $y(u)$ in the control θ denotes the *digital observations* or *read-out* given by the path u, described in the next sections.

22.3.7 The space $\mathcal{H}$

For this system the conditions on f, g will be strengthened slightly so that the solutions to (22.3.10) are strong, and thus the paths of solutions to the equation belong to the space $\mathcal{H}$ where

$$\mathcal{H} = C([0,T];\mathbf{H}) \cap L^{\infty}([0,T];\mathbf{V}).$$

Let $|u|$ denote the uniform norm on this space; that is

$$|u| = \sup_{0\le t\le T}|u(t)|$$

and note that with this norm $\mathcal{H}$ is separable.

22.3.8 The observations

Controls will be based on the information received from digital observations of the solution path $u \in \mathcal{H}$ made at **observation times**

$$0 < t_1 < \ldots < t_p < T$$

that are fixed for the problem. For completeness we write $t_0 = 0$ and $t_{p+1} = T$, but observations are not made at these times. The **digital observation** (taking its value in $\mathbb{N}$) at time t_i is denoted by y_i where

$$y_i : \mathcal{H} \to \mathbb{N} \quad (i = 1, \ldots, p)$$

The complete set of observations for a path u is recorded as

$$y(u) = (y_1(u), \ldots, y_p(u))$$

The ith observation y_i is assumed to be *non-anticipating* in the sense that $y_i(u)$ depends only on the past $u \restriction [0, t_i]$ of the path u up to time t_i.

The digital observation functions y_i are fixed for the whole problem.

22.3.9 Ordinary and relaxed feedback controls for digital observations

The feedback controls considered here are as follows. M is the fixed compact metric space as before.

Definition 22.3.11 An ***ordinary feedback control based on digital observations*** is a measurable function

$$\theta : \mathbb{N}^p \times [0, T] \to M$$

where θ is non-anticipating in the sense that if $t_k \leq t < t_{k+1}$ for $k = 1, \ldots, p$, then $\theta(y, t)$ depends on t and only the first k components of y, namely $(y_1, \ldots, y_k)$. For $t \in [0, t_1)$ a control θ depends only on t.

A ***relaxed feedback control based on digital observations*** is a measurable function

$$\varphi : \mathbb{N}^p \times [0, T] \to \mathcal{M}_1(M)$$

where $M_1(M)$ is the space of probability measures on M and φ is non-anticipating in the same sense as for an ordinary control.

Write $\hat{C}$ and $\hat{D}$ for these sets of ordinary and relaxed feedback controls.

For brevity, in the current context, by a *control* (ordinary or relaxed) we mean a feedback control based on digital observations. The interpretation of a relaxed control is as before.

Nonstandard (i.e. internal) controls have standard parts that are given as follows, using the compactness of non-feedback controls (Theorem 22.2.5). Let Φ be an internal control:

$$\Phi : {}^*\mathbb{N}^p \times {}^*[0,T] \to {}^*\mathcal{M}(M)$$

It is sufficient to construct the standard part of Φ only for $\mathbf{n} \in \mathbb{N}^p$. For $\tau \in {}^*[0, t_1)$ the control Φ is a function of time only (so it is a no-feedback control as in Section 22.2.3) and hence on the first subinterval $[0, t_1)$ we define its standard part as $\varphi = {}^\circ\Phi$ in the sense of the weak topology on $\mathfrak{D}$ (see the remark following Theorem 22.2.5). For $k = 1, \ldots, p$ and $\mathbf{n} \in \mathbb{N}^p$ write $\mathbf{n} \upharpoonright k = (n_1, \ldots, n_k)$ and notice that

$$\Phi(\mathbf{n}, \tau) = \Phi(\mathbf{n} \upharpoonright k, \tau)$$

for $t_k \leq \tau < t_{k+1}$. So define $\Phi_{\mathbf{n}\upharpoonright k} : {}^*[t_k, t_{k+1}[\to {}^*\mathcal{M}_1(M)$ by

$$\Phi_{\mathbf{n}\upharpoonright k}(\tau) = \Phi(\mathbf{n} \upharpoonright k, \tau).$$

Each $\Phi_{\mathbf{n}\upharpoonright k}$ is an internal relaxed control from ${}^*\mathfrak{D}$ (restricted to ${}^*[t_k, t_{k+1}[$), so, using Theorem 22.2.5 it has a standard part:

$$\varphi_{\mathbf{n}\upharpoonright k} := {}^\circ\Phi_{\mathbf{n}\upharpoonright k} : [t_k, t_{k+1}] \to \mathcal{M}_1(M)$$

From this we define the standard part $\varphi = {}^\circ\Phi$ to be the standard relaxed control given by

$$\varphi(\mathbf{n}, t) = \varphi_{\mathbf{n}\upharpoonright k}(t)$$

for $t \in [t_k, t_{k+1}[$. (Note that for fixed $\mathbf{n} \in \mathbb{N}^p$ if we write $\Psi(\tau) = \Phi(\mathbf{n}, \tau)$ and $\psi(\tau) = \varphi(\mathbf{n}, \tau)$ then ${}^\circ\Psi = \psi$.)

The weak or narrow topology on $\mathfrak{C}$ and $\mathfrak{D}$ is extended to $\hat{\mathfrak{C}}$ and $\hat{\mathfrak{D}}$ in the natural way (by regarding these as subsets of the products of $\mathfrak{C}^{\mathbb{N}^p}$ and $\mathfrak{D}^{\mathbb{N}^p}$ with the product topology) and then the standard part we have constructed is the standard part with respect to this topology — see [7] for details. The above together with the results of that paper give:

Theorem 22.3.12 *$\hat{\mathfrak{D}}$ is compact and the subset $\hat{\mathfrak{C}}^S$ of uniform step controls in $\mathfrak{C}$ is dense in $\hat{\mathfrak{D}}$.*

The following is the extension of Theorem 22.2.8 to show how the standard part of a digital feedback control acts.

Proposition 22.3.13 *Let Φ be an internal feedback control with $\varphi = {}^\circ\Phi$. If $z : [0,T] \times M \to \mathbb{R}$ is a bounded, measurable function continuous on M and $Z : {}^*[0,T] \times {}^*M \to {}^*\mathbb{R}$ is a bounded uniform lifting (according to Definition 22.2.7) then for any $\mathbf{n} \in \mathbb{N}^p$*

$$\int_0^T Z\big(\tau, \Phi(\mathbf{n}, \tau)\big) d\tau \approx \int_0^T z\big(t, \varphi(\mathbf{n}, t)\big) dt.$$

22.3.10 Costs for digitally observed controls

The cost of a control is defined in the expected way: for a control φ (ordinary or relaxed) define

$$\mathcal{J}(\varphi, u) = \mathbb{E}\left(\int_0^T h\left(t, u, \varphi(y(u), t)\right) dt + \bar{h}(u)\right) \tag{22.3.11}$$

where u is any solution to the equation with control φ. The cost functions h and $\bar{h}$ are given, with properties set out below. As usual the aim is to find a control which minimizes the cost over the set of controls. We will see that the solution to the equation is unique for a given control so we can define the cost $\mathcal{J}(\varphi) = \mathcal{J}(\varphi, u^\varphi)$ (where u^φ is the solution for the control φ). Then the optimal control problem is reduced to finding a control such that

$$\mathcal{J}_0 := \inf\{\mathcal{J}(\theta) : \ \theta \text{ an ordinary feedback control}\}$$

is obtained. As is to be expected, in general the set of ordinary feedback controls does not have sufficient closure properties and we prove existence of an optimal *relaxed* control.

22.3.11 Solution of the equations

As noted above the conditions on the functions f, g, h, $\bar{h}$ need to be strengthened (and modified to take account of the form of feedback). We fix the following set of assumptions for the current discussion.

Conditions 22.3.14 *There exist constants $a > 0$ and $L > 0$ such that:*

(a) $f : [0,T] \times \mathcal{H} \times M \to \mathbf{H}$ *is jointly measurable, bounded, non-anticipating, Lipschitz continuous in the second variable and continuous in the third variable. Specifically*

(i) $|f(t,u,m)| \leq a$

(ii) $u \restriction t = v \restriction t \Longrightarrow f(t,u,m) = f(t,v,m)$

(iii) $|f(t,u,m) - f(t,v,m)|_{\mathbf{H}} \leq L|u-v|$

(iv) *$f(t, u, \cdot)$ continuous for a.a. $t \in [0, T]$, for all $u, v \in \mathcal{H}$ and $m \in M$.*

(b) *$g : [0, T] \times \mathcal{H} \to L(\mathbf{H}, \mathbf{V})$ is jointly measurable, bounded, non-anticipating and Lipschitz continuous in the second variable. Specifically*

(i) $|g(t, u)|_{\mathbf{H},\mathbf{V}} \leq a$

(ii) $u \restriction t = v \restriction t \Longrightarrow g(t, u) = g(t, v)$

(iii) $|g(t, u) - g(t, v)|_{\mathbf{H},\mathbf{H}} \leq L|u - v|$ *for a.a. $t \in [0, T]$ and all $u, v \in \mathcal{H}$. For convenience we also assume that g is diagonal, i.e. $\forall n \in \mathbb{N}$ $g_n = \mathrm{Pr}_n(g \restriction \mathbf{H}_n) : [0, T] \times \mathcal{H} \longrightarrow S_d(n)$ where $S_d(n)$ denotes the space of $n \times n$ diagonal matrices (that is, $a_{ij} \neq 0$ iff $i \neq j$).*

(c) *$g(t, u)$ is invertible and*

$$|g^{-1}(t, u)f(t, u, m)| \leq a$$

for a.a. $t \in [0, T]$, all $u \in \mathcal{H}$ and $m \in M$.

(d) *$h : [0, T] \times \mathcal{H} \times M \to \mathbb{R}$ is jointly measurable, non-negative, bounded, non-anticipating in the second variable and continuous in the second and third variables. Specifically*

(i) $|h(t, u, m)| \leq a$

(ii) $u \restriction t = v \restriction t \Longrightarrow h(t, u, m) = h(t, v, m)$

(iii) *$h(t, \cdot, \cdot)$ is continuous for a.a. $t \in [0, T]$, all $u, v \in \mathcal{H}$ and $m \in M$.*

(e) *$\overline{h} : \mathcal{H} \to \mathbb{R}$ is non-negative, continuous and bounded. Specifically*

$$|\overline{h}(u)| \leq a$$

for all $u \in \mathcal{H}$.

Existence and uniqueness of solutions to the equation With the information, control and cost structures now defined, we return to the equations (22.3.10) under consideration. Since these equations involve feedback of the entire past of the process both directly in the coefficients f, g and through the control φ there are new considerations concerning existence and uniqueness. The first task therefore is to show how the basic existence theorem of [6] can be extended to prove the following.

Theorem 22.3.15 *There is an adapted Loeb space $\mathbf{\Omega} = (\Omega, \mathcal{F}, (\mathcal{F}_t)_{t \in [0,T]}, P)$ carrying a Wiener process w of covariance Q, with the following property. For any f, g satisfying Conditions 22.3.14(a,b,c), any initial condition $u_0 \in \mathbf{H}$ and any feedback control φ (ordinary or relaxed) the equation (22.3.10) has a unique strong solution $u = u^{\varphi}$ on $\mathbf{\Omega}$.*

The proof of this is rather long and somewhat complicated technically. The essence is to solve the internal hyperfinite dimensional Galerkin approximation in $\mathbf{H}_N$ with the internal control ${}^*\varphi$ giving an internal solution $U^{{}^*\varphi}$, and then take the standard parts. The latter involves transferring the Girsanov theorem to the hyperfinite setting, and the use of special topologies defined on the spaces $C(\mathcal{H},\mathbf{H})$ and $C(\mathcal{H}\times M,\mathbf{H})$ to make them separable so that Anderson's Luzin Theorem can be applied.

22.3.12 Optimal control

For the rest of this section the space $\mathbf{\Omega} = \big(\Omega, \mathcal{F}, (\mathcal{F}_t)_{t\in[0,T]}, P\big)$ and Wiener process w are fixed as those given by Theorem 22.3.15. The following is proved by taking the internal control in the proof of the previous theorem to be an arbitrary internal control Φ instead of ${}^*\varphi$.

Theorem 22.3.16 *Suppose that Φ is an internal control and U^Φ is the solution to the following equation in $\mathbf{H}_N$ on the internal space $\bar{\mathbf{\Omega}} = \big(\Omega, \mathcal{A}, (\mathcal{A}_\tau)_{\tau\in{}^*[0,T]}, \Pi\big)$ of the previous theorem:*

$$\begin{aligned} &dU(\tau) \\ &= \Big\{-\nu^*AU(\tau) - B_N\big(U(\tau),U(\tau)\big) + F_N(\tau,U,\Phi(Y(U),\tau),U)\Big\}d\tau \quad (22.3.12)\\ &+ G_N(\tau,U)dW(\tau) \end{aligned}$$

with $V(0)=U(0)=U_0=\mathrm{Pr_N}{}^*u_0$ and B_N, F_N, G_N as before, and $Y=(Y_i)$ a lifting of y with respect to μ. Then $u={}^\circ U^\Phi$ is the unique strong solution to the controlled equation (22.3.10) with $\varphi={}^\circ\Phi$ and

$$J(\varphi) = {}^\circ\hat{J}(\Phi)$$

where $\hat{J}(\Phi)$ is defined by

$$\hat{J}(\Phi) = {}^*\mathbb{E}\left(\int_0^T {}^*h\left(\tau, U^\Phi, \Phi(Y(U^\Phi),\tau)\right)dt + {}^*\bar{h}(U^\Phi)\right)$$

The existence of an optimal relaxed control now follows easily:

Theorem 22.3.17 *The cost function J is continuous on $\mathfrak{D}$*

Proof. Suppose that $\varphi_n \to \varphi$ in $\mathfrak{D}$. Theorem 22.3.16 shows that $\hat{J}({}^*\varphi_n) \approx J(\varphi_n)$ and so $\hat{J}(\varphi_M) \approx {}^*J(\varphi_M)$ for small infinite M.

Again using Theorem 22.3.16 we have $\hat{J}(\varphi_M) \approx J({}^\circ\varphi_M) = J(\varphi)$. Thus ${}^*J(\varphi_M) \approx J(\varphi)$ for small infinite M and so $J(\varphi_n) \to J(\varphi)$ as required. □

The compactness of $\mathfrak{D}$ and the density of step-controls now gives:

Theorem 22.3.18 *For feedback controls based on digital observations $J_0 = \hat{J}_0$ (the infimum for relaxed controls) and there is an optimal relaxed control φ_0 with $J(\varphi_0) = J_0 = \hat{J}_0$.*

22.4 Optimal control for $d = 3$

The extra complications of the 3D stochastic Navier-Stokes equations mean that we are only able to consider the simplest kind of controls. The controlled equation considered takes the form

$$\begin{aligned} u(t) = u_0 + \int_0^t \{-\nu A u(s) - B(u(s)) + f(s, u(s), \theta(s))\} \, ds + \\ + \int_0^t g(s, u(s)) \, dw(s) \end{aligned} \tag{22.4.1}$$

which has the same appearance as equation (22.3.1) for no-feedback controls in dimension $d = 2$.

The possible non-uniqueness of solutions means that the very definition of cost and optimality has to be refined — see below — and this requires a space that is rich enough to support essentially all possible solutions for any given control. For this it is necessary to employ a Loeb space.

First we fix the conditions on the coefficients f, g in the equation. These are the natural extension of the basic conditions for the fundamental existence of solutions in Theorem 22.2.3.

Conditions 22.4.1 *There exists an L^2 function $a(t) \geq 0$ such that*

(a) $f : [0,T] \times \mathbf{H} \times M \longrightarrow \mathbf{V}'$ *is jointly measurable with linear growth, continuous in the second variable on each K_n and continuous in the third variable, i.e.*

- **(i)** $|f(t,u,m)|_{\mathbf{V}'} \leq a(t)(1 + |u|)$,
- **(ii)** $f(t,\cdot,m) \in C(K_n; \mathbf{V}'_{\mathrm{weak}})$ *for all finite n, where $K_n = \{u \in \mathbf{V} : \|u\| \leq n\}$ as before (with the strong $\mathbf{H}$-topology).*
- **(iii)** $f(t,u,\cdot)$ *continuous*

for a.a. $t \in [0,T]$, all $u \in \mathbf{H}$ and all $m \in M$.

(b) $g : [0,T] \times \mathbf{H} \longrightarrow L(\mathbf{H},\mathbf{H})$ *is jointly measurable with of linear growth, and continuous in the second variable on each K_n, i.e.*

- **(i)** $|g(t,u)|_{\mathbf{H},\mathbf{H}} \leq a(t)(1 + |u|)$,
- **(ii)** $g(t,\cdot) \in C(K_n; L(\mathbf{H},\mathbf{H})_{\mathrm{weak}})$ *for all finite n,*

for a.a. $t \in [0,T]$ and all $u \in \mathbf{H}$.

22.4.1 Existence of solutions for any control

Theorem 22.2.3 then gives the following, where as before the key feature to note is the uniform energy bound.

Theorem 22.4.2 *There is a filtered probability space* $\mathbf{\Omega} = (\Omega, \mathcal{F}, (\mathcal{F}_t)_{t\geq 0}, P)$ *with the universal property that for any* f, g *satisfying Conditions 22.4.1(a,b) and any ordinary or relaxed control, the equation (22.4.1) has a solution satisfying the energy inequality*

$$\mathbb{E}\left(\sup_{t\in[0,T]} |u(t)|^2 + \int_0^T \|u(t)\|^2 dt \right) < E \tag{22.4.2}$$

with a constant $E = E\,(u_0, a)$ *uniform over the set of controls.*

The space fulfilling this result in [6, Theorem 6.4.1] is a particular Loeb space, which we need to specify later. Note that in the current 3D setting solutions may not be unique (this is an open problem).

22.4.2 The control problem for 3D stochastic Navier-Stokes equations

In order to formulate the optimal control problem for the 3D stochastic Navier-Stokes equations it is necessary to have a space $\mathbf{\Omega}$ as in Theorem 22.4.2 that has solutions for all controls. Fix such a space; then bearing in mind the possible non-uniqueness of solutions, the optimal control problem is formulated as follows. For a given control φ define

$$\begin{aligned} \mathcal{U}^\varphi := \{u :\ & u \text{ is weak solution on } \mathbf{\Omega} \\ & \text{to (22.4.1) for } \varphi \text{ and satisfying (22.4.2)}\,\} \end{aligned} \tag{22.4.3}$$

Then, taking the functions $h, \bar{h}$ satisfying the conditions of Definition 22.3.2 let

$$\mathcal{J}(\varphi, u) = \mathbb{E}\left(\int_0^T h\big(t, u(t), \varphi(t)\big)dt + \bar{h}(u(T))\right). \tag{22.4.4}$$

The cost for φ is then defined by

$$\mathcal{J}(\varphi) := \inf\{\mathcal{J}(\varphi, u) :\ u \in \mathcal{U}^\varphi\}. \tag{22.4.5}$$

Now set

$$\mathcal{J}_0 = \inf\{\mathcal{J}(\theta, u) :\ \theta \in \mathfrak{C}, u \in \mathcal{U}^\theta\} = \inf\{\mathcal{J}(\theta) :\ \theta \in \mathfrak{C}\}$$

which is the optimal cost for ordinary controls. Likewise

$$\hat{\mathcal{J}}_0 = \inf\{\mathcal{J}(\varphi, u) : \varphi \in \mathfrak{D}, u \in \mathcal{U}^\varphi\} = \inf\{\mathcal{J}(\varphi) : \varphi \in \mathfrak{D}\}$$

is the optimal cost for relaxed controls. The optimal control problem is to find if possible an optimal control θ and optimal solution $u \in \mathcal{U}^\theta$ such that

$$\mathcal{J}(\theta, u) = \mathcal{J}(\theta) = \mathcal{J}_0$$

and similarly for relaxed controls

$$\mathcal{J}(\varphi, u) = \mathcal{J}(\varphi) = \hat{\mathcal{J}}_0.$$

Remark The above definitions are of course relative to the space $\mathbf{\Omega}$ — so strictly this should be acknowledged by writing $\mathcal{J}_0^{\mathbf{\Omega}}$, etc. However, the space $\mathbf{\Omega}$ defined in the next section will be fixed for the rest of the paper.

22.4.3 The space $\mathbf{\Omega}$

In order to obtain an optimal control and optimal solution to the 3D stochastic Navier-Stokes equations the space must be rich enough to carry a good supply of solutions for each given control; we will see that the Loeb space used in [4] (see also [6]) to solve the general existence problem for the stochastic Navier-Stokes equations in dimensions ≤ 4 is just what is needed. For the rest of this paper we take $\mathbf{\Omega}$ to be this space, defined as follows.

Fix an infinite natural number N and take the internal space

$$\bar{\mathbf{\Omega}} = \big(\Omega, \mathcal{A}, (\mathcal{A}_\tau)_{\tau \in {}^*[0,T]}, \Pi\big)$$

where Ω is the canonical space of continuous functions ${}^*C\big({}^*[0,T]; \mathbf{H}_N\big)$, and Π is the Wiener measure induced by the canonical Wiener process $W(\tau) \in \mathbf{H}_N$ having covariance operator Q_N $(= \mathrm{Pr}_N\, {}^*Q\, \mathrm{Pr}_N)$. The filtration $(\mathcal{A}_\tau)_{\tau \in {}^*[0,T]}$ is that generated by the internal process $W(\tau)$.

Take the Loeb measure and Loeb algebra $P = L(\Pi)$, $\mathcal{F} = L(\mathcal{A})$ and put

$$\mathcal{F}_t = \bigcap_{t < {}^\circ\tau} \sigma(\mathcal{A}_\tau) \vee \mathcal{N}$$

where $\mathcal{N}$ denotes the family of P-null sets. This gives the adapted Loeb space

$$\mathbf{\Omega} = \big(\Omega, \mathcal{F}, (\mathcal{F}_t)_{t \in [0,T]}, P\big)$$

which carries the process $w = {}^\circ W$, which is a Wiener process in $\mathbf{H}$ with covariance Q.

The fundamental existence result for the 3D stochastic Navier-Stokes equations on the space $\mathbf{\Omega}$ in [4] or [6] is proved by taking the internal solution $U(\tau,\omega)$ to the Galerkin approximation of dimension N, so $U(\tau,\omega)$ is an internal process with values in $\mathbf{H}_N$. The existence of a uniform *finite* energy bound allows the definition of the standard part process $u = {}^\circ U$ in $\mathbf{H}$, which is the required solution. We do not need to repeat this here in order to establish existence, but the proof of the key theorem below concerning approximate solutions uses similar techniques.

22.4.4 Approximate solutions

Recall the main idea for using nonstandard methods in deterministic optimal control theory used in 2-dimensions. Take a minimizing sequence of controls θ_n and consider a nonstandard control $\Theta = {}^*\theta_K$ for infinite K and a solution $U = U^\Theta$ to the equation for this control. Then $u = {}^\circ U$ is a standard solution for the control $\varphi = {}^\circ\Theta$ which is thus optimal. The problem with this approach in the 3D stochastic setting is that the solution U associated with Θ would live in ${}^*\mathbf{H}$ and be carried by the nonstandard space ${}^*\mathbf{\Omega}$ (provided we can make sense of that). We showed in Section 22.3 how the uniqueness of solutions for $d = 2$ allows us to move back to the space $\mathbf{\Omega}$ to complete the story, but here an alternative approach is needed. This is provided by the notion of *approximate solutions*, first introduced in [11] in order to establish results on attractors for 3D stochastic Navier-Stokes equations. The rather technical notion required in [11] has been adjusted below to suit the current needs — and is somewhat simpler than the corresponding notion in [11].

Definition 22.4.3 (Approximate solutions) Fix an initial condition $u_0 \in H$ and let $E = E(u_0, \bar{a})$. Let $\Phi \in {}^*D$ be an internal control (ordinary or relaxed) and define sets X_j^Φ and X^Φ as follows.

(a) For each $j \in {}^*N$ denote by X_j^Φ the internal class of internal processes $U : {}^*[0,T] \times \Omega \to H_N$ that are *adapted to the filtration (A_τ), with the following properties:

(i) With Π-probability $\geq 1 - \frac{1}{j}$ on Ω, for all $\tau \in {}^*[0,T]$ and all $k \leq j$:

$$\begin{aligned} &\Big| U_k(\tau) - U_k(0) - \int_0^\tau \Big\{ -\nu A U_k(\sigma) - B_k\big(U(\sigma), U(\sigma)\big) \Big\} d\sigma \\ &- \int_0^\tau F_k^\Phi\big(\sigma, U(\sigma)\big) d\sigma - \int_0^\tau G_k\big(\sigma, U(\sigma)\big) dW(\sigma) \Big| \leq 2^{-j} \end{aligned} \tag{22.4.6}$$

where $B_k = ({}^*B, {}^*e_k)$, $F_k^\Phi\big(\sigma, U(\sigma)\big) = \Big({}^*f\big(\sigma, U(\sigma), \Phi(\sigma)\big), {}^*e_k\Big)$ and $G_k = ({}^*g, {}^*e_k)$.

(ii)

$$\mathrm{E}_\Pi\left(\sup_{\sigma\in{}^*[0,T]}|U(\sigma)|^2+\nu\int_0^T\|U(\sigma)\|^2d\sigma\right)\le E+\frac{1}{j} \tag{22.4.7}$$

(iii)

$$|U(0)-u_0|\le\frac{1}{j} \tag{22.4.8}$$

(b) Define $X^\Phi=\bigcap_{j\in\mathbb{N}}X_j^\Phi$. This is the set of ***approximate solutions*** to the controlled stochastic Navier-Stokes equations for control Φ.

Remark The sequence of internal sets $\mathcal{X}_j^\Phi$ is decreasing with j and since the set $\mathcal{X}^\Phi$ involves $\mathcal{X}_j^\Phi$ only for finite j then we have $\mathcal{X}_K^\Phi\subseteq\mathcal{X}^\Phi$ for all $K\in{}^*\mathbb{N}\backslash\mathbb{N}$.

The theory of [11], adapted to the modified definition of approximate solution here gives the following key result. First we must define the internal cost $\bar{\mathcal{J}}(\Phi,U)$ for an internal process $U\in\mathbf{H}_N$ on $\mathbf{\Omega}$ and control Φ by

$$\bar{\mathcal{J}}(\Phi,U)=\mathbb{E}_\Pi\left(\int_0^T {}^*h\big(\tau,U(\tau),\Phi(\tau)\big)d\tau+{}^*\bar{h}\big(U(\tau)\big)\right)$$

which is different from ${}^*\mathcal{J}_{{}^*\Omega}(\Phi,U)$ (which doesn't actually make sense).

Theorem 22.4.4

(a) *Let Φ be an internal control (that is, $\Phi\in{}^*\mathfrak{D}$) and $U\in\mathcal{X}^\Phi$. Then for P-a.a. $\omega\in\Omega$, $|U(\tau,\omega)|$ is finite for all $\tau\in{}^*[0,T]$ and $U(\cdot,\omega)$ is weakly S-continuous. The process defined by*

$$u(t,\omega)={}^\circ U(\tau,\omega)$$

for $\tau\approx t$ belongs to $\mathcal{U}^{{}^\circ\Phi}$ (where ${}^\circ\Phi$ is the standard part of Φ in the weak topology on $\mathfrak{D}$ as described in Section 22.2.3) and

$$\bar{\mathcal{J}}(\Phi,U)\approx\mathcal{J}({}^\circ\Phi,u).$$

(b) *Let $\varphi\in\mathfrak{D}$ be a standard control and $u\in\mathcal{U}^\varphi$. Then there exists $U\in\mathcal{X}^{{}^*\varphi}$ with $u={}^\circ U$ as defined in part (a), and hence*

$$\bar{\mathcal{J}}({}^*\varphi,U)\approx\mathcal{J}(\varphi,u).$$

Proof (Sketch). The proof of (a) follows quite closely the proof of Theorem 6.4.1 in [6]. The main difference here compared with that proof is (i) the presence of the control in the drift term f, which is dealt with routinely using

Anderson's Luzin Theorem, and (ii) the fact that internally we only have an approximate equation. That is, we have $\approx$ instead of $=$ in the internal equation for each finite co-ordinate of the solution. That is, from the approximate equality (22.4.6) with $j = K$ for some infinite K, using overspill, we have

$$U_k(\tau) \approx U_k(0) + \int_0^T \{-\nu AU_k(\sigma) - B_k(U(\sigma), U(\sigma))\}d\sigma + \int_0^T F_k^\Phi(\sigma, U(\sigma))d\sigma + \int_0^T G_k(\sigma, U(\sigma))dW(\sigma)$$

for finite k. This gives equality after taking standard parts.

For (b) the theory in [11] can be adapted easily to obtain suitable liftings of the solution u, giving an internal process $U \in \mathcal{X}^{*\varphi}$ such that $u = {}^\circ U$. The remaining fact about the cost follows from part **(a)** with $\Phi = {}^*\varphi$ and using the fact that ${}^\circ({}^*\theta) = \theta$. □

22.4.5 Optimal control

The results concerning optimal control of the stochastic 3D equation follow from the next more general theorem.

Theorem 22.4.5 *Let f, g, h and $\bar{h}$ satisfy Conditions 22.4.1.*

(a) For every control φ there is an optimal solution $u^\varphi \in \mathcal{U}^\varphi$ with $\mathcal{J}(\varphi, u^\varphi) = \mathcal{J}(\varphi)$

(b) Let φ_n be a sequence of controls and $u_n \in \mathcal{U}^{\varphi_n}$, with $\mathcal{J}(\varphi_n, u_n)$ converging. Then there is a control φ and $u \in \mathcal{U}^\varphi$ with $\mathcal{J}(\varphi, u) = \lim_{n\to\infty} \mathcal{J}(\varphi_n, u_n)$.

Proof. (a) For a control $\varphi \in \mathfrak{D}$ take a minimizing sequence $(u_k)_{k\in\mathbb{N}} \subset \mathcal{U}^\varphi$ such that $\mathcal{J}(\varphi, u_k) \searrow \mathcal{J}(\varphi)$. Using Theorem 22.4.4 (b), for each k take an approximate solution $U_k \in \mathcal{X}^{*\varphi}$ such that $u_k = {}^\circ U_k$ and $\bar{\mathcal{J}}({}^*\varphi, U_k) \approx \mathcal{J}(\varphi, u_k)$. Weakening this gives

$$U_k \in \mathcal{X}_k^{*\varphi}$$

and

$$|\bar{\mathcal{J}}({}^*\varphi, U_k) - \mathcal{J}(\varphi, u_k)| < \frac{1}{k}$$

for each finite k so by $\aleph_1$-saturation there is an infinite K with $U_K \in \mathcal{X}_K^{*\varphi} \subseteq \mathcal{X}^{*\varphi}$ and $\bar{\mathcal{J}}({}^*\varphi, U_K) \approx \mathcal{J}(\varphi, u_K)$. Then Theorem 22.4.4 (a) gives ${}^\circ U_K \in \mathcal{U}^\varphi$ with $\mathcal{J}(\varphi, {}^\circ U_K) = \mathcal{J}(\varphi)$, so we may take $u^\varphi = {}^\circ U_K$.

(b) The proof is similar to the proof of (a). For each k take an approximate solution $U_k \in \mathcal{X}^{*\varphi_k}$ such that $u_k = {}^\circ U_k$ and $\bar{\mathcal{J}}({}^*\varphi_k, U_k) \approx \mathcal{J}(\varphi_k, u_k)$. Using $\aleph_1$-saturation there is an infinite K with $U_K \in \mathcal{X}_K^{*\varphi_K} \subseteq \mathcal{X}^{*\varphi_K}$ and $\bar{\mathcal{J}}({}^*\varphi_K, U_K) \approx \mathcal{J}_1 = \lim_{n\to\infty} \mathcal{J}(\varphi_n, u_n)$.

Now let $\varphi = {}^\circ({}^*\varphi_K)$ and $u = {}^\circ U_K$; Theorem 22.4.4 (a) gives $u \in \mathcal{U}^\varphi$ and $\mathcal{J}(\varphi, u) = {}^\circ\bar{\mathcal{J}}({}^*\varphi_K, U_K) = \mathcal{J}_1$ □

Corollary 22.4.6

(a) There exists a relaxed control φ_0 (and solution $u^{\varphi_0} \in \mathcal{U}^{\varphi_0}$) that achieves the minimum cost for ordinary controls (i.e. such that $\mathcal{J}(\varphi_0, u^{\varphi_0}) = \mathcal{J}(\varphi_0) = \mathcal{J}_0$);

(b) There is an optimal relaxed control $\hat{\varphi}_0$ and optimal solution $u^{\hat{\varphi}_0} \in \mathcal{U}^{\hat{\varphi}_0}$ with $\mathcal{J}(\hat{\varphi}_0, u^{\hat{\varphi}_0}) = \mathcal{J}(\hat{\varphi}_0) = \hat{\mathcal{J}}_0$.

Proof. For (a) take a minimizing sequence of ordinary controls (and solutions) and for (b) take a minimizing sequence of relaxed controls and solutions. Then apply (b) of the previous theorem. □

Remark 22.4.7 One consequence of the possible non-uniqueness of solutions is that unlike in dimension 2 we cannot prove that $\hat{J}_0 = J_0$, so this theory is rather less satisfactory than that for 2D.

22.4.6 Hölder continuous feedback controls ($d = 3$)

Recall that in dimension $d = 2$, the existence of optimal ordinary feedback controls in a general setting was ensured by taking *Hölder continuous feedback controls* (Section 22.3.5). The technique of approximate solutions allows the extension of this idea to the 3D stochastic setting. Controls take the form $\theta(t, u(t))$, and the controlled equation (22.4.1) becomes

$$\begin{aligned} u(t) = u_0 + \int_0^t \{ -\nu A u(s) - B(u(s)) + f(s, u(s), \theta(s, u(s))) \} ds + \\ + \int_0^t g(s, u(s))\, dw(s) \end{aligned} \tag{22.4.9}$$

with cost function

$$\mathcal{J}(\theta, u) = \mathbb{E}\left(\int_0^T h(t, u(t), \theta(t, u(t)))dt + \bar{h}(u(T)) \right) \tag{22.4.10}$$

for $u \in \mathcal{U}^\varphi$

We maintain the Conditions 22.4.1 on the coefficients $f, g, h, \bar{h}$.

The controls considered are the Hölder continuous feedback controls $\mathfrak{C}(\alpha,\beta,\gamma,c_1,c_2)$ given by Definition 22.3.7.

The controls we consider are *Hölder continuous feedback controls* defined as follows. So we fix constants $(\alpha,\beta,\gamma,c_1,c_2)$ and as before write $\mathfrak{C}^H = \mathfrak{C}(\alpha,\beta,\gamma,c_1,c_2)$. We continue to work with the space $\mathbf{\Omega}$ as defined above (Section 22.4.3); it is routine to see that for any control $\theta \in \mathfrak{C}^H$ there is a solution to (22.4.9) having the energy bound (22.4.2). As before write $\mathcal{U}^\theta$ for the set of all such solutions, and then the minimal cost is defined as expected: for $\theta \in \mathfrak{C}^H$ extend the earlier definition to give:

$$\mathcal{J}(\theta) = \inf\{\mathcal{J}(\theta,u) : \ u \in \mathcal{U}^\theta\}$$

and then set

$$\mathcal{J}_0^H = \inf\{\mathcal{J}(\theta) : \ \theta \in \mathfrak{C}^H\}.$$

The standard part of an internal control is defined in the natural way:

Definition 22.4.8 Suppose that $\Theta \in {}^*\mathfrak{C}^H$, so $\Theta : {}^*[0,T] \times {}^*\mathbf{H} \to {}^*M$. Define ${}^\circ\Theta = \theta \in \mathfrak{C}^H$ by

$$\theta(t,u) = {}^\circ\Theta(t,u) = {}^\circ\Theta(\tau,U)$$

for any $\tau \approx t$ and $U \approx u$ in $\mathbf{H}$ (this makes sense because of the condition (22.3.7)).

22.4.7 Approximate solutions for Hölder continuous controls

For an initial condition $u_0 \in \mathbf{H}$ and an internal control $\Theta \in {}^*\mathfrak{C}^H$ the sets $\mathcal{X}_j^\Theta$ and $\mathcal{X}^\Theta$ are defined as in Definition 22.4.3 with F_k^Θ modified in the obvious way:

$$F_k^\Theta(\sigma,U(\sigma)) = \left({}^*f(\sigma,U(\sigma),\Theta(\sigma,U(\sigma))),{}^*e_k\right)$$

Then we have the counterpart of Theorem 22.4.4 for Hölder continuous controls.

Theorem 22.4.9

(a) Let Θ be an internal Hölder continuous control (that is, $\Theta \in {}^*\mathfrak{C}^H$) and $U \in \mathcal{X}^\Theta$. Then for a.a. $\omega \in \Omega$, $|U(\tau,\omega)|$ is finite for all $\tau \in {}^*[0,T]$ and $U(\cdot,\omega)$ is weakly S-continuous. The process defined by

$$u(t,\omega) = {}^\circ U(\tau,\omega)$$

for $\tau \approx t$ belongs to $\mathcal{U}^{\circ\Theta}$ and

$$\bar{\mathcal{J}}(\Theta,U) \approx \mathcal{J}({}^\circ\Theta,u).$$

(b) Let $\theta \in \mathfrak{C}^H$ *be a standard Hölder continuous control and* $u \in \mathcal{U}^\theta$*. Then there exists* $U \in \mathcal{X}^{*\theta}$ *with* $u = {}^\circ U$ *as defined in part (a), and hence*

$$\bar{\mathcal{J}}({}^*\theta, U) \approx \mathcal{J}(\theta, u).$$

The proof is similar to the proof of Theorem 22.4.4 except for dealing with the terms that contain the control, and this is routine using the Hölder continuity property.

Finally we have the counterpart of Theorem 22.4.5.

Theorem 22.4.10 *Let* f*,* g*,* h *and* $\bar{h}$ *satisfy Conditions 22.4.1.*

(a) For every Hölder continuous control θ *there is an optimal solution* $u^\theta \in \mathcal{U}^\theta$ *with* $\mathcal{J}(\theta, u^\theta) = \mathcal{J}(\theta)$*.*

(b) Let θ_n *be a sequence of controls in* $\mathfrak{C}^H$ *and* $u_n \in \mathcal{U}^{\theta_n}$*, with* $\mathcal{J}(\theta_n, u_n)$ *converging. Then there is a control* $\theta \in \mathfrak{C}^H$ *and* $u \in \mathcal{U}^\theta$ *with* $\mathcal{J}(\theta, u) = \lim_{n\to\infty} \mathcal{J}(\theta_n, u_n)$*.*

Proof. Proved from Theorem 22.4.9 just as Theorem 22.4.5 followed from Theorem 22.4.4. □

The existence of an optimal control now follows.

Corollary 22.4.11 *There is an optimal Hölder continuous control* θ_0 *and optimal solution* $u^{\theta_0} \in \mathcal{U}^{\theta_0}$ *with* $\mathcal{J}(\theta_0, u^{\theta_0}) = \mathcal{J}(\theta_0) = \mathcal{J}_0^H$*.*

Proof. Take a minimizing sequence θ_n of controls in $\mathfrak{C}^H$ and solutions $u_n \in \mathcal{U}^\theta$ with $\mathcal{J}(\theta_n, u^{\theta_n}) \to \mathcal{J}_0^H$ and apply (b) of the previous theorem. □

Appendix: Nonstandard representations of the spaces $\mathbf{H}^r$

For ease of reference the most important facts about the nonstandard representation of members of the spaces $\mathbf{H}^r$ are summarized here, together with other crucial nonstandard matters. For full details consult [6].

The space ${}^*\mathbf{H}$ has a basis $({}^*e_n)_{n\in{}^*\mathbb{N}} = (E_n)_{n\in{}^*\mathbb{N}}$ given by the nonstandard extension ${}^*e$ of the function $e : \mathbb{N} \to \mathbf{H}$. For each $U \in {}^*\mathbf{H}$ there is a unique internal sequence of hyperreals $(U_n)_{n\in{}^*\mathbb{N}}$ such that

$$U = \sum_{n=1}^{{}^*\infty} U_n E_n.$$

and the subspace $\mathbf{H}_N$ of ${}^*\mathbf{H}$ defined by

$$\mathbf{H}_N = \Big\{ \sum_{n=1}^{N} U_n E_n : U \in {}^*\mathbb{R}^N,\ U \text{ internal} \Big\}$$

On $\mathbf{H}$ (and also $\mathbf{V}$ or any of the spaces $\mathbf{H}^r$) there are both weak and strong topologies. Here are the most important nonstandard facts.

Proposition 22.4.12 *Let $U \in {}^*\mathbf{H}$ (or $\mathbf{H}_N$) Then*

(a) *U is (strongly) nearstandard to u in $\mathbf{H}$ (denoted $U \approx u$) if $|U - {}^*u| \approx 0$ (and then $|U| \approx |u|$).*

(a) *U is weakly nearstandard to u in $\mathbf{H}$ (denoted $U \approx_w u$) if $(U, {}^*v) \approx (u, v)$ for all $v \in \mathbf{H}$ and $|u| \leq {}^\circ|U|$ (allowing this to be ∞).*

(b) *If U is strongly nearstandard then it is weakly nearstandard and the standard parts agree, so we write ${}^\circ U$ for the standard part in whichever topology it may be nearstandard.*

(c) *If $|U|$ is finite then U is weakly nearstandard and $({}^\circ U)_n = {}^\circ(U_n)$ for finite n.*

(d) *If $\|U\|$ is finite then U is strongly nearstandard in $\mathbf{H}$.*

(e) *If $|U|, |V| < \infty$ then*

$$U \approx_w V \iff U_i \approx V_i \quad \forall i \in \mathbb{N}.$$

Similar facts obtain for other spaces in the spectrum $\mathbf{H}^r$.

From (22.2.2) we have the following lemma (a slight extension of the Crucial Lemma (Lemma 2.7.7) of [6]), which is crucial for taking standard parts of the quadratic term $B(U)$, which is the source of many of the difficulties when discussing the Navier-Stokes equations.

Lemma 22.4.13 (Crucial Lemma) *If $U, V \in {}^*\mathbf{V}$ with $\|U\|$ and $\|V\|$ both finite, and $z \in \mathbf{V}$ then*

$${}^*b(U, V, {}^*z) \approx b(u, v, z)$$

where $u = {}^\circ U$ and $v = {}^\circ V$ (with $u, v \in \mathbf{V}$.) Hence, if $\|U\| < \infty$

$${}^*B(U) \approx b(u) \text{ in } \mathbf{V}' \text{ (weakly)}$$

For the proof consult any of [6, 9, 10].

Taking standard parts of the term AU needs the following observation.

Lemma 22.4.14 *If $U \in {}^*\mathbf{V}$ with $\|U\|$ finite with $u = {}^\circ U$ then*

$${}^*AU \approx Au \text{ in } \mathbf{V}' \text{ (weakly)}$$

References

[1] S. ALBEVERIO, J.E. FENSTAD, R. HOEGH-KROHN and T. LINDSTROM, *Nonstandard Methods in Stochastic Analysis and Mathematical Physics*, Academic Press, New York, 1986.

[2] E.J. BALDER, New fundamentals of Young measure convergence, in *Calculus of Variations and Differential Equations* (Eds. A. Ioffe,S. Reich & I. Shafrir), CRC Research Notes **410**, CRC Press, Boca Raton, 1999.

[3] H.I. BRECKNER, *Approximation and Optimal control of the Stochastic Navier-Stokes Equation*, PhD Thesis, Halle (Saale), 1999. `http://webdoc.sub.gwdg.de/ebook/e/2002/pub/mathe/99H101/of_index.htm`

[4] M. CAPIŃSKI and N.J. CUTLAND, "Stochastic Navier-Stokes equations", Acta Applicanda Mathematicae, **25** (1991) 59–85.

[5] M. CAPIŃSKI and N.J. CUTLAND, "Existence of global stochastic flow and attractors for Navier-Stokes equations", Probability Theory and Related Fields, **115** (1999) 121–151.

[6] M. CAPIŃSKI and N.J. CUTLAND, *Nonstandard Methods for Stochastic Fluid Mechanics*, World Scientific, 1997.

[7] N.J. CUTLAND, "Infinitesimal methods in control theory: deterministic and stochastic", Acta Appl. Math., **5** (1986) 105–135.

[8] N.J. CUTLAND, Loeb measure theory, in *Developments in nonstandard mathematics*, Eds. N.J. Cutland, F. Oliveira, V. Neves, J. Sousa-Pinto, Pitman Research Notes in Mathematics Vol. 336, Longman, 1995.

[9] N.J. CUTLAND and K. GRZESIAK, "Optimal control for 2-dimensional stochastic Navier-Stokes equations", to appear in Applied Mathematics and Optimization.

[10] N.J. CUTLAND and K. GRZESIAK, "Optimal control for 3-dimensional stochastic Navier-Stokes equations", Stochastics and Stochastic Reports, **77** (2005) 437–454.

[11] N.J. CUTLAND and H.J. KEISLER, "Global attractors for 3-dimensional stochastic Navier-Stokes equations", Journal of Dynamics and Differential Equations", **16** (2004) 205–266.

[12] N.J. Cutland and H.J. Keisler, "Attractors and neoattractors for 3D stochastic Navier-Stokes equations", Stochastics and Dynamics, **5** (2005) 487-533.

[13] G. DaPrato and A. Debussche, "Dynamic programming for the stochastic Navier-Stokes equations", Mathematical Modelling and Numerical Analysis, **34** (2000) 459–475.

[14] G. DaPrato and J. Zabczyk, *Stochastic Equations in Infinite Dimensions*, Cambridge University Press, Cambridge, 1992.

[15] R.V. Gamkrelidze, *Principles of Optimal Control*, Plenum, New York, 1978.

[16] K. Grzesiak, *Optimal Control for Navier-Stokes Equations using Nonstandard Analysis*, PhD Thesis, University of Hull, UK, 2003.

[17] A. Ichikawa, "Stability of semilinear stochastic evolution equations", Journal of Mathematical Analysis and Applications, **90** (1982) 12–44.

[18] S.S. Sritharan (ed.), *Optimal Control of Viscous Flow*, SIAM Frontiers in Applied Mathematics, Philadelphia, 1998.

[19] R. Temam, *Navier-Stokes Equations*, North-Holland, Amsterdam, third edition, 1984.

[20] M. Valadier, Young measures, in *Methods of Nonconvex Analysis* (A. Cellina, ed.), Lecture Notes in Mathematics 1446, Springer-Verlag, Berlin, 1990.

23

Local-in-time existence of strong solutions of the n-dimensional Burgers equation via discretizations

João Paulo Teixeira*

Abstract

Consider the equation:

$$u_t = \nu\Delta u - (u \cdot \nabla)u + f \quad \text{for} \quad x \in [0,1]^n \ \text{ and } \ t \in (0,\infty),$$

together with periodic boundary conditions and initial condition $u(t,0) = g(x)$. This corresponds a Navier-Stokes problem where the incompressibility condition has been dropped. The major difficulty in existence proofs for this simplified problem is the unbounded advection term, $(u \cdot \nabla)u$.

We present a proof of local-in-time existence of a smooth solution based on a discretization by a suitable Euler scheme. It will be shown that this solution exists in an interval $[0,T)$, where $T \leq \frac{1}{C}$, with C depending only on n and the values of the Lipschitz constants of f and u at time 0. The argument given is based directly on local estimates of the solutions of the discretized problem.

23.1 Introduction

The Burgers equation

$$u_t - \nu\Delta u + (u \cdot \nabla)u = f \qquad x \in D \subset \mathbb{R}^n,\ t > 0$$

provides an example of a model for flows that takes into account the interaction between diffusion and (nonlinear) advection. This is probably the simplest nonlinear physical model for turbulence.

*Instituto Superior Técnico, Lisbon, Portugal.
jteix@math.ist.utl.pt

This equation is a simplification of the Navier-Stokes equations where the incompressibility constraint on the flow has been dropped. The Navier-Stokes equations are usually studied by taking projected solutions of the Burgers equations onto a subspace of divergence free functions. The main difficulty in the analysis of nonlinear flows, the advection term, $(u \cdot \nabla)u$, is present in both equations.

It is a consequence of the incompressibility condition that there are only trivial Navier-Stokes flows for $n = 1$. This is not the case in Burgers flows, where the case $n = 1$ is already nontrivial. Cole [6] and Hopf [10] have studied the problem in the real line:

$$\begin{cases} u_t - \nu u_{xx} + u u_x = 0 & x \in \mathbb{R},\ t > 0 \\ u(x,t) = u_0(x) & x \in \mathbb{R}. \end{cases} \tag{23.1.1}$$

Using the transformation of variables

$$u = -2\nu \frac{v}{v_x}, \tag{23.1.2}$$

(23.1.1) was reduced to and initial value problem for the heat equation:

$$\begin{cases} v_t - \nu v_{xx} = 0 & x \in \mathbb{R},\ t > 0 \\ v(x,t) = v_0(x) & x \in \mathbb{R}. \end{cases} \tag{23.1.3}$$

(Where $u_0' = -2\nu \frac{v}{v_0'}$). This enabled them to get a formula for a solution of problem (23.1.1), in terms of a Gaussian integral. Thus, and under very mild conditions on u_0, a solution, u, exists for all $x \in \mathbb{R}$ and $t > 0$, and is smooth.

It would be reasonable to expect that the $n > 1$ case should behave in a similar way. The usual standard theory for this type of equations uses a weak formulation of the problem in Sobolev spaces, a Galerkin approximation to show existence of weak solutions and, finally, regularity estimates. However, these methods only lead to partial results. For well-posed problems with generic initial and boundary data, the best that can be obtained is local in-time existence of regular solutions. It can also be shown that regular (strong) solutions are unique (if they exist). See [4, 13, 9, 12, 16] for accounts on these methods.

To gain some new insight into this problem, we develop a hyperfinite approach to the following model problem on a compact domain.

Model Problem: *Let* $\mathbb{T}^n = \mathbb{R}^n/\mathbb{Z}^n$ *be an n-dimensional torus. Assume that* f *is locally Lipschitz continuous on* $\mathbb{T}^n \times [0,\infty)$ *and* $u_0 \in C^{2,1}(\mathbb{T}^n)$ *(that is,* u_0 *is twice differentiable and all its second partial derivatives are Lipschitz continuous on* $\mathbb{T}^n$*). Let* $\mathcal{D} = \mathbb{T}^n \times (0,\infty)$*. Let* $\nu \in \mathbb{R}^+$*. Let* $f : \mathbb{R}^n/\mathbb{Z}^n \to \mathbb{R}$,

$u_0 : \mathbb{R}^n/\mathbb{Z}^n \to \mathbb{R}^n$. *Our task is to study the initial value problem for the Burgers equations:*

$$\begin{cases} u_t - \nu\Delta u + (u \cdot \nabla)u = f & \text{in } \mathcal{D} \\ u = u_0 & \text{on } \mathbb{T}^n \times \{0\}. \end{cases} \tag{23.1.4}$$

A (strong) solution of this problem is a sufficiently smooth u satisfying (23.1.4). By "sufficiently smooth u" we mean that $u : \mathbb{T}^n \times [0, \infty) \to \mathbb{R}$ is such that $u(\cdot, t) \in C^2(\mathbb{T}^n)$ for all $t \in [0, \infty)$ and $u(x, \cdot) \in C^1([0, \infty))$ for all $x \in \mathbb{T}^n$.

23.2 A discretization for the diffusion-advection equations in the torus

We now look for a discretized version of problem (23.1.4). Since we are mainly interested in the existence result, we will do this in the simplest way possible. First we work in the standard universe. To discretize $\mathbb{T}^n$, we introduce an h-spaced grid on $\mathbb{T}^n$. Choose $M \in \mathbb{N}_1$, and let $h = \frac{1}{M}$. Then, let:

$$\mathbb{T}^n_M = \Big\{0, h, 2h, \ldots, (M-1)h, 1\Big\}^n = h\ (\mathbb{Z} \bmod M)^n.$$

Consistently with our interpretation of $\mathbb{T}^n_M$ as a discrete version of the torus, $\mathbb{T}^n$, we define addition in $\mathbb{T}^n_M$ as follows: given any $x = (m_1, m_2, \ldots, m_n)h$ and $y = (l_1, l_2, \ldots, l_n)h$ in $\mathbb{T}^n_M$, let:

$$x + y = \Big((m_1 + l_1) \bmod M, (m_2 + l_2) \bmod M, \ldots, (m_n + l_n) \bmod M\Big)h$$

This makes addition well-defined in $\mathbb{T}^n_M$; furthermore, it will behave in a similar way to addition in $\mathbb{T}^n$. In particular, the set of grid-neighbors of any $x \in \mathbb{T}^n_M$,

$$\{x \pm he_i : i = 1, 2, \ldots, n\}$$

is well-defined.

As for the discretization of time, consider $T \in \mathbb{R}^+$ and $K \in \mathbb{N}_1$. Let $k = \frac{T}{K}$, and define:

$$I^T_K = \Big\{0, k, 2k, \ldots, (K-1)k, T\Big\} = k\ (\mathbb{N} \cap [0, K));$$

To each triple $d = (M, K, T)$, with $M, N \in \mathbb{N}_1$ and $T \in \mathbb{R}^+$, we associate a discretization as defined above. We will, later on, introduce some restrictions on the set of admissible d. For now, given any $d = (M, K, T)$, with $M, N \in \mathbb{N}_1$ and $T \in \mathbb{R}^+$, we let:

$$\mathcal{D}_d = \mathbb{T}^n_M \times I^T_K$$

$$\overline{\mathcal{D}}_d = \mathbb{T}_M^n \times (I_K^T \cup \{T\})$$

The elements of $\overline{\mathcal{D}}_d$ are called gridpoints. Any function U whose domain is a subset of $\overline{\mathcal{D}}_d$ is called a gridfunction.

The discrete Laplacian can be defined as a map $\Delta_d : \mathbb{R}^{\overline{\mathcal{D}}_d} \to \mathbb{R}^{\overline{\mathcal{D}}_d}$ given by:

$$\Delta_d U(x,t) = \frac{1}{h^2} \sum_{i=1}^{n} \Big(U(x+he_i,t) - 2U(x,t) + U(x-he_i,t) \Big). \qquad (23.2.1)$$

The definition of addition in $\mathbb{T}_M^n$ makes this well-defined.

The discrete version of the nonlinear parabolic operator, P, that occurs in the diffusion-advection equations is defined as the map $P_d : \mathbb{R}^{\overline{\mathcal{D}}_d} \to \mathbb{R}^{\mathcal{D}_d}$ given by:

$$\begin{aligned} P_d U(x,t) &= \frac{U(x,t+k) - U(x,t)}{k} - \nu \Delta_d U(x,t) \\ &\quad + \sum_{i=1}^{n} U_i(x,t) \frac{U(x+he_i,t) - U(x-he_i,t)}{2h}. \end{aligned}$$

With

$$\lambda = \frac{k}{h^2} = \frac{TM^2}{K} \in \mathbb{R}^+,$$

we get:

$$\begin{aligned} &P_d U(x,t) = \\ &= \frac{1}{\lambda h^2} \Bigg(U(x,t+k) - (1 - 2n\nu\lambda) U(x,t) - \\ &- \lambda \sum_{i=1}^{n} \Big(\Big(\nu - \frac{h}{2} U_i(x,t)\Big) U(x+he_i,t) + \Big(\nu + \frac{h}{2} U_i(x,t)\Big) U(x-he_i,t) \Big) \Bigg) \end{aligned} \qquad (23.2.2)$$

The discretized version of problem (23.1.4) is, then:

$$\begin{cases} P_d U(x,t) = f(x,t) & \text{if } (x,t) \in \mathcal{D}_d \\ U(x,0) = u_0(x,0) & \text{if } x \in \mathbb{T}_M^n. \end{cases} \qquad (23.2.3)$$

Let us look more closely at the finite difference equation in (23.2.3). If we solve for $U(x,t+k)$, we get:

$$\begin{aligned} &U(x,t+k) = \big(1 - 2n\nu\lambda\big) U(x,t) \\ &+ \lambda \sum_{i=1}^{n} \Big(\Big(\nu - \frac{h}{2} U_i(x,t)\Big) U(x+he_i,t) + \Big(\nu + \frac{h}{2} U_i(x,t)\Big) U(x-he_i,t) \Big) \\ &+ \lambda h^2 f(x,t) \end{aligned} \qquad (23.2.4)$$

Equation (23.2.4), together with the initial condition in (23.2.3), gives a recursive formula for the unique solution of problem (23.2.3).

Define a map

$$(\mathbb{R}^n)^{\overline{\mathcal{D}}_d} \times (\mathbb{R}^n)^{\overline{\mathcal{D}}_d} \ni (U,V) \stackrel{\Phi}{\longmapsto} \Phi(U,V) \in (\mathbb{R}^n)^{\overline{\mathcal{D}}_d},$$

by:

$$\begin{aligned}\Phi(U,V)(x,t) &= \left(1-2n\nu\lambda\right)U(x,t)+\\ +\lambda\sum_{i=1}^{n}&\left(\left(\nu-\frac{h}{2}V_i(x,t)\right)U(x+he_i,t)+\left(\nu+\frac{h}{2}V_i(x,t)\right)U(x-he_i,t)\right),\end{aligned} \tag{23.2.5}$$

whenever $(x,t) \in \overline{\mathcal{D}}_d$. Equation (23.2.4) can now be written as:

$$U(x,t+k) = \Phi(U,U)(x,t) + \lambda h^2 f(x,t)$$

In what follows, we will require the right-hand side of (23.2.5) to be in the form of a weighted average, that is, all coefficients multiplying $U(x,t)$ and $U(x \pm he_i,t)$ must be positive. For this, we will always assume that:

$$\lambda < \frac{1}{2n\nu} \tag{23.2.6}$$

We also require that:

$$|U_d(x,t)| < \frac{2\nu}{h} \qquad \text{for all } (x,t) \in \overline{\mathcal{D}}_d, \tag{23.2.7}$$

where U_d is the solution of (23.2.3), relative to d.

This means that the set of admissible discretizations is:

$$\left\{(M,K,T) \in \mathbb{N}_1 \times \mathbb{N}_1 \times \mathbb{R}^+ : \lambda < \frac{1}{2n\nu} \wedge \forall (x,t) \in \overline{\mathcal{D}}_d \; |U_d(x,t)| < \frac{2\nu}{h}\right\}$$

Note that condition (23.2.7) is easily satisfied when we work in the nonstandard universe. Whenever h is infinitesimal, $\frac{2\nu}{h}$ will be infinitely large; so, if U is kept finite, then condition (23.2.7) holds.

23.3 Some standard estimates for the solution of the discrete problem

In this section, we derive estimates that will not require a nonstandard discretization.

Consider any (standard) discretization, d. Given $U : \overline{\mathcal{D}}_d \to \mathbb{R}^n$ and $A \subset \overline{\mathcal{D}}_d$, let:

$$\begin{aligned}
\|U\|_{L^\infty_d(A)} &= \max_{(x,t)\in A} |U(x,t)| \\
[U]_{L^\infty_d(A)} &= \max_{(x,t),(y,t)\in A\; x\neq y} \frac{|U(x,t)-U(y,t)|}{|x-y|} \\
[[U]]_{L^\infty_d(A)} &= \max_{(x,t),(x,t+k)\in A} \frac{|U(x,t+k)-U(x,t)|}{k}
\end{aligned}$$

Here, $|\cdot|$ denotes the Euclidean norm on $\mathbb{R}^j$.

We begin by showing some properties of Φ.

Lemma 1 *Let $U, V \in (\mathbb{R}^n)^{\overline{\mathcal{D}}_d}$. Let $h > 0$ be such that:*

$$h < \frac{2\nu}{\|V\|_{L^\infty_d(\overline{\mathcal{D}}_d)}}. \tag{23.3.1}$$

Then, for all $(x,t) \in \overline{\mathcal{D}}_d$:

$$|\Phi(U,V)(x,t)| \leq \|U\|_{L^\infty_d(\overline{\mathcal{D}}_d)}.$$

Proof. For any $(x,t) \in \overline{\mathcal{D}}_d$:

$$\begin{aligned}
&\big|\Phi(U,V)(x,t)\big| \leq \big|\left(1-2n\nu\lambda\right)U(x,t)\big| + \\
&+\lambda\sum_{i=1}^n\left(\left|\left(\nu-\frac{h}{2}V_i(x,t)\right)U(x+he_i,t)\right|+\left|\left(\nu+\frac{h}{2}V_i(x,t)\right)U(x-he_i,t)\right|\right).
\end{aligned}$$

Let $M = \|U\|_{L^\infty_d(\overline{\mathcal{D}}_d)}$. By condition (23.3.1):

$$\nu \pm \frac{h}{2}V(x,t) > 0.$$

Thus:

$$\begin{aligned}
&\big|\Phi(U,V)(x,t)\big| \leq \left(1-2n\nu\lambda\right)|U(x,t)| + \\
&+\lambda\sum_{i=1}^n\left(\left(\nu-\frac{h}{2}V_i(x,t)\right)|U(x+he_i,t)|+\left(\nu+\frac{h}{2}V_i(x,t)\right)|U(x-he_i,t)|\right).
\end{aligned}$$

Since $|U(x,t)| \leq M$, we get:

$$\begin{aligned}
|\Phi(U,V)(x,t)| &\leq \left(1-2n\nu\lambda\right)M + \lambda M\sum_{i=1}^n\left(\nu-\frac{h}{2}V_i(x,t)+\nu+\frac{h}{2}V_i(x,t)\right) \\
&\leq \left(1-2n\nu\lambda\right)M + \lambda M 2\nu n = M.
\end{aligned}$$

□

Lemma 2 *Let* $U, V, W, Z \in (\mathbb{R}^n)^{\overline{\mathcal{D}}_d}$. *Then, for all* $(x,t) \in \overline{\mathcal{D}}_d$:

$$\Phi(U,W)(x,t) - \Phi(V,Z)(x,t) = \Phi(U-V,W)(x,t) + \\ + \frac{\lambda h}{2}\sum_{i=1}^{n}\Big(Z_i(x,t) - W_i(x,t)\Big)\Big(V(x+he_i,t) - V(x-he_i,t)\Big).$$

Proof. For all $(x,t) \in \overline{\mathcal{D}}_d$:

$$\Phi(U,W)(x,t) - \Phi(V,Z)(x,t) = (1-2n\lambda\nu)\Big(U(x,t) - V(x,t)\Big) + \\ + \lambda\sum_{i=1}^{n}\bigg(\Big(\nu - \frac{h}{2}W_i(x,t)\Big)U(x+he_i,t) - \Big(\nu - \frac{h}{2}Z_i(x,t)\Big)V(x+he_i,t) + \\ + \Big(\nu + \frac{h}{2}W_i(x,t)\Big)U(x-he_i,t) - \Big(\nu + \frac{h}{2}Z_i(x,t)\Big)V(x-he_i,t)\bigg)$$

$$= (1-2n\lambda\nu)\Big(U(x,t) - V(x,t)\Big) + \\ + \lambda\sum_{i=1}^{n}\bigg(\Big(\nu - \frac{h}{2}W_i(x,t)\Big)\Big(U(x+he_i,t) - V(x+he_i,t)\Big) + \\ + \Big(\nu + \frac{h}{2}W_i(x,t)\Big)\Big(U(x-he_i,t) - V(x-he_i,t)\Big)\bigg) + \\ + \frac{\lambda h}{2}\sum_{i=1}^{n}\bigg((Z_i(x,t) - W_i(x,t))V(x+he_i,t) - \\ - (Z_i(x,t) - W_i(x,t))V(x-he_i,t)\bigg)$$

$$= \Phi(U-V,W)(x,t) + \\ \frac{\lambda h}{2}\sum_{i=1}^{n}\Big(Z_i(x,t) - W_i(x,t)\Big)\Big(V(x+he_i,t) - V(x-he_i,t)\Big).$$

□

Lemma 3 *Let d be a discretization such that:*

$$h < \frac{2\nu}{\|u_0\|_{L^\infty} + T\|f\|_{L^\infty}}$$

If U is the solution of the discrete problem, (23.2.3), then:

$$\|U\|_{L^\infty_d(\overline{\mathcal{D}}_d)} \le \|u_0\|_{L^\infty} + T\|f\|_{L^\infty}$$

In particular, d is an admissible discretization.

Proof. Let $M = \|u_0\|_{L^\infty}$ and $L = \|f\|_{L^\infty}$. We show, by induction on $t = 0, k, \ldots, T$, that for all $\tau = 0, k, \ldots, t$ and for all $x \in \mathbb{T}^n_M$:

$$|U(x,t)| \leq M + tL.$$

By the initial condition in problem (23.2.3), the result holds for $t = 0$. Assuming it is valid for some $t \in I^T_K$, we have, by the hypothesis on h and the induction hypothesis, that:

$$h < \frac{2\nu}{M+TL} < \frac{2\nu}{M+tL} < \frac{2\nu}{|U(x,t)|},$$

for all $(x,t) \in \mathbb{T}^n_M \times \{0, k, 2k, \ldots, t\}$. Hence, by Lemma 2:

$$|U(x,t+k)| \leq |\Phi(U,U)(x,t)| + k|f(x,t)| \leq M + tL + kL = M + (t+k)L. \quad \square$$

23.4 Main estimates on the hyperfinite discrete problem

The nonstandard analytical setup we now use is as follows. We work in a superstructure $\langle V(\mathbb{R}), {}^*V(\mathbb{R}), * \rangle$. We will omit the stars on all standard functions of one or several variables and usual binary relations. Any $x \in {}^*\mathbb{R}^n$ is called finite iff there exists $m \in \mathbb{N}$ such that $|x| < m$. Each finite $x \in {}^*\mathbb{R}$ can be uniquely decomposed as $x = r + \varepsilon$, where $r \in \mathbb{R}$ and ε is an infinitesimal; r is called the standard part of x, and denoted by $\operatorname{st} x$. If $x, y \in {}^*\mathbb{R}$ are such that $x - y$ is infinitesimal, then we say that x is infinitely close to y, and write $x \approx y$. Similarly, if $x, y \in {}^*\mathbb{R}^n$:

$$x \approx y \quad \text{iff} \quad |x - y| \approx 0 \quad \text{iff} \quad x_i \approx y_i, \quad \text{for each } i = 1, \ldots, n.$$

Let $j, l \in \mathbb{N}$. If $x \in {}^*\mathbb{R}^j$ is finite then let:

$$\operatorname{st} x = \operatorname{st}(x_1, \ldots, x_j) = (\operatorname{st} x_1, \ldots, \operatorname{st} x_j).$$

If $F : A \subset {}^*\mathbb{R}^j \to {}^*\mathbb{R}^l$ is an S-continuous function, define ${}^\circ F$ by:

$$^\circ F(\operatorname{st} x) = \operatorname{st}(F(x)) \quad \forall x \in A. \tag{23.4.1}$$

For sets $A \in \mathbb{R}^j$, let:

$$^\circ A = \Big\{ \operatorname{st} x : \text{ "x is finite" and } x \in A \Big\}.$$

Each "circle" map as introduced above is sometimes called a standard part map.

In this, and other similar definitions of this work, we consider ${}^*\mathcal{V}$, where $\mathcal{V} = \{\Delta_d\}$. ${}^*\mathcal{V}$ is an internal *family of linear maps indexed in the internal set of all admissible d; each $\Delta_d \in {}^*\mathcal{U}$ is a *linear map acting on the vector space of internal gridfunctions $U : \overline{\mathcal{D}}_d \to {}^*\mathbb{R}$.

Similarly, we can consider ${}^*\mathcal{U}$, where

$$\mathcal{U} = \left\{P_d : \ M \in {}^*\mathbb{N}, T \in {}^*\mathbb{R}^+, K \in {}^*\mathbb{N} \text{ and } \frac{TM^2}{K} < \frac{1}{2n\nu}\right\},$$

${}^*\mathcal{U}$ is now an internal family of linear maps, indexed on an internal set. Each $H^T_{MK} \in {}^*\mathcal{U}$ is then a *linear map acting on the vector space of internal gridfunctions.

For the norms and seminorms introduced in the previous Section, we can proceed similarly. We get families of internal norms and seminorms which, by transfer, satisfy the following. Given $U : \overline{\mathcal{D}}_d \to {}^*\mathbb{R}^n$ and $A \subset \overline{\mathcal{D}}_d$:

$$\|U\|_{L^\infty_d(A)} = {}^*\max\left\{|U(x,t)| : \ (x,t) \in A\right\},$$

$$[U]_{L^\infty_d(A)} = {}^*\max\left\{\frac{|U(x,t) - U(y,t)|}{|x-y|} : \ (x,t), (y,t) \in A \text{ and } x \neq y\right\},$$

$$[[U]]_{L^\infty_d(A)} = {}^*\max\left\{\frac{|U(x,t+k) - U(x,t)|}{k} : \ (x,t), (x,t+k) \in A\right\}.$$

Lemma 4 *Let d be an admissible discretization such that h is infinitesimal. Let*

$$L_0 = \max\left([u_0]_{L^\infty_d(\mathbb{T}^n_M)}, \frac{1}{n}([f]_{L^\infty_d(\overline{\mathcal{D}}_d)})^{1/2}\right).$$

If U is the solution of the discrete problem (23.2.3) then, for $T < \frac{1}{2nL_0}$:

$$[U]_{L^\infty_d(\overline{\mathcal{D}}_d)} \lesssim \frac{L_0}{1 - 2nL_0T}$$

Proof. To work internally, we will begin by requiring a weaker condition on h, namely:

$$h < \frac{2\nu}{\|u_0\|_{L^\infty} + T\|f\|_{L^\infty}}. \tag{23.4.2}$$

Lemma 3 shows that this is a sufficient condition for admissibility of d, which is all that is needed for now.

Fix $z \in \mathbb{T}^n_M$ and write $V(x,t) = U(x+z,t)$. Then, by Lemma 2:

$$U(x+z,t+k) - U(x,t+k) = \Phi(V,V)(x,t) - \Phi(U,U)(x,t) + \\ + \lambda h^2(f(x+z,t) - f(x,t))$$

$$= \Phi(V-U,V)(x,t) + \\ + \frac{\lambda h}{2}\sum_{i=1}^{n}\Big(U_i(x,t) - V_i(x,t)\Big)\Big(U(x+he_i,t) - U(x-he_i,t)\Big) + \\ + \lambda h^2(f(x+z,t) - f(x,t))$$

By *recursion, we will construct a function $L : \{0,k,2k,\ldots,T\} \to {}^*\mathbb{R}$ such that, for all $t = 0,k,\ldots T$ and $x,z \in \mathbb{T}^n_M$:

$$|U(x+z,t) - U(x,t)| \le L(t)|z| \tag{23.4.3}$$

Let $L(t)$ be given by:

$$\begin{cases} L(0) &= L_0 \\ L(t+k) &= L(t) + nk(L(t))^2 + nkL_0^2 \end{cases}$$

For $t = 0$, we get:

$$|U(x+z,0) - U(x,0)| = |u_0(x+z) - u_0(x)| \le [u_0]_{L^\infty_d(\mathbb{T}^n_M)}\,|z| \le L_0|z|.$$

Now, we assume that for all $\tau = 0,k,\ldots,t$ and $x \in \mathbb{T}^n_M$

$$|U(x+z,\tau) - U(x,\tau)| \le L(t)|z|.$$

Since d is admissible (and so $|V(x,\tau)| = |U(x+z,\tau)| < \frac{2\nu}{h}$), Lemma 1 implies that $|\Phi(V-U,V)| \le L(t)|z|$. Hence:

$$|U(x+z,t+k) - U(x,t+k)| \le L(t)|z| \\ + \frac{\lambda h}{2}\sum_{i=1}^{n} L(t)|z|L(t)2h + \lambda h^2[f]_{L^\infty}|z| \\ \le \Big(L(t) + nk(L(t))^2 + nkL_0^2\Big)|z|.$$

This shows inequality (23.4.3) by *induction.

Note that $L(t)$ defines the Euler iterates for the standard ODE initial value problem:

$$\begin{cases} y' &= ny^2 + nL_0^2 \\ y(0) &= L_0 \end{cases} \tag{23.4.4}$$

Since $y' \geq 0$, $y(t) \geq L_0$. Hence, $y' \leq 2ny^2$, and thus:

$$y(t) \leq \frac{L_0}{1 - 2nL_0 t}$$

Now, use the fact that h is infinitesimal. By the convergence of the Euler iterates to the solutions of problem (23.4.4), we conclude that:

$$L(t) \approx y(t) \leq \frac{L_0}{1 - 2nL_0 t}$$

This estimate is valid for t in any compact interval where problem (23.4.4) has solution; in particular, the estimate will be valid for $t < \frac{1}{2nL_0}$. □

Lemma 5 *Let d be an admissible discretization such that h is infinitesimal. Let L_0 be as in Lemma 4 and*

$$M_0 = \max\left([[u_0]]_{L^\infty_d(\mathbb{T}^n_M)}, \frac{1}{n}\left([[f]]_{L^\infty_d(\overline{\mathcal{D}}_d)}\right)^{1/2}, L_0\right), \quad \text{with}$$

$$[[u_0]]_{L^\infty_d(\mathbb{T}^n_M)} = {}^*\max\left\{\nu\Delta u_0(x) - (u\cdot\nabla)u_0(x) + f(x,0) \,:\, x \in \mathbb{T}^n_M\right\}$$

If U is the solution of the discrete problem (23.2.3) then, for $T < \frac{1}{2nL_0}$:

$$[[U]]_{L^\infty_d(\overline{\mathcal{D}}_d)} \lesssim \frac{M_0}{1 - 2nL_0 T}$$

Proof. Again we begin by requiring that h satisfies the weaker statement:

$$h < \frac{2\nu}{\|u_0\|_{L^\infty} + T\|f\|_{L^\infty}} \tag{23.4.5}$$

Write $V(x,t) = U(x,t+k)$. Then, by Lemma 2:

$$U(x,t+2k) - U(x,t+k) = \Phi(V,V)(x,t) - \Phi(U,U)(x,t) + \\ + \lambda h^2(f(x,t+k) - f(x,t))$$

$$= \Phi(V-U,V)(x,t) + \\ + \frac{\lambda h}{2}\sum_{i=1}^{n}\Big(U_i(x,t) - V_i(x,t)\Big)\Big(U(x+he_i,t) - U(x-he_i,t)\Big) \\ + \lambda h^2(f(x,t+k) - f(x,t))$$

By *recursion, we again construct a function $M : \{0,k,2k,\ldots,T\} \to {}^*\mathbb{R}$ such that, for all $t = 0,k,\ldots T$ and $x \in \mathbb{T}^n_M$:

$$|U(x,t+k) - U(x,t)| \leq M(t)k. \tag{23.4.6}$$

Let $M(t)$ be given by:

$$\begin{cases} M(0) &= M_0 \\ M(t+k) &= M(t) + nkM(t)L(t) + nkM_0^2 \end{cases}$$

Here, L is the function introduced in the proof of Lemma 4.

For $t = 0$, we get:

$$\begin{aligned} \frac{U(x,k) - U(x,0)}{k} &= \nu \sum_{i=1}^{n} \delta^0_{h,i,i} U - \sum_{i=1}^{n} U_i \delta^0_{h,i} U + f(x,0) \\ &\approx \nu \Delta u_0(x) - (u_0 \cdot \nabla) u_0(x) + f(x,0) \\ &\leq [[u_0]]_{L^\infty_d(\mathbb{T}^n_M)} \\ &\leq M_0. \end{aligned}$$

Now, we assume that for all $\tau = 0, k, \ldots, t$ and $x \in \mathbb{T}^n_M$

$$|U(x, \tau + k) - U(x, \tau)| \leq M(t)k.$$

Since d is admissible (and so $|V(x,\tau)| = |U(x, \tau + k)| < \frac{2\nu}{h}$), Lemma 1 implies that $|\Phi(V - U, V)| \leq M(t)k$. Hence:

$$\begin{aligned} |U(x, t+2k) - U(x, t+k)| &\leq M(t)k + \frac{\lambda h}{2} \sum_{i=1}^{n} M(t)kL(t)2h + \lambda h^2 [[f]]_{L^\infty} h \\ &\leq \left(M(t) + nkM(t)L(t) + nkM_0^2 \right) k \\ &= M(t+k)k. \end{aligned}$$

This shows inequality (23.4.6) by *induction.

Now, we note that $M(t)$ defines the Euler iterates for the standard ODE initial value problem:

$$\begin{cases} z' &= n\, z\, y + nM_0^2 \\ z(0) &= M_0. \end{cases} \tag{23.4.7}$$

Here, $y(t)$ is the solution of problem (23.4.7), as given in the proof of Lemma 4.

Since $z' \geq 0$, $z(t) \geq M_0$, so $nz(t)y(t) \geq nM_0L_0 \geq nM_0^2$. Hence, $z'(t) \leq 2nz(t)y(t)$, and thus:

$$z(t) \leq \frac{M_0}{1 - 2nL_0 t}$$

Now, use the fact that h is infinitesimal. By the convergence of the Euler iterates to the solutions of problem (23.4.4), we conclude that:

$$M(t) \approx z(t) \leq \frac{M_0}{1 - 2nL_0 t}$$

This estimate is valid for t in any compact interval where problem (23.4.4) has solution; in particular, the estimate will be valid for $t < \frac{1}{2nL_0}$. □

23.5 Existence and uniqueness of solution

By the results of section 23.4, the function $U(x,t)$ is S-continuous for $t \in \{0, k, \dots, T\}$, with $T < \frac{1}{2nL_0}$. Note that the set of admissible values for T depends only on L_0, that is, on the size of the Lipschitz constants of u_0 and f.

Therefore,

$$u(\operatorname{st} x, \operatorname{st} t) = \operatorname{st} U(x,t) \tag{23.5.1}$$

is well-defined and continuous on $\mathbb{T}^n \times [0,T]$. Also, Lemmas 4 and 5 imply that u is globally Lipschitz on $\mathbb{T}^n \times [0,T]$.

To show that u solves (locally in time) the problem and also to prove a uniqueness result, we need the following:

Lemma 6 *Let $a : \mathbb{T}^n \times [0,T] \to \mathbb{R}$ be Lipschitz continuous. Then, the problem,*

$$\begin{cases} v_t - \nu\Delta v + (a \cdot \nabla)v = f & \text{in } \mathcal{D} \\ v = u_0 & \text{on } \mathbb{T}^n \times \{0\}. \end{cases} \tag{23.5.2}$$

has a unique solution $v \in C^{2,1}(\mathbb{T}^n \times [0,T])$. Furthermore, if v is a solution of problem (23.5.2) then (in the nonstandard universe) v satisfies: for all $(x,t) \in \overline{\mathcal{D}}_d$, there exists an $\varepsilon \approx 0$ such that

$$v(x, t+k) = \Phi(v,a)(x,t) + \lambda h^2 f(x,t) + \varepsilon\lambda h^2.$$

Proof. Consider problem (23.5.2). Since a is given, the vectorial differential equation $v_t - \nu\Delta v + (a \cdot \nabla)v = f$ is just an uncoupled system of n scalar second order parabolic equations. Since a is a Lipschitz continuous function on $\mathbb{T}^n \times [0,T]$ and $u_0 \in C^{2,1}(\mathbb{T}^n)$, by the theory of parabolic equations (e.g. Friedman, [8]), the problem has a unique (strong) solution, $v \in C^{2,1}(\mathbb{T}^n \times [0,T])$. By the regularity of v:

$$\begin{aligned}
f(x,t) &= v_t(x,t) - \nu\Delta v(x,t) + (a(x,t)\cdot\nabla)v(x,t) \approx P_d v(x,t) \\
&= \frac{v(x,t+k) - v(x,t)}{k} - \nu\Delta_d v(x,t) + \sum_{i=1}^{n} a_i(x,t)\frac{v(x+he_i,t) - v(x-he_i,t)}{2h} \\
&= \frac{1}{\lambda h^2}\Bigg(v(x,t+k) - (1 - 2n\nu\lambda)v(x,t) - \\
&\quad - \lambda\sum_{i=1}^{n}\left(\left(\nu - \frac{h}{2}a_i(x,t)\right)v(x+he_i,t) + \left(\nu + \frac{h}{2}a_i(x,t)\right)v(x-he_i,t)\right)\Bigg)
\end{aligned}$$

Consequently, for each $(x,t) \in \overline{\mathcal{D}}_d$, there exists an $\varepsilon \approx 0$ such that:

$$\begin{aligned} v(x,t+k) &= \big(1 - 2n\nu\lambda\big)\, v(x,t) + \\ &+ \lambda \sum_{i=1}^{n} \left(\left(\nu - \frac{h}{2} a_i(x,t)\right) v(x+he_i,t) + \left(\nu + \frac{h}{2} a_i(x,t)\right) v(x-he_i,t) \right) + \\ &+ \lambda h^2 f(x,t) + \varepsilon\lambda h^2 \\ &= \Phi(v,a)(x,t) + \lambda h^2 f(x,t) + \varepsilon\lambda h^2. \end{aligned}$$

□

Here is our existence lemma:

Lemma 7 *Let u and T be as given by equation (23.5.1). Then $u \in C^{(2;1)}(\mathbb{T}^n \times [0,T])$ and u is a (strong) solution of the Burgers equation problem (23.1.4).*

Proof. Consider the problem:

$$\begin{cases} v_t - \nu\Delta v + (u\cdot\nabla)v = f & \text{in } \mathcal{D} \\ v = u_0 & \text{on } \mathbb{T}^n \times \{0\}. \end{cases}$$

By Lemma 6 the problem has a unique (strong) solution, $u \in C^{2,1}(\mathbb{T}^n \times [0,T])$ and, in the nonstandard universe, for all $(x,t) \in \overline{\mathcal{D}}_d$, there exists an $\varepsilon \approx 0$ such that:

$$v(x,t+k) = \Phi(v,a)(x,t) + \lambda h^2 f(x,t) + \varepsilon\lambda h^2.$$

Now, we show that $u = v$.

Let $r \in \mathbb{R}^+$ be arbitrary. We will show by induction on $t = 0, k, 2k, \ldots, T$ that for all $\tau = 0, k, 2k, \ldots, t$ and $x \in \mathbb{T}^n_M$:

$$|U(x,t) - {}^*v(x,t)| \le rt.$$

This implies that, for all $(x,t) \in \overline{\mathcal{D}}_d$:

$$u(\operatorname{st} x, \operatorname{st} t) = \operatorname{st} U(x,t) = \operatorname{st} {}^*v(x,t) = v(\operatorname{st} x, \operatorname{st} t),$$

as wanted.

(As usual, from this point on we omit stars on ${}^*v$). For $t = 0$ the statement follows from the initial conditions. Now, assuming the statement holds true for some $t = jk$, $j \in {}^*\mathbb{N}$:

$$\begin{aligned} U(x,t+k) - v(x,t+k) &= \Phi(U,U)(x,t) - \Phi(v,u)(x,t) - \varepsilon\lambda h^2 \\ &= \Phi(U-v,u)(x,t) - \varepsilon k + \\ &+ \frac{\lambda h}{2} \sum_{i=1}^{n} \Big(u_i(x,t) - U_i(x,t)\Big)\Big(v(x+he_i,t) - v(x-he_i,t)\Big). \end{aligned}$$

Since $v(\cdot,t)$ is a C^2 function on the compact set $\mathbb{T}^n$, there exists a finite C so that $|v(x+he_i,t)-v(x-he_i,t)| \leq 2hC$. Also, by the definition and continuity of u, $u_i(x,t) - U_i(x,t) = \delta \approx 0$. By the induction hypothesis and Lemma 3, $|\Phi(U-v,u)(x,t)| \leq rt$. Therefore:

$$\begin{aligned}\Big|U(x,t+k) - v(x,t+k)\Big| &\leq rt + \varepsilon k + \frac{\lambda h}{2}\, n\delta\, 2hC \\ &= rt + k\big(\varepsilon + nC\delta\big) \\ &\leq r(t+k).\end{aligned}$$

□

As consequence of the above Lemma, we get:

Theorem 1 *Let f be locally Lipschitz continuous on $\mathbb{T}^n \times [0,\infty)$ and $u_0 \in C^{2,1}(\mathbb{T}^n)$. Let*

$$L = \max\left([u_0]_{L_d^\infty(\mathbb{T}_M^n)}, \frac{1}{n}([f]_{L_d^\infty(\overline{\mathcal{D}}_d)})^{1/2}, [[u_0]]_{L_d^\infty(\mathbb{T}_M^n)}, \frac{1}{n}([[f]]_{L_d^\infty(\overline{\mathcal{D}}_d)})^{1/2} \right).$$

Here, $[[u_0]]_{L_d^\infty(\mathbb{T}_M^n)}$ is an upper-bound for the Lipschitz constant of the function $\Delta u_0(x) + (u_0\cdot\nabla)u_0(x) + f(x,0)$. Then, for any $T < \frac{1}{2nL}$, the problem

$$\begin{cases} u_t = \nu\Delta u - (u\cdot\nabla)u + f & \text{in } \mathcal{D} \\ u = u_0 & \text{on } \mathbb{T}^n \times \{0\}. \end{cases}$$

has a strong solution, $u \in C^{2,1}(\mathbb{T}^n \times [0,T])$.

The uniqueness theorem follows from:

Lemma 8 *Let u and v be solutions of the Burgers equation problem (23.1.4) on $\mathbb{T}^n \times [0,T]$, for some $T > 0$. Let $L_v = [v]_{L_d^\infty(\mathbb{T}^n\times[0,T])}$. Then:*

$$u(x,t) = v(x,t) \qquad \text{for all } x \in \mathbb{T}^n,\ \ 0 \leq t < \min\left\{\frac{1}{2nL_v}, T\right\}$$

Proof. Using the nonstandard statement of Lemma 6, (note that u and v are given functions), we conclude that for all $(x,t) \in \overline{\mathcal{D}}_d$, there exist infinitesimal δ_1 and δ_2 such that:

$$u(x,t+k) = \Phi(u,u)(x,t) + \lambda h^2 f(x,t) + \delta_1\lambda h^2$$

and

$$v(x,t+k) = \Phi(v,v)(x,t) + \lambda h^2 f(x,t) + \delta_2\lambda h^2.$$

Let Θ be the largest ik, $i \in {}^*\mathbb{N}$, such that $ik \leq \min\left\{\frac{1}{2nL_v}, T\right\}$. Let $r \in \mathbb{R}^+$ be arbitrary. We will show by induction on $t = 0, k, 2k, \ldots, \Theta$ that for all $\tau = 0, k, 2k, \ldots, t$ and $x \in \mathbb{T}_M^n$:

$$|{}^*u(x,t) - {}^*v(x,t)| \leq rt.$$

This implies that, for all $(x,t) \in \overline{\mathcal{D}}_d$:

$$u(\operatorname{st} x, \operatorname{st} t) = v(\operatorname{st} x, \operatorname{st} t),$$

as wanted.

(From hereon, we omit stars on ${}^*u$ and ${}^*v$). For $t = 0$ the statement follows from the initial conditions. Now, assuming the statement holds true for some $t = jk$, $j \in {}^*\mathbb{N}$:

$$u(x,t+k) - v(x,t+k) = \Phi(u,u)(x,t) - \Phi(v,v)(x,t) + (\delta_1 - \delta_2)\lambda h^2$$

Let $\varepsilon = \delta_1 - \delta_2 \approx 0$. Using Lemma 2:

$$\begin{aligned} u(x,t+k) - v(x,t+k) &= \Phi(u-v,v)(x,t) + \varepsilon k \\ &+ \frac{\lambda h}{2} \sum_{i=1}^{n} \Big(v_i(x,t) - u_i(x,t)\Big)\Big(v(x+he_i,t) - v(x-he_i,t)\Big) \end{aligned}$$

Note that $|v(x+he_i,t) - v(x-he_i,t)| \leq 2hL_v$. Using the induction hypothesis $|u_i(x,t) - v_i(x,t)| \leq rt$. Also, by Lemma 3 and the induction hypothesis, $|\Phi(u-v,v)(x,t)| \leq rt$. Therefore:

$$\begin{aligned} \Big|u(x,t+k) - v(x,t+k)\Big| &\leq rt + \varepsilon k + \frac{\lambda h}{2} nrt\, 2hL_v \\ &= rt + k\Big(\varepsilon + rntL_v\Big). \end{aligned}$$

Since $t \leq \Theta \leq \frac{1}{2nL_v}$, it follows that:

$$|u(x,t+k) - v(x,t+k)| \leq r(t+k). \qquad \Box$$

Theorem 2 *Let u and v be solutions of our Model Problem (23.1.4) on $\mathbb{T}^n \times [0,T]$, for some $T > 0$. Then $u = v$.*

Proof. Assume there exists $t \in [0,T]$ such that, for some $x \in \mathbb{T}^n$, $u(x,t) \neq v(x,t)$; let θ be the infimum of such t.

Note that:

$$u(x,\theta) = v(x,\theta), \quad \text{for all } x \in \mathbb{T}^n. \tag{23.5.3}$$

If $\theta = 0$ then equality (23.5.3) follows from the initial conditions; otherwise, it follows from continuity of u and v. Consider the problem:

$$\begin{cases} w_t = \nu \Delta w - (w \cdot \nabla) w + g & \text{in } \mathcal{D} \\ w = u(x, \theta) & \text{on } \mathbb{T}^n \times \{0\}. \end{cases} \tag{23.5.4}$$

where $g(x,t) = f(x, \theta + t)$, for all $(x,t) \in \mathbb{T}^n \times [0, T - \theta]$. Consider $w_1, w_2 : \mathbb{T}^n \times [0, T - \theta] \to \mathbb{R}$ given by $w_1(x,t) = u(x, \theta + t)$ and $w_1(x,t) = v(x, \theta + t)$. But w_1 and w_2 are solutions of problem (23.5.4) such that, for arbitrarily small $t > 0$:

$$w_1(x,t) \neq w_2(x,t) \quad \text{for some } x \in \mathbb{T}^n.$$

This contradicts Lemma 8. □

References

[1] J.M. Burgers, "A Mathematical Model Illustrating the Theory of Turbulence", Advances in Applied Mechanics, **1** (1948) 171–199.

[2] M. Capinski and N.J. Cutland, "A Simple Proof of Existence of Weak and Statistical Solutions of Navier-Stokes Equations", Proceedings of the Royal Society, London, Ser. A, **436** (1992) 1–11.

[3] R. Courant and D. Hilbert, *Methods of mathematical physics*, Interscience Publishers, New York, 1953–1962.

[4] P. Ciarlet, *The finite element method for elliptic problems*, North Holland, New York, 1978.

[5] C. Chang and H.J. Keisler, *Model Theory*, North Holland, Amsterdam, 1990.

[6] J.D. Cole, "On a Quasilinear Parabolic Equation Occuring in Aerodynamics", Quant. Appl. Math., **9** (1951) 225–236.

[7] N. Cutland, *Nonstandard Analysis and its Applications*, London Mathematical Society, Cambridge, 1988.

[8] A. Friedman, *Partial differential equations of parabolic type*, Prentice Hall, Inc., Englewood Cliffs, N.J., 1964.

[9] D. Henry, *Geometric theory of semilinear parabolic equations*, Springer-Verlag, New York, 1981.

[10] E. HOPF, "The Partial Differential Equation $u_t + uu_x = \nu u_{xx}$", Communications Pure and Applied Math., **3** (1950) 201–230.

[11] H.J. KEISLER, *Foundations of Infinitesimal Calculus*, Prindle, Weber and Schmidt, 1976.

[12] J.L. LIONS, *Quelques méthodes de résolution des problèms aux limites nonlinéaires*, Dunod, Gauthier-Villars, Paris, 1969.

[13] J.L. LIONS, *Control of systems governed by partial differencial equations*, Springer-Verlag, New York, 1972.

[14] A.J. MAJDA and A.L. BARTOZZI, *Vorticity and Incompressible Flows*, Cambridge Texts in Applied Math., Cambridge, 2002.

[15] K.D. STROYAN, *Introduction to the Theory of Infinitesimals*, Academic Press, New York, 1976.

[16] LUTHER W. WHITE, "A Study of Uniqueness for the Initialization Problem in Burgers' Equations", J. of Math. Analysis and Applications, **172** (1993) 412–431.

Part V

Infinitesimals and education

24

Calculus with infinitesimals

Keith D. Stroyan*

24.1 Intuitive proofs with "small" quantities

Abraham Robinson discovered a rigorous approach to calculus with infinitesimals in 1960 and published it in [9]. This solved a 300 year old problem dating to Leibniz and Newton. Extending the ordered field of (Dedekind) "real" numbers to include infinitesimals is not difficult algebraically, but calculus depends on approximations with transcendental functions. Robinson used mathematical logic to show how to extend all real functions in a way that preserves their properties in a precise sense. These properties can be used to develop calculus with infinitesimals. Infinitesimal numbers have always fit basic intuitive approximation when certain quantities are "small enough," but Leibniz, Euler, and many others could not make the approach free of contradiction. Section 1 of this article uses some intuitive approximations to derive a few fundamental results of analysis. We use approximate equality, $x \approx y$, only in an intuitive sense that "x is sufficiently close to y".

H. Jerome Keisler developed simpler approaches to Robinson's logic and began using infinitesimals in beginning U.S. calculus courses in 1969. The experimental and first edition of his book were used widely in the 1970's. Section 2 of this article completes the intuitive proofs of Section 1 using Keisler's approach to infinitesimals from [6].

24.1.1 Continuity and extreme values

Theorem 1 ***The Extreme Value Theorem***

Suppose a function $f[x]$ is continuous on a compact interval $[a,b]$. Then $f[x]$ attains both a maximum and minimum, that is, there are points x_{MAX} and $x_{\min}$ in $[a,b]$, so that for every other x in $[a,b]$, $f[x_{\min}] \leq f[x] \leq f[x_{\mathrm{MAX}}]$.

*University of Iowa, USA.
keith-stroyan@uiowa.edu

Formulating the meaning of "continuous" is a large part of making this result precise. We will take the intuitive definition that $f[x]$ is continuous means that if an input value x_1 is close to another, x_2, then the output values are close. We summarize this as: $f[x]$ is continuous if and only if $a \le x_1 \approx x_2 \le b \Longrightarrow f[x_1] \approx f[x_2]$.

Given this property of $f[x]$, if we partition $[a, b]$ into tiny increments,

$$a < a + \frac{1(b-a)}{H} < a + \frac{2(b-a)}{H} < \cdots < a + \frac{k(b-a)}{H} < \cdots < b$$

the maximum of the function on the finite partition occurs at one (or more) of the points $x_M = a + \frac{k(b-a)}{H}$. This means that for any other partition point $x_1 = a + \frac{j(b-a)}{H}$, $f[x_M] \ge f[x_1]$.

Any point $a \le x \le b$ is within $\frac{1}{H}$ of a partition point $x_1 = a + \frac{j(b-a)}{H}$, so if H is very large, $x \approx x_1$ and

$$f[x_M] \ge f[x_1] \approx f[x]$$

and we have found the approximate maximum.

It is not hard to make this idea into a sequential argument where $x_{M[H]}$ depends on H, but there is quite some trouble to make the sequence $x_{M[H]}$ converge (using some form of compactness of $[a, b]$.) Robinson's theory simply shows that the hyperreal x_M chosen when $1/H$ is infinitesimal, is infinitely near an ordinary real number where the maximum occurs. We complete this proof as a simple example of Keisler's Axioms in Section 2.

24.1.2 Microscopic tangency in one variable

In beginning calculus you learned that the derivative measures the slope of the line tangent to a curve $y = f[x]$ at a particular point, $(x, f[x])$. We begin by setting up convenient "local variables" to use to discuss this problem. If we fix a particular $(x, f[x])$ in the x-y-coordinates, we can define new parallel coordinates (dx, dy) through this point. The (dx, dy)-origin is the point of tangency to the curve.

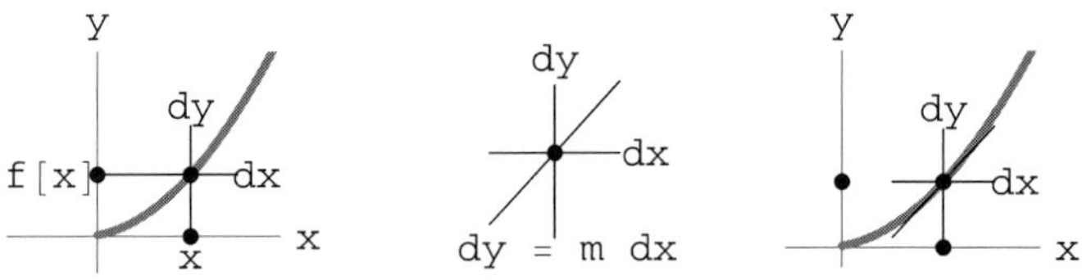

Figure 24.1.1: Microscopic Tangency

A line in the local coordinates through the local origin has equation $dy = mdx$ for some slope m. Of course we seek the proper value of m to make $dy = mdx$ tangent to $y = f[x]$.

You probably learned the derivative from the approximation

$$\lim_{\Delta x \longrightarrow 0} \frac{f[x + \Delta x] - f[x]}{\Delta x} = f'[x].$$

If we write the error in this limit explicitly, the approximation can be expressed as $\frac{f[x+\Delta x]-f[x]}{\Delta x} = f'[x] + \varepsilon$ or $f[x + \Delta x] - f[x] = f'[x] \cdot \Delta x + \varepsilon \cdot \Delta x$. Intuitively we say $f[x]$ is smooth if there is a function $f'[x]$ that makes the error small, $\varepsilon \approx 0$, in the formula

$$f[x + \delta x] - f[x] = f'[x] \cdot \delta + \varepsilon \cdot \delta x \tag{24.1.1}$$

when the change in input is small, $\delta x \approx 0$. The main point of this article is to show that requiring ε to be infinitesimal whenever δx is infinitesimal and x is near standard gives an intuitive and simple direct meaning to $\mathcal{C}^1$ smooth. Requiring only that ε tend to zero only for fixed real x, or pointwise, is not sufficient to capture the intuitive approximations of our proofs.

The nonlinear change on the left side of (24.1.1) equals a linear change, $m \cdot \delta x$, with $m = f'[x]$, plus a term that is small compared with the input change.

The error ε has a direct graphical interpretation as the error measured above $x + \delta x$ after magnification by $1/\delta x$. This magnification makes the small change δx appear unit size and the term $\varepsilon \cdot \delta x$ measures ε after magnification.

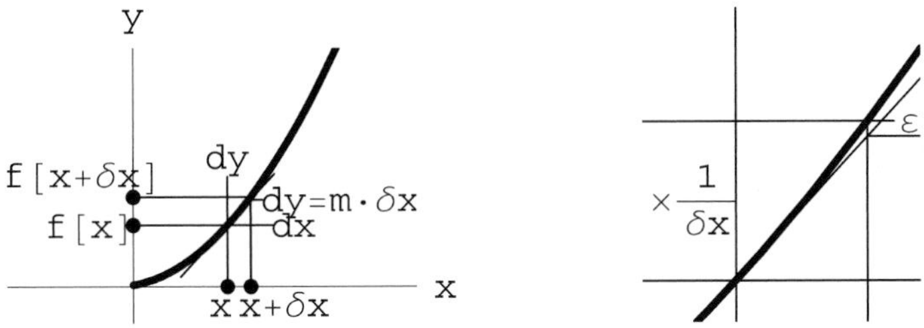

Figure 24.1.2: Magnified Error

When we focus a powerful microscope at the point $(x, f[x])$ we only see the linear curve $dy = m \cdot dx$, because $\varepsilon \approx 0$ is smaller than the thickness of the line. The figure below shows a small box magnified on the right.

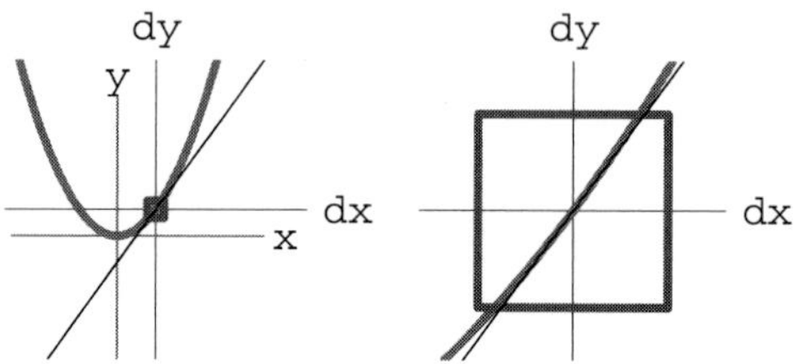

Figure 24.1.3: A Magnified Tangent

24.1.3 The Fundamental Theorem of Integral Calculus

Now we use the intuitive microscope approximation (24.1.1) to prove:

Theorem 2 ***The Fundamental Theorem of Integral Calculus: Part I***
Suppose we want to find $\int_a^b f[x]\,dx$*. If we can find another function* $F[x]$ *so that the derivative satisfies* $F'[x] = f[x]$ *for every* x*,* $a \leq x \leq b$*, then*

$$\int_a^b f[x]\,dx = F[b] - F[a]$$

The definition of the integral we use is the real number approximated by a sum of small slices,

$$\int_a^b f[x]\,dx \approx \sum_{\substack{x=a \\ \text{step } \delta x}}^{b-\delta x} f[x] \cdot \delta x, \text{ when } \delta x \approx 0$$

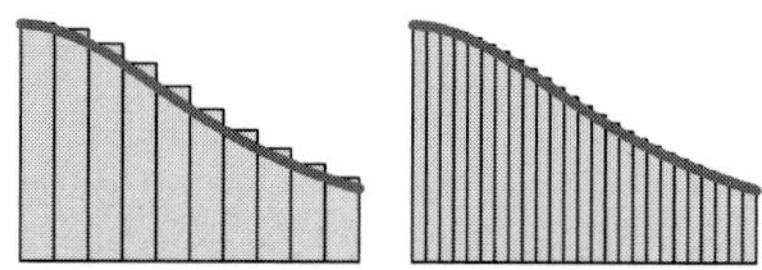

Figure 24.1.4: Sum Approximations

24.1.4 Telescoping sums and derivatives

We know that if $F[x]$ has derivative $F'[x] = f[x]$, the differential approximation above says,

$$F[x + \delta x] - F[x] = f[x] \cdot \delta x + \varepsilon \cdot \delta x$$

so we can sum both sides

$$\sum_{\substack{x=a \\ \text{step } \delta x}}^{b-\delta x} F[x+\delta x] - F[x] = \sum_{\substack{x=a \\ \text{step } \delta x}}^{b-\delta} f[x] \cdot \delta x + \sum_{\substack{x=a \\ \text{step } \delta x}}^{b-\delta x} \varepsilon \cdot \delta x$$

The telescoping sum satisfies,

$$\sum_{\substack{x=a \\ \text{step } \delta x}}^{b-\delta x} F[x+\delta x] - F[x] = F[b'] - F[a]$$

so we obtain the approximation,

$$\int_a^b f[x]\,dx \approx \sum_{\substack{x=a \\ \text{step } \delta x}}^{b-\delta x} f[x] \cdot \delta x = F[b'] - F[a] - \sum_{\substack{x=a \\ \text{step } \delta x}}^{b-\delta x} \varepsilon \cdot \delta x$$

This gives,

$$\begin{aligned}\left| \sum_{\substack{x=a \\ \text{step } \delta x}}^{b-\delta x} f[x] \cdot \delta x - (F[b'] - F[a]) \right| &\leq \left| \sum_{\substack{x=a \\ \text{step } \delta x}}^{b-\delta x} \varepsilon \cdot \delta x \right| \leq \sum_{\substack{x=a \\ \text{step } \delta x}}^{b-\delta x} |\varepsilon| \cdot \delta x \\ &\leq \max[|\varepsilon|] \cdot \sum_{\substack{x=a \\ \text{step } \delta x}}^{b-\delta x} \delta x = \max[|\varepsilon|] \cdot (b' - a) \\ &\approx 0\end{aligned}$$

or $\int_a^b f[x]\,dx \approx \sum_{\substack{x=a \\ \text{step } \delta x}}^{b-\delta x} f[x] \cdot \delta x \approx F[b'] - F[a]$. Since $F[x]$ is continuous, $F[b'] \approx F[b]$, so $\int_a^b f[x]\,dx = F[b] - F[a]$.

We need to know that all the epsilons above are small when the step size is small, $\varepsilon \approx 0$, when $\delta x \approx 0$ for all $x = a, a + \delta x, a + 2\delta, \cdots$. This is a uniform condition that has a simple appearance in Robinson's theory. There is something to explain here because the theorem stated above is false if we take the usual pointwise notion of derivative and the Riemann integral. (There are pointwise differentiable functions whose derivative is not Riemann integrable. See [5] Example 35, Chapter 8, p.107 ff.)

The condition needed to make this proof complete is natural geometrically and plays a role in the intuitive proof of the inverse function theorem in the next example.

24.1.5 Continuity of the derivative

We show now that the differential approximation

$$f[x+\delta x] - f[x] = f'[x] \cdot \delta + \varepsilon \cdot \delta x$$

forces the derivative function $f'[x]$ to be continuous,

$$x_1 \approx x_2 \Longrightarrow f'[x_1] \approx f'[x_2]$$

Let $x_1 \approx x_2$, but $x_1 \neq x_2$. Use the differential approximation with $x = x_1$ and $\delta x = x_2 - x_1$ and also with $x = x_2$ and $\delta x = x_1 - x_2$, geometrically looking at the tangent approximation from both endpoints.

$$\begin{aligned} f[x_2] - f[x_1] &= f'[x_1] \cdot (x_2 - x_1) + \varepsilon_1 \cdot (x_2 - x_1) \\ f[x_1] - f[x_2] &= f'[x_2] \cdot (x_1 - x_2) + \varepsilon_2 \cdot (x_1 - x_2) \end{aligned}$$

Adding these equations, we obtain

$$0 = \big((f'[x_1] - f'[x_2]) + (\varepsilon_1 - \varepsilon_2)\big) \cdot (x_2 - x_1)$$

Dividing by the nonzero term $(x_2 - x_1)$ and adding $f'[x_2]$ to both sides, we obtain, $f'[x_2] = f'[x_1] + (\varepsilon_1 - \varepsilon_2)$ or $f'[x_2] \approx f'[x_1]$, since the difference between two small errors is small.

This fact can be used to prove:

Theorem 3 ***The Inverse Function Theorem***

If $f'[x_0] \neq 0$ then $f[x]$ has an inverse function in a small neighborhood of x_0, that is, if $y \approx y_0 = f[x_0]$, then there is a unique $x \approx x_0$ so that $y = f[x]$.

We saw above that the differential approximation makes a microscopic view of the graph look linear. If $y \approx y_1$ the linear equation $dy = m \cdot dx$ with $m = f'[x_1]$ can be inverted to find a first approximation to the inverse,

$$\begin{aligned} y - y_0 &= m \cdot (x_1 - x_0) \\ x_1 &= x_0 + \tfrac{1}{m}(y - y_0) \end{aligned}$$

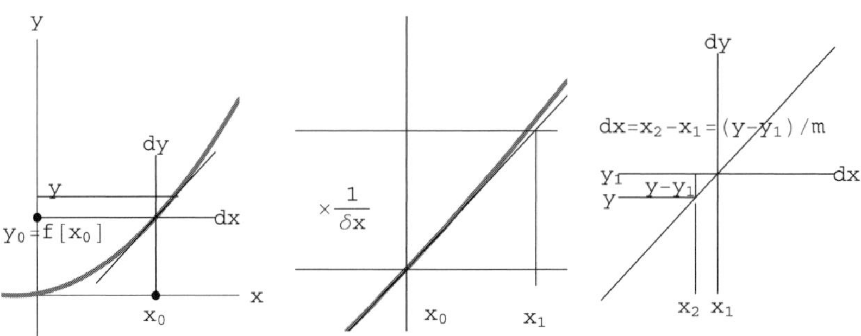

Figure 24.1.5: Approximate Inverse

We test to see if $f[x_1] = y$. If not, examine the graph microscopically at $(x_1, y_1) = (x_1, f[x_1])$. Since the graph appears the same as its tangent to within ε and since $m = f'[x_1] \approx f'[x_2]$, the local coordinates at (x_1, y_1) look like a line of slope m. Solving for the linear x-value which gives output y, we get

$$\begin{aligned} y - y_1 &= m \cdot (x_2 - x_1) \\ x_2 &= x_1 + \tfrac{1}{m}(y - f[x_1]) \end{aligned}$$

Continue in this way generating a sequence of approximations, $x_1 = x_0 + \frac{1}{m}(y - y_0)$, $x_{n+1} = G[x_n]$, where the recursion function is $G[\xi] = x + \frac{1}{m}(y - f[\xi])$. The distance between successive approximations is

$$|x_2 - x_1| = \big|G[x_1] - G[x_0]\big| \le \frac{1}{2} \cdot |x_1 - x_0|$$

$$|x_3 - x_2| = \big|G[x_2] - G[x_1]\big| \le \frac{1}{2} \cdot |x_2 - x_1| \le \frac{1}{2} \cdot \frac{1}{2} \cdot |x_1 - x_0|$$

by the Differential Approximation for $G[x]$. Notice that $G'[\xi] = 1 - f'[\xi]/m \approx 0$, for $\xi \approx x_0$, so $|G'[\xi]| < 1/2$ in particular, and

$$\begin{aligned} |x_2 - x_1| &= \big|G[x_1] - G[x_0]\big| \le \frac{1}{2} \cdot |x_1 - x_0| \\ |x_3 - x_2| &= \big|G[x_2] - G[x_1]\big| \le \frac{1}{2} \cdot |x_2 - x_1| \le \frac{1}{2^2} \cdot |x_1 - x_0| \\ &\vdots \\ |x_{n+1} - x_n| &\le \frac{1}{2^n} \cdot |x_1 - x_0| \\ |x_{n+1} - x_0| &\le |x_{n+1} - x_n| + |x_n - x_{n-1}| + \cdots + |x_1 - x_0| \\ &\le \left(1 + \frac{1}{2} + \cdots + \frac{1}{2^n}\right) \cdot |x_1 - x_0| \end{aligned}$$

A geometric series estimate shows that the series converges, $x_n \to x \approx x_0$ and $f[x] = y$.

To complete this proof we need to show that $G[\xi]$ is a contraction on some noninfinitesimal interval. The precise definition of the derivative matters because the result is false if $f'[x]$ is defined by a pointwise limit. The function $f[x] = x + x^2 \sin[\pi/x]$ with $f[0] = 0$ has pointwise derivative 1 at zero, but is not increasing in any neighborhood of zero.

24.1.6 Trig, polar coordinates, and Holditch's formula

Calculus depends on small approximations with transcendental functions like sine, cosine, and the natural logarithm. Following are some intuitive approximations with non-algebraic functions.

Sine and cosine in radian measure give the x and y location of a point on the unit circle measured a distance θ along the circle. Now make a small change in the angle and magnify by $1/\delta\theta$ to make the change appear unit size.

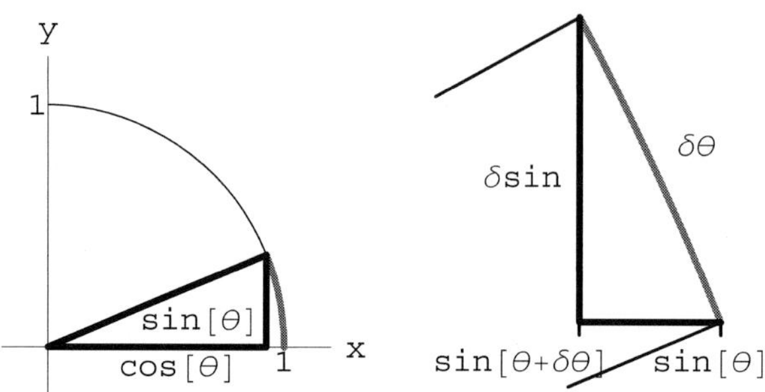

Figure 24.1.6: Sine increments

Since magnification does not change lines, the radial segments from the origin to the tiny increment of the circle meet the circle at right angles and appear under magnification to be parallel lines. Smoothness of the circle means that under powerful magnification, the circular increment appears to be a line. The difference in values of the sine is the long vertical leg of the increment "triangle" above on the right. The apparent hypotenuse with length $\delta\theta$ is the circular increment.

Since the radial lines meet the circle at right angles the large triangle on the unit circle at the left with long leg $\cos[\theta]$ and hypotenuse 1 is similar to the increment triangle, giving

$$\frac{\cos[\theta]}{1} \approx \frac{\delta \sin}{\delta\theta}$$

We write approximate similarity because the increment "triangle" actually has one circular side that is $\approx$-straight. In any case, this is a convincing argument that $\frac{d\sin}{d\theta}[\theta] = \cos[\theta]$. A similar geometric argument on the increment triangle shows that $\frac{d\cos}{d\theta}[\theta] = -\sin[\theta]$.

24.1.7 The polar area differential

The derivation of sine and cosine is related to the area differential in polar coordinates. If we take an angle increment of $\delta\theta$ and magnify a view of the circular arc on a circle of radius r, the length of the circular increment is $r \cdot \delta\theta$, by similarity.

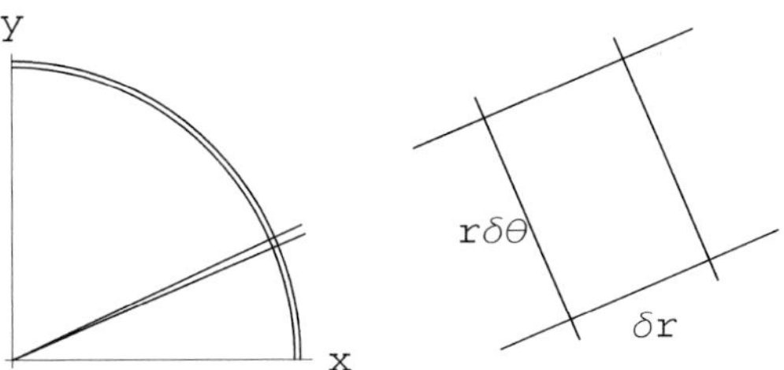

Figure 24.1.7: Polar increment

A magnified view of circles of radii r and $r + \delta r$ between the rays at angles θ and $\theta + \delta\theta$ appears to be a rectangle with sides of lengths δr and $r \cdot \delta\theta$. If this were a true rectangle, its area would be $r \cdot \delta\theta \cdot \delta r$, but it is only an approximate rectangle. Technically, we can show that the area of this region is $r \cdot \delta\theta \cdot \delta r$ plus a term that is small compared with this infinitesimal,

$$\delta A = r\delta\theta\delta r + \varepsilon \cdot \delta\theta\delta r, \ \varepsilon \approx 0.$$

Keisler's Infinite Sum Theorem [6] assures us that we can neglect this size error and integrate with respect to $rd\theta dr$.

Theorem 4 ***Holditch's formula***

The area swept out by a tangent of length R as it traverses an arbitrary convex curve in the plane is $A = \pi R^2$.

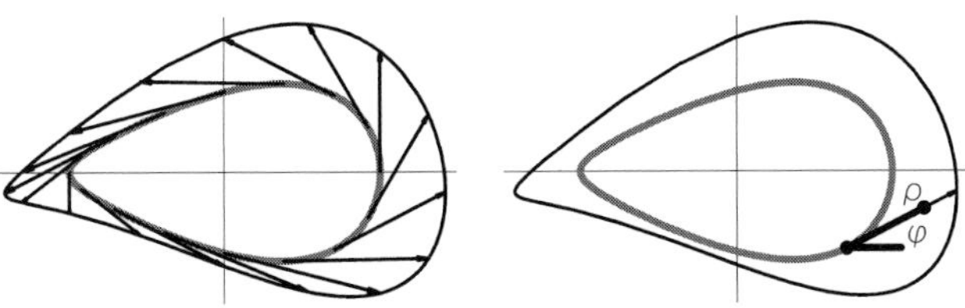

Figure 24.1.8: ρ-φ-coordinates

We can see this interesting result by using a variation of polar coordinates and the infinitesimal polar area increment above. Since the curve is convex, each tangent meets the curve at a unique angle, φ, and each point in the region swept out by the tangents is a distance ρ along that tangent.

We look at an infinitesimal increment of the region in ρ-φ-coordinates, first holding the φ-base point on the curve and changing φ. Microscopically this produces an increment like the polar increment:

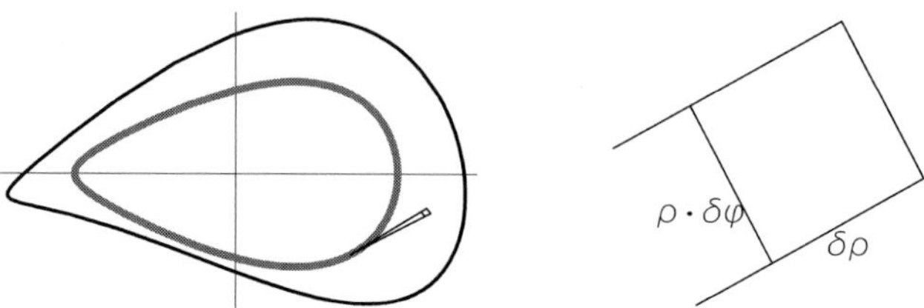

Figure 24.1.9: ρ-only increment

Next, looking at the base point of the tangent on the curve, moving to the correct $\varphi + \delta\varphi$-base point, moves along the curve. Microscopically this looks like translation along the tangent line (by smoothness) as shown on the left. Including this near-translation in the infinitesimal area increment produces a parallelogram:

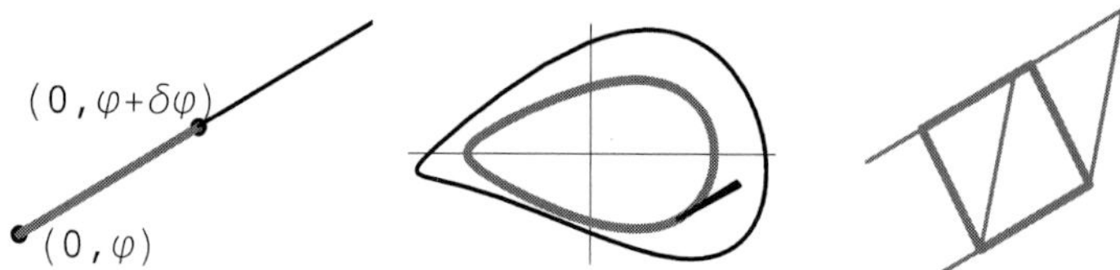

Figure 24.1.10: φ and ρ increment

of height $\rho \cdot \delta\varphi$ and base $\delta\rho$, or area $\delta A = \rho \cdot \delta\varphi \cdot \delta\rho$:

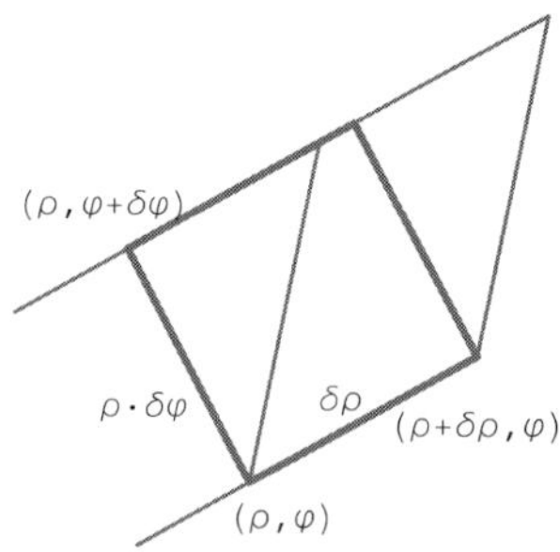

Figure 24.1.11: $da = \rho \cdot \delta\varphi \cdot \delta\rho$

Integrating gives the total area of the region

$$\int_0^R \int_0^{2\pi} \rho d\varphi d\rho = \pi R^2.$$

24.1.8 Leibniz's formula for radius of curvature

The radius r of a circle drawn through three infinitely nearby points on a curve in the (x, y)-plane satisfies

$$\frac{1}{r} = -\frac{d}{dx}\left(\frac{dy}{ds}\right)$$

where s denotes the arclength.

For example, if $y = f[x]$, so $ds = \sqrt{1 + (f'[x])^2}dx$, then

$$\frac{1}{r} = -\frac{d}{dx}\left(\frac{f'[x]}{\sqrt{1 + (f'[x])^2}}\right) = -\frac{f''[x]}{\left(1 + f'[x]^2\right)^{3/2}}$$

If the curve is given parametrically, $y = y[t]$ and $x = x[t]$, so $ds = \sqrt{x'[t]^2 + y'[t]^2}dt$, then

$$\frac{1}{r} = -\frac{d\left(\frac{dy}{ds}\right)}{dx} = \frac{y'[t]x''[t] - x'[t]y''[t]}{\left(x'[t]^2 + y'[t]^2\right)^{3/2}}$$

24.1.9 Changes

Consider three points on a curve $\mathbb{C}$ with equal distances Δs between the points. Let α_I and α_{II} denote the angles between the horizontal and the segments connecting the points as shown. We have the relation between the changes in y and α:

$$\sin[\alpha] = \frac{\Delta y}{\Delta s} \tag{24.1.2}$$

The difference between these angles, $\Delta\alpha$, is shown near p_{III} (figure 24.1.12).

The angle between the perpendicular bisectors of the connecting segments is also $\Delta\alpha$, because they meet the connecting segments at right angles.

These bisectors meet at the center of a circle through the three points on the curve whose radius we denote r. The small triangle with hypotenuse r gives

$$\sin\left[\frac{\Delta\alpha}{2}\right] = \frac{\Delta s/2}{r} \tag{24.1.3}$$

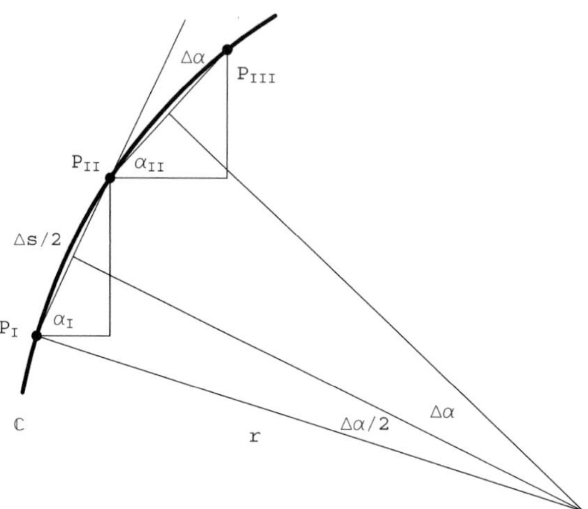

Figure 24.1.12: Changes in s and a

24.1.10 Small changes

Now we apply these relations when the distance between the successive points is an infinitesimal δs. The change

$$-\delta \sin[\alpha] = -\delta\left(\frac{\delta y}{\delta s}\right) = \sin[\alpha] - \sin[\alpha - \delta\alpha] = \cos[\alpha]\cdot\delta\alpha + \vartheta\cdot\delta\alpha, \quad (24.1.4)$$

with $\vartheta \approx 0$, by smoothness of sine (see above). Smoothness of sine also gives,

$$\sin\left[\frac{\delta\alpha}{2}\right] = \frac{\delta\alpha}{2} + \eta\cdot\delta\alpha, \text{ with } \eta \approx 0.$$

Combining this with formula (24.1.3) for the infinitesimal case (assuming $r \neq 0$), we get

$$\delta\alpha = \frac{\delta s}{r} + \iota\cdot\delta\alpha, \text{ with } \iota \approx 0.$$

Now substitute this in (24.1.4) to obtain

$$-\delta\left(\frac{\delta y}{\delta s}\right) = \cos[\alpha]\frac{\delta s}{r} + \zeta\cdot\delta s, \text{ with } \zeta \approx 0.$$

By trigonometry, $\cos[\alpha] = \delta x/\delta s$, so

$$-\frac{\delta\left(\frac{\delta y}{\delta s}\right)}{\delta x} = \frac{1}{r} + \zeta\cdot\frac{\delta s}{\delta x} \approx \frac{1}{r},$$

as long as $\frac{\delta s}{\delta x}$ is not infinitely large.

Keisler's Function Extension Axiom allows us to apply formulas (24.1.3) and (24.1.4) when the change is infinitesimal, as we shall see. We still have a gap to fill in order to know that we may replace infinitesimal differences with differentials (or derivatives), especially because we have a difference of a quotient of differences.

First differences and derivatives have a fairly simple rigorous version in Robinson's theory, just using the differential approximation (24.1.1). This can be used to derive many classical differential equations like the tractrix, catenary, and isochrone, see: Chapter 5, Differential Equations from Increment Geometry in [11].

Second differences and second derivatives have a complicated history. See [2]. This is a very interesting paper that begins with a course in calculus as Leibniz might have presented it.

24.1.11 The natural exponential

The natural exponential function satisfies

$$\begin{aligned} y[0] &= 1 \\ \frac{dy}{dx} &= y \end{aligned}$$

We can use (24.1.1) to find an approximate solution,

$$y[\delta x] = y[0] + y'[0] \cdot \delta = 1 + \delta x$$

Recursively,

$$\begin{aligned} y[2\delta x] &= y[\delta x] + y'[\delta x] \cdot \delta x = y[\delta] \cdot (1 + \delta x) = (1 + \delta x)^2 \\ y[3\delta x] &= y[2\delta x] + y'[2\delta x] \cdot \delta x = y[2\delta x] \cdot (1 + \delta x) = (1 + \delta x)^3 \\ &\vdots \\ y[x] &= (1 + \delta x)^{x/\delta x}, \text{ for } x = 0, \delta x, 2\delta x, 3\delta x, \cdots \end{aligned}$$

This is the product expansion $e \approx (1 + \delta)^{1/\delta x}$, for $\delta x \approx 0$.

No introduction to calculus is complete without mention of this sort of "infinite algebra" as championed by Euler as in [3]. A wonderful modern interpretation of these sorts of computations is in [7]. W. A. J. Luxemburg's reformulation of the proof of one of Euler's central formulas $\sin[z] = z \prod_{k=1}^{\infty} \left(1 - \left(\frac{z}{k\pi}\right)^2\right)$ appears in our monograph, [13].

24.1.12 Concerning the history of the calculus

Chapter X of Robinson's monograph [10] begins:

The history of a subject is usually written in the light of later developments. For over half a century now, accounts of the history of the Differential and Integral Calculus have been based on the belief that even though the idea of a number system containing infinitely small and infinitely large elements might be consistent, it is useless for the development of Mathematical Analysis. In consequence, there is in the writings of this period a noticeable contrast between the severity with which the ideas of Leibniz and his successors are treated and the leniency accorded to the lapses of the early proponents of the doctrine of limits. We do not propose here to subject any of these works to a detailed criticism. However, it will serve as a starting point for our discussion to try to give a fair summary of the contemporary impression of the history of the Calculus...

I recommend that you read Robinson's Chapter X. I have often wondered if mathematicians in the time of Weierstrass said things like, 'Karl's epsilon-delta stuff isn't very interesting. All he does is re-prove old formulas of Euler.'

I have a non-standard interest in the history of infinitesimal calculus. It really is not historical. Some of the old derivations like Bernoulli's derivation of Leibniz' formula for the radius of curvature seem to me to have a compelling clarity. Robinson's theory of infinitesimals offers me an opportunity to see what is needed to complete these arguments with a contemporary standard of rigor.

Working on such problems has led me to believe that the best theory to use to underly calculus when we present it to beginners is one based on the kind of derivatives described in Section 2 and not the pointwise approach that is the current custom in the U.S. I believe we want a theory that supports intuitive reasoning like the examples above and pointwise defined derivatives do not.

24.2 Keisler's axioms

The following presentation of Keisler's foundations for Robinson's Theory of Infinitesimals is explained in more detail in either of the (free .pdf) files [12] and the Epilog to Keisler's text [6].

24.2.1 Small, medium, and large hyperreal numbers

A field of numbers is a system that satisfies the associative, commutative and distributive laws and has additive inverses and multiplicative inverses for nonzero elements. (See the above references for details.) Essentially this means the laws of high school algebra apply. The binomial expansion that follows is a consequence of the field axioms. Hence this formula holds for any pair of numbers x and Δx in a field.

$$(x + \Delta x)^3 = x^3 + 3x^2\Delta x + ((3x + \Delta x) \cdot \Delta x) \cdot \Delta x$$

To compare sizes of numbers we need an ordering. An ordered field has a transitive order that is compatible with the field operations in the sense that if $a < b$, then $a+c < b+c$ and if $0 < a$ and $0 < b$, then $0 < a \cdot b$. (The complex numbers can't be ordered compatibly because $i^2 = -1$.)

An ***infinitesimal*** is a number satisfying $|\delta| < 1/m$ for any ordinary natural counting number, $m = 1, 2, 3, \cdots$. Archimedes' Axiom is precisely the statement that the (Dedekind) "real" numbers have no positive infinitesimals. (We take 0 as infinitesimal.) Keisler's Algebra Axiom is the following:

24.2.2 Keisler's algebra axiom

Axiom 5 *The hyperreal numbers are an ordered field extension of the real numbers. In particular, there is a positive hyperreal infinitesimal, δ.*

Any ordered field extending the reals has an infinitesimal, but we just include this fact in the axiom. There are many different infinitesimals. For example, the law $a < b \Rightarrow a + c < b + c$ applied to $a = 0$ and $b = c = \delta$ says $\delta < 2\delta$. If k is a natural number, $k\delta < \frac{1}{m}$, for any natural m, because $\delta < \frac{1}{k \cdot m}$ when δ is infinitesimal. All the ordinary integer multiples of δ are distinct infinitesimals,

$$\cdots < -3\delta < -2\delta < -\delta < 0 < \delta < 2\delta < 3\delta < \cdots$$

Magnification at center c with power $1/\delta$ is simply the transformation $x \to (x-c)/\delta$, so by laws of algebra, integer multiples of δ end up the same integers apart after magnification by $1/\delta$ centered at zero.

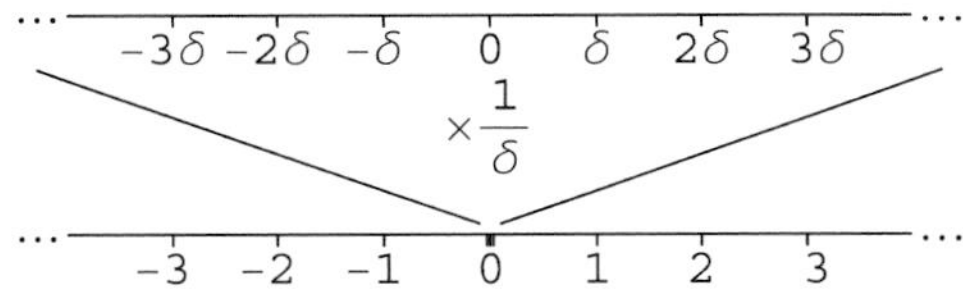

Figure 24.2.1: Magnification by $\frac{1}{\delta}$

Similar reasoning lets us place $\frac{\delta}{2}, \frac{\delta}{3}, \cdots$ on a magnified line at one half the distance to δ, one third the distance, etc.

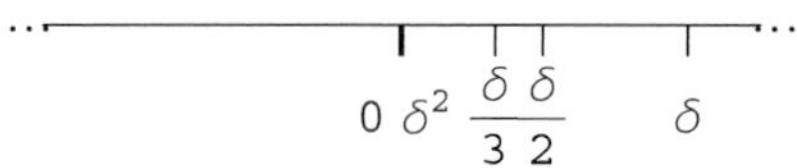

Figure 24.2.2: Fractions of δ

Where should we place the numbers δ^2, $\delta^3 \cdots$? On a scale of δ, they are infinitely near zero, $\frac{1}{\delta}(\delta^2 - 0) = \delta \approx 0$. Magnification by $1/\delta^2$ reveals δ^2, but moves δ infinitely far to the right, $\frac{1}{\delta^2}(\delta - 0) = \frac{1}{\delta} > m$ for all natural $m = 1, 2, 3, \cdots$

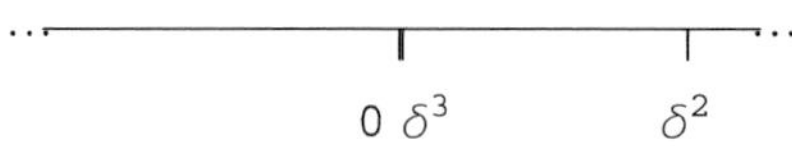

Figure 24.2.3: δ^3 on a δ^2 scale

Laws of algebra dictate many "orders of infinitesimal" such as $0 < \cdots < \delta^3 < \delta^2 < \delta$. The laws of algebra also show that near every real number there are many hyperreals, say near $\pi = 3.14159 \cdots$

$$\cdots < \pi - 3\delta < \pi - 2\delta < \pi - \delta < \pi < \pi + \delta < \pi + 2\delta < \pi + 3\delta < \cdots$$

A hyperreal number x is called ***limited*** (or "finite in magnitude") if there is a natural number m so that $|x| < m$. If there is no natural bound for a hyperreal number it is called ***unlimited*** (or "infinite"). Infinitesimal numbers are limited, being bounded by 1.

Theorem 6 ***Standard Parts of Limited Hyperreal Numbers***

Every limited hyperreal number x differs from some real number by an infinitesimal, that is, there is a real r so that $x \approx r$. This number is called the "standard part" of x, $r = st[x]$.

Proof. Define a Dedekind cut in the real numbers by $A = \{s : s \leq x\}$ and $B = \{s : x < s\}$. st$[x]$ is the real number defined by this cut. □

The fancy way to state the next theorem is to say the limited numbers are an ordered ring with the infinitesimals as a maximal ideal. This amounts to simple rules like "infsml × limited = infsml."

Theorem 7 ***Computation rules for small, medium, and large***

(*a*) *If p and q are limited, so are $p + q$ and $p \cdot q$*

(*b*) *If* ε *and* δ *are infinitesimal, so is* $\varepsilon + \delta$.

(*c*) *If* $\delta \approx 0$ *and* q *is limited, then* $q \cdot \delta \approx 0$.

(*d*) $1/0$ *is still undefined and* $1/x$ *is unlimited only when* $x \approx 0$.

Proof. These rules are easy to prove as we illustrate with (c). If q is limited, there is a natural number with $|q| < k$. the condition $\delta \approx 0$ means $|\delta| < \frac{1}{k \cdot m}$, so $|q \cdot \delta| < \frac{1}{m}$ proving that $q \cdot \delta \approx 0$. □

24.2.3 The uniform derivative of x^3

Let's apply these rules to show that $f[x] = x^3$ satisfies the differential approximation with $f'[x] = 3x^2$ when x is limited. We know by laws of algebra that

$$(x + \delta x)^3 - x^3 = 3x^2 \delta x + \varepsilon \cdot \delta x, \text{ with } \varepsilon = ((3x + \delta x) \cdot \delta x)$$

If x is limited and $\delta x \approx 0$, (a) shows that $3x$ is limited and that $3x + \delta x$ is also limited. Condition (c) then shows that $\varepsilon = ((3x + \delta x) \cdot \delta x) \approx 0$ proving that for all limited x

$$f[x + \delta x] - f[x] = f'[x] \cdot \delta x + \varepsilon \cdot \delta x$$

with $\varepsilon \approx 0$ whenever $\delta x \approx 0$.

Below we will see that this computation is logically equivalent to the statement that $\lim_{\Delta x \longrightarrow 0} \frac{(x+\Delta x)^3 - x^3}{\Delta x} = 3x^2$, uniformly on compact sets of the real line. It is really no surprise that we can differentiate algebraic functions using algebraic properties of numbers. You should try this yourself with $f[x] = x^n$, $f[x] = 1/x$, $f[x] = \sqrt{x}$, etc. This does not solve the problem of finding sound foundations for calculus using infinitesimals because we need to treat transcendental functions like sine, cosine, log.

24.2.4 Keisler's function extension axiom

Keisler's Function Extension Axiom says that all real functions have extensions to the hyperreal numbers and these "natural" extensions obey the same identities and inequalities as the original function. Some familiar identities are

$$\sin[\alpha + \beta] = \sin[\alpha] \cos[\beta] + \cos[\alpha] \sin[\beta]$$
$$\log[x \cdot y] = \log[x] + \log[y]$$

The log identity only holds when x and y are positive. Keisler's Function Extension Axiom is formulated so that we can apply it to the Log identity in the form of the implication

$$
\begin{aligned}
&(x > 0 \text{ and } y > 0) \Rightarrow \log[x] \text{ and } \log[y] \text{ are defined}\\
&\quad\text{and}\\
&\log[x \cdot y] = \log[x] + \log[y]
\end{aligned}
$$

The Function Extension Axiom guarantees that the natural extension of $\log[\cdot]$ is defined for all positive hyperreals and its identities hold for hyperreal numbers satisfying $x > 0$ and $y > 0$. We can state the addition formula for sine as the implication

$$
\begin{aligned}
&(\alpha = \alpha \text{ and } \beta = \beta) \Rightarrow \sin[\alpha], \sin[\beta], \sin[\alpha + \beta], \cos[\alpha], \cos[\beta] \text{ are defined}\\
&\quad\text{and}\\
&\sin[\alpha + \beta] = \sin[\alpha]\cos[\beta] + \cos[\alpha]\sin[\beta]
\end{aligned}
$$

The addition formula is always true so we make the logical real statement (see 24.2.7 below) begin with $(\alpha = \alpha$ and $\beta = \beta)$.

24.2.5 Logical real expressions

Logical real expressions are built up from numbers and variables using functions.

(a) A real number is a real expression.

(b) A variable standing alone is a real expression.

(c) If $e_1, e_2, \cdots, e_n$ are a real expressions and $f[x_1, x_2, \cdots, x_n]$ is a real function of n variables, then $f[e_1, e_2, \cdots, e_n]$ is a real expression.

24.2.6 Logical real formulas

A logical real formula is one of the following:

(i) An equation between real expressions, $E_1 = E_2$.

(ii) An inequality between real expressions, $E_1 < E_2$, $E_1 \le E_2$, $E_1 > E_2$, $E_1 \ge E_2$, or $E_1 \ne E_2$.

(iii) A statement of the form "E *is defined*" or of the form "E *is undefined.*"

24.2.7 Logical real statements

Let S and T be finite sets of real formulas. A logical real statement is an implication of the form,

$$S \Rightarrow T.$$

The functional identities for sine and log given above are logical real statements.

Axiom 8 ***Keisler's Function Extension Axiom***
Every real function $f[x_1, x_2, \cdots, x_n]$ has a "natural" extension to the hyperreals such that every logical real statement that holds for all real numbers also holds for all hyperreal numbers when the real functions in the statement are replaced by their natural extensions.

There are two general uses of the Function Extension Axiom that underlie most of the theoretical problems in calculus. These involve extension of the discrete maximum and extension of finite summation. The proof of the Extreme Value Theorem below uses a hyperfinite maximum, while the proof of the Fundamental Theorem of Integral Calculus uses hyperfinite summation and a maximum.

Theorem 9 ***Simple Equivalency of Limits and Infinitesimals***
Let $f[x]$ be a real valued function defined for $0 < |x-a| < \Delta$ with Δ a fixed positive real number. Let b be a real number. Then the following are equivalent:

(a) *Whenever the hyperreal number x satisfies $a \neq x \approx a$, the natural extension function satisfies*

$$f[x] \approx b.$$

(b) *For every real accuracy tolerance θ there is a sufficiently small positive real number γ such that if the real number x satisfies $0 < |x-a| < \gamma$, then*

$$|f[x] - b| < \theta.$$

Condition (b) is the familiar Weierstrass "epsilon-delta" condition (written with θ and γ.) Notice that the condition $f[x] \approx b$ is NOT a logical *real* statement because the infinitesimal relation is NOT included in the formation rules for forming logical *real* statements.

Proof. We show that (a) $\Rightarrow$ (b) by proving that not (b) implies not (a), the contrapositive. Assume (b) fails. Then there is a real $\theta > 0$ such that for every real $\gamma > 0$ there is a real x satisfying $0 < |x - a| < \gamma$ and $|f[x] - b| \geq \theta$. Let $X[\gamma] = x$ be a real function that chooses such an x for a particular γ. Then we have the equivalence

$$\gamma > 0 \Leftrightarrow (X[\gamma] \text{ is defined}, 0 < |X[\gamma] - a| < \gamma, |f[X[\gamma]] - b| \geq \theta)$$

By the Function Extension Axiom this equivalence holds for hyperreal numbers and the natural extensions of the real functions $X[\cdot]$ and $f[\cdot]$. In particular, choose a positive infinitesimal γ and apply the equivalence. We have $0 < |X[\gamma] - a| < \gamma$ and $|f[X[\gamma]] - b| > \theta$ and θ is a positive real number. Hence, $f[X[\gamma]]$ is not infinitely close to b, proving not (a) and completing the proof that (a) implies (b).

Conversely, suppose that (b) holds. Then for every positive real θ, there is a positive real γ such that $0 < |x - a| < \gamma$ implies $|f[x] - b| < \theta$. By the Function Extension Axiom, this implication holds for hyperreal numbers. If $\xi \approx a$, then $0 < |\xi - a| < \gamma$ for every real γ, so $|f[\xi] - b| < \theta$ for every real positive θ. In other words, $f[\xi] \approx b$, showing that (b) implies (a) and completing the proof of the theorem. □

Other examples of uses of the Function Extension Axiom to complete the proofs of the basic results of Section 1 follow.

24.2.8 Continuity and extreme values

We follow the idea of the proof in Section 1 for a real function $f[x]$ on a real interval $[a, b]$. Coding our proof in terms of real functions.

There is a real function $x_M[h]$ so that for each natural number h the maximum of the values $f[x]$ for $x = a + k\Delta x$, $k = 1, 2, \cdots, h$ and $\Delta x = (b - a)/h$ occurs at $x_M[h]$. We can express this in terms of real functions using a real function indicating whether a real number is a natural number,

$$I[x] = \begin{cases} 0, & \text{if } x \neq 1, 2, 3, \cdots \\ 1, & \text{if } x = 1, 2, 3, \cdots \end{cases}$$

When the natural extension of the indicator function satisfies $I[k] = 1$, we say that k is a hyperinteger. (Every limited hyperinteger is an ordinary positive integer. As you can show with these functions.) The maximum of the partition can be described by

$$\left(a \leq x \leq b \;\&\; I\left[h\frac{x-a}{b-a}\right] = 1\right) \Rightarrow f[x] \leq f[x_M[h]]$$

We want to extend this function to unlimited "hypernatural" numbers. The greatest integer function Floor$[x]$ satisfies, $I[\text{Floor}[x]] = 1$, $0 \leq x - \text{Floor}[x] \leq 1$. The unlimited number $1/\delta$, for $\delta \approx 0$ gives an unlimited $H = \text{Floor}[x]$ with $I[H] = 1$ and

$$\left(a \leq x \leq b \;\&\; I\left[H\frac{x-a}{b-a}\right] = 1\right) \Rightarrow f[x] \leq f[x_M[H]]$$

There is a greatest partition point of any number in $[a,b]$, $P[h,x] = a + \text{Floor}\left[h\frac{x-a}{b-a}\right]\frac{b-a}{h}$ with $a \le P[h,x] \le b$ & $I\left[h\frac{P[h,x]-a}{b-a}\right] = 1$ and $0 \le x - P[h,x] \le 1/h$. When we take the unlimited hypernatural number H we have $x - P[H,x] \le 1/H \approx 0$ and $P[H,x]$ a partition point in the sense that $(a \le x \le b$ & $I\left[H\frac{P[H,x]-a}{b-a}\right] = 1)$, so we have

$$f[P[H,x]] \le f[x_M[H]].$$

Let $r_M = \text{st}[x_M[H]]$, the standard part. Since $a \le x_M[H] \le b$, $a \le r_M \le b$. Continuity of the function in the sense $x_1 \approx x_2 \Rightarrow f[x_1] \approx f[x_2]$ gives

$$f[x] \approx f[P[H,x]] \le f[x_M[H]] \approx f[r_M], \text{ so } f[x] \le f[r_M]$$

for any real x in$[a,b]$.

One important comment about the proof of the Extreme Value Theorem is this. The simple fact that the standard part of every hyperreal x satisfying $a \le x \le b$ is in the original real interval $[a,b]$ is the form that topological compactness takes in Robinson's theory: A standard topological space is compact if and only if every point in its extension is near a standard point, that is, has a standard part and that standard part is in the original space.

24.2.9 Microscopic tangency in one variable

Suppose $f[x]$ and $f'[x]$ are real functions defined on the interval (a,b), if we know that for all hyperreal numbers x with $a < x < b$ and $a \not\approx x \not\approx b$, $f[x+\delta x] - f[x] = f'[x]\cdot\delta x + \varepsilon\cdot\delta x$ with $\varepsilon \approx 0$ whenever $\delta x \approx 0$, then arguments like the proof of the simple equivalency of limits and infinitesimals above show that $f'[x]$ is a uniform limit of the difference quotient functions on compact subintervals $[\alpha,\beta] \subset (a,b)$. More generally, in [12] we show:

Theorem 10 ***Uniform Differentiability***

Suppose $f[x]$ and $f'[x]$ are real functions defined on the open real interval (a,b). The following are equivalent definitions of, "The function $f[x]$ is smooth with continuous derivative $f'[x]$ on (a,b)."

(*a*) *Whenever a hyperreal x satisfies $a < x < b$ and x is not infinitely near a or b, then an infinitesimal increment of the extended dependent variable is approximately linear on a scale of the change, that is, whenever $\delta x \approx 0$*

$$f[x+\delta x] - f[x] = f'[x]\cdot\delta x + \varepsilon\cdot\delta x \textit{ with } \varepsilon \approx 0.$$

(*b*) *For every compact subinterval $[\alpha,\beta] \subset (a,b)$, the real limit*

$$\lim_{\Delta x \longrightarrow 0} \frac{f[x+\Delta x] - f[x]}{\Delta} = f'[x] \textit{ uniformly for } \alpha \le x \le \beta.$$

(c) *For every pair of hyperreal* $x_1 \approx x_2$ *with* $a < \mathrm{st}[x_i] = c < b$, $\frac{f[x_2]-f[x_1]}{x_2-x_1} \approx f'[c]$.

(d) *For every* c *in* (a,b), *the real double limit,* $\lim_{x_1 \longrightarrow c, x_2 \longrightarrow c} \frac{f[x_2]-f[x_1]}{x_2-x_1} = f'[c]$.

(e) *The traditional pointwise defined derivative* $D_x f = \lim_{\Delta x \longrightarrow 0} \frac{f[x+\Delta x]-f[x]}{\Delta x}$ *is continuous on* (a,b).

Continuity of the derivative follows rigorously from the argument of Section 1, approximating the increment $f[x_1]-f[x_2]$ from both ends of the interval $[x_1, x_2]$. It certainly is geometrically natural to treat both endpoints equally, but this is a "locally uniform" approximation in real-only terms because the x values are hyperreal.

In his *General Investigations of Curved Surfaces* (original in draft of 1825, published in Latin 1827, English translation by Morehead and Hiltebeitel, Princeton, NJ, 1902 and reprinted by the University of Michigan Library), Gauss begins as follows:

A curved surface is said to possess continuous curvature at one of its points A, if the directions of all straight lines drawn from A to points of the surface at an infinitely small distance from A are deflected infinitely little from one and the same plane passing through A. This plane is said to touch the surface at the point A.

In [4], Chapter A6, we show that this can be interpreted as C^1-*embedded* if we apply the condition to all points in the natural extension of the surface.

24.2.10 The Fundamental Theorem of Integral Calculus

The definite integral $\int_a^b f[x]\, dx$ is approximated in real terms by taking sums of slices of the form

$$f[a] \cdot \Delta x + f[a + \Delta x] \cdot \Delta x + f[a + 2\Delta x] \cdot \Delta x + \cdots + f[b'] \cdot \Delta x,$$

where $b' = a + h \cdot \Delta x$ and $a + (h+1) \cdot \Delta x > b$

Given a real function $f[x]$ defined on $[a, b]$ we can define a new real function $S[a, b, \Delta x]$ by

$$S[a, b, \Delta x] = f[a] \cdot \Delta x + f[a + \Delta x] \cdot \Delta x + f[a + 2\Delta x] \cdot \Delta x + \cdots + f[b'] \cdot \Delta x,$$

where $b' = a+h \cdot \Delta x$ and $a+(h+1) \cdot \Delta x > b$. This function has the properties of summation such as

$$\left|S[a, b, \Delta x]\right| \leq |f[a]| \cdot \Delta x + |f[a+\Delta x]| \cdot \Delta x + |f[a+2\Delta x]| \cdot \Delta x + \cdots + |f[b']| \cdot \Delta x$$

$$\left|S[a, b, \Delta x]\right| \leq \mathrm{Max}\Big[\left|f[x]\right| : x = a, a + \Delta x, a + 2\Delta x, \cdots, b'\Big] \cdot (b - a)$$

We can say we have a sum of infinitesimal slices when we apply this function to an infinitesimal δx,

$$\int_a^b f[x]\,dx \approx \sum_{\substack{x=a \\ \text{step } \delta x}}^{b-\delta} f[x] \cdot \delta x$$

or

$$\int_a^b f[x]\,dx = \operatorname{st}\left[\sum_{\substack{x=a \\ \text{step } \delta x}}^{b-\delta x} f[x] \cdot \delta x\right], \text{ when } \delta x \approx 0$$

Officially, we code the various summations with the functions like $S[a, b, \delta x]$ (in order to remove the function $f[x]$ as a variable.) We need to show that this is well-defined, that is, gives the same real standard part for every infinitesimal, $S[a, b, \delta x] \approx S[a, b, \iota]$ and both are limited (so they have a common standard part.)

When $f[x]$ is continuous, we can show this "existence," but in the case of the Fundamental Theorem, if we know a real function $F[x]$ with $F'[x] = f[x]$ for all $a \leq x \leq b$, the proof in Section 1 interpreted with the extended summation functions and extended maximum functions proves this "existence" at the same time it shows that the value is $F[b] - F[a]$. The only ingredient needed to make this work is that

$$\max\left[|\varepsilon[x, \delta x]| : x = a, a + \delta x, a + 2\delta x, \cdots, b'\right] = \varepsilon[a + k\,\delta x, \delta x] \approx 0$$

This follows from the Uniform Differentiability Theorem above when we take one of the equivalent conditions as the definition of "$F'[x] = f[x]$ for all $a \leq x \leq b$."

Notice that $\varepsilon[x, \Delta x]$ is the real function $\frac{f[x+\Delta x]-f[x]}{\Delta x} - f'[x]$, so we can define an infinite sum by extending the real function

$$S_\varepsilon[a, b, \Delta x] = \sum_{\substack{x=a \\ \text{step } \delta x}}^{b-\delta x} |\varepsilon| \cdot \delta x$$

24.2.11 The Local Inverse Function Theorem

In [1] Michael Behrens noticed that the inverse function theorem is true for a function with a uniform derivative even just at one point. (It is NOT true for a pointwise derivative.) Specifically, condition (d) of the Uniform Differentiability Theorem makes the intuitive proof of Section 1 work.

Theorem 11 ***The Inverse Function Theorem***

If m is a nonzero real number and the real function $f[x]$ is defined for all $x \approx x_0$, a real x_0 with $y_0 = f[x_0]$ and $f[x]$ satisfies

$$\frac{f[x_2] - f[x_1]}{x_2 - x_1} \approx m \text{ whenever } x_1 \approx x_2 \approx x_0$$

then $f[x]$ has an inverse function in a small neighborhood of x_0, that is, there is a real number $\Delta > 0$ and a smooth real function $g[y]$ defined when $|y - y_0| < \Delta$ with $f[g[y]] = y$ and there is a real $\varepsilon > 0$ such that if $|x - x_0| < \varepsilon$, then $|f[x] - y_0| < \Delta$ and $g[f[x]] = x$.

Proof. This proof introduces a "permanence principle." When a logical real formula is true for all infinitesimals, it must remain true out to some positive real number. We know that the statement "$|x - x_0| < \delta \Rightarrow f[x]$ is defined" is true whenever $\delta \approx 0$. Suppose that for every positive real number Δ there was a real point r with $|r - x_0| < \Delta$ where $f[r]$ was not defined. We could define a real function $U[\Delta] = r$. Then the logical real statement

$$\Delta > 0 \Rightarrow \big(r = U[\Delta],\ |r - x_0| < \Delta,\ f[r] \text{ is undefined}\big)$$

is true. The Function Extension Axiom means it must also be true with $\Delta = \delta \approx 0$, a contradiction, hence, there is a positive real Δ so that $f[x]$ is defined whenever $|x - x_0| < \Delta$.

We complete the proof of the Inverse Function Theorem by a permanence principle on the domain of y-values where we can invert $f[x]$. The intuitive proof of Section 1 shows that whenever $|y - y_0| < \delta \approx 0$, we have $|x_1 - x_0| \approx 0$, and for every natural n and k,

$$|x_n - x_0| < 2|x_1 - x_0|,\ |x_{n+k} - x_k| < \tfrac{1}{2^{k-1}}\,|x_1 - x_0|\,,\ f[x_n] \text{ is defined,}$$
$$\big|y - f[x_{n+1}]\big| < \tfrac{1}{2}\big|y - f[x_n]\big|$$

Recall that we re-focus our infinitesimal microscope after each step in the recursion. This is where the uniform condition is used.

Now by the permanence principle, there is a real $\Delta > 0$ so that whenever $|y - y_0| < \Delta$, the properties above hold, making the sequence x_n convergent. Define $g[y] = \underset{n \longrightarrow \infty}{\text{Lim}}\ x_n$. □

24.2.12 Second differences and higher order smoothness

In Section 1 we derived Leibniz' second derivative formula for the radius of curvature of a curve. We actually used infinitesimal second differences, rather than second derivatives and a complete justification requires some more work.

One way to re-state the Uniform First Derivative Theorem above is: The curve $y = f[x]$ is smooth if and only if the line through any two pairs of infinitely close points on the curve is near the same real line,

$$x_1 \approx x_2 \Rightarrow \frac{f[x_1] - f[x_2]}{x_1 - x_2} \approx m$$

A natural way to extend this is to ask: What is the parabola through three infinitely close points? Is the (standard part) of it independent of the choice of the triple? In [8], Vítor Neves and I show:

Theorem 12 ***Theorem on Higher Order Smoothness***

Let $f[x]$ be a real function defined on a real open interval (α, ω). Then $f[x]$ is n-times continuously differentiable on (α, ω) if and only if the nth-order differences $\delta^n f$ are S-continuous on (α, ω). In this case, the coefficients of the interpolating polynomial are near the coefficients of the Taylor polynomial,

$$\delta^n f[x_0, ..., x_n] \approx \frac{1}{n!} f^{(n)}[b]$$

whenever the interpolating points satisfy $x_1 \approx \cdots \approx x_n \approx b$.

References

[1] Michael Behrens, A Local Inverse Function Theorem, in *Victoria Symposium on Nonstandard Analysis*, Lecture Notes in Math. 369, Springer-Verlag, 1974.

[2] H.J.M. Bos, "Differentials, Higher-Order Differentials and the Derivative in the Leibnizian Calculus", Archive for History of Exact Sciences, **14** 1974.

[3] L. Euler, *Introductio in Analysin Infinitorum, Tomus Primus*, Lausanne, 1748. Reprinted as L. Euler, Opera Omnia, ser. 1, vol. 8. Translated from the Latin by J. D. Blanton, *Introduction to Analysis of the Infinite*, Book I, Springer-Verlag, New York, 1988.

[4] Jon Barwise (editor), *The Handbook of Mathematical Logic*, North Holland Studies in Logic 90, Amsterdam, 1977.

[5] Bernard R. Gelbaum and John M. H. Olmsted, *Counterexamples in Analysis*, Holden-Day Inc., San Francisco, 1964.

[6] H. Jerome Keisler, *Elementary Calculus: An Infinitesimal Approach*, 2nd edition, PWS Publishers, 1986. Now available free at `http://www.math.wisc.edu/~keisler/calc.html`

[7] MARK MCKINZIE and CURTIS TUCKEY, "Higher Trigonometry, Hyperreal Numbers and Euler's Analysis of Infinities", Math. Magazine, **74** (2001) 339-368.

[8] VITOR NEVES and K. D. STROYAN, "A Discrete Condition for Higher-Order Smoothness", Boletim da Sociedade Portugesa de Matematica, **35** (1996) 81–94.

[9] ABRAHAM ROBINSON, "Non-standard Analysis", Proceedings of the Royal Academy of Sciences, ser A, **64** (1961) 432–440

[10] ABRAHAM ROBINSON, *Non-standard Analysis*, North-Holland Publishing Co., Amsterdam, 1966. Revised edition by Princeton University Press, Princeton, 1996.

[11] K.D. STROYAN, Projects for Calculus: The Language of Change, on my website at
`http://www.math.uiowa.edu/%7Estroyan/ProjectsCD/estroyan/indexok.htm`

[12] K.D. STROYAN, Foundations of Infinitesimal Calculus,
`http://www.math.uiowa.edu/%7Estroyan/backgndctlc.htm`

[13] K.D. STROYAN and W.A.J. LUXEMBURG, *Introduction to the Theory of Infinitesimals*, Academic Press Series on Pure and Applied Math. 72, Academic Press, New York, 1976.

25

Pre-University Analysis

Richard O'Donovan*

Abstract

This paper is a follow-up of K. Hrbacek's article showing how his approach can be pedagogically helpful when introducing analysis at pre-university level.

Conceptual difficulties arise in elementary pedagogical approaches. In most cases it remains difficult to explain at pre-university level how the derivative is calculated at nonstandard values or how an internal function is defined. Hrbacek provides a modified version of IST [8] (rather Péraire's RIST) which seems to reduce all these difficulties. This system is briefly presented here in its pedagogical form with an application to the derivative. It must be understood as a state-of-the-art report[1].

25.1 Introduction

Infinitesimals are interesting when teaching analysis because they give meaning to symbols such as dx, dy or $df(x)$ and all related formulae. Also, symbolic manipulations are usually simpler than with limits.

In secondary school, real numbers are introduced with no formal justification. It is sometimes shown that $\sqrt{2} \notin \mathbb{Q}$. It is traditional to quote that π is not a rational either (with no proof)... and therefore the set of real numbers exists. The algebraic rules of the rationals are applied to the new numbers — students would never imagine that it could be otherwise.

On the one hand, it is possible to use the students' intuition of infinity and infinitesimals to make them find out most of the computational rules of a system containing infinitesimals, as shown in [10]. On the other hand,

*Collège André-Chavanne, Genève, Switzerland.
richard.o-donovan@edu.ge.ch

[1]The author has been teaching analysis with infinitesimals at pre-university level for several years. This paper has been possible thanks to many exchanges with Karel Hrbacek and also many helpful remarks by Keith Stroyan.

Hoskins writes: "the logical basis of NSA needs to be made clear before it can be used safely" [6]. Most published books on the subject seem to confirm this view. Di Nasso, Benci and Forti state: "Roughly, nonstandard analysis consists of two fundamental tools: the star-map and the transfer principle" [2]. Foundational aspects are fundamental but they should not be a prerequisite to study infinitesimals. Frege and Wittgenstein are not studied in kindergarten prior to learning how to add natural numbers.

Our motivations may have a more pedagogical origin than those exposed by Hrbacek in [8], but globally the reasons why we feel unsatisfied with the available textbooks are the same. Because of these difficulties, some colleagues have been tempted to work with infinitesimals by ignoring the question of transfer altogether. But without transfer, analysis with infinitesimals is analysis in a non-archimedean field, and these fields are known to have very special and unwanted properties as shown in [4] and [13]. Nonstandard analysis ensures that the properties of completeness are transferred to a certain class of objects.

25.2 Standard part

A major difficulty in nonstandard analysis is the existence of external functions. Consider the "updown" function:

$$f : x \mapsto 2 \cdot \mathbf{st}(x) - x$$

At standard scale it is indistinguishable from the identity function but at the infinitesimal scale the function is everywhere decreasing with a rate of change (slope?) equal to -1 *and* it is S-continuous and satisfies the intermediate value property! Why does it not satisfy that if a continuous function has a negative slope everywhere on an interval, then the function is decreasing on that interval?

It appears difficult to explain to students that statements which use the standardness predicate are "external" but that we can nonetheless use it safely for the definition of the derivative. How can a student understand when it is acceptable to use $\mathbf{st}[\ ]$ and when is it not? Some students have pointed out that if the derivative is given by $\mathbf{st}(\frac{\Delta f(x)}{h})$ for all x, then $f'(x+\delta) = f'(x)$. The classical textbooks such as [9], [14] or [5] use an indirect definition for the derivative at nonstandard points which makes a drastic distinction in nature between standard and nonstandard numbers: at standard points, the derivative is calculated using that point and other nonstandard neighbours but at nonstandard points some form of transfer is used. How can we be convincing when explaining why there are two different definitions?

Even Stroyan's uniform derivative [14] is not totally satisfying here because it requires that we recognise a "real function", which in turn means that we need a definition of what such a function is, and what an extension is.

25.3 Stratified analysis

Stratified analysis may be an answer. An interesting aspect is that the theory is adapted for elementary teaching at the same time as it is being developed. The interactions are mutual.

In the adaptation for high-school, we have deliberately avoided references to the words "standard" and "nonstandard" for reasons already discussed in [10].

Instead of the binary relation $\sqsubseteq$ introduced by Hrbacek, we have suggested to use a concept of level, symbolised by $\mathbf{v}(x)$. If $a \sqsubseteq b$ then $a \in \mathbf{v}(b)$: a is relatively standard to b means that a is at the level of b. For high-school, we feel that this is a simpler concept. All of the adaptations have been discussed with Hrbacek so that they satisfy both the educational requirements and the theoretical consistency.

The following is a possible description given to the students:

Levels

We get to know the familiar natural numbers $1, 2, 3, \dots$ when we learn to count. But it would be a rare child that actually counts (in steps of 1) to more than a thousand or so. It is not necessary to keep going further, because somewhere at this point one gets the idea that the counting process never ends (this is called: potential infinity), and one is capable of forming the set of natural numbers $\mathbb{N}$, i.e. it is possible to consider the collection of all natural numbers as one "thing" (this is called: actual infinity). It is neither necessary nor possible to count all natural numbers to do it!

Also, sums, products, differences, quotients, powers, roots, etc. of these familiar numbers are or can be known at this stage.

All these familiar numbers are at the same **level**.

If x is at the **level** of y, it means that x can be known when y can be known; written

$$x \in \mathbf{v}(y)$$

Familiar numbers are all at the **coarsest** level.

$$\mathbf{v}(0) = \mathbf{v}(1) = \mathbf{v}\left(\frac{3}{2}\right) = \mathbf{v}(\sqrt{3})$$

But then there are natural numbers unknown at this level, numbers that are so large (we will say "unlimited") that they do not belong to $\mathsf{v}(0)$.

Let K be such an unlimited natural number, i.e. $K \notin \mathsf{v}(0)$. It takes a higher level of knowledge to know K, but once we do, we also know $2K, K+1\ldots$ and many other objects which all belong to $\mathsf{v}(K)$. In increasing our knowledge about numbers, we do not lose former knowledge, so $1, 2, 3, \ldots$ are also at $\mathsf{v}(K)$: they remain known. Also, the reciprocal: $\frac{1}{K}$ is an "infinitesimal" that becomes known at the level of K. Hence a number such as, say, $3 + \frac{5}{31 \cdot K}$ is at $\mathsf{v}(K)$ (a **finer** level) and is infinitely close to 3.

But again, this level of knowledge does not exhaust the set $\mathbb{N}$, so there are numbers not at $\mathsf{v}(K)$ — hence not at $\mathsf{v}(0)$, etc. The levels potentially expand forever and never fully exhaust $\mathbb{N}$.

- There are many levels, each making new numbers known.
- At $\mathsf{v}(0)$ there are no infinitesimals.

Infinitesimals are defined relatively to a level. α-infinitesimals and α-unlimited are numbers which are not at $\mathbf{v}(\alpha)$ and which are, respectively, less in modulus, greater in modulus, than any nonzero number at $\mathbf{v}(\alpha)$. For an α-limited number ξ, the α-shadow (noted $\mathbf{sh}_\alpha(\xi)$) is the unique number at $\mathbf{v}(\alpha)$ which is α-infinitesimally close to ξ. The x-shadow replaces the concept of standard part in this relativistic framework[2].

We give a more restricted definition of levels than Hrbacek's. We only include numbers in the levels, not sets. This slight difference can be considered as a restriction of the definition given by Hrbacek not a contradiction: the only sets we use explicitly in our course are intervals, $\mathbb{N}$ or $\mathbb{R}$.

One of the immediate advantages of stratified analysis is the possibility to easily identify "acceptable" statements — those that will eventually transfer. A statement about x is acceptable if it makes either no reference to levels or only to the level of x. The same holds if instead of x there is a list $\overline{x} = x_1, x_2, \ldots$

A function f is defined by a rule that tells us what $f(x)$ is when x is in the domain of f. The rule may depend on some parameters, $p_1, p_2, \ldots$ A function is acceptable if the rule either does not refer to levels or refers only to $\mathbf{v}(x, p_1, p_2, \ldots)$. (Although probably not necessary at this school level, transfer ensures that if there are two different rules using parameters from different levels which give the same value for all values of the variable, then they define

[2] Again, for reasons related to politics in the educational realm, shadow has been preferred to "standard part". It also yields a fairly intuitive image that numbers cast a shadow on coarser levels (provided they are finite with respect to that level); shadows contain less information.

the same function and its level is the coarsest one where the function can be defined).

$f : x \mapsto 2 \cdot \mathbf{sh}_0(x) - x$ is not an acceptable function because there is an absolute reference to $\mathbf{v}(0)$ in the shadow predicate. If adapted as $2 \cdot \mathbf{sh}_x(x) - x$ then, as $\mathbf{sh}_x(x) = x$, it simplifies to the identity function. Theorems about continuity are for (acceptable) continuous functions and this function does not satisfy the conditions. The student can see this with no "external" knowledge.

It seems reasonable to state that from then on, acceptable functions will be simply called functions (so the statements of theorems will look very much like the classical ones).

25.4 Derivative

The derivative of a function is defined by:

Let $f :]a; b[\to \mathbb{R}$ be a function and $x \in]a; b[$

f is differentiable at x iff there is an $L \in \mathbf{v}(x, f)$ such that, for any $< x, f >$-infinitesimal h, with $x + h \in]a; b[$,

$$\mathbf{sh}_{<x,f>}\left(\frac{f(x+h) - f(x)}{h}\right) = L$$

then the derivative is

$$f'(x) = L$$

It is a direct observation that the derivative is "acceptable".

For $f : x \mapsto x^2$ at $x = 2$ the derivative is simple and works exactly as in any other nonstandard method. For x in general, the quotient simplifies to $2x + h$. The only parameter of f is $2 \in \mathbf{v}(0)$ hence $\mathbf{v}(x, f) = \mathbf{v}(x)$. As h is x-infinitesimal, $\mathbf{sh}_x(2x + h) = 2x$. For a direct calculation of $f'(2 + \delta)$, we have $2 + \delta \in \mathbf{v}(\delta)$ hence a $[2 + \delta]$-infinitesimal is also a δ-infinitesimal. Let h be a δ-infinitesimal.

$$\mathbf{sh}_\delta\left(\frac{(2+\delta+h)^2-(2+\delta)^2}{h}\right) = \mathbf{sh}_\delta\left(\frac{4h+2\delta h+h^2}{h}\right) = \mathbf{sh}_\delta(4 + 2\delta + h) = 4 + 2\delta.$$

Thus stratified analysis satisfies one of our major requirements: a single definition of the derivative which applies to all numbers.

In [10] it is shown that using the definition of infinitesimals, the students could work out the rules of computation as exercises and find that for standard a and infinitesimal δ, $a \cdot \delta$ is infinitesimal. The same exercises adapted to acceptable statements yield that if δ is a-infinitesimal, then $a \cdot \delta$ is a-infinitesimal.

25.5 Transfer and closure

Some definitions have been re-written several times and the definitions we use today may not be final. All the proofs of theorems in the syllabus must be checked and cross-checked.

The only principles that are needed are Hrbacek's closure principle: If f is an acceptable function, then $f(x) \in \mathbf{v}(x, f)$ for all x in the domain of f. This extends the observation that operations with "familiar" numbers do not yield infinitesimals. The other is a simple form of transfer which states that an acceptable statement is true for $\mathbf{v}(\alpha)$ iff it is true for all $\mathbf{v}(\beta)$ with $\alpha \in \mathbf{v}(\beta)$.

This adaptation to pre-university level is not completed yet. We hope to finish a first version of a handout, with proofs, within a year or so. Our goal is still the same: for most people, infinitesimals "are there", but can they be used to make maths easier to teach and learn and still remain rigorous? We hope to be able to contribute an answer to this question in a not too distant future.

References

[1] Occam's razor, In *Encyclopaedia Britannica*, 1998. CD version.

[2] Vieri Benci, Marc Forti and Mauro Di Nasso, The eightfoldpath to nonstandard analysis, Quaderni del Dipartimento di Mathematica Applicata "U. Dini", Pisa, 2004.

[3] Martin Berz, "Non-archimedean analysis and rigorous computation", International Journal of Applied Mathematics, **2** (2000) 889–930.

[4] Martin Berz, "Cauchy theory on Levi-Civita fields", Contemporary Mathematics, **319** (2003) 39–52.

[5] F. Diener and G. Reeb, *Analyse Non Standard*, Hermann, 1989.

[6] Roy Hoskins, *Standard and Nonstandard Analysis*, Ellis Horwood, 1990.

[7] Roy Hoskins, 2004. personal letter.

[8] Karel Hrbacek, Stratified analysis?, in this volume.

[9] Jerome Keisler, *Elementary Calculus*, University of Wisconsin, 2000. `www.math.wisc.edu/~keisler/calc.html`

[10] John Kimber and Richard O'Donovan, Non standard analysis at pre-university level, magnitude analysis. In D. A. Ross N. J. Cutland, M. Di Nasso (eds.), *Nonstandard Methods and Applications in Mathematics*, Lecture Notes in Logic 25, Association for Symbolic Logic, 2006.

[11] Mauro Di Nasso and Vieri Benci, Alpha theory, an elementary axiomatics for nonstandard analysis, Quaderni del Dipartimento di Mathematica Applicata "U. Dini", Pisa, 2001.

[12] Carol Schumacher, *Chapter Zero*, Addison-Wesley, 2000.

[13] Khodr Shamseddine and Martin Berz, "Intermediate values and inverse functions on non-archimedean fields", *IJMMS*, (2002) 165–176.

[14] K. D. Stroyan, *Mathematical Background: Foundations of Infinitesimal Calculus*, Academic Press, 1997.
`www.math.uiowa.edu/%7Estroyan/backgndctlc.html`

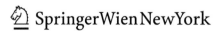

Palaeosurfaces: Recognition, Reconstruction and Palaeoenvironmental Interpretation

Geological Society Special Publications

Series Editor A. J. FLEET

GEOLOGICAL SOCIETY SPECIAL PUBLICATION NO. 120

Palaeosurfaces: Recognition, Reconstruction and Palaeoenvironmental Interpretation

EDITED BY

M. WIDDOWSON

Department of Earth Sciences
The Open University
Milton Keynes
UK

1997
Published by
The Geological Society
London

THE GEOLOGICAL SOCIETY

The Society was founded in 1807 as The Geological Society of London and is the oldest geological society in the world. It received its Royal Charter in 1825 for the purpose of 'investigating the mineral structure of the Earth'. The Society is Britain's national society for geology with a membership of around 8000. It has countrywide coverage and approximately 1000 members reside overseas. The Society is responsible for all aspects of the geological sciences including professional matters. The Society has its own publishing house, which produces the Society's international journals, books and maps, and which acts as the European distributor for publications of the American Association of Petroleum Geologists, SEPM and the Geological Society of America.

Fellowship is open to those holding a recognized honours degree in geology or cognate subject and who have at least two years' relevant postgraduate experience, or who have not less than six years' relevant experience in geology or a cognate subject. A Fellow who has not less than five years' relevant postgraduate experience in the practice of geology may apply for validation and, subject to approval, may be able to use the designatory letters C Geol (Chartered Geologist).

Further information about the Society is available from the Membership Manager, The Geological Society, Burlington House, Piccadilly, London W1V 0JU, UK. The Society is a Registered Charity, No. 210161.

Published by the Geological Society from:
The Geological Society Publishing House
Unit 7
Brassmill Enterprise Centre
Brassmill Lane
Bath BA1 3JN
UK
(*Orders*: Tel 01225 445046
Fax 01225 442836)

First published 1997

The publisher makes no representation, express or implied, with regard to the accuracy of the information contained in this book and cannot accept any legal responsibility for any errors or omissions that may be made.

British Library Cataloguing in Publication Data

A catalogue record for this book is available from the British Library

ISBN 1-897799-57-8
ISSN 0305-8719

Typeset by E & M Graphics,
Midsomer Norton, Bath BA3 2NS, UK

Printed by The Alden Press, Osney Mead, Oxford, UK

Distributors

USA
AAPG Bookstore
PO Box 979
Tulsa
OK 74101-0979
USA
(*Orders*: Tel (918) 584-2555
Fax (918) 560-2652)

Australia
Australian Mineral Foundation
63 Conyngham Street
Glenside
South Australia 5065
Australia
(*Orders*: Tel (08) 379-0444
Fax (08) 379-4634)

India
Affiliated East-West Press PVT Ltd
G-1/16 Ansari Road
New Delhi 110 002
India
(*Orders*: Tel (11) 327-9113
Fax (11) 326-0538)

Japan
Kanda Book Trading Co.
Tanikawa Building
3–2 Kanda Surugadai
Chiyoda-Ku
Tokyo 101
Japan
(*Orders*: Tel (03) 3255-3497
Fax (03) 3255-3495)

Contents

Africa

South America

The geomorphological and geological importance of palaeosurfaces

M. WIDDOWSON

Department of Earth Sciences, The Open University, Walton Hall, Milton Keynes, MK7 6AA, UK

This Special Publication is a collection of papers which identifies, describes, and interprets the occurrence of palaeosurfaces in the geological record. The concept of a palaeosurface is one which is generally understood across the geomorphological and geological sciences but, according to discipline and application, often differs in the emphasis and detail of its interpretation. In order to encompass the widest range of contributions and views upon the subject of palaeosurfaces, a broad remit was deliberately adopted during the preparation of this volume.

Background and objectives

Many working groups within the geological and geomorphological sciences share a common link through the Geological Society, and it is hoped that this volume will further illustrate the potential of cross-disciplinary study. Geology and geomorphology, though often approached by different research schools, are both expressions of complex Earth systems and as such can rarely be treated independently. Since the investigation of palaeosurfaces requires an understanding of both geomorphological and geological processes, their study is clearly a theme which demands this type of cross-disciplinary approach. Therefore, the aim of this volume is to bring together expertise from a variety of fields by researchers in both the geological and geomorphological disciplines who have adopted such a combined approach to palaeosurface research.

Contributions have been encouraged from the fields of geomorphology, geology, geochemistry, palynology and palaeoenvironmental studies. The papers include studies of geomorphological evolution, reconstruction of palaeolandscapes, lateritization and bauxitization, palaeo-karstification, sequence stratigraphy, geochemistry of rock alteration, the preservation of palaeosurface elements in both glaciated and sub-tropical regions, and the use of palaeosurfaces as an indicator of regional-scale neotectonic deformation. Moreover, many incorporate discussion regarding fundamental aspects of evolution, interpretation, reconstruction, and palaeoenvironmental implications of ancient palaeosurfaces. The research areas represented are geographically widespread and include examples from Australia, Africa, South America, India, and Europe.

The importance of palaeosurfaces

This Special Publication is timely for two main reasons. First, in recent years geologists, geophysicists, and geomorphologists have begun to look anew at landscape evolution because it has become increasingly apparent that the characteristics of major landscape components, such as palaeosurfaces, can potentially provide important clues to fundamental questions regarding the nature of macro-scale (i.e. tectonic) processes, and the rate and timing of uplift and erosion. Such an approach is evident from many recent papers which have sought to define both uplift and erosion in terms of the surface expression of crustal and mantle dynamics (e.g. McKenzie 1984; England & Houseman 1988; England & Molnar 1990; Cox 1989; Watts & Cox 1989; Gilchrist & Summerfield 1994; Brown *et al.* 1994*b*; Povey *et al.* 1994; Stüwe *et al.* 1994; Burbank *et al.* 1996): Second, there has been common recognition of the benefit of a more unified treatment of both landscape evolution and sediment deposition systems. The development of this latter approach has been greatly assisted by the advent of advanced computer modelling which has allowed calculation of macro-scale erosion and sedimentation budgets. As a consequence, the dynamics of landscape erosion, fluvial systems, and associated stratigraphical evolution of sedimentary basins, so long treated as independent research themes within geomorphology and geology, are now being considered in a more unified approach (e.g. Brown *et al.* 1990; Schlager 1993; Whiting *et al.* 1994; Burgess & Allen 1996; Widdowson & Cox 1996). Clearly, future dialogue and exchange between the different working groups will provide a basis for fashioning an increasingly holistic treatment of the processes acting both upon and beneath the Earth's surface. However, there remains great potential for exploring the interaction between tectonics and long-term landscape evolution; the great antiquity and lateral extent of

From Widdowson, M. (ed.), 1997, *Palaeosurfaces: Recognition, Reconstruction and Palaeoenvironmental Interpretation,* Geological Society Special Publication No. 120, pp. 1–12.

many palaeosurfaces can provide a suitable spatial and temporal framework for evaluating the link between tectonic processes and their effect upon the development of geomorphological components and associated erosional-depositional systems. Such a treatment is desirable since it has the potential to reveal information about fundamental mechanisms acting deep within the Earth, and to make a valuable contribution to our understanding of how the landscapes, and the processes acting upon them, function over long time scales.

What is a palaeosurface? Some ideas

It is evident from the papers submitted for this volume that the concept of palaeosurfaces encompasses a wide range of geological and geomorphological phenomena. This makes difficult any attempt to provide a palaeosurface definition which readily encompasses all areas of study. One fundamental distinction which can be made, however, is the difference between exogenic and endogenic surfaces: Exogenic processes act upon the Earth's exposed surface and include the action of ice, wind, water, and biota which result in denudation and a net reduction in elevation. Clearly, many identified palaeosurfaces have evolved largely as the result the exogenic geomorphological agents of weathering and erosion. However, there exist other types of surfaces which are demonstrably of an endogenic origin. These include constructional processes, often of geological origin, such as those resulting from crustal uplift (i.e. orogenic and epeirogenic), and igneous activity (e.g. surface volcanic structures and surface doming by magmatic injection). Since endogenic examples are typically constructional in nature they tend to be associated with an increase in elevation of relief. Good examples include those resulting from epeirogenesis (a broad uplift of the crust resulting from thermal or dynamic processes within the Earth), examples arising from the structural effects of faulting and folding in regions of tectonic convergence, or those of a volcanogenic derivation such as lava fields. Nevertheless, it is important to realize that in the ancient record pure endogenic surfaces may be uncommon, for unless these endogenic surfaces rapidly become covered and effectively fossilized by younger materials an increasingly important exogenic contribution to surface morphology will result from continued exposure.

In the broadest sense it can be argued that the term palaeosurface should be used without any genetic connotation to simply mean 'an ancient surface'. Accepting such a view, the term then adequately describes any identifiable surface or horizon of demonstrable antiquity. Within a geological context, the importance of identifying ancient surfaces in the rock record has long been recognized since they are often employed to divide stratigraphies (or other lithological or pedological units) into genetically and temporally related packages. Such *geological* surfaces often result from fundamental changes or hiatuses in deposition processes influenced by factors including tectonism or environmental change. Alternatively, they may be related both directly or indirectly to changes in the nature or rate of geomorphological processes which have the potential to alter or interrupt the established pattern of stratigraphical development.

Historically there has existed an often subtle, but nevertheless fundamental difference in the way in which the term has been employed by geologists and geomorphologists. It can be held that this difference of research perspective arises from the manner in which the two disciplines view the ancient record. Within the geological sciences the term appears to have attained a rather wide interpretation because geologists tend to view Earth surface systems in a depositional or constructive (endogenic) framework and therefore consider a whole variety of surface types, whether they be formed by erosion or other forms of depositional gaps (i.e. unconformities), as an integral and fundamental part of the stratigraphical record. Consequently, the term has been employed to describe surfaces preserved in both sub-aerial and sub-aqueous sequences and it therefore encompasses a variety of features originating from a range of erosional and depositional processes (e.g. unconformity surfaces arising from the erosion and removal of strata, breaks in sedimentation, or other fundamental changes in deposition or deposition rates, for instance). However, it is evident geomorphologists often consider the ancient record in an essentially destructive (exogenic) framework, since their work is primarily concerned with the processes of weathering and erosion; these being the two main factors which determine the evolution of landforms. *Geomorphologically* speaking, the palaeosurface concept involves features, typically within a palaeolandscape framework, which are characteristically erosional and usually, though not exclusively, sub-aerial in origin. Consequently, the term is commonly accepted as being something much more specific and complex than the rather generic 'ancient surface' definition suggested above. With respect to the papers presented in this volume, it is clear that both the geological and geomorphological viewpoints are valid methods of conducting palaeosurface research. Since both geomorphological and geological approaches are represented in the volume, it is worth further exploring these two viewpoints a little further.

Palaeosurfaces in geomorphology

Geomorphologically, the term palaeosurface has often been used to describe ancient 'planation surfaces' that have evolved in response to particular combinations of geomorphological processes which existed during the geological past. In addition, the idea of antiquity implicit in the palaeosurface term has, in the past, led to a preoccupation with flat, or near-flat, erosion plains. This is perhaps a legacy of the pedagogues of denudational chronology which suggest that an end-product of landform evolution is attained (i.e. a peneplain) if denudational cycles are left uninterrupted long enough to approach completion. Such a view is not necessarily valid because the generation of near-flat surfaces is not simply a function of age; clearly, other factors such as the nature of the landforming processes, process rates, lithostructural control, and tectonic regime should also be considered. Moreover, examples of palaeosurfaces, displaying significant relief are now recognized, for instance the etchsurfaces of the Sudetes described by Migoń (1997). Similarly considerable relief may exist wherever an assemblage of ancient landforms is identified, though in these cases there remains the question of whether these together should more strictly be termed *palaeolandscapes* (see below). The research presented in this Special Publication adopts a considered viewpoint since the term palaeosurface is employed to encompass a variety of ancient landform phenomena. These have may different genetic origins and include peneplains, panplains, pediplains, etchplains, etchsurfaces, palaeoland-scapes, as well as other surfaces, such as lava plains (Widdowson *et al.* 1997) which may more accurately be described as fundamentally endogenic in origin

A question of scale also appears to be implicit in the palaeosurface term. For example, whilst all identifiable geomorphological surfaces ranging in size from localized examples of limited extent (e.g. river terraces within an individual river catchment) through to regional or continental scale erosion surfaces could legitimately be called palaeosurfaces, the former small- to meso-scale features are not normally the subject of palaeosurface research. This bias may have arisen largely as a function of a more transient nature, limited durability, and poor preservation potential of areally restricted surfaces, as well as obvious difficulties presented in terms of their recognition in the ancient record. By contrast, the large-scale features which have required very long periods of time to mature and which have given rise to planation surfaces of immense lateral extent, are more commonly considered as palaeosurfaces presumably because the term stresses a temporal importance (i.e. sufficient time required for them to evolve) and because, once formed, features of this nature prove robust, thereby ensuring their long-term survival in the geological record.

Extensive palaeosurfaces of this nature have long been recognized as an important component of the landscape and their evolution, development, and subsequent preservation comprises a crucial aspect of many of the great models of landscape evolution formulated during the past century. These models sought to explain the formation of laterally extensive surfaces by considering differing types of geomorphological processes and concepts such as peneplanation, pedimentation, and etchplanation have been forwarded (e.g. Davis 1899; Penck 1953; King 1953; Thomas 1974). Many have subsequently been much developed and amended to explicate the existence of surfaces in different tectonic and climatic environments (e.g. morphotectonics, cratonic regimes, and climatic geomorphology; Hills 1961; Fairbridge & Finkl 1980; Büdel 1982; Bremer 1985; Ollier 1985). Interestingly, the fact that these types of palaeosurfaces are characteristically sub-aerial in origin has resulted in a close association with residual deposits, and palaeosurface research is, as a result, often closely allied with studies examining the evolution of laterites, silcretes, and bauxites (McFarlane 1976; Bowden 1997). This association is no accident since the preservation of geomorphological palaeosurfaces and the development of lateritic deposits often share the common prerequisites of protracted sub-aerial exposure, relative tectonic quiescence, and a degree of climatic stability. However, it should be noted that bauxites and thick, indurated laterites often appear to be dependant upon a later, slow incision of low-relief palaeosurfaces (e.g. resulting from epeirogeny), and therefore should not always be considered coeval with palaeosurface formation (McFarlane 1983; Thomas 1994).

Most importantly, the formation of large-scale palaeosurfaces will have often influenced sedimentation in adjacent marine basins (i.e. continental shelf and deep ocean), and the offshore coastal environment for example, or in lakes through changes in sediment supply, fluvial regime, or landscape rejuvenation. As a result, the development of many large-scale sub-aerial palaeosurfaces should be recognisable from changes in the pattern of sedimentation in associated basins.

Palaeolandscapes

Another dimension to understanding palaeosurfaces which needs to be addressed and clarified is the question of palaeolandscapes. Although a palaeosurface is commonly used to describe a single genetic entity (i.e. a surface resulting from a particular combination of endogenic or exogenic processes), it has also been employed to describe the surface of an ancient *landscape* which has either become exhumed or else otherwise made apparent. These palaeosurface landscapes may incorporate a collection of genetically unrelated landforms often of differing ages and origins. It is clear that the recognition of palaeolandscapes represent a special and complex aspect of palaeosurface research requiring both detailed study and extreme care in order to differentiate the various landscape components into an understandable evolutionary chronology (Gunnell 1997). In many cases this is a comparatively recent development to palaeosurface research since it can only be successfully achieved by combining both geomorphological and geological data sources, a task made easier with the advent and availability of sophisticated GIS-based techniques. Nevertheless it is also evident that effort directed toward the reconstruction and interpretation of these landscape palaeosurfaces can be most rewarding since it offers immense potential in terms of assessing the effects of environmental change on the landscape, understanding palaeoenvironmental controls, and determining the patterns and processes of landscape evolution which have occurred in the geological past.

Palaeolandscapes are explored in a number of papers within this volume. For example, the reconstruction and environmental interpretation of a complex, partially buried karst landscape in southern Ireland forms the subject of a palynological investigation by Coxon & Coxon (1997). Palynological evidence also forms the basis of the research by Jolley (1997) who utilizes palaeosurfaces preserved within the inter-trappean beds of the Skye lava field to provide a series of time slices which help constrain both the chronology of eruption and highlight fundamental elevational and topographical variations occurring during the Palaeocene construction of the Skye volcanic edifice. This theme of identifying different palaeosurfaces and their utilization as time slices, is similarly adopted by Lacika (1997) who, through the recognition of specific combinations of exogenic and endogenic landforms, explores the evolution and development of a Neogene volcanic landscape in Central Slovakia.

Palaeosurfaces in geology

Palaeosurfaces form an integral and important part of the geological record since they represent fundamental geological, geomorphological, and climatic events which aid in dividing stratigraphies into beds or members or formations. For example, they may occur as erosion surfaces truncating structure and stratigraphy in both the sub-aerial and sub-aqueous sedimentary records, or may be preserved as rock-weathering horizons and associated residual deposits which themselves contain important palaeo-environmental information regarding climate, flora, soil and rock chemistry. Alternatively, they may occur as stratigraphical gaps which may be associated with eustatic or tectonic influences upon sedimentation as evinced by sequence stratigraphy models. In this context, it is evident that palaeosurface investigation in the geological record should not be confined only to surfaces of sub-aerial origin.

In the geological sense a palaeosurface could be any recognizable surface which is preserved in the rock record. A whole hierarchy of such surfaces may be identified ranging from small-scale stratal bounding surfaces between rapidly deposited ripple or dune sets for example, to dateable hiatuses in deposition which, at the regional scale, occur as major unconformities representing surface exposure of immense duration. Again, it is the larger, typically regional scale features, which have been the focus of palaeosurface research because it is these which represent the important climatically, or tectonically controlled events.

Surfaces of erosion and/or non-deposition within the geological record have recently taken on extra importance with the advent of sequence stratigraphy (e.g. Vail *et al.* 1991, and references therein). This concept represents a relatively new, but rapidly developing technique in sedimentology which offers a new perspective to palaeosurface studies. It represents an holistic approach which seeks to describe sediment packages bounded by genetically related surfaces in both the sub-aerial and sub-aqueous environments of the geological record. Sequence stratigraphy may be used to divide the sedimentary record into packages of strata bounded by unconformities (marked by non-deposition and/or erosion) and their laterally equivalent correlative conformites found within the stratigraphically complete sections. Subaerially, these sequence boundaries may be characterized by subaerial erosion, the development of palaeosols, or karstification. Importantly, all of these surfaces may be traced laterally into complex marine unconformities: it is these latter which are more commonly preserved in the geological record (e.g. Coe 1996; Gale 1996). With such an approach there is exciting potential for integrating the marine and non-marine record as a method of obtaining a better understanding of the dynamics of onshore palaeosurface evolution both in terms of the relationship between erosion and the sediment budget, and the associated weathering regimes, and climatic controls. Such studies integrating the marine stratigraphical record and continental denudation are currently in their infancy but represent an obvious area where geological and geomorphological expertise can readily combine.

Toward a working definition

The term palaeosurface has hitherto been employed in a non-specific fashion and within this volume its usage has largely been left to the interpretation and emphasis of individual authors. However, whilst a generic term is attractive because it avoids all the complexities and special case scenarios which inevitably arrive with the constraint of definition, it suffers from that same lack of specification by a vagueness of common understanding. For instance, the question of what exactly constitutes 'ancient' is a moot point; the word is non-specific and is often used to represent differing degrees of antiquity in different countries. This confusion arises largely because its meaning is inevitably influenced by particular research traditions and by researchers' familiarity with the geological history of his or her own region of interest. Consequently, in Europe ancient palaeosurfaces are often considered to be those of Tertiary age, those of Australia being Mesozoic age and older, whilst an ancient palaeosurface in recently glaciated regions of northern Europe may be any feature which predates the last ice age. Therefore, in the longer term it may be an idea to evolve a more useful description other than to simply specify 'an ancient surface'. This description should aim to tackle the questions of age, scale, and those processes and features which are considered as being important to the recognition and preservation of palaeosurfaces. It is also arguable that there is now a need to distinguish more precisely between palaeosurfaces comprising a single genetic entity (i.e. one formed by a specific combination of processes), and those which incorporate a variety of ancient landform components which would, perhaps, be better served by the term palaeolandscapes. However, caution should be exercised before resorting to a complex hierarchical system of palaeosurface types, for with such systems there is often a compulsion to extend the system to specify the particulars of all eventualities. For instance, classification schemes according to age, the various types or combinations of endogenic and exogenic processes, the subsequent evolution and modification of the surface, its mode of preservation, and whether it currently constitutes an exposed landsurface or exists only as a unconformity surface within the stratigraphical record, are all valid methods. In most instances I believe these approaches would prove unwieldy and, in the long term, counterproductive.

Therefore, it is not the intention here to formulate any formal definition of palaeosurface. What is offered instead, is an attempt at a description which satisfies the range of examples presented in this volume and which, it is hoped, will provide a starting point and stimulus for further discussion. The key points in its design must include reference to the antiquity of the surface and an attempt to combine the geological and geomorphological criteria discussed earlier. The question of scale should also be addressed since in both the geological and geomorphological usage there exists the potential to include all features from the smallest river terrace, erosional bench, or stratigraphical bounding surface through to continental-scale planation surfaces and regional unconformities. Attention should therefore be restricted to those surfaces which have a regional significance whether this be in terms of tectonics, climate, or geomorphological regime. The term palaeosurface should, therefore, indicate *an identifiable topographic surface of either endogenic or exogenic origin, recognizable as part of the geological record or otherwise of demonstrable antiquity, which is, or was, originally of regional significance, and which as a consequence of its evolution, displays the effects of surface alteration resulting from a prolonged period of weathering, erosion, or non-deposition.* This is by no means an exhaustive definition, but importantly is not restricted to sub-aerial surfaces since it *does not* exclude the many examples of geological importance which develop in the sub-aqueous environment.

Establishing landscape antiquity : a problem area

Two of the fundamental aims of this collection of papers are to look at the evidence for the existence of palaeosurfaces, and to document their long term evolution over geological time. However, one of the fundamental problems facing palaeosurface research is that of determining the age of these ancient surfaces. In those cases where the surfaces are preserved within a stratigraphy an estimation of their age can be made by dating the rocks immediately above or below and, providing the lithology is suitable, this may be achieved via a range of palaeontological or geochronological methods. Successful applications of the stratigraphical technique may be realized where the palaeosurfaces immediately post-dates the lithology below or was covered by a dateable lithology shortly after its formation (e.g. Walsh *et al.* 1987). Unfortunately, such instances are not common, and are often restricted to those localities which lie at an interface between erosional and depositional environments (for example, the margins of sedimentary basins which are particularly susceptible to periodic sedimentation resulting from sea-level changes caused by eustatic or isostatic influences). However, many palaeosurfaces are remote from such situations and dateable cover is typically absent. In these latter cases this lithological approach can, at best, only provide a maximum age as yielded by the bedrock upon which the surface has developed. These limitations are best demonstrated by those palaeosurfaces developed in cratonic interiors comprising Archaean and Proterozoic rocks where a maximum lithological age is often of little value. Nevertheless, much of the research of recent decades which has focused upon identifying the processes of landscape evolution and establishing the rates at which they act, relies upon some estimation of landform age; a problem further compounded by the fact that many landscapes are demonstrably polycyclic; that is they display elements of different ages corresponding to particular combinations of factors, or else relating to changes in tectonic and climatic conditions during the past. Nevertheless, it should also be remembered that whilst polycyclic landscapes may be the geomorphological norm, different elements within a given landscape evolve at vastly differing rates, and hence have different 'lifetimes' thus providing the opportunity of landforms of great antiquity to exist within a modern dynamic landscape (Brunsden 1993). If some landforms are by nature long lived and therefore resistant to later climatic and tectonic changes, then the modern landscape may contain many and, perhaps, we are currently failing to recognize them.

The endurance of palaeolandforms and palaeosurfaces over geological time can be in little doubt. The literature cites many examples; the evidence for the antiquity of Australian palaeolandsurfaces, some of which may be Mesozoic and older, is examined in this volume by Twidale in the second of the introductory papers. Survival of very old palaeosurfaces is attributed to several factors, but particularly to unequal erosion, reinforcement mechanisms, and the prevalence of vertical (epeirogenic or isostatic), rather than orogenic, earth movements. However, despite the undoubted recognition of ancient surface remnants, their documentation as an integral part of the geological history has not always achieved unequivocal acceptance. There appears to be two main reasons for this apparent reluctance to accept geomorphological antiquity which will now be discussed.

First there is the historical precedent of the great evolutionary theories of landscape development, the earliest of which was forwarded nearly a century ago by Davis (1899). In such models it was assumed that over geological time successive uplift initiated cycles of erosion led to base-levelling and effectively destroyed any of the pre-existing landforms which may have formed in the earlier stages of landscape evolution. It was soon realized, however, that in practice, this palimpsest created by universal base-levelling rarely, if ever, occurred since most regions of the Earth's surface are not sufficiently tectonically quiescent over the time periods required for such a widespread base-levelling to be completed. In effect, many regions are geomorphologically polycyclic with remnants of many different evolutionary stages being preserved. Nevertheless, despite being demonstrably incorrect, the idea of such base-levelling and associated destruction of the earlier landforms as the landscape evolves through different stages seems to have persisted, apparently as a form of conventional wisdom: this type of misconception has had a very strong influence upon, and has provided a major obstacle against, a general acceptance of geomorphological components of geological antiquity. This reluctance to accept landform antiquity may, to some extent, have also been reinforced in recent decades through the detailed quantification of modern erosional and weathering regimes and process rates which became popular during the 1950s through to the 1970s. In many cases these studies served to highlight and emphasize the dynamism of the modern landscape and, most importantly, extrapolation of these modern rates back over geological time made the survival of ancient landforms an apparently unlikely scenario within such obviously dynamic systems. More lately, research has demonstrated that process rates actually vary considerably both spatially and temporally in response to tectonic, climatic, and geomorphological conditions. As a result, recognition of the influence of tectonic regime and the importance of climate change have come to the fore (for a detailed review see chapters 8 & 11, Thomas 1994). These more modern approaches suggest that whilst many regions of geomorphological interest do exhibit rapid rates of erosion and weathering, equally there are those which do not. For instance, across the fundamental geomorphological boundaries represented by the great escarpments such as the Western Ghats, India, Drakensberg of South Africa, and Great Dividing range of Australia (Pain 1985) a major spatial variation in process types and rates exist; other convincing examples of such variation are similarly provided by recent detailed studies of the geomorphology of the Australian continent (e.g. Gale 1992). Though there are a number of reasons why estimates of current and Quaternary process rates may be more rapid than those experienced in the geological past (e.g. human impact, isostatic and eustatic effects of Pleistocene deglaciaton), the fact remains that wherever the existence of surfaces of Palaeozoic, Mesozoic, or even Tertiary age have been demonstrated they are invariably offered as something very unusual. It may be time to re-examine whether preservation of ancient surfaces is, after all, such an extraordinary phenomenon given that a host of other factors (e.g. climate, lithology, tectonics, and geomorphological environment), and not simply time alone, are now known play the crucial roles in determining whether landforms are preserved. Accordingly, it is interesting to speculate whether the paucity of recognized geologically ancient landforms is, in fact, more apparent rather than real, and instead simply a function of the current limitations in our ability to recognize them in the first place, and then adequately demonstrate their antiquity in the second. With respect to future palaeosurface research there is clearly an argument for directing investigation to those regions in which conditions are most conducive for the preservation of ancient surfaces. But how do we identify such regions? In the past it has been generally accepted that the best potential for their preservation exists in the tectonically stable interiors of the continents (i.e. cratons), and certainly it is within these cratonic cores, where geomorphological and geological processes tend to be slower, that many of the 'classic' examples have been described. Consequently, it comes as some surprise to find ancient surfaces preserved in geomorphologically and geologically active regions. For instance, Kennan *et al.* (1997) describe surfaces of Mio-Pliocene age within the Andes orogenic belt, whilst Battiau-Queney (1997) cites Triassic surfaces in the Alps that have not only survived glacial effects but also which retain much valuable information regarding the nature of pre-glacial environments. Similarly, examples of saprolites developed upon Palaeozoic, Mesozoic, and Tertiary palaeosurfaces which have survived the Pleistocene glaciation of Scandinavia are discussed in detail by Lidmar-Bergström *et al.* (1997), and those of Tertiary age by Whalley *et al.* (1997).

The second major obstacle to a widespread acceptance of landform antiquity which continues to hinder progress severely, is the problem of establishing accurately, via independent dating techniques, the age of geomorphological features. Determining absolute ages for landforms has been elusive for a variety of reasons and, with regard to this problem, it is perhaps worth briefly examining this question of landscape dating with respect to the development of geochronology within the geological sciences.

In the geological record the longevity and durability of the rock components which make up the various units and formations comprising the stratigraphy are now rarely questioned. It is the appliance of radiometric dating methods which are largely responsible for dating the rocks and giving absolute ages to the boundaries which define the great epochs that build up the Earth's history. Technological advances in the science continue to improve absolute dating of rocks and the events which have affected them, yet an examination of the older literature reveals that such acceptance of the extreme antiquity of rocks and fossils was the subject of great debate during the last century. At that time it must have seemed inconceivable that definitive absolute dating of inert rock would ever be achieved. The discovery of the radioactive decay series only a few decades later helped set the foundations for one of the most fundamental aspects of modern geology - the science of geochronology. By contrast, the ages of geomorphological elements within the landscape, of which the rocks are an integral part, are not yet, as it were, set in stone.

Determining the absolute ages of geomorphological elements has seemed, by contrast, notoriously difficult. This is no failure of scientific ingenuity, but rather the fact that most exposed landscapes are often subject to continued modification through dynamic factors and variation in rates resulting from climatic and tectonic environment. This makes it very difficult to give precise ages to features of ancient origin which may yet remain in the modern dynamic landscape and thus be subject to more recent modification. Nevertheless, there is overwhelming evidence for the antiquity of entire surfaces and of those surface elements which comprise parts of the modern landscape. As mentioned earlier, protracted periods of time allow geomorphological processes to result in the development of widespread surfaces which may be recognized in both the geological record and in forming elements in the modern landscape. Therefore claims to their antiquity seem inescapable but difficult to quantify, and unless such features are somehow preserved or fossilized, for instance by the deposition of sediments or covering by lava flows, accurate dating of these surfaces often remains the source of debate.

Where ancient surfaces have escaped modification by more recent geomorphological processes, however, new analytical techniques such as those involving cosmogenically formed nuclides have become possible through modern technologies and promise one solution to the problem of surface dating. Although the detailed systematics of this technique are currently still being refined (Kurz *et al.* 1990; Lal 1991), it provides a means of measuring the duration of exposure from the presence of minute quantities of these radionuclides (e.g. ^{10}Be, ^{26}Al) and stable nuclides (e.g. ^{3}He, ^{21}Ne) which have formed by the interaction of cosmic radiation with atoms exposed at the surface of geomorphological features (e.g. Nishiizumi *et al.* 1991; Brown *et al.* 1994*a*). Similarly, surface exposure of minerals, particularly quartz and feldspar, provides the basis for thermoluminescene (TL) and optically stimulated dating (OD) techniques (for review see Aitken 1992). These have proved particularly useful in archaeological applications and are now becoming increasingly used for establishing the date of exposure of Pleistocene-Recent sediment surfaces. The technique requires that the sediment has received at least a short exposure to sunlight during its transport and deposition, and has subsequently been covered by the deposition of younger sedimentary layers (e.g. Huntley *et al.* 1985; Murray 1996).

Other isotopically-based approaches may prove useful for determining the age of weathering profiles which have developed in association with many palaeosurfaces. Foremost amongst these is Uranium-series disequilibria which are generated during rock weathering by the incorporation of the uranyl ion by iron oxides and oxyhydroxides which develop at differing stages during alteration. This makes the technique particularly useful for the study of weathering rinds, pedogenic weathering profiles, and laterites (e.g. Short *et al.* 1989; Mathieu *et al.* 1995). Variations in oxygen isotope ($\delta^{18}O$) have been used as an indirect method of dating regoliths based upon changing composition of meteoric waters in response to climatic variation (Bird & Chivas 1988; Bird *et al.* 1989). The development of iron oxides and oxyhydroxides during the evolution of residual deposits such as laterite can also provide substance for a palaeomagnetic record, and hence a further method for establishing the age of deeply weathered palaeosurfaces (e.g. Schmidt *et al.* 1984; Kumar 1985; Schmidt & Ollier 1988). Another technique which is becoming increasingly popular in the fields of geomorphology and tectonics is that of apatite fission track analysis (AFTA). Accurately measured AFT data provide the information required enabling calculation of the time-temperature path of rock through the upper crust (Gallagher *et al.* 1991). It can, therefore, be employed as a method of evaluating rates of uplift (Hurford 1991, and references therein). Moreover, since upward movement of rock is often associated with surface denudation, AFTA can provide quantified data regarding rates of surface erosion. This approach has been successfully applied to large-scale denudation and geomorphological evolution of passive continental margins (e.g. Gilchrist *et al.* 1994; Gallagher *et al.* 1995).

Many of the techniques described above are relatively recent developments, or have only become practical through the development and refinement of modern technologies. Clearly, if these techniques continue to develop they may yet provide the geomorphologist with dating tools comparable to the absolute dating methods employed by geologists.

Palaeosurfaces - wider applications

Palaeosurface reconstruction is clearly important as a method of assessing neotectonic effects since they can provide important time horizons upon which the effects of deformation can be readily discerned (Widdowson 1997; Molina-Ballesteros *et al.* 1997). In addition, there has been much recent emphasis on the role of climate change with regard to the environmental aspects of landscape evolution and the importance of changes which are seen in the geological record. Palaeosurface study can offer much information regarding palaeoenvironments, and the nature of subsequent modification by changing erosional and weathering regimes can reveal key aspects of subsequent environmental change (Borger 1997; Gutierrez & Javier-Gracia 1997).

One of the most significant developments in recent years have been the rapid advance in computing techniques to evaluate and process large amounts of data which are now been collected and collated. In the forefront, techniques such as remote sensing and GIS allow new and rapid appraisal of spatial data (Ringrose & Migoń 1997) and have become an important part of the Earth Scientist's tool kit. Remote sensing applications provide immense potential for geographically extending current regions of palaeosurface research (White *et al.* 1997). Similarly, the widespread availability of geochemical analytical techniques has also resulted in the generation of much data and a rigorous quantitative treatment of this information is clearly desirable (McAlister & Smith 1997).

In the longer term these advances in techniques and understanding of the relationships between geomorphological processes, including the evolution of the sedimentary record, and tectonic processes will provide the means by which we can begin to re-examine, evaluate, and perhaps test the 'classic' models of landscape evolution. Moreover, their continued development will permit consideration of a whole range of variables notably those of a climatic and tectonic nature.

Conclusions and outstanding problems

In summary, it is generally accepted that further refinements of methods for establishing the age of palaeosurfaces is required if they are to be understood within their geological and geomorphological context. Clearly, this type of study will provide another area of common ground which requires fundamental contribution from both geological and geomorphological fields. It is also evident that future investigation and identification of palaeosurfaces will need to adopt an increasingly holistic approach if our understanding is to improve. This will mean greater amounts of data of different types requiring sophisticated computing for its integration and interpretation. Clearly, it is necessary that GIS-based systems be more widely applied if we are to identify new examples of palaeosurfaces and determine the relative importance of different factors instrumental in their evolution.

It is clear too, that many different types of palaeosurfaces exist and whilst these can currently be broadly grouped in to exogenic and endogenic forms, more work will required if we are to evolve a more informative, robust, and readily workable system of description. However, before embarking upon such a task a degree of caution should be exercised and consideration given as to whether it will prove ultimately worthwhile or even desirable to instigate a complex classification system of palaeosurfaces. Until this problem is resolved, it remains incumbent on individual researchers to outline the nature of the palaeosurface which they are discussing. What is perhaps more urgently required is the discussion of palaeosurfaces as expressions of the larger interlinked erosional-depositional system. It should always be emphasized, for instance, that the action of geomorphological agents resulting in development of a new surface at one locality must result in fundamental changes of a geological nature which are expressed in the depositional history and stratigraphical record at another.

The core of the research included in this volume was originally presented at a workshop-style meeting held at the Department of Earth Sciences, Sheffield University on the 15th - 16th April 1994 which itself formed part of the UK contribution to I.G.C.P. 317 *Palaeoweathering Records and Palaeosurfaces* (Widdowson 1995). The object of this meeting was to provide a UK venue for presentation of current Palaeosurface research and, most importantly, a forum to foster cross-disciplinary discussion. A number of important inter-disciplinary issues were identified during the meeting and these have provided a thematic framework for work presented in this volume. The success of the palaeosurfaces theme was immediately apparent, and was made further evident from the lively discussions during the closing sessions. This exchange of ideas and information has continued in subsequent correspondence and, consequently, the original core material has been substantially expanded during preparation of this volume both in the form of extension of work originally presented at the meeting and as additional papers. As may be expected from such a cross-disciplinary theme there is a wide range of conceptual and analytical approaches adopted by the different authors. Therefore, for ease of organization, this volume has been simply divided in terms of the geographical region. The result is a collection of research papers drawing upon a wide range of expertise and knowledge, and which is presented in a manner which will be of direct interest to both geologists and geomorphologists alike.

I wish to thank all the contributors to this volume, for their lively discussions, interest, and support during the preparation of this interdisciplinary promotion of palaeosurface research. I remain indebted to the many scientists, many of whom are listed below, who have given freely of their time and expertise and, through their constructive comments and reviews have helped maintain the high standard of research and presentation within this volume: Phil Allen, Robert Allison, Y. Battiau-Queney, Karna Lidmar-Bergström, Hardy Borger, Des Bowden, Chris Clark, Angela Coe, Keith Cox, N. J. Cox, Pete Coxon, C. H. Emeleus, J. P. G. Fenton, Steve Flint, Andrew Goudie, Yanni Gunnell, Dave Jolley, John McAlister, Marty McFarlane, Poitr Migoń, Dave Nash, Philip Ringrose, Dave Rothery, Torsten Schwarz, Bernie Smith, Mike Summerfield, Mike Thomas, C. R. Twidale, John Walden, Nick Walsh and Brian Whalley. I also thank those additional reviewers who wish to remain anonymous.

I am grateful to Andrew Goudie, Piotr Migoń, Des Bowden and Andy Fleet for their constructive comment and suggestions on an earlier version of this paper. I also thank the many research colleagues who have greatly influenced my thoughts regarding geomorphological and geological issues past and present, and record my particular thanks to Marty McFarlane for encouraging an early involvement in the palaeosurfaces topic. Finally, IGCP 317 and the Royal Society are gratefully acknowledged for providing grants which aided in subsidizing the Palaeosurfaces workshop, and last but not least, the staff of the Department of Earth Sciences, Sheffield University for providing a convivial venue for the initial meeting.

References

AITKEN, M. J. 1992. Optical dating. *Quaternary Science Reviews*, **11**, 127–131.

BATTIAU-QUENEY, Y. 1997. Preservation of old palaeosurfaces in glaciated areas: examples from the French western Alps. *This volume.*

BIRD, M. I. & CHIVAS, A. R. 1988. Oxygen isotope dating of the Australian regolith. *Nature,* **331**, 513–516.

——, CHIVAS, A. R. & ANDREW, A. S. 1989. A stable isotope study of lateritic bauxites. *Geochimca et Cosmochemica Acta*, **53**, 1411–1420.

BORGER, H. 1997. Environmental changes during the Tertiary: the example of palaeoweathering residues in central Spain. *This volume.*

BOWDEN, D. J. 1997. The geochemistry and development of lateritized footslope benches: The Kasewe Hills, Sierra Leone. *This volume.*

BREMER, H. 1985. Randschwellen: a link between plate tectonics and climatic geomorphology. *Zeitscrift für Geomorphologie N.F., Supplementband,* **54**, 11–21.

BROWN, E. T., BOURLÈS, D. L., COLIN, F., SANFO, Z., RAISEBECK, G. M. & YIOU, F. 1994*a*. The development of iron crust laterite systems in Bukina Faso, West Africa examined with *in situ* produced cosmogenic nuclides. *Earth and Planetary Science Letters*, **124**, 19–33.

BROWN, R. W., SUMMERFIELD, M. A., & GLEADOW, A. J. W. 1994*b*. Apatite fission track analysis: Its potential for the estimation of denudation rates and implications for models of long-term landscape development. *In*: KIRKBY, M. J. (ed.) *Process Models and Theoretical Geomorphology*, 24–53.

——, RUST, D. J., SUMMERFIELD, M. A., & GLEADOW, A. J. W. & DE WIT, M. C. J. 1990. An early Cretaceous phase of accelerated erosion on the south-western margin of Africa: Evidence from fission-track analysis and the offshore sedimentary record. *Nuclear Tracks and Radiation Measurements*, **17**, 339–351.

BUDEL, J. 1982. *Climatic Geomorphology* (translated by FISCHER L. & BUSCHE D.). Princeton University Press, Princeton.

BURBANK, D. W., LELAND, J., FIELDING, E., ANDERSON, R. S., BROZOVIC, N. REID, M. R. & DUNCAN, C. 1996. Bedrock incision, rock uplift and threshold hillslopes in the northwestern Himalayas. *Nature*, **379**, 505–510.

BRUNSDEN, D. 1993. The persistence of land-forms. *Zeitscrift für Geomorphologie N.F., Supplementband,* **93**, 13–28.

BURGESS, P. M. & ALLEN, P. A. A forward modelling analysis of the controls on sequence stratigraphical geometries. *In*: HESSELBO, S. P. & PARKINSON, D. N. (eds) *Sequence Stratigraphy in British Geology.* Geological Society, London, Special Publication, **103**, 9–24.

COE, A. L. 1996. Unconformities within the Portlandian Stage of the Wessex Basin and their sequence stratigraphical significance. *In*: HESSELBO, S. P. & PARKINSON, D. N. (eds) *Sequence Stratigraphy in British Geology.* Geological Society, London, Special Publication, **103**, 109–143.

COX, K. G. 1989. The role of mantle plumes in the development of continental drainage patterns. *Nature*, **342,** 873–877.

COXON, P. & COXON, C. 1997. A pre-Pliocene or Pliocene land surface in County Galway, Ireland. *This volume.*

DAVIS, W. M. 1899. The geographical cycle. *Geographical Journal*, **14**, 481–504.

ENGLAND, P. C & HOUSEMAN, G. A. 1988. The Mechanics of the Tibetan Plateau. *Philosophical Transactions of the Royal Society of London, Series A-Mathematical and Physical Sciences*, **321**, 1557, 3–22.

——, & MOLNAR, P. 1990. Surface uplift, uplift of rocks, and exhumation of rocks. *Geology*, **18**, 1173–1177.

FAIRBRIDGE, R. W. & FINKL, C. W. 1980. Cratonic erosonal unconformities and peneplains. *Journal of Geology*, **88**, 69–86.

GALE, A. 1996. Turonian correlation and sequence statigraphy of the Chalk in southern England. *In*: HESSELBO, S. P. & PARKINSON, D. N. (eds) *Sequence Stratigraphy and its Applications to British Geology.* Geological Society, London, Special Publication, **103**, 177–195.

GALE, S. J. 1992. Long-term landscape evolution in Australia. *Earth Surface Processes and Landforms,* **17**, 323–343.

GALLAGHER, K, SAMBRIDGE, M., & DRIJKONINGEN, G. 1991. Genetic algorithms: an evolution from Monte Carlo methods for highly non-linear geophysical optimisation. *Geophysical Research Letters,* **18**, 2177–2180.

——, HAWKESWORTH, C. & MANTOVANI, M. S. M. 1995. Denudation, fission track analysis and the long-term evolution of passive margin topography: application to the southeast Brazil margin. *Journal of South American Earth Sciences,* **8**(1), 65–77.

GILCHRIST, A. R. & SUMMERFIELD, M. A., 1994. Tectonic models of passive margin evolution and their implications for theories of long-term landscape development. *In*: KIRKBY, M. J. (ed.) *Process Models and Theoretical Geomorphology.* Wiley, Chichester, 55–84.

——, A. R., KOOI, H. & BEAUMONT, C. 1994. Post Gondwana geomorphic evolution of southwestern Africa: implications for the controls of landscape development from observations and numerical experiments. *Journal of Geophysical Research,* **99**, 12 211–12 228.

GUNNELL, Y. 1997. Topography, palaeosurfaces and denudation over the Karnataka uplands, southern India. *This volume.*

GUTIERREZ-ELORZA, M. & GRACIA, F. J. 1997. Environmental interpretation and evolution of the Tertiary erosion surfaces in the Iberian Range (Spain). *This volume.*

HILLS, E. S. 1961 Morphotectonics and the geomorphologhical sciences with special reference to Australia. *Quarterly Journal of the Geological Society of London,* **117**, 77–89.

HUNTLEY, D. J., GODFREY-SMITH, D .I., & THEWALT, M. L. W. 1985. Optical dating of sediments. *Nature*, **313**, 105–107.

HURFORD, A. J. 1991. Uplift and cooling pathways derived from fission track analysis and mica dating: a review. *Sonderuck aus Geologische Rundschau.* Band **80**/2, 349–368.

JOLLEY, D. W. 1997. Palaeosurface palynology of the Skye Lava Field, and the age of the British Tertiary volcanic province. *This volume.*

KING, L. C. 1953 Canons of landscape evolution. *Bulletin of the Geological Society of America*, **64**, 721–752.

KUMAR, A. 1985. Palaeolatitudes and age of Indian laterites, *Palaeogeography, palaeoclimatology, Palaeoecology*, **53**, 231–237.

KURZ, M. D., COLODNER, D., TRULL, T. W., MOORE, R. B. & O'BRIAN, K. 1990. Cosmic ray exposure dating with in situ produced cosmogenic ^{3}He: results from young Hawaiian lava flows. *Earth and Planetary Science Letters*, **97**, 177–189.

LACIKA, J. 1997. Neogene palaeosurfaces in the volcanic area of Central Slovakia. *This volume.*

LAL, D. 1991. Cosmic ray labeling of erosion surfaces: *in situ* nuclide production rates and erosion models. *Earth and Planetary Science Letters*, **104**, 424–439.

LIDMAR-BERGSTRÖM, K., OLSSON, S. & OLVMO, M. 1997. Palaeosurfaces and associated saprolites in southern Sweden. *This volume.*

MCALISTER, J. J. & SMITH, B. J. 1997. Geochemical trends in Early Tertiary palaeosols from northeast Ireland: a statistical approach to assess element behaviour during weathering. *This volume.*

MCFARLANE, M. J. 1976. *Laterite and Landscape.* Academic, London.

—— 1983. The temporal distribution of bauxitisation and its genetic implications. *In*: MELFI, A. J. & CARAVALHO, A. (eds) *Lateritisation Processes: Proceedings II International Seminar on Lateritisation Processes, Sao Paulo, Brazil, July 1982*, 197–207.

MCKENZIE, D. 1984 A possible model for epeirogenic uplift. *Nature*, **307**, 616–618.

MATHIEU, D., BERNAT, M., & NAHON., D. 1995. Short-lived U and Th isotope distribution in a tropical laterite derived from granite (Pitinga river basin, Amazonia, Brazil): Application to assessment of weathering rate. *Earth and Planetary Science Letters*, **136**, 703-714.

MIGOŃ, P. 1997. Tertiary etchsurfaces in the Sudetes Mountains, SW Poland: a contribution to the pre-Quaternary morphology of Central Europe. *This volume.*

MOLINA-BALLESTEROS, E., TALEGÓN, J. G. & HERNÁNDEZ, M. A. 1997. Palaeoweathering profiles developed on the Iberian hercynian basement and their relationship to the oldest Tertiary surface in central and western Spain. *This volume.*

MURRAY, A. S. 1996. Development in optically stimulated luminescence and photo-transferred luminescence dating of young sedimiments: Applications to a 2000 year sequence of flood plain deposits. *Geochimca and Cosmochimica Acta*, **60**, No.4, 565– 576.

NISHIIZUMI K., KOHL, C. P., ARNOLD, J. R., KLEIN, J., FINK, D. & MIDDLETON, R. 1991. Cosmic ray produced ^{10}Be and ^{26}Al in Antarctic rocks: exposure and erosion history. *Earth and Planetary Science Letters*, **104**, 440–454.

OLLIER, C. D. 1985. The Morpotectonics of Passive Continental Margins: Introduction. *Zeitscrift für Geomorphologie N.F., Supplementband*, **54**, 1–9.

PAIN, C. F. 1985. Morphotectonics of the continental margins of Australia. *Zeitschrift für Geomorphologie N.F.*, Supplementband, **54**, 23–35.

PENCK, W. 1953. *Morphological Analysis of Landforms* (translated by H. CZECH & K. C. BOSWELL). Macmillan, London.

POVEY, D. A. R., SPICER, R. A., & ENGLAND, P. C. 1994. Palaeobotanical investigations of early Tertiary palaeoelevations in northeastern Nevada - Initial results. *Review of Palaeobotany and Palynology*, **81**, 1–10.

RINGROSE, P. S. & MIGOŃ, P. 1997. Analysis of digital elevation data from the Scottish Highlands and recognition of pre-Quaternary elevated surfaces. *This volume.*

SCHMIDT, P. W., PRASAD, V. & RAMAN, P. K., 1983. Magnetic ages of some Indian laterites. *Palaeogeography, Palaeoclimatology, Palaeoecology*, **44**, 185–202.

—— & OLLIER, C. D. 1988. Palaeomagnetic dating of Late Cretaceous to Early tertiary weathering in New England, N.S.W., Australia. *Earth Science Reviews*, **25**, 363–371.

SCHLAGER, W. 1993. Accommodation and supply - a dual control on stratigraphic sequences. *Sedimentary Geology*, **86**, 111–136.

SHORT, S. A., LAWSON, R. T., ELLIS, J. & PRICE, D. M. 1989. Thorium-Uranium disequilibrium dating of late Quaternary ferruginous concretions and rinds. *Geochimca et Cosmochimica Acta*, **530**, 1379–1389.

STÜWE, K., WHITE, L., & BROWN., R. 1994. The influence of eroding topography on steady-state isotherms. Application to fission track analysis. *Earth and Planetary Science Letters*, **124**, 63–74.

THOMAS, M. F. 1974. *Tropical Gemorphology.* Macmillan, London and Halstead, New York.

—— 1994. *Gemorphology in the Tropics: a Study of Weathering and Denudation in Low Latitudes.* Wiley, Chichester.

VAIL. P. R., AUDEMARD, F., BOWMAN, S. A., EINSER, P. N. & PEREZ-CRUZ, C. 1991. The Stratigraphic signatures of Tectonics, Eustasy and Sedimentology - An Overview. *In*: EINSELE, G., RICKEN, W. & SEILACHER, A. (eds) *Cycles and Events in Stratigraphy*. Springe, New York, 617–659.

WALSH, P. T, ATKINSON, K. BOULTER, M. C. & SHAKESBY R. A. 1987. The Oligocene and Miocene outliers of West Cornwall and their bearing on the geomorpological evolution of Oldland Britain. *Philosophical Transactions of the Royal Society of London, Series A*, **323**, 211–245.

WATTS, A. B. & COX, K. G. 1989. The Deccan Traps: an interpretation in terms of progressive lithospheric flexure in response to a migrating load. *Earth and Planetary Science Letters*, **93**, 85–97.

WHALLEY, W. B. REA, B. R., RAINEY, M. M. & MCALISTER, J. J. 1997. Rock weathering in blockfields: some preliminarty data from mountain plateaus in North Norway. *This volume.*

WHITE, K., DRAKE, N. & WALDEN, J. 1997. Remote sensing for mapping palaeosurfaces on the basis of surficial chemistry: a mixed pixel approach. *This volume.*

WHITING, B. M, KARNER, G. D. & DRISCOLL, N. W. 1994. Flexural and stratigraphic development of the west Indian continental margin. *Journal of Geophysical Research*, **99**, 13 791–13 811.

WIDDOWSON, M. 1995. Conference Report. Tertiary and pre-Tertiary palaeosurfaces: recognition, reconstruction and environmental interpretation. *Journal of the Geological Society, London*, **152**, 193–195.

—— 1997. Tertiary palaeosurfaces of the SW Deccan, Western India: implications for passive margin uplift. *This volume.*

—— & COX, K. G. 1996. Uplift and erosional history of the Deccan traps, India: Evidence from laterites and drainage patterns of the Western Ghats and Konkan

Coast. *Earth and Planetary Science Letters*, **137**, 57–69.
——, WALSH, J. N. & SUBBARAO, K. V. 1997. The geochemistry of Indian bole horizons: palaeoenvironmental implications of Deccan intravolcanic palaeosurfaces. *This volume*.

The great age of some Australian landforms: examples of, and possible explanations for, landscape longevity

C. R. TWIDALE

Department of Geology and Geophysics, University of Adelaide, South Australia 5005, Australia

Abstract: According to the major conventional models of landscape evolution there ought to be no land surfaces older than Oligocene, save some that have been exhumed. In reality, substantial areas of the Australian continent (as well as other parts of Gondwana and of Laurasia) are of Mesozoic or greater ages. Surfaces of Late Jurassic or Early Cretaceous ages are especially widespread. Many are of exhumed origin, but several notable features are of epigene-etch type. Their age ranges have been determined stratigraphically. In some instances (e.g. Gawler Ranges, Hamersley Surface, Eastern Uplands), the stratigraphic evidence is incontrovertible. Elsewhere the position and elevation of massifs, such as Arnhem Land, relative to Cretaceous shorelines demonstrate their essential antiquity. The survival of such very old palaeosurfaces is attributed to several factors, but particularly to unequal erosion, reinforcement mechanisms and the prevalence of vertical (epeirogenic or isostatic), rather than orogenic, earth movements.

Australia is well known as an arid continent and as a land of sweeping plains. It is also old in respect of some of its rocks and parts of its land surface. For at least 500 Ma Australia, Antarctica, Peninsula India, Africa south of the Atlas, the Arabian Peninsula, Malagassy and parts of eastern South America together constituted a supercontinent called Gondwana (Suess 1885). Named after a geological formation that in turn took its name from a local Indian tribe – the word Gondwana means 'land of the Gonds' – the supercontinent began to fragment *c.* 200 Ma ago, in the Early Mesozoic, but Australia and Antarctica did not separate until *c.* 60 Ma ago. Thus, in Australia, any landforms that are more than *c.* 60 Ma old are relics of Gondwanan landscapes and date from a time when Australia was part of the southern supercontinent.

Theoretically, in terms of the conventional geomorphological wisdom, it is impossible for any such remnants to have survived to the present day, apart, that is, from any fragments that have been preserved by burial by sediments or volcanic materials and subsequently exhumed. The lay view of the world, and that of some early geologists such as James Hutton (1788, 1795, p. 236), is one of constant disintegration, decay and wearing away under the influence of the Sun's heat, of moisture, of rivers, waves, glaciers, and wind. The conventional models of landscape evolution reflect this common-sense perception and argue against the protracted survival of landforms. Steady state development (e.g. Hack 1960), and base-levelling by Davisian peneplanation (Davis 1899, 1909) both imply a virtually continuous reshaping of the land surface, so that all is essentially youthful. Base-levelling by scarp retreat (King 1942), though allowing remnants of the 'original' surface to persist for some time, nevertheless implies that none can survive longer than the time it takes to reduce a large land mass to base level. For a land mass of continental or subcontinental size, such North America, Australia or Africa south of the Sahara, and assuming tectonic stability, but making allowance for isostatic compensations for erosion and deposition, this is, according to Schumm (1963), of the order of 33 Ma. Thus, even in the most favourable circumstances there ought to be no landforms (except those that have been exhumed) of an age > 33 Ma, i.e. in terms of the geological time-scale, no older than Oligocene. The reality is very different, for although duricrusted remnants of various Cainozoic ages, and later dunefields and riverine plains, dominate the landscape, at least 10% and possibly as much as 20% of the contemporary Australian land surface is of essentially Mesozoic or greater age, i.e. at least 65 Ma old (for reviews see Hills 1955; Twidale 1976; Young 1983; Wyrwoll 1988; Twidale & Campbell 1988). In Australia, the dating of Gondwanan surfaces and forms has been facilitated by their relationship to sedimentary sequences laid down during extensive repeated Cretaceous marine transgressions and to Early Cainozoic volcanic deposits. Landscapes of comparable antiquity are preserved not only in other components of Gondwana but also in remnants of Laurasia, which together with the southern supercontinent constituted the great land mass known as Pangaea.

From Widdowson, M. (ed.), 1997, *Palaeosurfaces: Recognition, Reconstruction and Palaeoenvironmental Interpretation,* Geological Society Special Publication No. 120, pp. 13–23.

Exhumed remnants

Some of the ancient remnants that constitute part of the contemporary landscape have been buried and later exhumed. They range in age from Archaean to Late Pleistocene, though substantial areas date from the Late Proterozoic, the Early Cambrian and, especially, the Late Jurassic or earliest Cretaceous. The latter occur at the margins of sedimentary basins occupied by the Early Cretaceous seas (Fig. 1), which inundated almost half of the present Australian continent and reduced it to an archipelago consisting of several very large islands, so that at that time Australia was truly an island continent (Frakes *et al.* 1988). Such exhumed pre-Cretaceous surfaces are prominent along the eastern margin of the Australian Craton (Twidale & Campbell 1992; see Fig. 2 for locations), for instance, in the Isa Highlands (Twidale 1956; see Fig. 3). They are also well represented in the Pine Creek and Katherine areas of the Northern Territory. The former region is underlain by Proterozoic granite, the latter by Cambrian limestones, but both were buried by sandstones laid down in the Early Cretaceous epicontinental seas (Stuart-Smith *et al.* 1987). Stripping of the Cretaceous beds has re-exposed the surface over which the seas advanced: a granite plain with occasional low whalebacks and boulders south of Pine Creek and, near Tindal, south of Katherine, a karst plain with groups of pinnacles and low, flat-crested hills. The extraordinary feature of the karst region is that even minor forms like solution hollows, flutings and bedding planes widened by chemical attack predate the marine transgression, for residual patches show that they were covered and infilled by Cretaceous sandstone (Twidale 1984).

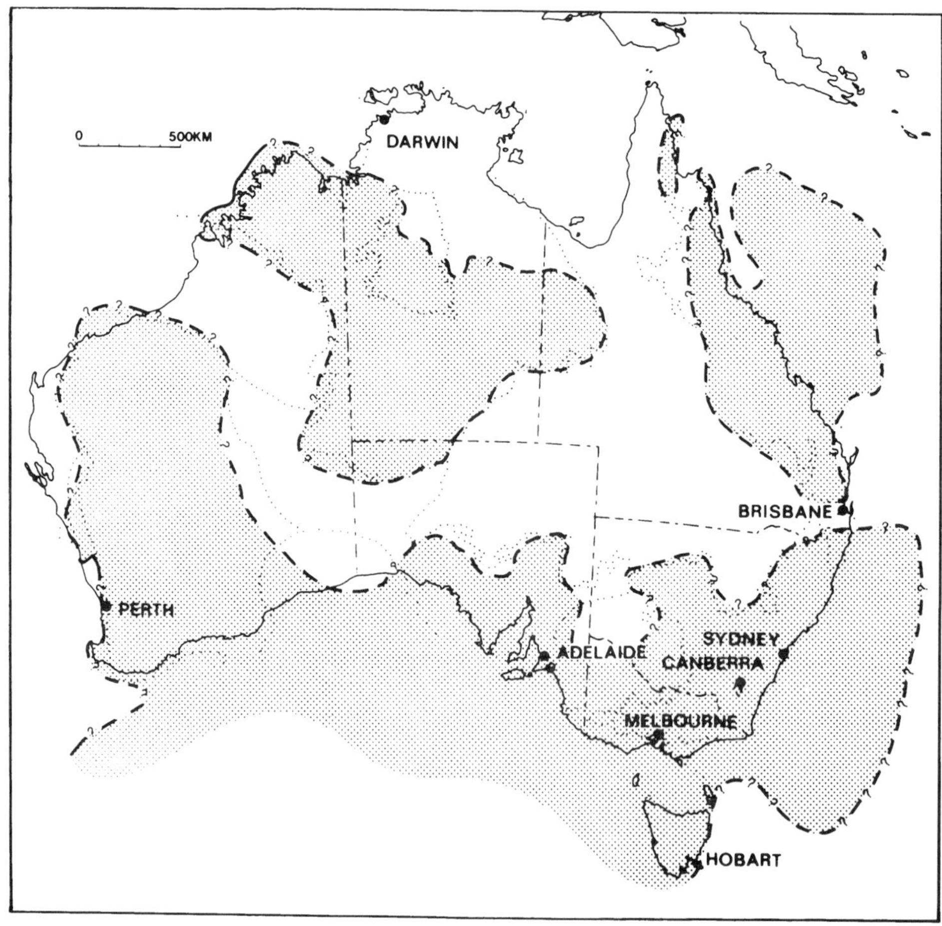

Fig. 1. Extent of Early Cretaceous marine transgression. Land masses shaded. (After Frakes *et al.* 1988).

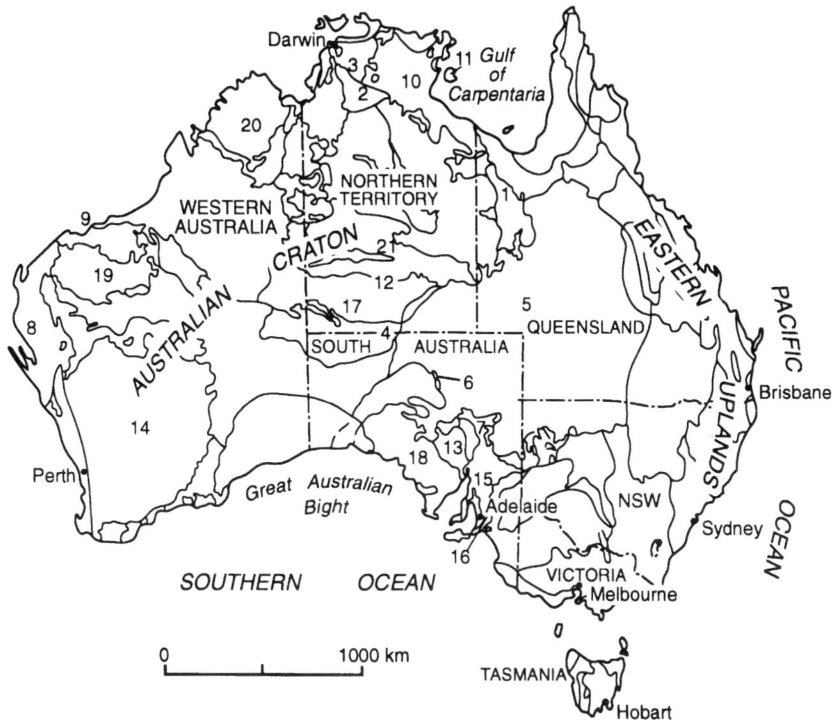

Fig. 2. Location map. Key: 1, Isa Block and Highlands; 2, Katherine; 3, Pine Creek; 4, Kulgera and Everard Ranges; 5, Eromanga Basin; 6, Peake and Davenport Ranges; 7, Flinders Ranges; 8, Carnarvon Basin; 9, Port Hedland; 10, Arnhem Land; 11, Groote Eylandt; 12, MacDonnell Ranges; 13, Arcoona Plateau; 14, Yilgarn Block; 15, Mid North; 16, Mt Lofty Ranges; 17, Ayers Rock; 18, Gawler Ranges; 19, Hamersley Ranges; 20, Kimberley Block; 21, Barrow Creek; 22, Pilbara Block.

The pre-Cretaceous landscape is also resurrected in the Kulgera–Everard Ranges area, where scattered remnants of sediments of probable Jurassic age are preserved between the granitic domes (Fig. 4); south and southwest of the Eromanga Basin (Jack 1931; Wopfner 1964); in the Peake and Davenport Ranges (Wopfner 1968); and in the northernmost Flinders Ranges (Woodard 1955), where Mt Babbage is a remnant of Neocomian marine sediments resting unconformably on a planate surface eroded in various disturbed Precambrian rocks. With the single exception of those preserved in Mt Babbage, the Cretaceous sediments have been stripped to expose the unconformity in an exhumed surface of pre-Cretaceous age (Fig. 5).

To the west, exhumed pre-Cretaceous landscapes are prominent in and adjacent to the Carnarvon Basin. Even the Murchison Gorge, inset into the lateritized Victoria Plateau, demonstrably and, in part, predates the Cretaceous transgression, for patches of the Cretaceous cover are preserved perched unconformably on the underlying Silurian sediments in the upper sections of the gorge (Hocking *et al.* 1987). To the north, east of Port Hedland, granite inselbergs are in process of re-exposure from beneath a cover of Early Cretaceous marine strata (Fig. 6a). The Cretaceous strata occur, albeit discontinuously, on the plains between the hills and also form mesas which stand higher than the granitic remnants, showing that the entire inselberg landscape was submerged beneath the Cretaceous seas and sediments. A Jurassic surface is partly re-exposed around Shay Gap, still further to the east of Port Hedland (Twidale & Campbell 1988). Regional stratigraphy suggests that the entire inselberg landscape of the western Pilbara (Fig. 6b), between the Carnarvon Basin and Port Hedland, is exhumed from beneath Cretaceous strata.

Thus, exhumed forms feature prominently in the old elements exposed in the contemporary Australian landscape. They have been preserved in unconformity by burial, though some, such as the

(a)

(b)

(c)

Fig. 3. (**a**) and (**b**) Bevelled crests of quartzite ridges, Isa Highlands near Mt Isa, northwest Queensland; (**c**) partly exhumed unconfomity between Proterozoic granite and lateritized Upper Cretaceous beds; south of Cloncurry, Isa Highlands.

Fig. 4. These granite domes near Kulgera, and others like them in the eastern Musgrave Block, on the Northern Territory–South Australia border, have been exhumed from beneath (?)Jurassic sediments. This is demonstrated by outcrops of the latter, like the one forming the mesa shown here, rising from the plains eroded in granite and standing higher than the hills sculpted in the crystalline rocks.

summit surface of the Isa Highlands, have, in part at least, been exposed for several scores of millions of years.

Epigene and etch forms

Parts of the large islands isolated by the Early Cretaceous seas were eroded by rivers graded to the then shorelines. Many such epigene or subaerial surfaces survive in the present landscape. Some, preserved by silcrete and other siliceous accumulations, remain intrinsically intact, but most have been modified, for most regoliths developed on them are readily eroded with the result that these epigene surfaces have been changed in some degree. In some instances, as for example with laterite- or calcrete-capped surfaces, only the friable sandy or silty A-horizon has been stripped (though in the case of lateritic regoliths the rest of the profile has either continued to develop – see e.g. McFarlane 1986 – or has been degraded). Despite such modifications, the essential morphology of the original surface is simulated or preserved (see e.g. Hills 1975, p. 300). Elsewhere, however, the entire mantle of weathered material has been removed exposing the base of the regolith or lower limit of effective weathering, which is called the *weathering front* (Mabbutt 1961*a*). Exposed fronts are called *etch* or etched surfaces (Falconer 1911; Jutson 1914; the term is due to Wayland 1934*a*).

Examples of old subaerial or *epigene* landscapes are preserved in the Mt Lofty Ranges (Hossfeld 1926; Miles 1952; Campana 1958) and adjacent areas (e.g. Daily *et al.* 1974); the central and southern Flinders Ranges (Twidale 1980); the Arcoona Plateau; the western part of the Yilgarn Block, in the Darling Ranges; and throughout the Eastern Uplands (see e.g. Craft 1932 ; Hills 1934; Twidale 1956; de Keyser 1964; Young 1981; Ruxton & Taylor 1982).

The evidence for the antiquity of the regions named is mainly stratigraphic and topographic. Thus, in the Eastern Uplands, the summit-high plains are deeply incised by valleys (Fig. 7a and b),

(a)

(b)

Fig. 5. (**a**) Mt Babbage, a mesa remnant of Early Cretaceous marine sediments in the northern Flinders Ranges, South Australia. The unconformity between the Mesozoic and the Precambrian is coincident with the level of the high plain, an example of which, cut in granitic rocks, is shown in (**b**).

of which the floors of some are occupied by lava flows of Tertiary age and in some instances of Eocene age (Fig. 7c). Thus, the valleys must be older than the volcanic flows and the upland surfaces into which they are incised are most probably of even greater age (Hills 1934; Young & McDougall 1985; Taylor *et al.* 1985). The dating of the volcanic extrusions also shows that the rate of headward erosion of even major rivers has been very slow.

Similarly, the Arnhem Land Plateau (Kakadu) was a large island in the Early Cretaceous seas. A Cretaceous shoreline has been identified at the base of the Kakadu escarpment, at the western edge of the upland (Needham 1982), and to the east, on Groote Eylandt, the highest Cretaceous shoreline stood only 75 m above present sea level (Frakes & Bolton 1984), whereas the Plateau rises 125–175 m higher. To the south, the MacDonnell Ranges also bordered the Cretaceous seas, and the fragmentary summit surface, preserved mainly on quartzite ridges, may well have been eroded by rivers graded to the then shorelines, as suggested by Mabbutt (1966); certainly it stands much higher than duricrusted valley floors of Miocene age. Still further to the south, the summit-high plain surface of the Arcoona Plateau is at least of Early Cretaceous age, for the Early Cretaceous marine transgression (Fig. 1) extended into valleys already incised below the high plain level.

Examples of etch surfaces, identified by surviving remnants of regolith, beneath which the level of the weathering front is coincident with that of the adjacent plains, are well represented in the Australian landscape. An extensive etch surface is reported from the Yilgarn Block of Western Australia (Jutson 1914; Mabbutt 1961*b*; Finkl & Churchward 1973). The original regolithic surface (Jutson's Old Plateau) was probably eroded and weathered during the Cretaceous, for sediments of this age are preserved in valleys cut in to its western margin (e.g. Playford *et al.* 1976), but the regolith

(a)

(b)

Fig. 6. (**a**) Granite inselbergs and intervening plains partly exhumed from beneath Early Cretaceous marine strata, near Port Hedland, in the northwest of Western Australia. (**b**) This inselberg landscape in the western Pilbara, northern Western Australia, is probably exhumed from beneath Cretaceous marine strata.

(a)

(b)

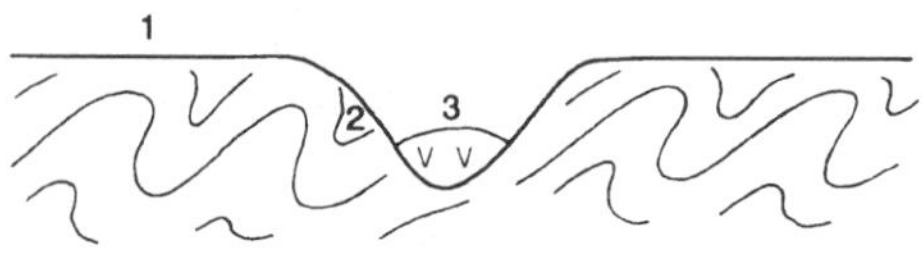

(c)

Fig. 7. (**a**) High plain with monadnock, southern Eastern Uplands, near the New South Wales–Victoria border. (**b**) High plain preserved on Late Palaeozoic felsite, Newcastle Ranges, north Queensland; the surface may be epigene-etch type or is possibly an extension of the exhumed pre-Cretaceous surface preserved in the Isa Highlands (Fig. 3). (**c**) Diagrammatic cross-section showing valley occupied by lava (3): clearly the valley (2) and the high plain (1) predate the lava.

had been stripped to expose the etch plain, Jutson's New Plateau, by the Eocene (Van de Graaff *et al* 1977). Etch surfaces are also well developed in central Australia where they are widely preserved (Mabbutt 1965), the Mid North region of South Australia (Alley 1973) and the eastern Mt Lofty Ranges (Twidale & Bourne 1975). The bevelled bedrock crest of Ayers Rock is probably an etch surface and was shaped beneath a regolith in Late Cretaceous times, though when it was exposed is not known. The steep bounding scarps of the residual, including flared sidewalls (Twidale 1962), are also of etch type but are of more recent (later Cainozoic) derivation (Twidale 1978; Harris & Twidale 1991).

Many granitic domes and boulders are also of etch character, as are many of the intervening valley floors and plains, so that the entire inselberg landscapes can be interpreted as exposed complex weathering fronts (e.g. Hassenfratz, 1791; Falconer 1911; Twidale 1982). In some instances the stripping of the regolith took place only recently, so that the survival of the forms poses no problems, but in others the exposure of the weathering front evidently occurred several scores of millions of years ago. The Eocene age of the West Australian etch plain known as the New Plateau has already been mentioned. In the Gawler Ranges of South Australia, developed on dacitic–rhyolitic ignimbrites of Lower Mesoproterozoic age, there is strong suggestion that the regolith was stripped from the present summit surface during the Early Cretaceous (Campbell & Twidale 1991; see also Wopfner 1969), implying that the planation and weathering of the surface took place even earlier (Fig. 8). The massif appears to have been little changed during the past 60 Ma, for not only are silcreted remnants of putative Eocene age preserved in the piedmont of the ranges and in present valley floors, but riverine sediments derived from the upland are few and thin (Campbell & Twidale 1991; also Firman 1983).

Again, the summit-high plain of the Hamersley Ranges once carried a cover of ferruginous regolith. Some remains but much has been translocated and redeposited in the floors of valleys running radially off the upland (Fig. 9). The relocated material became the Robe River Pisolite which protected the old valley floors to such an extent that the adjacent divides were preferentially eroded, leaving the elongate sinuous valley floors in positive relief. The Pisolite is in places underlain by fossiliferous riverine silts which show that the ferruginous cover was deposited and the upland surface stripped in the Eocene (Twidale *et al.* 1985). The planation and weathering of the Hamersley Surface must have taken place in earliest Tertiary or Mesozoic times, and a Cretaceous age is suggested by regional stratigraphy and palaeogeography (e.g. Hocking *et al.* 1987); although this is a minimum age and it may be older.

Complexity of the Gondwana landscape

Just as the present landscape is a palimpsest consisting of parts of many land surfaces, so was

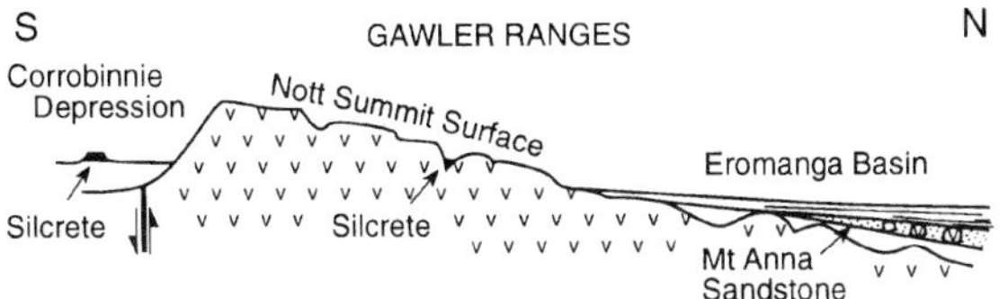

Fig. 8. Diagrammatic cross-section through the Gawler Ranges, South Australia, showing high plain of etch-type and correlation with Early Cretaceous Mt Anna Sandstone, containing corestones from the volcanic upland.

the Gondwana landscape already complex. For example, an exhumed Permian towerkarst has been reported from the Carnarvon Basin (Hocking *et al.* 1987) and early Cambrian pediments from the southern Isa Highlands (Öpik 1961). Exhumed glaciated pavements of Permian age have long been recognized from southeastern Australia and forms of similar origin but of various Precambrian ages have been reported from northwestern Australia (e.g. Perry & Roberts 1968). Much of the ridge and valley topography of the east Kimberley region of the north of Western Australia is exhumed from beneath Cambrian and Proterozoic rocks (Young 1992). Many minor exhumed Proterozoic surfaces are exposed, for example near Barrow Creek, in the Northern Territory, and near the eastern shore of Lake Gairdner, in the Gawler Ranges, in the arid interior of South Australia (Twidale *et al.* 1976). In the central Pilbara, granitic inselbergs are in process of resurrection from beneath a mixed sedimentary–volcanic sequence of either uppermost Archaean or basal Proterozoic age (Twidale 1986).

Fig. 9. On the skyline, the Hamersley Surface, northwest of Western Australia, with Mt Wall on the left and, in the foreground, sinuous mesa capped by Robe River Pisolite.

Other remnants of Gondwana

Surfaces of similar types and antiquity are reported from other parts of the original Gondwana. Thus, the classical inselberg landscapes described from east Africa (Tanzania) by Bornhardt (1900) are of exhumed type and pre-Cretaceous age (Willis 1936), and can be correlated with the earlier mentioned (Fig. 6) granitic inselberg landscapes from the northwest of Western Australia. Though there is debate concerning the precise character and age of the summit-high plains of the Drakensberg, there can be little doubt they are erosional and of Late Mesozoic age, though they are probably etched (see King 1962; Partridge & Maud 1987; Twidale 1990). Stratigraphically-dated lateritized surfaces are reported from West Africa (Michel 1978), East Africa (see e.g. Wayland 1934*b*; Pallister 1960; McFarlane 1991) and peninsular India, where exhumed sub-basaltic, as well as epigene, landscapes which predate the Deccan Traps, are also preserved (Choubey 1969; Demangeot 1978). The upper surface of the Roraima Plateau, in the Guyanan Shield, is evidently a Cretaceous feature (Briceño & Schubert 1990) and Gondwanan surfaces have recently been reported from Patagonia (M. Zarate, pers. comm., 1993).

Discussion

Australia is not unique in its ancient landscapes, though the presence of extensive Early Cretaceous marine transgressions and widespread Early Tertiary volcanicity in eastern Australia has greatly facilitated the dating of these very old palaeoforms (Twidale 1991*a, b*, 1994). Moreover, direct effects of Late Cainozoic glaciation were confined to the southeastern uplands and Tasmania so that the dating evidence has not been materially destroyed or buried. Nonetheless, it is increasingly clear that even in those parts of the Laurasian components of Pangaea that were glaciated very old palaeoforms of an age comparable to those described from Gondwana have survived (for review see Twidale & Vidal Romani 1994). Erosion and deposition associated with Late Cainozoic glaciations have blurred the picture but where a preglacial sedimentary cover remains, exhumed forms have been located: Proterozoic (sub-Torridonian) in northwest Scotland (Williams 1969), Proterozoic in Greenland, Proterozoic and Early Palaeozoic in North America (e.g. Cowie 1960; Ambrose 1964), Cambrian in Sweden (Rudberg 1970), Cretaceous or Jurassic in southern Sweden (Lidmar-Bergström 1989). In the latter region, re-exposure of minor forms is exemplified in the Borrås skåra, a gorge of Cretaceous age (Lidmar-Bergström & Åkesson 1987). The epigene and etch forms inherited from preglacial landscapes and preserved in glaciated terrains are increasingly recognized (see e.g. Fogelberg 1985).

Such old epigene or etch surfaces pose fundamental problems for general geomorphological theory and call to question several articles of faith concerning landscape evolution. The survival of such forms (Twidale 1976; Young 1983) can, in some measure, be explained first in terms of the inherent toughness of the materials in which they are shaped because many, though not all, are eroded into quartzite, sandstone or other resistant siliceous rocks; second, their standing high on and therefore dry, in the landscape (Twidale 1991*a,b*; Twidale & Campbell 1992), for most weathering is due to reactions with water; and third, by the fact that river erosion does not operate over the entire landscape, as suggested by Davis (1909, pp. 266–267) but only in and near stream channels, as urged by Crickmay (1932, 1976; see also Knopf 1924; Horton 1945). The retreat of escarpments may also be unequal in time and space, with some sectors of the scarp being worn back great distances where major rivers have penetrated deeply into uplands, but with intervening sectors virtually stable (Crickmay 1974; Partridge & Maud 1987). The high plains or plateaux behind such stable zones scarp remain untouched by river erosion and thus persist. Reinforcement or positive feedback mechanisms such as those implied, for example, in unequal stream activity, may also contribute to survival, for all river systems are enhanced by use: the more run-off, the more incision, the more surface and subsurface drainage flows into the channel, and so on.

The operation of several of these mechanisms, notably unequal river erosion, is facilitated by anorogenic tectonic regimes. There are exceptions, but very old palaeosurfaces are not, by and large, well preserved in tectonically-active orogenic zones or at convergent plate margins. On stable cratons and old orogens, however, epeirogenic and isostatic activities allow preservation of surfaces in unconformity beneath basins and, combined with unequal river activity, on exposed, uplifted, but otherwise undisturbed, uplands. Mantle plumes (e.g. Cox 1989) could produce similar essentially vertical uplift and may be germane to the preservation of palaeosurfaces in peninsular India, and southern and eastern Africa, for example. Such plume-generated uplifts can presumably affect craton or shield, orogen or platform, though their impacts may be more readily recognized in anorogenic regions. Nevertheless, it is for good reason that most palaeolandscapes have been recognized in cratonic and in very old orogenic regimes in Australia, Africa, peninsular India, northern Europe and the Americas, and that they are especially prominent in Australia and Africa south of the Sahara, both of which land masses are distant from active plate margins.

Factors like aridity have also been invoked as favouring the persistence of land forms and surfaces, and rates of weathering and erosion are indeed low on desert plains (see e.g. Corbel 1959), but many very old palaeoforms occur, as Young (1983) has pointed out, in areas dominated by warm humid climates through the relevant geological ages (e.g Kemp 1978). Low rates of headward erosion have also been cited as facilitating preservation of surfaces (e.g. Young 1983; Taylor *et al.* 1985), as have minor mechanisms such as gully gravure (Bryan 1941). But even taken together, such factors relieve, rather than resolve, the difficulties raised by Gondwanan landforms and landscapes. Little wonder that many, both in Australia and elsewhere, cannot bring themselves to accept the evidence of their antiquity. But if such very old palaeosurfaces exist they must be possible, and as Cardinal de Retz remarked more than 300 years ago, albeit in a rather different context, 'Not everything that is incredible is untrue'.

A summary of this paper was presented at an IGCP 317 meeting held in Sheffield, UK, April 1994. I was unable to travel to England to deliver it personally and I wish to thank Dr Robin Scott for presenting it on my behalf. I express my appreciation to Dr M. J. McFarlane, University of Botswana, Gaborone, an anonymous reviewer and Dr M. Widdowson for constructive comments.

References

ALLEY, N. F. 1973. Landscape development in the Mid–North of South Australia. *Transactions of the Royal Society of South Australia,* **97**,1–17.

AMBROSE, J. W. 1964. Exhumed palaeoplains of the Precambrian shield of North America. *American Journal of Science,* **262**, 817–857.

BORNHARDT, W. 1900. *Zur Oberflächengestaltung und Geologie Deutsch Ostafrikas.* Reimer, Berlin.

BRICEÑO, H. O. & SCHUBERT, C. 1990. Geomorphology of the Gran Sabana, Guyana Shield, southeastern Venezuela. *Geomorphology,* **3**, 125–141.

BRYAN, K. 1941. Gully gravure – a method of scarp retreat. *Journal of Geomorphology*, **3**, 89–107.

CAMPANA, B. 1958. The Mt Lofty region and Kangaroo Island. *In* : GLAESSNER, M. F. & PARKIN, L. W. (eds) *The Geology of South Australia*. Melbourne University Press/Geological Society of Australia, Melbourne, 3–27.

CAMPBELL, E. M. & TWIDALE, C. R. 1991. The evolution of bornhardts in silicic volcanic rocks, Gawler Ranges, South Australia. *Australian Journal of Earth Sciences*, **38**, 79–93.

CHOUBEY, V. D. 1969. Study of pre-Deccan Trap erosion surface in central India. *Geological Society of India Bulletin*, **6**, 79–82.

CORBEL, J. 1959. Vitesse d'érosion. *Zeitschrift für Geomorphologie*, **3**, 1–28.

COWIE, J. W. 1960. Contributions to the geology of north Greenland. *Meddelelser om Groenland*, **164.**

COX, K. G. 1989 The role of mantle plumes in continental drainage patterns. *Nature*, **342**, 873–877.

CRAFT, F. A. 1932. The physiography of the Shoalhaven valley. *Proceedings of the Linnaean Society of New South Wales*, **57**, 245–260.

CRICKMAY, C. H. 1932. The significance of the physiography of the Cypress Hills. *Canadian Field Naturalist*, **46**, 185–186.

—— 1974. *The Work of the River*. Macmillan, London.

—— 1976. The hypothesis of unequal activity. *In* : MELHORN, W. N. & FLEMAL, R. C. (eds). *Theories of Landform Development*. State University of New York, Binghamton, 103–109.

DAILY, B., TWIDALE, C. R. & MILNES, A. R. 1974. The age of the lateritized land surface on Kangaroo Island and the adjacent areas of South Australia. *Journal of the Geological Society of Australia*, **21**, 387–392.

DAVIS, W. M. 1899. The geographical cycle. *Geographical Journal*, **14**, 481–504.

—— 1909. *Geographical Essays*. Dover, Boston.

DE KEYSER, F. 1964. *Innisfail, Queensland. 1:250 000 Geological Series, Explanatory Notes*. Bureau of Mineral Resources, Geology and Geophysics, Canberra.

DEMANGEOT, J. 1978. Les reliefs cuirassés de l'Inde du Sud. *Travaux et Documents de Géographie Tropicale*, **33**, 97–111.

FALCONER, J. D. 1911. *The Geology and Geography of Northern Nigeria*. Macmillan, London.

FINKL, C. W. & CHURCHWARD, H. M. 1973. The etched land surfaces of southwestern Australia. *Journal of the Geological Society of Australia*, **20**, 295–307.

FIRMAN, J. B. 1983. Silcrete near Chundie Swamps: the stratigraphic setting. *Quarterly Geological Notes. Geological Survey of South Australia*, **85**, 2–5.

FOGELBERG, P. (ed.) 1985. Preglacial weathering and planation. *Fennia*, **163**, 283–383.

FRAKES, L. A. & BOLTON, B. R. 1984. Origin of manganese giants: sealevel change and anoxic-oxic history. *Geology*, **12**, 83–86.

——, BURGER, D., APTHORPE, M. *et al.* 1988. Australian Cretaceous shorelines, stage by stage. *Palaeogeography, Palaeoclimatology, Palaeoecology*, **59**, 31–48.

HACK, J. T. 1960. Interpretation of erosional topography in humid temperate regions. *American Journal of Science*, **238A**, 80–97.

HARRIS, W. K. & TWIDALE, C. R. 1991. Revised age for Ayers Rock and the Olgas, central Australia. *Transactions of the Royal Society of South Australia*, **115**, 109.

HASSENFRATZ, J.-H. 1791. Sur l'arrangement de plusieurs gros blocs de différentes pierres que l'on observe dans les montagnes. *Annales de Chimie*, **11**, 95–107.

HILLS, E. S. 1934. Some fundamental concepts in Victorian physiography. *Proceedings of the Royal Society of Victoria*, **47**, 158–174.

—— 1955. Die Landoberflächen Australiens. *Die Erde*, **7**, 195–205.

—— 1975. *Physiography of Victoria*. Whitcombe and Tombs, Melbourne.

HOCKING, R. M., MOORS, H. T. & VAN DE GRAAFF, W. J. E. 1987. Geology of the Carnarvon Basin, Western Australia. *Geological Survey of Western Australia Bulletin*, **133**.

HORTON, R. E. 1945. Erosional development of streams and their drainage basins. *Geological Society of America Bulletin*, **56**, 275–370.

HOSSFELD, P. S. 1926. *The geology of portions of the Counties Light, Eyre, Sturt and Adelaide*. MSc Thesis, University of Adelaide.

HUTTON, J. 1788. Theory of the Earth; or an investigation of the laws observable in the composition, dissolution, and restoration of land upon the globe. *Transactions of the Royal Society of Edinburgh*, **1**, 209–304.

—— 1795. *Theory of the Earth, with Illustrations*. Cadell, Junior and Davies, London; Creech, Edinburgh, two vols.

JACK, R. L. 1931. Report on the geology of the region north and northwest of Tarcoola. *Geological Survey of South Australia Bulletin*, **15**.

JUTSON, J. T. 1914. An outline of the physiographical geology (physiography) of Western Australia. *Geological Survey of Western Australia Bulletin*, **61**.

KEMP, E. M. 1978. Tertiary climatic evolution and vegetation history in the southeastern Indian Ocean region. *Palaeogeography, Palaeoclimatology and Palaeoecology*, **24**, 169–208.

KING, L. C. 1942. *South African Scenery*. Oliver and Boyd, Edinburgh.

—— 1962. *Morphology of the Earth*. Oliver and Boyd, Edinburgh.

KNOPF, E. B. 1924. Correlation of residual erosion surfaces in the eastern Appalachians. *Geological Society of America Bulletin*, **35**, 633–668.

LIDMAR-BERGSTRÖM, K. 1989. Exhumed Cretaceous landforms in southern Sweden. *Zeitschrift für Geomorphologie Supplement-Band*, **72**, 21–40.

—— & ÅKESSON, G. 1987. Borrås skåra – a gorge of Cretaceous age. *Geologiska Föreningens i Stockholm Förhandlingar*, **109**, 327–330.

MABBUTT, J. A. 1961*a*. 'Basal surface' or 'weathering front'. *Proceedings of the Geologists' Association of London*, **72**, 357–358.

—— 1961*b*. A stripped land surface in western Australia. *Transactions of the Institute of British Geographers*, **29**, 101–114.

—— 1965. The weathered land surface in central Australia. *Zeitschrift für Geomorphologie,* **9**, 82–114.

—— 1966. Landforms of the western MacDonnell Ranges. *In* : Dury, G. H. (ed.) *Essays in Geomorphology.* Heinemann, London, 83–119.

McFarlane, M. J. 1986. Geomorphological analysis of laterites and its role in prospecting. *Geological Survey of India Memoir,* **120**, 29–40.

—— 1991. Some sedimentary aspects of lateritic weathering profile development in the major bioclimatic zones of tropical Africa. *Journal of African Earth Sciences,* **12**, 267–282.

Michel, P. 1978. Cuirasses bauxitique et ferrugineuses d'Afrique occidentale. Aperçu chronologique. *Travaux et Documents de Géographie Tropicale,* **33**, 11–32.

Miles, K. R. 1952. Geology and underground water resources of the Adelaide Plains area. *Geological Survey of South Australia Bulletin,* **27**.

Needham, R. S. 1982. *East Alligator, Northern Territory. 1:100 000 Geological Map Commentary.* Bureau of Mineral Resources, Geology and Geophysics, Canberra.

Öpik, A. A. 1961. The geology and palaeontology of the headwaters of the Burke River, Queensland. *Bureau of Mineral Resources, Geology and Geophysics Bulletin,* **53**.

Pallister, J. W. 1960. Erosion cycles and associated surfaces of the Mengo district, Buganda. *Overseas Geology and Mineral Resources,* **8**, 26–36.

Partridge, T. C. & Maud, R. R. 1987. Geomorphic evolution of southern Africa since the Mesozoic. *Transactions of the Geological Society of South Africa,* **90**, 179–208.

Perry, W. J. & Roberts, H. G. 1968. Late Precambrian glaciated pavements in the Kimberley region, Western Australia. *Journal of the Geological Society of Australia,* **15**, 51–56.

Playford, P. E., Cockbain, A. E. & Low, G. H. 1976. Geology of the Perth Basin, Western Australia. *Geological Survey of Western Australia Bulletin,* **124**.

Rudberg, S. 1970. The sub-Cambrian peneplain in Sweden and its slope gradient. *Zeitschrift für Geomorphologie Supplement-Band,* **9**, 157–167.

Ruxton, B. P. & Taylor, G. 1982. The Cainozoic geology of the Middle Shoalhaven Plain. *Journal of the Geological Society of Australia,* **29**, 239–246.

Schumm, S. A. 1963. Disparity betwen present rates of denudation and orogeny. *United States Geological Survey Professional Paper,* **454**.

Stuart-Smith, P., Needham, R. S. & Wallace, D. A. 1987. *Pine Creek, Northern Territory. 1:100 000 Geological Map Series Commentary.* Bureau of Mineral Resources, Geology and Geophysics, Canberra.

Suess, E. 1885. *Der Antlitz der Erde.* Tempsky, Vienna.

Taylor, G., Taylor, G. R., Bink, *et al.* 1985. Prebasaltic topography of the northern Monaro and its implications. *Australian Journal of Earth Sciences,* **32**, 65–71.

Twidale, C. R. 1956. Chronology of denudation in northwest Queensland. *Geological Society of America Bulletin,* **67**, 867–882.

—— 1962. Steepened margins of inselbergs from northwestern Eyre Peninsula, South Australia. *Zeitschrift für Geomorphologie,* **6**, 51–69.

—— 1976. On the survival of palaeoforms. *American Journal of Science,* **276**, 1138–1176.

—— 1978. On the origin of Ayers Rock, central Australia. *Zeitschrift für Geomorphologie Supplement-Band,* **31**, 177–206.

—— 1980. Landforms. *In* : Corbett, D. W. P. (ed.) *A Field Guide to the Flinders Ranges.* Rigby, Adelaide, 13–41.

—— 1982. The evolution of bornhardts. *American Scientist,* **70**, 268–276.

—— 1984. The enigma of the Tindal Plain, Northern Territory. *Transactions of the Royal Society of South Australia,* **108**, 95–103.

—— 1986. Granite platforms and low domes: newly exposed compartments or degraded remnants? *Geografiska Annaler (Series A),* **68**, 399–411.

—— 1990. The origin and implications of some erosional landforms. *Journal of Geology,* **98**, 343–364.

—— 1991*a*. Gondwana landscapes: definition, dating and implications. *In* : Radnakrishna, B. P. (ed.) *The World of Martin F.Glaessner* Memoir of the Geological Society of India, **20**, 225–263.

—— 1991*b*. A model of landscape evolution involving increased and increasing relief amplitude. *Zeitschrift für Geomorphologie,* **35**, 85–109.

—— 1994. Gondwanan (Late Jurassic and Cretaceous) palaeosurfaces of the Australian craton. *Palaeogeography, Palaeoclimatology, Palaeoecology,* **112**, 157–186.

—— & Bourne, J. A. 1975. Geomorphological evolution of the eastern Mt Lofty Ranges, South Australia. *Transcations of the Royal Society of South Australia,* **99**, 197–209.

—— & Campbell, E. M. 1988. Ancient Australia. *GeoJournal,* **16**, 339–354.

—— & —— 1992. Geomorphological development of the eastern margin of the Australian Shield. *Earth Surface Processes and Landforms,* **17**, 319–331.

—— & Vidal Romani, J. R. 1994. The Pangaean inheritance. *Cuadernos Laboratorio Xeolóxico de Laxe,* **19**, 7–36.

——, Bourne, J. A. & Smith, D. M. 1976. Age and origin of palaeosurfaces on Eyre Peninsula and in the southern Gawler Ranges, South Australia. *Zeitschrift für Geomorphologie,* **20**, 28–55.

——, Horwitz, R. C. & Campbell, E. M. 1985. Hamersley landscapes of the northwest of Western Australia. *Revue de Géographie Physique et Géologie Dynamique,* **26**, 173–186.

Van De Graaff, W. J. E., Crowe, R. W. A., Bunting, J. A. & Jackson, M. J. 1977. Relict Early Cainozoic drainages in arid Western Australia. *Zeitschrift für Geomorphologie,* **21**, 379–400.

Wayland, E. J. 1934*a*. Peneplains and some erosional landforms. *Geological Survey of Uganda, Annual Report and Bulletin,* **1**, 77–79.

—— 1934*b*. Peneplains of east Africa. *Geographical Journal,* **83**, 79.

Williams, G. E. 1969. Characteristics and origin of a

Precambrian pediment. *Journal of Geology*, **77**,183–207.

WILLIS, B. 1936. *East African Plateaus and Rift Valleys. Studies in Comparative Seismology*. Carnegie Institute Publication Washington DC, **470**.

WOODARD, G. D. 1955. The stratigraphic succession in the vicinity of Mount Babbage Station, South Australia. *Transactions of the Royal Society of South Australia*, **78**, 8–17.

WOPFNER, H. 1964. Permian–Jurassic history of the western Great Artesian Basin. *Transactions of the Royal Society of South Australia*, **88**, 117–128.

—— 1968. Cretaceous sediments on the Mt Margaret Plateau and the evidence for neo-tectonism. *Quarterly Geological Notes. Geological Survey of South Australia*, **28**, 7–11.

—— 1969. Mesozoic Era. *In:* PARKIN, L. W. (ed.) *Handbook of the Geology of South Australia*. Geological Survey of South Australia, Adelaide, 133–171.

WYRWOLL, K.-H. 1988. Time in the geomorphology of Western Australia. *Progress in Physical Geography*, **12**, 237–263.

YOUNG, R. W. 1981. Denudational history of the south-central uplands of New South Wales. *Australian Geographer*, **15**, 77–88.

—— 1983. The tempo of geomorphological change. Evidence from southeastern Australia. *Journal of Geology*, **91**, 221–230.

—— 1992. Structural heritage and planation in the evolution of landforms in the East Kimberley. *Australian Journal of Earth Sciences*, **39**, 1421–1512.

—— & MCDOUGALL, I. 1985. The age, extent and geomorphological significance of the Sassafras basalt, southeastern New South Wales. *Australian Journal of Earth Sciences*, **32**, 323–331.

Analysis of digital elevation data for the Scottish Highlands and recognition of pre-Quaternary elevated surfaces

PHILIP S. RINGROSE[1] & PIOTR MIGOŃ[2]

[1] *Department of Petroleum Engineering, Heriot-Watt University, Riccarton, Edinburgh EH14 4AS, UK*

[2] *Department of Geography, University of Wrocław, pl. Uniwersytecki 1, 50-137 Wrocław, Poland*

Abstract: Remnants of elevated palaeosurfaces in the Scottish Highlands, proposed in numerous earlier studies, have been re-evaluated using digital elevation data from the Ordnance Survey. Histograms of elevation frequency, for areas 5, 10 and 20 km^2, display multi-modal distributions which can be interpreted in terms of plateau and base-level components. The effects of glaciation, lithology and faulting can be inferred from patterns in the elevation distribution. The histogram modes reveal a series of elevated surfaces which broadly confirm previous, qualitative models of inclined Tertiary palaeosurfaces, but indicate a much more complex pattern. Many of the elevated surfaces are seen to be inclined towards two major faults, the Great Glen Fault and the Ericht-Laidon Fault. However, the highest surfaces may have crossed these fault zones uninterrupted. A model in which an episode of tectonic reactivation imposes a major change on pre-Quaternary geomorphological evolution in this area is proposed.

The occurrence of elevated surfaces of low relief and the alignment of summits in different parts of the Scottish Highlands have long been recognized (see Sissons 1976 for a review). These features have been interpreted as remnants of planation surfaces, subsequently uplifted and dissected by fluvial and glacial erosion (e.g. Linton 1951; Godard 1965; George 1966; Sissons 1976). Several attempts have been made to correlate summit surfaces in the Highlands and to use them as indicators of geomorphic developments in this area in pre-Quaternary times. The common model adopted was a general tilting of the Highlands to the east, with the main preglacial watershed being located in the Western Highlands (Linton 1951).

Several objections to this early model, based on altitudinal correlations alone, have been raised, most recently by Hall (1991). The influence of differential tectonic movements during the Tertiary period has been generally neglected; the importance of deep selective weathering and differential erosion in the formation and maintenance of such surfaces has been increasingly appreciated; and the existence of Tertiary etch-type surfaces has been proposed. Pre-Quaternary etchsurfaces, which retain remnants of weathering mantles, have been described from Northeast Scotland (the Buchan surface) and some parts of the Grampians (Hall 1986, 1987; Hall & Mellor 1988). Such surfaces are likely to have had complex, lithologically-controlled relief and are unlikely to have formed very flat surfaces. Finally, most previous studies relied on a subjective interpretation of contour maps, and elevated surfaces of low relief were identified in the field 'by eye'. A clearer understanding of elevated palaeosurfaces in the Scottish Highlands is thus desirable.

The uplift history of this area also has important implications for understanding tectonic history and basin development in northern Europe in general. The Scottish massif has been an important source of sediment supply for the surrounding continental shelf during the Tertiary (e.g. Galloway *et al.* 1993; Turner & Scrutton 1993), and Tertiary uplift and subsidence patterns are important to understanding the neotectonics of the region (Muir Wood 1989; Davenport *et al.* 1989). Tertiary geomorphology is also important as it provides one of the few insights into geological conditions prior to the Quaternary glaciations, whose effects dominate so much of the current surface environment.

We have therefore set out to test and evaluate previous interpretations of elevated surfaces in the Scottish Highlands using recently available Ordnance Survey digital elevation data (1:50 000 scale). This dataset comprises surface elevations on a 50 m spaced grid with an accuracy in elevation determination of *c.* 2 m (Ordnance Survey 1992). For the purposes of this type of study, it can be regarded as an exhaustive database of onshore surface elevations. Each 20 × 20 km datafile (termed a 'tile') comprises 160 801 points – 46 tiles

From Widdowson, M. (ed.), 1997, *Palaeosurfaces: Recognition, Reconstruction and Palaeoenvironmental Interpretation,* Geological Society Special Publication No. 120, pp. 25–35.

were acquired for this study. Our main objectives were:

1. to establish whether or not unequivocal elevated surfaces exist;
2. to discover if regional correlation of such surfaces can be established, in terms of elevation statistics;
3. to improve our understanding of the geomorphological evolution of the pre-Quaternary in this area.

The area studied comprises an 80 × 220 km block across the central part of the Scottish Highlands (Fig. 1). It extends from the Inner Hebrides in the west to the North Sea coast in the east and includes several geomorphologically distinct regions: part of the Western Highlands, the Monadhliath Mountains, the Cairngorms, part of the Grampian Mountains and the Buchan area. The main aspects of the pre-Quaternary and Quaternary development of these areas have been outlined in several papers, including Godard (1965), Sugden (1968), Hall (1986, 1987, 1991), Hall & Mellor (1988) and Le Coeur (1988).

Most of the area is underlain by metamorphic rocks; the Precambrian Moine in the west and the Dalradian supergroup in the east, both intruded by a number of Caledonian granitoids (Fig. 1). The Highlands are also crossed by several major NW–SE faults, including the Great Glen Fault. These faults have had a considerable influence on

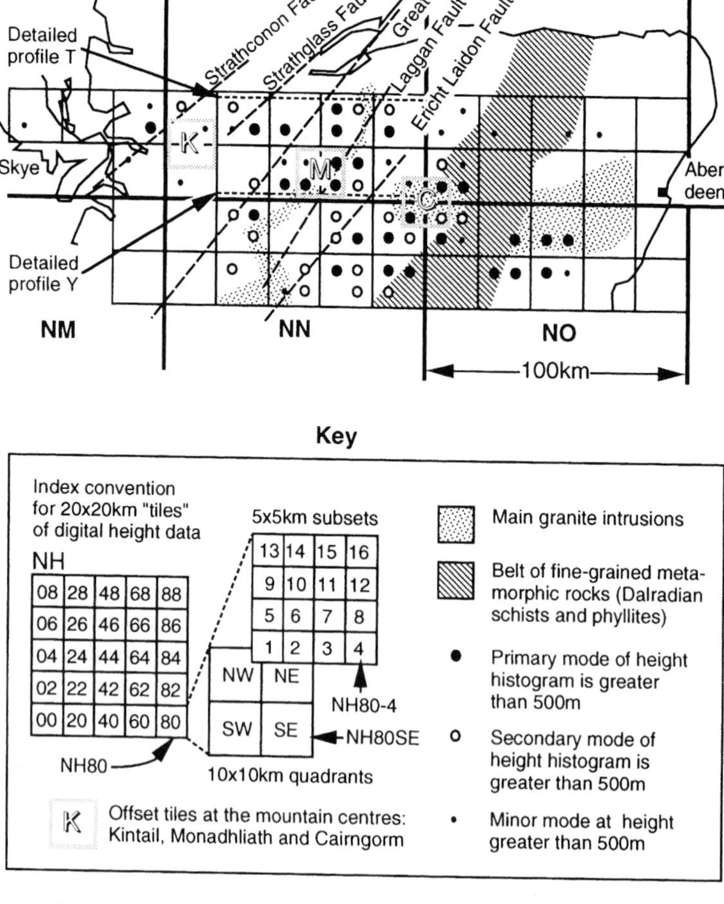

Fig. 1. Map showing locations for digital elevation data study, with the main geological features and the distribution of principal elevated surfaces indicated.

the development of the Scottish basement throughout the Palaeozoic (e.g. Rogers *et al.* 1989) and their reactivation during the Tertiary has also been suggested (Holgate 1969). Reactivation of basement faults during post-glacial times has also been proposed (Ringrose 1989; Ringrose *et al.* 1991; Fenton 1992), but the relation of these late Quaternary movements to possible Tertiary displacements is not yet understood.

Methods and concepts

The digital elevation data were first displayed on contour plots to identify the main geomorphic features. Figure 2 shows an example contour display of four 20 × 20 km tiles, with a contour interval of 100 m. These displays can be used to identify the main surfaces and surface boundaries in much the same way as with a traditional topographic map except that the density of elevation data allows one to study a broad range of map scale and elevation resolution when evaluating features of interest (as illustrated in the Fig. 2 insert). Histograms of elevations for areas of varying size were then evaluated for the main regions of interest. By varying the sample area for a histogram plot, different degrees of resolution of elevation variability could be assessed. Three scales were used: (1) 20 × 20 km data tiles (this being the primary form of the data); (2) 10 × 10 km quadrants of each tile; (3) 5 × 5 km subsets from each tile.

The indexing convention is illustrated in Fig. 1. In each case, 20 m 'bins' for elevation values were

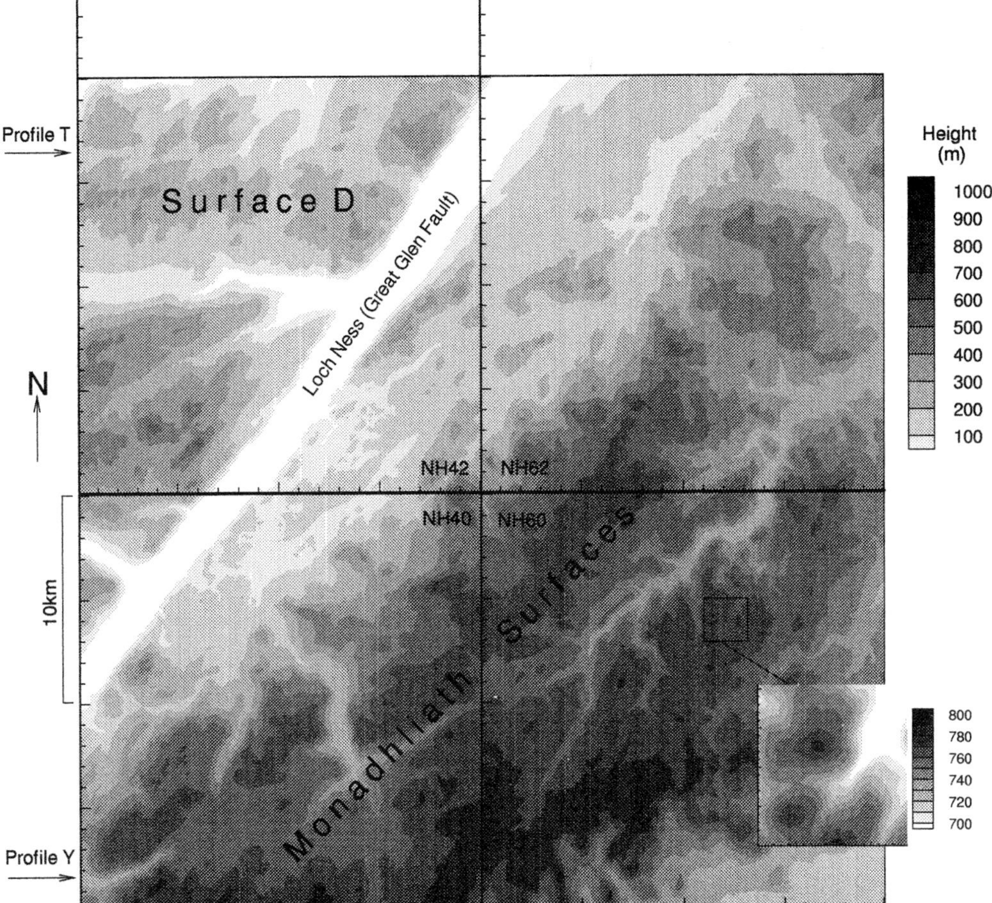

Fig. 2. Contour map of a portion of the digital data. Surface D and the Monadhliath Surfaces refer to prominant elevation modes identified in the histogram study. The insert illustrates the smaller-scale topography of the regional surfaces.

used, as this was found to be practical and of sufficient resolution for this study. (Bins as small as 5–10 m could have been used, given the accuracy of elevation determination at *c.* 2 m, but would have been impractical for this study involving an elevation range of 1200 m.)

In order to establish a conceptual setting for the interpretation of these histograms, theoretical slope profiles and their corresponding elevation histograms were generated. Each of these models relates to an overall postulate that an elevated surface has undergone some degree of denudation towards a base level, related to a river system or sea level. Three hypothetical slope forms were considered (Fig. 3a): (1) base-level dominant, representing a portion of a mature basin with only small remnants of upland terrain; (2) plateau dominant, representing an upland terrain with relatively underdeveloped incision to a new base level; (3) a transitional case, where some upland is preserved but erosion to base level is fairly well developed.

The elevation histograms for these profiles (Fig. 3b) are clearly diagnostic of the slope type, having modes of the distribution at the bottom for 'base level', the top for 'plateau' and at top and bottom for 'transitional'. The effect of adding random noise (i.e. stochastic variation in the absolute elevation of a number of profiles) was evaluated for the transitional case and a broad, bimodal histogram was produced (Fig. 3c). The relative strength of the two modes can be tuned by choosing different slope functions to give weight to the base level or plateau end of a spectrum of possible slope profiles. The main conclusion drawn from this

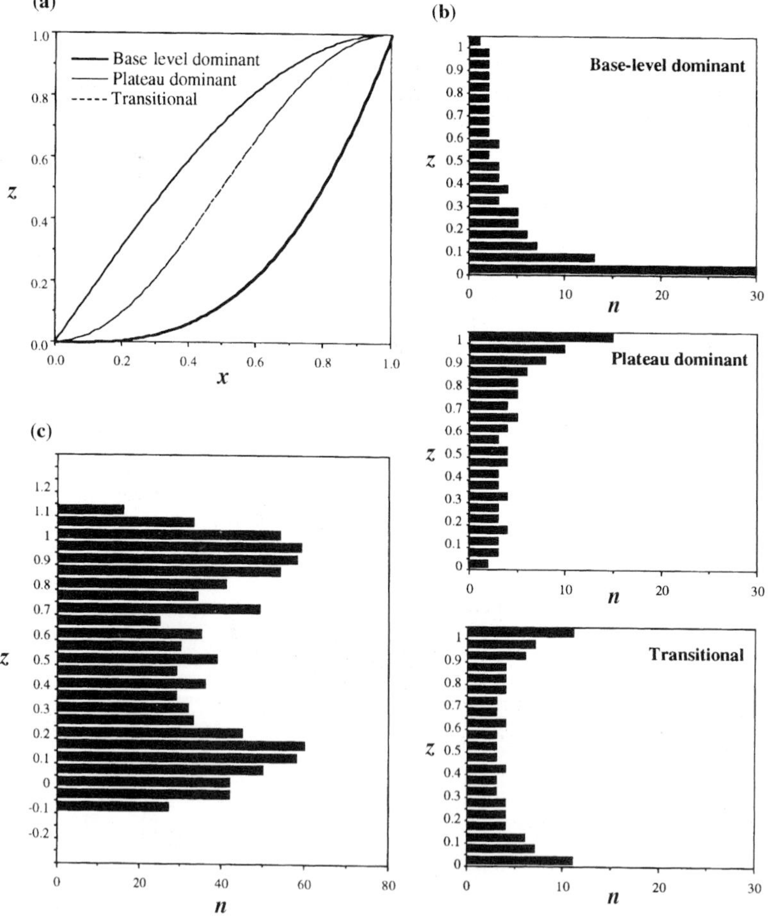

Fig. 3. Theoretical slope profiles (**a**) and corresponding elevation distributions (**b**). Random noise (uniform pdf, $z = \pm 0.25$)was added to the transitional profile to generate the distribution in (**c**). The functions used for the profiles are: base level dominant, $z = x^3$; plateau dominant, $z = 1 - \cos(x + 1)$; transitional, $z = \sin(x/2)$.

theoretical study is that dissected upland terrains should be characterized by *bimodal* distributions of the elevation histogram. The addition of 'noise' could represent a number of morphological processes occurring in the real world, causing loss of the underlying signal, and suggests that in real systems the expected bimodal distributions should be detectable.

Several factors mean that real elevation distributions are more complex: (1) most natural surfaces are slightly tilted, which will cause some broadening of the corresponding mode for a histogram for any sizeable area; (2) natural surfaces are not strictly planar but are likely to have some inherent relief (this could be as much as 100 m or more where strongly selective weathering and denudation has occurred, as described by Thomas 1989 and Hall 1991), this will also cause a broadening of the histogram mode; (3) complex uplift and denudation history may well result in a multiplicity of surfaces which may or may not be resolved on a elevation histogram. We have, however, found that the simple concepts of a local base level and dissected plateau provide a powerful tool for the interpretation of elevation histograms in this area.

Evaluation of digital elevation data

Histograms of the main mountain centres (20 × 20 km)

Figure 4 shows histograms for 20 × 20 km tiles positioned over the main mountain centres in the study area (for locations see Fig. 1). Several general characteristics are evident. The bell-shape of the upper part of Kintail histogram is probably due to severe erosion by mountain glaciers during the Quaternary. There is little evidence for any upland plateaux, although the peaks may relate to some ancient high level plateau above 1200 m. This pattern corresponds with a general paucity of high level planar surfaces in the western parts of the study area, almost certainly because of the effects of glaciation. The Monadhliath histogram shows a very pronounced high level mode at *c.* 750 m. This corresponds with the Monadhliath surface identified by Hall (1991) and others. Dissection of this plateau is evident in the low-side tail of the distribution and in a secondary mode at 200 m, which corresponds to the local base level. In the Cairngorms, a complex multimodal histogram is observed. Modes at *c.* 920 and 690 m appear to reflect remnants of high level plateaux but no clear patterns are evident at this scale.

Regional evaluation of quadrants (10 × 10 km)

A regional evaluation of the characteristics of the digital elevation data was done using histograms of 10 × 10 km quadrants. In an effort to classify the diverse nature of these multimodal distributions and identify the main high-level surfaces, three classes of histogram were established: (1) histograms with the primary mode at a elevation > 500 m; (2) histograms with a prominent

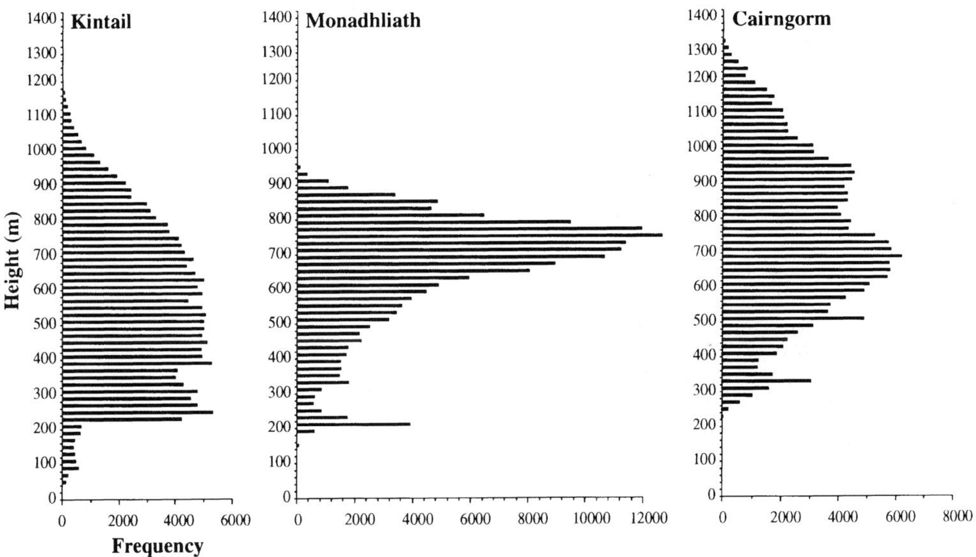

Fig. 4. Elevation histograms for three (offset) 20 × 20 km tiles at selected mountain centres (for locations see Fig. 1).

secondary mode at a elevation > 500 m; (3) histograms with a minor mode at a elevation > 500 m.

Examples of these classes of histogram are shown in Fig. 5a and their distribution is shown in Fig. 1. Histograms with only a 'base-level' signature (such as in the lower half of NH62SW, Fig. 5a) and no evidence for a significant higher level surface were excluded. The cut-off value of 500 m is quite arbitrary (significant modes do occur below this) but serves as a completely objective screening exercise for identifying high-level surfaces without reference to their origin or age. The main high-level modes occur in the central portion of the study area. In the west, whatever high-level surfaces may have existed have largely been destroyed by the effects of glaciation and in the east the modes are mostly below 500 m. The main influence of subsurface lithology is a lack of prominent surfaces over the Dalradian outcrop (Fig. 1). It appears that the finer-grained metamorphic rocks of this belt give less potential for upland surface preservation than either the coarser-grained Moine series (to the west) or the granite intrusions. These differences might also be explained by structural (rather than lithological) phenomena, such as the density of faults and

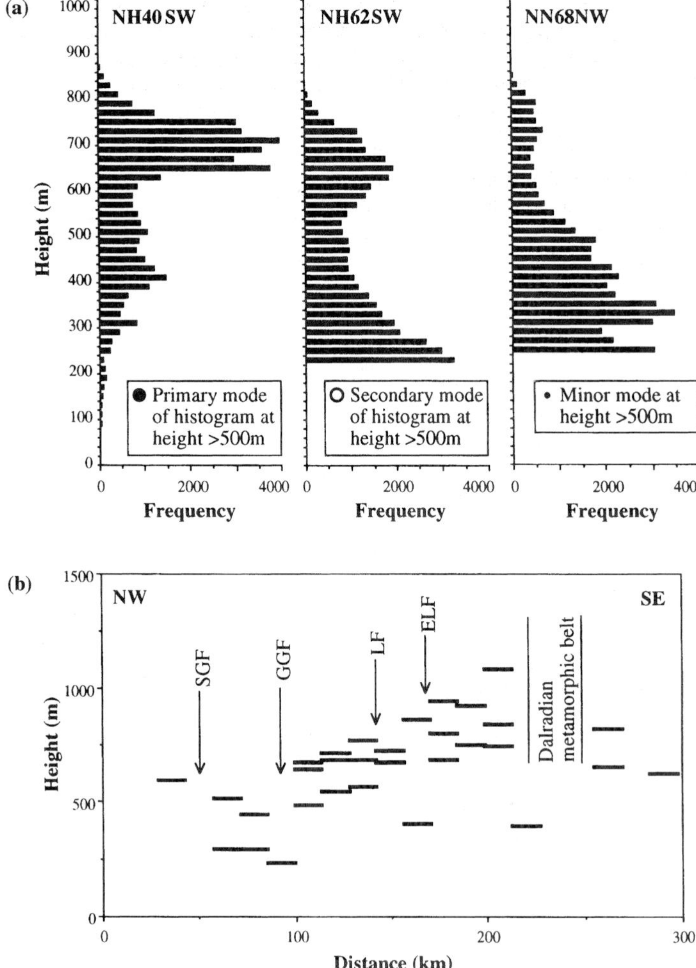

Fig. 5. (**a**) Examples of elevation histograms of tile quadrants (10 × 10 km) in the Central Highlands, showing examples of the three classes of 'upper modes' in the distributions. (**b**) Distribution of upper modes from tile quadrants projected onto a NW–SE line, with location of major faults indicated (cf. Fig. 1). Dataset incorporates all tiles in the rectangle between corner indices NH22 and NO40.

fractures or the degree of folding (the Dalradian province being more deformed).

The vertical distribution of 'upper modes' of these histograms was evaluated by projecting values from the central region (between NH22 and NO40) onto a NW–SE line (Fig. 5b). This projection is perpendicular to the main faults, which appear to have had some influence of their distribution. In particular, the elevation of modes appears to decrease towards the Great Glen Fault on both sides. No trends were evident on a projection of the same data onto an E–W line.

Detailed profiles of subsets (5 × 5 km)

The central area was then evaluated at a still higher resolution, using 5 × 5 km subsets along two lines, T and Y (Fig. 6; for location see Fig. 1). At this scale, the influence of faults on the distribution of histogram modes is more apparent. A possible correlation scheme (based only on elevation information) is indicated on Fig. 6 by the letters B–H. Some of these interpretations are quite tentative (as indicated by dashed lines) and other schemes might well be composed, but our aim at this stage is to pose a preliminary model to be tested by further detailed analyses. The following trends are apparent:

1. The highest surfaces B and C may be remnants of an extensive surface at *c.* 700–900 m. This corresponds with the Monadhliath surface (Fig. 4). However, at this higher resolution the Monadhliath 'surface' appears to comprise at least two distinct surfaces (B and C), or alternatively a complex surface with local topography.
2. The Great Glen and the Ericht-Laidon Faults are the focus of several surfaces which dip towards them: surfaces marked D and E dip towards the Great Glen, and surfaces marked F and G dip towards the Ericht-Laidon axis. Surface D to the west of the Great Glen Fault is the best developed and most laterally extensive; the other 'surfaces' of this group could be steeply inclined surface segments or could represent a more complex stepped topography.
3. The lowest surfaces, marked H, are all related to the current base level of drainage, focused down Strath Glass, the Great Glen and Strath Spey (the Ericht-Laidon axis).

In this area, there is thus evidence for: (1) an extensive regional surface at *c.* 800 m; (2) intermediate surfaces between 200 and 600 m, but above current base level, which dip towards to major fault-controlled axes; and (3) low-level surfaces related to current drainage. A further, more extensive analysis of this high resolution dataset is needed to resolve some of the uncertainties in this interpretation.

Discussion

Spatial patterns in elevation distributions

Elevated surfaces occur dominantly in the central part of the Highlands, mostly in the area between the Strath Glass Fault in the west and the Dalradian belt in the east. In the west, the absence of surfaces probably reflects very intense glacial erosion and deep dissection of preglacial uplands (as suggested by Linton 1959). Alternatively, extensive surfaces may never have formed in this area if it was persistently tectonically unstable. The results of the histogram analysis favour the first hypothesis, since elevated uplands have been identified in glacially-eroded areas (Glen Roy, Gaick Forest and the Cairngorms) as well as less dissected areas (e.g. the Monadhliath). Moreover, there is growing evidence that glacial erosion of Scottish uplands was highly selective and many uplands may have survived glaciation relatively intact, leaving distinct traces of pre-Quaternary morphogenesis (Sugden 1968; Hall & Sugden 1987; Hall & Mellor 1988).

The absence of elevated surfaces in the east may be due to a number of factors. It may simply be that palaeosurfaces were always less elevated in the east (as suggested originally by Linton 1951). It is also clear that the lithology of Dalradian metamorphic belt is less favourable for the preservation of high-altitude surfaces. However, this zone does separate major regions of upland (over 600 m) and lowland (100–200 m), and a complex (non-planar) palaeosurface may have existed, as suggested by Hall (1991) and indicated by the presence of pre-Quaternary sediments and weathering mantles dating back to the Miocene (Fitzpatrick; 1963, McMillan & Merritt 1980; Hall 1986, 1987, 1991; Hall & Mellor 1988). The highest altitudes in the Grampians, exceeding 1000 m, occur close to this Dalradian belt (Cairngorms, Ben Avon) and even within it (Lochnagar). This suggests that considerable uplift of the central Highlands relative to the Northeast lowland has occurred (Hall 1991). Since there are no mountain fronts or flights of tectonic steps in between, large-scale warping is likely. We suggest that the belt of varied, strongly-folded Dalradian rocks was the most prone to warping by slow tectonic movements, intensive enough to cause a predominance of erosion over the formation of extensive surfaces.

Influence of faults

Four major faults cross the central part of the Highlands: the Strath Glass Fault (SGF), the Great

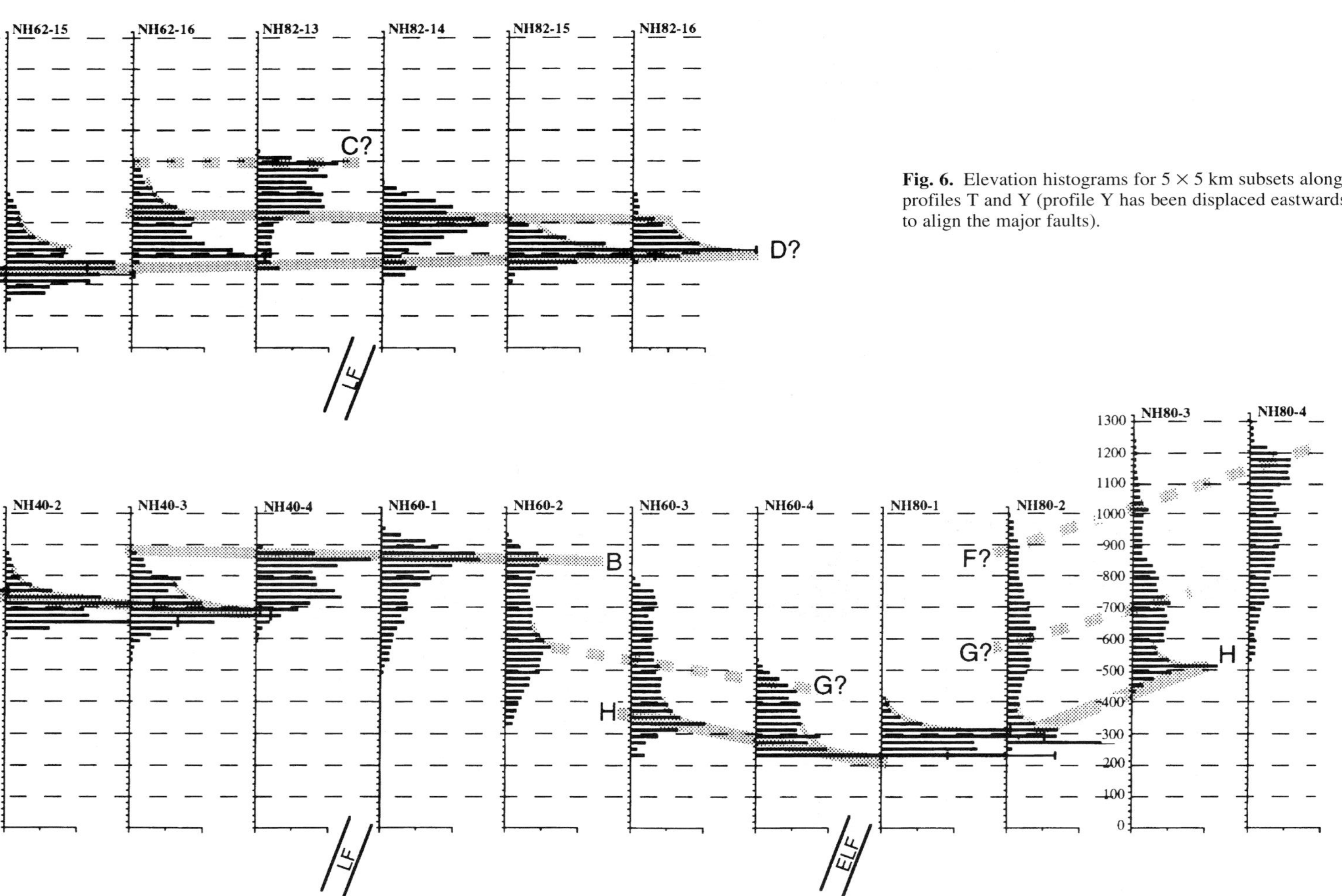

Fig. 6. Elevation histograms for 5 × 5 km subsets along profiles T and Y (profile Y has been displaced eastwards to align the major faults).

Figure 6 continued →

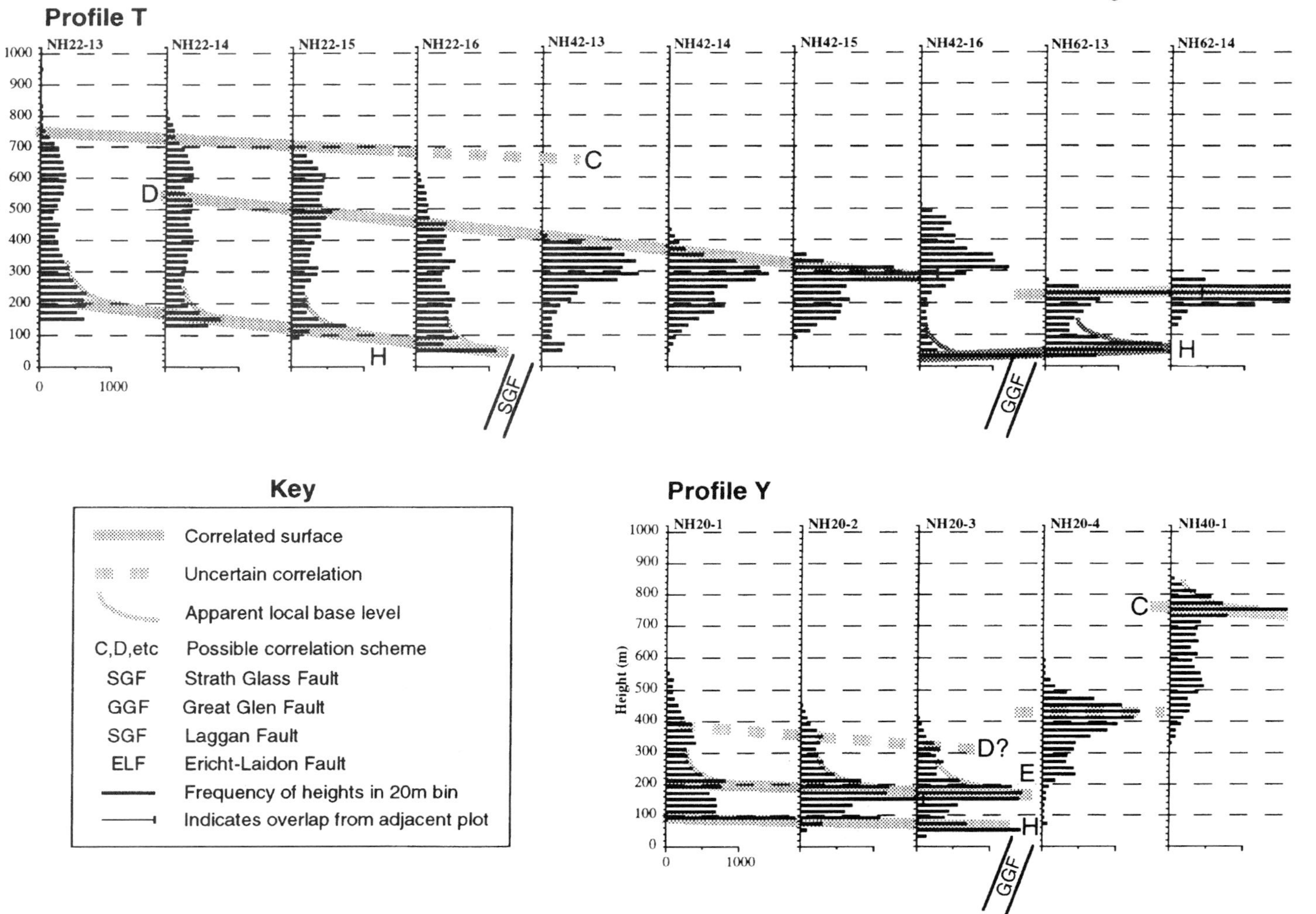

Glen Fault (GGF), the Laggan Fault (LF) and the Ericht-Laidon Fault (ELF). As all of them cross areas where elevated pre-Quaternary surfaces are relatively common, the potential exists to assess their influence on surface development and to infer their degree of tectonic activity during the Tertiary.

The GGF and ELF appear to have exerted a major influence on the development and preservation of elevated surfaces, many of which slope towards these fault zones. The highest preserved surfaces (Fig. 6) may have crossed the GGF relatively uninterrupted, but lower surfaces clearly slope towards the GGF. This suggests a reactivation of the GGF at some time after the formation of the highest preserved surfaces. If we assume that the highest surfaces must be post-Cretaceous (Linton 1951; Hall 1991), this implies fault reactivation in the early or mid Tertiary. This is consistent with a study in the Moray Firth area of the GGF, by Underhill & Brodie (1993), who infer 'post-Early Cretaceous reactivation of the GGF'. Without a more rigorous basis for correlation we are not able to establish or quantify displacements of these palaeosurfaces across faults; however, several possible displacements could be inferred across the GGF (Fig. 6):

- surface D appears to higher on the west side of the GGF in profile T;
- in profile Y, surface D, on the west, is significantly lower than any possible matching surface on the east side of the GGF;
- in both profiles anomalous modes in the elevation distribution occur immediately adjacent to the GGF (west of the fault in profile T and east of the fault in profile Y). These could represent fault blocks or fault-related erosion surfaces.

Thus, several possible instances of tectonic disruption of palaeosurfaces can be identified. Some of these observations may be consistent with a downthrow to the northwest across the GGF (of *c.* 180 m) suggested by Holgate (1969) from his reconstruction of ancient drainage patterns in the Highlands. However, the true picture appears to be a lot more complex than a simple downthrow to the west. Lateral fault movements producing apparent offsets of inclined palaeosurfaces are also possible. We hope that further work on these data will clarify these inferences. It is, however, already clear that the GGF has had a major influence on palaeosurface development in this area. This could be due to block movement of pre-existing surfaces or due to the creation of new erosional surfaces due to an introduction of new base levels along reactivated fault axes. A similar influence of the ELF is also apparent but here the effects are less extensive and less clear. The SGF and LF do not appear to have had any appreciable effect on palaeosurface development.

Structural geomorphology of the central Highlands

The early model of a uniform tilting of one major surface to the east (Linton 1951) is clearly too simplistic in the light of this dataset. A possible high level surface (*c.* 800 m) gently inclined to the east across the area can be inferred, but lower (younger) surfaces are strongly influenced by the GGF and ELF, and slope towards them both westwards and eastwards. The occurrence of structurally-delineated blocks with different geomorphic expression suggests a strong control of geomorphology by contemporaneous tectonic movements. The two areas of maximum uplift in the Highlands appear to be the Cairngorms in the east and the Western Highlands in the west. A zone of less intensive uplift or erosion (or both) is situated in between, with the GGF zone acting as the main focus of depression or erosion. The causes of uplift are beyond the scope of this discussion but we speculate that the Cairngorm region owes its elevated history to processes of glacial rebound, whereas the Western Highlands uplift is more likely to relate to tectonic and magmatic processes associated with Tertiary volcanism.

The dating of these surfaces is difficult. They are almost certainly older than the Miocene gravels of the Buchan area (McMillan & Merritt 1980) and younger than Cretaceous. They probably relate to, and post-date, the Palaeocene uplift in the west of Scotland. However, precise dating of these morphological features is likely to be problematic due to the paucity of datable residues. The most promising way forward is an integrated interpretation of tectonics and morphology, incorporating onshore and offshore data.

Conclusions

The surfaces we have identified in this study do not differ significantly from those described qualitatively by previous workers (Sissons 1976; Hall 1991). However, using this digital dataset we have uncovered a much more complex distribution of surfaces than previously supposed. Some qualitatively mapped 'erosion surfaces' (e.g. Hall 1991, p. 21) are not evident in our histogram datasets. The results of this analysis substantiate objections raised by Hall (1991) as to the basic assumptions in former reconstructions. Differential movements implied by Hall (1991) are more than likely and some of the tectonic zones have been identified. Major faults have clearly influenced geomorphic evolution of the Scottish Highlands in pre-Quaternary times. The pattern of surface development is much more complicated than has been assumed hitherto and the concept of a large-scale uniform tilting cannot be sustained.

We have identified a possible tectonic event where a fairly extensive elevated surface was interrupted by dissection or block movement focused on two major fault zones, the Great Glen Fault and the Ericht-Laidon Fault. Further work, using the higher resolution dataset, is needed to improve our model of geomorphic evolution of this area of upland terrain. We have also shown how digital topographic data can contribute to geomorphological analysis and be used to quantitatively evaluate previous, more subjective, geomorphological models.

We would particularly like to thank Michael Thomas (University of Stirling) for discussions on the etching concept, Gillian Pickup (Heriot-Watt University) for computer programs to manipulate the digital data and Mike Widdowson and two anonymous referees for valuable advice on the manuscript. The work was done with the support of a Royal Society Research Grant.

References

DAVENPORT, C. A., RINGROSE, P. S., BECKER, A., HANCOCK, P, & FENTON, C. 1989. Geological investigations of late and post glacial earthquake activity in Scotland. *In*: GREGERSEN, S. & BASHAM, P. W. (eds), *Earthquakes at North-Atlantic Passive Margins: Neotectonics and Postglacial Rebound,* Kluwer, Dordrecht, 175–194.

FENTON, C. 1992. Postglacial faulting in Scotland: An overview. *In*: FENTON, C. (ed.) *Neotectonics in Scotland A Field Guide.* University of Glasgow, Glasgow, 4–15.

FITZPATRICK, E. A. 1963. Deeply weathered rock in north-east Scotland, its occurrence, age and contribution to the soils. *Journal of Soil Science,* **14**, 33–42.

GALLOWAY, W. E., GARBER, J. L., XIJIN LIU & SLOAN, B. J. 1993. Sequence stratigraphic and depositional framework of the Cenozoic fill, Central and Northern North Sea Basin. *In*: PARKER, J. R. (ed.) *Petroleum Geology of Northwest Europe, Proceedings of the 4th Conference.* The Geological Society, London, 33–43.

GEORGE, T. N. 1966. Geomorphic evolution in Hebridean Scotland. *Scottish Journal of Geology,* **2**, 1–34.

GODARD, A. 1965. *Recherches en Géomorphologie en Écosse de Nord-ouest,* Masson et Cie, Paris.

HALL, A. M. 1986. Deep weathering patterns in north-east Scotland and their geomorphological significance. *Z.eitschrift Für Geomorphologie, N.F.,* **30**, 407–422.

—— 1987. Weathering and relief development in Buchan, Scotland. *In*: GARDINER, V. (ed.) *International Gemorphology 1986 Part II.* John Wiley & Sons, London, 991–1005.

—— 1991. Pre-Quaternary landscape evolution in the Scottish Highlands. *Transactions of the Royal Society, Edinburgh: Earth Sciences,* **82**, 1–26.

—— & MELLOR, A. 1988. The characteristics and significance of deep weathering in the Gaick area, Grampian Highlands, Scotland. *Geografiska Annaler,* **70A(4)**, 309–314.

—— & SUGDEN, D. E. 1987. Limited modification of mid-latitude landscapes by ice-sheets: the case of northeast Scotland. *Earth Surface Proceedings Landforms,* **12**, 531–542.

HOLGATE, N. 1969. Palaeozoic and Tertiary transcurrent movements on the Great Glen fault. *Scottish Journal of Geology*, **5(2)**, 97–139.

LE COEUR, C. 1988. Late Tertiary warping and erosion in Western Scotland. *Geografiska Annaler,* **70A(4)**, 361–367.

LINTON, D. L. 1951. Problems of the Scottish scenery. *Scottish Geographical Magazine,* **67**, 65–85.

—— 1959. Morphological contrasts between eastern and western Scotland *In*: MILLER, R. & WATSON, J. W. (eds) *Geographical Essays in Honour of Alan Ogilvie.* Nelson, Edinburgh, 16–45.

MCMILLAN, A. A. & MERRITT, J. 1980. A reappraisal of the Tertiary deposits of Buchan, Grampian region. *Report of the Institute Geological Science*, **80/**1.

MUIR WOOD, R. 1989. Fifty million years of 'passive margin' deformation in North West Europe. *In*: GREGERSEN, S. & BASHAM, P. W. (eds), *Earthquakes at North-Atlantic Passive Margins: Neotectonics and Postglacial Rebound.* Kluwer, Dordrecht, 7–36.

ORDNANCE SURVEY 1992. *Digital Elevation Data: 1:50 000 scale Elevation Data User Manual.* Ordnance Survey, 1992.

RINGROSE, P. S. 1989. Recent fault movement and palaeoseismicity in western Scotland. *Tectonophysics,* **163**, 305–314.

——, HANCOCK, P., FENTON, C. & DAVENPORT, C. A. 1991. Quaternary tectonic activity in Scotland. *In*: FORSTER, A. *et al.* (eds) *Quaternary Engineering Geology.* Geological Society, London, Engineering Geology Special Publication **7**, 679–686.

ROGERS, D. A., MARSHALL, J. A. E. & ASTIN, T. R. 1989. Short Paper: Devonian and later movements on the Great Glen fault system, Scotland. *Journal of the Geological Society, London,* **146**, 369–372.

SISSONS, B. J. 1976. *Scotland.* Methuen, London.

SUGDEN, D. E. 1968. The selectivity of glacial erosion in the Cairngorm Mountains, Scotland. *Institute of British Geographical Transactions,* **45,** 79– 92.

THOMAS, M. F. 1989. The role of etch processes in landform development. *Z.eitschrift Für Geomorphologie, N.F.,* **33,** 129–142 and 257–274.

TURNER, J. D. & SCRUTTON, R. A. 1993. Subsidence patterns in western margin basins: evidence from the Faeroe–Shetland Basin. *In*: PARKER, J. R. (ed.) *Petroleum Geology of Northwest Europe, Proceedings of the 4th Conference.* Geological Society, London, 975–983.

UNDERHILL, J. R. & BRODIE, J. A. 1993. Structural geology of Easter Ross, Scotland: implications for movement on the Great Glen Fault zone. *Journal of the Geological Society, London,* **150,** 515–527.

A pre-Pliocene or Pliocene land surface in County Galway, Ireland

PETER COXON[1] & CATHERINE COXON[2]

[1] *Department of Geography and* [2] *Environmental Sciences Unit, Trinity College Dublin, Dublin 2, Ireland.*

Abstract: This paper describes a site on the Carboniferous limestone of County Galway, Ireland, where a complex of gorges, cave passages and shallow surface depressions is filled with organic silt and clay overlain by white quartz sand. The dating of the biogenic deposits to the Late Pliocene by biostratigraphical means provides a record of this largely undocumented period of Irish geological history. However, the particular importance of this site is that unlike other Irish karst infills, it represents not just the localized preservation of material in a closed depression but evidence of a more widespread cover of sediments suggesting the preservation of a Pliocene or pre-Pliocene land surface. This implies that glacial action throughout the Pleistocene has resulted in relatively little bedrock erosion in this region and raises the possibility that the present day landscape of the western Irish limestone lowlands may retain influences of preglacial karstification.

Tertiary deposits in Ireland range in type from the very extensive (*c.* 390 000 ha) basalt plateau and associated palaeosurfaces (Interbasaltic Beds) found in the northeast of the country, to the very localized fills or weathering residues within karstic depressions which are found exclusively within areas of Carboniferous limestone (for authorities see references on Table 1 and localities on Fig. 1). The age of the Tertiary deposits, like their geomorphology, is variable and includes the impressive basalts of Early Tertiary age (Palaeocene) of northeastern Ireland and the thick Oligocene infill of the Lough Neagh Basin. Less extensive sediments of Oligocene age have also been found as karstic infills (i.e. Ballymacadam, Table 1), as have younger materials (e.g. the Miocene deposits at Hollymount). The latter two examples are well documented (see references on Table 1) whilst other deposits of possible Tertiary age are less so.

References in the literature to (as yet) undated landforms of possible Tertiary age are numerous (Davies 1970; Mitchell 1980; Davies & Stephens 1978), as are unpublished reports (due principally to mineral exploration activities) of deeply weathered limestone and associated infills. Phreatic passages and extensive cave systems now far above (and below) current water table levels have also been widely commented upon. Some of the recorded geomorphological features are large-scale landforms that are thought to be of considerable antiquity, most probably Tertiary in age (e.g. the area of The Doons in County Sligo which may be an area of 'cockpit karst' Davies & Stephens 1978). The papers by Mitchell (1980, 1985) discuss many of the landscape elements of possible Tertiary origin and these, and other sources, suggest that a great deal remains to be discovered regarding Ireland's Tertiary geomorphology.

As shown on Table 1, records of Tertiary deposits younger than Miocene are rare with only two known localities of Pliocene age: a tentative age accorded to weathering residues at Ballyegan, County Kerry, and the Pliocene material from the area around Pollnahallia, County Galway. The geomorphological and geological contexts of the Pliocene sediments around the vicinity of Pollnahallia were described in detail in an earlier paper (Coxon & Flegg 1987). This paper includes new information on the extent of these deposits, and hence of a Tertiary palaeosurface, and it evaluates the significance of this surface to the landscape history of the region as a whole.

The regional geomorphology

The area between Headford and Tuam (County Galway) described in this paper lies at the heart of an extensive limestone lowland stretching from Co. Clare through east Galway to southeast Mayo, where pure Visean limestone susceptible to karstification is overlain by a thin, patchy cover of glacial deposits. The discontinuous cover of glacigenic sediments includes areas of lodgement till (up to 2–3 m thick), glacially transported rafts of the local bedrock and patches of sand and gravel, the latter forming localized kame topography. While the influence of glaciation means that the area does not have the appearance of a classic karst landscape, features such as caves and swallow

From Widdowson, M. (ed.), 1997, *Palaeosurfaces: Recognition, Reconstruction and Palaeoenvironmental Interpretation*, Geological Society Special Publication No. 120, pp. 37–55.

Fig. 1. Location map of some of the more important Tertiary sites referred to in the text.

holes are found, and the hydrologic system is karstic in nature. The natural surface drainage network is highly fragmented, few of the major rivers having a natural surface outflow to the sea or to Loughs Mask and Corrib; the present integrated network is the result of drainage operations carried out from the mid-nineteenth century onwards, and much of it is a storm run-off system inactive in summer (see Coxon & Drew 1986 for a comparison of natural and present-day drainage networks).

Water from swallow holes and influent reaches of rivers generally drains westwards along lines of high permeability with underground flow velocities of the order of 100–150 m h^{-1} determined by water tracing experiments (documented in Drew & Daly 1993). The karstic nature of the aquifer is also indicated by the high proportion of failed wells (comparable to adjacent areas of poor or non-aquifer) combined with a high proportion of wells with excellent yields and a large number of major springs (Daly 1985). Within the area shown in Figure 3B, groundwater flows southwestwards to a line of springs situated 1–3 km east of the shore of Lough Corrib (Drew & Daly 1993).

Figure 2 is a detailed geomorphological map of the region compiled from aerial photographs, from a field survey by Corcoran and Flegg (in Coxon & Flegg 1987) and from a field survey by Coxon (1986). It shows the distribution of surface depressions, small scarps, bare limestone and superficial deposits in a 200 km^2 area centred on Pollnahallia. Some of the closed depressions marked on Fig. 2 may be irregularities in the glacial drift, but many are of karstic origin. Some of these latter depressions are shallow features but there are also deeper collapse features, e.g. at Pollaturk (Fig. 2) and near Knockmaa (D1 on Fig. 2 and Fig. 11)

A characteristic feature of the western Irish limestone lowland are *turloughs*. These are seasonal lakes which generally contain one or more swallow holes and do not have a natural surface

Table 1. *Tertiary sites in Ireland (see Fig.1 for location)*

Site	Nature of evidence/Age	Publication(s)
Interbasaltic beds, e.g. Giant's Causeway, Craigahulliar and Ballypalady, Co. Antrim, and Washing Bay, Co. Tyrone	Extensive and numerous organic horizons (some blanket peats or peaty soils) lying on weathered basalt surfaces. The palaeobotany of these organic sediments tentatively suggests an Early Palaeocene age	Watts 1962, 1970 (review); Curry *et al.* 1978
Lough Neagh Clays, southern part of Lough Neagh, e.g. Washing Bay, County Tyrone and Thistleborough, Co. Antrim	350 m of predominantly lacustrine and swamp sediments deposited in a large subsiding basin. Although the palaeobotany (like that of the Interbasaltic Beds) requires modern research an Oligocene age (Chattian or Rupelian) can be implied	Watts 1970; Curry *et al.* 1978; Boulter 1980; Wilkinson *et al.* 1980
Ballymacadam, Co. Tipperary	A solution pipe in Carboniferous limestone with an infill including clay sediments rich in biogenic material. The palaeobotany suggests an Oligocene age	Watts, 1970; Boulter & Wilkinson 1977; Curry *et al.* 1978; Boulter 1980
Aughinish Island, Co. Limerick	Deep karstic hollows encountered during site investigations for an aluminium smelting plant. Limited palaeobotanical information implied a Middle Tertiary age	Clark *et al.* 1981; Mitchell 1985
Tynagh, Co. Galway	Altered sulphide ore lying in karstic hollows developed along faults. Authors suggest that a log of *Cupressus* wood may imply a middle Tertiary age	Mitchell 1980; Monaghan & Scannell 1991
Galmoy and Lisheen, Cos Kilkenny and Tipperary	Deep (40 m +) karstification of limestone identified in mineral exploration boreholes. Palynology of organic clays in the depressions suggests a middle Tertiary age	Unpublished reports submitted to exploration companies
Hollymount, Co. Laois	20 m of quartz sand, lignite and other organic sediments infilling a karst solution hole. Detailed palynology indicates a Miocene age	Hayes 1978; Watts 1985
Ballyegan, Co. Kerry	Palaeosols containing gibbsite and quartz are exposed in a limestone quarry capped with glacigenic sediments. The authors suggest the weathering is preglacial (possibly 'Pliocene?' but noting that they recovered no palynomorphs to prove this conclusion)	Battiau-Queney & Saucerotte 1985; Battiau-Queney 1987
Ballygaddy Townland, Co. Offaly	Karst depression (possibly 40 m deep) containing weathered stony clays. A possible Tertiary age was inferred by the authors	Beese *et al.* 1983
Pollnahallia, Co. Galway	Shallow karst depressions, gorges and caves in limestone overlain by organic sediments including lignite. Palynology indicates a Pliocene age	Mitchell 1980; Coxon & Flegg 1987; this paper

outlet. Typically, they fill from groundwater (or in some instances also from surface water) over a few days in the autumn and empty over several weeks in the springtime. Thus, in the winter they are lakes several metres deep, while in summer they are pasture land used for grazing (see Coxon 1986 for a detailed discussion of turlough hydrology and geomorphology). Several turloughs exist in the vicinity of Pollnahallia; both active turloughs and drained turloughs (whose flooding regime has been seriously affected by drainage schemes) are shown in Fig. 2.

The area around Kilwullaun and Pollnahallia

The general geomorphology

Extensive surveys in this local area (Figs 3 and 4) confirmed in detail the karstic nature of the limestone bedrock in the region as a whole. A resistivity survey (Mullen 1981) and a comprehensive drilling programme were carried out in the early 1980s in and around Kilwullaun and Pollnahallia. The location of these boreholes and the results of the resistivity survey are summarized on Fig. 3c and the details are discussed in Coxon & Flegg (1987). The geomorphology of this small area of 3.3 km^2 is very complex and includes large shallow depressions (up to 5 m deep), deeper depressions and fractured limestone (up to 15 m deep), deep gorges (up to 20 m deep) and cave passages (identified in borehole A, Figs 3 and 8) within the limestone. One such gorge, associated with deep depressions and caves in the bedrock, occurs at Pollnahallia and it was recognized from sand-quarrying operations, from a resistivity survey and from borehole data. The form of the sediments lying in this gorge can be seen in Figs 4, 9, and 10. Figure 4 shows the local geology of the sand-pit area depicting the resistivity lows, the distribution of the important geological units, the location of

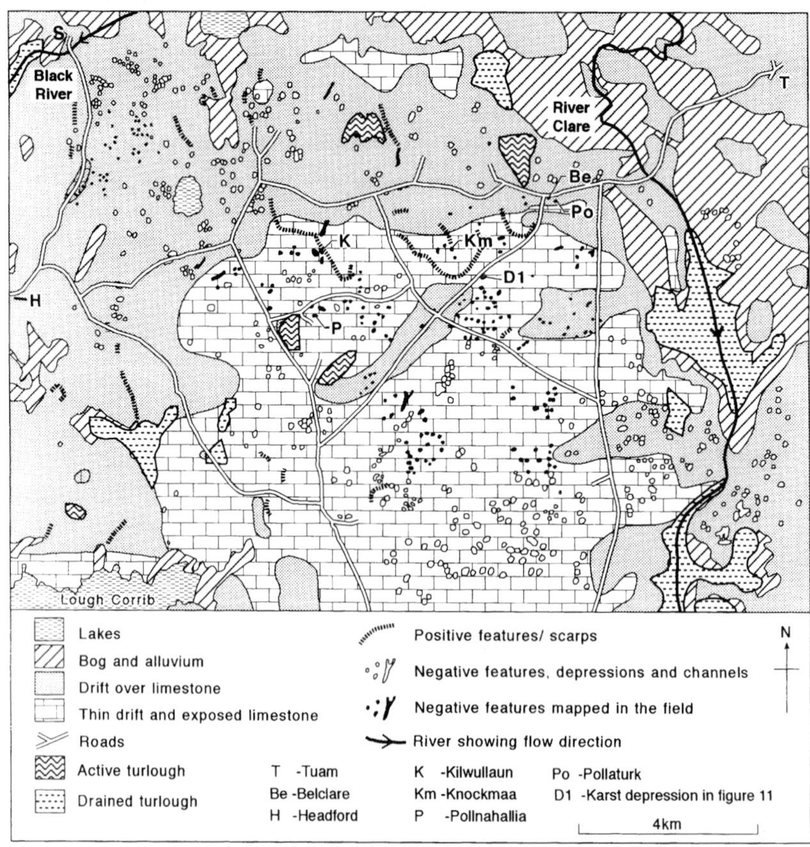

Fig. 2. Geomorphological map of the area between Tuam and Headford [after Corcoran and Flegg in Coxon & Flegg (1987) with some alterations].

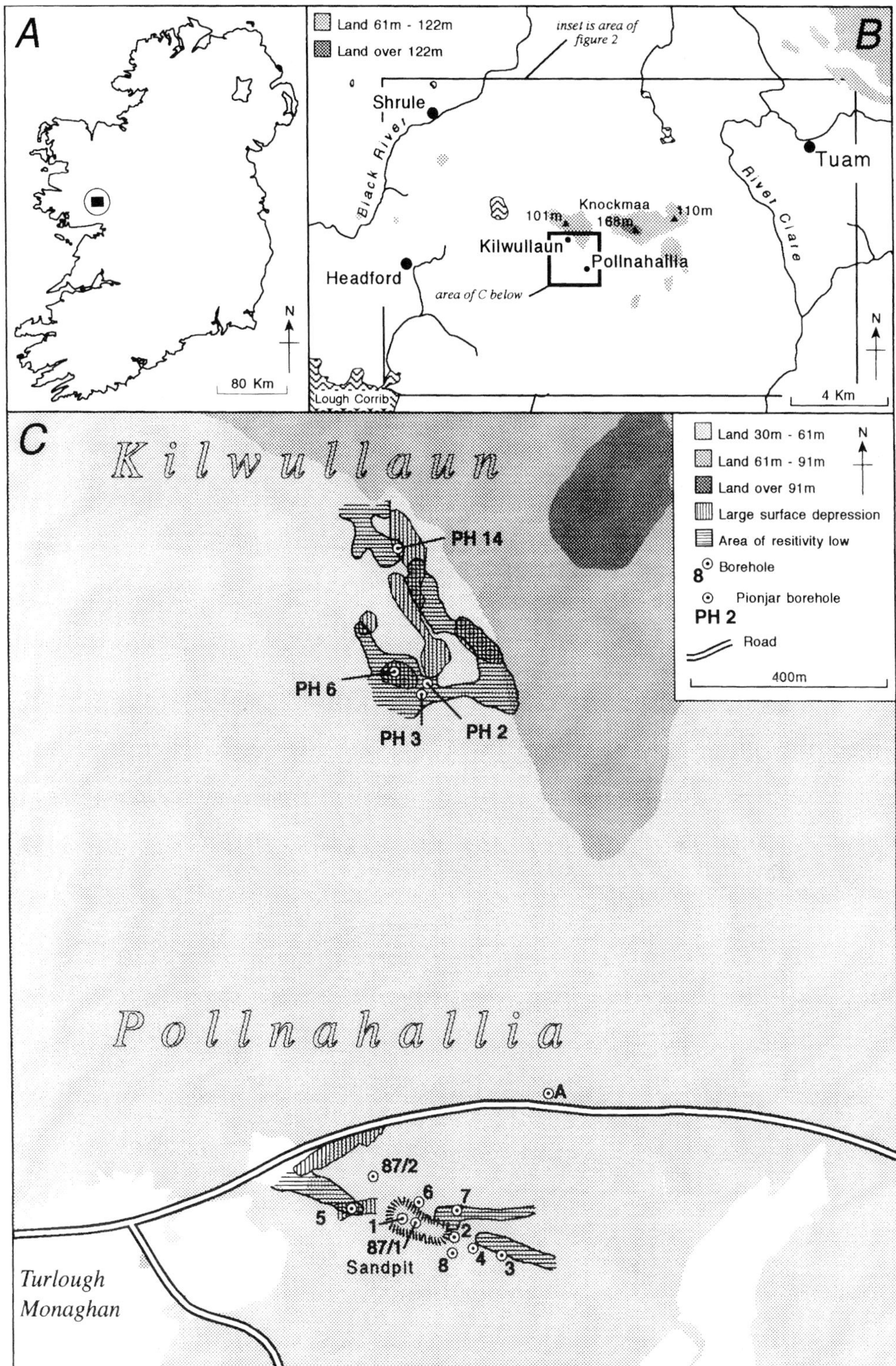

Fig. 3. The location of the Kilwullaun and Pollnahallia areas and the borehole locations and results of the resisitivity survey carried out in the area. A and 1-8 are Geological Survey of Ireland boreholes referred to in Coxon & Flegg (1987). The Pionjar holes are also referred to in that publication. Boreholes 87/1 and 87/2 were drilled with funding from the Royal Society in 1987 (see also Figs 5 and 9).

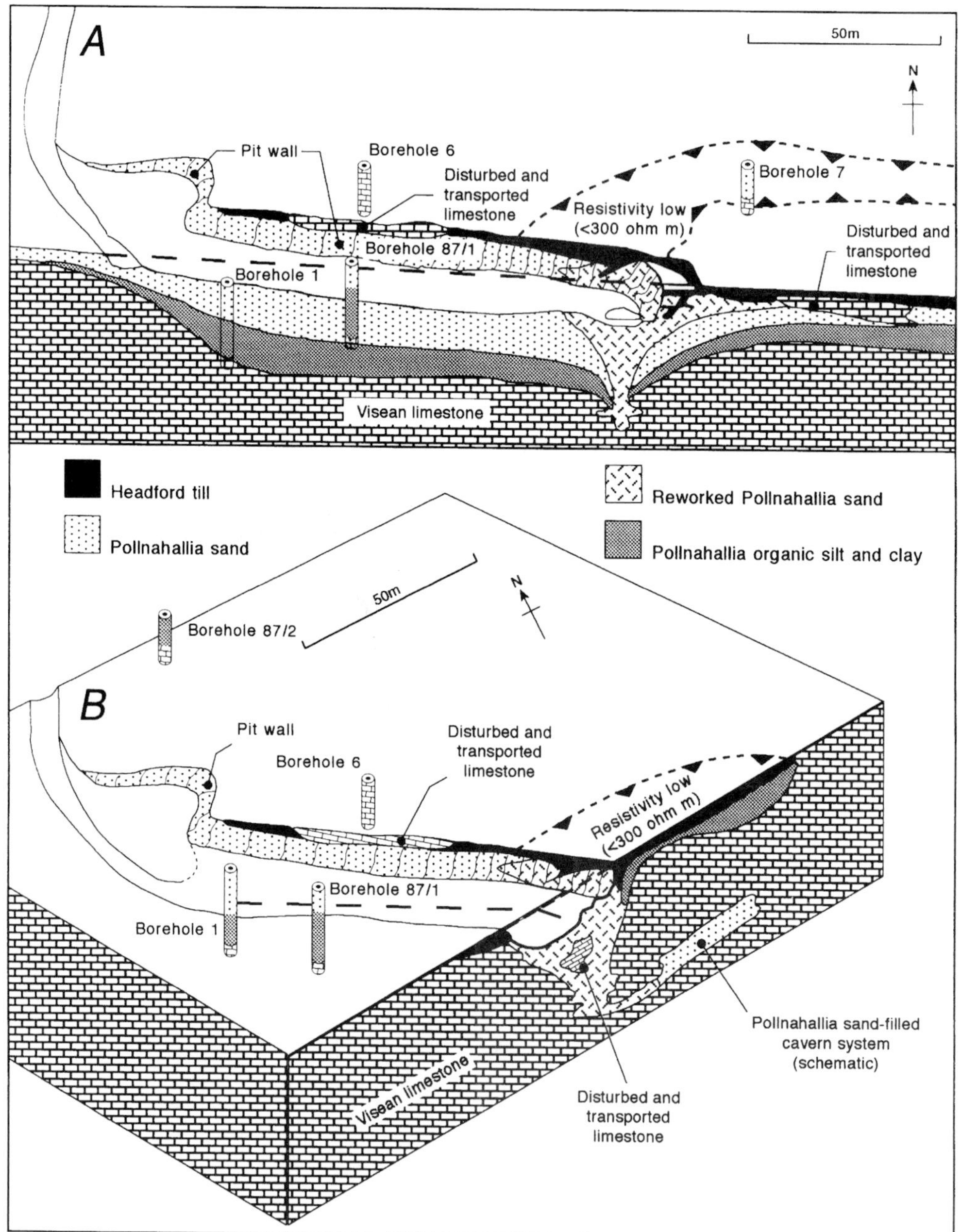

Fig. 4. Schematic cross-section of the deposit at Pollnahallia based on the original surveys (Coxon & Flegg 1987) and on unpublished boreholes carried out late in 1987 with funding from the Royal Society (boreholes 87/1 and 87/2).

boreholes and the reconstructed nature of the surface of the limestone bedrock.

The nature and origin of the deposits lying on the limestone

Several distinct lithological units lying on the limestone were identified and described by Coxon & Flegg (1987) and were classified within an informal lithostratigraphy.

Pollnahallia organic silt and clay. Directly overlying the limestone in both the Kilwullaun and Pollnahallia areas are organic silts and clays containing compressed biogenic material. The extent and nature of these organic materials is variable and ranges from thin black silty clays in shallow depressions in the Kilwullaun area to thicker accumulations exposed at (or near) the surface as at borehole 87/2, and to thick lignite accumulations as recorded in boreholes 1 and 87/1 (Figs 5 and 12). In the shallow depressions and in surface exposures the sediments are primarily yellow, white or dark grey and black silty clays containing organic debris. Much of the organic material from these shallow sites is heavily oxidized. These sediments appear to represent organic sedimentation in wet conditions, in shallow

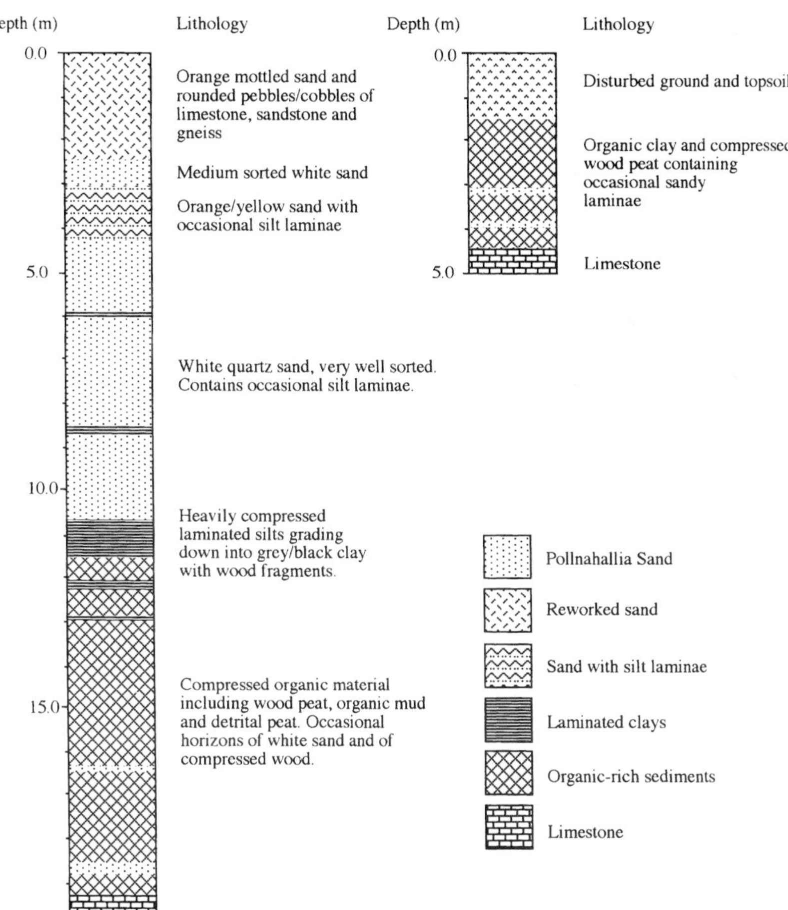

Fig. 5. Borehole logs of the 1987 drilling programme.

depressions on the limestone surface, possibly as soils, blanket peats or in shallow pools.

The continuous core obtained in 1987 (borehole 87/1; Figs 3–5 and 9) has allowed a more detailed analysis of the long organic record from the gorge at Pollnahallia. This core shows the sediment lining the base of the gorge to be predominantly composed of organic material (including woody detritus), with laminated clays and occasional sand horizons becoming more frequent towards the top of the core (Fig. 5). The alternating nature of the sedimentation suggests that the organic detritus in the gorge was deposited in slow moving water with the sandy horizons indicating episodes of faster flow whilst the finer sediments probably represent standing water.

Pollnahallia sand. The Pollnahallia organic silt and clay unit can be seen to coarsen upwards within the gorge sequence where increasingly thick horizons of sand occur (Fig. 5). Eventually, pure white sand overlies the organic sediments. This sand is very-well sorted (inclusive graphic mean = 0.2 mm; Fig. 6) and silica-rich (99%). Sections in the quarry showed large-scale cross-bedding of the white sands and it has been inferred that these sediments represent wind-blown material infilling the gorge. The source of the silica sand is unknown but it is possibly the product of a weathered residue of any of a number of rock types that lie to the west. The most likely source (Coxon & Flegg 1987) are the quartzites of Connemara, found 30 km to the west of Headford. Deeply weathered quartzites can be found in a number of localities in Connemara (see Fig. 13 for an example) and such weathering could have occurred throughout the Tertiary (Mitchell 1980, 1985).

Subsequent reworking of the sands by glacifluvial melt water is apparent in the sections as are shear planes, large injections of the overlying till and rafts of glacially-transported bedrock (Figs 4, 9 and 10). Some of the sections showed evidence of post-depositional subsidence (faulting) indicating that limestone solution and subsequent collapse was also occurring after the sands had been deposited.

Headford till. The till capping the sequence at Pollnahallia (Figs 9 and 10) is a lodgement till containing a strongly preferred orientation of clasts and numerous shear structures suggesting ice movement from the northwest. The till has incorporated large limestone rafts [one is $16 \times 4 \times 20(+)$ m in dimension], and it has been injected down into the underlying sand and has extensively sheared and disturbed the underlying

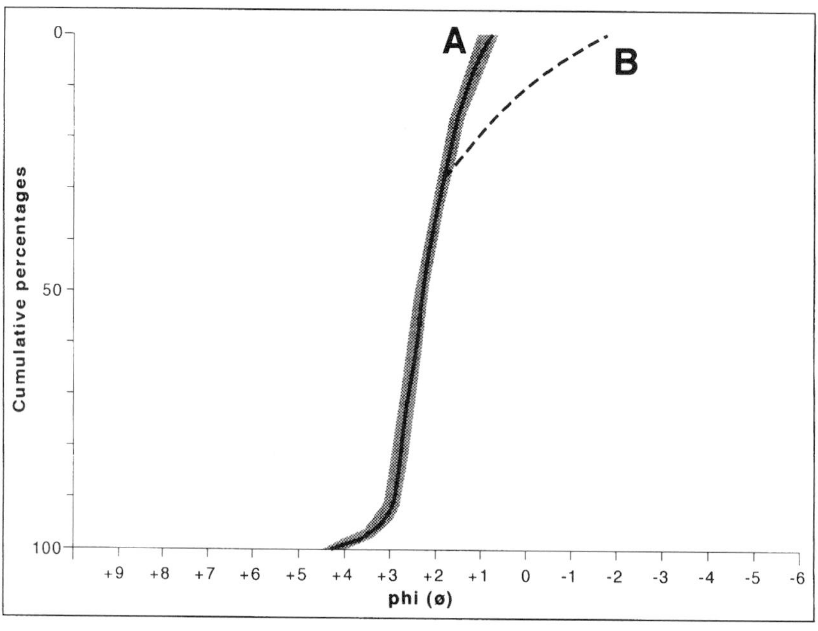

Fig. 6. Particle size analysis curves for the Pollnahallia sand. (**A**) line, average – envelope is for 15 samples. (**B**) Reworked sand, $n = 5$).

deposits in parts of the pit. Till of this type is found smeared in patches across the whole of the area and its age is unknown, although the lack of decalcification suggests that it probably dates from the Midlandian (Last) Glaciation.

The age of the Pollnahallia organic silt and clay

Methodology

Detailed palynological analyses of Geological Survey of Ireland's borehole 1 (Fig. 4) in the Pollnahallia pit and of bagged samples from GSI Pionjar boreholes in the Kilwullaun area suggested that the biogenic material in both areas was probably Pliocene in age (Coxon & Flegg 1987). The pollen samples were treated using standard techniques (Moore *et al.* 1991), and the pollen types recognized follow the schemes of Faegri & Iversen (1975), Birks (1973) and Zagwijn (1960), whilst the tetrads of Ericales were grouped on size criteria. The nomenclature follows Clapham *et al.* (1962) and the percentage sums are of the total terrestrial pollen sum (*P*). Pollen types of taxa less frequently encountered in Irish Quaternary sequences were identified to modern taxa using modern reference material in the Subdepartment of Quaternary Research, Cambridge, and slide material from the southeastern United States made available by W. A. Watts. Material collected by one of us (P.C.) from Late Tertiary (Reuverian) and Early Pleistocene (Tiglian) sites in the Netherlands (Obel Pit and Pit Russel-Tiglian Egypte – see Zagwijn & deJong 1982) was also used for comparative purposes.

The pollen count data are reproduced on a relative abundance diagram (Fig. 7). The percentages used in the diagram and in the text are calculated from the total pollen sum (*P*), which is given on the diagram. Percentages of indeterminables, i/d (indeterminable) bisaccates, aquatics and lower plants are calculated in their own sums [total pollen (*P*) + indeterminables . . .]. Some taxa occur only sporadically and these do not have their own curves. Values of less than 1% *P* for these taxa are represented on the pollen diagram by a '+', and where the percentage representation is 1% *P* or over the value is shown in parentheses after the taxon abbreviation; the pollen morphology and specific identifications of important palynomorphs are discussed at length in Coxon & Flegg (1987).

Pollen assemblage biozones

Although there is not a lot of internal variability in the pollen diagram (Fig. 7) it has been subdivided

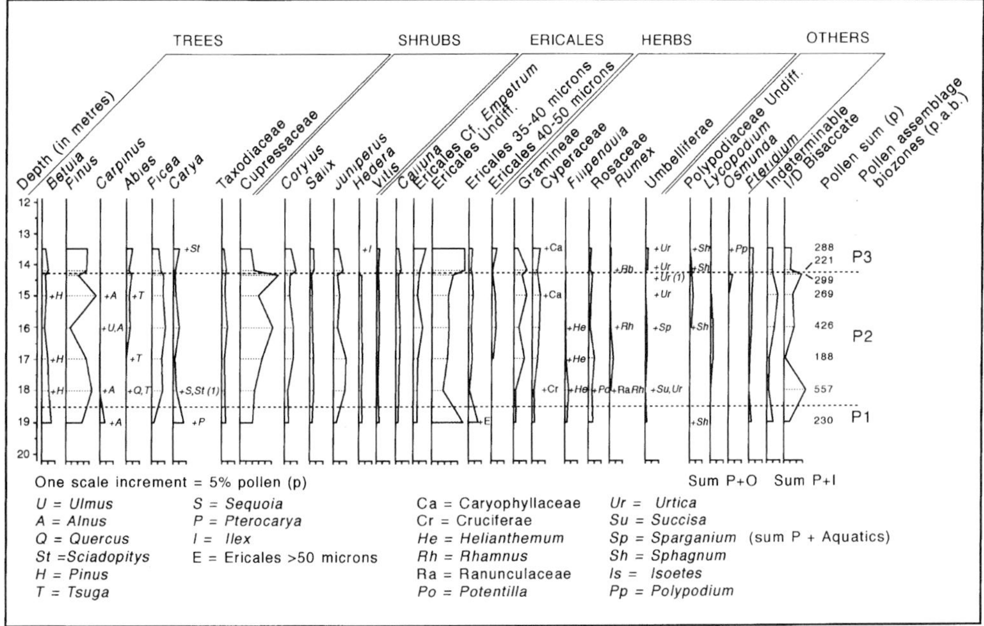

Fig. 7. A percentage pollen diagram from GSI borehole 1 at Pollnahallia (from Coxon & Flegg 1987).

into the following pollen assemblage biozones (pab) to allow description and correlation.

Pollen assemblage biozone P1.
Ericales–*Pinus*–*Carpinus* pab (basal sample below 18.50 m).

This lowermost sample contains dominant percentages of *Pinus* (12.5%) and Ericales (25%) pollen. In this zone *Carpinus* and Ericales (35–40 μm) both produce peaks (7% *P*). A single grain of *Pterocarya* was also recorded from this zone. Other taxa present in significant amounts include *Betula*, *Alnus*, Taxodiaceae, Cupressaceae (9% *P*), *Corylus*, *Salix*, Rosaceae and numerous herb types.

Pollen assemblage biozone P2.
Cupressaceae–Ericales pab (18.50–14.25 m).

The lower boundary of P2 is drawn where pollen of Ericales falls below 15%, *Carpinus* disappears and *Carya* falls to below 5% *P*. The zone is uniform showing only the following minor variations: *Betula* pollen declines throughout, *Pinus* pollen shows a distinct trough (3% *P* from values of over 15% *P*) at 16.00 m, *Abies* first appears at 16.00 m, *Juniperus* is more important earlier in the zone (over 8% *P*, falling to below 2% *P*) and Ericales (40–50 μm) appears at 16.00 m. Other taxa present at significant levels include *Picea*, *Carya*, *Corylus*, *Salix*, *Calluna*, Ericales cf. *Empetrum*, Gramineae, Cyperaceae and a number of herb types. Sporadic or single occurrences of *Quercus*, *Ulmus*, *Alnus*, *Sequoia*, *Sciadopitys*-type, *Tsuga*, *Hedera* and *Vitis* are also note worthy.

Pollen assemblage biozone P3.
Ericales–*Pinus* pab (14.25–13.50 m).

The lower boundary of this zone has been placed at the only point of considerable change in the pollen diagram, i.e. where the percentage representation of Cupressaceae falls (from 29 to 9%) whilst that of *Corylus*, *Salix*, *Juniperus* and Ericales rises. The overall taxa represented are as in P2 with the exception of *Ilex* and *Polypodium* found only in the sample from 13.50 m.

Vegetational history

The pollen diagram from Pollnahallia represents a vegetation cover dominated by ericaceous, cupressaceous and coniferous trees and assorted shrubs. Little can be said regarding the exact nature of the more important elements of the vegetation, beyond that the landscape certainly had a cover of heath associated with *Pinus* and *Picea*. *Abies* appeared in the vegetation by the mid-point of P2. The tree and/or shrub species contributing to the pollen of the Cupressaceae are unknown.

Other trees contributing pollen include *Betula*, *Ulmus*, *Quercus*, *Tsuga*, *Alnus*, *Carpinus* and *Pterocarya* (the latter two taxa only during P1), *Carya*, Taxodiaceae, *Sequoia*, and *Sciadopitys*. This represents a wide range of taxa, possible species and habitats. Many of the (now) exotic taxa are only present at background levels, possibly indicating a regional rather than local presence. Some change in forest composition is indicated by the early disappearance of *Carpinus* and *Pterocarya* and the appearance of *Abies* later in P2.

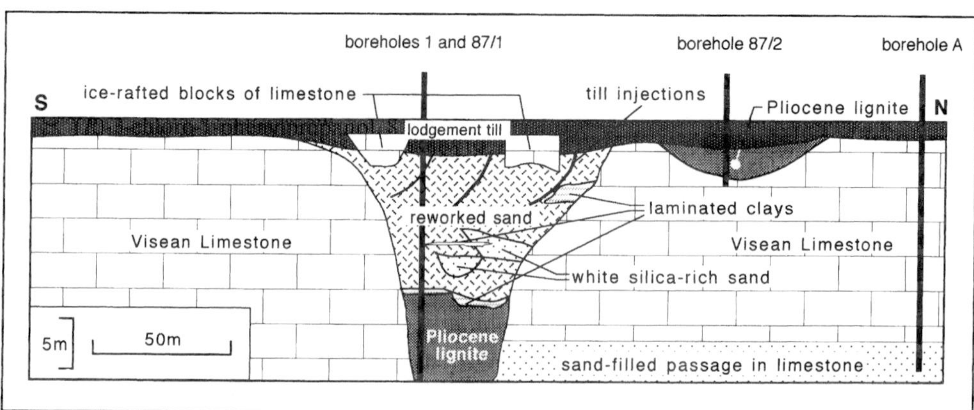

Fig. 8. A schematic cross-section of the deposit at Pollnahallia based on the original surveys (Coxon & Flegg 1987) and on unpublished boreholes carried out late in 1987 with funding from the Royal Society. The borehole locations are shown on Fig. 3.

New palynological information, from borehole 87/1, shows a marked assemblage change at the top of the organic deposit associated with the facies change to laminated clays with sand horizons (at *c.* 10.50–13.00 m in borehole 87/1; Fig. 5). The change involves an increase in taxa indicating climatic deterioration (e.g. the rising values of Ericales and *Juniperus*) and the disappearance of thermophilous taxa. The lithological and palynological changes suggest that the top of the organic sedimentation may represent climatic deterioration at the end of the Pliocene and the beginning of the Pleistocene (see correlation below).

Samples taken for pollen analysis from the thin clays and biogenic materials preserved in the shallower limestone depressions contain mostly weathered, degraded and crumpled pollen due to oxidation of these thinner sequences. However, preserved palynomorphs (e.g. samples from borehole 87/2, Figs 3–5) that contain well-preserved material and from PH 6 (a locality sampled with a Pionjar) from Kilwullaun (Fig. 3; details in Coxon & Flegg 1987) confirm a correlation to the thicker sequences in the Pollnahallia gorge.

The vegetation cover providing pollen during the deposition of the organic sediments at Pollnahallia was clearly diverse, with coniferous trees and heath dominant, but with many other tree species growing regionally. Areas of shrubs and open ground were also present, adding further to the diversity. This variation in vegetation cover may represent the range of available habitats in an area with open limestone topography interspersed with wet, sheltered gorges. Most of the taxa for which there are specific enough identifications can grow successfully today in Ireland.

Biostratigraphic correlation

Dating the organic sediments at Pollnahallia is difficult as there is a lack of a biostratigraphic framework within the Tertiary and Quaternary outside of the Middle Pleistocene and Holocene (Coxon 1993). The correlation of Tertiary deposits (especially those of Neogene age) is equally, if not more, difficult. Many of the palynomorphs recorded at Pollnahallia (e.g. Taxodiaceae, *Tsuga*, *Carya* and *Sequoia*) are known to be indicative of the Late Tertiary in northwestern Europe. In Ireland, although there are deposits of Oligocene age at Lough Neagh (Watts 1970; Wilkinson *et al.* 1980) and Ballymacadam (Watts 1957, 1970; Boulter & Wilkinson 1977; Boulter 1980), there is only one Irish site of Neogene age available for comparison and that is at Hollymount, Co. Laois, 153 km southeast of Pollnahallia (Burgess, pers. comm.; Hayes 1978; Boulter 1980; Watts 1985).

The 20 m of sediment covered by till at Hollymount occupy a closed depression in limestone. Hayes (1978) identified a pollen assemblage listed by Watts (1985) as being predominantly composed of *Pinus*, *Quercus*, *Corylus*, *Myrica* and Ericales. Also present at Hollymount were the pollen of *Taxodium*-type, *Symplocos*, *Tsuga*, *Sciadopitys*, *Liquidambar* and *Palmae*-type. Boulter (1980) lists the taxa from Hollymount as form-genera and suggests an Upper Pliocene age, whilst Watts (1985) comments that 'The flora resembles Miocene or earliest Pliocene assemblages from northwest Europe described by van der Hammen *et al.* (1971)'. The Pollnahallia assemblages differ from the Hollymount ones in the following respects: (1) Cupressaceae, *Abies* and *Picea* are important at Pollnahallia; (2) *Symplocos*, *Liquidambar* and *Palmae*-type are not recorded at Pollnahallia.

The nearest Neogene sites with a reported palaeobotanical record are those described from the 'Pocket Deposits' of Derbyshire, 475 km east of Pollnahallia (Ford & King 1969; Boulter & Chaloner 1970; Boulter 1971; Boulter *et al.* 1971; Ford 1972). Detailed palynological analyses of two fossiliferous beds below glacial till and filling enclosed depressions in the limestone have been described from sites in Derbyshire at Bee's Nest pit, Brassington and Kenslow Top pit, Friden (Boulter 1971). The geomorphological location of Pollnahallia and these Derbyshire sites is similar and the palaeobotanical evidence provides a useful comparison.

Boulter (1971) describes the palynological evidence from the Derbyshire sites in some detail recording two distinct floristic assemblages. The first (Kenslow) has an important ericaceous element along with woodland taxa and the second (Bee's Nest) has an absence of heathland taxa and lower values of spores and triporate palynomorphs.

The Pollnahallia assemblages differ markedly from both of the Derbyshire assemblages in that the former contain high levels of Cupressaceae and do not contain *Symplocos*, *Tricolpopollenites* -types, *Cedrus*, *Cryptomeria*-type, *Podocarpus* -type or high percentages of *Tsuga* and *Sciadopitys* (the latter two taxa being rare at Pollnahallia). The pollen diagram from Kenslow Top pit interestingly contains high levels of Ericaceae pollen and is similar in that respect to the assemblages from Pollnahallia. This similarity may reflect open heathland on sandy soils being present in both environments during the deposition of the organic sediments.

The presence of both Miocene and Pliocene elements in the flora from the Derbyshire material

Fig. 9. The drilling of borehole 87/1. The elongate sand pit in the Pollnahallia sand can be seen as can the sand infill and injections of Headford till. A large raft of limestone bedrock can be seen in the section behind the quarry machinery.

allowed the material to be assigned to the Miocene–Pliocene boundary by Boulter (1971), Boulter & Chaloner (1970) and Curry *et al.* (1978). More complete palynological records for the Late Tertiary and Early Pleistocene are found in the Netherlands (*c.* 1000 km to the east–southeast of Pollnahallia) and most correlations from the British Isles have been made to these type areas (e.g. Gibbard *et al.* 1991). In the Netherlands extensive deposits of Neogene and Early Pleistocene age have been carefully mapped and their pollen content systematically analysed (e.g. Zagwijn 1960, 1975, 1985; van der Hammen *et al.* 1971) to produce a biostratigraphic framework. The most notable similarities between the Pollnahallia pollen assemblages and those recorded in the Netherlands are found in the Late Pliocene (Reuverian) and Early Pleistocene (Tiglian) of the latter.

The pollen assemblages representative of the Middle Pliocene (Brunsummian) contain marked differences from later deposits and include Miocene relics such as *Glyptostrobus*, Palm-type, *Spirematospermum*, *Engelhardtia* and *Symplocos* (Zagwijn 1960; van der Hammen et al 1971). The Late Pliocene (Reuverian) is characterized by *Sequoia*, *Taxodium*, *Sciadopitys*, *Nyssa*, *Liquidambar*, *Aesculus*, *Carya*, *Pterocarya*, *Tsuga* and *Eucommia*, whilst the Tiglian, which follows the cold conditions prevalent at the onset of the Pleistocene (in the Praetiglian), has much lower values of some of the Tertiary tree types, notably *Sequoia*, *Sciadopitys*, *Taxodium*, *Nyssa* and *Liquidambar*. During the Tiglian, *Carya*, *Pterocarya* and *Tsuga* [Zagwijn's (1960) 'Early Pleistocene'-types] are more important than the previous 'Pliocene group'.

The Pollnahallia assemblages can be compared to those from the Reuverian and Tiglian in the following manner.

Similar characteristics of Pollnahallia assemblages to those from Reuverian deposits.

1. Presence of Tertiary types including *Sequoia*, Taxodiaceae, *Sciadopitys*, *Carya*, and *Pterocarya* – although some of these taxa are recorded in low percentages at Pollnahallia the levels of *Carya* and Taxodiaceae are significant;
2. Presence of Cupressaceae pollen and *Pinus*-haploxylon type.

Similar characteristics of Pollnahallia assemblages to those from Tiglian deposits:

1. Presence of Ericales pollen in high percentages;
2. Low values of some Tertiary types, e.g. *Sequoia*, *Sciadopitys*, *Pterocarya* and *Tsuga*; low values of *Ulmus* and *Quercus*;
4. Presence of a continuous curve for *Corylus*;
5. Presence of *Picea* and *Abies* in significant amounts;
6. Absence of *Aesculus*, *Eucommia*, *Nyssa* and *Liquidambar* at Pollnahallia.

There are many problems involved in making long-distance correlations between the Netherlands and western Ireland. Among these are the difficulties in assigning importance to, for example, the absence of mixed oak forest types at Pollnahallia and the high Ericales pollen percentages recorded. Can such characteristics be used in correlation when the climate and habitat availability in western Ireland in the past were probably as distinctly different from those of the Netherlands as they are at the present time? The organic infill of the gorge at Pollnahallia was deposited in a limestone landscape situated on the western edge of Europe. The Late Tertiary and Early Pleistocene sediments described above from the Netherlands are predominantly fluviatile (floodplain, overbank and swamp) sediments deposited by very large rivers on the margin of continental Europe. Biostratigraphic correlations between two such different depositional and climatic environments must be made with caution.

Accepting that the flora history of Ireland is likely to differ from that of nearby countries (the differences between Britain, neighbouring continental Europe and Ireland during the Holocene are slight but that is not to say that such differences have necessarily always been small), certain elements in the floral record can still be useful in attempting a correlation. The most important taxa in this respect are the Tertiary tree types found in the 'Pliocene group' – *Sequoia*, Taxodiaceae, *Sciadopitys*, *Carya*, and *Pterocarya*—which are present at Pollnahallia. Although some were rare, others were an important part of the assemblage. These taxa suggest a Late Pliocene age for the Pollnahallia organic silt and clay. The low percentages of some types and the absence of key elements of the 'Pliocene group' (*Aesculus*, *Eucommia*, *Nyssa* and *Liquidambar* at Pollnahallia) may be explained by habitat and climatic differences, but there is also the

Fig. 10. The Pollnahallia sand and a till injection at the eastern end of the sand pit. The overlying Headford till can be recognized.

Fig. 11. A karst depression in Visean Limestone labelled D1 on Fig. 2.

possibility that the Pollnahallia deposits are Early Pleistocene in age. The survival of tree types into the Pleistocene should be similar in Ireland and the Netherlands assuming that the climatic deterioration of the Praetiglian affected both countries.

In conclusion, the important biostratigraphical elements of the pollen diagram, Fig. 7, (and of the individual samples from Pollnahallia and Kilwullaun) include the presence of typical Late Tertiary taxa; e.g. *Sequoia*, Taxodiaceae, *Sciadopitys*, *Carya* and *Pterocarya* . Such taxa are frequently found in Pliocene deposits in the Netherlands (Zagwijn 1960) and this, the absence of pre-Pliocene marker taxa and the apparent climatic deterioration recorded in the upper part of the sequence allows a probable correlation to be made to the Reuverian of the Netherlands (Coxon 1993), possibly Reuverian C.

The age of the limestone landscape

From the geomorphological and sedimentological investigations described above, it can be concluded that a complex system of gorges, depressions and enclosed passages in limestone exists in the area around Pollnahallia and Kilwullaun, and that dateable organic material has been found both in the base of a gorge at Pollnahallia and in shallow surface depressions on the limestone at Pollnahallia and Kilwullaun. This finding is of considerable geomorphological significance.

Firstly, the approximate dating of the organic-rich silt and clay by biostratigraphical means to the Late Pliocene or Early Pleistocene implies that the limestone surface underlying the deposits is at least of this age. Thus, karstification of the limestone of the area must have taken place before the Late Tertiary. However, karstification of Carboniferous limestone during the Tertiary Period has already been established at a number of sites, as outlined in the Introduction, and what is unique about the Pollnahallia/Kilwullaun site is that, unlike the other Irish karst infills, it represents not just the localized preservation of biogenic Pliocene material in karstic depressions but a more widespread cover of Tertiary sediments suggesting the preservation of a surface that is Pliocene or pre-Pliocene in age.

Several old, complex, partially unblocked cave passages of presumed preglacial or interglacial age occur on the western Irish limestone lowlands [e.g. Ballyglunin Cave, 13 km east–southeast of Pollnahallia, described by Drew (1973), and the Gort river caves in south county Galway – see Farr

1984]. The dating of the sand-filled cave passage at Pollnahallia to pre-Pleistocene times gives strength to the argument for a preglacial origin of other passages. Drew (1973) suggests five stages in the development of Ballyglunin cave, placing its origin at least prior to the last glacial, and possibly much earlier, and a pre-glacial origin is clearly a possibility.

Furthermore, the fact that surface as well as subterranean features at Pollnahallia appear to have survived the Pleistocene glaciations gives rise to the possibility that other surface landscape features of the western Irish limestone lowlands may retain a preglacial influence. In this context, the origin of turloughs is of interest.

The turloughs are one of the most distinctive features of the limestone lowland, but their origin is not yet firmly established. Williams (1964) envisaged them as glacial erosional and depositional features – many of them simply hollows in the glacial drift – which developed a karstic function postglacially. However, more recent work (Coxon 1986) suggests that although the turlough shape and extent is often determined by glacial deposition, actual bedrock hollows are present and an alternative possibility to a glacial erosional origin is a solutional one. The depressions may originally have been single closed depressions such as dolines or cockpits, or more complex forms such as uvalas, which were subsequently modified by glacial erosion and deposition. Given that a well-developed network of closed depressions has not had time to develop in karstic areas of Ireland since the last glaciation, interglacial solutional processes may not have been prolonged enough to do more than reactivate and modify existing karst landforms, so if karstic processes are to be invoked this is likely to place the origin of turloughs in preglacial time. The evidence for a remnant pre-Pleistocene land surface at Pollnahallia, implying that there has been virtually no bedrock erosion in this area since the Pliocene, is clearly of great significance in this context. Equally, rather than having a postglacial origin, the lines of high permeability in the aquifer associated with the turloughs may represent the re-use of remnants of a subterranean drainage network created by extensive solution during the Tertiary, which is partially clogged by glacial drift and therefore inefficient at coping with high flows, resulting in the expulsion of water under pressure and surface ponding in the turloughs.

Fig. 12. Pliocene lignite exposed in a shallow trench. Here the Pollnahallia organic silts and clays lie just below the modern soil and on an irregular limestone surface. This photograph was taken at the site of borehole 87/2 (Figs 3 and 4).

Fig. 13. Heavily weathered quartzite from near Bangor Erris in the Nephin Beg Range. Similarly weathered bedrock may have provided the source of the Pollnahallia Sand.

It is possible that turloughs are polygenetic features: some may be glacial hollows with post-glacial flow routes, while others may have a more complex history, involving earlier phases of solution. The importance of the Pollnahallia site lies in the fact that it adds credibility to the hypothesis of a long, complex history for these characteristic landforms of the western Irish limestone lowlands.

The finding that the limestone surface over a considerable area (at least 3 km^2) and probably a considerably larger area) was already in its current form by the Early Pleistocene not only provides evidence of the landscape configuration at this time, but also brings into question the efficacy of the Pleistocene glaciations in modifying all of the elements of the Irish landscape. The glacial episode resulting in the deposition of the Headford till is possibly just the most recent in a sequence of glacial events to have affected the area over the last 2 Ma, yet relatively shallow depressions contain organic clays and silt that have apparently survived these events. In addition to depositing till, the ice has apparently had a streamlining effect on local rock protuberances, and it has transported sizeable limestone rafts, but it appears to have eroded very little bedrock in this low lying area of Galway. Identifying such selective glacial modification of the landscape is important when considering the age and development of Ireland's surface.

While any comment on the age of geomorphological features beyond the few square kilometres where the Pollnahallia organic silt and clay is found must be highly speculative, the fact that a fragment of Pliocene or pre-Pliocene surface has survived in this area gives rise to the possibility that the landscape retains some influences from this early period of karstification over a wider area.

In this latter respect it is interesting to compare the region around Pollnahallia with that of southern Derbyshire where the limestone surface contains numerous sand-filled karst depressions or Pocket Deposits (Ford & King 1969; Cox & Harrison 1979; Cox & Bridge 1977; Harrison & Adlam 1984). Some of the Derbyshire Pocket Deposits contain organic materials dated to the Miocene–Pliocene boundary (see above) and are, like Pollnahallia, infilled depressions within limestone that contain sand, clay and organic deposits. The Derbyshire Pocket Deposits are believed to have had a complex history, forming as a sheet of fluviatile sediments (fans) laid down in front of a retreating (Triassic) escarpment (Ford & King 1969; Ford 1972). Contemporaneous and ensuing collapse of the underlying limestone lowered and preserved patches of the fan sediments in protected hollows which were subsequently capped by glacial sediments. The Pliocene sediments at Pollnahallia also appear to have been preserved within karstic depressions in the limestone surface and as such they may have, in part, a similar history to the Miocene–Pliocene fills of Derbyshire. Indeed, Ford & King (1969 p. 65) suggest the existence of gorges within the limestone allowing subaerial water courses to fill with sediment as at Pollnahallia. Such similarity, both in age and geomorphological setting, suggests that widespread mantles of weathered residues draped the limestone surfaces of parts of the British Isles by the Late Tertiary and were preserved within depressions in the limestone or across the surface of the bedrock (as at Pollnahallia) where subsequent erosion was ineffectual.

Conclusions

The dating of the organic sediments at Pollnahallia by biostratigraphical means to the Late Pliocene is of particular interest because these deposits are found not only in gorges and deep depressions but also in shallow depressions in the limestone surface over an area of *c*. 3 km^2, indicating the preservation of a Pliocene or pre-Pliocene land surface in this area. Laminated silts that overlie the organic sediments grade up into widespread sand deposits and the latter appear to represent environmental change, possibly climatic deterioration, at the onset of the Pleistocene. Such a change in environmental conditions led to the mobilization of weathered materials which can be identified in the aeolian silica sand deposits infilling the gorge at Pollnahallia. The general geomorphological context of the Pliocene sediments, their associated Pleistocene cover and examples of the borehole evidence have been schematically summarized on Fig. 8.

In addition to giving an exciting glimpse of the Tertiary landscape, the site has provided important evidence concerning Pleistocene glacial activity in the locality, as the existence of Tertiary sediments at the ground surface implies that minimal glacial erosion has taken place in this area. This opens the possibility that the limestone landscape over a wider area of the western Irish limestone lowlands may retain influences from Tertiary karstification. The discovery of this Tertiary palaeosurface certainly adds credence to the theory that Ireland's surface retains many geomorphological elements inherited from the Tertiary Period.

The authors would like to thank Aubrey Flegg (Geological Survey of Ireland), David Mullen and the Applied Geophysics Unit, University College Galway, for access to unpublished material. P. Burgess is acknowledged for access to unpublished undergraduate thesis work. Funding from the Royal Society for the drilling of boreholes 87/1 and 87/2 is gratefully acknowledged. The comments of the referees greatly assisted the redrafting of this paper.

References

BATTIAU-QUENEY, Y. 1987. Tertiary inheritance in the present landscape of the British Isles (examples from Wales, the Mendip Hills and south-west Ireland). *In*: GARDINER, V. (ed.) *International Geomorphology 1986 Part II*, Wiley Chichester, 979–989.

—— & SAUCEROTTE, M. 1985. Paléosols pré-glaciaires de la carrière de Ballyegan (Co. Kerry, Irlande). *Hommes et Terres du Nord,* **1985** (3), 234–237.

BEESE, A. P., BRÜCK, P. M., FEEHAN, J. & MURPHY, T. 1983. A silica deposit of possible Tertiary age in the Carboniferous Limestone near Birr, county Offaly, Ireland. *Geological Magazine*, **120**, 331–340.

BIRKS, H. J. B. 1973. *Past and present vegetation of the Isle of Skye – a Palaeoecological Study.* Cambridge University Press, Cambridge.

BOULTER, M. C. 1971. A palynological study of two of the Neogene plant beds in Derbyshire. *Bulletin of the British Museum (Natural History) Geology,* **19** (7) 359–410.

—— 1980. Irish Tertiary plant fossils in a European context. *Journal of Earth Sciences Royal Dublin Society,* **3**, 1–11.

—— & CHALONER, W. G. 1970. Neogene fossil plants from Derbyshire (England). *Review of Palaeobotany and Palynology,* **10**, 61–78.

—— & WILKINSON, G. C. 1977. A system of group names for some Tertiary pollen. *Palaeontology*, **20**, 559–79.

——, FORD, T. D., IJTABA, M. & WALSH, P. T. 1971. Brassington Formation: a newly recognised Tertiary formation in the southern Pennines. *Nature*, **231**, 134–136.

CLAPHAM, A. R., TUTIN, T. G. & WARBURG, E. G. 1962. *Flora of the British Isles* 2nd edition. Cambridge University Press, Cambridge.

CLARK, R. G., GUTMANIS, J. C., FURLEY, A. E. & JORDAN, P. G. 1981. Engineering geology for a major industrial complex at Aughinish Island, Co. Limerick, Ireland. *Quarterly Journal of Engineering Geology*, **14**, 231–239.

COX, F. C. & BRIDGE, D. MCC. 1977. *The limestone and dolomite resources of the country around Monyash, Derbyshire: Description of 1:25,000 resource sheet SK 16.* Mineral Assessment Report of the Institute of Geological Sciences, **26**.

—— & HARRISON, D. J. 1979. *The limestone and dolomite resources of the country around Wirksworth, Derbyshire: Description of parts of sheets SK25 and 35.* Mineral Assessment Report of the Institute of Geological Sciences, **47**.

COXON, C. E. 1986. *A study of the hydrology and geomorphology of turloughs.* PhD Thesis, University of Dublin, Trinity College.

—— & DREW, D. P. 1986. Groundwater flow in the lowland limestone aquifer of eastern Co. Galway and eastern Co. Mayo, western Ireland. *In*: PATERSON, K. & SWEETING, M. (eds) *New Directions in Karst.* Geo Books, Norwich, 259–280.

COXON, P. 1993. Irish Pleistocene biostratigraphy. *Irish Journal of Earth Sciences,* **12**, 83–105.

—— & FLEGG, A. M. 1987. A Late Pliocene/Early Pleistocene deposit at Pollnahallia, near Headford, Co. Galway. *Proceedings of the Royal Irish Academy,* **87B**, 15–42.

CURRY, D. ADAMS, C. G., BOULTER, M. C., DILLEY, F. C., EAMES, F. E., FUNNELL, B. M. & WELLS, M. K. 1978. *A Correlation of the Tertiary Rocks of the British Isles.* Geological Society London, Special Report, **12**.

DALY, D. 1985. *Groundwater in Co. Galway, with particular reference to its protection from pollution.* Geological Survey of Ireland, Report to Galway County Council.

DAVIES, G. L. 1970. The enigma of the Irish Tertiary. *In*: STEPHENS, N. & GLASSCOCK, R. E. (eds) *Irish Geographical Studies.* Queen's University, Belfast, 1–16.

—— & STEPHENS, N. 1978. *Ireland.* Methuen London.

DREW, D. P. 1973. Ballyglunin cave, Co. Galway, and the hydrology of the surrounding area. *Irish Geography*, **6**(5), 610–617.

—— & DALY, D. 1993. *Groundwater and karstification in mid-Galway, south Mayo and north Clare.* Geological Survey of Ireland Report Series, RS 93/3 (Groundwater).

FAEGRI, K. & IVERSEN, J. (revised by FAEGRI, K.) 1975. *Textbook of Pollen Analysis,* Blackwell, Oxford.

FARR, M. 1984. The Churn, Gort, Co. Galway. *Cave Diving Group Newsletter*, **70**, 31.

FORD, T. D. 1972. Field Meeting in the Peak District. 11–13 June 1971. *Proceedings of the Geologists' Association.* **83**, 231–236.

—— & KING, R. J. 1969. The origin of the silica sand pockets in the Derbyshire limestone. *Mercian Geologist,* **3**, 51–69.

GIBBARD, P. L., WEST, R. G., ZAGWIJN, W. H. *et al.* 1991. Early and Early Middle Pleistocene correlations in the southern North Sea basin. *Quaternary Science Reviews*, **10**, 23–52.

HAMMEN, T. VAN DER, WIJMSTRA, T. A. & ZAGWIJN, W. H. 1971. The floral record of the Late Cenozoic of Europe. *In*: TEUREKIAN, K. K. (ed.), *Late Cenozoic Glacial Ages.* Yale University Press, New Haven, CT, 391–424.

HARRISON, D. J. & ADLAM, K. A. MCL. 1984. *The limestone and dolomite resources of the Peak District of Derbyshire and Staffordshire. Description of parts of 1:50,000 geological sheets 99, 111, 112, 124 and 125.* Mineral Assessment Report of the British Geological Survey, **144**.

HAYES, F. L. 1978. *Palynological studies in the south-eastern United States, Bermuda and south-east Ireland.* MSc Thesis, University of Dublin, Trinity College.

MITCHELL, G. F. 1980. The search for Tertiary Ireland. *Journal of Earth Sciences Royal Dublin Society*, **3**, 13–33.

—— 1985. The Preglacial landscape. *In*: EDWARDS, K. J. & WARREN, W. P. (eds) *The Quaternary History of Ireland.* Academic, London, 17–37.

MONAGHAN, N. T. & SCANNELL, M. J. P. 1991. Fossil cypress wood from Tynagh Mine, Loughrea, Co. Galway. *Irish Naturalist's Journal*, **23**(9), 377–378.

MOORE, P. D., WEBB, J. A. & COLLINSON, M. E. 1991. *Pollen Analysis,* 2nd Edition. Blackwell, Oxford.

MULLEN, D. 1981. *A resistivity survey on karstic limestone.* MSc Thesis, Applied Geophysics Unit, University College Galway.

WATTS, W. A. 1957. A Tertiary deposit in County Tipperary. *Scientific Proceedings of the Royal Dublin Society,* **27**, 309–311.

—— 1962. Early Tertiary pollen deposits in Ireland. *Nature,* **193**, 600.

—— 1970 Tertiary and interglacial floras in Ireland. *In*: STEPHENS, N. & GLASSCOCK, R. E. (eds) *Irish Geographical Studies.* Queen's University, Belfast, 17–33.

—— 1985. Quaternary vegetation cycles. *In*: EDWARDS, K. J. & WARREN, W. P. (eds) *The Quaternary History of Ireland.* Academic, London, 155–185.

WILKINSON, G. C., BAZLEY, R. A. B. & BOULTER, M. C. 1980. The geology and palynology of the Oligocene Lough Neagh clays of Northern Ireland. *Journal of the Geological Society,* **137**, 1–11.

WILLIAMS, P. W. 1964. *Aspects of the limestone physiography of parts of counties Clare and Galway, western Ireland.* PhD Thesis, University of Cambridge.

ZAGWIJN, W. H. 1960. Aspects of the Pliocene and Early Pleistocene vegetation in the Netherlands. *Mededelingen Geologie Stichting CIII,* **5**, 1–78.

—— 1975. Variations in climate as shown by pollen analysis, especially in the Lower Pleistocene of Europe. *In*: WRIGHT, A. E. & MOSELEY, F. (eds) *Ice*

Ages: Ancient and Modern. Seel House Press, Liverpool, 137–152.

—— 1985. An outline of the Quaternary stratigraphy of The Netherlands. *Geologie en Mijnbouw,* **64**, 17–24.

—— & DE JONG, J. 1982. Tegelen–Reuver area. *In*: BRYANT, R. H. (ed.) *Quaternary Research Association field guide, Easter Field Meeting 1982, Soesterberg, The Netherlands*. Quaternary Research Association, London, 52–92.

Geochemical trends in Early Tertiary palaeosols from northeast Ireland: a statistical approach to assess element behaviour during weathering

JOHN J. McALISTER & BERNARD J. SMITH
School of Geosciences, The Queens University Belfast, Belfast BT7 1NN, UK

Abstract: Geochemical and statistical analyses (principle components and Cluster analysis) are used to study changes in element concentration in a range of chemically complex palaeosols from northeast Ireland produced under humid tropical conditions. Chemical analyses characterize the samples and statistical techniques show trends in elemental concentration due to chemical weathering processes. Simple ratios allow the concentration of each element to be examined in individual samples and a ternary diagram classifies these palaeosols with respect to their degree of lateritization. Results demonstrate the general value of statistical analyses in grouping chemically complex palaeosols in terms of element enrichment and depletion as a result of chemical weathering. This approach is independent of any preferred mechanism in the process of rock weathering. It also points out that the theoretical behaviour of elements does not follow those determined by statistical analysis. This technique could also be applied to the classification of present-day lateritic soils.

The landscape of northeast Ireland is dominated by plateau basalts and intervening palaeosols which date back to the Tertiary (62–65 Ma) and are best seen in extensive exposures around the margins of the basalt province (Wilson 1972; Smith & McAlister 1986). The basalt series has traditionally been divided into three major components – lower, middle and upper, – although each of these is comprised of numerous individual lava flows. Between the major flows are two interbasaltic palaeosols, developed under subaerial weathering conditions in a humid tropical type climate (Cole *et al.* 1912; Eyles *et al.* 1952; Patterson 1952, 1955; Charlesworth 1953; Montford 1970; Davies & Stephens 1978; Mitchell 1981). Previous research has characterized these palaeosols geochemically and mineralogically (Eyles *et al.* 1952; Charlesworth 1953; McAlister & McGreal 1983; McAlister *et al.* 1984, 1988). This information is important since it reflects the long-term significance of chemical and leaching environments on the parent rock. Previous research has suggested hydrolysis of silicate and aluminosilicate minerals as an important factor during weathering of volcanic rocks and lavas (Wilson 1975; Nahan 1977; Ollier 1984; Nahan *et al.* 1985; Valeton 1994). However, recent research has highlighted the importance of other mechanisms such as microbial action (Wilson & Jones 1983; McFarlane & Heydeman 1984; Eckhardt 1985; Karavaiko 1988; McFarlane *et al.* 1994). Solubility related to pH and Eh conditions dictates whether elements are enriched or depleted at the site of release. Some elements can exist in several oxidation states, for example, iron may be in any one of three. The stability of this element in any one oxidation state will depend on the energy involved in adding or removing electrons. Eh varies with the concentrations of the reacting substances and if H^+ or OH^- is involved, Eh varies with pH of the solution. As pH increases Eh becomes lower and so oxidation proceeds more rapidly in alkaline solution (Ollier 1984).

The aim of this study is to examine enrichment and depletion on a purely statistical basis, thereby avoiding preconcieved models and ideas of element behaviour. The principal intention is to demonstrate the value of statistical analysis as a tool for identifying relationships between the concentrations of a wide range of elements during the weathering process. Enrichment and depletion is examined and the aim is to ascertain degrees of interdependence between different elements in terms of their mobility as a result of weathering.

Pearson's product moment correlation coefficients show the existence of correlations between two variables and is denoted by the correlation coefficient, r. Perfect and positive correlation occurs when one variable increases by precisely the same proportion as the other, and a perfect and negative correlation occurs when one variable decreases by precisely the same proportion. If r^2 is considered, then $100 \times r^2\%$ of the variation in the values of the variable Y may be accounted for by the linear relationship with the variable X. A correlation of 0.55 means that 30% of the variation

From Widdowson, M. (ed.), 1997, *Palaeosurfaces: Recognition, Reconstruction and Palaeoenvironmental Interpretation*, Geological Society Special Publication No. 120, pp. 57–65.

of the random variable Y is accounted for by difference in the variable X. (Walpole 1968; Rollinson 1993). Correlated variables were reduced to four main factors using principle components analysis (Le Maitre 1982; Rollinson 1993) and these four factors were accepted as holding the maximum variation in the data. Cluster analysis produces a dendrogram from this data, which divides the samples into groups with respect to their element concentrations and movement (enrichment or depletion).

Enrichment and depletion properties of single elements in individual samples are studied by calculating concenration ratios. This is carried out by dividing element concentrations in the samples by those in the basalt parent rock from the interbasaltic study areas (Wolfenden 1965). Analyses of basalt from other locations in surrounding areas are given in Table 5 so that comparison studies may be carried out. The degree of lateritization in the palaeosols is indicated by plotting SiO_2, Al_2O_3 and Fe_2O_3 concentrations on a ternary diagram (Schellman 1981). Using this method some samples, including basalt, are shown to be kaolinized, the remainder lie in the zones characteristic of weak to moderate lateritization (Fig. 3).

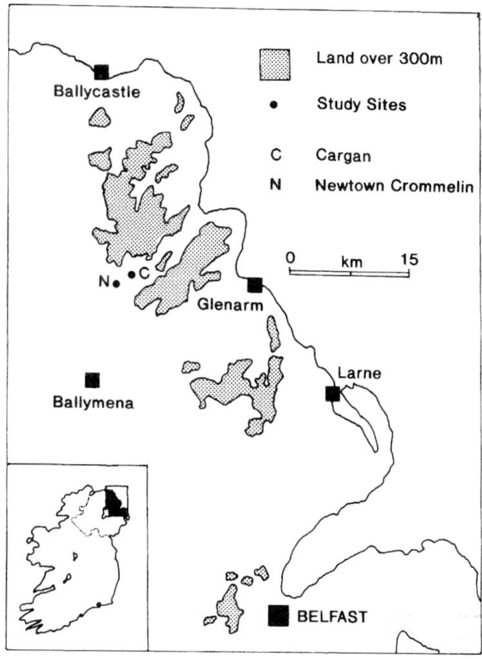

Fig. 1. Location map showing the study sites.

Site areas and samples

Study areas included an exposed audit at Skerry mine, an underground site at Solomons Drift mine (both located near Newtown Crommelin) and a cutting along a low ridge near Cargan, County Antrim (Fig. 1). Outcrops of ore up to 30 cm occur at Skerry mine with a deposit of lithomarge underneath, outcrops of basalt are present in thick layers above the bed. Salmons Drift mine is an underground mine, which consists of pisolitic iron ore and bauxite, rests on an uneven floor of 'pavement'. At Cargan, bauxite occurs as masses of lenticular bands under poor pisolitic iron ore. Samples from exposed sites were selected from different coloured horizons in the profiles. Underground sampling in the mine presented problems and representative samples were collected as conditions allowed. Samples L1–L6 were collected from Skerry mine, L7–L11 from Cargan and L13–L23 from Salmons Drift mine. Sample L12 is basalt from the areas studied since the original parent material was destroyed to produce the profiles.

Sample pre-treatment

Samples were air dried at 30–35°C and those more intensely weathered were gently broken down, mixed and reduced to laboratory size using a riffle box technique (Hesse 1971). Other samples were chipped using a stainless steel jaw crusher and representative portions of all the samples were ground for 60 s in a Tema grinding mill (tungsten–carbide rings), dried overnight at 105°C and stored for XRF analysis. Both pellet and glass disc samples were prepared (Hutchison 1974) and analysed using a Phillips PW14-1020 X-ray fluorescence spectrometer

Results, statistical analysis and discussion

Results of element analysis are shown in Table 1. Standard deviation (S) and mean (X) values were calculated for each element and these were normalized using a 'Z-score' technique:

$$Z_i = (X_i - X)/S$$

X_i = sample concentration, X = mean concentration and S = standard deviation.

This technique transformed all the data to a mean value of zero and a standard deviation of 1.0 and allowed a more meaningful comparison of the various elements to be made. Statistical analysis was carried out on a VAX computer using an SPSS programme package. A significant correlation is achieved if the absolute value is >0.55. Critical

Table 1. *Total element analysis*

	Sample number																						
	L1	L2	L3	L4	L5	L6	L7	L8	L9	L10	L11	L12	L13	L14	L15	L16	L17	L18	L19	L20	L21	L22	L23
%																							
Fe_2O_3	17.3	20.0	14.0	19.4	31.5	24.3	25.7	32.9	22.9	25.1	21.4	10.9	40.0	44.3	42.0	12.9	10.7	7.9	24.7	30.9	34.6	44.3	29.1
Al_2O_3	17.7	20.1	15.1	18.9	41.6	30.4	28.3	32.1	37.8	39.0	45.3	15.5	24.6	22.7	20.8	22.7	18.9	43.5	32.1	24.6	24.5	20.8	28.3
SiO_2	45.3	41.2	43.9	40.6	8.8	28.5	30.0	20.0	14.0	24.2	9.0	49.2	14.0	23.0	23.2	41.8	44.2	31.0	28.0	26.9	23.0	26.9	25.5
TiO_2	2.3	2.04	1.0	1.1	4.23	2.8	2.9	2.9	3.2	3.4	3.1	0.9	9.3	5.4	4.7	1.7	1.1	4.4	3.1	7.7	4.2	4.6	4.7
CaO	11.5	11.6	9.1	9.9	0.25	0.4	0.1	0.3	0.3	0.18	0.11	9.5	0.6	0.6	0.6	1.5	3.0	0.5	0.5	0.8	0.8	0.8	0.8
MgO	7.6	7.5	10.6	11.2	0.17	0.3	0.2	0.3	0.3	0.45	0.18	11.4	0.7	0.5	0.4	4.9	8.8	0.4	0.3	0.5	0.5	0.5	0.6
NaO	2.3	2.2	0.8	0.4	0.13	0.2	0.26	0.2	0.16	0.21	0.13	1.5	0.3	0.3	0.25	0.2	0.16	0.3	0.24	0.3	0.3	0.2	0.3
K_2O	0.15	0.08	0.4	0.16	0.09	0.09	0.09	0.04	0.04	0.04	0.04	0.4	0.06	0.05	0.08	0.05	0.05	0.05	0.02	0.03	0.03	0.02	0.05
MnO	0.18	0.19	0.2	0.18	0.05	0.05	0.2	0.18	0.12	0.19	0.14	0.13	0.05	0.04	0.04	0.15	0.2	0.1	0.1	0.07	0.07	0.06	0.06
P_2O_5	0.2	0.18	0.1	0.08	0.04	0.04	0.13	0.1	0.04	0.16	0.11	0.08	0.05	0.06	0.06	0.25	0.14	0.13	0.2	0.05	0.05	0.12	0.08
H_2O	4.6	6.9	9.4	10.1	19.4	13.1	12.8	11.7	18.6	15.3	22.6	2.6	8.1	6.3	8.8	15.3	14.1	12.2	12.1	10.4	10.9	5.0	10.9
p.p.m																							
Cu	30	20	120	90	90	150	140	100	60	130	50	90	40	80	130	140	230	120	240	200	120	140	150
Cr	30	20	530	550	300	130	400	410	390	320	430	450	300	200	270	580	570	260	300	240	210	120	290
Ni	180	190	600	650	400	1300	900	590	400	700	400	790	140	300	400	1000	600	400	900	300	200	300	300
Zn	160	120	130	110	51	120	130	160	110	100	100	150	50	40	50	150	150	300	310	100	80	80	50
V	279	220	213	236	410	377	339	398	436	411	344	163	387	480	242	309	176	392	589	368	550	427	546
Sr	254	246	732	539	0.0	0.0	0.0	0.0	0.0	0.0	0.0	158	7.0	33	71	0.0	0.0	1.0	42	0.0	42	0.0	43
Rb	4.0	4.0	8.0	9.0	4.0	2.0	2.0	2.0	2.0	2.0	2.0	4.0	4.0	4.0	4.0	5.0	4.0	2.0	3.0	3.0	3.0	4.0	3.0
Nb	3.0	5.0	3.0	5.0	5.0	3.0	1.0	9.0	15	3.0	4.0	4.0	58	33	58	10	7.0	14	33	7.0	59	55	35
Y	32	37	27	22	4.0	4.0	5.0	3.0	4.0	3.0	2.0	19	14	33	7.0	59	16	0.0	1.0	2.0	5.0	6.0	3.0
Ba	64	82	474	360	53	73	51	54	75	61	12	141	287	379	526	36	34	90	358	177	368	91	383

Sum totals are not recorded due to large variations in moisture content.

values of pearson's product moment correlation coefficients for 20 degrees of freedom is 0.537 and in this study values > 0.55 are used to indicate correlations between the elements analysed with respect to enrichment and depletion during the weathering process (Walpole 1968; Ebdon 1981). Positive correlation values between two elements means that they undergo enrichment or depletion together, negative correlation values indicate that as one element increases the other decreases and vice versa.

Elements with significant correlations were tabulated (Table 2) and results show that as iron and aluminium are enriched; magnesium, manganese, silicon, zinc, calcium, rubidium and yttrium are significantly depleted. Calcium, magnesium, potassium and silicon are correlated with vanadium which is enriched, as these elements are depleted. Manganese is depleted as iron, titanium and niobium are enriched. Copper shows no correlation with any of the other elements and chromium and barium are correlated with the depletion and enrichment of nickel and niobium, respectively.

Loadings show relationships between the original elements and the new variables (Table 3). This table shows that component 1 (38.2% of total variation) has high positive loading on silicon, calcium, magnesium, sodium, potassium, strontium, rubidium and yttrium, and high negative loading on iron, alluminium, titanium, vanadium and niobium, and suggests that component 1 was interpretable in terms of enrichment and depletion properties of these elements. Loadings on the first three components were related to enrichment and depletion of the various elements present.

The influence of the original samples on these new components was assessed by calculating standardized component scores. Scores on the first four components were subjected to cluster analysis which resulted in a dendrogram (Fig. 2). Samples are placed in three groups by this statistical technique, with respect to their element concentrations. Samples L1–L4 and L12 are shown in Group 1. Large variations in element enrichment and depletion are observed in the chemically-weathered basalts, which are placed in Groups 2 and 3. (Fig. 2).

Concentration ratios between the samples and basalt are shown in Table 4. Values > 1.0 represent element enrichment and those < 1.0, depletion. Enrichment of iron, aluminium, titanium and vanadium correlates with the depletion of silicon, calcium, magnesium, sodium and potassium. Manganese, strontium and yttrium are enriched in samples L1–L4 and this may be due to depletion of more mobile elements, however, these elements are depleted in the more intensely weathered samples. Isolated cases of enrichment are observed in samples L6 and L16 for chromium and nickel (Table 4). Some minor elements show a much greater and more significant degree of enrichment than other major constituents. Concentration ratios of 3.78 and 2.88 for chromium are observed in

Table 2. *Schematic representation of patterns showing relative enrichment and depletion of constituent elements*

Element		Enrichment ↑ Depletion ↓	
Fe	↑	Ti, Nb Mg, Mn, Zn	↑ ↓
Al	↑	Si, Ca, Mg, Rb, Y	↓
Si	↓	Al, V Y, Ca, Mg, Na, Sr	↑ ↓
Ti	↑	Fe, Nb Mn	↑ ↓
Ca	↓	Al, V Si, Mg, Na, K, Rb, Y, Sr	↑ ↓
Mg	↓	Fe, Ti, V, Al V, Ca, K, Sr, Rb, Si, Mn, Na, Y	↑ ↓
Na	↓	Si, Ca, Mg	↓
K	↓	V Ca, Mg, Sr, Rb	↓
Mn	↓	Fe, Ti, Nb Mg, P	↑ ↓
P	↓	Mn	↓
Cu	↓		
Cr	↓	Ni	↓
Zn	↓	Fe	↑
V	↑	Si, Ca, Mg, K	↓
Sr	↓	Ca, Mg, K, Rb, Si	↓
Rb	↓	Al Ca, Mg, K, Sr, Y	↑ ↓
Nb	↑	Fe, Ti Mn, Ba	↑ ↓
Y	↓	Al Ca, Rb, Si, Mg	↑ ↓
Ba	↓	Nb	↑
Ni	↓	Cr	↓

Table 3. *Element loadings reduced to four main factors using Principle Components Analysis*

Element	Factor			
	1	2	3	4
Fe	*–0.63563	*–0.56356	0.00234	0.18389
Al	*–0.66873	–0.41149	–0.13669	–0.42069
Si	*0.85688	0.00600	0.07737	0.36974
Ti	*–0.60452	*–0.51385	–0.01875	0.01568
Ca	*0.90916	0.25080	–0.19076	–0.08951
Mg	*0.95004	0.14079	0.10653	–0.01645
Na	*0.68148	0.18141	–0.52624	–0.03970
K	*0.67449	0.32095	0.29515	–0.41447
Mn	0.40538	–0.44243	–0.40925	*0.55407
Cu	–0.13880	–0.18604	*0.61512	*0.54994
Cr	0.08838	–0.42578	*0.76642	–0.20668
Ni	0.18236	*–0.62439	*0.63845	0.12661
Zn	0.23646	*–0.59535	–0.10327	0.27788
V	*–0.74470	–0.03196	–0.03791	0.23067
Sr	*0.76689	0.38962	0.19879	–0.16807
Rb	*0.66095	0.51498	0.32842	0.03402
Nb	*–0.54654	*0.61643	–0.00970	0.39928
Y	*0.66129	0.16505	–0.13841	0.29806
Ba	–0.03947	*0.73812	0.37406	0.20567
Eigenxvalue	7.64187	3.66393	2.34119	1.64568
Variation (%)	38.2	18.3	11.7	8.2
Cumulative variation (%)	38.2	56.5	68.2	76.5

* Significant values.

samples L3 and, L6 respectively, and a significant enrichment is shown for zinc in samples L18 and L19.

Concentration ratios reach values of 14.5,14.5 14.75 and 13.75 for niobium in samples L13, L15, L21 and L22, respectively. Minor element enrichment and depletion has also been observed in other studies (McFarlane 1976; McFarlane *et al.* 1994). From a chemical point of view, both valancy and size of the cation are important in weathering.

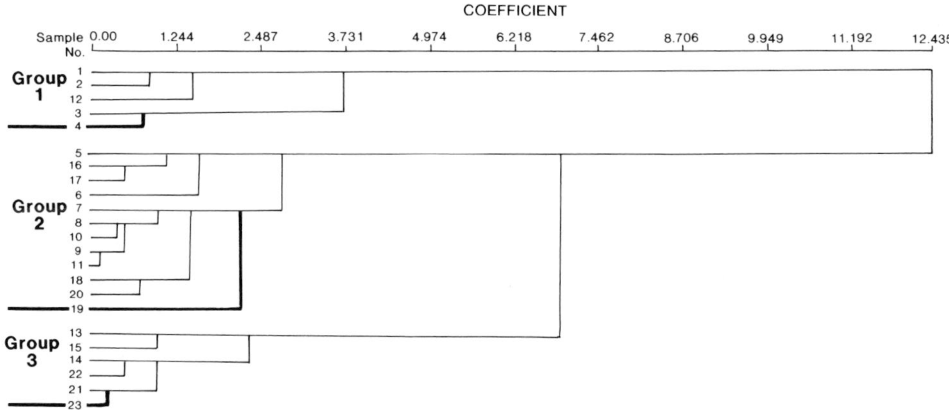

Fig. 2. Dendrogram showing the three sample groups.

Table 4. *Concentration ratios with respect to basalt*

Element	Sample																					Basalt
	L1	L2	L3	L4	L5	L6	L7	L8	L9	L10	L11	L13	L14	L15	L16	L17	L18	L19	L20	L21	L22	L23	L12
Fe	1.61	1.87	1.31	1.76	1.82	2.27	2.40	3.07	1.91	2.34	2.00	3.74	5.29	4.14	1.20	1.00	0.73	2.40	2.98	3.23	5.26	2.78	1.00
Al	1.09	1.34	0.97	1.22	1.34	1.96	1.83	2.07	3.17	2.20	2.93	1.58	1.46	3.78	1.46	1.22	2.81	2.07	1.59	1.58	1.34	1.83	1.00
Si	0.87	0.84	0.86	0.83	0.18	0.53	0.16	0.41	0.29	0.41	0.16	0.49	0.47	0.47	0.85	0.90	0.63	0.57	0.55	0.47	0.55	0.52	1.00
Ti	2.56	2.30	1.08	1.28	4.83	3.18	3.28	3.26	6.07	3.79	3.51	10.49	6.49	5.32	1.92	1.26	4.98	3.45	8.72	4.79	5.25	5.26	1.00
Ca	1.21	1.22	0.95	1.05	0.03	0.04	0.01	0.03	0.03	0.02	0.01	0.06	0.06	0.07	0.16	0.31	0.05	0.05	0.08	0.08	0.07	0.08	1.00
Mg	0.61	0.65	0.92	1.04	0.01	0.03	0.02	0.02	0.03	0.04	0.02	0.06	0.05	0.03	0.43	0.77	0.03	0.02	0.06	0.04	0.04	0.05	1.00
Na	1.53	1.48	0.54	0.27	0.09	.014	0.17	0.13	0.11	0.14	0.09	0.22	0.18	0.17	0.10	0.10	0.18	0.16	0.16	0.20	0.12	0.17	1.00
K	0.60	0.35	1.55	0.65	0.80	0.35	0.15	0.15	0.15	0.15	0.15	0.25	0.20	0.35	0.20	0.20	0.10	0.05	0.10	0.10	0.10	0.02	1.00
Mn	1.40	1.50	1.40	1.40	0.40	0.40	1.50	1.40	1.00	1.50	1.10	0.40	0.40	0.05	1.20	1.50	0.70	0.80	0.70	0.08	0.50	0.50	1.00
P	3.00	2.67	1.33	1.00	0.33	0.67	2.00	1.33	0.67	2.33	1.67	0.67	1.00	1.00	3.67	2.00	2.00	2.67	0.67	0.67	1.67	1.00	1.00
Cu	0.33	0.22	1.33	1.00	1.00	1.67	1.56	1.11	0.67	1.44	0.50	0.44	0.93	1.44	1.56	2.54	1.33	2.67	2.22	1.33	1.36	1.67	1.00
Ni	0.22	0.24	0.46	0.82	0.51	1.64	1.13	0.75	0.51	0.89	0.51	0.17	0.42	0.51	1.27	0.80	0.50	1.13	0.38	0.25	0.38	0.38	1.00
Zn	1.06	0.80	0.87	0.73	0.34	0.80	0.87	1.06	0.73	0.67	0.67	0.33	0.27	0.33	1.00	1.00	2.00	2.06	0.67	0.53	0.53	0.33	1.00
V	1.71	1.35	1.31	1.45	2.52	2.32	2.08	2.44	2.67	2.52	2.11	2.37	2.94	1.48	1.90	1.08	2.40	3.61	2.26	3.37	2.62	3.35	1.00
Sr	1.60	1.56	4.63	3.41	0.00	0.00	0.00	0.00	0.00	0.00	0.00	0.04	0.21	0.45	0.00	0.00	0.01	0.27	0.00	0.27	0.00	0.27	1.00
Rb	1.00	1.00	2.00	2.50	2.00	0.50	0.50	0.50	0.50	0.50	0.50	1.00	1.00	1.00	1.25	1.00	0.50	0.75	0.75	0.75	1.00	0.75	1.00
Nb	0.75	1.25	0.75	1.25	1.25	0.75	0.25	2.22	3.75	0.75	1.00	14.5	8.25	14.5	2.50	1.75	3.50	8.25	1.75	14.8	13.8	8.75	1.00
Y	1.68	1.95	1.42	1.16	0.21	0.21	0.26	0.16	0.21	0.46	0.11	0.74	1.74	0.37	3.11	0.84	0.00	0.05	0.11	0.26	0.32	0.16	1.00
Ba	0.45	0.58	3.36	2.55	0.38	0.52	0.36	0.38	0.53	0.43	0.09	2.04	2.69	3.37	0.25	0.24	0.64	2.54	1.26	2.61	0.65	2.72	1.00
Cr	0.07	0.04	3.78	1.22	0.67	2.88	1.13	0.91	0.86	0.71	0.85	0.67	0.44	0.60	1.29	1.27	0.58	0.67	0.53	0.47	0.27	0.64	1.00

Table 5. *Element analysis of the basalt from three other locations in the interbasaltic area*

Location	SiO_2	Al_2O_3	Fe_2O_3	TiO_2	CaO	MgO	Na_2O	K_2O	MnO
Glenarm	43.6	11.6	14.1	1.5	7.3	15.4	1.4	0.3	0.2
Ballymena	45.2	14.8	11.9	0.9	9.9	11.5	1.9	0.3	0.2
Giant's Causeway	44.5	15.3	13.1	1.9	8.5	9.7	2.2	0.5	0.3
Study site (L 12)	49.2	15.5	10.9	0.9	9.5	11.4	1.5	0.4	0.1

Ionic potential provides a measure of the behaviour of ions towards water. If attraction between a positive ion (Me) and oxygen is weak compared to that between hydrogen and oxygen, Me remains free in solution during weathering. If the two bonds are of comparable strength, the structure forms an insoluble hydroxide. Me—O bond strength should be greatest for ions with high positive charge and small ionic radius. Therefore, ionic potential should be a measure of the tendancy of an ion to remain free to form an anion with oxygen or to precipitate as an hydroxide. There is good agreement when ionic properties of elements at both sides of the periodic table are expressed as functions of geometric quantities, but not for those situated in the middle. The latter elements have a tendancy to form covalent bonds and this distorts the large O^{2-} anion. As a result the Me—O and O—H bonds are no longer a simple matter of charge and radius. Ionic potential is defined as the ratio of ionic charge

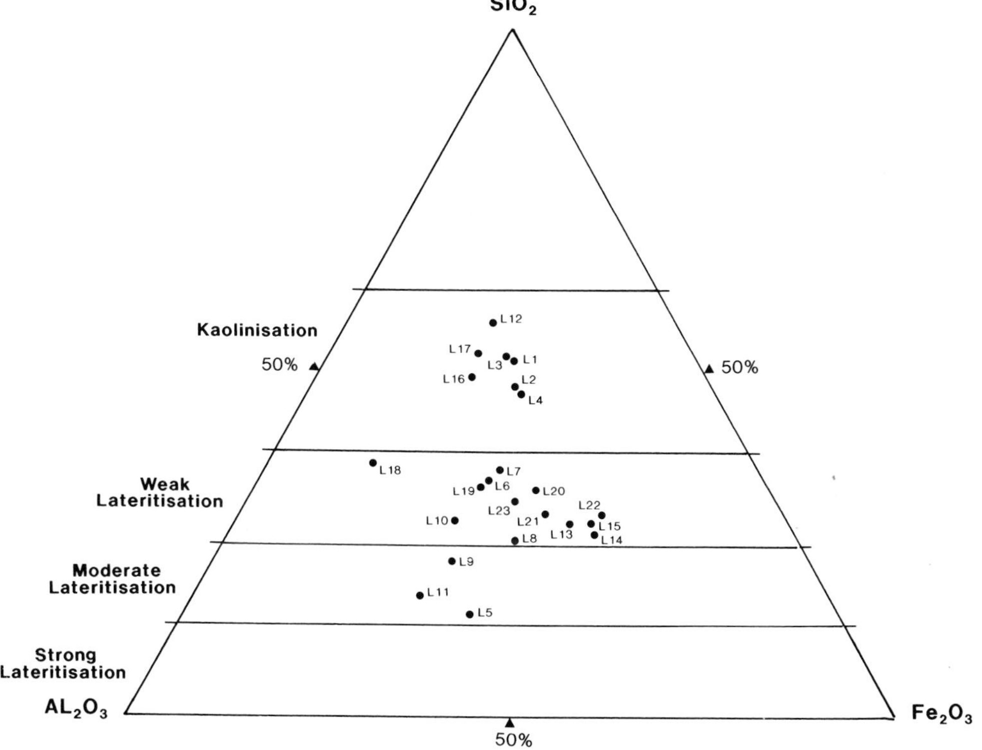

Fig. 3. Sample classification with respect to lateritization (after Schellman 1981).

to ionic radius and subdivides elements into groups of small, highly-charged cations with high field strength (HFS.) and large cations with low charge and low field strength (LFS.). The latter are also known as large ion lithophile elements (LILE) (Goldschmidt 1954; Zeissinki 1965; Bell & Lott 1966; Krauskopf 1967; Shannon 1976; Rollinson 1993).

Element enrichment and depletion in this study show similarities to the concept of ionic potential. Group 1 elements in the ionic potential table match those found in this study that are soluble and easily lost. Group 2 elements match those that are relatively insoluble and become precipitated as hydroxides. Enrichment properties of these elements, found in the samples studied, correlate with those in Group 2 with respect to ionic potential, with the exceptions of nickel, chromium and phosphorus which were enriched in isolated cases only (Table 4). Although there are similarties between enrichment and depletion of the elements in ionic potential Groups 1 and 2 and those found in the samples studied, there is no complete agreement between the theoretical behaviour of these elements, as predicted by their ionic potentials, and their actual distribution. Ionic potential can be used to predict general chemical behaviour of the material and the overall process of weathering. It is inadequate in explaining the chemical behaviour of many ions since some cations in the same group may be more strongly retained by the weathered matrix than others. Generalizations taken from inorganic chemistry may be useful in weathering studies but they must be applied with care. Copper is an example of this, since depletion would be expected for this element during the weathering process due to the value of it's ionic potential. In this study copper is enriched in a number of samples in the weathered profile and shows no correlation with any of the other other elements with respect to it's weathering pattern (Table 4).

Element concentration ratios with respect to basalt allow enrichment and depletion patterns of individual elements to be studied. Results plotted on a ternary diagram for basalt, which follows Schellman's research on laterite classification (Schellman 1981), indicates kaolinization of primary silicates and classifies the more intensely weathered palaeosols as being weakly to moderately laterized. Results would indicate that the statistical methods studied provide a means of combining the concentrations of a wide range of major and minor elements to give an overall pattern of element enrichment and depletion due to chemical weathering. A ternary diagram serves to classify the palaeosols with respect to their degree of lateritization.

Conclusions

Statistical analysis provides information regarding enrichment and depletion of a wide range of elements during the chemical weathering process. This technique also demonstrates that theoretical element enrichment and depletion patterns are not followed. Cluster analysis is used as an approach rather than a model and is independent of any mechanism or process of rock weathering. This approach therefore provides a useful tool for any researcher in the field of rock weathering where the important established concepts may be assessed.

The authors would like to thank the Geochemical, Cartographic and Secretarial staff of the School of Geosciences for preparation of the manuscript.

References

BELL, C. F. & LOTT, K. A. K. 1966. *Modern approach to Inorganic Chemistry*. Butterworths, London.

CHARLESWORTH, J. K. 1953. *The Geology of Ireland. An introduction.* Oliver and Boyd, Edinburgh.

COLE, G. A. J., WILKINSON, S. B., MCHENRY, A., KILROE, J. R., SEYMOUR, H. J. & MOSS, C. E. 1912. *Memoirs of the Geological Survey of Ireland.* Deptartment of Agriculture Technical Instruction for Ireland.

DAVIES, G. L. & STEPHENS, N. 1978. *The Geomorphology of the British Isles: Ireland.* Methuen, London.

EBDON, D. 1981. *Statistics in Geography. A practical approach.* Basil, Blackwell, Oxford, 75–80.

ECKHARDT, F. E. W. 1985. Solublization, transport and deposition of mineral cations by micro-organisms – Efficient rock weathering agents. *In*: DREVER, J. I. (ed.) *The Chemistry of Weathering.* Reidel, Dordrecht, 161–173.

EYLES, V. A., BANNISTER, F. A., BRINDLEY, G. W. & GOODYEAR, J. 1952. *The Composition and origin of the Antrim Laterites and Bauxites*. Memoirs of The Geological Survey, HMSO, Belfast.

GOLDSHMIDT, V. M. 1954. *Geochemistry*. Clarendon, Oxford.

HARMAN, H. 1960. *Introduction to Statistics.* Macmillan, New York.

HESSE, P. R. (ed.) 1971. *A Textbook of Soil Chemical Analysis,* John Murray, London.

HUTCHISON, C. S. 1974. *Laboratory Handbook of Petrographic Techniques.* Wiley, Chichester.

KARAVAIKO, G. I. 1988. Micro-organisms and their significance for biotechnology of metals. *In*: KARAVAIKO, G. I., AGATE, A. D., GROUDER, S. N. & AVAKYAM, Z. A. (eds) *Biotechnology of Metals.* United Nations Environment Programme, Moscow.

KRAUSKOPF, K. B. 1967. *Introduction to Geochemstry. International Series in Earth and Planetary Sciences.* McGraw–Hill, Maidenhead, 593–595.

LE MAITRE, R. W. 1982. *Numerical Petrology: statistical interpretation of geochemical data.* Elsevier, Amsterdam.

LURLEY, D. N. & MAXWELL, A. E. 1963. *Factor Analysis*

as a Statistical Method. Butterworth, London.

McALISTER, J. J. & McGREAL, W. S. 1983. An investigation of deep weathering products from a fossil laterite horizon in Central Antrim, N. Ireland. *Proceedings of the 11 International Seminar on Lateritisation Processes, Sao Paulo,* July, 4-12, 1982, 345–357.

——, —— & WHALLEY, W. B. 1984. The application of X-ray and thermoanalytical techniques to the mineralogical analysis of an interbasaltic horizon. *Microchemical Journal*, **29**, 267–274.

——, SVEHLA, G. & WHALLEY, W. B. 1988. A comparison of various pretreatment and instrumental techniques for the mineralogical characterisation of chemically weathered basalt. *Microchemical Journal,* **38,** 211–231.

McFARLANE, M. J. 1976. *Laterite and Landscape*. Academic, London.

—— & HEYDEMAN, M. T. 1984. Some aspects of kaolinite dissolution by a laterite-indigenous micro-organism. *Geo. Eco. Trop.* **8,** 73–91.

——, BOWDEN, D. J. & GIUSTI, L. 1994. The behaviour of chromium in weathering profiles associated with the African surface in parts of Malawi, *In*: ROBINSON, D. A. & WILLIAMS, R.G. (eds) *Rockweathering and Landform Evolution.* Wiley, Chichester.

MITCHELL, G. F. 1981. Other Tertiary events. *In*: HOLLAND, C. H. (ed.) *The Geology of Ireland.* Scottish Academic Press, Edinburgh, 231–234.

MONTFORD, G. F. 1970. The terrestial environment during the upper Cretaceous and Tertiary times. *Proceedings of the Geologists' Association*, **81**, 181–204.

NAHAN, D. 1977. Time factor in iron crusts genesis. *Catena,* **4,** 249–254.

——, BEAUVAB, A. & TRESCASES, J. 1985. Manganese concentration through chemical weathering of metamorphic rocks under lateritic conditions. *In*: DREVER, J. I. (ed.) *The Chemistry of Weathering.* Reidel, Dordrecht, 161–173.

OLLIER, C. 1984. *In*: CLAYTON, K. M. (ed.) *Weathering.* Longman, London.

PATTERSON, E. M. 1952. A petrochemical study of the Tertiary lavas of northeast Ireland. *Geochimica Cosmochimica Acta*, **2**, 283–299.

—— 1955. The Tertiary lava succession in the northeast part of the Antrim Plateau. *Proceedings of the Royal Irish Academy*, **57B**, 112–1798.

ROLLINSON, H. R. 1993. *Using Geochemical Data: Evaluation, Presentation , Interpretation.* Longman, London.

SCHELLMAN, W. 1981. Consideration on the definition and classification of laterites. *Proceedings of the International Seminar on Lateritisation Processes*, 1–10.

SHANNON, R. D. 1976. Revised effective ionic radii and systematic studies of interatomic distances in halides and chalogenides. *Acta Crystallographica Section A*, **32,** 751–767.

SMITH, B. J. & McALISTER, J. J. 1986. Tertiary weathering environments and products in northeast Ireland. *In*: GARDINER, V. (ed.) *International Geomorphology, Part 11*. 1007–1031. Wiley, Chichester.

VALETON, I. 1994. Element concentration and formation of ore deposits by weathering. *Catena,* **21,** 99–129.

WALPOLE, R. E. 1968. *Introduction to Statistics*. Macmillan, New York.

WILSON, H. E. 1972. *Regional Geology of N. Ireland.* Geological Survey of N. Ireland, HMSO, Belfast.

WILSON, M. J. 1975. Chemical weathering of some primary rock forming minerals. *Soil Science,* **119,** 349–355.

—— & JONES, D. 1983. Lichen weathering of minerals: implications for pedogenesis. *In*: WILSON, R. C. L. (ed.) *Residual Deposits: Surface Related Weathering Processes and Materials.* Geological Society, London, Special Publication, **11**, 5–12.

WOLFENDEN, E. B. 1965. Geochemical behaviour of trace element during bauxite formation in Savawak, Malaysia. *Geochimica et Cosmochimica Acta,* **29**, 1051.

ZEISSINKI, H. E. 1965. The mineralogy and geochemistry of nickeliferrous lateritie profile, Greenvale, Queensland, Australia. *Mineral Deposita, Berlin*, **4**, 132–152.

Palaeosurface palynofloras of the Skye lava field and the age of the British Tertiary volcanic province

DAVID W. JOLLEY

Centre for Palynological Studies, University of Sheffield, Mappin Street, Sheffield S1 3JD, UK

Abstract: Collection and palynological analysis of intratrappean boles, mudstones, sands, conglomerates and coals from the Skye lava field has allowed the reconstruction of geomorphological features and vegetation distribution on five palaeosurfaces. These palaeosurfaces show a changing pattern of streamside, swamp and upland vegetation which fall into three forest types: mixed mesophytic, upland Taxodiaceae and montane conifer forest. Comparisons of the Skye palynofloras with others analysed from the intra and intertrappean beds of the Antrim, Mull and Small Isles lavas, gives evidence of three (possibly four) main phases of British Tertiary volcanic province extrusive activity. In comparison to complete palynological records from the Faeroe–Shetland Basin, Skye and Antrim palynofloras can be dated as ranging from 58.23 Ma to 57.99 Ma, while those of the Mull Lavas, and coals below the Faeroe Islands Middle/Upper Basalts indicate initiation and resumption of extrusion after 55.00 Ma. Evidence of the age of palynofloras from the intratrappean beds of Eigg and Muck is limited, and does not dispute the Chron 27r age suggested by isotopic analysis (*c.* 62 Ma). Finally palynological evidence is presented showing two major subsidence phases, and one uplift phase during the 0.24 Ma existence of the Skye Lava field. It is suggested that altitudes in excess of 1200 m were experienced during thermal doming related to the emplacement of the Cuillin centre. The relative timing of this and other volcanic events is compared to subsidence patterns in the marine sedimentary record.

During the Palaeocene, the Isle of Skye was on the eastern margin of the North Atlantic rift at around 62°N latitude. Active volcanism was a prominent feature of the landscape of the Hebrides area, with lava fields developed in Antrim, Mull, Ardnamurchan, Rum and Skye, forming the British Tertiary volcanic province (BTVP). Further to the north and west, extensive volcanism was taking place along the rifting North Atlantic margin, with deep marine sedimentation in the intervening Rockall–Faeroe Basin. The vegetation of this period was dominated by humid forests growing in a subtropical to warm temperate climate, abundant evidence of which is preserved in the form of pollen, both in deep marine sediments and in the intra and intertrappean sediments of the lava fields. Thin red-brown boles (the term bole is used here as indicating all red-brown intratrappean weathering products, regardless of origin, see Widdowson *et al.* 1997), poor coals, fluvio-lacustrine silts, sands and conglomerates throughout the lava piles provide evidence of fossil land surfaces which mark periods of prolonged erosion, pedogenesis and sedimentation between successive lava outpourings. The current work represents an examination of the palynofloras (pollen, spores and algae) preserved in the varied sediments of these surfaces, principally from the Skye Lava Field with additional material from Antrim, Mull, Muck, Eigg and Rum.

The recovery of palynomorphs from the diverse sediments of the intra and intertrappean beds allows comparisons to be made between the stratigraphical record of the eruptive centres and the complete offshore marine record. The rapid turnover of palynofloras recorded in the offshore record is evidence of significant climatic, evolutionary and migratory trends in the parent vegetation. First order calibration of these and equivalent events to magnetostratigraphy (Ali & Jolley 1996), has allowed accurate dating of volcanic activity in the BTVP with reference to a chronostratigraphical framework.

Because of the varied nature of the lithologies that yielded palynomorphs, the palaeoecology and palaeogeography of the fossil land surfaces represented by the sediments of the Skye lava pile can be reconstructed. The geomorphological reconstruction of these palaeosurfaces provides significant information regarding the duration of the quiescent phases represented by the sediments. These data are used to provide key information regarding the elevation of successive palaeosurfaces in the west central Skye lava field and the fields' subsidence and uplift pattens in reponse to major magmatic body emplacement.

From Widdowson, M. (ed.), 1997, *Palaeosurfaces: Recognition, Reconstruction and Palaeoenvironmental Interpretation,* Geological Society Special Publication No. 120, pp. 67–94

Age of the BTVP: previous work

The ages of the principal igneous complexes of the BTVP have been recently summarized by Mussett *et al.* (1988), Hitchen & Ritchie (1993) and Ritchie & Hitchen (1996). These authors have concentrated on a number of dating methods as the basis for their stratigraphy, placing particular emphasis on K-Ar and Ar-Ar whole rock analysis. In addition, magnetostratigraphy has indicated the dominantly reversed polarity of the BTVP lavas (Mussett *et al.* 1988) restricting the isotope derived ages to Chrons 26r and 24r. These ages, together with the similarity of lava composition between the sequences of Antrim, Mull, Ardnamurchan and Skye have been taken to indicate contemporaneous activity for the Hebridean centres (for example see Hitchen & Ritchie 1993).

The fact that biostratigraphy has played a minor role in the comparative age dating of these BTVP centres is largely due to the lack of availability of suitable regional pollen sequences for the 60–54 Ma period, a factor redressed in this study. With few exceptions, (e.g. Boulter & Kvacek 1989), little previous biostratigraphical information is available for the intra and intertrappean sediments associated with the lavas of the BTVP and the wider (Fig. 1) North Atlantic Igneous Province (NAIP). Micropalaeontological analysis of the intrabasaltic sediments of the East Erland volcano (Ridd 1983) has indicated a Late Cretaceous, Campanian age, while palynology conducted on the exposures in East Greenland (Soper *et al.* 1976*a*; Soper *et al.* 1976*b*) has demonstrated a range of Late Cretaceous to Eocene ages. More recent palynological investigation of the Mull and Ardnamurchan intrabasaltics by Boulter & Manum (1989) and Boulter & Kvacek (1989), followed up earlier studies of these intrabasaltic beds by Simpson (1961) and Srivastava (1975). Previously dated as Miocene or Maastrichtian, Boulter & Kvacek (1989) proposed a Palaeocene age (nannofossil zones NP9/10), and a new floral province, the BIP (Brito-Arctic Igneous Province), to incorporate all similar floras within the NAIP. Boulter & Kvacek (1989) also reported extensive macrofloral remains from the leaf beds of Mull and Ardnamurchan. Similar records of macrofloras have also been made by Anderson & Dunham (1966) from Glen Osdale, Skye and compared by these authors to the Mull macroflora.

Further to the north, a palynological study of the coaly strata between the Lower Basalts and the Middle–Upper Basalts of the Faeroe Islands, was made by Lund (1983, 1988) who suggested a Late Palaeocene age, whilst on the eastern margin of Greenland, cores taken by the ODP have enabled some dating of the sediments immediately overlying the seaward dipping reflector series (SDRS) of basalts. Boulter & Manum (1989) examined a series of cores taken from the Outer Vøring Plataeu Hole 642E, demonstrating that intra Lower Basalt sediments were of latest Palaeocene age. Sediments immediately overlying the SDRS in the Vøring Plataeu and in the Irminger Basin are of mid-Early Eocene age (51–52.5 Ma; Jolley, unpublished data).

Materials

The database for this study comprises a series of field samples and other well sample material from the Hebrides and west Shetland areas. Samples from the Isle of Skye were collected from outcrops of intrabasaltic sediments throughout the lava pile. Careful selection of localities ensured a good stratigraphical and geographical coverage of the lava sequence. While the stratigraphical position of the Palagonite Tuff plant beds at the base of the lava pile (Anderson & Dunham 1966) is not in doubt, the interrelationship of the intrabasaltic sediments in central and northern Skye appears to be confused. One of the most prominent north Skye intrabasaltic localities, in Glen Osdale (NG 230 440) was reported to occur at the boundary between the Ramascaig and Osdale Lava Groups (Anderson & Dunham 1966). Recently, England (1994) recorded the obvious change in basalt type occurring some 250 m higher up Healabhal Mor (NG 220 445) and Healabhal Beag (NG 224 422), questioning the validity of the stratigraphical approach of Anderson & Dunham (1966). Problems concerning the correlation of the intra-basaltics in the north and central area are currently unavoidable; fortunately in west central Skye, detailed mapping of lavas and intrabasaltic sediments by Williamson & Bell (1994) has provided a reliable local stratigraphy. It is the mapping of these boles and other sedimentary horizons that has made practical the recognition of four continuous sedimentary horizons within the west central Skye basalt pile. By regarding these intercalations as a 'type sequence' termed erosion surfaces E2-E5 (Fig. 2), it has proven possible using palynofloral and lithological evidence to attempt correlations to other sedimentary units which occur in the area studied by Anderson & Dunham (1966).

Samples of the oldest sedimentary horizon (E1), the Palagonite Tuff plant beds of eastern and north Skye, where taken from Camas Ban (NG 491 423), Camas Tianavaig (NG 513 389) and Glen Uig (NG 420 636). These beds do not occur in west central Skye (see below), but occur immediately above the basal tuffs of the Palaeocene sequence (Fig. 3). The

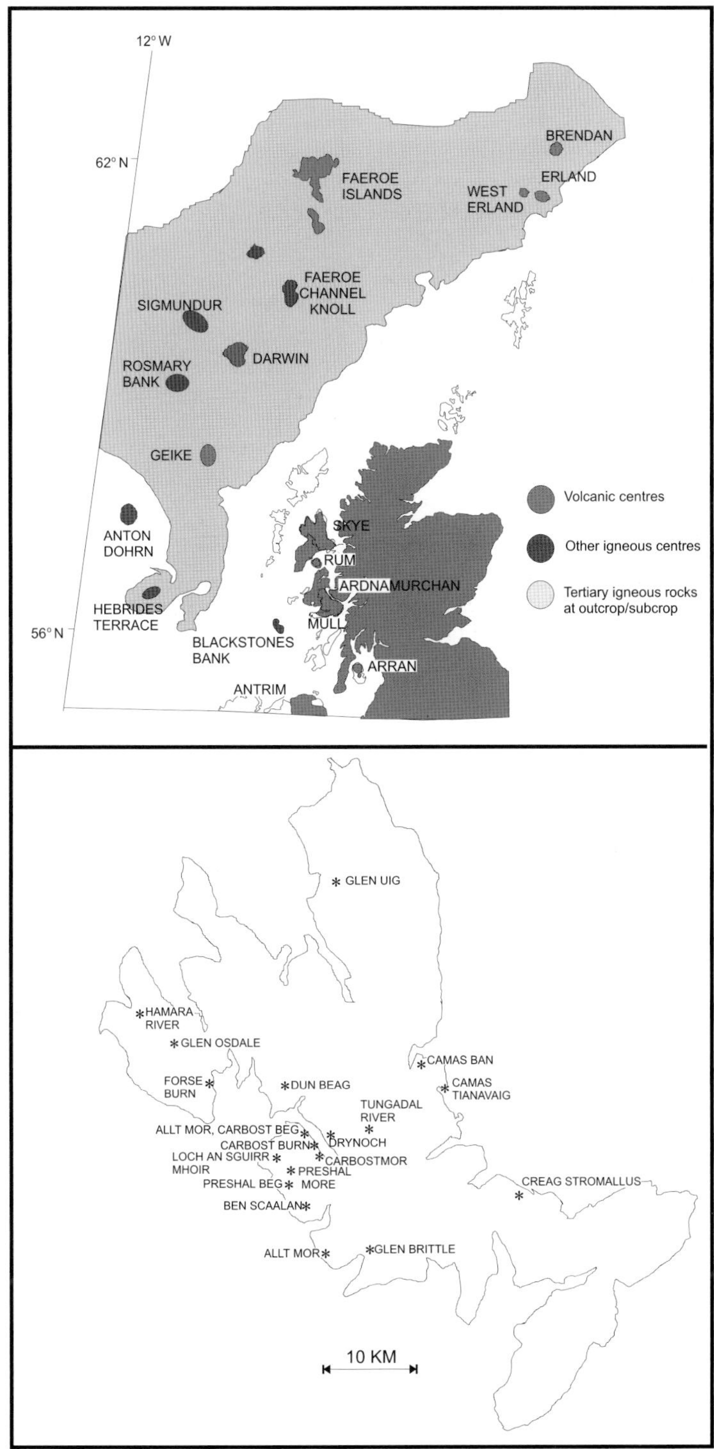

Fig. 1. Map of the main igneous centres in the BTVP and Skye sampling localities.

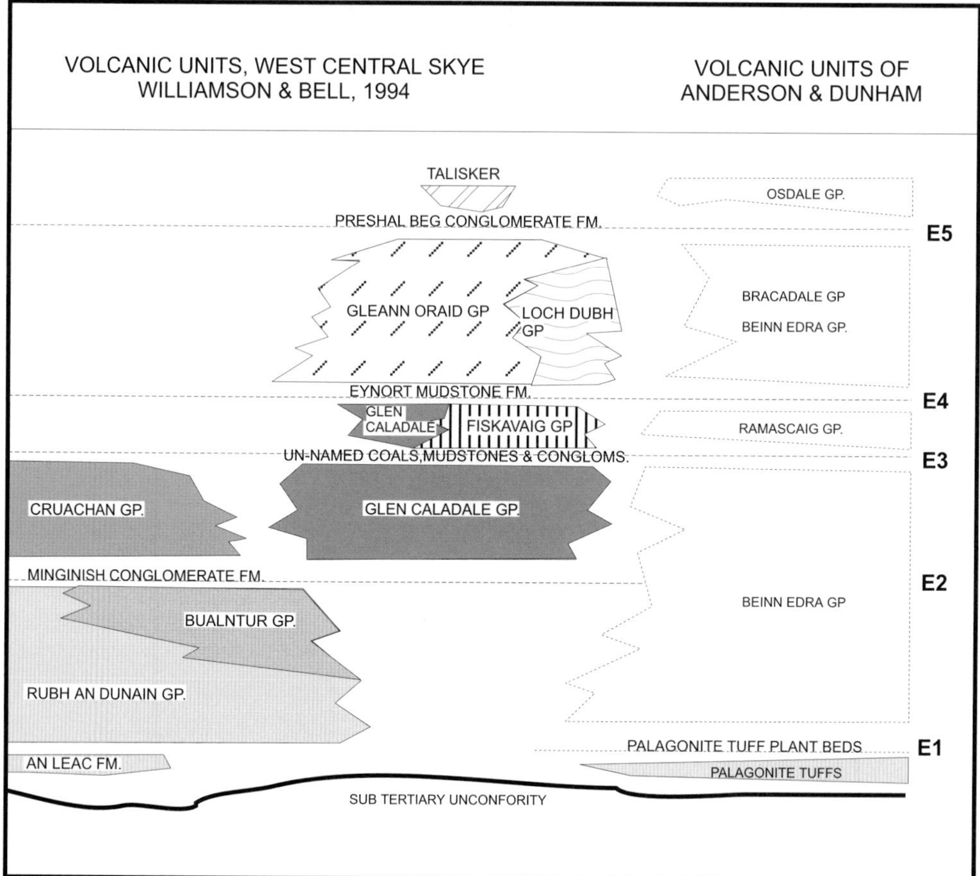

Fig. 2. Schematic relationship of intrabasaltic surfaces with the Formations of Williamson & Bell (1994) and possible correlation with those of Anderson & Dunham (1966). The palaeosurfaces proposed here are labelled E1–E5.

succeeding sedimentary horizon (E2) lies between the Bualintur and Cruachan Lava Groups of Williamson & Bell. Termed the Minginish Conglomerate Formation by these authors (Fig. 2), these polymictic conglomerates, sandstones, silts and poor coals were sampled on the slopes of Glen Brittle (NG 415 215) and on the sea cliffs between Loch Brittle and Loch Eynort (NG 375 196; NG 366 205). A younger sedimentary horizon (E3), is found further to the north between the Glen Caladale and Fiskavaig Lava Groups, consisting of a series of purple and reddened shales with some coaly strata. Good exposures of this horizon occur at Carbost Burn (NG 374 310) and at Allt Mor, Carbostbeg (NG 366 326). Above the Fiskavaig Lava Group, surface E4 is represented by extensive boles and red brown mudstones, widely exposed in the country between Loch Eynort and Loch Harport. Overlying this sedimentary horizon are the lavas of the Gleann Oraid and Loch Dubh Lava Groups. These are succeeded by the Talisker Lava Group, comprising the MORB type lavas of Preshal Mor and Preshal Beg (Thompson *et al.* 1972). Below the Talisker Lava Group is surface E5, the Preshal Beg Conglomerate Formation (Williamson & Bell 1994). In places the more silty matrix of this coarse conglomerate was collected for palynological analysis.

Correlation of the upper four of these five sedimentary horizons with those identified in central and northern Skye by Anderson & Dunham (1966) is problematical. In the absence of detailed re-mapping of these areas, the palynological and sedimentological comparison of localities, combined with a knowledge of the possible relationships between the west central lava groups

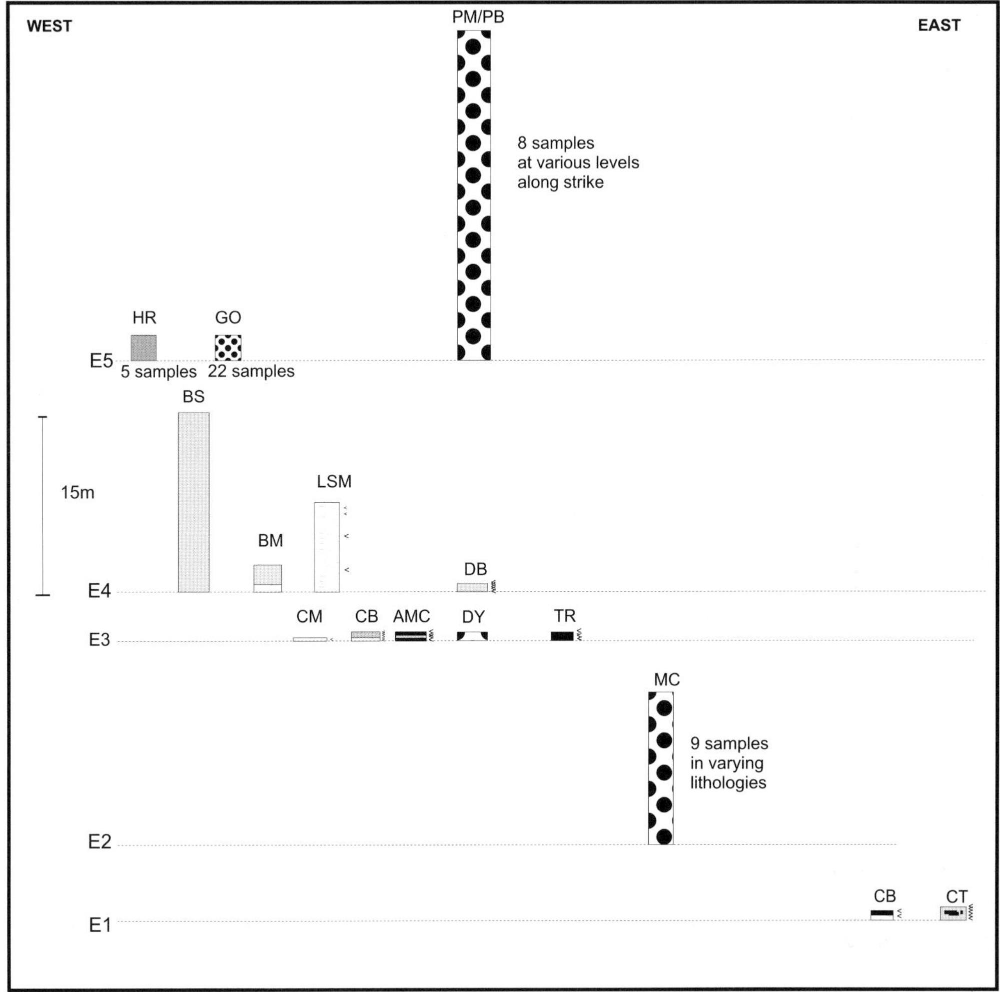

Fig. 3. Schematic representation of the lithologies and sampling regime of the five palaeosurfaces identified here. The lithological key is as for Fig. 7b.

and those of the north (Williamson & Bell 1994), provides a tentative comparison. In the central and northern area, congolmerates and coals in Tungadal River (NG 429 353), reddened organic shales and sands at Dun Beag, Bracadale (NG 333 301) and conglomerates, sandstones and shales at Glen Osdale and in the Hamara River (NG 468 196) were sampled, in addition to numerous bole horizons.

All materials collected where subjected to processing for palynomorphs involving dissolution of silicates in hydrofluoric acid with minimal oxidation in nitric acid. In the case of the coals, a short treatment with 5% sodium hyperchlorite was required to remove the abundant amorphous organic matter. During processing a 7 μm sieve was used, with the resultant residues being prepared as strew mounts for examination under an optical microscope.

The palynofloras

Palynofloral assemblages recovered from individual samples of intra or intertrappean sediments of the BTVP provide a picture of the vegetation at the time of deposition. This supplies information as to the age of the specific erosion surface, and detail of the distribution of vegetation on that surface. Changes in palynofloral associations recovered from the intratrappean sediments are attributable not only to variation in altitude and local environment in the lava fields, but also to factors which affect the regional British palynoflora of this period, evolution, migration and climate. It is regional factors which allow us to correlate the palynofloras of the BTVP with coeval associations in the Faeroe–Shetland Basin and the North Sea Basin (including eastern England).

The palynofloras recovered from the Skye, Mull and Antrim intra and intertrappean beds were of moderate diversity, and shared a common feature, that they are moderately to poorly preserved. Broadly, three types of autochthonous palynofloral assemblages were recorded. Rare assemblages from Skye are dominated by *Pityosporites* spp. (pines) and *Inaperturopollenites dubius* (*Taxodium* types) with low frequencies (Fig. 4) of *Leiotriletes adriennis* (ferns) and *Retitricopites retiformis* (floodplain shrubs or trees). These are interpreted as being derived from a humid montane conifer & fern forest with a low frequency of streamside taxa. They are most closely similar to a more common assemblage dominated by *Inaperturopollenites hiatus* (swamp cypresses), with fewer *Pityosporites* spp. and some streamside taxa that probably represent vegetation from an upland Taxodiaceae forest and riparian community (Fig. 5). Some macrofossil evidence exists for this community within the Minginish Conglomerate Formation. Both of these assemblage types are of low diversity, but the latter is found in several localities where it is admixed with pollen and spores from a lowland flora. Pollen of the Juglandaceae (hickories and

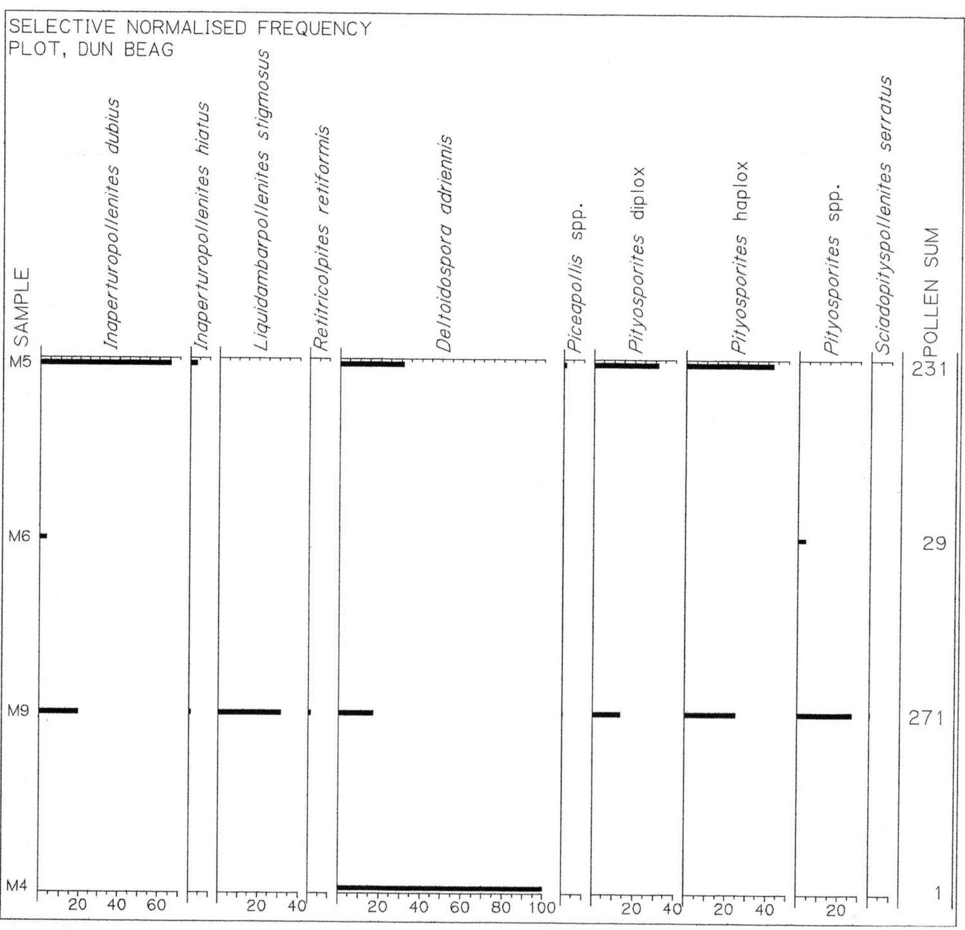

Fig. 4. Selective normalized frequency plot, Dun Beag.

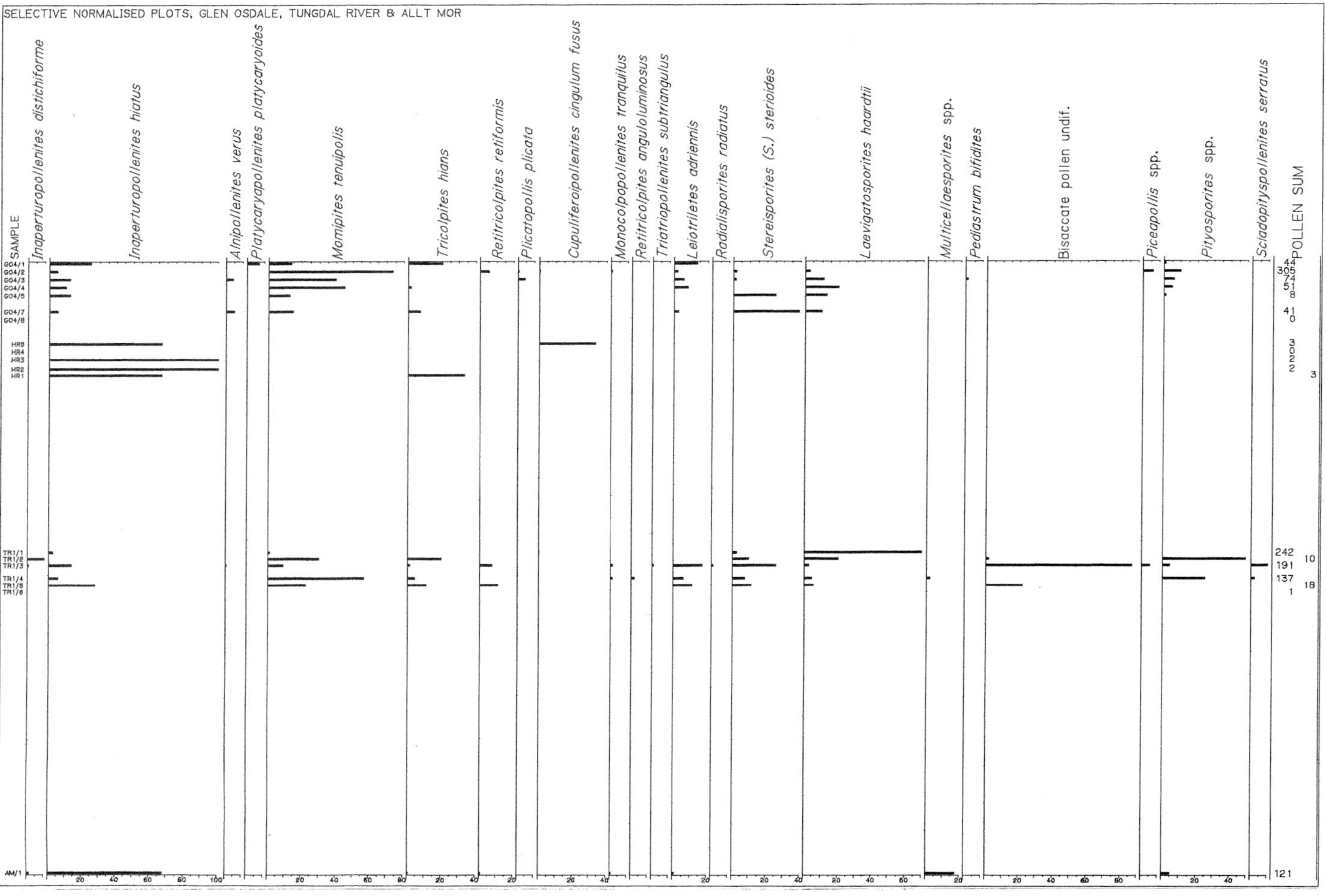

Fig. 5. Selective normalized frequency plot, Allt Mor (E2), Tungadal River (E3) & Glen Osdale (E5).

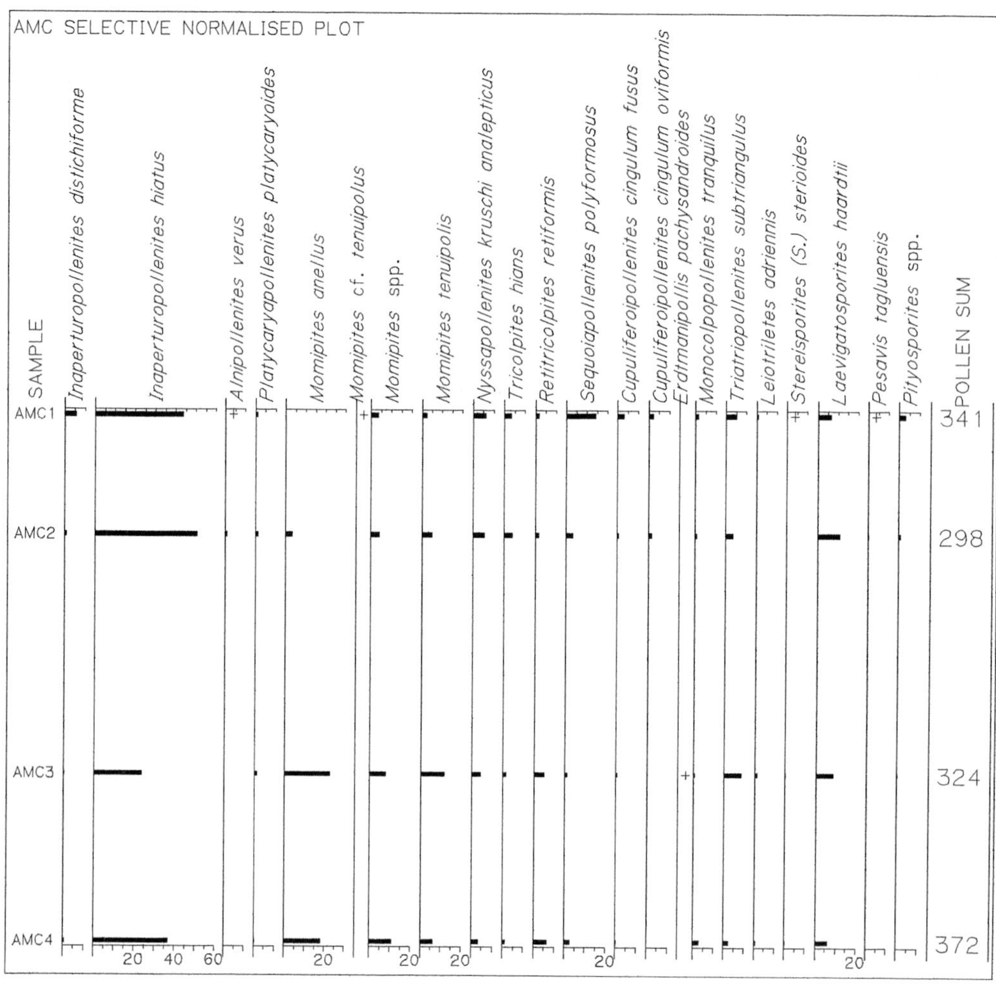

Fig. 6. Selective normalized frequency plot, Allt Mor, Carbostbeg.

walnuts) is usually the most common component in lowland assemblages, high frequencies of *Momipites* species implying extensive swampy areas probably around ephemeral intraflow lakes and on watercourses. Frequencies of pollen from streamside angiosperm taxa are usually high in lowland assemblages (*Retitricolpites retiformis, Tricolpites hians* and *T.* cf. *hians*), accompanied by pollen from more mature forest trees (Fig. 6) and considerable numbers of spores. These spores may have been derived mainly from epiphytic ferns, although the often common presence of *Stereisporites* species (mosses) suggests nutrient poor iron-rich soil conditions on the colonized lava surface (Fig. 5). True lowland pollen assemblages contain abundant streamside taxa with more common representatives derived from climax forest trees such as *Cupuliferoipollenites cingulum* subspecies and *Quercoidites microhenrici* (?oaks & chestnuts).

The distribution of these palynofloral assemblages and their relationship to the lithologies in which they are preserved provides information regarding the age of the Skye lava field, the duration of the sedimentary interruptions between major lava groups and to the altitude of the lava surface at different times in the fields' evolution. Additional information as to the behaviour of plant communities under stress imposed by local volcanism and the evidence for climatic variation is also available from the analysis of the palynological assemblages.

Stratigraphy of the lava fields

Intratrappean sediments on Skye

The formalized nomenclature for the intratrappean beds of Skye proposed by Williamson & Bell (1994) recognized three intervals worthy of formational status (Fig. 2), along with a number of other sedimentary intercalations not incorporated in the scheme. As this palynological study incorporates localities in the central and north Skye areas, the number of significant sedimentary intercalations is increased to five. To avoid imposing a further nomenclature, which is believed to be unjustified until the whole of the Skye intratrappean geology is reviewed, these sedimentary intercalations are referred to as erosion surfaces E1 - E5 in stratigraphical order. This reflects the significance, not only of the clastic sediments themselves, but also the importance of the bole horizons and the areas where sediments are absent.

Surface E1. The oldest of the significant erosion surfaces in the lava pile, the Palagonite Tuff plant beds, occurs only around Portree and Uig in north and east Skye. At Camas Ban and Glen Uig, the sequence consists of thin red boles on the Palagonite Tuffs, overlain by thin shales, or coals, as at Camas Ban. At Camas Tianavaig, the sequence is somewhat thicker, with around 0.80m of reddened carbonaceous shales, which are recorded as bearing plant fossils (Anderson & Dunham 1966). The correlation of this unit with the stratigraphy of the west central Skye area is assisted by the lithological comparability of the Palagonite Tuffs with the An Leac Formation exposed around Soay Sound (Williamson & Bell 1994), although there is no evidence of the sedimentary units seen in the north and east. The sediments appear to represent localized accumulation of muds and sandy muds in low-lying lacustrine areas, with abundant clastic material draining from the lava flow surfaces. At Camas Tianavaig, reworked amygdales are apparent in the shales, which also have an overall red cast that suggests the transportation of lateritic soil material into the lake from near-by flows. Disseminated plant material is also present in the middle of this unit, where recycled amygdales are absent. Lateritic soils of this type are seen at the base of the Palagonite Tuff plant beds at Camas Ban, where a thin red bole (0.30 m) is succeeded by a thin coal (approximately 0.30 m), reportedly mined during the 17th Century (Anderson & Dunham 1966).

The microscopic organic content of these Palagonite Tuff plant beds is dominated by small fragments of angular, equidimensional black wood. Pollen and spores are extremely rare, the sections at Glen Uig and Camas Ban proving to be barren. However, the section at Camas Tianavaig contained high frequencies of amorphous humic kerogen, together with some structured brown wood ('vitrinite'). In this interval, a low-diversity pollen and spore flora was recovered, dominated by *Inaperturopollenites hiatus*, but including *Momipites*, *Retitricolpites retiformis* and *Tricolpites hians*. A single specimen of the chlorophycean alga, *Botryococcus braunii* at the lower limit of these beds confirms a freshwater depositional environment. However, amorphous humic kerogen is precipitated when humic acid in freshwater solution comes into contact with a saline water mass. This process suggests that erruption of the Palagonite Tuffs took place in a marginal marine setting, rather than in the freshwater lakes proposed by Anderson & Dunham (1966).

The lack of recovery from localities other than Camas Tianavaig, prevents a full interpretation of the ecosystem of the Skye lava field during the interval of surface E1. However, the assemblages recovered at Camas Tianavaig suggests a lowland flora dominated by *Taxodium* and Juglandaceae in a high water table swamp community, with floodplain angiosperm taxa along watercourses. The lowland nature of this flora is confirmed by the petrology of the Palagonite Tuffs, which were erupted into shallow (saline?) waters (Anderson & Dunham 1966). The palynofloras on the upper surface of these beds represent initial colonising of the newly created land surface. The occurrence of lacustrine muds and sands with the development of boles and poor, dull coals suggests that a significant period of time is represented by this horizon below the lowermost lavas. The pollen present are derived from gymnosperm and angiosperm taxa (*Taxodium* & ?Platanaceae) indicating development of primary swamp forests in the area during a extended period of time, perhaps in the order of several hundred years. The abundance of subangular equidimentional black wood is typical of the palynofacies seen in the lower and mid part of Hebridean bole horizons where extensive oxidation of organic matter has occurred. However, the abundance of this material in all the samples may indicate the occurrence of regular forest fires, perhaps caused by lava flows or pyroclastic activity, resulting in an extensive charcoal supply to the lacustrine sediments. Although siltstones comparable to those of north and east Skye appear to be lacking above the possibly correlative An Leac Formation of west central Skye, the limited nature of the exposure does not rule out their presence on the west side of the island.

Surface E2. The oldest major intrabasaltic sedimentary units exposed in west central Skye at this level were termed the Minginish Conglomerate Formation by Williamson & Bell (1994). The formation is subdivided into three members, corresponding to outcrops around Glen Brittle. These sediments represent polymictic conglomerates, sandstones, siltstones and some coals deposited by a braided river system (see Williamson & Bell for details of lithologies) which flowed across the surface of the Bualintur and Rubha' an Dunain Lava Groups. These fluviatile deposits are correlative with nearby thin boles and to the north with the Carbost Pier borehole sandstone bed (Williamson & Bell 1994). A further exposure occurs on the eastern slopes of Creag Stromallus in east Skye.

Of the three locations exposing the conglomerates around Glen Brittle, the Allt Mor section has yielded the richest palynofloras, containing the most argillaceous sediments. Shales within the conglomerates at Allt Mor yielded a rich assemblages dominated by *Inaperturopollenites hiatus* and *I. dubius*, with common *Pityosporites* spp., *Leiotriletes adriennis* and some *Retitricolpites retiformis*, *Tricolpites hians* and *Monocolpopollenites tranquilus* (Fig. 6). This assemblage is suggestive of an upland Taxodiaceae forest with an understorey of cyathacean ferns, and streamside vegetation dominated by floodplain angiosperm taxa, in particular the Platanaceae. Similar, but impoverished assemblages are recorded from the alternating sands and conglomerates of the Culnamean Member in Glen Brittle.

The intrabasaltic conglomerates of Rum (Meighan *et al.* 1981; Emeleus 1973, 1985) have been suggested as the proximal equivalents of the Minginish Conglomerates (Williamson & Bell 1994), both being derived from the elevated area over the Rum Central Complex which sourced the fluvial system which flowed north westwards over Skye. The presence of Jurassic rocks in the source area for the Minginish Conglomerates fluvial system is confirmed by the occurrence of a reworked palynological assemblage in the Allt Mor section. The abundance of *Classopollis* spp. and the common occurrence of *Micrhystridium fragile* suggests an Early Jurassic age consistent with the Hettangian–Sinemurian Broadford Beds of southeast Rum (Smith, 1988). The north easterly direction of the flow of the Rum-sourced fluvial system (Fig. 7a) is indicated by the geometry of the conglomerates of the southwest central area (Williamson & Bell 1994), but the lack of exposure in the north limits the geographical reconstruction of this surface. The floras of E2 appear to have been dominated by an upland Taxodiaceae forest, perhaps with some pine stands in the highest altitude areas. This community was dissected by braiding fluvial systems, flanked by floodplain angiosperm communities dominated by the members of the Platanaceae.

Surface E3. The upland Taxodiaceae forest and braided fluvial system of Surface E2 times was destroyed by the lava flows of the Cruachan and Glen Caladale Lava Groups. However, similar lithologies to the Minginish Conglomerate Formation exist on the north shores of Loch Harport (NG 320 390) where elongate lenses of conglomerate outcrop at Drynoch (Bell, pers. comm.). These pass laterally into boles, shales and coals, but the position of at least one of the fluvial channels along the Loch Harport axis during E3 times suggests a continuation of the trend established during the deposition of the Minginish Conglomerates. This particular trend may relate to faulting along the axis of Loch Harport, or represent a persistent topographic low in the area. On the southwest flank of Loch Harport, beds equivalent to the Drynoch conglomerates occur at Carbost Burn, Carbostmor and Allt Mor, Carbostbeg. At Carbostbeg in the Allt Mor burn, the basalts of the Glen Caladale Lava Group are overlain by a dominantly dull coal which passes upwards into carbonaceous shales, further dull coals and an uppermost bright coal totalling 1.10 m in thickness. A similar sedimentary feature is seen in the waterfall at Carbost Burn, where a 0.30 m bole resting on amygdaloidal basalts is overlain by 1.18 m of siltstones, in turn capped by a thick mugearite of the McFarlane's Rock Formation. The siltstones are reddened in the basal 30 cm and become increasingly well laminated towards the top, with red mud filled (dessication?) cracks in the middle part. Further to the southwest at Carbostmor, the boundary between the Glen Caladale Lava Group and the McFarlane's Rock Formation is apparently marked by a prominent red bole.

In the far west of the west central area, Surface E3 is marked by boles at McFarlane's Rock, Beinn nan Cuithean and Preshal Beg (see Williamson & Bell 1994) the first two localities including thin siltstones and tuffs. On the northwest side of Loch Harport, the rising slopes of Roineval are dominantly bog covered with poor exposure. At the western foot of this mountain, the Tungadal River exposes well laminated fissile shales with thin coals and gritty 'conglomerates' overlain by basalts. The beds total around 2.20 m in thickness and contain recognisable seeds and leaf fragments. Palynological evidence suggests these are attributable to Surface E3, appearing to represent swamp deposition on the western margin of the Drynoch conglomerate fluvial system.

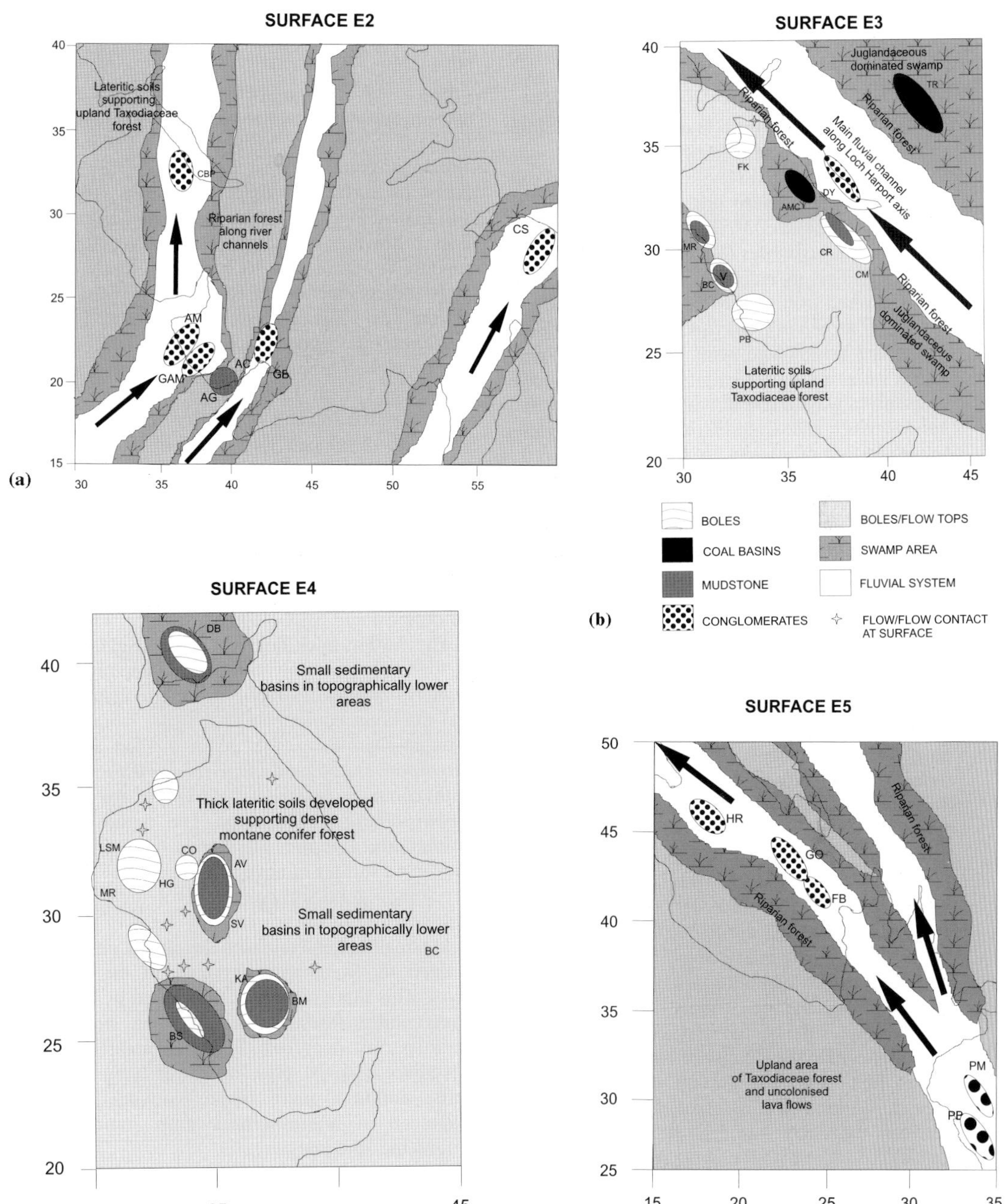

Fig. 7a–d. Schematic maps of vegetation and sedimentation patterns on palaeosurfaces E2–E5. Abbreviated localities are: GB, Glen Brittle; AG, Allt a Ghaid; AC, An Crocan; GAM, Allt Geodh'a Ghamhna; AM, Allt Mor; CBP, Carbost peir borhole; CS, Creag Stromallus; PB, Preshal Beg; BC, Beinn nan Cuithean; MR, McFarlane's Rock; FK, Fiskavaig Bay; AMC, Allt Mor, Carbostbeg; DY, Drynoch; CR, Carbost Burn; CM, Carbostmor; TR, Tungadal River; BS, Bem Scaalan; BM, Biod Mor; KA, Kearra; SV, Stockval; AV, Arnaval; LSM, Loch an Sguirr Mhoir; CO, Creag Omain; DB, Dun Beag; PM, Preshal More; FB, Forse Burn; GO, Glen Osdale; HR, Hamara River.

The sediments attributed to this erosion surface contain the most diverse and abundant palynofloras recovered from the intrabasaltics of Skye. The bole at Carbostmor contains only the equidimentional black wood fragments associated with mid - lower profile lateritic soils. This is repeated in the bole at the base of the Carbost Burn Section with the exception of rare specimens of *Inaperturopollenites hiatus*, and surprisingly continued into the overlying shales. These are rich in black wood, with only *I. hiatus, Momipites* spp., *Tricolpites hians* and *Stereisporites* (*Stereisporites*) *stereioides* recovered. This assemblage is in total contrast to that recovered from Allt Mor, Carbostbeg. Here an *I. hiatus, Momipites anellus*, and *M. tenuipolus* dominated assemblage with common *Laevigatosporites haardti* and *Tricolpites hians* in the lower coals, gives way to one in which *Sequoiapollenites polyformosus* and *Cupuliferoipollenites cingulum* subspecies are prominent. A similar palynoflora is recovered from the Tungadal River section. Here, *I. hiatus* is less prominent than *Momipites* spp., but common *T. hians* and *Retitricolpites retiformis* occur. The frequency of fern spores is higher at Tungadal River, with very high frequencies of *L. haardti* in some samples and common *Leiotriletes adriennis* throughout. *S.*(*S*) *stereioides* is also more common and *Pityosporites* spp. is sporadically present in large numbers.

Surface E3 contains in its sediments, a diverse group of palynofloras and palynofacies. The higher altitude lateritic soils may have supported a montane conifer forest sourcing the pine pollen in Tungadal River (Fig. 7b), with upland Taxodiaceae forest at lower levels. The localised swamp at Carbostbeg, presumably on the margin of the Drynoch conglomerate fluvial system was vegetated by a Juglandaceae dominated swamp community, although the increase in frequency of pollen from a secondary angiosperm forest with large forest trees including *Sequoia*, suggests that the swamp became increasingly dry and stable. On the opposite side of the Drynoch conglomerate fluvial system in the locality of Tungadal River, the clastic input into the swamp environment was greater, resulting in conglomeratic and silty bands between the coaly laminae. Here again, a juglandaceous dominated angiosperm swamp community with a rich (epiphytic?) fern component dominated, but with considerable contribution from a mixed floodplain angiosperm and fern community. In addition, occasional fluvial input into the area brought in not only coarser clastic sediment, but pollen from a montane conifer forest.

At Tungadal River some pollen and spores of Jurassic origin are recorded, recycled possibly from the Rum area where the fluvial systems appear to have originated (Williamson & Bell 1994). Only one reworked grain is encountered at Carbostbeg, reflecting the lowland, quiescent environment.

Surface E4. This is perhaps the most unusual erosion surface of the five recorded as major breaks in the lava pile. The complex of lowland swamps, floodplains and surrounding upland Taxodiaceae forest of Surface E3 times was destroyed by the subsequent extrusion of the Fiskavaig Lava Group. The following quiescent phase gave rise to a topography dominated by bare flow tops, lateritic soils and ephemeral lake basins in low lying areas. On the west coast of the west Central area (Fig. 7c) thick boles developed at Loch an Sguirr Mhoir and nearby at McFarlane's Rock, Beinn nan Cuithean and Creag Omain. In the area of Ben Scaalan, and nearby at Kearra and Biod Mor, shales and thin coals are developed with boles in the sequence. Similar deposits are seen in the area of Arnaval and Stockval (Williamson & Bell 1994). North of Loch Harport, near Bracadale, roadside exposures of reddened shales and a thin sand are seen at Dun Beag, Bracadale, which are possibly correlative with the more southerly exposures of Surface E4 on lithological and palynological evidence.

The 7.00 m thick bole at Loch an Sguirr Mhoir yielded only poor palynofloras with occurrences of *Inaperturopollenites hiatus* with *Tricolpites hians* and common equidimentional black wood. However, at Dun Beag, the shales yielded abundant *Pityosporites* spp. and *Inaperturopollenites dubius* with common *Leiotriletes adriennis* and *Liquidambarpollenites stigmosus*. The nature of the dominantly lateritic lithologies of Surface E4 suggests a surface of unvegetated lava flows, others developing good soil profiles involving prolonged pedogenesis. In the lower lying areas of the lava field, small ephemeral lakes developed into which drained the weathering products of the eroding lavas. This topography was dominated by a humid pine conifer forest, in which some streamside angiosperms occurred. In other areas, upland Taxodiaceae forest grew on lower lying lateritic soils, although there is no evidence for the existence of angiosperm swamp communities around the lake margins.

Surface E5. The sediments of this erosive surface represent some of the thickest and most spectacular of the Skye intrabasaltics. Termed the Preshal Beg Conglomerate Formation by Williamson & Bell (1994) these are coarse, angular pebble to cobble size clast conglomerates with little sorting and a dominantly sandy matrix, reaching around 40 m in thickness. At both Preshal Beg and Preshal More, the upper part of this bed is intensely lateritized and overlain by the columnar jointed basalts of the Talisker Formation. Williamson & Bell compared the structure of these deposits to older conglomerates recorded by Emeleus (1972; 1985) and Emeleus & Gyopari (1992) from Rum, Canna and Sanday, suggesting an origin as debris flow deposits in a proximal fan setting. Further to the north, palynological evidence suggests that the sediments exposed in Glen Osdale may approximate in age to this erosion surface. Exposed in Glen Osdale, the sediments are a strike section through matrix supported conglomerates with rounded clasts and well bedded fine to medium sands, with some shales. These are lithologically similar to the sediments of the Minginish Conglomerate Formation, suggesting a mid-fan deposit with dilute debris flows, channel deposits and infill. Further north at Hamara River, the deposits are wholly medium to fine sands, these lower energy sediments lacking the well rounded pebbles seen at Glen Osdale.

Palynological data from the Preshal More and Preshal Beg conglomerates is limited by problems of recovery from coarse-grained facies. However, the presence of numbers of *Inaperturopollenites hiatus* and *I. distichiforme* suggests a source area rich in upland Taxodiaceae forest, *Tricolpites hians* and *Momipites* spp. perhaps representing streamside and angiosperm swamp vegetation. Further north at Glen Osdale, a rich palynoflora is recovered. *Momipites* spp. dominates the palynoflora, with *I. hiatus* and some *Pityosporites* spp. The presence of an abundance of fern spores, including *Laevigatosporites haardti* and *Stereisporites* (*Stereisporites*) *stereioides*, indicates a humid climate.

The presence of this unit of very coarse sediments of a type similar to the older proximal deposits of Rum, Canna and Sanday, suggests that the source of deposition had shifted to the area of the Cuillin Hills from the south. This is reflected in the coarse conglomerates in the Preshal Beg and Preshal More area, with the finer grained, more mature conglomerates and sandstones at Glen Osdale and Hamara River. The vegetation represented in the sediments seems to suggests the presence of a montane conifer forest on the highest intraflow areas (Fig. 7d), but with a upland Taxodiaceae forest dominating elsewhere. In lower lying area on the fluvial systems, a juglandaceous primary community occured in swampy areas, with floodplain angiosperm forest dominated by platanaceous trees. The abundance of *Stereisporites* spp. implies the presence of low nutrient iron-rich soil bog areas on the erosion surface of the Gleann Oraid and Loch Dubh Lava Groups.

Intratrappean sediments of Rum, Canna & Sanday

The occurrence of coarse conglomeratic horizons on Rum, Canna and Sandy have been reported by Emeleus (1973, 1985), and Emeleus & Gyopari (1992). The small number of samples analysed here were provided by C. H. Emeleus, and are from the silty sands in conglomeratic sequences on the north slopes of Fionchra (NG 3365 0068) belonging to the base of the Upper Fionchra Member, Canna Lava Formation (Emeleus 1985). The palynofloras recovered (Fig. 8) are typical of an upland Taxodiaceae forest, with some input from a juglandaceous dominated swamp and streamside angiosperm forest, a flora characteristic of uplands in this part of the BTVP. Occurrences of *Plicapollis pseudoexcelsus* and *Sequoiapollenites polyformosus*, together with the calculated palaeoaltitude of the flora (see below) suggest a correlation of this interval with erosion surface E3 in the Skye Lava field. Similar sediments from the base of the Lower Fionchra Member of the Canna Lava Formation in Maternity Hollow (NG 3469 0061), exhibit a similar, but lower diversity palynoflora to those from Fionchra. Their stratigraphical position implies that they are probably better attributed to Surface E2; this awaits further work for confirmation.

Intratrappean sediments of Muck and Eigg

A small number of samples where provided by C. H. Emeleus from the intratrappean horizons of the Eigg Lava Formation from Eigg and Muck. These are largely tuffs and boles, and yielded few palynomorphs. However, a sample from Druim an Lochain, Eigg (NM 4716 8741) yielded *Inaperturopollenites hiatus* and *Nyssapollenites kruschi* subsp. *analepticus*, suggesting that an upland Taxodiaceae forest may have occurred on the lateritic soil of the flow surface.

Intratrappean sediments of Mull and Ardnamurchan

A small number of samples from Ardtun Head in the collection of the Centre for Palynological Studies were examined for palynomorphs. These samples were taken from the bottom and middle leaf beds of the intratrappean sediments (NM 3775 2520) exposed east of the ravine described by Bailey *et al.* (1923). Bottom leaf bed samples (Fig. 9) yielded a flora dominated by *Inaperturopollenites hiatus* and *Pityosporites* spp. with

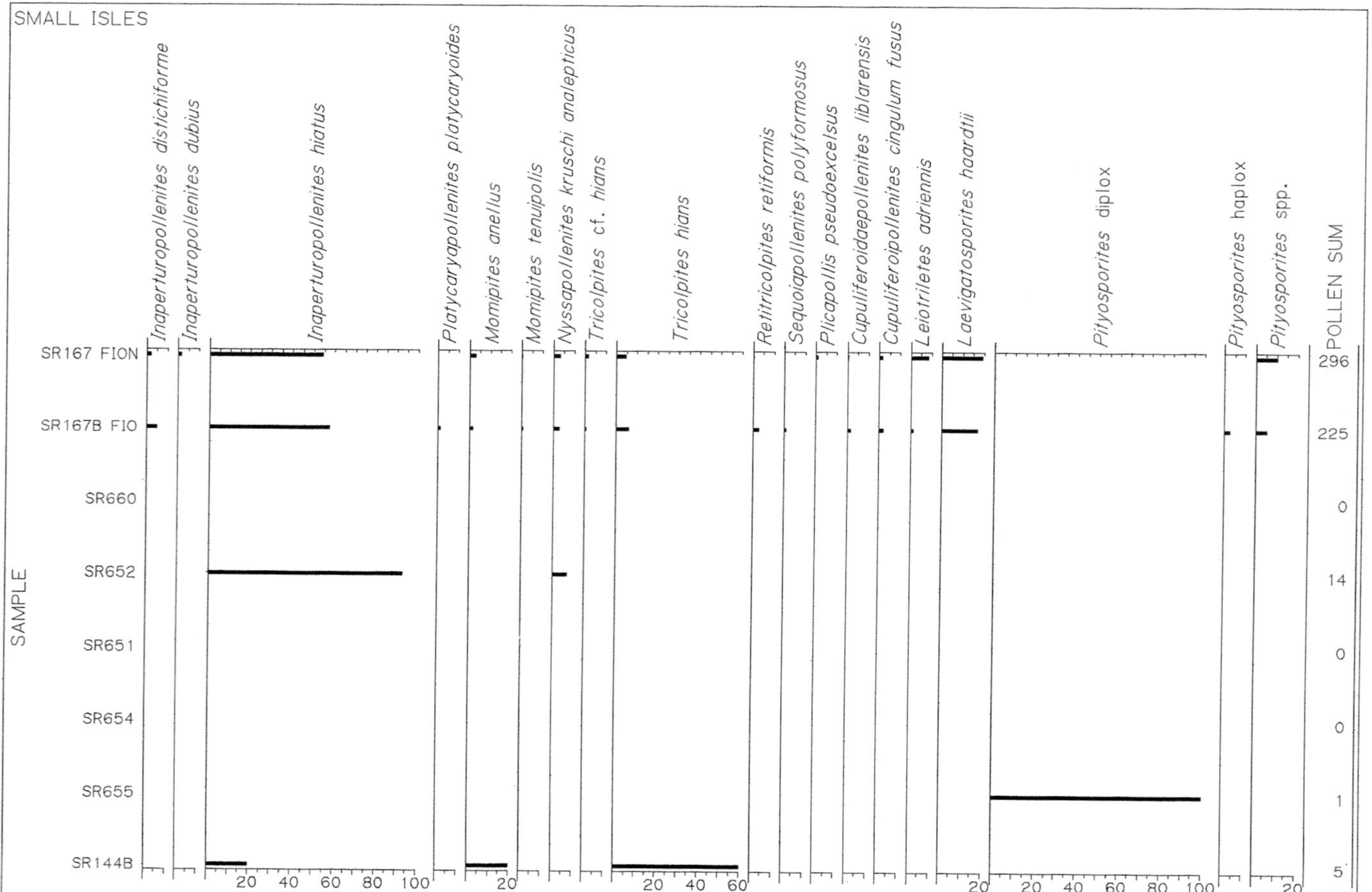

Fig. 8. Selective normalized frequency plot, Small Isles. Samples SR167and 167B are from the Upper Fionchra Member, SR144B from the Lower Fionchra Member, samples SR660 and SR652 are from Eigg, Eigg Lava Formation and samples SR651, 654, 655 are from Muck, Eigg Lava Formation.

common *Tricolpites hians* and frequent occurrences of *Cupuliferoidaepollenites liblarensis* and *Cupuliferoipollenites cingulum* subsp. *fusus*. Middle leaf bed samples yielded slightly different assemblages, with common to abundant occurrences of *Monocolpopollenites tranquilus,* and fewer *I. hiatus*. This palynoflora suggest the occurrence of a lowland *Taxodium* swamp forest in a localized lake basin with fluvial input bringing in pollen from a floodplain angiosperm forest and a (*Ginkgo* dominated?) mixed mesophytic forest. The presence of *Pityosporites* spp. emphasizes the allocthonous palynofloral components, which were probably being derived from a montane conifer forest. Interestingly, the extensive macrofossil records of Boulter & Kvacek (1989) from these leaf beds did not contain any Piniaceae-type foliage, leading these authors to conclude a upland source area for the bisaccate pollen prevalent in the palynofloras. A greater diversity of other taxa have been recorded by other authors from these deposits, some of which are discussed below.

Intra and intertrappean sediments of Antrim

A sequence of samples from boles and intra and intertrappean mudstones collected by M. Widdowson were treated for palynological analysis. The red mudstones recovered from between the Chalk surface and Lower Basalt Formation at Glenarm Quarry, northwest of Larne (D 305 145) proved to be barren of structured palynomorphs, containing small, angular fragments of black wood. This is unusual in view of the macrofossil evidence recorded from this location by Gardener (1883–1886) and Boulter & Kvacek (1989) which pointed to rapid fluctuation in organic fossil preservation in the sediments, possibly related to differential contemporaneous oxidation. However, samples from Craigahuilliar Quarry, near Portrush (C 880 390) where volcaniclastic silts and lignitic clays rest on the lowermost flow of the Tholeiitic Member of the Interbasaltic Formation, yielded a rich palynoflora. Here, *Pityosporites* spp., *Inaperturopollenites hiatus, Momipites tenuipolus*

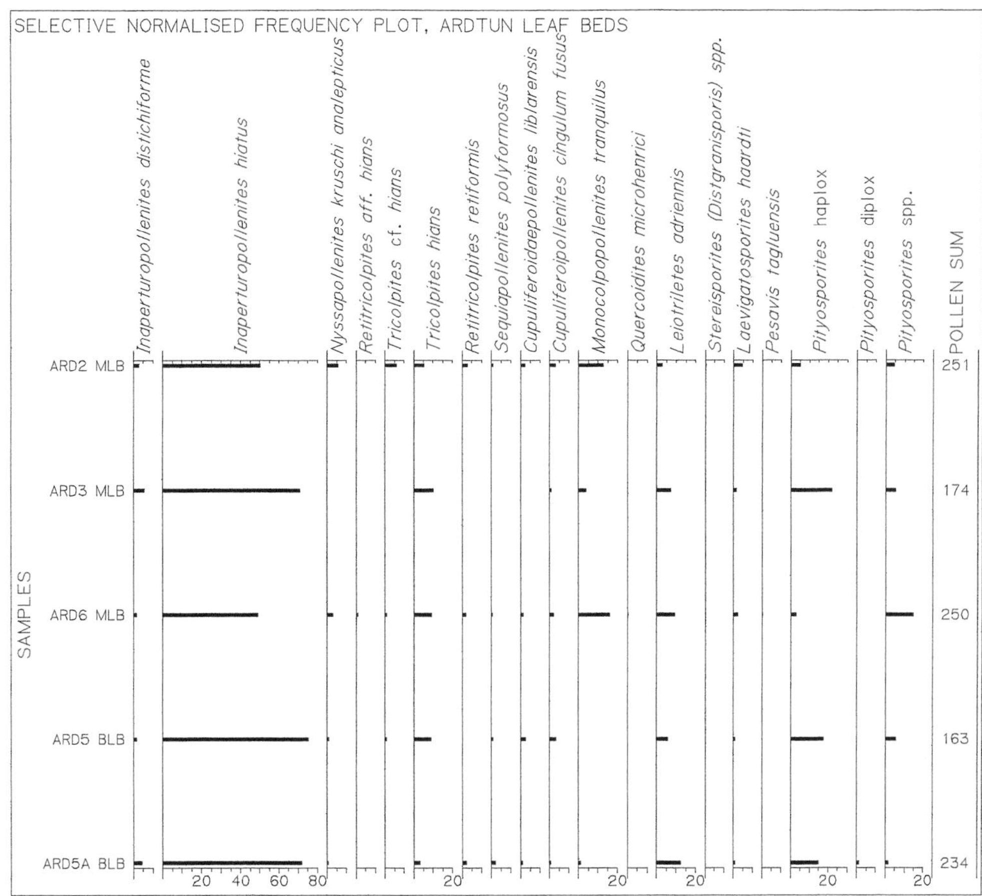

Fig. 9. Selective normalized frequency plot, Ardtun, Mull. BLB, bottom leaf bed, MLB, middle leaf bed.

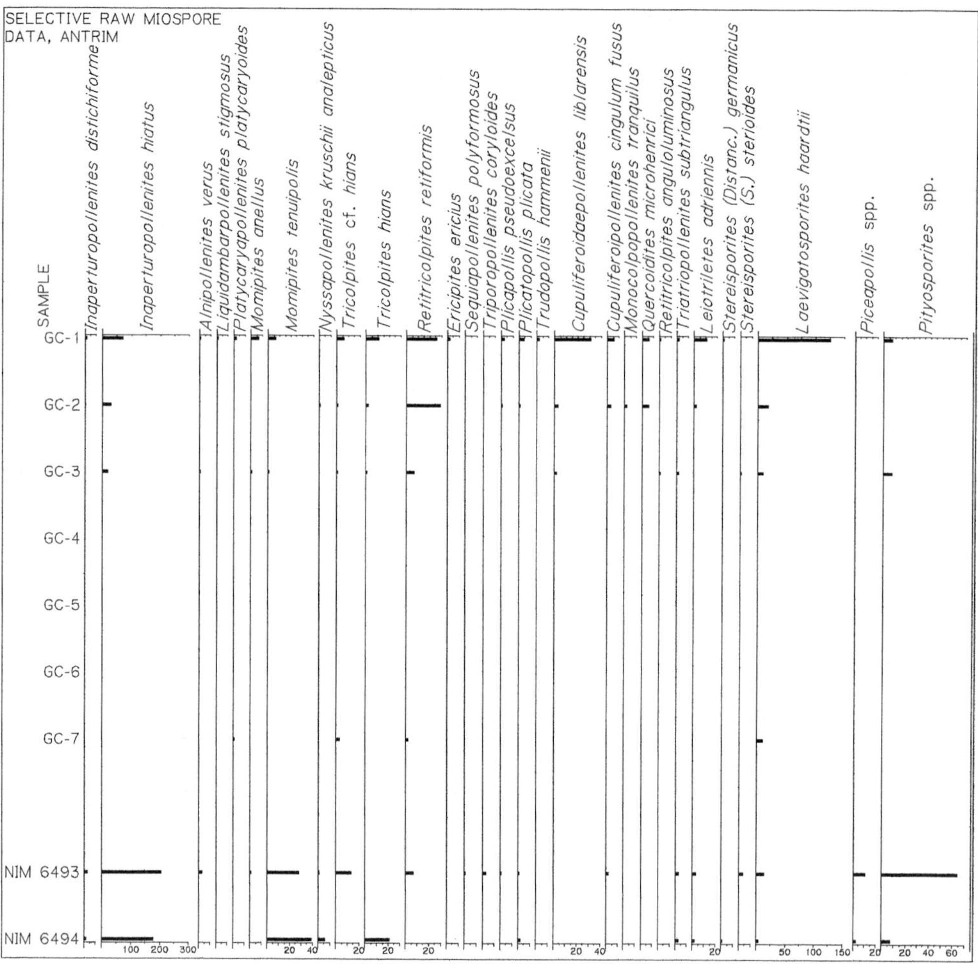

Fig. 10. Selective normalized frequency plot, Antrim. NIM 6494 is taken from the tuffaceous mudstones, NIM 6493 from the overlying lignites at Craigahuilliar quarry. Samples GC-1 to GC-7 are from the Interbasaltic Formation, Giants Causeway.

and *Tricolpites* cf. *hians* dominate diverse assemblages. The sediments are lacustrine in origin, the palynofloras (Fig. 10) implying a dominant *Nyssa–Taxodium*–Juglandaceae swamp forest vegetation, with associated floodplain angiosperm forest along fluvial channels. The diversity of these assemblages, and the presence of taxa from a mixed mesophytic forest suggests a lowland setting.

Higher up the sequence, prominent boles exposed in the cliff sections lying northeast of the Giant's Causeway (e.g. D 950 445), Lower Basalt Formation and Interbasaltic Member (Wilson & Manning 1978) yield similar assemblages to those seen in the Craigahuilliar section. However, they show a lesser dominance by the *Nyssa–Taxodium*–Juglandaceae swamp forest with greater representation by a floodplain angiosperm forest and by an increase in taxa which occur in paratropical rain forest palynofloras (Boulter & Hubbard 1982). The diversity of this flora, and the presence of taxa seen in paratropical assemblages is concordant with the mixed mesophytic forest suggested for the macrofloras of the Interbasaltic Member by Boulter & Kvacek (1989). The thickness of this member and the character of the subtropical vegetation that grew on its lateritic soils imply lengthy periods separating extrusion of the Tholeiitic Member from the underlying Lower Basalt Formation.

The age of the British Tertiary volcanic province

In assessing the age of the BTVP by the palynofloral content of the various intrabasaltic and interbasaltic sediments, it is necessary to consider the relatively complete record of vegetational change offered by the pollen record of sediments in the West Shetland Basin. A number of sections of the Palaeocene–Early Eocene interval from along the strike of Quadrants 214 to 204 have been examined and used to produce the composite pollen record shown in Fig. 11. Although there is some variation present across the strike of the southwest–northeast palaeoshelf, the patterns of palynomorph abundance in the Selandian, Thanetian and Ypresian of this area are remarkably consistent, showing similarities to those of the North Sea Basin (including southeastern England), but with a greater abundance of terrestrial palynomorphs. Variations in the frequency of certain taxa throughout this period have proven to be of significant stratigraphical value, although some first and last occurrences of specific taxa are also useful. By comparing this pollen record of the West Shetland Basin (Fig. 10) and contemporary floras in the North Sea Basin (Jolley 1992, 1996, in press) with those of the BTVP, it has proven possible to determine accurate dates for the intrabasaltic sediments.

The dating of events in the composite pollen record for the West Shetland Basin is achieved by utilising the similarity of these floras with those of the southern North Sea Basin. In this latter region a directly calibrated, high-resolution comparative palyno-magnetostratigraphy has been established using the same sample sets (Jolley 1992, 1996, in press; Ali & Jolley 1996; Ellison *et al.* 1996), with magnetic and palynological events being tied to the time scale of Cande & Kent (1992, 1995). Of particular relavance to this study is the work conducted on the Ormesby/Thanet depositional sequence in East Anglia. Here, the identification of Chron 26n (57.554–57.911 Ma) in the Ormesby Clay Formation spans the interval of palynological association sequences O/Th C2–intra C6, giving each an average duration of 0.08 Ma. Extrapolation of six more association sequences of this duration from the upper limit of Chron 26r, downsection to the base of the section in East Anglia and western Europe (Jolley in press) indicates an age of around 58.40 Ma for the oldest sediments. This age is concordant with that suggested by Knox (in press), and spans the time interval during which some of the significant erruptive events took place in the BTVP. Similar calculations have been undertaken for the interval between the Palaeocene/Eocene boundary and the upper limit of the Balder Formation and its equivalents (Jolley 1996). In this interval magnetostratigraphical information is supplimented by Ar/Ar dates for distinctive ashes from the Ølst Formation, Denmark. The ages inferred for the BTVP intratrappean beds studied are detailed below.

The Eigg Lava Formation

The work of Emeleus (1985) has indicated that the Eigg Lava Formation on Eigg and Muck is older than the nearby Canna Lava Formation because of its geological relationship with the Rum central complex by which it is penetrated. The limited sample database, together with the nature of the dominantly thin bole intrabasaltic horizons present mean that insufficient evidence is available to accurately date the Eigg Lava Formation.

The Canna Lava Formation and Skye Lava Groups

The conglomerates and mudstones of the Canna Lava Formation sampled on Rum are correlative with those sampled on Skye. The dominance of *Inaperturopollenites hiatus, Tricolpites* and *Retitricopites* species with common to abundant *Momipites*, *Laevigatosporites haardti* and *Sequoiapollenites polyformosus* is characteristic of the latest Selandian and early Thanetian in the West Shetland Basin. By comparison this gives ages from 57.99 Ma for the youngest possible palyno-floras, to 58.23 Ma for the oldest (Fig. 11) occurrence of this palynoflora, a total of 0.24 Ma for the eruption of the lava field. Interestingly, this period falls within Chron 26r, a factor confirmed by Ali & Jolley (1996), which corresponds to the reversed polarities obtained from the Skye Lava Field by Wilson *et al.* (1972) and to the ages of around 58–60 Ma provided by isotopic analysis of Rum lavas and the Western Granophyre (Mussett 1984). Given the rate of development of the plant communities on fresh lava flows, from initial colonisation from isolated refuges, to development of climax forest with large forest trees, the maximum time for lava eruption may have been less than 0.18 Ma. A simple calculation demonstrates that eruption rates would average a minimum of around 1 m of lava every 9 years. This is a minimum estimation since it takes no account of contemporaneous, or subsequent erosion of the lava pile. However, the short time period of the extrusive activity, and the abundance of bole horizons in the lava field may indicate that many 'red beds' are altered tephras (Bell *et al.* 1996), rather than soil horizons.

The timing of the final stage of this eruptive phase is coincident with the occurrence of tuffaceous sands, comparable in age to the Balmoral Tuffite of the North Sea (Knox & Holloway 1992), in the West Shetland Basin. The petrology and palynology of the Balmoral Tuffite was reported by Jolley & Morton (1992) from the mid-Moray Firth. Its composition was found to be similar to the petrology of the MORB type Preshal

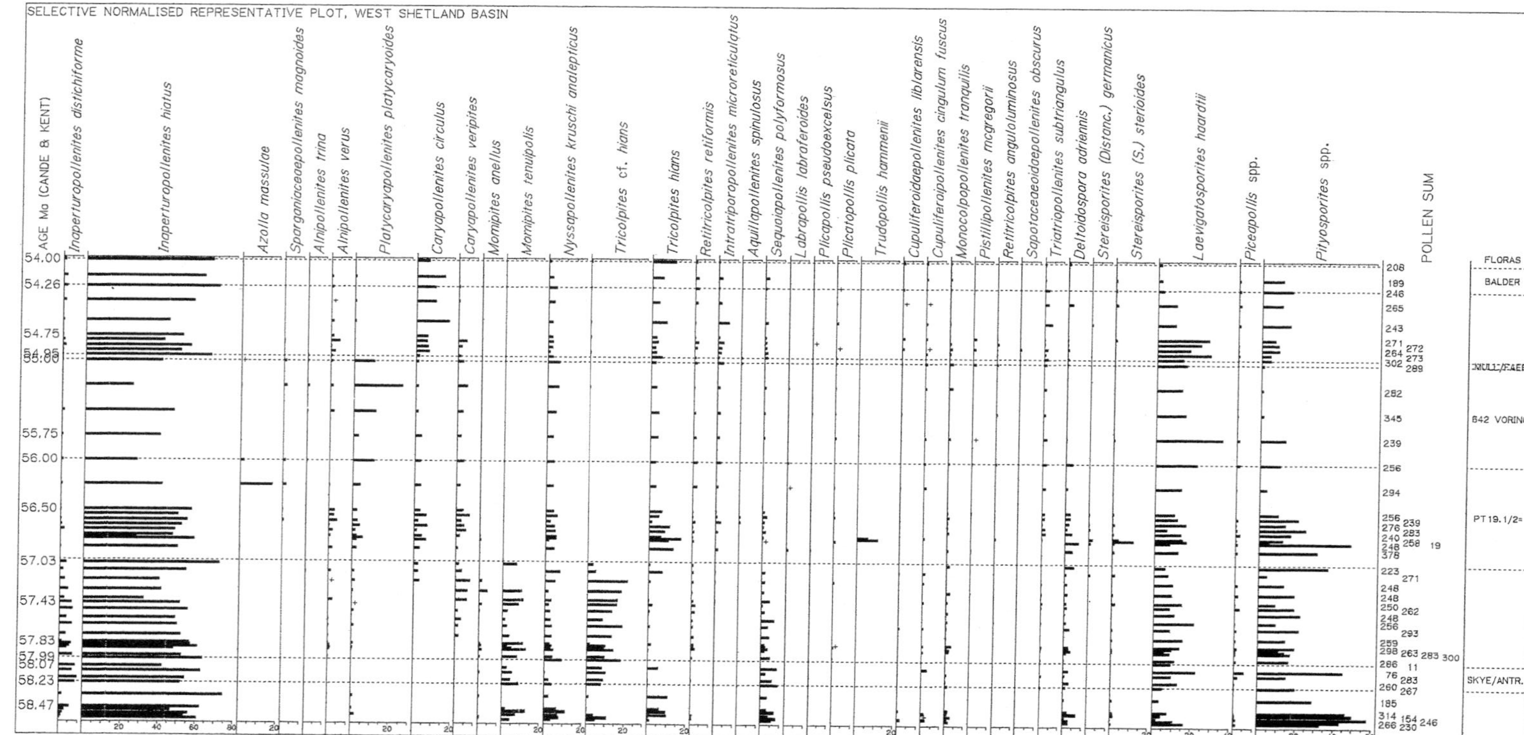

Fig. 11. Selective normalized frequency plot, composite West Shetland Basin section. The right hand column shows the stratigraphical relationships of the Skye/Antrim lavas to the Mull and Faeroes intrabasaltics. The age of the palynofloras in the intrabasaltics in ODP 642 from the Vøring Plateau and the Balder Formation are also shown.

More basalts (Talisker Lava Group), a similar tholeiitic composition characterising the West Shetland Basin tuffite (Morton *et al.* 1988). However, it is not concluded that the Skye Centre was responsible for this pyroclastic deposit, since it may equally be attributable to the Erland volcanic centre to the north, or to the Greenland–Faeroe province as suggested by the high Ti - tholeiitic composition (Morton *et al.* 1988).

The Antrim Basalts

The palynofloras of the intra and intertrappean sediments of the uppermost Lower Basalt Formation and Intrabasaltic Formations (including the lowermost Tholeiitic Member at Craigahuilliar) of Antrim contain *Inaperturopollenites hiatus, Momipites* and *Tricolpites* species dominated palynofloras of close similarity to those seen from Skye and Rum (Fig. 9). This indicates a comparable age of between 57.99–58.23 Ma, which corresponds to the reversed polarities for the whole of the lava series (Mussett *et al.* 1988) and with the Ar^{40}–Ar^{39} dates of 61–58 Ma (Thomson 1986) and 59 Ma (Thomson *et al.* 1984) for the lavas and volcanics of the Interbasaltic Formation respectively. This confirms the comparability of the Skye and Antrim Lavas suggested by Hitchen & Ritchie (1993) and Ritchie & Hitchen (1996). The occurrences of *Aquilapollenites* spp. in the museum samples of the Interbasaltic Formation utilized by Boulter & Kvacek (1989) are not associated with the palynofloral assemblages characteristic of the earliest Eocene Mull and Ardnamurchan intratrappean beds (see below), but rather suggests that this relict Late Cretaceous taxon may have found a refuge in the lava fields of the BTVP. However, the upper part of the Tholeitic Member and Upper Basalt Formation remain undated, and may concievably be of a significantly younger age.

The Mull and Ardnamurchan Basalts

The palynofloras recovered from the base of the Plateau Lava Group are distinctly different from those seen in the Canna Lava Formation, Skye Lava Series or in the Antrim Basalts. In the leaf beds, the dominance of *Monocolpopollenites tranquilus* (?*Ginkgo*) is pronounced, and points to a comparison with the earliest Ypresian assemblages of the West Shetland Basin (Fig. 11) where this taxon becomes common for the first time. This correlation is reinforced by the occurrence of other pollen taxa recorded in the more extensive studies of Mull and Ardnamurchan intrabasaltics, in particular those of Simpson (1961) and Srivastava (1975). Simpson presented the first detailed work on the intrabasaltic leaf beds of Mull and Ardnamurchan, and despite the lack of quantitative data, and the authors belief that his material was of Miocene age, this still provides a valuable reference.

The presence of *Aquilapollenites* spp. in the records of Simpson from the Mull intrabasaltics (as *Taurocephalus proteus*) and *Labrapollis labraferoides* (as *Engelhardtia granulata*) from the intrabasaltics of Ardnamurchan clearly indicate a comparison with a short lived palynoflora seen in the earliest Ypresian of the West Shetland Basin. Simpson also confirms the common occurrence of *Monocolpopollenites tranquilus* in these sediments and the occurrence of *Triatriopollenites subtriangulus* (as *Euryale spinosa*) and *Trudopollis variabilis* (as *Haloragis scotia* and *H. bremanoirensis*) in the prebaslatics of Ardnamurchan and the intrabasaltics of Mull, further components of the earliest Ypresian flora (Fig. 11). Similar records were provided by Srivastava (1975), who suggested a Maastrichtian age for the sediments based on the occurrence of *Dinogymnium,* and *Aquilapollenites* spp.

The former dinoflagellate thecate taxon is reworked from the Late Cretaceous, but the occurrence of *Aquilapollenites* spp. confirms the reports of Simpson. However, it does contribute to the debate that the presence of *Aquilapollenites* spp. does not represent pollen of a plant of the BTVP flora, but is a direct result of rejuvenation of sources at times of lowered sea level and increased reworking of palynomorphs. In addition to these taxa, Srivastava reported occurrences of *Azolla cretacea*, the spore massulae of a freshwater aquatic fern seen in the latest Palaeocene in the West Shetland Basin (Fig. 11).

The correlative palynoflora in the West Shetland Basin is dated as being from 55.00 Ma at the oldest (i.e. at the Palaeocene/Eocene boundary), to around 54.92 Ma at the youngest. This palynoflora has a particular ecological significance as a pioneer vegetation on the upper surface of the Hebrides terrace lavas, coinciding with a major depositional sequence boundary.

The Faeroe Islands basalts

Although no material has been examined from the Faeroe Islands intrabasaltics, the work of Lund (1983, 1988) provides some quantitative detail of the palynomorphs present between the Lower Basalt and Middle–Upper Basalt Series. The occurrence of abundant *Inaperturopollenites hiatus, Momipites* spp. with common *Sequoiapollenites polyformosus*, *Caryapollenites* spp. and *Alnipollenites verus* gives a palynoflora closely similar to that in the earliest Ypresian of the West Shetland Basin. This comparison is strengthened by the occurrences of *Alnipollenites trina*, a taxon limited to the uppermost part of Sele Formation Unit 1b and basal 2a (Knox & Holloway 1992) in the North Sea, and by *Montanapollis* sp. This latter taxon is rare in the North Sea Basin, but occurs in the lowermost part of Sele Formation Unit 2a. It is also rare in Mull, where its occurrence in the intrabasaltics was recorded by Srivastava (1975) and is unknown from the rest of the Palaeocene or Eocene of the UK.

In summary, palynofloras of an earliest Ypresian age have been identified in the intrabasaltics of the Faeroes, Mull and Ardnamurchan. They are clearly different in palynofloral composition to those of the early Thanetian and indicate the concentration of extrusive igneous activity on this margin of the north Atlantic rift into three, possibly four periods. The earliest lavas appear to be those of the Eigg Lava Formation, with isotope dates corresponding to Chron 26r (Ritchie & Hitchen, 1996). Approximately four million years separates these extrusives from the Late Palaeocene Antrim/Canna/Skye Lavas of 58.23–57.99 Ma, which are in turn separated from the initiation of activity in the Mull Lavas from 55.00, possibly lasting to around 54.5 Ma (Fig. 12). Finally, the Sgurr of Eigg Pitchstone Formation has provided isotopic dates of around 52 Ma (Dickin & Jones 1983), which if correct, indicate a last phase of BTVP volcanism separated from the Mull activity by a quiescent period of approximately 2.5–3 Ma.

The onset of activity in the Mull and Ardnamurchan centres, corresponds to the resumption of lava extrusion in the Faeroes. To the north of the NAIP, other intrabasaltic sediments reported by Boulter & Manum in ODP Hole 642E on the

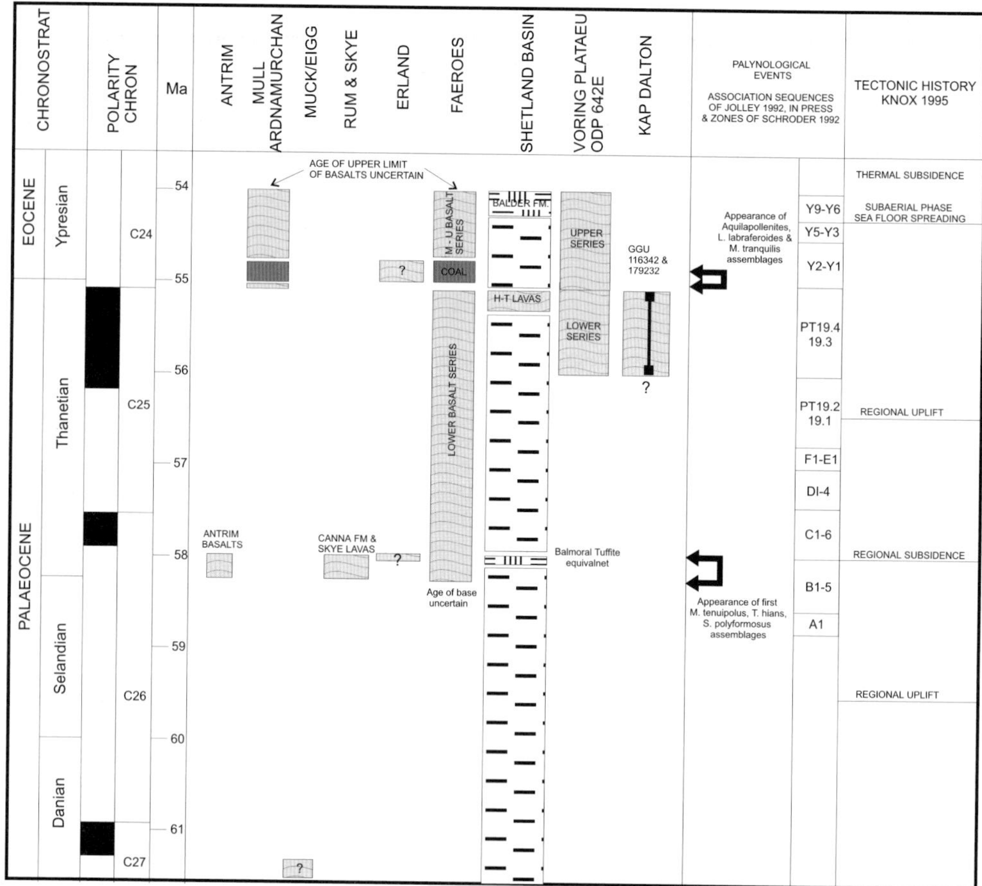

Fig. 12. A correlation of intra and intertrappean sediments in the NAIP. Undulating lines indicate lavas, dashed lines indicate marine sediments, alternating vertical and horizontal lines show tuffs and heavy stipple, thicker intrabasaltic sediments.

Vøring Plataeu yielded palynofloras containing *Apectodinium* species, with a pollen assemblage typical of the last 1 Ma period of the Palaeocene. The position of these intrabasaltic estuarine sediments below the Upper/Lower Basalt series boundary (Viereck *et al.* 1988) in Hole 642E suggests that the boundary may approximate to the Palaeocene–Eocene boundary, also marked by the stratigraphical position of the palynofloras between the Lower and Middle Basalts of the Faeroes and at the base of the lavas in Mull.

A re-examination of archive material in the Centre for Palynological Studies from the work of Soper & Costa (1976) in east Greenland was also undertaken. Sample GGU 179232, taken from the 'basalts without dykes' in the Kap Dalton graben, yielded an assemblage with common specimens of the dinoflagellate cyst *Apectodinium*, similar to that recorded from Hole 642E by Boulter & Manum (1989). A Late Palaeocene, 55–56 Ma age is also suggested by the pollen floras of samples GGU 179232 and 116342 which are rich in *Intratriporopollenites microreticulatus*. The Bopladsdalen Formation sediment samples studied by Soper & Costa were also re-examined, containing a dinoflagellate cyst palynoflora characteristic of the mid-Early Eocene, the dating of this and the underlying basalts suggesting that extrusive igneous activity ceased in this area around the Palaeocene/Eocene boundary. This is concomitant with the date suggested for the boundary between the Lower Basalts and Middle–Upper Basalts of the Faeroes and the age of the lowermost Mull intrabasaltics.

Tectonism

Tectonic and erosive phases on the Skye Lava Field

The presence of the Palagonite Tuffs and the An Leac Formation at the base of the Skye lava pile has been used to indicate the eruption of the initial activity into a shallow saline water environment (Anderson & Dunham 1966; Williamson & Bell 1994). Subsequent activity on the Skye Lava Field has been assumed to have been balanced by subsidence, resulting in a low-lying, but terrestrial, series of eruptions. The work of Meighan *et al.* (1981) and Emeleus (1973, 1985) on the conglomerates of the Canna Lava Formation has indicated that the Rum Centre was an uplifted area, which sourced the fluvial systems draining across the Skye Lava Field. Recently, the reconstructions of Skye and Rum topography proposed by Williamson & Bell (1994, Fig. 28) have followed this, envisaging Rum as being part of an upland area, with low-lying land in the Skye region to the north. A system of fissure–vent volcanoes which parallel the Skye west coast was also proposed, perhaps being related to the dyke-fed volcanoes suggested by England (1992) for the Skye Lava Field.

The acquisition of palynological data from the Skye and Rum intrabasaltics has allowed consideration of the tectonic processes and erosional responses involved in the fields formation. Each palynofloral association recovered, contains a record of the floras that produced it. Such floras are sensitive in their spatial distribution to temperature, showing this by declines with increasing latitude and altitude. It is possible to compare the composition of the fossil floras with those of modern forests, using details of botanical affinity and comparisons of diversity. The difference in composition between the fossil forest type present at sea level in the region at 58 Ma (a mixed mesophytic forest, see Fig. 11) and the forests represented by the intrabasaltic palynofloras, (Fig. 13) is related to altitude. The degree of this difference varies between surfaces E1–E5, dependant upon the change in elevation. The three forest types present in the intratrappean sediments and coeval deposits of the Faeroe–Shetland Basin (mixed mesophytic forest, upland Taxodiaceae forest, montane conifer forest) can be compared to modern forests in China (Wang 1961; Wolfe 1979; see Fig. 13). Using modern sites where a mixed deciduous and evergreen forest occurs as the lowland vegetation, the altitudinal limits of the modern forest types can be applied to the equivalent fossil forest to give a thoretical altitudinal distribution by analogy. Resultant estimations of altitude for the floras of the five erosion surfaces in the west central Skye Lava Field were combined with the thicknesses of the lava groups in west central Skye given by Williamson & Bell (1994) to give a history of the tectonic activity in the field. Altitudinal estimations from the erosion surfaces E1–E5 were taken from the west central Skye area only, to ensure minimization of the effects of local variations in geomorphology.

Accepting that initial eruption of the An Leac Formation and Palagonite Tuffs took place in shallow saline water (?marine) conditions, the fact that the composition of the palynofloral assemblages in the Palagonite Tuff plant beds resembles equivalent palynofloras in West Shetland Basin sections is not surprising. The compositional change in the palynoflora from those derived from a mixed mesophytic forest to those from an upland Taxodiaceae forest between surfaces E1 to E2, suggests a considerable increase in altitude for the area. This altitudinal gain is estimated to be close to that derived from the combined thicknesses of the Rubah' an Dunain and Bualintur Lava Groups

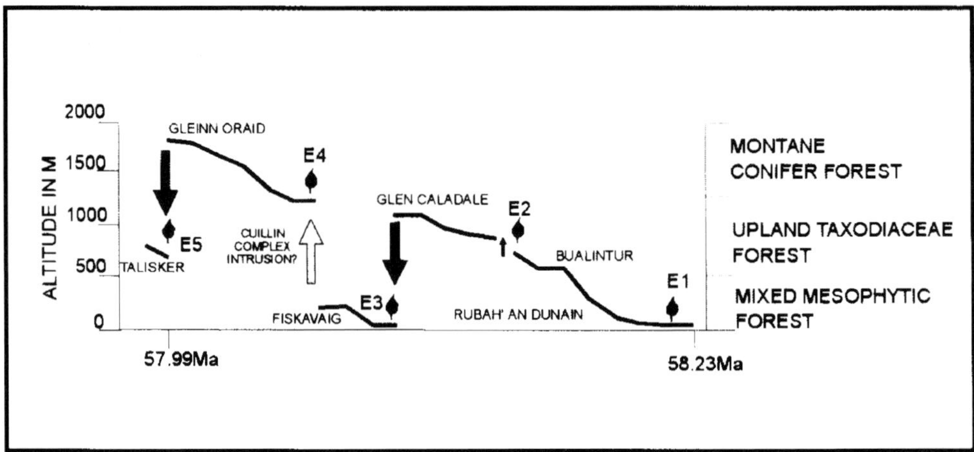

Fig. 13. Subsidence and uplift plot for the west central Skye area during the eruption of the main lavas. Leaf symbols indicate palynofloral sample positions plotted at the repective altitude they indicate, solid black lines indicate the cumulative thickness of lavas between erosion surfaces.

which lie between them, the deficit of 128 m is probably best attributed to erosive processes (Fig. 13). Hence E2 can be interpreted as a constructional palaeosurface produced by the accumulation of lava not balanced by subsidence.

The succeeding palynofloras of erosion surface E3 recorded on Skye and possibly on Rum, are exceptional in their diversity, representing mixed mesophytic swamp forest. This implies a lowland habitat with an elevation below 500 m. As the Glen Caladale Lava Group which underlies this is some 395 m in thickness, a cumulative loss of around 1100 m in altitude over the period between the E2 and E3 surfaces must be accounted for. This is the first major subsidence period identified in the west central Skye record.

Eruption of the 185 m thick Fiskavaig Lava Group prior to the establishment of erosion surface E4 cannot account for the estimated elevation of 1200 m indicated by the palynofloras, implying a major uplift of around 1000 m during this period. This uplift phase appears to have been short-lived, because the 555 m thickness of the succeeding Gleann Oraid Lava Group was erupted during a subsiding phase, prior to the development of the E5 erosion surface. Erosion was undoubtably extensive during this period, suggesting irregular upland topography. Extensive erosion cannot explain the 1000 m+ discrepancy between the altitude suggested by the E4 palynoflora with the additional thickness of the Gleann Oraid Lava Group and that suggested by the palynoflora of surface E5. Erosion would have led to extensive denudation of the uppermost lavas of the Gleann Oraid Lava Group, which is instead preserved by rapid subsidence prior to E5. Above surface E5, there is no evidence as to the altitude of the area during eruption of the Talisker Lava Group.

The stratigraphical position of the Canna Lava Formation above the Rum Central Complex makes it unlikely that emplacement of this complex was responsible for the period of uplift between E1 and E2. The initiation of uplift in the Rum–Canna area appears to have beeen between eruption of the Eigg

Fig. 14. Taxa typical of intratrappean sediments on Skye. All taxa are photographed in transmitted light, phase contrast and are approximately ×1000 unless otherwise stated. (1) Pityosporites labdacus (P. diploxylonoid type) Dun Beag M9 V29/3 ×400. (2) *Pityosporites microalatus* (P. haploxylonoid type) Dun Beag M9 J30/1 ×400. (3) *Inaperturopollenites distichiforme* Allt Mor, Carbostbeg AMC3, W39/1. (4) *Laevigatosporites haardti* Allt Mor, Carbostbeg AMC3, K32. (5)*Inaperturopollenties dubius* Dun Beag M9 O37/2. (6) *Sequoiapollenites polyformosus* Allt Mor, Carbostbeg AMC1, V34. (7) *Pesavis tagluensis* Allt Mor, Carbostbeg AMC1 P34/1 ×600. (8) *Lycopodiumsporites* spp. Allt Mor, Carbostbeg AMC3 J25. (9) *Monocolpopollenites tranquilus* Ardtun Head, ARD6 M46. (10) *Momipites tenuipolus* Allt Mor, Carbostbeg AMC3 P29/3 (11) *Momipites anellus* Allt Mor, Carbostbeg AMC1 P30/4. (12) *Momipites* cf. *tenuipolus* Allt Mor, Carbostbeg AMC1 N27.

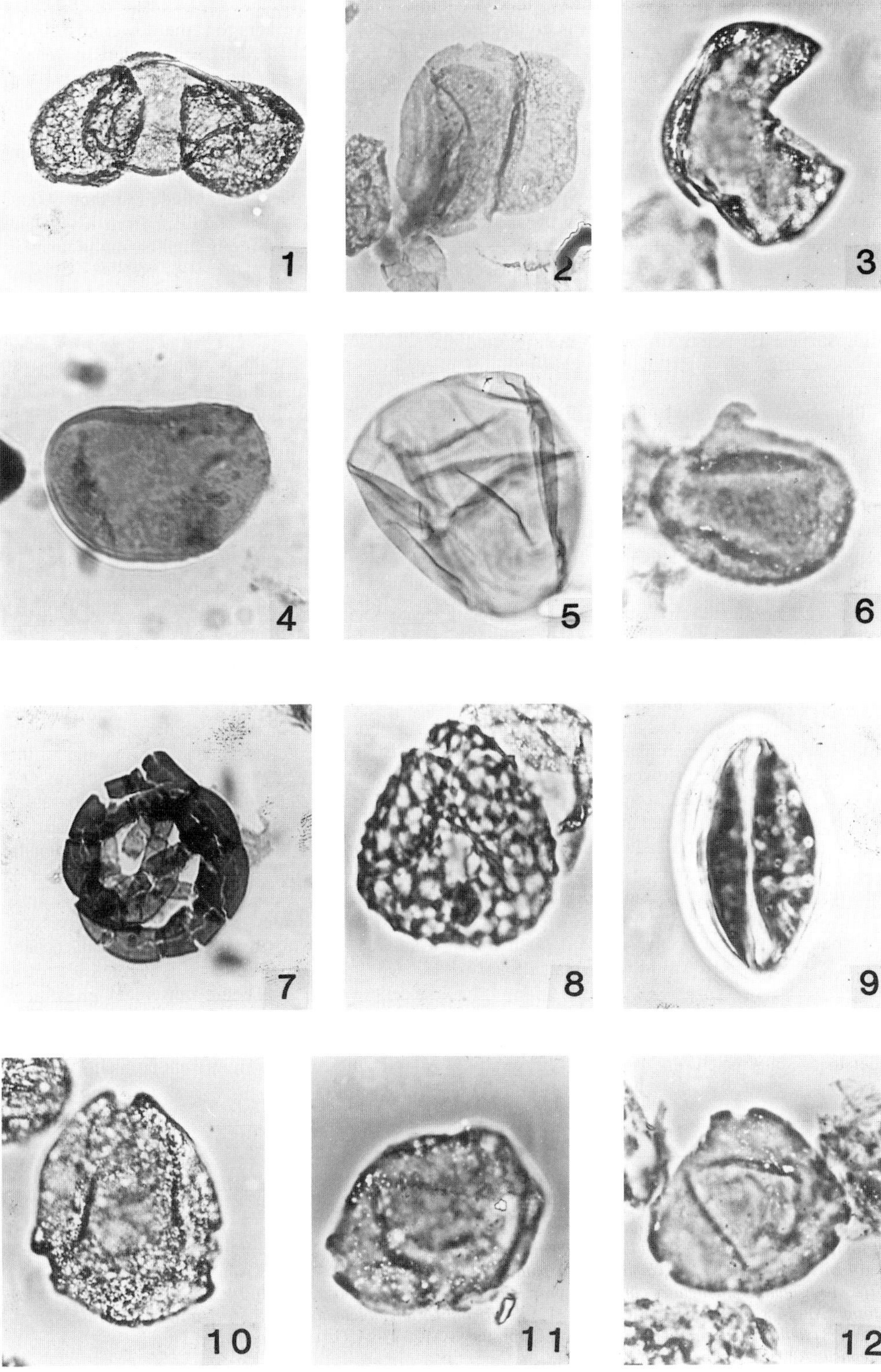
1
2
3
4
5
6
7
8
9
10
11
12

Lava Formation and the eruption of the Canna Lava Formation–Skye Lava Series, and was connected with intrusion of the Rum Centre. Denudation or sustained uplift across the Rum–Sleat area, later sourced the Minginish Conglomerate Formation of E2, as evidenced by sedimentary and lava thicknesses (England 1994). The phase of rapid subsidence to surface E3, followed by subsequent uplift prior to surface E4 is perhaps explained as a slowing down in lava production along the nearby fissures, resulting in thermal subsidence, followed by a change to uplift related to the initial emplacement of the Cuillin Centre at depth. The switch to a more southerly clastic source from the Cuillin area during surface E5 points to emplacement related thermal uplift, although the constant control exercised on lava distribution by active faults (Williamson & Bell 1994) cannot be discounted as a contributory factor.

All of the palynofloras derived from the intra and intertrappean beds of Antrim yield palynofloras of lowland character, implying altitudes of less than 500 m during development of the Lower Basalt and Interbasaltic Formations. This is in direct contrast to the Skye–Rum area, and reflects a more even subsidence rate, creating accommodation space for the thick flows seen in the Antrim area. Cumulative thicknesses of around 800 m have been given for the Tertiary lavas of Antrim (Wilson & Manning 1978), in comparison to the 2000 m for the Skye field. If the two areas are accepted to have been active simultaneously then the eruptive rate of Antrim was considerably slower than that of Skye, with accumulation rates of around 1 m every 237 years, a level of activity that may have allowed compensating subsidence, leading to the development of the lowland palaeosurfaces currently preserved. If the Upper Basalt Formation, and possibly part of the Tholeiitic Member prove to be of a younger age (i.e. around 55 Ma), then this accumulation figure would be considerably reduced.

Tectonic activity associated with the younger Mull and Ardnamurchan lavas is less well understood. Since the current database is limited by the concentration of the efforts of previous palynological workers on the lowermost intrabasaltics (the leaf beds), their remains a general lack of knowledge of the long term subsidence and uplift history of this area. The palynofloras of the leaf beds of Mull analysed for this study were derived from a lowland deciduous forest. This, by comparison to equivalent West Shetland Basin palynofloras, suggests an altitude less than 500 m, although the presence of pollen from a montane conifer forest would imply high elevation areas in the catchment.

Conclusions

Previous studies of the pollen and spore floras of the BTVP intrabasaltic sediments have suffered from the lack of a complete palynofloral record for the Palaeocene in the region, against which to constrain their stratigraphy. Studies of commercial West Shetland Basin sections of terrestrial, shallow marine, and basinal sediments have allowed a composite pollen record to be presented, against which the intrabasaltic floras can be compared. In conjunction with isotopic evidence, this has lead to recognition of four phases of eruptive activity in the BTVP at approximately 52 Ma (Sgurr of Eigg Pitchstone Formation), ?54.50–55.00 Ma (Mull and Ardnamurchan), 57.99–58.23 Ma (Skye and Antrim ?pars.) and approximately 62 Ma (Eigg Lava Formation), dates comparable to those derived from isotopic studies of lavas and to the record of palaeomagnetism. The palynofloras and macrofloras of the Mull and Antrim intrabasaltics have been previously termed the 'BIP flora' (Brito-Arctic Igneous Province) by Boulter & Manum (1989, 1993) and Boulter & Kvacek (1989), who regarded them as mostly pioneer forest, dramatically affected by vegetation destruction from volcaniclastic activity around the Palaeocene/Eocene boundary. While it is agreed that the concept of the BIP as being a discrete high latitude flora, separated from the broadleaved Arctic floras of Alaska by the barrier of the rifting North Atlantic, there remain significant problems with the

Fig. 15. (1) *Plicapollis pseudoexcelsus* Fionchra, 167A, R37. (2) *Plicapollis pseudoexcelsus* Antrim GC1, U27. (3) *Triatriopollenites subtriangulus* Allt Mor, Carbostbeg AMC1 V43. (4) *Erdtmanipollis pachysandroides* Allt Mor, Carbostbeg AMC3 K26/2. (5) *Liquidambarpollenites stigmosus* Allt Mor, Carbostbeg AMC3 R27. (6) *L. stigmosus* Allt Mor, Carbostbeg AMC3 J25. (7) *Alnipollenites verus* Antrim GC1 O32/3. (8) *Tricolpites* cf. *hians* Craigahuilliar Quarry, MPA40386, H41/2. (9) *Cupuliferoidaepollenites liblarensis* subsp. *liblarensis* Allt Mor, Carbostbeg AMC1 D27/2. (10) *Cupuliferoipollenites cingulum* subsp. *pusillus* Allt Mor, Carbostbeg AMC1 P31. (11) *C. cingulum* subsp. *oviformis* Allt Mor, Carbostbeg AMC1 R40. (12) *Retitricolpites anguloluminosus* Allt Mor, Carbostbeg AMC3 V31/3. (13) *Retitricolpites retiformis* Allt Mor, Carbostbeg AMC3 J36/4. (14) *Nyssapollenites kruschi* subsp. *analepticus* Allt Mor, Carbostbeg AMC3 H34.

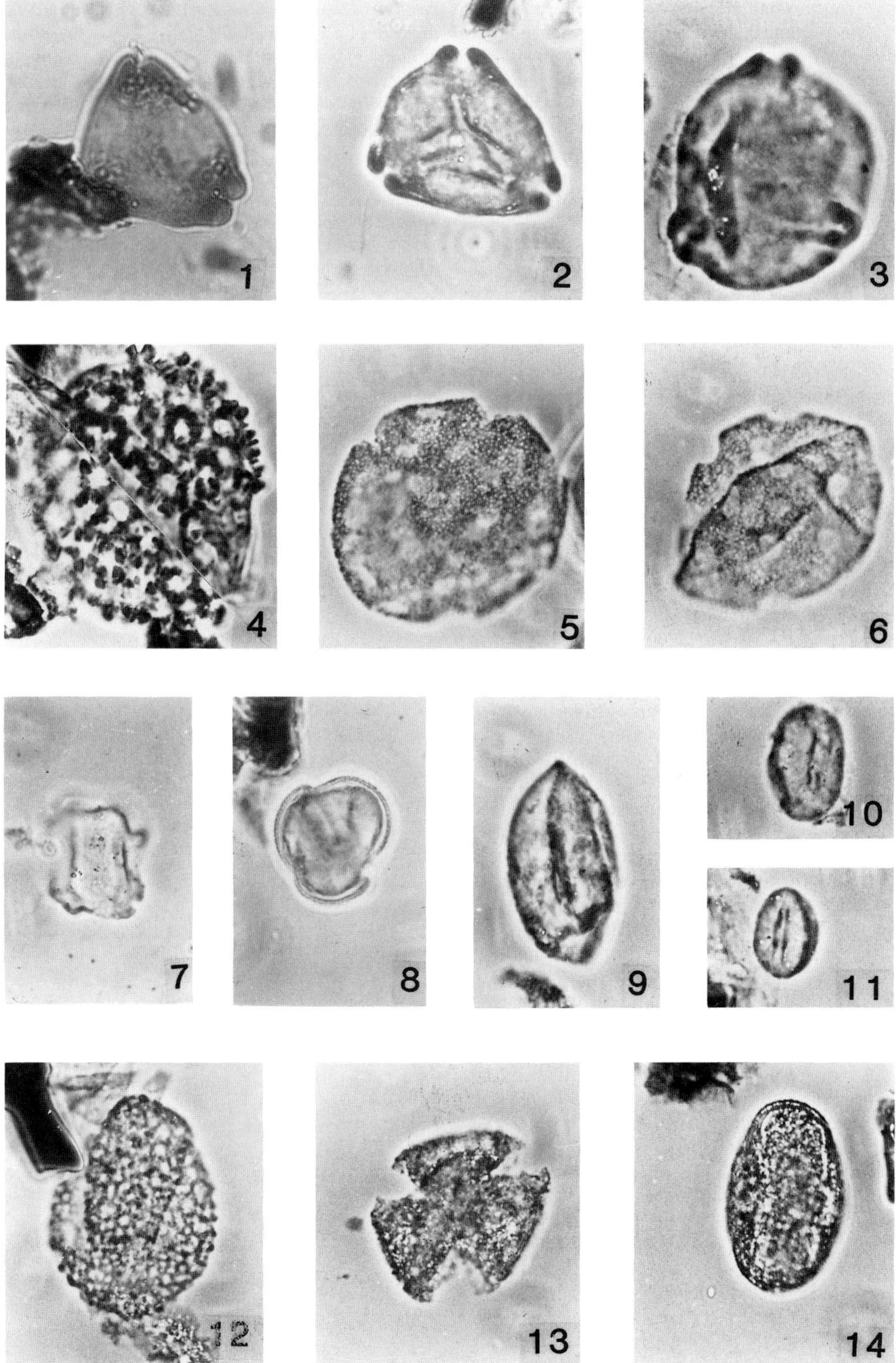
1
2
3
4
5
6
7
8
9
10
11
12
13
14

pioneer forest interpretation. The pollen records of the West Shetland Basin and equivalent records in the North Sea indicate errors of varying magnitude in the current stratigraphical comparison of BIP floras from Antrim, Mull, Ardnamurchan, the Faeroes, and the Vøring Plateau presented by Boulter & Manum (1989, 1993).

Confirmation of the disparate ages of the BTVP intrabasaltics could be taken to indicate that this pioneer vegetation existed throughout the 58.5–54 Ma period, in response to repeated volcanic activity, rather than to the limited Palaeocene/Eocene (NP9/NP10) boundary age. Examination of the pollen profiles from the West Shetland Basin indicates that the floras of this region were not static, repeated pioneer forests, but varied widely in composition with time. They remain broadly similar in composition to the pollen profiles from southeastern England and western continental Europe, since they show similar taxa and variation in response to changing temperatures as seen in equivalent North Sea sections (Schroder 1992). Re-evaluation of dating of BIP floras into a 7.5 Ma time span, which predates the 54.26–54.00 pyroclastic activity of the Balder Formation and its equivalents (see Jolley 1996), removes this supposed major influence on BIP forest structure. Instead, the BIP forests seem to be preserved in areas of mainly extrusive activity in major igneous centres, where they acted as pioneer vegetation. Here, their taxonomic composition depended upon altitude, latitude, prevailing mean annual temperature, soil type, and importantly on the length of quiescent periods between flows. Accordingly, not all the palynofloras of the BTVP intrabasaltics can be regarded as pioneer communities, because the current work demonstrates that (e.g. Allt Mor, Carbostbeg, later phase) secondary or climax forests occured within the lava fields as well as in the margins of the West Shetland Basin. The regional similarity of the palynofloras of the BTVP in the 62.00–54.26 Ma period indicates that while the effects of repeated eruptions within the lava fields and their tectonic evolution (i.e. uplift and subsidence) changed the local forest structure, mixed mesophytic to mixed angiosperm and conifer forests dominated the lowland margin. These stable Palaeocene floras fall within the BIP concept, but do not differ significantly in their taxonomic composition from the mid USA palynofloras of Wyoming (Fredericksen & Christopher 1978; Nichols & Ott 1978; Pocknall 1987). The principal distinction of the BIP lies in the lower diversity (even outside the lava fields) of the palynofloras and macrofloras in comparison to those of souteastern England and the mid USA. This difference is possibly attributable to lower land surface temperatures as a result of the higher latitude of the BIP floras, combined with the western margin isolation of the region. However, the role of elevated atmospheric CO_2 on the floras around volcanic centres is not yet fully understood; in view of the amounts of CO_2 produced at such centres, this may prove to have been a further contributory factor to the structure of the BIP forests.

The author is greatly indebted to Brian Bell for his assistance in locating field sections in west central Skye, for providing material from the Allt Mor section and for informative and enjoyable discussions. Sarah Wilson is thanked for her help in field collecting and for helpful comments. C. H. Emeleus is thanked for sample material and comments on the geology of the Small Isles. Samples from Antrim were provided by Mike Widdowson, who also contributed helpful comments on the manuscript. Additional samples from Antrim were provided by Ian Mitchell of the Geological Survey of Northern Ireland. C. H. Emeleus and J. P. G. Fenton are thanked for constructive reviews of the manuscript.

Taxonomic Note: The author citations of the taxa discussed in the text are as those contained in papers by Jolley (1992, 1996, in press). Illustrations of many of the most frequently occurring taxa, and their typical preservation are given in Figs 14 and 15.

References

ALI, J. R. & JOLLEY, D. W. 1996. Chronostratigraphic framework for the Thanetian and lower Ypresian deposits of Southern England. *In:* KNOX, R. W. O'B. CORFIELD, R. & DUNAY, R. E. (eds) *Correlation of the Early Palaeogene in Northwest Europe*. Geological Society, London, Special Publication, **101**, 129–144.

ANDERSON, F. W. & DUNHAM, K. C. 1966. *The Geology of Northern Skye*. Memoirs of the Geological Survey, Great Britain.

BAILEY, E. B., CLOUGH, C. T., WRIGHT, W. B., RICHEY, J. E. & WILSON, G. V. 1924. *Tertiary and post Tertiary Geology of Mull, Loch Aline and Oban*. Memoirs of the Geological Survey, Great Britain.

BELL, B. R., WILLIAMSON, I. T., HEAD, F. E. & JOLLEY, D. W. 1996. On the origin of a reddened interflow bed within the Palaeocene lava field of North Skye. *Scottish Journal of Geology*, **32**, 117–126.

BOULTER, M. C. & HUBBARD, R. N. L. B. 1982. Objective paleoecological and biostratigraphic interpretation of Tertiary palynological data by multivariate statistical analysis. *Palynology* **6**, 55–68.

—— & KVACEK, Z. 1989. The Paleocene Flora of the Isle of Mull. *Special Papers in Palaeontology*, **42**, 1–149.

—— & MANUM, S. B. 1989. The Brito-Arctic igneous province flora around the Paleocene/Eocene boundary. *In:* ELDHOLM, O., THIEDE, J., TAYLOR, E. *et al. Proceedings of the Ocean Drilling Programme*, Scientific Results, **104**, 663–680.

—— & —— 1993. Further comments on a geological map of the southern Deccan Traps, India and its structural implications. *Journal of the Geological Society, London*, **150**, 791–793.

CANDE, S. C. & KENT, D. V. 1992. A new geomagnetic polarity time scale for the Late Cretaceous and Cenozoic. *Journal of Geophysical Research*, **97**,B10, 13917–13951.

—— & —— 1995. Revised calibration of the geomagnetic polarity time scale for the Late Cretaceous and Cenozoic. *Journal of Geophysical Research*, **100**, 6093–95.

DICKIN, A. P. & JONES, N. W. 1983. Isotope evidence for the age and origin of pitchstones and felsites, Isle of Eigg, NW Scotland. *Journal of the Geological Society, London*, **140**, 691–700.

ELLISON, R. A., ALI, J., HINE, N. & JOLLEY, D. W. 1996. Recognition of Chron C25n in the upper Palaeocene Upnor Formation. *In:* KNOX, R. W. O'B., CORFIELD, R. & DUNAY, R. E., (eds) *Correlation of the Early Palaeogene in Northwest Europe*. Geological Society, London, Special Publication, **101**, 185–194.

EMELEUS, C. H. 1973. Granophyre pebbles in Tertiary conglomerates on the Isle of Canna, Inverness-shire. *Scottish Journal of Geology*, **9**, 157–159.

—— 1985. The Tertiary lavas and sediments of northwest Rum, Inner Hebrides. *Geological Magazine*, **112**, 419–437.

—— & GYOPARI, M. C. 1992. *British Tertiary Volcanic Province*. London, Chapman & Hall.

ENGLAND, R. W. 1992. The role of Paleocene magmatism in the tectonic evolution of the Sea of the Hebrides Basin: implications for basin evolution on the NW Seaboard. *In*: PARNELL, J. (ed.) *Basins on the Atlantic Seaboard: Petroleum Geology, Sedimentology and Basin Evolution*. Geological Society, London, Special Publication, **62**, 163–174.

—— 1994. The structure of the Skye lava field. *Scottish Journal of Geology*, **30**, 33–37.

FREDERIKSEN, N. O. & CHRISTOPHER, R. A. 1978. Taxonomy and biostratigraphy of Late Cretaceous and Paleogene triatriate pollen from South Carolina. *Palynology*, **2**, 113–145.

GARDNER, J. S. 1883–1886. *A monograph on the British Eocene Flora 2*. Palaeontographical Society (Monograph).

HITCHEN, K., & RITCHIE, D. K. 1993. New K-Ar ages, and a provisional chronology, for the offshore part of the British Tertiary Igneous Province. *Scottish Journal of Geology*, **29**, 73–85.

JOLLEY, D. W. 1992. Palynofloral association sequence stratigraphy of the Paleocene Thanet Beds and equivalent sediments in eastern England. *Review of Palaeobotany and Palynology*, **74**, 207–237.

—— 1996. The earliest Eocene sediments of eastern England; an ultra-high resolution palynological correlation. *In:* KNOX, R. W. O'B., CORFIELD, R. & DUNAY, R. E. (eds) *Correlation of the Early Palaeogene in Northwest Europe*, Geological Society, London, Special Publication, **101**, 219–254.

—— in press. Palynostratigraphy and depositional history of the Palaeocene Ormesby/Thanet depositional sequence set in southeastern England, and its correlation with continental west Europe and the North Sea. *Review of Palaeobotany & Palynology.*,

—— & MORTON A. C. 1992. Palynological and petrological characterization of a North Sea Palaeocene volcaniclastic sequence. *Proceedings of the Geologists' Association*, **103**, 119–127.

KNOX, R. W. O'B. 1996. Tectonic controls on sequence development in the Palaeocene and earliest Eocene of SE England: implications for North Sea stratigraphy. *In*: HESSELBO, S. P. & PARKINSON, D. N. (eds) *Sequence Stratigraphy in British Geology.* Geological Society, London, Special Publication, **103**, 209–230.

—— & HOLLOWAY, S. 1992. *Lithostratigraphic nomenclature of the UK North Sea. 1. Palaeogene of the Central and Northern North Sea.* UKOOA, 1–133.

LUND, J. 1983. Biostratigraphy of interbasaltic coals from the Faeroe Isalnds. *In* BOTT, M. H. P., SAXOV, S., TALWANI, M. & THIEDE, J. (eds) *Structure and development of the Greenland - Scotland Ridge*, Plenum, New York, 417–423.

—— 1988. A late Paleocene non-marine microflora from the interbasaltic coals of the Faeroe Islands, North Atlantic. *Bulletin of the Geological Society of Denmark*, **37**, 181–203.

MEIGHAN, I. G., HUTCHISON, R., WILLIAMSON, I. T. & MACINTYRE, R. M. 1981. Geological evidence for the different relative ages of the Rum and Skye Tertiary central complexes (abs). *Journal of the Geological Society, London*, **139**, 659.

MORTON, A. C., EVANS, D., HARLAND, R., KIING, C. & RITCHIE, D. K. 1988. Volcanic ash in a cored borehole W of the Shetland Islands: evidence for Selandian (late Palaeocene) volcanism in the Faeroes region. *In*: MORTON, A. C. & PARSON, L. M. (eds) *Early Tertiary Volcanism and the Opening of the NE Atlantic*. Geological Society, London, Special Publication, **39**, 263–269.

MUSSETT, A. E. 1984. Time and duration of igneous activity on Rhum and adjacent areas. *Scottish Journal of Geology*, **20**, 273–279.

——, DAGLEY, P. & SKELHORN, R. R. 1988. Time and duration of activity in the British Tertiary Igneous Province. *In*: MORTON, A. C. & PARSON, L. M. (eds) *Early Tertiary Volcanism and the Opening of the NE Atlantic*. Geological Society, London, Special Publication, **39**, 337–348.

NICHOLS, D. J. & OTT, H. L. 1978. Biostratigraphy and evolution of the *Momipites - Caryapollenites* lineage in the Wind River Basin, Wyoming. *Palynology*, **2**, 93–112.

POCKNALL, D. 1987. Palynomorph biozones for the Fort Union and Wasatch Formations (Upper Paleocene - Lower Eocene), Powder River Basin, Wyoming and Montana, USA. *Palynology*, **11**, 23–36.

RIDD, M. F. 1983. Aspects of the Tertiary geology of the Faeroe - Shetland channel. *In:* BOTT, M. H. P., SAXOV, S., TALWANI, M. & THIEDE, J. (eds) *Structure and development of the Greenland–Scotland Ridge: New Methods and Concepts*. Plenum, New York, 91–108.

RITCHIE, D. K. & HITCHEN, K. 1996. Early Palaeogene offshore igneous activity to the NW of the UK and its relationship to the North Atlantic Igneous Province. *In:* KNOX, R. W. O'B., CORFIELD, R. & DUNAY, R. E. (eds) *Correlation of the Early Palaeogene in Northwest Europe*, Geological Society, London, Special Publication, **101**, 63–78.

SCHRODER, T. 1992. A palynological Zonation for the Paleocene of the North Sea basin. *Journal of Micropalaeontology,* **11**, 113–126.

SIMPSON, J. B. 1961. The Tertiary pollen flora of Mull and Ardnamurchan. *Transactions of the Royal Society of Edinburgh,* **64**, 421–468.

SMITH, N. J. 1988. The age and structural setting of limestone and basalt on the Main Ring Fault in southeast Rhum. *Geological Magazine,* **122**, 439–445.

SOPER, N. J. & COSTA, L. I. 1976. Palynological evidence for the age of Tertiary Basalts and post-basaltic sediments at Kap Dalton, central east Greenland. *Rapport Grondlands Geoliske Undersøgelse*, 123–127.

——, HIGGINS, A. C., DOWNIE, C., MATTHEWS, D. W. & BROWN, P. E. 1976*a*. Late Cretaceous-early Tertiary stratigraphy of the Kangerdulugssuaq area, east Greenland, and the age of opening of the north-east Atlantic. *Journal of the Geological Society, London,* **132**, 85–104.

——, DOWNIE, C., HIGGINS, A. C. & COSTA, L. I. 1976*b*. Biostratigraphic ages of Tertiary basalts on the east Greenland continental margin and their relationship to plate separation in the northeast Atlantic. *Earth and Planetary Science Letters,* **32**, 149–157.

SRIVASTAVA, S. K. 1975. Maastrichtian microspore assemblages from the interbasaltic lignites of Mull, Scotland. *Palaeontographica Abt.B,* **150**, 125–156.

THOMPSON, P. 1986. *Dating the British Tertiary Igneous Province in Ireland by the* ^{40}Ar–^{39}Ar *stepwise degassing method.* PhD Thesis, University of Liverpool.

——, WATTS, S. & DURRANI, S. A. 1984. Concordant fission track and Ar–Ar age determination for the Interbasaltic Formation of Northern Ireland. *Geophysical Journal of the Royal Astronomical Society,* **77**, 326.

THOMPSON, R. N., ESSON, J. & DUNHAM, A. C. 1972. Major element chemical variation in the Eocene lavas of the Isle of Skye, NW Scotland. *Journal of Petrology,* **21**, 265–293.

VIERECK, L. G., TAYLOR, P. N., PARSON, L. M., MORTON, A. C., HERTOGEN, J., GIBSON, I. L. & THE ODP LEG 104 SCIENTIFIC PARTY 1988. Origin of the Vøring Plataeu volcanic sequence. *In*: MORTON, A. C. & PARSON, L. M. (eds) *Early Tertiary Volcanism and the Opening of the NE Atlantic.* Geological Society, London, Special Publication, **39**, 69–83.

WANG, C. W. 1961. *The Forests of China.* Maria Moors Cabot Foundation, **5**.

WIDDOWSON, M., WALSH, J. N. & SUBBARAO, K. V., 1997. The geochemisty of Indian bole horizons: palaeoenvironmental implications of Deccan intravolcanic palaeosurfaces. *This volume.*

WILLIAMSON, I. T. & BELL, B. R. 1994. The Palaeocene lava field of west-central Skye, Scotland: Stratigraphy, palaeogeography and structure. *Transactions of the Royal Society of Edinburgh: Earth Sciences,* **85**, 39–75.

WILSON, H. E. & MANNING, P. I. 1978. *Geology of the Causeway Coast.* Memoir of the Geological Survey of Northern Ireland. HMSO, Belfast.

WILSON, R. L., DAGLEY, P. & ADE-HALL, J. M. 1972. Palaeomagnetism of the British Tertiary igneous province: the Skye lavas. *Geophysical Journal of the Royal Astronomical Society,* **28**, 285–293.

WOLFE, J. A. 1979. Temperature parameters of humid to mesic forests of eastern Asia and relation to forests of other regions of the northern hemisphere and Australasia. *U.S. Geological Survey Professional Paper,* **1106**, 1–37.

Palaeosurfaces and associated saprolites in southern Sweden

KARNA LIDMAR-BERGSTRÖM[1], SIV OLSSON[2] & MATS OLVMO[3]

[1] *Department of Physical Geography, Stockholm University, S-10691 Stockholm, Sweden*

[2] *Department of Quaternary Geology, Tornavägen 13, S-22363 Lund, Sweden*

[3] *Department of Physical Geography, Earth Science Centre, Göteborg University, S-41381 Göteborg, Sweden*

Abstract: Saprolite remnants from different palaeosurfaces in southern Sweden have been analysed by XRD and SEM analyses. They represent two clearly different types. The first is a clay- and silt-rich saprolite with a kaolinite-dominated clay mineral association representative of mature saprolites and with chemically altered quartz grains. This saprolite type is associated with sub-Cambrian, sub-Jurassic and sub-Cretaceous denudation surfaces. The second type is gravelly and in a youthful stage of alteration with mainly vermiculitic clay minerals in the fine fractions. Microtextures developed on quartz grains indicate a mainly mechanical breakdown. Compared to overlying till beds this saprolite is in a more advanced stage of chemical alteration. It is not associated with any specific denudation surface. Its characteristics and thickness indicate a Plio-Pleistocene age.

The saprolites represent deep weathering of the bedrock surface at different times. The weathering resulted in thin kaolinitc saprolites during the Late Proterozoic, thick kaolinitic saprolites from the latest Triassic through the Jurassic and Cretaceous, and medium thick immature saprolites from the Pliocene and onwards. The depth of the deep weathering has been decisive for the shape of the present relief and thus etch processes have been of fundamental importance in shaping the relief, even in a formerly glaciated area.

Due to movements in the Earth's crust different parts of the Precambrian basement surface of Fennoscandia have been exposed to denudation at different times (Lidmar-Bergström 1995). At the end of the Proterozoic, the Baltic Shield was denuded to a surface of low relief which subsequently was covered by Lower Palaeozoic rocks. Parts of this primary peneplain (mainly sub-Cambrian) have not been re-exposed until recent geological times and so retain almost their original form over large areas in eastern Sweden and in the Middle Swedish Lowlands. Other parts were already re-exposed in the Mesozoic, and deep weathering and subsequent stripping caused a hilly relief, which was partly covered by Jurassic and Cretaceous sediments. Some of these areas have been exhumed late in the geologic history and still have their Mesozoic forms and remnants of the old saprolites, while other areas were exhumed earlier or never had any Mesozoic cover. Here the relief continued to develop. The relief of these latter areas is characterized by plains with residual hills. In contrast to the exhumed forms this relief is labelled 'new'. It is interpreted to be of Tertiary age (Lidmar-Bergström 1982, 1988).

Saprolite remnants occur all over the country (Lundqvist 1985; Elvhage & Lidmar-Bergström 1987). Where they occur in direct contact with cover rocks it is possible to roughly date them. Thin kaolinitic saprolites, which are solidified by diagenetic processes, are encountered below Cambrian cover rocks, and thick kaolinitic saprolites occur beneath and around Jurassic and Cretaceous cover rocks. In addition, there is an abundance of reports on unspecified saprolites.

The aim here is to elucidate further the etch processes during different geological periods. The characteristics of saprolites overlain by cover rocks have been ascertained and compared to the characteristics of other saprolites, enabling distinction between saprolites produced at different times and in different environments, and also to use the saprolites to further confirm extensions of the old denudation surfaces. A second purpose of the paper is to once again stress that the main landforms in southern Sweden are formed by etch processes (Lidmar-Bergström 1982, 1989, 1995) during particular time intervals in the geologic history. Only to a lesser degree do the current landforms depend on glacial erosion.

Study area

The South Swedish Dome (Fig. 1) is an appropriate study area since cover rocks of different ages onlap directly the Precambrian basement over a comparatively small area. Lower Palaeozoic cover onlaps the dome in the north and east, and

From Widdowson, M. (ed.), 1997, *Palaeosurfaces: Recognition, Reconstruction and Palaeoenvironmental Interpretation,* Geological Society Special Publication No. 120, pp. 95–124.

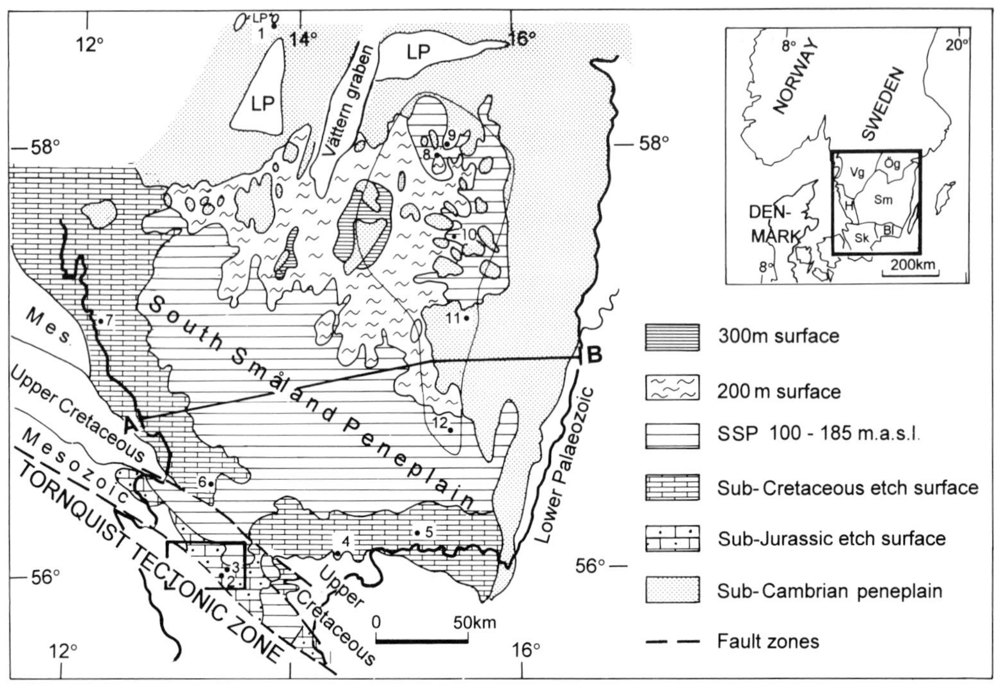

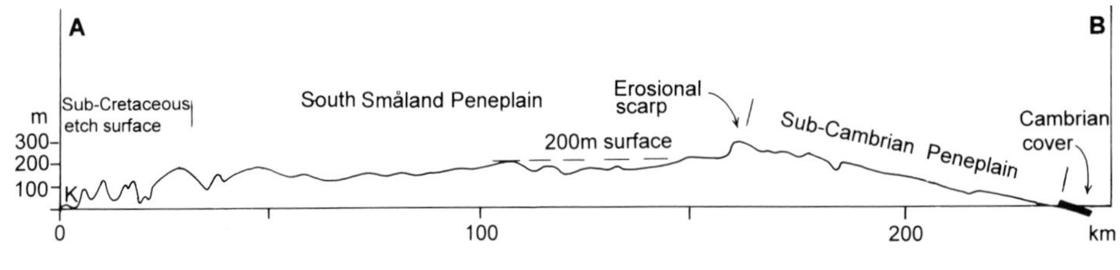

Fig. 1. Denudation surfaces in the Precambrian basement of southern Sweden and associated cover rocks (without symbols), simplified from Lidmar-Bergström (1988). LP, Lower Palaeozoic. Investigated sites with palaeoweathering are marked and numbered. Detailed map of rectangular area around sites 2 and 3 in Fig. 2. The extension of the area with known ocurrences of gravelly saprolites are marked with a thin line around sites 8–12. Provinces: Vg, Västergötland; Ög, Östergötland; Sm, Småland; H, Halland; Sk, Skåne; Bl, Blekinge.

Mesozoic in the southwest. The extent of exhumed (sub-Cambrian, sub-Jurassic, sub-Cretaceous) surfaces as well as younger Tertiary surfaces, have been mapped previously (Lidmar-Bergström 1988, 1994; Fig. 1). The landforms of the exhumed bedrock surfaces and their associated saprolites were identified where they were in contact with the cover rocks, and then followed outwards from this cover with the aid of closely spaced profiles on contour maps. The exhumed surfaces are inclined and are cut by three near-horizontal surfaces of much younger age. These levels may be recognized as follows: the South Småland Peneplain (SSP) with its main surface at 125–185 m above sea level (a.s.l.) and a lower part down to 100 m a.s.l; a 200 m surface with frequent residual hills; and a 300 m surface.

The present climate is temperate with the lowest temperatures in February with means of –1°–3°C and the highest temperatures in July (mean of 17°C). The precipitation varies between 500 mm a^{-1} in the east to over 1000 mm in the western parts (Ångström 1958).

Methods

Sites with saprolite remants were located on the different denudation surfaces and a number of them selected for detailed analyses (Fig. 1). The relationship between the bedrock forms and associated saprolites was studied in the field and associated stratigraphies ascertained. Samples from the saprolites and overlying till beds were studied by X-ray diffraction analysis (XRD). Scanning electron microscopy (SEM) of quartz grains was also used to complement investigation of saprolite material from sites 2, 4, 6, 8 and 12. Grain size analyses were carried out according to standard sieving and pipette methods (Stål 1972). The mineralogical composition of the parent rock from sites 8–12 was determined by standard microscopic analyses of thin sections.

XRD analyses

Samples with high amounts of 'free iron oxides' were treated according to the citrate–bicarbonate–dithionite (CBD) method of Mehra & Jackson (1960). The samples were deflocculated in distilled water by ultrasonic treatment. The clay fraction (< 2 μm) was collected during sedimentation. From some of the samples an additional size fraction 10–2 μm was collected. 'Oriented' mounts were prepared according to the filter-membrane peel-off technique (Drever 1973). Diffractograms were recorded on a Philips diffractometer using the K_{α} radiation of a fine-focus copper tube. The scanning speed was $1° \, 2\theta \, \mathrm{min}^{-1}$. All samples were X-ray scanned and air-dried; (1), after solvation with ethylene glycol (EG) (2), and after saturation with potassium or heating at 400°C (3). Selected samples were also X-ray scanned after: (4) saturation with magnesium and treatment with glycerol, and (5) after heating at 550°C or digestion in hot 2 M HCl. Mineral identifications rest on criteria given by Brindley & Brown (1980).

Semi-quantative estimates of clay minerals are based on peak areas, while for non-clay minerals peak heights have been used. The term 'kaolinite' is used to refer to minerals of the kaolin group, as no differentiation between polymorphs has been attempted. Diffractograms of the samples are shown in Figs 12 and 13.

SEM analyses

The samples were studied using SEM in order to illustrate the microtextural characteristics of quartz grains. Samples were prepared according to standard methods (Krinsley & Doornkamp 1973). Thirty quartz grains (0.25–0.5 mm) from each sample were randomly picked out, mounted on an aluminium specimen plug and coated with gold. In each case all samples were analysed in the SEM (Zeiss DSM940) by noting the existence or non-existence of 30 diagnostically valid surface textures and shape characteristics compiled from Culver *et al.* (1983) and, with some modification, from Bull *et al.* (1987). Finally, the frequency of different textures etc. were calculated.

Exhumed landforms and associated saprolites

Sub-Cambrian landforms and associated saprolites

Where the sub-Cambrian surface extends from beneath the Cambrian cover it is commonly an extremely flat surface which continues in some areas for > 100 km. Its relative relief over these areas is < 20 m (Rudberg 1954; Lidmar-Bergström 1995), but it is occasionally separated into bedrock blocks by low fault-line scarps. Hills rising above the plain in Västergötland (Fig.1) occur as mesas and buttes of Lower Palaeozoic rocks, and the basement below the bottom Cambrian strata may be kaolinized to a maximum depth of 5 m (Elvhage & Lidmar-Bergström 1987 and refs therein). Beyond the Cambrian cover the shallow weathered upper part of the basement surface is never preserved. However, below one small butte of Cambrian strata it was possible to take a sample of this old saprolite in an abandoned underground quarry.

Lugnås (site 1; Table 1). The Proterozoic gneiss, softened by weathering, is somewhat porous but the pores have, as a rule, been filled with quartz and calcite and less frequently with iron oxides, principally haematite (Hadding 1929). The gneiss is directly overlain by an arkose with small inclusions of kaolin. The cement in the arkose consists mainly of calcite. The constituent grains are distinctly worn and indicates aeolian transport. Dreikanter also occurs both in the arkose and in an associated conglomerate. According to Hadding (1929) there is evidence of cementation of the weathered rock prior to the Cambrian transgression since iron oxide cement commonly occurs in the gneiss. In the arkose the oxide is present only in the form of loose grains detached by denudation. A quartz cement has also been deposited before the transgression. It is this cement alone that has prevented the gneiss pebbles in the conglomerate, as well as kaolin inclusions in the arkose, from disintegration. These observations suggest a phase of silcrete and ferricrete formation prior to the Cambrian transgression.

Table 1. *Sub-Cambrian, sub-Jurassic and sub-Cretaceous saprolites and their characteristics*

Sites/location h.a.s.l.(m)	Parent rock composition	Stratigraphy	Mineralogy of < 2 μm fraction
1. **Lugnås** 100 m sub-Cambrian	medium–coarse biotite-gneiss	Cambrian strata arkose/conglomerate	
		1 m cemented saprolite (sample just below cover)	**ka** (il, exp. M-L)
2. **Djupadal** 60 m sub-Jurassic	veined gneiss (Table 2)	2 m fluvioglacial deposits	
		1 m volcanic tuff (L/M Jurassic) 3 m sandstone and clay (L Jurassic) 0.06 m conglomerate	
		44 m saprolite (sample 10 cm below cover)	**ka** (il, I/S)
3. **Snälleröd** 65 m exhumed sub-Jurassic	veined gneiss (Table 2)	glacial deposits	
		Saprolite	**sm,** ka (il, fsp, qz)
4. **Ivön** 10-30 m sub-Cretaceous	coarse granite with biotite (Table 3)	10 m glacial deposits	
		15 m calcarenite (Campan.)	
		>30 m saprolite (sample 3–5 m below cover)	**ka**, il
5. **Dalhejaberg** 15 m exhumed sub-Cretaceous	medium-coarse granite with K-fsp augen (Table 3)	no cover	
		joints	**ka** (il, M-L)
6. **Kåphult** 100 m exhumed sub-Cretaceous/new relief	granitoid gneiss, granitic–granodioritic	*c.* 2 m glacial deposits	
		> 0.5 m saprolite	**sm,** ka (il)
7. **Trönninge** 25–30 m exhumed sub-Cretaceous	gneiss, granodioritic with sillimanite	glacial deposits	
		0.5 m saprolite	**ka**, I/S (ca, qz, chl)

ka, kaolinite; sm, smectite; il, illite; M-L, mixed layer (undefinable); I/S, interstratified illite–smectite; chl, chlorite; fsp, feldspar; qz, quartz; ca, calcite.
Bold type, dominant; normal type, abundant; (parentheses), minor (i.e. < 10 %) component.

Sub-Jurassic landforms and associated saprolites

In central Skåne remnants of Jurassic cover rocks are common (Figs 1 and 2). The exhumed sub-Jurassic landscape is characterized by an undulating hilly relief with a relative relief of 20–50 m and this landscape lies mainly below 100 m a.s.l. At this level it merges with a near horizontal plain of Tertiary age (Lidmar-Bergström *et al.* 1991). The basement lying below or in close proximity of the Jurassic sediments is often kaolinized and these kaolins have been investigated as an economic prospect by The Geological Survey of Sweden (Shaikh 1987). Generally, the thickness of the weathered basement fluctuates between 10 and 30 m. The deposits are extensive and various rock types of Precambrian age are kaolinized. The weathered rocks show clearly the original structure of the parent rock. The depth of the kaolinization

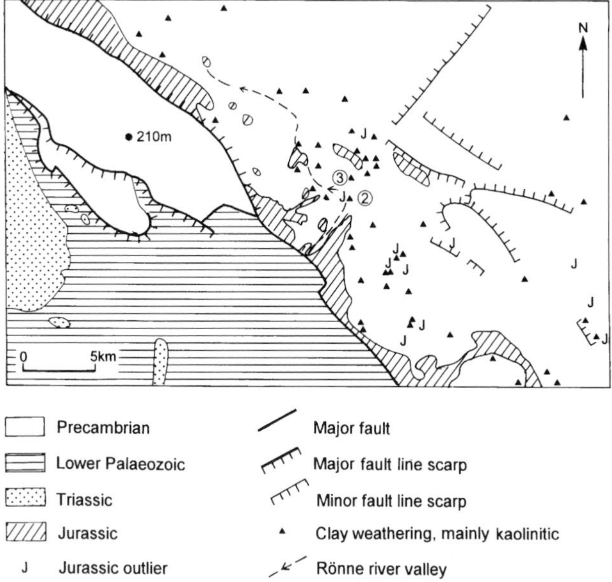

Fig. 2. Kaolinitic saprolite remnants and cover rocks in central Skåne, compiled from several sources. Investigated sites are marked. For location see Fig. 1.

can be up to 50 m. There is a regular vertical zoning with decreasing intensity of kaolinization downwards. The kaolinization probably occurred continously and it is generally assumed that epigene weathering played the dominant role in the genesis of these kaolins. Two sites have been chosen for sampling, the one immediately beneath the cover rocks and the other close by.

Djupadal and Snälleröd (sites 2 and 3; Fig. 2, Table 1). At Djupadal the Rönne River has cut a 20 m deep valley mainly in the Precambrian basement. However, Lower Jurassic strata (including volcanic tuff) form part of the southern valley side, and also occur in the valley northwestwards, indicating that the valley is partly an exhumed sub-Jurassic depression. The continuation of the valley to the southwest is along a fracture line. The present valley is a Late Weichselian melt-water channel and it also contains eroded fluvioglacial deposits (Ringberg 1984). The relative relief in the fresh basement amounts to at least 90 m.

A road cut shows the kaolinized basement beneath the Lower to Middle Jurassic sediments (Norling *et al.* 1993; Tralau 1973) and a nearby boring has penetrated 44 m of kaolinized gneiss. Fresh bedrock is exposed in the valley bottom in the form of small ridges and cupolas (in gneiss) and tors (mainly in amphibolite). The stripping of the saprolite in the valley seems to be largely a Late Weichselian phenomenon (Lidmar-Bergström *et al.* 1991).

Two samples have been collected from the kaolinized basement (originally a veined gneiss), one from the bottom of the road cut at Djupadal (Fig. 2, site 2) and the second along a rivulet at Snälleröd 1.5 km NNE of Djupadal (Fig. 2, site 3). The sample from Djupadal consists of massive, soft lumps of a white, fine-grained material lacking any visually detectable grains of primary minerals. Chemical data on this argillized material and its parent rock collected from borings in the Djupadal–Snälleröd area (Table 2) show that the weathered material is highly depleted in cations (calcium, potassium and sodium) and that kaolinization is significant. The modal composition of the kaolinized rock (material < 0.025 mm) corresponds to a kaolinite content exceeding 85%, which agrees with the result of the XRD analysis of our sample from Djupadal.

The material from Snälleröd, site 3, is less weathered . The rock is certainly disintegrated, but the angular, gravel-sized material contains less altered granules of gneiss. Some remnant feldspar and quartz are found in the clay fraction, which is, in contrast to the Djupadal sample, dominated by smectite, although kaolinite is also abundant.

Sub-Cretaceous landforms and associated saprolites

An undulating hilly relief emerges from below a cover of Cretaceous rocks in NE Skåne and southern Halland (Fig. 1; Lidmar-Bergström 1982,

Table 2. *Chemical composition of parent rocks and washed kaolin (< 0.025 mm) in the Djupadal–Snälleröd area*

	parent rock	washed kaolin	Modal composition (%)	(washed kaolin)
SiO_2	66.7	45.1	Kaolinite	86
TiO_2	0.62	0.86	Illite	3
Al_2O_3	14.4	35.8	Berthierine	5
Fe_2O_3	3.4	3.7	Smectite	–
MnO	0.07	0.01	Quartz	1
CaO	1.7	0.1	K-feldspar	1
MgO	0.64	0.28	Carbonate	–
Na_2O	3.5	–	Anatase	0.9
K_2O	5.1	0.4	Others	3
H_2O^+	0.6	12.8		
H_2O^-	0.3	1.2		
CO_2	1.1	< 0.1		
BaO	0.19	0.03		

Source: Lidmar-Bergström *et al.* 1988.

1985, 1989). The relative relief varies between 20 and 200 m, but to the east of Skåne, in Blekinge, it is replaced by a joint-aligned valley landscape with a relative relief of up to 50 m, whilst to the north, along the west coast, it is replaced by a joint-aligned valley landscape with a relative relief between 20 and 100 m.

The undulating hilly relief in northeast Skåne is associated with frequent occurrences of remnants of a kaolinitic saprolite up to 60 m thick. The shaping of the relief by kaolinitic deep weathering and subsequent stripping is best observed in the abandoned quarry on Ivö island in Ivösjön (Fig. 3; Lidmar-Bergström 1989). Remnants of the Cretaceous cover are also found within the joint valleys in Blekinge (Mattsson 1962; Kornfält & Bergström 1986; Kornfält 1993). The northernmost Cretaceous outlier on the west coast is encountered at Särdal (Bergström *et al.* 1973; Fig. 1, at A). The Cretaceous cover rocks rest both on fresh and kaolinized bedrock. Kaolinization has also been reported from several drillings along the west coast and below Cretaceous deposits in southern Halland (Fig. 1; Lidmar-Bergström 1982).

Ivön (site 4; Table 1). At the northern border of the Cretaceous cover in NE Skåne several hills of Precambrian bedrock protrude through the cover. Similar hills with characteristic steep slopes make up the undulating hilly relief to the north of the border of the cover. At the foot of the northern end of the hill, Ivöklack, remnants of the Cretaceous cover of Campanian age (Christensen 1973) have been quarried. Before removal, the Cretaceous strata rested on both unaltered and deeply kaolinized granite, the latter reaching a maximum depth of at least 30 m. Quarrying of the kaolin has exposed parts of the weathering front, which is distinct and gives rise to a steep bedrock wall (Fig. 3). Thus, it is here clearly demonstrated how the steep bedrock walls of the surrounding hills have been mainly formed by deep weathering and subsequent stripping. The rock is sheeted and the weathering has widened the fractures. Large corestones are common. A Cretaceous conglomerate overlies the basement along the hill slope. The conglomerate consists of large boulders of unaltered local granite and a matrix of a calcarenite. The granite boulders were concentrated after erosion of the saprolite but before the Campanian transgression. In contrast, the corestones of the saprolite are weathered and thus some weathering has continued in the saprolite after its covering by the Upper Cretaceous strata.

Samples were taken of a reddish and a grey variant of the kaolinized granite. The $< 2\,\mu m$ fraction of both samples has a similar mineralogical composition with *c.* 90% kaolinite and 10% mica. The results conform with chemical data on the kaolinized rock (cf. Table 3) and the low cation content (Ca, Na and K), which suggets that both feldspars (originally 68%) have been more or less completely replaced by kaolinite (potassium in the analysis is derived mainly from the mica).

Saprolite remnants within the sub-Mesozoic hilly relief

*Dalhejaberg (*site 5; Table 1). Dalhejaberg is a hill situated in Blekinge within the joint valley

Fig. 3. The kaolin clay pit at Ivön, Skåne. The uppermost part of the rock has been exhumed by quarrying of the Upper Cretaceous limestone at the end of the last century; the lower steep part of the bedrock surface is the weathering front, exposed by exploiting the kaolinitc saprolite. The form of the fresh bedrock slope is identical with slope forms of other residual hills in the surrounding area.

Table 3. *Mineralogical and chemical composition of the unweathered Vånga granite (Vå)* and chemical composition of the overlying kaolinitic saprolite†, Ivön, site 4*

Minerals (% of volume)	Granite	Chemical composition of samples < 5 µm	Grey	Red
Quartz	25	SiO_2	48	46.3
Plag. (incl. sericite	17	Al_2O_3	37	35.0
K-feldspar	51	Fe_2O_3	1.10	2.73
Biotite	5	TiO_2	0.03	0.20
Muscovite	–	CaO	0.22	0.16
Chlorite	–	MgO	0.08	0.49
Epidote	+	K_2O	0.79	0.88
Phrenite	+	Na_2O	0.07	0.12
Pumpellyite	–	L.O.I.	13.10	14.10
Titanite	–			
Zircon	+	Modal composition		
Apatite	+			
Fluorite	1	Koalinite	88	88
Topaz	+	Mica	11	10
Calcite	–	Quartz	–	–
Opaques	+	Montmorillonite	1	2

* *Source*: Kornfält & Bergström 1990; C. Bristow pers. comm.
† *Source*: Kornfält 1993.

landscape. The hill has the same steep-sided shape as the hills in NE Skåne and is thus thought to have formed by Mesozoic deep weathering and subsequent stripping (Lidmar-Bergström 1989). The hill is elongated in a NNW–SSE direction and is 50 m a.s.l. It is below the Late Weichselian highest shoreline, which here ranges between 55 and 65 m a.s.l. (Ringberg 1971). The bedrock surface is exposed along the steep slopes (Fig.4). The hillside shows no signs of glacial erosion and it would have been to the lee of earlier ice movements which ranged from N20–40E to N5–10W (Ringberg 1971). Joints widened by weathering occur along sections of the southwestern side (Fig. 5).

The sample was taken from weathered Karlshamns granite in one of the joints. It consists mainly of gravel and sand in equal proportions. Its clay fraction is almost pure kaolinite.

Kåphult (site 6; Table 1). Kåphult is situated within the undulating hilly land in southern Halland with a local relative relief of 50 m. The site is also close to the lower level (100 m a.s.l.) of the SSP. In the area there are frequent occurrences of Upper Cretaceous flints in both till and fluvioglacial deposits. They are often weathered and interpreted as residues from a former Cretaceous cover (Lidmar-Bergström 1982). The closest occurrence of Cretaceous cover rocks (Albian–Cenomanian) *in situ* is 13 km to the SW. The site is on a SW tilted tectonic block. Detached Cretaceous flints 1 km SW of the site might be derived from a preglacial occurrence in the depression between two tilted blocks (Fig. 6a). To the southwest of a small hill, rising 30 m above the immediate surroundings, a gravel pit had penetrated down through the till cover and exposed the weathered basement, now a sandy saprolite. The boundary between the weathered rock and the till is slightly undulating and the uppermost part of the saprolite is, in places, slightly folded. The saprolite is clearly truncated. The site is in a lee side position in relation to the main direction of the Pleistocene ice movements from the NE, and well above the highest Late Weichselian shoreline (Robison 1983).

A sample was taken *c.* 0.5 m below the boundary with the overlying till. Its clay fraction is smectite-dominated (> 75%) and has approximately equal proportions of kaolinite and illite. Kaolinite is, however, dominant in the 10–2 µm fraction, thus reflecting the contrast in the preferential size distributions of kaolinite and smectite. It is also

Fig. 4. Site 5, Dalhejaberg. The steep slope is characteristic of the residual hills formed in granites by the Mesozoic deep weathering and subsequent stripping of the saprolite.

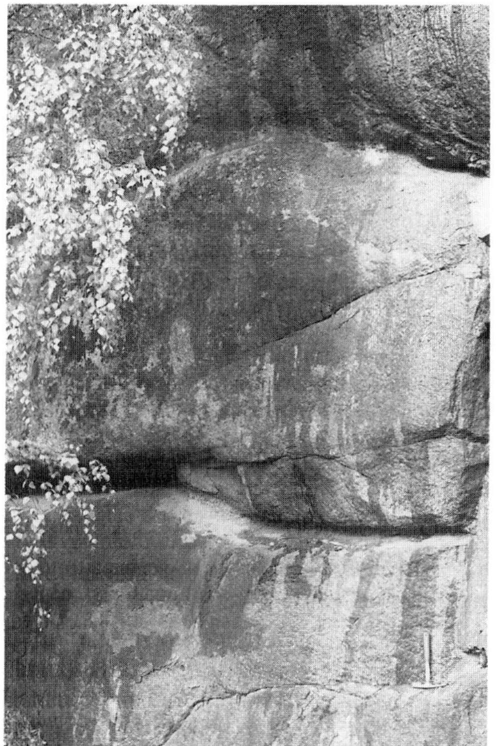

Fig. 5. Detail of the hillside at Dalhejaberg, located immediately to the right of the hill side shown in Fig. 4. The sample of weathered rock was taken from the innermost part of the weathered joint. Note the forms of the hillside, which lack glacial overprinting.

noticeable that the smectite does not dissolve in HCl, a fact that suggests that it is a type displaying high chemical stability.

Trönninge (site 7; Table 1). The site is situated in central Halland on bedrock block tilted to the southwest (Fig. 6b), the summit surface of which coincides with the sub-Cambrian peneplain since Cambrian fissure fillings occur in the area (Samuelsson 1982; I. Lundqvist, Geol. Survey, Göteborg, pers. comm.). The relative relief in the topography of the bedrock block increases from *c.* 50 m at the coast to > 135 m 15 km inland.

In connection with road constructions, a weakly undulating bedrock surface was exposed below the cover of Weichselian glacial deposits. A weathering mantle, *c.* 0.5 m thick, was preserved in depressions. The relative relief on a macroscale is here *c.* 90 m. The sampling site is situated on a horizontal bedrock surface 500 m NE of a residual hill reaching 71 m a.s.l. The main direction of the Weichselian ice movements in the area was from NE, and the highest Late Weichselian shoreline is at 70–75 m a.s.l. Pre-Weichselian interglacial and glacial deposits have been found in the area (Påsse 1990).

The gneiss has become disintegrated into a white, poorly-sorted gravelly sand. The clay fraction is kaolinite-dominated with smectite as the second most abundant phyllosilicate. Chlorite is present, as are non-clay minerals, mainly quartz and calcite.

Location of sites with gravelly saprolites (gruss)

The gruss sites are distributed over an area in eastern Småland and southern Östergötland within the areas of 'new relief' on the South Swedish Dome (Fig. 1).The sites can be grouped as follows (Fig. 7). (1) Locations in close proximity to the dissected sub-Cambrian peneplain, which constitutes the summits in large parts: (a) close below the almost intact peneplain; (b) in joint valleys incised in the peneplain; and (c) at the sides of hills, which have been isolated by denudation of the peneplain. (2) Locations on the SSP at: (a) hill sides; (b) hill tops; and (c) shallow valleys. Five out of 35 sites have been selected for more detailed studies.

Knasekärret (site 8; Table 4). The site is on the SE side of a residual hill reaching 30 m above the adjacent area. A large gravel pit has been opened in the decomposed granite. Corestones in all stages of development are frequent along the entire section (Figs 8 and 9), and large rounded boulders of the local rock type are abundant in the surrounding terrain. The latter are interpreted to have been detached from the saprolite by glacial erosion.

The samples taken *c.* 2.5 m below the ground surface, just below and above the saprolite/till contact, demonstrate a discontinous mineralogical trend across the boundary. The clay fraction of the saprolite material contains, besides unaltered primary minerals of the granite, a mixture of initial (e.g. vermiculite) and interim-end products of weathering (i.e. kaolin minerals). The major constituent, a kaolin mineral, produces very broad basal reflections and a diffraction band at 20–23° 2θ, suggesting that the mineral has a highly disordered structure and is probably halloysite. The chemical stability of the 14 Å phase is low, which is shown by complete dissolution with acid treatment.

The overlying till has a mineralogy which best can be described as youthful, since several minerals occur that are highly susceptible to dissolution or

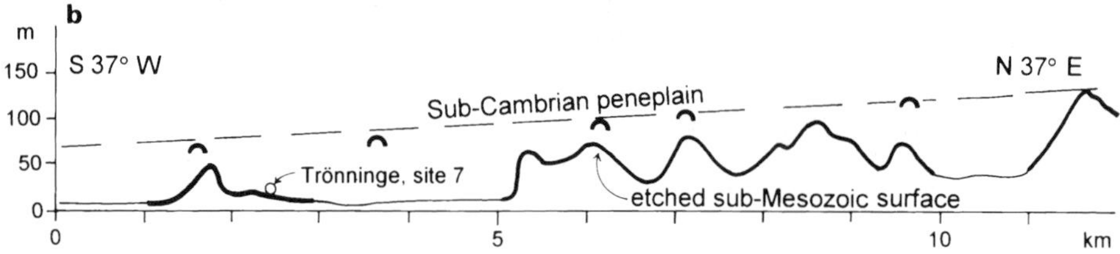

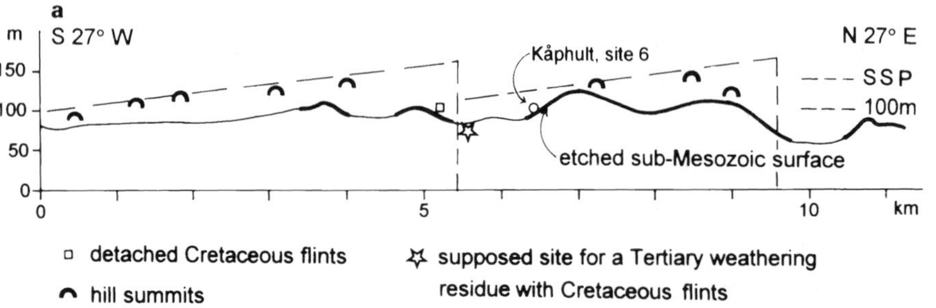

Fig. 6. Profiles showing the locations of sites 6 (**a**) and 7 (**b**) within an undulating hilly terrain. The thick line reflects where the bedrock is close to the surface.

vermiculitization during pedogenesis (e.g. amphibole, chlorite, biotite-like illite).

Malexander (site 9; Table 4). A residual hill, reaching 220 m a.s.l. (approximately the peneplain level) and 40 m above the adjacent area has been cut through for a new road. The bedrock is heavily weathered to gruss along orthogonal fracture systems (Fig. 10). The weathered parts of the rock are easily distinguishable from the fresh rock due to a red–brown staining and darkening by moisture. The contact with fresh rock is often very sharp. In the central part of the section weathering along vertical joints extends from the surface to below the road surface, i.e. > 15 m. Till can be found in 'pockets' at the surface, elsewhere unaltered rock is exposed. The width of weathering joints vary from centimetres to zones *c.* 5 m wide. Large rounded to semi-rounded boulders of the local rock type are abundant in the surrounding terrain and some of them, now detached from the residual hill, can easily be traced back to their original positions. The weathered rock normally consists of gravel but a higher clay content has been observed in some fractures. Three samples were taken for analysis at depths varying from 10.5 to 12 m below the surface.

The grain size of the samples varies from gravel to moderately sorted silty clay. The clay fraction of all samples is dominated by smectite and illite, but differs with respect to their quartz content; the clay-rich sample has no quartz. All samples also contain secondary iron minerals, such as haematite (peaks at 3.67 and 2.69 Å) and lepidocrocite (δ-FeOOH; peak at 6.26 Å), which are dissolved by the CBD treatment (cf. Fig. 13*b*).

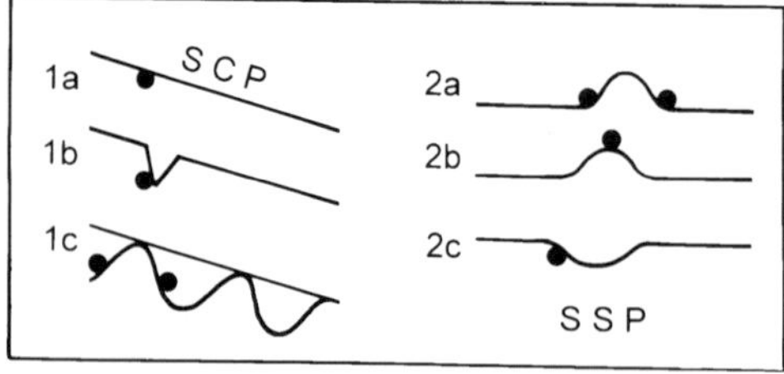

Fig. 7. Location of sites with gravelly saprolites in relation to: (1) the sub-Cambrian peneplain (SCP): (a) close below; (b) in joint aligned valleys incised in the peneplain; (c) at the sides of hills isolated by denudation below the peneplain surface; and (2) the South Småland peneplain (SSP): (a) at hillsides; (b) at hill tops; (c) in shallow valleys.

Duvedal (site 10; Table 4). A gravel pit is located on the northwest slope of a large residual hill reaching 242 m a.s.l. (approximately the peneplain level) and 77 m above the surrounding area. The bedrock has disintegrated to a depth exceeding 3 m. Corestones in *in situ* positions are frequent in the central part of the gravel pit. They may originally have been more common in other parts of the pit judging by the abundance of rounded boulders left on the site. Besides the overlying till bed, a varved glacial clay was encountered in a depression in the southeast part of the gravel pit resting on the fractured and partly disintegrated bedrock surface. The clay has been deformed and appears, in places, to be injected into the saprolite structure.

Also these samples demonstrate a discontinuity in the clay mineral evolution across the boundary between the saprolite and the cover units. For

Table 4. *Sites with gravelly saprolites*

Sites/location a.s.l. (m)	Parent rock composition	Stratigraphy at sampling point	Mineralogy of < 2 μm fraction
8. Knasekärret 180 hillside (1c) 25 m below sub-Cambrian peneplain	Quartz 24% Plagioclase (sericite) 38% Alkali feldspar 31% Biotite (chlorite) 6% Ass. min.: apatite, sphene, opaques	0.5–2.5 m Sandy till (0.5 m Redeposited gravel) >9 m Saprolite with corestones	**chl, il, qz,** fsp, (amph, ver, M-L, ka) **ka (halloysite?)**, verm,il, (qz + fsp < 20%)
9. Malexander 200–220 summit (1a) close below sub-Cambrian peneplain	Quartz 23% Plagioclase (sericite) 48% Alkali feldspar 19% Biotite (chlorite) 7% Ass. min.: apatite, sphene, zircon, opaques	Orthogonal fracture systems; rock compartments changing to corestones in the upper part; fine laminaton of saprolite in close contact with unaltered rock	**sm, il** (lep, hae)
10. Duvedal 185 hillside (1c) 60 m below sub-Cambrian peneplain	Quartz 18% Plagioclase (sericite) 37% Alkali feldspar 32% Biotite (chlor., epid.) 8% Ass. min.: apatite, sphene, epidote, hornblende, opaques (common)	2.5 m Sandy till; glacial clay Redeposited gravel 3 m Saprolite	**qz, fsp,** verm; **verm, il, fsp**, qz, (chl, hae) **verm, V/S,** ka, (qz, fsp)
11. Hulu-Triabo 190 valley (1b) 30 m below sub-Cambrian peneplain	Quartz 16% Plagioclase (sericite) 47% Alkali feldspar 22% Biotite (chlorite) 13% Ass. min.: apatite, sphene, epidote, rutile, opaques (common)	Up to 15 cm wide steeply-dipping fractures; early stage of corestone development	**qz, fsp, verm***, il, chl, M-L
12. Skruv 145 valley (2a) in SSP	Quartz 26% Plagioclase (sericite) 49% Alkali feldspar (13% Biotite (chlorite, epidote, muscovite) 9%	0.75 m Ablation till 2.5 m Basal till 0.5 m Redeposited gravel > 5 m Saprolite	 **qz, fsp,** il, M-L[†] **verm*, il,** M-L, qz, fsp

ka, kaolin mineral; sm, smectite; il, illite; M-L, mixed layer (undefineable); verm, vermiculite; [*], low charge; [†], Al-hydroxi interlayered; V/S, interstratified vermiculite-smectite; chl, chlorite; fsp, feldspar; qz, quartz; amph, amphibole; lep, lepidocrocite; hae, haematite.
Bold, dominant; normal, abundant; (normal), minor (< 10%) component.

Fig. 8. Site 8, Knasekärret. Saprolite with corestones.

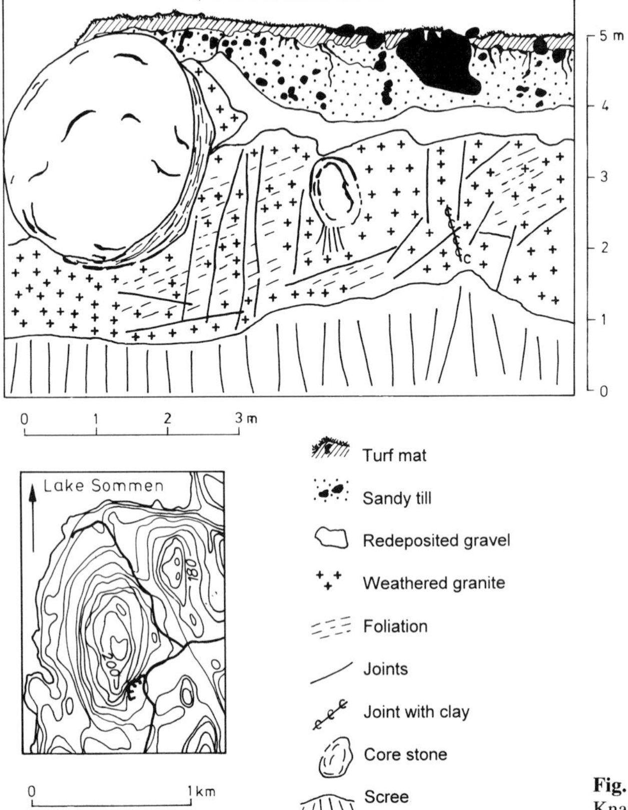

Fig. 9. Sketch of the section at site 8, Knasekärret.

Fig. 10. Site 9, Malexander. Deep weathering along joints and incipient corestones.

instance, kaolin, which is relatively abundant in the clay fraction of the saprolite, cannot be detected in the overlying units. Mineralogical differences between the overlying sediments, such as the high vermiculite content of the clay, which contrasts with a low abundance and/or very poor crystallinity of the phyllosilicates in the coarse diamictite, may partly reflect mineralogical differences inherent with the genesis of the sediments. Consequent contrasts in permeability etc. can also explain that the susceptibility to weathering differs and has determined the evolution of clay minerals in the sediments.

Hulu-Triabo (site 11; Table 4). In a shallow valley incised just 30 m below the sub-Cambrian peneplain, weathered rock was encountered in an excavated rock outcrop, *c.* 30 m long and 4–5 m high. Weathering has taken place, mainly along steeply WSW-dipping fractures, to a depth exceeding the height of the excavated section. Till can be found in 'pockets' at the surface, otherwise unaltered rock outcrops at the ground surface. The sample was taken in a fracture 2 m below the bare bedrock surface.

The clay mineral association, with approximately equal proportions of vermiculite (including a low-charge type), illite, chlorite and interstratifications of 10 and 14 Å minerals, is essentially the same as in the samples from sites 10 and 8, with the exception that the kaolin mineral is not present.

Skruv (site 12; Table 4). In a shallow, *c.* 10 m deep valley, in the South Småland Peneplain a gravel pit has been open for many years in the decomposed bedrock. The saprolite is developed both in granodiorite, dolerite and the more metamorphosed facies of the granitoid. Corestones in all stages of development are found along the section and some fine examples were found in an earlier excavation in the central part of the gravel pit (Fig. 11). Glacially detached, well-rounded boulders of the local bedrock, interpreted as corestones from the local saprolite, are typical for this area.

The clay mineral suite of the saprolite, consisting of approximately equal proportions of vermiculite

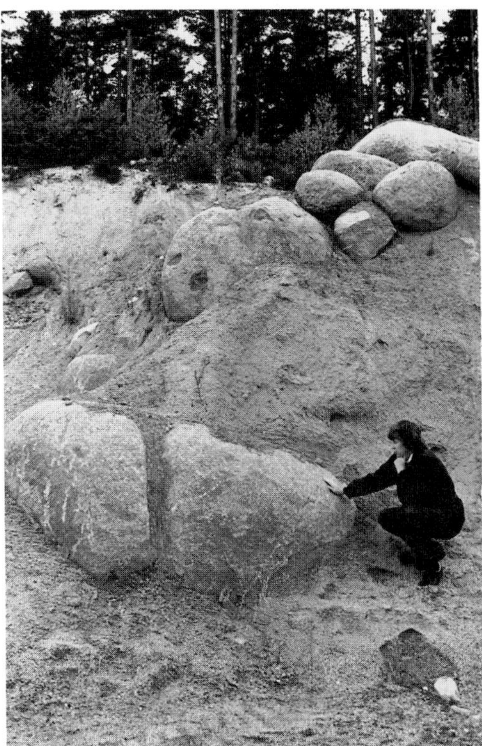

Fig. 11. Site 12, Skruv. Corestones *in situ*.

(including a low-charge type), illite, and interstratifications of illite and 14 Å minerals, is similar to that of site 11. The mineralogy of the till at a depth of *c.* 1 m is, in part, a result of processes connected to pedogenesis. The phyllosilicates are low in abundance and/or have a poor crystallinity. Various poorly-defined interstratified minerals occur, including a vermiculite phase with chloritic or aluminium hydroxide interlayers, which typically forms under slightly acid conditions during pedogenesis.

Interpretation of XRD and SEM data

XRD analyses

The results of the XRD analyses are given in Tables 1 and 4, and is also shown with diffractograms in Figs 12 and 13.

Argillaceous saprolites from stratigraphically controlled sub-Cambrian and sub-Mesozoic sites. The clay mineralogy of the sub-Cambrian (site 1), sub-Jurassic (site 2), and sub-Cretaceous (site 4) saprolites from southern Sweden is characterized by very high ratios of kaolinite to 2 : 1 phyllosilicates. The latter minerals are mainly illite/mica and expansible mixed-layered clay minerals. Hence, all these samples are approaching a single-phase assemblage of kaolinite, which is characteristic of intensely altered saprolites in the 'final' stage of weathering irrespective of parent rock compositions (see review in Weaver 1989; Hall *et al.* 1989; Störr 1993). Site 3 (exhumed sub-Jurassic) is stratigraphically and geographically closely connected to site 2, but the samples from the former site are different, having two major phases, smectite and kaolinite. A vertical zoning with decreasing kaolinization, associated with downwards increasing 2 : 1 clay mineral content, is frequently encountered in bodies of residual kaolin (Velde & Meunier 1987; Störr 1993) and was also observed in this sub-Jurassic saprolite. It is therefore likely that sample 3 simply reflects the weathering regime prevailing in the relatively rigid structure of the relict rock close to the weathering front, where smectite is stable due to impeded drainage conditions.

Other argillaceous saprolites. Sites 5–7 are not stratigraphically controlled by overlying cover rocks, but are situated within undulating hilly sub-Cretaceous relief. Site 6 is also close to the 100 m level of the SSP. High kaolinite dominance characterizes samples 5 (Dalhejaberg) and 7 (Trönninge), whereas sample 6 (Kåphult) has two major phases, smectite and kaolinite. The smectite has a high chemical stability since it is not severely attacked by hot 2 M HCl. These latter samples may be interpreted as representatives of different weathering regimes at varying depths in a mature saprolite body. The routine XRD analysis gives little support for differentiating them from the previous group of samples on clay mineralogical grounds.

Gravelly saprolites (gruss). The sample population is fairly heterogeneous with respect to mineralogical composition. All samples are multiphase associations with easily weatherable, trioctahedral 14 Å minerals being ubiquitous constituents. Heterogeneity in composition, both on a large and small scale, can be expected in granitoid rock in its early stage of decomposition. In fracture systems, where water fluxes are high, argillization may be intense, whereas the unfractured rock may persist almost unaltered. On a small scale, multiphase aggregates form due to the transformation of primary mineral grains along cleavage planes and at grain contacts. The nature of the clay minerals formed is dependent on host crystal compositions and on the variety of active geochemical micro-environments which may co-exist in the parent rock (Meunier & Velde 1979; Velde & Meunier 1987). Thus, rock composition has a fundamental control on clay mineral evolution during the initial stage of weathering, as has been shown, for instance, by Hall *et al.* (1989).

The mineral association in sample 11 (Hulu-Triabo) includes plagioclase and potassium feldspars together with illite, and iron and magnesium phyllosilicates such as chlorite, vermiculite/illite and vermiculite, including a low-charge type. Except that chlorite is absent, the sample from site 12 (Skruv) has essentially the same clay mineralogy. These samples represent a stage when breakdown of the rock has merely involved exfoliation and physical fragmentation. Mineralogical transformations are restricted to release of potassium and a decrease in layer charge due to oxidation and some loss of structural iron in biotite, resulting in vermiculitization and precipitation of iron oxides. The low-charge type of vermiculite is often found to form when the oxidation of biotite occurs at neutral pH. Under acid conditions, oxidation is balanced by the release of octahedral iron, which maintains a high surface charge and promotes the formation of 'normal' vermiculite (Wilson *et al.* 1984). However, some low-charge vermiculites may require a dioctahedral mica as a precursor (Aragoneses & García-Gonzáles 1991). The mica to vermiculite to smectite 'continous' sequence seems to be the most frequent pathway for biotite weathering in temperate/subtropical regions (e.g. Jackson 1965; Gjems 1967; Loveland 1984; Wilson *et al.* 1984; Righi & Meunier 1991).

The sample from site 9 (Malexander) is smectite–illite dominated. Iron released from weathered minerals has accumulated in the fine fraction as the iron oxides, lepidocrocite and haematite. Reducing conditions are required for lepidocrocite to form and its occurrences are almost exclusively restricted to hydromorphic soils. However, once formed, lepidocrocite may persist for a long time under soil conditions, since the rates of dissolution and transformation are very slow (Schwertmann & Taylor 1977; Taylor 1987). The mineral suite of sample 9 could easily be fitted into the postulated alteration sequence as a representative of a poorly-drained leaching environment prevailing, for instance, in fractures in the coherent rock at the base of a weathering profile. However, the mineral distribution in this clay-rich sample may reflect the superimposed effect of sorting in water penetrating the fractures, by which the finest-grained minerals became concentrated.

The occurrence of a kaolin mineral in sample 8 (Knasekärret) and 10 (Duvedal) distinguishes these samples from other gruss samples; the remaining clay mineral suites are similar (vermiculite and illite in samples 8; vermiculite and mixed-layered vermiculite/smectite in sample 10). The ease with which all 14 Å minerals decomposed on acid treatment suggests that they are trioctahedral, possibly derived from biotite.

Progressive leaching of cations eventually leads to kaolin formation, which initially occurs mainly at the expense of clay minerals of low stability and/or plagioclase. Kaolin minerals are generally most extensively developed in humid tropical climates, but kaolin formation is in no way restricted to tropical soils (see review in Weaver 1989). In their studies of the initial stages of weathering of a granite, Meunier & Velde (1979) and Velde & Meunier (1987) found that illite was generally formed at muscovite/orthoclase grain contacts, but in fractures, where water movement was rapid, kaolinization could be intense. Similarly, Pruett & Murray (1991), investigating Cretaceous palaeosols in Canada, suggested that kaolinite formation was intensified during periods of a depressed groundwater table allowing more intense leaching of cations. It cannot be excluded that even small differences in parent rock composition between the sites (granites at site 8 and 10, granodiorites at the other sites) had a fundamental control on clay mineral evolution during initial alteration. However, the overall high permeability of the gruss would favour kaolinite formation and also at particularly free-draining positions in the terrain. Site 8 and 10 are positioned on hillsides, whereas the other sites are situated on valley bottoms.

Overlying glacial deposits. The saprolites at sites 8, 10 and 12 are covered by Weichselian glacial deposits. The mineralogy of the latter can best be interpreted in terms of the effects of Holocene pedogenesis on minerals derived from granitoid bedrock mainly by mechanical comminution. The amount of quartz and feldspars in the clay fraction of the tills is significantly higher than in the saprolite samples. The phyllosilicate suite consists of easily weatherable trioctahedral Fe- and Mg-phyllosilicates, such as biotite and chlorite, and poorly-defined interstratifications of illite and vermiculite, which form from biotite and chlorite during pedogenesis. Occasionally, chloritized or aluminium hydroxide-interlayered vermiculite also occurs (site 12), i.e. minerals which typically form under acid leaching conditions in humid temperate regions (e.g. Loveland 1984). Sampling depth, permeability and topography-related drainage status determine the extent of vermiculitization and other secondary alterations and, hence, the proportions between the inherited and secondary phyllosilicates.

Comparison of the gravelly saprolites and overlying till beds. Mineralogical trends are discontinuous over the saprolite/till boundaries, indicating that saprolite formation was a process distinct from Holocene soil formation. This is best illustrated at site 8, where the stratigraphic distribution of kaolin minerals and iron chlorite can be described as a reversed 'evolutionary trend'. Iron chlorite, which is highly susceptible to weathering, is one of the major phyllosilicates in the till and ubiquitous in the fresh bedrock, but has not been detected in the gruss. Kaolin minerals, which form under conditions of intense leaching, have a contrasting distribution. Trace amounts of kaolin minerals probably occur in the till, but are most likely inherited from the ancient weathering cover since they probably do not form under the present pedogenic regime. According to Wilson *et al.* (1984), there is no evidence that kaolinite is forming at present in soils in Scotland. The frequent kaolinite occurrence in Scottish soils is suggested as evidence that ancient weathering covers have had a great influence on present soil clay mineralogy.

SEM analyses of quartz grains

Samples from five of the sites were analysed by SEM (Fig. 14). The results of the analyses are summarized in Table 5. Some of the characters,

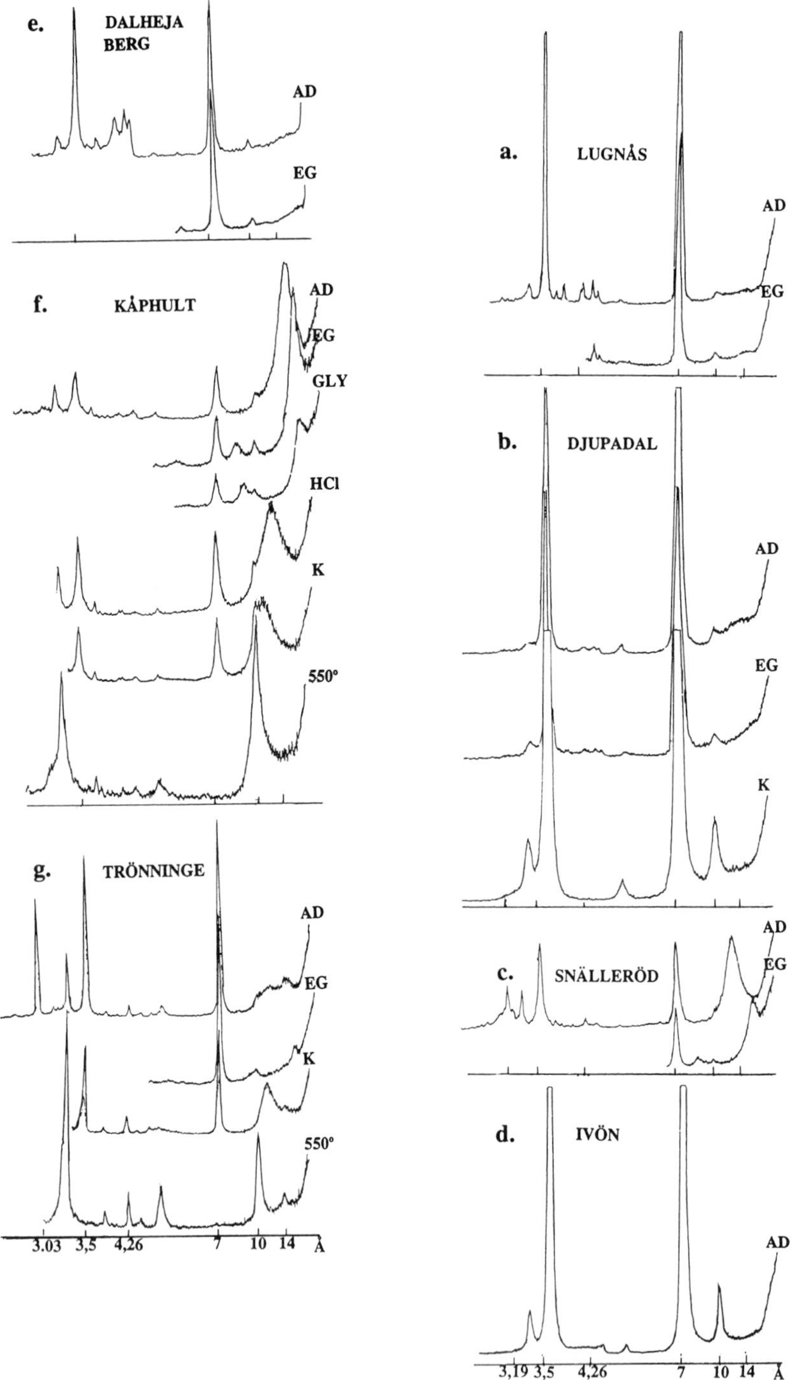

Fig. 12. X-ray diffractograms of the clay fraction of samples of the mature saprolites at: (**a**) site 1, Lugnås (sub-Cambrian); (**b**) site 2, Djupadal (sub-Jurassic); (**c**) site 3, Snälleröd (exhumed sub-Jurassic); (**d**) site 4, Ivön (sub-Cretaceous); (**e**) site 5, Dalhejaberg (exhumed sub-Cretaceous); (**f**) site 6, Kåphult (exhumed sub-Cretaceous/new relief); (**g**) site 7, Trönninge (exhumed sub-Cretaceous). AD, air dried; EG, ethylene-glycol solvated; GLY, glycerol solvated after Mg saturation; K, potassium saturated; HCl, digested in 2 M HCl; 550°, heated at 550°C. CuK_{α} radiation.

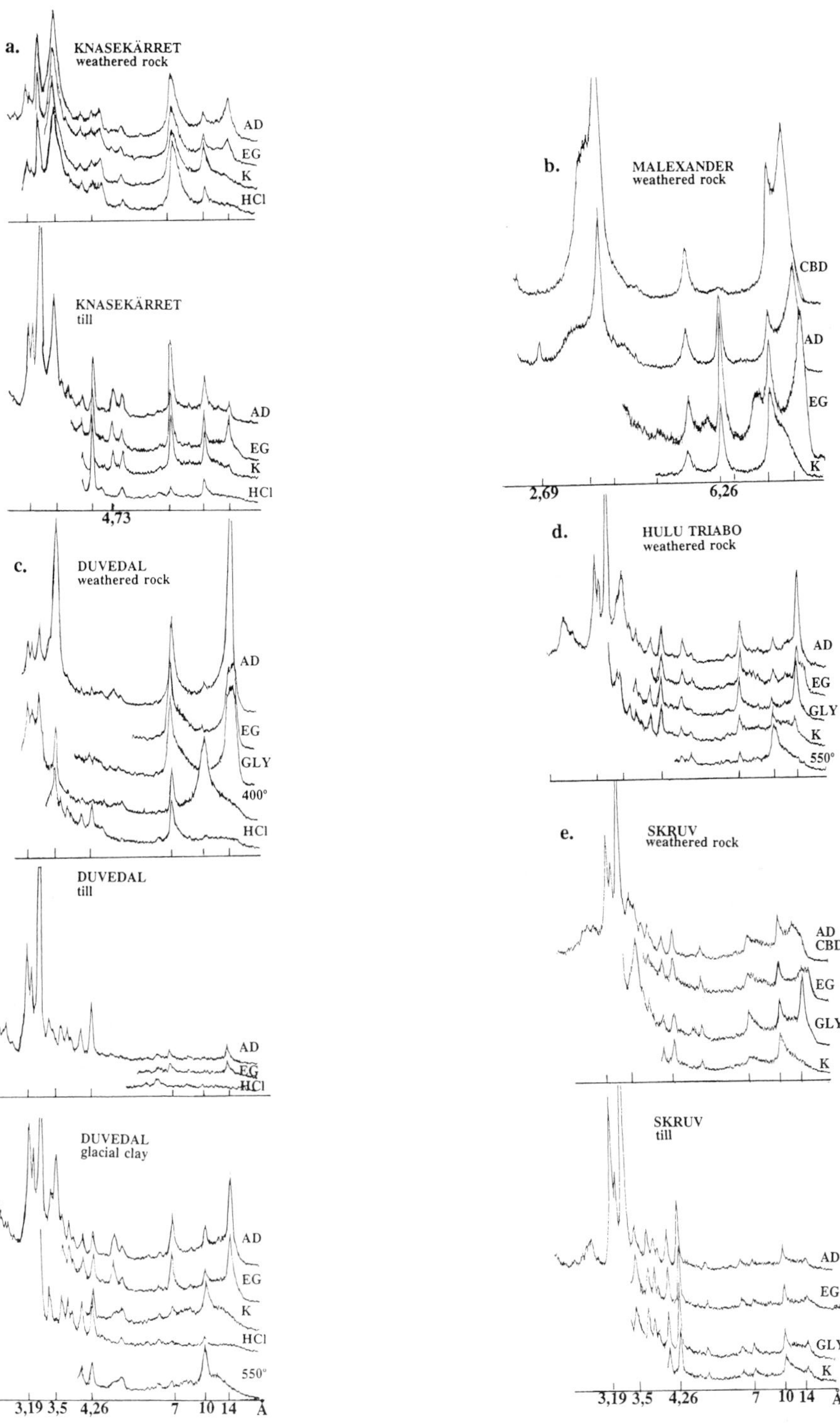

Fig. 13. X-ray diffractograms of the clay fraction of samples of gruss saprolites and overlying glacial deposits at: (**a**) site 8, Knasekärret (weathered rock and overlying till); (**b**) site 9, Malexander; (**c**) site 10, Duvedal (weathered rock and overlying till/glacial clay); (**d**) site 11, Hulu-Triabo; (**e**) site 12, Skruv (weathered rock and overlying till). Abbreviations – see explanations to Fig. 12. Note the basic difference between the diffractograms. Figure 12 shows mature saprolites dominated by kaolinite and smectite while Fig. 13 shows immature saprolites with a spectrum of clay minerals.

(a)

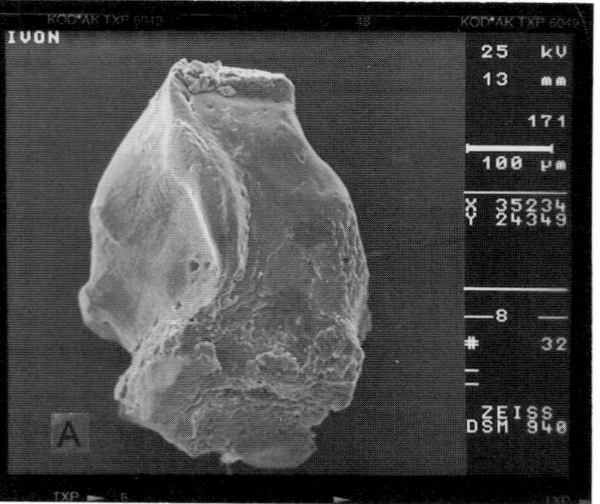

(b)

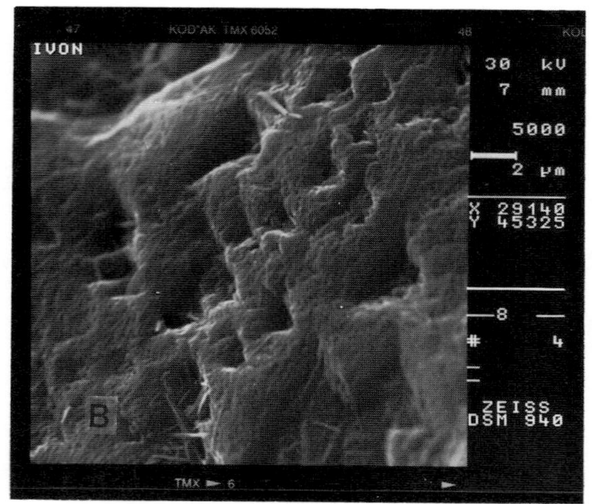

(c)

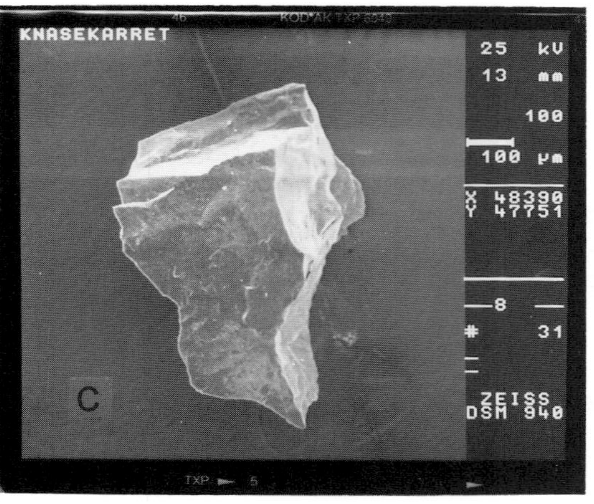

Fig. 14. SEM photos of microtextures on sand quartz grains. (**a**) Sub-rounded quartz grain from site 4, Ivön, sub-Cretaceous. Note the ocurrence of quartz crystal terminations to the right and the large solution pit, approximately in the centre of the grain. Rounding of grains by chemical weathering is typical for the mature saprolites. (**b**) Oriented V-pits on grain from site 4,Ivön, indicating intense chemical etching of the grain surface. (**c**) Angular grain formed by mechanical breakage, from site 8, Knasekärret, a gravelly, immature saprolite. Note the sharp edges and broken surfaces.

Table 5. *Results of scanning electron microscopy (SEM) on quartz grains*

Texture categories	Djupadal (2)	Ivön (4)	Kåphult (6)	Knasekärret (8)	Skruv (12)
Large conchoidal fractures	20	20	40	87	87
Small conchoidal fractures	40	23	77	97	87
Large breakage blocks	63	50	56	50	23
Small breakage blocks	87	73	73	67	93
Arc-shaped steps	37	23	77	90	90
Random scratches and grooves	0	3	3	0	0
Oriented scratches and grooves	3	17	0	1	3
Parallel steps	43	57	43	23	63
Non-oriented V-shaped pits	0	3	0	0	0
Meandering ridges	0	3	3	0	0
Dish-shaped concavities	30	13	33	6	0
Upturned plates	100	87	90	77	57
Microblocks (chemical or mechanical)	97	33	60	23	20
Roundness – rounded	0	0	0	0	0
Roundness – sub-rounded	33	27	37	3	3
Roundness – sub-angular	60	43	43	27	60
Roundness – angular	7	30	20	70	37
Cleavage flake	93	43	63	67	67
Precipitation platelet	10	0	7	0	17
Cracked surface	63	17	23	47	40
Chemically formed V-shaped oriented pits	13	63	30	3	10
Cleavage plane	90	87	77	67	57
Silica precipitation	100	97	100	100	90
Solution pits and hollows	100	100	93	93	100
Dulled surface from solution of silica	43	30	33	13	17
Chattermarks	0	0	0	0	0
Star cracking	0	0	0	0	0
Low relief	3	3	0	0	0
Medium relief	73	36	63	60	53
High relief	23	43	37	40	47

Frequency of grains with different microtextures in % of total number of grains (30) examined.

which have been used for differentiation and grouping of the samples, are commented on here. The information is based mainly on Krinsley & Doornkamp (1973), Doornkamp (1974), Goudie (1981), Söderman *et al.* (1983), Williams *et al.* (1986), Bull *et al.* (1987), Borger (1993) and others referred to below.

Conchoidal fractures and breakage blocks of different sizes, as well as arc-shaped steps, are well known textures formed in environments promoting mechanical breakage. Among the observed textures chemically-formed,V-shaped oriented pits are maybe the most evident texture indicating very intense chemical etching. Surface cracking has been reported in many studies. Their diagnostic value is, however, doubtful since the interpretation of these features is uncertain. They might indicate chemical action in hot desert environments (Goudie *et al.* 1979; Krinsley & Doornkamp 1973 p. 17).

The texture categorized as microblocks is here defined as blocky cleavage flakes (*c.* 5 μm) often found in depressions in the grain surface. Krinsley & Doornkamp (1973) consider these blocks as being chemically produced by rapid disintegration of the grain surface. They are not obliterated due to their protected position.

Rounding of grains may be caused either by mechanical abrasion or chemically by solution *in situ,* as shown by Crook (1968). As the samples compared here are all saprolites, the degree of rounding must reflect solution rounding.

Dulled surfaces produced by the action of solution result in a surface or capping layer of amorphous silica giving a dull appearance to the grain surface. This seems to be most typical of diagenetic effects and indicates solution and reprecipitation of silica. In this investigation it may be used as a measure of the magnitude of chemical activity which has affected the saprolites.

Some basic differences between the samples. The samples of gravelly saprolites from Knasekärret (site 8) and Skruv (site 12) have a high percentage of characters indicating mechanical breakages. These characters, especially conchoidal fractures and arc-shaped steps, are infrequent in the samples from Ivön (site 4, sub-Cretaceous) and Djupadal (site 2, sub-Jurassic), whilst the frequency is intermediate for the sample from Kåphult (site 6, exhumed sub-Cretaceous/100 m level of SSP). On the other hand, sub-rounded grains are common in the samples from Ivön, Djupadal and Kåphult, but they are almost absent at Knasekärret and Skruv. Dulled surfaces from solution of silica are also more frequent at Ivön, Djupadal and Kåphult. The two latter characters indicate chemical alteration. Thus, there is a clear difference between the samples from Ivön, Djupadal and Kåphult, and the samples from Knasekärret and Skruv, the former having been more affected by chemical weathering than the latter.

V-pits are most common in the samples from Ivön (63%) and Kåphult (30%) but are rare in the Djupadal (13%) and Skruv (10%) samples and very rare in the sample from Knasekärret. Thus the most intense chemical etching of the quartz grains has occurred at Ivön and Kåphult.

Surface cracks are common in the Djupadal, Knasekärret and Skruv samples while infrequent in the samples from Kåphult and Ivön. They might indicate that, at times, hot and arid conditions have been prevalent at the former sites.

The sample from Djupadal is distinguished from the Ivön and Kåphult samples in other characteristics too. Microblocks, cleavage flakes, and subangular grains are very frequent at Djupadal. Thus, it seems that Djupadal might have experienced environments slightly different from those that produced the saprolites at Ivön and Kåphult.

Reprecipitation in the form of euhedral quartz crystals are frequent in the Ivön, Kåphult and Djupadal samples, which indicates very slow precipitation (Krinsley & Doornkamp 1973).

Summary of characteristics of the saprolites and covering tills

The results of the field investigation and analyses of samples (grain size, XRD, SEM) are summarized in Table 6. A summary of grain size analyses from eight of the sites is also presented in Fig. 15.

Table 6. *Summary of characteristics of saprolites and glacial deposits at investigated sites*

Sites/location	Depth of saprolite (m)	Corestones	Major clay minerals	SEM Q.gr.	Fine fraction < 60μm (%)
1. Lugnås sub-Cambrian	1	No	ka		
2. Djupadal sub-Jurassic	44	?	ka	ch	65
3. Snälleröd E. sub-Jurassic	?	?	sm, ka		
4. Ivön sub-Cretaceous	> 30	Yes	ka	ch	49
5. Dalhejaberg E. sub-Cretaceous	In joint	–	ka		
6. Kåphult E. sub-Cretaceous/N.r.	> 0.5	No	sm, ka	ch	29
7. Trönninge E. sub-Cretaceous	0.5	No	ka		
8. Knasekärret N.r. Hs *Till*	> 9	Yes	ka, verm, il *chl, il*	m	3-6 *29*
9. Malexander N.r. Su	In joints > 15 m	Incip.	sm, il		6
10. Duvedal N.r. Hs *Till* *Clay*	> 3	Yes	verm, V/S, ka *verm* *verm, il*		5–6 *28*
11. Hulu-Triassic N.r. Va	In joints	Incip.	v, il, chl.		6
12. Skruv N.r. Va *Till* *Redeposited gravel*	> 5	Yes	v, il, ML *il, ML*	m	5–8 *24* *13*

Bold stratigraphiclly controlled key sites; *italics, glacial deposits.* E, exhumed; Ca, Cambrian; Ju, Jurassic; Cr, Cretaceous; N.r., new relief; Hs, hillside; Su, summit; Va, valley; m/ch depicts dominance for mechanical or chemical alteration of quartz grains.

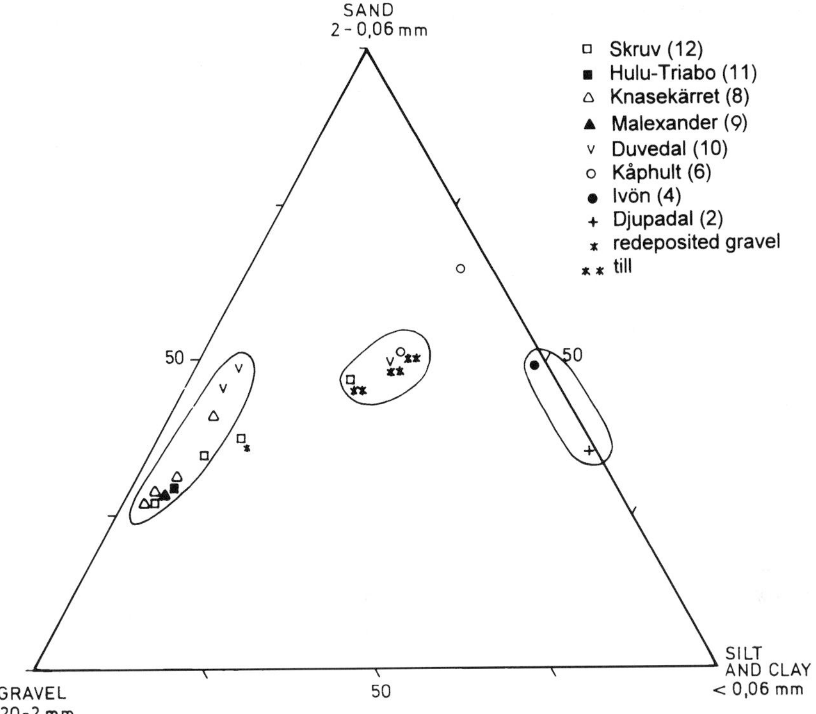

Fig. 15. Grain-size distribution from eight of the investigated sites.

The two most fine-grained saprolites are those from Ivön (4, sub-Cretaceous) and Djupadal (2, sub-Jurassic). The Kåphult (6, exhumed sub-Cretaceous/100 m level of SSP) sample can be classified as a sandy saprolite. The samples from Dalhejaberg (5, exhumed sub-Cretaceous) and Trönninge (7, exhumed sub-Cretaceous) should also be classified as gravelly sand, but granulometric data for these samples are considered less informative. The samples represent disintegration products formed *in situ* in a shallow zone in contact with coherent but kaolinized rock. They are not necessarily representative of the saprolite, but are indicative of its former existence. The saprolites from the sites 8–12 are characterized by a high gravel content and very small amounts of silt and clay. The tills overlying these latter sites are sandy and have a more mixed grain size distribution than the saprolites. The redeposited gravel found immediately overlying some of the saprolites has a grain size distribution fairly similar to the gravelly saprolites.

The SEM analyses of the quartz grains from five samples led to a similar grouping of the saprolites. The most chemically affected samples are those from Ivön, Djupadal and Kåphult, with the sample from Ivön as the most typical of a high-energy chemical environment, i.e. hot and humid. The samples from Knasekärret and Skruv are characterized by very limited chemical decomposition and the saprolites must be regarded as mainly mechanically produced.

The diffractograms (Figs 12 & 13) clearly demonstrate the clay mineralogical difference between the saprolite samples. Mature kaolinite/illite or smectite/kaolinite/illite assemblages characterize the samples from Lugnås, Djupadal, Snälleröd, Ivön, Dalhejaberg, Kåphult and Trönninge (sites 1–7). It is evident that argillization is most extensive at Djupadal and Ivön, where primary minerals, except quartz, are rare even in coarse grain-size fractions. However, all the samples can be interpreted as representatives of different weathering regimes at varying depths within mature saprolite bodies. The clay mineral assemblages of the gravelly saprolites (sites 8–12) are in marked contrast to the previous group. They are heterogeneous multi-phase associations dominated by various vermiculite minerals typical of those found in granitoid rocks during their initial stage of alteration. The frequency of corestones indicates that they developed in an active phase of alteration (Thomas 1994). The kaolin mineral found in the samples from Knasekärret and

Duvedal may reflect better drained sites on hillsides compared with those at Skruv and Hulu-Triabo. The latter sites are situated in shallow valleys.

The clay mineralogy of the tills is distinctly different from that of the underlying saprolites and interpreted as the result of mechanical comminution of granitoid rocks and are modified to varying extent by Holocene pedogenesis.

Thus the present investigation clearly separates the immature, gravelly saprolites from the mature clay-and silt-rich saprolites. Further the gravelly saprolites show a more advanced weathering than the overlying till beds and so they must have had a much longer time for their formation. These saprolites probably just lay dormant during colder phases and became reactivated during warmer phases, as at present.

Saprolites below Cambrian cover rocks have a maximum thickness of 5 m. Fine-grained kaolinitic saprolites from the sub-Mesozoic surfaces can be over 50 m deep. The maximum thickness of the gravelly saprolites is not known, but it certainly exceeds 10 m.

Age of the saprolites

The minimum age of the saprolite at Lugnås, site 1, is stratigraphically controlled by its overlying cover of Lower Cambrian rocks, the one at Djupadal, site 2, by its Lower Jurassic cover rocks, and the one at Ivön, site 4, by its Upper Cretaceous (Campanian) cover rocks.

Recent investigations of Upper Triassic–Lower Jurassic sedimentary strata in Skåne (Ahlberg 1990; Arndorff 1994) confirm and refine earlier conclusions regarding environmental conditions (summary in Lidmar-Bergström 1982) at the Jurassic–Triassic boundary. In the Rhaetian (Latest Triassic) the climate changed from arid to humid and the fossil flora indicates a tropical climate. Saprolite thicknesses of 50 m, as encountered in Central Skåne, may have been produced during a time interval of 1 to 25 Ma according to calculated figures based on saprolite production from different parts of the world (i.e. 2–48 m Ma^{-1}; Thomas 1994). In a humid tropical climate the time required is comparatively short. Thus there was enough time between the climate change to humid conditions and deposition of the covering sedimentary strata for the observed saprolites to develop. The Djupadal sample has all the characteristics of a mature saprolite formed in a chemically active environment. It is, however, possible that some of the characters found in the quartz grains of this saprolite are inherited from earlier times (Norian and early Rhaetian) which had an arid climate. The weathering mantle at Snälleröd is interpreted to belong to the same deep weathering event as caused the saprolite formation at Djupadal on the basis of it clay mineralogy and the position of the site close to remnants of the Lower Jurassic cover rocks.

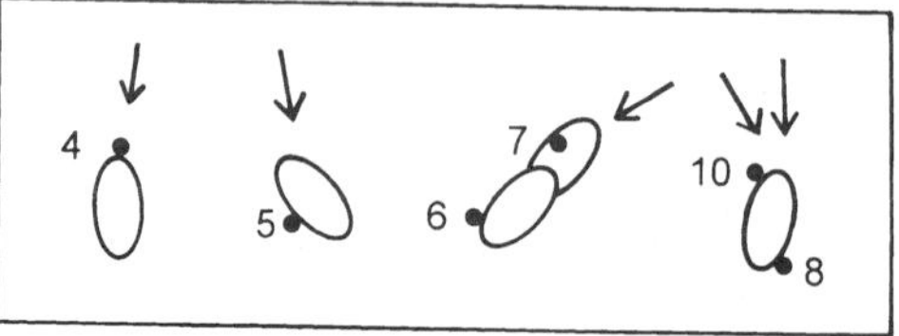

Fig. 16. Location of sites in relation to ice movements.

Northeastern Skåne and southern Halland stayed in an uplifted position and were not flexed down and transgressed by the sea until the Cretaceous (Norling & Bergström 1987). Thus there was a long time during which the basement here was exposed to subaerial denudation including saprolite formation and etching of the Precambrian bedrock surface. Apart from cooler periods in the Aptian and Cenomanian the Cretaceous temperatures stayed high until the Campanian (references in Lidmar-Bergström 1982). The lowermost strata on the west coast belong to the upper Lower Cretaceous–lower Upper Cretaceous interval, and the Upper Cretaceous (Campanian) in northeast Skåne and Blekinge. Kaolinitic clays began to be deposited in the Late Rhaetian, and kaolinitic clays as well as pure quartz sand are typical in Mesozoic sediments (summary in Lidmar-Bergström 1982). Thus the weathering remnants (sites 5, 6, 7), found in close proximity to the Cretaceous cover rocks, and with several characters in common with the saprolites found below the Jurassic and Cretaceous cover rocks, are interpreted as having formed in the Jurassic–Late Cretaceous time interval. On the west coast this is likely to have been not later than the Early Cretaceous. The possibility that the sandy saprolite at site 7 has evolved during warm humid periods in the Miocene, cannot be excluded though (Fig. 17). The prerequisitse in this case would be that the exhumation of the sub-Cretaceous surface occurred in the Miocene, a question which is still open (Lidmar-Bergström 1993). An alternative explanation can also be given for site 6 (Kåphult). This site is situated in an area where the sub-Cretaceous relief merges with the SSP, and the saprolite might, therefore, belong to the time of its formation (see below). Samples from stratigraphically controlled sites in southern Halland (if these can be obtained) should shed more light on this question.

The gravelly saprolites are found below Weichselian deposits. Typical mineralogical evolutionary trends are discontinuous or reversed at the saprolite/till boundaries, indicating that saprolite formation is a process distinct from Holocene

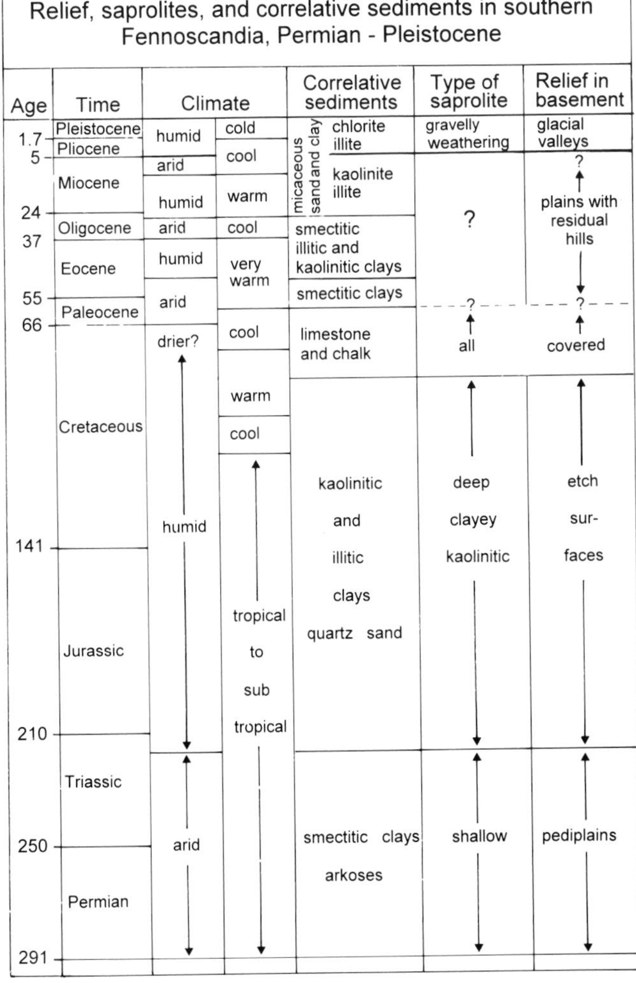

Fig. 17. Correlation of relief evolution, saprolite formation, sedimentation and climate. Summary from different sources.

pedogenesis. The saprolites are clearly pre-Weichselian, but saprolite formation may have been reactivated during later periods of ameliorated climate since it is well known that the mechanical disintegration of rocks becomes the active process in cold climates. Moreover chemical leaching was probably not even during the Pleistocene cold phases a totally inactive process, as suggested by evidences presented by Ugolini (1986) and Locke (1986) of weathering in arctic climates. If, however, we were to assume that the encountered 10 m thick saprolites were produced entirely during the Eemian interglacial, then the resulting estimate of rate of formation would be *c.* 1000 m Ma^{-1} (Table 7); this is totally unrealisitc when compared with the weathering rates quoted by Thomas (1994). To get within the cited rate of 2–48 m Ma^{-1}, the whole Pleistocene, or the Pleistocene *and* the Pliocene, are needed for a saprolite of 10 m to form. However, the lack of correspondance between these saprolites and the mineralogy of the Tertiary clays found in surrounding basins (Fig. 17) strengthens the idea of a Pliocene–Pleistocene age for the gravelly saprolites.

The gravelly saprolites are more or less similar to other saprolites found in Fennoscandia (Roaldset *et al.* 1982; Peulvast 1985; Kejonen 1985; Lahti 1985; Sørensen 1988; Lundqvist 1988). Several of these saprolites are clearly recognized as pre-Weichselian on stratigraphical grounds. As Peulvast (1985) claims, their thickness also indicates a long time for formation. Similar saprolites are also abundant in Scotland (Hall *et al.* 1989) and interpreted to have formed during warm temperate conditions in the Pliocene and continued to develop during interglacial Pleistocene times. Outside the former

Table 7. *Calculated time for production of 10 m saprolite*

Eem	10 m/10 000 years	1000 m ma^{-1}
Eem + Holstein	10 m/30 000 years	330 m ma^{-1}
Last million years with climates suitable for saprolite production	10 m/140 000 years 10 m/200 000 years	70 m ma^{-1} 50 m ma^{-1}
Quaternary with climates suitable for saprolite production	10 m/280 000 years 10 m/400 000 years	35 m ma^{-1} 25 m ma^{-1}
Quaternary + Pliocene with climates suitable for saprolite production	10 m/5 280 000 years 10 m/5 400 000 years	1.9 m ma^{-1} 1.85 m ma^{-1}

Quaternary climates suitable for saprolite production are roughly calculated with two alternatives from ‰ $\delta^{18}O$ benthic foraminifers by Ruddiman *et al.* (1989).

glaciated areas sandy saprolites displaying limited chemical alteration are of common occurrence (Bakker 1967; Millot 1970).

Saprolite and relief relationships

The sub-Cambrian peneplain and its associated saprolites

As there was no vegetation during the formation of the sub-Cambrian saprolite it is thought that the etching of the bedrock was accompanied by stripping, and consequently thick saprolites never developed. A prolonged period of time with stable tectonic conditions resulted in a peneplain where deep weathering, accompanied by stripping and pedimentation, produced extremely flat surfaces. The process might best be described with the terminology of Mabbut (1966), as mantle-controlled planation. The saprolites were often incrusted by silcrete and ferricrete. Aeolian processes were also active, though they probably had no importance during the development of the very flat plain.

In eastern Småland and southern Östergötland the present relief reveals incision of up to 100 m in the uplifted sub-Cambrian peneplain surface. If this lowering was caused by deep weathering and subsequent stripping, a period of between 2 and 50 Ma would have been required once the cover rocks had been removed. The uplift has been interpreted to have occurred during the Tertiary but whether this happened Early or Late Tertiary times has not yet been established (Lidmar-Bergström 1993).

The sub-Mesozoic etchsurfaces and associated saprolites

The kaolinitic saprolites in central Skåne were produced during a period of 1–10 Ma in the Latest Triassic–Early Jurassic in a tropical and humid climate. Partial stripping of the saprolite had probably already occurred during the weathering period which had resulted in the slightly undulating terrain below the Jurassic cover rocks. The present topography has been accentuated by further stripping, but the unevenness of the fresh bedrock is much greater than that seen at surface. Thus it can be labelled a partly stripped etchsurface (Thomas 1974, 1994).

At the site at Ivön the relationship between deep weathering and subsequent stripping on the one hand, and the hilly relief on the other, is clearly demonstrated (Fig. 3). The same kind of hills with steep slopes can be followed both westwards, northwards and eastwards. Dalhejaberg (Fig. 4) is a typical representative of a hill produced by Mesozoic deepweathering and subsequent stripping. Most of the stripping had probably already occurred in the Mesozoic because the Cretaceous remnants in the area are often found lying directly upon fresh bedrock and in pits on freshly exposed surfaces.

On the west coast the profiles of the relief in connection with sites 6 and 7 show an undulating hilly landscape interpreted as the sub-Cretaceous etchsurface. The summit surfaces might very well indicate a Permian–Triassic surface, which is interpreted to be the starting surface for the Mesozoic etching (Lidmar-Bergström 1993). However, in northern Halland this surface is close to or identical with the sub-Cambrian peneplain since Cambrian fissure fillings are common. The maximum relief along the profile at site 7 was calculated to be 135 m. If this relief is caused by deep weathering and subsequent stripping it must correspond to a saprolite production of similar thickness, which could have been produced within time intervals of 3 to 67.5 million years duration in the Jurassic–Early Cretaceous.

Remnants of the kaolinitic saprolite are fewer on

the west coast than in NE Skåne. This might be explained by a more complete stripping of the saprolites already before the Cretaceous transgressions because the west coast probably was more exposed to erosion both subaerially and by wave action during the Cretaceous transgressions.

The Tertiary surfaces

The South Småland Peneplain (SSP) including the 100 m level, the 200 m surface, and the 300 m surface are thought to have been produced during the Palaeogene or Miocene by short periods of deep weathering and subsequent stripping and pedimentation, which mainly promoted the formation of plains with residual hills (Lidmar-Bergström 1982, 1995). However, no saprolites correlatable with the argillaceous Palaeogene - Miocene sediments in Denmark and the North Sea (Fig. 17) are found upon these surfaces, maybe with the exception of the saprolite from site 6 (though this is probably Mesozoic). Thus it is thought that the saprolites associated with the formation of these surfaces have either been totally stripped or else are now hidden below Quaternary deposits. This stripping ought to have occurred during the arid phases of the Tertiary, of which the latest occurred in the Late Miocene.

The gravelly saprolites and the bedrock forms.

The gravelly saprolites are found within the areas of 'new relief' and often close below the uplifted sub-Cambrian surface. Their loose character and clay mineralogy, which is totally different from the cemented saprolite at Lugnås (site 1, sub-Cambrian), clearly shows that these gravelly saprolites formed after exhumation of the sub-Cambrian surface. Saprolites at sites close to the sub-Cambrian peneplain might have developed directly from this surface after erosion of the cover rocks. This denudation may be a comparatively late stage phenomenon. These circumstances imply that the term SSP is not appropriate in the northeastern part of the South Swedish Dome and the current representation of the SSP in these areas really only relates to the height above sea level for the valley systems. The only site from the SSP proper is number 12 (Skruv) and this is situated in a shallow valley, which itself might postdate the formation of the SSP.

The scale of the forms associated with the saprolite remnants depend on the depth of the latter. Minor bedrock forms ought to have been produced by stripping of this saprolite. The bedrock forms around the sites 8 and 9 have features, which are closely connected with the fractures of the bedrock. Site 9, Malexander, with its weathered joints, is similar to many sites outside the glaciated areas with weathering following joints and making hillsides unstable (Thomas 1994 p. 56, pl 3-V). The deep weathering has also produced many of the well rounded boulders, now incorporated in the till cover.

Saprolites, denudation surfaces, and glacial erosion

Saprolites are rare phenomena in glaciated landscapes. The common situation is that the ice sheets have totally stripped the saprolites and that there has been no neoformation since the last glacial event. Exceptions are found on particular rock types which have inherited properties that make them weather more easily after exposure (Samuelsson 1973, 1985).

Exhumed relief, glacial erosion, and postglacial wave erosion.

The saprolites at sites 1, 2 and 4 are situated below protecting cover rocks. The exhumed sub-Cambrian peneplain has no preserved saprolite remnants since the very flat relief has facilitated glacial stripping. In central Skåne the frequent occurrences of deep saprolite remnants and the frequent remnants of the Jurassic cover rocks show that the glacial erosion has only partly exposed the Mesozoic weathering front. Protruding parts of the fresh basement are often found forming cores of drumlins, which are common in the area (Lidmar-Bergström *et al.* 1991). On exposed surfaces the basement has only been slightly scoured by eroding ice sheets, but in the intervening depressions the fresh bedrock has not been reached by this erosion. Site 2 with its covering Jurassic sequence has been exposed by glacio-fluvial erosion in the Late Weichselian.

The frequent remnants of kaolinitic saprolites and Cretaceous cover rocks within the exhumed sub-Cretaceous relief show that glacial erosion has not everywhere exposed the Mesozoic weathering front. Site 4 is situated at the northern base of a residual hill. Here the Cretaceous cover and underlying saprolite have been eroded on both the eastern and western flanks of the hill but are preserved at the northern end, in what can be called the stoss side and lee positions (Fig. 16). Also in this area the protruding hills of Precambrian crystalline rocks similarly form the cores of drumlins.

Thus, even if a slight glacial scouring is normal for the areas with exhumed sub-Mesozoic relief in south Sweden, the glacial erosion has not been able to remove completely the thick saprolites nor to remove totally the later cover rocks. Even detailed surface forms such as weathering pits of Cretaceous age are preserved on glacially striated surfaces (Magnusson & Lidmar-Bergström 1983; Lidmar-Bergström 1989). Of the investigated sites besides Ivön (site 4), typical stoss side and lee position preservation also occur at Trönninge (7). The Mesozoic saprolite remnants at Dalhejaberg (5) and Kåphult (6) are both in more typical lee side positions (Fig. 17).

The sites Dalhejaberg and Trönninge are below the highest Late Glacial shoreline, which means that they have been exposed to erosion by wave action. However, both sites are in positions that were sheltered during the Late Weichselian regression by an archipelago emerging due to isostatic uplift.

The saprolite remnants within the 'new' relief and glacial erosion

The remnants of the gravelly saprolites in eastern Småland and southern Östergötland are all found in a lee position with respect to glacial erosion (Fig. 16): in fractures (sites 9 and 11), lee side (site 8), stoss side lee position (site 10), and in a shallow valley (site 12). At several of the sites in eastern Småland and Östergötland large well rounded boulders of the local bedrock are abundant in the terrain and can often be seen utilised in the local stone walling. In many cases it is clear that these boulders are detached core stones from the local saprolite. It can thus be concluded that before the Weichselian ice advance the saprolites must have been more widespread and as commented on above their thickness indicates that they are much older.

The areal extent of the encountered saprolite remnants in eastern Småland and Östergötland is delimited in Fig. 1. The area is probably larger, but from what we know, saprolites of the investigated type are absent in the western part of the South Småland Peneplain and the 200 m surface, although weathering is encountered in syenites. These observations might indicate a more effective glacial stripping of the saprolites in the western parts of Småland.

Summary - Etching and long term landform evolution within a cratonic environment

Etchprocesses have acted on the Precambrian basement during specific periods and produced characteristic relief and saprolites. The Late Precambrian etchprocesses with immediate stripping of produced saprolites formed the extremely flat sub-Cambrian peneplain. During the Mesozoic deep weathering event the saprolites accumulated and by intermittent stripping and renewed deepweathering a differentiation of the relief was achieved (cf. Thomas 1980; Kroonenberg & Melitz 1983), which resulted in a hilly relief (Lidmar-Bergström 1995).

The Mesozoic deep weathering event is evidenced in Europe both as preserved saprolites below and adjacent to cover rocks, and in the Mesozoic sedimentary record (Almeborg *et al.* 1969; Vachtl 1969; Störr 1975; Störr *et al.* 1977; Shaw 1981; Dupui 1992), and also in Greenland (Pulvertaft 1979) and North America (Grout 1919; Bergquist 1944). This event is interpreted to have been of fundamental importance for the evolution of the relief in the exposed parts of Baltic Shield (Lidmar-Bergström 1982, 1995).

Remnants of Early Tertiary saprolites are widespread in Europe (Bakker 1967; Millot 1970; Borger *et al.* 1993). The saprolite remnants in northern Fennoscandia (Hirvas *et al.*1988) might, in part, date back to the Early Tertiary since redeposited fossils of this age are found in the area (Hirvas & Tynni 1976), but in southern Scandinavia saprolite remnants from the Early Tertiary have not yet been identified. These latter were probably stripped during arid events in the Tertiary.

The gravelly saprolites are witnesses of etch-processes continuing to operate since the Late Tertiary to the present, although the weathering rate most certainly fluctuated with the climatic changes during this time span. For Scotland and Norway with its present high precipitation, Thomas (1994) concludes that: 'renewal of any saprolite removed by glaciation has hardly begun'. Thomas continues: 'In former glaciated areas, there may be a problem of getting the system started, because of the near total removal of moisture-retaining regolith.' The gravelly saprolites in southern Sweden which have escaped glacial erosion are, therefore, regarded as reactivated and not fossilized saprolites, but it has apparently been much more difficult than perhaps anticipated, for saprolite production to start again on these glacially scoured surfaces, with the possible exception for particular rock types.

Concluding remarks

Etch processes have resulted in saprolites with different characters and thickness during different times, and obviously etching is an important factor for the long term evolution of the observed relief (Thomas 1980, 1994; Lidmar-Bergström 1982, 1995; Lundqvist 1985; Söderman 1985; Hall 1986). Crustal movements cause different parts of a cratonic basement to be exposed at different times with characteristic forms and saprolites being the result. The transgressions of the sea with subsequent deposition of sediments protects the craton surface from further weathering for long periods of time (Elvhage & Lidmar-Bergström 1987). The cratonic regime (Fairbridge & Finkl 1980) can, therefore, be revealed by the study of the relationship between relief, saprolites, and cover rocks of different age.

These relationships seem to be a new field of research from which much knowledge of relief evolution might benefit during coming years. However, a knowledge of the characteristics of different types of saprolites alone is often diagnostically insufficient, and consequently two basic misconceptions arise regarding saprolite remnants in formerly glaciated areas. The first and most common misconception is that the occurrence of minerals such as kaolinite and gibbsite is exclusively connected to, and indicative of, saprolite formation in warm, more or less humid climate regardless of (1) the amount of these clay minerals, (2) the extent of argillization, (3) the drainage status and so on (discussion in Bouchard *et al.* 1995). Our study demonstrates, however, that when the entire spectrum of clay minerals (and non-clay minerals) is considered, together with other independant parameters, it is obvious that the sub-Mesozoic, mature saprolites have a basically different character compared to the young, gravelly saprolites, irrespective of kaolin minerals being present or not in the latter. The other misconception is that glacial erosion has been of such efficacy that neither relief nor saprolites have survived from preglacial times. The rising awarness among glacial geomorphologists of the limited glacial erosion on the northern continents resulting from long lasting conditions of non-erosive cold based ice sheets (Sugden 1978; Hall & Sugden 1987; Dyke 1993; Kleman 1994), together with an appreciation of the non-glacial cause of relief within shield areas as demonstrated in several studies (Lidmar-Bergström 1982, 1987, 1988, 1989, 1994, 1995; Peulvast 1985; Söderman 1985; Hall 1987) contributes to make this type of research of great interest. Precambrian basement areas of the different continents, are suitable for comparisons of exhumed surfaces and their associated saprolites as well as 'new relief' and younger saprolites, regardless of glaciations.

This study was supported by a grant awarded by the Swedish Natural Science research Council.

References

AHLBERG, A. 1990. *Provenance, stratigraphy, paleoenvironments and diagenesis of the Lower Jurassic strata in the Helsingborg railway tunnel, southern Sweden.* Licentiate of Philosophy, Historical Geology. Thesis 2. Institute of Geology. Department of Historical Geology and Palaeontology. University of Lund.

ALMEBORG, J., BONDAM, J. & HELLER, E. 1969. Kaolin deposits of Denmark. International Geological Congress Report of the twenty-third session Czechoslovakia 1968. *Proceedings Symposium I. Kaolin deposits of the world A – Europe,* 75–84.

ÅNGSTRÖM, A. 1958. *Sveriges klimat. Generalstabens litografiska anstalts förlag.* Stockholm.

ARAGONESES, F. J. & GARCÍA-GONZÁLES, M. T. 1991. High-charge smectite in Spanish 'Rana' soils. *Clays and Clay Minerals,* **39**(2), 211–218.

ARNDORFF, L. 1994. *Upper Triassic and Lower Jurassic Palaeosols from Southern Scandinavia.* Lund Publications in Geology, **116**.

BAKKER, J. P. 1967. Weathering of granites in different climates, particularly in Europe. *In*: P. MACAR (ed.) *L'Evolution Des Versants.* Les Congrès de et Colloques de l'Université de Liége, **40**, 51–68.

BERGQUIST, H. R. 1944. Cretaceous of the Mesabi Iron Range, Minnesota *Journal of Paleontology,* **18**, 1–30.

BERGSTRÖM, L., CHRISTENSEN, W. K., JOHANSSON, C. & NORLING, E. 1973. An extension of Upper Cretaceous rocks to the Swedish West coast at Särdal. *Bulletin Geological Society of Denmark,* **22**, 84–154.

BORGER, H. 1993. Quartzkornanalyse mittels Rasterelektonenmikroskop und Dünnschliff unter besonderer Berücksichtigung tropischer Vervitterungsresiduen. *Zeitschrift für Geomorphologie, Neu Folge,* **37**(3), 351–375.

——, BURGER, D. & KUBINOK, J. 1993. Verwitterungsprozesse und deren Wandel im Zeitraum Tertiär – Qvartär. *Zeitschrift für Geomorphologie, Neue Folge,* **37**(2), 129–143.

BOUCHARD, M., JOLICEUR, S. & PIERRE, G. 1995. Characteristics and significance of two pre–late-Wisconsinan weathering profiles (Adirondacks, USA and Miramichi Highlands, Canada). *Geomorphology,* **12**, 75–89.

BRINDLEY, G. W. & BROWN, G. (eds) 1980. *Crystal Structures of Clay Minerals and their X-Ray Identification.* Mineralogical Society Monograph, London, **5**.

BULL, P. A., GOUDIE, A. S., PRICE WILLIAMS, D. & WATSON, A. 1987. Colluvium: A scanning electron microscope analysis of a neglected sediment type. *In*: MARSHALL, JOHN R. (ed.) *Clastic Particles. Scanning electron microscopy and shape analysis of sedimentary and volcanic clasts.* Van Nostrand Reinhold, New York, 16–35.

CHRISTENSEN, W. K. 1973. Upper Cretaceous belemnites from the Kristianstad area in Scania. *Fossils and Strata,* **7**.

CROOK, K. A. W. 1968. Weathering and the roundness of quartz sand grains. *Sedimentology,* **11**, 171–182.

CULVER, S. J., BULL, P. A., CAMPBELL, S., SHAKESBY, R. A. & WHALLEY, W. B. 1983. Environmental discrimination based on quartz grain surface textures: a statistical investigation. *Sedimentology,* **30,** 129–136.

DOORNKAMP, J. C. 1974. Tropical weathering and the

ultramicroscopic characteristics of regolith on Dartmoor. *Geografiska Annaler,* **56A**(1–2), 73–82.

DREVER, J. I. 1973. The preparation of oriented clay mineral specimen for X-ray diffraction analysis by a filter-membrane peel technique. *American Mineralogist,* **58,** 553–554.

DUPUIS, C. 1992. Mesozoic kaolinized giant regoliths and Neogene halloysitic cryptokarsts: two striking paleoweathering types in Belgium. *In*: SCHMITT, J. M. & GALL, Q. (eds) *Mineralogical and Geochemical Records of Paleoweathering.* IGCP 317. Ecole de mines de Paris. Mémoires des sciences de la terre 1992, **18**.

DYKE, A. S. 1993. Landscapes of cold-centred Late Wisconsinan ice caps, Arctic Canada. *Progress in Physical Geography,* **17**, 223–247.

ELVHAGE, C. & LIDMAR-BERGSTRÖM, K. 1987. Some working hypothesis on the geomorphology of Sweden in the light of a new relief map. *Geografiska Annaler,* **69A**(2), 343–358.

FAIRBRIDGE, R. W. & FINKL, C. W. 1980. Cratonic erosional unconformities and peneplains. *Journal of Geology,* **88,** 69–86.

GJEMS, O. 1967. Studies on Clay Minerals and Clay-mineral Transformation in Soil Profiles in Scandinavia. *Meddelelser fra Det Norske Skogsforsöksvesen,* **81**, **Bd XXI,** 303–415.

GOUDIE, A. S. (ed.) 1981. *Geomorphological Techniques.* George & Unwin, London.

——, COOKE, R. U. & DOORNKAMP, J. C. 1979. The formation of silt from quartz dune sand by salt-weathering processes in deserts. *Journal of Arid Environments,* **2**, 105–112.

GROUT, F. F. 1919. Clays and shales of Minnesota. *United States Geological Survey, Bulletin*, **678**.

HADDING, A. 1929. *The Pre-Quaternary Sedimentary Rocks of Sweden III. The Palaeozoic and Mesozoic Sandstones of Sweden.* Lunds Universitets Årsskrift NF Avd. 2 bd 25:3 (Kungliga Fysiografiska Sällskapets Handlingar NF bd 40:3).

HALL, A. M. 1986. Deep weathering patterns in north-east Scotland and their geomorphological significance. *Zeitschrift für Geomorphologie NF,* **30**, 407–422.

—— 1987. Weathering and relief development in Buchan, Scotland. *In*: GARDINER, V. (ed.) *International Geomorphology 1986: 11.* Wiley, London, 991–1005.

—— & SUGDEN, D. E. 1987. Limited modification of midlatitude landscapes by ice sheets: The case of north-east Scotland. *Earth Surface Processes and Landforms,* **12**, 531–542.

——, MELLOR, A. & WILSON, M. J. 1989. The clay mineralogy and age of deeply weathered rock in north-east Scotland. *Zeitschrift für Geomorphologie NF*, Suppl. **72**, 97–108.

HIRVAS, H. & TYNNI, R. 1976. Tertiary clay deposits at Savukoski, Finnish Lappland, and observations of Tertiary microfossils, preliminary report. (In Finnish, with English summary.) *Geologi,* **28**, 33–40.

——, LAGERBÄCK, R., MÄKINEN, K., NENONEN, K., OLSEN, L., RODHE, L. & THORESEN, M. 1988. The Nordkalott Project: studies of Quaternary geology in northern Fennoscandia. *Boreas*, **17**, 431–437.

JACKSON, M. L. 1965. Clay transformation in soil genesis during the Quaternary. *Soil Science,* **99**, 15–22.

KEJONEN, A. 1985. Weathering in the Wyborg rapakivi area, southeastern Finland. *Fennia,* **163**(2), 309–313.

KLEMAN, J. 1994. Preservation of landforms under ice sheets and ice caps. *Geomorphology*, **9**, 19–32

KORNFÄLT, K. A. 1993. Beskrivning till berggrundskartan Karlskrona NV/SV (Description to the map of solid rocks Karslkrona NV/SV). *Sveriges Geologiska Undersökning Af,* **179**.

—— & BERGSTRÖM, J. 1986. Beskrivning till berggrundskartan Karlshamn NO. (Description to the map of solid rocks Karlshamn NO). *Sveriges Geologiska Undersökning Af,* **154**.

—— & —— 1990. Beskriuning till berggrunds - Kartorna Karlshamn SV och SO (Description to maps of solid rocks Karlshamn SV och SO). *Sveriges Geologiska Undersökning Af,* **167**.

KRINSLEY D. H. & DOORNKAMP, J. C. 1973. *Atlas of Quartz Sand Surface Textures.* Cambridge University Press, Cambridge.

KROONENBERG, S. B. & MELITZ, P. J. 1983. Summit levels, bedrock control and the etchplain concept in the basement of Suriname. *Geologie en Mijnbouw*, **62**, 389–399.

LAHTI, S. I., 1985. Porphyritic pyroxene-bearing granitoids – a strongly weathered rock group in central Finland. *Fennia,* **163**(2), 315–321.

LIDMAR-BERGSTRÖM, K. 1982. *Pre-Quaternary geomorphological evolution in southern Fennoscandia.* Meddelanden från Lunds Universitets Geografiska Institution. Avhandlingar 91/Sveriges Geologiska Undersökning **C 785**.

—— 1985. Regional analysis of erosion surfaces in south Sweden. *Fennia,* **163**(2), 341–346.

—— 1988. Denudation surfaces of a shield area in south Sweden. *Geografiska Annaler,* **70A**(4), 337–350.

—— 1989. Exhumed Cretaceous landforms in south Sweden. *Zeitschrift für Geomorphologie, NF,* Suppl. **72**, 21–40.

—— 1993. Denudation surfaces and tectonics in the southernmost part of the Baltic Shield. *Precambrian Research,* **64**, 337–345.

—— 1994. Morphology of the bedrock surface. *In*: FREDÉN, C. (ED.) *Geology*. National Atlas of Sweden, 44–54.

—— 1995. Relief and saprolites through time on the Baltic Shield. *Geomorphology,* **12**.

——, BERGSTRÖM, L., HILLFORS, A., NYBERG, R., SAMUELSSON, L., SHAIKH, N. A., STRÖMBERG, B. & SWANTESSON, L. 1988. Preglacial weathering and landform evolution in Fennoscandia. *Guide to excursions.* Lunds universitets naturgeografiska institution.

——, ELVHAGE, C. & RINGBERG, B. 1991. Landforms in Skåne, south Sweden. Preglacial and glacial landforms analysed from two relief maps. *Geografiska Annaler,* **73A**(2), 61–91.

LOCKE, W. W. 1986. Rates of hornblende etching in soils on glacial deposits, Baffin Island, Canada. *In*: COLEMAN, S. & DETHIES, D. (eds) *Rates of Chemical Weathering of Rocks and Minerals.* Academic, London, 129–146.

LOVELAND, P. J. 1984. The soil clays of Great Britain: II England and Wales. *Clay Minerals*, **19**(5), 681–707.

LUNDQVIST, J. 1985. Deep weathering in Sweden. *Fennia*, **163** (2), 287–292.

—— 1988. The Revsund area, Central Jämtland – an example of preglacial weathering and landscape formation. *Geografiska Annaler*, **70A**(4), 291–298.

MABBUTT, J. A. 1966. The mantle controlled planation of pediments. *American Journal of Science*, **264**, 78–91.

MAGNUSSON, S.-E. & LIDMAR-BERGSTRÖM, K. 1983. Fossila vittringsformer från krittiden på Kjugekull (Fossil Cretaceous weathering forms on Kjugekull, south Sweden). *Svensk Geografisk Årsbok*, **59**, 124–137.

MATTSSON, Å. 1962. Morphologishe Studien in Südschweden und auf Bornholm über die nichtglaziale Formenwelt der Felsenskulptur. Meddelanden från Lunds Universitets Geografiska Institution. *Avhandlingar*, **39**.

MEHRA, O. P. & JACKSON, M. L. 1960. Iron oxide removal from soils and clays by a dithionite–citrate system buffered with sodium bicarbonate. *Clays and Clay Minerals*, 7th National Conference. Pergamon Press, London. 317–327.

MEUNIER, A. & VELDE, B. 1979. Weathering mineral facies in altered granites. The importance of local small-scale equilibria. *Mineralogical Magazine*, **43**, 261–268.

MILLOT, G. 1970. *Geology of Clays*. Springer, New York.

NORLING, E. & BERGSTRÖM, J. 1987. Mesozoic and Cenozoic tectonic evolution of Scania, southern Sweden. *Tectonophysics*, **137**, 7–19.

——, AHLBERG, A., ERLSTRÖM, M. & SIVHED, U. 1993. *Guide to the Upper Triassic and Jurassic Geology of Sweden*. Research Papers, Sveriges Geologiska Undersökning **Ca 82**.

PÅSSE, T. 1990. Beskrivning till jordartskartan Varberg NO. (Descripton to the Quaternary map Varberg NO.) *Sveriges Geologiska Undersökning*, **Ae 102**.

PEULVAST, J.-P. 1985. *In situ* weathered rocks on plateaus, slopes and strandflat areas of the Lofoten-Vesterålen, North Norway. *Fennia*, **163**(2), 333–340.

PRUETT, R. J. & MURRAY, H. H. 1991. Clay mineralogy, alteration history, and economic geology of the Whitemud Formation, Southern Saskatchewan, Canada. *Clays and Clay Minerals*, **39**(6), 586–596.

PULVERTAFT, T. C. R. 1979. Lower Cretaceous fluvial–deltaic sediments at Kuk, Nugssuaq, West Greenland. *Bulletin Geological Society of Denmark*, **28**, 57–72.

RIGHI, D. & MEUNIER, A. 1991. Characterization and genetic interpretation of clays in an acid brown soil (Dystrochrept) developed in a 'granitic saprolite'. *Clays and Clay Minerals*, **39**(5), 519–530.

RINGBERG, B. 1971. Glacialgeologi och isavsmältning i östra Blekinge. *Sveriges Geologiska Undersökning*, **C 661**.

—— 1984. Beskrivning till jordartskartan Helsingborg SO (Description to the Quaternary map Helsingborg SO). *Sveriges Geologiska Undersökning*, **Ae 51**.

ROALDSET, E., PETTERSEN, E., LONGVA, O. & MANGERUD, J. 1982. Remnants of preglacial weathering in western Norway. *Norsk Geologisk Tidskrift*, **62**, 169–178.

ROBISON, J. M. 1983. Glaciofluvial sedimentation: a key to the deglaciation of the Laholm area, southern Sweden. *Lundqua Thesis*, **13**.

RUDBERG, S. 1954. *Västerbottens berggrundsmsorfologi*. Geographica 25, Uppsala.

RUDDIMAN, W. F., RAYMO, M. E., MARTINSON, D. G., CLEMENT, B. M. & BACKMAN, J. 1989. Pleistocene evolution: Northern hemisphere ice sheets and North Atlancic Ocean. *Paleoceanography*, **4**(4), 353–412.

SAMUELSSON, L. 1973. Selective weathering of igneous rocks. *Sveriges Geologiska Undersökning*, **C690**.

——1982. Beskrivning till berggrundskartan Kungsbacka NO (Description to the map of solid rocks Kungsbacka NO). *Sveriges Geologiska Undersökning*, **Af 124**.

——1985. Beskrivning till berggrundskartan Göteborg NO (Description to the map of solid rocks Göteborg NO). *Sveriges Geologiska Undersökning*, **Af 136**.

SCHWERTMANN, U. & TAYLOR, R. M. 1977. Iron oxides. *In*: DIXON, J. B. & WEED, S. B. (eds) *Minerals in Soil Environments*. Soil Science Society of America. Madison, Wisconsin, 145–180.

SHAIKH, N. A. 1987. Mineralogical and chemical characteristics of Swedish kaolins. Summaries – *Proceedings of Sixth Meeting European Clay Groups*, Seville, Spain 1987. Sociedad Espanola de Arcillas.

SHAW, H. F. 1981. Mineralogy and petrology of the argillaceous sedimentary rocks of the U.K. *Quarterly Journal of Engineering Geology, London*, **14**, 277–290.

SÖDERMAN, G. 1985. Planation and weathering in eastern Fennoscandia. *Fennia*, **163**(2), 347–352.

——, Kejonen, A. & Kujansuu, R. 1983. The riddle of the tors at Lauhavuori, western Finland. *Fennia*, **161**(1), 91–144.

SØRENSEN, R. 1988. *In-situ* rock weathering in Vestfold, southeastern Norway. *Geografiska Annaler*, **70A**(4), 299–308.

STÅL, T. 1972. Kornfördelning. Förslag till geotekniska laboratorieanvisningar, del 4. Byggforskningens Informationsblad B2. Statens Institut för byggnadsforskning. Stockholm.

STÖRR,M. (ed.) 1975. Kaolin deposits of the GDR in the northern region of the Bohemian massif. Ernst-Moritz-Arndt-Universität, Greifswald.

—— 1993. Lagerstätten von Tonrohstoffen. *In*: JASMUND, K. & LAGALY, G. (eds) *Tonminerale und Tone*. Steinkopff Verlag, Darmstadt, 193–211.

——, KÖSTER, H. M., KUZVART, M., SZPILA, K. & WIEDEN, P. 1977. Kaolin deposits of central Europe. *In*: GALAU, E. (ed.) *Proceedings 8th International Kaolin Symposium and Meeting on Alunite*, Madrid–Rome, **K-20**.

SUGDEN, D. E. 1978. Glacial erosion by the Laurentide ice sheet. *Journal of Glaciology*, **20**(83), 367–391.

TAYLOR, R. M. 1987. Non-silicate oxides and hydroxides. *In*: NEWMAN, A. C. D. (ed.) *Chemistry of Clays and*

Clay Minerals. Mineralogical Society Mon. 6, Longman, Harlow, 129–201.

THOMAS, M. F. 1974. *Tropical Geomorphology*. Macmillan, London.

—— 1980. Timescales of landform development on tropical shields – a study from Sierra Leone. *In*: CULLINGFORD, R. A., DAVIDSON, D. A. & LEWIN, J. (eds) *Timescales in Geomorphology*. Wiley, Chichester, 333–354.

——1994. *Geomorphology in the Tropics. A study of weathering and denudation in low latitudes*. Wiley, Chichester.

TRALAU, H. 1973. *En palynologisk åldersbestämning av vulkanisk aktivitet i Skåne*. Fauna och Flora 68, Stockholm, 121–176.

UGOLINI, F. C. 1986. Processes and rates of weathering in cold and polar desert environments. *In*: COLEMAN, S. & DETHIER, D. (eds) *Rates of Chemical Weathering of Rocks and Minerals*. Academy Press, London, 193–225.

VACHTL, J. 1969. Review of kaolin deposits of Europe. International Geological Congress. Report of twenty-third session. Czechoslovakia 1968. *Proceedings. Symposium I. Kaolin deposits of the world A – Europe*, 13–24.

VELDE, B. & MEUNIER, A. 1987. Petrologic phase equilibria in natural clay systems. *In*: NEWMAN, A. C. D. (ed.) *Chemistry of Clays and Clay Minerals*. Mineralogical Society Mon. 6, Longman, Harlow, 423–458.

WEAVER, C. H. 1989. *Clays, Muds, and Shales. Developments in Sedimentology 44*. Elsevier, Amsterdam.

WILLIAMS, A. G., TERNAN, L. & KENT, M. 1986. Some observations on the chemical weathering of the Dartmoor granite. *Earth Surface Processes and Landforms,* **11**, 557–574.

WILSON, M. J., BAIN, D. C. & DUTHIE, D. M. L. 1984. The soil clays in Great Britain: II Scotland. *Clay Minerals,* **19**, 709–735.

Preservation of old palaeosurfaces in glaciated areas: examples from the French western Alps

YVONNE BATTIAU-QUENEY

CNRS (URA 1688) and IGCP 317, University of Sciences and Technology of Lille, 59655 Villeneuve d'Ascq Cedex, France

Abstract: Remnants of deeply-weathered rock have been recorded in many regions which were ice-covered during the Pleistocene. They generally belong to shields and Palaeozoic massifs. This paper presents an example from the French Alps which testifies to the good preservation of very old (> 200 Ma) landforms in a high relief area situated on the western side of the 'Grandes Rousses' which was entirely covered by ice 30 000 years ago. Sedimentary structures, such as ripples on sandstone bedding planes and other erosional near-shore features, dating back to the Late Triassic before the development of the north, passive, margin of the Tethys ocean, are found in this high relief area. These are remarkably well preserved and occur in close proximity to typical glacial landforms. More generally, widespread elements of a pre-Triassic erosional surface are still recognized in the present landscape. Evidently, ice was unable to obliterate these 200 Ma old features in a high mountain area which was strongly glaciated 30 000 years ago.

Palaeoweathering remnants have been recorded for many years in regions which were severely glaciated during the Pleistocene. In most cases these regions belong to shields or Palaeozoic massifs where ice sheets and ice-caps have covered a low or moderate pre-glacial relief (Battiau-Queney 1981, 1984; Bouchard 1985; Bouchard & Pavich 1989; Fogelberg 1985; Godard 1989; Hall 1986; Lidmar-Bergström 1988; Lundqvist 1985; Peulvast 1985; Roaldset *et al.* 1982). Several authors (e.g. Lidmar-Bergström 1989) claim that pre-glacial features which formed at or near the palaeoweathering front could have initiated the supposed glacial 'roches moutonnées'.

Descriptions of pre-glacial landforms which survived after the retreat of ice are less frequent in high mountainous regions. This paper presents a case of remarkable preservation of Triassic palaeofeatures in the French western Alps. The survival of 200 Ma old features further constrains the effects of the Late Pleistocene ice dynamics in this region.

Geomorphological and geological setting of the western slopes of the Grandes Rousses

The study area is located in Dauphiné, near Alpe d'Huez, on the western slopes of the Grandes Rousses massif (Fig. 1). It belongs to the external crystalline massifs of the western Alps (Battiau-Queney 1993). The relief is high with differences exceeding 2500 m between the Romanche valley of Bourg d'Oisans (710–730 m) and the ice-covered summits of the Grandes Rousses (Etendard peak, 3468 m; Bayle peak, 3466 m). Cirque and short valley glaciers are present above 2600 m. Steep slopes exceeding 30° are common but some areas display a more subdued relief. This is the case of the studied area which is locally called 'the lake plateau' (Fig. 2) and forms an intermediate level at 2000–2200 m between the Grandes Rousses ice-field and the steep lower slopes of the deep valleys of the Eau d'Olle and the Romanche.

This region is a classical area in which to study the development of the continental margin of the Mesozoic Tethys which preceded the Cenozoic Alpine compression (Amaudric du Chaffaut & Fudral 1986; Lemoine *et al.* 1986; Gillcrist *et al.* 1987; Lemoine & Graciansky 1988; Grand 1988).

Four main phases characterize the palaeotectonic evolution of this region: (1) The pre-rift period, in Late Triassic, when sandstones and carbonates were deposited on the post-Hercynian surface in a tidal flat, arid environment; (2) the Tethyan rifting phase in Early and Middle Jurassic: the Palaeozoic basement and its Triassic cover were faulted in tilted blocks and half-grabens; (3) the post-rift subsidence phase of the new passive margin, in relation to the opening and spreading of the Ligurian ocean in Late Jurassic and Early Cretaceous; (4) the closure of the ocean and the plate collision which initiated the Alpine orogeny in Late Cretaceous.

The fault geometry (Fig. 3) is typical of extensional structures inherited from phases 2 and 3. These faults were reactivated and sometimes

From Widdowson, M. (ed.), 1997, *Palaeosurfaces: Recognition, Reconstruction and Palaeoenvironmental Interpretation,* Geological Society Special Publication No. 120, pp. 125–132.

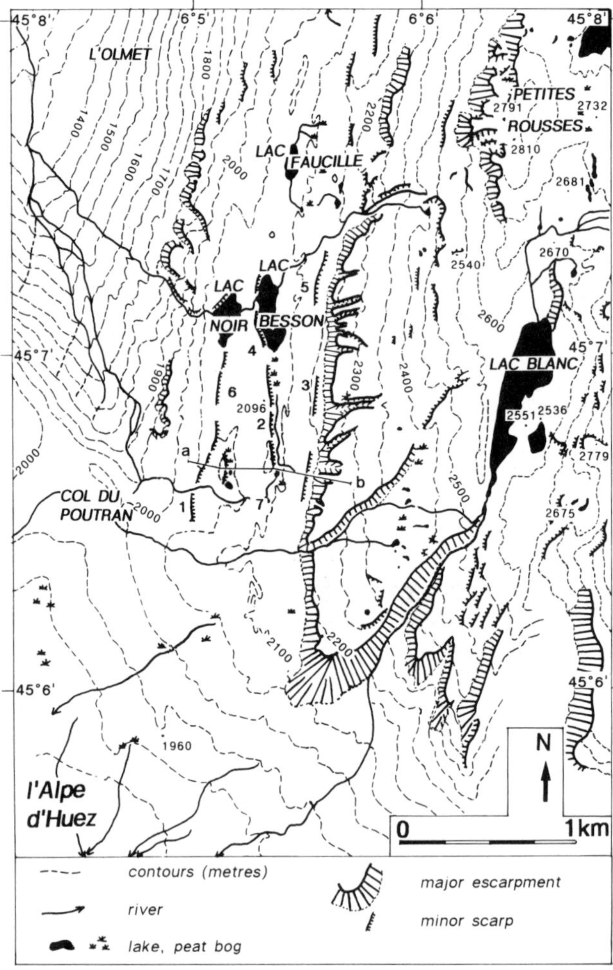

Fig. 1. Topographic map of the studied area. Contour interval: 10 m. 1, Site of ice-polished Palaeozoic gneiss (Fig. 7); 2, site of Triassic ripples on a 22° slope; 3, site of Triassic cliff (MC on Fig. 4); 4, site of Figs 5 and 6 and point of view of Figs 2–7; 5–7, ice-polished crystalline rock. Line a–b is line of section of Fig. 3.

inverted during phase 4 (Grand 1988). All the faults have significant geomorphological expression and produce fault scarps which are easily recognized in the present landscape (Fig. 4) despite the fact that the vertical throw varies from 10 m to a few tens of metres (Grand 1988). Features inherited from the first phase of the regional palaeotectonic evolution are remarkably well preserved in the study area, proving that only slight Alpine Late Cretaceous and Tertiary deformations were superimposed on the Tethyan rifting structures in this area (Amaudric du Chaffaut & Fudral 1986). The geometry of the faulted basement can easily be observed because the western slope of the Grandes Rousses massif mostly lacks its Triassic and Jurassic sedimentary cover, contrary to the Bourg d'Oisans graben in the Romanche valley where > 1000 m of Liassic sediments are still present. The pre-Triassic (Hercynian) basement, which crops out in the Grandes Rousses and the other crystalline massifs of the Alpine External Zone, has been stripped of its Mesozoic sedimentary cover in Late Miocene and Pliocene, in response to a strong regional uplift (Debelmas 1974; Battiau-Queney 1993). We don't know the exact thickness and extent of the remaining Triassic cover on the lake plateau of Alpe d'Huez at the onset of Pleistocene glaciation but it was probably, like today, thicker on the downfaulted side of the tilted blocks and thin or absent on their upper edge.

Post-Hercynian palaeosurface and Triassic palaeofeatures

In spite of post-Triassic faulting, it is relatively easy to reconstruct the pre-Triassic (or post-Hercynian) surface in this area. The Palaeozoic basement,

Fig. 2. A general view of the lakes, northwards from point 4 of Fig. 1. In the background, the Sept-Laux massif. To the right, the 'Petites Rousses' (2810 m). Lake Besson (2070 m) and Lake Noir (2045 m) are located in structurally-controlled basins. The west shore of Lake Besson (F) is a fault scarp a few metres high. Triassic sandstones crop out on the 20–22° slope dipping from lake Besson to Lake Noir. Widespread ripples (R) are observed. Palaeozoic gneiss crops out just above S. It is typically ice-polished and locally carved with grooves.

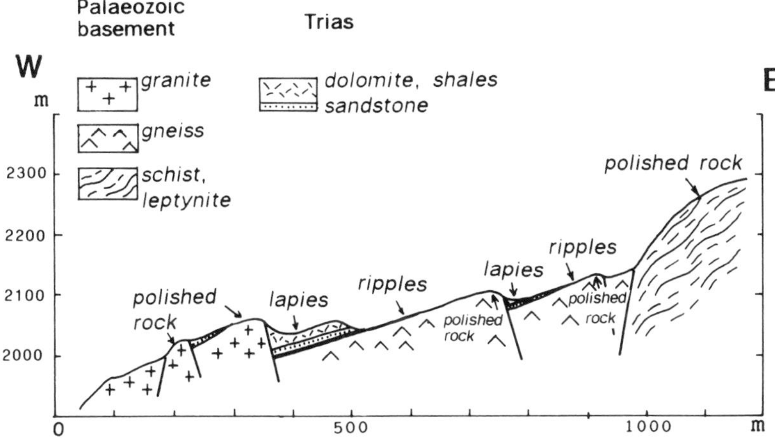

Fig. 3. Cross section in the south part of the studied area (line a–b on Fig. 1). The fault system is typical of structural extension related to the development of a passive margin (here the Tethyan margin). Faults have a direct expression in the landscape. Ice flow was from east to west during the maximum ice extent 30 000 years ago. Glacial erosion was strictly controlled by the faulted relief. Note: 'lapies' is the French word for 'Karren' (small-scale solution clints and grikes).

Fig. 4. The western slopes of the Grandes Rousses massif (view towards the ENE). PB, Pic Blanc (3327 m); PR, Petites Rousses (2810 m). The photograph was taken in July: small glaciers and firns can be seen below the summit ridge. Palaeozoic crystalline and metamorphic rocks crop out on these upper slopes. FS, a 50 m high fault scarp; RM, widespread ripples on Triassic sandstone steeply dipping westwards. Reddish dolomite crops out in the same area. MC, Triassic dolomite abuts on a small gneissic palaeocliff. In the same area, Palaeozoic gneiss has been ice-polished and grooved. L, emplacement of Lakes Besson and Noir.

mainly gneiss and granite, crops out at several places, especially on the eastern edges of tilted blocks (Fig. 3). Towards the west, each block is progressively covered by a thin layer of Middle to Late Triassic marine orthoquartzites which are themselves covered by Late Triassic reddish dolomites (Lemoine *et al.* 1986). Locally, for example at location 3 of Fig. 1, the reddish dolomite lies directly on the crystalline basement without any fault. It abuts against what seems to be a fossilized cliff a few metres high or the steep edge of a basin.

In contact with overlying Triassic sediments, the crystalline rocks have been weathered. At location 5 of Fig. 1, it is possible to recognize an upper bleached zone over a pink coloured rock, a type of profile which could be similar to those described in the southern Massif Central and characterized by low-temperature albitization phenomena (Schmitt 1992). Nearby, the upper surface of the Triassic basal sandstone exhibits well-preserved ripples (Fig. 5) typical of a tidal flat environment (Mégard-Galli & Baud 1977). The ripple-bearing sandstone is only preserved where the slope dips 20° or more (Fig. 3). It disappears on the upper edge of the tilted blocks where the crystalline basement has been ice-moulded.

Small-scale solution features, clints and grikes (the German 'Karren' or the French 'lapiés'), are commonly observed at the surface of the Triassic reddish dolomite which overlies the ripple-bearing sandstone (Fig. 6). Other examples of limestone pavement (or 'Karrenfeld') are frequent in the Alps in the same climatic environment. According to most authors, they could develop in a few thousand years after the retreat of ice and are still active (Nicod 1972; Maire 1990; Salomon & Maire 1992). Contrary to ripple marks they have no palaeomorphological significance.

To summarize, in this area, very old features,

Fig. 5. Ripples on upper bedding surface of Triassic sandstones (location 4 of Fig. 1). Slope 22° westwards. These sedimentary structures reflect a tidal flat environment. On the right, an ice-deposited gneissic boulder. Gneiss crops out on the top left.

ripples of Triassic age (> 200 Ma) which were exhumed at least before the Late Pleistocene Glaciation, are perfectly preserved at the surface of steeply dipping sandstone beds (Fig. 5). Such preservation is surprising given the close proximity of features which testify to strong glacial erosion (Fig.7). It brings into question the efficacy of Pleistocene glacial action in this type of high relief area.

Present and past glacial environment

It is important to consider the dynamics of the regional glaciers in the Late Pleistocene (Fig. 3). On the western slopes of the Grandes Rousses, glaciers are currently restricted to above 2800 m, but in the Late Pleistocene (Würm or Weichselian), the whole region was ice-capped, except a few summits ('nunataks') (Battiau-Queney 1993). The Würmian glaciers reached their maximum extent between 30 000 and 25 000 years ago. They retreated rapidly thereafter. From pollen analysis and ^{14}C dating, we know that the main valleys (Romanche, Vénéon, Eau d'Olle) were free of ice 14 000 years ago.

The main valley glacier (the Romanche glacier) was fed by the Vénéon and upper Romanche glaciers. Ice flow was channelled in the deep trough of Bourg-d'Oisans, which is now partially filled with post-glacial deposits. Watersheds were carved by ice into glacier cirques which fed extensive ice-caps below. The studied area forms an intermediate level between 1800 and 2200 m, which was above the main ice flows. Several times in the Late Würmian and once more during the Little Ice Age, the bowl-shaped depression of Alpe d'Huez was partially covered with ice flowing down from the Grandes Rousses, but during the same cold stages the area around Lakes Noir and Besson probably remained free of ice (Chardon 1991).

During the maximum glacial expansion on the lake plateau of Alpe d'Huez, ice flow was strongly controlled by the pre-existent relief and locally slowed down by a series of small scarps, fault scarps and dolomite cuestas (Fig. 3). In the case of

Fig. 6. Location 4 on Fig. 1. An ancient Triassic shore platform. The view is to the northeast. Triassic sandstones dip 20–22° westwards. Ripples are widespread on this slope (see also Fig. 5). In the background, reddish dolomite (D) crops out, overlying the sandstones. Lapies ('Karren') are well developed at the surface. On this steep dip-slope, neither dolomite nor sandstone presents any typical glacial feature.

the east-facing scarp, the steep front produced a transverse obstacle to the main direction of ice flow coming from the Grandes Rousses, according to the general slope.

As a main consequence of the effect of the pre-glacial underlying relief, ice flow was alternatively compressive on the upper edge of east-facing scarps and extensive on the back slopes (Chorley *et al.* 1984). Pressure melting was high on the 'upstream' side and upper edge of the scarps, providing the high contact forces required by abrasion. Ice pressure was considerably reduced on the lee side of scarps, leading to the separation of ice from the underlying rock in the case of steeper slopes >20°. The resulting glacial erosion was efficient on top of hills but completely ineffective on back slopes. On the upper edge of tilted blocks, the Triassic sediments have disappeared and crystalline rocks have been smoothed, polished and grooved. On back slopes, the previously described ripples remain untouched. The field study suggests that a minimum gradient of the back slopes (20–22°) is required to produce a good preservation of these Triassic palaeostructures.

Conclusions

Very old palaeofeatures have survived in a high mountainous environment which was severely glaciated at least in the Late Pleistocene. These small-scale ripples marked the final development of the post-Hercynian planation surface with the onset of deposition of transgressive Triassic sediments in a tidal flat environment more than 200 Ma ago. They were partially exhumed from a thick Jurassic cover in Late Miocene–Pliocene, in response to a strong uplift of the crystalline external massifs of the Western Alps.

Several additional factors explain the preservation of these fragile features in the area: (1) a system of tilted blocks with fault strike transverse to the local ice flow; (2) a relatively low pre-glacial relief (less than a few tens of metres) but sufficient to produce strong lateral differences of ice pressure at the contact with the bedrock.

The case of the 'lake plateau' of Alpe d'Huez is certainly not unique in the western French Alps. Other areas, for example the Taillefer plateau (Fig. 8), deserve similar types of study and, in addition,

Fig. 7. Ice-polished Palaeozoic gneiss (location 1 on Fig. 1) on the upper edge of a tilted block. Abraded surface testifies to the effectiveness of glacial erosion in a place where ice pressure was high.

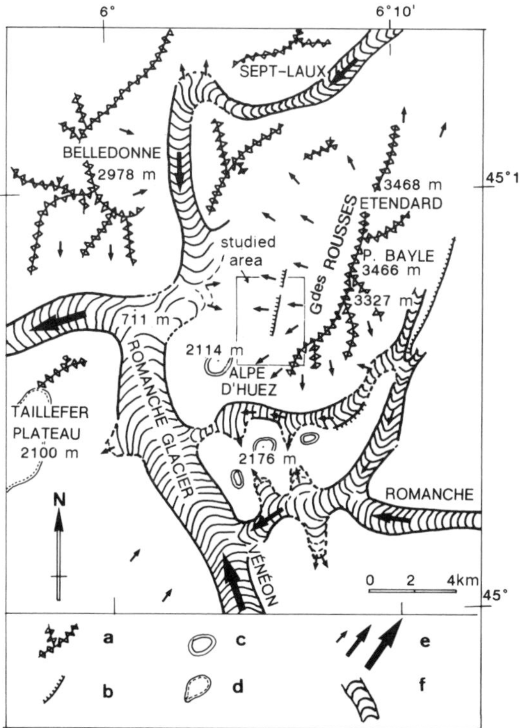

further research is needed on the observed pre-Triassic palaeoweathering profiles which have developed on Palaeozoic gneiss. Finally, the uneven distribution of the effects of glacial scouring is demonstrated: it was essentially concentrated in the valleys, for example the deep Romanche trough. However, glacial erosion was at a minimum in areas which were above and away from the main ice flows. For that reason, high mountains show (surprisingly) good palaeosurface records.

I wish to thank two anonymous referees and Mike Widdowson who greatly helped me to improve the earlier version of this paper with fruitful comments and language corrections. This work is a contribution to the IGCP 317 project.

Fig. 8. The studied area in its regional glacial environment (tentative reconstruction of the Würmian palaeoenvironment). a, Arêtes and cirque glaciers; b, scarp, asymmetric ridge; c, flat or rounded summit (Mesozoic shales); d, ice-scoured, relatively flat, plateau; e, ice flow (the size of arrow is proportional to ice thickness and estimated velocity); f, main valley glaciers.

References

AMAUDRIC DU CHAFFAUT, S. & FUDRAL, S. 1986. De la marge océanique à la chaîne de collision dans les Alpes du Dauphiné. Réunion extraordinaire de la Société Géologique de France, 4–8 Septembre 1984. *Bulletin de la Société Géologique de France*, **8**, 197–231.

BATTIAU-QUENEY, Y. 1981. Les effets géomorphologiques des glaciations quaternaires au Pays-de-Galles (Grande-Bretagne). *Revue de Gémorphologie Dynamique*, **30**, 63–73.

—— 1984. The pre-glacial evolution of Wales. *Earth Surface Processes and Landforms*, **9**, 229–252.

—— 1993. *Le relief de la France. Coupes et croquis*. Masson, Paris.

BOUCHARD, M. 1985. Weathering and weathering residuals on the Canadian shield. *Fennia*, **163**, 327–332.

—— & PAVICH, M. 1989. Characteristics and significance of pre-Wisconsinan saprolites in the northern Appalachians. *Zeitschrift für Geomorphologie, NF, Suppl. Bd*, **72**, 125–137.

CHARDON, M. 1991. L'évolution tardiglaciaire et holocène des glaciers et de la végétation autour de l'Alpe d'Huez (Oisans, Alpes Françaises). *Revue de Géographie Alpine*, **2**, 39–53.

CHORLEY, R., SCHUMM, S. A., SUGDEN, D. E. 1984. *Geomorphology*, Methuen, New York.

DEBELMAS, J. 1974. Les Alpes Franco-italiennes. *In*: DEBELMAS, J. (ed.) *Géologie de la France*. Doin, Paris, **t.z**, 387–442.

FOGELBERG, P. (ed.) 1985. Preglacial weathering and planation. *Fennia*, **163**, 283–383.

GILLCRIST, R., COWARD, M. & MUGNIER, J. L. 1987. Structural inversion and its controls: examples from the Alpine foreland and the French Alps. *Geodinamica Acta (Paris)*, **1**, 5–34.

GODARD, A. 1989. Les vestiges des manteaux d'altération sur les socles des Hautes Latitudes: identification, signification. *Zeitschrift für Geomorphologie*, Suppl. Bd. **72**, 1–20.

GRAND, T. 1988. Mesozoic extensional inherited structures on the European margin of the Ligurian Tethys. The example of the Bourg d'Oisans half-graben, western Alps. *Bulletin de la Société Géologique de France*, **8**, 613–621.

HALL, A. M. 1986. Deep weathering patterns in north-east Scotland and their geomorphological significance. *Zeitschrift für Geomorphologie, NF*, **30**, 407–422.

LEMOINE, M. & GRACIANSKY, P. C. DE 1988. Histoire d'une marge continentale passive: les Alpes occidentales au Mésozoïque. Introduction. *Bulletin de la Société Géologique de France*, **8**, 597–600.

——, BAS, T., ARNAUD-VANNEAU, A. *ET AL*. 1986. The continental margin of the Mesozoic Tethys in the western Alps. *Marine and Petroleum Geology*, **3**, 179–199.

LIDMAR-BERGSTRÖM, K. (ed.) 1988. *Preglacial weathering and landform evolution in Fenno-scandia*. Field symposium in southern Sweden, May 16–20, 1988. Guide to excursions, Lund.

—— 1989. Exhumed Cretaceous landforms in south Sweden. *Zeitschrift für Geomorphologie, NF*, Suppl. Bd. **72**, 21–40.

LUNDQVIST, J. 1985. Deep weathering in Sweden. *Fennia*, **163**, 287–292.

MAIRE, R. 1990. Recherches géomorphologiques et spéléologiques sur les karsts de haute montagne. *Karstologia, Mém.*, **3**.

MÉGARD-GALLI, J. & BAUD, A. 1977. Le Trias moyen et supérieur des Alpes Nord-Occidentales: données nouvelles et corrélations stratigraphiques. *Bulletin du BRGM*, **2**, 233–250.

NICOD, J. 1972. *Pays et paysages du calcaire*. PUF, Paris.

PEULVAST, J. P. 1985. *In situ* weathered rocks on plateaux, slopes and strandflat areas of the Lofoten-Vesterålen, North Norway. *Fennia*, **163**, 333–340.

ROALDSET, E., PETTERSEN, E., LONGVA, O. & MANGERUD, J. 1982. Remnants of preglacial weathering in western Norway. *Norsk Geologisk Tiddskrift*, **62**, 169–178.

SALOMON, J. N. & MAIRE, R. (eds) 1992. *Karst et évolutions climatiques*. Presses Univ., Bordeaux.

SCHMITT, J. M. 1992. Triassic albitization in southern France: an unusual mineralogical record from a major continental paleosurface. *In*: SCHMITT, J. M. & GALL, Q. (eds) *Mineralogical and Geochemical Records of Paleoweathering*. Ecole Nationale Supérieure des Mines de Paris, Mémoire des Sciences de la Terre, **18**, 115–131.

Rock weathering in blockfields: some preliminary data from mountain plateaus in North Norway

W. BRIAN WHALLEY, BRICE R. REA, MICHELLE M. RAINEY & JOHN J. McALISTER

School of Geosciences, The Queen's University of Belfast, Belfast BT7 1NN, UK

Abstract: The formation of blockfields is a process usually attributed to weathering. In mountain areas this is generally assumed to be mechanical weathering (frost shattering). Evidence from two high plateaus [900 and 1350 m above sea level (a.s.l.)] in North Norway (*c.* 70° N) suggests that chemical action is at least as important as mechanical activity in blockfield formation. The bedrock in both areas consists of complex banded gabbros. Blockfields circumscribe ice masses and are generally > 1 m thick. They contain high percentages of material in the silt and clay sized fractions, including a variety of clay minerals: gibbsite, chlorite, vermiculite and kaolinite, as well as magnetite/maghemite. The blockfield thickness and presence of these weathering products suggests both a considerable (pre-Pleistocene) length of time required for development as well as warmer conditions than are found now (mean annual air temperature *c.* 0°C) or in the period since deglaciation. It is suggested that these blockfields represent a preglacial palaeosurface which formed initially under warmer conditions and has survived, largely intact, beneath all the Pleistocene ice sheets.

The review paper by White (1976) provides a basis for an appreciation of the 'blockfield problem'. However, there are still major gaps in our knowledge of mountain-top blockfields; specifically, the rate of formation and the climatic conditions involved. The review of cryoplanation by Priesnitz (1988) further highlights the lack of detailed knowledge about cryoplanation terraces and cryopediments on mountains as well as the mechanisms involved in the formation of all of these landforms. Recent work in eastern USA (e.g. Clark & Hedges 1992) relates to the interpretation of past conditions rather than current formation and mechanisms of formation. Investigations on extant blockfields thus provide a useful means of placing past events in a 'process' framework.

The formation of mountain-top blockfields in northern Norway has been of interest for many years following early work by Svenonius (1909), Högbom (1914) and, more recently, by Lundqvist (1948) and Dahl (1966). All these authors suggest that blockfields have evolved very recently, being landforms produced by rapid freeze–thaw weathering of bedrock which has been exposed since the end of the last glaciation, i.e. they are *c.* < 10 000 years old. Such a scenario suggests very rapid weathering of what, certainly in many Norwegian instances, are weathering-resistant rocks. Alternatively, blockfield formation throughout the Pleistocene during phases of deglaciation has been postulated by Nesje *et al.* (1988). Freeze–thaw is still thought to be mainly responsible for formation but the longer time period allows for the production of deep weathering profiles and rounded debris. A third possibility is that blockfields have been forming since pre-Pleistocene times, initially under warm temperate conditions, with later exposure to periglacial conditions (Roaldset *et al.* 1982; Dahl 1987, 1987; Nesje *et al.* 1988; Nesje 1989). Most deep weathering would thus be due to chemical action during pre-Pleistocene periods when the climate was warm temperate, even at high altitudes.

The interpretation of long-term geomorphology of a region may be significantly related to its long-term denudation. Recent research on Scandinavian saprolites (e.g. Lidmar-Bergström 1988; Fogelberg 1985; Goddard 1989) has highlighted the general issue of pre-Pleistocene weathering in Norway. Dahl (1966) investigated blockfields in the Narvik area (*c.* 65° N) and, more recently, Kleman & Borgström (1990) discussed mountain blockfields in western Sweden (62° N). However, to date, none have addressed the issue of North Norwegian mountain blockfields, although Malmström (1991) has identified the widespread nature of these deposits. Appreciation of mountain-top blockfields also relates to debates about interglacial refugia, for example, the discussion by Ives (1966) of Ragnar Dahl's Narvik investigations. Blockfields may represent palaeosurfaces which have been forming since Tertiary times or earlier. Reconstruction of the preglacial land surface using these blockfields is relatively easy compared to working with buried

From Widdowson, M. (ed.), 1997, *Palaeosurfaces: Recognition, Reconstruction and Palaeoenvironmental Interpretation,* Geological Society Special Publication No. 120, pp. 133–145.

palaeosurfaces as the blockfields are still exposed on mountain and plateau tops and can be mapped from air photos. Further investigations are however required in order to interpret their origin, age and weathering environment during their development.

In this paper we examine selected blockfield sites on parts of plateaus at Øksfjordjøkelen (*c.* 900 m a.s.l.) and Bredalsfjellet (Lyngen Alps, *c.* 1350 m a.s.l.), north Norway, to show how the weathering products of these two high mountain blockfields may relate to the three basic models of formation outlined above.

Site areas, description and samples

Field work was carried out on two plateau summits in North Norway (Fig. 1). Øksfjordjøkelen is a large (40 km^2) plateau glacier with a surrounding blockfield at *c.* 900 m. Bredalsfjellet is a rather smaller plateau, but here with a thin glacier (< 1 km^2), at 1350 m in the Lyngen Alps. There are other examples of plateau summits with blockfields in both areas. Further details of the glaciers and plateaus are given in Gordon *et al.* (1987) and Gellatly *et al.* (1988, 1989). Rocks of both areas are mainly banded gabbroic rocks of Silurian age (Krauskopf 1954). They are variable in composition but typically contain amphibole, pyroxene and feldspar; rock types include troctolite (i.e. olivine gabbro), anorthosite gabbro, garnet–biotite-bearing gneisses, pyroxene–plagioclase hornfels and syenites.

The blockfield surface shows sorted stone circles of various sizes; in some cases these may be revealed by the decaying ice margin (Whalley *et al.* 1981). Pits were dug in blockfields at a number of

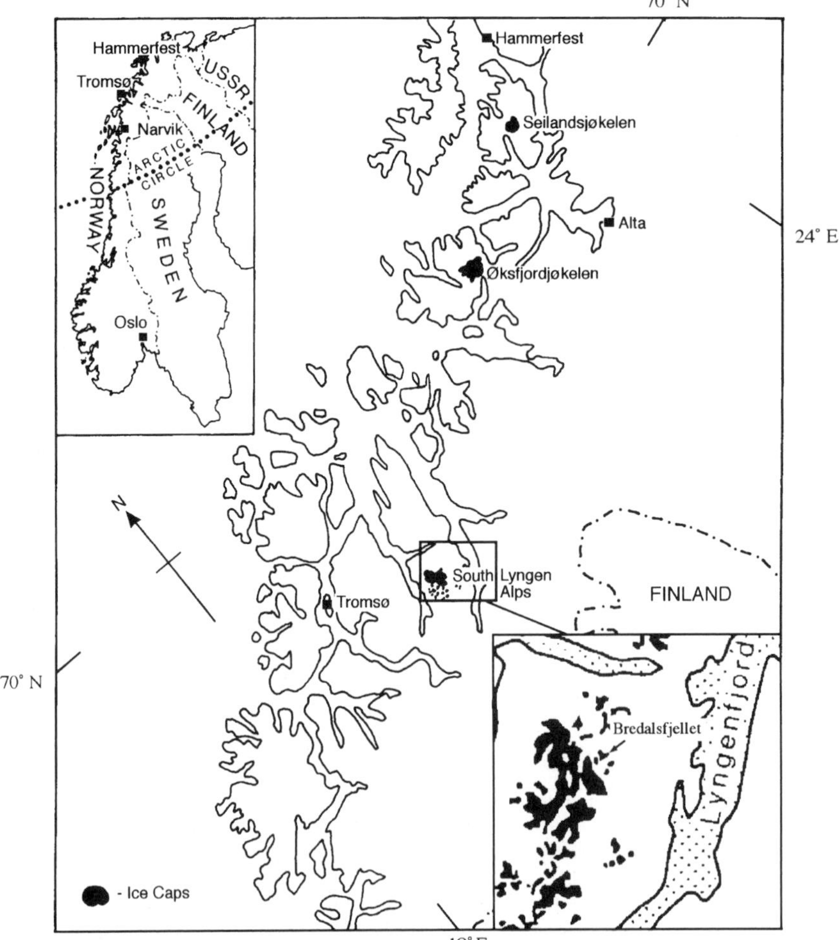

Fig. 1. Location map showing the location of Øksfjordjøkelen and Bredalsfjellet.

locations around Øksfjordjøkelen and on Bredalsfjellet to investigate depth of weathering and profile characteristics. Depths varied from 0.5 m to 1 m on both plateaus. In many cases intact bedrock was not reached because of large boulders towards the base of the profile. Profiles generally had a cobble/boulder surface cover with angular cobbles and occasional boulders of varying size found within a matrix of fines. In some cases (Fig. 2), virtually the whole profile was composed of fines. It is important to note that the residual soils are typically non-structured saprolites. Frost-churning has destroyed any structure which may have been produced during formation. Thus, it is only profile depth and mineralogy that can provide an insight into the weathering processes which formed them. Mixing of profiles by frost-churning removes the need for a vertical sampling strategy. Sampling is in many places limited by the exposure of blockfield, i.e. on nunataks, on the plateau beyond the ice margins and in areas free of perennial snow. Thus, samples comprising interstitial fines and a number of larger pieces of rock were collected from accessible locations.

On the Øksfjordjøkelen plateau it is possible to identify areas of sparse blockfield development (although bedrock and boulders may be severely weathered, Fig. 3) which are relatively close to present ice limits. The thickness of blockfield generally increases away from the ice towards plateau edges, although, in some cases, ice is retreating from a position directly over thick blockfield. No traces of erratics/till were found on these plateau surfaces and both plateaus may be considered to have developed autochthonous blockfields.

Laboratory analyses and results

Samples were wet sieved (British Standard 410 1969) to provide basic size information with Coulter Counter analysis being used for the $< 63\ \mu m$ fraction (Table 1). Figure 4 shows a sample of the particle size ranges found. It should be noted that all of the remaining samples would plot somewhere inside the 'envelope' (Fig. 4). The most important result shown in Table 1 and Fig. 4 is enrichment of the samples in the silt and clay size fractions.

Clay minerals were identified from both XRD (CuK_α) and DTA analyses of the $< 63\ \mu m$ and $< 2\ \mu m$ fractions. Pre-treatment was applied for both types of analysis. Samples were ultrasonically dispersed and fine fractions separated by centrifuge

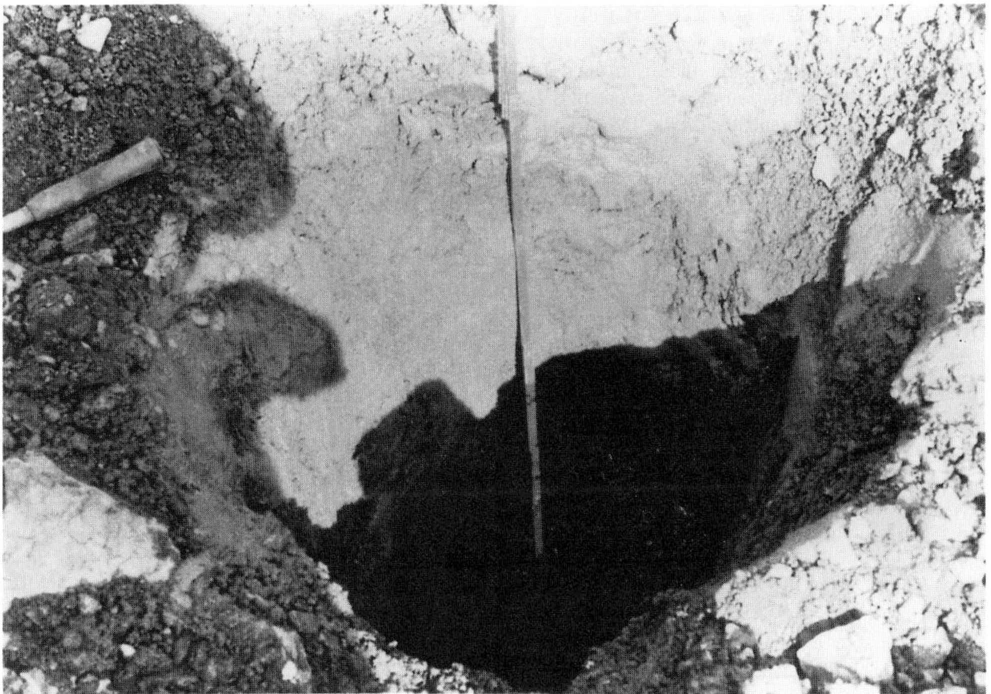

Fig. 2. Pit on Bredalsfjellet, Lyngen, containing almost all silty fines (photo: D. J. Tate).

Fig. 3. Weathered boulder and bedrock on the Øksfjordjøkelen plateau.

(750 r.p.m. for 3 min.) and then saturated with Mg^{2+}. Preferred orientation mounts were prepared using a rapid membrane technique (McAlister & Smith 1996). XRD slides were also heat treated (610°C for 1 h), glycolated and saturated with K^+, Mg^{2+} and NH^{4+}. Powdered samples < 63 μm were diluted for DTA at a 1 : 1 ratio with calcined alumina and run isothermally from ambient to 1000°C at 10°C min^{-1}. Cation saturation, glycolation and heat treatment of the clay complex was carried out in order to enhance mineralogical interpretation. A chlorite/vermiculite peak is observed at 14–15 Å and glycolation results in the expansion of vermiculite which shows a shift in d-spacing to 15–16 Å. Heat treatment results in dehydration of the clay complex and results in complete contraction to an anhydrous spacing of 9.4–9.6 Å (e.g. Fig. 5a–c). Chlorite is unaffected by

Table 1. *Particle size distributions for samples BM1–BM6 and BM9–BM 15*

Size Fraction	BM1	BM2	BM3	BM4	BM5	BM6	BM9	BM10	BM11	BM12	BM13	BM14	BM15
2–1 mm	2.39	8.36	30.83	45.00	30.79	20.74	5.95	17.32	2.66	7.10	7.24	8.60	3.17
1–500 μm	1.59	10.17	17.10	19.53	24.60	14.52	7.83	16.57	3.95	6.82	9.89	7.47	2.81
500–250 μm	2.60	11.99	7.46	9.00	14.36	12.46	7.95	12.92	8.99	6.19	15.06	10.71	7.18
250–180 μm	4.66	8.42	2.52	3.35	5.26	6.23	4.58	6.05	9.71	4.44	9.63	8.88	7.60
180–125 μm	10.09	9.38	2.35	2.75	4.50	6.01	4.64	6.04	10.50	7.76	8.38	9.29	9.10
125–90 μm	14.19	8.38	2.23	2.15	3.55	5.25	4.48	4.94	10.59	9.47	6.77	8.68	8.82
90–63 μm	17.69	7.32	2.35	2.18	2.90	5.45	5.37	4.80	9.36	12.76	6.16	6.14	9.47
< 63 μm	46.79	36.08	35.16	16.04	14.04	29.34	59.20	31.36	44.24	45.46	36.78	40.23	51.85

No analysis was carried out for samples BM7 and BM8 as the initial samples were too small to allow all tests to be carried out. It is important to note the enrichment in the fine < 63 μm size fraction.

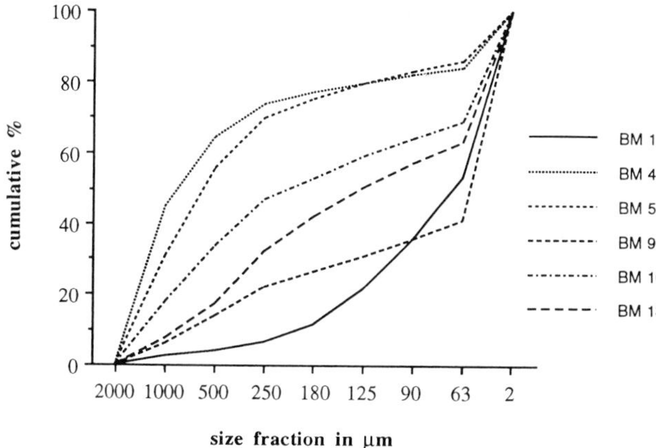

Fig. 4. Cumulative percentage curves for a representative number of samples from Table 1. The remaining samples not shown all plot inside the envelope between BM1, BM4, BM5 and BM9.

any of these treatments and heating is therefore used as a diagnostic feature.

Although vermiculite was found in some samples, saturation, heat and glycolation treatment suggested that 2.1 chlorite was also present in several samples. The analysis of two samples of plateau bedrock, likely to represent parent materials of the blockfields, showed somewhat variable mineralogy and chemistry. These samples were assumed unweathered, having been taken from a recently deglaciated foreland where the glacier had removed the blockfield and scoured the bedrock. Clay mineralogy is given in summary form for all samples (Table 2). Representative XRD traces and a DTA trace are shown in Figs 5–7.

Particle size and clay minerals: Discussion

Particle size analyses suggested that there was an important contribution to breakdown from mechanisms that could not be attributable to frost shattering. The effectiveness of frost shattering is believed to be best applied to wedging of blocks along joints. These joint-sized blocks are then broken down by chemical weathering processes which exploit small cracks and intergranular microfractures (Dredge 1992; Rainey 1994). The parent rocks contain very few interconnecting cracks or microfractures (< 10 μm wide) and thus inhibit breakdown through mechanical processes. Lautridou & Seppälä (1986) carried out extensive freeze–thaw cycling of metamorphic rocks from Finnish Lapland. The rocks were found to be highly resistant to this form of weathering. Chemical weathering, on the other hand, tends to exploit grain boundaries and intragranular microfractures to produce material characteristically high in silt and clay size fractions (Wambeke 1962; Hall *et al.* 1989). The high proportion of silt size material, shown in Table 1 and Fig. 4, is not commensurate with the crystal/grain size of the gabbros. This suggests that chemical weathering has been an important process in the production of fines below the grain size of the parent material.

Rock surface summer temperatures at Øksfjord-jøkelen may often exceed 25°C, as in other high mountains (Whalley *et al.* 1984). The number of days per year when rock surface temperatures are high enough to promote chemical weathering are few. Thus, present-day chemical action may be sufficient to produce both weathering rinds and clay minerals. However, it cannot easily account for the whole thickness of the blockfield which may in places exceed 1 m. Further, the ice caps on both summits have been much more extensive in the past and, indeed, may only now be retreating from Little Ice Age (*c.* 1600–1880 AD) maxima. Even the full length of the (uninterrupted) Holocene is unlikely to have produced such thicknesses of weathered material. The type of clay minerals found do suggest that the environment for formation may have been substantially different from those pertaining today and, indeed, the whole of the Holocene. The following observations remark on the possible origin of the clay minerals.

1. Vermiculite was found as an important component in some of the blockfield specimens. It is typically found in gabbroic saprolites (Kodama *et al.* 1988) and is thought to be indicative of long-term weathering (Wilson 1967; Basham 1974).
2. Chlorite is present in varying amounts in many specimens and is the direct result of weathering of olivine gabbro but may also occur in parent material. Serpentized chlorite is also found in small amounts. Chlorite has been suggested as a product of physical weathering of rocks

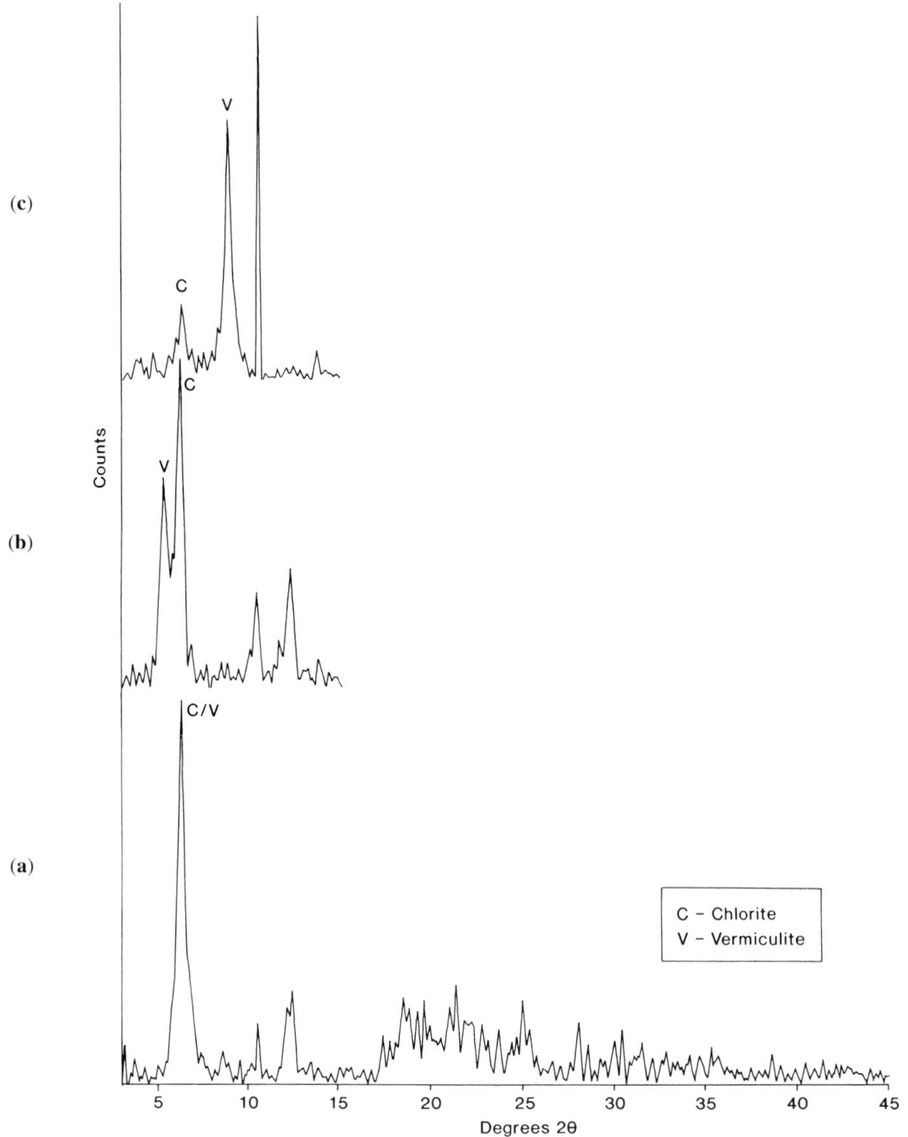

Fig. 5. XRD traces from sample BM2 (pit on Øksfjordjøkelen plateau) showing peak shifts and disappearance using: (**a**) Mg^{2+} saturation; (**b**) Mg^{2+} saturation + glycolation; (**c**) Mg^{2+} saturation + heated at 610°C.

(Bockheim 1982), although it can be converted from biotite (Jackson *et al.* 1977).

3. Kaolinite was common to nearly all samples (with the exception of a parent material). It is most probably the product of plagioclase weathering in the gabbro and occurs even around coarser particles in the blockfield material (e.g. Michailidis *et al.* 1993). Plagioclases were the commonest components of the < 2 μm fraction and present in all samples (not shown in Table 2).
4. Gibbsite was found in several samples, often in considerable amounts. It is commonly associated with kaolinite and is the product of biotite weathering. Basham (1974) found vermiculite/kaolinite/gibbsite as indicative of deep weathering in Aberdeenshire gabbro formed under a warm, humid climate, possibly

Table 2. *Results of comparative XRD examination of the $< 2\ \mu m$ fraction of samples from blockfield and related sites*

Sample	Description/location	Vermiculite	Chlorite	Kaolinite	Gibbsite	Magnetite	Amphiboles	Quartz	Mica
BM1	Moraine Øks	xx	xxx	xxx	xx	x	xx	x	x
BM2	Blockfield Øks	xx	xxx	xx	xxx	x	x	xx	xx
BM3	Blockfield Øks	x	xxx	xx	xxx		x	xx	x
BM4	Blockfield Øks			xxx	x				
BM5	Subglacial material Øks	xxx	xxx	x			x		
BM6	Subglacial material Øks	xxx	xxx	xx			xx		
BM7	Blockfield interjoint Øks			xxx			xx	x	
BM8	Debris deglaciated foreland Øks			x					
BM9	Blockfield Bredalsfjellet		xxx	xx	xxx		xx	xx	xx
BM10	Blockfield Øks	xxx	xxx	xxx	x		x	xx	xx
BM11	Blockfield Øks	xxx	xxx	xx	xxx		x	xx	xx
BM12	Blockfield Øks		xxx		xxx			x	
BM13	Blockfield Øks		xxx	x	xxx		x	xx	x
BM14	Blockfield Øks		xx	x	xx		x	xx	x
BM15	Blockfield Øks		xxx	x		xxx		xx	x
BM16	Parent material Øks						x		
BM17	Parent material Øks		xxx				x		x

xxx, Major; xx, Minor; x, Trace. Øks, Øksfjordjøkelen.

The deglaciated foreland, moraine and subglacial samples were collected to show that the blockfield cover extends over the plateau beneath the present icefield.

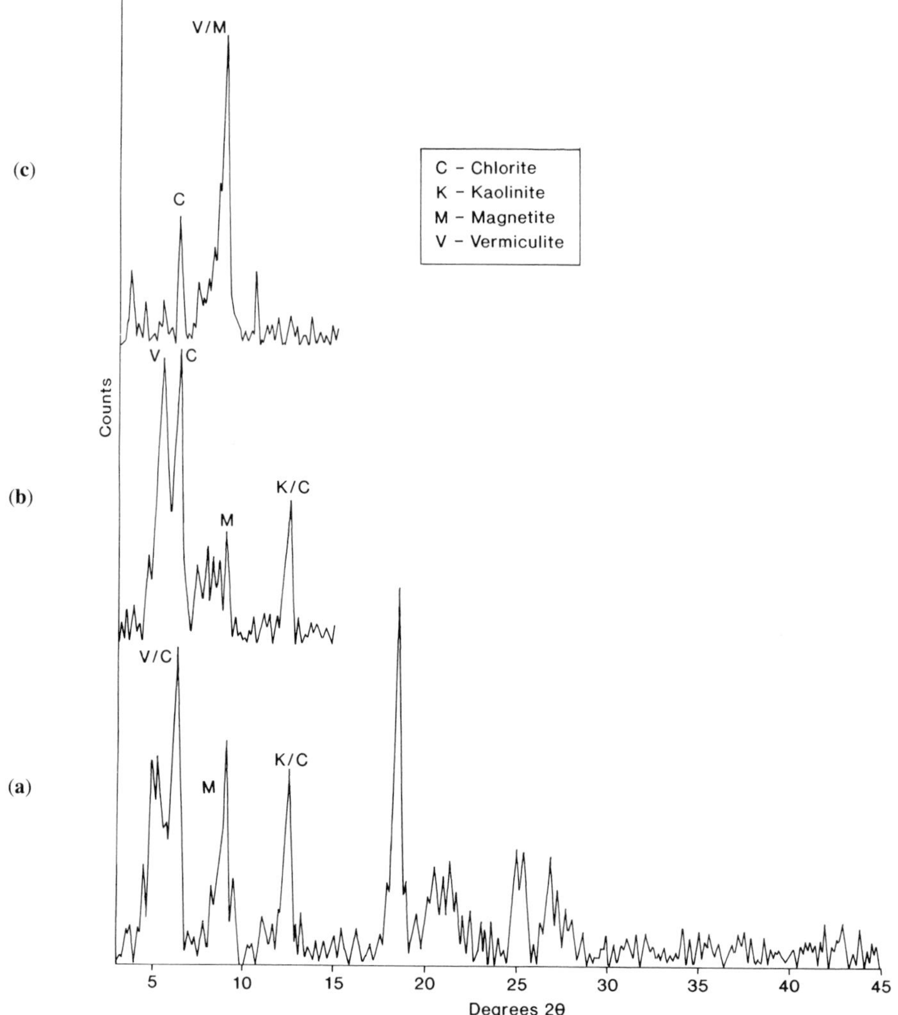

Fig. 6. XRD traces from sample BM6 (subglacial sample from an outlet glacier of Øksfjordjøkelen) showing peak shifts and disappearance using: (**a**) Mg^{2+} saturation; (**b**) Mg^{2+} saturation + glycolation; (**c**) Mg^{2+} saturation + heated at 610°C.

from mid-Tertiary times. Although it is possible that some gibbsite may also be partly formed under present-day conditions the vermiculite/kaolinite/gibbsite presence is suggestive of long-term weathering.

5. Amphiboles are again found commonly, as the complex parent material has some amphibolite facies (Ball *et al.* 1963). Amphiboles can give rise to vermiculite seen, for example, in the sandy saprolite described by Abreau & Vairnho (1990). The weathering of gabbro at Mount Mégantic described by Clement & De Klimpe (1977) shows similar results.

Samples also show that clay minerals were present in the subglacial material (BM5 and 6) as well as the moraine on the edge of the blockfield (BM1). The smooth but striated nature of the glacier bed and foreland shows that the glacier has removed blockfield subglacially from upstream on the plateau and that some has been deposited in the moraine. Parent material (BM16 and 17) from the striated foreland shows no vermiculite/kaolinite/gibbsite. Similarly, the glacier foreland (BM8) material contains little clay.

The findings by Basham (1974) and Wilson (1967), as well as Hall & Mellor (1988), are all

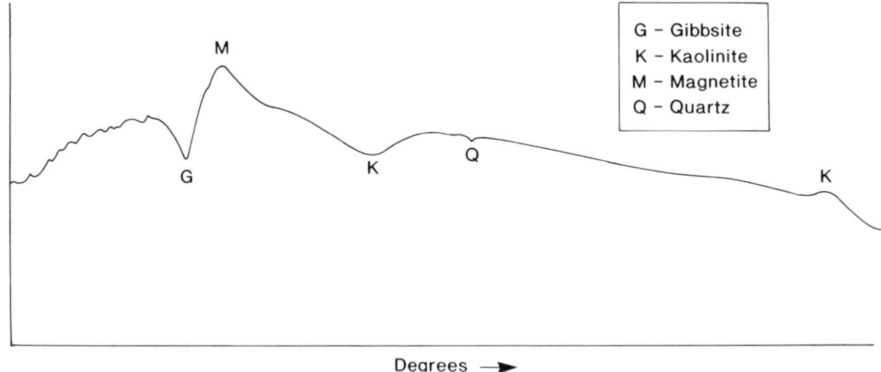

Fig. 7. DTA trace (sample BM2) from pit on Øksfjordjøkelen plateau showing main clay minerals.

similar to the situation in North Norway and provide a similar interpretation, i.e. of pre-Pleistocene weathering having taken place under warmer, possibly wetter, conditions than at present. The thicknesses of saprolites in Scotland, e.g. Hall *et al.* (1989), similarly suggest that they cannot have had time to develop during the interglacial periods. This must be especially true of North Norway where conditions have been much more severe than in Scotland in the post-Glacial. Further, the length of ice-free time available will have been severely reduced in North Norway. We conclude, therefore, that the weathering products on the blockfields are probably Tertiary in age and cannot have formed substantially at any stage during the Pleistocene. However, some Pleistocene weathering might have taken place on bedrock surfaces left uncovered by ice masses during the last glaciation (e.g. Fig. 3). Unfortunately, there is no independent age control available for this supposition.

SEM observations

Frost shattering is essentially a mechanical weathering process which comminutes material into smaller components by splitting and fracturing along weaknesses. Chemical weathering involves removal and alteration of minerals. The two processes are inherently different and evidence for each may be imprinted on grain surfaces. Mechanical action (stress-induced breakdown) produces a variety of patterns characterized by fracture surfaces and sharp edges (Gallagher 1987; Whalley 1996). Ultimately, chemical weathering may tend to produce more rounded and smooth grains; solution, evident in the form of pitting and etched Vs on surfaces, provide excellent evidence of chemical attack. Grains from a number of samples were investigated using SEM. Very angular and broken grains, typical of mechanical breakdown, are rare; instead, evidence of chemical action is found on many grains. Figure 8a–c shows three examples of typical surface textures from chemically weathered grains, while Fig. 8d shows that typical of freeze–thaw mechanical breakdown. Grains in Fig. 8a, a plagioclase feldspar, show cleavage-controlled breakdown. Grains in Fig. 8b, another feldspar, have been weathered to a much greater degree. The whole surface of the grain is covered in solution pits. Grains in Fig. 8c, a calcium-rich feldspar, shows classic V-in-V etching, indicative of chemical weathering. These grains are typical examples of the surface textures found on grains from the blockfield fines. In some instances the chemical weathering is severe while in others it may be less pronounced. These differences may be the result of mineralogical differences or variations in the weathering environment at a local or microscale. The profiles have been mixed thoroughly by frost heaving, and so grains which were originally relatively fresh and only newly released from the parent rock may now be at the surface. SEM investigations show that evidence of chemical weathering is present on most grains examined from the fines.

The importance of micro-crack structure and water retention

Although clay mineralogy and SEM provides some basic information about ages and environmental conditions during weathering, they do not fully explain the mechanisms involved in blockfield formation. Observations made on stones excavated from pits help to explain some of the weathering features.

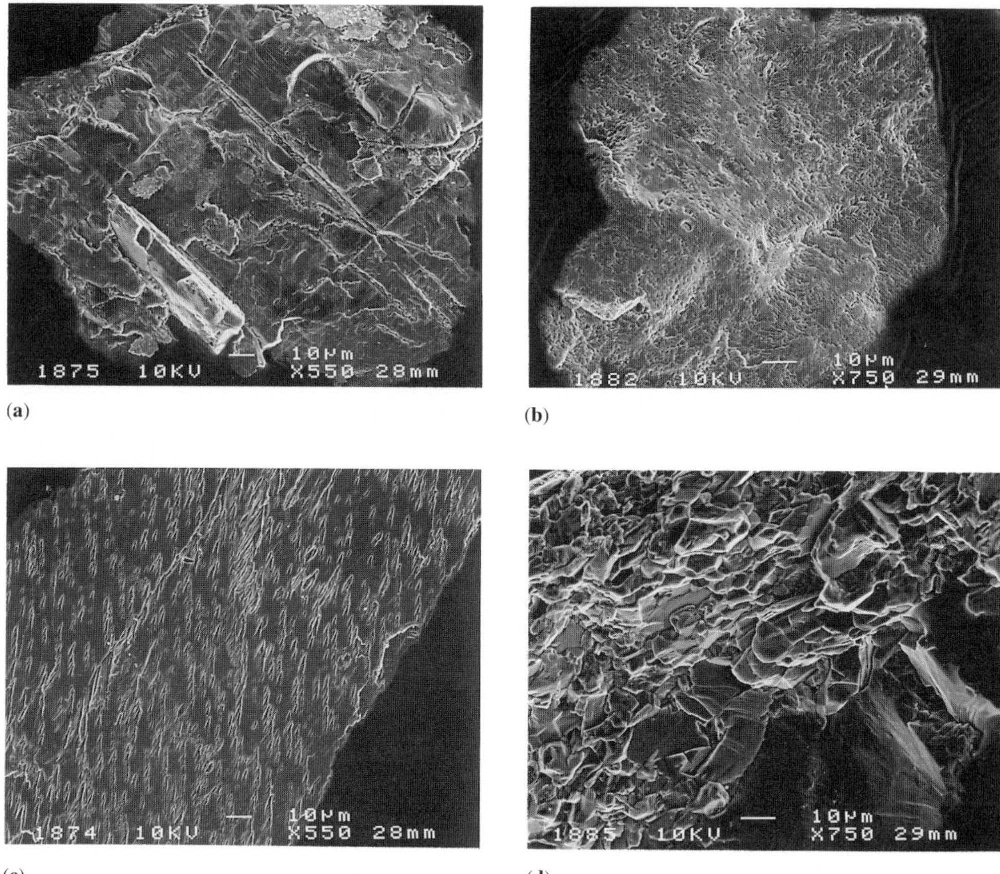

(a) (b) (c) (d)

Fig. 8. (**a**) A feldspar grain showing cleavage-controlled breakdown; the cleavages control the lines of chemical attack. (**b**) A magnesium silicate grain (probbaly olivine). Here cleavage is less pronounced and the upper surface is covered in solution pits. (**c**) This calcium-rich feldspar shows classic V-in-V etching indicative of chemical weathering. Note the extended microfracture from the top middle of the micrograph. Splitting of the grain is most likely to occur along such weaknesses. (**d**) This micrograph is of a 'fresh' aluminium silicate grain taken from supra-glacial material on Glacier de Tsidjiore Nouve, Switzerland. Breakdown is again cleavage-controlled but is mechanical, shown by the fresh, unweathered fracture surfaces. The surface texture is in obvious contrast to those shown in (**a**)–(**c**).

If weathering is to take place, by frost action or chemical alteration, then ingress of water into cracks will be important but there are very few data on this problem. Several types or conditions might apply.

1. Rocks exposed on the surface. These would be exposed to short (seasonal) and very short (days) term temperature variations near the rock surface. Water would gain access to cracks after rain or during snow melt. Frost shattering might result if cracks are available in 'intact rock' while chemical alteration is possible with sufficiently high temperatures and water presence. Weathering rind formation would be possible together with alteration of near-surface grains of susceptible minerals (especially, plagioclase and mafic minerals).
2. (a) Rocks below the surface.
   Perhaps < 1 m – short-term (seasonal) variations near the rock surface would apply. Water would be present for much of the year but temperature fluctuations would be damped and unlikely to be much above the mean annual air temperature. Chemical alteration is unlikely under present climatic conditions. Frost shattering might result

if the severity of the cold wave penetration allowed, but again, only near the surface.

2. (b) Below 1 m – long-term (years) moisture variations within crack/microfracture systems might occur but temperature variations would be highly damped. Neither mechanical nor physical weathering are likely under these conditions.

The wetting process is rather different than the drying process; the latter being a two stage mechanism. Capillary water can penetrate even thin cracks under low pressure gradients. These cracks will remain wet unless exposed at, or very near, the blockfield surface. There seem reasonable grounds for justifying the above classification, even though the actual amounts of interaction and times are unknown. However, there is some qualitative observational evidence for this division. In several pits on Øksfjordjøkelen, stones found in the fines were broken open. Although the outer surfaces were damp, the broken surfaces were almost always dry. The rock only breaks easily when there are cracks present and it was usually the case that fracturing occurred wholly or partly down these cracks. The surfaces of the cracks appeared wet, indicating moisture in the crack. However, no chemical alteration could be seen on such crack surfaces. This suggests that, away from the outer surface, there is little present-day chemical alteration. Condition (1) is thus the only mode of alteration presently operative. However, physical rock breakdown, although slow (Lautridou & Seppälä 1986) might allow an increase in surface area and, ultimately, enhanced chemical alteration if conditions became suitable. The large quantities of fines do indicate that chemical, rather than purely physical, break-up is the most important component of blockfield formation. Physico-chemical conditions are most suitable at, or very near, the blockfield surface.

Implications for blockfield development

Rates of fines production (associated with freeze–thaw) from hard metamorphic and igneous rocks under this climatic regime are very low (Lautridou & Seppälä 1986). As the blockfield depth increased the rate of weathering would be reduced. However, this does not take into account the 'churning' effect produced by frost action in the blockfield. In general, large blocks will come to the surface and fines will move downwards. If water has access to cracks in near-surface blocks then frost shattering might occur, increasing surface area and chemical alteration. Pre-Pleistocene (or perhaps during interglacials) the climate was certainly conducive to chemical activity which could have produced the depth of weathering seen on plateau surfaces. As climate deteriorated, rates of chemical weathering reduce until finally limited to only a few days per year and only on exposed surfaces where summer insolation can produce sufficient heating to allow chemical action to proceed. Thus, it is difficult to evaluate the overall response of weathering and weathering rates to climate change.

Some blockfields and related landscapes in the USA have been attributed to purely (or at least mainly) to periglacial processes (e.g. Clark & Ciolkosz 1988). However, some high arctic landscapes have been interpreted, as here, to having inherited a substantial pre-Pleistocene component (Watts 1983). It is possible that ancient chemical weathering might be revealed by excavating to the bottom of the blockfield material. As yet, however, there is no timescale for blockfield development. In other, extra-glacial locations, non-tropical rates of saprolite formation have varied between 2–10 and 38 m Ma^{-1} (Thomas 1994). It is not yet known how such rates might relate to those on the high plateau surfaces in North Norway considered here; especially where Mediterranean, or even tropical, rates of weathering might be appropriate.

The presence of blockfield and its emergence, apparently uneroded, from under ice-cap and glacier (Gordon *et al.* 1987) is significant in terms of both glacier erosion and landscape development. This has been considered by Sugden (1989) in terms of landscape development in Scotland. In North Norway, the proximity of both eroded glacier foreland next to uneroded, deeply-weathered blockfield has far-reaching implications for both the effectiveness of glacial erosion and thermal history.

Conclusions

The depth and mineralogy of blockfield deposits found on both plateaus suggest non-contemporaneous and long-term weathering. Preservation of pre-glacial landforms and surfaces beneath Pleistocene ice sheets has been suggested by various authors (e.g. Sugden & Watts 1977). Lidmar-Bergström (1988) and Lundqvist (1988) suggest that, for certain regions of Sweden, the extent of vertical glacial erosion has been limited, comprising mainly the stripping of a regolith cover from the land surface. Kleman (1994) suggests that landforms can be preserved beneath large and small ice bodies provided the ice is cold-based. Roaldset *et al.* (1982) identified a number of deeply-weathered profiles in western Norway which they concluded to be the direct result of 'lateritic/bauxitic weathering in a warm humid climate'. Elsewhere, preservation of pre-glacial landforms have been reported by Ives (1974, 1975) and

Sugden & Watts (1977) from Baffin Island; Ives (1958) and Ives *et al.* (1976) in Labrador and by Hall & Mellor (1988) and Hall *et al.* (1989) from Scotland. In all these, the site-related evidence suggests a pre-Weischelian glaciation (> 120 000 years) for the age of the blockfields in the various study regions – with which the situation in North Norway appears to agree. A blockfield origin in Tertiary times under different climatic conditions is certainly plausible. During the Quaternary, the profiles produced through the weathering history of the blockfield have been destroyed. Continual frost-heaving processes active during subaerial exposure in interglacials and interstadials has brought corestones to the surface and removed any traces of a 'normal' weathering profile. The clay mineralogy provides the only key to the weathering history.

These blockfields may well represent a palaeosurface pre-dating the Pleistocene glaciations. The mineralogies of clay fines reflect a Mediterranean, or even tropical, type weathering environment. This palaeosurface is presently exposed subaerially and highlights the longevity of some landsurfaces even in high latitude regions where extensive glacial modification of the landscape has occurred. Finally, it is suggested that site-specific investigations from a wide geographical spread now need to be linked to 'process' studies and a development of techniques to examine rates and environmental conditions of weathering.

We thank the various sponsors and assistants for QUB expeditions to Lyngen and Øksfjordjøkelen, especially the Earthwatch 1992 volunteers; Stephen McCarron and Derrick Tate for sample collection in Lyngen. We also thank Jeff Wilson for discussion about our data. Also thanks to Gill Alexander and Moira Pringle for cartographic assistance. This paper was much improved through the comments of two anonymous referees and the editor.

References

ABREAU, M. M & VAIRINHO, M. 1990. Amphibolite alteration to vermiculite in a weathering profile of gabbro-diorite. *In*:: DOUGLAS L. A. (ed.) *Soil Micromorphology: A Basic and Applied Science*. Developments in Soil Science, **19**, Elsevier, Amsterdam, 493–500.

BALL, T. K., GUNN, C. B., HOOPER, P. R. & LEWIS, D. 1963. A preliminary geological survey of Loppen district, West Finnmark. *Norsk Geologisk Tidsskrift*, **43**, 215–246.

BASHAM, I. R. 1974. Mineralogical changes associated with deep weathering of gabbro in Aberdeenshire. *Clay Minerals*, **10**, 189–202.

BOCKHEIM, J. G. 1982. Properties of a chronosequence of ultraxerous soils in the Trans-Antarctic Mountains. *Geoderma*, **28**, 239–255.

CLARK, G. M. & CIOLKOSZ, E. J. 1988. Periglacial geomorphology of the Appalachian highlands and interior highlands south of the glacial border – a review. *Geomorphology*, **1**, 191–200.

—— & HEDGES, J. 1992. Origin of certain high-elevation local broad uplands in the Central Appalachians south of the glacial border. *In*: DIXON, J. C. & ABRAHAMS, A. D. (eds) *Periglacial Geomorphology*. Wiley, Chichester, 31–61.

CLEMENT, P. & DE KIMPE, C. R. 1977. Geomorphological conditions of gabbro weathering at Mount Mégantic, Quebec. *Canadian Journal of Earth Sciences*, **14**, 2262–2273.

DAHL, R. 1966. Blockfields, weathering pits and tor-like forms in the Narvik mountains, Nordland, Norway. *Geografiska Annaler*, **48A**, 55–85.

—— 1987. The nunatak theory reconsidered. *Ecological Bulletins*, **38**, 77-94.

DREDGE L. A. 1992. Breakup of limestone bedrock by frost shattering and chemical weathering, eastern Canadian. *Arctic: Arctic and Alpine Research*, **24**, 314–323.

FOGELBERG, P. 1985. A field symposium on preglacial weathering and planation held in Finland, May 1985. *Fennia*, **163**, 283–286.

GALLAGHER, J. J. Jr 1987. Fractography of sand grains broken by uniaxial compression. *In*: MARSHALL, J. R. (ed.) *Clastic Particles*. Van Nostrand Reinhold, New York, 189–228.

GELLATLY, A. F., GORDON, J. E., WHALLEY, W. B. & HANSOM J. D. 1988. Thermal regime and geomorphology of plateau ice caps in northern Norway: Observations and implications. *Geology*, **16**, 983–986.

——, WHALLEY, W. B., GORDON, J. E., HANSOM J. D. & TWIGG D. S. 1989. Recent glacial history and climatic change, Bergsfjord, Troms-Finmark, Norway. *Norsk Geografisk Tidsskrift*, **43**, 19–30.

GODDARD, A. (ed.)1989. Weathered mantles (saprolites) over basement rocks of High Latitudes. *Zeitschrift für Geomorphologie*, Suppl. Bd, **72**.

GORDON, J. E., WHALLEY, W. B., GELLATLY, A. F. & FERGUSON, R. I. 1987. Glaciers of the Southen Lyngen Peninsula, Norway. *In*: GARDINER, V. (ed.) *International Geomorphology*: Part 1. Wiley, Chichester, 743–758.

HALL, A. M. & MELLOR, A. 1988. The characteristics and significance of deep weathering in the Gaick area, Grampian Highlands, Scotland. *Geografiska Annaler*, **70A**, 309–314.

——, MELLOR, A. & WILSON, M. J. 1989. The clay mineralogy and age of deeply weathered rock in north-east Scotland. *Zeitschrift für Geomorphologie*, Suppl. Bd, **72**, 97–108.

HÖGBOM, B. 1914. Über die geologische Bedeutung des Frostes. *Uppsala Universitet. Geologisk Institut. Bulletin*, **12**, 257–390.

IVES, J. D. 1958. Glacial geomorphology of the Torngat Mountains, northern Labrador. *Geographical Bulletin*, **12**, 47–75.

—— 1966. Block fields, associated weathering forms on mountain tops and the nunatak hypothesis. *Geografiska Annaler*, **48A**, 220–223.

—— 1974. Biological refugia and the nunatak hypothesis.

In: IVES, J. D. & BARRY, R. G. (eds) *Arctic and Alpine Environments*. Methuen, London, 604–636.

—— 1975. Delimitation of surface weathering zones in eastern Baffin Island, northern Labrador and Arctic Norway: A discussion. *Geological Society of America Bulletin*, **86**, 1096–1100.

——, NICHOLS, H. & SHORT, S. 1976. Glacial history and palaeoecology of North-eastern Nouveau-Québec and Northern Labrador. *Arctic*, **29**, 48–52.

JACKSON, M. L., LEE, S. Y., UGOLINI, F. C. & HELMKE, P. A. 1977. Age and uranium content of soil micas from Antarctica by the fission particle track replica method. *Soil Science*, **123**, 241–248.

KLEMAN, J. 1994. Preservation of landforms under ice sheets and ice caps. *Geomorphology*, **9**, 19–32.

—— & BORGSTRÖM, I. 1990. The boulder fields of Mt. Fulufjället, West-Central Sweden. *Geografiska Annaler*, **72A**, 63–78.

KODAMA, H., DE KIMPE, C. R. & DEJOU, J. 1988. Ferrian saponite in a gabbro saprolite at Mont Mégantic, Quebec. *Clays and Clay Minerals*, **36**, 102–110.

KRAUSKOPF, K. B. 1954. Igneous and metamorphic rocks of the Øksfjord area, Vest-Finnmark. *Norges geologiske Undersökelse, Arbok*, **188**, 29–50.

LAUTRIDOU, J. P. & SEPPÄLÄ, M. 1986. Experimental frost shattering of some Pre-Cambrian rocks, Finland. *Geografiska Annaler*, **68A**, 89–100.

LIDMAR-BERGSTRÖM, K. 1988. Preglacial weathering and landform evolution in Fennoscandia. *Geografiska Annaler*, **70A**, 273–276.

LUNDQVIST, G. 1948. De svenska fjallens natur. Svenska Turistf. Forl., Stockholm.

—— 1988. The Revsund area, central Jämtland – An example of preglacial weathering and landscape formation. *Geografiska Annaler*, **70A**, 291–298.

MCALISTER, J. J. & SMITH B. J. 1995. A rapid preparation technique for X-Ray Diffraction analyses of clay minerals in weathered rocks. *Microchemical Journal*, **52**, 53–61.

MALMSTRÖM, B. 1991. Blockhavens utbredning i Skandinavien. *Svensk Geografisk Arsbok*, **67**, 110–113.

MICHAILIDIS, K., TSIRAMBIDES, A. & TSAMANTOURIDIS, P. 1993. Kaolin weathering crusts on gabbroic rocks at Griva, Macedonia, Greece. *Applied Clay Science*, **8**, 19–36.

NESJE, A. 1989. The geographical and altitudinal distribution of blockfields in southern Norway and its significance to the Pleistocene ice sheets. *Zeitschrift für Geomorphologie,* Suppl. Bd, **72**, 41–53.

——, DAHL, S. O., ANDA, E. & RYE, N. 1988. Blockfields in southern Norway: significance for the Late Weichselian ice sheet. *Norsk Geologisk Tidsskrift*, **68**, 149–169.

——, ANDA, E., RYE, N., LIEN, R., HOLE, P. A. & BLIKRA, L. H. 1987. The vertical extent of the Late Weichselian ice sheet in the Nordfjord-Møre area western Norway. *Norsk Geologisk Tidsskrift*, **67**, 125–141.

PRIESNITZ, K. 1988. Cryoplanation. *In:* CLARK, M. J. (ed.) *Advances in Periglacial Geomorphology*. Wiley, Chichester, 49–67.

RAINEY, M. M. 1994. *Microfractures in the weathering of igneous rock*. PhD Thesis, Queen's University of Belfast.

ROALDSET, E., PETTERSON, E., LONGVA, D. & MANGERUD, J. 1982. Remnants of preglacial weathering in western Norway. *Norsk Geologisk Tidsskrift*, **3**, 169–178.

SUGDEN, D. E. 1989. Modification of old land surfaces by ice sheets. *Zeitschrift für Geomorphologie,* Suppl. Bd, **72**, 163–172.

—— & WATTS, S. H. 1977. Tors, felsenmeer, and glaciations in northern Cumberland Peninsula, Baffin Island. *Canadian Journal of Earth Science*, **14**, 2817–2823.

SVENONIOUS, F. 1909. Om skarf eller blockhaven på vara hogfjall. *Geologiska Foren.ings Forhandlingar*, **31**.

THOMAS, M. F. 1994. Ages and geomorphic relationships of saprolite mantles. *In*: ROBINSON, D. A. & WILLIAMS, R. B. G. (eds) *Rock Weathering and Landform Evolution,* Wiley, Chichester, 287–301.

WAMBEKE, A. R. VAN, 1962. Criteria for classifying tropical soils by age. *Journal of Soil Science*, **13**, 124–132.

WATTS, S. H. 1983. Weathering processes and products under arid arctic conditions. *Geografiska Annaler*, **65A**, 85–98.

WHALLEY, W. B. 1996. Scanning Electron Microscopy. *In*: MENZIES, J. (ed.) *Glacial Environments – Processes, Sediments and Landscape. Vol. 2. Sediments, Forms and Techniques,* Methuen/ Chapman & Hall, London, 357–375.

——, GORDON, J. E. & THOMPSON, D. L. 1981. Periglacial features on the margins of a receding plateau ice cap, Lyngen, north Norway. *Journal of Glaciology*, **27**, 492–496.

——, MCGREEVY, J. P. & FERGUSON, R. I. 1984. Rock temperature observations and chemical weathering in the Hunza region, Karakoram: preliminary data. *In:* MILLER, K. J. (ed.) *The International Karakoram Project*, *2*. Cambridge University Press, Cambridge, 616–633.

WHITE, S. E. 1976. Rock glaciers and block fields, review and new data. *Quaternary Research*, **6**, 77–97.

WILSON, M. J. 1967. The clay mineralogy of some soils derived from a biotite-rich quartz gabbro in the Strathdon area, Aberdeenshire. *Clay Minerals*, **7**, 91–100.

Environmental interpretation and evolution of the Tertiary erosion surfaces in the Iberian Range (Spain)

M. GUTIÉRREZ-ELORZA[1] & F. J. GRACIA[2]

[1]*Dept. Ciencias de la Tierra, Universidad de Zaragoza, 50009 Zaragoza, Spain*

[2]*Dept. Geologia, Facultad de Ciencias del Mar, Universidad de Cádiz, 11510 Puerto Real, Cádiz, Spain*

Abstract: The geomorphological analysis of regional Neogene erosion surfaces is of great geological interest as it helps in the reconstruction of the environmental evolution of continental areas. Moreover, the structural analysis of their deformation is a good approach to the identification of the main neotectonic trends at a regional scale. The Iberian Range is an excellent example of the application of this kind of study. It shows four stepped Neogene erosion surfaces, well preserved in general and whose ages range from the middle Miocene to the Villafranchian. Their formation is related to planation processes in semi-arid conditions. The stepping of surfaces is related to distensive Neogene tectonic pulses of sufficient magnitude for the planation processes to be interrupted.

Planation or erosion surfaces can be defined as plains formed under conditions of prolonged and complete equilibrium between tectonic movements and surficial exogenic processes. Endogenic forces increase the relief energy, whilst exogenic agents, mainly through the action of gravity, erode rocks, and transport and accumulate debris in sedimentary basins. Today these surfaces appear as fragmentary remnants which represent the record of ancient denuded landforms.

The importance of reconstructing the long-term development and evolution of landforms has been recognized for a long time. This work is done through the study of sedimentary formations and their relationship to erosive events, especially to the genesis of erosion surfaces (Thomas & Summerfield 1987). These kinds of investigation can supply valuable data about ancient climates, deformational episodes, karstification periods, time necessary for their formation and age of the planation surfaces.

The origin of erosion surfaces is linked to the classic theories about the evolution of landforms by Davis (1899), Penck (1924) and King (1953). In order to obtain a correct knowledge and interpretation of such forms, detailed studies are needed of different rock resistances, palaeoclimatological interpretation of correlative sediments, analysis of palaeosols related to planation episodes, the type of erosive surface (buried, exposed, etc.; following Adams 1975) and existing base level during its formation.

These detailed investigations are important for establishing correlations among cycles of planation at a continental scale (Melhorn & Edgar 1975), which may be interpreted within the scope of geodynamics. These studies are also important in economic geology: some mineral deposits of secondary origin are located in deeply eroded areas, while other residual ore minerals such as bauxite, iron, nickel and manganese are linked to deposits correlated with planation processes (Pecsi 1970; Ollier 1991). Furthermore, the present geometry of deformed surfaces can give data about the tectonic genesis of the relief and provide criteria for the definition and dating of tectonic episodes.

The Iberian Range constitutes an excellent example of an alpine chain with a general development of polycyclic Neogene erosion surfaces. The development of surfaces clearly contrasts with the almost total absence of such forms in the other two important, and currently active, alpine ranges of the Iberian Peninsula: the Pyrenees and Betics (north and south of the Iberian Range, respectively). This fact, as well as the present distribution of surfaces in the Iberian Chain, are indicators of the great importance that endogenic activity has had in the genesis and evolution of these palaeoforms.

The Iberian Range

The Iberian Range is located in the northeastern part of the Iberian Peninsula (Fig. 1). It is characterized by the existence of large planated areas, generally over 1000 m, above which appear several rounded mountains, sometimes reaching > 2000 m. The Iberian Range has a NW–SE orientation, with a Palaeozoic basement of quartzites and shales, and a mainly calcareous Mesozoic cover. Both are folded and follow a

From Widdowson, M. (ed.), 1997, *Palaeosurfaces: Recognition, Reconstruction and Palaeoenvironmental Interpretation,* Geological Society Special Publication No. 120, pp. 147–158.

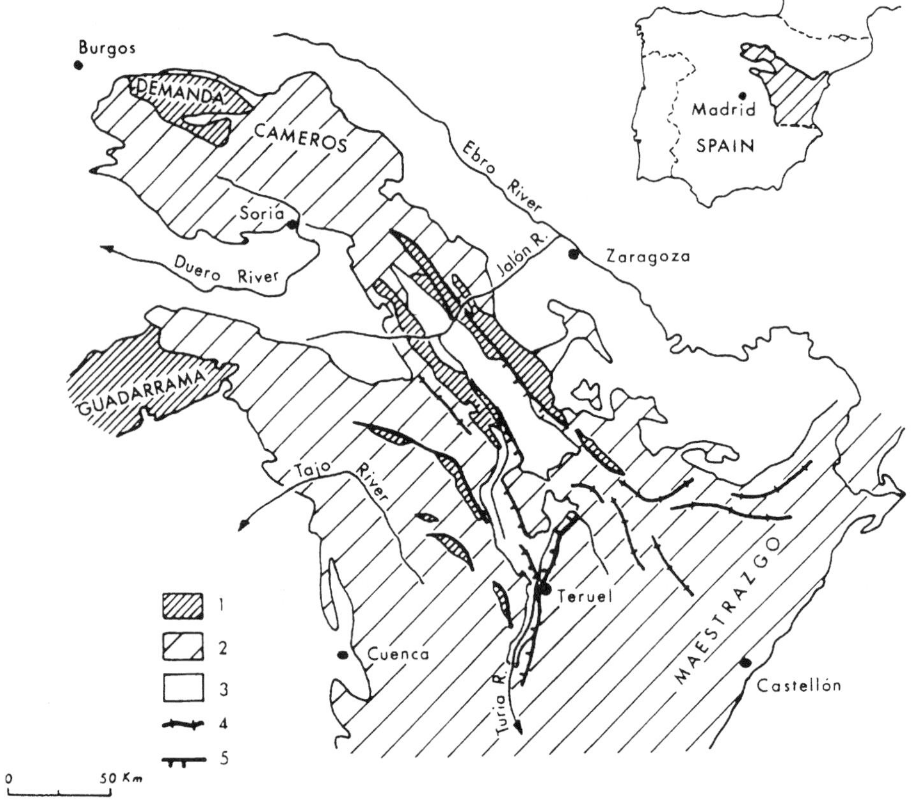

Fig. 1. Location map and geological sketch of the Iberian Range. Key: 1, Palaeozoic basement; 2, Mesozoic cover; 3, post-orogenic tertiary depressions; 4, main fold trends; 5, main faults (normal/general).

NW–SE direction, in a Germanic or Saxonian tectonic style (Stille 1931). The origin of the range can be considered as an evolutionary result of the so-called 'Iberian Aulacogen' (Alvaro *et al.* 1978), created during the Triassic and deformed during the Palaeogene. The main tectonic phases affecting the range took place in the Upper Oligocene and Lower Miocene through a compressive NE–SW stress field, related to the convergence of the Iberian and African plates. Since then, several extensional episodes have occurred.

The tectonic phases that followed the main folding of the chain led to the formation of a horst and graben relief, with uplifted blocks and tectonic basins. The depressions were filled by continental sediments mainly during the Neogene. The sedimentation in the central zones of the depressions was of lacustrine type, bordered by alluvial fans at the edges. The tectonically uplifted reliefs are commonly represented by Palaeozoic materials. Finally, the last important tectonic pulse took place at the end of the Pliocene, creating new grabens, which have been infilled by Plio-Quaternary and Quaternary alluvial fans. This tectonic activity has continued until the Holocene.

The erosion surfaces

This synthesis includes the conclusions made by a number of previous studies carried out by several geomorphologists and geologists during the last 20 years. The methods and techniques used include photogeomorphological mapping, sedimentological analysis of correlative deposits, palaeontological dating of such deposits, microstructural neotectonic analysis, etc. Figure 2 is an example of cartographic synthesis of one of these regional studies.

In the Iberian Range four Neogene surfaces can be distinguished, all of them mostly erosive in nature. They appear as stepped plains starting from the areas of Palaeozoic relief with a gentle slope towards the Neogene continental basins, showing

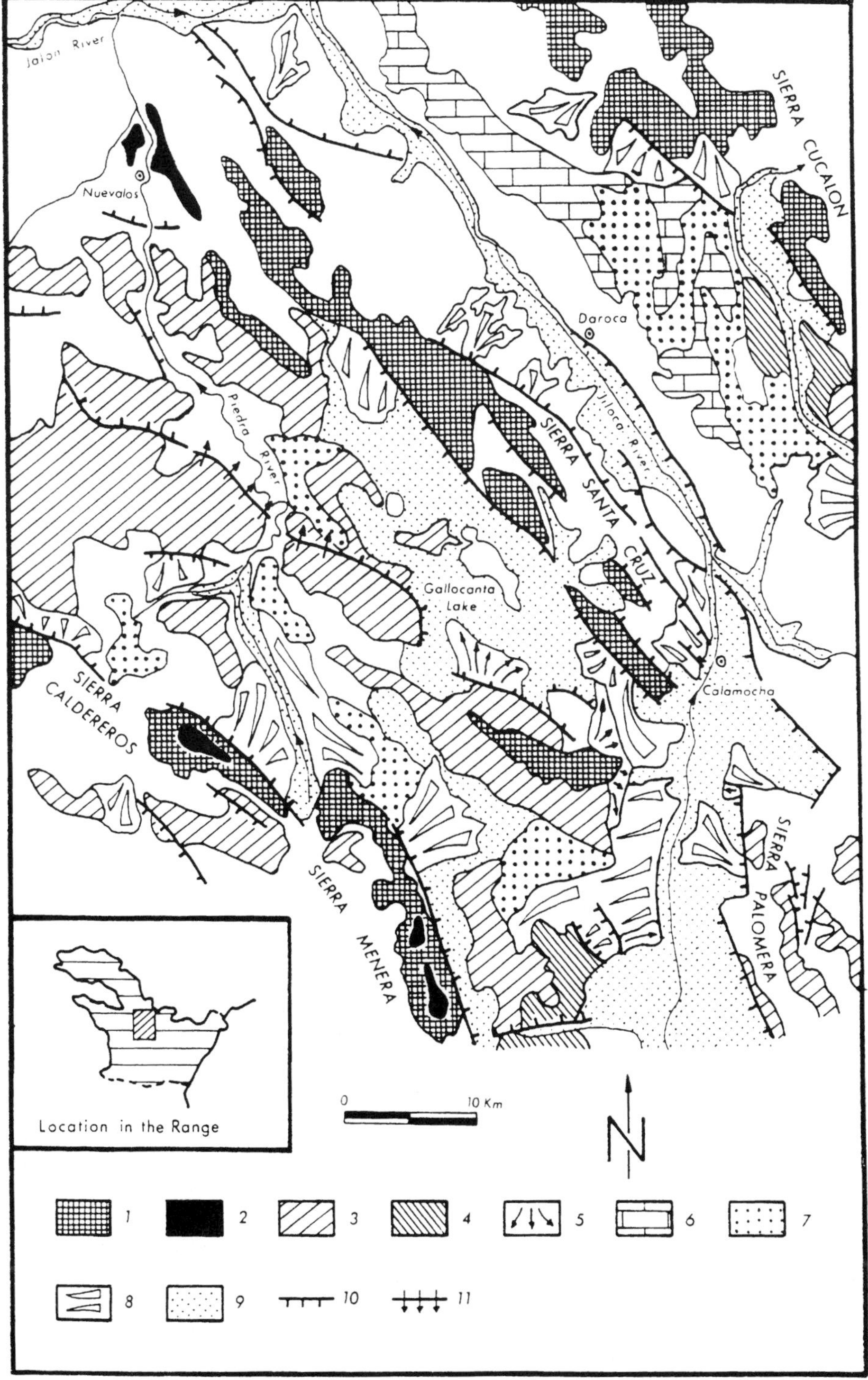

Fig. 2. Geomorphological map of the central zone of the Iberian Range (after Gracia *et al.* 1988). This map is a synthesis of a previous cartography at a 1 : 50 000 scale. 1, Residual reliefs (with respect to surfaces S_2, S_3 and S_4); 2, erosion surface S_1; 3, erosion surface S_2; 4, erosion surface S_3; 5, erosion surface S_4; 6, Upper Neogene limestones ('Paramo' units); 7, Pliocene detritic units; 8, Plio-Quaternary pediments; 9, Quaternary deposits; 10, normal faults; 11, flexures.

a typical 'piedmonttreppen' outline. These continental basins constituted the ancient regional base levels during the development of the surfaces.

The oldest surface (S_1) consists of a 'Gipfelflur', or a planated summit, in the main mountain ranges. Its present preservation is poor, since it is highly eroded. In some localities it appears to be partly buried by Neogene deposits but may be identified as an exposed unconformity which separates a folded basal unit from a upper horizontal formation Neogene (Fig. 3). Such outcrops are quite frequent along the chain (Loranca Depression, Almazan Basin, etc.). The age of the sedimentary units helps to identify both situations (summit and exhumed surfaces) as belonging to the same planation level, which has suffered later tectonic disruptions. Palaeontological data (Daams *et al.* 1987) from the Neogene sedimentary units (some of them folded and truncated by the planation level, and others overlying them unconformably) bracket the age of the surface at the Lower–Middle Miocene transition (Aragonian).

At a lower level the so-called 'Main Erosion Surface of the Iberian Chain' appears (Peña *et al.* 1984), S_2; the name refers to its great development along the Range (Fig. 4). It starts from the areas of Palaeozoic relief, sometimes smoothly connecting with the summit level S_1. In places where this connection is not preserved, geometric extrapolation of the surface using computer allows its reconstruction. It seems that the formation of the Main Erosion Surface began upon the summit surface by a process of progressive reworking and lowering, and then adapted itself to the new base level imposed by basin subsidence. In other cases this connection is disrupted by the later activity of normal faults. This surface mainly affects the Mesozoic calcareous outcrops of the range. Although it is somewhat deformed by later tectonic activity, in several locations its connection with the last sedimentary episode of the Neogene basins ('Paramo Lacustrine Limestones') of Upper Miocene–Pliocene age can be observed, (Turolian–Ruscinian; Adrover *et al.* 1982).

A new surface, here named the 'Border Surface of Neogene Basins' (S_3) (Gracia *et al.* 1988), appears, poorly developed in some places, some tens of metres below the Main Erosion Surface though starting from it through a very smooth waning slope unit. It appears at the edges of some Neogene basins and connects with the very last sedimentary episode ('Upper Paramo' lacustrine

Fig. 3. Correlative unconformity (partially exhumed) to erosion surface S_1, near Teruel. In the foreground are a series of folded, dipping calcareous Jurassic units. These have been eroded and bevelled by the erosion surface S_1 which is here presented as an unconformity beneath the overlying (horizontal) Upper Miocene sediments which form the hills in the background of the photograph.

Fig. 4. Main erosion surface of the Iberian Range at Aliaga, 30 km north of Teruel. The Guadalope river valley can be seen in the foreground, on the far side of the valley a series of folded Lower Cretaceous rocks. Folding is related to Alpine deformation during Oligocene–Lower Miocene times. The most important and widespread Tertiary erosion surface bevels all these tectonic structures across the skyline of the photograph.

limestones), of Lower Pliocene age (Adrover *et al.* 1982). When both lacustrine 'Paramo' units (lower and upper) are clearly distinguishable, an intensively rubefacted detritic formation (Middle Red Unit) appears between them, which can be interpreted as a deposit formed by transported regoliths (terra rossa).

This surface also appears at the edges of the basins that have suffered a renewed tectonic subsidence during the Miocene–Pliocene transition. In other basins unaffected by this tectonic pulse, the S_3 surface did not develop and the S_2 surface was continuously reworked until the Upper Pliocene (Fig. 5). This process explains the connections existing between the S_2 surface and several 'Paramo Lacustrine Limestone' levels of differing ages.

Finally, a Plio-Quaternary Erosion Surface (S_4) appears at the edges of some Neogene basins and Plio-Quaternary grabens. It is represented by pediments at the head of Plio-Quaternary alluvial fans and is very poorly developed. The palaeontological dating of these fans (Crusafont *et al.* 1964; Adrover 1974) indicates a Pliocene–Quaternary (Villafranchian) age. Table 1 represents a general scheme of the erosion surfaces, their correlative sedimentary units and the tectonic episodes during the Neogene in the Iberian Range.

Palaeoenvironmental reconstruction and genesis of the surfaces

The development of the Neogene erosion surfaces of the Iberian Range is linked to a number of geological and environmental factors such as time, climate, lithology and tectonics.

Time is the first factor which must be considered since the generation of a planation level requires a protracted time period for the erosive processes to act. The greater the duration of these processes, the wider extension and the better the development of the planation surface is seen. In this manner, the summit surface (S_1) was developed in the period from the final Oligocene until the middle Miocene, which represents a time of *c.* 10 Ma. Hence, its current poor preservation is due to its antiquity and because later tectonic and erosive processes have severely degraded it. The Main Erosion Surface of the Iberian Range, the widest and best preserved in

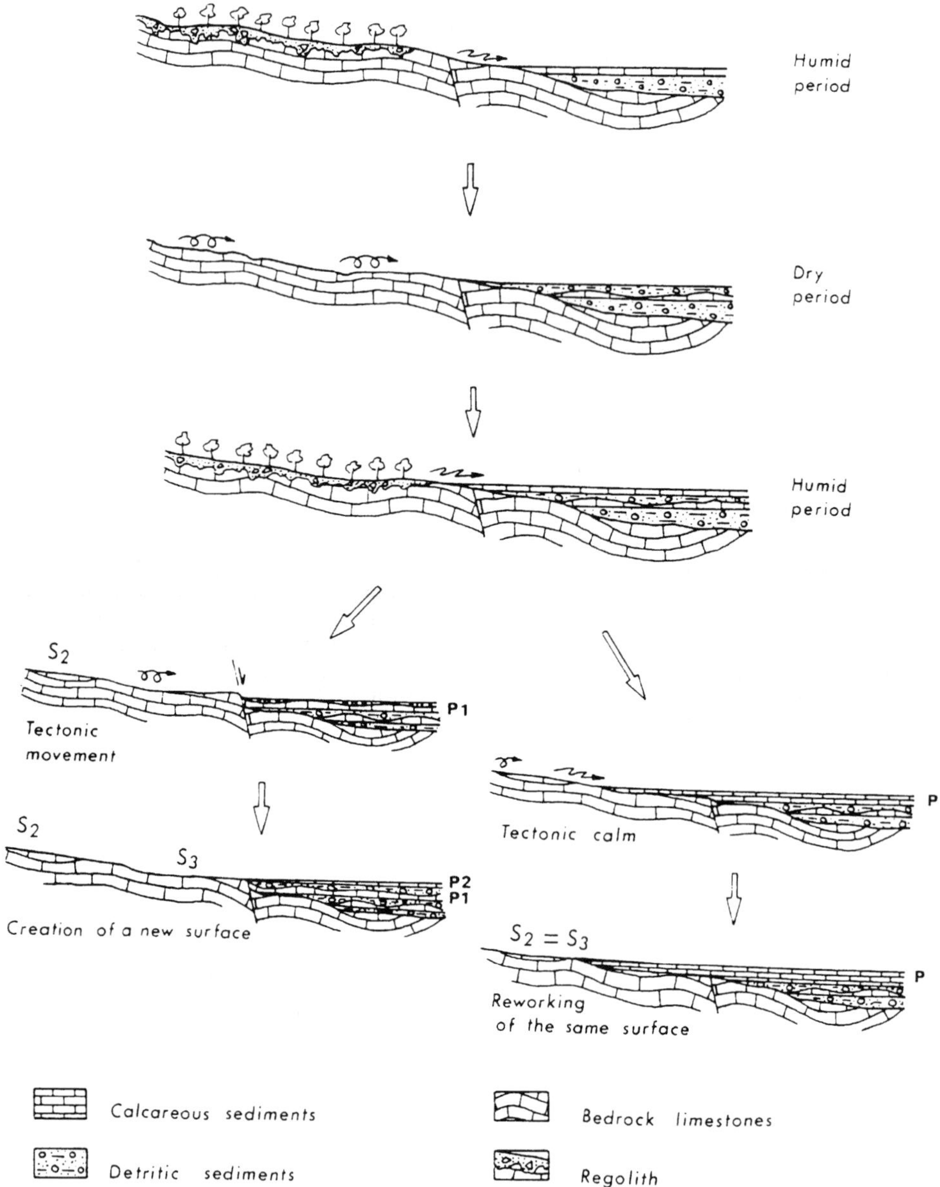

Fig. 5. Theoretical model for the genesis of stepped erosion surfaces at the margin of sedimentary basins, related to changing climatic conditions and margin tectonic activity. This model explains the genesis of surface S_3. P1 and P2, lower and upper Paramo units, respectively.

the chain, took the longest period of time for its development: between 12 and 18 Ma (from the middle Miocene until the end Miocene, or even until the middle Pliocene, depending on the sites). This long period of time allowed for good development of the surface. The Border Surface of Neogene basins, S_3, shows a more reduced development, resulting from a shorter period of formation (*c.* 2 Ma, equivalent to the Lower Pliocene). Finally, the Plio-Quaternary surface shows the most limited extent, reflecting a very short period of development (only *c.* 1 Ma), corresponding to the Villafranchian.

The second factor to consider is the climate. The analysis of Tertiary deposits of the sedimentary basins indicate that the climate was mainly semi-

Table 1. *Evolution of the Neogene erosion surfaces in the Iberian Range*

Epoch	Erosion surfaces	Tectonic episodes	Sedimentary units
Quaternary		Late distension	Drift deposits
	S_4	Distension	Plio-Quaternary fans
	S_3 S_2	Distension	Upper Red Unit
Pliocene			Upper Paramo Unit
	S_2	Distension	Middle Red Unit
			Lower Paramo Unit
Miocene	S_1	Compression–Distension	Lower Red Unit
Palaeogene		Main compressive episodes	Basal Unit and pre-Neogene materials

Relationships are shown with tectonic phases and correlative sedimentary units.

arid. These climatic conditions are reflected by the existence of thick evaporitic formations, typically gypsum and anhydrite, and by the presence of many calcrete levels. Nevertheless, several minor more humid fluctuations have existed (Daams *et al.* 1988). However, the impact of the periods of more arid conditions are reduced towards the mountains which surround the Tertiary basins, since these retained more humid climates throughout.

The third factor to be taken into account in the genesis of the erosion surfaces is the lithology. Many silicic rocks are strongly resistent to erosion; hence special climatic and environmental conditions are needed in order to cause such rocks to be weathered and transported.

The widest and best developed erosion surfaces appear in areas of calcareous rock, particularly in the case of the Main Erosion Surface. On the areas of high relief surrounding the Neogene basins a general karstification of all the calcareous Mesozoic outcrops of the chain took place (Gutiérrez & Peña 1989). These solution processes were favoured by the planated topography together with the generation of residual red clays (terras rossas). This cryptokarstic dissolution of limestones caused a carbonate enrichment of the surficial waters which, when they flowed into the basins, led to the precipitation of lacustrine carbonatic muds rich in organic fragments, travertines, etc. The transition to a drier period, or alternatively a weak tectonic pulse, would have led to the removal of the red clays and their subsequent accumulation in the basins. In this way, the correlative sedimentary deposits of the basins (Upper Miocene) consist of an alternating sequence of red detritic sediments, mainly clays, and lacustrine limestones which are sometimes travertinic. Such interpretation is consistent with the work of previous authors (Moissenet 1985; Lozano 1988) who have pointed out that these sediments appear to reflect alternating climatic conditions.

Tectonic activity is the third and final factor which must be considered in the development of the Iberian erosion surfaces. On the one hand, the tectonic pulses during the Neogene maintained a topographic gradient between the relief and the basins. This situation favoured the transport of debris towards base level with a continuous lowering of the erosion surfaces. Nevertheless, at certain times, the vertical tectonic movements exceeded a certain threshold, creating a sudden break in the hypsometric profile and resulting in a topographic disconnection between areas of relief and basins. In this way the planation processes were completely interrupted, leading to the generation of a new surface starting at a lower position, and adapted to the new tectonically established base level.

Thus, whilst the equilibrium between tectonics and denudation was being maintained, each surface was forming without significant interruption. When this equilibrium was disrupted, due to more intense endogenic activity, the development of the surface was effectively halted, beginning the generation of a new one at a lower position. In this sense the case of surfaces S_2 and S_3 is very clear (Fig. 5); this duplication of surfaces was only produced in those basins that suffered a tectonic lowering at the end of the Miocene (Gracia 1989, 1990). In these cases the Pliocene sedimentary sequence records this evolution: the last lacustrine carbonate episode (Paramo Limestones) is duplicated in two levels: Lower Paramo and Upper Paramo, separated by an intermediate detritic level with red clays (Middle Red Unit).

This model explains the 'staircasing' of Neogene erosive surfaces in the Iberian Range and their relationships with the main Miocene–Pliocene tectonic episodes. To a large extent this interpretation is equivalent to Penck's model of the generation of a piedmont staircase related to tectonic events ('piedmonttreppen'; Penck 1924). Parallel slope retreat occurred during the development of surfaces S_3 and S_4, as can be deduced from their geometries. The contacts between surface S_2 and the areas of residual relief, whether these contain the summit level S_1 or not, are unclear and distorted because they almost always coincide with active Neogene faults. This fact suggests an accelerated tectonic uplift between the final formation of surface S_1 and the onset of the formation of surface S_2, leading to mechanical contacts and waxing valleyside slopes ('Steilrelief' following Penck's model).

Deformation of the Main Erosion Surface

As the S_2 surface is the most extensive and best developed, it is an excellent marker horizon for determining the effects of the tectonic activity which has taken place after its final formation during the middle Pliocene. In this study a structural–contour map or tectono–morpho–isohipses map (Bashenina 1978) of the S_2 surface has been constructed for the whole range, using 200 m contour intervals (Fig. 6). This kind of map needs a pre-existing cartographic map of the erosion surface and clear identification of the tectonic incidents that have affected it – the contour interval must be in accordance with the magnitude of the deformation affecting the surface. We have considered as an initial assumption, that the planation was originally horizontal, although obviously its slope would have probably been higher in the proximity of residual relief. These maps permit a visualization of the relief generated by tectonics *after* the formation of the surface, as well as geometrical analysis of the deformation.

From the map we can infer the existence of domes and the presence of tilted and tectonically stepped surfaces towards the neogene grabens and the Mediterranean coast (Valencia Trough). It can be deduced that the Neogene grabens suffered renewed subsidence during the Upper Pliocene, caused by reactivation of the bordering faults. This represented an important morphogenetic control upon the Quaternary fluvial network, as well as on the creation of Quaternary lacustrine tectonic basins (Gallocanta Lake; Gracia 1990).

The map also allows us to develop a morphostructural subdivision of the range. In this sense, the central-eastern sector is characterized by a relief consisting of domes and grabens. The tectonic depressions are more common in the central part and the coastal zone. The tectonic deformations are mainly brittle, made up of normal faults and flexures, indicating important vertical movements. The geometry of these faults is related to a radial distensive tectonic regime, which reactivate previous Alpine structures (Simón 1984).

The southwestern sector shows much less deformation. The relief is formed by wide plains gently declining towards the inner Tertiary basins of the Iberian Peninsula (Duero and Tajo Basins). In this way, deformation diminishes while approaching the most stable zones of the Iberian Meseta.

The northwestern sector is characterized by great residual relief (Demanda–Cameros Ranges) where the erosion surfaces are hardly developed. Only erosion surfaces upon limestones can be recognized. This has been one of the most active zones of the chain during the Tertiary, with thrusting of the Palaeozoic basement over Tertiary series occurring. This marked activity prevented development of planation surfaces during the Neogene, as also happened in the Pyrenees and in the Betic Ranges. Today, this zone is the most seismically active in the whole range; several historic and recent earthquakes have shown intensities of up to VIII on the MSK scale (Alfaro *et al.* 1987).

Geomorphological and neotectonic evolution

After the main compressive episode, which took place at the Upper Oligocene–Lower Miocene boundary, a long period of relative tectonic quiescence allowed the development of the erosion surface S_1 (Fig. 7a). Its final age can be located at the Lower–Middle Miocene boundary. This surface was deformed by a tectonic episode in the middle Miocene (Table 1), during which the landscape was broken into uplifted relief areas and intramountain basins (Fig. 7b). The erosive products of such an episode were deposited into the basins in alluvial fan and playa-lake environments under a more or less semi-arid climate.

In the surrounding mountain zones, intense denudation took place, during which time the Main Erosion Surface of the Iberian Range (S_2) was developed through pediplanation processes, at least during its final stages. In many areas of residual relief a summit planation level can be recognized, representing the S_1 erosion surface (Fig. 7c). At different points the surface S_2 connects with the last episode of the limestone sedimentation of the basins, which culminates as an end-filling sedimentary surface, developed during the Late

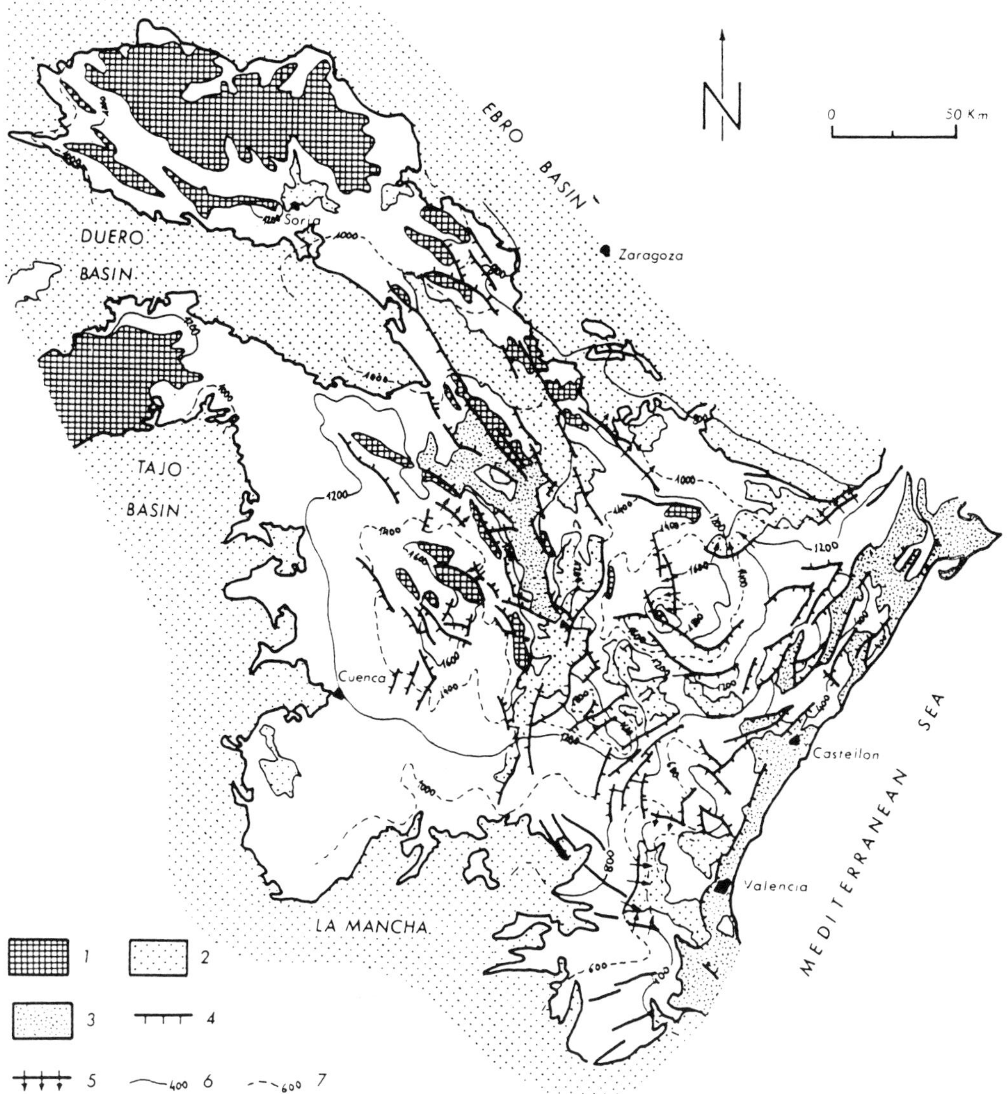

Fig. 6. Structural–contour map of the Main Erosion Surface of the Iberian Range. 1, Residual reliefs; 2, post-orogenic tertiary deposits; 3, Plio-Quaternary and Quaternary deposits; 4, normal faults; 5, flexures; 6, isohipses (equidistance of curves, 200 m); 7, auxiliary curves.

Miocene (Turolian). Planation subsequently favoured the infiltration of overland flow and, consequently, the karstic corrosion of the Mesozoic limestones, although it is possible that this process may have already started slightly before the development of surface S_2.

The long period of tectonic stability was broken by a weak extensional episode at the Miocene–Pliocene boundary, with a tectonic reactivation of relief. The resultant erosive products were deposited in the basins, unconformably fossilizing the Lower Paramo Unit. These deposits (Middle Red Unit), with detritic channels and calcareous crusts (calcretes), were the result of the removal of residual clays formed by the karstic corrosion that occurred during the planation period. Finally, new lacustrine environments developed in the basins, leading to an accumulation of calcareous materials

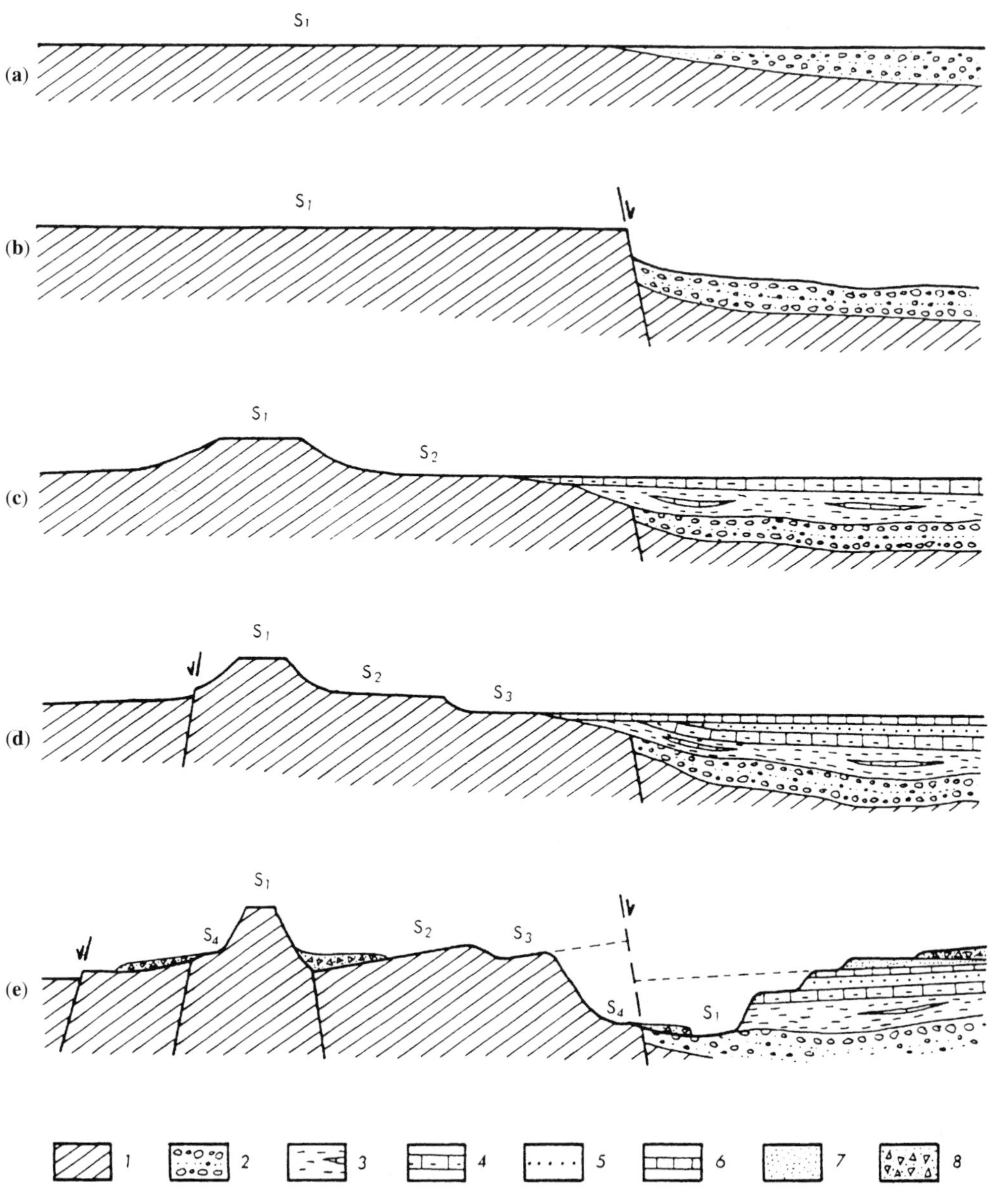

Fig. 7. Evolution model for the erosion surfaces of the Iberian Range. (**a**) Lower Miocene; (**b**) Middle Miocene; (**c**) Upper Miocene; (**d**) Middle Pliocene; (**e**) Quaternary. 1, Pre-Neogene materials; 2, Basal Unit; 3, Lower Red Unit; 4, Lower Paramo Unit; 5, Middle Red Unit; 6, Upper Paramo Unit; 7, Upper Red Unit; 8, Plio-Quaternary fans.

(Upper Paramo Unit) during the Lower Pliocene (Ruscinian).

The results of this tectonic episode differed from place to place, generating distinct relief gradients. In areas where the resulting topographic rupture was important, the creation of a new erosion surface (S_3) began, which can be identified mainly at the borders of the Neogene basins (Fig. 7d). If the topographic rupture was of small magnitude, then surface S_2 suffered only a slight reworking and development of surface S_3 was not initiated. Thus, in such areas, the reworked surface S_2 connects with the final sedimentary Upper Paramo Unit of the Neogene basins. A later but very weak tectonic pulse then affected the region, leaving a detritic layer, the Upper Red Unit, which unconformably overlies the Upper Paramo Unit. As in the former case, it is made up of transported 'terras rossas' with interbedded calcareous crusts of Upper Pliocene age.

Subsequently, an important tectonic event occurred in the range. It notably transformed the previously planated landscape, creating the main morphological features of the range as can be seen today. The erosional surfaces were deformed and displaced from their original positions, showing a doming and rifting relief, declining towards the newly created grabens. The higher topographic gradients led to the generation of wide alluvial sheets, forming pediments (surface S_4) both on the uplifted zones and inside the grabens (Fig. 7e), during the middle Villafranchian. The tectonic pulses continued, although with less intensity, until the present day (Fig. 8). Exogenic processes modified the landforms during the Quaternary and the fluvial network dissected the Neogene erosion surfaces. As a result of these processes we can now find erosion surface S_1 in an exhumed situation (Fig. 7e).

Conclusions

The study of Neogene erosion surfaces is of great geomorphological and geological interest since it permits the reconstruction of the morphological and environmental evolution of continental areas which otherwise lack sufficient sedimentary records. Furthermore, the structural analysis of the tectonic deformation of erosion surfaces gives data about the tectonic evolution following their formation, the relief generated by the endogenic activity and the structural geometry of the neotectonic phases.

The Iberian Range represents an excellent example of the application of this kind of study. Four Neogene erosion surfaces can be identified whose genesis is related to moderate tectonic activity and to several denudation stages that have operated over long periods and upon several different lithologies (mainly limestones). The successive stepping of surfaces is related to specific tectonic pulses of greater magnitude which interrupted the planation processes leading to the development of new surfaces at lower positions. This model is very close to that proposed by Walter Penck.

Fig. 8. Normal fault affecting Holocene slope deposits and a rendzina soil. Road from Caminreal to Rubielos de la Cérida (Teruel).

The development of wide, stepped erosion surfaces in the Iberian Range needs to be properly located within the geotectonic framework of the Western Mediterranean, where Neogene evolution is characterized by several compressive and distensive tectonic phases that are related to the convergence of the Iberian and African plates. The later tectonic deformation of the erosion surfaces is represented by widespread doming in the central-eastern sector of the range, as well as the creation of grabens and tectonic depressions, especially towards the Valencia Trough.

References

ADAMS, J. 1975. Introduction. *In*: ADAMS, J. (ed.) *Planation Surfaces*. Dowden, Huchinson & Ross, Stroudsburg, 1–13.

ADROVER, R. 1974. Un relleno kárstico plio-pleistocénico en el Cerro de Los Espejos en Sarrión (prov. de Teruel, España). Nota preliminar. *Acta Geol. Hisp.*, **4**, 142–143.

—— 1986. *Nuevas faunas de roedores en el Mio-Plioceno continental de la región de Teruel (España). Interés biostratigráfico y paleoecológico.* Doct. Thesis, Publicaciones del Instituto de Estudios Turolenses, Teruel.

——, FEIST, M., GINSBURG, L., GUERIN, C., HUGUENEY, M. & MOISSENET, E. 1982. L'âge et la mise en relief de la formation détritique culminante de la Sierra Pelarda (province de Teruel, Espagne). *Comptes Rendus de l'Académie des Sciences du Paris*, **295**, 231–236.

ALFARO, J. A., CASAS, A. M. & SIMÓN, J. L. 1987. Ensayo de zonación sismotectónica en la Cordillera Ibérica, Depresión del Ebro y su borde sur-pirenaico. *Estudios Geologicos*, **43**, 445–457.

ALVARO, M., CAPOTE, R. & VEGAS, R. 1978. Un modelo de evolución geotectónica para la Cadena Celtibérica. *Acta Geológica Hispánica*, **14**, 172–177.

BASHENINA, N. V. 1978. Map of morpho and tectono-morpho-isohypses. *In*: DEMEK, J. & EMBLETON, C. (eds) *Guide to Medium-scale Geomorphological Mapping*. E. Schwerzerbart'sche Verlagsbuchhandlung, Stuttgart, 68–80.

CRUSAFONT, M., HARTENBERGE, J. L. & HEINTZ, E. 1964. Un nouveau gissement de mammifères d'âge Villafranchien de la Puebla de Valverde (prov. de Teruel), Espagne. *C.R. Acad. Sci. Paris*, **258**, 2869–2871.

DAAMS, R., FREUDENTHAL, M. & ALVAREZ, M. 1987. Ramblian – a new stage for continental deposits of early Miocene age. *Geologie en Mijnbouw*, **65**, 297–308.

——, FREUDENTHAL, M. & VAN DER MEULEN, A. J. 1988. Ecostratigraphy of micromammal faunas from the Neogene of Spain. *In*: FREUDENTHAL, M. (ed.) *Biostratigraphy and Paleoecology of the Neogene Micromammalian Faunas from the Calatayud–Teruel Basin (Spain)*. Scripta Geologica, Special Issue 1, Leiden, 287–302.

DAVIS, W. M. 1899. The geographical cycle. *Geographical Journal*, **XIV**, 481–504.

GRACIA, F. J. 1989. A model of the genesis and evolution of erosion surfaces in a mediterranean context. Examples from the Iberian Chain (Spain). *In*: SEUFFERT, O. (ed.) *Abstracts of the 2nd International Conference on Geomorphology*, Frankfurt, Geoöko plus, **1**, 110–111.

—— 1990. *Geomorfología de la región de Gallocanta (Cordillera Ibérica central)*. Tesis Doctoral, Universidad de Zaragoza.

——, GUTIÉRREZ, M. & LERÁNOZ, B. 1988. Las superficies de erosión neógenas en el sector central de la Cordillera Ibérica. *Revista de la Sociedad Geológica España*, **1**, 135–142.

GUTIÉRREZ, M. & PEÑA, J. L. 1989. El karst de la Cordillera Ibérica. *In*: DURÁN, J. J. & LÓPEZ, J. (eds) *El Karst en España*, Monografía Sociedad Española de Geomorfología, **4**, 151–162.

KING, L. C. 1953. Canons of landscape evolution. *Bulletin of the Geological Society of America*, **64** (7), 721, 742–751.

LOZANO, M. V. 1988. *Estudio geomorfológico de las sierras de Gúdar (prov. de Teruel)*. Doct. Thesis, Universidad de Zeragoza.

MELHORN, W. N. & EDGAR, D. E. 1975. The case for episodic, continental-scale erosion surfaces: a tentative geodynamic model. *In*: MELHORN, W. N. & FLEMAL, R. C. (eds) *Theories of Landform Development*. Allen & Unwin, London, 243–276.

MOISSENET, E. 1985. Les aplanissements partiels sur les calcaires de la Chaine Ibérique orientale. In: *Cônes Rocheux*. Mémories et Documents de Geographie. Edition du Centre National de la Recherche Scientifique, Paris, 127–139.

OLLIER, C. 1991. *Ancient Landforms*. Belhaven, London.

PECSI, M. 1970. Introduction. *In*: PECSI, M. (ed.). *Problems of Relief Planation*. Académiai Kiadó, Budapest, 11–12.

PENCK, W. 1924. *Die Morphologische Analyse: Ein Kapital der Physikalischen Geologie*. Geographische Abhandlungen, 2, Reihe, Heft 2, Stuttgart.

PEÑA, J. L., GUTIÉRREZ, M., IBÁÑEZ, M. J. *ET AL.* 1984. *Geomorfología de la provincia de Teruel*. Instituto de Estudios Turolenses, Teruel.

SIMÓN, J. L. 1984. *Compresión y distensión alpinas en la Cadena Ibérica oriental*. Instituto de Estudios Turolenses, Teruel.

STILLE, H. 1931. Die Keltiberische Scheitalung. Nachr. v. d. Ges. d. Wiss. z. Göttingen. Mat.-Phy., Kl. Fachg., IV (Geol. und Min.), Nr. 10. Berlin (Translated by San Miguel, M., 1948. La Divisoria Ibérica. *Publicaciones Extranjeras sobre Geología de España*, IV, 297–303. Madrid

THOMAS, M. F. & SUMMERFIELD, M. A. 1987. Long-term landform development: key themes and research problems. *In*: GARDINER, V. (ed.) *International Geomorphology*, 1987, Part II. Wiley, Chichester, 935–956.

Environmental changes during the Tertiary: the example of palaeoweathering residues in central Spain

HARALD BORGER

Geographisches Institut, Universität Tübingen, Hölderlinstr. 12, D-72074 Tübingen, Germany

Abstract: In central Spain three main weathering phases are found which can be put in a chronological sequence with tectonic and sedimentation phases. Weathering intensity, as recorded by micromorphological analysis, has plainly decreased during the course of the Tertiary reflecting the prevailing environmental conditions.

During the Palaeogene intensive chemical activities produced, on the mineral grains of the deeply decomposed rocks, features typical of humid tropical climate.

In the Miocene the Palaeogene regolith of the southern Tajo Basin was truncated and partly incorporated into sediments which comprise reworked weathering residua of this basin and of the adjoining Montes de Toledo. After sedimentation another weathering phase started. During this phase the feldspars were heavily corroded, but in contrast to the Palaeogene residues, the quartz shows hardly any features caused by etching.

In the Pliocene the rañas were deposited during a second main transportation phase. Particularly in the SE of the Montes de Toledo, these sediments were subjected to intense weathering, but only where conditions were favourable. Due to sufficient soil humidity in intramontane basins, high temperatures, and presence of iron, quartz corrosion was locally possible once more during Pliocene times. The rañas of the Tajo Basin, however, were substantially less attacked.

The Montes de Toledo are part of the Variscan Massif in central Spain and stretch from the La Mancha in the east, including the Sierra de Guadalupe, to the Estremadura in the west. In the north they are bounded by the Tajo Basin and in the south by the Guadiana Basin. For the present investigation an area of *c.* 6000 km^2 was mapped and sampled, stretching from the Río Tajo in the north up to the southern spurs of the Montes de Toledo, as shown in Fig. 1. The mountains of the research area represent an updoming within the Iberian Massif and are mainly composed of Palaeozoic quartzites, sandstones and schists. In the northern foreland the geology comprises granites and granodiorites.

Gehrenkemper (1978) documented evidence for several phases of palaeomorphodynamic processes in the area. An etchplain formation (base-levelling), which probably lasted until the end of the Oligocene, was followed by an uplift of the Variscan folded rock formations by the northward thrusting of the Betic Cordillera during the Miocene. Both the granites outcropping in the southern Tajo Basin and the intramontane basins of the Montes de Toledo were buried by reworked Palaeogene weathering residues of the mountains during this upheaval phase. The resulting loamy–sandy Miocene sediments reached an overall thickness of 60–100 m. Further uplift occurring during the Rhodanic phase of folding and semi-arid to arid climatic conditions in the mid-Pliocene led to the formation of a traditional glacis developed upon remnants of the older etchplain. After their formation, the flat-to-undulose surface of this glacis became the basal surface upon which the raña developed (cf. Stäblein 1973; Wenzens 1977). Under more humid conditions the older coarse detritic weathering residues of the quartzites of the uplifted mountains were destabilized, transported and deposited as the raña sediments. Gehrenkemper (1978, p. 67) shows that this took place in three phases: 'Raña formation began with a mudflow-like movement of the masses ... limited to the near parts of the mountain fringe ... The overlying deposits were affected by torrential dynamics, mainly in the middle part of the glacis. In the end-phase of raña formation, run-off processes were more regular, leading to the formation of the raña plain.'

The detritic raña formations appear associated only with quartzitic mountain ranges. 'The term 'raña' has both a stratigraphic and a morphological meaning ...' (Espejo 1987, p. 399). Stratigraphically rañas are Upper Pliocene sediment formations and accumulation glacis in the geomorphological sense of the term.

The glacis and the Miocene sands were deeply dissected by the southern tributary rivers of the Río

From Widdowson, M. (ed.), 1997, *Palaeosurfaces: Recognition, Reconstruction and Palaeoenvironmental Interpretation,* Geological Society Special Publication No. 120, pp. 159–173.

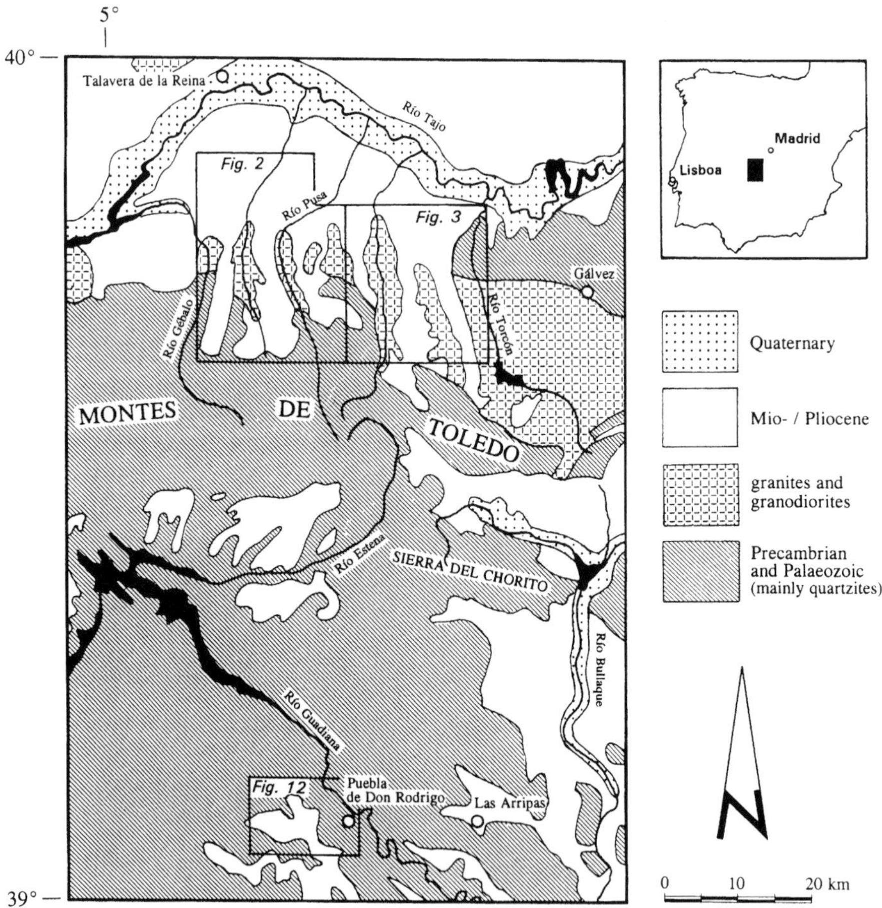

Fig. 1. Location and geology of the study area (after: 'Mapa Geologico de la Peninsula Ibérica, Baleares y Canarias', 1 : 1 000 000, Instituto Geologico y Minero de España, 1981).

Tajo and today, between Río Gébalo and Río Torcón (Figs 2 and 3) the relief is characterized by distinctive raña-covered crests on interfluves which slope gently northwards at an inclination of *c*. ± 1°. The Quaternary erosion has meant that the rivers have reached the decomposed granite and exhumed the old Lower Tertiary weathering base at the valley floors. The rivers have now partly cut some metres into the unweathered granite, for example, the Río Pusa west and northwest of Los Navalmorales. As a result, a near complete sampling of the entire sequence from the unweathered granite, its regolith and the Miocene cover, up to the raña surface, was possible in the field. A generalized profile is shown in Fig. 4. However, between Río Torcón and Galvez the weathered granite forms the whole of the recent land surface without any overlying Miocene sands and raña sediments. The alteration intensity of the lower parts of regolith profiles is characterized by decomposed granite with spheroidal weathering and deep kaolinization of the granodiorite, as exposed in the quarry 'La Cantera' in the valley of Río Torcón (Fig. 3). The upper parts of the Palaeogene weathering residues are not commonly preserved. The profiles were truncated by the Miocene erosion and the eroded material became incorporated into the Miocene sediments.

Methods

Mechanical and chemical influences leave a number of different microfeatures on the surfaces of quartz grains, and the environment which is responsible for the characteristic features can thereby be determined quite precisely by scanning electron microscopy (SEM) (Margolis & Kennet

1971; Krinsley & Doornkamp 1973; Margolis & Krinsley 1974). Mechanically derived microfeatures, such as those caused by different transportation media, can be distinguished from microfeatures caused by chemical processes which develop during tropical weathering phases. In addition, single grains of quartz can often show features caused by a number of different processes. In such cases, the chronology of weathering phases and phases of transport, which has influenced a grain at different times in the course of its history, can be deduced, as demonstrated with the example of the 'Monheimer Höhensande' in south Germany (Borger 1993*b*). With regard to investigation of the

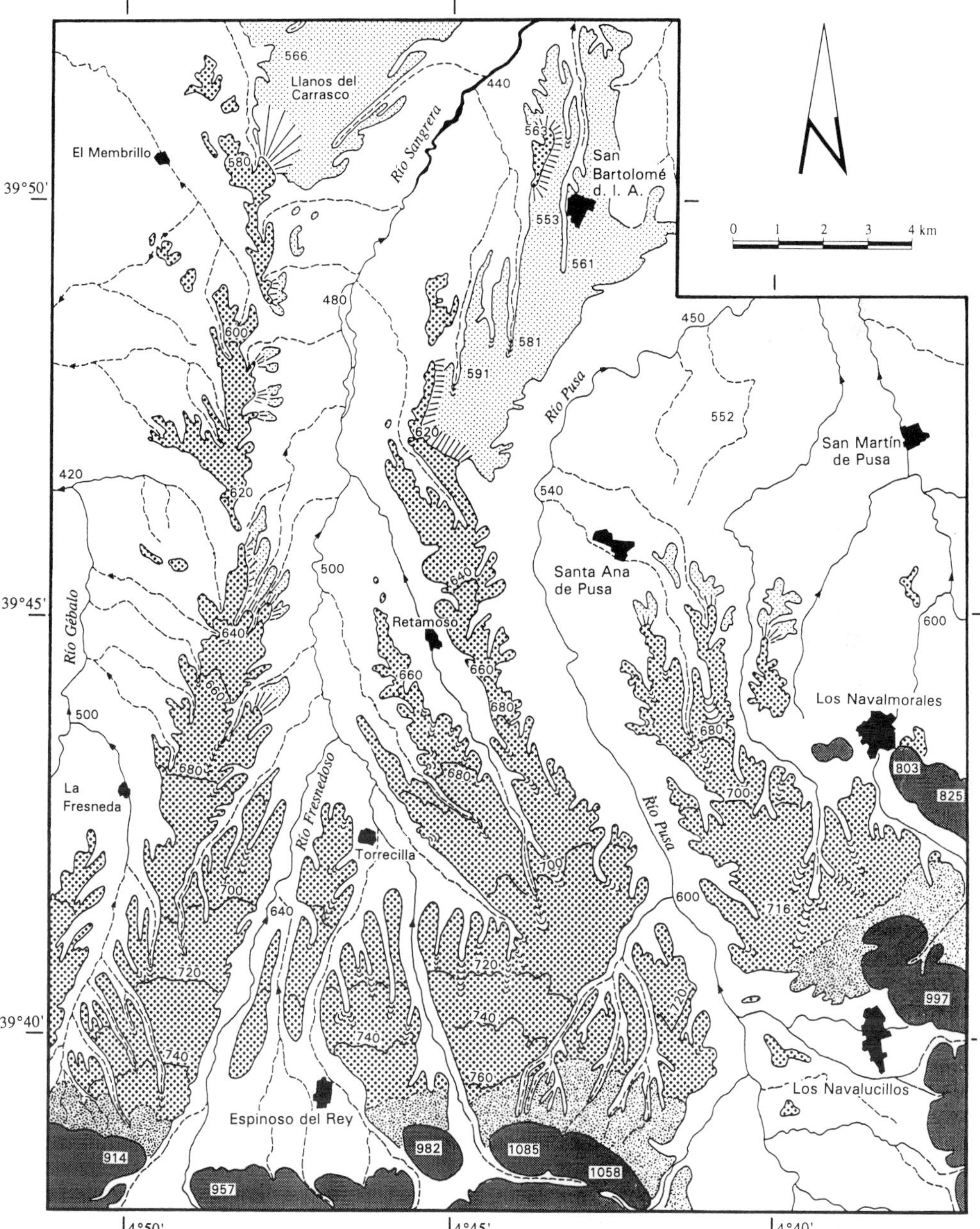

Fig. 2. Geomorphology of the rañas in the northern foreland of the Montes des Toledo west of Los Navalmorales.

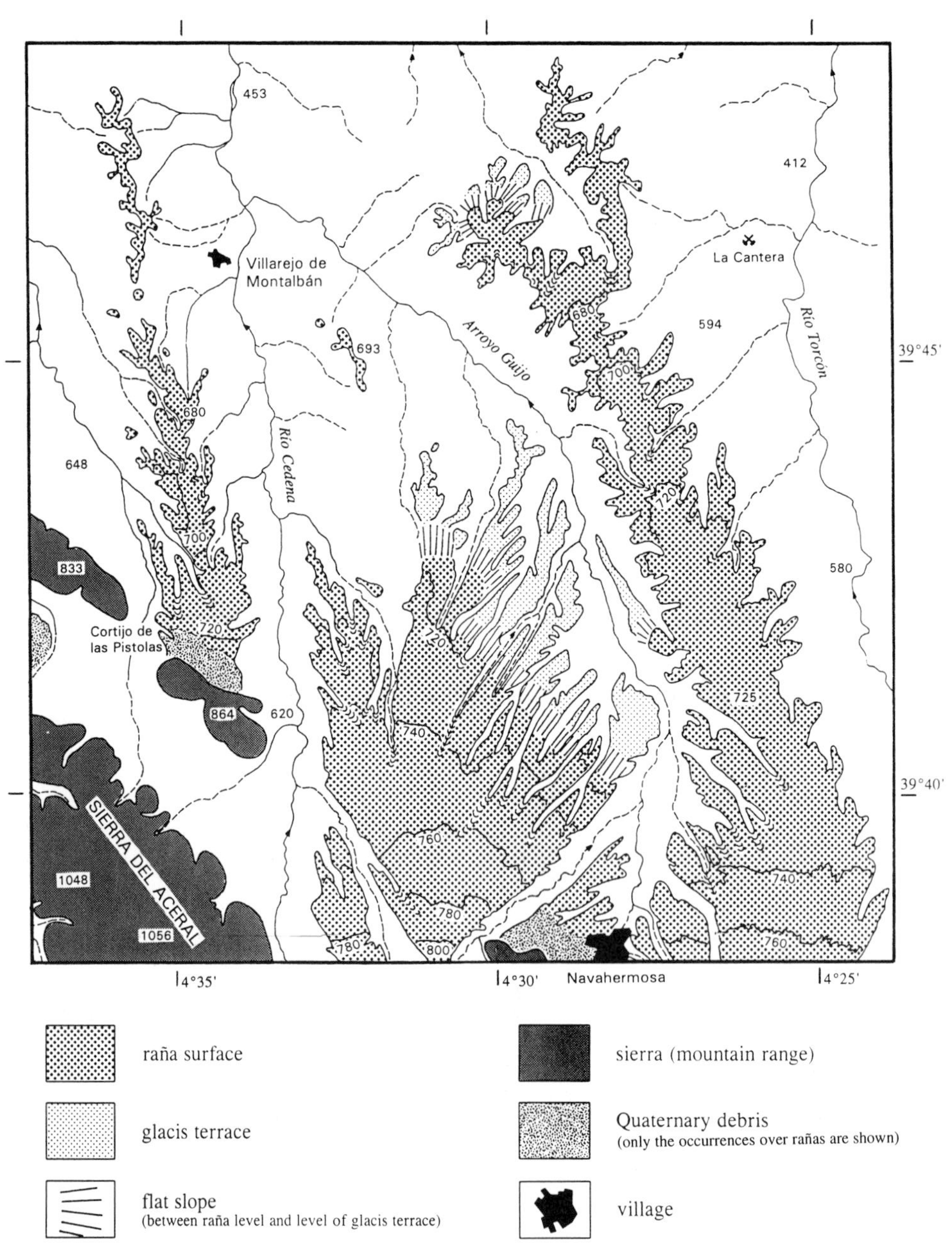

Fig. 3. Geomorphology of the rañas in the northern foreland of the Montes des Toledo north of Navahermosa.

chemical weathering processes, the interpretation of thin sections (by SEM and optical microscope) is also particularly useful. In humid hot climates the weathering of quartz produces the same result regardless the source material: typically visible in thin sections is an intense splitting of the quartz grains, followed by corrosion and dissolution (Eswaran *et al.* 1975; Schnütgen & Späth 1983; Borger 1992). Detailed investigation employing SEM and optical microscopy can provide a method which enables a scale of quartz weathering intensity to be set up (Borger 1993*a*, 371f).

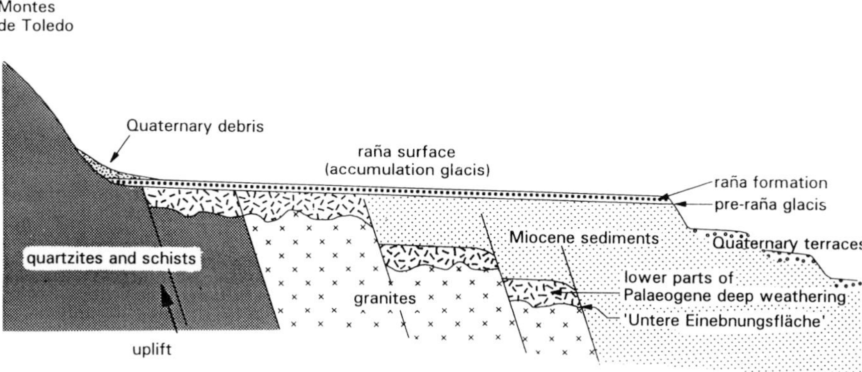

Fig. 4. Generalized profile of the northern foreland of the Montes de Toledo, not to scale (modified after Gehrenkemper 1978).

Using this scale, the environmental conditions which caused a decomposition of samples with an unknown history can be inferred. In this scale, unweathered quartz grains are placed as stage 1. The initial splitting of quartz grains represent the second stage with only very minor quartz corrosion. In stage 3 the etching of the quartz surfaces is clearly increased and a more irregular shape of formerly smooth-faced fractures begins to develop. The distinctive feature of the quartz grains at stage 4 is the penetration of corrosion embayments, which then become very deeply cavernous. In stage 5, the quartz grains are fragmented by numerous deeply etched micro-caverns which are in contact with each other (cf. 'runiquartz', Eswaran *et al.* 1975). The quartz fragments are forced apart (mostly by penetrating secondary iron and/or kaolinitic matrix), but their relative positions remain unchanged. Stage 6 represents the most extreme weathering with near to complete quartz pseudomorphs. Examples of each weathering stage are given in Borger (1993*a*, p. 368).

A total of 81 samples from central Spain were investigated to ascertain which geomorphological conclusions can be made by using the above methodology. Further laboratory methods were employed to support the microscopic results, particularly analyses by X-ray diffraction and electron microprobe (energy dispersive system). In addition to the quartz weathering study, the effect of weathering upon feldspar and mica minerals has also received attention.

Weathering of the crystalline material (southern Tajo Basin)

The micromorphological processes of granite weathering are verified by thin section in particular (cf. Kubiniok 1988; Schnütgen 1992; Borger 1992). Unweathered granite shows an undisturbed fabric composed of fresh quartz, feldspar and mica. The first signs of weathering are characterized by a splitting of quartz grains and feldspar minerals. Often the fractures follow lines of inherent weakness within the mineral grains (tectonic structures and/or structural characteristics) which were already present in the unweathered rock (cf. Schnütgen & Späth 1983). The fractures are typically characterized by two smooth-faced surfaces, running parallel to each other through the quartz grains (weathering stage 2). With few exceptions, the splitting does not appear to depend upon the grain boundaries and hence granular disintegration of the granite does not proceed by disaggregation of constituent mineral grains (Fig. 5). In this distinctive way the fractures cross nearly all mineral grains irrespective of composition. However, in the first stages, the mica of the investigated Spanish granite was more resistant to this splitting as a result of its physical properties. Kubiniok (1988) made the same observations in granites of the 'Odenwald' (Germany).

During the initial stages of the weathering the sodium feldspars altered to kaolinite. The fractures within split plagioclase widened irregularly by solution: an irregular shape developed and the

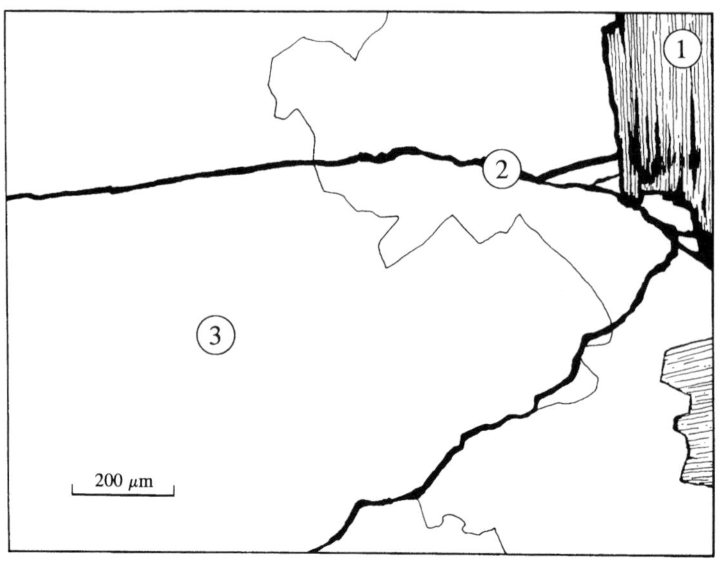

Fig. 5. Weathered biotite, ①, and quartz from granite near Molina de Bodegas (Río Pusa southwest of Los Navalmorales). Mobilized iron, ②, invaded the fractures of the quartz, ③. The quartz grains are mostly fractured across to their boundaries.

mineral grains became heavily fragmented. The fragments are forced apart by the invasion of kaolinic matrix fillings. However, their relative position and orientation remains unchanged during *in situ* weathering, since they show similar extinctions in thin section. The matrix resulting from the feldspar weathering also penetrated into the fractures of the quartz grains. In this fashion, the fragments of the quartz grains which had also been subject to splitting are similarly pushed apart by crystallizing pressure of the penetrating matrix. This widening is a feature typical of tropical weathering as described by Schnütgen & Späth (1983) at comparable samples from Sri Lanka.

At an advanced weathering stage the micas (initially biotite, then muscovite) become subject to alteration, the biotite having been bleached and widened at its edges. Firstly, the iron of the biotite becomes mobilized, causing the bleaching of the crystal, and invading the fractures developed in the other minerals (quartz and feldspar) as an amorphous iron-gel, partly along extended paths (Fig. 5). With the increase of the iron content within the matrix the solubility of the quartz also rises. In addition to the splitting, the etching of the quartz surfaces clearly increases and features such as corrosion embayments and etched microcaverns begin to develop (weathering stage 3; Fig. 6).

In thin section, stacks of crystallized kaolinite are visible within the clayey matrix which originated from the weathered plagioclase. In addition, smaller solution features can be seen by SEM on the quartz surfaces both inside and outside the cavernous penetrations. These are mainly chemically etched V-patterns and aligned solution crevasses, both controlled by crystal lattice orientation.

In Spain the weathering under a tropical environment during the Palaeogene led to heavily roughened grain surfaces, thereby offering an increased surface area to the subsequent weathering processes. To an increasing degree the quartz grains were fragmented and reduced by chemical etching whilst feldspar minerals were completely replaced by kaolinite by this stage. Nevertheless, some small remnants of former large homogeneous feldspars are preserved (Fig. 7). Using this, among other things, the original mineralogical constituents can be reconstructed. The amount of matrix has increased from almost zero in the unweathered rocks to $>20\%$ in the weathering residues of corresponding samples. Molina *et al.* (1991, p. 349) observed that, in comparable residues in which nearly half of the feldspar minerals are weathered, the silt content clearly decreases upwards in the profiles.

In particular, the complete dissolution of the feldspars is obvious in samples with myrmekite (Fig. 8). Here, the plagioclase has been weathered to kaolinite and merely the hieroglyphic quartz

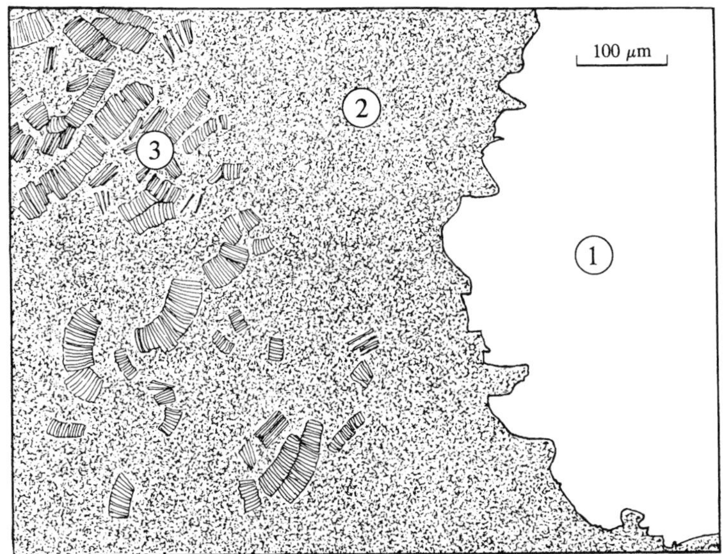

Fig. 6. Quartz grain, ①, from a kaolinized crystalline rock from the 'La Cantera' quarry west of the Río Torcón. Stacks of well-crystallized kaolinite minerals, ③, occur inside the clayey matrix, ②.

strips are preserved. At high magnification (500 times), well-crystallized kaolinite becomes visible by optical microscope (Fig. 6).

These described residues, which can only result from humid and hot climates, represent the lower parts of tropical deeply-weathered profiles in which the structure of the granite and granodiorite is partly still preserved. In this regolith, quartz grains of weathering stages 2–5 occur. The bases of the investigated profiles is identical with the weathering front ('Untere Einebnungsfläche' according to Büdel 1957) of the Lower Tertiary when a vast etchplain was developed on the Hercynian basement. These micromorphological

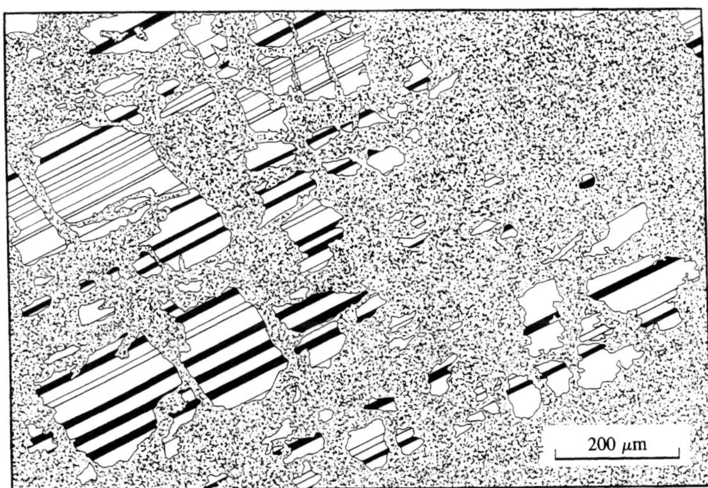

Fig. 7. Corroded plagioclase from weathered granite, taken from a road cut along C 401 west of Urbanización Río Cedena. The feldspar mineral has disintegrated into several small fragments but these fragments remain in their original positions.

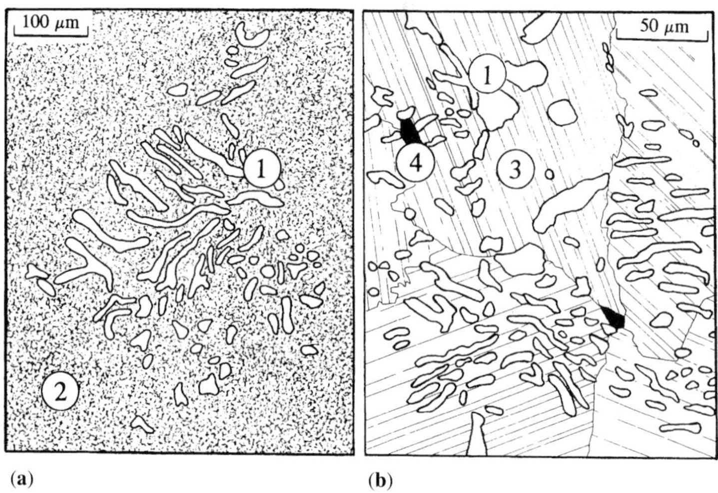

Fig. 8. (**a**) Weathered myrmekite from 'La Cantera', west of the Río Torcon. Only the quartz, ①, remains, whilst the feldspar is completely kaolinized, ②. (**b**) In comparison, unweathered myrmekite from gneiss west of Stammbach (Frankenwald, south Germany) with quartz, ①, plagioclase, ③, and biotite, ④.

results coincide with the analyses of the relief (cf. Gehrenkemper 1978). In one and the same level the etchplain truncated very different lock structures independently of their lithology; for instance the granites and granodiorites of the Tajo Basin as well as the quartzites and schists of the Montes de Toledo.

The Miocene sediments

The intense chemical weathering in the Lower Tertiary provided the source of the detrital material which was then deposited in the Miocene as loamy sands and then finally as raña sediments of the Pliocene. During the Miocene phase of transportation the upper parts of the granite weathering profiles had been removed and incorporated into the Miocene sands. In thin section this can be demonstrated because of a high number of quartz grains with consertal texture. Another portion of the Miocene sediments derive from the quartzites of the Montes de Toledo. In thin section, corresponding rounded quartzite grains can be identified and hence the chronological order of the weathering and transport phases can be determined by SEM work. Older features resulting from chemical weathering are still clearly evident on quartz grains in spite of the mechanical overprint caused by the younger transportation phase. The presence of corroded fragmented feldspar grains, also observed in thin section shows that after the redeposition of the weathering residues of the Lower Tertiary, a phase of chemical weathering took place again in the Miocene. The original position of each single fragment of corroded feldspar grains has remained unchanged, indicating their *in situ* character (in a similar fashion to the example of granite weathering described earlier (cf. Fig. 7). Even the lamellar twinning of the plagioclase incorporated in these Miocene sediments is preserved intact. Clearly, if this weathering had taken place before the redeposition, these feldspars would inevitably have fallen apart during transport and the original grain shape would have been entirely destroyed. However, the quartz shows few post-sedimentary Miocene corrosion features and, apart from traditional corrosion features of older weathering phases, the Miocene weathering appears to have left quartz grains at weathering stages 1–3. The older corrosion features are overprinted by the mechanically-generated microfeatures of the Miocene transportation, whilst post-sedimentary corrosion features would have overprinted the mechanically-generated features. From this a differentiation and evaluation of the weathering intensity of different phases becomes possible by SEM. Obviously, the environmental conditions during the Miocene alteration of these sediments had been enough to dissolve plagioclase, but the climatic conditions required for quartz weathering had not been present. The weathering was clearly less intense in the Miocene than in the Lower Tertiary. This environmental change can be identified in different areas of Europe (cf. Borger *et al.* 1993).

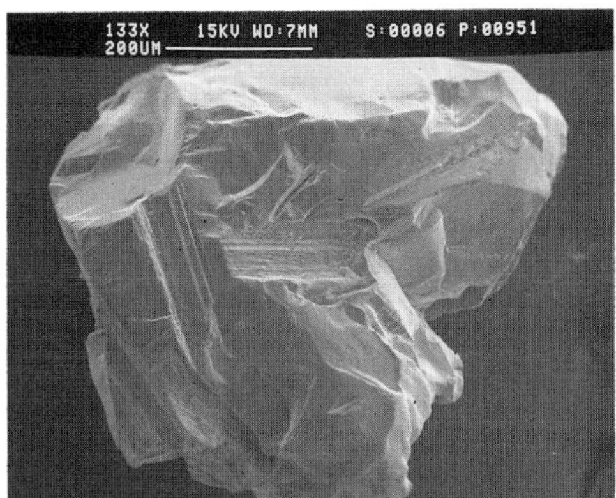

Fig. 9. Quartz with impressions of former existing micas (Miocene sand south of Talavera de la Reina).

SEM similarly demonstrates that a large proportion of the quartz grains are of granitic origin. Figure 9 shows a quartz grain with a negative impression of a former existing mica. The preservation of such an impression, together with the angular grain shape, demonstrates that this grain was not transported over long distances and that the transportational effect was not energetic. The mainly mechanical V-shaped patterns and curved grooves as crescentic impact scars on the quartz grains of the Miocene sands gives reason to believe that the transport was predominantly fluviatile. The impact scars occur at the edges of the grains, but the grains' surfaces remain unaffected by mechanical stress, which also suggests that this transport did not have an intense effect upon the grains.

The raña sediments of the Montes de Toledo

Typical appearance

The flat raña surfaces, which slope downwards at angles of only *c.* 0.5–2.0°, dominate the relief of extensive parts of the mountains and their surroundings. The extremely distinctive morphological appearance of the rañas is emphasized by the later Quaternary dissection. The preserved raña strips on the interfluves between the Quaternary valleys range in size from some tens of metres to a few of kilometres in width, but often extend > 25 km in length.

A characteristic of the raña sediments is a chaotic bedding of poorly-sorted pebbles and fine material, and the fact that they do not show any layering. Normally, the raña sediments are *c.* 4–6 m, but sometimes reach > 10 m in depth. The maximum of thickness of the fanglomeratic raña sediments is *c.* 25 m, as described by Stäblein & Gehrenkemper (1977). The coarse material of the rañas mainly consists of subangular to rounded quartzite debris

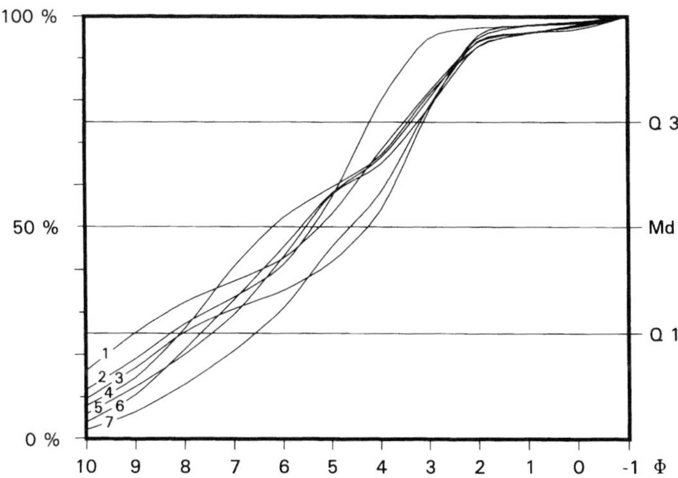

Fig. 10. Granulometric curves of raña sediments. 1, Raña west of Puebla de Don Rodrigo (4326,96-357,14; 580 m NN); 2, Raña west of Puebla de Don Rodrigo (4326,84-357,00; 600 m NN); 3, Raña east of Santa Maria (4319,32-377,46; 610 m NN); 4, Raña south of Río Estenilla (4371,0-351,16; 670 m NN); 5, Raña south of Río Estenilla (4371,0-351,16; 670 m NN) 6, Raña de las Puercas (4327,24-354,92; 610 m NN); 7, Raña El Caserejo, northeast of Las Arripas (4331,92-380,92; 655 m NN).

derived from the mountain areas. The fine material between the coarse raña scree has a poor sorting similar to all raña sediments, as visible from the granulation curves of Fig. 10. The coefficient of sorting (*So*), after Trask (1932), is *c.* 4 to >5 on average. This generally poor sorting suggests mud flow-like and torrential conditions of sedimentation and, therefore, a semi-arid climate at the time of deposition (Muñoz Jimenez & Asensio Amor 1975; Gehrenkemper 1978). Like the quartz grains of the Miocene sands, the quartz grains of the fine material of the raña sediments show pre-sedimentary features of chemical weathering which have survived the overprint caused by the later transportation. By this means the derivation of the raña sediments from the residues of an older weathering phase is recognizable.

Kaolinite is the dominant clay mineral of all raña sediments. In most of the samples the kaolinite proportion of the fine material is *c.* 25%. The brown hue of the raña sediments is caused by the presence of goethite, as determined by X-ray diffraction analysis (XRD; Fig. 11a).

In the north of the Montes de Toledo the quartzite debris is, in general, coarser than that found in the south of the mountains. After sedimentation, the quartzite pebbles of the northern glacis of the Montes de Toledo were subjected to only a small degree of decomposition because traces of intense chemical weathering from the period subsequent to the raña sedimentation rarely occur. Pebbles with incrustations of high iron content occur only in locally restricted areas (for example, on the raña surface west of Río Cedena near to the farmyard 'Cortijo de las Pistolas'; Fig. 3). By contrast, on the raña surfaces and in the upper soils in the south, and especially in the southeast of the mountains an enrichment of small pebbles (*c.* 2–3 cm in diameter) with dark brown to black iron incrustations occurs in many places. These pebble accumulations are described in the literature as being typical of raña surfaces, for instance by Espejo (1987). Such pebbles are heavily weathered and leached through to their centres. Very often they are completely impregnated by secondary iron. In thin section many of the quartz grains which compose these pebbles show an intense dissolution with deeply corroded embayments and etched caverns. The corrosion features are mostly filled with opaque iron oxides and hydroxides.

The raña west of Puebla de Don Rodrigo

In the southeast of the Montes de Toledo several intramontane basins have been filled by raña sediments. A distinctive example of this is the basin near Las Arripas (Fig. 1). The sediments of this basin have not been extensively dissected during the Quaternary, and the drainage furrows are alveolate and only slightly deepened. Even the main drainage, the Arroyo del Puerto de las Tinajas

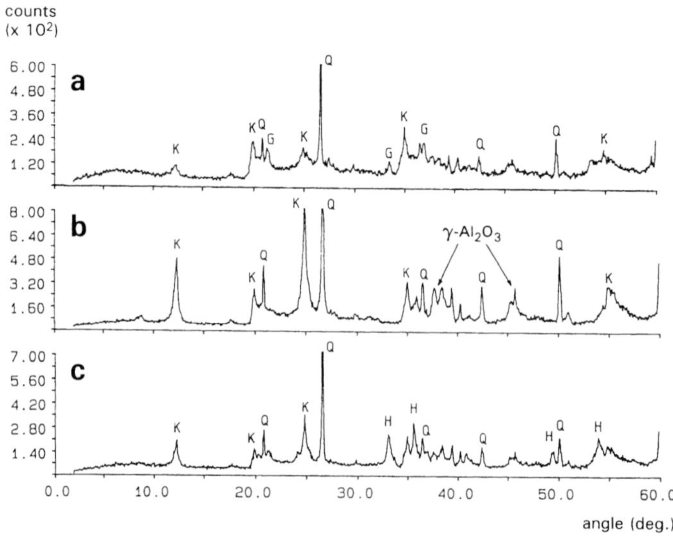

Fig. 11. XRD traces from raña sediments. (**a**) Raña east of Alcaudete de la Sierra (4404,00-345,24; 615 m NN); (**b**) Raña west of Puebla de Don Rodrigo (4326,96-357,14; 580 m NN); (**c**) Raña Maleta, west of Casa del Coto (4329, 64-349,28; 605 m NN). G, goethite; H, hematite; K, kaolinite; Q, quartz.

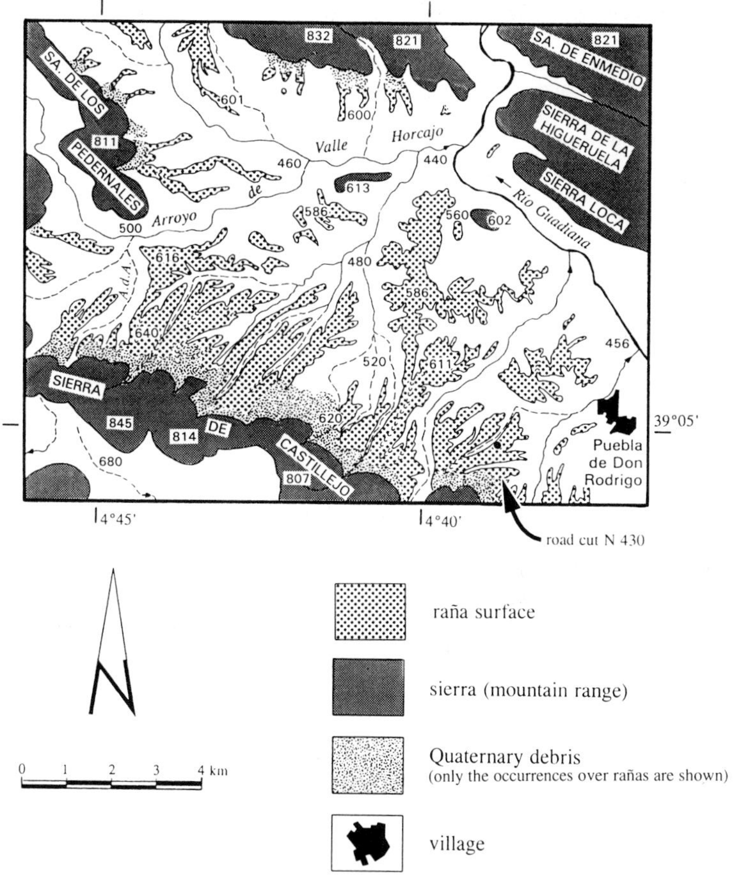

Fig. 12. Geomorphology of the intramontane basin west of Puebla de Don Rodrigo (SE Montes des Toledo).

(arroyo means torrential creek), has cut its wide valley floor little more than 20 m deeper than the raña surface. A comparable situation occurs west of Puebla de Don Rodrigo (Fig. 12). Like a bay, the basin near Puebla de Don Rodrigo opens out to the valley of Río Guadiana in the east, where the raña sediments have been completely eroded by the river. In contrast with the raña near Las Arripas, the rañas west of Puebla de Don Rodrigo are heavily dissected by deeply entrenched, steep-sided arroyo valleys due to the proximity of the Río Guadiana as the receiving stream. After a very short upper stream course the arroyos quickly cut deep into the raña level. The Arroyo de Alconcillo (Fig. 12), for example, flows a total distance of just 2 km into the Arroyo de Valle Horcajo yet cuts a valley 100 m below the level of the raña surface. The Arroyo de Valle Horcajo runs through the basin from west to east – as main drainage of this basin – to the Río Guadiana.

At a road cut along N 430, west of Puebla de Don Rodrigo (Fig. 12), raña sediments occur with a different appearance compared to most of those in the north of the mountains. Here, 8–10 m under surface, the profile shows a striking mottled pattern of intense red and white colours. The sediment is completely interspersed with this distinctive mottled coloration and is similar to that described by Espejo (1987, p. 410) with respect to the raña sediments developed in the south of the Sierra de Guadalupe. The mottling affects both the fine material between the quartzite pebbles as well as the quartzite pebbles themselves and indicates that this detrital material has again been weathered *in situ* after deposition as raña sediment. From this it is obvious that, after the raña sedimentation, intense chemical weathering was again possible, but only in localized moist locations of Pliocene basins. The red colouring caused by hematite and the white colouring caused by quartz and kaolinite can be identified in thin section as well as by XRD (Fig. 11c). In addition, the post-raña weathered sediments also show violet colours. The corresponding XRD analyses indicate traces of an aluminium oxide (not determinable in detail; Fig. 11b). Hematite and the existence of independent aluminium minerals are further indications of a post-sedimentary weathering to which raña material was exposed locally.

In thin sections it is visible that inside the red iron-bearing patches of the profile, the quartz weathering is substantially stronger than inside the white iron-poor patches. Many quartz grains of the hematite-saturated parts of the pebbles are intensively corroded and show deeply etched microcaverns and microembayments (Fig. 13) corresponding to quartz weathering stages 1–4. SEM images reveal that the surfaces of those embayed quartz grains often show etched V-patterns (Fig. 14). When developed on the 1 0 1 1 lateral faces of quartz crystals these V-patterns take on trapezium-like outlines, showing the dependence of solution features to crystalline textures. The solution processes can be demonstrated by laboratory experiments as shown by Dove & Crerar (1990) and Dove (1994), for example. Dense development of the V-patterns cause extensive etching of the grain surface with promontories occasionally remaining. Continued corrosion results in the disruption of the quartz surfaces leading to the development of the deeply etched microcaverns within the quartz grains (Fig. 15). In addition to these solution caverns, small corrosion features also occur on the quartz surfaces inside the caverns. These small features (< 2–10 μm) are visible by SEM but are not recognizable in thin section because of their size. Particularly because of the corrosion features, the cavernous penetrations become considerably extended and the quartz grains become smaller and smaller. The activation energy for quartz dissolution in pure water at 25°C is *c.* 89 ± 5 kJ/mol^{-1} SiO_2 (Tester *et al.* 1994). Further experiments indicate a sixteenfold increase of the dissolution rate for an increase in temperature from 25 to 50°C (Tester *et al.* 1994), showing the importance of temperature on quartz dissolution.

In addition to the corrosion features, reprecipitation of silica is visible, an effect also typical of an environment of strong chemical weathering. As a result of the reprecipitation, small idiomorphic quartz crystallization of 2–3 μm in size may be created but, more commonly, the constituent grains become coated with an amorphous silica layer of just nanometre thickness which can also coat corrosion features. These observations indicate a close interplay of dissolution and limited reprecipitation during weathering of the quartz

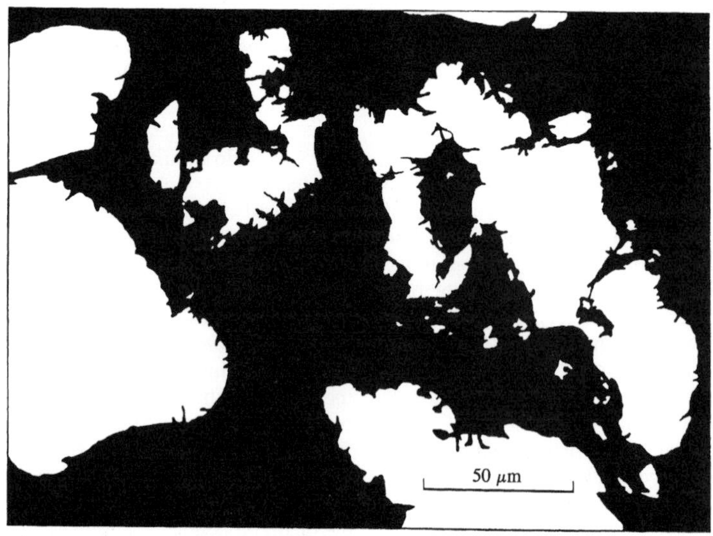

Fig. 13. Part of an Upper Pliocene corroded quartzite pebble from the raña sediment west of Puebla de Don Rodrigo. The hematite matrix (black) fills the etched microcaverns and corrosion embayments of the quartz grains (white). Parts of the quartz are replaced by the hematite.

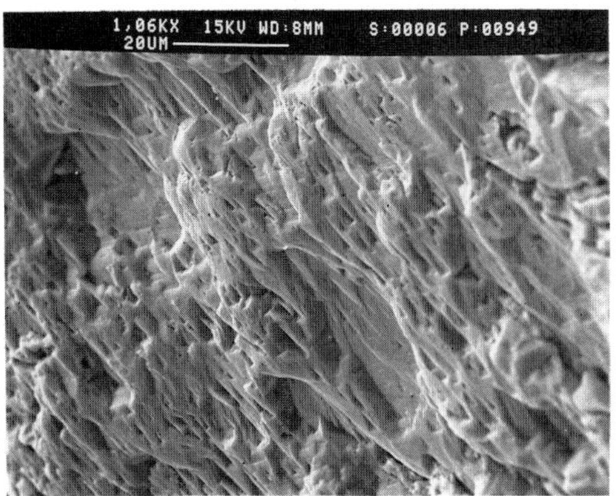

Fig. 14. Oriented V-patterns caused by chemical etching.

grains. Furthermore, etched caverns and dissolution features can be 'healed' in part by secondary quartz crystal growth that is typically characterized by sharp-edged domains.

In contrast to the iron-bearing pebbles, the white pebbles fall apart even under relatively minor mechanical influence (for instance, during sampling). After deposition, all the sediment was firstly bleached causing the quartzite pebbles to become friable and permeable. At a later stage a hematitic gel was then able to penetrate unhindered into the quartzite pebbles, where the iron strengthened the effect of ongoing quartz corrosion. The quartz grains inside the iron-poor patches are comparatively less weathered. After crystallization of the iron, only the iron-cemented pebbles were rehardened.

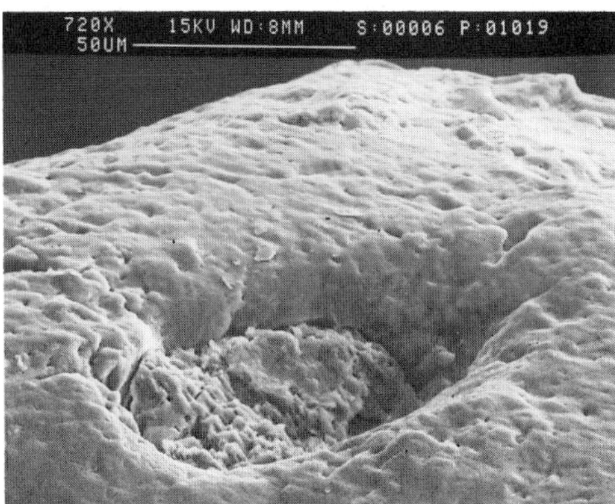

Fig. 15. Quartz grain showing the entrance of a deep corrosion cavern. The microcavern is filled with kaolinic clay.

The weathering of quartzite pebbles at the end of the Pliocene could only take place locally and under favourable conditions where both moisture and temperature were ideal. Most of the other raña sediments throughout the whole investigated area, especially those in the north, do not show weathering to such a degree. However, it should not be presumed that at this time there had been large differences in temperature between the north and the south of the Montes de Toledo. Rather, the raña weathering in the southeast of the Montes de Toledo was possible due to sufficient *soil humidity* in almost self-contained basins during a time which the temperature was probably high enough in the whole investigated area. In the basins of the southeast of the mountains, a groundwater table was probably present close to the raña surface before dissection of the sediments started in the Quaternary. In addition, an increase in the strength of the quartz weathering was caused by the massive penetration of iron (cf. Schnütgen & Späth 1983; Borger 1992).

Conclusions

The micromorphological analyses indicate that effects of intense weathering in central Spain decreased from the Palaeogene to the Plio-Pleistocene boundary. In the region of the Montes de Toledo, the Palaeogene weathering of the basement was an essential prerequisite for the genesis of the later developed glacis. According to micromorphological investigations and relief analyses, all rocks of the study area became deeply weathered under a hot and humid Palaeogene climate and an extensive etchplain evolved. Remains of this denudation surface are still preserved. The rock weathering during this stage of Palaeogene etchplain development was fundamental to the later geomorphological events which resulted in the raña sedimentation.

The analyses show that the granite weathering began with an intense splitting of the mineral grains, which was largely independent of inherent grain boundaries. At the same time, the plagioclase minerals were subject to the onset of corrosion and kaolinization and the resulting neogenic matrix penetrated into widening fractures. In the course of weathering, the fractures, and then with an increasing degree the kaolinization processes, led to the granular disintegration of the rock. As weathering continued, the feldspar minerals were eventually completely dissolved and even the quartz grains heavily corroded. In addition, iron originating from the biotite was mobilized and increased quartz solubility.

The etched features on the quartz grains of the Miocene sands are overprinted by features indicative of mechanical transport. From this observation it is clear that most of these quartz sands are redeposited weathering residues that were transported over short distances only. During this phase of transportation, the profiles of the decomposed granites had been truncated and partly incorporated into the Miocene sediments. Both the Miocene sands and the raña sediments show pre-sedimentary traces of weathering. In both cases the mechanical overprint did not eliminate the Palaeogene (and/or even older) weathering features completely. The sedimentation of the raña cover happened during environmental conditions of a semi-arid climate. However, for the main part, the evolution of the detrital material which later formed the Neogene sediments' itself dates back to the period of tropical rock decomposition during the Palaeogene and perhaps older periods. The flat landscapes of the Neogene were not initiated by the glacis development but, rather, represent the decomposition zone of a pre-existing etchplain which had been buried by the debris derived from the weathered rocks of the uplifted mountains and their surroundings.

The post-sedimentary weathered quartzite pebbles of the raña profiles to the west of Puebla de Don Rodrigo demonstrate that intense weathering processes continued locally even until the upper Pliocene. The current variation in temperatures in both the north and the south of the Montes de Toledo are negligible: it seems reasonable to assume that also during the Pliocene the variation was similarly very minor though generally throughout the whole area the temperature must had been high enough for intense quartz dissolution. The regional differences of the environmental conditions, with its different effects on the weathering during the late Pliocene cannot be explained by climatic limits. Also, iron enrichments, which could increase quartz solubility, are observable in all raña sediments. Therefore, the only mechanism left which could enable the late-stage reoccurrence of quartz corrosion experienced in the southeast Montes de Toledo is an increased soil humidity resulting from an elevated water table in intramontane basins. Here, sufficient humidity and temperature, as well as iron enrichment, were together able to give a short interlude with weathering conditions comparable with those under humid tropical climates. Similar weathering intensities are unlikely to have developed contemporaneously in central Europe since here the elevated temperatures required were no longer present at the turn of the Plio-Pleistocene.

For financial support I am very grateful to the DFG (Deutsche Forschungsgemeinschaft; project Bo 967/1-1). The laboratory investigations mainly were carried out at the Geographisches Institut, Universität zu Köln, Germany, and at the School of Geography, University of Oxford, UK. In addition, analyses were done at the Mineralogisch–Petrographisches Institut, Universität zu Köln (XRD) and at the Department of Earth Sciences, University of Oxford (energy dispersive system). For their energetic support and helpful discussions I thank Hanna Bremer, Andrew Goudie, Achim Schnütgen, Mike Widdowson and Thomas Dortmann.

References

BORGER, H. 1992. Paleotropical weathering on different rocks in Southern Germany. *Zeitschrift für Geomorphologie NF*, Suppl. **91**, 95–108.

—— 1993*a*. Quarzkornanalyse mittels Rasterelektronenmikroskop und Dünnschliff unter besonderer Berücksichtigung tropischer Verwitterungsresiduen. *Zeitschrift für Geomorphologie NF*, **37**(3), 351–375.

—— 1993*b*. Monheimer Höhensande, Transport- und Verwitterungsphasen im Dünnschliff und Elektronenmikroskop. *Geologische Blätter für NO-Bayern und angrenzende Gebiete*, **43**(4), 247–270.

——, BURGER, D. & KUBINIOK, J. 1993. Verwitterungsprozesse und deren Wandel im Zeitraum Tertiär-Quartär. *Zeitschrift für Geomorphologie NF*, **37**(2), 129–143.

BÜDEL, J. 1957. Die 'Doppelten Einebnungsflächen' in den feuchten Tropen. *Zeitschrift für Geomorphologie NF*, **1**, 201–228.

DOVE, P. M. 1994. The dissolution kinetics of quartz in sodium chloride solutions at 25° to 300°C. *American Journal of Science*, **294**, 665–712.

—— & CRERAR, D. A. 1990. Kinetics of quartz dissolution in electrolyte solutions using a hydrothermal mixed flow reactor. *Geochimica et Cosmochimica Acta*, **54**, 955–969.

ESPEJO, R. 1987. The soils and ages of the 'raña' surfaces related to the villuercas and altamira mountain ranges (Western Spain). *Catena*, **14**, 399–418.

ESWARAN, H., SYS, C. & SOUSA, E. C. 1975. Plasma infusion – a pedological process of significance in the humid tropics. *Anales de Edafología y Agrobiología*, **34**, 665–673.

GEHRENKEMPER, J. 1978. Rañas und Reliefgenerationen der Montes de Toledo in Zentralspanien. *Berliner Geographische Abhandlungen*, **29**.

KRINSLEY, D. H. & DOORNKAMP, J. C. 1973. *Atlas of Quartz Sand Surface Textures*. Cambridge University Press, Cambridge.

KUBINIOK, J. 1988. Kristallinvergrusung an Beispielen aus Südostaustralien und Deutschen Mittelgebirgen. *Kölner Geographische Arbeiten*, **48**.

MARGOLIS, S. V. & KENNET, J. P. 1971. Cenozoic glacial history of Antarctica recorded in sub antarctic deep-sea cores. *American Journal of Science*, **271**, 1–36.

—— & KRISNLEY, D. H. 1974. Processes of formation and environmental occurrence of microfeatures on detrital quartz grains. *American Journal of Science*, **274**, 449–464.

MOLINA, E., GARCÍA GONZÁLEZ, M. T. & ESPEJO, R. 1991. Study of paleoweathering on the Spanish Hercynian basement. Montes de Toledo (Central Spain). *Catena,* **18**, 345–354.

MUÑOZ JIMENEZ, J. & ASENSIO AMOR, I. 1975. Los depositos de raña en el borde noroccidental de los Montes de Toledo. *Estudios Geograficos,* **XXXVI**, 140–141, 779–806.

SCHNÜTGEN, A. 1992. Speroidal weathering, granular disintegration and loamification of compact rock under different climatic conditions. *Zeitschrift für Geomorphologie NF*, Suppl. **91**, 79–94.

—— & SPÄTH, H. 1983. Mikromorphologische Sprengung von Quarzkörnern durch Eisenverbindungen in tropischen Böden. *Zeitschrift für Geomorphologie NF*, Suppl. **48**, 17–34.

STÄBLEIN, G. 1973. Rezente und fossile Spuren der Morphodynamik in Gebirgsrandzonen des Kastilischen Scheidegebirges. *Zeitschrift für Geomorphologie NF*, Suppl. **17**, 177–194.

—— & GEHRENKEMPER, H. 1977. Rañas der Sierra de Guadalupe, Untersuchungen zu Gebirgsrandformationen. *Zeitschrift für Geomorphologie NF*, **21**(4), 411–430.

TESTER, J. W., WORLEY, W. G., ROBINSON, B. A. GRIGSBY, C. O. & FEERER, J. L. 1994. Correlating quartz dissolution kinetics in pure water from 25 to 625°C. *Geochimica et Cosmochimica Acta*, **58**, 2407–2420.

TRASK, P. D. 1932. *Origin and Environment of Source of Sediments of Petroleum.* Gulf Publishing Co., Houston.

WENZENS, G. 1977. Zur Flächengenese auf der Iberischen Halbinsel. *Karlsruher Geographische Hefte,* **8,** 63–87.

Palaeoweathering profiles developed on the Iberian Hercynian Basement and their relationship to the oldest Tertiary surface in central and western Spain

E. MOLINA BALLESTEROS[1], J. GARCÍA TALEGÓN[2] & M. A. VICENTE HERNÁNDEZ[2]

[1] *Dpto. de Geología, Universidad de Salamanca, Spain*

[2] *Insto. de Recursos Naturales y Agrobiología, CSIC, Salamanca, Spain*

Abstract: On the southwest border of the Duero Basin the Hercynian basement of the Iberian Peninsula is fossilized by a siderolithic sedimentary cover whose thickness increases progressively eastward. Two different palaeoweathering mantles have developed in this zone. From west to east, these mantles are progressively separated in the vertical sense. Whereas the lower mantle affects the Hercynian basement to a depth of > 18–20 m, being fossilized eastward by the siderolithic cover, the upper mantle affects both the Hercynian basement to the west and the siderolithic cover to the east to a depth of *c*. 3–5 m.

Mineralogical, petrographic and geochemical techniques have shown that the lower mantle shows an upward destruction of the parent minerals with the development of new 2 : 1 phyllosilicates (smectite-like minerals) in the middle levels of weathering and an enrichment in 1 : 1 phyllosilicates in the upper levels. By contrast, the higher weathering mantle shows important palaeosol features, an enrichment in CT opal, oxyhydroxides mobilizations and occasional concentrations of alunite dated at 58–67 Ma.

A planation surface appears related to the uppermost of these mantles, its remnants now uneven owing to the Alpine tectonic phases. Eastward, the Tertiary sediments of the Duero Basin unconformably fossilize the remnants of this palaeosurface.

The Iberian Hercynian Massif (Solé & Llopis 1952) forms the greater part of the western Iberian Peninsula and is its oldest geological entity (Fig. 1). In the provinces of Zamora, Salamanca and Avila, located on the southwestern border of the Duero Basin, this basement is fossilized by a siderolithic sedimentary cover whose thickness increases progressively eastward, becoming several tens of metres thick. These sediments are of alluvial origin and appear organized in a superposition of banks, each of which is no more than 1–2 m thick. In each bank the facies change from 'channel lag' and/ or 'channel bar' at the bottom to 'alluvial plain' facies, at the top (Jiménez 1972; Corrochano 1977; Alonso Gavilán 1981). Thus, the grain size changes from pebbles and gravels to fine sand, quartz being the dominant mineral (> 80%).

The upper levels of this cover show an enrichment in CT opal (Blanco & Cantano 1983; Blanco 1991) and an irregular redistribution of iron oxyhydroxides, leading to drastic changes in the colours of these parts of the sediments. In opal-rich profiles, it is common to find some alunite $[(SO_4)_2KAl_3(OH)_6]$ concentrations filling cracks and fissures. These concentrations have been dated by radiometric methods at between 58.4 and 67.7 Ma. (Blanco *et al.* 1982). Likewise, this cover is unconformably fossilized eastward by the Tertiary sediments of the Duero Basin

Studies performed by different authors since the beginning of the eighties (e.g. García Abbad & Martín Serrano 1980; Molina & Blanco 1980) have shown that the Hercynian basement has a weathering mantle of > 18–20 m in depth, whose remnants outcrop in the landscape or are fossilized by the siderolithic cover. From bottom to top, three main levels, or horizons, can be distinguished in this weathering mantle (Molina *et al.* 1990; Vicente *et al.* 1994): (1) a lower level in which the parent minerals are dominant, although some of them (i.e. plagioclases) are already weathered; (2) a middle level where chlorites are altered and a new 2 : 1 phyllosilicate at 14 Å is formed; and (3) an upper level in which only resistant parent minerals (i.e. quartz and some micas) occur, encompassed within a mass of 1 : 1 phyllosilicates with an upward enrichment in CT opal.

The enrichment in opal is a general fact in both the Hercynian basement westward and the siderolithic sedimentary cover eastward. However, if the sedimentary cover is thick enough (commonly > 25–30 m thick) it does not affect the fossilized weathered basement.

This paper addresses this silicification and its

From Widdowson, M. (ed.), 1997, *Palaeosurfaces: Recognition, Reconstruction and Palaeoenvironmental Interpretation*, Geological Society Special Publication No. 120, pp. 175–185.

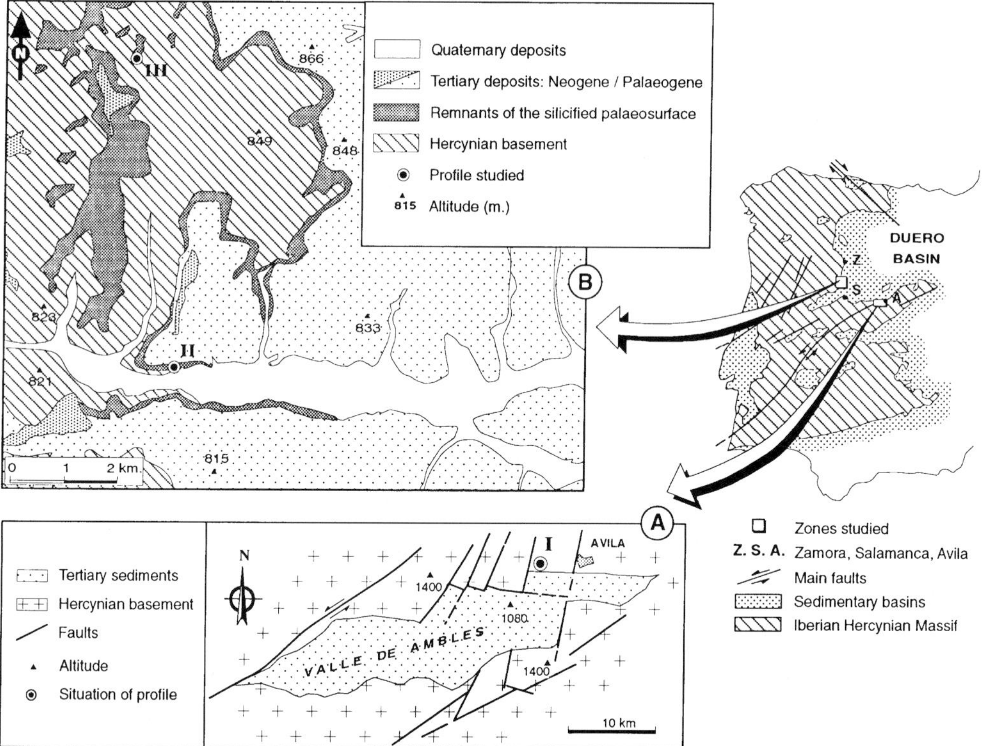

Fig. 1. Geographical and geological situation of the profiles studied.

role in the development of the regional landscape. To do so, several profiles were studied on both the basement and the sedimentary cover. Three of these profiles were chosen as the most characteristic. Profile I is located in the surroundings of the city of Avila (central Spain) (40°39′05″ N; 4°47′15″ W); it has no sedimentary cover at the top and has developed over a granitoid parent rock. The other two profiles are located in Salamanca province (western Spain). Profile II (40°50′45″ N; 5°51′40″ W) has a leucogranite as parent rock and profile III (40°52′55″ N; 5°49′00″ W) is developed over orthogneisses, having a siderolithic cover of *c.* 3–4 m thick.

All these weathering profiles display strong degrees of silicification at the top. Moreover, alunite concentrations have been detected in profile II in the form of fissure fillings.

Methodology

Micromorphology, X-ray diffraction (XRD) techniques and chemical analyses were used in this study.

Since most samples are soft and break down easily, they were placed in special boxes to preserve the original structure of the different levels of the profiles. In the laboratory they were dried at room temperature, subjected to a vacuum and impregnated with a mixture of polyester (cronolithe) and different proportions of solvent, catalyser and activator in order to achieve a more or less rapid polymerization (Brewer 1964, 1976; Benayas 1984). After 3–4 weeks of drying, hard blocks were formed. As water is a polar liquid it cannot be used either as lubricant and/or coolant; thus, special oils were used to cut the blocks. The slices obtained were fixed to a glass slide and, after polishing, thin sections were obtained for study under a Leitz Laborlux 12 Pol S microscope.

XRD diagrams of the $< 2\,\mu m$ fraction were obtained using a Philips PW 1730 instrument with a Ni-filter and CuK_α radiation.

For the geochemical study, once the samples had been ground to a powder, they were dissolved under pressure in an acid solution ($NO_3H + FH$) in a Milestone MLS-1200 microwave digestor. The final solutions were buffered with boric acid. Identification of the main elements was performed in a Plasma ICP, model II from Perkin-Elmer.

Profile description

In complete profiles, on both the siderolithic cover and the Hercynian basement, a bleached level or horizon, no more than 2 m deep, has developed at the top. The uppermost part of this level shows dissolution hollows and cavities some decimetres in depth and breadth, giving rise to a mesokarstic relief. Some of these cavities have concentrations of resistant minerals collapsed from their walls, resulting from the dissolution of opal. In some cases, a column-like structure has developed in these parts of the profiles.

Downward, another level of ochre, red and purple spots occurs due to the different re-distribution of iron oxyhydroxides. This level is especially well developed on schists; in these cases, red and purple colours are dominant, crossed by some ochre and white longitudinal spots related to fissures. In some cases, this mottled level may reach > 5 m in thickness. At the contact between the weathered schists and the siderolithic cover ferruginous concentrations occur.

Where the parent rocks are granitoids *sensu lata* or gneisses, this mottled level is not so well developed; in these cases, the red and purple spots are more or less discontinuous, affecting no more than 1–2 m of the weathered rock, whose dominant colour then becomes ochre downward.

In most of the profiles studied the bleached horizon and the upper parts of the coloured level are rich in opal, which then forms the cement of the weathered rock. Downward, opal contributions appear concentrated in certain zones; they fill fissures and veins in the weathered basement whilst in the siderolithic cover this mineral appears mainly in levels of coarse grains.

Profile I

The landscape of the area where the profile outcrops is a platform 1.100–1.200 m above sea level (a.s.l.). Its geological structure is defined by several small Hercynian blocks limited by two trends of faults: E–W and NNE–SSW (Fig. 1A; after Garzón *et al.* 1981). The lithology of these blocks is granitoid (granodiorite and adamellite), crossed by some episyenite aplite, porphyroid dykes and quartz veins. The top of these blocks is a weathering mantle *c.* 18–20 m thick which is more or less preserved beneath the silicified saprolite. Both the blocks and the weathering mantle slope slightly southward and are overlapped by the Oligocene arkosic sandstones of the 'Valle de Amblés', a tectonic basin of alpine origin (Garzón & López 1978; Garzón *et al.* 1981). From bottom to top in the weathering mantle it is possible to distinguish the following levels, or horizons (García Talegón *et al.* 1991; Vicente *et al.* 1994) (Fig. 2):

1. **Lower level**: the parent rock is a granodiorite crossed by a system of fissures defining polyhedra of different sizes and shapes. These polyhedra display the typical 'onion skin' weathering, grading from the corestones of unweathered rock to saprolite of preserved structures.
2. **Middle level**: the parent rock is totally disintegrated, giving rise to a generalized saprolite. While below the structure of the parent rock is still preserved, upward it grades into a subhorizontal laminated structure with a concomitant change in colour from greyish to the ochre of the weathered rock.
3. **Upper level**: this level is silicified by CY opal with drastic changes in colours from ochre to dark red and/or white. Its structure changes from massive to laminated or columnar, with dissolution features at the top.

Profiles II and III

These comprise the weathering profiles developed on a platform which slopes slightly downward to the southeast at an altitude between 870 and 800 m. This platform forms the top of another block in which, from bottom to top, two different lithologies can be distinguished (Fig. 1B; after Alonso Gavilán 1981): (1) the Hercynian basement formed by granitoids and highly metamorphosed rocks that are weathered to a depth of *c.* 20–30 m; and (2) a siderolithic sedimentary cover whose thickness and extent increase progressively eastward. Eastward, both the Hercynian basement and the siderolithic cover have been fossilized by the unconformable cover of Tertiary sediments of the Duero Basin. The upper 2–4 m of the saprolite in the weathered basement and of the sedimentary cover are strongly silicified.

At the top of profile II (Fig. 2), dissolution processes have resulted in different degrees of opal mobilization, giving rise to a mesokarstic relief. The bleached horizon developed on the saprolite is very thin or may be absent. This saprolite displays fissures and cracks, some of them filled by CT opal and, in some cases, by alunite, as detected in thin sections. In the upper parts of the silicified saprolite, where the changes in colour are more significant, it is possible to find remnants of pedological features (e.g. bioturbation). These form pipes and burrows < 2 mm wide and several centimetres long, featuring a progressive change in colour from yellow and ochre, in the outer parts, to red in the inner parts.

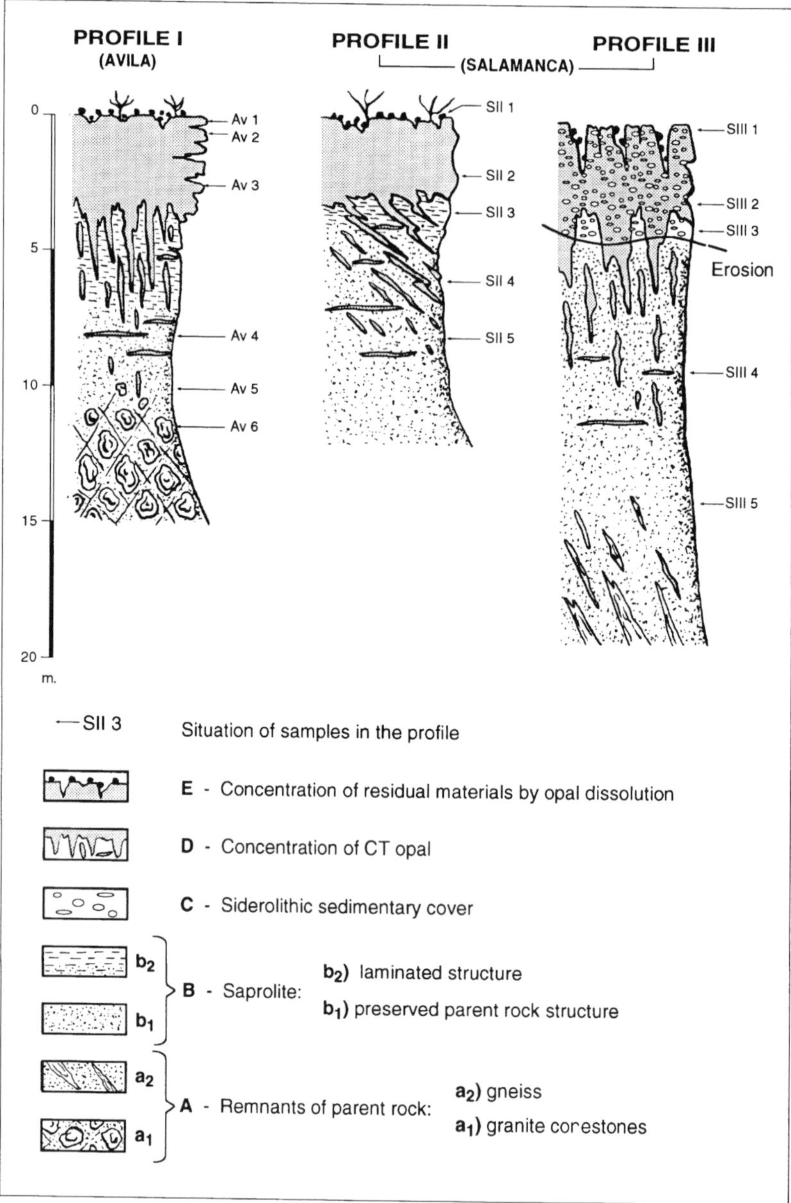

Fig. 2. Sketch of the main processes occurring in the profiles.

In profile III (Fig. 2) the silica, although concentrated in the sedimentary cover, has been illuviated into some fissures of the weathered basement. This cover displays a column-like structure, variable in depth, in which old cracks and fissures were, and still are, the leaching drains. This process carried down silica and the finest fractions, leading to the release of coarse grains which have collapsed into the cavities below. Repetition of these phenomena is responsible for the column structure and most of the dissolution hollows.

Discussion

Microscope studies of samples from the different levels of the weathered basement in profiles I and II point to a progressive upward reduction in primary minerals (plagioclases, biotites, chlorites and some potassium feldspars) and the appearance of an isotropic mass quite similar to the 'weathering plasma' described by Nahon (1991). The resistant minerals (mainly quartz and muscovite) become more evident, although their edges are corroded.

In the siderolithic cover of profile III, the most salient feature is likewise an upward reduction in the content of the more alterable minerals (some biotites and k-feldspars), giving rise to a texture formed of resistant minerals (quartz grains) encompassed within a brown and more or less anisotropic mass composed of clay, iron oxy-hydroxides and CT opal. However, in samples from the upper parts of the three profiles a contribution of colloform material also composed of opal, clays and oxyhydroxides appears; this covers the walls of the voids or fills them.

Study of thin samples from profile II shows that some voids have ovoid and/or circular sections, these being interpreted as the traces of roots of old palaeosols. They concentrate the oxyhydroxides around their walls (Fig. 3) and different types of coatings have developed as described by Brewer (1964, 1976) and Bullock *et al.* (1985). The white and ochre zones are distant from the voids and cavities, while dark red zones appear just on their borders accompanied by the development of 'complex neocutans' in Brewer's terminology. According to Dorronsoro *et al.* (1988) and Schwertmann & Fitzpatrick (1992), this suggests that during the ancient soil-forming processes these parts of the profiles were under low Eh conditions, these pipes and burrows being the areas where oxidation has occurred.

Fieldwork and micromorphological studies have also shown that the bleaching process affecting the upper parts of the profiles is related to the cracks and fissures crossing most of the above features. The presence of alunite concentrations, which are not generally common, appears to be related to this process of bleaching. Only in one thin section of profile II were some concentrations of small crystals of alunite found (Fig. 4).

The mineralogical and geochemical data appear in Tables 1a–3a and 1b–3b. Whereas the mineralogical data (a) refer to the $< 2\ \mu m$ fraction, in order to obtain an accurate study of the weathering products, the geochemical data (b) refer to the

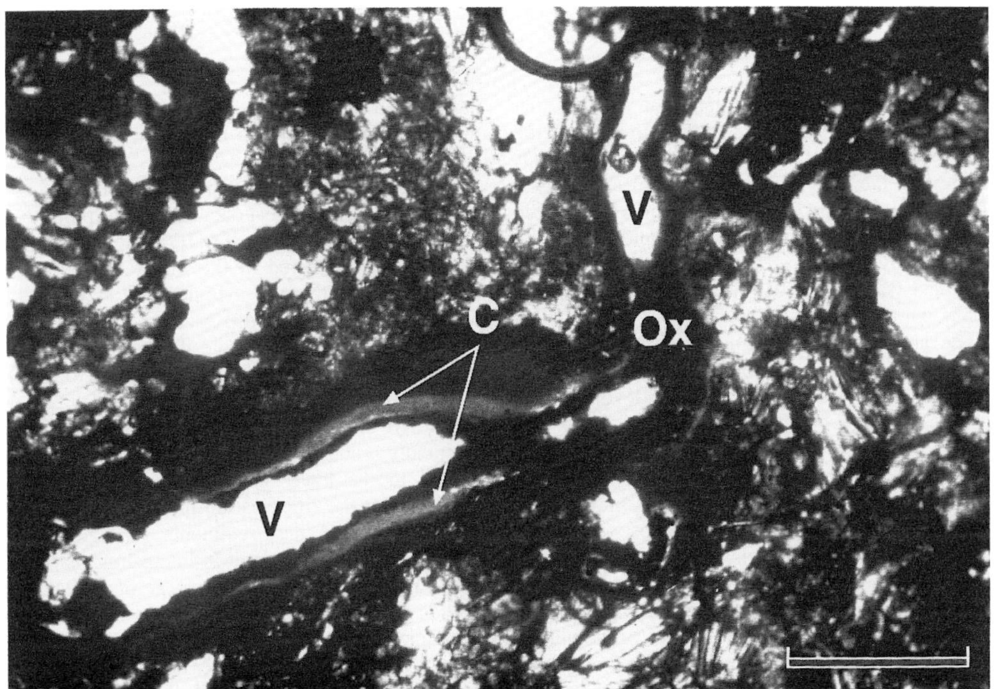

Fig. 3. Photomicrograph (one nicol) of a sample from the mottled level of profile II. Voids (V) are interpreted as being of pedological origin (tracks of old root and/or worm activity). Oxyhydroxide concentrations (Ox) and clay orientations (C) appear related to voids. Scale: 0.5 mm.

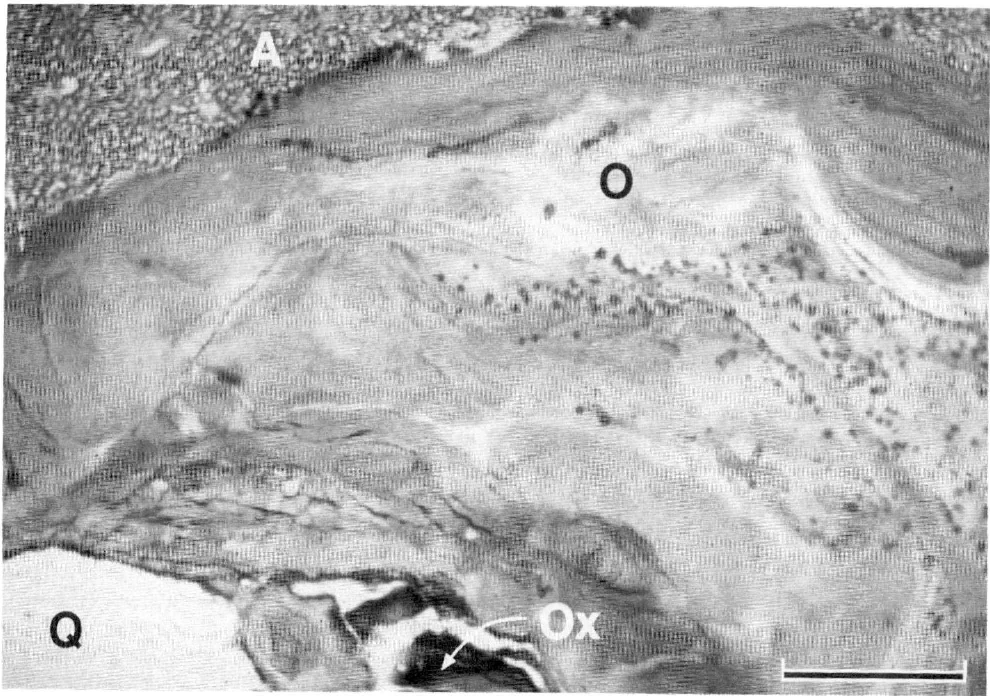

Fig. 4. Photomicrograph (plane polarized) of opal concentration (O) filling a crack from profile II. Small crystals of alunite (A) appear in the upper left corner of the feature. Quartz (Q) and oxyhydroxides (Ox) are in the lower left border of the feature. Scale: 0.5 mm.

whole sample. From these tables it is clear that there is not always a direct relationship between the appearance of opal and the increase in the SiO_2 content of the samples. Only in samples Av I, SIII 1 and SIII 2 is there a significant increase in the SiO_2/Al_2O_3 ratio, this being 6.79, 17.13 and 11.00, respectively. In the unweathered parent rock of profile I (sample Av 6), the value of this ratio is 4.53. All this suggests that the geochemical data alone are insufficient to interpret the weathering behaviour and that the opal contribution was a process mainly affecting the upper levels of the profiles.

The mineralogical data from profile I (Table 1a) show that CT opal, kaolinite and oxyhydroxides appear commonly in the lower levels of weathering. However, oxyhydroxides are not well represented in the upper levels; this may be due to leaching processes under acid conditions and/or organic-rich environments (Thiry & Milnes 1991; Rayot 1994). Moreover, thin sections of the uppermost parts of this profile show that some voids are filled by microquartz and/or length-fast chalcedony (Fig. 5). According to Thiry & Millot (1987) the existence of these types of quartz is consistent with leaching processes in these parts of the profile and has led to a reduction in the concentration of solutions which gives rise to more ordered forms of silica.

In profile II the unweathered parent rock is not visible. The geochemical data (Table 2b) point to a progressive increase in Fe_2O_3 content upward, which is not reflected in the mineralogical data, suggesting that there is a high Fe_2O_3 content in poorly crystalline oxyhydroxides.

It is also important to note that the siderolithic sedimentary cover appearing at the top of profile III (Table 3b) displays an upper part (samples SIII 1 and SIII 2) in which some components are drastically reduced (e.g. Al_2O_3, Fe_2O_3, K_2O) with respect to the lower part (sample SIII 3). This suggests that the silicification process is related to the weathering of these sediments. As regards this silicification phenomena, it is interesting that in a previous paper (García Talegón *et al.* 1994) we studied a profile similar to that of profile I, located quite close to it, the only difference being the existence of a siderolithic cover no more than 0.5 m thick at the top. The aim of the paper was to determine the effects of silicification on the upper levels of the saprolite. The main conclusion was that silicification leads to a drastic reduction (*c.*

Table 1a. *Mineralogical data profile I (< 2 µm fraction)*

Samples	Depth (m)	Quartz	Felspar	Mica	Chlorite	Kaolin	Smectite	Inters. I–Sm	Oxyhyd.	Opal
Av 1	0.20	xxx				xx				x
Av 2	0.50	xx				xxx				x
Av 3	2.50	x				xx			x	xx
Av 4	8.00	xx				xx	T		x	x
Av 5	10.00	xxx	T	T	T	x	T	T	x	T
Av 6*	11.00	xxx	xx	xx	x					

* Sample from a corestone of the parent rock.
xxx, Dominant; xx, abundant; x, scarce; T, traces.

Table 1b. *Geochemical data* (wt%) *of profile I*

Samples	Depth (m)	SiO_2	AlO_2O_3	TiO_2	Fe_2O_3	MnO	MgO	CaO	Na_2O	K_2O	P_2O_5	H_2O
Av 1	0.20	79.35	11.69	0.35	0.30	0.00	0.10	0.20	0.05	0.18	0.07	7.21
Av 2	0.50	72.54	17.94	0.52	0.08	0.01	0.04	0.13	0.03	0.06	0.06	8.46
Av 3	2.50	62.89	13.69	0.38	11.83	0.02	0.12	0.29	0.04	0.21	0.30	10.58
Av 4	8.00	65.43	18.22	0.42	6.65	0.01	0.10	0.09	0.04	0.27	0.15	8.43
Av 5	10.00	71.55	16.34	0.45	2.80	0.00	0.23	0.25	0.06	0.40	0.05	7.36
Av 6*	11.00	68.90	15.21	0.51	2.76	0.05	1.29	2.59	3.41	3.71	0.15	1.11

* Sample from a corestone of the parent rock.

Table 2a. *Mineralogical data profile II (< 2 µm fraction)*

Samples	Depth (m)	Quartz	Felspar	Mica	Chlorite	Kaolin	Smectite	Inters. I–Sm	Oxyhyd.	Opal
SII 1*	0.30	xx	T	x		xxx			x	x
SII 2	2.00	x	x	xx		xxx			x	x
SII 3	3.50	x	T	xxx		x	x	T	xx	x
SII 4	6.00	x	x	xxx		xxx	T	T	x	xx
SII 5	8.00	x	xx	xx	T	xxx	xx	T		x

* Leaching hole.
xxx, Dominant; xx, abundant; x, scarce; T, traces.

Table 2b. *Geochemical data* (wt%) *of profile II*

Samples	Depth (m)	SiO_2	AlO_2O_3	TiO_2	Fe_2O_3	MnO	MgO	CaO	Na_2O	K_2O	P_2O_5	H_2O
SII 1*	0.30	58.61	17.74	0.35	11.54	0.01	0.69	0.57	0.38	1.27	0.18	8.32
SII 2	2.00	68.52	16.50	0.30	6.80	0.02	0.76	0.29	0.47	2.97	0.13	2.97
SII 3	3.50	65.20	18.62	0.47	4.26	0.03	0.80	0.86	0.39	1.07	0.10	7.73
SII 4	6.00	64.77	19.65	0.49	2.04	0.02	0.68	0.79	0.29	1.24	0.77	8.75
SII 5	8.00	68.40	15.64	0.46	3.34	0.03	1.11	0.77	0.33	4.24	0.13	5.61

* Leaching hole.

Table 3a. *Mineralogical data profile III (< 2 μm fraction)*

Samples	Depth (m)	Quartz	Felspar	Mica	Kaolin	Smectite	Inters. I–S	Oxyhyd.	Opal
SIII 1	0.20	x		x	xx		T		xx
SIII 2	3.00	x		xx	xxx			x	x
SIII 3	4.00	x		xx	xxx			x	x
SIII 4	9.00	x		x	xx	T	xx	x	xx
SIII 5	14.00	xx	T	xx	xx	xxx	x	T	T

xxx, Dominant; xx, abundant; x, scarce; T, traces.

Table 3b. *Geochemical data* (wt%) *of profile III*

Samples	Depth (m)	SiO_2	AlO_2O_3	TiO_2	Fe_2O_3	MnO	MgO	CaO	Na_2O	K_2O	P_2O_5	H_2O
SIII 1	0.20	90.48	5.28	0.13	0.76	0.00	0.09	0.04	0.03	0.43	0.05	2.69
SIII 2	3.00	85.65	7.78	0.19	1.00	0.00	0.15	0.07	0.05	0.59	0.03	4.62
SIII 3	4.00	67.19	19.01	0.25	2.78	0.01	0.49	0.05	0.07	1.34	0.02	8.41
SIII 4	9.00	71.30	16.47	0.14	1.75	0.01	0.60	0.09	0.06	1.10	0.04	8.14
SIII 5	14.00	62.64	17.01	0.81	5.47	0.04	1.88	0.16	0.20	4.24	0.08	6.82

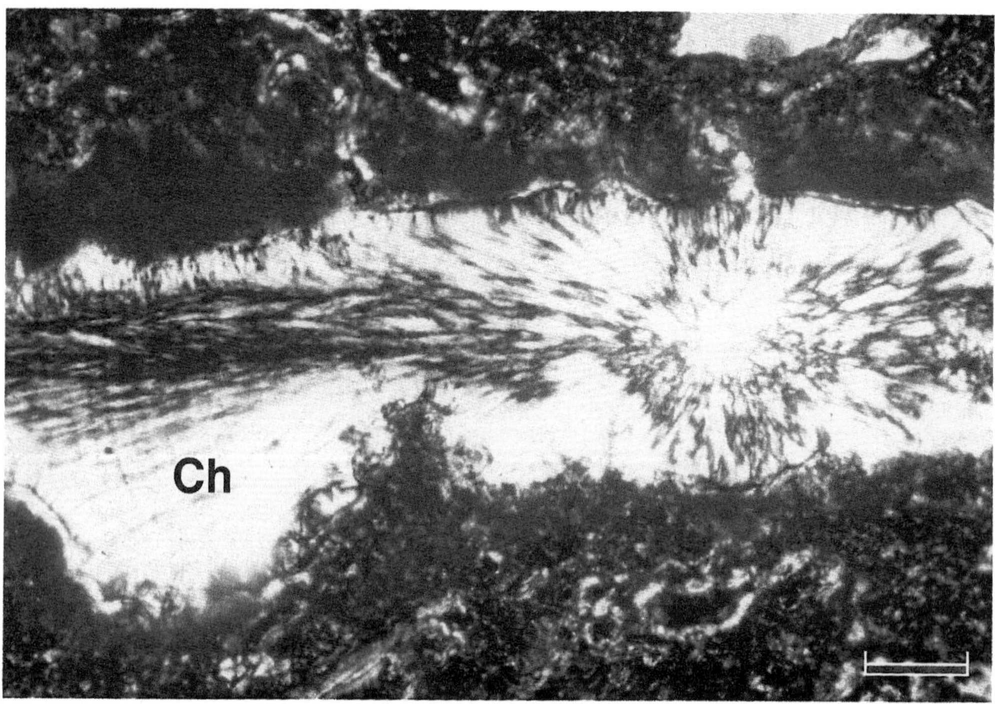

Fig. 5. Photomicrograph (cross polarized) of length-fast chalcedony (Ch) filling a void of sample from the upper level of profile I. Scale: 0.1 mm.

30%) of the 1 : 1 layer silicates inherited from the saprolite and a reduction in the crystallinity of these inherited minerals.

The data on profile III (Tables 3a and b) show that the weathered basement outcropping below the silicified cover represents the upper part of the weathering mantle. This suggests that the unweathered basement would lie several metres beneath the current landscape.

On comparing both the mineralogical and geochemical data from the three profiles, it is clear that there is a direct relationship between feldspar, mica and the K_2O content. This suggests that profile II is truncated at a weathering level slightly lower than the other two profiles. In this profile there is no bleached horizon and the coloured level with bioturbations appears at the top.

Progressing eastward in the region, where profiles II and III were chosen, the siderolithic cover becomes thicker and silicification affects only its upper levels. However, the Hercynian basement fossilized beneath is still deeply weathered but not silicified.

Apart from profile I, in which the unweathered parent rock has been studied, in the other two profiles this basement has not been found. Nevertheless, profiles II and III show a lower part relatively rich in MgO, CaO, Na_2O and K_2O (Tables 2b and 3b), in agreement with the presence of chlorites, micas and feldspars as primary minerals.

Special attention should be paid to the occurrence of a smectite-like mineral. This phyllosilicate appears in the middle levels of the three profiles studied and is common in most of the palaeoweathering profiles developed on the Hercynian basement. Its origin is related to the weathering of primary minerals, namely chlorites and illites (Vicente *et al.* 1991; Espejo *et al.* 1992).

Conclusions

The processes studied here are widespread throughout the western border of the Duero Basin. Since the dated alunite fills cracks and fissures in both the weathered basement and the sedimentary cover, the ages of these materials must be older than 58–67 Ma.

All the evidence points to the existence of two weathering mantles, the lower one showing an upward increase in kaolinite contents and the upper one a reduction in the inherited kaolinite content and a mobilization of silica and oxyhydroxides. Alunite concentration appears to have been one of the last processes to occur.

The question that arises here is whether the two weathering mantles are of different ages or whether they simply represent different weathering conditions (e.g. groundwater conditions or surface conditions). Studies performed in different parts of the Iberian Hercynian Massif (Saavedra & Camazano 1981; Corrochano & Pena dos Reis 1986) and in other Hercynian basement areas of Western Europe, e.g. Massif Central of France (Lapparent 1930; Simon-Coiçon 1989), have pointed to the existence of deep weathering mantles of Mesozoic age, probably Cretaceous. Stratigraphic, mineralogical and radiometric data suggest that, in the region studied, the oldest of these two mantles is Mesozoic *sensu lata* and the younger is of very early Tertiary age.

The profiles chosen are examples of many which can be related to one other by a planation surface, their remnants today being identified on both the Hercynian basement and the siderolithic cover. Thus, this surface could be of early Palaeocene age, in agreement with previous works referring to palaeosurfaces developed in other parts of Western Europe and displaying similar characteristics (Grandin & Thiry 1983; Blanc-Valleron & Thiry 1993). Once developed, it was then rendered uneven by Alpine tectonism during the Tertiary (Molina *et al.* 1987; Martín Serrano 1988), its remnants becoming either exhumed or else fossilized at the landsurface beneath younger Tertiary sediments. This palaeosurface is the result of a set of biogeochemical processes acting together, in the sense of the French authors (Erhart 1967; Millot 1980; Blanc-Valleton & Thiry 1993). The main source of the siderolithic sediments was the weathered Hercynian basement, which supplied resistant minerals and kaolinites. The source of silica seems to have been both the weathered Hercynian basement and the superficial weathering of the siderolithic cover. Given the situation of the mottled level, located just beneath the bleached horizon, it seems that during the surface development the general water table would have always been a few metres beneath the landsurface. According to the micromorphological data, lateral migration of solutions within the regional groundwater would have been important. Study of the upper parts of the bleached horizon shows that silica and fine fractions (silt and clays) were leached downward and concentrated a few metres below, this process being repeated many times. The resistant and unleached materials were concentrated at the top of the profiles and later removed by erosion. Alunite concentrations seem to be related to the last stages of this landscape evolution and originated by the action of very acid solutions passing through the profiles.

All these processes, working together and repeated over millions of years on a relatively stable basement, produced a progressive lowering of the relief, giving rise to the planation surface in which resistant lithologies (quartz dykes and quartzites) were thrown into relief, forming inselbergs, hills and mountains with rather regular slopes.

The level rich in CT opal is hard and forms a kind of 'hardpan' or 'duripan', in the sense of Thiry (1993), which protects the lower levels from erosion. Where this protective cap has disappeared, an 'etchplain' has developed some 20–30 m below the old surface. This is the origin of some 'mesas, in the typography of this region.

Eastward, both the etch surface and the old planation surface are fossilized by the Tertiary sediments of the Duero Basin.

We thank to Dr M. Widowson for his suggestions and critical revision of this paper. We also thank to Professor C. R. Twidale and Dr L. Campbell for their explanations and suggestions during the first draft of this work.

This research has been supported by the CICYT (Spain) under project no. PAT91-1570-CO3-O3 CE, and by the European Community under project no. STEP 90-0101.

References

ALONSO GAVILÁN, G. 1981. *Estratigrafía y sedimentología del Paleógeno del borde suroccidental de la Cuenca del Duero (Provincia de Salamanca)*. Tesis Doctoral, Universidad de Salamanca.

BENAYAS, J. 1984. *Atlas de Micromorfología de Suelos e Introducción a la Micromorfología*. Monografía 84, Ed. E.T.S. Ingenieros Agrónomos, Universidad Politécnica, Madrid.

BLANC-VALLETON, M. M. & THIRY, M. 1993. Mineraux argilleux, paléoalterations, paléopaysages et sequence climatique. Exemple du Paléogène continental de France. *In*: PAQUET H. & CLAUER, N. (eds) *Sédimentologie et Géochimie de la Surface. Colloque á la memoire de George Millot* , Acad. des Sciences, Paris, 199–216.

BLANCO, J. A. 1991. Primera parada. Los procesos de slilicificación asociados al Paleógeno basal del borde SW de la Cuenca del Duero: Sobre alteritas pre-terciarias del zócalo hercínico. *In*: BLANCO, J. A., MOLINA, E. & MARTIN SERRANO, A. (eds) *Alteraciones y paleoalteraciones en la morfología del oeste peninsular*, 199–209, Publ. Instituto Tecnológico Geominero de España (ITGE)., monografía n° 6, Madrid.

—— & CANTANO, M. 1983. Silicification contémporaine à la sédimentation dans l'unité basale du Paléogène du bassin du Duero (Espagne). *Sciences Géologiques Mémoire*, Strasbourg, **72**, 7–18.

——, CORROCHANO, A., MONTIGNY, R. & THUIZAT, R. 1982. Sur l'âge du debut de la sédimentation dans le bassin tertiaire du Duero (Espagne). Atribution au Paléocène par datation isotopic des alunites de l'unité inferiure. *C.R. Academie Sciences Paris*, **295**, 259–262.

BOULET, R., BOCQUIER, G. & MILLOT, G. 1977. Géochimie de la surface et formes du relef. I. Déséquilibre pédoclimatique dans les couvertures pédologiques de l'Afrique tropical de l'Ouest et son rôle dans l'aplanisement des reliefs. *Sciences Géologiques Bulletin, Strasbourg*, **30**, 235–243.

BREWER, R. 1964. *Fabric and Mineralogical Analysis of Soils*. Wiley, New York.

—— 1976. *Fabric and Mineralogical Analysis of Soils*. Krieger, Huntinton, New York.

BULLOCK, P., FEDOROFF, N., JONGERIUS, A., STOOB, G., TURSINA, T. & BABEL, U. 1985. *Handbook for Soil Thin Section Description*. Waine Research, Mount Pleasant, Albrighton.

BUSTILLO, M. A. & MARTIN SERRANO, A. 1980. Caracterización y significado de las rocas silíceas y ferruginosas del Paleoceno de Zamora. *Tecniterrae*, **36**, 1–16.

CORROCHANO, A. 1977. *Estratugrafía y sedimentología del Paleógeno de la provincia de Zamora*. Tesis Doctoral, Universidad de Salamanca.

—— & PENA DOS REIS, R. 1986. Analogías y diferencias en la evolución sedimentaria de las Cuencas del Duero, Occidental portuguesa y Lousa (Península Ibérica). *Studia Geológica Salmanticensia*, **22**, 309–326.

DORRONSORO, C., ALONSO, P. & RODRIGUEZ, T. 1988. La hidromorfía y sus rasgos micromorfológicos. *Anales de Edafología y Agrobiología*, Número Homenaje Prof. Kubiena, **47**, 243–278.

ERHART, H. 1967. *La génèse des sols en tant que phénomène géologique* 2nd Edition. Masson et Cie. Paris.

ESPEJO, R., MOLINA, E. & VICENTE, M. A. 1992. Mecanismos fundamentales de alteración sobre el Macizo Hercínico Ibérico. III Congreso Geológico de España y VIII Congreso Latinoamericano de Geología, Salamanca 1992, *Simposios*, **1**, 216–224.

GARCÍA ABBAD, F. J. & MARTÍN SERRANO, A. 1980. Precisiones sobre la génesis y cronología de los relieves apalachianos del Macizo Hespérico (Meseta Central española). *Estudios Geológicos*, **36**, 391–401.

GARCÍA TALEGÓN, J., MOLINA, E. & VICENTE, M. A. 1991. Weathering processes in granites. *7th Euroclay Conference, Dresden, Proceedings*, **2**, 405–409.

——, —— & —— 1994. Nature and characteristics of 1 : 1 phyllosilicates from weathered granite, Central Spain. *Clay Minerals*, **29**, 727–734.

GARZÓN, G. & LOPEZ, N. 1978. Los roedores fósiles de Los Barros (Avila). Datación del Paleógeno continental en el Sistema Central. *Estudios Geológicos*, **34**, 571–575.

——. UBANELI, A. G. & ROSALES, F. 1981. Morfoestructura y sedimentación terciarias en el valle de Ambés (Sistema Central español). *Cuadernos de Geología Ibérica* **7**, 655–665.

GOLBERY, R. 1980. Early dyagenetic Na-alunite in Miocene algal mats intertidal facies. Ras Sudar, Sinai. *Sedimentology*, **27**, 189–198.

GRANDIN, G. & THIRY, M. 1983. Les grandes surfaces continentales tertiaires des régions chaudes. Successions des types d'alteration. *Cahiers ORSTOM. Série Géologie*, **13**, 3–18.

JIMÉNEZ, E. 1972. El Paleógeno del borde S.O. de la

Cuenca del Duero. Los escarpes del Tormes. *Studia Geológica Salmanticensia*, **3**, 67–111.

LAPPARENT, J. DE 1930. Comportement minéralogique et chimiques des produits d'alteration élaborés aux dépens des gneiss du Massif Central français avant l'établissement des dépôts sédimentaires de l'Oligoène. *C.R. Academie des Sciences, Paris*, **190**, 1062–1064.

MARTIN SERRANO, A. (Ed.) 1988. *El relieve de la región occiental zamorana. La evolución geomorfológica de un borde del Macizo Hespérico*. Instituto de Estudios Zamoranos 'Florian de Campos', Zamora.

MILLOT, G. 1980. Les grandes aplanissements des socles continentaux dans les pays subtropicaux, tropicaux et desertiques. *In*: *Livre Jubilaire de la Societé Géologique de France*, Mém. H. Sér., **10**, 295–305.

——, BOCQUIER, G. & PAQUET, H. 1976. Géochimie des paysages tropicaux. *La Recherche*, **7**, 236–244.

MOLINA, E. & BLANCO, J. A. 1980. Quelques précisions sur l'altération du Massif Hercynien espagnol. *C.R. Academie Sciences, Paris*, **290**, 1293–1296.

——, ——, PELLITERO, E. & CANTANO, M. 1987. Weathering processes and morphological evolution of the Spanish Hercynian Massif. *In*: GARDINER, V. (ed.) *International Geomorphology 1986*. Wiley, Chichester, Part II, 957–977.

——, GARCIA GONZALEZ, M. T. & ESPEJO, R. 1991. Study of paleoweathering on the Spanish Hercynian basement. Montes de Toledo (Central Spain). *Catena*, **18**, 345–354.

NAHON, D. B. 1991. *Introduction to the Petrology of Soils and Chemical Weathering*. Wiley, New York.

RAYOT, V. 1994. Altération du centre de l'Australie: Role des solutions salines dans la genèse des silcrètes et des profiles blanchis. *Ecole de Mine de Paris, Memiore Sciences de la Terre*, **22**.

SAAVEDRA, J. & CAMAZANO, M. S. 1981. Origen de niveles continentales silicificados con alunita en el pre-Luteciense de Salamanca, España. *Clay Minerals*, **16**, 163–171.

SATO, M. 1960. Oxidation of sulfide ore bodies. 2. Mechanisms of oxidation of sulfide minerals at 25°C. *Economic Geology*, **55**, 1202–1231.

SIMON-COINÇON, R. 1989. Le rôle des paléoaltérations et des paléoformes dans les zocles: l'exemple du Rouergues (Mssif Central fançais). *Ecole de Mine de Paris, Memiore Sciences de la Terre*, **9**.

SOLÉ, L. & LLOPIS, N. 1952. *Geografía de España y Portugal, Vol. I: España. Geografía Física*. Montaner y Simon, Barcelona.

SCHWERTMANN, V. & FITZPATRICK, R. W. 1992. Iron minerals in surface environments. *Catena Suplement*, **21**, 7–30.

THIRY, M. 1993. Silicifications continentales. *In*: PAQUET, H. & CLAUER, N. (eds) *Sédimentologie et Géochimie de la Surface. Colloque á la memoire de George Millot*. Acad. des Sciences, Paris, 177–198.

—— & MILLOT, G. 1987. Mineralogical forms of silica and their sequence of formation in silcretes. *Journal of Sedimentary Petrology*, **57** (2), 343–352.

—— & MILNES, A. R. 1991. Pedogenic and groundwater silcretes at Stuart Creek opal field. South Australia. *Journal of Sedimentary Petrology*, **61** (1), 111–127.

VICENTE, M. A., MOLINA, E. & ESPEJO, R. 1991. Clays in paleoweathering processes: Study of a typical weathering profile in the Hercynian basement in the montes de Toledo (Spain). *Clay Minerals*, **26**, 81–90.

——, GARCIA TALEGON, J. IÑIGO, A. C., MOLINA, E. & RIVES, V. 1994. Weathering mechanisms of silicated rocks in continental environments. *In*: THIEL, M. J. (ed.) *Conservation of Stone and Other Materials. I*. Spon, London, 320–328.

Tertiary etchsurfaces in the Sudetes Mountains, SW Poland: a contribution to the pre-Quaternary morphology of Central Europe

PIOTR MIGOŃ

Department of Geography, University of Wrocław, pl. Uniwersytecki 1, 50-137 Wrocław, Poland

Abstract: This paper presents the evidence of the major part played by deep weathering accompanied, or followed, by stripping of a weathering mantle in the formation of Tertiary surfaces in the Sudetes. It also highlights the complicated nature of the topography of surfaces developed upon differentiated bedrock over protracted time-scales. The evidence is gained from the available data about weathering mantles and Tertiary sediments, and from the analysis of pre-Quaternary landforms such as rolling plains, scarps and inselbergs. A classification of the Tertiary surfaces, recognized here as etchsurfaces, is proposed for the Sudetes. The palaeosurfaces in the Sudetes are regarded as a part of an extensive etch surface which dominated the Tertiary landscape of Central and Northern Europe.

The Sudetes Mountains comprise the northeastern part of the Bohemian Massif, which itself is a very distinctive geological and morphological unit in the eastern part of the European Variscides. The massif encompasses a large part of the Czech Republic, together with parts of Poland and Germany (Fig. 1).

Old surfaces, presumed to be pre-Quaternary in age, have been recognized in the Sudetes since the

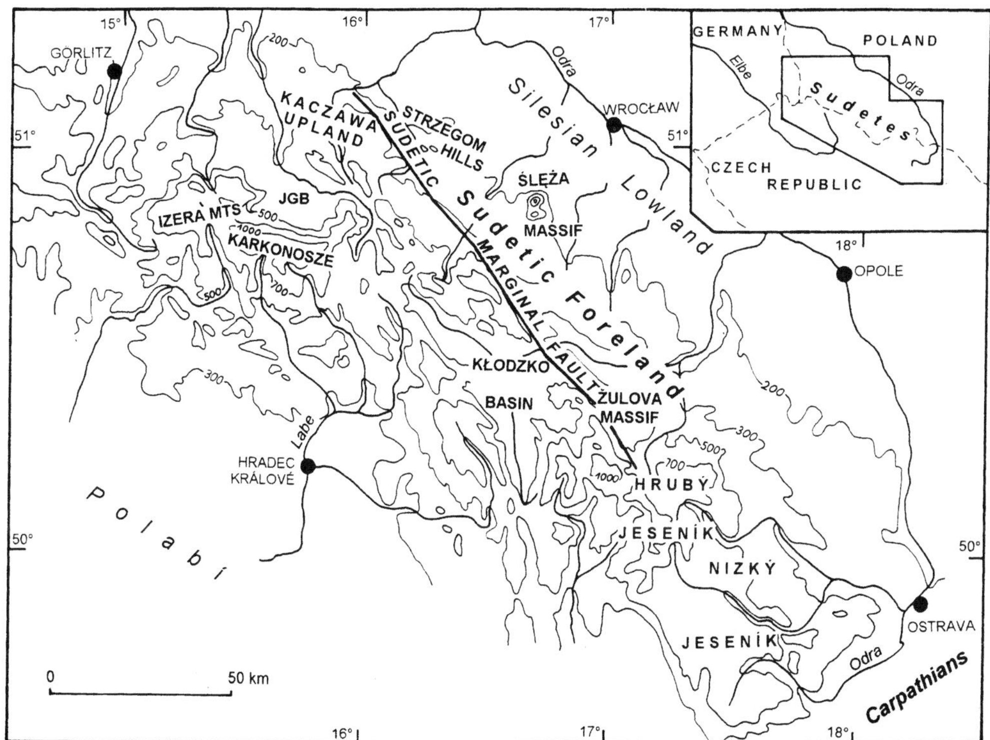

Fig. 1. General hypsometry of the Sudetes and the Sudetic Foreland and the key locations. JGB, Jelenia Góra Basin.

From Widdowson, M. (ed.), 1997, *Palaeosurfaces: Recognition, Reconstruction and Palaeoenvironmental Interpretation,* Geological Society Special Publication No. 120, pp. 187–202.

early twentieth century (Kral 1985). Major work has concentrated on recognition of the age of planation surfaces, which have subsequently been used to infer a scheme of denudation chronology (Jahn 1953; Szczepankiewicz 1954; Pernarowski 1963; Walczak 1968), rather than on the origin and environmental significance of these ancient landscapes. An important paper by Czudek & Demek (1970) directed attention towards geomorphic processes operating in Tertiary and pre-Tertiary times, and has emphasized the role of deep weathering, yet, surprisingly, these new ideas have been incorporated into relatively little modern work, the exceptions being Czudek (1977), Jahn (1980) and Ivan (1983).

The aim of this paper is twofold: firstly, it aims to provide comprehensive evidence of the deep weathering origin of pre-Quaternary morphological surfaces in the Sudetes; secondly, it aims to highlight the complicated nature of the topography of old surfaces, which is a direct result of structurally-controlled weathering and denudation over protracted time-scales.

Geological and geomorphological context

The Sudetes may be considered as a part of the Variscan basement of Central Europe. The main orogenic phase, which involved intensive faulting, folding, thrusting and metamorphic processes, occurred in late Devonian to early Carboniferous times (Don & Żelaźniewicz 1990). This was followed by a very long period of relative tectonic quiescence, lasting throughout the entire Mesozoic and Palaeogene, yet interrupted by minor faulting and warping at the turn of the Cretaceous and Palaeogene (Laramian phase). However, during the Neogene and Quaternary, many faults of Palaeozoic age underwent a considerable reactivation and this has led to differential uplift of individual parts of the Sudetes (Oberc 1977; Dyjor 1986). Consequently, the Sudetes today consist of numerous horsts, domes and grabens of different sizes and altitudes. The Sudetic Foreland, situated to the northeast of the Sudetes (Fig. 1), is of special geomorphological interest, since the Cenozoic uplift affecting this region has been relatively minor and hence older topography may have been preserved.

Lithologically, the Sudetes are a very complex area (Fig. 2). Crystalline rocks of Precambrian and Palaeozoic age comprise the major geological structure of the Sudetes. These are mostly gneisses, mica schists, greenschists, phyllites and gabbro, into which several granitoid bodies were intruded in the Variscan post-orogenic stage; the latter are strongly diversified in terms of their petrology, geochemistry and joint pattern. Large intramontane depressions generated during the Variscan movements, i.e. the Intrasudetic Trough and the North Sudetic Trough (Fig. 2), were subsequently infilled by terrestrial Carboniferous–Triassic conglomerates, sandstones and mudstones. The youngest rock formation occurring within the Sudetes region consists of Upper Cretaceous shallow marine sandstones and marls, though these sediments were again confined to pre-existing Variscan troughs and never occupied the whole Sudetes area (Teisseyre 1960) (Fig. 2). The Upper Cretaceous transgression was characteristic for large areas of the Bohemian Massif and forms part of the global sea level rise which occurred at this time. Since the Upper Cretaceous, the Sudetes have undergone an entirely subaerial evolution and today comprise a denuded area from which the older rocks have been progressively eroded away (Teisseyre 1960; Jahn 1980; Gilewska 1987).

Tertiary climate data indicates that the climate of the Sudetes was probably warmer and more humid than it is today (Gilewska 1987; Głazek & Szynkiewicz 1987; Sadowska 1977; Stuchlik 1980); a view which accords well with the general palaeoenvironmental reconstructions made for Central Europe (Buchardt 1978) and the Northern Hemisphere (Wolfe 1980). Although many details of European Tertiary palaeoclimates are not yet fully known, or else remain controversial, it is difficult to determine precisely climate conditions at this time, however it seems likely that for most of the Palaeogene it was characterized by alternating wet and dry seasons (i.e. subtropical savanna type). A slight deterioration of climate in the Miocene resulted in more Mediterranean-type climates, with longer period of seasonal dryness possibly verging toward semi-arid conditions during the Late Miocene (Głazek & Szynkiewicz 1987). In the Pliocene, the climate was warm-temperate, not dissimilar to that of the present-day (Sadowska 1987).

Palaeosurfaces in the Sudetes

The oldest palaeosurface identified in the Sudetes is that onto which the Upper Cretaceous sea transgressed during Cenomanian times. This surface has been found preserved in its original form beneath the Cenomanian sediments and investigation shows (Don 1989) that its relief was extremely low (not exceeding 20 m). However, exhumed inselbergs have been reported rising above this surface in the vicinity of Prague (Kral 1985). The sub-Cretaceous surface is a part of an extensive pre-Upper Cretaceous surface which characterizes the northern part of the Bohemian Massif (Demek 1982; Kral 1985). Since numerous post-Cretaceous faults occur, the original extent of

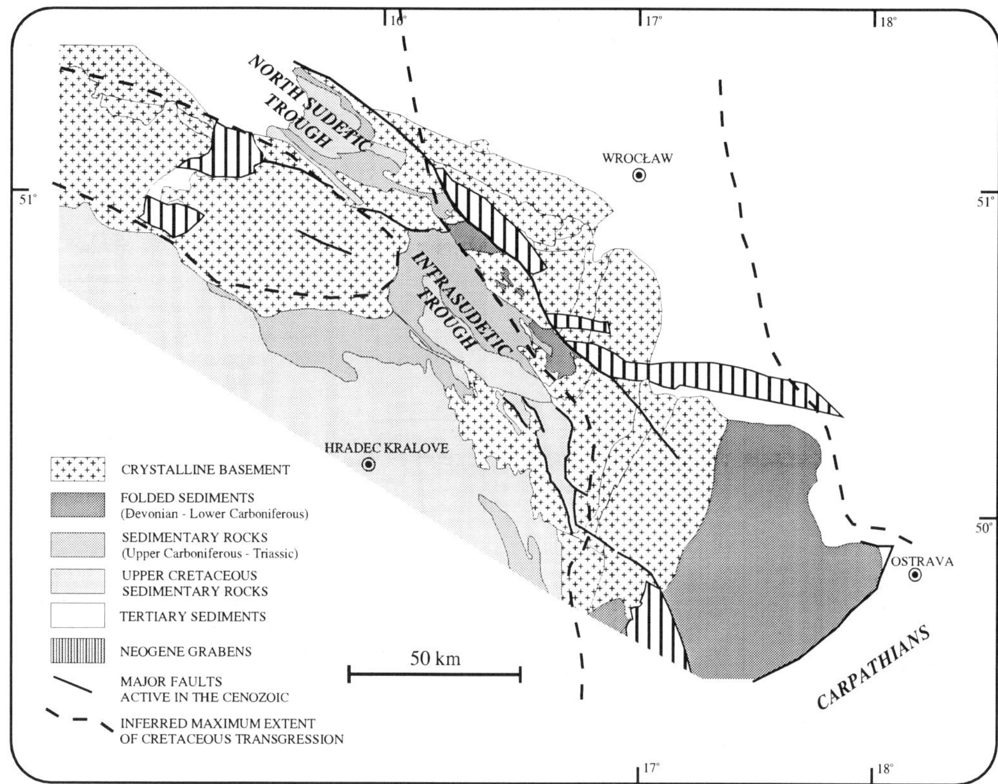

Fig. 2. Simplified geological map of the Sudetes and the Sudetic Foreland.

this surface cannot reliably be projected or reconstructed from the current limited outcrop of the Cretaceous sediments.

The main palaeosurface in the region is the Tertiary surface, often referred to as the ‘Palaeogene planation surface’ (Jahn 1953, 1980; Klimaszewski 1958; Walczak 1968). This, again, can be found throughout the whole Bohemian Massif (Demek 1982; Kral 1985). Since a very advanced stage of planation has been assumed for this surface, the more differentiated relief developed upon it has been assumed to be the result of subsequent erosion. Consequently, only flat or almost flat areas have been regarded as relics of Tertiary relief. Several surfaces at different altitudes have been recognized in various parts of the Sudetes and different ages for them have been proposed (Szczepankiewicz 1954; Klimaszewski 1958; Dumanowski 1961; Pernarowski 1963; Walczak 1968; Don 1989). However, the investigations which have recognized this variety of surfaces have failed to take into account the complexity of Neogene tectonics, the changing nature of the geomorphic processes operating during the Tertiary and especially geomorphological consequences of deep weathering in a warm climate.

Several surfaces of local significance have been also reported: these include horizontal surfaces related to lithology and structural planes within the Upper Cretaceous clastic sediments which give rise to a stepped topography of sandstone plateaux (Pulinowa 1989). These surfaces are obviously of Tertiary–Quaternary age, but correlation with equivalent denudational surfaces developed upon the older crystalline rocks is unclear. This may be due to the operation of structurally-controlled processes such as scarp retreat and large-scale piping preventing any correlation being established (Pulinowa 1989).

Subslope benches up to 3 km wide commonly occur along major valleys and, although the importance of Pleistocene cryopedimentation in their formation has been stressed (Czudek & Demek 1971; Demek 1980; Czudek 1988), the occurrence of Pliocene sediments and weathering profiles clearly point to a Late Tertiary age for these landforms (Jahn *et al.* 1984). In addition, since these benches are apparently cut into the main surface (surfaces), the Tertiary age of the latter is further confirmed.

Tertiary surface in the Sudetes

Age

In the absence of datable sediments any discrimination between Tertiary surfaces of different ages remains highly speculative. Thus, by contrast to previous studies, no attempt is made here to establish strict time constraints for the formation of the Tertiary surfaces. Nevertheless, it can be strongly argued that extensive Tertiary surfaces did form during Palaeogene and Early/Middle Miocene times because they are certainly older than the residual tectonic relief. This relief consists of mountain fronts, hundreds of metres high, grabens and horsts, the formation of which began in the Middle Miocene and continued during the Pliocene and Early Quaternary. Ensuing phases of endogenic activity are well documented in the sedimentary record of the Neogene (Dyjor 1986) and features such as elevated surfaces of low relief, truncated valleys and drainage pattern changes provide good morphological evidence for the preferential development of tectonic rather than denudational relief in post-Middle Miocene times (Czudek 1977; Migoń 1993*b*).

In the Sudetic Foreland the denudational surface is mantled by clastic deposits of Miocene and Pliocene age (Sadowska 1977; Ciuk & Piwocki 1979; Dyjor 1986). The Miocene age of sediments is demonstrated by the occurrence of palaeobotanically-constrained brown coal seams which are intercalated with more widespread sands, silts and clays (Sadowska 1977). Similarly, the 'Lower Silesian Basaltic Formation' was erupted onto the main denudational surface; dating of this formation by the K–Ar method at 28–2 Ma (Birkenmajer *et al.* 1977) provides an upper age bracket of Oligocene–Pliocene for the surface.

Weathering mantles

Weathering products of pre-Quaternary age are well known in the Sudetic Foreland and, to a lesser extent, in the Sudetes. They developed on a variety of lithologies, including granites (Kužvart 1965; Franz 1969; Borkowska & Czerwiński 1973; Budkiewicz 1974; Kural 1979; Migoń & Czerwiński 1994), gneisses (Kościówko 1982), serpentinites (Niśkiewicz 1967) and basalts (Kozłowski & Parachoniak 1960; Stoch *et al.* 1977; Dyjor & Kościówko 1986). The thickness of these weathering profiles may attain 60–100 m, but the depth of weathering is often highly variable even over short distances, as proven from borehole data (Kural 1979; Niśkiewicz 1982).

The most advanced alteration, irrespective of bedrock lithology, is reported from the Sudetic Foreland (Niśkiewicz 1967; Budkiewicz 1974; Kural 1979; Kościówko 1982) where granite has been decomposed to a highly kaolinitic clay (kaolinite content up to 96%), 30–60 m thick, whilst weathered basic rocks give rise to nickel-bearing red earths in which hydrosilicates and clay minerals are a significant component. Tertiary volcanics and sediments provide a significant age constraint for advanced alteration because they are Miocene in age and, hence, deep weathering mantles must be mostly of Palaeogene age. This age is further demonstrated by the fact that the depth of weathering profiles developed upon the Middle and Late Miocene basalts is much less than profiles reported from older rocks (Stoch *et al.* 1977), and the post-depositional kaolinization of Miocene clastic sediments around the Strzegom Hills in the western part of the Sudetic Foreland is relatively insignificant (Kural 1979).

Preservation of Palaeogene weathering mantles is largely due to the protective cover afforded by overlying Neogene sediments and volcanic rocks; as a consequence these mantles are only occasionally exposed on the topographic surface (Fig. 3). Generally, the Palaeogene mantles are best preserved in those areas where either Late Cenozoic uplift has been relatively small or in regions which have undergone a relative subsidence. As a result, the greatest reported depths of clay weathering (> 50 m) have been reported from downfaulted blocks (Kužvart 1965; Kural 1979) (Fig. 3).

In the more mountainous areas, preservation of advanced chemical alteration is exceptional; instead, relatively shallow profiles of gruss-type weathering are the norm (Borkowska & Czerwiński 1973; Migoń &Czerwiński 1994). These latter are usually 5–10 m deep, display little sign of chemical changes and a predominance of the gravel- and sand-size fraction (up to 98%). Granitic grusses are more or less evenly disintegrated and features like corestones and abrupt boundaries between weathered and non-weathered rock are rare. The geochemistry of grusses is strongly controlled by the primary chemical composition of the parent rock (Migoń & Czerwiński 1994).

The gruss weathering pattern is apparently controlled by topography, since profiles tend to thicken in places where the availability of water is greater (e.g. at the footslope and along valley sides). The scarcity of colluvial deposits indicates that little of the original profiles have been destroyed due to slope processes.

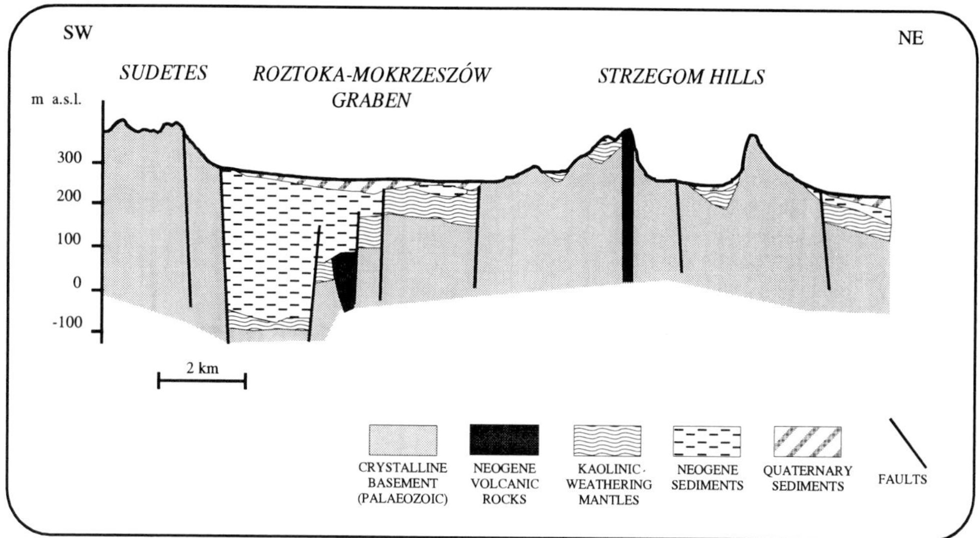

Fig. 3. Cross-section across the Strzegom Hills in the Sudetic Foreland, to illustrate relationships between bedrock, weathering mantles, Neogene sediments and volcanic rocks (after Kural 1979, simplified).

It is difficult to place the gruss weathering in the context of geologic time. Clearly, some grusses must predate the Pleistocene since they are overlain by sediments left behind by the Pleistocene (Elsterian, Saalian) ice sheets which advanced from Scandinavia (Jahn 1960). However, in other locations the sediments resting upon grusses are Late Glacial slope deposits and Holocene peat bogs which could indicate the grusses continued to form during Pleistocene times (Jahn 1968; Migoń & Czerwiński 1994). The thickness of grusses in those localities affected by ice-sheet erosion is, however, very limited (e.g. 1–2 m). Therefore, the depth and intensity of gruss weathering is regarded to be a function of time, topography and bedrock lithology rather than it is of any palaeoclimatic significance (Migoń, Czerwiński 1994).

Thus, it appears that weathering mantles in the Sudetes and the Sudetic Foreland may be subdivided into two main groups. Firstly, the older weathering mantles which are directly related to the formation of the Tertiary surface: these continued to form until the Miocene and are products of deep, chemical and selective weathering over a long timespan. Secondly, the gruss type which began to evolve in response to changes in the weathering regime to predominantly physical weathering: these are typical of the Pliocene and Quaternary and their occurrence is controlled by topography, hence, gruss weathering appears to be younger than the main features of regional relief.

The spatial restriction of advanced weathering to the lowland areas, graben and other areas less prone to denudation, suggests that such deep alteration was much more widespread in the past. However, equivalent weathering profiles in the more elevated areas have either been significantly truncated or else completely stripped off due to tectonics and erosion. Features of Neogene sediments around the Sudetes, treated in more detail below, provide the confirmation of the statement above.

Neogene sediments

Neogene deposits are widespread around the Sudetes, attaining a maximum thickness of 300–400 m in the Tertiary tectonic grabens (Dyjor, Kuszell 1977; Dyjor *et al.* 1978; Kural 1979; Kasiński & Panasiuk 1987; Růžička 1989; Ciuk *et al.* 1992). These sediments rest directly on weathered bedrock and older Palaeogene deposits are not known around the Sudetes. In addition, there are also some scattered outcrops of Tertiary sediments inside the Sudetes (Zimmermann 1937; Jahn *et al.* 1984), regarded as Pliocene in age.

The Neogene sediments are predominantly terrestrial clastic deposits, with some organic accumulation in the Miocene (Sadowska 1977). Their source was the elevated block of the Sudetes which had been subject to uplift since the turn of the Palaeogene/Neogene and, consequently, the pattern of sedimentation and the composition of sediments was directly influenced by the topography and the geological structure of particular areas. This relationship is well documented in the Strzegom Hills, in the western part of the Sudetic Foreland (Dyjor & Kuszell 1977; Kural 1979) (Fig. 3). The area is separated from the margin of the Sudetes by the deep tectonic Roztoka–Mokrzeszów graben and, close to the margins of the graben, coarse sedimentation took place resulting in a system of alluvial fans and debris flows. However, the predominating sediments were kaolin clays and loams which accumulated in shallow lakes or on alluvial plains and may be well dated palynologically (e.g. from brown coal/lignite) and from the occurrence of intercalated volcanic tuffs. These age data demonstrate that the main redeposition of clay weathering products in the area of the Strzegom Hills took place in the Middle Miocene (Kural 1979). However, further to the east the main redeposition is dated on the basis of similar evidence as early Late Miocene (Ciuk & Piwocki 1979) or even Pliocene (Růžička 1989).

Sudetes Neogene sediments typically consist of allogenic clay minerals, kaolinized feldspars and quartz (Kural 1979; Kościówko 1982; Dyjor 1986), which together indicate an advanced stage of alteration in the source areas. By contrast, Pliocene sediments are mostly quartzose gravels and sands (Dyjor 1966) with relatively little silt and clay, and this composition is interpreted as being the result of deep bedrock incision which followed uplift of the whole area. In general, the sedimentary record of the Tertiary testifies to little denudation during the Palaeogene which was then replaced by a phase of intensive stripping of weathering products and bedrock exposure in the Neogene.

Landforms

Although the whole area of the Sudetes is regarded as an area of inherited Tertiary relief (Jahn 1980), the Late Neogene tectonics and exogeneous processes operating during the Quaternary may have significantly changed Tertiary landform assemblages. Therefore, it is probable that the relief of palaeosurfaces can be recognized in detail in relatively few areas. Observation reveals that Tertiary landscape elements are best preserved within tectonically subsiding blocks, such as intramontane basins and the Sudetic Foreland, and upon those elevated surfaces which have been subject to little subsequent incision. The latter dominate over the western and easternmost part of the Sudetes. Typical pre-Quaternary landforms of the Sudetes are undulating plains, escarpments and inselberg-like hills.

Undulating plains. True plains are very rare features in the Sudetes, and old landscape remnants tend to be either undulating (rolling) plains or hilly areas. Undulating plains are typically developed upon bedrock lithologies such as fine-grained granite, gneisses and mica schists, phyllites and greenschists. In the easternmost part of the Sudetes (i.e. Nizky Jesenik Massif), there occur extensive surfaces of low relief developed upon Lower Carboniferous conglomerates, sandstones and mudstones (Czudek 1977). Although elevations within these plains have gentle slopes, normally < 10°, they may attain a height of 80–120 m. They are separated by broad trough-like valleys, the pattern of which often does not accord with the present-day drainage pattern. The relief of undulating plains when looked at in very great detail displays lithological control; e.g. the rolling landscape of the Izera Foreland consists of hills built of granitogneisses, whilst footslopes and valleys are underlain by laminated gneisses and mica schists.

The occurrence of remnants of Oligocene/Miocene basalt plateaux, resting on undulating plains in the Western Sudetes and in the Sudetic Foreland, suggests a pre-Neogene age for those surfaces of low relief (Pernarowski 1963; Walczak 1968). Indeed, deep dissection of basaltic plateaux and neighbouring plains, clearly related to Neogene uplift of the Sudetes (Pernarowski 1963; Oberc & Dyjor 1973), confirms an essentially Early Tertiary age for the plains. In numerous places, deep chemical weathering of pre-basaltic bedrock or of the oldest generation of lavas has been revealed, while younger basalts have been only slightly weathered (Berg 1935; Kozłowski, Parachoniak 1960; Kural 1979).

Escarpments. Escarpments of various heights, lengths and geomorphic expression occur in all parts of the Sudetes, though they are much less developed in the Sudetic Foreland. These escarpments separate different lithologies and as such are typical structure-related landforms testifying to the differing resistance of particular rock complexes. Perhaps the most typical settings for their development are at the margins of intrusive bodies, especially granitoid intrusions (Figs 4 and 5). In most cases, marginal escarpments face inward toward the granitoid masses (Dumanowski 1963). The best developed marginal escarpments occur where the country rock has been subject to contact metamorphism and has been altered to highly resistant hornfelses and andalusite schists: such scarps reach 200–400 m in height. Another characteristic setting for escarpments are margins of Laramian grabens, which preserve inliers of Permian rocks. The Permian sandstones and conglomerates typically form the lower ground whilst the margins, built of Early Palaeozoic metavolcanic rocks, are 50–100 m high. Since there are no

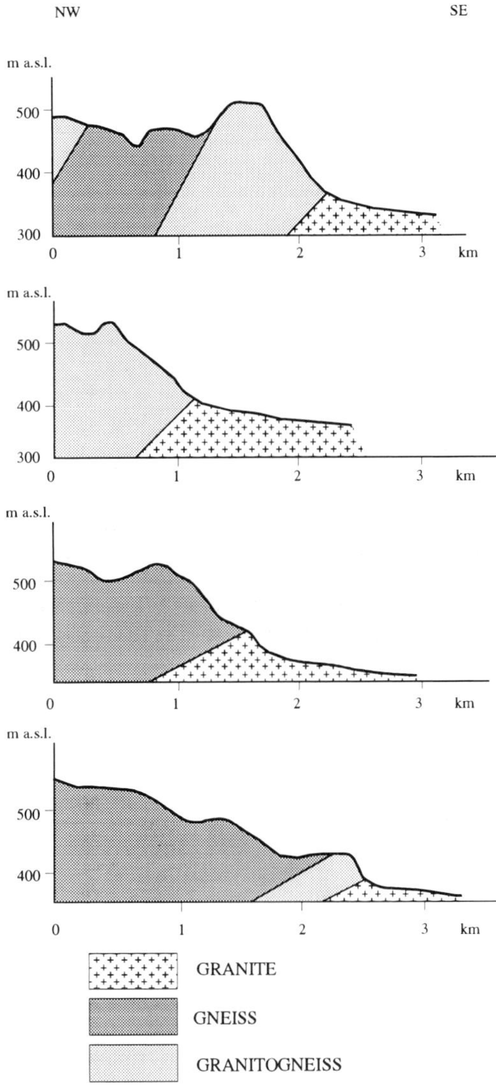

Fig. 4. Cross-sections of the lithologically-controlled western margin of the Jelenia Góra Basin.

obvious signs of young tectonics along these graben margins, the escarpments themselves are apparently the result of selective denudation and thus essentially fault-line scarps.

A very characteristic feature of the escarpments is their close adjustment to lithological boundaries (Fig. 4) which results in the different appearance of different scarps. Some scarps are sinuous while others retain straight courses over very long distances. The absence of features such as embayments and outliers, which are characteristic of retreating scarps, indicates that little, if any, scarp retreat has occurred in this region. Thus, scarps in crystalline areas may be regarded as features due to lithologically-controlled downwearing and indicate significant selectivity of geomorphic processes during the Tertiary. However, there are distinct signs of back-wearing in areas developed upon Permo-Mesozoic sedimentary rocks, which suggests a continuous development of the surfaces underlain by the Permian and Mesozoic sediments and their diachronous age (Pulinowa 1989; Tułaczyk 1992). Consequently, it is impossible to propose any strict correlation with the principal Tertiary surface developed upon basement rocks.

Inselbergs. Isolated hills (inselbergs) characterize Variscan granitoid massifs in both the Sudetes as well as in the Sudetic Foreland (Gellert 1931; Demek 1964; Ivan 1983; Migoń 1992, 1993*a*) (Fig. 6). They are also known to occur in gabbro, serpentinites and gneisses of the Sudetic Foreland (Gellert 1931; Pernarowski 1963). Most inselbergs are apparently lithologically or structurally controlled, being related to either petrologic differences or joint pattern; the latter is especially the case with granitic inselbergs. However, in the case of some isolated hills in the Sudetic Foreland there appear to be hardly any differences in rock characteristics, yet inselbergs have formed nevertheless.

Granitic inselbergs in the Jelenia Góra Basin and Žulova Massif provide the best examples of lithological control on inselberg formation. These striking features are rocky hills built of fine-grained granite, rising to 10–200 m in height, surrounded by a rolling plain which is itself underlain by a more porphyritic variant of the granite. Similarly, protrusions of equigrained granite within porphyritic granite give rise to hills typically 50–100 m high (Fig. 6).

Structural control is exerted by the dome-like pattern of joints and this results in dome-shaped hills up to 100 m high, which may be regarded as bornhardts (Migoń 1993*a*). Sheeting, which consists of widely spaced arcuate joints, is clearly visible on hillslopes. However, in the present state, these inselbergs are block-strewn rather than domical inselbergs, which indicates a considerable subaerial degradation over protracted periods of time (Fig. 7). Low elevations, 20–30 m high, termed whalebacks or ruwares (Thomas 1974*a*,*b*; Twidale 1982), are even more common in the granitoid areas. They again appear to be structurally controlled, being related to either a lower density of the orthogonal joint set or minor domes. Outcrops of finer granite amidst a coarser variant tend to form low hills too.

Fig. 5. Western margin of the Jelenia Góra Basin. The flat area in the foreground is underlain by granite whilst the scarp face is built of gneisses and granitogneisses (cf. Fig. 4).

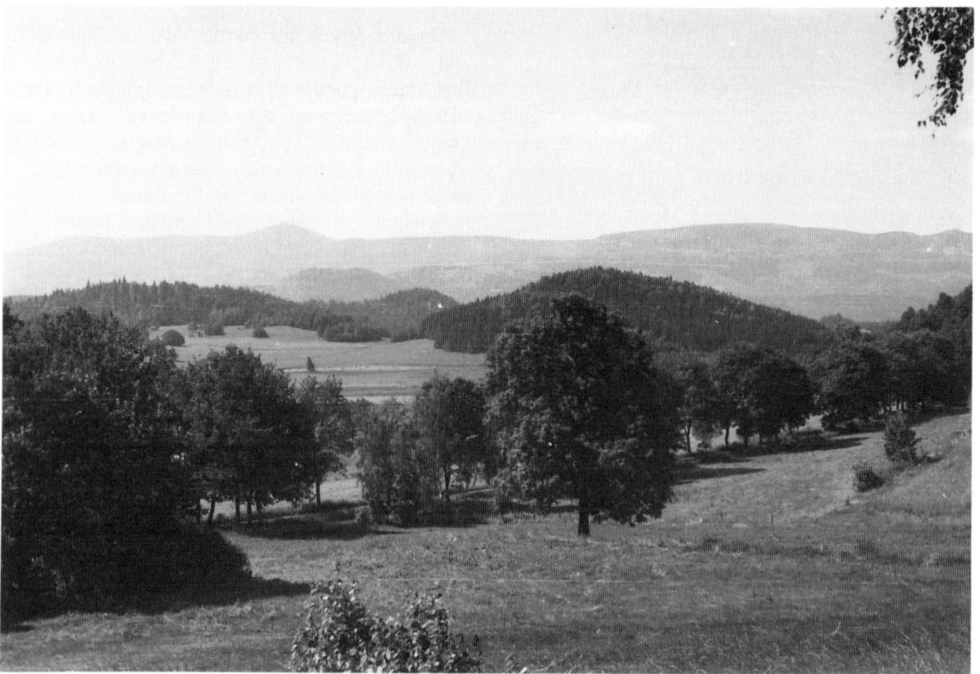

Fig. 6. Hilly granite landscape of the Jelenia Góra Basin, individual hills being joint or lithologically controlled. On the skyline the rolling summit plateau of the Karkonosze Mountains can be seen.

Fig. 7. One of the block-strewn inselberg-like hills in the Jelenia Góra Basin. Extensive boulder fields testify to considerable remodelling of a primary dome.

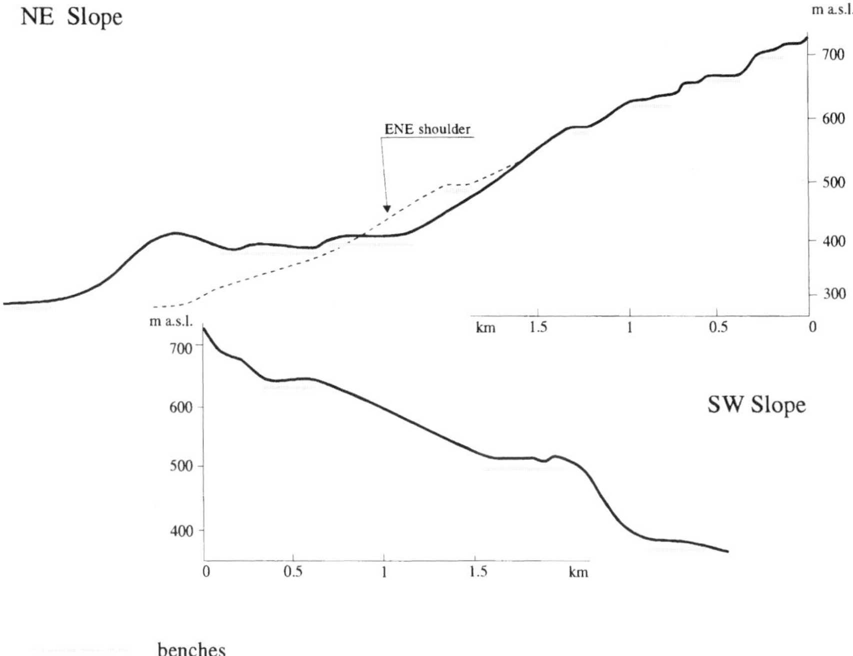

Fig. 8. Vertical distribution of benches on the slopes of Mt Ślęża inselberg, Sudetic Foreland.

Gabbro and serpentinite inselbergs in the Sudetic Foreland are characterized by considerable height, rising to 500 m high in the case of Mt Ślęża (Fig. 8). Since they are neither horsts nor constructive features related to past volcanism, they must represent landforms due entirely to denudation. Furthermore, they must be landforms of considerable antiquity, as demonstrated by the occurrence of Miocene sediments, including brown coal, at their base (Gellert 1931). A remarkable feature of their hillslopes is a stepped profile, best exemplified on the slopes of Mt Ślęża (Fig. 8). Hillslopes consist of alternating benches and steeper, often rocky, segments 30–100 m high. The most distinct benches are arranged concentrically around the summit. Although these benches are covered by Pleistocene frost-shattered debris, thick chemical weathering has been reported to occur beneath this debris (Szczepankiewicz 1958).

Close adjustments of inselbergs and related hills to characteristics of the underlying bedrock is an important indicator of selectivity of geomorphic processes. The presence of inselbergs and their associated features in the Sudetes indicates downwearing as the dominant mode of landscape evolution in this region.

The origin of the Tertiary surface in the Sudetes

The wide occurrence of Palaeogene–Miocene weathering mantles, features of Neogene sediments, and strong structural control of landforms provide the evidence that the main Tertiary surface in the Sudetes has been formed as a result of protracted deep weathering and subsequent removal of a weathering mantle. Thus, its origin is viewed here in the context of etch planation which applies to 'landforms which may have been produced by one or more distinct episodes of stripping of a relict saprolite mantle' and 'landforms continuing to evolve by surface erosion acting on pre-weathered material' (Thomas & Thorp 1985). These landforms are termed *etchplains* or *etchsurfaces* and this descriptive distinction simply reflects the range of altitudinal differences within the etched surfaces.

The idea of the significant part played by deep weathering was first expressed in the 1930s (Berg 1935) but the modern terminology of surfaces originating due to etch planation (Thomas 1974*a*, 1989) has been rarely adopted to the Sudetes subsequently (Demek 1976), despite the fact that

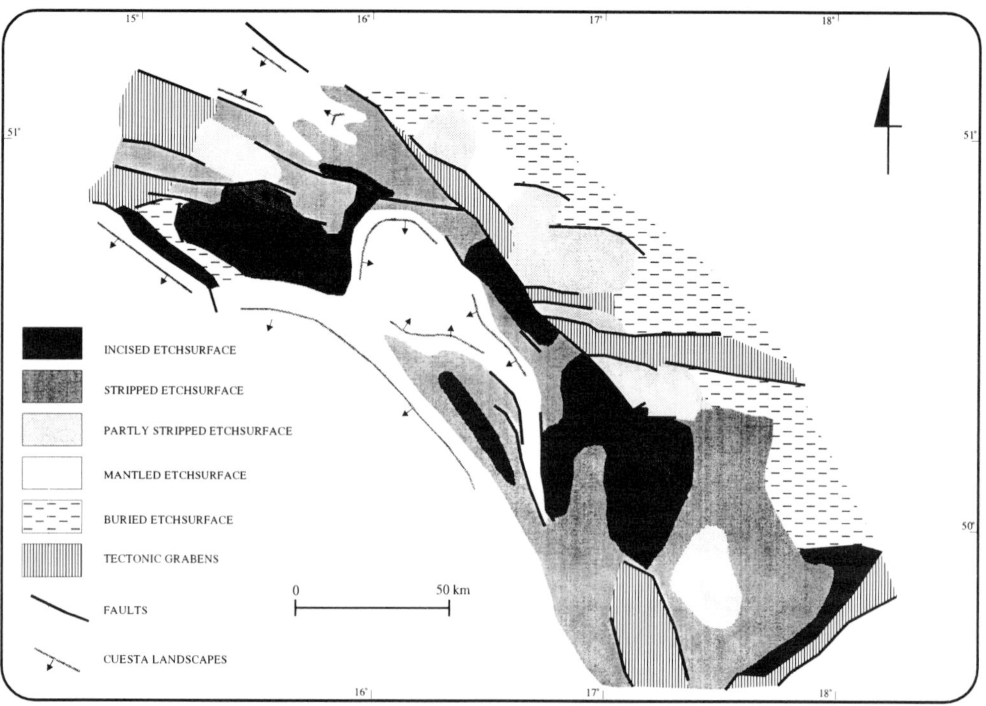

Fig. 9. Regional distribution of etch surfaces in the Sudetes and the Sudetic Foreland. Compare with Figs 1 and 2 to see relations to hypsometry and geology (lithology).

the great geomorphic role of deep weathering has indeed been acknowledged (Jahn 1980).

The variety of landscapes and the differing degree of preservation of ancient weathering mantles allows a more detailed classification of Tertiary surfaces to be proposed for the Sudetes. The proposal classification basically follows that for etchplains proposed by Thomas (1989). However, most of the old surfaces in the Sudetes are not etchplains but etchsurfaces, since they have considerable relief and are hardly real plains. The spatial distribution of etchsurfaces is directly related to the history of Late Tertiary tectonics (Fig. 9), since it has been uplift that has caused stripping of older regoliths and the dissection of surfaces of low relief.

In the most elevated parts of the Sudetes, such as the Karkonosze Mountains, Izera Mountains or Hruby Jesenik, incised etchsurfaces form elevated plateaux of low relief (Figs 9 and 10A). They are deeply dissected by valleys, undercut by glacial cirques and broken up by faults. The major part of the Sudetes is occupied by stripped etchsurfaces from which the whole thickness of Tertiary regolith has been eroded away (Fig. 10B and C). As an effect of stripping the configuration of the weathering front has been exposed and this may be seen to consist of structurally or lithologically

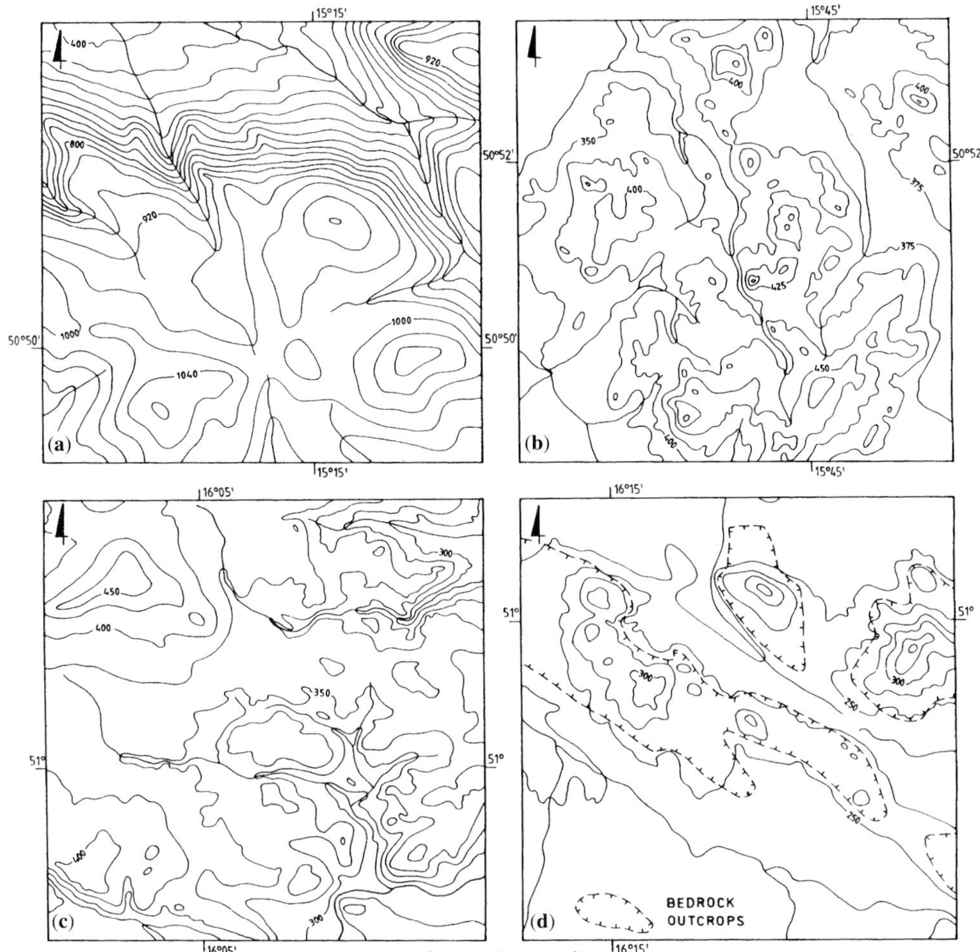

Fig. 10. Contour maps of selected etch surfaces of the Sudetes and the Sudetic Foreland. (**a**) Incised etch surface of the Izera Mts (fine- to medium-grained granite); (**b**) stripped etch surface of the Jelenia Góra Basin (coarse-grained granite), (**c**) stripped etch surface of the Kaczawa Upland (greenschists); (**d**) partly stripped etch surface with scattered inselbergs of the Strzegom Hills (medium-grained granite).

controlled hills, intervening trough valleys and basins and only rarely plains (Fig. 10B). Classic examples of stripped etchsurfaces are granite multiconvex cupola landscapes (the name proposed by Thomas 1974*b*), which are characteristic for the Žulova Massif in the Sudetic Foreland (Demek 1964, 1976; Ivan 1983) and for the Jelenia Góra Basin in the Western Sudetes (Migoń 1992, 1993*a*). Those illustrate well the complex bedrock control upon landscape development. There is a certain hierarchy in this control, since the pattern of higher and lower grounds relates to petrology and geochemistry of granite masses, while location, shape and relief details of individual hills are directly influenced by joint pattern (Demek 1964; Migoń 1993*a*).

The Nizky Jesenik in the Eastern Sudetes (Figs 1 and 9) provides an example of a partly stripped etchsurface which has a very low relief consisting of broad, flat-topped elevations, shallow basins and wide valleys. However, basins and valleys are known to be underlain by deep sandy weathering of local Carboniferous sandstones and slates (Czudek 1983). Partly stripped etchsurfaces are typical for the Sudetic Foreland, and for inselberg landscapes especially (Fig. 10D). Between inselbergs there are pockets of chemically-weathered rocks which are tens of metres deep. Mantled etchsurfaces occupy relatively small areas and their occurrence is confined to the subsiding parts of the Sudetic Foreland (Fig. 9). The existence of mantled etchsurfaces with an irregular weathering front has been proved by boreholes (Kural 1979, Niśkiewicz 1982). This weathering mantle is only occasionally pierced by fresh rock outcrops.

Wider implications for palaeomorphological research

Relief of Tertiary etchsurfaces

The main contribution of etch planation theory for palaeomorphological reconstructions is that it allows reappraisal of old landscapes which are neither plains or planation surfaces, but instead may exhibit considerable relief. The Sudetic examples illustrate that only a few lithological settings favour formation of true plains. In most cases, selective weathering followed by a period of mantle stripping has resulted in hilly or inselberg-like areas. Higher elevations indicate locations where, for different reasons (i.e. lithology, geochemistry, and topographic setting), the rocks have been more resistant to weathering. Differential etching also accounts for the existence of lithological escarpments. The Neogene surface comprised different types of etchsurfaces, including mantled, partly stripped, stripped and incised etchsurfaces because landform changes through time have consisted of continuous transformations of one type of etchsurface into another. However, climatic deterioration and increasing tectonic activity towards the end of the Tertiary have resulted in extensively stripped etchsurfaces developing. Consequently, Tertiary relief is, in most cases, an exposed weathering front which may not have form any flat surface (Fig. 10).

Alternating phases of etching and tectonically- or climatically-induced stripping may result in an increase of relief through time (e.g. Twidale 1991; Thomas 1994*a*,*b*). Indeed, the height of inselbergs and scarps in the Sudetes, which occasionally rise to 300–500 m, cannot be explained in terms of simple replacement of the etching regime by a phase of removal of the weathering products. Altitudinal differences testify to a very long period of etch planation which has perhaps led to plain formation in one place yet to considerable relief development in another. Since inselbergs and scarps are inherent and coeval elements of the Tertiary surface, there is no reason to consider them separately in palaeoenvironmental research and it is argued here that inselberg landscapes and true planation surfaces are compartments of the same palaeosurface. Similarly, it may be argued that scarps do not necessarily separate surfaces of different ages, but rather surfaces formed in response to etching/stripping processes acting upon varying lithologies and structures. In this manner, ideas of dynamic etch planation, developed in low latitudes (Thomas 1989), may also be successfully applied to explain Tertiary relief in mid and high latitudes.

Significance of inselbergs

Inselbergs and related hills, such as whalebacks, ruwares and castle koppies, are useful indicators of structural control upon development of landscape. Moreover, since they are essentially denudational landforms, they may indicate the degree and, perhaps, mode of surface lowering. The location of inselbergs in the Sudetes is clearly influenced by contrasting lithologies or else spatial variation in joint pattern. The most characteristic lithological settings for inselberg formation are the inliers of fine-grained granite within a mass of porphyritic granite (e.g. Jelenia Góra Basin), serpentinite outcrops (e.g. Sudetic Foreland) and scattered outcrops of contact-metamorphosed country rock within granitoid bodies (e.g. Eastern Sudetes). Jointing, in turn, exerts its influence on the shape of inselbergs and causes them to develop as bornhardts, whalebacks or castle koppies. A domical pattern of fractures not only controls the appearance of hills but also influences the pattern of subaerial degradation and, so, the majority of inselbergs in the Sudetes may be considered to be 'inselbergs due to resistance' (Büdel 1957).

However, there are also isolated hills which appear to be neither lithologically nor structurally controlled. Furthermore, they are not associated with any escarpments and so cannot be regarded as 'inselbergs due to position' (Büdel 1957). Hillslope benches are relatively common minor features on these isolated hills and so it seems plausible that these particular inselbergs have been developing according to the model of episodic exposure as outlined by Twidale & Bourne (1975). The gradual exposure of inselbergs is facilitated by the juxtaposition of two contrasting environments: the bare, and thus dry, slopes of bedrock elevations and the moist regolith around then. This model is supported by the fact that in the Sudetic Foreland deep weathering profiles are still observed on the footslopes of these inselbergs.

In general, inselbergs in the Sudetes are either more resistant compartments in former deep weathering profiles (as described above) or have come into being due to diversified weathering controlled by topography and related moisture availability. They prove that the dominant mode of landscape evolution has been downwearing, consisting of alternating episodes of etching and stripping, and that relief may have increased following etchsurface development. Contrasts between hills and plains have become accentuated through time. The considerable height of some inselbergs, which greatly exceed typical depths of deep weathering, clearly suggest a multistage development. Since inselbergs have been proved to be at least pre-Pliocene features (Gellert 1931; Pernarowski 1963; Migoń 1992), they may bear testament to a very long subaerial development, perhaps even reaching beyond the Mesozoic/ Tertiary boundary.

Palaeomorphological context

The identification of etchsurfaces in the Sudetes fits well to the known palaeoenvironmental reconstructions of the Tertiary in Central and Northern Europe. The theoretical basis for etchplain development was first outlined by Büdel (1957) and subsequently by Bakker & Levelt (1964), though few detailed studies of Tertiary etchplains followed these conceptual studies. The reasons for this might be the assumed scarcity of older weathering mantles together with a common belief that pre-Quaternary morphology was largely obliterated by powerful Scandinavian ice sheets during the Pleistocene.

However, numerous detailed studies of perfectly preserved stripped etchsurfaces in higher latitudes have appeared in the last decade. etchsurfaces, and more locally etchplains, have been described in Sweden (Lidmar-Bergström 1982, 1988; Lundqvist 1988), Finland (Söderman 1985), Scotland (Hall 1986) and Wales (Battiau-Queney 1984). An at least Early to Mid-Tertiary age of deep weathering has been argued for the above examples, whilst stripping of regolith is assumed to have dominated towards the end of the Tertiary. For the Fennoscandian Shield a continuous development of etchsurfaces since Palaeozoic times has been suggested (Lidmar-Bergström 1982; Söderman 1985). The early ideas of Büdel (1957) have also been confirmed in Germany where various remnants of Tertiary deep weathering have been found (Kubiniok 1988; Borger 1992).

There are some differences in relief between etchsurfaces found in the Sudetes and those described from Fennoscandia, which are described as being very flat over long distances (Lidmar-Bergström 1982). This major difference may be related to lithological differences since availability of rock to differential weathering is a very important factor, though one which is often difficult to quantify. However, it may also be suggested that it is climate-related weathering which is responsible for relief differentiation rather than simply lithological or structural variation. It has been demonstrated that remnants of grusses, and subordinately clayey grusses, dominate in high latitudes (Lidmar-Bergström 1982; Hillefors 1985; Kejonen 1985; Smith & McAlister 1987; Sorensen 1988; Hall *et al.* 1989; Le Coeur 1989) and only rarely give rise to deep diversified profiles. These types of weathering are believed to indicate a temperate, or at most a Mediterranean, climate. By contrast, very advanced deep weathering is widespread around 50° N (i.e. the Sudetes) and may account for the higher relief of etchsurfaces in the crystalline areas of Central Europe.

Conclusions

Despite the intensive Late Tertiary faulting which resulted in the horst-and-graben gross morphology of the Sudetes, Tertiary surfaces are well preserved in both the uplifted Sudetic block and in the downthrown Sudetic Foreland. Genetically, these surfaces represent etchsurfaces of various types. Although stripped etchsurfaces are the most common, partly stripped and mantled etchsurfaces have also been preserved in the Sudetic Foreland. The significant part played by deep weathering in the formation of the Tertiary surface is confirmed by the occurrence of weathering mantles, the features of Neogene deposits, and the occurrence of landforms such as inselbergs, plains, basins and escarpments. The latter group of features display distinct signs of bedrock control and are commonly interpreted as features of a weathering front.

Etch planation in the Sudetes has certainly occurred over a very long period of time, and may have begun in the Mesozoic, and has consisted of alternating phases of etching and stripping. Changing geomorphic processes through time account for considerable differentiation of relief in favourable places. Thus, the Tertiary palaeosurface comprises areas of high and low relief which are, however, of the same origin and age. Although etch planation is a continuous process, periods of predominant etching in the Palaeogene and predominant stripping in the Neogene may be recognized.

The Tertiary etchsurface is not unique to the Sudetes but forms part of an extensive etchsurface characteristic of Central and Northern Europe during the Tertiary. Its existence has long been postulated (Büdel 1957; Bakker & Levelt 1964; Czudek & Demek 1970) and there is now growing evidence that it dominated the Tertiary landscape of Europe.

I wish to thank Mike Widdowson, Andrew Goudie and an anonymous referee for the careful review of this paper and numerous valuable suggestions for improvement. I am also indebted to Karna Lidmar-Bergström, Cliff Ollier and Michael Thomas for discussions about etching concept and long-term landform development of crystalline areas. Waldemar Sroka and Barbara Bierońska kindly helped in the preparation of the figures. The conference presentation and the very first draft of this paper were prepared during the author's stay in the Department of Geography, London Guildhall University. The generous help of Peter Allen is here greatly appreciated.

References

BAKKER, J. P. & LEVELT, TH. W. M. 1964. An inquiry into probability of a polyclimatic development of peneplains and pediments (etchplains) in Europe during the Senonian and Tertiary period. *Publications Service Geologique du Luxembourg*, **14**, 27–75.

BATTIAU-QUENEY, Y. 1984. The pre-glacial evolution of Wales. *Earth Surface Processes and Landforms*, **9**, 229–252.

BERG, G. 1935. *Erläuterungen zu Blatt Alt-Kemnitz.* Geologische Karte von Preußen, Lieferung, 276.

BIRKENMAJER, K., JELEŃSKA, M., KĄDZIOŁKO-HOFMOKL, M. & KRUCZYK, J. 1977. Age of deep-seated fracture zones in Lower Silesia (Poland), based on K–Ar and palaeomagnetic dating of Tertiary basalts. *Rocznik Polskiego Towarzystwa Geologicznego*, **47**, 545–552.

BORGER, H. 1992. Paleotropical weathering on different rocks in Southern Germany. *Zeitschrift für Geomorphologie, NF*, Suppl. Bd., **91**, 95–108.

BORKOWSKA, I. & CZERWIŃSKI, J. 1973. On some mineralogical and textural features of granite regoliths in the Karkonosze Mts. *Studia Geographica*, **33**, 43–56.

BUCHARDT, B. 1978. Oxygene isotope palaeotemperatures from the Tertiary period in the North Sea area. *Nature*, **275**, 121–123.

BUDKIEWICZ, M. 1974. Niektóre złoża kaolinu okolic Świdnicy na Dolnym Śląsku. *Prace Geologiczne PAN*, **87**, 1–60.

BÜDEL, J. 1957. Die "Doppelten Einebnungsflächen" in den feuchten Tropen. *Zeitschrift für Geomorphologie, NF*, **1**, 201–228.

CIUK, E. & PIWOCKI, M. 1979. Trzeciorzed w rejonie Zabkowic Śląskich. *Biuletyn Instytutu Geologicznego*, **320**, 27–56.

——, DOKTOR, M. ET AL. 1992. *Litologia utworów trzeciorzedu w polskiej części Niecki Żytawskiej i ich własności fizyko-mechaniczne.* Prace Geologiczne PAN, p. 137.

CZUDEK, T. 1977. Reliefgenerationen im Ostteil des Nizky Jesenik (Gesenke) in der Tschechoslovakei. *In:* BÜDEL, J. (ed.) *Beiträge zur Reliefgenese in verschiedenen Klimazonen.* Würzburger Geographische Arbeiten, **45**, 39–68.

—— 1983. Relief a fosilni zvětraliny v okoli Bruntalu. *Sbornik Československe Geograficke Společnosti*, **88**, 289–297.

—— 1988. Kryopedimente – wichtige Reliefformen der rezenten und pleistozänen Permafrostgebiete. *Petermanns Geographische Mitteilungen*, **132**, 161–173.

—— & DEMEK, J. 1970. Některe problemy interpretace povrchovych tvarů Česke Vysočiny. *Zpravy Geografickeho Ustavu ČSAV*, **7**, (1), 9–28.

—— & —— 1971. Pleistocene cryoplanation in the Česka Vysočina highlands, Czechoslovakia. *Institute of British Geographers Transactions*, **52**, 95–112.

DEMEK, J. 1964. Slope development in granite areas of Bohemian Massif (Czechoslovakia). *Zeitschrift für Geomorphologie*, Suppl. Bd., **5**, 82–106.

—— 1976. Pleistocene continental glaciation and its effects on the relief of the northeastern part of the Bohemian Highlands. *Studia Societatis Scientiarum Torunensis, Toruń – Polonia*, **8C**, 4–6, 63–74.

—— 1980. Kryopedimenty: jejich vznik a vyvoj. *Scripta Facultatis Scientiarum Naturalium Universitatis Purkynianae Brunensis*, **10**, (5), 221–232.

—— 1982. Zarovnane povrchy Česke vysočiny. *In: Geomorfologicka konference,* Univerzita Karlova, Praha, 37–46.

DON, J. 1989. Jaskinia na tle ewolucji geologicznej Masywu Śnieżnika. *In*: Jahn, A., KOZŁOWSKI, S., WISZNIOWSKA, T. (eds) *Jaskinia Niedźwiedzia w Kletnie.* Ossolineum, Wrocław, 58–79.

—— & ŻELAŹNIEWICZ, A. 1990. The Sudetes – boundaries, subdivision and tectonic position. *Neues Jahrbuch für Geologie, Paläontologie usw., Abhandlungen*, **179**, 121–127.

DUMANOWSKI, B. 1961. Krawedź Sudetów na odcinku Gór Sowich. *Zeszyty Naukowe Uniwersytetu Wrocławskiego*, **B**, 7.

—— 1963. Stosunek rzeźby do struktury w granicie Karkonoszy. *Acta Universitatis Wratislaviensis*, **9**, 27–35.

DYJOR, S. 1966. Młodotrzeciorzedowa sieć rzeczna

zachodniej cześci Dolnego Śląska. *In*: *Z geologii Ziem Zachodnich*. PWN, Warszawa, 287–318.

—— 1986. Evolution of sedimentation and palaeogeography of nearfrontiers areas of the Silesian part of the Paratethys and the Tertiary Polish–German Basin. *Geologia, Kwartalnik AGH*, **12**, (3), 7–23.

—— & Kościówko, H. 1986. Rozwój wulkanizmu i zwietrzelin bazaltowych Dolnego Śląska. *Archiwum Mineralogiczne*, **41**, 111–122.

—— & Kuszell, T. 1977. Neogeńska i czwartorzedowa ewolucja rowu tektonicznego Roztoki – Mokrzeszowa. *Geologia Sudetica*, **12**, (2), 113–131.

——, Dendewicz, A., Sadowska, A. & Grodzicki, A. 1978. Neogeńska i staroplejstoceńska sedymentacja w obrebie stref zapadliskowych rowów Paczkowa i Kedzierzyna. *Geologia Sudetica*, **13**, (1), 31–65.

Franz, H.-J. 1969. Die geomorphologische Bedeutung der Granitverwitterung in der Oberlausitz. *Petermanns Geographische Mitteilungen*, **113**, 249–254.

Gellert, J. F. 1931. Geomorphologie des mittelschlesischen Inselberglandes. *Zeitschrift der Deutschen Geologischen Gesellschaft*, **83**, 431–447.

Gilewska, S. 1987. The Tertiary environment in Poland. *Geographia Polonica*, **53**, 19–41.

Głazek, J. & Szynkiewicz, A. 1987. Stratygrafia młodotrzeciorzedowych i staroczwartorzedowych osadów krasowych oraz ich znaczenie paleogeograficzne. *In: Problemy młodszego neogenu i eoplejstocenu w Polsce*. Ossolineum, Wrocław, 113–130.

Hall, A. M. 1986. Deep weathering patterns in north-east Scotland and their geomorphological significance. *Zeitschrift für Geomorphologie, NF*, **30**, 407–422.

——, Mellor, A. & Wilson, M. J. 1989. The clay mineralogy and age of deeply weathered rock in north-east Scotland. *Zeitschrift für Geomorphologie, NF*, Suppl. Bd., **72**, 97–108.

Hillefors, A. 1985. Deep-weathered rock in western Sweden. *Fennia*, **163**, 293–301.

Ivan, A. 1983. Geomorfologicke poměry Žulovske pahorkatiny. *Zpravy Geografickeho Ustavu ČSAV*, **20**, (4), 49–69.

Jahn, A. 1953. Morfologiczna problematyka Sudetów Zachodnich. *Przeglad Geograficzny*, **25**, 51–59.

—— 1960. The oldest periglacial period in Poland. *Biuletyn Peryglacjalny*, 9, 159-162.

—— 1968. Peryglacjalne pokrywy stokowe Karkonoszy i Gór Izerskich. *Opera Corcontica*, **5**, 9–25.

—— 1980. Main features of the Tertiary relief of the Sudetes Mountains. *Geographia Polonica*, **43**, 5–23.

——, Łańcucka-Środoniowa, M. & Sadowska, A. 1984. Stanowisko utworów plioceńskich w Kotlinie Kłodzkiej. *Geologia Sudetica*, **18**, (2), 7–43.

Kasiński, J. R. & Panasiuk, M. 1987. Geneza i ewolucja strukturalna Niecki Żytawskiej. *Biuletyn Instytutu Geologicznego*, **357**, 5–35.

Kejonen, A. 1985. Weathering in the Wyborg rapakivi area, southeastern Finland. *Fennia*, **163**, 309–313.

Klimaszewski, M. 1958. Rozwój rzeźby terytorium Polski w okresie przedczwartorzedowym. *Przeglad Geograficzny*, **30**, 3–43.

Kościówko, H. 1982. Rozwój zwietrzelin kaolinowych na przedpolu Sudetów Wschodnich. *Biuletyn Instytutu Geologicznego*, **336**, 7–59.

Kozłoski, S. & Parachoniak, S. 1960. Produkty wietrzenia bazaltów w rejonie Lubania na Dolnym Śląsku. *Acta Geologica Polonica*, **10**, 285–324.

Kral, V. 1985. Zarovnane povrchy Česke vysočiny. *Studie ČSAV*, **10**.

Kubiniok, J. 1988. Kristallinvergrusung an Beispielen aus Südostaustralien und deutschen Mittlegebirgen. *Kölner Geographische Arbeiten*, **48**.

Kural, S. 1979. Geologiczne warunki wystepowania kaolinów w zachodniej cześci masywu strzegomskiego. *Biuletyn Instytutu Geologicznego*, **313**, 9–68.

Kužvart, M. 1965. Geologicke poměry moravoslezskych kaolinu. *Sbornik geologickych věd, LG*, **6**, 87–146.

Le Coeur, C. 1989. La question des alterites profondes dans la region des Hebrides internes (Ecosse occidentale). *Zeitschrift für Geomorphologie, NF*, Suppl. Bd., **72**, 109–124.

Lidmar-Bergström, K. 1982. Pre-Quaternary geomorphological evolution in southern Fennoscandia. *Sveriges Geologiska Undersökning*, **C**, 785.

—— 1988. Denudation surfaces of a shield area in south Sweden. *Geografiska Annaler*, **70A**, 337–350.

Lundqvist, J. 1988. The Revsund area, Central Jämtland – an example of preglacial weathering and landscape formation. *Geografiska Annaler*, **70A**, 291–298.

Migoń, P. 1992. Inherited landforms in the crystalline areas of the Sudetes Mts. A case study from the Jelenia Góra Basin, SW Poland. *Geographia Polonica*, **60**, 123–136.

—— 1993*a*. Kopułowe wzgórza granitowe w Kotlinie Jeleniogórskiej. *Czasopismo Geograficzne*, **64**, 3–23.

—— 1993*b*. Geomorphological characteristics of mature fault-generated range fronts, Sudetes Mts., southwestern Poland. *Zeitschrift für Geomorphologie, NF*, Suppl.-Bd., **94**, 223–241.

—— & Czerwiński, J. 1994. Problem wieku zwietrzelin granitowych masywu karkonosko-izerskiego. *Acta Universitatis Wratislaviensis*, **1702**, 19–26.

Niśkiewicz, J. 1967. Budowa geologiczna Masywu Szklar. *Rocznik Polskiego Towarzystwa Geologicznego*, **37**, 387–415.

—— 1982. Geologiczne warunki wystepowania chryzoprazu i pokrewnych kamieni ozdobnych w masywie Szklar (Dolny Śląsk). *Geologia Sudetica*, **17**, 125–138.

Oberc, J. 1977. The Late Alpine Epoch in South-west Poland. *In:* Pożaryski, W. (ed.) *Geology of Poland, Vol. IV, Tectonics*. Wydawnictwa Geologiczne, Warszawa, 451–475.

—— & Dyjor, S. 1973. Postepy erozji młodotrzeciorzedowej w okolicy Leśnej na Pogórzu Izerskim. *Przeglad Geologiczny*, **21**, 177–182.

Pernarowski, L. 1963. Morfogeneza północnej krawedzi Wzgórz Niemczańskich. *Acta Universitatis Wratislaviensis*, **10**.

Pulinowa, M. Z. 1989. Rzeźba Gór Stołowych. *Prace Naukowe Uniwersytetu Śląskiego w Katowicach*, 1008.

RŮŽIČKA, M. 1989. Pliocen Hornomoravskeho uvalu a Mohelnicke brazdy. *Sbornik geologickych věd, A*, **19**, 129–151.

SADOWSKA, A. 1977. Roślinność i stratygrafia górnomioceńskich pokładów wegla Polski południowo-zachodniej. *Acta Palaeobotanica*, **18**, (1), 87–122.

—— 1987. Plioceńskie flory południowo-zachodniej Polski. *In*: Problemy *młodszego neogenu i eoplejstocenu w Polsce*. Ossolineum, Wrocław, 43–52.

SMITH, B. J. & MCALISTER, J. J. 1987. Tertiary weathering environments and products in northeast Ireland. *In:* GARDINER, V. (ed.) *International Geomorphology 1986*, Part II. Wiley, Chichester, 1007–1031.

SÖDERMAN, G. 1985. Planation and weathering in eastern Fennoscandia. *Fennia*, **163**, 347–352.

SORENSEN, R. 1988. *In-situ* rock weathering in Vestfold, southeastern Norway. *Geografiska Annaler*, **70A**, 299–308.

STOCH, L., DYJOR, S., KALMUS, M. & SIKORA, W. 1977. *Zwietrzeliny bazaltowe Dolnego Śląska*. Prace Mineralogiczne PAN, **56**.

STUCHLIK, L. 1980. Chronostratygrafia neogenu Polski południowej (północna cześć Paratetydy Centralnej) na podstawie badań paleobotanicznych. *Przeglad Geologiczny*, **28**, 443–448.

SZCZEPANKIEWICZ, S. 1954. Morfologia Sudetów Wałbrzyskich. *Prace Wrocławskiego Towarzystwa Naukowego*, **B**, 65.

—— 1958. Peryglacjalny rozwój stoku Masywu Ślęży. *Biuletyn Peryglacjany*, **6**, 81–92.

TEISSEYRE, H. 1960. Rozwój budowy geologicznej od prekambru po trzeciorzed. *In*: TEISSEYRE, H. (ed.) *Regionalna Geologia Polski, t. III, Sudety, z. 2.* Polskie Towarzystwo Geologiczne, Kraków, 335–357.

THOMAS, M. F. 1974*a*. *Tropical Geomorphology. A Study of Weathering and Landform Development in Warm Climates.* MacMillan, London.

—— 1974*b*. *Granite Landforms: A Review of Some Recurrent Problems of Interpretation.* The Institute of British Geographers, Special Publication, **7**, 13–37.

—— 1989. The role of etch processes in landform development. *Zeitschrift für Geomorphologie, NF*, **33**, 129–142, 257–274.

—— 1994. Applicable models for preglacial landform development. *Geomorphology*, **12**, 3–15.

—— & THORP, M. B. 1985. Environmental change and episodic etchplanation in the humid tropics of Sierra Leone: the Koidu etchplain. *In*: DOUGLAS, I. & SPENCER, T. (eds) *Environmental Change and Tropical Geomorphology.* Allen and Unwin, London, 239–268.

TUŁACZYK, S. 1992. Cuesta landscape in the middle part of the Sudetes Mts. *Geographia Polonica*, **60**, 137–150.

TWIDALE, C. R. 1982. *Granite Landforms.* Elsevier, Amsterdam.

—— 1991. A model of landscape evolution involving increased and increasing relief amplitude. *Zeitschrift für Geomorphologie, NF*, **35**, 85–109.

—— & BOURNE, J. A. 1975. Episodic exposure of inselbergs. *Geological Society of America Bulletin*, **86**, 1473–1481.

WALCZAK, W. 1968. *Sudety.* PWN, Warszawa.

WOLFE, J. 1980. Tertiary climates and floristic relationships at high latitudes in the Northern Hemisphere. *Palaeogeography, Palaeoclimatology, Palaeoecology*, **30**, 313–323.

ZIMMERMANN, E. 1937. *Erläuterungen zu Blatt Hirschberg*. Geologische Karte von Preußen, Lieferung, 276.

Neogene palaeosurfaces in the volcanic area of Central Slovakia

JÁN LACIKA

Institute of Geography, Slovak Academy of Sciences, Štefánikova 49, 814 73 Bratislava, Slovakia

Abstract: Two types of Neogene palaeosurfaces occur in the Central Slovakia Volcanic Area, these being volcanic and planational in origin. The volcanic palaeosurfaces are currently represented as topographic remnants which were originally created as a result of Neogene–Pleistocene andesite–rhyolite (Badenian–Panonian) and basaltic (Pontian–Upper Pleistocene) volcanic activity in the region. During the intervening and subsequent periods of relative volcanic and tectonic quiescence, geomorphological processes of landform evolution became dominant and, consequently, a total of four generations of Neogene planation surfaces may be identified in the study area. From the topmost, or oldest surface, these are: a high-level Middle Miocene surface; an Upper Miocene middle or intramontane level; an intermediate level of Pontian age; and lowermost, a sub-montane level dated as Upper Pliocene–Lower Pleistocene, found at the present river level. All of the Neogene palaeosurfaces in the Central Slovakia Volcanic Area, of both volcanic and erosional origins, currently display an advanced state of erosion and dissection as a result of Alpine uplift effects: a total of six main types of relief may be recognized.

Rapid erosion of palaeosurfaces has taken place in the tectonically active areas of the Alpine orogen, especially in regions where block tectonic movements have accelerated the rate of erosional dissection of any pre-existing, flatter relief. Despite this, remnants of the pre-Quaternary landforms may be recognized in the West Carpathian Mountains of Central Slovakia, although it is important to stress that they are typically in a very advanced state of erosion and dissection. Together, the landforms constitute an often complex palaeosurface topography. Two main groups of these palaeosurfaces can be identified: those of volcanic origin, including cone and caldera remnants; and those which are planational or erosional in origin.

The current work is based largely upon a synthesis of previous geomorphological and geological research in the Central Slovakia Volcanic Area, and aims to provide an overview of the development and evolution of the palaeosurfaces within the region. This appoach is supported by a combination of cartographic analyses and field observations in which the topographically flat elements identified in the landscape are recognized as 'potential planation surfaces'. These features are then interpreted within the context of the established sequence of palaeosurface development. Elements of both the Neogene volcanic landscapes, together with planation surfaces which developed in the intervening and post-volcanic quiescent phases, may be recognized in the modern landscape. The main themes of the collated data regarding Neogene palaeosurfaces in the study area are given in Table 1. The information is divided into general, regional and local scales.

General outline of the study area

The Central Slovakia Volcanic Area is located in the central part of the eastern European Alpine orogen, and is situated in the southwestern margin of the West Carpathians uplift (Figs 1 and 2). This uplifted region slopes towards both the Intra-Carpathian Danube Lowland in the southwest and the South Slovakia intermontane basins in the south, and is chiefly drained by a northeasterly flowing tributary of the Danube (the Hron River).

Volcanic activity in the Central Slovakia Volcanic Area is linked with the tectonic evolution of the inner margin of the Carpathian arc and is associated with the subduction zone which sloped northwards towards the Panonian basins: Kalinčiak *et al.* (1989) considers that volcanism was largely a consequence of the penetration of a mantle diapir associated with this subduction, and a summary of absolute dating and fission track results for the Central Slovakia Volcanic Area are given in Table 2. The volcanism comprised a complex of Neogene volcanoes which evolved through a series of active phases occurring predominently during the Miocene and Pliocene, though reduced basaltic eruption continued into the Pleistocene (e.g. Tables 3 and 4). During the eruptions, the existent network of then active faults became major conduits for the

From Widdowson, M. (ed.), 1997, *Palaeosurfaces: Recognition, Reconstruction and Palaeoenvironmental Interpretation*, Geological Society Special Publication No. 120, pp. 203–219.

Table 1. *Summary of scales of data collection used the interpretation of Central Slovakian palaeosurfaces*

Scale of observation	Approaches or techniques of palaeosurface interpretation
General	* Theories of planation * Volcanological methods †‡ Identification of 'potential planation surfaces'
Regional	* Interpretation and reconstruction of volcanic landforms * Interpretation and reconstruction of planation surfaces * Sedimentary correlation between neighbouring basins * recognition and identification of weathering profiles * Geological properties – structural and lithological aspects * Absolute dating of constituent volcanic rocks * Palaeogeographical reconstructions †‡ Areal composition and interpretation of 'potential planation surfaces' † Morphostructural properties of elements of relief
Local	* Structural and lithological properties of different rock types * Results of radiometric dating †‡ Identification of preserved elements of 'potential planation surfaces' † Mesoscale morphostructural relief analyses † Morphogenesis of landforms

* Approaches used by previous authors; † observational techniques developed during the current study; ‡ 'potential planation surfaces' are interpreted as planation surface remnants.

rising magma and, in many cases, these faults were later associated with the subsequent uplift of the various volcanic landforms which characterize the region. As a consequence, subsequent Alpine tectonic movements have transformed the Central Slovakia Volcanic Area into a mosaic-like system of horsts (mountainous areas) and grabens or basins (Fig. 2). The distribution of these morphotectonical

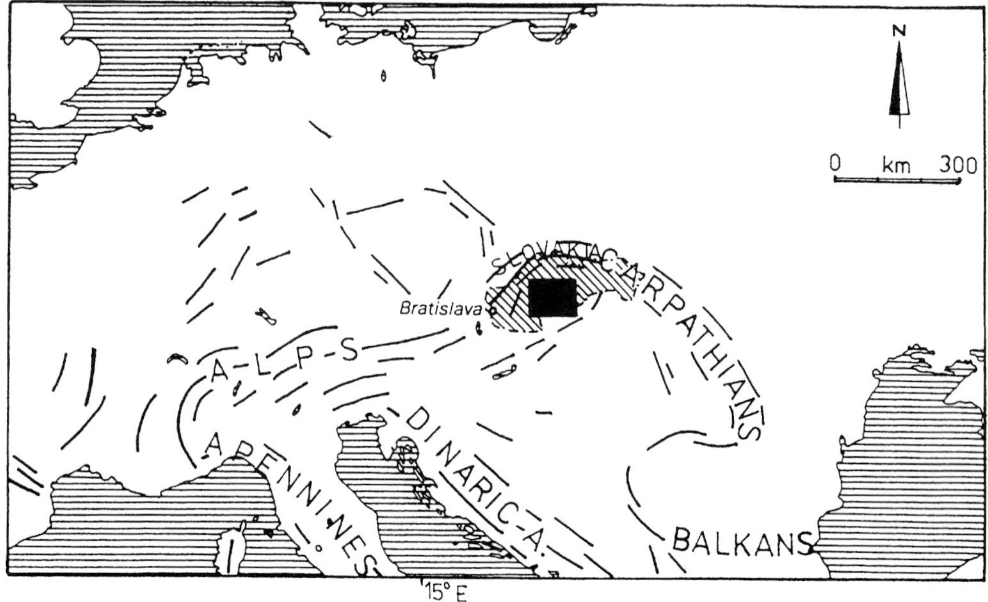

Fig. 1. General location of study area. Shaded rectangle denotes area of Fig. 2.

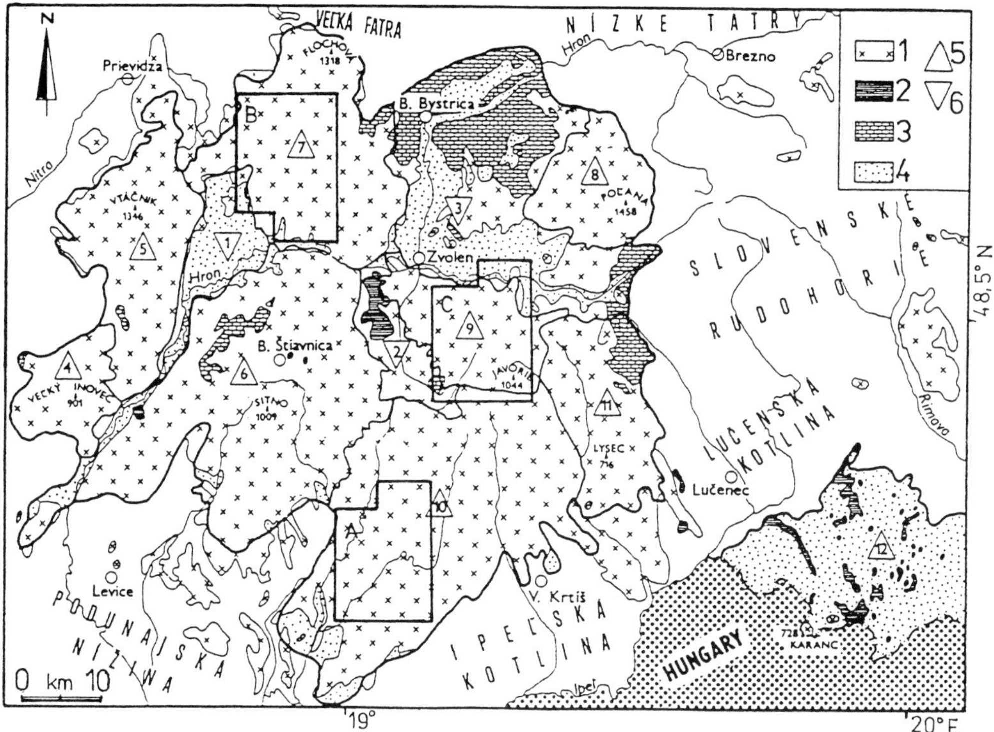

Fig. 2. Geological and geomorphological division of study area (after Lacika 1992). 1, Andesites and rhyolites; 2, basalts; 3, basement; 4, Tertiary sediments, geomorphological division: 5, elevated areas (i.e. mountains and plateaux); 6, topographic depressions and basins. Localities are given as numbered figures in triangles: 1, Žiarska Kotlina Basin; 2, Pliešovská Kotlina Basin; 3, Zvolenská Kotlina Basin; 4, Pohronský Inovec Mts; 5, Vtáčnik Mts; 6, Štiavnické Vrchy Mts; 7, Kremnické Vrchy Mts; 8, Polana Mts; 9, Javorie Mts; 10, Krupinská Planina Mts; 11, Ostrôžky Mts; 12, Cerová Vrchovina Mts (not included in current study). Inset rectangles A–C give positions of Figs 6–8 respectively.

units is currently expressed as a series of eight mountain ranges, each corresponding to one or more major volcanic centres and three major intermontane basins (Fig. 2). In addition, to the southeast of the main volcanic area, there are also the smaller volcanic areas of the Cerová Vrchovina Mts, Lučenská Kotlina Basin and Slovenské Rudohorie Mts, which are separated from the Central Slovakia Volcanic Area and hence lie beyond the limits of the current study. The distribution of the main stratovolcano centres with respect to the current mountain ranges of the Central Slovakia Volcanic Area is given in Fig. 3.

Mountain ranges

West of the Hron River valley there are three mountain ranges (Fig. 2), these are, from south to north: the Pohronský Inovec Mts (901 m), the Vtáčnik Mts (1348 m) and the Kremnické Vrchy Mts (1318 m). Geologically, the Pohronský Inovec Mts is a large horst and comprises part of the Štiavnické Vrchy volcanic structure which is itself divided by the Hron River valley. A similar, but even larger, horst forms the Vtáčnik Mts, whilst the Northern Kremnické Vrchy Mts, which comprise the highest and most uplifted horst in the eastern margin of the mountains and whose steep easterly scarp slopes into the Zvolenská Kotlina Basin, are significantly modified by landslides. By contrast to the Pohronský Inovec and Vtáčnik Mts, the Northern Kremnické Vrchy Mts display a more complicated structure in which geomorphologically distinct regions can be identified.

To the east of the Hron River valley lie the Štiavnické Vrchy Mts (1009 m) which are composed of the remnants of a large Neogene stratovolcano that has been uplifted into another complicated horst structure. The horst contains

Table 2. *Summary of radiometric and fission track (F/T) dating results from volcanic rocks in the study area*

Locality	Rock type	Method	Age (Ma)
Krupinská Planina and Ostrôžky Mts			
Hrušov*	Pyroxene andesite	K–Ar	18.5 ± 0.9
Lysec Hill*	Amphibole andesite	K–Ar	18.2 ± 0.8
Cerovo*	Andesitic breccia	K–Ar	17.3 ± 0.8
Šiavnické Vrchy and Pohronský Inovec Mts			
Banská Štiavnica†	Granodiorite	F/T	17.1 ± 0.4
Tanád Hill*	Pyroxene andesite	K–Ar	17.0 ± 0.5
Sabová Skala Hill†	Rhyolitic glass	F/T	15.3 ± 0.3
Rudno Nad Hronom†	Rhyolite	F/T	12.3 ± 1.0
Kremnické Vrchy Mts			
Kordíky†	Amphibole andesite	F/T	16.2 ± 0.6
Horný Turčok*	Pyroxene andesite	K–Ar	15.2 ± 1.0
Stará Kremnička*	Rhyolite	K–Ar	11.3 ± 0.3
Polana and Javorie Mts			
Breziny*	Pyroxene andesite	K–Ar	16.7 ± 1.2
Zvolenská Slatina†	Amphibole andesite	F/T	16.6 ± 0.3
Detva†	Pyxroxene-biotite andesite	F/T	13.6 ± 0.4
Strelníky†	Rhyodacite tuffs	F/T	12.9 ± 0.4

* Bagdasarjan *et al.* 1970; † Repcok 1981.

several geomorphologically different parts controlled by tectonic block movements. Northeast of the Zvolenská Kotlina Basin there lie the Polana Mts, which are in area the smallest, but highest in elevation (1458 m), and represent the best preserved of the Neogene volcanic landforms in the study area. South of the Polana Mts, beyond the Slatina River valley, lie the Javorie Mts. In their northernmost part they are relatively low and dissected, but become higher and less dissected in their southernmost part where they reach an altitude of 1004 m. The southernmost ridge of the Javorie Mts slopes gently sothwards and southeastwards towards the Krupinská Planina plateau (775 m) and Osrtrôžky Mts (887 m), respectively. In its western margins the large Krupinská Planina plateau is divided by deep canyon-like valleys displaying a radial drainage pattern which divides the plateau into several flat-topped interfluve ridges. Futher to the east, the Ostrôžky Mts have a similiar, but more dissected, relief cut by a dense network of deep valleys.

Intermontane basins

There are three major intermontane basins recognized in the study area (Fig. 2). The Žiarska Kotlina Basin, which lies between the Kremnické Vrchy, Štiavnické Vrchy and Vtáčnik Mts in the northwest region of the volcanic area, is the youngest and is of post-volcanic origin. Its broadly triangular shape is delimited by well-developed marginal fault scarps which border the adjoining horst regions. By contrast, the Pliešovská Kotlina Basin, which lies in the centre of the volcanic region and formed as a depression between two stratovolcanoes, was later modified by tectonic block movements. An absence of marine and lacustrine sediments in the structure of the Pliešovská Kotlina Basin is the result of the palaeodrainage and valley network being altered by an extensive Lower Pleistocene basalt lava flow (Halouzka 1986). The third and largest basin in the Central Slovakia Volcanic Area is the tectonically complicated Zvolenská Kotlina Basin; this lies to the north of the Javorie Mts, and between the Polana and Kremnické Vrchy Mts, which lie to the east and west, respectively. This basin is divided by faults into several small grabens and horsts which expose rocks of pre-volcanic basement, together with the Neogene volcanic products and Neogene–Quaternary post-volcanic sediments.

Volcanic palaeosurfaces

The volcanically-generated palaeosurfaces may be considered as surface geomorphology which was constructed as a direct result of the volcano–tectonic phases affecting the region (i.e. a volcanic topography). These surfaces were dominated by a series of stratovolcanoes and neighbouring caldera collapses which formed basins later filled with a combination of volcanogenic sediments and lavas.

Table 3. *Age of the Neogene and Quaternary volcanic features in the study area*

Volcanic region	Phases of activity
Southern volcanic zone	
(Krupinská Planina plateau, Ostrôžky Mts)	
Submarine volcanoes	Lower Badenian
Lysec Volcano	Lower–Middle Badenian
Štiavnica volcanic centre	
(Štiavnické Vrchy Mts, Pohronský Inovec Mts, NE Danube Lowland, W Krupinská Planina plateau, S Vtáčnik Mts, S Kremnické Vrchy Mts)	
Lower stratovolcano	Lower–Middle Badenian
Caldera	Upper Badenian–Lower Sarmatian
Upper stratovolcano	Sarmatian
Basalt intrusions	Pontian
Basalt extrusions	Upper Pleistocene
Javorie volcanic centre	
(Javorie Mts, N Ostrôžky Mts, N Krupinská Planina plateau, E Zvolenská Kotlina Basin, Pliešovská Kotlina Basin)	
Lower stratovolcano	Lower–Middle Badenian
Volcanotectonic depression	Middle Badenian–Lower Sarmatian
Upper stratovolcano	Sarmatian
Basalt extrusion	Lower Pleistocene
Polana volcanic centre	
(Polana Mts, N Zvolenská, Kotlina Basin)	
Lower stratovolcano	Middle–Upper Badenian
Upper stratovolcano	Sarmatian
Kremnica volcanic centre	
(Kremnické Vrchy Mts, Žiarska Kotlina Basin)	
Lower stratovolcano	Lower–Upper Badenian
Volcanotectonic depression	Upper Badenian–Lower Sarmatian
Middle stratovolcanoes	Sarmatian
Upper volcanoes and intrusions	Upper Sarmatian–Panonian
Vtácnik Volcanic Centre	
(Vtáčnik Mts, W Ziarska Kotlina Basin)	
Upper volcanoes and intrusions	Upper Sarmatian–Panonian

After Konečný *et al.* (1984*a*).

This volcanogenic terrain has subsequently been modified by a complex interplay of post-volcanic processes and associated evolution of planation surfaces during the intervening and subsequent periods of volcanic and tectonic quiescence.

Development of the volcanic palaeosurfaces

The following genetic interpretations of the original volcanic palaeosurfaces are based upon a synthesis of structural–lithological and biostratigraphical analyses originally published by Konečný *et al.* (1984*a*,*b*) and Kalinčiak *et al.* (1989).

The remnants of the stratovolcanoes are perhaps the most typical landforms in the Central Slovakia Volcanic Area but, with the exception of the single stage evolution of the volcanic structure in the Vtačnik volcanic centre, most appear to be the result of more than one volcanic constructional phase (Fig. 3). For example, the Štiavnica, Javorie and Polana volcanic centres consist of two generations of stratovolcanoes, whilst a total of three generations of stratovolcanoes may be identified in the construction of the Kremnica volcanic centre. A brief outline of the evolution of the important volcanic centres is given below (see also Table 3).

The Lower Štiavnica Stratovolcano represents the largest volcanic edifice generated during the volcanic history of the study area. The distance between the centre and margin of the cone is in excess of 45 km. After the early stages of volcanic

Table 4. *Correlative scheme for chronological evolution of the volcanic topography and development of the recognised plantation surfaces in the Central Slovakian Volcanic area. (N.B. Column of volcanic activity includes phases of basalt volcanic activity identified in the Lučenská Kotlina Basin and Cerová Vrchovina Mts).*

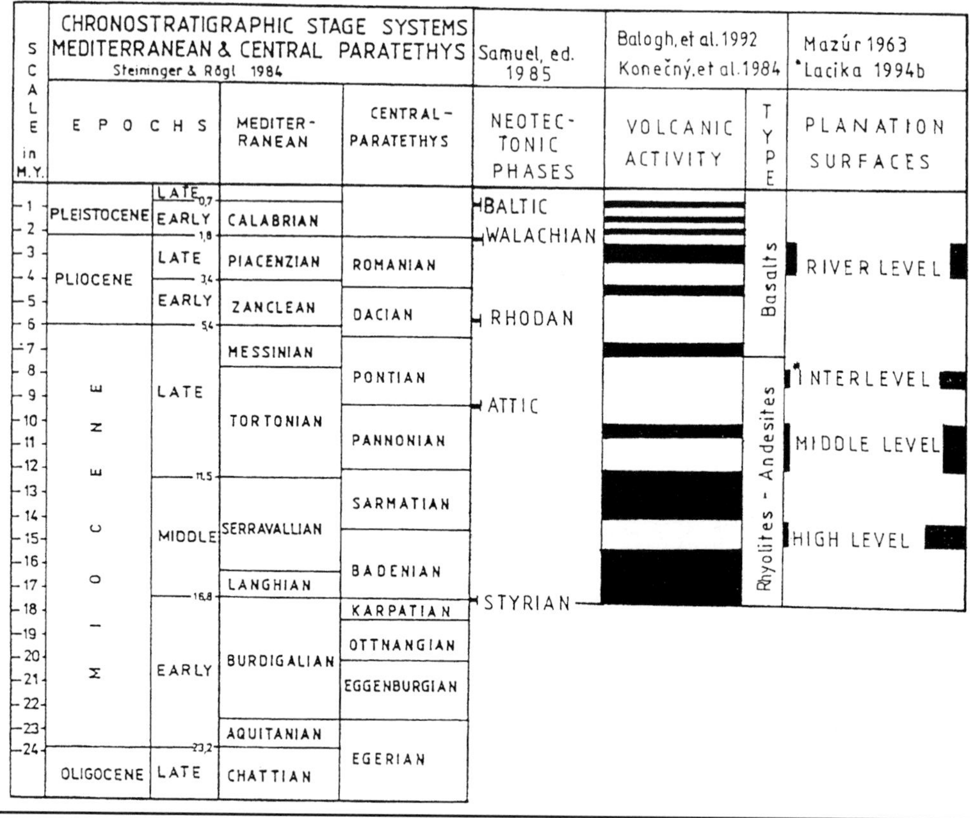

activity at Stiavnica, a large caldera (> 20 km diameter) developed in the centre of the strato-volcano. Palaeoenvironmental interpretation of different vegetational zones identified upon the slope of this caldera margin indicate that it was originally 600–700 m in elevation (Planderová in Konečný *et al.* 1984*a*). The caldera has subsequently been filled by later intrusive and extrusive products to such an extent that the lavas eventually breached the caldera margin and flooded and filled peripheral radial palaeovalleys. The Upper Štiavnica Stratovolcano was then built upon these tectonically disturbed and eroded remnants of the lower stratovolcanic structure. By contrast to the single-vent structure of the earlier edifice, the Upper Štiavnica Stratovolcano developed through the activity of several dispersed volcanic centres. Both the Upper Štiavnica Stratovolcano and neighbouring Upper Javorie Stratovolcano were originally extremely large structures. However, the volcanic eruptions of the Upper Javorie Strato-volcano tended to preferentially develop towards the south and, as a result, an extensive raised periphery with a large proluvial plain has developed on the margin of this structure (Lukniš 1972).

SCALE in M.Y.
CHRONOSTRATIGRAPHIC STAGE SYSTEMS MEDITERRANEAN & CENTRAL PARATETHYS
Steininger & Rögl 1984
Samuel, ed. 1985
Balogh, et al. 1992
Konečný, et al. 1984
Mazúr 1963
•Lacika 1994b
EPOCHS
MEDITER-RANEAN
CENTRAL-PARATETHYS
NEOTEC-TONIC PHASES
VOLCANIC ACTIVITY
TYPE
PLANATION SURFACES
1
2
3
4
5
6
7
8
9
10
11
12
13
14
15
16
17
18
19
20
21
22
23
24
PLEISTOCENE
LATE
0,7
EARLY
1,8
CALABRIAN
PLIOCENE
LATE
3,4
EARLY
5,4
PIACENZIAN
ZANCLEAN
ROMANIAN
DACIAN
MIOCENE
LATE
11,5
MIDDLE
16,8
EARLY
MESSINIAN
TORTONIAN
SERRAVALLIAN
LANGHIAN
BURDIGALIAN
AQUITANIAN
PONTIAN
PANNONIAN
SARMATIAN
BADENIAN
KARPATIAN
OTTNANGIAN
EGGENBURGIAN
23,2
OLIGOCENE
LATE
CHATTIAN
EGERIAN
BALTIC
WALACHIAN
RHODAN
ATTIC
STYRIAN
Basalts
Rhyolites - Andesites
RIVER LEVEL
INTERLEVEL
MIDDLE LEVEL
HIGH LEVEL

The Upper Polana Stratovolcano is of Sarmatian age and was constructed upon same centre of an earlier strongly eroded Badenian vent structure. Explosive eruption characterized the earlier periods of its volcanic activity resulting in lahars which developed upon the southwest margin of the cone. The eruptions which dominated the final stages of volcanic activity occurred mainly on the southern flanks of the Polana Volcano.

The Kremnica volcanic region has, perhaps, had the most complicated development (Table 3) because, subsequent to the creation of the Lower Kremnica Stratovolcano, a very large portion of it subsided creating a large volcanotectonic

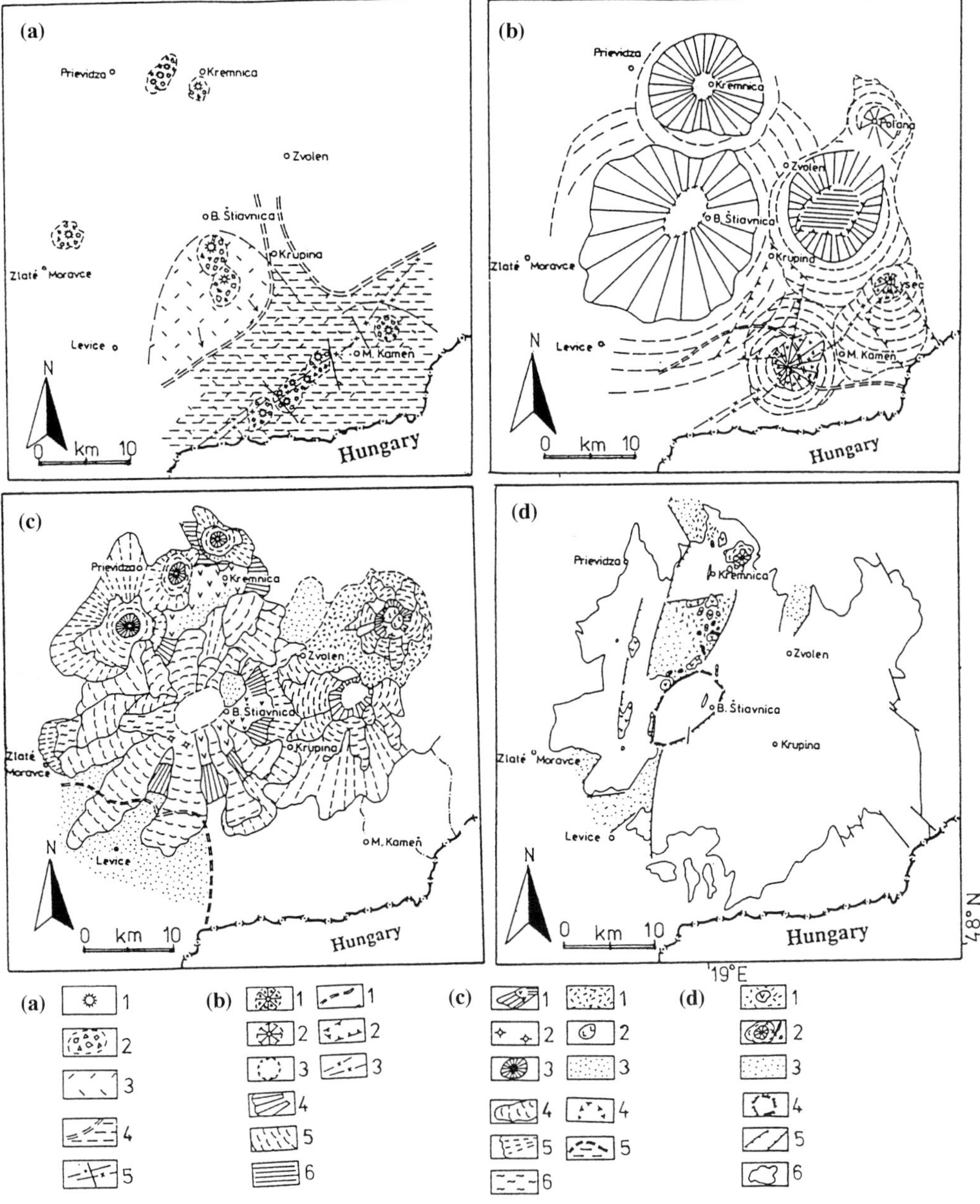

Fig. 3. Palaeogeographic scheme of study area during the phase of andesite–rhyolite volcanic activity (after Konečný *et al.* 1984*a*). (**a**) Lower Badenian: 1, volcanic centres and extrusive domes; 2, accumulations of coarse volcanoclastic rocks around volcanic centres; 3, areas of finer, reworked volcanoclastic rocks; 4, maximum extent of the marine sediments; 5, Šahy–Lysec volcano–tectonic zone and associated syn-volcanic faulting. (**b**) Late Lower Badenian to Midle Badenian: 1, pyroclastic volcanic cones; 2, stratovolcanoes; 3, central volcanic zone; 4, proximal stratovolcanic debris; 5, distal accumulations of reworked volcanoclastic rocks; 6, hyaloclastites in the Javorie volcanotectonic depression; 7, limit of maximum northward extent of marine sediments; 8, syn-volcanic depressions and grabens; 9, Sahy–Lysec volcanotectonic zone. (**c**) Sarmatian: 1, exposed remnants of Badenian stratovolcanoes; 2, vents and dykes; 3, volcanic cones; 4, effusive complexes; 5, reworked volcanoclastic rocks laid down as prolluvial and alluvial fans; 6, ignimbrite sheet; 7, rhyodacite pyroclastic rocks; 8, rhyodacite extrusions; 9, areas of sedimentation; 10, Polana caldera; 11, extent of the marine transgression towards the north. (**d**) Panonian: 1, extrusive domes and volcanoclastic rocks of rhyolite composition (i.e. of late Sarmatian to early Panonian age); 2, stratovolcano lava flows and outcrops of basaltic andesites; 3, areas of sedimentation; 4, marginal faults of Štiavnica horst; 5, marginal faults around structural depressions; 6, extent of recently denuded volcanic rocks.

depression. A similar graben-like form of this nature occurs in the northern Javorie Mts (Kalinčiak *et al.* 1989). In both cases the subsidence of these depressions has been closely associated with subsequent infilling of these depressions by local intrusions and extrusions from the nearby volcanic centres.

By contrast to the sub-aerial volcanism which characterizes the northern stratovolcanoes, a different form of activity evolved in the southernmost part of the study area. Here, the location of the volcanic centres was controlled by the presence of SW–NE faults, which, in their earliest phases, were manifest as submarine volcanoes typically exhibiting phreatic activity. With continued explosive activity, the Lysec Volcano eventually emerged from beneath the Neogene sea. Other nearby volcanic centres, however, appear to have developed predominantly in a terrestrial environment since limited local occurrence of hyaloclastite breccias indicates that only in some cases did the associated lava flows actually reach the sea.

Age of volcanic palaeosurfaces

Table 2 gives a summary of the available K–Ar results (Bagdasarjan *et al.* 1970, 1977), together with the fission track results (Repčok 1981). These results have been correlated with the lithological and biostratigraphical interpretation of the rock ages (Konečny *et al.* 1984*a*,*b*); the latter authors have used the correlation as a basis for a scheme outlining the palaeogeographical and volcanotectonic evolution of the Central Slovakia Volcanic Area (Fig. 3). These results are compared with other schemes in Table 4. The onset of volcanic activity in the Central Slovakia Volcanic Area occurred *c.* 17–18 Ma and, during subsequent eruptive phases, changed in both its extent and intensity. Eruptions had begun in the south of the Krupinská Planina plateau and Ostrôžky Mts during Lower Badenian times (Table 3), after which there followed two main phases of volcanic activity (typically generating andesites and rhyolites) during the Badenian and Sarmatian. The Badenian and Sarmatian eruptive phases are separated by periods of relative quiescence at 15–14 and 12–11 Ma respectively, each lasting *c.* 1 Ma (Table 4). The Štiavnica, Kremnica, Polana and Javorie centres were active during the Badenian activity, whilst the Vtáčnik volcanic centre became active during the Sarmatian. During the Panonian volcanic activity was restricted to the Kremnica and Vtáčnik volcanic centres.

A second major burst of volcanism in the Central Slovakia Volcanic Area, characterized by locally-restricted basaltic eruptions, took place over a period of *c.* 7 Ma, from the Pontian to Lower Pleistocene (Table 3). These types of eruption are chiefly associated with localized extension (e.g. in the Štiavnické Vrchy Mts and Pliešovská Kotlina Basin areas).

Post-volcanic development of the volcanic palaeosurfaces

Mechanisms of dynamic morphotectonic development, together with long-term destructive morphoclimatic influences, have caused much destruction of the Neogene volcanic palaeosurfaces in the Central Slovakia Volcanic Area. The original volcanic relief has been transformed to a much modified form, often as a result of controls related to the structural and lithological properties of the constituent volcanic rocks.

There is an absence of volcanic palaeosurfaces older than Sarmatian age in the study area because the existing Badenian volcanoes were already deeply eroded before the onset of the ensuing Sarmatian activity. The conical Lysec Hill (716 m) was built on the ridge plateaux of the Ostrôžky Mts, whilst the dome-like features of the Rohy Hills (675 m) developed in the eastern part of the Zvolenská Kotlina Basin. Both the Ostrôžky Mts and the Rohy Hills are constructed from Badenian volcanic rocks but cannot be readily recognized as erosional remnants of the Badenian volcanic landforms since they seem to largely consist of differentially eroded intrusions and extrusions which must originally have lain beneath the original Badenian topography (Konečny *et al.* 1984*a*).

There are several examples of Sarmatian volcanic palaeosurfaces preserved in the present-day relief. For example, the geomorphology of the former Sarmatian Polana Stratovolcano may be divided into two distinct parts (Lacika 1993). Firstly, there is the caldera in the centre of the Polana Mts (Fig. 4), which has been deeply eroded and disturbed along the associated faults; in addition, a radial, centripetally-draining valley network has developed within the caldera. This inwardly-draining river system is itself drained by the Hrochot River valley which cuts through caldera structure on the western periphery of the former volcano. Secondly, the remaining slopes comprising the lower flanks of the old stratovolcano have been divided into a series of well-developed centrifugally-draining valleys and crests. However, later tectonic movements in the north-west margin of the Polana Mts have modified this radial drainage network into a sub-parallel pattern of tributaries. Elsewhere in the Polana Mts, the relatively well-preserved volcanic landforms of the stratovolcano are largely retained due to the fact

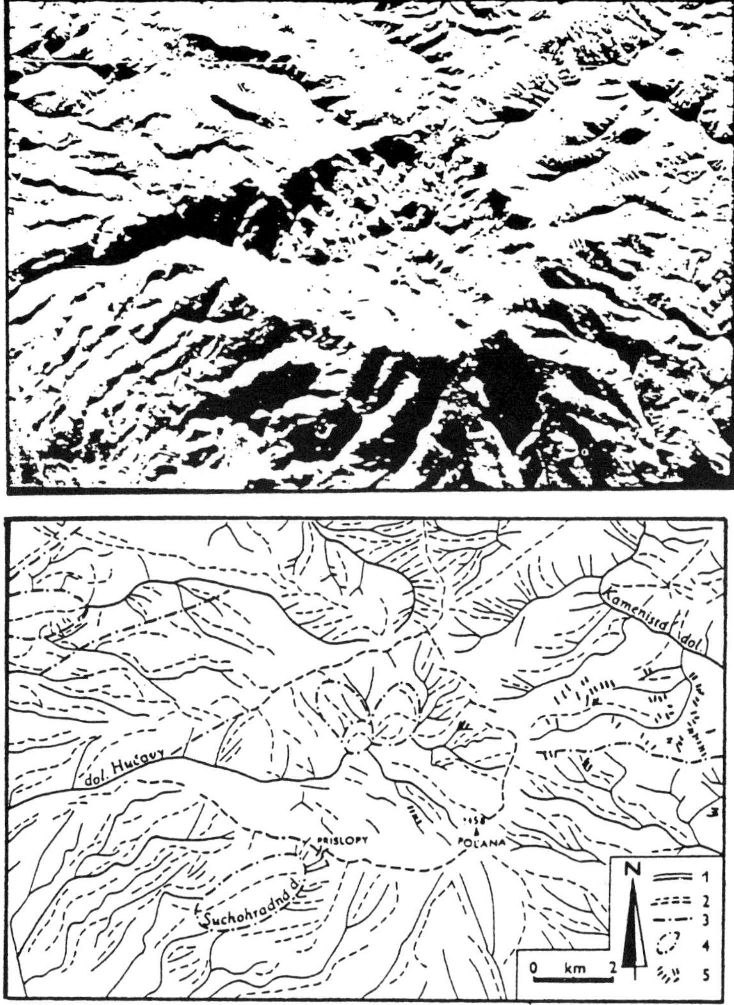

Fig. 4. Radar image of Polana Mts. This may be geomorphologically interpreted (Lacika 1992) as follows:1, valley lines; 2, crest lines: 3, other linear features; 4, non-linear topographic features; 5, small-scale topographic dividing features.

that the Polana Volcano has remained isolated from the major uplifts which have affected the other parts of the Central Volcanic Region, and also by the fact that the thick, extensive complex of resistant lava flows have, in effect, acted as a shield against rapid erosion (Lacika 1993).

The Sarmatian Javorie Stratovolcano is less well-preserved than that of Polana and is currently in a stage of near-complete destruction. Nevertheless, there are several features which record the presence of former volcanic landforms. The two main ridges of the Javorie Mts follow two geologically identified structural lines. The northern ridge (Fig. 5) is situated close the former southern margin of the central caldera, whilst the southern ridge is located at the southern margin of the Badenian volcanotectonic depression. The best indication of the original Sarmatian Javorie Stratovolcano is seen in the region of the Krupinská Planina plateau (Figs 5 and 6) where there exists a typical radial valley network which developed upon the last remnants of the Javorie volcanic cone.

In other parts of the study area the Neogene volcanic palaeosurfaces have been completely destroyed. Their relief have been transformed from the original volcanic relief to a lithologically or

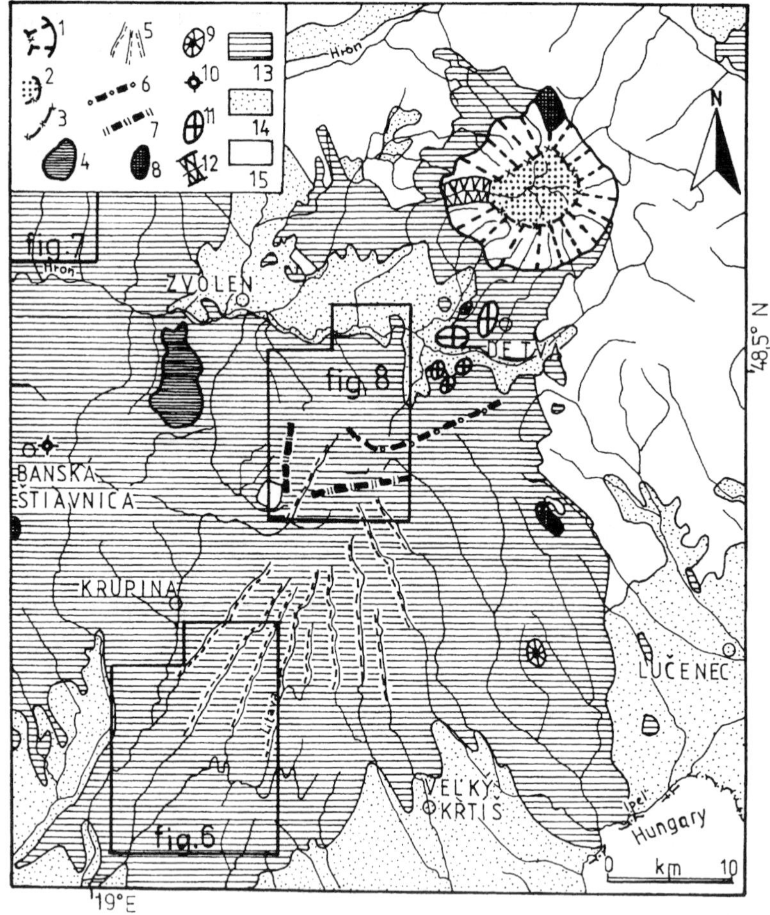

Fig. 5. Interpretation of the volcanic palaeosurfaces within the study area. **Volcanic landforms**: 1, periphery of highly eroded stratovolcano; 2, caldera remnants; 3, marginal crest of caldera; 4, structurally-related footslope plateaux developed on the surface of eroded basalt lava flows. **Post-volcanic landforms**: 5, radial valley network; 6, mountain crests following geologically recognizable caldera margin; 7, mountain crest following geologically identified margin of highly eroded volcanotectonic depressions. **Landforms representing more advanced stages of denudation**: these are typically developed by selective erosion of the different volcanic rocks and are the result of inherent variation in lithology or stucture: 8, large table mountains formed on andesite–rhyolite lava flows by selective erosion; 9, conical elevations formed upon central extrusions or vent plugs; 10, conical elevations developed upon basaltic dykes; 11, dome-like elevations formed on intrusions and extrusions, 12; major valley cutting through remnants of the peripheral flank of the former volcano; 13, undivided relief on volcanic rocks. **Landforms developed upon non-volcanic rocks**: 14, undivided relief on Tertiary sediments; 15, undivided relief on basement rocks.

sturcturally-controlled topography. These landforms, which include the Sarmatian and Panonian Volcanoes of Štiavnické Vrchy Mts, Vtáčnik Mts and Kremnické Vrchy Mts, which subsequently developed on the strongly uplifted horsts, have been described by earlier authors (e.g. Lukniš 1972). The erosion and denudation of the Upper Štiavnica Stratovolcano, for example, has been so intense in its centre that rocks of the pre-volcanic basement have now become exhumed .

Flat-topped 'table mountains' and stepped or 'trappean' topography are common landforms in the areas characterized by the remnants of the stratovolcanoes. These features have been created by selective erosion of softer volcanoclastics between the more resistant lava flows. The largest

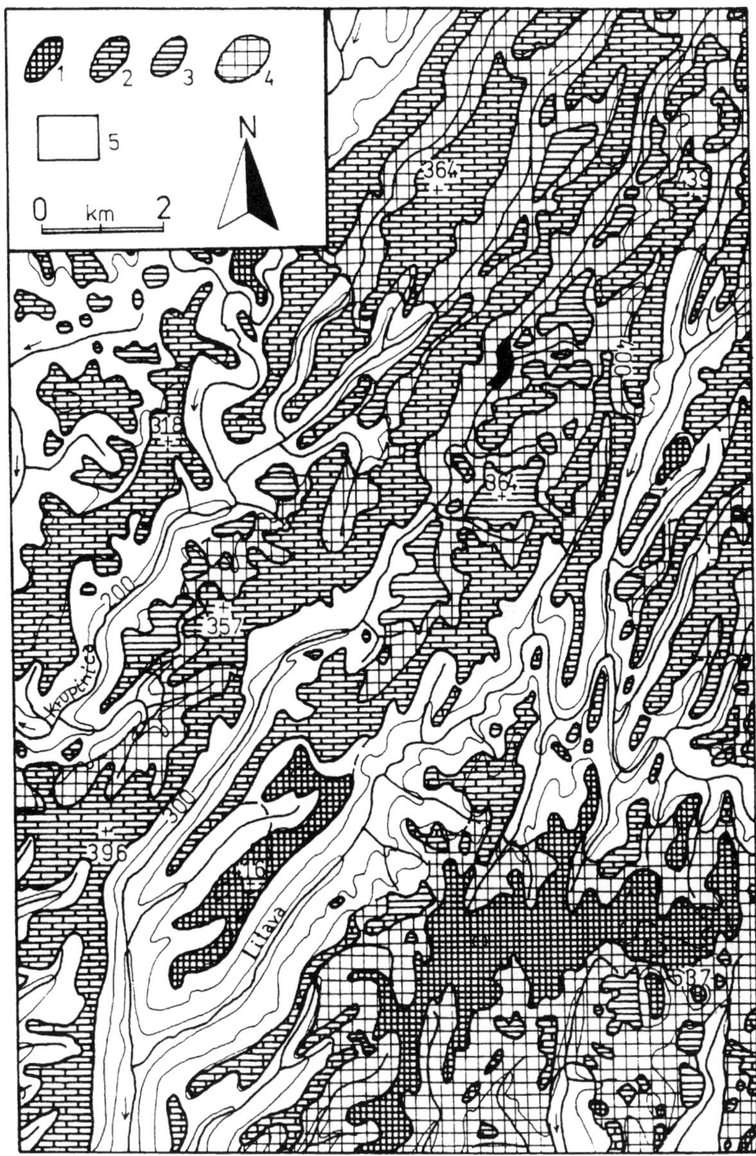

Fig. 6. Potential planation surfaces pattern in Krupinská Planina Plateau. 1, Upper plateau; 2, ridge crest plateaux; 3, footslope plateaux (including stepped slopes and saddle plateaux); 4, relatively undissected relief; 5, strongly dissected relief.

examples of these features are located in the Lower Pleistocene basaltic lava flows (Halouzka 1986) of the northern Pliešovská Kotlina Basin (Fig. 5).

Planational palaeosurfaces

The present interpretations of the Neogene planation surfaces in the West Carpathians is largely based upon the work of Lukniš and Mazúr which was conducted in the early 1960s. These authors presented a general scheme for the evolution of relief in the West Carpathians during the Neogene and Quaternary periods, and subsequent researchers have accepted and applied this scheme into their own regional work with little, if any, further modification (e.g. Činčura 1969).

Lukniš and Mazúr suggest that the evolution of relief of the area as a whole was controlled by periodic changes in the dynamics of the tectonic movements. They assumed that during periods of relative tectonic and volcanic quiescence the formation of planation surfaces became prevalent, whilst erosional processes and associated dissection of these planation surfaces became dominant during the periods of tectonic rejuvenation. By adopting this general model of periodic uplift interspersed by periods of tectonic quiescence, Mazúr (1963, 1964, 1965) identified three generations of Neogene planation surfaces in the West Carpathians: a top or high level (Upper Miocene); a middle or intramontane level (Upper Miocene); and a river or sub-montane level (Upper Pliocene-Lower Pleistocene). However, Lukniš (1964) recognized only two Neogene surfaces, since he excluded Mazúr's top, or high, level which was assumed to represent exhumed remnants of a much earlier (Upper Cretaceous–Paleaogene) planation surface. Lacika (1994) identified four generations of planation surfaces, an extra surface evolving during the Pontian period, termed the interlevel, being recognized between Mazúr's middle and river levels (Table 4).

Evolution of the planational palaeosurfaces

Over the past century, many different types of planation surface have been distinguished and a variety of theories regarding the dominant evolutionary processes forwarded. Those theories which recognize a genetically specific type of surface according to the dominant planation process include, for example: peneplain (Davis 1898, 1899; Penck 1924), pediment (Gilbert 1880), pediplains (King 1962), etchplain (Wayland 1933) and cryoplanation surface (Bryan 1946).

In the present case of the palaeosurfaces of Central Slovakia, more than one process seems to have been responsible for palaeosurface evolution. Indeed, the changing climatic and tectonic controls have apparently favoured different mechanisms and hence genetically distinct types of palaeosurface at different times during the Neogene. Mazúr (1965) and Lukniš (1964) interpreted the middle level as a pediment or pediplain: a model for the genesis of the top level (e.g. Mazúr 1963, 1964, 1965) was not suggested. The biostratigraphical analyses of Planderová (1975, 1978) indicate that during the Panonian the climate was more arid and, as such, would be more conducive to mechanisms of pediplanation.

By contrast, the relatively warm and humid climate of the Middle Miocene (Planderová 1975), together with an associated reduction of tectonic effects, as indicated by reduced volcanic activity (Konečný *et al.* 1984*a*), would have been more favourable to the processes of peneplanation. This particular mechanism is considered the most probable origin of the top or high level, an opinion supported by Kraus (1989) who, in addition, recognized that such Miocene climatic conditions would also be suitable for extensive kaolinization.

Similarly, the interlevel could be interpreted as a surface formed by peneplanation since Planderová (1975) identified increasing climate humidity during the Pontian period. However, considering the extent and nature of the remnants of this surface, it seems likely that this planation phase was not long-lived and consequently did give rise to a fully-developed peneplain. The ensuing semi-arid climate during the Upper Pliocene was favourable for the formation of the river level by pedimentation processes .

It is interesting to note that Slovak geomorphologists have often discussed the existence of an etchplain in the West Carpathians relief but, to date, relatively little data has been published on this subject. This particular model of palaeosurface evolution, together with that of cryoplanation, may provide a suitable avenues for future geomorphological investigation in the area.

Age of the planational palaeosurfaces

Since planation surfaces must be younger than the rocks they truncate, the ages determined for the local volcanic and sedimentary rocks provide a maximum age limit for such surfaces. However, volcanic materials extensively covered the entire region from early Badenian times onwards, so providing little chance of recognizing any pre-Badenian planation surfaces in the Central Slovakia Volcanic Area (with the possible exception where profiles or sections are exhumed from beneath these volcanic rocks).

Remnants of the late Badenian top or high level (Table 4) must necessarily be sought beneath the Sarmatian and younger volcanic rocks, or alternatively beneath the sediments of syngenetic marine transgressions, or those which filled the subsiding calderas and volcanotectonic depressions formed during the Upper Badenian and Lower Sarmatian phases. Given the geological environment of the Krupinská Planina plateau and south Ostrôžky Mts, such surface remnants probably have the maximum potential for preservation in the southern volcanic region.

Extensive 'potential planation surfaces' (for futher explanation of this term see below) occur on the Sarmatian volcanic rocks and typically belong to Mazúr's middle level or younger surfaces.

Additionally, there are many indications of the probable existence of the interlevel in the Kremnické Vrchy Mts since a series of step-like plateaux may be identified developed here upon the Panonian volcanic structures. This interpretation (Table 4) is consistent with an Upper Panonian or younger age (i.e. Pontian), and the fact that the Pontian period is characterized by a significant reduction of tectonic and volcanic activity (Samuel 1985).

Mazúr's and Lukniš's age interpretations of the river level as the Upper Pliocene planation surface can be readily verified from field relationships. Plateaux areas, which are thought to be remnants of river level surfaces, are situated slightly above the highest of the Lower Pleistocene river terraces, or else are found developed upon Pliocene sediments filling the intermontane basins (e.g. within the Zvolenská Kotlina Basin).

Late-stage development of the planational palaeosurfaces

Regional tectonic effects and associated uplift has been the main causative factor of the rapid dissection of the Neogene planation surfaces, whilst morphoclimatic conditions have accelerated or slowed the destructive erosional processes on the developing horsts and grabens. As may be expected, the older palaeosurfaces display this destruction to the greatest extent: the top level surface has been subject to much more degradation than the younger middle, inter or river levels, and consequently the remnants of the top level surface are very rare in the study area. The remnants of the middle level surfaces has probably been better preserved in the present-day relief because they were originally the best developed and most extensive planation surfaces developed in the West Carpathians (Mazúr 1965; Lukniš 1964). Similarly, the preservation of remnants of the river level surface is relatively common, especially in the basin areas. Although this particular surface is not particularly extensively developed, its potential for preservation in the modern landscape is a consequence of its relative youth. Information regarding the extent and preservation of the interlevel palaeosurface is currently scant.

Identifying patterns of potential planation surfaces

The term *potential planation surface* requires further explanation with respect to the current study: it is here considered to be that part of the relief which may be readily distinguished from its surroundings by virtue of its near horizontality. This explanation accepts that such apparent flatness materially aids the distinction of such components from the more dynamic and more dissected younger landforms which have been subsequently created. Mapping of potential planation surfaces onto the topographical maps, scale 1 : 25 000, has been employed in the current work and the accuracy of the resulting maps verified through fieldwork.

Four different patterns of potential planation surfaces have been distinguished in the study area on the basis of analyses of their areal composition and size: (1) a series of extensive sub-parallel or radially arranged ridges with plateau-like summits – these surfaces generally display relatively little dissection in the form of step-like topography around their margins; (2) areas of extensive foot-slope plateaux, again affected by little step-like dissection; (3) Areally restricted upper-level, ridge-top, footslope or saddle plateaux, typically displaying more advanced effects of dissection: these currently occur at a variety of levels due to the effects of extensive faulting; (4) Rare, isolated, and small step-like plateaux.

The first of these patterns may be identified on Fig. 6, which is an example of one of the most extensive and best developed of the potential planation surfaces in the study area. It may be recognized forming the Krupinská Planina plateau, in the Ostrôžky Mts area, and in the southern part of the Štiavnické Vrchy Mts. Typically, this pattern of topography is indicative of an originally extensive planation surface subsequently cut by deeply eroded river valleys and thus giving rise to a series of extensive plateaux-topped interfluves.

The second pattern type of areally extensive footslope plateaux is similarly common and well represented in the study area. It can be readily identified as forming Kunešov plateau in the Kremnické Vrchy Mts (e.g. in the upper left-hand quadrant of Fig. 7) and similar patterns can be recognized in the Pohronský Inovec Mts (Lehota plateau), Štiavnické Vrchy Mts (Studenec plateau) and Zvolenská Kotlina Basin (Strelníky plateau). This pattern tends to be the result of a more advanced stage of erosion and dissection than observed in those areas characterized by the plateaux-topped interfluves descibed above; this is largely because these extensive footslope plateaux have typically been subject to attack by agents of lateral erosion.

The third pattern type occurs in the south Kremnické Vrchy Mts and is the most frequent in the Central Slovakia Volcanic Area (e.g. lower part of Fig. 7). The morphotectonic evolution of the study area forming the horst–graben mosaic has caused frequent dissection of originally extensive planational palaeosurfaces into a series of step-like

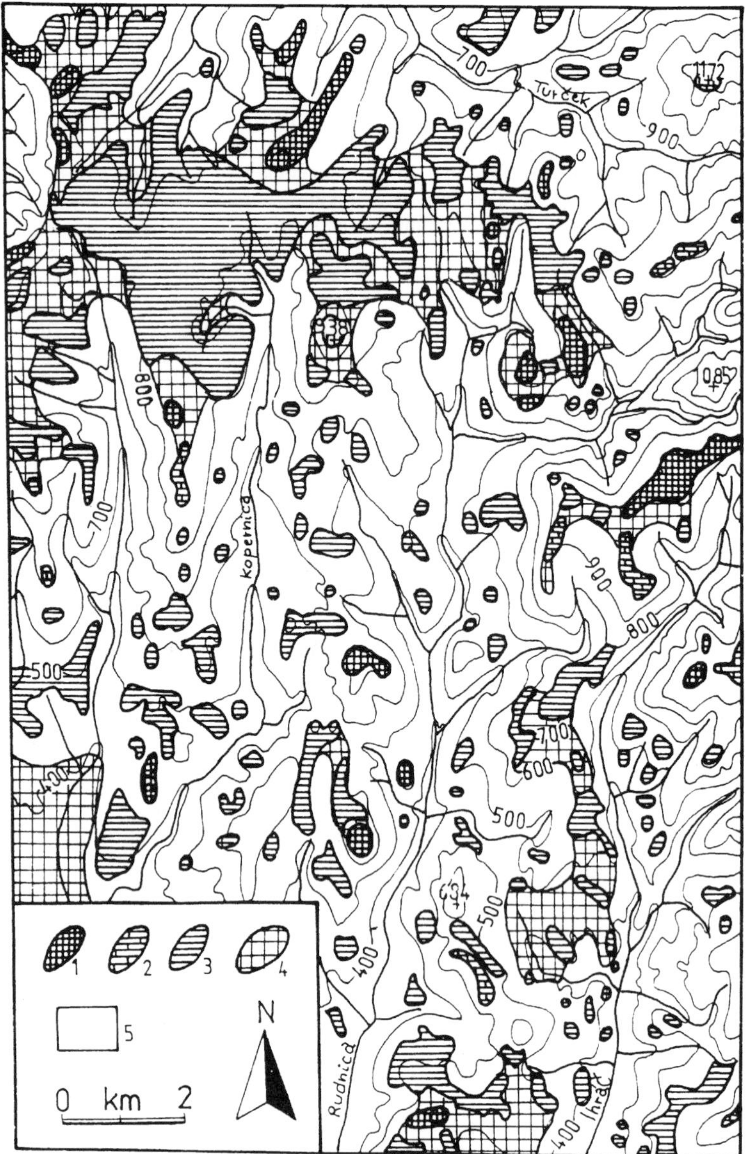

Fig. 7. Potential planation surfaces pattern in Kremnické Vrchy Mts. 1, Upper plateau; 2, ridge crest plateaux; 3, footslope plateaux (inluding stepped slopes and saddle plateaux); 4, less dissected relief; 5, more dissected relief.

plateaux controlled by local faulting. The size of dissected surfaces is dependent upon the density of faults which have affected the originally flat palaeosurface.

There are five regions in the study area which exhibit the fourth pattern, these are: the Polana Mts, Javorie Mts (Fig. 8), Vtácnik Mts, western Štiavnické Vrchy Mts and northeast Kremnické Vrchy Mts (middle part of Fig. 7). These areas exhibit a paucity of potential planation surfaces, or else display only isolated, very small step-like plateaux. Such areas typically display a particularly heterogeneous geological structure and the small plateaux features often occur quite frequently in combination with numerous rocky landforms. The occurrence of this pattern of small plateaux as a topographic component may be interpreted in two ways. Firstly, they may represent remnants of former planation surfaces which have been severely eroded after uplift in the horsts regions; this appears to be the case for the small plateaux features of the Štiavnické Vrchy Mts, Kremnické Vrchy Mts and Vtáčnik Mts; secondly, they may simply be structurally-controlled features which have developed in regions where prolonged planational phases have been unable to act effectively; this is their most probable origin in the case of the Polana and Javorie Mts.

Conclusions

Two types of palaeosurface remnants may be identified in the Central Slovakia Volcanic Area; these being volcanic and planational in origin. Three stages can be distinguished during the subsequent erosional modification of volcanic palaeosurfaces; these are manifest as the different patterns of the potential planation surfaces which are observed in the morphological analyses of the

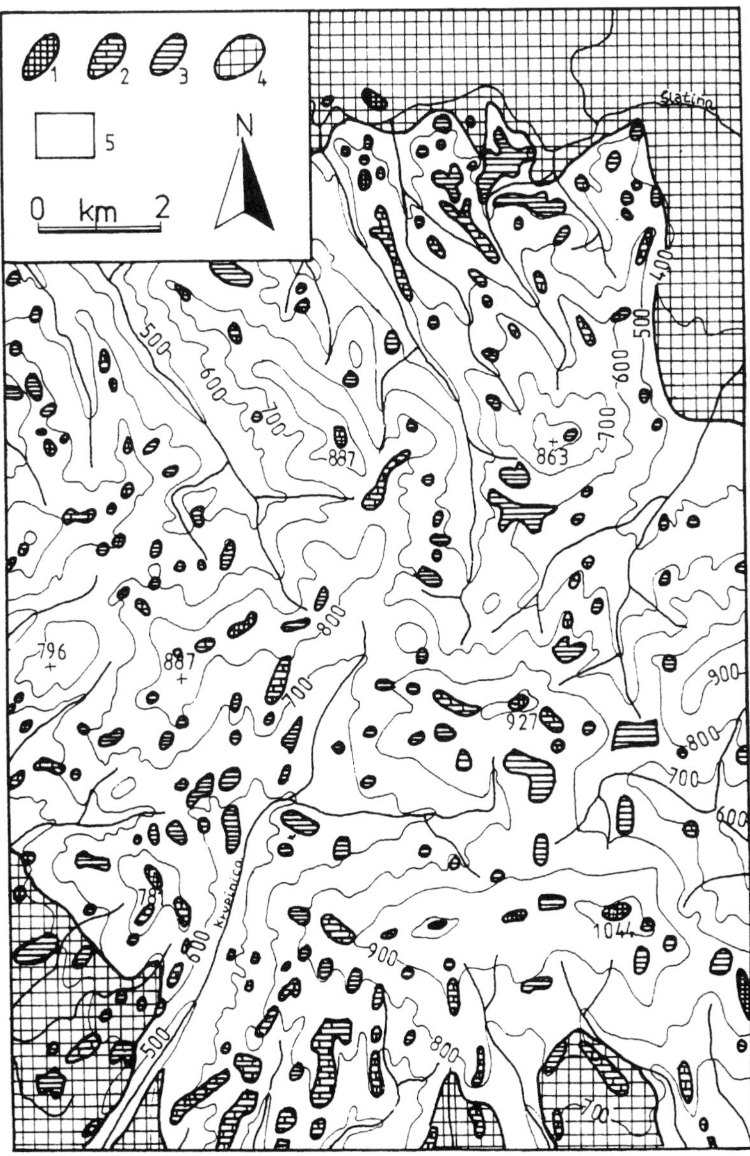

Fig. 8. Patterns of potential planation surfaces recognized in the Javorie Mts. Upper plateau; 2, ridge crest plateaux; 3, footslope plateaux (including stepped slopes and saddle plateaux); 4, less dissected relief; 5, more dissected relief.

study area. As a result, a total of six types of relief may be recognized.

1. Strongly modified volcanic relief unaffected by post-volcanic planation. This type of relief has been identified in the Polana Mts and is largely the result of the preservation of the Sarmatian Upper Polana Stratovolcano. In addition, the prevalence of the fourth pattern type of potential planation surfaces in the Polana Mts indicates the absence of a post-Sarmatian planation phase in this region.
2. Highly modified volcanic relief, probably unaffected by post-volcanic planation processes. This type of relief has been formed in the Javorie Mts: the erosional transformation of the Upper Javorie Stratovolcano area has been so strong that a total absence of post-Sarmatian planation is more difficult to prove than in the Polana Mts.
3. Relief developed upon volcanic rocks derived from volcanic and planational palaeosurface elements which have been highly modified as a result of subsequent uplift and erosion. This type of relief has developed in the Vtáčnik Mts, northeast Kremnické Vrchy Mts and the west Štiavnické Vrchy Mts. Here, the original Neogene volcanic landforms have been completely transformed by erosion and tectonic dissection on the most uplifted horsts of the Central Slovakia Volcanic Area. These regions also display the fourth pattern type of the potential planation surfaces. The reasons for the evolution of this pattern are not entirely clear, but it is either the result of an absence of planation phases during the evolution of the relief, or else all traces of any former planation surfaces have been entirely destroyed by erosion.
4. Highly modified planational palaeosurfaces, usually apparent as a series of step-like plateaux. The occurrence of this type of landform is, perhaps, most frequent in the study area. The best examples exist in the south Kremnické Vrchy Mts and in the south Pohronský Inovec Mts. Here, the Neogene volcanic palaeosurfaces have been completely destroyed by erosion, denudation and planation. The planation palaeosurfaces have subsequently been dissected into a series of step-like features by later tectonic movements along faults.
5. Extensive footslope plateaux. This relief type is similar to the preceding one but generally less tectonically disturbed. The Kunešov plateau (the Kremnické Vrchy Mts) is a good example of this type of relief.
6. Plateau-topped ridges derived from the deep dissection and modification of originally extensive planational palaeosurfaces. These occur in the regions exhibiting the best preserved Neogene planation surfaces in the study area. This type of relief has formed in the Krupinská Planina plateau, Ostrôžky Mts and south Štiavnické Vrchy Mts.

Two anonymous referees are gratefully thanked for their detailed and useful comments during the preparation of this manuscript. M. Widdowson is acknowlegded for improvments in the style and presentation of the final version.

References

BAGDASARJAN, G. P., KONEČNÝ, V. & VASS, D. 1970. Príspevok absolutných vekov k vyvojovej schéme neogénneho vulkanizmu stredného Slovenska. *Geologické práce,* **51**, 47–61.

——, ——, DUBLAN, L., KONEČNÝ, V. & PLANDEROVÁ, E. 1977. Príspevok k stratigrafickej pozícii stratovulkánov Javoria a Polany. *Geologické práce*, **68**, 141–151.

BRYAN, K. 1946. Cryopedology, the study of frozen ground and intensive frost action with suggestion on nomenclature. *American Journal of Science,* **244**, 622–642.

ČINČURA, J. 1969. Morfogenéza južnej časti Turčianskej kotliny a severnej časti Kremnických vrchov. *Náuka o Zemi, Goegraphica*, **2**, IV, Bratislava.

DAVIS, W. M. 1898. The peneplain. *American Geologist*, **22**(B), 207–239.

—— 1899. The geographical cycle. *Geographical Journal,* **14**(A), 481–503.

GILBERT, G. K. 1880. Contributions to the history of Lake Bonneville. *US Geological Survey, 2nd Annual Report*, 167–200.

HALOUZKA, R. 1986. Z nových poznatkov o stratigrafii kvartéru terasových náplavov riek Západných Karpát [stredné Pohronie, Orava, Turiec]. *Regionálna geológia Západnych Karpát,* **21**, 167–175.

KALINČIAK, M., KONEČNÝ, V. & LEXA, J. 1989. Štruktúra a vyvoj neogenných vulkanitov Slovenska vo vztahu k blokovej tektonike. *Geologické práce,* **88**, 79–105.

KING, L. C. 1962. *The Morphology of the Earth.* Oliver & Boyd, Edinburgh.

KONEČNÝ, V., LEXA, J. & PLANDEROVÁ, E. 1984*a*. Stratigrafické členenie neovulkanitov stredného Slovenska. *Západné Karpaty,* Séria geologia **9**, Bratislava.

——, PLANDEROVÁ, E. & LEXA, V. 1984*b*. *Geologická mapa stredoslovenských neovulkanitov v mierke 1:100 000.* Geologický ústav Dionyza Štúra, Bratislava.

KRAUS, I. 1989. Kaolíny a kaolinitové íly Západných Karpát, *Západné Karpaty*, Seria min. petrogr. geochem. metalogen. **13**, Bratislava.

LACIKA, J. 1992. The best preserved stratovolcano in the West Carpathians. *In:* STANKOVIANSKY, M. & LACIKA, J. (eds) *Excursion Guide-book,*

International Symposium: Time, Frequency and Dating in Geomorphology. Tatranská Lomnica-Stará Lesná, June 16–21, 1992, Bratislava, 65–68.

—— 1993. Morfoštrukturná analýza Polany. *Geografický Časopis*, **45**, (2–3), 233–250.

—— 1994. Contribution to the knowledge of the age of planation surfaces in the Slovenské stredohorie Mts. *Geographia Slovaca,* **7**, 81–102.

LUKNIŠ, M. 1964. Pozostatky starších povrchov zarovnavania reliefu v ceskoslovenských Karpatoch. *Geografický Časopis,* **XVI**, (3), 289–298.

—— 1972. Zvysky tretohornych zarovnanych povrchov. *In:* LUKNIŠ M. *et al.* (eds) *Slovensko II–Príroda.* Obzor, Bratislava.

MAZÚR, E. 1963. *Žilinská kotlina a prilahlé pohoria.* SAV, Bratislava.

—— 1964. Intemountain basins a characteristic element in the relief of Slovakia. *Geografický Časopis*, **XVI**, (2), 105–126.

—— 1965. Major features of the West Carpathians in Slovakia as a result of young tectonic movements. *In:* MAZÚR. E. & STEHLÍK, O. (eds) *Geomorphological Problems of Carpathians.* SAV, Bratislava, 9–54.

PENCK, W. 1924. *Die morphologische Analyse.* Engelhorn, Stuttgart.

PLANDEROVÁ, E. 1975. Data on climatic changes in the Neogene of the Central Parathetys on the basis of palinology. 6th Congr. Reg. Commit. Mediter. Neogene Stratigr., Bratislava.

—— 1978. Microflorizones in Neogene of Central Parathetys. *Západné Karpaty*, Seria geologia **3**, 7–34.

REPČOK, I. 1981. Datovanie niektorých stredoslovenskych neovulkanitov metódou po delení uránu [fission track]. *Západné Karpaty*, Séria min. petr. geochem. metalog. **8**, 59–104.

SAMUEL, O. (ed.) 1985. *Chronostratigrafická a synoptická tabulka.* Geologický ústav Dionyza Štúra, Bratislava.

WAYLAND, E. J. 1933. *Peneplains and Some Other Erosional Platforms.* Annual Report Bulletin, Protectorate of Uganda Geological Survey Department of Mines, Note 1, 77–79.

Tertiary palaeosurfaces of the SW Deccan, Western India: implications for passive margin uplift

M. WIDDOWSON

Department of Earth Sciences, Parks Road, Oxford OX1 3PR, UK
(Current address: Department of Earth Sciences, The Open University, Walton Hall, Milton Keynes MK7 6AA, UK)

Abstract: Two genetically distinct lateritized palaeosurfaces of different ages are recognized in the southwest Deccan Traps region of Western India using a combination of geochemical, topographical, and satellite image data. The Deccan Traps were erupted at the Cretaceous–Tertiary boundary (*c.* 65 Ma), and comprise a huge area of originally near-horizontal basalt lavas covering much of northwest Peninsular India, and topographically forming the coast-parallel Western Ghats escarpment and elevated Maharashtra plateau to the east.

Remnants of the older, palaeosurface currently exist as a series of isolated, laterite-capped plateaux forming the highest elevations along the Western Ghats (15°30′–18°15′ N). This surface is of late Cretaceous–early Tertiary age, and originally developed upon flows which lay at, or near to, the top of the lava sequence. This lateritization phase was terminated by a period of uplift and extensive erosion in lower- to mid-Tertiary times during which the low-lying, low-relief coastal (Konkan) plain developed through the eastward recession of the Ghats scarpline. A second phase of lateritisation occurred upon this coastal pediplain during mid- to late Tertiary times. Since the earlier uplift had gently deformed the lava pile prior to the development of the pedimented surface, the low-level Konkian laterite lies with marked angular unconformity upon the lava stratigraphy. Both surfaces have been subject to further large-scale distortion resulting from continuing uplift effects.

Development and evolution of these Deccan palaeosurfaces is important since together they provide a record of uplift effects in western India. Moreover, they offer a datum against which the uplift erosional history may be further constrained and demonstrate that uplift effects have acted upon the Indian margin throughout the Tertiary. Since such longevity of uplift is difficult to reconcile with the commonly cited thermal and dynamic post-rift mechanisms known to act upon passive margins, the morphological and structural evolution of the rifted Deccan margin is better described in terms of denudational isostasy.

Understanding the evolution of volcanic rifted margins (VRM) is of fundamental geological importance because their structural and geomorphological development represents the interplay of geological and geophysical mechanisms at the plate tectonic scale. Continental rifting is often intimately associated with extensive and rapid volcanic events, namely the generation of continental flood basalt (CFB) provinces; these place important constraints upon the dynamics of the rifting process. Many VRM clearly share similar structural and geomorphological characteristics indicating a common pattern of evolution. These include monoclinal flexuring of the margin, and an immense laterally continuous escarpment separating the coastal zone from an elevated inland plateau. Nevertheless, most studies have tended to concentrate on the depositional and structural record of the offshore sedimentary basins as a method of constraining models of passive margin evolution. Yet the key issue remains regarding the relationship of the onshore post-rift history to evolution of the coastal plain–escarpment–plateau morphology, which ultimately regulates offshore sediment fluxes. Although sediment supply is known to vary temporally due to climatic changes, maturation of the drainage system, and continued flexural adjustment of the margin itself, it is the development and evolution of the 'great escarpments' that appear to exert a fundamental control.

Evolution of the Deccan lavas, the construction of the volcanic sequence, and the associated palaeo-environmental effects of the eruptions have received much attention, but the subsequent morphotectonic evolution of the Deccan margin has, with a few notable exceptions (e.g. Kumar 1975; Kailasam 1979; Ollier & Powar 1985), been largely neglected. Clearly, the post-eruptive evolution of this particular VRM is of crucial importance if its continental-scale geomorphology, drainage development, and associated uplift and erosional history are to be more fully understood.

From Widdowson, M. (ed.), 1997, *Palaeosurfaces: Recognition, Reconstruction and Palaeoenvironmental Interpretation,* Geological Society Special Publication No. 120, pp. 221–248.

With this in mind, this paper is divided into four sections. The first describes the general geological and geomorphological setting which led to the formation of the lateritized palaeosurfaces; the second examines the nature of the laterites and the reconstructed palaeosurfaces with respect to the described geological structure; the third looks at the subsequent uplift and associated deformation of the palaeosurfaces; in the fourth, the uplift and erosional history of the Western Ghats is discussed with respect to both the long-term evolution of the Indian continental margin and the key morphotectonic factors which have controlled both palaeosurface formation and their subsequent uplift and deformation.

Geological and geomorphological setting

The studied area lies in the southwest corner of the Deccan Traps continental flood basalt (CFB) province of western peninsular India (Fig. 1). Geologically speaking, the Deccan lavas were rapidly erupted over a period of 0.5–5 Ma at, or near to, the Cretaceous–Tertiary (K/T) boundary (e.g. Courtillot *et al.* 1986; Duncan & Pyle 1988; Baksi 1994). These lavas formed a broadly lenticular mass consisting largely of sub-horizontal basaltic lavas which today cover an area of approximately 500 000 km^2. Thicknesses of > 1200–1700 m are exposed along the Western Ghats, though total thicknesses probably exceed 2000 m along the margin of the continent (see Mahoney 1988 for general review). The original extent of the Traps may have exceeded 1 500 000 km^2 (Krishnan 1956) but because the lava sequences were rifted apart during the separation of the Seychelles microcontinent, the present Deccan region only represents the easterly, albeit larger, fragment of the original CFB province. More westerly extremes appear to have foundered either as offshore fault blocks (Naini & Talwani 1983; Chandrasekharam 1985), or else comprise part of the submerged Seychelles–Mascarene plateau (Devey & Stephens 1991).

The Indian continental margin is geomorphologically interesting since it displays many features endemic to other rifted CFB provinces (e.g. Paraná of Brazil; Karoo of SE Africa; and Etendeka of SW Africa). These features include an elevated inland plateau; a huge, erosionally controlled escarpment; a coast-parallel monoclinal flexure; and a low-lying coastal plain. This morphological similarity clearly indicates a shared pattern of tectonic and geomorphological evolution for such margins. More uniquely, the Indian margin has, through conducive climatic and tectonic factors, been the site of extensive laterite development during the Tertiary. It is these laterites, and in particular the surfaces upon which they formed, that provide a substantive record of the post-rift evolution of the Deccan margin.

The two lateritized palaeosurfaces identified within the southwest Deccan lava field differ markedly with regard to their ages and the type of geomorphological surface upon which they evolved. The older, high-level Ghats crest laterite began to develop upon the post-Deccan lava plain shortly after eruptive phase had ended; the low-level Konkan laterite developed much later upon an low-lying erosion surface, or pediment, formed by eastward recession of the western Ghats escarpment. A combination of extensive field-based studies and LANDSAT imagery enables mapping of the remnants of these two lateritized palaeosurfaces. These image data reveal that remnants of the high-level, Ghat crest laterite currently only occur as a capping layer upon isolated mesas, whereas the low-level, Konkan material forms laterally extensive, but deeply dissected plateaux. This distribution reflects their different ages and consequently the degree of erosion they have suffered. Reconstruction and height contouring of these lateritized remnants demonstrates that large-scale warping has affected both the upper and lower-level surfaces and, moreover, the variation in the degree of deformation records the style of epeirogenic uplift throughout the Tertiary. It is important to note that such long-lived uplift experienced at this and other passive margins is difficult to reconcile with thermally driven rift-related processes, and is better explained in terms of isostatic uplift in response to denudational unloading.

The Deccan Flood Basalts

The Deccan has been the focus of much recent research and debate because it has been used as a prime example of the interaction between continental rifting and hotspot plumes which, when their effects are combined, provide a mechanism for generating the huge volumes of lavas associated with CFBs (e.g. White & McKenzie 1989; Richards *et al.* 1989; Kent *et al.* 1992).

During eruption of the Deccan CFB province the chemical composition of the lavas changed. The causes of this secular variation were a combination of different petrogenetical processes which profoundly affected lava composition (e.g. variations in mantle source, fractional crystallization, and degree of crustal contamination). Consequently, the chemical compositions of the lavas vary in any given section through the lava pile (e.g. Najafi 1981; Cox & Hawkesworth 1985) thus providing the basis for a lava stratigraphy (Table 1). Such stratigraphical studies were first carried out along

Table 1. *Deccan basalt stratigraphy. Each formation is defined on the basis of the characteristic chemical signatures of the constituent lava packages. Chemical variation between different lava packages is the result of petrogenetic processes affecting lava composition which have resulted in shifts in Sr and Ba contents, Zr/Nb and Ba/Y ratios, and the* $^{87}Sr/^{86}Sr$ *isotope ratio (for further details see Mitchell & Widdowson 1991).*

Group	Sub-Group	Formation	Thickness (m)
		Panhala	> 175
	Wai	Mahabaleshwar	280
		Ambenali	500
		Poladpur	400
Deccan Basalt	Lonavala	Bushe	325
		Khandala	100–180
		Bhimashankar	< 140
		Thakurvadi (upper)	
	Kalsubai	Thakurvadi (middle)	210–400
		Thakurvadi (lower)	
		Neral	080–145
		Igatpuri	> 150
		Jawhar	n/d

n/d, not determined.

the Western Ghats, where relief is strongest and the exposure along road sections is excellent. Combining the results of several such geochemical studies, the stratigraphy and structure of the lava pile has since been determined on a regional scale (e.g. Subbarao & Hooper 1988). Huge tracts have been mapped using the chemostratigraphical technique and the structure of the southern Deccan lava pile is now well established and accepted (e.g. Devey & Lightfoot 1986; Beane *et al.* 1986; Khadri *et al.* 1988; Mitchell & Widdowson 1991; Khadri & Nagar 1994; Subbarao et al. 1994).

At the broadest scale, the structure of the Deccan province forms the easterly fragment of a lensoid structure with the thickest sequences occurring near its original centre in the Nasik area (Fig. 1). This central area appears to have undergone the most erosion exposing the oldest formations of the Kalsubai subgroup (Fig. 2), with progressively younger lavas preserved towards the periphery (Subbarao *et al.* 1994). The present study area represents the southern peripheral region and comprises lavas of the uppermost, Wai Subgroup. Within the study area the stratigraphy reveals a southward stratigraphical overstep by successively younger lava units (Fig. 3a). This overstep is considered to be a primary stratigraphical feature consistent with a southerly migration of the volcanic centre during eruption, and may be explained in terms of the movement of the Indian plate over the reunion hotspot *c.* 60–65 Ma. (e.g. Cox 1983; Mitchell & Widdowson 1991). Detailed stratigraphical investigation also reveals that the formation geometries define a large, southerly dipping, anticlinal–monoclinal feature which may be traced 10–30 km inland (east) of the Western

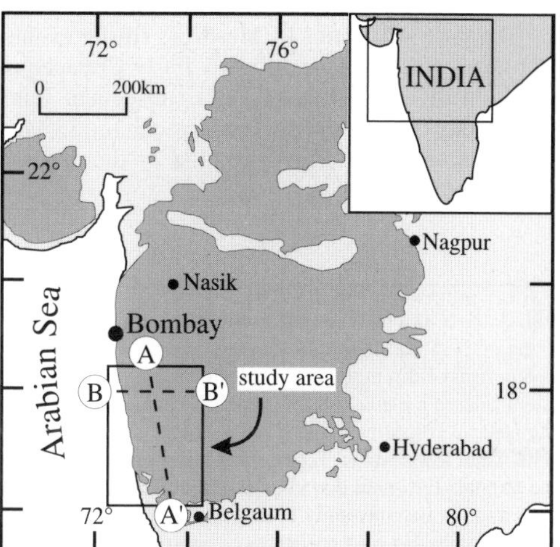

Fig. 1. Areal extent of the Deccan Traps continental flood basalt (CFB) province in northwest peninsular India (dark shading). Rectangle outlines the studied area in the southwest of the province. For details of lines of section A–A' and B–B' see Figs 3a and 3b.

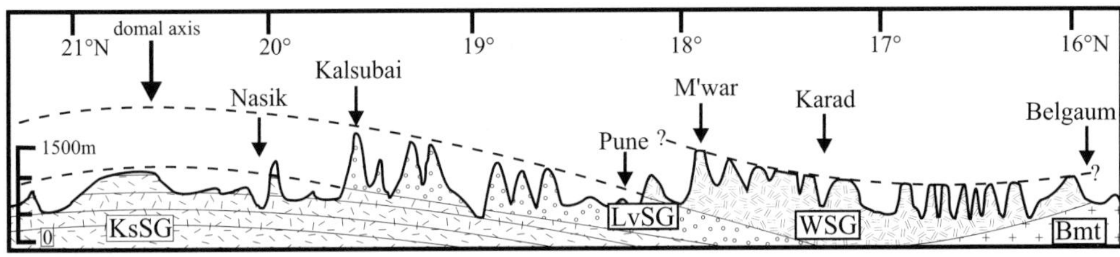

Fig. 2. Simplified N–S sketch section between 16° N and 21° N running along the Western Ghats though Nasik and Belgaum (see Fig. 1), showing geometries of main stratigraphical Sub-Groups of the Deccan lava sequence. Note central 'dome' structure developed north of the Nasik region. Bmt, pre-Deccan basement; KsSG, Kalsubai Sub-Group; LvSG, Lonavala Sub-Group; WSG, Wai Sub-Group.

Ghats escarpment between 19°30′N–15°30′ N (Figs 3b & 4). This is interpreted as a post-eruptive modification to the lavas (Widdowson 1990).

It is important to note that these stratigraphical and structural studies unequivocally demonstrate an absence of any large-scale faulting within the study area. The lack of faulting in vicinity of the escarpment and Konkan Plain, together with an absence of any large-scale, seismically-determined structural offset of the base of the volcanic sequence across the line of the Western Ghats (Kaila *et al.* 1981) confirms that the present escarpment should be considered as a purely erosional feature.

Geomorphology of the Deccan Volcanic Rifted Margin (VRM)

The evolution of the west coast of India is the consequence of a series of Mesozoic rifting events beginning with the Late Jurassic–Early Cretaceous break-up of Gondwanaland, and subsequent Late Cretaceous detachment of Madagascar (*c.* 88 Ma; Storey *et al.* 1995). The formation of the present Indian margin was the result of a ridge-jump which detached India from the Seychelles bank at the time of the Deccan volcanism (*c.* 65 Ma; Norton & Sclater 1979; Chandrasekharam 1985). Today, the most striking geomorphological feature of western India is the Western Ghats escarpment running the entire length of peninsular India. The origins and nature of this huge feature have been the subject of morphological study and debate for over a century (e.g. Foote 1876; Medlicott & Blandford 1879; Auden 1975; Ollier & Powar 1985; Subrahmanya 1987, 1994; Radhakrishna 1991). Although important, the escarpment forms only one component of the geomorphology of the western Indian margin. From the present-day coastline eastwards four broad geomorphological zones may be recognized (all elevations are given above current mean sea level).

(*i*) *The Konkan plain.* A low-lying coastal strip, 50–60 km wide, lying between the coastline and Ghats foothills. Within the study area, it comprises a coastal belt of dissected, laterite-capped plateaux which have a maximum elevation of 250 m in the north (e.g. Srivardhan) and 70 m in the south (e.g. Devgarh). The inland tract adjacent to the Ghats is

Fig. 3. N–S and E–W sketch sections through the southwest Deccan (section lines A-A' and B-B', Fig. 1). (**a**) Detailed N–S sketch section of the formation geometries (Wai subgroup) within the study area. The southerly component of dip in the Mahabaleshwar area is *c.* 0.3°. Southerly *overstep* by successively younger basalt lava units is considered a *primary* stratigraphical feature consistent with the southerly migration of the volcanic centre during the Deccan eruptions (Cox 1983; Mitchell & Widdowson 1991). Cross-hatched ornament shows position of high-level (H.L.) and low-level (L.L.) laterites upon the lava stratigraphy. High-level laterites are developed upon basalt flows of the Panhala Fm (note that laterite-capped mesas only occur south of latitude 18°09′ N), and the low-level laterites lie with angular unconformity upon exposed basalts of the Poladpur and Ambenali Fms which form the Konkan (coastal) plain lying to the west of the Ghats escarpment. (**b**) Detailed W–E sketch section at 18°00′ N showing detail of the coast-parallel assymetric anticlinal structure developed across the Ghats crest zone. The westerly component of dip across the outer Konkan is *c.* 0.5°–0.7° whilst generally a much lower easterly components (e.g. < 0.1°) exist across the Maharashtra plateau. West of the Ghats escarpment erosion has removed *c.* 1–1.5 km of basalt exposing progressively older formations towards the anticlinal core. High-level laterites are restricted to the Ghats crest zone; low-level (Konkan) laterites are developed upon lavas of the Ambenali and Poladpur Fms exposed in the anticlinal core.

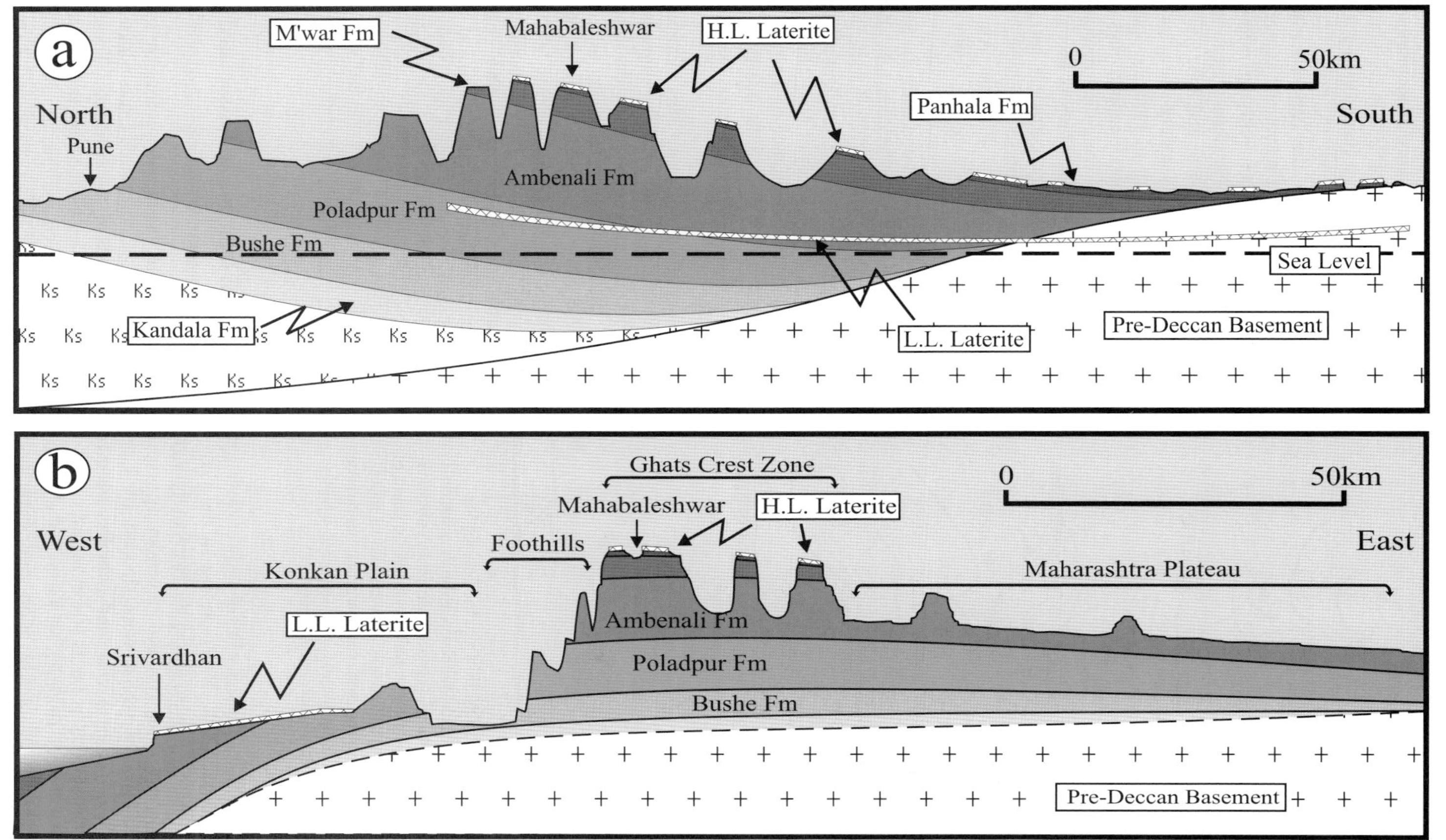
a
North
Pune
M'war Fm
Mahabaleshwar
H.L. Laterite
Ambenali Fm
Panhala Fm
0
50km
South
Poladpur Fm
Bushe Fm
Sea Level
L.L. Laterite
Pre-Deccan Basement
Kandala Fm
Ks
b
West
Srivardhan
Konkan Plain
L.L. Laterite
Foothills
Mahabaleshwar
Ghats Crest Zone
H.L. Laterite
Ambenali Fm
Poladpur Fm
Bushe Fm
Maharashtra Plateau
0
50km
East
Pre-Deccan Basement

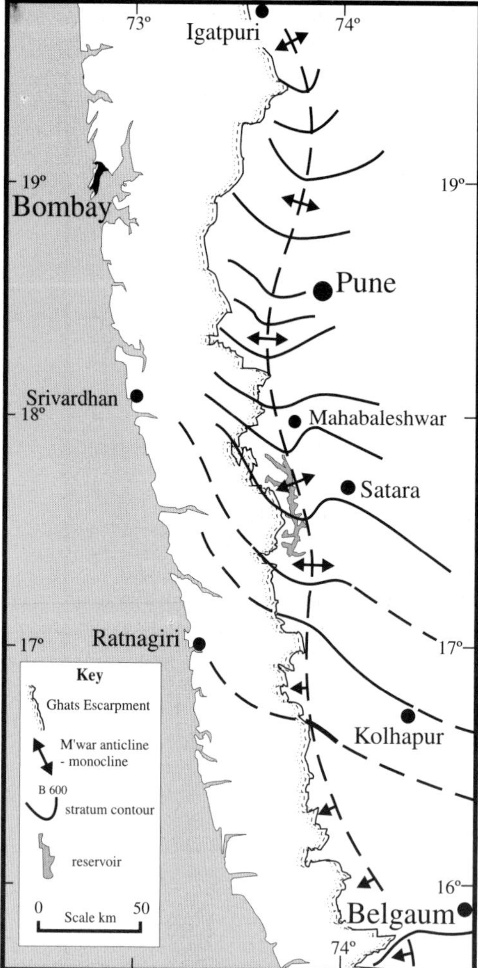

Fig. 4. Stratum contours drawn to the base of the Thakurvadi, Bushe, and Mahabaleshwar Fms which together define the large, southerly-dipping, anticlinal feature which becomes essentially monoclinal south of 17°15'N. It is important to note that the axis of the feature is not coincident with the scarpline, but between 19°30′ N–15°30′ N commonly lies 10–30 km inland (east) of the Western Ghats escarpment.

not lateritized and typically lies at a lower elevation than the coastal belt. Rivers rising in the Ghats foothills drain westward across the Konkan through a series of deeply incised meandering river systems, many of which were flooded by Late Pleistocene sea-level rise and which currently display tidal effects as far as 40 km inland.

(*ii*) *The Ghats foothills.* A series of elongate spurs 300–600 m in elevation and extending westward from the main Ghats escarpment. These steep-sided features have been carved by the fast-flowing, short river courses which drain the western flank of the Ghats.

(*iii*) *The Western Ghats.* Topographically the highest zone, reaching 1450 m in the northern part of the study area, and incorporating the main escarpment and associated isolated plateaux and ridges immediately to the east. Laterite-capped mesas form the highest elevations adjacent to the escarpment, whilst lower, outlying ridges project eastward across the Maharashtra plateau so that the boundary between this zone and zone (iv) is highly intricate in plan. This boundary is marked in most places by a combination of elongate spurs, interfluves, and isolated basalt-capped mesas which together comprise an irregular, discontinuous, inland-facing escarpment which is less steep than the main seaward-facing escarpment, and generally less than half the elevation.

(*iv*) *The Maharashtra Plateau.* An elevated inland plateau with little relief, comprising the flood plains of major easterly draining river systems (e.g. the Krishna and Bhima). Toward the north of the study area the Maharashtra plateau lies at *c.* 800 m, falling to 640 m near Satara, and 550 m near Kolhapur; an average gradient of less than 1.5 m per km. At distances of 10–20 km east of the Western Ghats the coalesence of the river systems has removed virtually all of the interfluves leaving only occasional basalt mesas as the only topography.

The edge of the Western Ghats currently forms the main drainage divide of Peninsular India. Eastwardly draining tributary rivers feeding the Krishna and Bhima rise adjacent to the Ghats escarpment, often within 50 km of the Arabian Sea, yet discharge into the Bay of Bengal. Such extreme asymmetry of continental-scale drainage pattern is not unique to western India, and comparable examples in southern Brazil (Paraná) and southeast Africa (Karoo) were used by Cox (1989) to argue that the initiation of such drainage patterns was a result of the dynamic plume uplift prior to continental rifting. If this is the case, the Deccan drainage on the Maharashtra plateau has its origins *c.* 65 Ma.

The Deccan laterites and associated palaeosurfaces

In southwest Deccan laterite profiles are capped by erosionally resistant iron-rich crusts which form the uppermost levels of the residuum. These laterite weathering profiles formed by the *in situ* breakdown of basalt lava and thus can be considered as primary laterites. Preservation and exposure of entire profiles is uncommon, but in some localities erosion has provided examples of complete progressions from basalt through to the indurated vermiform laterite. These occur either in the form of precipitous cliffs around the margins of the high-level mesas of the Western Ghats, or along the tops of the incised Konkan river valleys. However, access to the mesa-top profiles is notoriously difficult, and whilst access to the valley sections is generally easier, degradation of Konkan profiles by mass movement and dense vegetation often mean that detail is partially masked.

Where fully exposed, the nature of these profiles agrees with the model of groundwater laterite formation described by McFarlane (1976). They typically comprise a top layer of highly indurated vermiform laterite which passes down into softer vermiform and pisolitic layers, then into unindurated saprolitic and weathered basalt horizons below. The structure of individual laterite profiles is similar throughout the southwest Deccan and indicates a common pattern of evolution for primary laterite development in the region.

In addition to the primary laterite profiles described above, allocthonous accumulations of lateritic material may be also be recognized. These are restricted to the Konkan where they typically occur in locally restricted topographical depressions. Such material is clearly secondary in origin because it often incorporates a variety of lithic clasts, older laterite debris and, importantly, it does not display the typical alteration profile since it lies directly upon unaltered basalt bedrock. As a consequence, this more modern lateritic material can be readily identified on the basis its physical appearance, mineralogy, and chemical composition. These minor occurrences are, therefore, excluded from the following discussion which is concerned with development, distribution and destruction of the widespread primary laterites.

Mapping Deccan laterite

Figure 5 is a map of laterite distribution in the southwest Deccan based upon a combination of extensive fieldwork and satellite image processing. The size of the region studied, large number of lateritized plateaux, together with their remote location and logistical difficulties in reaching individual mesas prevents practical survey of many sites. To produce a comprehensive map of the extent of Deccan laterite, and details of its elevation throughout the studied region, the laterite occurences in remote areas were identified using satellite imagery (LANDSAT MSS) and the resulting location information combined with elevation data from 1:63 360 (1 inch: 1 mile) topographical maps.

Many satellite-based lithological discrimination techniques rely on the presence of characteristic absorption features within the reflection spectra of different lithologies. For geological purposes the most useful absorption features are the electronic transitions and charge transfer in the transition metals (e.g. Fe^{3+}, Ti^{3+}, Cr^{3+}, etc.). These can be used to determine the composition of surfaces comprising metallic silicates and metallic oxides which are typical constituents of rock materials (see White 1997). Furthermore, the processing technique may be further refined to differentiate surfaces with dissimilar iron oxide content (White *et al.* 1992). Consequently, the contrasting chemistry of basalt and indurated laterite surfaces in the study area, coupled with differences in vegetation cover, has permitted accurate demarcation of lateritized areas using a combination of suitable LANDSAT MSS scanner bands.

The particulars of the image processing technique are discussed in detail elsewhere (Widdowson 1990), but the technique is based primarily upon detailed field survey of accessible sites. This ground control information was then used to design and refine image processing techniques aimed at enhancing other areas of laterite. There are two main factors which permit laterite mapping from satellite. First, expanses of indurated laterite support only sparse vegetation and this, together with severe die-back during the dry season, results in semi-continuous exposure of the laterite surface over very large areas; by contrast, the modern, non-indurated weathering products and soils which accumulate locally around plateaux footslopes or in depressions, tend to retain some degree of vegetation cover throughout the year thus providing an effective masking to this material. Second, exposed indurated laterite surfaces have a characteristically high iron content (i.e. 50–80% Fe_2O_3), which contrasts with both the modern, unconsolidated weathering products (i.e. 20–30% Fe_2O_3) and the unaltered basalt (11–18% Fe_2O_3) beneath.

The resulting satellite-generated map is demonstrably accurate because it recognizes not only those areas verified during subsequent field observation, but also effectively identifies all of the laterite occurrences listed in the detailed survey of Fox (1923) who achieved his survey by enlisting the help of local peoples. Most importantly, the technique identifies many more inaccesible and hitherto undocumented occurences of laterite-capped mesas high in the Western Ghats and delimits accurately the large expanses of Konkan laterite.

Laterite distribution

Widespread occurrence of indurated laterite has long been recognized in the southwest Deccan

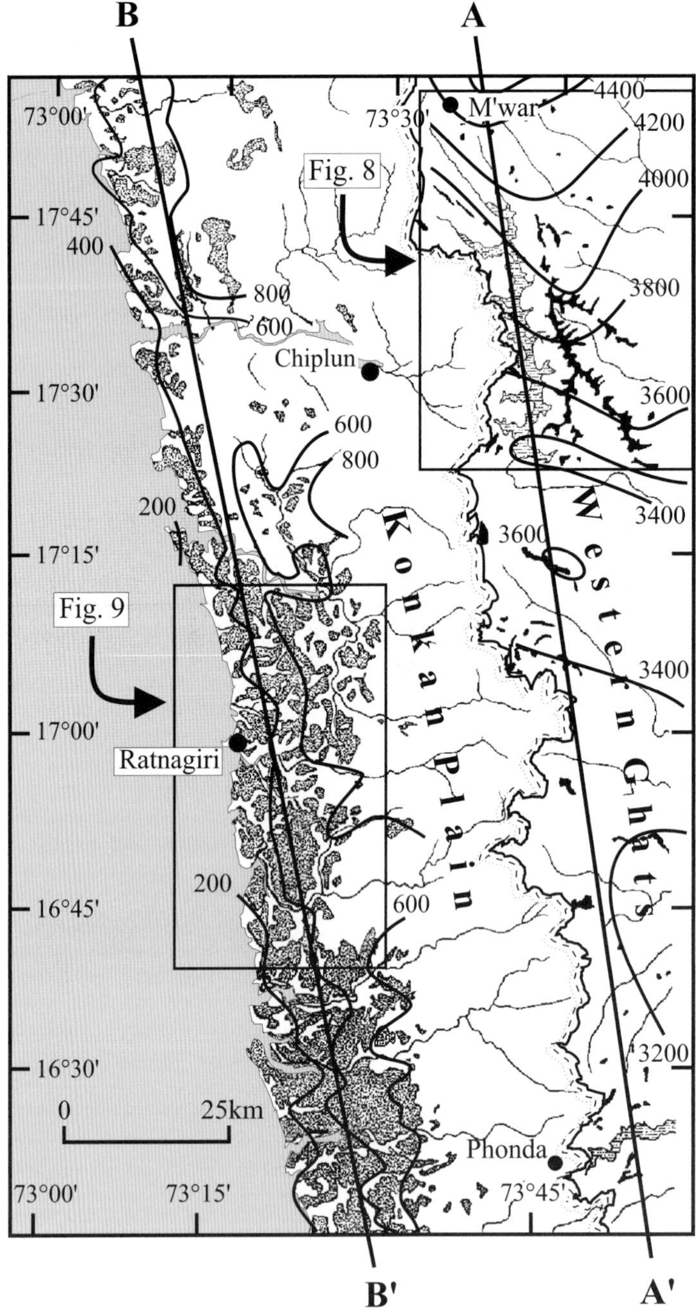

Fig. 5. Map of laterite distribution in the southwest Deccan. The mapped outcrop has been determined using a combination of suitably processed LANDSAT images and extensive field survey. Distribution of high-level laterites (black shading) is restricted to the highest mesas of the Ghats crest zone. Low-level (Konkan) laterite (dark stipple) forms a wide, coast-parallel belt most extensively preserved in the south. Contours revealing the current form of both lateritized palaeosurfaces are drawn using elevation data from the areas of preserved laterites. Note that since height data is taken from original Indian 1 inch:1 mile maps, all contour elevations are given in feet; where appropriate text coverts elevations to metres. Inset rectangles show areas of detailed laterite maps Figs 8 and 9. Detail of section lines A-A' and B-B' drawn along the lateritized palaeosurfaces is given in Fig. 11.

Fig. 6. Precipitous cliffs around the margins of the high level mesas of the Western Ghats along the Bamnoli Hills, near Patan (73°57′ 17°23′; elevation *c.* 1070 m). The laterite cliff section is 10–15 m in height and represents an entire weathering profile from unaltered basalt (Panhala Fm) through to the capping layers of indurated, vermiform laterite. Laterite breakaways occur around the edges of these mesas, and the basalt slopes below the cliffs are littered with blocky debris derived from the indurated layers capping the profile.

(Foote 1876; Medlicott & Blandford 1879; Fermor 1909), and Fox (1923) gives detailed reference to many individual occurrences of laterite plateaux. However, the distribution of Deccan laterite is by no means uniform. This investigation confirms that indurated laterite is found in two distinct topographical situations, these being: (i) capping the high mesas of the Western Ghats (zone 2; Fig. 6) and (ii) a wide belt developed upon the outer Konkan plain (zone 1; Fig. 7). King (1967) identified these two topographically distinct occurrences as the 'High-level' and 'Low-level laterites'.

High- or upper-level (Ghats crest) Laterites. The image data reveal that the high-level laterites are entirely restricted to the flat mesa tops of the highest peaks (900–1500 m) along the Ghats ridge (Fig. 5). However, north of Bhor (73°52′, 18°09′) there is little evidence of further mesa-cappings along the Ghats crest zone, and near Nasik (73°43′, 19°35′) the highest point of the Ghats (Kalsubai peak, 1650 m), is composed entirely of eroded basalt masses. South of 18°00 N, isolated laterite-capped basalt mesas are common along the Ghats ridge to the southern limit of the lavas near Belgaum (74°26′, 15°50′). Beyond Belgaum the laterites pass onto Archaean and Proterozoic basement rocks which form the southern continuation of Western Ghats, though in the Karnataka uplands laterite occurs as a discontinuous rubbly crust (Gunnell 1997) rather than the well-defined mesa cappings of the Deccan.

The Deccan high-level laterites typically form a distinctive 10–20 m high cliff around the margins of the basalt mesas often displaying a basal 'notch' where the profile passes from the indurated upper layers into the saprolite and altered basalt below. Surface relief across the mesa cappings is virtually nil. The presence of cliff breakaways and raft-like blocks of indurated laterite debris on lower slopes testify to the ongoing destruction of these profiles and, in some locations, areas of less than a few hundred square metres remain. The most extensive occurrences (> 200 km^2) lie between the Koyna reservoir and Vena river valleys capping the Bamnoli hill range (Fig. 8). These comprise a series of long sinuous laterite-capped ridges or interfluves extending 70 km SSE from the Mahabaleshwar

Fig. 7. View across the Konkan laterite plateau inland of Ratnagiri (73°24′ 17°04′; elevation 200 m). Note the gently undulating form of this lateritized plateau. The patchy, semi-continuous exposure of the Fe-rich indurated laterite and vegetation die-back during the dry season are the main factors which permit accurate mapping from satellite images.

massif. Elsewhere, at numerous localities immediately east of the escarpment, and at elevations just below the undisturbed laterite cappings, occur extensive undulose areas littered with 3–4 m thick laterite rafts of up to 10 m^2. These are indicative of a general topographical lowering where the broken fragments and sheets of the resistant topmost layers of the profile remain as block fields following erosion and removal of the lower, less indurated layers. More commonly, the hills which occur below the preserved lateritic level display a ridge-like morphology despite the layered nature of the constituent basalt. Field observation indicates that this is the pattern typical of later morphological evolution of the basalt ridges which acts once the laterite cap is removed and a direct chemical and physical attack can begin on the lavas beneath.

Most importantly, in addition to the flatness of the undisturbed mesa surfaces, there is a striking concordance of elevation between adjacent lateritized summits within mesa groups and, on a regional scale, between the mesa groups themselves. Such remarkable similarity in elevation cannot be coincidental since it is most unlikely that individual mesas evolved laterite cappings independantly. More probably, this concordancy indicates that they originally formed part of a much more continuous surface which has been subsequently dissected and, for the most part, destroyed by later erosion.

Low-level (*Konkan*) *Laterites*. Much of the outer Konkan between Srivadhan (73°01′, 18°03′) and Devgarh (73°23′, 16°22′) comprises dissected lateritized plateaux forming a coast-parallel belt typically 20–30 km wide but which narrows to about 15 km in the north (Fig. 5). In detail, Konkan laterite often forms a characteristic capping to the coastal cliffs (e.g. at Ratnigiri), and is particularly common immediately inland. It occurs most extensively in the region between 16°15′ N and 17°15′ N but the Konkan laterite plateaux generally extend no further inland than 73°40′ E, and have a quite marked easterly boundary with the inner Konkan. The inner Konkan comprises a low (i.e. 30–100 m elevation), 20–30 km wide, unlateritized corridor lying between the coastal plateaux and the Ghats escarpment. Isolated laterite-capped plateaux are documented as far north as the Bombay area, for example at Matheran and Tungar Hills (73°16′, 18°59′; 72°50′, 19°24′) (Wilkins *et al*. 1994), whilst southwards, beyond 16°30′ N, they are widely developed upon the pre-Deccan basement rocks of Goa and Karnataka states.

The low-level laterites also form flat, poorly

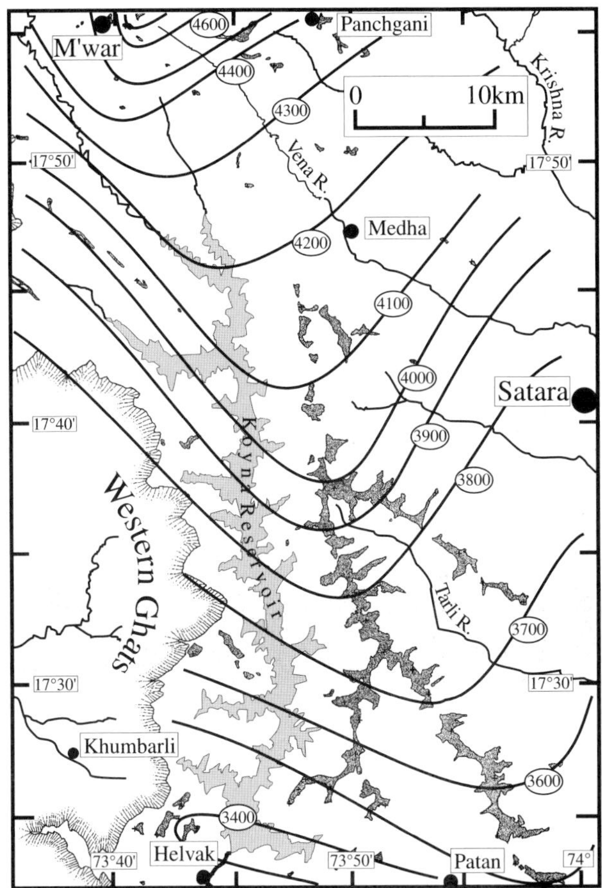

Fig. 8. Detailed map of high-level laterite plateaux (dark stipple) comprising the summits of the Bamnoli Hill range which lie between the Koyna reservoir and the Vena and Tarli rivers. Other discontinuous and isolated lateritized mesa summits lie immediately inland of the Western Ghats escarpment and locally upon the Mahabaleshwar (M'war) - Panchgani plateau in the north. Note that contours (in feet) drawn using these remaining high-level laterite localities reveal an anticline structure exactly similar to that developed in the basalt stratigraphy beneath (see Fig. 4).

vegetated expanses, but the surface topography of these low-level lateritized plateaux is gently undulose in character (Fig. 7). South of 17°15′ N huge tracts of semi-continuous outcrop (*c.* 2000 km^2) can be easily discerned on the satellite imagery, broken only by the deep meandering gorges of the main westward-flowing river systems of the Vashishti, Shashtri, Kajvi, Muchkundi, and Vaghotan (Fig. 9). Such deep entrenchment of the Konkan drainage indicates a relative base-level fall subsequent to the lateritization of the outer Konkan. As observed with the upper-level plateaux, the laterites of the coastal belt also lie at concordant elevations indicating that they originally comprised a lateritized palaeosurface of regional extent prior to the river incision.

Interestingly, whilst the low-level laterite typically covers the more elevated regions of the outer Konkan plain, it does not always form the highest topography because, in some instances, unaltered basalt spurs rise above the general level of the Konkan laterite. Field investigation reveals no evidence of eroded laterite blocks or other lateritic detritus upon the flanks of these ridges which might otherwise indicate that they were the remnants of another lateritized surface intermediate between those of the high- and low-levels. Instead, it seems that Konkan laterite developed around the footslopes of these ridges leaving them as unaltered basaltic ‘islands’ upon the Konkan.

Development of the Deccan Laterites

It has long been recognized that topography and climate are fundamental controls on laterite formation. Sub-tropical regions with wet, or seasonally wet climates (Tardy *et al.* 1991), and exhibiting only minor topographical relief are commonly cited requisite conditions (McFarlane 1976, and references therein). Time is also a crucial factor since prolonged periods are required to produce the residual accumulations of the least mobile elements, principally iron and aluminium, which together constitute the lateritic profile. The

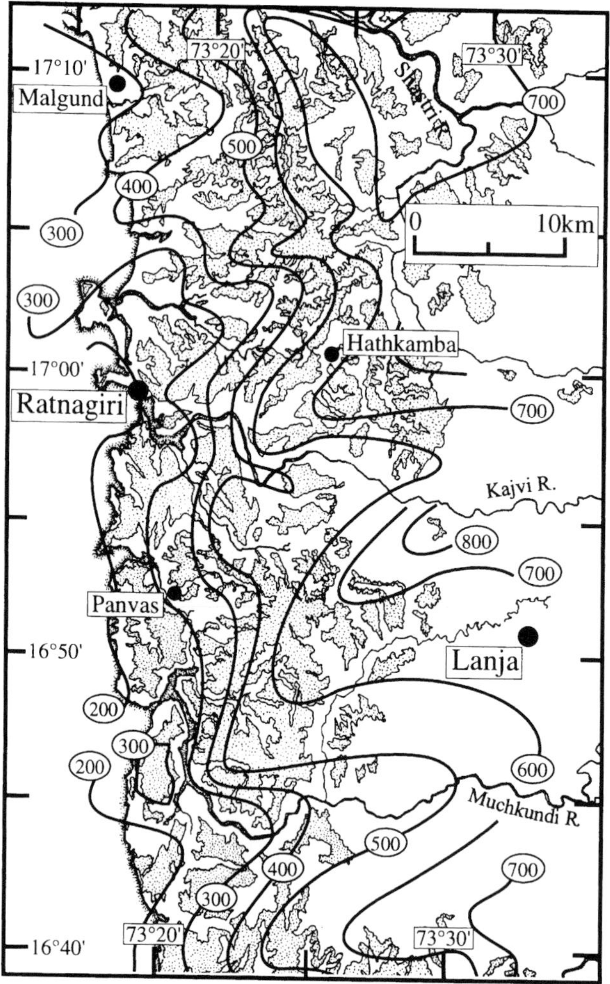

Fig. 9. Detailed map showing the distribution of the low-level (Konkan) laterite plateaux inland of Ratnagiri (stippled regions). This is one of the most extensive regions of indurated laterite, and forms a semi-continuous blanket across the gently undulose topography. Deeply entrenched (80–120 m), westward-flowing river meanders (e.g. Shastri, Kajvi, and Muchkundi river systems) have now dissected this laterite belt. Contours (in feet) display the general westward slope of the lateritized surface. Note also inland deflection of the contours along the courses of these major rivers.

existence of low-relief topography is considered favourable to laterite formation because it permits ingress of water and prevents active run-off and erosion of the resulting weathering mantle (e.g. Goudie 1983; Summerfield 1991; Thomas 1994). Remnants of regional laterites, similar in character to those of the Deccan, are recognized in the continents of South America and Africa (e.g. Aleva 1981; McFarlane 1983; Späth 1987), and many of these laterites are associated with peneplains of great antiquity, a connection that prompted past authors to consider lateritization as fundamental to emerging models of landscape evolution and denudation chronology (e.g. Campbell 1917). It should be noted, however, that regional peneplanation is not necessarily a pre-requisite for lateritization since studies in Uganda (McFarlane 1976) raise questions regading the use of tiered duricrusts as chronological marker horizons. Of more relevance to the current study are conditions of prolonged tectonic stability (Bardossy 1981) which assist in the evolution and preservation of topographical surfaces, and which result in the limited drainage incision and run-off conditions conducive to extensive laterite development.

It is now held that laterites evolve with the landscape because a thick laterite profile is a consequence of water table lowering associated with limited vertical erosion; protection by the resulting duricrust development then exerts a fundamental control upon later morphological development (McFarlane 1976). However, in the case of the Deccan it is clear that the elevated plateaux upon which the upper and lower-level Deccan laterites currently exist would have been inimical to laterite formation. The fact that these Deccan laterites are in disequilibrium with their present topographical environment is further demonstrated by their advanced and ongoing destruction. However, it would be wrong to suggest that elevated laterites become fossilized (Woolnough 1918), because in the studied area solute-charged springs occur at the base of some mesa profiles attesting to continuing limited element mobilization. Nevertheless, in the Deccan it is clear that the widespread development of laterite relates to periods in the geological past when tectonic, topographic and climatic conditions were favourable. Moreover, any recent changes to the structure of the laterite profiles appear to have been very limited in their effects since both upper- and lower-level laterites are known to retain ancient palaeomagnetic signatures (see below). Thus, given the widespread distribution and the height concordancy of mesa groups, it is reasonable to conclude that both the upper- and lower-level laterites developed upon palaeosurfaces of regional extent and, importantly, that evolution of these palaeosurfaces considerably predates the development of the present topographical environment.

Since the remnants of two lateritized surfaces are recognized, a geomorphological model is required which adequately explains their existence within the geological context outlined earlier. The first step in constructing such a model, is to determine the relative ages of the upper- and lower-level laterites. The second step is to construct contours upon the two lateritized palaeosurfaces using a combination of elevation data and the laterite distribution data derived from the satellite imagery; an exercise which effectively aids reconstruction of the ancient lateritized palaeosurface within the study area and, importantly, reveals the current palaeosurface morphology. This latter provides a record of subsequent neotectonic effects which have affected the margin. Establishing both the relative age of Deccan laterites and the morphology of the reconstructed palaeosurfaces are crucially important because their development and subsequent deformation and destruction places fundamental contraints upon the geomorphological evolution of the Indian continental margin.

Age of the Upper and Lower-Level regional laterites

Determining the age of laterite is problematic. Clearly, since both upper and lower-level laterites are developed upon the Deccan basalts, the age of these lavas (*c.* 65 Ma) puts a maximum constraint upon their antiquity. However, more accurate dating proves difficult for two main reasons: First, unlike instances elsewhere in India (e.g. the coastal laterites of Gujurat, Valeton 1983), there is an absence of dateable sediments overlying the laterites of the southwest Deccan which could provide the basis for a minimum age. Second, since evolution of thick laterite profiles requires protracted periods of subaerial alteration, ascribing a precise geological age may be impracticable. Fortunately, other lines of evidence which clarify the matter are available in the form of geomorphological observation, chemostratigraphical studies illustrating basalt-laterite relationships, and palaeomagnetic data.

Early investigations (e.g. Medlicott & Blandford 1879) had suggested a fundamental age difference between upper- and lower-level laterites on the basis of geomorphological observation. However, because there was no way of determining the precise relationship between basalt stratigraphy and the overlying laterites, the idea could not be explored further. The recent chemostratigraphical studies (Table 1) reveal that west of the Ghats scarpline the process of scarp recession has removed 1–1.5 km of lavas which, within the studied area, comprise much of the Wai Subgroup (i.e. Poladpur to Panhala Formations; Fig.3b). The Konkan coastal plain therefore represents a pedimented tract resulting from this scarp recession. Since the lower-level laterite developed upon this pediment it is highly discordant with the basalt stratigraphy and can only have begun its development after significant erosion of the lava pile had already occurred. The nature of this unconformity between the lower-level Konkan laterite and the structure and stratigraphy of the lavas may is explored by Widdowson (1990) and Widdowson & Cox (1996) using detailed geochemical investigation of Deccan laterites (i.e. trace element signatures and ratios). Of importance is these authors' conclusion that no unconformity exists between the topmost (Panhala) Formation and the upper-level laterite. This latter observation is crucial since it suggests that no significant erosion of the lavas occurred prior to the development of the upper-level laterites. On this basis of this evidence alone, the upper-level laterite must considerably predate development of lower-level laterite upon the Konkan pediment.

Interestingly, a major difference in the palaeomagnetic ages of the upper- and lower-level laterites was demonstrated by Schmidt *et al.* (1983). These data demonstrate a late Cretaceous–early Tertiary age for the upper-level laterites, and a mid- to late Tertiary age for the low-level laterites. However, in the absence of the detailed morphological and structural information now available, these authors were unable to forward an evolutionary model for the upper- and lower-level laterites other than to describe the observed height-age relationship. Moreover, even with modern analytical techniques it may be difficult to demonstrate more precise palaeomagnetic ages because it is debatable whether an exposed laterite can ever become truly 'fossilized' in the geological sense (Bourman *et al.* 1987). Nevertheless, the fact that these Indian laterites data do record an ancient magnetic pole, together with periods of magnetic reversal (Kumar 1985), illustrates that any subsequent magnetic resetting by later alteration to the laterite profile has not masked the earliest signatures. Most importantly, these palaeomagnetic data indicate that development of the upper laterite must have begun almost immediately after the lava eruptions had ended; a conclusion consistent with the observed conformable relationship between the upper-level laterite and Panhala formation described above. The fact that pervasive subaerial weathering had already become a common phenomenon toward the close of the Deccan volcanic episode is further evidenced by an increasing frequency of bole horizons toward the top of the basalt stratigraphy (Widdowson *et al.* 1997). In this context, the upper laterite should perhaps be considered as a final 'super-bole' which, uninterrupted by subsequent lava eruptions, was allowed to mature into a thick laterite profile. With regard to the low-level laterite, the palaeomagnetic data confirm that its evolution began much later, which is consistent with the time lapse required for development of the Konkan pediplain by scarp recession.

A model for the development of the upper and lower-level lateritized palaeosurfaces

Toward the end of the Deccan volcanic episode climatic conditions suitable for lateritization were already well established. Long hiatuses in eruption (*c.* 1000–5000 years) permitted deep weathering of the lavas evinced by the presence of thick red boles and associated palaeosol horizons. Once volcanism had ceased completely, the extensive lavas of western India presented an endogenic plain of low relief and immense lateral extent where alteration of the surface flows could proceed uninterrupted. This combination of climatic and topographical factors provided an ideal site for extensive laterite development; protracted alteration, coupled with restricted run-off, slow rates of incision, and associated ingress of ground waters, resulted in the development of thick laterite profiles. It is the remnants of this lateritized, 65 Ma lava-plain palaeosurface (henceforth termed the post-Deccan surface) which currently cap the highest mesas of the Ghats.

Although the low-relief nature of the post-Deccan palaeosurface can be reliably assumed by analogy with modern lava field topography, determining its original elevation is more problematic. If it is assumed that elevations were primarily the result of crustal thickening resulting from the volcanism, then given the currently observed maximum thickness of 1.5 km for the lavas, and assuming similar thicknesses of associated magmatic underplating (Cox 1980), a very simple calculation using Airey isostasy suggests that the lava plateau may have originally lain at an elevation of 500–700 m shortly after eruptions ended (Widdowson 1990). However, what is certain, and geomorphologically speaking more important, is the fact that this lava plain palaeosurface apparently exhibited a low easterly tilt (Subramanian 1987) away from the newly-rifted continental margin since it was drained by rivers which flowed eastwards into the Bay of Bengal (Cox 1989). Accordingly, its westernmost reaches (i.e. the regions of the lava surface adjacent to the newly rifted margin) must have lain at some altitude above sea level, yet the remoteness of the effective base level where the rivers discharged (i.e. the Bay of Bengal), meant that erosional processes must have been extremely slow on the continental backslope. This characteristic is common to other continental drainage divides of this type (e.g. Great Divide, SE Australia, Pain 1985, and southwestern Africa, Gilchrist *et al.* 1994*a*).

Lateritization of the post-Deccan lava palaeosurface continued until it was terminated by a widespread, regional (epeirogenic) uplift during the lower Tertiary which instigated the erosional régime and began the destruction of the lateritized palaeosurface and deep dissection of the lavas beneath. Moreover, this uplift ended the long-term ingress of groundwater and hence any further widespread modification or maturation of the laterite profiles thereby preserving the lower Tertiary palaeomagnetic ages in the laterites. This chronology of events is further supported by additional geochronological and geomorphological investigations (i.e. Bruckner 1989), which confirm that erosion of the upper-level laterite having already commenced by the middle–upper Palaeogene (i.e. late Eocene–Oligocene), with more advanced stages of destruction following in the Miocene. The mechanisms triggering such epeirogenic uplift along the passive margin can only be speculated upon, but commonly cited causes include post-rift thermal and magmatic effects. Though, as will be later discussed, problems remain with the time scales over which such mechanisms are known operate and the longevity of uplift effects observed at many VRM.

Throughout development of the high-level laterite the edge of the nascent continental margin formed the western limit of the palaeosurface. It is here that the western Ghats scarp had its origin, probably as an emergent seaward-facing fault-scarp generated as a consequence of the rifting process. Initially the proto-escarpment would have been adjacent, or near to, the locus of rifting, but once geomorphological (i.e. exogenetic) processes became dominant, its subsequent evolution resulted in an easterly recession of the scarpline across the continental margin. In this manner the recession not only removed large thicknesses of basalt but also effectively began consuming the western edge of the lateritized lava surface (Fig. 10a). Given the current relative positions of the southwest Deccan Ghats scarpline and the offshore edge of the

a) Morphological evolution of the S.W. Deccan during mid-upper Tertiary times

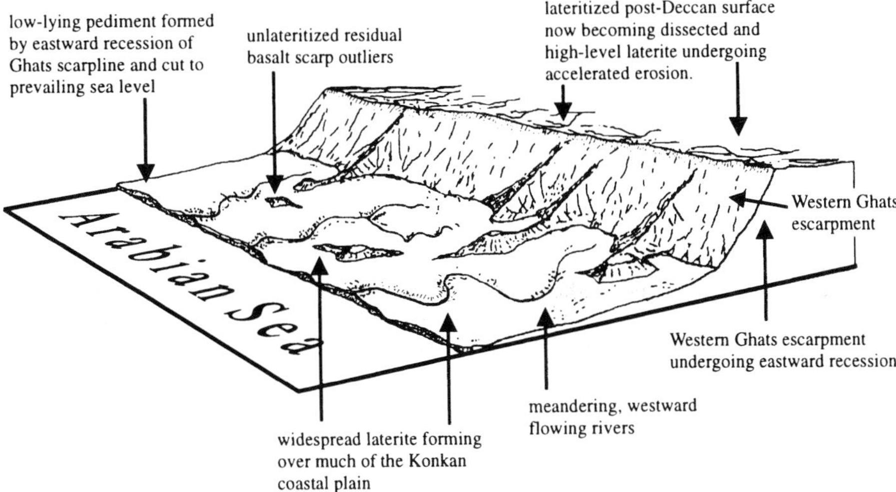

b) Morphological evolution of the S.W. Deccan during upper Tertiary to Recent times

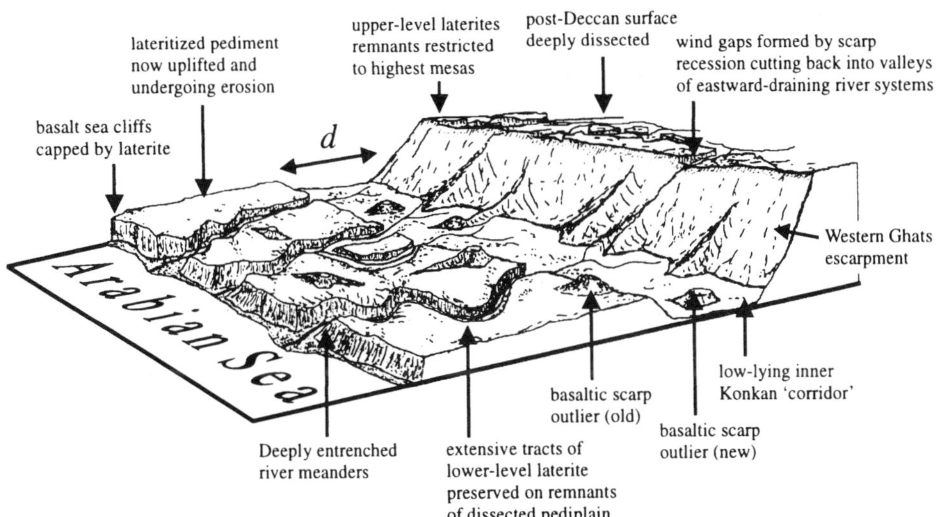

Fig. 10. Block diagrams showing the main morphological elements comprising the southwest Deccan during (**a**) mid-upper Tertiary, and (**b**) upper Tertiary-Recent times. Distance *d* is the extent of eastward suppression since mid-Tertiary times.

continental margin (i.e. 200 m isobath; Naini & Talwani 1983), an estimated 120–180 km of scarp recession has occurred since the Seychelles rifting event *c.* 65 Ma. Simple calculation gives a recession rate of 1.8–2.8 m per 1000 years; a value consistent with estimates of scarp recession determined for the Australian Great Escarpment (Pain 1985), and the Drakensburg of South Africa (King 1967).

On the continental backslope east of the escarpment, the processes vertical and lateral erosion by the easterly draining rivers systems progressed at a much slower rate. Nevertheless, over time these systems have effectively etched into, and largely consumed the lateritized lava plateau ahead of the receding Ghats escarpment. This situtation has continued to the present day so that rivers draining the Maharashtra plateau currently lie 300–400 m below the Ghats crest laterites indicating the total amount of downcutting since this vertical erosion began to take effect on this backslope during the late Eocene–Oligocene. Consequently, upper-level laterites can now be only be found immediately east of the escarpment upon a narrow zone of elevated interfluves which have escaped both the erosion resulting from the coalescence of the plateaux valleys to the east, and that caused by scarp recession to the west. These remaining high-level laterite plateaux represent the last vestiges of the post-Deccan lava surface.

Apart from the partial destruction of the Deccan lava-plain surface, the main geomorphological consequence of this recession was the formation of the Konkan pediplain: a laterally extensive, low-relief shelf cut into the basalt stratigraphy along the emergent continental margin (figs 3b & 10a). Besides being much younger, the lower-level palaeosurface (henceforth termed the Konkan surface) differs fundamentally from the upper-level lava-plain palaeosurface in two other important respects. First, it is the product of exogenic rather than endogenic processes; therefore, the undulose nature of this low-level lateritized palaeosurface (Fig. 7) and presence of residual basaltic hills (i.e. inselbergs) can be explained as a consequence of its origin as an erosional rather than constructional surface. Second, its primary elevation can be better constrained since its evolution would have been dependent upon local base level which, in this case, was sea-level along the nearby coast. Such pediments are characterized by low-amplitude topography which, coupled with sluggish drainage and associated low rates of erosion is ideal for laterite development (Mabutt 1966). Evolution of the Konkan, together with continuing favourable climatic conditions, thus provided a suitable environment for the second major phase of lateritization to affect the southwest Deccan.

A phase of renewed regional uplift eventually brought to an end the period of apparent tectonic stasis which had permitted the development of the Konkan pediplain. This was reflected in a fundamental change in the geomorphological systems of the Konkan (Bruckner 1989) resulting in renewed incision of the drainage system, terminating widespread lateritization of the pediplain, and thereby preserving the mid-upper Tertiary (Late Miocene?) palaeomagnetic signatures in the laterite. Today, the short, westerly draining river systems which rise at the Ghats escarpment, traverse the outer Konkan via a series of deeply entrenched meanders (*c.* 100–200 m). It is these river systems that have effectively dissected the low-level palaeosurface into numerous, areally extensive laterite-capped plateaux (Fig. 10b).

It should be noted that the southern Konkan which is cut into the Archaean and Proterozoic rock of Kanataka and Kerala, displays a more complex history of lateritization. Here, a series of stepped lateritized surfaces may be recognized (Bruckner 1989), probably indicating periodic uplift punctuated by subsequent geomorphological adjustment to base level. However, within the current study area, such periodic adjustment may explain the existence of the low-lying inner Konkan 'corridor'. This corridor feature may represent a more recent eastward extension of the pediplain which has since been cut to a lower relative base level as a consequence of the last major phase of margin uplift (Fig. 10b). Moreover, using the scarp recession rates suggested above, its width (*c.* 20–30 km) is consistent with an estimated distance of scarp recession since uplift and commencement of low-level laterite erosion during the mid-upper Tertiary.

Modification and morphological evolution of the upper and lower level palaeosurfaces

Generally speaking, the current form of ancient palaeosurface is unlikely to reflect original surface morphology unless palaeosurfaces occur in cratonic interiors remote from active tectonics regions; more commonly their gross morphology will have been modified through neotectonic effects. The degree of deformation experienced will depend upon the age of the palaeosurface and its tectonic setting. As discussed earlier, western India is known to have been subject to two major rifting phases during the Mesozoic, and regional or epeirogenic uplift is known to have subsequently affected the margin during the Tertiary (Kailasam 1975, 1979). With respect to this uplift, it is most unlikely that the post-Deccan lava surface and the low-level Konkan pediment will have entirely escaped these neotectonic effects because, unless this uplift has been entirely uniform across the Deccan, variations in the degree of vertical movement might be expected to have caused significant warping or distortion of the surfaces. Moreover, an increased degree of deformation ought to have affected the upper-level (post-Deccan) palaeosurface, given its greater age. In this context, contouring the lateritized palaeosurfaces in the southwest Deccan proves very useful since it not only reveals these neotectonic effects, which are a consequence of the morphotectonic evolution of the western Indian margin, but also the differences in the degree of deformation suffered by the two palaeosurfaces provides a record of this uplift. Reconstruction of the upper- and lower-level palaeosurfaces may be achieved using the available laterite height and distribution data. The results of contouring the upper and lower-level laterite surfaces are shown in Fig. 5.

The upper-level palaeosurface

The contours are best constrained east of the Koyna reservoir between lats 17°20′–17°55′ N where the contours define a low-amplitude, southerly dipping antiform (Fig. 8). The southerly dip component of the contoured surface is on average 0.25°–0.3° (i.e. an elevational difference of less than 350 m over 70 km) and is very similar to that displayed by the lava structure of this region. The axis of the antiform trends approximately NNW–SSE and lies approximately 10–15 km east of the Ghats scarpline coincident with the elongate spur which constitutes the Bamnoli Range. Dips on the limbs of this feature appear to be greatest in the north (~1°) but become less steep towards the south (< 0.3°) broadening the antiform until south of 17° N, the general structure becomes more monoclinal in form.

Comparison of the structural contours of the lava pile (see again, Fig. 4) and those drawn on the upper-level palaeosurface reveal a remarkable degree of similarity indicating that development of the laterite antiform and lava pile anticline are closely related. Shared similarities with the lava pile structure continue in the southernmost regions where the height of the laterite mesas adjacent to the escarpment define a northerly dip of the laterite surface mimicking exactly the dip reversal documented in the lavas of this area. Moreover, the contour pattern is truncated by the Ghats escarpment, lending support to the argument that high-level lateritized palaeosurface once extended much further westwards and has since been destroyed by eastward retreat of the Ghats scarpline.

It is clear that the anticlinal structure is not a primary constructional feature developed within the Deccan lava pile (Widdowson & Cox 1996). This can be readily demonstrated by the fact it affects both lavas and the post-Deccan laterites alike, and by the fact that headwaters of the eastward flowing drainage system (i.e. Bhima, Krishna, and Koyna rivers) rise adjacent to the present scarp edge and only cross the anticlinal axis after flowing eastward for some 10–25 km; thus the original rift-flank

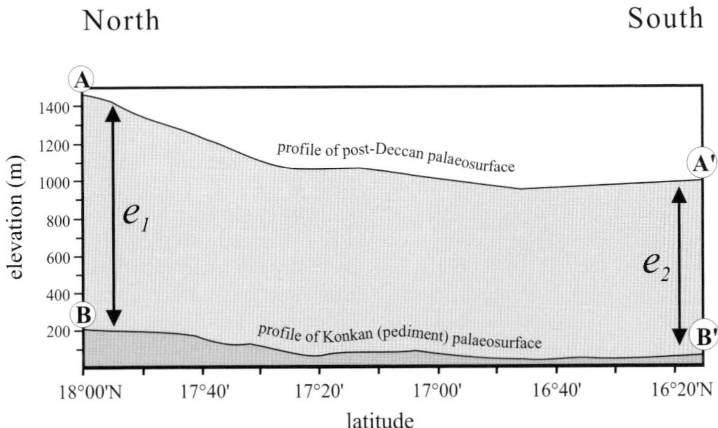

Fig. 11. Coast-parallel (approximately N–S) profiles along the upper-level (post-Deccan) and lower-level (Konkan) lateritized surfaces showing the height relationship between the two palaeosurfaces. The general form of the surface profiles is similar since both indicate relative uplift in the northern regions. However, measurements of elevation difference between the two palaeosurfaces varies from north to south of the studied region since e_1 = 1210 m and e_2 = 935 m. This indicates that the effects of this differential uplift are more accentuated (i.e. $e_1 - e_2$ = 280 m) in the older, upper-level post-Deccan palaeosurface.

drainage (Cox 1989) is antecedent with respect to this structure. In effect, the Western Ghats anticline–monocline represents a Tertiary deformation of the continental margin and, therefore, the westerly dip component affecting the coastal (Konkan) tract is a secondary feature.

The lower-level palaeosurface

The extensive lateral exposure of Konkan laterite allows the contours of the lower surface to be well constrained. Inspection of laterite height distribution reveals systematic east–west and north–south variations in elevation. The most prominent feature of the Konkan contours is the westerly dip of 0.4°–0.7° though, importantly, this is less than that displayed by underlying the basalt and consequently the laterite lies on progressively younger formations towards the coast (Fig. 3b). This westwards slope can be recognized across many of the larger Konkan plateaux since their more easterly margins often lie 100–200 feet (30–60 m) higher than those adjacent to the coast. This coastwards dip is consistent along the entire length of the field area and culminates in laterites at a maximum height of 800 feet (240 m) some 20–25 km inland. East of these highest Konkan laterites the topography descends sharply to low-lying inner Konkan corridor.

The height data also suggest a southerly decline in elevation because the laterite belt in the north lies, on average, at 750 feet (230 m) near Srivadhan, 350 feet (100 m) inland of Ratnagiri, and descends to 200 feet (60 m) in the southern Konkan, just inland of Devgarh. Most of the contours appear to be essentially coast-parallel, but in detail actually run at a slight angle to the coastline thus bringing progressively more elevated laterite plateaux adjacent to the northern coastline (Figs 5 & 9).

Closer inspection reveals a series of shallow east–west 'synclines' in the laterite surface which deflect the contours inland. Most of these features correspond with the esturaries and lower courses of present-day rivers such as the Muchkundi, Kajvi, and Shashtri (Fig. 9); the most prominent palaeosurface syncline is developed around the Vashishti the largest Konkan river system within the studied area and its associated tributaries. The origin of these synclines is unclear, but they may represent a large-scale marginal cambering of the Konkan plateaux in response to river incision; an as yet unidentified structural control; or alternatively, a succession of stepped laterite terraces as seen in coastal basement terrain to the south. Bruckner (1989) suggests these later lateritized terraces developed upon the basement rocks may be the result of continuing periodic uplift across the Konkan. Clarification of these features in the Deccan region awaits further inspection and investigation to determine whether the laterite is continuous in these synclines, or else occurs as a series of discrete terraces which cannot be distinguished from the widespread laterite coppings of the Konkan plateaux using the current satellite imagery.

North–South (coast-parallel) palaeosurface profiles

Figure 11 is a comparison of coast-parallel (north–south) profiles drawn from the contours of the upper- and lower-level surfaces. It is immediately apparent that both the Konkan and post-Deccan surfaces are most elevated in the north of the study area and, moreover, that both display similar longitudinal topographical profiles, including the dip reversal in the south. The fundamental difference between the two is the *degree* of this elevational variation; for instance over a distance of 160 km (i.e. between Mahableshwar 17°55′ N and Bavda 16°30′ N) the elevation of the upper laterite surface descends by approximately 460 m, whilst over the same distance, the elevational difference of the Konkan surface is only about 180 m. From stratigraphical relationships it is clear that the upper level exactly mimics the southerly dipping structure developed in the lavas. Beane *et al.* (1986) and Devey & Lightfoot (1986) argue that this southerly dip component within the lava pile is a primary constructional feature produced during the development of the volcanic edifice (i.e. the flank of a migrating shield volcano) but given the north–south dip of the later palaeosurfaces this interpretation appears to be incorrect. This error is self-evident because the Konkan surface is erosional in origin and its development therefore controlled by local base level (i.e. sea level); consequently it was probably near horizontal in the north–south sense when originally formed. This being the case, the currently observed variation in elevation along the Konkan palaeosurface must be the result of a later differential uplift, the effects of which are most accentuated in the north. Moreover, if differential uplift along the Deccan margin is the underlying cause of these southerly palaeosurface dips, then the older upper-level post-Deccan palaeosurface should have experienced significantly more of these uplift effects than the younger Konkan surface. This relationship is clearly indicated by Fig. 11, and the increased southerly slope developed in the post-Deccan palaeosurface can be simply explained by its greater antiquity and the certainty that it must have been subject to these effects for a much longer period of time. The fact that this differential uplift has been operating over a considerable period of time is further evidenced by the fact that the Konkan laterites lie on progressively older basalts toward the north (i.e. an angular unconformity; Fig. 3a); this indicates that the basalts of these northerly regions had already become more elevated prior to erosion of the coastal tract and associated development of the Konkan pediplain.

Discussion–Neotectonics of the Deccan margin

From the preceding sections describing palaeosurface morphology three fundamental observations are made which require further explanation. First, the upper level palaeosurface is dominated by an anticlinal–monoclinal structure virtually identical to that developed in the basalt stratigraphy and, most importantly, evolution of this structure post-dates the development of the upper-level laterite. Second, the main feature of the low-level, Konkan laterite surface is that it displays a consistent coastward dip which is less than that developed in the basalt stratigraphy upon which it lies. There are two possible causes for this coastward slope; either the slope represents the original pediment slope away from the foot of the late Tertiary Ghats scarpline or, alternatively, it is a deformational feature caused by continued coast-parallel uplift and associated coastward cambering of the entire Deccan margin. Third, coast-parallel (north–south) profiles along the upper- and lower-level palaeosurfaces reveal both are more elevated toward the north but, importantly, this difference in elevation is more accentuated in the older, upper-level palaeosurface. The most probable explanation for this observation is that the Deccan margin has experienced differential uplift throughout the Tertiary which has resulted in greater relative uplift in the older, upper-level palaeosurface. Such differential uplift may explain why remnants of the two palaeosurfaces become less common, and eventually absent, north of the study area. The underlying causes of this long-term, large-scale palaeosurface deformation are addressed below.

Uplift mechanisms

Marginal uplift and the presence of elevated regions adjacent to passive rifts (also termed 'randschwellen' in the geomorphological literature, e.g. Bremer 1985) are certainly a common geological phenomenon along nascent and juvenile continental margins, though the precise mechanism causing these uplifts has been the source of much debate (for a geomorphological review see Ollier 1985). Historically, the most commonly cited processes are predominently endogenic phenomena which include rift-related mechanisms of crustal thinning (McKenzie 1978), magmatic underplating (Cox 1980), transient thermal effects (Cochran 1983) and secondary convective effects associated with extension and flexural unloading (Buck 1986; Wiessel & Karner 1989). In those examples where rifting is associated with hotspot volcanism, as in the case of the Deccan CFB (White & McKenzie 1989), the dynamic and thermal effects of the hotspot plume are considered a fundamental controlling factor of passive margin uplift. The common characteristic shared by these latter mechanisms is that the predicted dynamic uplift should begin to decline shortly after completion of the rifting event (i.e. either once seafloor spreading moves the margin away from the hotspot locus, or when both the thermal and magmatic acmes have begun to wane; Richards *et al.* 1989). However, geological and geomorphological observations along the Deccan, and other similar rifted CFB margins (e.g. E Greenland, Karoo, Paranà) clearly demonstrate that such margins remain elevated and, even more surprisingly, often continue to rise over geological time (Brooks 1985; Cox 1988*a*; Gallagher *et al.* 1995; Widdowson & Cox 1996). Clearly, this phenomenon of continued uplift cannot be readily explained in terms of the above mechanisms. Instead, the persistence of elevated inland regions (e.g. Maharashtra Plateau), its associated escarpment (e.g. Western Ghats), and low-lying coastal plain (e.g. Konkan) indicate a significant erosional contribution to the post rift modification of passive margin landscapes (Gilchrist & Summerfield 1994). In recent years numerical modelling has done much to illustrate the potential of denudational unloading in controlling long-term uplift along continental margins (Summerfield 1989; Brown *et al.* 1990; Summerfield 1991; Gilchrist *et al.* 1994*a*). Such models are attractive because they are the result of exogenic process which are unaffected by the temporal restrictions that control the themal and magmatic mechanisms outlined above. Given that the nature of the deformation affecting both the post-Deccan and Konkan palaeosurfaces demonstrates that uplift has occurred throughout the Tertiary (*c.* 60 Ma), the best explanation for this uplift lies in a long-term driving mechanism, namely, continued isostatic adjustment in response to denudation. Models of denudational isostasy have been outlined to account for the geomorphological evolution of other passive margins (Thomas & Summerfield 1987; Gilchrist and Summerfield 1994); in essence, these models propose a history of continued adjustment through coast-parallel uplift and cambering of passive margins in response to the inland recession of continental-scale escarpments and associated offshore loading by sediment deposition. The relevance of this mechanism to the morphotectonic evolution of the Indian margin will now be discussed in detail.

Denudational isostasy: a model for the uplift and deformation of the Deccan palaeosurfaces

The Deccan margin is an ideal candidate the denudational model because the eastward recession of the Western Ghats is demonstrably responsible for the removal of 1–1.5 km of basalt over the region now exposed as the Konkan Plain (Figs 3b & 12) and, furthermore, much of the eroded material has been deposited offshore in a series of half-grabens comprising the outer continental margin and continental slope. According to the model, the net result of this onshore unloading and offshore loading is manifest as an inland, coast-parallel axis of uplift resulting in a large-scale seaward cambering of the continental margin and associated coastal plain (figs 13a, b). Both these effects are apparent over the entire SW Deccan in the form of the anticlinal-monoclinal feature and the westward dip of the basalt stratigraphy across the Konkan.

To better understand the magnitude of erosion effects caused by the Western Ghats scarp recession, and hence the potential for isostatic response, it is helpful to make estimates based on a representative region of the SW Deccan where the geomorphological, stratigraphical, and structural relationships are well constrained. Since it is the amount of material eroded (i.e. depth of dissection, *d*) which ultimately controls the isostatic response of a given region (Gilchrist *et al.* 1994*b*), we can consider the east–west tract of basalts between Ratnagiri and Srivardhan (i.e. 17–18° N) in order to make an estimate of the volume of material which has been removed by scarp retreat; this particular region is chosen because the top of the original lava sequence can be readily determined from the elevations of the post-Deccan surface (i.e. high-level laterites). Over the currently exposed Konkan (i.e. from the current Deccan coastline to the foot of the Western Ghats escarpment) the volume of material which has been removed by scarp retreat between Ratnagiri and Srivardhan is at least 8.0×10^3 km^3 since values for *d* in this region are of the order of 1300 m (Fig. 12). In effect, volumes greater than twice this amount can be assumed if it is accepted that the Western Ghats escarpment had its origins at, or near to, the edge of the continental shelf (i.e. 200 m isobath; Naini & Talwani 1983) and has since recessed to its current position during the Tertiary (Widdowson 1990). This large-scale removal of dense basaltic overburden is augmented further inland by the erosion of basalt across the Maharashtra plateau thus providing additional potential for isostatic adjustment eastward of the Western Ghats escarpment. However, although this erosion has occurred over a large surface area, the thicknesses of basalt removed east of the Ghats are much smaller since dissection (*d*) between latitudes 17°–18° N is of the order of 600 m when measured from the remnants of the lateritized remnants of the post-Deccan surface down to the level of the current Maharastra Plateau (Fig. 12). Moreover, this eroded material has been transported to the Bay of Bengal via the eastward-draining rivers (a

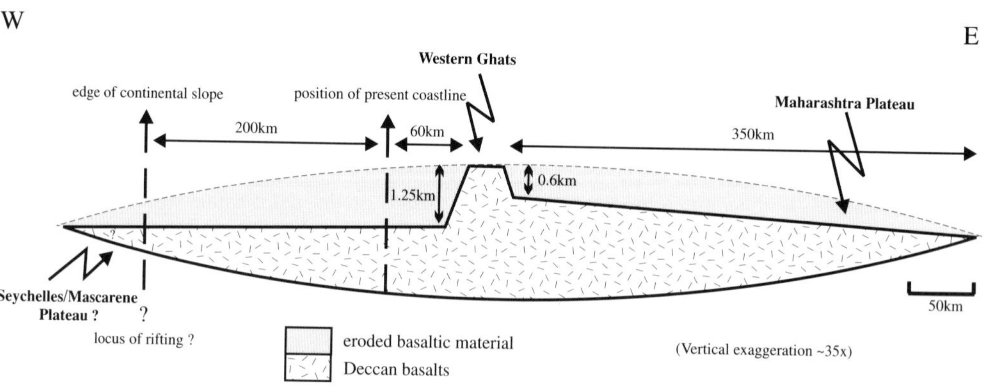

Fig. 12. Simplified schematic figure indicating the general lensoid form of Deccan lava pile at ~18°00′ N. Estimates are given of basalt thicknesses which have been removed to the west of the current escarpment by scarp recession (1.25 km), and thicknesses removed across the elevated Maharashtra plateau inland of the escarpment (0.6 km).

distance of 500–700 km), and is therefore unlikely to have contributed significantly to Deccan structural evolution through its subsequent sedimentation and associated loading effects.

These estimates are designed to demonstrate the maginitude of erosion, but it is important to realize that they only consider one small segment of the continental margin (i.e. that comprising the southern Deccan); in the wider context, the scarp-line of the Western Ghats extends well beyond the basalt terrain to the southernmost tip of India (Cape Comorin), and provides testimony to the scale of denudational unloading which has affected this particular passive margin. Clearly, redistribution of crustal loads at this scale along a rifted continental margin provide immense potential for fundamental isostatic compensation along the entire coastline and, most importantly, predicts the observed pattern of structural and geomorphological features endemic to such margins. In the particular case of the Deccan, denudational isostasy adequately satisfies the earlier assertion that the coast-parallel anticlinal-monoclinal axis is entirely a post-eruptive structure, and that the morphology of the reconstructed post-Deccan palaeosurface is a deformational feature resulting from coast-parallel uplift (Figs 13a, b). In a similar fashion, the westward dip recorded in the basalt stratigraphy of the Konkan also becomes a consequence of rotation of the margin in response to denudational unloading and offshore sedimentary loading. Moreover, arguments that the westward dip of the Konkan palaeosurface (see again, Fig. 9) represents the original pediment slope seem untenable given the angular disconformity between the Konkan laterite and the more steeply dipping coastal lava sequence (Fig. 3b). Clearly, this disconformity demonstrates that the effects of westward cambering of the lavas was already advanced prior to erosion and lateritization of the Konkan pediplain; it is therefore reasonable to assume that this cambering continued to affect the Deccan margin and Konkan pediplain as a consequence of ongoing scarp recession throughout the late Tertiary and to the present day.

The nature of Deccan uplift - some geomorphological and seismologial considerations

It is clear that the denudational model accounts for many of the geomorphological and structural observations in the studied region, and satisfies many of the outlined spatial and temporal constraints regarding the timing and nature of the uplift. However, acceptance of this model as a

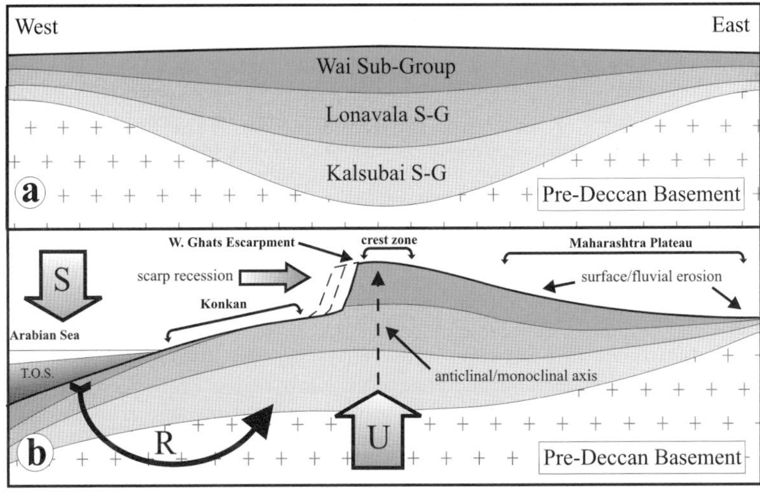

Fig. 13. Schematic diagram illustrating the uplift and associated geomorphological and structural evolution of the Deccan margin (cross section at ~18°00′ N) resulting from denudational isostasy (general model modified after Thomas and Summerfield, 1987). (**a**) Pre-rift lensoid structure of lava pile at the end of the Deccan eruptive phase; (**b**) modified structure resulting from Tertiary uplift effects along the Indian VRM. S, subsidence of offshore continental margin in response to sediment loading; U, uplift resulting from isostatic adjustment caused by removal of lava overburden through scarp recession and denudational mechanisms; R, rotation and associated cambering of continental margin; TOS, Tertiary offshore sediments. For details of basalt stratigraphical nomenclature see Table 1.

viable explanation for the evolution of the Deccan margin still leaves a number of important questions unanswered. The first of these concerns the nature and frequency of isostatic compensation since the resulting uplift will provide fundamental controls upon geomorphological evolution and seismicity along the Indian margin. For instance, the presence of incised meanders across the Deccan Konkan, together with development of river terraces in the basement terrain of the coastal lowlands to the south, may represent discrete uplift events. Moreover, if adjustment occurs as a series of small, but frequent flexural compensations in response to the changing isostatic régime, then relatively weak seismic events would result. If, however, isostatic compensation becomes impeded and stresses are only released periodically, then infrequent earthquakes of much greater magnitude will occur. The style of seismicity in the SW Deccan indicates that the latter might be the normal mechanism of adjustment of the region as erosion progresses (Widdowson & Mitchell in press).

A second point which requires addressing is the nature of the tectonic conditions controlling the formation of the Konkan pediplain. The development of such a laterally continuous and topographically uniform erosion plain would seem to indicate a period of relative tectonic quiescence along the Indian margin during which base-level (i.e. sea-level) remained at a standstill relative to the landmass. Yet such quiescence is difficult to reconcile with the evidence for continued uplift evinced by the deformation of the coastal lava stratigraphy and post-Deccan surface prior to development of the Konkan pediment. The reason why the such an extensive coastal plain should have evolved can only be speculated upon, but the denudational model accommodates a number of possibilities. First the pediplain could have been generated during a prolonged period of stasis during which isostatic adjustment was impeded. The causes of this interlude in adjustment could be associated with a slowing in the rate of scarp recession, perhaps the result of mid-Tertiary climatic variation for example. Alternatively, the fact that by mid-Tertiary times scarp recession must have progressed from attenuated continental crust adjacent to the rift locus, and into the thicker and structurally stronger crustal material further inland, may have resulted in progressively less frequent, but seismically more potent isostatic adjustment of the margin. The second possibility is that the pediplain developed during a period of ‘dynamic stasis’, possibly marking a transitional period between the conflicting effects of post-rift thermal subsidence and later erosionally induced isostatic rebound along the continental margin (Widdowson & Cox 1996). The evolution of the Konkan pediplain currently remains a fascinating aspect of the evolution of the Indian margin, and the answers as to its true origin may well lie in future investigation of the offshore sedimentary record.

The third point is that whilst the denudational model readily explains the observed east–west structures (i.e. coastal cambering and the anticlinal–monoclinal feature), it does not account for the differential uplift observed in the north–south sense along both the post-Deccan and Konkan palaeosurfaces (Fig. 11). In terms of erosion, this differential uplift has been most effective because to the northward of approximately 18° N the upper level laterite surface is missing, and the basalts exposed at the top of the highest point of the Ghats near Igatpuri (19°40′ N) are stratigraphically well below those in the study area. The amount of material removed can be estimated by simple calculation if the post-Deccan surface is extrapolated northward and it is also assumed that the Wai Subgroup extended with its present thickness into the Igatpuri area (it seems unlikely that it was thicker than this, and it might well have been thinner). This maximum estimate implies the erosion of an extra 1.5 km of the sequence in this northern area. One possible cause of this increased uplift and erosion in the Nasik region was suggested by Watts & Cox (1989) in terms of flexure of the Indian margin in response to volcanic loading by the basalt pile. However, since these authors adopted a relatively high value for the elastic thickness of the Indian margin (i.e. $T_e = 100$ km), their preferred model predicted the removal of 5.5km in the Nasik region and a thinner, but nevertheless significant thickness from those sequences in the southwest Deccan (i.e. above the Panhala Fm). Such a conclusion is clearly incorrect given that the post-Deccan palaeosurface, which lies directly upon the Panhala flows, represents the preservation of the original lava sequence top in the southwest Deccan. Interestingly, if lower T_e values are adopted (i.e. $T_e < 50$ km), which are perhaps more consistent with rifted and structurally weakened continental crust (e.g. Naini & Talwani 1983; Barton & Wood 1985), then this flexural model can still allow for the preferential erosion in the Nasik region whilst permitting preservation of the entire sequence further southward.

Flexural loading clearly provides one possibility for the observed N–S variations in uplift, but other possible contributing factors should also be considered. These largely concern the wider tectonic environment of the Indian sub-continent and include, for instance, the continued northward movement of the Indian subcontinent since hard collision with Eurasia (*c.* 40 Ma). This northward movement largely resulted in deformation and crustal thickening associated with the Himalayan orogen, but underthrusting and flexuring of the Indian plate may provide a likely explanation for the evolution of the Ganges basin (Lyon-Caen & Molnar 1985). In addition, there is evidence for associated reactivation and reversal of normal faults in the Deccan basalts of the Narmada and Tapti rift zones (Deshmukh & Sehgal 1988), but it remains an open question whether such compressive forces may have been transmitted even further southward thereby affecting uplift in the Deccan. Regional compressional stresses are also known to have affected the sea floor and associated marine sediments around the Indian continent largely as a result of ridge-push forces (e.g. Karner & Wiessel 1990). However, it remains to be demonstrated whether these oceanic crustal stresses are capable of being transmitted to the continental crust thereby producing a coupled deformation in the adjacent continental margin. Nevertheless, Whiting *et al.* (1994) argue that loading effects generated by the Indus fan in the Arabian Sea may be capable of producing limited flexural uplift along the western Indian margin. To summarize, collisional and flexural mechanisms do have the potential to affect uplift along the western Indian margin, but much more work is required to determine their contribution, if any, to the patterns of deformation observed in the lava sequence and lateritized palaeosurfaces.

Impact of uplift on drainage patterns

A final point of discussion concerns the impact of the geomorphological evolution of the Deccan margin and associated uplift upon drainage patterns. It has already been mentioned that the Western Ghats represent the main drainage divide for Peninsular India, and that the headwaters of eastward-draining systems (i.e. Krishna and Godavari rivers) are antecedent with respect to the coast-parallel upwarping of the margin. What remains to be examined is the origin of this eastward-draining system and the consequences of long-term scarp recession.

Where CFB are developed on a relatively stable platform inshore of a rifted margin, as in the case of the Deccan, there is the possibility that the drainage pattern was initiated first by the dynamic/thermal plume effects, and then subsequently augmented by the surface uplift generated by the emplacement of the lavas and associated underplating (i.e. the constructional topography of the lava sequence). If this is the case, the subsequent differential uplift effects have left this dome-flank drainage pattern largely unaffected and, since the initial triggering of erosion during the lower Tertiary, it has simply continued to slowly etch itself into the post-Deccan surface thus forming the current Maharashtra plateau. In the study area the total surface lowering achieved during this time is, on average, 300–400 m (i.e. the elevation between the upper-level laterites and the present-day river plains) giving an estimated average incision rate of only 5–10 m per Ma. A recent collation of river data worldwide (Hovius in press) demonstrates that the Krishna and Godavari river systems which drain much of the Maharashta plateau, currently carry a sediment load which translates to a modern incision rate of 94 m and 219 m per Ma respectively. Clearly this is much faster than the estimated long-term rate of fluvial incision based on the elevation differences given above. The underlying causes for this discrepancy remain to be addressed, but since rapid erosion rates similarly occur inland of escarpments located at other rifted margins, it must indicate a fundamental geomorphological or tectonic control. One possible explanation is that the uplift axis generated inland of the escarpment effectively produces a 'wave of uplift' which proceeds ahead of the recessing escarpment. The effects of this uplift wave are currently manifest as the anticlinal–monoclinal axis and antecedent relationship described above, and must produce an ongoing rejuvenation along the upper courses of the eastward draining system. Such rejuvenation could, potentially, provide a mechanism for the increased sediment loads. Another possibility lies in the idea of iterative uplift mentioned earlier, since this could cause periodic regional rejuvenation of the inland drainage system and supply pulses of sediment into the rivers. If the latter is correct, then the current situation of high sediment load may represent an abnormally high value following just such an uplift episode.

Another consequence of inland scarp recession is the fact that it effectively consumes the catchment of the plateau drainage. In other words, expansion of the coastal plain drainage system effectively proceeds at the expense of the headwaters of the plateau system (Fig. 14). However, the Konkan drainage may never have become greatly extended because, if the isostatic model is correct, westward cambering of the margin should progressively drown the river mouths along the Arabian sea coast whilst the Konkan river headwaters are simultaneously being extended eastward. Moreover, the long-term evolution of the Konkan river systems may be further complicated by Tertiary eustatic variation, and more recently by the sea-level rise associated with the Pleistocene deglaciation. Therefore, although there is evidence from offshore sediments suggesting that coastal drowning has occurred in the past, more detailed work is required to better document the evolution of the Konkan drainage.

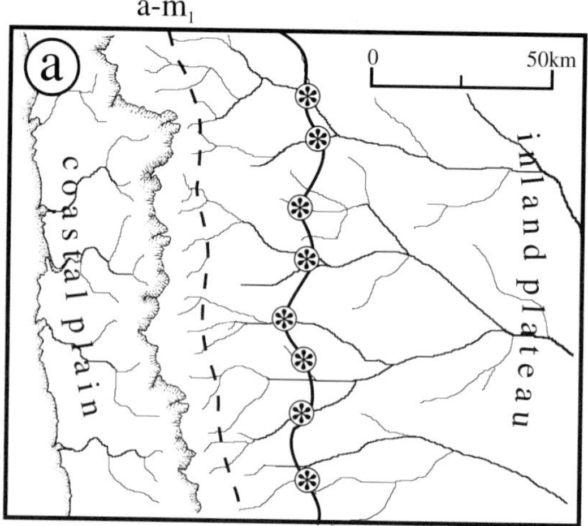

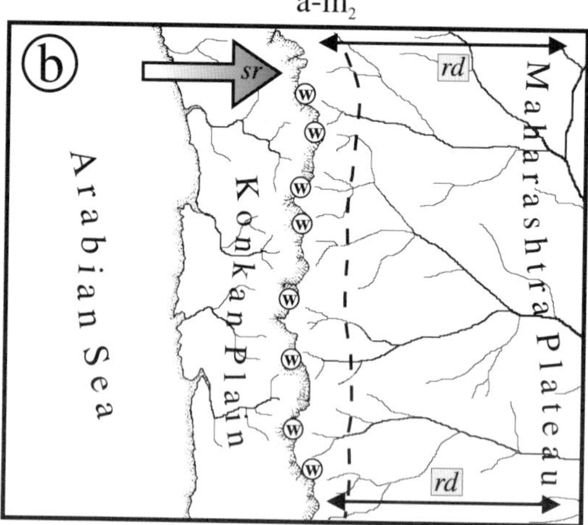

Fig. 14. Evolution of drainage patterns. (**a**) Past (e.g. mid-Tertiary) situation; scarpline (toothed ornament) has already recessed some distance eastward producing a low-lying coastal plain to the west drained by short, meandering rivers. On the elevated plateau inland (east) of the escarpment the headwaters of the plateau drainage systems are antecedent with respect to the migrating uplift axis (dotted line a-m_1) which has generated an anticlinal-monoclinal structure and produced limited rejuvenation and downcutting. Solid line indicates position of scarpline following continued recession as shown in Fig. 14b. Asterisks mark localities where the later scarpline will intersect the major tributaries of the established plateau river valley systems (**b**) Present situation; continued scarp recession (*sr*) has extended the coastal (Konkan) plain eastwards, but cambering of the margin in response to offshore sediment loading has now drowned the outer coastal plain together with the lower reaches of the original coastal drainage system. During this scarp recession 20-30 km of the headwater catchment of the plateau drainage system has been consumed. Large 'wind gaps' (w) have developed where the new escarpment intersects the old river valley systems. Loss of catchment area results in a large tract characterized by rivers with reduced discharge (*rd*), however, inland migration of the uplift axis (a-m_2) produces limited rejuvenation and renewed downcutting in the remaining headwater valleys.

The effects of scarp recession on the plateau drainage are more readily appreciated since this eastward recession clearly results in the beheading of the easterly draining plateau river system; the presence of huge wind gaps along the Ghats crest (e.g.) and wide valleys with small misfit streams rising at the escarpment edge immediately testify to erosion of the headwater catchment and an associated reduction of fluvial capacity (see again Fig. 14). In the long term this destruction of the upper courses of the plateau drainage can only be accompanied by a progressive reduction in downstream discharge. However, this long-term reduction in the fluvial capacity is notwithstanding other climatic and tectonic factors which may work in an opposing manner during the shorter term to increase discharge, rate of incision, and sediment load (Kale & Rajaguru 1987). Although much work remains to be done regarding the evolution of the Konkan and that of the Maharashtra plateau, what is apparent is the fundamental importance of the Western Ghats morphotectonic régime with regard to the dynamic complexity of the drainage system and the development of the offshore sedimentary record.

Summary and conclusions

Coastal regions and inland escarpments of many VRM appear to have undergone uplift but the timing and underlying causes of this phenomenon have remained the matter of some debate. Dynamic and thermal effects of hotspot plume activity may be the cause of initial surface uplift along the flanks of newly-rifted continental margins, but it is evident from both geological and geomorphological observation that uplift continues long after the plume effects have decayed. At many of these VRM (e.g. SE Africa, E. Brazil, and E. Greenland) the pattern of uplift is often difficult to determine but, perhaps uniquely, two lateritized palaeosurfaces provide a record of post-rift uplift along the southwest Deccan margin. The geomorphological environment in which these two surfaces are currently located, together with the regional deformation which they have suffered, leaves little doubt that a major and permanent uplift affected the western Indian margin throughout the Tertiary. The clue to the mechanism driving this uplift comes from the detailed chemostratigraphical studies which reveal the structure and outcrop pattern of the remaining lava sequence. These demonstrate that the lavas have undergone considerable erosion during the 65Ma since their eruption. Thicknesses of 1–1.5 km have been removed from the western edge of the rifted Deccan CFB province, thus creating the coastal (Konkan) plain which today forms a lowland fringing much of northwest peninsular India. This erosion is the result of eastward retreat of the Western Ghats escarpment, and the observed uplift is therefore interpreted as a consequence of the isostatic response to this onshore denudational unloading and concommitant offshore sedimentary loading, which has produced a lithospheric flexuring of the entire margin. Moreover, because denudational unloading is independent of plume effects, it provides a long-term mechanism which permits the generation of permanent and continuing uplift over geological time. In detail, the erosionally controlled Western Ghats escarpment may have had its earliest origins in the form of seaward-facing normal fault scarps generated during the rifting phase 65 Ma. During the Tertiary, the position and geological relationship of these early scarps has changed fundamentally through geomorphological processes such as scarp retreat which has caused significant inland recession of the feature across the evolving continental margin.

During the evolution of the continental margin, two phases of lateritization can be recognized which developed independently upon two palaeosurfaces of genetically different origins. Each phase corresponds to a period during which the topographic, tectonic, and climatic conditions were conducive to the development of widespread laterite. The first phase began during the early Tertiary and is the result of pervasive, long-term alteration of the newly formed Deccan lava plain which lay inland of the nascent continental margin. A later uplift and associated erosion effectively terminated lateritization upon this post-Deccan surface, but subsequent development of a coastal pediment cut by scarp recession into the lava sequence to the west provided a topographical site suitable for a second lateritization phase to occur during mid-upper-Tertiary times.

The importance of the Deccan palaeosurfaces is threefold. First, reconstruction of the palaeosurfaces indicates that uplift is not uniform along the Indian margin and that the pattern of uplift is consistent with that expected from denudational unloading resulting from scarp recession. Second, the differing degrees of deformation suffered by the two palaeosurfaces provides a record of Tertiary uplift effects, and by virtue of their differing ages the current form of these surfaces reveals important information regarding the post-rift evolution of this continental margin. Third, identification of the original top of the lava sequence provides a datum against which the erosional history of the region may be further constrained. Such protracted uplift of a passive continental margin is beyond that expected by transient thermal effects and therefore challenges some of the accepted ideas of VRM development.

In conclusion, the Deccan continental margin should be viewed within the outlined framework of passive margin evolution and erosion. Comparison with other examples of similar VRM at later stages of their erosional development (e.g. Karoo 190 Ma, and Paraná, 120 Ma; Cox 1988*b*) indicate that continuation of the current erosion processes in Western India will lead to further exhumation, associated isostatic uplift, and seismicity. In the future it will become necessary to provide a more robust description of the evolution of rift margins which discriminates between the tectonic uplift component and that attributable to erosionally induced isostatic rebound. Clearly, the onshore signature of passive margin evolution is complex, but the similarities of different margins of different ages clearly indicate a common pattern of evolution. Future work should aim to contribute significantly to a global model of the rift process, and thereby to an understanding of the subsequent continental-scale drainage patterns and the rates of offshore sedimentation.

This research, including fieldwork and acquisition of satellite images, was supported by the NERC under research grants GT4/85/GS/83 and GR9/963. The research has greatly benefited from discussion with contemporary colleagues at the Department of Earth Sciences, Oxford University, and in particular I record my thanks to Keith Cox, Andrew Goudie, Marty McFarlane, and Clive Mitchell for providing constructive criticism and comment during the preparation of this work. Angela Coe is also thanked for providing a thorough review of the manuscript. I am grateful for logistical help provided in India by Professor K.V. Subbarao, Professor F.B. Antao and O. A. Fernandez, and for additional computing and drafting facilities offered by the Department of Geological Sciences, Durham University.

References

ALEVA, G. J. J. 1981. Bauxites and other duricrusts on the Guiana Shield, South Americ. *In*: *Lateritisation Processes*, Proceedings of an International Seminar on Lateritisation processes, Trivandrum, 1979. Balkema, Rotterdam, 261–269.

AUDEN, J. B., 1975. *Seismicity Associated with the Koyna Reservoir*, Maharashtra. UNESCO Technical Report, Paris. RP/1975-76/2.2223.

BAKSI, A. K. 1994. Geochronological studies o whole-rock basalts, Deccan Traps, India: evaluation of the timing of volcanism relative to the K-T boundary. *Earth and Planetary Science Letters*, **121**, 43–56.

BARDOSSY, G. 1981. Palaeoenvironments of laterites and lateritic bauxites - effect of global tectonism on bauxite formation. *In*: *Lateritisation Processes*, Proceedings of an International Seminar on Lateritisation processes, Trivandrum, 1979. Balkema, Rotterdam, 287–294.

BARTON, P. J. & WOOD, R. J. 1984. Tectonic evolution of the North Sea basin: crustal stretching and subsidence. *Geophysical Journal of the Royal Astronomical Society*, **79**, 987–1022

BEANE, J. E., TURNER, C. A., HOOPER, P. R., SUBBARAO, K. V. & WALSH, J. N. 1986. Stratigraphy, composition, and form of the Deccan basalts, Western Ghats, India. *Bulletin of Volcanology*, **48**, 61–83.

BOURMAN, R. P., MILNES, A. R. & OADES, J. M. 1987. Investigations of ferricretes and related surficial ferruginous materials in parts of southern and eastern Australia. *Zeitschrift für Geomorphologie N.F., Supplementband,* **64**, 1–24.

BREMER, H. 1985. Randschwellen: a link between plate tectonics and climatic geomorphology. *Zeitscrift für Geomorphologie N.F., Supplementband,* **54**, 11–21.

BROOKS, C. K. 1985. Vertical crustal movements in the tertiaty of central East Greenland: a continental margin at a hot-spot. *Zeitschrift für Geomorphologie N.F., Supplementband,* **54**, 101–117.

BROWN, R. W., RUST, D. J., SUMMERFIELD, M. A., GLEADOW, A. J. W. & DE WIT, M. C. J. 1990. An early Cretaceous phase of accelerated erosion on the southwestern margin of Africa: Evidence from fission-track analysis and the offshore sedimentary record. *Nuclear Tracks and Radiation Measurements*, **17**, 339–351.

BRUCKNER. H. 1989. Kustennahe Tiefelander in Indien - ein Beitrag zur Geomorphologie der Tropen. *Dusseldorfer Geographische Schriften* Heft, **28**. Heinrich-Heine-Universitat, Dusseldorf.

BUCK, W. R. 1986. Small-scale convection induced by passive rifting : The cause for uplift of rift shoulders. *Earth and Planetary Science Letters*, **77**, 362–372.

CAMPBELL, J. M. 1917. Laterite. *Mineralogical Magazine*, **17**, 27–77, 120–128, 171–179, 220–229.

CHANDRASEKHARAN, D. 1985. Structure and evolution of the western continental margin of India deduced from gravity, seismic, geomagnetic and geochronological studies. *Physics of the Earth and Planetary Interiors*, **41**, 186–198.

COCHRAN, J. R. 1983. Effects of finite extension times on the development of sedimentary basins. *Earth and Planetary Science Letters*, **66**, 289–302.

COURTILLOT, V. E., BESSE. J., VANDAMME, D., MONTIGNY, R. & CAPETTA, H. 1986. Deccan flood basalts and the Cretaceous/Tertiary boundary. *Earth and Planetary Science Letters*, **80**, 361–374.

COX, K. G. 1980. A model for flood basalt vulcanism. *Journal of Petrology*, **21**, 629–650.

—— 1983. Deccan Traps and the Karoo: stratigraphic implications of possible hot spot origins. *IAVCEI programme and abstracts of XVIII IUGG General Assembly, Hamburg*, 96.

—— 1988*a*. The Karoo Province. In: MACDOUGALL J.D (ed.) *Continental Flood Basalts*. Kluwer, Dordrecht,. 239–271.

—— 1988*b*. Inaugural Address. In: SUBBARAO, K.V (ed.) *Deccan Flood Basalts: Geological Society of India Memoir,* **10**, 15–22

—— 1989. The role of mantle plumes in the development of continental drainage patterns. *Nature*, **342,** 873–877.

—— & HAWKESWORTH, C. J. 1985. Geochemical Stratigraphy of the Deccan Traps at Mahabaleshwar, Western Ghats, India, with implications for Open system Magmatic Processes. *Journal of Petrology*, **26**, 355–377.

DESHMUKH, S. S. & SEHGAL, M. N. 1988. Mafic dyke swarms in the Deccan Volcanic Province of Madhya Pradesh and Maharashtra. *In*: SUBBARAO, K. V (ed.) *Deccan Flood Basalts: Geological Society of India Memoir,* **10**, 323–340.

DEVEY, C. W. & LIGHTFOOT, P. C. 1986. Volcanological and tectonic control of stratigraphy and structure in the western Deccan Traps. *Bulletin of Volcanology*, **48**, 195–207.

—— & STEPHENS, W. E. 1991. Tholeiitic dykes in the Seychelles and the original spatial extent of the Deccan. *Journal of the Geological Society, London*, **148,** 979–983.

DUNCAN, R. A. & PYLE, D. G. 1988. Rapid extrusion of the Deccan flood basalts at the Cretaceous/Tertiary boundary. *Nature*, **333**, 841–843.

FERMOR, L. L. 1909. Manganese in Laterite. *Memoir of the Geological Survey of India,* **37**, 370–389.

FOOTE, R. B. 1876. Geological features of the south Maratta County and adjacent districts. *Memoir of the Geological Survey of India*, **12**, 1–268.

FOX, C. S. 1923. The bauxite and aluminous laterite occurrences of India. *Memoir of the Geological Survey of India*, **49**, 1–287.

GALLAGHER, K., HAWKESWORTH, C. & MANTOVANI, M. S. M. 1995. Denudation, fission track analysis and the long-term evolution of passive margin topography: application to the southeast Brazil margin. *Journal of South American Earth Sciences,* **8**(1), 65–77.

GILCHRIST, A. R. & SUMMERFIELD, M. A. 1990 Differential denudation and the flexural isostasy in the formation of rifted-margin upwarps. *Nature*, **346**, 739–742.

—— & SUMMERFIELD, M. A., 1994. Tectonic models of passive margin evolution and their implications for theories of long-term landscape development. *In*: KIRKBY, M. J. (ed.) *Process Models and Theoretical Geomorphology*. Wiley, Chichester, 55–84.

——, KOOI, H. & BEAUMONT, C. 1994*a*. Post Gondwana geomorphic evolution of southwestern Africa: implications for the controls of landscape development from observations and numerical experiments. *Journal of Geophysical Research*, **99**, 12211–12228.

——, SUMMERFIELD, M. A. & COCKBURN, H. A. P. 1994*b*. Landscape dissection, isostatic uplift, and the morphotectonic development of orogens. *Geology*, **22**, 963–966.

GOUDIE, A. S. 1983. *Duricrusts in Tropical and Subtropical Landscapes*. Oxford University Clarendon Press.

GUNNELL, Y. 1997. Topography, palaeosurfaces and denudation over the Karnataka uplands, southern India. *This volume*.

HOVIUS, N. (in press). Controls on sediment supply by large rivers. *In*: SHANLEY, K. W. & MCCABE, P. J. (eds) *Relative Role of Eustasy, Climate and Tectonics in Continental Rock*. SEPM Special Publication.

KAILA, K. L., MURTY, P. R. K., DIXIT, M. M. & LAZARENKO, M. A. 1981. Crustal structure from deep sesimic sounding along the Koyna II (Kelsi-Loni) profile in the Deccan Trap area, India. *Tectonophysics*, **73**, 365–384.

KAILASAM, L. N. 1975. Epeirogenic studies in India with reference to recent vertical movements. *Tectonophysics*, **29**, 505–521.

—— 1979. Plateau uplift in Peninsula India. *Tectonophysics*, **61**, 243–269.

KALE, V. S. & RAJAEURU, S. N. 1987. Late Quaternary alluvial history of the northwestern Deccan upland region. *Nature*, **325**, 612–614.

KARNER, G. D. & WEISSEL, J. K. 1990. Compressional deformation in the central Indian Ocean: Why is it where it is. *Proceedings of the Ocean Drilling Program, Scientific Results*, **116**: *College Station, TX (Ocean Drilling Program)*, 279–289.

KENT, R. W., STOREY, M. & SAUNDERS, A. D. 1992. Large igneous provinces: sites of plume impact or plume incubation? *Geology*, **20,** 891–894.

KHADRI, S. F. R., SUBBARAO, K. V. & BODAS, M. S. 1988. Magnetic studies on a thick pile of Deccan flows at Kalsubai. *Memoirs of the Geological Society of India*, **10**, 163–189.

—— & NAGAR, R.G. 1994. Magnetostratigraphy of Malwa Deccan Traps near Mandu region, Madhya Pradesh, India. *Memoirs of the Geological Society of India*, **29**, 199–209.

KING, L. C. 1967. *The Morphology of the Earth* (2nd edn). Oliver and Boyd, Edinburgh.

KRISHNAN, M. S. 1956. *Geology of India and Burma. (3rd edition)*. Higginbothams, Madras.

KUMAR, A. 1985. Palaeolatitudes and age of Indian laterites. *Palaeogeography, Palaeoclimatology, Palaeoecology*, **53**, 231–237.

KUMAR, S. 1975. A model for the subsidence history of the west coast of India. *Tectonophysics*, **27**, 167–176.

LYON-CAEN, H. & MOLNAR, P. 1985. Gravity anomalies, flexure of the Indian plate, and the structure, support and evolution of the Himalaya and Ganga Basin. *Tectonics*, **4**, 513–538.

MABBUTT, J. A. 1966. Mantle-controlled planation of pediments. *American Journal of Science*, **264**, 78–91.

MAHONEY, J. J. 1988. Deccan Traps. *In*: MACDOUGALL, J. D. (ed.) *Continental flood Basalts*. Kluwer, Dordrecht, 151–194.

MCFARLANE, M. J. 1976. *Laterite and landscape*. Academic Press, London.

—— 1983. The temporal distribution of bauxitisation and its genetic implications. *In*: MELFI, A. J. & CARVALHO, A. (eds) *Lateritisation Processes*. Proceedings 2nd International Seminar on Lateritisation Processes, São Paulo, Brazil, 1982, 197–207.

MCKENZIE, D. 1978. Some remarks on the development of sedimentary basins. *Earth and Planetary Science Letters*, **40**, 25–32.

MEDLICOTT, H. B. & BLANDFORD, W. T. 1879. *A Manual of the Geology of India (Part I: Peninsular Area)*. Trubner, London.

MITCHELL, C. & WIDDOWSON, M. 1991. A Geological Map of the Southern Deccan Traps, India. *Journal of the Geological Society, London*, **148**, 495–505.

NAINI, B. R. & TALWANI, M. 1983. Structural Framework and the Evolutionary History of the Continental Margin of Western India. *In*: WATKINS J. S. & DRAKE C. L. (eds) *Studies in Continental Margin Geology*. AAPG Memoir, **34**, 167–191.

NAJAFI, S. J., COX, K. G. & SUKHESWALA, R. N. 1981. Geology and geochemistry of basalt flows (Deccan Traps) of the Mahad-Mahabaleshwar section, India. *In*: SUBBARAO, K. V. & SUKHESWALA, R. N. (eds) *Deccan volcanism and related provinces in other parts of the world*. Geological Society of India Memoir, **3**, 300–315.

NORTON, I. O. & SCLATER, J. G. 1979. A model for the evolution of the Indian Ocean and the breakup of Gondwanaland. *Journal of Geophysical Research*, **84**, 6803–6830.

OLLIER, C. D. 1985 The Morpotectonics of Passive Continental Margins: Introduction. *Zeitscrift für Geomorphologie NF, Supplementband*, **54**, 1–9.

—— & POWAR, K. B. 1985. The Western Ghats and the morphotectonics of peninusular India. *Zeitschrift für Geomorphologie NF, Supplementband*, **54**, 37–56.

PAIN, C. F. 1985. Morphotectonics of the continental margins of Australia. *Zeitschrift für Geomorphologie NF, Supplementband*, **54**, 23–35.

RADHAKRISHNA B. P. 1991. An excursion into the past – "the Deccan volcanic episode". *Current Science*, Special issue **61** (9 & 10), 641–647.

RICHARDS, M. A., DUNCAN, R. A. & COURTILLOT, V. E. 1989. Flood basalts and hot spot tracks: plume heads and tails. *Science*, **246,** 103–107.

SCHMIDT, P. W., PRASAD, V. & RAMAN, P. K. 1983. Magnetic ages of some Indian laterites. *Palaeogeography, Palaeoclimatology, Palaeoecology*, **44,** 185–202.

SPÄTH, H. 1987. Landform development and laterites in northwestern Australia. *Zeitscrift für Geomorphologie NF, Supplementband*, **64**, 25–32.

STOREY, M., MAHONEY, J. J., SAUNDERS, A. D., DUNCAN, R. A., KELLEY, S. P. & COFFIN, M. F. 1995. Timing of hot-spot related volcanism and the breakup of Madagscar and India. *Science*, **267**, 852–855.

SUBBARAO, K. V. & HOOPER, P. R. 1988. Reconnaissaance map of the Deccan Basalt Group in the Western Ghats, India, Scale 1:100000. *In*: SUBBARAO, K. V. (ed.). *Deccan Flood Basalts: Geological Society of India Memoir*, **10**.

——, CHANDRASEKHARAM, D., NAVANEETHAKRHISHNAN, & HOOPER, P. R. 1994. Stratigraphy and structure of parts of the central Deccan Province: eruptive models. *In*: SUBBARAO, K. V. (ed.) *Volcanism (Radhakrishna volume)*. Wiley, New Delhi, 321–332.

SUBRAHMANYA, K. R. 1987. Evolution of the Western Ghats, India - a simple model. *Journal of the Geological Society of India*, **29**, 446–449.

—— 1994. Post-Gondwana tectonics of the Indian peninsula. *Current Science*, **67**, 527–530.

SUBRAMANIAN, K. S. 1987. Possible cause of Easterly tilt of the Southern part of India. *Journal of the Geological Society of India*, **29**, 362–363.

SUMMERFIELD, M. A. 1989. Tectonic geomorphology: convergent plate boundaries, passive continental margins and supercontinent cycles. *Progress in Physical Geography*, **13**, 431–441.

—— 1991. Sub-aerial denudation of passive margins: regional elevation versus local relief models. *Earth and Planetary Science Letters*, **102**, 460–469.

TARDY, Y., KOBILSEK, B. & PAQUET, H. 1991. Mineralogical composition and geographical distribution of African and Brazilian periatlantic laterites. The influence of continental drift and tropical paleoclimates during the past 150 million years and implications for India and Australia. *Journal of African Earth Sciences*, **12 No.1/2**, 283–295.

THOMAS, M. F. 1994. *Gemorphology in the Tropics: a Study of Weathering and Denudation in Low Latitudes*. Wiley, Chichester.

—— & SUMMERFIELD, M. A. 1987. Long-term landform development: key themes and research problems. *In*: GARDINER, V. (ed.) *International Geomorphology 1986: Proceedings of the First International Conference on Geomorphology*. Part 2, 935–956. Wiley, Chichester.

VALETON, I. 1983. Palaeoenvironment of lateritic bauxites with vertical and lateral differentiation. *In*: WILSON, R. C. L. (ed.) *Residual Deposits: Surface Related Weathering Processes and Materials*. Geological Society, London, Special Publication, **11**, 77–90

WATTS, A. B. & COX, K. G. 1989. The Deccan Traps: an interpretation in terms of progressive lithospheric flexure in responce to a migrating load. *Earth and Planetary Science Letters*, **93**, 85–97.

WEISSEL, J. K. & KARNER, G. D. 1989. Flexural uplift of rift flanks due to mechanical unloading of the lithosphere during extension. *Journal of Geophysical Research*, **94**, 13919–13950.

WHITE, K., WALDEN, J. & ROLLIN, E. M. 1992. Remote sensing of pedogenic iron oxides using Landsat Thematic Mapper data of Southern Tunisia. *Remote Sensing from Research to Operation*. Proceedings of 18th Annual Conference of the Remote Sensing Society, Dundee, 179–187.

WHITE, R. S. & MCKENZIE, D. 1989. Magmatism at rift zones: the generation of volcanic continental margins and flood basalts. *Journal of Geophysical Research*, **94,** 7685–7729.

WHITING, B. M., KARNER, G. D. & DRISCOLL, N. W. 1994. Flexural and stratigraphic development of the west Indian continental margin. *Journal of Geophysical Research*, **99**, 13791–13811.

WILKINS. A., SUBBARAO, K. V., INGRAM, G. & WALSH, J. N. 1994. Weathering regimes within the Deccan Basalts. *In*: SUBBARAO, K. V. (ed.). *Volcanism (Radhakrishna Volume)*. Wiley, New Delhi, 217–232.

WIDDOWSON, M. 1990. *Uplift History of the Western Ghats, India*. DPhil thesis, Oxford University.

—— & COX, K. G. 1996. Uplift and erosional history of the Deccan traps, India: Evidence from laterites and drainage patterns of the Western Ghats and Konkan Coast. *Earth and Planetary Science Letters*, **137**, 57–69.

—— & MITCHELL, C. (in press). Large-scale stratigraphical, structural, and geomorphological constraints for earthquakes and seismicity in the southern Deccan Traps, India.

——, WALSH, N. & SUBBARAO, K. V. 1997. The geochemistry of Indian bole horizons: palaeoenvironmental implications of Deccan intravolcanic palaeosurfaces. *This volume*.

WOOLNOUGH, W. G. 1918. Physiographic significance of laterite in Western Australia. *Geological Magazine NS*, **6**, 385–393.

Topography, palaeosurfaces and denudation over the Karnataka Uplands, southern India

YANNI GUNNELL

Laboratoire de Géographie Physique, URA 1562–CNRS, Université Blaise-Pascal, 29 boulevard Gergovia, 63 037 Clermont-Ferrand cedex 1, France;

Present address: Départment de Géographie, Université Paris 7, Case 7001, 2 Place Jussieu, 75251, Paris cedex 05, France

Abstract: A synthesis of new information from extensive field reconnaissance and systematic map analyis from southern India is presented. A new interpretation for landscape development in the Karnataka upland region is given. Three phases of geomorphological evolution are identified.

A fundamental, pre-Deccan Trap palaeosurface is recognized. While becoming largely fossilized at 65 Ma by the eruption of the Deccan lavas, its southern continuation, studied here, remained exposed throughout the Tertiary and evolved in response to uplift and crustal flexuring of the Indian margin. Remnants of this surface now occur as the highest summits of the region (*c*. 1900 m above sea level).

A younger, intermediate or post-Deccan Trap palaeosurface may be recognized by a combination of superimposed drainage and morphological mapping techniques. It evolved in response to the loading of the Indian shield by the 1–2 km thick Deccan lava pile.

The third and final phase of landform development is complex since it has not resulted in a single surface of regional extent but rather a series of contemporaneous, but locally restricted, levels at varying altitudes. These are not interpreted as the result of cyclical erosion but as controlled in acyclical fashion by individual river catchments and lithological heterogeneity. The possible causes for this complexity are discussed and confronted with existing schemes of landscape development on other continents.

The current work represents a synthesis of new data and previous studies and sets out to draw attention to the tropical upland landscape of an Archean craton which has largely escaped geomorphological scrutiny. Whilst a few authors have repeatedly revisited the Tamilnadu Lowlands and the eastern flank of the Mysore Plateau, much less attention has been given to the Karnataka Plateau itself. No fewer than nine different interpretations of the landscape have nonetheless been compiled from these authors who have tended to restrict their focus to certain portions of the area (Table 1). An examination of Table 1 occasions the following comments. (1) There is inconsistency among the authors regarding what topographic unit should be called Mysore Plateau and what should be termed Karnataka Plateau (see Fig. 3). It is true that common understanding locates the Mysore Plateau to the south and the Karnataka Plateau to the north. The former pertains to the Cauvery catchment and is higher by 300 m than the latter, which pertains to the Tungabhadra catchment (itself a tributary of the Krishna, Figs 1 and 7). However, such a geographical distinction is no proof that these two parts of one large upland region are genetically distinct, yet this appears to be assumed by most authors except Petit (1982). (2) Studies have repeatedly tended to treat the Mysore Plateau and the neighbouring Tamilnadu Lowlands as two contrasting and competing geomorphic systems. There is no doubt that this bias has brought much conceptual wealth to geomorphological thinking (Büdel 1982). Yet, the northern two-thirds of the Karnataka Upland region have usually been left aside, giving the impression that the Tamilnadu Lowlands and the Mysore Plateau somehow go naturally together, while the Mysore Plateau and Karnataka Plateau are separate. This is not just a matter of scale of analysis and could be geographically misleading. (3) The different levels identified by the different authors are strongly disparate in their names as well as in their altitudes, leaving aside the additional disagreement concerning the age of a given level. This inconsistency poses the problem of how to identify a surface and whether it should be defined as a mean altitude, a modal altitude, or a bracket of altitudes which allows for a certain relief. Whilst a disunity apparently prevails in the matching of planation level altitudes, previous authors seem more agreed on the cyclical nature of these surfaces, invoking tiered levels of distinct age, generated by a sequence of uplift events, each followed by prolonged planation during tectonic quiescence. Each new cycle of planation entails the rejuvenation and areal contraction of the upper, older level by

From Widdowson, M. (ed.), 1997, *Palaeosurfaces: Recognition, Reconstruction and Palaeoenvironmental Interpretation*, Geological Society Special Publication No. 120, pp. 249–267.

Table 1. *Survey of planation surfaces identified by previous authors*

	Chatterjee 1961*	Vaidyanadhan 1967, 1977	King 1967†	Brunner 1968	Babu 1975
Quaternary		400 m (Eastern Ghats)	Rejuvenation Uplift	300250 m (Tamilnadu surface)	150 m (Palghat–Madurai)
Pliocene	400–300 m	700 m (Karnataka Plateau)	Dissection		300 m (Palani–Tirupattur)
			Uplift	450–400 m (Chitoor surface)	
	550–450 m		Dissection	750–650 m (Punganur surface)	
Miocene	850–600 m	1000 m (Mysore Plateau)	Uplift	900–750 m (Bangalore surface)	600 m (Manantody–Kolar regions)
Oligocene	1150–900 m		Extreme planation		
Eocene					
			High level laterite		
Cretaceous		2000 m (Bababudan)			900 m (Mysore Plateau)
Jurassic		2500 m (Nilgiri)	Gondwana surface (Nilgiri)		1500–2700 m (Shevaroy–Nilgiri)

* No ages given. Position in table is purely tentative and assumed by default.
† No altitudes given.

headward erosion. Petit however, once again, stands out with a fundamentally alternative approach which provides a platform for alternative interpretation (see Table 1).

In this context it seems appropriate to reassess the whole geometry of the landscape based upon recent extensive field expeditions across the area. This current work utilizes simple, classical geomorphological tools: a set of 1 : 50 000 topographical sheets and a 1 : 500 000 geological map (the only one available), and provides a preliminary frame of understanding into which fission track data from a recent series of sampling expeditions will eventually be incorporated.

General outline of the topography

In the current work the Mysore Plateau and Karnataka Plateau are grouped together as *Karnataka Uplands*, stressing that the entire backslope of the Western Ghats from the southern edge of the Deccan Traps to the northern foot of the Nilgiri Highlands is being considered as a single geomorphic unit (Fig. 1). A degree of geological unity also befits the Karnataka Uplands insofar as the core of the study area coincides with the Western Dharwar Craton, otherwise known as Karnataka Nucleus (Fig. 2) – one of the three most anciently consolidated crustal terranes of the Indian shield.

The Karnataka Uplands are a vast area perched to the west above the Tamilnadu plains via the scarp interpreted by Büdel (1982) as an etchscarp in the flank of a domal uplift, but mitigated by Brunner (1968) and later by Demangeot (1975) as partly due to a fault line. On their western flank the Karnataka Uplands are perched above the Konkan Lowlands via the Great Escarpment of the Western Ghats (Fig. 1). Consequently, major rivers leave the Karnataka Uplands in both directions via major waterfalls and gorges such as the Shimsha Falls and the sequence of Cauvery Falls to the east, as well as many dozens over the Ghats to the west. Owing to the sluggishness of stream erosion on cratonic interiors in tropical climates, it seems that the escarpments on both edges of the Karnataka Uplands have thus delayed the onset of any waves of cyclical headward erosion. As such, the uplands' perched position fully justifies considering them as a separate physiographical unit to be dealt with as a whole. It is important to stress at this point that in these distant upper reaches of the mainly east-flowing rivers, it would be unwise to look for correlations between detrital sediments in the Bengal offshore region and the planation levels found here some 500 km away from the Bay of Bengal base level. Furthermore, these rivers are effectively cut off from direct base-level influence by waterfalls and transverse mountain ridges (Kappat Hills, Sandur Hills, Fig. 3; Appalachian-type ridge-and-valley relief of the Eastern Ghats/Cuddapah formation, Figs 1 and 2). The range of methods available to study long-term landscape development in the Karnataka Uplands thus appears quite restricted.

Demangeot 1975	Naruse 1980*	Petit 1982, 1985	Radhakrishna 1965, 1991
Minor alterations to landscape			Cymatogenic uplifting Rejuvenation of plateaus
'Tamilnadu cycle'	< 500 m (Tamilnadu surface) 700 m (Mandya surface 800 m (Mysore surface)	Various local levels developed over the course of the Cenozoic and controlled by structural factors:	430–360 m (Newer Mysore Plateau, stage 2) 600–550 m (Newer Mysore Plateau, stage 1) Creation of Western Ghats scarp by faulting
900–600 m ('Mysore cycle')	900 m (Bangalore surface)	1200 m (Mercara), 1000 m (Chikmagalur), 850–950 m (Coorg), 700–750 m (upper Tungabhadra), 650–550 m (lower Tungabhadra) . . .	900 m (Older Mysore Plateau) Upwarp in Western Ghats
Shevaroy level		1800–1900 m ('infra-trap' surface)	(Nilgiri surface)
2600 m ('Nilgiri')			

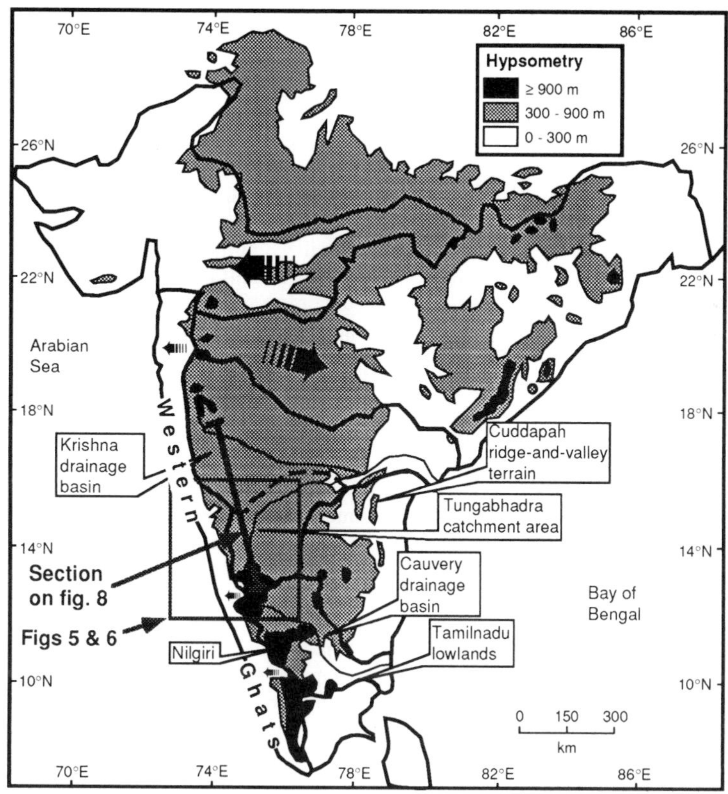

Fig. 1. General location map of peninsular India (i.e. south of the Ganges plains) showing highlands, upland plateau areas and lowlands. Major drainage divides (thick black lines), attending rivers (thin black lines) and relief features mentioned in text are shown here. Tungabhadra sub-unit of the Krishna Basin shown in pecked outline. Heavy arrows in drainage basins outside study area indicate, for reference, general direction of flow and base level dependence. Location of Figs 5, 6 and 8 also shown.

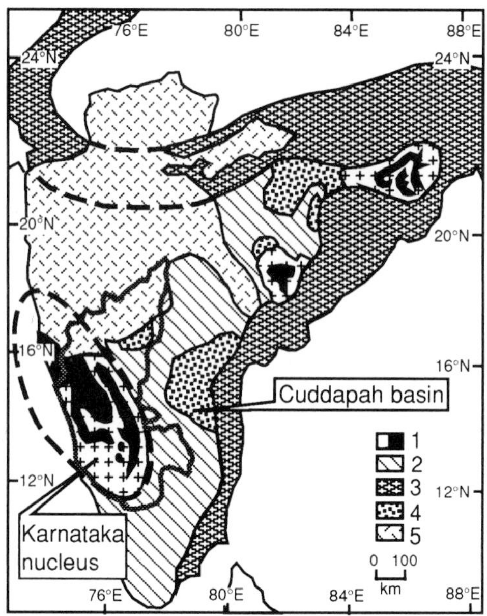

Fig. 2. Geological provinces of peninsular India (as interpreted by Radhakrishna & Naqvi 1986). Karnataka Nucleus is also known as Western Dharwar Craton (see Fig. 4). 1, Archean cratons (tonalitic gneiss with greenstone belts); 2, granodiorites, gneisses and granulites of Early Proterozoic Mobile Belt; 3, granulites and gneisses of Middle Proterozoic Mobile Belt; 4, Middle Proterozoic sedimentaty basins; 5, Mesozoic sediments and Deccan flood basalts. Outline of Karnataka state locates Figs 3 and 4.

The topography of the Karnataka Uplands was summarized by a synthetic map of modal altitudes (Fig. 5). This treatment of data is judged to produce a more faithful representation of planation levels than small-scale hypsometric atlas maps with unsuitable contour intervals (cf. Figs 3 and 5; see caption Fig. 5 for discussion). It appears clearly that below the 1000 m upper bracket no altitude bracket exceeds 100 m, which is an acceptable elevation range when attempting to define a continental-scale planation level (i.e. several hundreds of kilometres in the case of the Indian subcontinent); furthermore, no gap between the altitude classes appears to exceed 50 m on the Karnataka Uplands which, at the outset, makes it difficult to identify any clear-cut cycles of denudation in the later stages of landscape development. In a humid tropical environment, where weathering profiles are several tens of metres thick, it is indeed reasonable to expect in the conceptual framework of cyclical denudation, a minimum of 40–50 m separating two continental-scale planation surfaces. In addition to the surface lowering by stripping of the existing weathered mantle (effected by rhexistasis), cyclical models postulate an additional component of net crustal uplift which can, in theory, induce a supplementary increment of weathering and stripping. The pattern of modal altitude levels depicted on Fig. 5 therefore reveals a borderline situation: it allows hesitation between a series of low-magnitude cycles (known as epicycles, or interrupted cycles in Davisian geomorphology), or an alternative (acyclical) mode of differential lowering of the topography. This ambiguity will be discussed in the following sections.

The higher levels of the Western Ghats ridge and Bababudan Hills in Fig. 5 have been classed separately as 'highly dissected' since the distinction between modal altitude and mean altitude starts to lose its geographical significance due to mountainous and irregular relief. Maximum altitude (Fig. 6) was therefore brought in to provide a better idea of the geometry of accordant summit levels.

Starting point: the Pre-Deccan Trap planation surface

The geomorphological history of the Deccan flood basalt plateau has lasted at the most 60–65 Ma, but that of the Precambrian basement to the south of the Deccan Traps is much longer. In effect, the Karnataka Craton has evolved under subaerial conditions for 60–65 Ma more than its extension to the north beneath the Deccan Traps where it was fossilized at the K/T boundary by the eruptions. It is suggested here to define the trap/basement unconformity as the Pre-Deccan Trap palaeosurface. This implies that the surface was established during the Mesozoic, possibly before the breakup of Gondwanaland, but no assumptions are made here on whether it corresponds to the Gondwana surface of King (1967) or whether it is actually Jurassic rather than Cretaceous (see Table 1). Such affirmations appear to be more model dependent than objectively established.

In spite of having been partly fossilized by the Deccan lavas, there does not appear to be a striking difference in elevation between the Pre-Deccan Trap surface on the basement rocks and the adjacent basalt plateaus. Indeed, when travelling from the basement to the south onto the volcanics to the north, there is no overriding impression of 'climbing up' onto a higher level due to more recently superadded volcanic materials. This simple observation suggests that Tertiary planation has been operating in the same plane, on trap and basement alike.

In the vicinity of the trap/basement boundary, the Malprabha and Ghatprabha Rivers (Fig. 7) maintain their W–E courses regardless of the

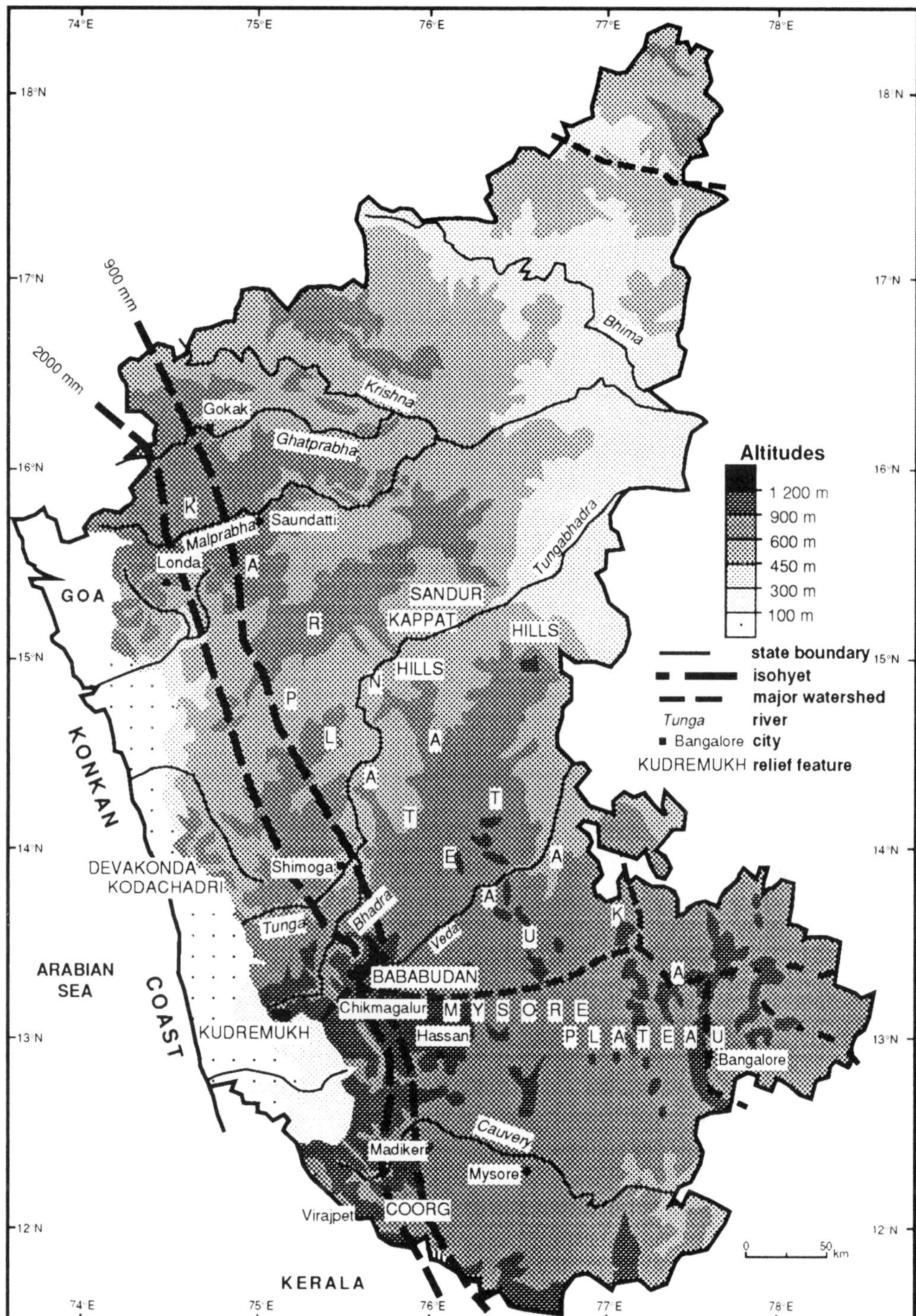

Fig. 3. General location map of Karnataka with place names, rivers and relief features referred to in text. Political state boundary was preserved to match Fig. 4. Main continental divide along Western Ghats is not shown; focus is restricted to the upland region second order watersheds.

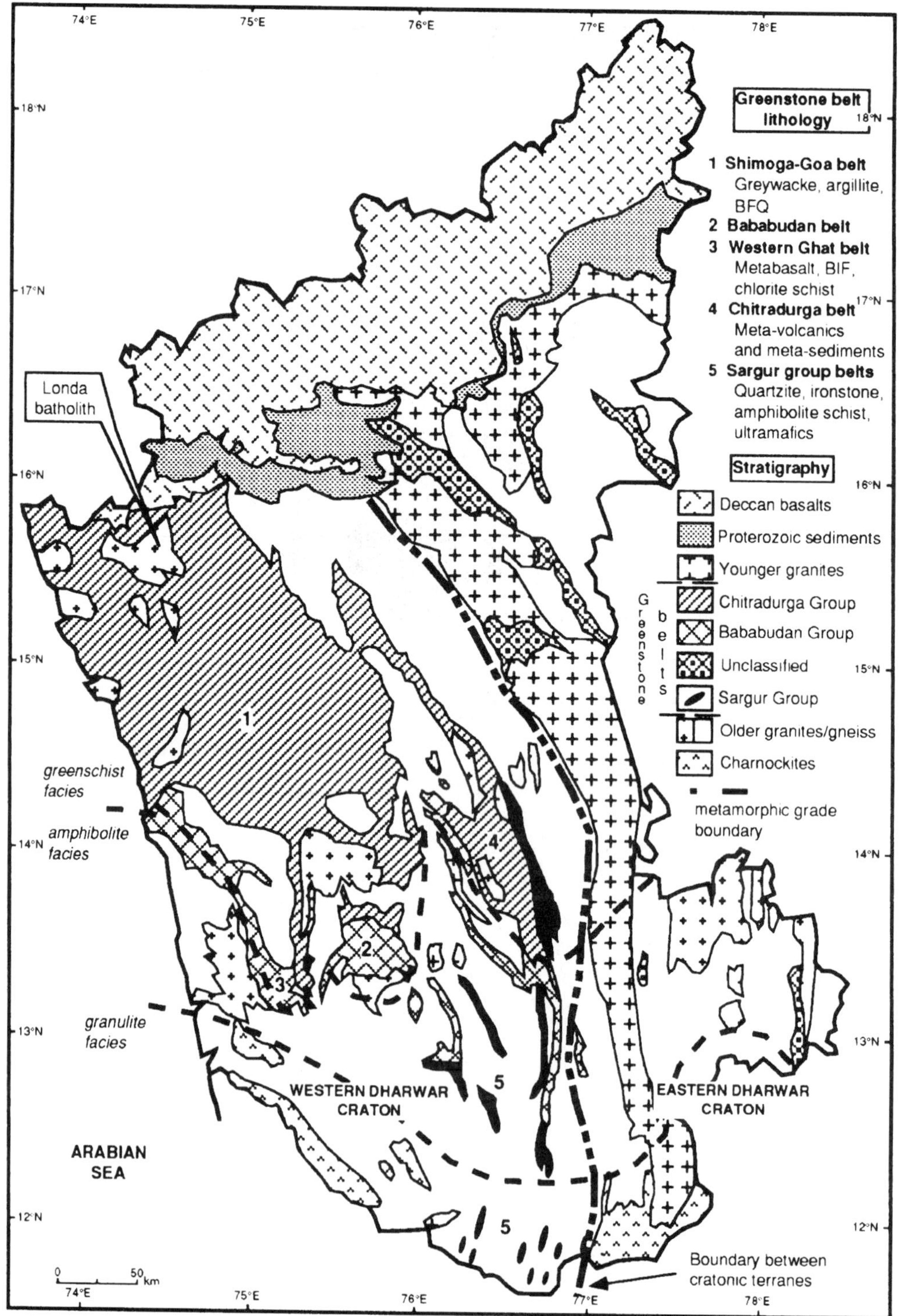

Fig. 4. Geological outline map of Karnataka. Political state boundary is the only available framework for mapping (modified after Swamy Nath & Ramakrishnan 1981).

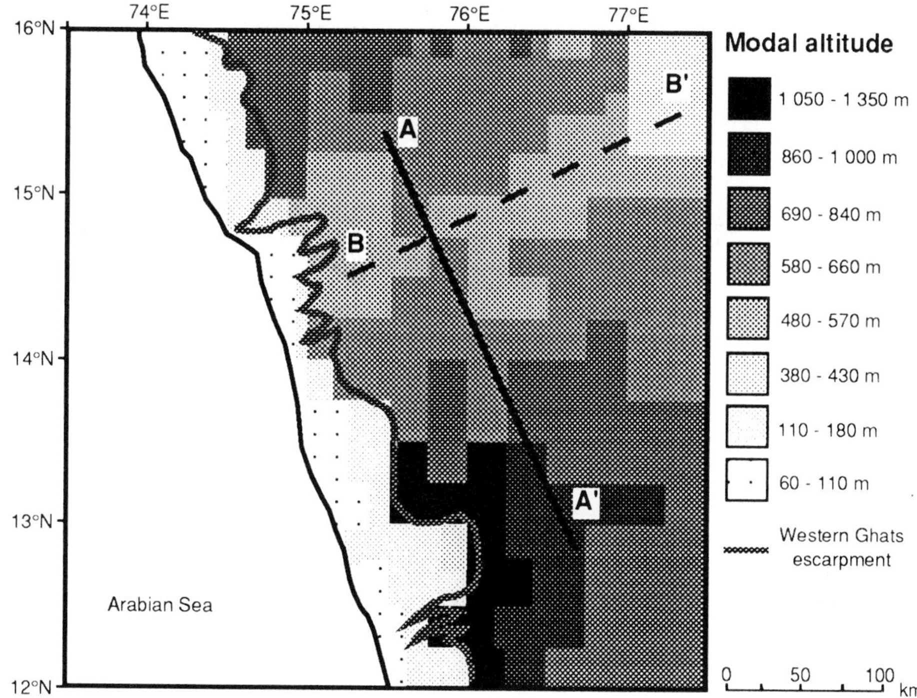

Fig. 5. Modal altitudes computed over the Karnataka Uplands. Each of the 174 squares of the grid represents one 1 : 50 000 topographic sheet, for which the modal altitude was established with a ± 20 m accuracy level by singling out, by definition, the most frequently occurring contour. The modal altitude classes were elaborated graphically from a dispersion diagram: the largest leaps between values in the series were chosen as 'natural' class intervals, understood to depict major breaks in slope over the Karnataka Uplands, i.e. possible cyclical knickpoints. See text for discussion of lines A–A′ and B–B′.

underlying geology: they cut through basement igneous rocks, Deccan basalt, and Proterozoic sandstones and quartzites in varying orders of succession. This preliminary fact leads to several corollary remarks. (1) These rivers show signs of having been established by superimposition onto a Post-Deccan Trap surface. Whether this surface was a true planation level, bevelling basalt and basement (at *c.* 850 m near Belgaum, for example), or the structural, post-eruptive lava surface might well call for debate (see Widdowson 1997): a review by Kale & Gupte (1986) tentatively concluded on the 'inadequacies and limitations in the available methodology' to ascertain the denudation chronology of the Deccan Traps; nevertheless, these authors do, however, claim to identify multiple denudational surfaces in each of the large Deccan river catchments (Godavari, Bhima and Krishna). Dikshit & Wirthmann (1992) do likewise. At this stage, it is therefore premature to assert whether the identifiable signs of denudation on the Deccan lavas (truncation of flows in particular) are of regional or merely local significance, and whether they correspond to inter- or post-eruptive events. The impending question is, in essence, one of timing: did sufficient time elapse between the eruption of the lavas and the superimposition of the rivers for a major planation surface to develop? It could be concluded, for example, that there was not much volcanic local relief to bevel in the first place and, furthermore, that the ongoing intense (tropical) chemical denudation of a low-relief, layered basalt structure may not yield clear-cut field signatures of denudation. It can at least be said that the summit levels of the Proterozoic sandstone and quartzite landforms found emerging from beneath the edge of the trappean flows lying at the same altitude are certainly polygenetic in nature. Hence, the Pre-Deccan Trap palaeosurface appears to be exhumed and reworked in the same plane by a Post-Deccan Trap denudational episode. (2) Near Saundatti (Fig. 7), the superimposed Malprabha drainage network has even separated an outlier of Proterozoic quartzite from its outcrop which lies a few kilometres further north, and thereby exhumed the Archean palaeosurface from beneath the Proterozoic rocks

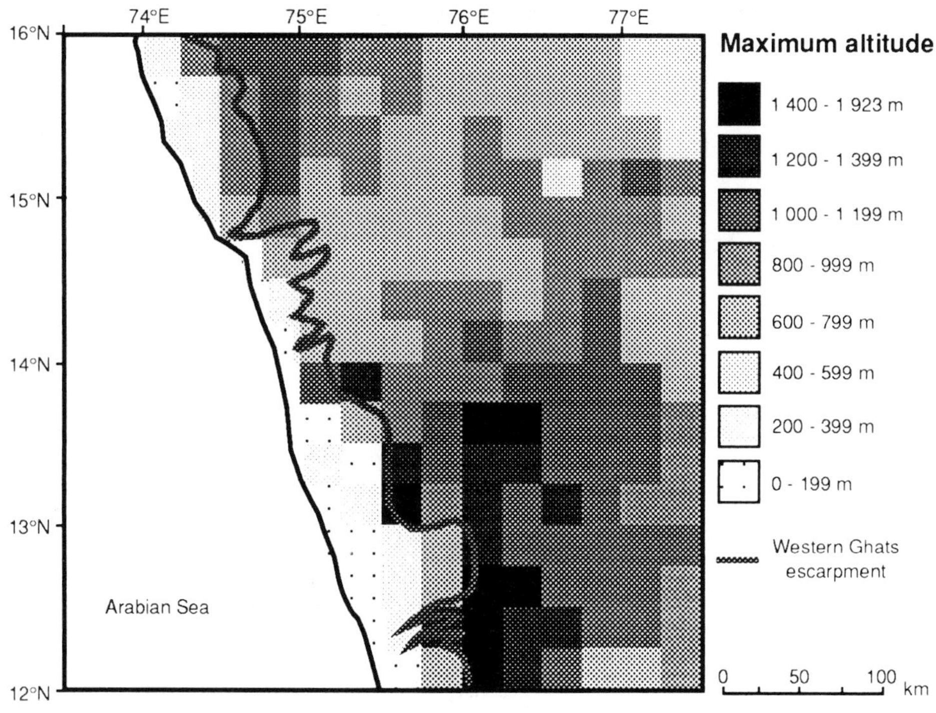

Fig. 6. Maximum altitude over the Karnataka Uplands, i.e. highest spot height value recorded on each 1 : 50 000 topographic sheet.

which lie unconformably on the Archean gneiss basement. In effect, this Archean palaeosurface corresponds to the exhumed heterolithic unconformity between the Archean gneiss and the overlying Proterozoic sediment pile. It seems likely that the exhumation of this corridor of Archean gneiss, which isolates the Saundatti outlier, was guided by a fracture in the quartzite (Gokhale & Joshi 1990). Although this example is, perhaps, local and anecdotal, it shows that the landscape in the vicinity of the trap–basement boundary is a juxtaposition of identifiable planation facets of at least three widely different generations: in this particular instance, the Pre-Deccan Trap surface at the top of the Saundatti table mountain, possibly reworked by a Post-Deccan Trap surface (see above), dominates the Archean surface by an average of 100 m. (3) According to field evidence so far, the basalt flows appear to cap the Proterozoic sediment cover rather than infill conspicuous pre-existing palaeovalleys. This leads to the conclusion that the Pre-Deccan Trap surface bevelled both the sediments and the crystalline basement rocks. In other words, the Proterozoic sediments were probably not forming a plateau perched above the Archean basement at the K/T boundary. From this follows that the structural landforms, developed in the Proterozoic sediment cover (Fig. 7), are presumably derived from the post-eruptive surface by differential erosion, and thus give evidence as to the degree of entrenchment of valleys since the early Tertiary. This scarp-and-vale landscape, with a relative wealth of cuestas and accordant tributaries to the otherwise often discordant Malprabha and Ghatprabha, is original on a Precambrian basement; it stands in contrast to the ridge-and-basin relief of the metamorphic terrain elsewhere on the craton (Fig. 7). A close examination of available large-scale topographic maps shows that rivers have cut into the initial surface, i.e. have rejuvenated the Pre-Deccan Trap palaeosurface, by only 120 m on average. This rejuvenation is usually less in the far interior of the region, though somewhat more in the vicinity of the uplifted Western Ghats. For example, the Ghatprabha cuts by at least 160 m through sandstone and quartzite ridges in the vicinity of the Gokak and Kaladgi basement windows (Fig. 7); stream incision values go up to 300 m for the uppermost reaches of the Malprabha, which slices into the basement rocks, which are themselves capped by bauxitized basalt ridges. The basalt

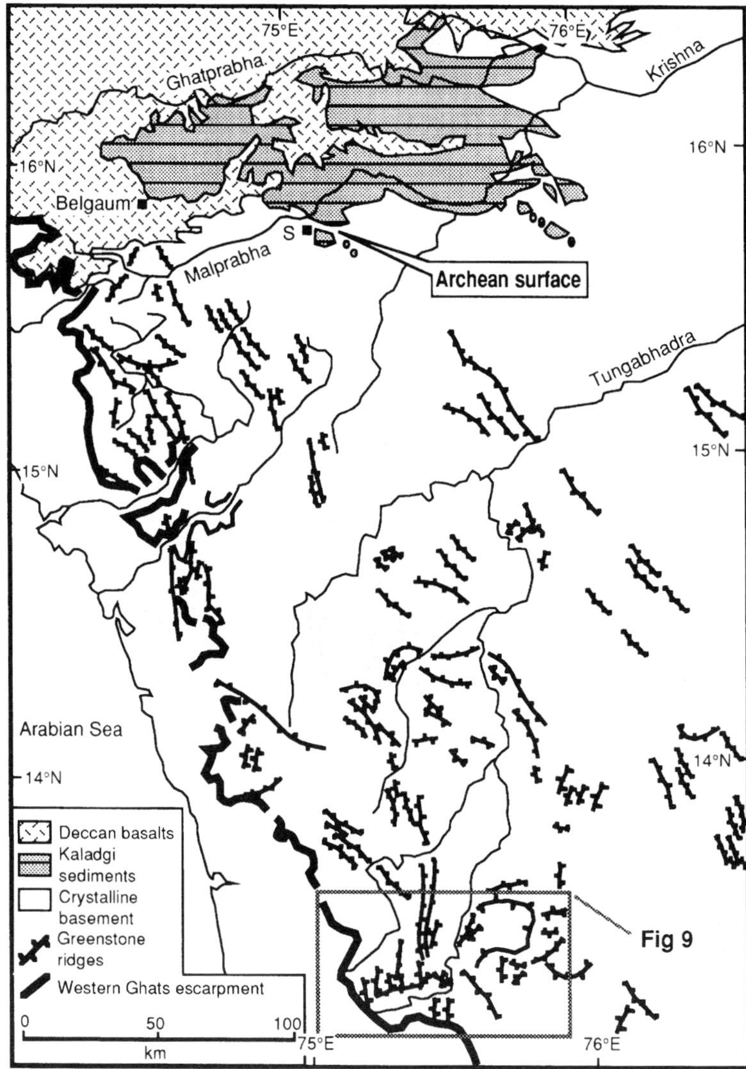

Fig. 7. Northwestern part of study area showing contrast between ridge-and-basin terrain on Dharwar crystalline basement (NW–SE Dharwar trend is noticeable from orientation of greenstone ridges) and Proterozoic Kaladgi sediments, dominated by scarp-and-vale relief. The Malprabha River runs off Deccan basalts onto the crystalline basement, then enters and leaves the sedimentary province several times before rejoining the Krishna River, which indicates a superimposed river course. Similarly, the Ghatprabha River flows across the three represented structural provinces. On the basement area to the south, most rivers flow across the structural grain of the craton and cut gorges into the greenstone ridges. Location of identified Archean surface near Saundatti (S) and neighbouring quartzite outlier also shown.

cappings serve as a good marker horizon in the landscape to measure the depth of post-eruptive incision. These values, which clearly increase westwards, give some indirect indication as to the degree of differential uplift of the Western Ghats rift shoulder: on a W–E traverse, the higher values of linear incision in the Ghats would reveal the greater degree of crustal uplift along the plateau rim – a geomorphological signature of flexural isostasy at the flank margin uplift. The lower values of incision inland are indicative of a relative downwarp of the lithosphere along the southern edge of the Deccan Traps (cf. line A–A′ on Fig. 5). There, Tertiary rejuvenation of the Pre-Deccan

Trap palaeosurface has accordingly remained modest and its origin may well be related to another downwarp described further north on the Deccan Traps themselves (Devey & Lightfoot 1986; Mitchell & Cox 1988). It is important to state, however, that these linear incision data cannot be considered without caution as reliable estimates of areal denudation for the region as a whole.

However, the heart of the current research is to establish what happens to the Pre-Deccan Trap surface further south towards the core area of the exposed Karnataka Craton and to constrain the geometry of this palaeosurface in the absence of any volcanic or biostratigraphical record. It is known from the lensoid structure of the southern Deccan lava pile (e.g. Watts & Cox 1989) that the Precambrian basement (i.e. Pre-Deccan Trap surface) is found *c.* 2000–3000 m below sea level (b.s.l.) at the latitude of Nasik but climbs to 500–1000 m (b.s.l.) at Mahabaleswar (Kaila & Krishna 1992), and finally emerges at an altitude of 800 m *c.* 250 km to the south of Mahabaleshwar (near Belgaum). Thus, the Pre-Deccan basement surface must have a slope of 0.52–0.72%. If this palaeosurface slope is extended 300 km further southwards from Belgaum, then its projected height at the corresponding latitude on the Karnataka plateau is 1500–2000 m (Fig. 8). Allowing for the initial uncertainty in the depth of the Deccan traps further north, this elevation bracket agrees exceptionally well with the highest summit planes found at that particular latitude: Kudremukh (1892 m) along the Western Ghats ridge and Bababudangiri (1923 m) further inland, both bevelling the Archean metabasalts of the Bababudan Greenstone Group (Fig. 8). Approached from a different angle, it could equally be said that these summit levels, by far the highest to be found on the Karnataka Uplands, can be age-bracketed since they cannot be younger than the present modal plateau surface developed at altitudes below, and cannot be older than the Archean surface on Peninsular gneiss upon which the greenstone series are known to lie unconformably (a basal conglomerate is found in several places at the base of the supracrustals): these summits therefore must represent an intermediate episode of planation, i.e. the Pre-Deccan Trap surface, brought to these considerable heights by a process of elastic upwarping of the lithosphere. Whether this upwarping is solely due to the flood basalt overburden to the north as in the flexural scheme described by Watts & Cox (1989), or by *motu proprio* uplift along the 13° N parallel, remains to be fully substantiated by devising a broader model inclusive of data all the way down the western margin of India. But the idea of a basement forebulge, analogous to topographic swells observed at the periphery of volcanic seamounts (e.g. Walcott 1970), is tempting. Such a scenario would imply that post-Cretaceous denudation is responsible for removing a large wedge of basement terrain between the trap edge and the Bababudan–Kudremukh line (Fig. 8). It would, however, be

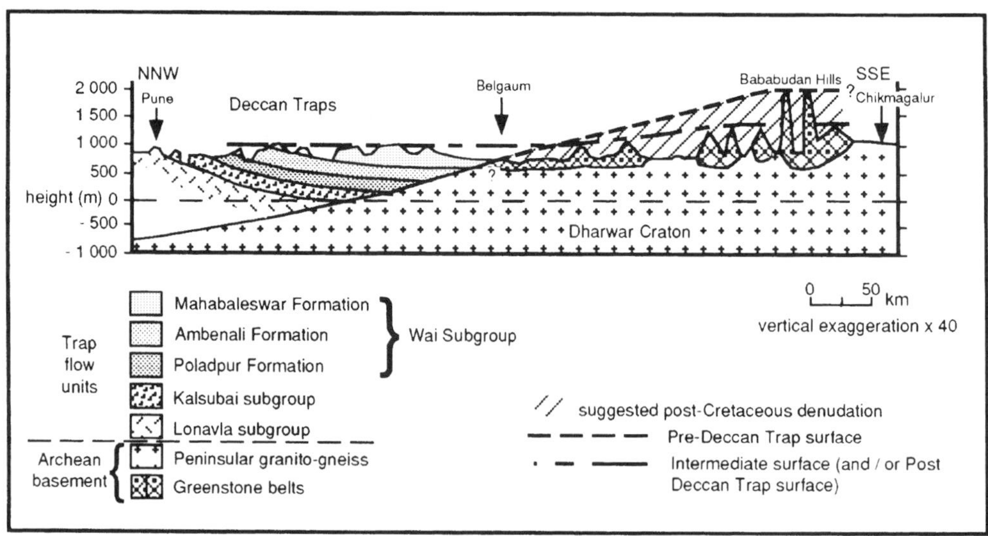

Fig. 8. Trap–basement contact and suggested warping of the Pre-Deccan Trap and Intermediate palaeosurfaces. See text for discussion (modified after Mitchell & Cox 1988).

necessary to distinguish between this first-order undulation of long wavelength, depicted on Fig. 8, and the second-order undulations, of shorter wavelength, nested in the former, the latter being: (1) The 16°45′ N synclinal corridor identified by Devey & Lightfoot (1986) and Mitchell & Widdowson (1991), which thus indicates that the Krishna River is adapted to the regional tectonics. (2) The 16°10′ N anticlinal structure (possibly to be equated with the so-called 'Vengurla basement arch' of Biswas 1988), which would explain the existence of the Gokak basement window (Figs 3 and 7) and account for the antecedent character of the Ghatprabha. The Pre-Deccan Trap surface on the northern edge of the Deccan Traps in Madhya Pradesh was also found to be upwarped using various geomorphological criteria (Choubey 1971; Venkatakrishnan 1984, 1987), which would lend even further support to the flexural mechanism suggested above. The wavelength of the basement undulations observed on and beyond the southern edge of the Deccan Trap load relies heavily on the estimated elastic thickness of the Dharwar Craton lithosphere. As previously discussed by Watts & Cox (1989), the elastic thickness best fitting all the structural and geometrical aspects of the region is debatable. In addition, 'buried basement structures', thought to be the result of crustal density heterogeneities owing to the greenstone/gneiss contrasts of the buried Dharwar Craton (Mitchell & Widdowson 1991), may have locally complicated the flexural response of the lithosphere to the flood basalt load.

What happens to the Pre-Deccan Trap Gipfelflur (i.e. accordant summit level) to the south of the Kudremukh–Bababudan line and as far as the Nilgiri Massif remains a matter of conjecture at this stage, but it seems reasonable to suppose that in Coorg district the charnockite domes which form the Western Ghats ridge at 1600–1700 m also derive from the Pre-Deccan Trap summit surface. If one follows the Western Ghats ridge from Coorg back north to the trap/basement contact, Kodachadri (13°54′ N–74°52′ E; 1343 m) and Devakonda (13°54′ N–74°44′ E; 1102 m) are the only summits apart from Kudremukh to recall this Gipfelflur: this evidence further substantiates the idea of a northward plunging palaeosurface and highlights the fact that such a palaeosurface is apparently poorly preserved when adequately resistant rocks are absent. This is indeed the case north of the Kudremukh Massif: Kodachadri and Devakonda represent the last bastions of the Bababudan Greenstone Group (see figs 3 and 4), after which the lithology is largely dominated by meta-greywackes and meta-argillites (Fig 4) and provides little potential for palaeosurface preservation.

Younger palaeosurface development: field evidence for a possible Intermediate surface notched into the Pre-Deccan Trap surface

Evidence from drainage pattern

Evidence for a surface cut into and developed below the Pre-Deccan Trap level must be sought where remains of the latter still occur, i.e. around the Bababudan Greenstone Massif. Here, a number of drainage anomalies can be observed, providing clues to the mode of landscape development which has led to the present relief: the Bhadra, the major river flowing to the Bay of Bengal in the northern two-thirds of the Karnataka Uplands (Fig. 1), is here only in its uppermost reaches; and yet, so close to its source, it cuts a close succession of five gorges (Fig. 9), *c.* 200 m deep, and largely ignoring structural and lithological guidance by ridges of Banded Ferruginous Quartzite (BFQ hereafter in text and figures), depressions of weaker gneiss and greenschist, and at least some of the faults apparent on satellite images and pointed out by Chadwick *et al.* (1985) or Petit (1985; Fig. 9). In classical geomorphological terms, this situation lends support to the hypothesis of a drainage pattern superimposed upon an initial surface of planation which was lower than the Pre-Deccan Trap palaeosurface but higher than the present modal altitude. This level of erosion is all the more apparent given that the lithology of the area is sufficiently diverse to preserve benches of differential erosion at various altitudes. The wave of erosion associated with the formation of this Intermediate surface has, in particular, reached up into the weaker core of greenschist rocks of the Bababudan metabasalt and BFQ horseshoe ramparts. There is conflicting evidence as to whether the flat-bottomed Jagar Basin (Fig. 7) is an example of differential weathering in a Precambrian ring structure (Petit 1985) or is predominantly due to the disposition of a complex syncline structure as mapped by Chadwick *et al.* (1985). The answer has to be that both factors are combined: the bottom of Jagar Basin lies at 800 m, i.e. 1000 m below the Bababudan summits, but its only outlet is via the Somvahini River, which enters the basin through a gorge in the southern limb of the horseshoe and leaves it via another gorge through the northern limb of the horseshoe. Here again, evidence for a superimposed drainage pattern is apparent, where the river course apparently ignores both the syncline and the weaker rocks. At least, if the aforementioned lineaments are correctly interpreted as Dharwar trend fractures, it certainly appears, from Fig. 9, that while some gorges are guided by structure others are not. An Intermediate

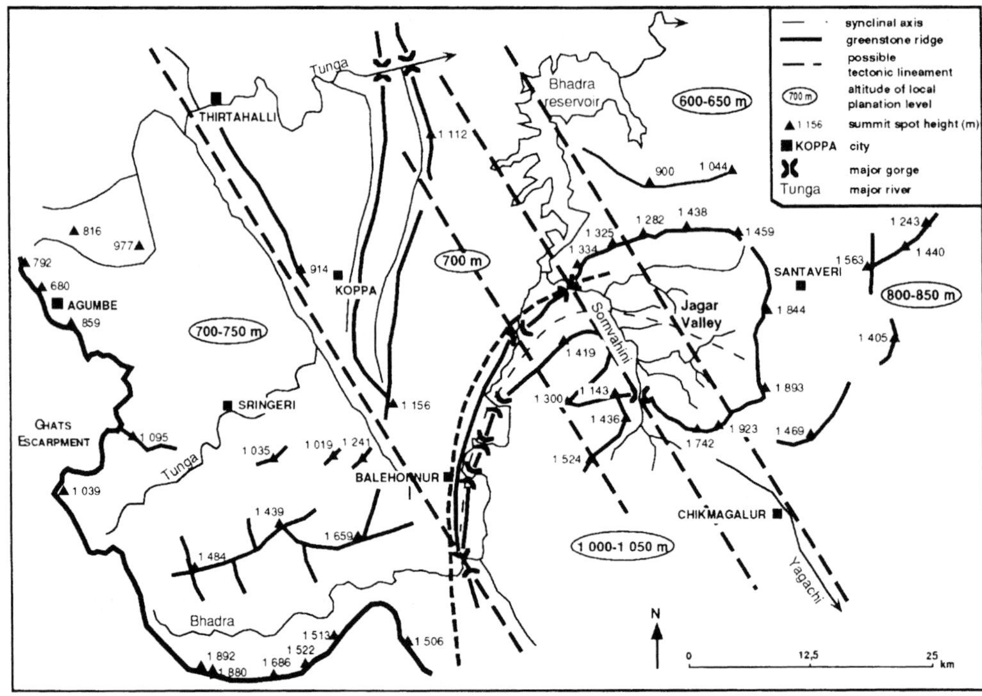

Fig. 9. Bababudan–Kudremukh area. Sketch map of local planation levels, possible superimposed drainage and structural greenstone ridges [partly after Petit (1985) and Chadwick *et al.* (1985)].

surface cut into the tapering spurs and lower stratigraphical outcrops of the Bababudan greenstone belt, occurring at *c.* 1150–1350 m, thus appears to have been the original surface of superimposition.

Evidence from ferricrete remains

If the fossilized and rubbly laterite crust found extensively though discontinuously across the Karnataka topography represents the signature of a fundamental planation level in the history of the uplands, it would have to be of this Intermediate level, indicating a climate a good deal wetter than presently found east of the 900 mm isohyet and conducive to monosiallitization (formation of kaolinite). Kaolinite is the only clay mineral able to provide a host structure for haematite (Tardy 1994) in order to produce the proper forms of induration which are preserved in the aforementioned laterite. Whether this upland laterite may be equated with the 'high-level' laterite of King (1967) and similarly so-called by all reviewed Indian authors (e.g. Vaidyanathan 1977) remains unclear, since no 'high-level' laterite was found described in the literature other than on the Deccan basalts (*c.* 1300 m elevation) which lie to the north of the present study area, or on the much higher (*c.* 2400 m elevation) massifs and their satellites in Kerala and Tamilnadu to the south (the author has seen several occurrences of bauxite and duricrust in the Nilgiri Massif). This, in itself, poses yet another element of confusion in the literature: when is a 'high level' a high level when an altitude difference of up to 1100 m may separate the aforementioned ferricrete occurrences over the entire southern Indian upland region? As to the age of this cyclical palaeosurface, it has to be Post-Deccan Trap and yet must be of sufficient antiquity to allow tropical rivers of weak erosive power to cut subsequently into 200 m of quartzite ridges (Fig. 7). A late Oligocene–early Miocene terminal age may be a probable candidate, planation having thus been achieved prior to the hard collision phase of Himalayan orogeny. This constraint would also satisfy the current assumption that downwearing removes *c.* 1 m of material in 60–80 ka in rocks of granitic composition (Tardy 1993). Interestingly, this rate translates as 500–700 m of denudation between the K/T boundary and the beginning of the Miocene, which matches surprisingly well the difference of altitude between the Bababudan summits and the surrounding greenstone ridge summits (Shimoga Belt) suggested as belonging to the Intermediate level .

If this 1150–1350 m Intermediate level is extended northwards from the Bababudan key study area to the rest of the Karnataka Uplands, it appears that most of the summit levels of the greenstone ridges north of 13°30′ N coincide with this altitude bracket (Fig. 6). To the south, where greenstone outcrops dwindle significantly (Fig. 4), Peninsular gneiss and charnockite dominate; here, a true 1200 m level, interpreted as the Intermediate surface, is remarkably well preserved at and around the hill station of Madikeri (otherwise known as Mercara) on the more resistant charnockites. This Madikeri level dominates the broad Upper Cauvery gneissic basin by a steep, 350 m high erosional escarpment. This is the opportunity to stress the necessity of having a sufficiently varied provision of rock types whose response to differential erosion ensures a more complete geomorphological record of denudational events and a richer display of structural landforms. This factor points to the relative poverty of the Dharwar Craton in this respect as compared with, for example, the exceptional array to be found in the crystalline uplands of Madagascar (Petit 1990). It is no oversimplification to point out that in the study region gneiss and meta-greywacke behave systematically as the weakest rocks, upon which the most perfect low-lying landscapes of planation are developed; all other lithologies comprising the crystalline basement of the craton tend to dominate as residual hills of varying amplitude. As landforms, these hills are difficult to classify, largely because a genetic classification of structural landforms in crystalline basement structures has yet to be developed to the same extent as that which already exists for sedimentary structures (Godard *et al.* 1994).

To summarize, the Intermediate level palaeosurface of the Karnataka Uplands survives more by recognition of the remains of its alleged by-products (i.e. buried ferricrete rubble) than by its spatial geometry in the landscape, though the latter is apparent in the accordant summits of residual ridges, localized benches on suitable rock types (periphery of Bababudan complex, Madikeri) and indications of superimposed drainage. This hypothesis assumes further that the upwarping of the Pre-Deccan Trap summit surface was an early Tertiary isostatic response to the sudden overburden of the high density, thick flood basalt pile, ensuring the necessary modification of base level for a new cycle of erosion to cut back into the Pre-Deccan Trap surface. It cannot be ruled out that the Intermediate level is in fact the continuation of the Post-Deccan Trap topography upon which Ghatprabha and Malprabha Rivers were superimposed further north (Figs 7 and 8); in which case this Post-Deccan Trap surface, as its Pre-Deccan ancestor, is also upwarped from its *c.* 900 m level on the Deccan Traps to the 1150–1350 m levels observed at the latitude of the Bababudan–Kudremukh watershed (Figs 3 and 8).

Present-day topography yielded by rejuvenation of Intermediate level: some suggestions

In the hypothesis of an acyclical degradation of the Intermediate level to the present modal altitudes (Fig. 5), it would appear that the aforementioned duricrust was downwasted, accompanying the differential denudation of the Intermediate palaeosurface in relation to lithological contrasts and intensity of dissection. Such a model does not exclude the possibility that a duricrust-forming climate did not prevail during part of the downwasting episode since the Oligocene; indeed, denudation (notably chemical denudation) and duricrust formation need not be mutually exclusive.

Observations substantiating an acyclical rejuvenation of the Intermediate surface may be summarized as follows. (1) West of the 2000 mm isohyet (Fig. 3), duricrust formation is still active, although potentially more so on the iron-rich rocks of the Western Ghats greenstone belt than on granito-gneisses such as the Londa batholith (Fig. 4) or the Peninsular gneisses. However, intensity of stream dissection due to drainage capture from west-flowing river networks is also dominant in this area and counter-productive to ferricrete preservation in any form other than rounded hills strewn with slumped blocks of ferricrete; these hills correspond to a 700 m level along the Western Ghats ridge. (2) East of the 2000 mm isohyet (Fig. 3), the ferricrete is essentially fossilized in polyphased soil profiles, being broken down into residual nodules and pisoliths. As a result, its remains have not been entirely removed from the landscape. These rubbly residues are found anywhere on the 600 m surface north of 14° N (developed largely on meta-greywackes), still *in situ* but buried, or else deposited in river terraces along the Tungabhadra and its tributaries; they are also found at a number of locations on the 900 m level around Hassan on gneiss, or in southern Coorg district around Virajpet at the same altitudes.

One possible clue to the acyclical development of the present-day topography is lent by the following points: the passage from the 600 m level in the north to the 900 m level in the south (cf. Fig. 5, line A–A′) is absolutely gradual and no cyclical evidence in the form of topographical notches or knickpoints can be determined either in the field or from maps. It can therefore be argued that ferricrete emanating from the Intermediate palaeosurface has been wasted down to one vast undulose topographic surface owing to: (1) external geodynamic factors: chiefly structural and petrographic controls on differential weathering and stripping; (2) internal geodynamic factors: among the more likely possibilities to be explored are differential isostatic upwarp of the passive margin, Deccan Trap overload, repercussions from the Himalayan collision, as well as possible in-plane stresses due to ridge push from the Carlsberg ridge and diffuse margin dynamics in the Central Indian Basin (Royer 1993).

Thus, the fact that the present-day Karnataka Uplands are an assemblage of different levels of modal altitude above which residual reliefs stand proud, must not lead to the conclusion that these levels belong to different generations or cycles. The mid-Tertiary Intermediate surface may instead have simply been rejuvenated to different levels for a number of possible reasons. (1) The broad ENE--WSW topographic low in the north-central part of the Karnataka Uplands corresponds to a relative tectonic downwarp of the crust; hence the 600 m low (Fig. 5, line B–B′). In effect, the Tungabhadra River virtually follows a northerly course before veering northeastwards and appearing to follow the aforementioned axis of downwarp. The course of the Tungabhadra seems therefore to be influenced primarily by the palaeoslopes of the Pre-Deccan Trap and Intermediate surfaces as depicted in Fig. 8, and only subsidiarily by the easterly tilt of the Deccan peninsula as a result of the uplift of the Western Ghats (Figs 1, 3 and 7). (2) In the region of the Western Ghats tectonic bulge, field evidence indicates that the 600 m level is tilted up to *c.* 750 m. The eastward tilt of the Karnataka Plateau is thus more acutely detectable in the immediate vicinity of the Western Ghats than further inside the craton. (3) The 900 m level in the south belongs to the Cauvery catchment (Figs 1, 3 and 5), and as such denudation had no reason to operate at the same rate as in the Tungabhadra–Krishna basin to the north of the Bababudan divide (Figs. 1, 3 and 9).

Going back to the key area of the Bababudan Hills, it can be clearly shown that lowering of topography has led to different local levels of planation according to the catchment involved. The vast pedestal of the Bababudan Massif (Fig. 9) is found at 1000–1050 m in the south on the Chikmagalur granodiorite (Yagachi–Cauvery catchment); *c.* 850 m to the east (Vedavati–Tungabhadra catchment); *c.* 700–750 m to the west (Upper Tunga Basin on the immediate backslope of the Ghats); finally *c.* 600–650 m to the north (Tunga–Bhadra confluence area of Shimoga). The drainage network of all these catchments eventually join up with the Bay of Bengal, their single base level, but this does not mean that the various planation levels reviewed are the result of different erosional cycles.

These factors are suggested in order to emphasize that a variety of regional and local geographical controls can influence the geometry of planation events. In this instance, two equally valid explanations are possible.

Firstly, an acyclical etching of the craton at different rates in different drainage basins; in which case the Karnataka Upland topography is to be considered as a mosaic of partly autonomous, regional-scale levels controlled by local and regional base levels as well as intensity of dissection by streams. These themselves are dictated by the structural configuration of the Dharwar craton (Figs 4 and 7). It may be recalled here that one of the more appealing theories of acyclical etching was developed from geomorphological studies on the Mysore plateau (Büdel 1982), although the acyclical model proposed here is somewhat more complex than the fundamental idea of an irregular weathering front being stripped of its regolith.

Secondly, a juxtaposition of local planation levels at different altitudes originating from one single cycle of erosion. In this hypothesis, the cycle also yields different tiered basins owing to local structural controls of base level across the path of the main rivers flowing to the Bay of Bengal, NNW–SSE trending greenstone ridges in particular (cf. Kappat and Sandur Hills and Eastern Ghats; Figs 1, 3 and 7). The assumptions are therefore the same as in the previous alternative, with strong structural control of surface lowering, but allows for a single major wave of headward erosion to initiate cyclical rejuvenation.

A third possible hypothesis, whereby each tier (i.e. each statistical class) of modal altitude identified on Fig. 5 could correspond to an autonomous epicycle, is rejected here. Such an interpretation would imply either that repeated (and hypothetical) eustatic variations in the Bay of Bengal propagated headward erosion far into the Karnataka Uplands; or that sharp and repeated pulses of tectonic uplift of the Karnataka region occurred during the Neogene. Either condition appears unrealistic in the context of a cratonic interior flanked by two passive margins.

It is clear that the validity of the two favoured scenarios will still depend heavily on the tectonic regime of the Dharwar Craton during the course of the Cenozoic. Evidence needs to be brought forward which might objectively discriminate between slow epeirogeny and rapid uplift followed by tectonic quiescence (see Discussion and Conclusions).

If it is accepted that Neogene and Quaternary times are sufficient to waste down the Karnataka Uplands topography from the 1150–1350 m surface to the present 600–900 m range, then the current elevation difference between maximum altitude (i.e. residual relief summits) and modal altitude is in keeping with rates of denudation for tropical regions suggested by Birot (1978). According to Birot, the exhumation of a 300–400 m inselberg by differential downwearing can be achieved in *c.* 10 Ma. This order of magnitude gives ample time in the period since the Miocene for a process of incremental etching and stripping to generate the 100–500 m amplitude bornhardts and ridges currently observed across the Karnataka Uplands. Birot's estimate implies 30 climatic oscillations between the dry and the more humid (wavelength of 300 ka), which could possibly fit West Africa. In southern India, such repeated climatic modifications have so far not been recorded, but a slow tendency to tectonic uplift could provide the same net effect of erosional stripping as dry-phase scouring. This tectonic uplift is also necessary to account for the fluvial downcutting of 200 m deep gorges by the Tungabhadra in a plateau area 500 km away from its Bay of Bengal base level. Moreover, soil profile evidence points to a steady average lengthening of the dry season over much of the interior Karnataka Uplands due to the rain-shadow effect of the Western Ghats being uplifted and providing a barrier to the monsoon circulation (Gunnell & Bourgeon 1997). Such uplift means that downwearing processes may have become more efficient over recent geological time, attendant aridification simply reinforcing the tectonically-induced denudation.

Comparing with the stepped surfaces on the other fragments of Gondwanaland: did India miss a step?

Given that the Deccan Traps provided such a valuable marker from which to work, it seems surprising that past authors have attempted to constrain the planation levels in southern India by resorting to presumed equivalents on distant fragments of Gondwanaland. Nonetheless, fitting India back into a world landscape development model ultimately has its merits. It is generally accepted that India does display remnants of a 'post-Gondwana' palaeosurface, in the sense of Dixey and King. This palaeosurface was controlled by the new base levels which resulted from the breakup of Gondwanaland and deposits from this worldwide episode of denudation are found elsewhere as the 'Continental intercalaire' in Africa or the Cenomanian detrital sediments of Madagascar. In southern India, this post-Gondwana surface can be tentatively equated here with the Pre-Deccan Trap surface, also described as the 'Cretaceous peneplain' in Madhya Pradesh (Dixey 1971), sealed by the Deccan flood basalt of uppermost Maastrichtian modal age.

However, unlike the other continental fragments of Gondwanaland, southern India appears to lack a late Cretaceous–early Tertiary cyclical surface notched into the previous higher one, which elsewhere usually constitutes the fundamental reference surface for subsequent geomorphological evolution (Petit 1990): it usually bears regional names, e.g. the Sul Americana level for the Brazilian shield, the African Surface in Africa, the Tampoketsa level in Madagascar, the Great Australian Pediplain in Australia. These are all capped to some extent by ferricrete, bauxite, calcrete or silcrete, and are remarkable by their smoothness and perfection. Compared to other Gondwana fragments, the Indian plate exhibited a much greater mobility reflected in rapid drift rates (Besse & Courtillot 1989) and this, together with the huge areas which became subject to taphrogenic processes at the K/T boundary, may have mitigated the Tertiary development of cyclical planation surfaces over the old Dharwar Craton. As mentioned, smooth and extensive remnants of the late Cretaceous Afro–American surface remain to be found over the Karnataka Uplands, and the Indian landscape does not fit readily in the broad scheme established by King (1967). At the same time, it cannot be ruled out that the Post-Deccan Trap Intermediate surface identified here, and tentatively attributed to late Palaeogene times, does correspond to a belated, and therefore much more imperfect, fundamental surface, equivalent to its African, Brazilian or Malagasy counterparts: in short, an Indian Surface. Combined with the important increase in sea-floor spreading and volcanic activity particularly affecting the Indian region (e.g. Larson & Olson 1991), with therefore higher production levels of CO_2 and an average temperature 10°C higher than today (Tardy 1986), weathering rates were indeed much higher than today (Bardossy 1981), thus preparing a deep weathered mantle to be removed in the drier phases of the Tertiary. Both Godard (1965) and Lageat (1989), working in the Scottish Highlands and in the South African Bushveld, respectively, concluded that younger planation surfaces tend to develop over smaller areas than the older fundamental surfaces of the geological past. Furthermore, the younger the planation surfaces are, the more they appear to be subject to lithological control. Such an observation is broadly in agreement with the evidence given above which suggests that the weathering potential of world climates has decreased over the course of the Cenozoic, thus resulting in a relative increase in the importance of lithological and structural control in geomorphic processes. Lageat (1989) extends this model by attempting to establish the relative ages of planation levels on igneous rocks in terms of the manner and degree by which each erosive episode has been influenced by lithological contrasts. Such a situation, albeit at a much lower level of resolution, is exemplified by the stacking of surfaces over the Karnartaka Uplands since these reveal the lithological hierarchy of the Dharwar Craton (i.e. metabasalts, BFQ and charnockite at the top, gneiss and meta-greywackes at the bottom of the scale of resistance).

The existence of later Miocene surfaces, usually recognized by global models of planation (e.g. King 1967), is more controversial than the aforementioned Afro–American surface. On the other continents, such levels as Post-African I or pediplain P2 in Brazil, both also reported in Madagascar or on the Guyana shield, have been postulated. However, it is difficult to recognize equivalents through geometric and cyclic reasoning on the Karnataka Uplands. Nevertheless, it cannot be ruled out that the local levels of planation identified, for example, on the Bababudan pedestal (Fig. 9), actually correspond to such benches developed during a late Tertiary cycle, as formerly hypothesized.

The presence of stepped palaeosurfaces is usually corroborated by the identification of stepped soil covers which display depletion in exchangeable cations and relative enrichment in iron and aluminium compounds. Such soils often take the form of massive duricrusts and typically characterize the high-level, sparsely-inhabited surfaces of the tropical world, namely the African and Sul Americana Surfaces. Yet in southern India, on the Karnataka Uplands, no such cyclical high levels of soil empoverishment are found. It seems likely that the barren, culminating summits of the Bababudan Hills and of Kudremukh along the Western Ghats ridge (*c.* 1900 m elevation, Figs 3, 6 and 7), are the result of duricrusting of the areally restricted Archean Banded Iron Formation (BIF) outcrops found there (see photographs in Radhakrishna *et al.* 1986), and as such cannot be unequivocally connected with palaeosurface formation. A similar case may be found in the itabirites of the Quadrilatero Ferrifero in Minas Gerais, Brazil, where Tricart (1961) prefers the term *chapeau de fer* to avoid confusion with the genetic connotations normally associated with the term *cuirasse*. If the outcrop pattern of BIF is the contributing factor of these highest level 'laterites', then they have no 'age' as such, especially in a climate which is still conducive to ferricrete formation. They are, therefore, of no use in reconstructing palaeosurfaces, since they simply develop in response to a bedrock outcrop which already contains economic-grade quantities of haematite.

Discussion: the model of acyclical evolution and its implications

The absence of extensive and clearly identifiable stepped planation surfaces below the Intermediate level on the Dharwar Craton calls for an alternative model of landscape evolution. It is suggested that acyclical evolution has operated on the Karnataka Uplands at least since the mid-Tertiary. This concept implies a continuous incremental removal of material from the previous topographic plane of erosion, in effect, wearing down the topography in response to low amplitude epeirogenic uplift.

In such a scheme, the cumulative effect of basal deepening of the weathering front (lowering of the water table due to uplift) and surface removal of the weathered mantle leads to the low angle amalgamation of one level of planation onto the other and therefore defeats the meaning of palaeosurface. At any time, the topography remains vast and retains low relief, but can equally be described as very ancient or quite juvenile, since few signatures of past processes remain recorded in the landscape. Such an evolution was proposed by Klein (1984) in the Armorican Massif of France along the western fringes of the Paris Basin.

It is indeed remarkable that, unlike in West Africa, not one single laterite-capped mesa is to be found on the Karnataka Uplands, save in a very localized major watershed position east of Bangalore. Unfortunately, much geomorphological effort has focused on the aforementioned case, which gives the impression that the Karnataka Uplands semi-arid area is a palaeosurface corresponding to this duricrusted level of inherited kaolinite-bearing profiles (e.g. Demangeot 1975). Although the present-day topography may have been derived from such a level, the aforementioned conclusion is misleading since there is little left of this conjectural ferricrete level with which to demonstrate its original extent. Furthermore, the present-day topographic surface is blanketed, in the area receiving < 800 mm rainfall, by non-kaolinite-bearing red soil profiles which are in equilibrium with present-day climatic conditions (Bourgeon 1992).

The acyclical model of topography is known, in French terminology, as *surface de dégradation lente* or *topographie de rajeunissement lent*; an equivalent concept has been coined *dynamic etchplain* by Thomas (1989). This is formed through a combination of chemical denudation and mechanical erosion operating on a seasonal wet–dry cycle repeated over geological time-scales. The important point is that the climate may remain hot and the length of the dry season can fluctuate, but the overall process is not fundamentally modified since the pace is largely set by the *tectonic* regime.

Conclusions

The salient features of a qualitative model for acyclical evolution are as outlined below.

Tectonic regime

For an acyclical evolution of the landscape to stand the test of fission track modelling: (1) Uplift has to be slow and weak but ongoing; i.e. on balance, periods of possible standstill must amount to less than periods of positive crustal uplift. Slow rejuvenation operates much better when the curvature of plateau upwarp is low and a situation of major continental divide, as in the Karnataka Uplands, considerably delays adjustment to new base levels. Furthermore, the more extensive the initial fundamental planated area, the better the chance of its preservation over long periods of time, insofar as the initial low relief ensures the perpetuation of lateral surface wash processes as opposed to linear dissection. This latter consideration may account for the remarkable smoothness of the late Cretaceous–early Tertiary fundamental surfaces preserved in Africa and Brazil; (2) The climate must be warm and have sufficient rainfall to ensure ongoing weathering; the rainfall regime can vary to some extent, but the total length of the dry season will mainly affect the degree of hardening of the laterite minerals, thus steering the landscape towards further downwearing (wetter climate), or conversely towards a greater immunisation to downwearing (the French term *cuirasse* expresses well the protective action of a duricrust). If the protective capping by duricrust is effective enough, inversion of relief can occur where differential weathering has worn down the topography to lower levels. If uplift encourages headward erosion and duricrust breakaway retreat across the upland plateau, then the mesas are gradually destroyed. In such a situation downwearing may then resume or accelerate and the duricrust rubble, if it is not exported out of the system by rivers, will either be fossilized or digested and reworked; iron, silica and aluminium will accordingly recombine to new minerals in equilibrium with the current soil forming conditions.

The tropical landscape

This can be considered as an open system which is subjected to both positive and negative feedback loops over time. However, only drier phases can preserve signatures of previous wetter episodes, while wetter phases of climate tend to obliterate any trace of previous dry phases. Thus, the environmental record of the open system is biased but, as pointed out by Tardy (1993), the thermodynamics responsible for the geochemical evolution of a tropical landscape operate within a limited range of possibilities. It is indeed the case that a relatively narrow range of temperatures and rainfall, and a handful of chemical elements (Si, Al, Fe, O, H), cooperate to form a finite number of soil types and landscape patterns. In spite of this alleged simplicity, however, it might be added that conflicting classification schemes of tropical soils have impeded attempts to comprehensively understand the dynamics of tropical landscapes. It is argued here that tectonic regime is of paramount importance in controlling those geochemical pathways, since the movement of elements in the landscape is a question of balance between soil forming processes (pedogenesis) and profile stripping processes (morphogenesis) just as much as one of climatic variation.

Palaeosurfaces

The concept of a palaeosurface loses much of its strength in the Indian context presented here, since the topography as one finds it today has existed through, and been modified by, a long history of changing climates and tectonic pulses. In cratonic interiors devoid of unconformable and datable sediments, the original geometry of the surface from which the present-day planation topography is derived is unknown, and can only be very crudely outlined from little more than small-scale mapping and field experience. Palaeosurfaces can only be dated with confidence when proved to be exhumed (see Twidale 1997).

In spite of the speculative model proposed above, there is no compelling evidence as to whether the evolution of the Karnataka Uplands was purely acyclical, the prevailing German viewpoint (Büdel 1982); or totally cyclical, the largely Davisian approach adopted by King (1967) as well as the reviewed Indian authors (Table 1). A more 'agnostic' view would be a combination of both, which the field evidence reported in this paper could support. Future fission track results should enable these options to be resolved, since acyclical episodes of denudation generally tend to be active during tectonic uplift, whereas cyclical surfaces are expected to be formed during periods of standstill. Fission track analysis should therefore constrain the occurrence or absence of post-Cretaceous pulses of uplift both along the passive margin upwarp and further inland. In view of the series of recent shallow (5–10 km focal depth) earthquakes in a region adjoining the study area (e.g. September 30, 1993 tremor in Latur, *c*. 18°15′ N, 76°45′ E), there seems little doubt that tectonic disturbance in the southern Indian shield is an ongoing process, although its dynamic causes and geomorphological expression at the surface need to be studied in greater detail.

The author is grateful to the URA 141 and URA 1562 of the CNRS for funding two field surveys in the Western Ghats and to the French Institute in Pondicherry for providing a grant towards a recently accomplished third expedition in Karnataka. Thanks are also extended to Y. Lageat, N. J. Cox and an anonymous referee for a helpful review of the manuscript and figures, and especially to M. Widdowson for some much needed streamlining of the English.

References

BABU, P. L. V. P. 1975. A study of cyclic erosion surfaces and sedimentary unconformities in the Cauvery basin, South India. *Journal of the Geological Society of India*, **6** (3), 349–353.

BARDOSSY, G. 1981. Paleoenvironments of laterites and lateritic bauxites – effect of global tectonism on bauxite formation. *In*: *Proceedings of the International Seminar on Lateritization Processes (Trivandrum, India, 11–14 Dec. 1979)*, 287–294.

BESSE, J. & COURTILLOT, V. 1989. Paleogeographic maps of the continents bordering the Indian Ocean since the early Jurassic. *Journal of Geophysical Research*, **93**, 11 791–11 808.

BIROT, P. 1978. Evolution des conceptions sur la genèse des inselbergs. *Zeitschrift für Geomorphologie*, Suppl. B, **31**, 1–41.

BISWAS, S. K. 1988. Structure of western continental margin of India and related igneous activity. *In*: SUBBARAO, K. V. (ed.) *Deccan Flood Basalts*. Memoir of the Geological Society of India, **10**, 371–390.

BOURGEON, G. 1992. *Les sols rouges de l'Inde péninsulaire méridionale*. Publications du département d'écologie, Institut Français de Pondichéry, Vol. 31.

—— & GUNNELL, Y. 1996. Present-day climatic influences on geomorphology and soil profiles at the western edge of the Karnataka Plateau, peninsular India, in preparation.

BRUNNER, H. 1968. Geomorphologische Karte des Mysore Plateaus (Südindien). Ein Beitrag zur Methodik der morphologischen Kartierung in der Tropen. *Wissenschaftliche Veröffentlichungen*, **25**, 1–17.

BÜDEL, J. 1982. *Climatic Geomorphology*. Princeton University Press, Princeton.

CHADWICK, B., RAMAKRISHNAN, M. & VISWANATHA, M. N. 1985. Bababudan – A Late Archean intracratonic volcanosedimentary basin, Karnataka, Southern India. *Journal of the Geological Society of India*, **26**(11), 769–801.

CHATTERJEE, S. P. 1961. *National Atlas of India*, Plate 36 (Madras), Ministry of Scientific Research & Cultural Affairs, Government of India.

CHOUBEY, V. D. 1971. Pre-Deccan trap topography in Central India and crustal warping in relation to Narmada rift structure and volcanic activity. *Bulletin Volcanologique*, **35**, 660–685.

DEMANGEOT, J. 1975. Recherches géomorphologiques en Inde du Sud. *Zeitschrift für Geomorphologie*, **19**(3), 229–272.

DEVEY, C. W & LIGHTFOOT, P. C. 1986. Volcanological and tectonic control of stratigraphy and structure in the western Deccan traps. *Bulletin of Volcanology*, **48**, 195–207.

DIKSHIT, K. R. & WIRTHMANN, A. 1992. Strip planation in laterite – a case study from western India. *Petermanns Geographische Mitteilungen*, **136**,(1), 27–40.

DIXEY, F. 1971. The geomorphology of Madhya Pradesh. *In*: MURTHY, T. V. & RAO S. S. (eds) *Studies in Earth Sciences, West Commemorative volume, Sagar University*. Today & Tomorrow, New Delhi, 195–224.

GODARD, A. 1965. *Recherches géomorphologiques en Ecosse du Nord-Ouest*. Thèse Etat, Université de Paris, Les Belles Lettres, Paris.

——, LAGASQUIE, J.-J. & LAGEAT, Y. (eds) 1994. *Les régions de socle, apports d'une école de géomorphologie française*. Faculté des Lettres et Sciences humaines de l'Université Blaise Pascal, Fascicule 43, Clermont-Ferrand.

GOKHALE, N. W. & JOSHI, V. S. 1990. Comparative structural study of four outliers of Kaladgi Formation and their mode of upliftment. *Bulletin of the Indian Geologists' Association*, **23**(1), 13–16.

GUNNELL, Y. & BOURGEON, G. In Press. Soils and climate geomorphology on the Kamataka plateau, peninsular India.

KAILA, K. L. & KRISHNA, V. G. 1992. Deep seismic sounding studies in India and major discoveries. *Current Science*, **62**(1 & 2), 117–154.

KALE, V. S. & GUPTE, S. C. 1986. Some observations on the denudational surfaces of the Upper Godavari and Krishna basins (Maharashtra). *Transactions of the Institute of Indian Geographers*, **8**(1), 51–58.

KING, L. 1967. *The Morphology of the Earth* 2nd Edition, Oliver and Boyd, Edinburgh.

KLEIN, C. 1984. Une notion centenaire, la notion de cycle en géomorphologie. *Physio-géo*, **11**, 61–101.

LAGEAT, Y. 1989. *Le relief du Bushveld. Une géomorphologie des roches basiques et ultrabasiques*. Association des Publications de la Faculté des Lettres de Clermont-Ferrand, nouvelle série, 30.

LARSON, R. L. & OLSON, P. 1991. Mantle plumes control magnetic reversal frequency. *Earth and Planetary Science Letters*, **107**, 437–447.

MITCHELL, C. & COX, K. G. 1988. A geological sketch

map of the southern part of the Deccan Province. *In*: SUBBARAO, K. V. (ed.) *Deccan Flood Basalts*. Memoir of the Geological Society of India, **10**, 27–33.

—— & WIDDOWSON, M. 1991. A geological map of the Southern Deccan Traps, India and its structural implications. *Journal of the Geological Society, London*, **148**, 495–505.

NARUSE, T. 1980. Physiographic features. *In*: FUJIWARA, K. (ed.) *Geographical field research in South India*, Department of Geography, University of Hiroshima, 28–34.

PETIT, M. 1982. Aspects morphologiques fondamentaux des traps méridionaux du Deccan. *Bulletin de l'Association des Géographes Français*, **485–486**, 91–96.

—— 1985. Deux bassins dans les socles tropicaux: le Mutara granitique (Rwanda), les Baba Budan Hills métavolcaniques (Inde péninsulaire méridionale). *Physio–géo*, **13**, 79–90.

—— 1990. *Géographie physique tropicale*. Karthala, Paris.

RADHAKRISHNA, B. P. 1965. Geomorphological evolution of the Mysore Plateau. *Reprinted in*: MISRA, R. P. (gen. ed.) *Contributions to Indian Geography, vol. 2: Geomorphology*. Heritage Publishers, New Delhi, 31–40.

—— 1991. An excursion into the past - "the Deccan volcanic episode". *In*: *Extinct Plants, Evolution and Earth's History*. Current Science, Special issue **61**(9 & 10), 641–647.

——, DEVARAJU, T. C. & MAHABALESWAR, B. 1986. Banded iron-formation of India. *Journal of the Geological Society of India*, **28**, 71–91.

—— & NAQVI, S. M. 1986. Precambrian continental crust of India and its evolution. *Journal of Geology*, **94**, 145–166.

ROYER, J.-Y. 1993. La tectonique de l'Océan Indien. *La Recherche*, **24**(251), 160–167.

SWAMY NATH, J. & RAMAKRISHNAN, M. (eds) 1981. *Early Precambrian supracrustals of southern India*. Geological Survey of India, Memoir, **112**.

TARDY, Y. 1986. *Le cycle de l'eau*. Masson, Paris.

—— 1993. Climats, paléoclimats et biogéodynamique du paysage tropical. *Colloque 'Sédimentologie et géochimie de la surface' à la mémoire de Georges Millot*. Académie des Sciences, 141–175.

—— 1994. *Pétrologie des latérites et des sols tropicaux*, Masson, Paris.

THOMAS, M. F. 1989. The role of etch processes in landform development II. Etching and the formation of relief. *Zeitschrift für Geomorphologie*, **33**, 257–274.

TRICART, J. 1961. Le modelé du Quadrilatero Ferrifero, au S de Belo Horizonte (Brésil). *Annales de Géographie*, **LXX**, 255–272.

TWIDALE, C. R. 1997. The great age of some Australian landforms: examples of, and possible explanations for, landscape longevity. *This volume*.

VAIDYANADHAN, R. 1967. An outline of the geomorphic history of India south of 18 N Latitude. *Proceedings of the Seminar on Geomorphological Studies in India*, Centre for Advanced Studies, University of Saugar, Sagar, 121–130.

—— 1977. Recent advances in geomorphic studies of Peninsular India: a review. *Indian Journal of Earth Sciences*, S. Ray Volume, 13–35.

VENKATAKRISHNAN, R. 1984. Parallel scarp retreat and drainage evolution, Pachmari area, Madhya Pradesh, Central India. *Journal of the Geological Society of India*, **25**(7), 401–413.

—— 1987. Correlation of cave levels and planation surfaces in the Pachmari area, Madhya Pradesh: a case for base level control. *Journal of the Geological Society of India*, **29**, 240–249.

WALCOTT, R. I. 1970. Flexure of the lithosphere at Hawaii. *Tectonophysics*, **9**, 435–456.

WATTS, A. B. & COX, K. G. 1989. The Deccan Traps: an interpretation in terms of progressive lithospheric flexure in response to a migrating load. *Earth and Planetary Science Letters*, **93**, 85–97.

WIDDOWSON, M. 1997. Tertiary palaeosurfaces of the SW Deccan, India: implications for passive margin uplift. *This volume*.

The geochemistry of Indian bole horizons: palaeoenvironmental implications of Deccan intravolcanic palaeosurfaces

M. WIDDOWSON[1], J. N. WALSH[2] & K. V. SUBBARAO[3]

[1] *Department of Earth Sciences, Parks Road, Oxford OX1 3PR, UK; Current address: Department of Earth Sciences, The Open University, Walton Hall, Milton Keynes MK7 6AA, UK*

[2] *Department of Geology, Royal Holloway, University of London, Egham, Surrey TW20 OEX, UK*

[3] *Department of Earth Sciences, Indian Institute of Technology, Bombay 400 076, India*

Abstract: Deccan intravolcanic bole horizons represent weathering products formed during major hiatuses of a major volcanic episode. In these quiescent periods weathering processes pervasively altered the newly formed volcanic landscape and subsequent flows covered and effectively fossilized the resulting weathered palaeosurfaces. The current work is a detailed geochemical study which examines patterns of element mobilization during these intravolcanic weathering events.

The bole horizons normally rest on top of altered basaltic lavas. Both boles and altered lavas represent a comparatively early stage in weathering because the content of chemically residual elements, such as aluminium and iron, are closer to fresh basalt than laterite. Nevertheless, there is clear evidence for significant chemical alteration because the more mobile elements such as calcium and sodium have been substantially removed. A chemically distinctive nature for some boles can be demonstrated from chondrite normalized REE plots and, in these instances, strontium and neodymium isotopic compositions demonstrate that the fine-grained bole material is derived from a chemically distinct source. In addition, thin sections reveal that these fine grained portions commonly contains glass shards and occasionally fresh phenocrysts. It is therefore suggested that many Deccan boles are in fact weathered pyroclastic material, and that the pyroclastic content of the basaltic succession may be greater than previously supposed. A significant pyroclastic input during the Deccan eruptions has important palaeoenvironmental implications for the fate of late Cretaceous flora and fauna in peninsular India.

The Deccan flood basalt province currently covers an area of *c.* 500 000 km^2 of West Central India, although the original extent was probably in excess of 1 500 000 km^2 (Krishnan 1956). The sequence comprises a great thickness (> 2000 m exposed) of flat lying, predominantly tholeiitic, lava flows. Individual flows can easily be identified and, on the basis of detailed geochemical work, the entire lava sequence has been divided stratigaphically (Table 1) into three major subgroups and 13 formations (Beane *et al.* 1986; Subbarao *et al.* 1994). Each formation comprises a package of flow units (Cox & Hawkesworth 1984) which may vary in thickness and comprise both simple and compound (i.e. including several flow units) types. Nevertheless, it is clear that there have been many breaks in eruption. Often these breaks are simply manifest as glassy or scoriaceous material between flow units and representing relatively short respites in eruption. However, in some cases, well preserved weathered horizons occur at the flow tops. These latter are the so-called bole horizons and are indicative of prolonged hiatuses in eruption. They are often readily recognized in the field by their distinctive red coloration, though detailed examination reveals a range of colours from red through to brown and, in addition, a few green boles. When exposed in Ghats sections and road cuttings, boles can provide detailed profiles through the ancient weathering front and into the basalts beneath. Within the Deccan lava sequence there are also examples of laterally equivalent, presumably contemporaneous, fossiliferous lacustrine and fluvio-lacustrine deposits (Prasad & Khajuria 1995) which, when considered with the boles, indicate that a highly complex palaeoenvironment existed between the major eruptive events.

It is clear that the bole horizons are widely distributed throughout the Deccan but that the distribution is somewhat uneven. For example, an increased frequency of boles, possibly related to a major eruptive focus, is reported from the Nasik area (Deshmukh 1988). The current work similarly demonstrates a geographical control on bole

From Widdowson, M. (ed.), 1997, *Palaeosurfaces: Recognition, Reconstruction and Palaeoenvironmental Interpretation,* Geological Society Special Publication No. 120, pp. 269–281.

Table 1. *Stratigraphy of the Deccan lavas*

Group	Sub-group	Formation
Topmost lavas		
	Wai	Panhala Mahabaleshwar Ambenali Poladpur
Deccan basalts	Lonavala	Bushe Khandala
	Kalsubai	Bhimashankar Lower Thakurvadi Middle Thakurvadi Upper Thakurvadi Neral Igatpuri Jawhar
Lowermost lavas		

Individual formations comprise packets of lava flows with similar geochemical characteristics. For further details of the established chemostratigraphy see Mitchell & Widdowson (1991).

distribution since in some areas, such as that around Pune, boles are common whilst in others, such as Malshej Ghat, there are none. This does not appear to relate directly to stratigraphy since, for example, the Bhimashanker formation exposed in the Khandala and Matheran sections has no boles, whereas when exposed at Bhimashanker, boles occur (Wilkins *et al.* 1994). Within the area of this study however boles become progressively more common as one proceeds up the stratigraphy and are especially prolific in the topmost formations of the Wai subgroup (i.e. Mahabeleshwar and Panhala Formations). This general prevalence toward the top of the lava pile probably reflects increasingly frequent quiescent periods toward the close of the Deccan volcanic episode. Widdowson (1997) argues that the high level regional laterite preserved in the Western Ghats represents a final 'super bole' which continued to develop uninterrupted during the early Tertiary after the eruptions had ceased completely.

Sampling

As part of a detailed study into the geochemistry of weathering regimes in the Deccan basalts, the bole horizons were extensively sampled and analysed. Figure 1 shows the location of several of the Ghat sections through the Deccan Traps where sampling of the weathered profiles was carried out, covering all the stratigraphical formations from Igatpuri to Mahabaleshwar (Subbarao & Hooper 1988). In the Deccan, the larger proportion of boles are developed on vesicular flow tops but this is not an exclusive relationship because boles occur on both pahoe-hoe and aa flows. Detailed geochemical sampling was undertaken on many profiles to provide an overview of weathering processes. Samples were collected from the fresh, unaltered basalt up through the various stages of altered lava, into the fine grained bole. Typically, at least 5 or 6 samples were collected from the well-exposed weathered profiles.

Boles

Bole horizons are commonly considered to be the result of *in situ* alteration of flow tops which have been exposed for many hundreds, or possibly even thousands, of years. This being the case, the weathered crust should typically consist of a portion of fine grained material sitting on top of an altered flow top, perhaps an incipient soil horizon, which then grades downward into fresh lava below. This type of alteration sequence represents a fossilized profile reflecting weathering effects which occurred at the time of the eruptions, 65 Ma. However, the fine grained material is not always present and it is clear that the thickness of a single bole commonly varies laterally. Such variation is presumably the result of differences in the effectiveness of palaeoweathering, partial removal by contemporaneous erosional effects or, alternatively, by the scouring effects of later flow units as they moved over the weathering profile.

Inspection of a number of bole profiles reveals important differences which provide the basis for dividing the bole profiles into two groups. In many cases complete, or near complete, profiles are preserved displaying a progression of weathering products from the deeply altered horizon lying immediately below the subsequent lava flow through to underlying unaltered basalt. In such examples, the topmost, pervasively altered materials, commonly retain physical or textural clues as to their origin as flow tops. These features include a rubbly composition, zeolite infilled vesicles and, in some cases, a recognizable primary basaltic crystal fabric pseudomorphed by later weathering products. In these profiles the weathered materials are generally poorly consolidated, soft and earthy in appearance. Given the observed weathering progression and retention of primary features, such bole materials could be accurately described as saprolitic and are henceforth termed saprolitic boles. However, there exists a second group of bole-like materials within the

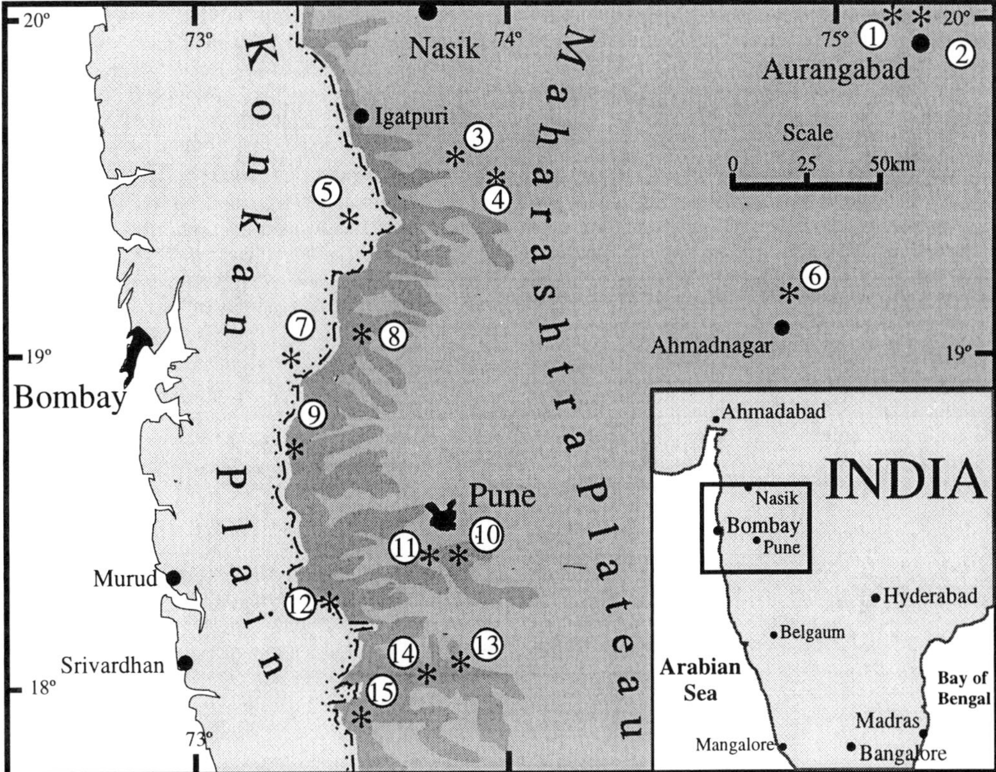

Fig. 1. Map of the western Deccan basalt province showing the location of the important Ghat sections investigated during the current study. Detailed chemical studies have been made on weathering profiles from these Ghat sections. 1, Outram Ghat; 2, Maishmal Ghat; 3, Bari Ghat; 4, Chandanpuri Ghat; 5, Malsej Ghat; 6, road section NE of Ahmednagar; 7, Matheran section; 8, Bhimashankar section; 9, Khandala section; 10, Sasvad Ghat; 11, Katraj Ghat; 12, Singahad section; 13, Kambatki Ghat; 14, Pasarni Ghat; 15, Poladpur–Mahabaleshwar section. Stippled ornament marks the Western Ghats escarpment; dark grey ornament represents the Western Ghats ridge zone. Inset shows geographical position of studied region.

lava sequence in which a recognizable weathering progression cannot readily be identified. These boles are usually laterally continuous, often apparently unaffected by partial erosion or contemporaneous removal and, in some examples, display a laminated structure. Such boles often lie with a sharp contact upon relatively unaltered flow tops, have a characteristically 'cherty' or brittle character and are typically indurated toward their top surfaces. This induration has hitherto been interpreted as baking of the bole surface by the subsequent flow. However, the lack of a weathering progression, together with this induration and presence of other unusual features, means that their origin as a flow-top alteration product cannot be made with confidence. The significance of this cherty group of boles is discussed in a later section.

Boles and laterites : chemical and physical differences

There exists some confusion regarding the usage of the term 'bole' and 'laterite' since the two have commonly been considered as interchangeable, especially with respect to the weathered horizons occurring within lava sequences. We prefer the term bole to describe those weathering horizons contemporaneous with the eruptive phase and which have subsequently become fossilized within the lava sequence. Such a distinction effectively avoids any misinterpretation resulting from the genetic implications which are implicit in the term laterite. Nevertheless, it is important to clarify the discrimination between boles and laterites more in the present context. The Deccan boles, and many similar intra-volcanic weathering horizons found in flood basalt provinces elsewhere, do not display the physical and textural characteristics or the residual element enrichment considered typical of lateritization. Laterites commonly display an advanced stage of physical and chemical maturity, since they are distinguished physically by the presence of pisolitic or vermiform textures (McFarlane 1976), and chemically by patterns of conspicuous silica depletion, advanced enrichment of iron and aluminium and extreme depletion of bases relative to the original parent composition (Schellmann 1986, 1994). Although inspection of the coloration and texture of the bole horizons clearly indicates that they are weathered, chemical analysis demonstrates that they are not 'lateritic' in composition. As Fig. 2 demonstrates, the boles have far less extreme weathering patterns than the laterites and are thus easily distinguished from them. The term bole is therefore useful to distinguish boles from laterites.

Geochemistry

Table 2 provides a comparison of analytical data for boles, laterites and unaltered Deccan basalt. The compositional range for unaltered Deccan basalt is given in Table 3. Major, trace and rare earth element analyses of the boles were made by ICP AES (atomic emission spectrometry) using the techniques described in Walsh *et al.* (1981) and Thompson & Walsh (1989). Those of the basalt and laterite were made by XRF utilizing fused beads for major elements and pressed pellets for trace elements (Potts 1987). Reliability of results was ensured by analysing and comparing a representative sample suite by both techniques. Loss on ignition was determined for a number of samples, and, relative to unaltered rock, increases in both the laterites and boles with levels between 10 and 25% were normal. The formation of hydrated minerals clearly indicates that chemical weathering is involved in the formation of both boles and laterites.

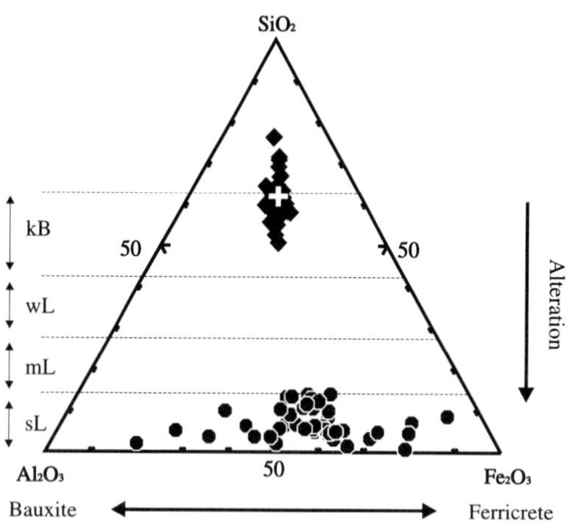

Fig. 2. Triangular plot showing composition of Deccan alteration products with respect to an average basaltic protolith. Details of the calculation of the field limits of kaolinitized basalt (kB), weakly (wL), moderately (mL) and strongly lateritized (sL) materials follow those set out in Schellmann (1986) using limits derived from the average Deccan basalt composition. Large cross, average Deccan protolith composition; diamonds, Deccan boles; circles, lateritic materials displaying the pisolitic and/or vermiform features considered typical lateritic textures.

In Table 2 the average Indian bole analysis is derived from all the collected bole materials and yields values broadly representative of the majority of samples. Bole analyses SB2 and WPB2 are individual samples representative of the saprolitic and cherty bole groups outlined above. The unusual composition of bole KCB1 is discussed in a later section. The laterite values are taken from an extensive suite of analyses representing indurated high- and low-level regional laterites (Widdowson 1997) derived from the *in situ* breakdown of Deccan basalt and displaying the key lateritic textures (i.e. presence of pisoliths and/or vermiform tubes).

Enrichment and depletion of the individual elements depends upon their relative mobility during the weathering process. Given an increasingly longer duration of sub-aerial exposure, the differences in concentration of constituent elements will become progressively more pronounced: less mobile elements, chiefly iron and aluminium, are typically considered as being residual since they become increasingly enriched during alteration. Figure 2 clearly shows that indurated laterites represent a highly advanced stage of alteration since, relative to the protolith composition, they display considerable enrichment in iron and aluminium with concentrations of these elements reaching 3.5 and 2.7 times, respectively, that of the average basalt (Table 2). Such enrichment is balanced by a rapid loss of the highly mobile elements sodium, calcium, magnesium and strontium associated with the breakdown of primary silicates in the earliest stages of weathering, and a concomitant decrease in silica content occurring throughout the lateritization process. The patterns of extreme concentration and depletion have been achieved as a result of a protracted sub-aerial exposure since the lateritization process began. By contrast, aluminium and iron levels in the average bole composition, and that of samples SB2 and WPB2, show only very minor enrichment from average basalt composition, indicating that they are

Table 2. *Comparison of analytical data for Deccan boles, laterites and unaltered Deccan basalt*

Element	Basalt (average)	Bole (average)	Laterite (average)	Bole SB2	Bole WPB2	Bole KCB1
Wt%						
SiO_2	48.83	52.54	7.11	48.37	52.68	67.73
TiO_2	2.50	2.24	3.09	3.18	2.68	1.18
Al_2O_3	13.72	16.65	36.76	14.18	15.67	10.85
Fe_2O_3	14.79	17.80	52.04	17.42	18.36	10.26
MnO	0.21	0.17	0.11	0.23	0.18	0.07
MgO	6.21	4.95	0.05	5.50	4.23	1.82
CaO	10.62	3.75	0.04	7.71	2.73	1.49
Na_2O	2.35	0.39	0.07	1.80	0.21	0.37
K_2O	0.30	1.14	0.14	0.51	1.59	5.89
P_2O_5	0.23	0.15	0.27	0.24	0.09	0.26
p.p.m.						
Ba	106	175	94.	82	86	404
Co	51	53	23	50	51	20
Cr	109	134	1424	64	133	97
Cu	217	289	61	166	575	101
Nb	11	9	20	14	15	9
Ni	85	95	46	65	99	62
Rb	10	42	19	24	46	228
Sr	227	117	16	171	64	74
V	358	196	1076	358	206	116
Y	36	38	17	47	26	44
Zn	105	94	42	105	108	73
Zr	150	122	247	156	155	114

Average bole composition is taken from current data set ($n = 47$ samples): average laterite composition is calculated from strongly lateritized samples (see Fig. 2) which display typical vermiform or pisolthic textures ($n = 46$ samples): average basalt composition is further detailed in Table 3. Individual samples: Bole SB2 is a weathered flow top (i.e. saprolitic bole) from within the Lower Ambenali Formation; boles WPB2 and KCB1 are cherty boles from within the Upper Ambenali and Lower Thakurvadi Formations, respectively. The pyroclastic nature of KCB1 is illustrated in Fig. 6.

much less weathered than the laterites. This fact is most clearly demonstrated by the differences in silica levels which in laterites are commonly present in quantities of < 15% that of the basalt protolith, whilst in the boles, silica levels generally remain within the basaltic range (Table 3). This behaviour is entirely consistent with geologically short periods of surface exposure and the idea of 'fossilization' of the bole profile after it became sealed by later lava flows.

Chemical variation in weathering profiles

The changes in bulk chemical composition during the weathering of Deccan basalts have been documented by Wilkins *et al.* (1994). These authors demonstrated that there are substantial changes in the major element compositions at the bole horizons. Detailed chemical profiles from unaltered basalt, through the weathered material into the boles above, were presented for a representative sequence from the Sinhagad section.

Table 4 gives a general representation of the chemical changes occurring during basalt weathering to produce a bole, in this case from the Sinhagad section, but the patterns illustrated are typical. The extent of the chemical alteration that has occurred in the bole horizons is clearly seen from the loss on ignition values (increasing from < 5% in the basaltic material to values in excess of 15% in the bole). Wilkins *et al.* (1994) also show that calcium and sodium are the two most readily removed major elements. CaO falls from 10% to *c.* 4% and Na_2O decreases from 2% to < 0.5% during bole formation. These changes are paralleled by other major element concentrations, but the changes are much less pronounced. MgO values fall from 5–6% to 4–5% and SiO_2 falls from 47 to 43%. Al_2O_3, Fe_2O_3 and TiO_2 concentrations show only small changes, varying only slightly through the profile and remaining within the basaltic limits (Table 3). The results presented by Wilkins *et al.* (1994) confirm that these latter three elements may be regarded as relatively immobile, at least in the

Table 3. *Compositional range of unaltered Deccan basalt*

Element	Average	Minimum	Maximum
Wt%			
SiO_2	48.83	45.64	52.45
Al_2O_3	13.72	11.78	18.44
TiO_2	2.50	1.27	4.04
Fe_2O_3	14.79	11.12	18.30
MnO	0.21	0.12	0.67
MgO	6.21	4.22	11.49
CaO	10.62	9.08	14.16
Na_2O	2.35	1.68	2.95
K_2O	0.30	0.00	0.87
P_2O_5	0.23	0.13	0.40
p.p.m.			
Ba	106	32	321
Co	51	39	70
Cr	109	31	443
Cu	217	76	425
Nb	11	2	31
Ni	85	41	308
Rb	10	0	32
Sr	227	106	442
V	358	251	477
Y	36	22	95
Zn	105	66	157
Zr	150	66	273

Data set $n > 400$ samples.

context of the chemical weathering regime associated with Deccan bole formation. The authors also comment on the behaviour of K_2O during weathering which, by contrast to that of Na_2O and CaO, appears to be complex. It is likely that potassium is also mobilized during chemical weathering but because it may also be easily incorporated into neo-formed clay minerals it can be retained in the developing weathering profile. Examination of the K_2O contents shown in Table 2 demonstrates that, despite initially low potassium contents in the basalt protolith, it remains, albeit partially depleted, even within the highly altered laterites whereas the other alkalis are almost entirely removed.

It is important to note that the chemical changes that occur during bole formation can be clearly distinguished from the changes that occur during secondary mineralization (zeolitization). The secondary minerals found include quartz, calcite and a range of zeolites, and these may be found either singly or as successive layers in vesicles. The fact that this mineralization post-dates bole formation is clear in many boles. However, since its presence could potentially overprint the element patterns produced by weathering, it is important in studying weathering profiles to avoid excessively zeolitized sequences or to make appropriate allowance for the zeolites. The relatively simple chemistry of the Deccan zeolites was demonstrated by Jeffery (1988), consequently it can be demonstrated that the effects of secondary mineralization are small when compared with the degree of alteration required to generate the average bole. As a result, the overall weathering trends in the bole profiles are clear.

In summary, the chemical changes during bole formation are significantly less than the chemical changes that occur during laterite formation. There are clear differences in chemistry between the Deccan boles and 'laterites'. The boles represent a chemical weathering event that is well defined but is substantially less intense than the weathering that formed the laterites. A further significant point emerging from the consideration of many bole profiles is that despite the fact the detailed structure of boles varies considerably both in the Deccan and elsewhere (e.g. Skye and Antrim), the most marked changes in element patterns occur at the point where altered lava changes to an essentially fine-grained material. This is especially apparent in the case of the cherty boles where the point of transition from lava to fine-grained material commonly coincides with unusual changes in element abundance. In these latter cases the fine-grained material is clearly weathered but could, in terms of its structure, be described as having sediment-like characteristics (e.g. presence of primary laminations).

Chemical comparison of bole types

The saprolitic and cherty bole groups also display important chemical differences. Compositions of three representative boles (SB2, WPB2 and KCB1) are given in Table 2. That of SB2 is consistent with the patterns of element enrichment and depletion expected from the alteration of an average basaltic precursor since both Fe_2O_3, TiO_2 and Al_2O_3 are slightly elevated, whilst MgO, CaO, Na_2O and Sr are noticeably depleted. The fact that SiO_2 concentrations show little change, together with the observation that the aforementioned element concentrations in SB2 remain within the basaltic range (Table 3), testifies to the limited degree of alteration represented by this sample. Compared with Deccan basalt, the cherty bole (WPB2) similarly displays elevated Fe_2O_3 and Al_2O_3 and comparable degrees of MgO, CaO, Na_2O and Sr depletion, thus clearly demonstrating that some alteration has affected the sample. However, the presence of elevated SiO_2 and K_2O, the concentrations of which lie beyond the basaltic range, are inconsistent with the weathering pattern observed in those bole materials

Table 4. *Chemical composition of samples taken through the Sinhagad bole and underlying altered basalt flow*

Sample	Shd-020 Bole	Shd-021 Bole	Shd-022 Altered basalt	Shd-023 Altered basalt	Shd-024 Altered basalt	Shd-026 Altered basalt	Shd-027 Protolith basalt
Depth (cm)	0	15	30	45	60	200	250
Wt%							
SiO_2	43.86	42.55	46.32	46.17	46.05	46	46.86
TiO_2	1.97	2.26	2.70	2.44	2.45	2.55	2.66
Al_2O_3	12.70	13.36	12.99	13.38	13.27	13.65	13.22
Fe_2O_3	15.11	16.11	15.78	15.00	14.96	14.65	15.92
MnO	0.15	0.17	0.20	0.19	0.19	0.21	0.21
MgO	4.57	4.77	5.34	5.61	5.61	5.96	5.74
CaO	4.18	4.97	9.07	9.18	8.79	10.53	10.44
Na_2O	0.33	0.45	1.69	1.63	1.48	1.89	2.09
K_2O	0.45	0.46	1.14	1.00	1.42	0.18	0.16
P_2O_5	0.13	0.11	0.27	0.23	0.24	0.24	0.23
LOI	16.55	14.79	4.50	5.17	5.54	4.14	2.47

Depth in centimetres is measured downwards from the top surface of the bole horizon (data supplied by A. Wilkins).

which are unequivocally derived from weathered flow surfaces. In the case of bole KCB1 these discrepancies in element behaviour become even more apparent. In the field this bole is a pervasively reddened, highly indurated material displaying a laminated structure and a sharp contact with the basalt beneath. Chemically, it is one of the most unusual examples because whilst it contains patterns of silica and potassium enrichment characteristic of the cherty bole types, the concentrations of these elements are abnormally elevated. Furthermore, many of the characteristic weathering trends are apparently reversed in this sample since Fe_2O_3, Al_2O_3 and TiO_2 are depleted whilst SiO_2, K_2O, Rb and Ba are highly enriched relative to the average basalt datum. In explanation, it could be argued that, due to their large ionic radius, K, Rb and Ba may be subject to preferential incorporation into clay minerals which would perhaps allow their retention during normal weathering conditions. However, the development of these clays could not explain the observed silica enrichment. It has been suggested that the conditions of basalt alteration during Deccan times (65 Ma) are not analogous those described by Schirrmeister & Storr (1994) as being characteristic of modern basaltic weathering phenomena. Yet, in order to account for these abnormal patterns of enrichment and depletion in KCB1, this alternative would require highly unusual palaeoenvironmental conditions at the time of bole formation. In the absence of supporting palaeoenvironmental evidence, and given the fact that the weathering pattern observed in the contemporaneous saprolitic boles (e.g. SB2) mitigates against the existence an extraordinary palaeoweathering regime, such an assumption can effectively be discounted. A second explanation of the cherty bole composition is that patterns of element mobility past and present have remained largely unchanged. In this case, the elevated concentrations of SiO_2 and K_2O could only be achieved if the concentrations currently observed in such boles are derived from protolith materials with initially higher concentrations of these elements. If changes in the nature of basalt weathering are discounted, then the only alternative explanation is that the cherty boles represent the alteration products of non-basaltic materials. If this latter interpretation is accepted, then the precursors of the cherty boles would have had to contain elevated SiO_2 (> 53%), elevated K_2O (> 1.59%), and elevated Ba, Rb and Sr, since these are depleted during alteration, and initially lower V, Cr, Fe_2O_3 and Al_2O_3 to be consistent with established patterns of element enrichment and depletion (Table 2). For a similar consistency in the case of bole KCB1, initially lower values of Fe_2O_3 (< 10%), Al_2O_3 (< 10%) and V (< 116 p.p.m.) together with elevated SiO_2 (> 68%), K_2O (> 6%), Rb (> 228 p.p.m.) and Ba (> 404 p.p.m.) would be required in the protolith composition: such values would have been well beyond the basaltic compositional range (Table 3). To summarize, relative to Deccan tholeiite, the cherty boles must, therefore, have been derived from a material of a more petrogenetically evolved nature.

Trace element signatures

Further evidence that some Deccan boles are compositionally inconsistent with derivation from the basalt is found within the trace element signatures. For instance, using diagnostic 'immobile' element ratios (Winchester & Floyd 1977) materials displaying relatively low degrees of chemical alteration may be related back to their protolith compositions. It should be noted, however, that in practice no element can be considered as being entirely immobile, especially where alteration is extreme (i.e. during lateritization), since a limited mobility can be demonstrated even for highly residual elements such as Ti, Cr and V (Widdowson 1990). However, the current bole data are suitable for such treatment since, compared with the highly altered lateritic materials, they clearly represent a relatively small degree of alteration. Using such an approach, Figs 3 and 4 demonstrate that whilst many of the Deccan bole data are consistent with derivation from a Deccan tholeiitic protolith, because they plot within the limits of the sub-alkaline basalt field, the remainder of the samples indicate derivation from igneous precursors of a more evolved nature, probably of an andesitic or even a dacitic composition. Further inspection of the bole field data reveal that these evolved samples typically correspond to the cherty bole examples, hence supporting the interpretation made earlier on the basis of the major element concentrations.

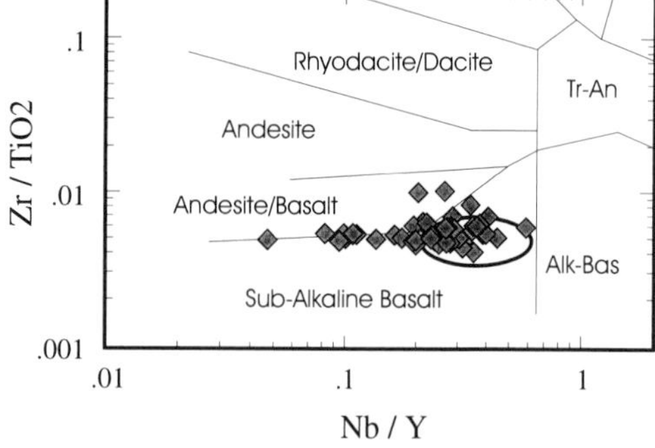

Fig. 4. Major groupings of volcanic rock types based upon characteristic proportions of minor and trace elements considered immobile during post-consolidation alteration (modified after Winchester & Floyd 1977). The composition of many bole samples fall beyond the limits of the typical basaltic compositions delimited by the elliptical area marked within the sub-alkaline basalt field. As demonstrated in Fig. 3, many boles have a more evolved character, here displaying an affinity to a basaltic andesite composition.

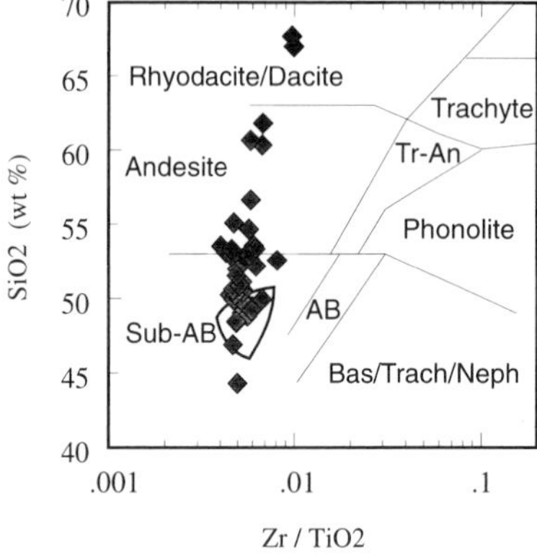

Fig. 3. Major groupings of volcanic rock types based upon proportions of SiO_2 and selected immobile and minor trace elements (modified after Winchester & Floyd 1977). The SiO_2 content and Zr/TiO_2 ratio may be used to determine the protolith compositions for alteration products which have been subject to relatively small degrees of weathering (i.e. kaolinitized field of Fig. 2). Note that despite some silica depletion resulting from the palaeoweathering effects, many bole samples plot within the andesitic field, with some sample compositions occurring in the rhyodacite/dacite field. Clearly, these more acidic compositions are not derived from a basaltic protolith. Irregular outlined area represents composition of unaltered Deccan basalts typical of the Wai subgroup.

Rare earth elements

This point is further demonstrated by the rare earth element plot, Fig. 5. The patterns for a modern Deccan soil show that modern weathering produces a series of sub-parallel chondrite normalized plots indicating the residual behaviour of the rare earth elements. These recent soils show a marked negative Ce anomaly which is typically produced during relatively early stages of alteration. Similar negative anomalies have been reported in some altered Deccan flow tops and weakly lateritized materials (Widdowson 1990). Development of these anomalies are associated with the double oxidation states for cerium (i.e. Ce^{3+} and Ce^{4+}) which result in profound differences in the mobility of the element during weathering (Braun *et al.* 1990). REEs are considered as relatively immobile under normal weathering conditions but detailed investigations demonstrate that relative enrichment of both LREE and HREE can occur at different stages of the weathering process (Nesbitt 1979; Marsh 1991). This is typically manifest as a relative HREE enrichment in the early stages of alteration (saprolitization), which then changes to extreme

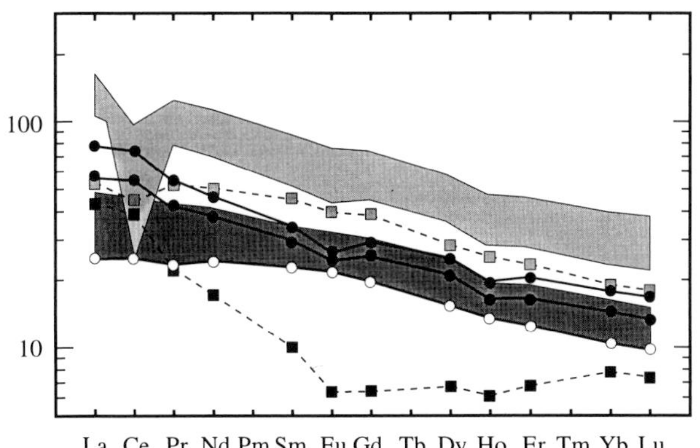

Fig. 5. Chondrite normalized REE plots for selected bole and lateritic materials. The light grey shaded region is the compositional range for a suite of samples from a modern Deccan soil weathering profile; the main feature is the development of a substantial negative Ce anomaly with increasing degrees of weathering. Typical compositions of basalt and altered basalt (Sinhagad profile) produce a series of sub-parallel plots shown as a dark grey shaded region. Light grey circles, unaltered Deccan basalt; light grey squares, weakly lateritized basalt; dark grey squares, strongly lateritized basalt; black circles, fine-grained ('cherty'-type) bole samples. Note that bole samples have plots distinguishable from basaltic compositions; a steeper chondrite normalized plot reflects their increased light to heavy rare earth ratio. There is also a small negative Eu anomaly. Clearly, these two bole samples have a different source material to the basalt material exposed immediately below them.

HREE depletion in residual deposits such as laterite (Duddy 1980). These effects are demonstrated in Fig. 5 by the REE plots for the two lateritized samples derived from alteration of Deccan basalt. The weakly lateritized example displays a small Ce anomaly, and slight LREE depletion when compared with the basaltic composition. By contrast, the strongly lateritized sample displays extreme HREE depletion.

The plots for the previously considered Sinhagad profile show a similar series of sub-parallel plots for the basalt samples from the lower portion of the profile (Fig. 5). The patterns for two fine-grained bole samples, while similar to each other, are markedly different from those typical of the Deccan lava since they are steeper (LREE enriched) and have a small negative europium anomaly. It could be argued that this steeper pattern is the result of the REE fractionation (i.e. HREE depletion), described above as being typical of the more advanced stages of alteration. However, given the absence of any negative Ce anomaly in these bole samples it would appear they have, in reality, undergone relatively little alteration and associated REE mobilization. To put it another way, the LREE enrichment observed in these boles is much greater than would be expected for the small degree of alteration they suffered and is therefore inconsistent with the pattern of alteration expected from a basaltic precursor. Moreover, the Eu anomalies they display are recognized as resulting from magmatic processes (i.e. plagioclase fractionation) and cannot be achieved during weathering. Since the patterns of REE behaviour and the presence of Eu anomalies in these boles cannot be fully explained in terms of weathering processes, a precursor of a more evolved composition is indicated for fine-grained, or 'cherty' bole materials.

There is now good reason to believe that this model can be widely applied. Detailed chemical analysis has shown that the fine-grained material at the top of many weathered profiles is distinctive and is from a source different to that of the altered basalt beneath it. This can not only be demonstrated in other Deccan boles, but also for bole profiles in other areas (e.g. British Tertiary igneous province) which display chemically distinct REE plots for the fine-grained material at the top of the profile, again indicating a separate source.

Strontium and neodymium isotopes

Strontium and neodymium isotopic compositions for bole materials were determined by Wilkins *et al.* (1994) for a restricted suite of samples. Whilst strontium isotopic compositions of these alteration products were considered to have been influenced by the local groundwater composition as a result of weathering, the preliminary neodymium isotope results in some instances indicated markedly different neodymium isotopic compositions between the weathered bole and unaltered basalt beneath. Since there is no known mechanism for producing this Nd isotopic shift by weathering processes, it must be concluded that these results lend further support to the idea that some of the weathered bole horizons are derived from sources other than those of basalt-derived weathering products.

Petrographic observations

Relatively few petrographic observations have been made due to the friable nature of many of the bole materials. However, the cherty types are more readily analysed due to their indurated nature. In many cases, pervasive alteration masks microstructure within the samples but investigations commonly reveal the presence of irregular, cuspate fragments of highly weathered volcanic glass together with pseudomorphed crystal forms within the matrix. These are interpreted as a relict pyroclastic fabric. For petrographic purposes the 'dacitic' bole (KCB1) reveals the most detail. This particular bole occurs within the lower Thakurvadi Formation at Bari Ghat. It is highly indurated and small feldspar phenocrysts can be readily observed within the fine-grained red matrix on cut surfaces. Most importantly, thin section examination reveals the presence of fresh glass shards (Fig. 6) and, more rarely, fragments of relatively unaltered plagioclase and pyroxene crystals embedded within the deeply weathered matrix.

The glass shards can be easily recognized both as large fragments (1–2 mm) embedded in the fine matrix and also comprising the bulk of the ground mass matrix itself. Given this prevalence of glassy material, KCB1 can be described as hyaloclastic tuff, especially since there is some evidence of layering and compaction of the finer shard fragments. The glass colour varies from pale yellow through to the common red–brown colour and occasionally black in the more altered fragments. Many of the larger glass shards show quench textures evidenced by the presence of radially-arranged bundles of elongate plagioclase

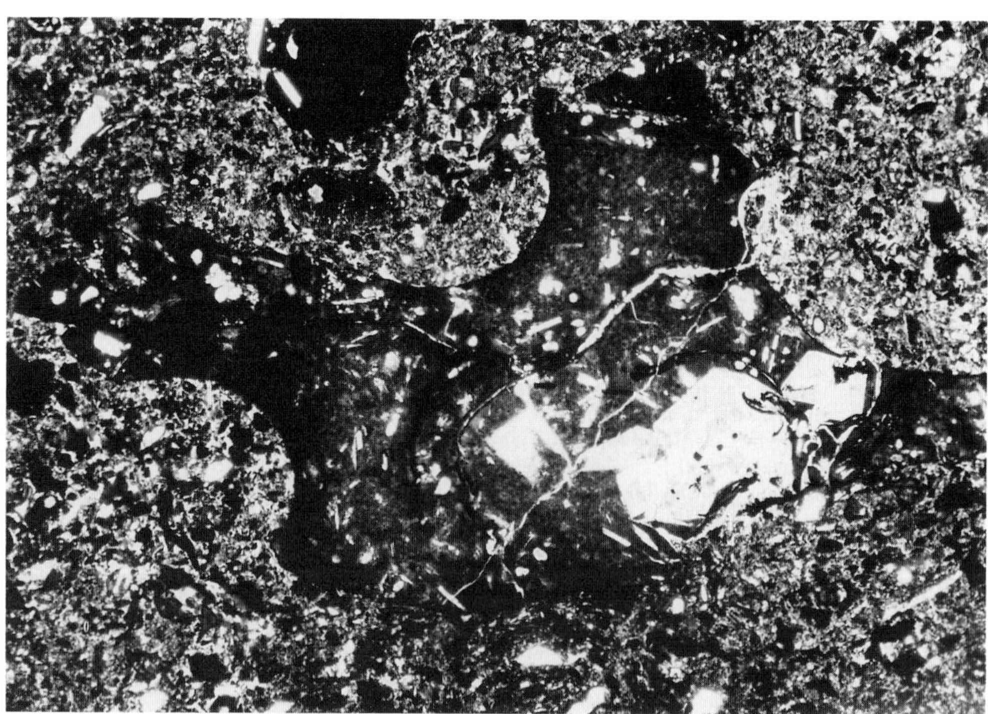

Fig. 6. Photomicrograph of typical cherty bole material (i.e. KCB1) in plane polarized light. Field of view is *c.* 3 × 2 mm and comprises a large cuspate glass shard (dark grey) across the middle region of the photomicrograph. Darker grey rim area around the periphery of the shard is the result of increased alteration. Euhedral feldspar (labradorite) crystals (0.5 mm) may be observed embedded in the right-centre of the shard (white polygonal features). Smaller, elongate feldspar crystals occur both within the shard and throughout the groundmass matrix. The matrix itself is composed of fine glass shards (max. *c.* 50 μm) together with (predominantly) feldspar microlites.

microlites. Larger crystals of plagioclase feldspar (*c.* An_{60}) occur either embedded in some of the larger shards or, more commonly, individually throughout the matrix. These often exhibit noticeable fracturing or undulose extinction which are features consistent with their derivation as volcanic ejecta. Bole KCB1 clearly represents a relatively unaltered example of a Deccan pyroclastic deposit. Given the above geochemical and petrographic evidence we conclude that the chemically distinct cherty boles represent the alteration products of petrogenetically evolved (i.e. andesitic–dacitic) pyroclastic materials. The presence of pyroclastic material of this nature within the Deccan lavas is of fundamental importance and their palaeoenvironmental implications are explored in the following section.

Discussion

We suggest that the most reasonable explanation for our results is that within the Deccan lava sequence there exist numerous pyroclastic horizons. This pyroclastic component has not been widely recognized previously due to the highly weathered nature of the materials and understandable confusion with boles derived from the weathering of basaltic flow tops. However, comparison with modern settings such as Hawaii or the Canary Islands indicate that scoria cones and more acid pyroclastic material are common features associated with basaltic eruption and their presence within the Deccan sequence is therefore not surprising. In addition, since the boles represent expressions of fossilized palaeosurfaces they have great potential as a source of palaeoenvironmental data recording conditions during the quiescent phases of these major volcanic epochs.

Duration of sub-aerial exposure is a key control to degree of weathering and is known to result in weathered horizons of differing thickness and degree of alteration. Pyroclastic materials are typically more porous to water penetration and hence more susceptible to weathering and element removal as compared with flow tops, and this may explain the poor state of preservation within many of the pyroclastic boles. As a consequence, weathering of the pyroclastics would have supplied a massive influx of the more mobile elements into the late Cretaceous Deccan drainage system which would eventually be passed into the coastal sea waters. Interestingly, variations in the nature of terrestrial drainage is one mechanism by which documented shifts in Sr isotopic content of ocean waters may be explained (Jones *et al.* 1994)

However, the unusual induration observed in many of the bole examples remains unexplained. This feature does not appear to be a weathering phenomenon, since disaggregation rather than consolidation might be expected as alteration progresses. Furthermore, lateritic induration can be discounted since these boles are clearly not laterites and this effect only occurs in the advanced stages of lateritization. One possibility is that the induration is a primary feature, and that the pyroclastic boles represent altered welded tuffs or a similar pyroclastic flow deposit. If this is so, then they record eruptive events which not only affected a very wide area, but which would have been particularly devastating to the extant flora and fauna. Further demonstration of such welding will require detailed examination of the preserved glass shard textures (Fisher & Schmincke 1984) and is beyond the scope of the current paper. Remaining questions concerning the timing of these pyroclastic eruptions will require additional isotopic investigation. For instance, it is uncertain whether the pyroclastics represent the initial stages of an eruptive phase which occurred immediately prior to massive, but relatively quiescent, basalt effusion or, alternatively, the end of such an eruptive phase during which more chemically evolved products were evacuated from an exhausted magma chamber.

The palaeoenvironmental implications for a substantial pyroclastic component in the Deccan are wide ranging. Many authors have cited the volcanic evolution of the province as a major contributor to the Cretaceous–Tertiary extinctions (McLean 1985; Officer *et al.* 1987). However, the quiescent nature of modern analogues of the basaltic eruptions are at odds with a postulated palaeoenvironmental disaster of global proportions (Sigurdsson 1990). If the pyroclastic boles of the Deccan are as widespread as we suspect, then clearly they provide evidence for a much more vigorous type of eruption together with a substantial ash component being generated during the Deccan episode. Ash fall-out would not only have been devastating to the existing flora and fauna within the fall-out zone, but injection of this ash into the higher levels of the atmosphere could have had a potentially damaging influence upon late Cretaceous climates. However, such links with global palaeoenvironmental disaster should be treated with caution because studies of infra- and intertrappean biota (Prasad & Khajuria 1995) indicate the eruptions had little immediate effect upon floral and faunal diversity. Furthermore, palynological investigation of intertrappean boles from the British Tertiary Igneous Province by Jolley (1997) demonstrate that significant floral recolonization and environmental regeneration occurred between major eruptive events. Preliminary palynological studies on the Indian boles shows similarly substantial floral regeneration between successive Deccan eruptions.

Summary

Red bole horizons are formed by pervasive sub-aerial weathering of previously exposed igneous materials and, together with locally associated lacustrine and fluvial sediments, are a common occurrence in continental flood basalt (CFB) sequences of the Deccan Trap (India) and British Tertiary Igneous Province (e.g. Skye, Mull and Antrim). Yet, unlike the lava sequences, the inter-trappean bole horizons have received little geological attention, although it is accepted that the formation of these weathering horizons temporally represents the greater part of the history of such CFB provinces.

Geochemical study of the weathered bole horizons demonstrates the role of element mobilization during chemical weathering, but indicates that the geochemical data cannot always be reconciled with established patterns of element depletion typical of alteration of a basaltic precursor. In addition, thin sections indicate that the fine-grained portions of some boles contain glass shards and fresh phenocrysts of a pyroclastic origin. Although the Deccan intertrappean surfaces all result from the same endogenetic process, namely the development of lava fields during flood basalt volcanism, the saprolitic and pyroclastic boles record two different types of palaeoenvironmental conditions. The former group are formed by protracted, *in situ* alteration of the lava surface during periods of volcanic quiescence. These quiescent periods, together with the development of saprolitic boles and associated incipient soil horizons upon the lava field, would have provided conditions suitable for the regeneration of indigenous biota. By contrast, the cherty boles are the result of a 'blanketing' of the lava landscape by extensive ash fall-out which would have led to the widespread destruction of established flora and fauna. Further study of such horizons may further our understanding regarding the efficacy of CFB eruptions in affecting ancient biota and climate.

We are grateful to the NERC (UK) for their support of the ICP–AES analytical facilities at Royal Holloway, University of London. Discussion with A. Wilkins, K. G. Cox and D. W. Jolley have provided useful additional insight during the preparation of this mauscript. Field work for MW was supported by NERC grant GT4/85/ GS/83.

References

BEANE, J. E., TURNER, C. A., HOOPER, P. R., SUBBARAO, K. V. & WALSH, J. N. 1986. Stratigraphy, composition, and form of the Deccan basalts, Western Ghats, India. *Bulletin of Volcanology*, **48**, 61–83.

BRAUN, J. -J., PAGEL, M., MULLER, J.-P., BILONG, P., MICHARD, A. & GUILLET, B. 1990. Cerium anomalies in lateritic profiles. *Geochimica et Cosmochemica Acta*, **54**, 781–795

COX, K. G. & HAWKESWORTH, C. J. 1984. Geochemical stratigraphy of the Deccan Traps at Mahabaleshwar, Western Ghats, India, with implications for open system magmatic processes. *Journal of Petrology*, **26**, 355–377.

DESHMUKH, S. S. 1988. Petrographic variations in compound flows of Deccan Traps and their significance. *In*: SUBBARAO, K. V. (ed.) *Deccan Flood Basalts*. Geological Society of India Memoir, **10**, 305–320.

DUDDY, I. R. 1980. Redistribution and fractionation of rare-earth and other elements in a weathering profile. *Chemical Geology*, **30**, 363–381

FISHER, R. V. & SCHMINCKE, H.-U. 1984. *Pyroclastic Rocks*. Springer, Berlin.

JEFFERY, K. 1988. *Mineral chemistry of zeolites from the Deccan basalts*. PhD Thesis, University of London.

JOLLEY, D. W. 1997. Palaeosurface palynofloras of the Skye Lava field, and the age of the British Tertiary volcanic province. *This volume*.

JONES, C. E., JENKYNS, H. C., COE, A. L. & HESSELBO, S. P. 1994. Sr isotopic variations in Jurassic and Cretaceous Seawater. *Geochimica et Cosmochimica Acta*, **58**(14), 3061–3074.

KRISHNAN, M. S. 1956. *Geology of India and Burma*, 3rd Edition. Higginbothams, Madras.

MARSH, J. S. 1991. REE fractionation and Ce anomalies in weathered Karoo dolerite. *Chemical Geology*, **90**, 189–194.

MCFARLANE, M. J. 1976. *Laterite and Landscape*. Academic, London.

MCLEAN, D. M. 1985. Deccan Traps mantle degassing in the terminal Cretaceous mass extinctions. *Cretaceous Research*, **6**, 235–259.

MITCHELL, C. & WIDDOWSON, M. 1991. A geological map of the southern Deccan Traps, India and its structural implications. *Journal of the Geological Society, London*, **148**, 495–505.

NESBITT, H. W. 1979. Mobility and fractionation of rare earth elements during weathering of a granodiorite. *Nature*, **279**, 206–210.

OFFICER, C. B., HALLAM, A., DRAKE, C. L. & DEVINE, J. D. 1987. Late Cretaceous and paroxysmal Cretaceous/Tertiary extinctions. *Nature*, **326**, 143–149

POTTS, P. J. 1987. *A Handbook of Silicate Rock Analysis*. Blackie, London.

PRASAD, G. V. R. & KHAJURIA, C. K. 1995. Implications of the infra- and inter-trappean biota from the Deccan, India, for the role of volcanism in Cretaceous–Tertiary boundary extinctions. *Journal of the Geological Society, London*, **152**, 289–296.

SCHELLMANN, W. 1986. A new definition of Laterite. *Geological Survey of India Memoir*, **120**, 1–7.

—— 1994 Geochemical differentiation in laterite and bauxite formation. *Catena*, **21**, 131–143.

SCHIRRMIESTER, L. & STORR, M. 1994. The weathering of basaltic rocks in Burundi and Vietnam. *Catena*, **21**, 243–256.

SIGURDSSON, H. 1990. Evidence for volcanic loading of the atmosphere and climatic response. *Palaeo-*

geography, Palaeoclimatology, Palaeoecology (Global and Planetary Change Section), **89**, 277–289.

SUBBARAO, K. V. & HOOPER, P. R. 1988. Reconnaissance map of the Deccan basalt group in the Western Ghats, India Scale 1 : 100 000. *In*: *Deccan Flood Basalts*. Geological Society of India Memoir, **10**, Bangalore.

——, CHANDRASEKHARAM, D., NAVANEETHAKRISHNAN, P., & HOOPER, P. R. 1994. Stratigraphy and structure of parts of the Central Deccan Basalt Province: eruptive models. *In*: SUBBARAO, K. V. (ed.) *Volcanism (Radhakrishna Volume)*. Wiley, New Delhi, 321–332.

THOMPSON, M. & WALSH, J. N. 1989. *Handbook of Inductively Coupled Plasma Spectrometry 2nd* Edition. Blackie, London.

WALSH, J. N., BUCKLEY, F. & BARKER, J. 1981. The simultaneous determination of the rare earth elements in rocks using inductively coupled plasma source spectrometry. *Chemical Geology*, **33**, 141–153.

WIDDOWSON, M. 1990. *Uplift history of the Western Ghats, India.* DPhil Thesis, University of Oxford.

—— 1997. Tertiary palaeosurfaces of the SW Deccan, India: implications for passive margin uplift. *This volume.*

WILKINS, A., SUBBARAO, K. V., INGRAM, G. & WALSH, J. N. 1994. Weathering regimes within the Deccan Basalts. *In*: SUBBARAO, K. V. (ed.) *Volcanism (Radhakrishna Volume)*. Wiley, New Delhi, 217–232.

WINCHESTER, J. A. & FLOYD, P. A. 1977. Geochemical discrimination of different magma series and their differentiation products using immobile elements. *Chemical Geology*, **20**, 325–343.

Remote sensing for mapping palaeosurfaces on the basis of surficial chemistry: a mixed pixel approach

KEVIN WHITE[1], NICK DRAKE[2] & JOHN WALDEN[3]

[1] *Department of Geography, University of Reading, Whiteknights, Reading RG6 2AB, UK*

[2] *Department of Geography, King's College London, Strand, London WC2R 2LS, UK*

[3] *School of Geography, University of Oxford, Mansfield Road, Oxford OX1 3TB, UK*

Abstract: Palaeosurfaces are often characterized by distinctive surficial material compositions; for example, iron oxide in lateritic surfaces in India or calcium carbonate in calcreted surfaces in Africa. The chemical constituents of interest often play a significant part in the induration, and hence the preservation, of palaeosurfaces. Furthermore, these components often have distinct spectral reflectance characteristics, enabling them to be detected from multispectral remote sensing systems. However, traditional image processing techniques, such as classification, are inappropriate for mapping such phenomena, which usually exhibit a gradually-varying distribution. In studies of palaeosurface development, the requirement is for data that provide quantitative estimates of the surficial material composition (i.e. the proportions of a given component, such as iron oxides, within an image pixel). Algorithms based on compositional analytical techniques can be applied to unmix the pixel spectral response into estimates of the proportions of the image components (such as vegetation, iron oxides, etc.) within each pixel of an image. The resulting fraction maps provide a powerful technique for mapping palaeosurfaces and analysing their development. This paper demonstrates how these techniques have been applied to studies of Quaternary piedmont surfaces in the Tunisian Southern Atlas.

The recognition, mapping and analysis of palaeosurfaces have played important roles in palaeoenvironmental reconstruction (Coxon & Flegg 1987) and landform evolution theories (Louis 1973; Melhorn & Edgar 1975). However, the study of palaeosurfaces presents severe logistical problems for Earth scientists. Specifically, palaeosurfaces often cover very large areas, large parts of which are often very inaccessible (Cotton 1973; Melhorn & Edgar 1975). This is where the use of satellite imagery can play a very useful role as an adjunct to conventional field mapping techniques, providing a synoptic view over large areas. For example, a single image from the Multispectral Scanner (MSS) or Thematic Mapper (TM), on board the Landsat series of Earth observation satellites, covers an area of 34 225 km^2 (Townshend *et al.* 1988). Even if accessibility were not a problem, the cost of detailed field survey of such an area would be prohibitively expensive for most applications.

A further factor promoting the application of remote sensing techniques in palaeosurface studies is that, in many instances, palaeosurfaces can be recognized by a distinct chemical or mineralogical composition of surficial materials; for example the regolith or soil cover developed on the surface (Gerrard 1992). For instance, Tertiary surfaces in Australia are characterized by calcretes (Milnes 1992) and silcretes (van de Graaff 1983). These chemical crusts often play a vital role in the preservation of the palaeosurfaces they cover (Goudie 1985). If these minerals have distinct spectral reflectance characteristics, then multispectral remote sensing systems can be used to identify the outcrops and map their extent (Drury 1994). For example, the extensive laterite deposits developed on Tertiary basalts in the Western Ghats of India (Singh 1988) can be detected and mapped due to the presence of iron oxide minerals (Widdowson 1997), which have very distinct reflectance properties in the visible part of the electromagnetic spectrum (Hunt *et al.* 1971).

Multispectral images consist of a raster array of measurements of radiance from the Earth's surface and atmosphere. Each cell in this array (known as a pixel) contains a digital number (DN) which is proportional to the satellite received radiance at a given wavelength. Each pixel will thus have a series of digital numbers, corresponding to the received radiance in the various wavelengths (or bands) being measured by the remote sensing system. The Landsat TM sensor has become a very widely used source of multispectral data: it combines relatively high spatial resolution (each pixel representing an area of *c.* 900 m^2, or 30 × 30 m, on the ground) with relatively high spectral resolution (six bands in the visible to short-wave infrared part of the electromagnetic spectrum, as well as one band operating in the thermal infrared

From Widdowson, M. (ed.), 1997, *Palaeosurfaces: Recognition, Reconstruction and Palaeoenvironmental Interpretation,* Geological Society Special Publication No. 120, pp. 283–293.

(which is not used in this project). Therefore, each pixel in a TM image has six spectral dimensions in the reflected solar part of the electromagnetic spectrum (Mather 1987).

When seeking to analyse types of land cover, the most common technique has been to seek to allocate pixels to different land cover classes according to their DN values in the different spectral bands. This is based on the fact that different land cover types (such as forest, soil, water, etc.) absorb and reflect solar radiation at different wavelengths and thus can be characterized by distinctive 'spectral signatures' in a multispectral image. Thus, an image can be used to map land cover based on the spectral information (Thomas *et al.* 1987). Although a powerful technique when dealing with discrete land cover types (such as different crops), the technique is inappropriate for gradually varying land cover types where individual pixels consist of a mixture of different scene components, such as natural vegetation communities. Furthermore, the classification procedure results in a loss of data because the interval radiance measurements are replaced by nominal (class) values (Haralick & Fu 1983).

The most common way of obtaining interval compositional data for image pixels is to use ratios (Mather 1987). This involves dividing the satellite radiance in one spectral band where reflectance is at a maximum for a given cover type by another band where reflectance is at a minimum due to absorption of radiation by that cover type, thereby exploiting information from the shape of the spectral curve (see Fig. 1). The problem here is that, in broad band satellite data such as TM imagery, these ratios are not unique to a given cover type, and the relationship between the proportion of the cover type within a pixel and the ratio value of that pixel is unknown. Furthermore, topographic relief, although subdued by the ratioing process, is not effectively removed. This means that shadows will also be controlling the ratio values in areas of variable relief, further degrading the relationship between the ground proportions and the ratio values (Mather 1987).

Multispectral data in the visible to short-wave infrared part of the electromagnetic spectrum are usually highly correlated between bands, a pixel with a high radiance value in one band will have high radiance values in the other bands and *vice versa* (Rothery 1987). In other words, in this part of the spectrum, surfaces tend to be generally bright (e.g. sand dunes) or generally dark (e.g. wet mudflats). As with other highly correlated data, the variation can be more efficiently described by principal components rather than the original spectral bands. These can be calculated using the standard Karhunen–Loeve transformation (Abrams *et al.* 1988, Sabins 1987). This enables the subtle spectral differences between surfaces to be detected (Sheffield 1985) and can be used to derive compositional information from mixed pixels, as outlined below.

Study area

The study area for this project is located in south-central Tunisia (Fig. 2). The geology consists of

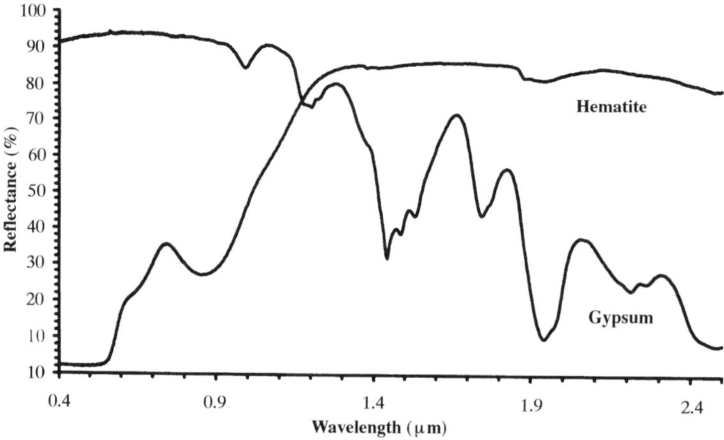

Fig. 1. Spectral reflectance curves of hematite and gypsum. Image ratioing seeks to exploit the shape of the reflectance spectra by dividing a waveband where reflectance is at a maximum (e.g. the hematite spectrum at 0.5 μm) by a waveband where most energy is absorbed (an absorption feature such as seen in the gypsum spectrum at 1.95 μm).

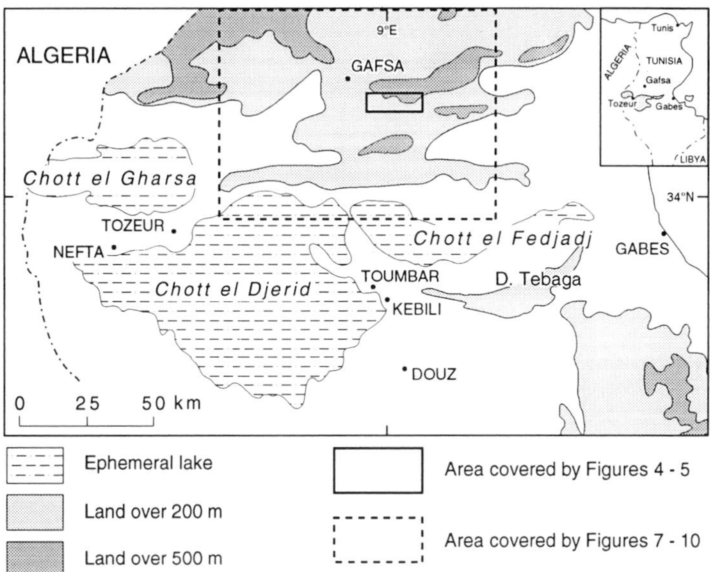

Fig. 2. Location of the study area.

Cretaceous dolomites, limestones, dolomitic limestones, marls, sandstones, clays, anhydrite and gypsum (Domerque *et al.* 1952), and Tertiary shales, sandstones, limestones, clays, marls, gypsum and phosphates (Burollet 1967). Quaternary landforms consist of alluvial fans, river terraces, gypsum-crusted *glacis d'erosion* (nested pediment surfaces), dunes, alluvial plains and playas.

The study area has a mean annual rainfall of 120–150 mm and a dry period of 7–8 months between April and November. Prevailing winds are northeasterlies, but hot, dry sirocco winds blow from the south for 20–60 days per year, often bringing large quantities of dust from the central Sahara. Mean monthly temperatures vary between 8°C in January to 33°C in August (data courtesy of Tunisian Meteorological Bureau, Tunis).

Directed principal components analysis for mapping piedmont surfaces on the basis of pedogenic iron oxide accumulation

The time-dependent accumulation of iron oxide minerals in dryland soils has been noted by several workers (e.g. Diaz & Torrent 1989; Torrent *et al.* 1980). This is thought to be due to the absence of organic matter in the soils, preventing the chelation and subsequent leaching of iron oxides (Schwertmann & Taylor 1989). Therefore, the amount of iron oxide in a dryland soil can be used as an indicator of relative age, if other factors such as parent lithology, elevation, etc., are held constant (White & Walden 1994). The ability to remotely sense the soil iron oxide content therefore holds potential for mapping surfaces of different ages.

Many alluvial fans in the Tunisian Pre-Sahara exhibit distinct slope segments (Blissenbach 1954) of different ages (Coque 1962). These represent old depositional surfaces, upon which pedogenesis and iron oxide enrichment begins when the locus of deposition on the fan shifts elsewhere. In section, three surfaces can be identified on these fans (Fig. 3), an upper surface nearest to the mountain front, an intermediate segment inset within this, and a contemporary depositional wash which is incised into the older surfaces. On one fan, the Oued es Seffaia fan, near Gafsa, the upper and intermediate surfaces have been dated at 50 000 and 5000 years respectively, by a combination of optically stimulated luminescence and AMS radiocarbon techniques (White *et al.* 1996). The upper surface of this fan has been shown to contain a greater concentration of canted antiferromagnetic minerals (e.g. hematite and goethite) and secondary minerals formed by weathering during soil formation (White & Walden 1994). Remote sensing was used to determine how widespread this phenomenon is (White *et al.* 1992).

Iron oxides have distinct spectral reflectance characteristics in the visible part of the electromagnetic spectrum (Fig. 1). This is dominated by the charge transfer band in the near ultraviolet, which is so intense that its tail extends into the

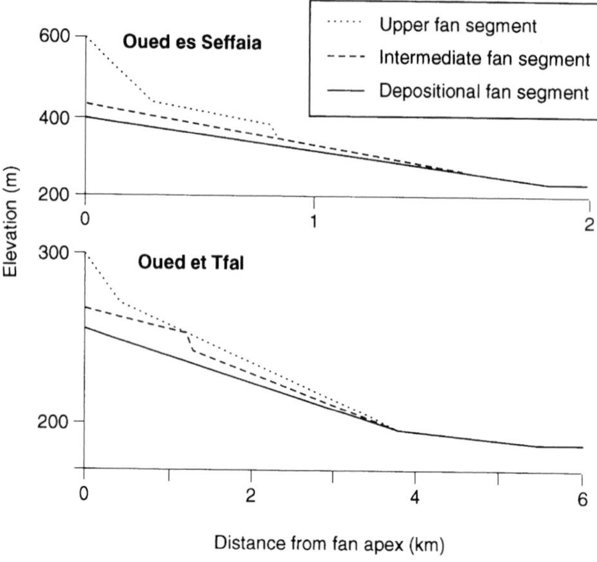

Fig. 3. Sections through two alluvial fans in the Tunisian Southern Atlas, showing the relationship between the segments of different ages. The upper (steeper) segments are thought to be of Mousterian age, the intermediate segments are thought to be of Capsian age. The depositional wash is the contemporary depositional

visible, giving a sharp absorption edge between 0.5 and 0.6 μm. This accounts for the trans-opaque behaviour of goethite and hematite, which imparts such strong coloration to these minerals (Hunt *et al.* 1971).

In order to see if remotely sensed imagery can be used to map the different fan surfaces on the basis of soil iron oxide content, a subscene from a Landsat TM image of 13 September 1987 (10:08 local solar time) was selected, covering the southern side of Djebel Orbata, just to the east of Gafsa, and processed as follows. (1) Atmospheric correction using the regression intersection method of Crippen (1987). (2) Calculation of ratios of TM band 3 divided by band 1 (a standard 'iron oxide' ratio) and band 4 divided by band 1 (a crude 'vegetation' ratio). TM band 1 (visible blue radiation) is strongly affected by atmospheric scattering, but the 3/1 and 4/1 ratios have been demonstrated to provide the best separability between soil iron oxides and vegetation using the technique outlined below (Fraser 1991). (3) Vegetation has strong spectral effects in the visible and near infrared (where iron oxide minerals also have distinctive features), which can therefore interfere with the interpretation of iron oxides in broad band data (Siegel and Goetz 1977; Segal 1983). The directed principal components approach uses a standardized transformation derived from the interband correlation rather than the covariance matrix (Singh & Harrison 1985), thereby normalising the variance on each component. If the two ratio images are used as input into this procedure, the vegetation signal is effectively decorrelated from the soil signal, thereby unmixing the effects of vegetation and pedogenic iron oxides within each pixel. Further details of these processing steps are given in Fraser (1991).

Figure 4 shows the first directed principal component image (containing the pedogenic iron oxide information) of the piedmont zone to the east of Gafsa. This has been calibrated by laboratory analysis (Mehra & Jackson 1960) of 14 field samples, selected to cover the full range of iron oxide concentrations encountered and located using Global Positioning System (GPS) equipment (Fig. 5); the correlation between remotely-sensed estimates and laboratory measurements of iron oxide content is highly significant ($r = 0.625$, $n = 14$). The image picks out the contemporary ephemeral depositional washes as areas of low iron oxide content (dark), because pedogenesis has not begun on these surfaces. The intermediate segments, dated to the Capsian (*c.* 5000 years ago), have a mean iron oxide content of 0.82 mg g^{-1} (standard deviation = 0.050 mg g^{-1}), whereas the upper segments, dated to the Mousterian (*c.* 50 000 years ago), are brightest on the image and have a mean iron oxide content of 0.91 mg g^{-1} (standard deviation = 0.003 mg g^{-1}). The intermediate fan segments can be clearly discriminated as distinct funnel-shaped segments around the contemporary depositional washes, thus the image products can be used qualitatively to map these old surfaces to improve our understanding of their morphology and mode of formation, as well as quantitatively, to

Fig. 4. The first directed principal component image of the alluvial fans forming on the south side of Djebel Orbata. This component contains the iron oxide information. Surfaces with low amounts of iron oxide (e.g. the depositional wash) appear dark. Surfaces with the highest iron oxide concentration (e.g. the upper fan segments) appear bright. Note that the technique does not remove the effects of relief, so iron oxide information cannot be retrieved from upland areas. The image covers an area of 7.5 × 15 km.

analyse the processes and rates of pedogenesis operating on these surfaces. However, it should be noted that, as the technique is based on ratios, relief remains a problem, as can be seen in Fig. 4; the shadows resulting from elevation differences within the upland parts of the image cannot easily be removed from these image products. Therefore, pixels in these upland areas cannot be calibrated against field measurements, thus presenting a severe limitation to the application of the technique in regions of high relief, or to images with low solar illumination angles.

Spectral mixture modelling for mapping surfaces on the basis of gypsum crust development

To overcome the problem of relief for proportions mapping, a technique is required that will isolate the shadows resulting from variable relief, so that the effect can be removed from the data, enabling proportions to be calculated as if the surface had uniform elevation. Spectral mixture modelling provides a method for doing this. The following example illustrates the technique.

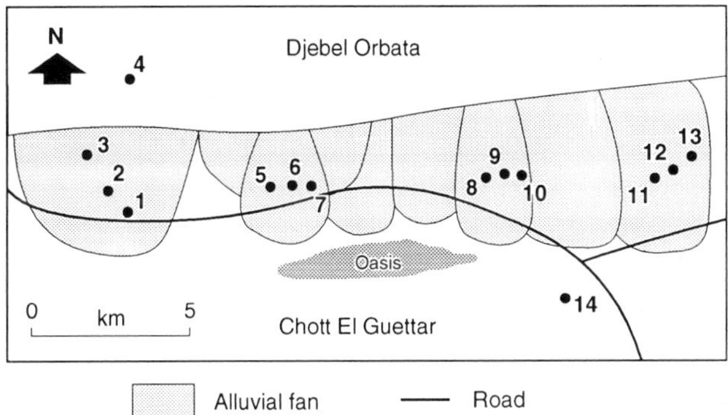

Fig. 5. Map of area covered by Fig. 4, showing the location of the 14 soil samples for laboratory determination of iron oxide content.

Apart from alluvial fans, the other type of piedmont surface found in the Tunisian Southern Atlas is the *glacis d'erosion*. Unlike the depositional fan surfaces, these are erosional surfaces formed at the mountain front, but morphologically they are similar in that they are nested, with lower surfaces inset within older surfaces. In the Tunisian Southern Atlas, the *glacis* surfaces are covered by gypsum crusts or gypcretes (White 1993).

Gypsum has very distinct spectral reflectance characteristics with absorption features due to vibrational combinations and overtones of molecular water, which forms individual clusters of molecules at specific sites in the gypsum crystal lattice (Fig. 1). The absorption features at 0.942, 1.135, 1.454, 1.9, 2.2 and 2.5 μm provide a highly diagnostic spectrum (Hunt 1977). The strong feature at 2.2 μm occurs in the centre of TM band 7.

In order to derive estimates of gypsum within each image pixel, a linear mixture model can be applied to the spectral data (Settle & Drake 1993). The technique assumes that pure examples of each scene component (i.e. endmembers) can be identified. The proportions of the endmember spectra are then varied to try and model the measured spectrum of each pixel. The proportions of endmembers which most closely resemble the observed pixel spectrum provide estimates of proportions which can be checked in the field. Smith *et al.* (1985) showed that principal components analysis can be used to determine the number of spectrally distinct components in the mixture and their respective endmembers. This empirical method involves plotting scattergrams of the significant principal components (i.e. those that are not simply noise). The endmembers lie at the scattergram extremities while the mixed pixels will lie within the region defined by the endmembers. These extremes can be located in the image and visited in the field for sample collection and identification using XRD analysis. In any image with significant relief, shadow will always form an endmember.

Six principal components were defined from the six TM spectral bands and enabled the identification of seven endmembers; marl, dolomitic limestone ('rock'), halite, gypsum, shadow, vegetation and phosphates. The locations of the marl, rock, halite and gypsum endmembers relative to principal components axes 2 and 3 are shown in Fig. 6. To estimate the proportions we have used the regularized estimator which Settle (1990) has shown to be more accurate than simple least squares. The method minimizes the difference between observed and modelled pixel spectra (Settle & Drake 1993):

$$(x - Mf)^{T}\, C^{-1}\, (- Mf) + v\, (f - g)^{T}(f - g) \quad (1)$$

where x is the pixel vector, M is a matrix whose columns are endmember spectra, ^{T} signifies the transpose of the matrix, f is a vector of unknown proportions, C is a matrix of errors and v is a smoothing parameter. The aim is not to make an exact fit to the data, as this means fitting to both spectral information and noise (errors in the pixel values) The fit is controlled by a smoothing parameter v; as this increases the pixels are forced to be

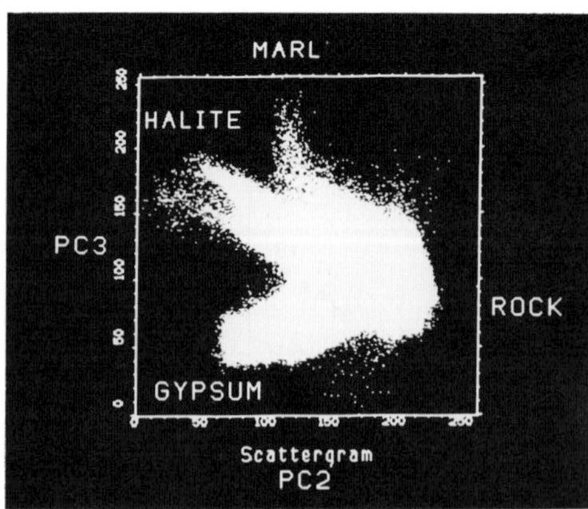

Fig. 6. An example scatterplot of principal components 2 and 3, showing the positions of some endmembers. See text for explanation.

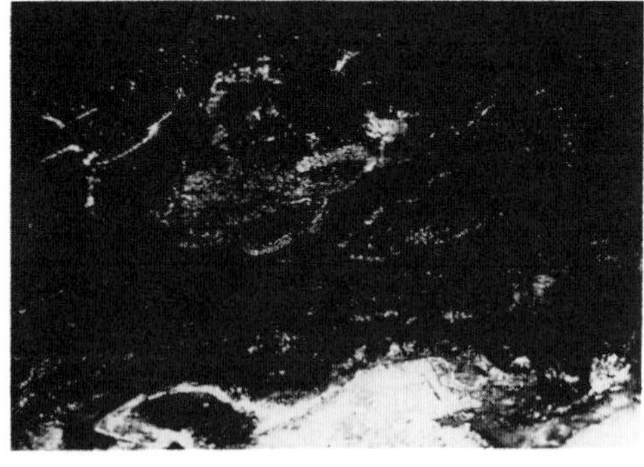

Fig. 7. The mixture map of gypsum distribution and abundance in the Tunisian Southern Atlas. When compared with a TM image (Fig. 8), the bright areas of highest gypsum concentration are found to be evaporitic crusts in the ephemeral lakes and gypsiferous sand dunes forming by deflation of these crusts. The pedogenic crusts characteristic of the *glacis d'erosion* show up as having medium concentrations of gypsum. The area of the image is 90 × 90 km.

more even mixtures. The approach is subject to two constraints, the first being:

$$f_1 > 0_{i=1...c} \quad (2)$$

where c is the number of components. This states that negative proportions are not allowed. The second constraint is:

$$f_1 + f_{2...} + f_c = 100\% \quad (3)$$

This states that the proportions in a pixel cannot add to more or less than 100%. It implies that we know all the spectral components in a pixel.

The technique yields a map of the proportion of shadow within each pixel, which can then be subtracted from the model, and all the other endmember proportions can be recalculated. The resulting gypsum fraction map is shown in Fig. 7, compared to a contrast stretched TM band 7 image (Fig. 8). The proportion estimates were calibrated against 24 field samples of varying gypsum content (determined using XRD), which were pinpointed in the image from GPS-derived locations (Table 1). The correlation between remotely-sensed estimates and laboratory measurements of gypsum content was highly significant ($r = 0.93$, $n = 24$). Further details of the validation are given in White & Drake (1993). The mixture map shows that the greatest concentrations of gypsum are found as evaporitic crust on playas and as aeolian sands derived from the playas and blown towards the southwest. The *glacis d'erosion* show up as patchy areas (5–30% gypsum) surrounding the upland areas and they can be clearly differentiated from the surrounding terrain. They can then be mapped with a high degree of accuracy (Fig. 9) and the distribution can be compared with the regional geology (Fig. 10). This simple analysis highlights the fact that the *glacis* are formed predominantly on outcrops of the

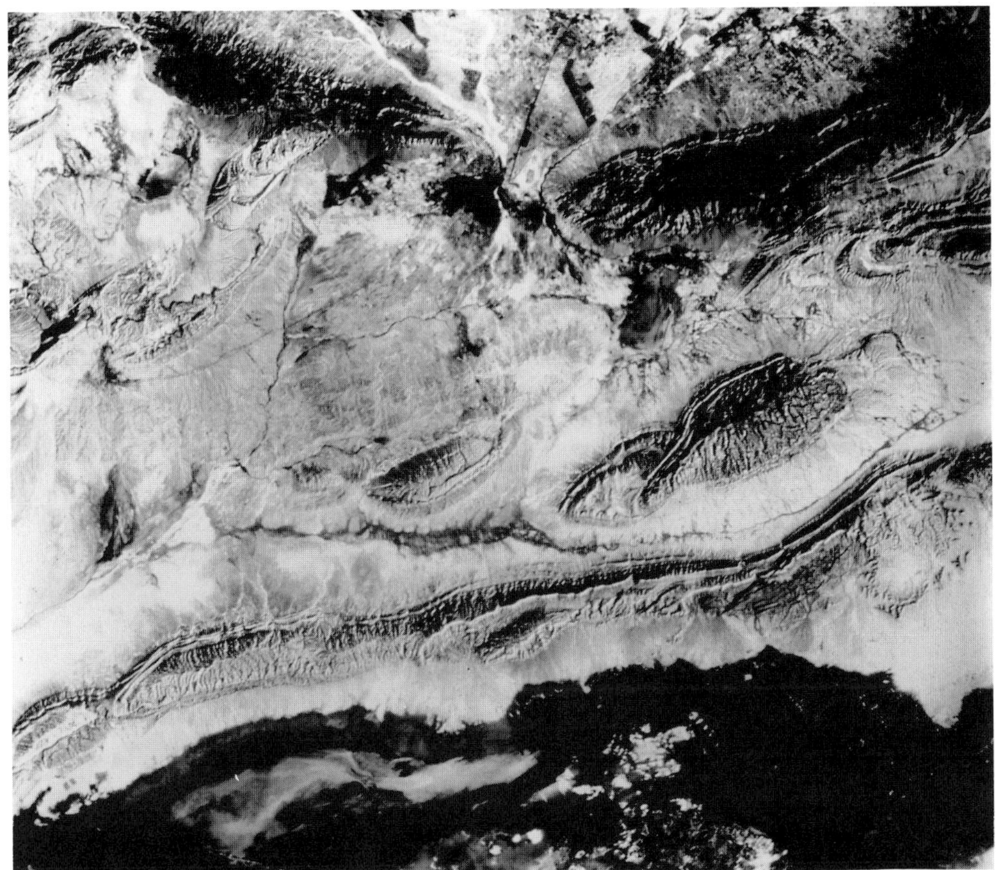

Fig. 8. Landsat TM band 7 image of the Tunisian Southern Atlas for comparison with Fig. 7. The area of the image is 90 × 90 km.

Table 1. *Ground (XRD) and satellite-derived estimates of gypsum content in the Tunisian Southern Atlas*

Site no.	Gypsum (%) (ground)	Gypsum (%) (imagery)
1	0.3	17.8
2	72.0	91.0
3	23.0	16.3
4	24.0	54.8
5	57.0	79.3
6	27.0	41.4
7	19.6	45.2
8	35.0	59.8
9	4.0	5.2
10	10.1	18.6
11	10.4	26.5
12	0.0	1.5
13	0.0	2.5
14	22.0	10.9
15	2.0	7.3
16	10.0	17.4
17	16.0	4.1
18	12.0	28.1
19	9.2	35.5
20	38.0	26.5
21	0.0	3.0

relatively poorly indurated Mio-Pliocene rocks in this area. This observation is significant because it highlights the importance of lithological characteristics in the process of pedimentation. The Mio-Pliocene rocks also contain evaporitic gypsum beds, which provides evidence for the importance of lithological composition in the process of gypsum crust formation (White & Drake 1993). These image products also permit analysis of the morphometric relationships between pediment surfaces and associated drainage basins and mountain fronts (White 1991). A major advantage of the technique is that shadow can be modelled as an endmember itself, and it can then be removed from the model, allowing the other components of the shaded pixels to be re-estimated on a pro rata basis, thereby removing the problem of relief.

Conclusions

The examples presented here have demonstrated that multispectral imagery from satellite sensors

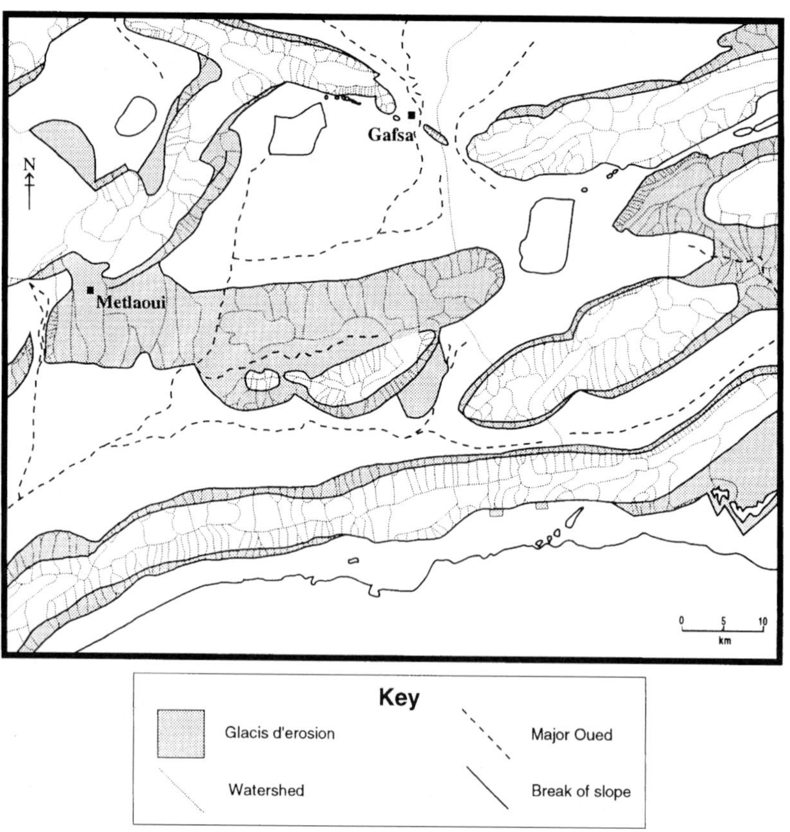

Fig. 9. Map showing the distribution of *glacis d'erosion* produced from Fig. 6 and checked in the field. Comparison with Fig. 10 indicates the close association of *glacis* with outcrops of poorly consolidated Mio-Pliocene sediments.

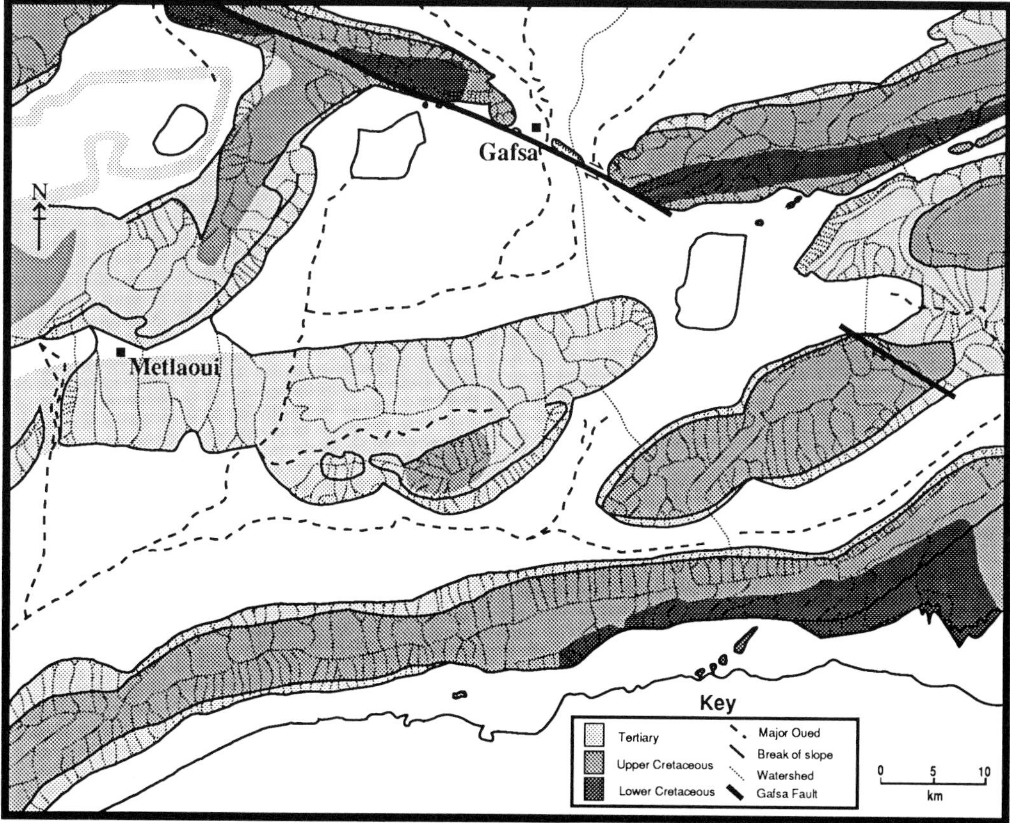

Fig. 10. Geological map of the Tunisian Southern Atlas after Castany (1951), for comparison with Fig. 9.

can be processed to yield proportions estimates of surficial material composition. Directed principal components analysis can unmix proportions from ratio data, but is still prone to problems caused by variable relief and illumination within an image. Spectral mixture modelling provides a technique of calculating proportions of image components within each pixel and enables the effects of relief to be removed from the image products. Although these examples have been drawn from the Quaternary, similar techniques can be applied to Tertiary and pre-Tertiary palaeosurfaces characterized by lateritic and bauxitic surficial materials. The results not only enable more detailed mapping of these surfaces over very large (often innaccessible) areas, but they also enable analyses of the processes of weathering and pedogenesis operating on these surfaces. If non-age dependent controls on surficial material composition (e.g. amount of secondary iron oxides) can be held constant (such as climate, source lithology, etc.), then it should prove possible to extrapolate palaeosurfaces from field mapped areas to unknown parts of the imagery. Detailed compositional estimates of surficial material composition also enable the testing of specific hypotheses of palaeosurface formation and preservation (see e.g. White *et al.* 1992, White & Drake 1993).

The authors are grateful to Erika Meller for help with hardcopy image production, Judith Fox for drawing the diagrams, and Drs David Rothery (The Open University) and Chris Clark (University of Sheffield) for their thoughtful and helpful comments on the first draft.

References

ABRAMS, M. J., ROTHERY, D. A. & PONTUAL, A. 1988. Mapping in the Oman ophiolite using enhanced Landsat Thematic Mapper images. *Tectonophysics*, **151**, 387–400.

BLISSENBACH, E. 1954. Geology of alluvial fans in semi-arid regions. *Geological Society of America Bulletin*, **65**, 175–190.

BUROLLET, P. F. 1967. Tertiary geology of Tunisia. *In*: MARTIN, L. (ed.) *Guidebook to the Geology and History of Tunisia*. Petroleum Exploration Society of Libya, Tripoli, 215–225.

CASTANY, G. 1951. *Carte Géologique de la Tunisie, echelle 1/500,000*. Services Géologique de la Tunisie, Tunis.

COQUE, R. 1962. *La Tunisie Présaharienne. Etude Géomorphologique*. Armand Colin, Paris.

COTTON, C. A. 1973. The theory of savanna planation. *In*: DERBYSHIRE, E. (ed.) *Climatic Geomorphology*. Macmillan, London, 171–185.

COXON, P. & FLEGG, A. M. 1987. A late Pliocene/early Pleistocene deposit at Pollnahallia, near Headford, Co. Galway. *Proceedings of the Royal Irish Academy*, **87b**, 15–42.

CRIPPEN, R. E. 1987. The regression intersection method of adjusting image data for band ratioing. *International Journal of Remote Sensing*, **8**, 137–155.

DIAZ, M. C. & TORRENT, J. 1989. Mineralogy of iron oxides in two soil chronosequences of central Spain. *Catena*, **16**, 291–299.

DOMERQUE, C., DUMON, E., LAPPARANT, A. F. DE & LOSSEL, P. 1952. *Sud et Extreme Sud Tunisiens*. Congrès Géologique International, Tunis.

DRURY, S. A. 1994. *Image Interpretation in Geology* 2nd Edition. Chapman & Hall, London.

FRASER, S. J. 1991. Discrimination and identification of ferric iron oxides using satellite Thematic mapper data: A Newman case study. *International Journal of Remote Sensing*, **12**, 635–641.

GERRARD, J. 1992. *Soil Geomorphology*. Chapman & Hall, London.

GOUDIE, A. S. 1985. Duricrusts and landforms. *In*: RICHARDS, K. S., ARNETT, R. R. & ELLIS, S. (eds) *Geomorphology and Soils*. George Allen and Unwin, London, 37–57.

HARALICK, R. M. & FU, K. S. 1983. Pattern recognition and classification. *In*: COLWELL, R. N. (ed.) *Manual of Remote Sensing* 2nd Edition. American Society of Photogrammetry, Falls Church, Virginia, 793–805.

HUNT, G. R. 1977. Spectral signatures of particulate minerals in the visible and near-infrared. *Geophysics*, **42**, 501–513.

——, SALISBURY, J. W. & LENHOFF, C. J. 1971. Visible and near-infrared spectra of minerals and rocks: 3, oxides and hydroxides. *Modern Geology*, **2**, 195–205.

LOUIS, H. 1973. The problem of erosion surfaces, cycles of erosion and climatic geomorphology. *In*: DERBYSHIRE, E. (ed.) *Climatic Geomorphology*. Macmillan, London, 153–170.

MATHER, P. M. 1987. *Computer Processing of Remotely Sensed Images*. Wiley, Chichester.

MEHRA, O. P. & JACKSON, M. L. 1960. Iron oxide removal from soils and clays by a dithionite–citrate system buffered with sodium bicarbonate. *In*: SWINFORD, A. (ed.) *Proceedings of 7th National Conference on Clays and Clay Minerals*, Washington DC, 1958. Pergamon, New York, 317–327.

MELHORN, W. N. & EDGAR, D. E. 1975. The case for episodic continental-scale erosion surfaces: a tentative geodynamic model. *In*: MELHORN, W. N. & FLEMAL, R. C. (eds) *Theories of Landform Development*. New York State University, New York, 243–276.

MILNES, A. R. 1992. Calcretes. *In*: MARTINI, I. P. & CHESWORTH, W. (eds) *Weathering, Soils and Paleosols*, Amsterdam. Elsevier, Amsterdam, 309–347.

ROTHERY, D. A. 1987. Improved discrimination of rock units using Landsat Thematic Mapper images of the Oman ophiolite. *Journal of the Geological Society, London*, **144**, 587–597.

SABINS, F. F. Jr. 1987. *Remote Sensing Principles and Interpretation* 2nd Edition. Freeman, New York.

SCHWERTMANN, U. & TAYLOR, R. M. 1989. Iron oxides. *In*: DIXON, J. B. & WEED, S. B. (eds) *Minerals in the Soil Environment* 2nd Edition. Soil Science Society of America, Madison, 379–438.

SEGAL, D. B. 1983. Use of Landsat multispectral data for definition of limonitic exposures in heavily vegetated areas. *Economic Geology*, **78**, 711–722.

SETTLE, J. J. 1990. Contextual models and the use of Dirichlet Priors for mixing problems. *Proceedings of 5th Australasian Remote Sensing Conference*, Perth, 314–323.

—— & DRAKE, N. A. 1993. Linear mixing and the estimation of ground cover proportions. *International Journal of Remote Sensing*, **14**, 1159–1171.

SHEFFIELD, C. 1985. Selecting band combinations from multispectral data. *Photogrammetric Engineering and Remote Sensing*, **51**, 681–687.

SIEGAL, B. S. & GOETZ, A. F. H. 1977. The effects of vegetation on rock and soil type discrimination. *Photogrammetric Engineering and Remote Sensing*, **43**, 191–196.

SINGH, A. & HARRISON, A. 1985. Standardised principal components. *International Journal of Remote Sensing*, **6**, 883-896.

SINGH, R. P. 1988. Tectonics and surfaces of Peninsular India. *In*: SINGH, S. & TIWARI, R. C. (eds) *Geomorphology and Environment*. Allahabad Geographical Society, Allahabad, 311–324.

SMITH, M. O., JOHNSON, P. E. & ADAMS, J. B. 1985. Quantitative determination of mineral types and abundances from reflective spectra using principal components analysis. *Journal of Geophysical Research*, **90**, C797–C804.

THOMAS, I. L., BENNING, V. M. & CHING, N. P. 1987. *Classification of Remotely Sensed Images*. Adam Hilger, Bristol.

TORRENT, J., SCHWERTMANN, U. & SCHULZE, D. G. 1980. Iron oxide mineralogy of some soils of two river terrace sequences in Spain. *Geoderma*, **23**, 191–208.

TOWNSHEND, J. R. G., CUSHNIE, J., HARDY, J. R. & WILSON, A. 1988. *Thematic Mapper Data; Characteristics and Use*. NERC Scientific Services, Swindon.

VAN DE GRAAFF, W. J. C. 1983. Silcretes in Western Australia: geomorphological settings, textures, structures and their genetic implications. *In*: WILSON, R. C. L. (ed.) *Residual Deposits: Surface Related Weathering Processes and Materials*. Blackwells, Oxford, 159–166.

WHITE, K. 1991. Geomorphological analysis of piedmont landforms in the Tunisian Southern Atlas using ground data and satellite imagery. *Geographical Journal*, **157**, 279–294.

—— 1993. Image processing of Thematic Mapper data for discriminating piedmont surficial materials in the Tunisian Southern Atlas. *International Journal of Remote Sensing*, **14**, 961–977.

—— & DRAKE, N. A. 1993. Mapping the distribution and abundance of gypsum in south-central Tunisia from Landsat Thematic Mapper data. *Zeitschrift für Geomorphologie*, **37**, 309–325.

—— & WALDEN, J. 1994. Mineral magnetic analysis of iron oxides in arid zone soils from the Tunisian Southern Atlas. *In*: MILLINGTON, A. C. & PYE, K. (eds) *Environmental Change in Drylands: Biogeographical and Geomorphological Perspectives*. Wiley, London, 43–65.

——, —— & ROLLIN, E. M. 1992. Remote sensing of pedogenic iron oxides using Landsat Thematic Mapper data of Southern Tunisia. *Remote Sensing from Research to Operation*, Proceedings of 18th Annual Conference of the Remote Sensing Society, Dundee, 179–187.

——, DRAKE, N. A., MILLINGTON, A. C., & STOKES, S. 1996. Late Quaternary evolution of a telescopically-segmented alluvial fan, southern Tunisia. *Palaeogeography, Palaeoclimatology, Palaeoecology*, in press.

WIDDOWSON, M. 1997. Tertiary palaeosurfaces of the SW Deccan, India: implications for passive margin uplift. *This volume*.

The geochemistry and development of lateritized footslope benches: The Kasewe Hills, Sierra Leone

D. J. BOWDEN

Newman College, Genners Lane, Bartley Green, Birmingham B32 3NT, UK

Abstract: A series of lateritized footslope benches are observed on the isolated hill masses which rise to 500 m above the interior plain of Sierra Leone. Field observations and chemical analyses of samples indicate that each bench has a characteristic geochemistry controlled by its own topography, hydrological environment and leaching history. The benches are related, each inheriting material from above and providing material for the bench below. The material is transported both mechanically and in solution. As a result, the parent material of each bench is not only the bedrock (metamorphosed volcanic and sedimentary rocks of Proterozoic age in the case of the Kasewe Hills), but also material which has undergone various degrees of lateritization. This cratonic region has been subjected to denudational processes since Mesozoic times at least. Denudational processes, especially etchplanation, are long continued, yet weathering, the development of breakaway slopes and subterranean flushing are patently continuing processes: these palaeosurfaces and palaeosoils may, therefore, be ancient in origin but they cannot be considered as being fossilized.

Isolated hill masses rise to 500 m above the interior plain of Sierra Leone (Fig.1). They are generally formed of bedrock with lateritized footslopes developed below the bedrock slopes. The Kasewe, Kosankoh and Malal Hills are formed of meta-volcanic rocks of the Rokel River Group (Upper Proterozoic), and the other hill masses are formed by schists and other metamorphic rocks of the older Port Loko Group (Archean/Lower Proterozoic) (Hawkes 1970; Morel 1979). Allen (1969) suggests that the main rock sequences present in the Kasewe Hills (i.e. the Kasewe Hills Formation) represent considerable late Precambrian volcanic activity with some evidence that some of the extrusions occurred under the sea. The main rocks are andesites, spillites and basalts which have all been affected by greenschist facies metamorphism. The composition of these bedrocks is highly variable, however, the analyses of four samples is presented (Table 1) as being indicative of the bulk bedrock geochemistry.

Footslopes extend to a distance between 2 and 3 km away from the bedrock slopes of the hills, and are mantled by lateritic duricrusts and related weathering products. These surficial rocks and weathering products have been given the name mantle rock (Brückner 1955). A series of benches with gentle slopes of between 1 and 7°, and downslope terminations between 20 and 90° (Fig. 2) have developed on the footslopes. These duricrusted benches are characterized by the development of pseudo-karst topography including pavements (bowe), caves, cambering, collapses and underground drainage (Bowden 1980*a*). The relationship between the lateritic mantle and the bedrock exemplifies Büdel's concept (1957,1982) of 'doppelten Einebnungsflächen' (double surfaces of lowering). Budel recognized that in many humid tropical terrains there were two zones of increased geomorphological activity, namely the ground surface where wash and other subaerial processes remove weathered material, and the subterranean weathering front where essentially geochemical and biochemical processes attack unweathered bedrock (Fig. 3).

Bench morphology

Field survey identified five benches, conventionally labelled A–E away from the Kasewe ridge (Fig. 2). The junction between the footslopes and the steep hillslope is of a variable nature, sometimes a junction between bench B and the bedrock slope, sometimes another bench, bench A, intervenes, and yet again there may be a junction obscured by the products of landslides and other mass wastage processes. The piedmont angle is always high. The general gradient of the footslope is between 1 and 2°, and never more than 7°, whereas the hillslope angles may be as high as 35°. Bench A is not always found. It is typically narrow never extending more than 500 m and commonly only 100 m, and composed of talus, a mixture of bedrock and regolith.Benches B–D are duricrusted. Bench D is the most consistent and well developed. Locally (for example, in the northeast of the hills) it has a mantle rock cover of 25 m, deep pseudo-karst

From Widdowson, M. (ed.), 1997, *Palaeosurfaces: Recognition, Reconstruction and Palaeoenvironmental Interpretation,* Geological Society Special Publication No. 120, pp. 295–305.

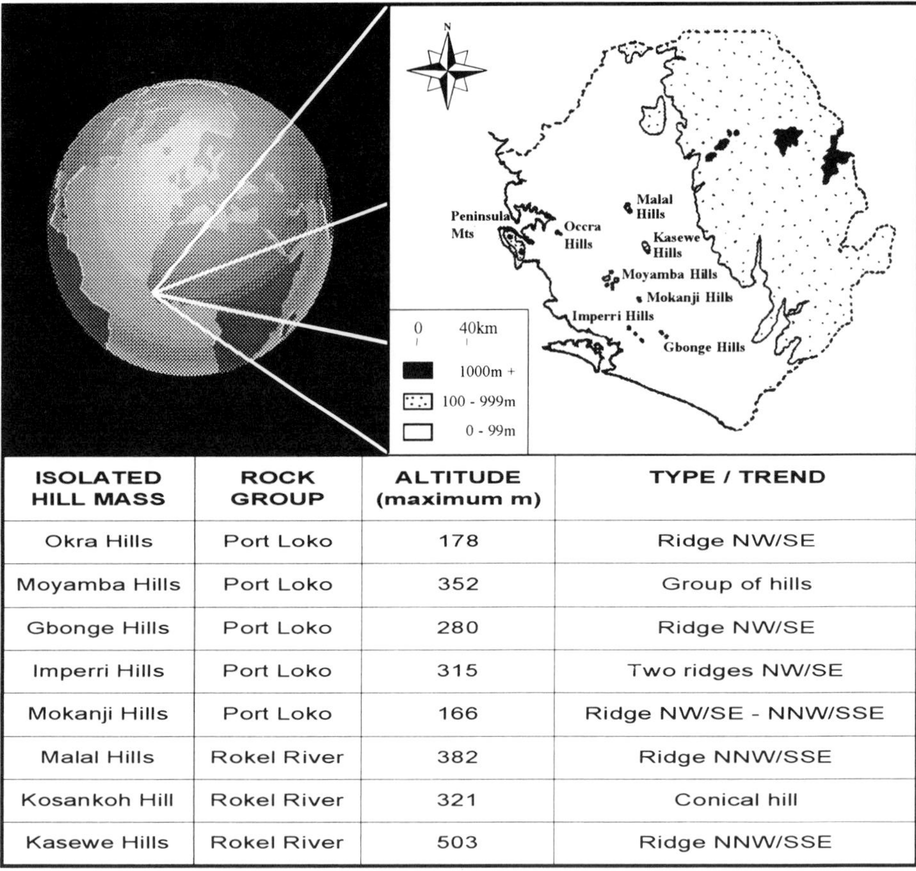

ISOLATED HILL MASS	ROCK GROUP	ALTITUDE (maximum m)	TYPE / TREND
Okra Hills	Port Loko	178	Ridge NW/SE
Moyamba Hills	Port Loko	352	Group of hills
Gbonge Hills	Port Loko	280	Ridge NW/SE
Imperri Hills	Port Loko	315	Two ridges NW/SE
Mokanji Hills	Port Loko	166	Ridge NW/SE - NNW/SSE
Malal Hills	Rokel River	382	Ridge NNW/SSE
Kosankoh Hill	Rokel River	321	Conical hill
Kasewe Hills	Rokel River	503	Ridge NNW/SSE

Fig. 1. Sierra Leone: isolated hill masses on the interior plain.

development and a precipitous downslope termination (Bowden 1987*a*). The overall width of the duricrust sheets varies from *c.* 2–3 km. Close to the downslope edges of the duricrust benches bowe surfaces are well developed and only locally disrupted by cambering and collapse. Bench E is less clearly defined, it is duricrusted but the degree of induration is less than that of the three main benches B–D. In some areas it forms a true bench, elsewhere it slopes down to a swamp or

Table 1. *Major element composition (%) of bedrock samples from the Kasewe Hills*

Sample	SiO_2	TiO_2	Al_2O_3	Fe_2O_3	MnO	MgO	CaO	K_2O	P_2O_5	Loss on ignition	Total (%)
Malempeh Quarry	43.95	1.24	14.77	15.45	bd	bd	9.87	bd	bd	14.72	85.28
Kasabere Summit	12.73	2.77	40.6	43.1	0.05	0.1	0.15	0.13	0.32	bd	99.95
004	9.86	2.15	31.38	33.35	0.04	0.08	0.12	0.1	0.25	22.66	77.33
084	50.48	1.69	14.52	12.84	0.25	0.1	3.03	0.39	0.2	16.5	83.5

The Malempeh Quarry is located at the northern end of the ridge [grid ref. (g.r.) 095270] and Kasabere (g.r. 096260) is the summit 400 m above the quarry. Sample 004 is also from northern end of ridge (g.r. 097260) and sample 084 is from the western slope of the hills (g.r. 101241). Grid references are from sheet 65 Yonibana 1 : 50 000 maps published by the Directorate of Overseas Surveys for the Sierra Leone Government (1965). bd, Concentration below detection limit.

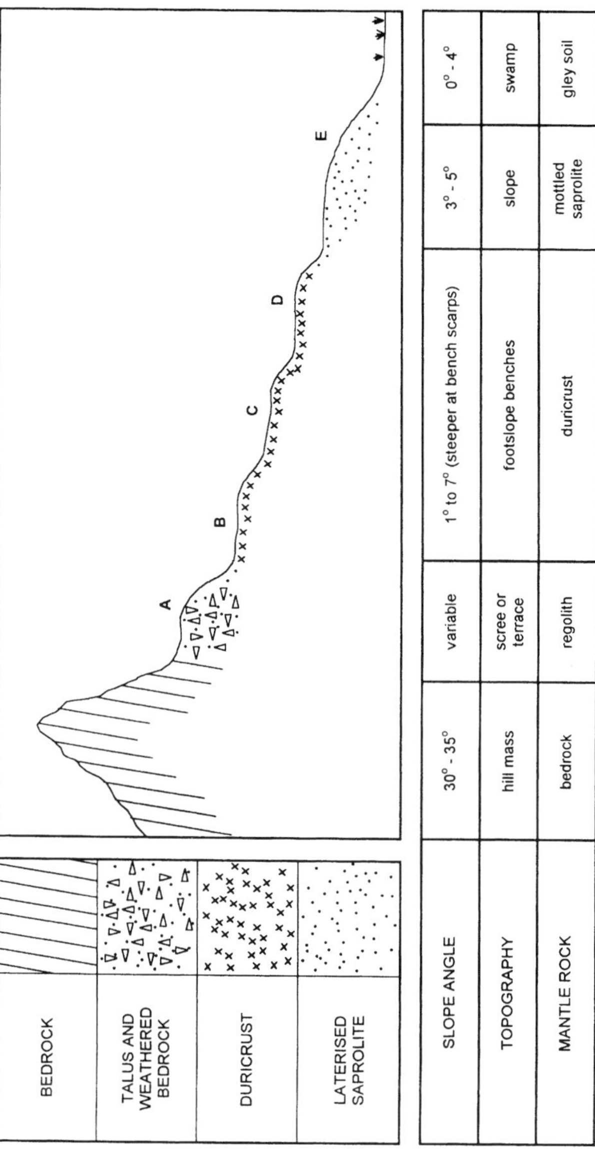

BEDROCK
TALUS AND WEATHERED BEDROCK
DURICRUST
LATERISED SAPROLITE
A
B
C
D
E
SLOPE ANGLE
30° - 35°
variable
1° to 7° (steeper at bench scarps)
3° - 5°
0° - 4°
TOPOGRAPHY
hill mass
scree or terrace
footslope benches
slope
swamp
MANTLE ROCK
bedrock
regolith
duricrust
mottled saprolite
gley soil

Fig. 2. A diagrammatic section across the Kasewe Hills to show major topographical features (not to scale). The horizontal distance is *c.* 3 km and the height some 300 m.

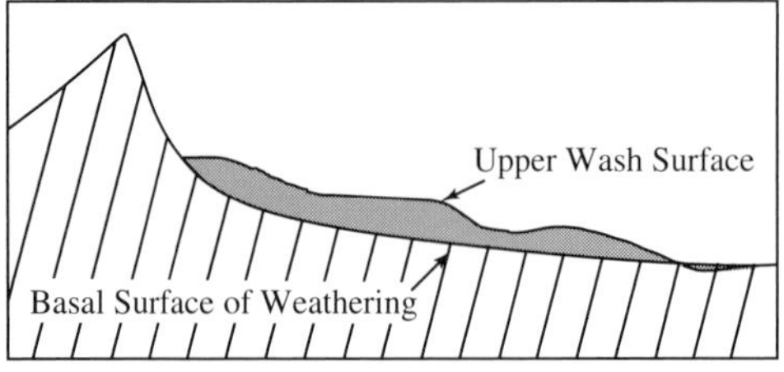

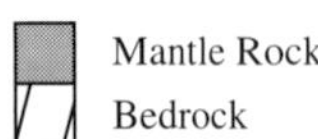

Fig. 3. The double surface of planation after Büdel(1982). The upper wash surface (*die Spül-Oberfläche*) and the basal weathering front (*die Verwitterungs-Basis Fläche*) together comprise the double surface of planation (*die doppelten Einebnungsflächen*).

watercourse. The lateral extent of the benches does vary and individual benches do not form a continuous apron around the hills.

Mantle rocks

The mantle rocks of the Kasewe Hills and the other hill masses display much textural and geochemical variation. Profiles through the mantle rock are deep and complex (Fig. 4). The material shows varying degrees of lateritization, a classification of the lateritic textures corresponds to a scheme developed by McFarlane (1976) from detailed work in Uganda. However, unlike the weathering sequences in Uganda, where clear endmembers may be identified, the laterites in the current Sierra Leonean research area often seem to display transitional forms (Fig. 5). The parent material of the laterite comprises both the bedrock, and lateritic and saprolitic materials inherited from higher benches (Bowden 1987*a*, *b*).

Mantle rock analysis

The mantle rocks were analysed by the author using the Phillips X-ray spectrometer the Geology Department at Bedford College (Regents Park) under the supervision of Dr G. Marriner. The standard lithium flux did not make satisfactory beads with the laterite samples and so a sodium tetraborate flux had to be used, thus rendering it impossible to analyse for sodium oxides. It is normal procedure for pulverized samples to be ignited before weighing to drive off moisture and volatiles. This was not attempted with these samples to avoid the possibility of further oxidation and because of the difficulties of estimating the proportion of water in the sample compared to that in the flux (Bowden 1987*a*). Analytical data given in this paper from other sources is duly acknowledged.

The chemistry of the laterites

If laterite is a residuum then it might be expected that its gross chemistry might reflect that of the parent rock (McFarlane 1976). On a macroscale it was indeed observed that laterites are preferentially developed on bedrocks which are richer in mafic minerals, since the rocks of the Kasewe Hills Formation give rise to well developed duricrusts on the footslopes of the Kasewe, Kosankoh and Malal Hills. By contrast, the intervening lowlands are underlain by metasediments together comprising the rocks of the Teye and Mabole Formations which Allen (1969) believes interdigitate with the lavas of the Kasewe Hills Formation. These sediments are mainly psammitic and pelitic and interpreted as deltaic deposits (Allen 1969). MacFarlane *et al.* (1981) demonstrate that the Mabole and Kasewe Hills Formation are contemporaneous and succeed the Teye Formation. The sediments which have also been metamorphosed have only poorly developed laterites with no duricrusting, although the low interfluves do have spaced pisolithic laterite (i.e. lateritic gravel) on them. Smaller scale examples of this phenomenon can be cited from along the course of the River Taia between Taiama and Njala (30 km SSE of the Kasewe Hills), where a series of dolerite dykes are observed with cappings of duricrust, whereas the metasediments into which they have been intruded are devoid of lateritic material (Bowden 1987*a*).

No clear relationship was found between fabric class and the major element chemistry of the laterite (Bowden 1987*b*). This seems to support Aleva's (1982) assertion that, 'the structures and textures now visible in lateritic rocks were imparted at different times and through different processes, with overlapping in time, space, and process being the rule not the exception'. Differences between bedrock, saprolite and pisolithic laterite can be largely explained in terms of desilicification, loss of aluminium, and a relative increase in iron. However, beyond the pisolithic stage no distinction can be made between the fabric (maturity sequence) and their major element concentrations (Fig. 6). The apparent loss of aluminium may be, at least partly, the result of organically-bound mobilization. In their work on aluminium leaching in Malawi, McFarlane & Bowden (1992) argue that microbially-mediated kaolinite dissolution and mobilization of aluminium appears to be the best option for explaining the removal of a high proportion of the aluminium as saprolite weathers to residuum. In the

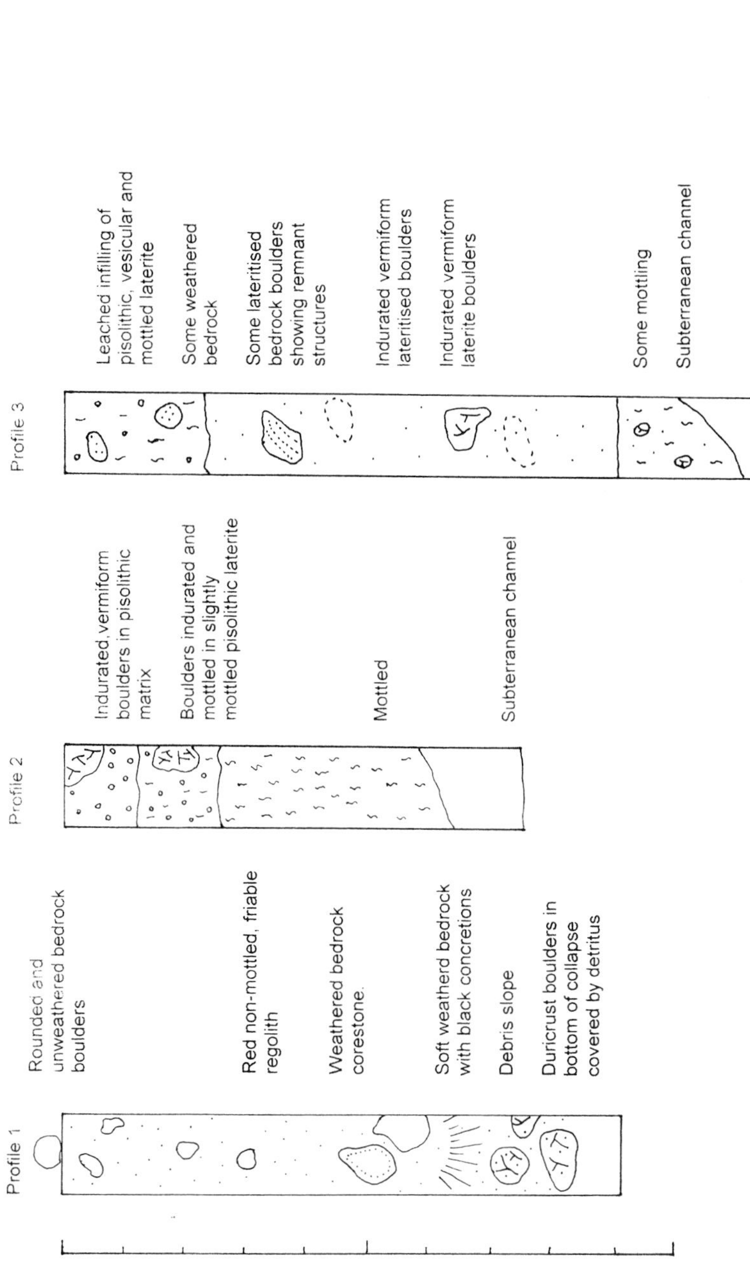

Fig. 4. Three mantle rock profiles from Kasewe Hills (scale in m). Profile 1 is from bench A at grid reference 127224; Profile 2 is from bench C at grid reference 131192; Profile 3 is from bench F at grid reference 130189. (Grid references are from sheet 65 Yonibana of the 1 : 50 000 map published by the Directorate of Overseas Survey for the Government of Sierra Leone 1965).

Type	Description	Type	Description
A	**Bedrock** Fresh unweathered rock.	D	**Pisolithic laterite** These contain rounded pisoliths of up to 3 cm in diameter with no internal differentiation, although the pisoliths may have yellow cutanes. The pisoliths may or may not be contiguous.
B	**Weathered bedrock** This is defined as material which has definite relict rock structures and textures. Pedologists would probably regard this material as parent material rather than as part of the soil profile. Here it is recognized as a stage in the development of the mantle rock.	E	**Pisolithic laterite developing into vermiform** Elongate structures begin to appear. Pisoliths are usually internally differentiated and become oval and even more attenuated. Tubular structures develop which may be lined with goethite.
		F	**Vermiform laterite** No or very few pisoliths present. Many tubular structures which display a great variation in form, colour and content.
C	**Mottled saprolite** Aggregation of iron minerals into mottles has occurred. The mottles may be irregular and poorly delimited or they may be more definite, hard with a dark colouring. The hard mottles may range up to 5 cm in diameter.	G	**Massive limonite** These are dark brown or ochre in colour. They may be uniform or banded, but lack the complexity of fabric displayed by the other lateritic material.

Fig. 5. Fabric classification of mantle rock.

Kasewe Hills the amount of organic matter that gives rise to humic and fulvic acids is much greater than in Malawi, so the amount of chemically-mobilized aluminium could potentially be greater; however, high Sierra Leonean rainfalls (Bowden 1980*b*) may dilute the concentrations of such acids and hence correspondingly reduce their mobilization properties; therefore leaving microbially-mediated mobilization as a sound hypothesis for the removal of aluminium from the saprolite.

Although similar textural variations and complexities are found on all of the benches, bulk chemistry of the major elements does indicate that each bench has a characteristic chemistry (Bowden 1987*a*, *b*) (Fig. 7 and Table 2). The trends show that the proportion of silica decreases downslope (with the exception of the lowest bench), whereas the proportion of iron oxide shows the opposite trend. This observation can be interpreted in terms of bench A being richer in fresh bedrock, with the lower surfaces having been subject to a longer period of weathering and leaching. The anomaly encountered at the lowest bench might reflect the bedrock change with the parent material since it is probable that this bench has developed on the Rokel River metasediments. The TiO_2 distribution could suggest that on each successive surface the leaching works on already depleted material derived from higher benches. The proportion of aluminium decreases steadily across the footslope benches. Only limited analysis of trace element variation are available (Table 3) but they do support the idea of differentiation in laterite chemistry between fabrics and benches.

In some laterites there is manganese enrichment (McFarlane 1976), in others, as here, the manganese is leached out; the reason for this manganese behaviour is not understood. It may be governed by biochemistry rather than geochemistry (*sensu stricto*), especially since manganese and iron

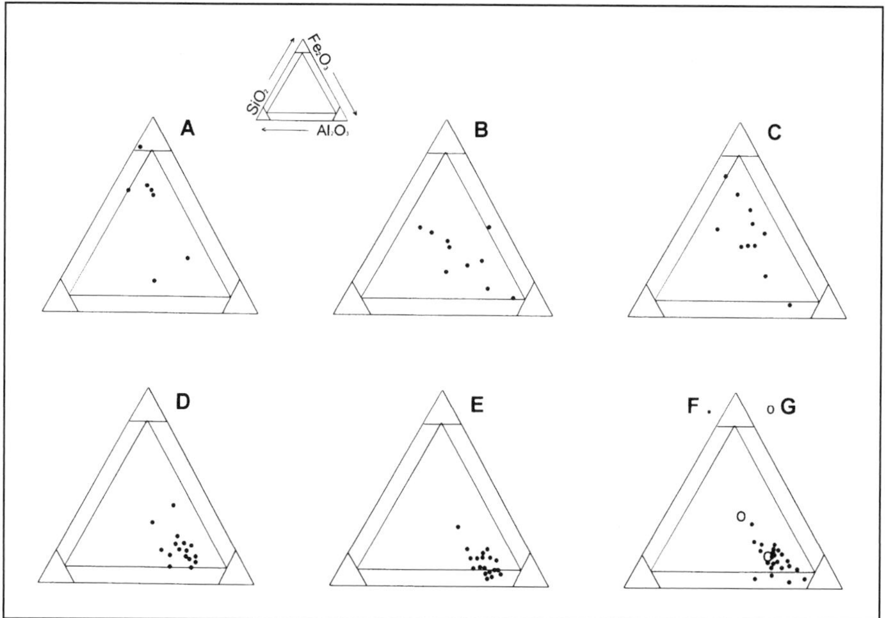

Fig. 6. Major oxide composition of mantle rock by fabric class (Kasewe Hills). Refer to Fig. 5 for explanation of fabric classes.

might be expected to otherwise behave similarly (McFarlane 1976). Chromium is expected to be geochemically immobile and to accumulate (Schellmann 1986). In Sierra Leone, Webb (1958) has demonstrated that chromium is residual and derives from the underlying parent rocks. However, trace element investigation of laterites developed on the African Surface in Malawi (McFarlane *et al.* 1992) do demonstrate a degree of chromium mobility and, as such, chromium can no longer be regarded as a resistant index element (McFarlane *et al.* 1994). In the Kasewe Hills it seems that its similar behaviour supports the Malawian observations, since it shows progressive depletion across the benches. If the weathering was wholly a geochemical process, then it might be expected that the chemical similarities of aluminium, iron and chromium would lead to similar patterns of behaviour during lateritization. However, the patterns are not similar so the evidence might indicate that some biochemical activity was encouraging the mobilization and leaching of chromium.

The chemistry and fabric of the laterites is complex, reflecting both the nature of the parent rocks, and the geomorphological and leaching history of the particular bench. To summarize, it thus appears that the nature of the laterite depends on a number of interrelated factors giving rise to this complexity including: (1) original (bedrock) geochemistry; (2) proximity of the water table; (3) the time available for leaching; (4) the extent to which the material already partially leached is inherited by successively lower surfaces.

A model of bench development

Although the benches are part of a suite of landforms, each bench is characterized by its geochemistry which is, in turn, controlled by its own topographical and hydrological environment. The benches are related, each inheriting material from above and also each providing material for the lower bench from its downslope termination (Fig. 8). This inheritance of material occurs by both geochemical and biochemical processes through mobilization and removal by surface and groundwaters, and mechanically through the development of scarp slopes (breakaways) (Moss 1965). Lateritization progresses in a fashion proposed by McFarlane (1976), so that induration caused either by welding of pisoliths, to produce a continuous phase variety, or by water table fall, which will initiate breakaway retreat with pseudo-karst development above the break and colluvial laterite gravel spreads below. Incorporated in the mantle rock on the edge of bench D on the northeast of the Kasewe Hills, rounded bedrock cobbles may be observed, which have been translocated by surface wash from the hill slopes now some 2 km away. The material inherited by the younger and lower surface will gradually be adjusted to the chemical and leaching conditions on that surface. Further falls in the water table would reinitiate pseudo-karst and regolith development above the initial break and set the processes in motion below. Repetitions of these events will lead to increasingly complex mantle rock profiles and successively lowering benches. Active etchplanation on the plains will reduce their level as the total relief increases (Thomas & Thorp 1985; Thomas 1989*a*, *b*). As the

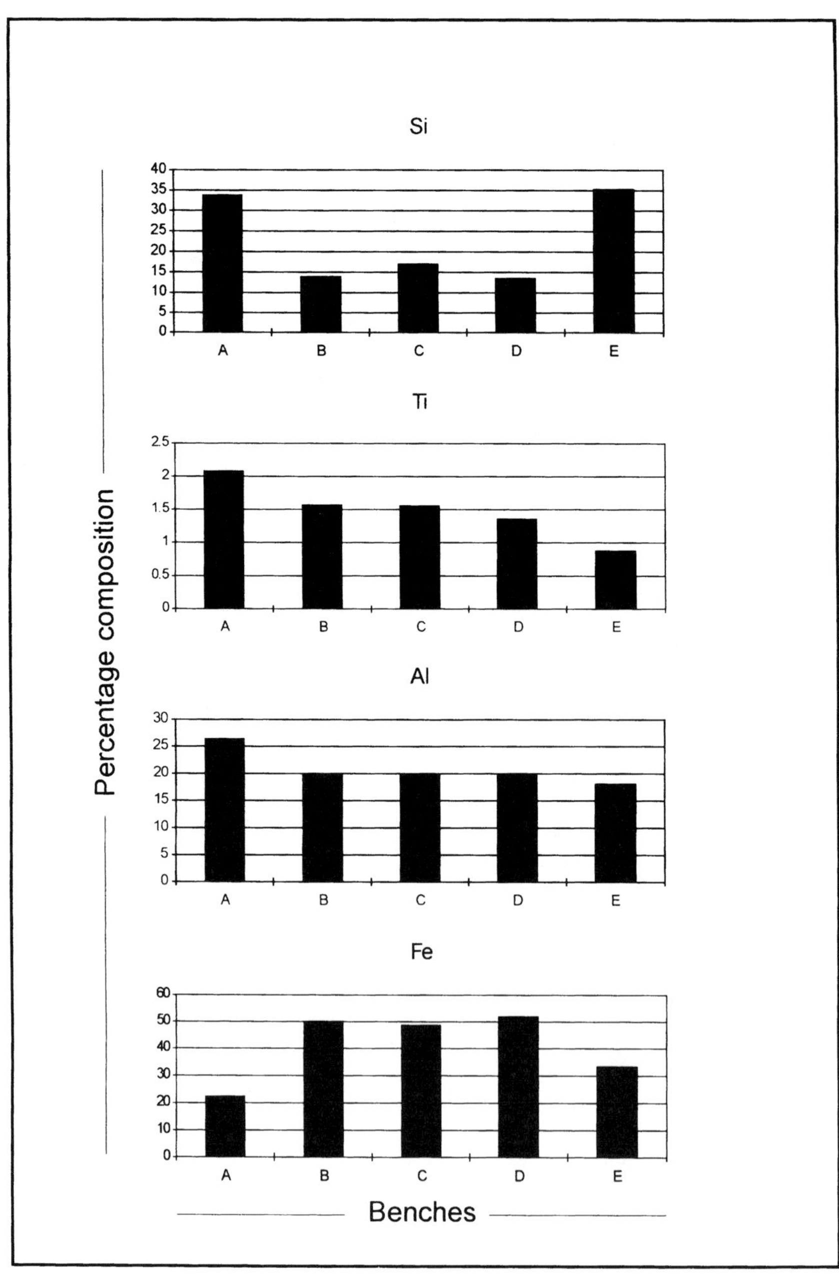

Fig. 7. Major oxide variation across footslopes (Kasewe Hills).

Table 2. *Mean mantle rock composition (&) by bench*

Bench	Number of samples	SiO_2	TiO_2	Al_2O_3	Fe_2O_3	MnO	MgO	CaO	K_2O	P_2O_5
A	3	33.83	2.08	26.33	22.50	0.15	0.19	0.06	0.17	0.19
B	21	13.90	1.57	19.91	50.10	0.15	0.57	0.19	0.13	0.40
C	12	17.05	1.56	19.66	48.60	0.08	0.08	0.03	0.15	0.17
D	24	13.56	1.36	19.62	51.90	0.12	0.09	0.15	0.17	0.31
E	12	35.30	0.88	18.07	33.40	0.04	1.86	0.02	0.86	0.17

base level intermittently falls so the laterites on the footslopes will be subject to phases of formation, destruction and reformation.

Age considerations

It is difficult to date either accurately or absolutely the mantle rock and the associated bench development. There are some time markers, but since it is maintained that the landscape has developed from itself by long continued etchplanation, albeit episodically, an absolute chronology is not possible. Certainly, the landscapes are post-Cambrian since this is the age of the Rokel Rivers metasediments (Allen 1969), and they are post- Ordovician since tillites from that period are locally eroded (Reid & Tucker 1972). The dolerite dykes which have been intruded into the plain are associated with the break up of Gondwanaland (Morel 1979) and this suggest that the plains may have a Mesozoic origin. Thomas (1980) recognized that in Sierra Leone there is '... the possibility of a continuity of landform development over timespans unparalleled in many of the classic areas for geomorphological study' and suggested that much of the landscape may date from the Cretaceous.

A model for the development of the interior plain of Sierra Leone seems to demand a dynamic approach which recognizes that over a long time span '... rough levelling down...' (Pitty 1973) occurs as the landscape undergoes long periods of denudation. This has not resulted in any predetermined sequence of cyclical or other landform evolution, but episodic denudation (Schumm 1979; Thomas 1980) acting on the 'transitory products of continual adaptation' (Büdel 1982) has resulted in the current dynamic landscape (Bowden 1987*a, b*).

Climatic change has no doubt influenced the rates of lateritization and denudation during the time since the upper Mesozoic. Quaternary climatic change is well documented in parts of West Africa (Thomas & Thorp 1980; Sowunmi 1981). For example, the Ogolian dry phase which began some 20 000 years BP resulted in widespread shrinkage of the forest cover. Yet two factors limit the effect of such short-term climatic changes in Sierra Leone. Sierra Leone's location has meant that it has kept close to the heartland of the humid tropics and not suffered such great climate variability as the Savanna zone. The climatic perturbations measured on timescales of 10 000 years or so are insignificant when compared to the 70 Ma years or so since the upper Mesozoic. This period corresponds to Büdel's (1982) 'tropicoid palaeo-Earth', a protracted period of tropical climate, allowing lateritization, and which is still continuing in Sierra Leone.

Table 3. *Mantle rock fabric and trace elements (p.p.m.)*

Mantle rock fabric	Ba	Co	Cr	Mn	Mo	Nb	Pb	Rb	Sr	Th	U	V	Y	Zn	Zr
Fabric B, weathered bedrock	69	11	180	675	3.9	16.3	24	16	19	17	< 3	586	29	6	213
Fabric E, pisolithic/vermiform	180	56	430	4150	7.7	8.5	60	11	15	9	12	1599	5	40	148
Fabric F, vermiform	35	< 5	316	203	20.2	8.9	7	8	14	17	14	1901	9	< 2	190

Minor oxide analysis by X-ray fluorescence, G. Hendry, University of Birmingham.

As the breakaway retreats so successively more complex mantle rock profiles develop. Both the profile and landscape develop in tandem.

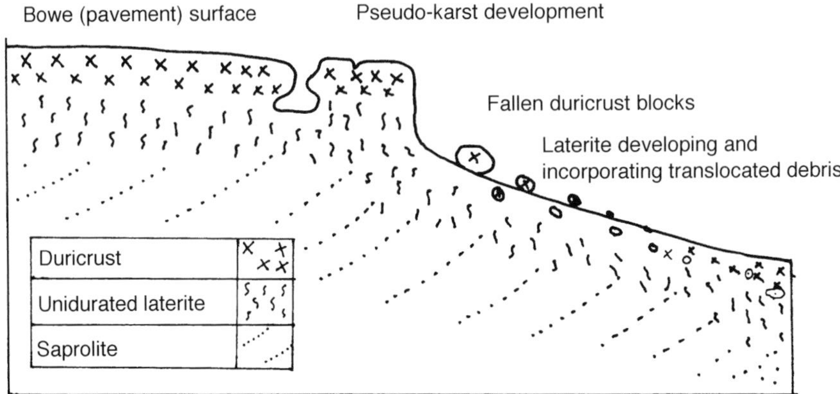

Fig. 8. The development of a footslope bench.

Palaeosurfaces and palaeosols?

Within these footslopes it has been demonstrated that denudational processes are still active. Under the present humid tropical climate of this region, which delivers some 2500 mm of rainfall per year (Bowden 1980*b*), it is likely that leaching and chemical and biochemical weathering are also active. The benches are diachronous landforms, their development and the development of the associated mantle must also be continuous and therefore diachronous. Hence, these 'palaeosurfaces' and 'palaeosols' may be ancient in origin but they are by no means fossilized.

I would like to thank the following: Dr McFarlane for her help in the preparation of this paper; Dr G. Hendry of the School of Earth Sciences, University of Birmingham, who carried out the X-ray fluorescence analysis of selected laterite samples for trace elements; R. J. McGill, University of Dundee, who carried out the X-ray fluorescence analysis of two of the bedrock samples (Kasabere and Malempeh samples, Table 1); and D. Harris, Newman College, who prepared the figures.

References

ALLEN, P. M. 1969. The geology of part of an orogenic belt in western Sierra Leone. *Geologische Rundschau Bel.* **58**.

ALEVA, G. J. J. 1982. Suggestions for a systematic structural and textural description of lateritic rocks. *Proceedings of the Second International Seminar on Laterisation Processes*, Sao Paulo, Brazil.

BOWDEN, D. J. 1980*a*. Sub-laterite cave systems and other pseudo-karst phenomena in the humid tropics: the example of the Kasewe Hills, Sierra Leone. *Zeitschrift fur Geomorphologie NF*, **24**(1), 77–90.

—— 1980*b*. Rainfall in Sierra Leone. *Singapore Journal of Tropical Geography*, **1**(2), 31–39.

—— 1987*a*. *Laterites and pseudo-karst: the geomorphology of the duricrusted footslopes surrounding the isolated hill masses rising above the interior plain of Sierra Leone*. PhD Thesis, University College London.

—— 1987*b*. On the composition and fabric of the footslope laterites (duricrust) of Sierra Leone, West Africa, and their geomorphological significance. *Zeitschrift fur Geomorphologie NF*, Suppl. Bd. **64**, 39–53

BRUCKNER, W. D. 1955. The mantle rock (Laterite) of the Gold Coast and its origin. *Geologische Rundschau*, **43**, 307–327.

BÜDEL, J. 1957. Die 'doppelten Einebnungsflächen in den feuchten Tropen. *Zeitschrift fur Geomorphologie NF*, **1**, 201–288.

—— 1982. *Climatic Geomorphology* (translated from German by L. Fisher and D. Busche). Princeton University Press, Princeton.

HAWKES, D. D. 1970. The Geology of Sierra Leone. *Proceedings from Conference on African Geology*, University of Ibadan, 471–482.

MACFARLANE, A., CROWE, M. J., ARTHURS, J. W., WILKINSON, A. F. & AVCOTT, J. W. 1981. *The Geology and Mineral Resources of Northern Sierra Leone*. Institute of Geological Sciences, Overseas Monograph, HMSO, London.

MACFARLANE, M. J. 1976. *Laterites in the Landscape*. Academic, London.

—— & BOWDEN, D. J. 1992. Mobilization of aluminium in the weathering profiles of the African surface in Malawi. *Earth Surface Processes and Landforms*, **17**, 789–805.

——, —— & Giusti, L. 1994. The behaviour of chromium in weathering profiles associated with African surface in parts of Malawi. *In*: Robinson, D. A. & Williams, R. B. G. (eds) *Rock Weathering and Landform Evolution*. 321–339.

——, ——, Watt, F. & Grime, G. W. 1992. Contemporary leaching of the African Erosion Surface in Malawi. *Memoires des Sciences de la Terre*. Ecole de Mines de Paris.

Morel, S. W. 1979. The geology and mineral resources of Sierra Leone. *Economic Geology*, **74**, 1536–1576.

Moss, R. P. 1965. Slope development and soil morphology in a part of S.W. Nigeria. *Journal of Soil Science*, **16**, 192–209.

Pitty, A. 1973. *Introduction to Geomorphology*, Metheun, London.

Reid, P. C. & Tucker, M. E. 1972. Probable late Ordovician marine sediments from Northern Sierra Leone. *Nature*, **238**, 38–40.

Schellmann, W. 1986. A new definition of laterite. *In*: *Laterisation Processes*, IGCP-127. Geological Survey of India, Memoirs, **120**, 1–7.

Schumm, S. A., 1979. Geomorphic thresholds: the concept and its application. *Institute of British Geographers Transactions, New Series*, **4**, 485–515.

Sowunmi, M. I. 1981. Aspects of late Quaternary vegetational changes in West Africa. *Journal of Biogeography*, **8**, 457–474.

Thomas, M. F. 1980. Timescales of landform development on tropical shields – a study from Sierra Leone. *In*: Cullingford, R. A., Davidson, D. A. & Levin, J. (eds) *Environmental Change in the Tropics*. Allen and Unwin, London, 333–354.

—— 1989*a*. The role of etch processes in landform development. I Etching concepts and their applications. *Zeitschrift fur Geomorphologie NF*, **33**(2), 129–142.

—— 1989*b*. The role of etch processes in landform development. II. Etching and the formation of relief. *Zeitschrift fur Geomorphologie. NF*, **33**(3), 257–274.

—— & Thorp, M. B. 1985. Environmental change and episodic etchplanation in the humid tropics of Sierra Leone: the Koidu etchplain. *In*: Douglas, I. & Spencer, T. (eds) *Environmental Change and Tropical Geomorpholgy*, Allen and Unwin, London, 239–267.

Webb, J. S. 1958. Observations of geochemical exploration in tropical terrains. *Symposium de Exploracion Geoquimica, 20th. International Geological Congress*, Mexico City, 1956, **1**, 143–173.

High-altitude palaeosurfaces in the Bolivian Andes: evidence for late Cenozoic surface uplift

L. KENNAN, S. H. LAMB & L. HOKE

Department of Earth Sciences, Parks Road, Oxford OX1 3PR, UK

Abstract: Low relief palaeosurfaces are widely preserved in the Cordillera Oriental of the Bolivian Andes at altitudes between 2000 m and 4000 m in a region up to 100 km wide and 600 km long. Remnants of these nearly horizontal or gently undulating erosion surfaces, from 5 to 2000 km^2 in area, cut across folded Palaeozoic to Cenozoic bedrock. They appear to be erosional pediments formed between 12 and 3 Ma along broad, low gradient, valleys, which are generally undeformed. In detail, the morphology of surface remnants can be used to construct two distinct palaeodrainage basins. Very little of the material eroded during surface cutting appears to have remained within the Cordillera Oriental and must have been carried into a foreland basin, now preserved in the Subandean fold and thrust belt. The palaeovalleys are generally parallel to the dominant north–south structural grain, forming a tortuous low gradient drainage system which flowed into the foreland basin in an east–west distance of < 150 km. Reconstructions of the drainage systems suggest that they have been uplifted *c.* 2 km since their formation. These estimates are consistent with plausible models of uplift of the Cordillera Oriental as a consequence of intense deformation in the Subandean fold and thrust belt since *c.* 10 Ma. However, only since 3 Ma have the surfaces been deeply dissected by drainage cutting more directly across the structural grain. The timing of deep dissection may be partly related to a climate change to wetter and colder conditions.

Regional planation surfaces provide an important datum for unravelling the tectonic evolution of a region. Those preserved at high altitudes in mountain belts are often interpreted as having formed at lower altitudes with subsequent uplift and dissection (Walker 1949; Hollingworth & Rutland 1968; Molnar & England 1990). Thus, planation surfaces may provide clues to the timing and amount of surface uplift, placing important constraints on dynamic processes in the lithosphere, if their origin and evolution are well understood. In this respect, it is important to understand both the processes that caused planation and those that lead to subsequent dissection.

In this paper we describe the geomorphology and geological context of high altitude erosion surfaces which are preserved in a vast region, up to 600 km long and 100 km wide, on the eastern flank of the Bolivian Andes. Geochronological data indicate that the surfaces formed in the interval 12–3 Ma, with most surface cutting probably pre-dating *c.* 9 Ma. They cut across regional shortening structures of Eocene to middle Miocene age and are now at altitudes between 2000 and 4000 m. The presence of widespread horizontal remnants suggests strongly that they have not been subsequently tilted and that the relative altitudes seen today are an original feature. Deep dissection of these surfaces only occurred in the last 3 Ma. We believe that the origin and dissection of these surfaces places important constraints on the timing, amount and nature of surface uplift in the Bolivian Andes.

Servant *et al.* (1989) have outlined the morphology of palaeosurfaces in the central part of the Bolivian Andes, identifying distinct phases of pediment formation and fluvial dissection. However, their work lacks precise chronological data. The morphology of palaeosurfaces in the southern part of the Cordillera Oriental has been described by Gubbels *et al.* (1993), who also present new geochronological data constraining the timing of surface formation and dissection. We have mapped palaeosurfaces over a much wider area of the Cordillera Oriental than previous workers. Here, we integrate our own observations of surface morphology with previous work (Gubbels *et al.* 1993) and with new geochronological data (Kennan *et al.* 1995) to reach a better understanding of how the surfaces developed through time. Finally, we discuss the links between the palaeosurfaces and tectonics and, possibly, climate change in this part of the Andes.

Geological setting

The Central Andes are the result of oblique subduction, since the Cretaceous, of the Nazca (or Farallon) plate beneath the western margin of South America (Pardo-Casas & Molnar 1987). The western edge of the South American plate has been

From Widdowson, M. (ed.), 1997, *Palaeosurfaces: Recognition, Reconstruction and Palaeoenvironmental Interpretation,* Geological Society Special Publication No. 120, pp. 307–323.

thickened from *c.* 35 km to up to *c.* 75 km (James 1971; Wigger *et al.* 1993; Dorbath *et al.* 1993), mainly as a consequence of tectonic shortening, with subordinate magmatic addition (Sheffels 1990; Kennan 1994; Lamb *et al.* 1996). Deformation in the Bolivian Andes has been active since the Eocene, accommodating overall *c.* 330 km of shortening (Kennan *et al.* 1995; Lamb *et al.* 1996). The Bolivian Andes are traditionally subdivided into physiographic provinces (Fig. 1) in which the geological evolution has generally been distinct throughout the Cenozoic.

The Cordillera Occidental is the active volcanic arc, consisting of spaced Miocene and Quaternary dacitic–andesitic volcanoes and thick ignimbrite sheets erupted through a sequence of poorly known older rocks (Avila 1991). Further east, the Altiplano forms a *c.* 200 km wide region of subdued topography at an average altitude of *c.* 3800 m, which has been essentially a region of internal drainage throughout the Cenozoic, confined by the Cordilleras Occidental and Oriental. Erosion of these two ranges has supplied 10 km or more of sediment to various depocentres within the Altiplano (Kennan *et al.* 1995, Lamb *et al.* 1996). East of the Altiplano, the Cordillera Oriental (Fig. 1) forms a rugged region up to 200 km wide, made up of Palaeozoic to Cenozoic strata and igneous rocks, with altitudes ranging between 2000 and 4500 m. The main phase of shortening in the Cordillera Oriental occurred prior to *c.* 12 Ma. Subsequent erosion resulted in the surfaces described in this paper. The eastern flank of the Cordillera Oriental drops abruptly towards the much lower Subandes. This is a thin-skinned fold and thrust belt which deforms foreland basin sediments deposited mainly after *c.* 10 Ma. There has been *c.* 140 km of shortening above a basal décollement in the last 10 Ma (unpublished oil company data, Roeder 1988; Hérail *et al.* 1990; Baby *et al.* 1993; Wigger *et al.* 1993).

Mapping of surface remnants

Scattered remnants of palaeosurfaces are found throughout the eastern part of the Cordillera Oriental (Figs 2 and 3), west of the Subandean fold and thrust belt, and south of the Cochabamba Lineament System (*c.* 17° S), extending as far south as the Argentinian border and beyond. Surface remnants are especially widespread south of Sucre (Fig. 2) where they have been previously studied (Gubbels *et al.* 1993). North of Sucre,

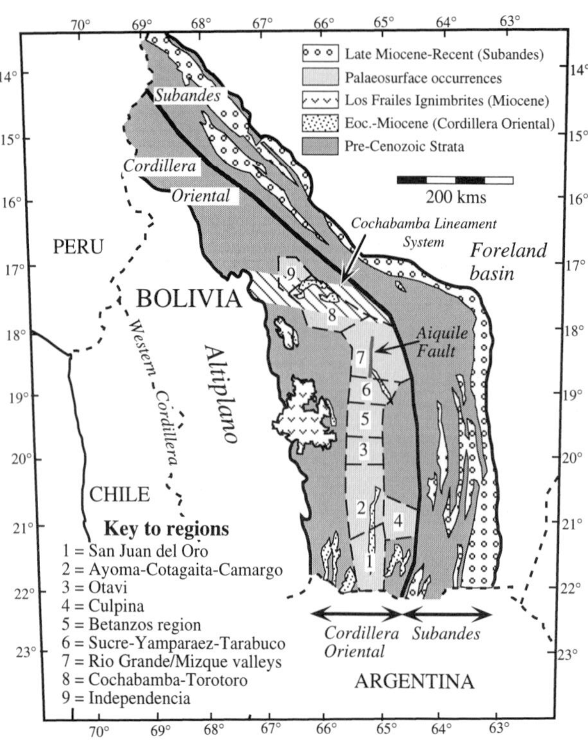

Fig. 1. Outline map of Bolivia showing the major physiographic provinces of the central Andes, outcrop of Pre-Cenozoic and Cenozoic rocks in the Cordillera Oriental and Subandes, and the locations of two major strike-slip zones; the Cochabamba Lineament System and the Aiquile Fault. The area in which high altitude palaeosurface remnants are found is shown in light stipple. The numbered regions are referred to in the text.

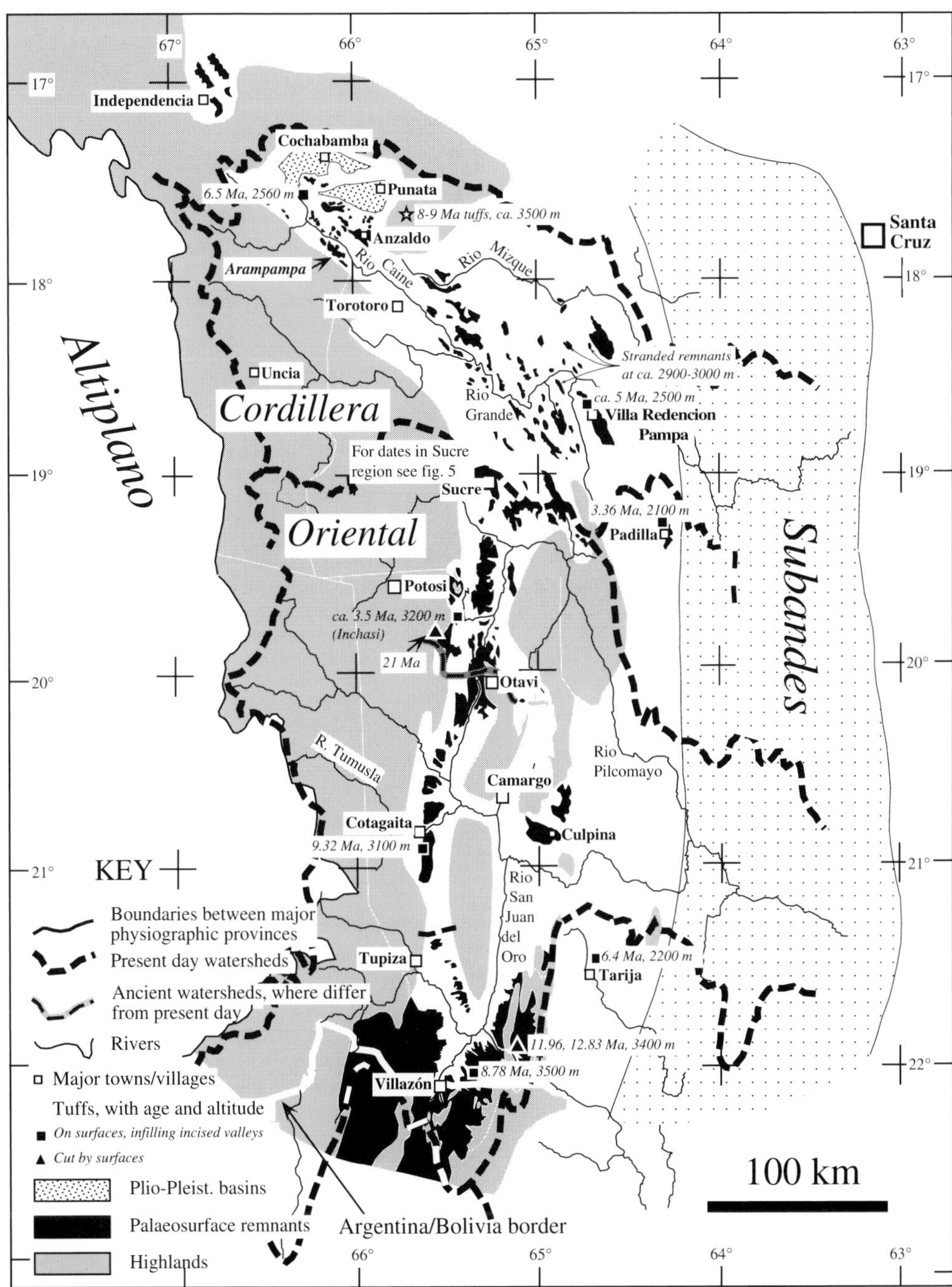

Fig. 2. Detailed map of the Cordillera Oriental showing the preserved palaeosurfaces (black). These clearly define north–south trending, flat-bottomed palaeovalleys, with regions in between up to 1000 m higher (grey). These surfaces drained towards the east into a foreland basin now deformed as the Subandes fold and thrust belt. Since *c.* 3 Ma they have been deeply dissected by rivers such as the Rio Pilcomayo and Rio Grande. The locations of dated tuffs or fossils, with altitude and age, are also shown (data from sources indicated in text).

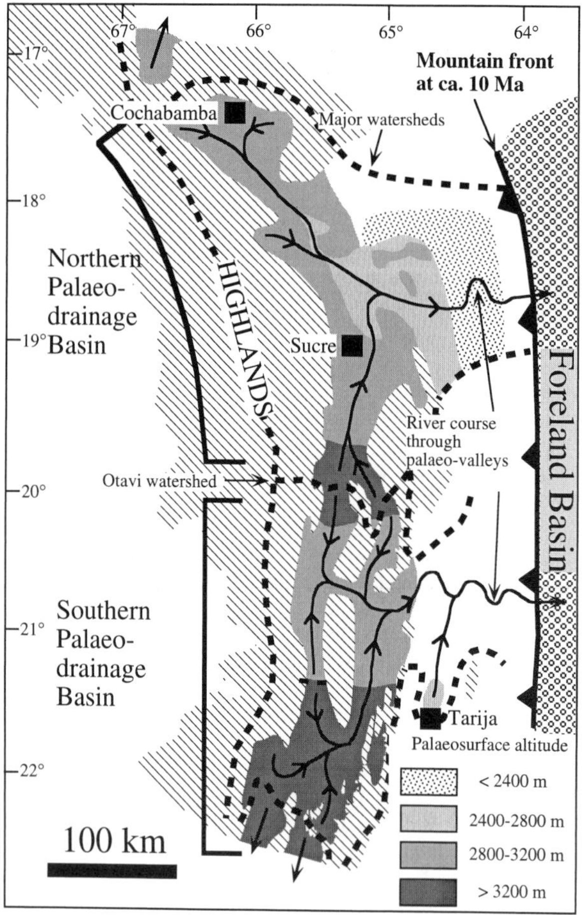

Fig. 3. Summary diagram showing proposed drainage paths (solid black) in the northern and southern palaeodrainage systems, the intervening highlands and watersheds, and general palaeosurface altitude. Present-day altitude clearly drops from *c.* 3600 m upstream to *c.* 2200 m downstream at typical gradients of 1 in 200. The ultimate sediment sink was a foreland basin, near sea level, at *c.* 10 Ma, now deformed as the Subandes. The downstream ends of the drainage paths are speculative because of Pliocene–recent dissection but their mouths are unlikely to lie much below 2100 m. Thus, we believe present-day palaeosurface altitude is the result of *c.* 2 km of uplift since 10 Ma. This estimate accords well with independent estimates from structural models and avoids assumptions associated with palaeobotanical altitude estimates.

individual surface remnants tend to be smaller. However, many of the key observations presented in this paper were made in this area.

We have mapped surface remnants and intervening higher areas using a combination of 1 : 250 000 Landsat Multispectral Scanner images, SPOT multispectral images, aerial photography and topographic maps. Most surface remnants have a very distinctive bright yellow colour on standard false-colour composite satellite images. This seems to correlate, in the field, with a marked reddening of both the bedrock and patchy, thin overlying gravel deposits and the 'bright' appearance is confined to surface remnants at altitudes of 2000–3800 m, providing an easy form of identification. Locally, the 'bright-coloured' covering has been removed by shallow, dendritic fluvial dissection. Surrounding slopes exposed during later dissection are less reddened in the field and much darker on all imagery, regardless of altitude.

The morphology and field relations of the surface were also examined during extensive fieldwork throughout the Cordillera Oriental, and tuffs where mantle surface remnants were collected and dated (Kennan 1994; Kennan *et al.* 1995). The surfaces appear in the field as benches, flat hill tops or plains cut directly across folded Palaeozoic to Cenozoic bedrock, either bare or with a thin covering of gravels, sands and interbedded tuffs. These cover sequences are generally much less than a few tens of metres thick and, in general, the bedrock planation surfaces and cover are considered together when discussing surface morphology. However, we must draw an important distinction between the time at which the surfaces were cut and the age of the cover. In many cases, several millions of years probably elapsed between surface cutting and cover deposition, and there may be no genetic link between the two. Locally, there are cover sequences up to 250 m thick, mostly as fans built up against highlands protruding above a surface.

Nomenclature of palaeosurface remnants

Previous workers have named two palaeosurfaces; the Chayanta Surface at *c.* 4000 m (Walker 1949; Servant *et al.* 1989) and the San Juan del Oro Surface at *c.* 3500–2850 m (Servant *et al.* 1989; Gubbels *et al.* 1993). This paper is concerned with the latter. The *c.* 4000 m Chayanta Surface is defined largely by a 100 km wide summit height accordance in the western part of the Cordillera Oriental (Servant *et al.* 1989). Its appearance in the field is often as low-gradient undulating ground, clearly cut by steeper, rugged gorges probably related to Pliocene and younger dissection (see below). Gubbels *et al.* (1993) considered these surfaces to be late Miocene. However, we believe they are considerably older. For instance, flat ignimbrite shields up to 15 km across overlie low-gradient, undulose topography at *c.* 4000 m in the Uncia–Challapata region (Fig. 2). These are as old as 21 Ma (Schneider 1985 in GEOBOL-SGAB 1992), suggesting the existence of regional early Miocene planation. The traditionally defined San Juan del Oro Surface is a combination of flat and gently undulose surfaces between *c.* 2850 and 3500 m, which cut across tightly folded Palaeozoic to Miocene rocks. However, our work indicates that, especially north of Sucre, there are also widespread and previously unmapped remnants of surfaces as low as *c.* 2000 m. These show the same general range of surface morphologies as the higher surfaces and appear to have formed in the same drainage system and yet they have not previously been mapped together with the San Juan del Oro Surface.

It is clear that a new terminology is required. We believe that the San Juan del Oro Surface, as defined by Gubbels *et al.* (1993), is part of a series of low-gradient fluvial drainage systems which we provisionally refer to under the umbrella term Cordillera Oriental Palaeodrainage Systems and which were responsible for planation. We assume that there has been no regional deformation of the palaeosurfaces and have therefore used the slope and height of the palaeosurfaces to reconstruct two separate palaeodrainage basins, separated by a watershed at *c.* 20° S (Fig. 3). The southern palaeodrainage basin seems to have had an outlet to the foreland at *c.* 21° S, while the much larger northern drainage basin had an outlet at *c.* 18° S. A major present day watershed between the Rio Grande system, draining into the Amazon, and the Pilcomayo system, draining into the Rio de la Plata, runs through the Sucre region, north of the ancient watershed. We suggest this is a result of the Pilcomayo cutting through a major highland (reaching over 4000 m) and capturing the southern part of the original northern drainage system. Geochronological data (see below) suggest that the two drainage basins have been distinct throughout their histories. We describe the palaeodrainage systems in terms of subareas (Fig. 1), which facilitate description and generally divide the systems into upstream and downstream parts.

Southern palaeodrainage basin

Morphology and field relationships

The remains of the southern drainage basin are found between the latitudes 19.75° S and 22.25° S in a region up to 100 km wide in the Bolivian Cordillera Oriental and northernmost part of the Argentinian Puna. It comprises north–south palaeovalleys, up to *c.* 30 km wide and at altitudes between 3800 and 3000 m, surrounded by steep-sided highlands reaching over 4000 m (Fig. 2). There are distinct 'palaeotributaries' which appear to follow the present river systems such as the Rio San Juan del Oro and also the tributaries and main stem of the Rio Tumusla. The heads of these palaeotributaries are separated by gentle watersheds from each other, from low-gradient drainage into salar basins in the Argentinian Puna to the south and from the northern palaeodrainage basin (Fig. 2).

All the palaeotributaries of this palaeodrainage system appear to drop in altitude and converge to the east, at the main outlet to the sedimentary basin in which eroded debris was deposited. In the following sections, the morphology and field relations of the palaeosurfaces are described in detail for various regions (see Fig. 1).

San Juan del Oro region (Fig. 1, region 1). The southernmost parts of the southern palaeodrainage basin, near the Argentinian border, are also the most extensive and best preserved (Fig. 4). Palaeosurfaces are found either side of the Rio San Juan del Oro (Fig. 2), mimicking its U-shaped course. Low gradient (often as low as 1 in 250) plains, up to *c.* 2000 km^2 in area, are broken by discontinuous, low strike-ridges. Surfaces are generally mantled by not more than a few tens of metres of gravel and sand cover except at their margins where gravel fans are banked as steep as *c.* 1 in 20 and up to *c.* 250 m thick. Surface altitude drops from *c.* 3800 m at the upstream margins to *c.* 3400 m at the edges of the deep gorges of the Rio San Juan del Oro and tributaries. The upper reaches of these gorges are marked by dendritic drainage networks. Away from the gorges, the drainage on the surfaces is largely unincised and some parts of the surface still drain into shallow lacustrine basins. In northernmost Argentina, the plains coalesce to

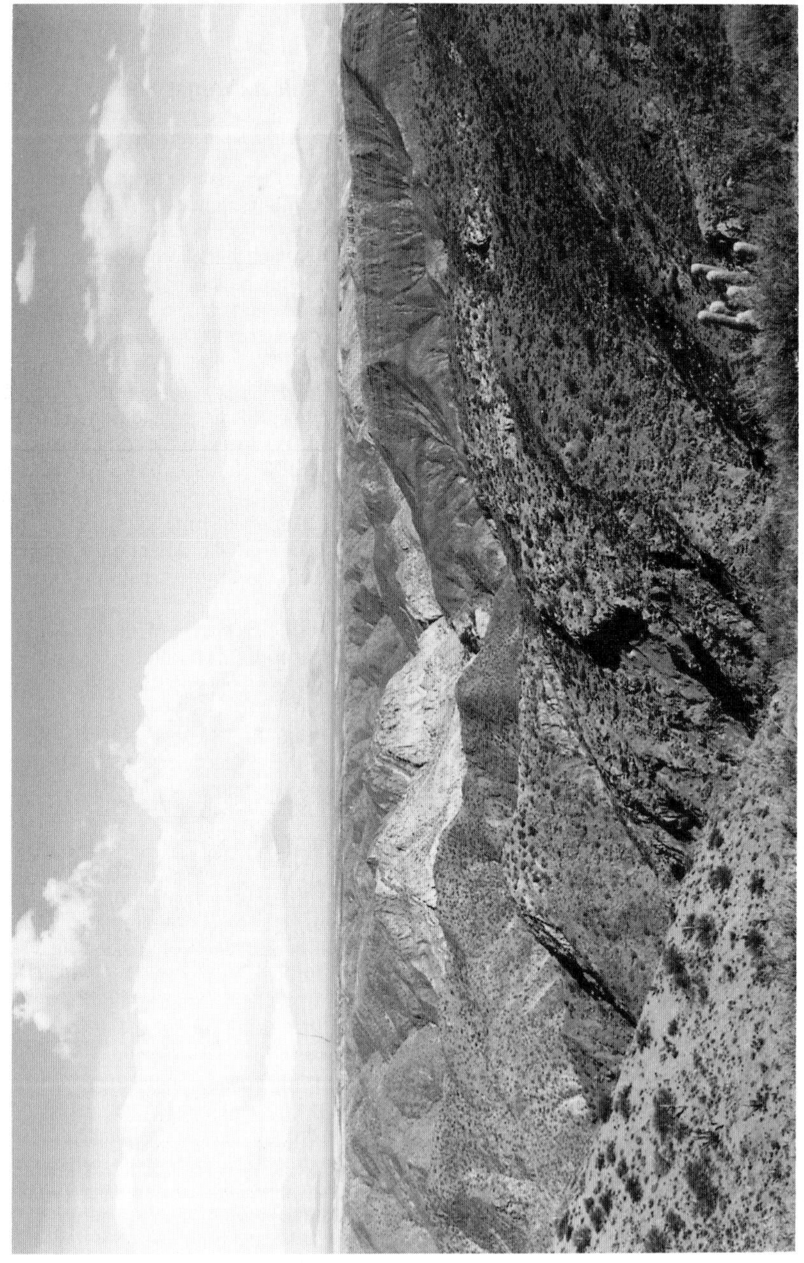

Fig. 4. Photograph looking south across gorge of Rio San Juan del Oro, near Villazón. Here, the palaeosurfaces lie at *c.* 3500 m, are cut nearly flat across underlying Palaeozoic and Cretaceous (bright) bedrock, with some upstanding strike ridges. There is a slight tilt down into the centre of wide palaeovalleys. They are either bare or mantled by only thin gravels (with some tuffs).

define a 100 km wide flat surface which merges with the Puna across a gentle watershed at *c*. 3800 m, near Villazón (Fig. 2).

Ayoma–Cotagaita–Camargo region (Fig. 1, region 2). Further north, in the central part of the southern palaeodrainage basin, erosion surface remnants at *c*. 3000–3200 m define a north–south palaeovalley running through Cotagaita, bounded by steep-sided strike-ridges of Palaeozoic to Cretaceous strata. Although relatively recent erosion has removed the few metres of brightly reflective surface (on satellite images and aerial photographs) material from many remnants, the underlying north–south structural grain has not yet had a significant influence on the developing drainage. Surface altitude drops progressively from both the northern and southern ends of this palaeovalley towards a low point near Cotagaita. A gentle watershed at *c*. 3500 m separates the southern part of this palaeovalley from surfaces along the Rio San Juan del Oro valley. In the northern part of the palaeovalley, now dissected by the Rio Ayoma (see Fig. 5), the surfaces cut across steeply-dipping bedrock, with little or no sediment cover over wide areas. There are no obvious sharp steps in surface height. On the upper slopes of the Rio Ayoma valley there are thin (< 50 m) and essentially flat-lying sequences of gravels which infill shallow pre-surface valleys. The tops of these sequences are planed off at the general surface level and they may represent the product of erosion during the very early stages of surface formation. Close to its upstream margins, gradients are relatively steep, reaching 1 in 40. Surface height then drops much more gradually south towards Cotagaita, with valley-parallel gradients as low as 1 in 250 on the bare bedrock surfaces. Drainage from the surfaces in the Cotagaita region must have continued east towards Camargo along the same general course as the present-day Rio Tumusla (Fig. 2), which cuts through a ridge rising to over 4000 m. There is no other possible outlet at or below 3000–3100 m.

Otavi region (Fig. 1; region 3). Just north of Otavi, the palaeosurfaces reach an altitude of *c*. 3450 m at the foot of a gentle east–west watershed which defines the limit between the southern and northern palaeodrainage basins (Fig. 5). Immediately south of the watershed, topographic contours drawn on the bare bedrock of the palaeosurface clearly show a moderate dip downstream and towards the centre of a palaeovalley, now occupied by the headwaters of the Rio Ayoma. North of the watershed the palaeosurfaces drop gently north from *c*. 3400 m with a gradient of *c*. 1 in 100. The watershed can be traced to the east of the Cretaceous strata of the Otavi syncline. The highest palaeosurfaces in the

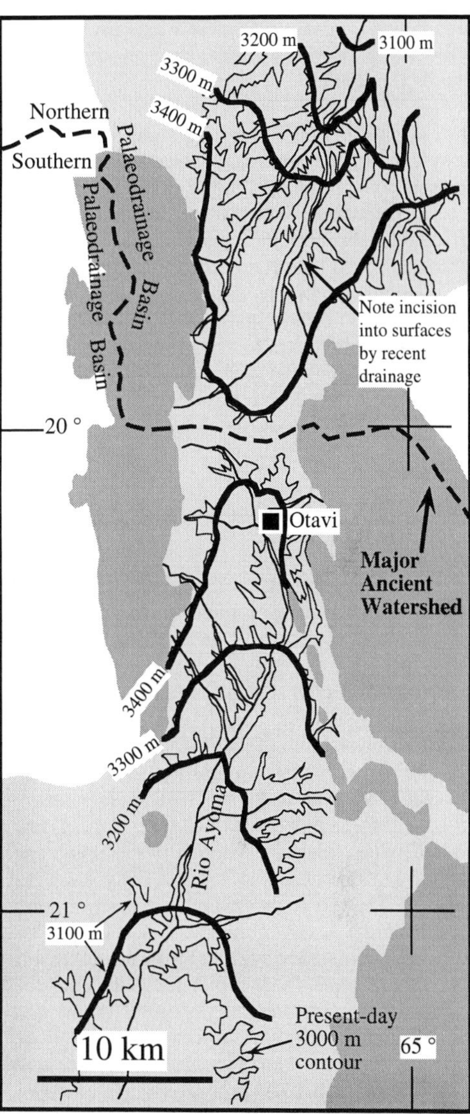

Fig. 5. Outline topographic map of a major ancient watershed, just north of Otavi, between the southern and northern palaeodrainage basins, showing detailed present-day contours and generalized contours of palaeosurface height (bold, black). Palaeovalleys are shown in light grey and intervening high areas, composed of folded Palaeozoic to Palaeogene strata, shown dark grey. Preserved surface remnants are almost entirely devoid of sediment cover. Note that palaeosurface altitude drops gently downstream and into the centre of major north–south palaeovalleys. Recent, finer topographic detail (thin contours) is the result of post-Pliocene incision of the palaeosurfaces by the present river system.

region (Avichuca Pampa, at *c.* 3620 m) are preserved along this part of the watershed, which never drops below *c.* 3500 m. Palaeovalleys drop gently to the north, towards Sucre, and to the south, towards Camargo.

Culpina region (Fig. 1; region 4). All the palaeovalleys within the southern palaeodrainage basin appear to have converged in the east near Culpina, east of Camargo. Here, there are widespread surface remnants at *c.* 3000 m, while narrow ranges rise to near 4000 m to the north and south (Fig. 2). The continuation of the drainage system beyond Culpina is not clear, though we believe that it ultimately continued into the foreland region at *c.* 21° S. Also, there may have been some link with drainage from the Tarija Basin, to the southeast, since at least 6.5 Ma (see below).

Age of palaeosurface remnants, aggradation and dissection

The age of palaeosurface formation is bracketed both by the age of deformed rocks below the surface and younger tuffs and sediments which mantle the surfaces. Tuffs in the Quebrada Honda lacustrine and fluvial sequences (east of Villazón), which infill shallow depressions immediately beneath surfaces south of the Rio San Juan del Oro, near the Argentinian border, have been dated between 11.96 and 12.83 Ma (MacFadden *et al.* 1990). These sediments, which are only a few hundred metres thick, are essentially flat-lying, and may have accumulated at a very early stage of planation. The top of the sediment sequence coincides with the level of nearby surfaces cut on bedrock. Tuffs dated at 9.32 and 8.78 Ma (Gubbels *et al.* 1993) come from sediments overlying the surfaces in the Cotagaita region (*c.* 3100 m) and near Villazón (*c.* 3700 m), respectively (locations shown on Fig. 2). Thus, the complete southern palaeodrainage basin, preserved at altitudes between 3800 and *c.* 3000 m, was formed between 12 and 9 Ma. The northern palaeodrainage basin also probably existed at this time (see below) and there must have been a major watershed between the two. Much of this watershed is still represented by steep-sided highlands rising over 1000 m above nearby surfaces. In the Otavi and Avichuca Pampa regions, however, flat surface remnants are present along the crest of the watershed suggesting that, locally, drainage cut through the highlands and one palaeodrainage system grew at the expense of the other.

The widespread presence of thin and generally undated gravel sequences, resting on the planation surfaces in the southern part of the southern palaeodrainage basin, indicates one or more limited phases of fluvial aggradation prior to dissection. This aggradation may be the result of both changes in the geometry of the drainage system and climate change, which resulted in some sediment being retained in the palaeodrainage basin after planation. There appears to have been little subsequent tectonic disturbance, although faults with vertical displacements generally < 10 m do cut the surface near the Argentinian border (Cladouhos *et al.* 1994).

Flat-lying tuffs from the Yesera Formation in the Tarija Basin, dated at 6.4 ± 0.4 Ma (M. Bonhomme in Troeng *et al.* 1993; GEOBOL-SGAB, Villazón sheet, 1 : 250 000 series) mantle a palaeosurface at *c.* 2200 m. This is *c.* 70 km southeast of Culpina, near where we believe the eastern outlet of the southern palaeodrainage basin may have flowed into the foreland basin. There does not appear to have been young faulting between the Tarija Basin and the San Juan de Oro region (separated horizontally by only *c.* 15 km across strike) suggesting that the downstream end of the southern drainage systems was *c.* 1600 m lower than the upstream end by *c.* 6.5 Ma and that drainage between Culpina and Tarija may have been very sinuous if the low gradients present in the upper part of the drainage system (e.g. Ayoma–Cotagaita region) were maintained. In this case, the palaeovalley downstream of Culpina would have to be at least 120 km long.

Northern palaeodrainage basin

Morphology and field relations

The northern palaeodrainage basin lies between 17.5° S and 19.75° S. Well-defined palaeovalleys at *c.* 2900–3400 m in the south and northwest of the basin converge on a broad area of low relief in the Rio Grande and Rio Mizque Valleys, where there are numerous small and scattered remnants of palaeosurfaces as low as *c.* 2100 m (Fig. 2). Although none of these have previously been mapped as part of the same drainage system as the higher palaeosurfaces, we believe we are justified in doing so. Their morphology and expression on satellite imagery and aerial photographs are similar to the higher surfaces and there are no topographic barriers or very steep steps between the two. The lowest surfaces are found immediately west of the Subandean fold and thrust belt, with higher regions to north and south, suggesting that this palaeodrainage basin had an outlet to the then foreland basin at *c.* 18.5° S. In the following sections, the morphology of surface remnants are described in detail for the regions shown on Fig. 1, starting with the higher upstream parts of the palaeodrainage basin before considering the scattered, lower remnants in the Rio Grande valley.

Betanzos region (*Fig. 1; region 5*). Near the Otavi watershed (Fig. 5), palaeosurfaces cut into Palaeozoic bedrock, with thin or no cover, and drop northwards from *c*. 3450 m to *c*. 3200 m over *c*. 15 km, before levelling out near Betanzos (Fig. 6). Nearby, there are some striking examples of the preserved palaeosurfaces which form extremely flat plains at *c*. 3200 m, with < 20 m relief over distances of > 10 km (Fig. 6). They define a north–south palaeovalley *c*. 60 km long and 30 km wide, bounded by higher regions rising to *c*. 4000 m, with no sign of regional tilt to the east or west. The surfaces cut across Palaeozoic and Cretaceous bedrock and are broken only locally by steep-sided protruding ridges of Cretaceous rocks. They are generally devoid of sediment cover except very locally, such as at Inchasi (Fig. 2) where *c*. 200 m of gravels, sands and tuffs have ponded against a protruding ridge (MacFadden *et al.* 1993).

Sucre–Yamparaez–Tarabuco region (*Fig. 1; region 6*). North of Betanzos, between Sucre and Tarabuco (Fig. 6), we have recognized several distinct surface types, which are described below. The highest palaeosurfaces are preserved east of Tarabuco, where gently undulose plains rise gradually from *c*. 3200 to 3300 m. Further east, there is a still-prominent watershed at *c*. 3600 m. This undulose surface merges to the west with near horizontal plains preserved near Yamparaez. Although similar to the Betanzos plains, here there are several distinct plain levels. The dominant level lies at 3160 m and shows < 10 m relief. There are also significant areas at *c*. 3040 m, and small, steep-sided remnants at 3200 and 3080 m. We have also noted small ($\ll 1$ km^2) very flat surface remnants at *c*. 3000 and 3100 m cut across Palaeozoic bedrock just north of Sucre on the road to Maragua. All these plains are bounded by short, very steep steps and are generally devoid of sediment cover. Sucre city lies at *c*. 2850 m, in a broad, shallow bowl cut into the surrounding flat plains. In places the bowl is steep-sided, but to the east it rises through gentle undulating slopes, mantled with thin gravels and tuffs, to merge with the flat plains near Yamparaez. The Sucre bowl contains a thin (10–50 m) gravel and sand fill with tuffs. The tuffs are locally very gently folded and cut by dextral strike-slip faults with displacements of several metres. This suggests that the surfaces in this region may be locally broken by small strike-slip faults. For instance, a major northtrending fault further north, referred to as the Aiquile fault, may have been active in the Plio-Pleistocene (Dewey & Lamb 1992). However, the lack of any fault scarps crossing preserved bare bedrock surfaces, and the general concordance of surface height across older fault lines, suggests there has not been significant dip-slip motion.

Cochabamba–Torotoro region (*Fig. 1; region 8*). Between Torotoro and Cochabamba, there are numerous palaeosurfaces on either side of the present-day Rio Caine–Rio Tapacari Gorge. This gorge is, in places, nearly 1000 m deep and the

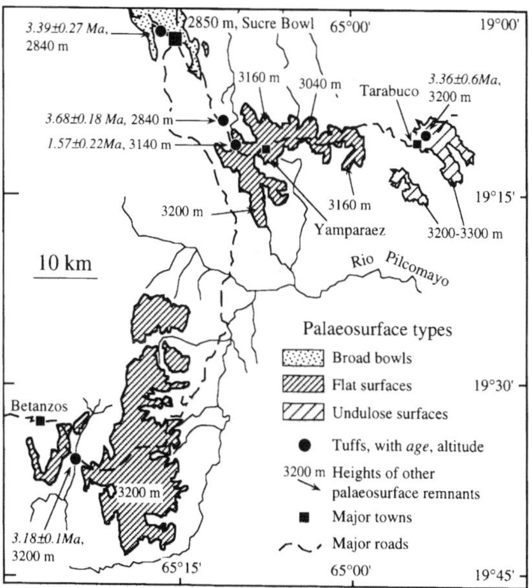

Fig. 6. Map of the preserved palaeosurfaces in the Betanzos and Sucre–Tarabuco regions. Note that higher, gently undulose regions (Tarabuco) grade downstream into flat plains which show distinct, steep-sided benches up to 5 km wide (altitudes of main benches shown). Altitudes and ages of dated tuffs are shown (data from Kennan *et al.* 1996). Near Betanzos, a 3.5 Ma tuff was deposited in a valley incised 150 m into an extensive 3200 m surface, on which deposition was occurring at the same time, clearly indicating the relatively local scale on which transport processes were operating. Over periods of 1–5 Ma, however, only a small fraction of the eroded sediment remained within the palaeodrainage systems, the bulk being carried into the foreland basin.

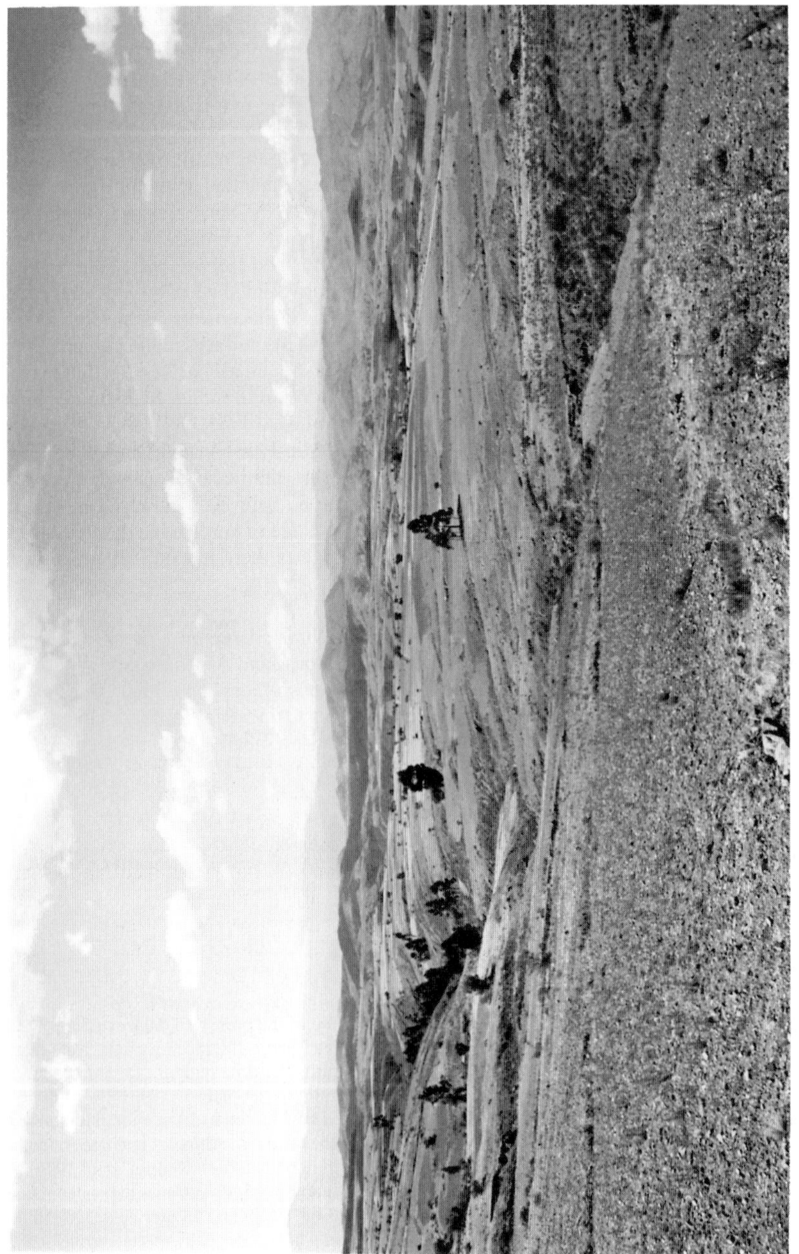

Fig. 7. Photograph looking down onto a large palaeosurface remnant at Arampampa and northwest towards the Cochabamba area. The gently undulose surface, at 3000–3100 m, cuts across lower Palaeozoic strata. Soils covering the surface are commonly only a few centimetres thick. Modest dissection (middle distance) has occurred in response to the deep downcutting of the Rio Caine Gorge, a few kilometres to the north. The surface cuts across the *c.* 19 Ma tuffs of the Parotani Basin but dissection started prior to 6.5 Ma when a tuff filled a palaeovalley cut 350 m into the surface.

surfaces are found close to the gorge, defining a northwest–southeast trending palaeovalley *c.* 20 km wide. The largest surface remnant, at Arampampa, has an area of *c.* 200 km^2 and lies at an altitude of *c.* 3000 m with < 100 m of gentle relief (Fig. 7). Like almost all surface remnants in this part of the palaeodrainage system, the surface is cut directly into folded bedrock, with no sediment cover (Fig. 8). North of Arampampa, palaeosurfaces rise through distinct steps towards an ancient watershed at *c.* 3400 m, preserved just south of Anzaldo. North of the watershed the palaeosurfaces drop towards the Plio-Pleistocene Punata and Cochabamba Basins (Fig. 2). Locally, there are well-preserved horizontal surfaces at 3150, 3050 and 2950 m, with steep steps between them. From here the palaeodrainage probably ran northwest through the area now occupied by the young basins, before swinging back to join the main palaeovalley (Figs 2 and 3). High ground to the east of Punata prevents any other exit. Palaeosurfaces are also patchily preserved to the west of Cochabamba. For instance, near Llavini there are small remnants of flat benches at 3150 and 2950 m which possibly once formed part of more extensive flat plains. Nearby, Cerro Llavini and numerous other steep-sided hills in the area have gently undulose tops at altitudes between 3000 and 3200 m, which show up as bright patches on satellite images.

Rio Grande and Rio Mizque Valleys (*Fig. 1; region 7*). The higher surfaces in both the Cochabamba and Sucre regions gradually drop downstream towards numerous lower surfaces, exposed along the courses of the Rio Grande and Rio Mizque where these rivers cut east across the structural grain of the Cordillera. These surfaces have not previously been described nor have they been related to the higher San Juan del Oro surfaces. Distinctive bevelled ridges and areas of low relief, all with a characteristic brightness on satellite images, occur at *c.* 2750, 2600–2500, 2400 and 2100–2200 m, with altitudes dropping consistently to the east (Figs 2 and 3). Even as far east as Villa Redención Pampa (Figs 2 and 3), there are some remnants of higher surfaces and bevelled strike-ridges at *c.* 2900 m. None of the height changes between adjacent palaeosurface remnants seems to be due to dip-slip faulting. No faults cut the surface remnants and remnant height remains constant across the major north–south Aiquile Fault. In general, dissection in this region has been more intense than further west.

Independencia region (*Fig. 1; region 9*). The most northerly palaeosurface remnants are found near Independencia, *c.* 70 km northwest of Cochabamba. They may be part of a third palaeodrainage system and are described in this section only for the sake of completeness. A marked watershed, rising to 5000 m, separates them from palaeodrainage in the Cochabamba region. Although there is significant Plio-Pleistocene displacement on the ESE trending Cochabamba Lineament System (Fig. 1) it is unlikely to have created a watershed of this height. The Independencia palaeosurface remnants are gently sloping valley shoulders, up to 1 km wide, at *c.* 3000 m, above the *c.* 1000 m deep gorges of the Rio Ayopaya and Rio Santa Rosa. The shoulders have the same characteristic bright reflectance on satellite images as palaeosurface remnants to the south. Substantial high late Cenozoic uplift and erosion north of the Cochabamba Lineament System, in the Cordillera Real and northern part of the Cordillera Oriental, may have obliterated most of the palaeodrainage basins in this region. Hérail *et al.* (1995) have recently described a *c.* 8 Ma valley filling sequence lying at *c.* 1000–2000 m on the eastern side of the Cordillera Real. The connection of these strata to the palaeosurfaces we have described remains unclear.

Age of surface remnants and dissection

There is little direct constraint on the timing of planation in the northern palaeodrainage basin. Clearly, the surfaces post-date 21 Ma volcanics southeast of Potosi (Grant *et al.* 1979). Immediately west of Potosi lies the 7–12 Ma Los Frailes ignimbrite shield (Fig. 1: dates in Kennan *et al.* 1995; Grant *et al.* 1979). Although possibly related 8–9 Ma tuffs are found far to the northeast, near Punata (Fig. 2), none are found on any of the surface remnants we have examined. This suggests that surface cutting and cleaning in this area may have continued beyond *c.* 7 Ma and is probably younger than in the southern palaeodrainage basin. A 6.5 Ma tuff from a valley cut 450 m into this surface near Parotani, places an upper age limit on surface planation near Cochabamba and indicates some downcutting somewhat earlier than in the Sucre region (Kennan *et al.* 1995).

Possible late Miocene (5–6 Ma) mammal fossils have been reported from thin deposits on a surface remnant at *c.* 2500 m (Muyu Huasi or Villa Redención Pampa; Marshall & Sempere 1991). Post-2.5 Ma fossils of cervidae or camellidae are absent so the deposit is certainly older than middle Pliocene. The same authors report an age of 3.36 ± 0.3 Ma for a tuff at Padilla, at 2100 m. Thus, by *c.* 5 Ma palaeosurfaces were being cut at *c.* 2500 m in the easternmost part of the northern palaeodrainage basin, and by *c.* 3 Ma they were

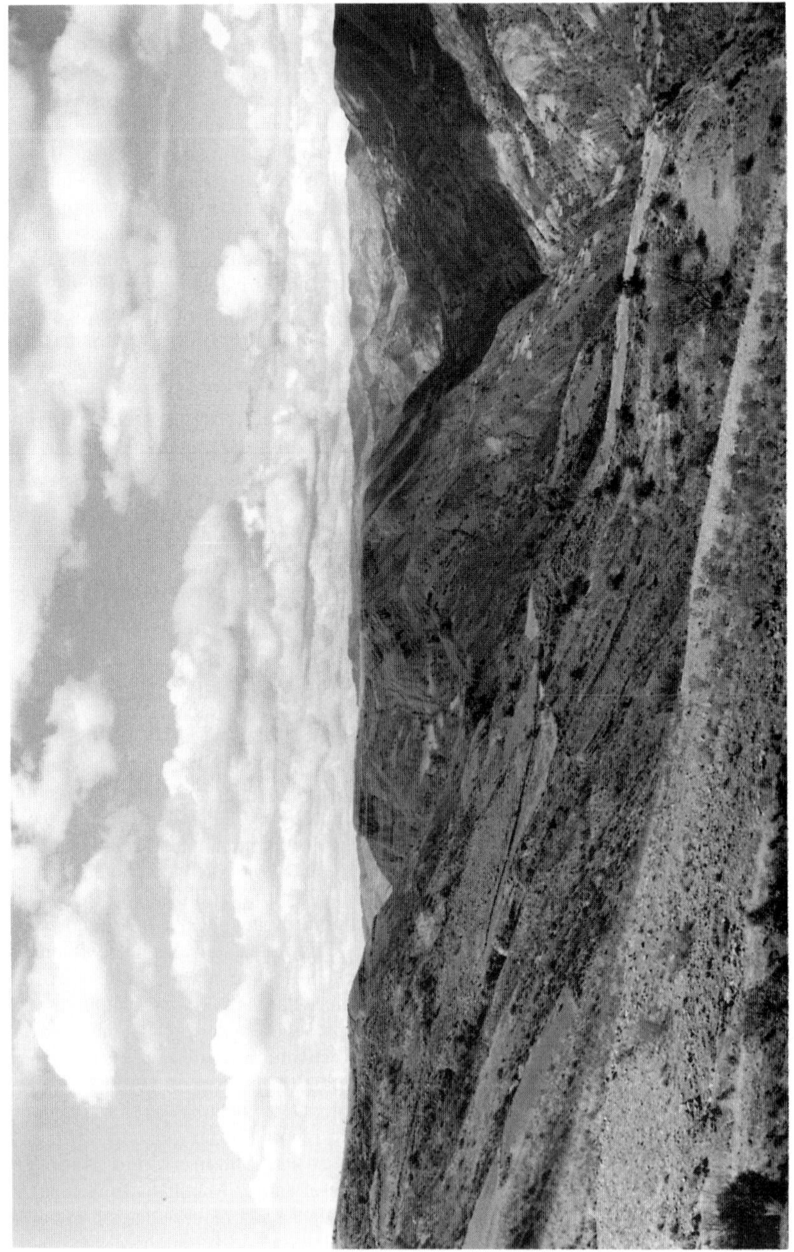

Fig. 8. Photograph looking southeast (approximately downstream along the palaeovalley) from Arampampa towards Limon Pampa, a small isolated palaeosurface remnant. Note that the surface is cut directly onto gently folded Palaeozoic strata. The extreme flatness of the surface can clearly be seen to be an erosive effect and not a result of aggradation of cover gravels. Note also that throughout this part of the Eastern Cordillera (see Fig. 2) there is little relief above the general palaeosurface level.

being cut as low as 2100 m, while nearby there are remnants of *c.* 2900 m palaeosurfaces (flat-topped hills and bevelled strike-ridges). This suggests that between *c.* 10 and 3 Ma the base level at the eastern end of the northern palaeodrainage system was lowered from *c.* 2900 to *c.* 2100 m (Fig. 9). Such base-level drop may account for the pre-6.5 Ma incision observed near Cochabamba. However, the rate of any incision prior to *c.* 3 Ma was much lower than that subsequently, when gorges up to 1000 m deep were cut.

There are numerous dates for sediments or tuffs which mantle the palaeosurfaces in the 1.5–4 Ma age range. These provide an upper limit for planation, as well as constraining the onset of deep dissection which must largely post-date these deposits. For instance, the shallow bowl around Sucre is mantled by a 3.5 Ma tuff which is very

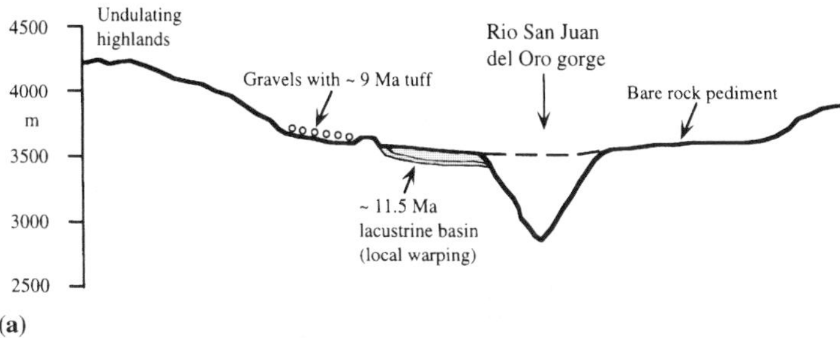

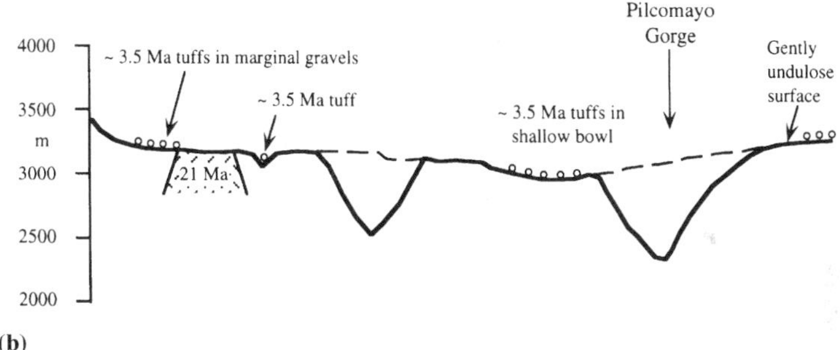

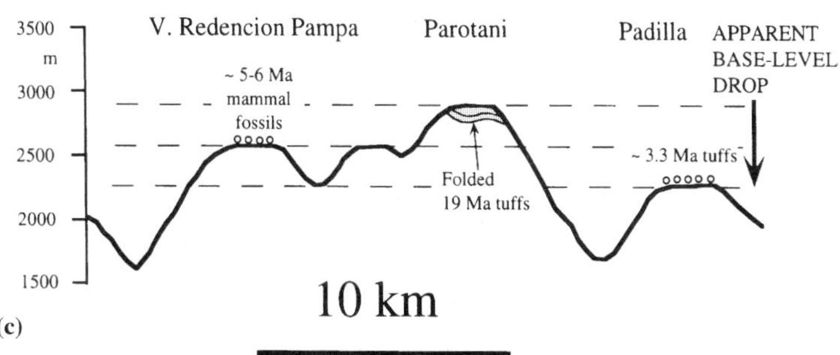

Fig. 9. Schematic approximately east–west cross-sections across the palaeodrainage systems in: (**a**) upstream (e.g. Rio San Juan del Oro); (**b**) midstream (e.g. Sucre–Tarabuco); and (**c**) downstream (e.g. Villa Redención Pampa–Padilla) regions. The near-flat surfaces define broad palaeovalleys cut by steep-sided Plio-Pleistocene gorges. Relationships to dated deposits are shown. Note that in the downstream region there may have been a base-level drop of *c.* 800 m between ?10 and 3 Ma. Subsequent dissection has been much more rapid and widespread. Small, *c.* 50–100 m, steps between discrete surface levels in midstream and upstream regions suggest that between 10 and 3 Ma base-level drops did not significantly modify these areas.

gently folded and faulted. Near Yamparaez, southeast of Sucre, a 1.5 Ma tuff (Kennan *et al.* 1995) in a shallow valley cut into the 3160 m plain indicates that large parts of the plains may have remained undissected until the Pleistocene. Further east, near Tarabuco, surfaces are cut by a moderately dense dendritic network of shallow channels which contain a *c.* 3.5 Ma tuff (Kennan *et al.* 1995). At Inchasi, east of Potosi (Fig. 2), 1–200 m of sands and gravels resting on the surface have yielded 3–4 Ma old mammal fossils (MacFadden *et al.* 1993). However, only 8 km to the north, we have found a 3.5 Ma tuff filling a valley incised 150 m into the 3200 m surface (Kennan *et al.* 1995), suggesting that by 3.5 Ma parts of the surface were being moderately dissected, probably related to drainage onto the lower surfaces nearer Sucre.

Tectonic controls on the northern palaeodrainage basin

The major ESE trending faults of the Cochabamba Lineament System have been active since at least the early Miocene (Kennan 1994). They may have directly controlled the line of the Cochabamba–Torotoro palaeovalley, and the position of the eastern outlet of the northern palaeodrainage basin, by providing a zone of more easily eroded rock. In contrast to the southern palaeodrainage system, where surfaces and sediment cover appear to have been largely 'fossilized' since *c.* 9 Ma, base level in the eastern part of the northern drainage basin seems to have slowly dropped after 9 Ma, resulting both in the continued cleaning of palaeosurfaces into the late Miocene and a greater range of palaeosurface heights by *c.* 3 Ma, ranging from *c.* 3400 to 2100 m. In this way, higher surfaces in the east became stranded. The base-level drop was possibly a result of the normal faulting on the Cochabamba Lineament System. This faulting was probably kinematically related to a marked divergence in shortening direction around the bend in the Bolivian Subandes, resulting in tangential extension in the Cochabamba region and further east (Kennan 1994). There is also some direct evidence for post-palaeosurface deformation. For instance, sequences mantling the surface near Sucre are gently folded and faulted (see above). Also, 15 km south of Cochabamba, a *c.* 10×4 km palaeosurface remnant is warped downwards into the Parotani Basin, which has been a local, fault-controlled depocentre since *c.* 20 Ma (Kennan 1994).

Discussion

Broad, low-gradient valleys lying between strike-parallel ranges 500–1000 m higher define at least two major palaeodrainage basins in the Cordillera Oriental (Fig. 3). These appear to have reached more or less their present form by the late Miocene and have not been subjected to later tectonic tilting – large areas are remain very near horizontal. Thus, we believe that we can use present-day observed gradients to reconstruct the drainage direction and intervening watersheds. In general, the landscape at this time probably looked very much like the northernmost part of the Puna along the Bolivia–Argentina border. Topography was subdued and, although there must have been some lowering of base level in the downstream parts of the northern palaeodrainage basin, there is no evidence for steep-sided, deep gorges such as those that exist today (Fig. 9). The early stages of this palaeodrainage pattern may have been slightly different to that which subsequently became established prior to Plio-Pleistocene deep erosion. Firstly, perfectly planar palaeosurfaces are found right on watersheds between the two palaeodrainage basins suggesting one grew at the expense of the other once major high areas between the two were breached. Secondly, the Rio Pilcomayo, part of the southern drainage system, has cut back through a major high region southeast of Sucre, capturing drainage in the Betanzos region that probably once drained north towards the Rio Grande as part of the northern palaeodrainage system.

We can crudely estimate the volume of sediment that must have been eroded during the cutting of the palaeovalleys. The *c.* 1000 m height of the intervening highlands suggests this thickness was eroded from the palaeodrainage system when it was being cut, giving a total sediment volume of *c.* $1–2 \times 10^4$ km^3 for both north and south palaeodrainage systems. We suggest that planation was the result of progressive widening of valleys as the low-gradient drainage system cut predominantly laterally, transporting weathered rock from the valley sides downstream in a low-gradient drainage system. However, a striking feature is the lack of any sediment sinks within both the northern and southern palaeodrainage basins sufficient to account for the estimated volume of material eroded during its formation. Only locally are there more than a few metres of sediment on surface remnants. Also, there is no indication of a major hiatus in deposition between *c.* 12 and 3 Ma (Coudert *et al.* 1993) in the Subandean zone, when sediment could only have been coming from the Cordillera Oriental. There has been *c.* 3 km of sediment deposited in the Subandean zone and foreland basin in the last 10 Ma (unpublished oil company data).

The above discussion suggests that the palaeodrainage basins could not have formed in a system of internal drainage. This stands in contrast to much of the Puna and Altiplano, where the present-day salar and lake basins are deep, long-lived sediment sinks, often structurally controlled. We can also rule out the possibility that sediment may have been removed into the Altiplano basins to the west. The intervening watersheds are too high and there is no evidence for significant vertical displacements between the Altiplano and palaeodrainage basins since the middle to late Miocene. In this respect, we believe it is significant that the palaeosurfaces do not appear to be tectonically tilted. Thus, the generally eastward gradient of the palaeodrainage systems strongly suggests that the sediment eroded during surface formation was transported out into a foreland basin, which is now part of the Subandean fold and thrust belt (Fig. 3). Our estimates of eroded sediment volume are sufficient to provide several hundred metres of debris to the basin adjacent to the mountain belt during the late Miocene.

The extensively preserved upstream parts of both northern and southern drainage basins clearly show gradients of *c.* 1 in 125 in their upper parts, dropping rapidly downstream to as low as 1 in 200–250. This is comparable with gradients of erosional reaches of many modern rivers draining ancient mountain belts and is much steeper than aggradational reaches of rivers such as the Rhine (Neils Hovius, pers. comm.). Also, in the Subandean zone, marine sediments were being deposited at *c.* 10 Ma (Yecua formation, Marshall & Sempere 1991). Thus, if the gradients in the palaeodrainage basins were maintained downstream towards the Subandes, which were at or near sea level, the palaeosurfaces must have been at significantly lower altitude than they are now. The downstream distance from the easternmost remnants to the Subandes was probably somewhat longer than the present cross-strike distance, taking into account palaeodrainage sinuousity and shortening at the back of the Subandean fold and thrust belt. Given reasonable estimates for this distance, we suggest that at *c.* 10 Ma the palaeodrainage basins in the Cordillera Oriental were probably between 2 and 2.5 km lower at *c.* 10 Ma than they are today (Fig. 10), and that they have been gradually uplifted since then.

We believe that the deformation in the Subandean zone since 10 Ma provides a plausible mechanism for the uplift described above. Thin-skinned shortening in the Subandes accommodated *c.* 140 km of underthrusting of the Brazilian Shield beneath the Cordillera Oriental (Roeder 1988; Sheffels 1990; Baby *et al.* 1993). The precise effect of this underthrusting on uplift in the Cordillera Oriental is unclear. However, we can crudely estimate the effect if we assume that the Cordillera Oriental behaved as a rigid block which was driven up a basal décollement inclined at a similar gradient to the basal décollement in the Subandean zone itself (unpublished oil company data; Wigger *et al.*

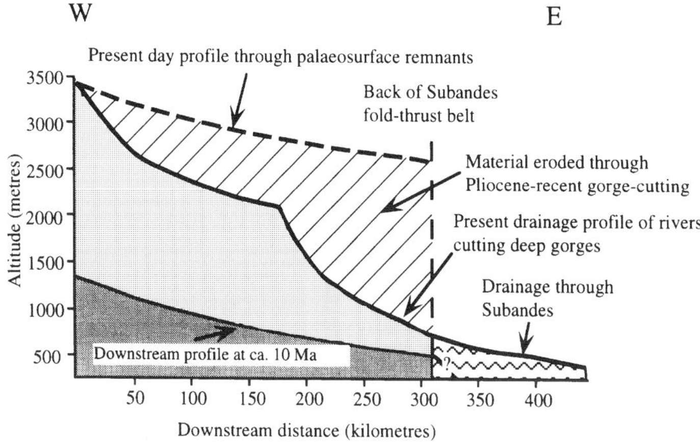

Fig. 10. Summary diagram showing proposed uplift of the palaeodrainage systems (based on real altitude data along the Rio Tumusla–Rio Pilcomayo Valleys). Palaeosurface remnants clearly define a drainage system with a downstream (back of Subandes) end now at *c.* 2000 m. This fed sediment into the foreland basin and must have been close to sea level at *c.* 10 Ma. As it was uplifted, the frontal portion flowing through the Subandes must have steepened but did not cause much upstream dissection. The present gradient of the floor of the Rio Tumusla–Rio Pilcomayo Valley is also shown. The prominent knickpoint may reflect upstream migration to the steep frontal portion of the palaeodrainage system.

1993). Watts *et al.* (1995) have shown that the Brazilian Shield in this region has high flexural rigidity. In this case, we might expect the surface uplift as a result of underthrusting to be more than that expected for full Airy isostatic compensation, possibly by a factor of two (Lamb & Vella 1987; Lamb & Bibby 1989). Thus, assuming a basal slope between 2 and 3° and ignoring the effects of erosion, 140 km of underthrusting would result in between 4 and 7 km of crustal thickening, and *c.* 2 km of surface uplift. This compares well with the estimated average surface uplift, deduced above from the elevations of the palaeodrainage basins. Other calculations, assuming homogeneous crustal thickening beneath the Cordillera Oriental as a consequence of underthrusting (cf. Isacks 1988), give similar estimates of surface uplift. Erosional unloading of the Cordillera Oriental by post-Pliocene dissection may contribute a further few hundred metres to the uplift of the palaeosurfaces.

The average shortening rate in the Subandes over *c.* 10 Ma, of *c.* 10–13 mm yr^{-1}, is about the same as present day rates deduced from foreland seismicity (Dewey & Lamb 1992). If this rate has been fairly constant, then the palaeosurfaces were probably uplifted to within a few hundred metres of their present altitude by 3 Ma (Fig. 10), and the downstream gradient from the easternmost surface remnants into the foreland basin was probably much steeper than that further upstream. Late drainage modification may reflect the effects of a late Cenozoic climate change or it may be that only in the last 3 Ma was the drainage system sufficiently perturbed for the entire system to be modified.

Conclusions

Between *c.* 12 and 3 Ma, an extensive system of valleys with marginal pediments developed within the Cordillera Oriental in a vast region 600 km long and 100 km wide. The lack of sediment sinks within the palaeodrainage system suggests that these valleys drained directly into the then foreland basin, now deformed as the Subandean fold and thrust belt. Projecting drainage gradients downstream indicates *c.* 2–2.5 km of uplift relative to the foreland since *c.* 10 Ma, providing a direct estimate of surface uplift which avoids many of the pitfalls of interpreting palaeoaltitudes from botanical data and compares favourably with uplift estimates from various structural models. We suggest that rapid dissection since *c.* 3 Ma most probably results from surface uplift related to Late Miocene and younger Subandean deformation and the effects of Pliocene climate change.

These palaeosurfaces also have important implications for reconstructing the tectonic evolution of the Cordillera Oriental. They can be used to bracket distinct early and late Miocene–Pliocene phases of movement on the Cochabamba Lineament System. Also, several recent discussions of Subandean structure (Roeder 1988; Coudert *et al.* 1993) have suggested that there may have been major out-of-sequence thrusting on faults at the back of the Subandes, and cross-sections commonly show ramps beneath these thrusts. However, the complete lack of regional palaeosurfaces tilt is inconsistent with late, ramp-related folding. The geomorphology of the Cordillera Oriental must be considered when attempting to balance cross-sections.

Palaeosurfaces like these in Bolivia may be relatively common features in many mountain belts. Unfortunately, they are likely to be short-lived. They may, however, record some features of the tectonic history of a mountain belt, such as the surface uplift history, which are not easy to deduce by other means. Examples in Bolivia, Argentina and elsewhere are clearly worthy of greater attention.

This work was supported by a grant from BP held by Professor J. F. Dewey, a Royal Society Research Fellowship (S.H.L.), Shell and British Council Studentships (L.K.) and an Austrian Academy of Sciences (APART) Research Fellowship (L.H.). We acknowledge helpful reviews by Phil Allen and Steve Flint and valuable discussion with Neils Hovius and Colin Stark.

References

AVILA, W. 1991. *Petrologic and Tectonic Evolution of the Cenozoic Volcanism in the Bolivian Western Andes.* Geological Society of America Special Paper, **265**, 245–257.

BABY, P., GUILLER, B., OLLER, J., HERAIL, G., MONTEMURRO, G. & ZUBIETA, D. 1993. Structural synthesis of the Bolivian Subandean Zone. *Third International Symposium on Andean Geodynamics, Oxford (extended abstracts),* 159–162.

CLADOUHOS, T. T., ALLMENDINGER, R. W., COIRA, B. & FARRAR, E. 1994. Late Cenozoic deformation in the Central Andes: fault kinematics from the northern Puna, northwestern Argentina and southwestern Bolivia. *Journal of South American Earth Sciences,* **7**, 209–228.

COUDERT, L., SEMPERE, T., FRAPPA, M., VIQUIER, C. & ARIAS, R. 1993. Subsidence and crustal flexure evolution of the Neogene Chaco foreland basin. *Third International Symposium on Andean Geodynamics, Oxford (extended abstracts),* 291–294.

DEWEY, J. F. & LAMB, S. H. 1992. Active tectonics of the Andes. *Tectonophysics,* **205**, 79–95.

DORBATH, C., GRANET, M., POUPINET, G. & MARTINEZ, C. 1993. A teleseismic study of the Altiplano and the Eastern Cordillera in Northern Bolivia: new constraints on a lithospheric model. *Journal of Geophysical Research*, **98**, 9825–9844.

GEOBOL-SGAB. 1992. Hoja Challapata (6237). Carta Geologica de Bolivia. 1 : 100 000.

GRANT, J. N., HALLS, C., AVILA SALINAS, W. & SNELLING, N. J. 1979. K–Ar ages of igneous rocks and mineralisation in part of the Bolivian tin belt. *Economic Geology*, **74**, 838–851.

GUBBELS, T. L., ISACKS, B. L. & FARRAR, E. 1993. High-level surfaces, plateau uplift, and foreland basin development, Bolivian central Andes. *Geology*, **21**, 695–698.

HÉRAIL, G., SHARP, W., VISCARRA, G. & FORNARI, M. 1995. La edad de la formacion Cangalli: Nuevos datos geocronologicos y su significado geologico. *Memorias del XI Congreso Geologico de Bolivia*, 361–365.

——, BABY, P., LOPEZ, M. *et al.* 1990. Structure and kinematic evolution of Subandean thrust system of Bolivia. *Second International Symposium on Andean Geodynamics, Grenoble (extended abstracts)*, 179–182.

HOLLINGWORTH, S. E. & RUTLAND, R. W. R. 1968. Studies of Andean uplift, part 1, Post-Cretaceous evolution of the San Bartolo area, north Chile. *Geological Journal*, **6**, 49–62.

ISACKS, B. L. 1988. Uplift of the Andean Plateau and Bending of the Bolivian Orocline. *Journal of Geophysical Research*, **93**, 3211–3231.

JAMES, D. E. 1971. Andean crustal and upper mantle structure. *Journal of Geophysical Research*, **76**, 3246–3271.

KENNAN, L. 1994. *Cenozoic tectonics of the central Bolivian Andes*. DPhil Thesis, Oxford.

——, LAMB, S. H. & RUNDLE, C. C. 1995. K–Ar dates from the Altiplano and Cordillera Oriental of Bolivia: Implications for the Cenozoic stratigraphy and tectonics. *Journal of South American Earth Sciences*, **8**, 163–186.

LAMB, S. H. & BIBBY, H. M. 1989. The last 25 Ma of rotational deformation in part of the New Zealand plate-boundary zone. *Journal of Structural Geology*, **11**, 473–492.

—— & VELLA, P. 1987. The last million years of deformation in part of the New Zealand plate-boundary zone. *Journal of Structural Geology*, **9**, 877–891.

——, HOKE, L. KENNAN, L. & DEWEY, J. F. 1997. Cenozoic evolution of the central Andes in Bolivia and northern Chile. *In:* BURG, J.-P. & FORD, M. (eds) *Orogeny Through Time*. Geological Society, London, Special Publication, **121**, 237–264.

MACFADDEN, B. J., ANAYA, F. & ARGOLLO, J. 1993. Magnetic polarity stratigraphy of Inchasi: a Pliocene mammal-bearing locality from the Bolivian Andes deposited just before the Great American Interchange. *Earth and Planetary Science Letters*, **114**, 229–241.

——, ——, PEREZ, H., NAESER, C. W., ZEITLER, P. K. & CAMPBELL, K. E. 1990. Late Cenozoic paleomagnetism and chronology of Andean basins of Bolivia: evidence for possible oroclinal bending. *Journal of Geology*, **98**, 541–555.

MARSHALL, L. G. & SEMPERE, T. 1991. The Eocene to Pleistocene vertebrates of Bolivia and their stratigraphic context: a review. *In:* SUAREZ, R. (ed.) *Fossiles y Facies de Bolivia, Volume 1. Revista Tecnica de Yacimientos Petroliferos Fiscales Bolivianos*, **12**, 631–652.

MOLNAR, P. & ENGLAND, P. 1990. Late Cenozoic uplift of mountain ranges and global climate change: chicken or egg? *Nature*, **346**, 29–34.

PARDO-CASAS, F. & MOLNAR, P. 1987. Relative motion of the Nazca (Farallon) and South American plates since late Cretaceous time. *Tectonics*, **6**, 233–248.

RAYMO, M. E. & RUDDIMAN, W. F. 1992. Tectonic forcing of late Cenozoic climate. *Nature*, **359**, 117–122.

ROEDER, D. 1988. Andean age structure of Eastern Cordillera (Province of La Paz, Bolivia). *Tectonics*, **7**, 23–39.

SERVANT, M., SEMPERE, T., ARGOLLO, J., BERNAT, M., FERAUD, G. & LO BELLO, P. 1989. Morphogenese et soulevement des Andes de Bolivie au Cenozoique. *Comptes Rendus a l'Academie des Sciences de Paris*, **309**, 417–422.

SHEFFELS, B. M. 1990. Lower bound on the amount of crustal shortening in the central Bolivian Andes. *Geology*, **18**, 812–815.

TROENG, B., CLAURE, H., OLIVEIRA, L., BALLÓN, R. & WALSER, G. 1993. Mapas Temáticos de Recursos Minerales de Bolivia: Hojas Tarija y Villazón. *Boletin del Servicio Geologico de Bolivia*, **3**.

WALKER, E. H. 1949. Andean uplift and erosion surfaces near Uncia, Bolivia. *American Journal of Science*, **247**, 646–663.

WATTS, A. B., LAMB, S. H., FAIRHEAD, J. D. & DEWEY, J. F. 1995. Lithospheric flexure and bending of the central Andes. *Earth and Planetary Science Letters*, **134**, 9–21.

WIGGER, P., SCHMITZ, M., ARANEDA, M. *ET AL.* 1993. Variation in crustal structure in the southern central Andes deduced from seismic investigations. *In:* REUTTER, K.-J., SCHEUBER, E. & WIGGER, P. (eds) *Tectonics of the Southern Central Andes*. Springer, Berlin, 23–48.

Index